GREATER SAGE-GROUSE

STUDIES IN AVIAN BIOLOGY

A Publication of the Cooper Ornithological Society

HTTP://WWW.UCPRESS.EDU/GO/SAB

Studies in Avian Biology is a series of works by the Cooper Ornithological Society since 1978. Volumes in the series address current topics in ornithology and can be organized as monographs or multi-authored collections of chapters. Authors are invited to contact the series editor to discuss project proposals and guidelines for preparation of manuscripts.

See complete series list on page 645.

GREATER SAGE-GROUSE

Ecology and Conservation
of a Landscape Species and Its Habitats

Steven T. Knick and John W. Connelly, *Editors*

Studies in Avian Biology No. 38

A PUBLICATION OF THE COOPER ORNITHOLOGICAL SOCIETY

University of California Press
Berkeley Los Angeles London

University of California Press, one of the most distinguished university presses in the United States, enriches lives around the world by advancing scholarship in the humanities, social sciences, and natural sciences. Its activities are supported by the UC Press Foundation and by philanthropic contributions from individuals and institutions. For more information, visit www.ucpress.edu.

Studies in Avian Biology, No. 38
For digital edition of this work, please see the UC Press website.

University of California Press
Berkeley and Los Angeles, California

University of California Press, Ltd.
London, England

Library of Congress Cataloging-in-Publication Data

Greater sage-grouse : ecology and conservation of a landscape species and its habitats /
Steven T. Knick and John W. Connelly, editors.
p. cm. — (Studies in avian biology ; no. 38)
"A Publication of the Cooper Ornithological Society."
Includes bibliographical references and index.
ISBN 978-0-520-26711-4 (cloth : alk. paper)
1. Sage grouse—Ecology. 2. Sage grouse—Habitat—Conservation.
3. Sagebrush—Ecology. I. Knick, Steven T. II. Connelly, John W. (John William), 1952–

QL696.G27E26 2011

639.9'78636—dc22 2010052102

19 18 17 16 15 14 13 12 11
10 9 8 7 6 5 4 3 2 1

The paper used in this publication meets the minimum requirements of ANSI/NISO Z39.48-1992 (R 1997)(*Permanence of Paper*).

Cover image: Greater Sage-Grouse. Photo by Terry R. Steele.

PERMISSION TO COPY

CONTENTS

CONTRIBUTORS

CAMERON L. ALDRIDGE
Natural Resource Ecology Laboratory
Colorado State University; and
U.S. Geological Survey
2150 Centre Avenue, Bldg. C
Fort Collins, CO 80526-8118
cameron_aldridge@usgs.gov

COURTNEY L. AMUNDSON
Department of Fisheries, Wildlife and
Conservation Biology
University of Minnesota
1980 Folwell Avenue
St, Paul, MN 55108
amun0113@umn.edu

WILLIAM L. BAKER
Ecology Program and Department of Geography
University of Wyoming
Laramie, WY 82071
bakerwl@uwyo.edu

ERIK A. BEEVER
U.S. Geological Survey
Forest and Rangeland Ecosystem Science Center
3200 SW Jefferson Way
Corvallis, OR 97331
(Current address: U.S. Geological Survey,
Northern Rocky Mountains Science Center, 2327
University Way, Suite 2, Bozeman, MT 59715)
ebeever@usgs.gov

CLAIT E. BRAUN
Grouse Inc.
5572 North Ventana Vista Road
Tucson, AZ 85750
sg-wtp@juno.com

THOMAS J. CHRISTIANSEN
Wyoming Game and Fish Department
351 Astle Avenue
Green River, WY 82935
tom_christiansen@wgf.state.wy.us

JOHN W. CONNELLY
Idaho Department of Fish and Game
1345 Barton Road
Pocatello, ID 83204
jcsagegrouse@aol.com

HOLLY E. COPELAND
The Nature Conservancy
Wyoming State Office
Lander, WY 82520
hcopeland@tnc.org

KEVIN E. DOHERTY
Wildlife Biology Program
University of Montana
Missoula, MT 59812
(Current address: U.S. Fish and Wildlife Service
3425 Miriam Avenue
Bismarck, ND 58501)
kevin_doherty@fws.gov

SEAN P. FINN
U.S. Geological Survey
Forest and Rangeland Ecosystem Science Center
970 Lusk Street
Boise, ID 83706
sfinn@usgs.gov

JOHN C. FREEMUTH
Cecil Andrus Center for Public Policy
Boise State University
Boise, ID 83725
jfreemu@boisestate.edu

EDWARD O. GARTON
Department of Fish and Wildlife Resources
University of Idaho
Moscow, ID 83844
ogarton@uidaho.edu

CHRISTIAN A. HAGEN
Oregon Department of Fish and Wildlife
63714 Parrell Road
Bend, OR 97702
christian.a.hagen@state.or.us

STEVEN E. HANSER
U.S. Geological Survey
Forest and Rangeland Ecosystem Science Center
970 Lusk Street
Boise, ID 83706
shanser@usgs.gov

CHARLES J. HENNY
U.S. Geological Survey
Forest and Rangeland Ecosystem Science Center
3200 SW Jefferson Way
Corvallis, OR 97331
hennyc@usgs.gov

ANN L. HILD
Department of Renewable Resources
University of Wyoming
Laramie, WY 82009
annhild@uwyo.edu

MATTHEW J. HOLLORAN
Wyoming Wildlife Consultants LLC
Laramie, WY 82072
matth@wyowildlife.com

JON S. HORNE
Department of Fish and Wildlife Resources
University of Idaho
Moscow, ID 83844
jhorne@uidaho.edu

DOUGLAS H. JOHNSON
U.S. Geological Survey
Northern Prairie Wildlife Research Center
204 Hodson Hall
1980 Folwell Avenue
St. Paul, MN 55108
douglas_h_johnson@usgs.gov

JOSEPH M. KIESECKER
The Nature Conservancy
Rocky Mountain Conservation Region
Fort Collins, CO 80524
jkiesecker@tnc.org

STEVEN T. KNICK
U.S. Geological Survey
Forest and Rangeland Ecosystem Science Center
970 Lusk Street
Boise, ID 83706
steve_knick@usgs.gov

MATTHIAS LEU
U.S. Geological Survey
Forest and Rangeland Ecosystem Research Center
Snake River Field Station
970 Lusk Street
Boise ID, 83706
(Current address: Biology Department, College of William and Mary, P.O. Box 8795, 540, Landrum Drive, Integrated Science Center (ISC) RM-2129, Williamsburg, VA 23185)
mleu@wm.edu

CARA W. MEINKE
U.S. Geological Survey
Forest and Rangeland Ecosystem Science Center
970 Lusk Street
Boise, ID 83706
(Current address: Stantec Consulting, 9400 SW Barnes Road, Suite 200, Portland, OR 97225)
cara.meinke@stantec.com

RICHARD F. MILLER
Eastern Oregon Agricultural Research Center
202 Strand Agricultural Research Center
Oregon State University
Corvallis, OR 97331
richard.miller@oregonstate.edu

ANN MOSER
Idaho Department of Fish and Game
600 S. Walnut
P.O. Box 25
Boise, ID 83707
ann.moser@idfg.idaho.gov

DAVID E. NAUGLE
Wildlife Biology Program
University of Montana
Missoula, MT 59812
david.naugle@umontana.edu

SARA J. OYLER-MCCANCE
U. S. Geological Survey
Fort Collins Science Center, 2150 Centre Avenue
Building C
Fort Collins, CO 80526
sara_oyler-mccance@usgs.gov

AMY POCEWICZ
The Nature Conservancy
Wyoming State Office
Lander, WY 82520
apocewicz@tnc.org

DAVID A. PYKE
U.S. Geological Survey
Forest and Rangeland Ecosystem Science Center
3200 SW Jefferson Way
Corvallis, OR 97331
david_pyke@usgs.gov

THOMAS W. QUINN
Rocky Mountain Center for Conservation Genetics and Systematics
Department of Biological Sciences
University of Denver
Denver, CO 80208
tquinn@du.edu

KERRY PAUL REESE
Department of Fish and Wildlife Resources
University of Idaho
Moscow, ID 83844
kreese@uidaho.edu

E. THOMAS RINKES
U.S. Bureau of Land Management
1387 Vinnell Way
Boise, ID 83709
tom_rinkes@blm.gov

MICHAEL A. SCHROEDER
Washington Department of Fish and Wildlife
P.O. Box 1077
Bridgeport, WA 98813
michael.schroeder@dfw.wa.gov

SAN J. STIVER
Western Association of Fish and Wildlife Agencies
2184 Richard St.
Prescott, AZ 86301
stiver@cableone.net

CYNTHIA M. TATE
Wyoming Game and Fish Department
Wyoming State Veterinary Laboratory
1174 Snowy Range Road
Laramie, WY 82070
ctate@uwyo.edu

W. MATTHEW VANDER HAEGEN
Washington Department of Fish and Wildlife
600 Capitol Way North
Olympia, WA 98501
matt.vanderhaegen@dfw.wa.gov

BRETT L. WALKER
Wildlife Biology Program
University of Montana
Missoula, MT 59812
(Current address: Colorado Division of Wildlife, 711 Independent Ave., Grand Junction, CO 81505)
brett.walker@state.co.us

MICHAEL J. WISDOM
U.S.D.A. Forest Service
Pacific Northwest Research Station
1401 Gekeler Lane
La Grande, OR 97850
mwisdom@fs.fed.us

PREFACE

Population declines of Greater Sage-Grouse (*Centrocercus urophasianus*) coupled with extensive loss of their sagebrush (*Artemisia* spp.) habitats prompted nine petitions from 1999 to 2005 for protection under the Endangered Species Act. As a result, the United States Fish and Wildlife Service (USFWS) faced the daunting task of determining status and trends for a species and its habitat currently distributed across 11 states and two provinces in western North America. The difficulty of this task was in large part due to complex interactions among invasive plants, altered fire and disturbance regimes, and land use. The Conservation Assessment of Greater Sage-Grouse and Sagebrush Habitats (Connelly et al. 2004) was written for the Western Association of Fish and Wildlife Agencies and was intended to aid the USFWS's range-wide listing decision issued in 2005. The assessment provided detailed information on Greater Sage-Grouse biology as well as analysis of population and habitat trajectories. It now provides the foundation for this volume.

The USFWS determined in March 2010 that Greater Sage-Grouse warranted listing under the Endangered Species Act, but that immediate action was precluded by higher priorities. Coalitions of private, state, federal, and nongovernmental entities have been formed, conservation strategies developed, and management actions to benefit sage-grouse and restore sagebrush have been implemented. Sage-grouse and sagebrush are priority concerns for industry, conservation organizations, and management agencies in the United States and Canada. Yet thousands of square kilometers of sagebrush steppe burn each year or are invaded by exotic plants, energy development has accelerated and influenced many more square kilometers of sage-grouse habitat, West Nile virus has infected local populations of sage-grouse, human densities in the western United States and their use of wildlands have increased, and sage-grouse populations continue to decline in many parts of their range. In this volume, we have documented the status of Greater Sage-Grouse populations and their sagebrush habitats and identified factors that influence their long-term persistence.

We have revised, updated, and reconfigured the original content of the 2004 assessment. Gaps caused by time limits in preparing the assessment have been addressed with new chapters that have greatly expanded the scope and depth of information contained in this volume. Management considerations were not provided in the 2004 assessment to avoid influencing policy direction. These now have been developed in a conservation implications section at the conclusion of each chapter.

This volume reflects a broad spectrum of expertise in sagebrush and sage-grouse ecology. Thirty-eight authors represented 20 organizations, including 6 state agencies, 3 federal agencies, 6 universities, and 5 nongovernmental organizations. In all, 63 reviewers represented 39 organizations, including 8 state agencies, 8 federal agencies, 18 universities, and 5 nongovernmental organizations. Only 11 individuals both authored and reviewed chapters.

We appreciate the support of the Cooper Ornithological Society in providing the forum for this collection. Carl Marti[1], series editor of *Studies in Avian Biology*, ensured the consistency and quality of the chapters. All chapters have been peer reviewed, most by both an expert in sage-grouse or sagebrush ecology and an outside expert in the ecological topic. At least one and often both associate editors reviewed each chapter in addition to the technical editor. Chapters authored by United States Geological Survey (USGS) scientists received an agency policy review, although other chapters and the volume as a whole do not necessarily reflect the views of the USGS, the Western Association of Fish and Wildlife Agencies, or any other state or federal agency. The volume has not been endorsed by any private, state, or federal organization but was developed solely as a scientific peer-reviewed publication by the Cooper Ornithological Society. Any use of trade, product, or firm names is for descriptive purposes only and does not imply endorsement by the United States government.

We thank the USGS Forest and Rangeland Ecosystem Science Center (STK) and Idaho Department of Fish and Game (JWC). Kathy Peter and Susan Haseltine of the United States Geological Survey provided financial and logistical support. Carol Schuler, Kate Kitchell, Sue Phillips, and Ruth Jacobs provided administrative support as well as much-needed encouragement. Elena Velasquez and Martina Zucchini translated abstracts into Spanish. The USGS funded page charges.

We dedicate this volume to the agency biologists who had the foresight to form the first Western Sage and Columbian Sharp-Tailed Grouse Technical Committee over 50 years ago for the purpose of developing the cooperation across administrative boundaries necessary to conserve prairie grouse. These individuals include Paul D. Dalke, Robert L. Eng, Clifton W. Greenhalph, Gordon W. Gullion, J. Burton Lauckhart, Levon Lee, A. Starker Leopold, Jessop B. Low, Donald D. MacLean, W. V. Masson, Otto C. Nelson, Robert L. Salter, and Charles F. Yocom. Their vision provides a model of collaboration that must be sustained if we are to conserve sage-grouse. Without their efforts, much of our work would not have been possible. We present the information in this volume in hopes that generations long into the future will be able to experience the sun rising on a vast open landscape, smell the pungent scent of sagebrush, and marvel at the centuries-old rite of Greater Sage-Grouse displaying on a lek.

STEVEN T. KNICK
JOHN W. CONNELLY

[1] Carl Marti died during the production of this volume. We will remember working with him as the series editor and also as a colleague and a friend.

FOREWORD

Thoughts on the Role of Science in Making Public Policy

John C. Freemuth

The Greater Sage-Grouse (*Centrocercus urophasianus*) has become a species whose listing status under the Endangered Species Act (ESA) is fraught with controversy. The recent past has seen court cases, high-level political intervention, and disputes over what constitutes the best available scientific information. Yet the conflict over the warranted listing of the sage-grouse presents an opportunity to engage the public about what is known about the science of the sage-grouse and its sagebrush habitat, and to promote public deliberation about possible solutions. That opportunity can perhaps raise public confidence in science and make the blatant politicizing of science more difficult.

SETTING THE STAGE

The role of science and expertise in making public policy may strike many readers as a relatively modern concern. Certainly, recent events surrounding the listing decisions for the sage-grouse have brought that role to the foreground and jump-started a new and intense discussion. But we have talked about that role for a long time. Woodrow Wilson, the only president to hold a Ph.D., noted in 1912,

> What I fear, therefore, is a government of experts. God forbid that in a democratic country we should resign the task and give the government over to experts. What are we for if we are to be scientifically taken care of by a small group of gentlemen who are the only men who understand the job? Because if we don't understand the job, then we are not a free people. We ought to resign our free institutions and go to school to somebody and find out what it is we are about. (Davidson 1956:83, as quoted in Smith 1991:2)

Wilson was not bashing experts but asking us to think about their appropriate role in a democracy. Today, the role of science in public land management is the question before us. It remains part of the context of decisions over whether to list the sage-grouse as an endangered species. I argue that that role can help us learn what we are about regarding our federal lands, while continuing to make decisions grounded in democratic practice.

I had the privilege of serving on, then chairing, the Science Advisory Board of the Bureau of Land Management (BLM) during the late 1990s. The board's purpose was to work with the BLM and its director on improving the communication of the bureau's science and research needs to other agencies, scientists, the U.S. Congress, and the public; helping get scientific findings to BLM field office staff; and coping with resource management issues. The board lasted until the early years of the Bush administration when its Federal Advisory Committee Act charter was not extended.

One of the most interesting and important projects of the board and BLM was writing the bureau's science strategy. That strategy was to delineate the role of science in BLM decision making and public land management. The strategy noted:

> Science is useful for evaluating alternatives and estimating outcomes. However, it is not the sole factor in making decisions because the state of natural resource science is often insufficient to give definitive cause-effect predictions. Unknowns and uncertainties will always be associated with predictions of decision outcomes. Science may reduce but can never completely eliminate the uncertainty regarding future events. However, the use of the best-available science—along with a consideration of political, social, and economic information—will result in the best-informed decisions. (United States Bureau of Land Management 2000:3–5)

In this document, science was considered a necessary but insufficient condition for making BLM management decisions. It was also my experience, both on the board and as someone who pays attention to the academic and actual worlds of federal land policy, that science can occupy other places in discussions over its role in public-land policy decisions. Indeed, it was likely that most of those involved with the writing of the statement above, as well as many other observers, were aware that there was quite a bit of debate and confusion over the role of science in natural resource and public lands decision making.

TWO ISSUES IN THE SCIENCE-POLICY RELATIONSHIP

Two regularly occurring issues influence any attempt to untangle the science-policy relationship. The first issue concerns viewing science as a process versus science as truth. Science as process refers to the use of the scientific method to produce knowledge, which is then subjected to peer review and published in professional journals. All chapters of this volume of *Studies in Avian Biology* were subjected to peer review, revised, and, in some cases, reviewed again, certainly meeting the process test.

The science-as-truth claim refers to the use of science as the proper way to solve policy conflicts over other ways of deciding. To use the BLM quote above as an example, the science-as-truth claim would thus disallow political, social, and economic concerns. Authors of this document did not make any such claim about science, nor were they even given the task of considering such concerns, but it nonetheless remains an important issue in the science-policy dialogue.

Science as Process

An extension of the process discussion is a question of particular concern to those who work on public-land management conflicts such as that of the sage-grouse. Is the knowledge produced helpful to land managers and other decision makers? This is a difference between applied and basic science. During my tenure with the BLM board, agency scientists often discussed being evaluated and rewarded for publication in journals, even though the information needed by land managers was of a different sort—more immediate and thus less able to go through long peer-review processes—but it had to be credible or it simply would not stand up in court or anywhere else. This volume is a good example. It has gone through a long and exhaustive review process both to ensure science credibility and to stand up in court. Beyond the immediate purpose of providing information for determining the listing status of Greater Sage-Grouse, how will it get used by managers and other decision makers to influence public-land policy?

It was also not clear whether land managers did a very good job at explaining to scientists both within and outside an agency (like BLM) how they made decisions. To put it differently, how did they use science to make a decision while still considering political, social, and economic information, especially under conflicting multiple-use laws? Having this kind of discussion would go far in allowing people to see the alternate perspectives in the science–public policy arena. Indeed, agency-based meetings or conferences should be designed to do exactly that.

Many suspect that more may be going on with the process question than how managers use science. Scientists always need to be transparent about how their own values did or did not inform their research choices. In 1994, a fascinating debate appeared in the journal *Conservation Biology* over whether conservation biologists should link arms with activists in efforts to reform grazing practices (Brussard et al. 1994). Other noted conservation biologists were concerned that conservation biologists might damage their credibility by openly advocating and supporting political positions. They suggested asking instead how livestock grazing can be managed to have the fewest impacts on biodiversity and ecosystem integrity (Brussard et al. 1994). They worried about the flaw of deductive reasoning which leads to a research question that states:

> "range management must be dramatically reformed." . . . Our work as scientists involves recognizing patterns based on data and only then formulating a general rule. More importantly, how can we hope to advance society's mission to preserve biological diversity if our audience of policymakers assumes that we intend to prove a presumed conclusion instead of attempting to falsify well-framed null hypotheses? (Brussard et al. 1994:920)

In this example, we see well-respected scientists having a polite discussion about the use of scientific information and how it informs policy. What the authors caution against is an approach that confuses scientific assertions with value judgments (grazing reduces biodiversity) and assumptions (grazing has a negative influence on ecosystem integrity). One begins with a supposedly neutral hypothesis arising from a presumed conclusion that grazing is bad or grazing must be reformed. It might then be a relatively easy to find the science in support of the hypothesis.

The more appropriate hypothesis in this argument ought to be along the lines that grazing has no effect on biological diversity. If it is falsified, then people can begin to speak of reform. Of course, as Brussard et al. (1994) also note, sometimes the effects of grazing could be positive. We also know that all of these hypotheses are very specific to place and perhaps not as useful in such a complex issue as the sage-grouse status. The authors remind us that good science starts with data, looks for patterns, then forms general rules that are subject to further testing through hypothesis development.

A second process-related issue concerns the influence of political appointees on the production and dissemination of science to be used in decision making. If science is indeed necessary (and the requirements and legal interpretations of many laws seem to make that a noncontroversial statement), then it can become strategic to influence what that science might be asserting. In May 2008, the United States Government Accountability Office issued a report on United States Fish and Wildlife Service reviews of ESA decisions (including Greater Sage-Grouse and Gunnison Sage-Grouse [*Centrocercus minimus*]) during the George W. Bush administration's tenure and concluded that political appointees indeed had intervened. Not surprisingly, there was debate over that role and whether or not there was undue pressure on scientists to alter documents and conclusions. As one former official, Craig Manson, put it, "the fact was the assistant secretary's office took a very active role in the ESA program, and that's perfectly proper for the assistant secretary's office to do so" (Manson 2008:A-7).

Manson is correct. Political appointees do take an active role in these sorts of decisions, and they should. They represent the president and the president's views. They are accountable to the president, and they are accountable to Congress and to all Americans. In the best of circumstances, they have to navigate a labyrinth of laws, values, and conflicting expectations. They are not given the task of simply following best science, because best science is not a trump. However, how they take an active role is important. As one reviewer of this introduction noted, changing data, altering reports, and attempting to intimidate scientists is unacceptable and destroys the integrity of the process.

We could decide as a society that best science ought to be the trump. We would have to do that through legislation, and in doing that, we open up a number of other issues regarding science and democracy. Such a solution might be thought extreme by many, though perhaps not by many of my friends working on the science needed for federal land-management decisions. More realistically, what is needed for this particular problem, based on the actions of political appointees in this case, is an administrative framework, a series of guidelines that make it difficult for the excesses reported by the Government Accountability Office to occur. Such a framework ought to provide an open and defensible accounting of why certain scientific information was edited, replaced, not allowed, and so on. After all, this *Studies in Avian Biology* volume, used by the United States Fish and Wildlife Service in its listing decision, must by law be made easily accessible to the public. There should not be a mysterious administrative process that only partially reveals and certainly does not explain why a decision was made. This is similar to the earlier call for scientists to be clear about their values and to disentangle those values from their science. Decision makers ought to be clear, too.

Science as Truth

The second issue, science as truth, involves the value one places on science and the knowledge it produces. A recent article by Lockwood (2008) on the mythic cowboy sets the stage for this perspective:

> From the first-hand perspective of a scientist—an insect ecologist, in particular—I can assure you that science cannot provide the answers to the deep questions concerning what we ought to do with western people and lands. Science is necessary, but it is not sufficient. And here's why.
>
> In the Myth of Scientism, the hero is an absolutely objective, rational discoverer of unchanging physical truths (like Isaac Newton and the apple incident, which of course never happened—but we know that myths don't collapse under the weight of facts). The scientist reveals how the world really is, not some fanciful tale replete with human desires. The problem, of course, is that according to this myth, science is value-free. So the hero can offer no moral lessons regarding our treatment of one another or the land.

If it claims to answer value questions, science, from this second perspective, moves from a way of knowing grounded in a process to a purveyor of truths. It has become a form of higher law. As Gregg Cawley and I have pointed out (Cawley and Freemuth 2007), one of the leading writers in public administration suggests that "the imperative of higher law is always conceived as derived from what is most valid, most powerful, most highly honored. Historically this has most frequently been God. But in the late nineteenth and early twentieth century America it has often been SCIENCE" (original emphasis; Waldo 1984:157). The science-as-truth claim suggests that it can answer questions not simply about what might or could be done but about what should be done.

Is there a path that leads away from the problems associated with these two issues? Perhaps, but it requires patience and dialogue. As Albert Teich, director of science and policy programs for the American Association for the Advancement of Science, recently wrote:

> The dominant mode of scientists engaging with the lay public in the past has been "public understanding of science"—an approach based on the belief that if people only knew more about science, they would see the world the way scientists do. This is a paternalistic posture, one that is based on a failure to recognize the legitimacy of people's values when they conflict with those of the scientist. The constituencies whose values are reflected in some of the political positions of the administration with which many scientists disagree are not necessarily uneducated or irrational. Rather, they have a different perspective on these issues, one that places a different value on the costs and risks versus the benefits of stem cell research, reductions in carbon dioxide emissions, or the preservation of an obscure species of aquatic plant that may block construction of a dam. They need to be engaged in genuine two-way conversation—a dialogue—rather than dismissed out of hand, or "educated." (Teich 2008:22)

Before scientists get upset at Teich's remarks, they might consider that some of the scientists who capture the public imagination are those who effectively communicate the excitement and challenges of modern science, such as the late Carl Sagan.

A POSSIBLE PATH

We should experiment with new forms of decision making and public discourse. Let us create a sage-grouse science and public policy deliberative forum, where scientists, the public, and interested decision makers might assemble to talk through the issues surrounding a possible listing of the sage-grouse. Scientists known for their skills in communication could present their best science in an accessible and understandable way, while letting us know how their values and assumptions guided and informed their work. This would mean distilling the diverse offerings that make up this volume, but that can be done, and scientists should be rewarded for doing it. Scientists could also debate and question each other so the public could see science at work. The public could ask questions and talk about the scientific information they think would be useful, but they should be prepared to put their own assumptions and values on the table as well, such as whether they have already decided that they are for or against listing. Conserving sage-grouse and sagebrush are very good goals, but some of the measures required to do that may not be accepted by an expanding human population in the western United States that has other priorities. We need to know if that is true.

We should hear as well the perspective of various sage-grouse working groups throughout the West who provide a useful way to link questions of science to questions of public and agency concerns and values. Managers and decision makers ought to come—indeed, ought to be compelled to come—and talk about the factors that govern their listing decisions and the role that science plays or does not play in those decisions. Together, perhaps all parties could develop the sort of guidelines that should be used by managers and decision makers in reviewing the scientific information needed for an effective listing decision. This process would take much work, but it might be more sustainable, and certainly would be more democratic, than our current one.

This volume is a hugely important part of this process. The work that has gone into it is extraordinary. But we all need to participate in developing a better way to link it to the other processes that will lead to a decision about the future of the sage-grouse and its habitats.

GREATER SAGE-GROUSE AND SAGEBRUSH

An Introduction to the Landscape

Steven T. Knick and John W. Connelly

The Greater Sage-Grouse (*Centrocercus urophasianus*) is often called an icon of the West because the species has become the symbol for conserving sagebrush (*Artemisia* spp.) ecosystems, one of the most difficult environmental challenges in North America. Sage-grouse have undergone long-term population declines and now are absent from almost half of their estimated distribution prior to Euro-American settlement (Schroeder et al. 2004) (Fig. I.1). Overall, population trends have been more stable in recent years, although sage-grouse numbers are still declining in some regions (Connelly et al. 2004). Proximate reasons for population declines differ across the sage-grouse distribution, but ultimately, the underlying cause is loss of suitable sagebrush habitat (Connelly and Braun 1997, Leonard et al. 2000, Aldridge et al. 2008). Some form and quantity of sagebrush within the landscape are necessary to meet seasonal requirements for food, cover, and nesting of sage-grouse (Patterson 1952, Connelly et al. 2000c). Thus, conserving and managing Greater Sage-Grouse is as much about the ecology of the bird as it is about understanding the dynamics of sagebrush ecosystems (Connelly et al. 2000c, Crawford et al. 2004).

Concluding that loss and degradation of sagebrush-dominated landscapes cause sage-grouse population declines is deceptively simple, much like the ecosystems themselves. Many species of conservation concern have either restricted ranges or a small number of factors contributing to their plight. In comparison, sage-grouse currently occupy 670,000 km^2 spanning parts of 11 western states and two Canadian provinces (Schroeder et al. 2004). This broad distribution encompasses highly diverse environments and an extensive array of ecological stressors. At least 11 species of sagebrush occur within the sage-grouse range, each differing in their specific plant community structure, productivity, resilience, and resistance to disturbance (West and Young 2000, Miller and Eddleman 2001).

The distribution and influence of multiple land uses also vary widely across the sage-grouse distribution, as summarized in Knick et al. (this volume, chapter 12). Conversion to croplands has eliminated or fragmented sagebrush in areas having deep fertile soils or irrigation potential. Sagebrush remaining in these areas has been reduced to agricultural edges or to relatively unproductive environments. Oil and gas resources are being developed primarily in the eastern portion of the sage-grouse range (Knick et al. 2003; Naugle et al., this volume, chapter 20), but exploration and development of wind and geothermal energy is increasing rapidly in many regions. Livestock grazing occurs throughout the sage-grouse range but has a

Knick, S. T. and J. W. Connelly. 2011. Greater Sage-Grouse and sagebrush: an introduction to the landscape. Pp. 1–9 *in* S. T. Knick and J. W. Connelly (editors). Greater Sage-Grouse: ecology and conservation of a landscape species and its habitats. Studies in Avian Biology (vol. 38), University of California Press, Berkeley, CA.

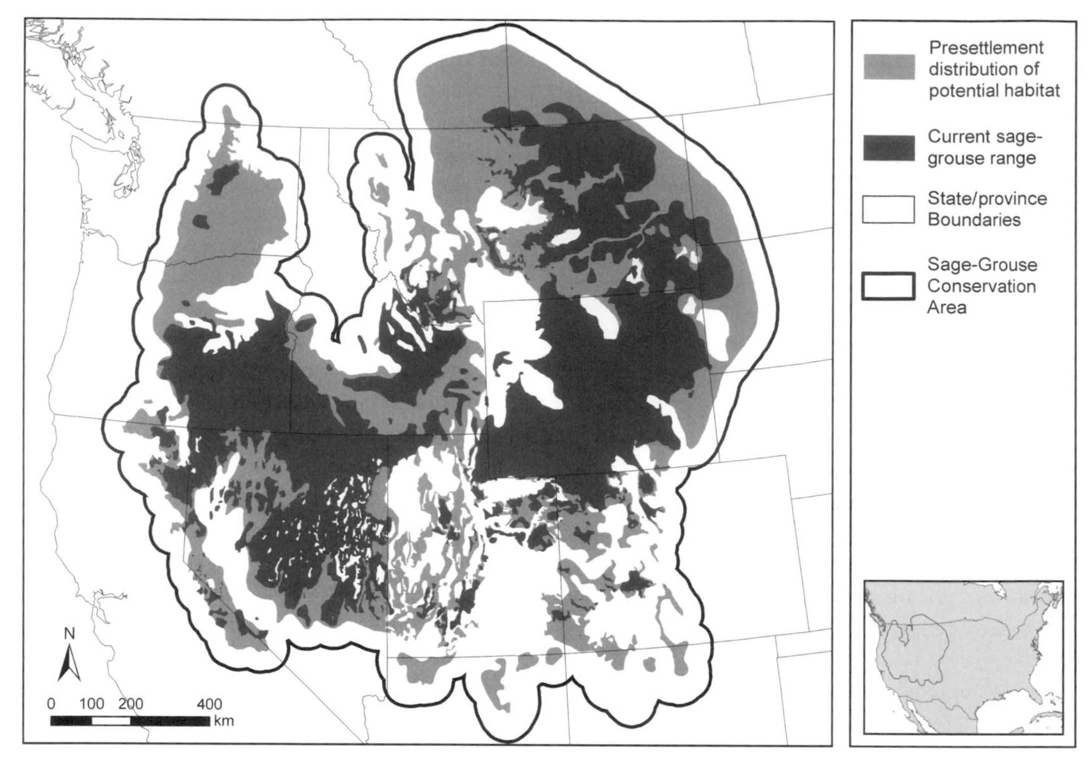

Figure I.1. Current and pre–Euro-American settlement distribution of sage-grouse in North America (Schroeder et al. 2004) within the Sage-Grouse Conservation Area (SGCA). Gunnison Sage-Grouse ranges also are included in southeastern Utah and southwestern Colorado. The SGCA was delineated by buffering the potential pre–Euro-American settlement distribution by 50 km.

more diffuse influence on soils and vegetation in contrast to land uses that remove or fragment habitat. Urbanization and human densities are increasing in the western United States as people choose to live near wilderness and recreation areas (Brown et al. 2005, Hansen et al. 2005). New corridors proposed for energy transmission would affect another 2% of the current sagebrush distribution. Less than 5% of the sage-grouse range presently is >2.5 km from a mapped road. Recreation, including off-road vehicles, is rapidly increasing on public lands. Leu and Hanser (this volume, chapter 13) describe how the collective effect of this human footprint influences the landscape structure of sagebrush-dominated habitats for sage-grouse.

The Greater Sage-Grouse is a landscape species; their large annual ranges can encompass >2,700 km^2 (Dalke et al. 1963, Schroeder et al. 1999, Leonard et al. 2000). Movements from lek sites used for breeding to nesting locations can exceed 25 km (Holloran and Anderson 2005), and seasonal ranges can be >80 km apart (Connelly et al. 1988). Sage-grouse use a variety of landscapes within their annual range that, at broad scales, range from expanses dominated by sagebrush to heterogeneous mosaics of grass and shrublands (Connelly et al. 2004). Large ranges and complex habitat selection challenge conservation because management focused on one area, usually centered around leks or habitat configuration and often emphasizing nesting and brood rearing, may not have the intended benefit when other seasonal ranges or habitat components may be at least as critical (Connelly et al. 2000c, Doherty et al. 2008). The large ranges also can encompass lands under multiple ownerships and uses. Knick (this volume, chapter 1) reviews the historical and legislative background that shaped the development of sagebrush landscapes and resulted in differing land characteristics among the mosaic of owners. Stiver (this volume, chapter 2) describes the efforts among public and private entities having diverse perspectives engaging in the collaborative use of sagebrush lands.

These conservation challenges are not unique to Greater Sage-Grouse and sagebrush. Fourteen of 18 grouse species receive national status of concern in at least one country, primarily because of habitat degradation and loss (Storch 2000). Globally, temperate grasslands, savannahs, and shrublands are the least protected biomes and have experienced extensive conversion to agriculture or invasion by exotic plant species (Brooks et al. 2004b). Extensive loss of grasslands and shrublands in the Western Hemisphere has led to population declines for many species of birds obligate to these systems (Peterjohn and Sauer 1999, Vickery et al. 1999, Brennan and Kuvlesky 2005, Askins et al. 2007).

This volume presents a multifaceted view of the ecology of Greater Sage-Grouse and sagebrush from wildlife biologists, landscape ecologists, and shrubland biologists. Authors and reviewers represented a broad expertise in research and management from across the sage-grouse range and were drawn from agency, academic, and private sectors. Thus, the syntheses of published literature combined with new analyses for this volume provide a foundation to develop effective approaches to conservation. History and our current use of the vast landscapes dominated by sagebrush can tell us much about land use, priorities, values, and resource management. The future will tell others about the effectiveness of conservation actions we implement today.

DISTRIBUTION OF GREATER SAGE-GROUSE

Sage-Grouse Conservation Area

The Sage-Grouse Conservation Area (SGCA) used throughout this volume to define the spatial extent of analyses was delineated from the estimated pre-settlement distribution of sage-grouse (Connelly et al. 2004, Schroeder et al. 2004). We added a 50-km buffer to the pre-settlement distribution to include adjacent factors, such as urban centers or agricultural regions, which may have contributed to previously extirpated populations (Aldridge et al. 2008) or to current trends (Fig. I.1). The SGCA includes parts of 14 U.S. states and three Canadian provinces encompassing 2,063,000 km^2. We first described this region as the boundary for conducting the range-wide conservation assessment (Connelly et al. 2004) and have carried it forward to this volume. No legal basis or agency endorsement exists for this region or our designation of the area.

Sagebrush is the dominant land cover on 500,000 km^2 within the SGCA (Fig. I.2). Of the sagebrush species, three subspecies of big sagebrush (Wyoming big sagebrush [*Artemisia tridentata* ssp. *wyomingensis*], basin big sagebrush [*A. t.* ssp. *tridentata*], and mountain big sagebrush [*A. t.* ssp. *vaseyana*]), two low forms (little sagebrush [*A. arbuscula*] and black sagebrush [*A. nova*]), and silver sagebrush (*A. cana*) are the most important to Greater Sage-Grouse (Connelly et al. 2000c). Connelly et al. (this volume, chapter 4) provide a summary of habitat requirements by Greater Sage-Grouse in a synthesis of published literature. Miller et al. (this volume, chapter 10) describe the characteristics of these primary sagebrush communities within the sage-grouse range and the challenges that exotic plant species, altered fire regimes, and climate change present to long-term conservation. Pyke (this volume, chapter 23) discusses the difficulties in restoring and rehabilitating these arid-land systems.

Sage-Grouse Management Zones

The Western Association of Fish and Wildlife Agencies defined seven Sage-Grouse Management Zones for assessing population and habitat trends independent of administrative and jurisdictional boundaries (Fig. I.3; Stiver et al. 2006). Management zones originally were delineated from floristic provinces, within which similar environmental factors influence vegetation communities (West 1983b, Miller and Eddleman 2001). Boundaries of management zones subsequently have been redefined, particularly in Montana and Wyoming, to better reflect potential linkages among populations and to include known leks outside the original zones (S.J. Stiver, pers. comm.).

HIERARCHICAL ORGANIZATION OF SAGEBRUSH SYSTEMS

This volume emphasizes regional and range-wide themes that describe sagebrush systems. Authors synthesized results from site-specific studies into broader patterns. Other analyses aggregated local- or small-scale events, such as individual fires or counts of sage-grouse at a lek, to provide regional or range-wide patterns of sagebrush communities, disturbance regimes, land use, and sage-grouse behavior and population dynamics. This broad-scale perspective was emphasized because monitoring is most effective when conducted and interpreted at population or range-wide scales (Connelly et al. 2003b, Reese and Bowyer 2007).

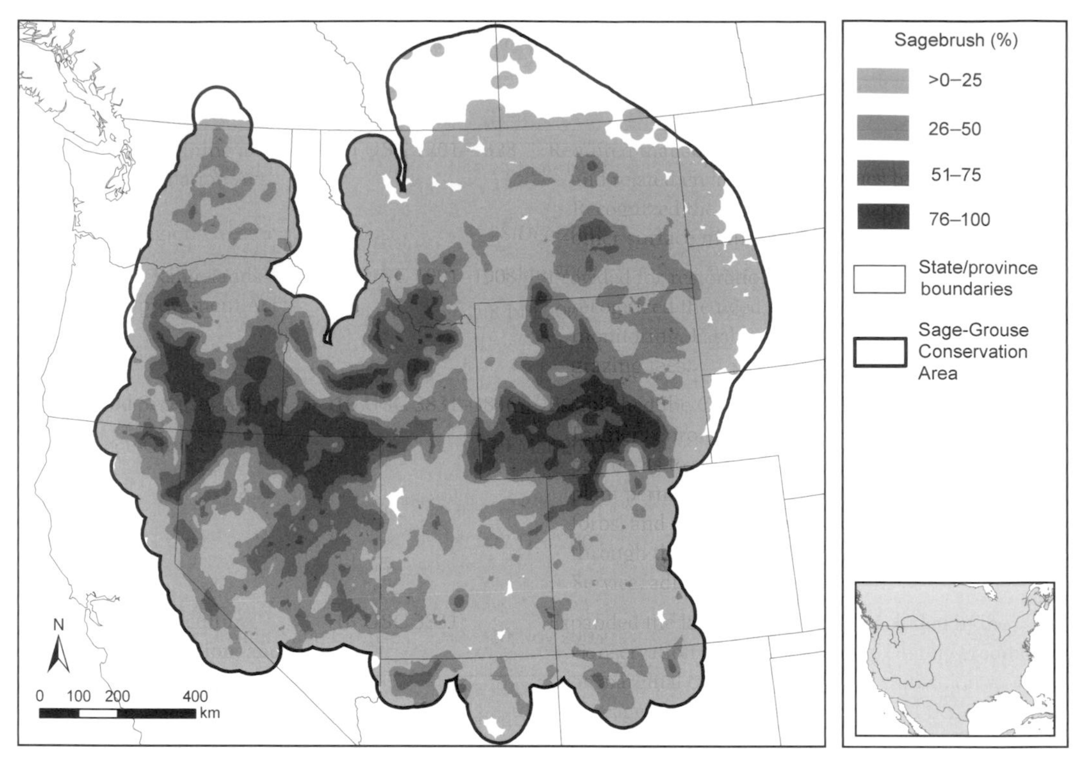

Figure I.2. Distribution of sagebrush land cover within the Sage-Grouse Conservation Area based on land cover mapped in the early 2000s in western North America. The map represents the percent of the landscape dominated by sagebrush land cover (Miller et al., this volume, chapter 10), not site-specific percentages of ground cover. As such, the map delineates a general representation of sagebrush distribution.

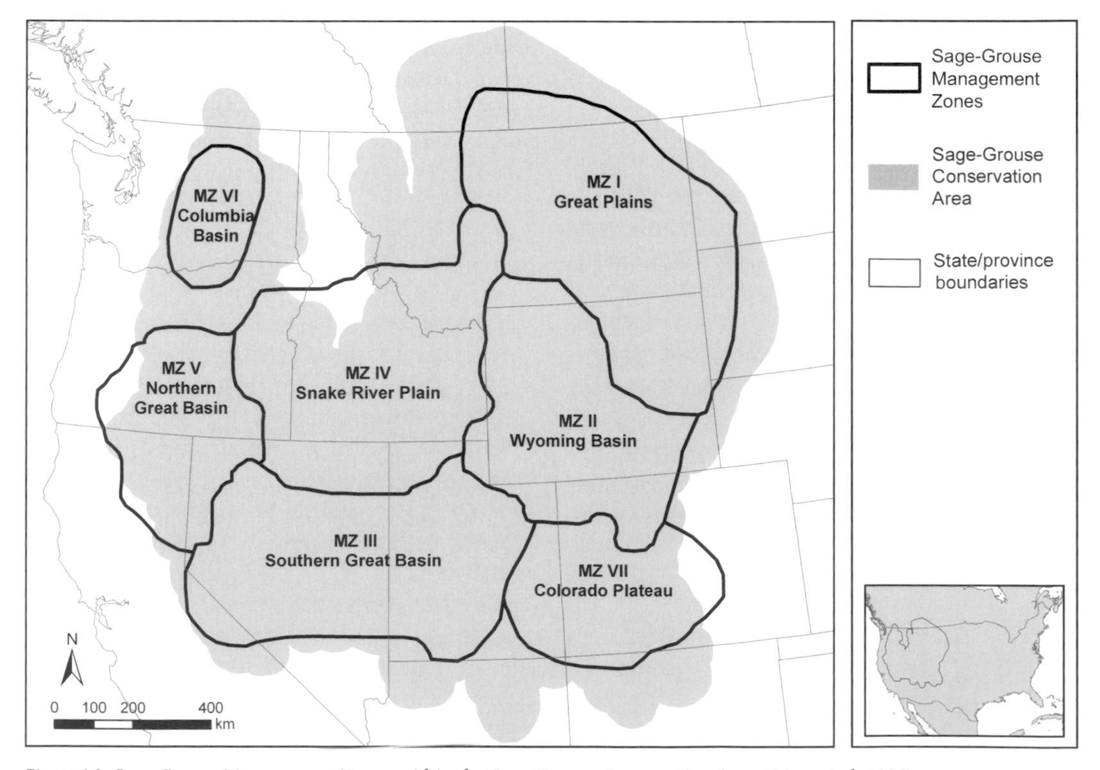

Figure I.3. Sage-Grouse Management Zones within the Sage-Grouse Conservation Area (Stiver et al. 2006).

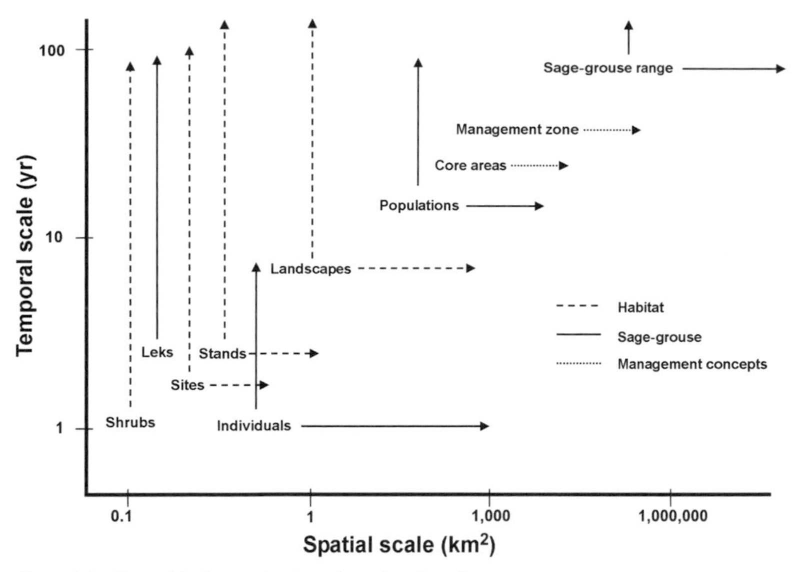

Figure I.4. Hierarchical organization of sagebrush and sage-grouse ecosystems.

Conservation planning and actions also have the greatest benefit for Greater Sage-Grouse when local actions are coordinated at regional or range-wide scales (Hemstrom et al. 2002; Wisdom et al. 2002c, 2005a). However, the validity of this paradigm rests on understanding the organizational structure of sagebrush systems.

The model of sagebrush systems as a hierarchical organization arranged along spatial and temporal scales is one of the unifying concepts underlying the information presented in this volume (Fig. I.4). This model presents ecological systems as an integrated assemblage of patterns and processes at smaller scales enclosed within successive levels at larger scales (Allen and Starr 1982, O'Neill et al. 1986, Urban et al. 1987). Levels that are adjacent in this hierarchy are most likely to share similar characteristics (Kotliar and Wiens 1990). Events in widely separated levels may be uncorrelated, and extrapolation of results across scales may lead to incorrect conclusions if organizational structure is not considered (Wiens 1981, 1989b).

Sagebrush systems have been studied most often in relatively small spatial and short temporal scales (Brown et al. 2002). Plant communities at sites functioned through time but were not considered as interacting parts of a larger landscape (West 2003b). Modeling sagebrush as a multiscale hierarchy can help illuminate landscape dynamics and long-term changes. Landscapes are mosaics of smaller patches shaped by disturbance patterns that have characteristic spatial and temporal scales (Delcourt et al. 1983, Levin 1992). Small-scale disturbances are most frequent, but rapid recovery periods maintain relatively stable patterns at larger scales (Urban et al. 1987). Unbalanced dynamics of disturbance relative to recovery at smaller scales can change patterns observed at larger scales. Miller et al. (this volume, chapter 10) describe fire characteristics for each of the Sage-Grouse Management Zones based on fire polygons mapped since the 1980s. Baker (this volume, chapter 11), using a landscape approach to estimate pre–Euro-American and current fire rotations, discusses how changes in size and frequency of fires relative to recovery have changed in sagebrush systems. Both chapters conclude that recovery periods are insufficient relative to the characteristics of the new disturbance regime. Consequently, sagebrush landscapes are being converted to exotic grasslands at scales sufficiently large to influence future disturbance dynamics and sage-grouse distributions.

A hierarchical perspective may also better describe habitat selection by sage-grouse, as presented by Connelly et al. (this volume, chapter 4). Conceptual models of hierarchical selection predict that coarse features at large spatial scales are important for initial location but finer features in the environment become primary factors in selecting individual sites (Johnson 1980, Kristan 2006). Site-specific characteristics are the most commonly studied components of habitat selection by sage-grouse (Connelly et al. 2000c, Hagen et al. 2007). Most research reviewed by Connelly et al. (this volume, chapter 4) was focused on fine-scale features, such as percent canopy cover and height of sagebrush selected by sage-grouse for nesting or brood rearing. Yet large-scale (2–50 km^2) features are important influences on seasonal use areas, population trends, and persistence (Holloran and Anderson 2005, Doherty et al. 2008). The presence of relatively large (thousands of km^2) expanses of sagebrush-dominated shrub steppe at broad spatial scales is necessary to support sage-grouse populations (Connelly et al. 2000c, Aldridge et al. 2008). However, the appropriate patch size or landscape configuration and spatial scale(s) are uncertain. Habitat features and spatial scale of selection by sage-grouse are explored by Hanser and Knick (this volume, chapter 19), Connelly et al. (this volume, chapter 4), and Johnson et al. (this volume, chapter 17).

Anthropogenic land use has altered landscapes used by Greater Sage-Grouse in most parts of their range (Knick et al. 2003, Connelly et al. 2004). However, land uses differ in the multiscale signature they impose (Knick and Rotenberry 1997). The Columbia Basin consists primarily of a matrix dominated by agriculture at large scales and interspersed sagebrush patches at small scales. In contrast, oil and gas development in the Wyoming Basin creates fragmentation in landscapes dominated by sagebrush at larger scales. Sage-grouse use each region, although the commingled effect of fragmentation and habitat loss at different scales has influenced each of these populations (Schroeder et al. 2000, Doherty et al. 2008). The diversity of areas used by sage-grouse across their range suggests that we have not identified either the appropriate scales or the relative importance of habitat arrangements across multiple scales. Consequently, our ability to fully understand effects of land uses and alterations often is limited to correlation rather than cause and effect.

The hierarchical scaling of sage-grouse as individuals, populations, and metapopulations (Fig. I.4) provides perspective for understanding different sources of mortality. Predation (Hagen, this volume, chapter 6), harvest (Reese and Connelly, this volume, chapter 7), and disease (Christiansen and Tate, this volume, chapter 8) are significant to individuals or local groups but are not significant factors influencing population trends. Similarly, West Nile virus (Walker and Naugle, this volume, chapter 9) has the potential to significantly decrease sage-grouse numbers or eliminate relatively small peripheral populations, but the effect on range-wide trends is less clear.

The range-wide distribution of sage-grouse, although encompassing a broad spatial extent, is composed primarily of small, mostly independent units. Forty-one populations within the Greater Sage-Grouse range previously were identified using geographic or physical barriers to movements to guide delineation; five spatially large populations were further divided into 24 subpopulations (Connelly et al. 2004). Dispersal distances to new breeding areas by juvenile sage-grouse from the leks where their mother was marked are not well documented but suggest that most distances are <27 km and average between 7 and 13 km (Dunn and Braun 1985, Schroeder and Robb 2003). Knick and Hanser (this volume, chapter 16) estimated that approximately 200 population units (components) exist within the range-wide distribution of mapped leks, using connectivity analysis to identify unique assemblages of leks linked by dispersal distances <18 km.

Conserving an area sufficiently large to contain an intact sagebrush system complete with disturbance and recovery dynamics is especially challenging because of the large spatial area and long temporal scale that will be required. Available financial and logistic resources limit the extent of conservation and management actions. However, core areas (Doherty et al., this volume, chapter 21), population units (Garton et al., this volume, chapter 15), or population components (Knick and Hanser, this volume, chapter 16) can help focus planning on large areas needed to sustain populations, thus avoiding a spatial checkerboard of unrelated actions that have less benefit to long-term conservation.

CHANGES IN GREATER SAGE-GROUSE POPULATIONS

Annual rates of change suggest a long-term decline for Greater Sage-Grouse throughout their range at an overall rate of 2.0% per year from 1965 to 2003 (Connelly et al. 2004). From 1965 to 1985, the range-wide population declined at an average rate of 3.5% per year. Populations declined from 1986 to 2003 at a lower annual rate of 0.4% and fluctuated around a level that was 5% lower than the 2003 population (Connelly et al. 2004). Garton et al. (this volume, chapter 15) update these trends for the range-wide distribution, management zones, and population levels and estimate decreases in carrying capacity that have influenced sage-grouse populations.

Sage-grouse populations at the edge of the current range are increasingly isolated from larger core areas (Schroeder et al. 1999, 2004). Dispersal rates may be adequate to maintain gene flow among individual populations or subpopulations, although two, the Columbia Basin and Mono Lake populations, are genetically isolated (Benedict et al. 2003, Oyler-McCance et al. 2005b). Knick and Hanser (this volume, chapter 16) assess connectivity within the network of sage-grouse leks to identify the contribution of spatial relationships to persistence of sage-grouse. Their analysis suggests that environmental and stochastic events, rather than genetic constraints (Oyler-McCance and Quinn, this volume, chapter 5), may have a greater influence on viability of populations.

CHANGES IN SAGEBRUSH LANDSCAPES

Individual and interacting components of quantity, composition, and configuration drive processes within sagebrush systems and influence suitability for sage-grouse. These fundamental characteristics of sagebrush landscapes have changed from what early North American explorers once described as a vast sea of wormwood (sagebrush; Frémont 1845). First, the quantity of area dominated by sagebrush land cover has been reduced. Conversion to cropland is the primary large-scale agent of reduction, although other land uses also remove sagebrush at smaller spatial scales (Knick et al. 2003). Sage-grouse persist in agricultural landscapes within the Great Plains and Columbia Basin management zones, although populations in both zones are declining (Schroeder et al. 2000, Connelly et al. 2004). Increases in agriculture have been correlated with declines in sage-grouse populations (Swenson et al. 1987, Leonard et al. 2000), and thresholds may exist beyond which sage-grouse may be extirpated from a region (Aldridge et al. 2008). Schroeder and Vander Haegen (this volume, chapter 22) describe how the Conservation Reserve Program may provide critical habitat that allows sage-grouse to persist in regions dominated by agricultural croplands.

Second, composition of sagebrush communities has changed and resulted in fire regimes that are profoundly altered from historic patterns. The role of fire has been altered across much of the western sage-grouse range in low-elevation sagebrush systems that have been invaded by cheatgrass (*Bromus tectorum*) and other exotic plant species (Young et al. 1972, Billings 1990, West and Young 2000). In contrast, fire that has been reduced or eliminated at higher elevations has facilitated juniper (*Juniperus* spp.) and pinyon (*Pinus* spp.) woodland expansion into sagebrush communities (Miller and Rose 1999). Increased resistance to fire in these systems ultimately results in successional dynamics that leave woodlands with little shrub or herbaceous understory. Suitable habitat is lost because Greater Sage-Grouse do not use cheatgrass-dominated areas or woodlands (Connelly et al. 2000c, Crawford et al. 2004).

Composition is also changed when structures such as communication towers and power lines decrease the landscape suitability for sage-grouse (Connelly et al. 2000c, Beck et al. 2006). Johnson et al. (this volume, chapter 17) estimate broad-scale relationships between anthropogenic features and numbers of sage-grouse counted at leks. Wisdom et al. (this volume, chapter 18) identify the importance of anthropogenic features in contributing to the extirpation of sage-grouse from previously occupied regions. The human footprint also may favor increases in generalist predators (Leu et al. 2008) that prey on sage-grouse (Coates et al. 2008; Hagen, this volume, chapter 6).

Third, configuration of sagebrush within the landscape mosaic also has changed. The increased edge in landscapes fragmented by the infrastructure network also directs or increases predator movements (Tewksbury et al. 2002) and isolates wildlife populations (Saunders et al. 1991, Trombulak and Frissell 2000). Spread of exotic species along these artificial corridors (Gelbard and Belnap 2003) alters the composition of the surrounding plant community and can subsequently increase fire and eliminate sagebrush. Leu and Hanser (this volume, chapter 13) demonstrate that increasing levels of human footprint create landscapes that are unsuitable for sage-grouse because sagebrush is reduced to a few widely dispersed patches.

CONSERVATION IMPLICATIONS

This volume is focused on Greater Sage-Grouse and sagebrush, but more than 350 other species depend on sagebrush for all or part of their

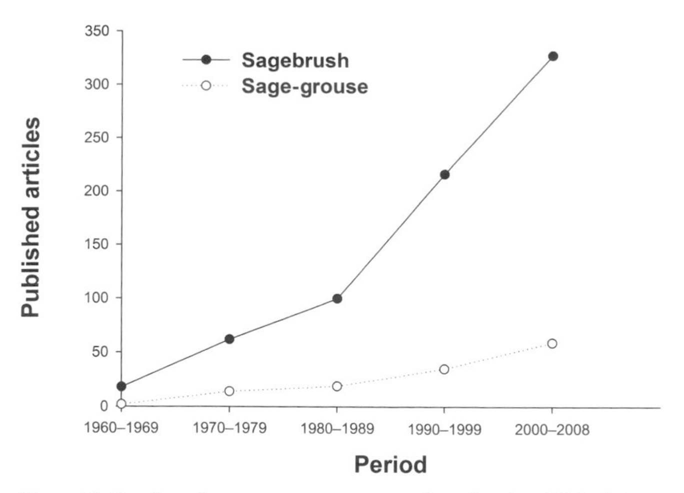

Figure I.5. Number of papers on sage-grouse and sagebrush published since 1960 in 33 refereed journals. Article titles and topic words were searched for "sage-grouse," "sage grouse," and "sagebrush."

existence (Suring et al. 2005a). A high proportion of the endemic and imperiled species in the western United States occur within sagebrush-dominated habitats. The Great Basin ecoregion contains 207 species of conservation concern (Suring et al. 2005a) and the second highest number of imperiled endemic species in the United States (Chaplin et al. 2000). Pygmy rabbits (*Brachylagus idahoensis*) within the Columbia Basin and Gunnison Sage-Grouse (*Centrocercus minimus*) within the Colorado Plateau either are or have been candidate species for federal listing under the Endangered Species Act (United States Department of the Interior 2000b, 2003a, 2005a, 2006a). The current range occupied by Gunnison Sage-Grouse has been reduced to 5,000 km^2 from its estimated pre-settlement distribution of 45,000 km^2, primarily because of habitat loss and alteration (Oyler-McCance et al. 2001, Schroeder et al. 2004). The different species of sagebrush in the intermountain region collectively make up the third most-threatened bird habitat in the United States (American Bird Conservancy 2007). Hanser and Knick (this volume, chapter 19) assess the efficacy of management focused on Greater Sage-Grouse to benefit other species of passerine birds that use sagebrush landscapes.

We are encouraged that the volume of research and number of papers published on sage-grouse and sagebrush have increased greatly in recent years (Fig. I.5). Those papers, many of which are summarized in this volume, have significantly advanced our understanding of the ecology of sagebrush-dominated landscapes and species that depend upon them. In addition to sage-grouse and sagebrush, a broader understanding of other aspects of this ecosystem are receiving interest and funding. This volume contains information on the role and management of invasive plant species and restoration (Pyke, this volume, chapter 23), wild equids (Beever and Aldridge, this volume, chapter 14), and disease (Christiansen and Tate, this volume, chapter 8; Walker and Naugle, this volume, chapter 9).

The first comprehensive evaluation of sage-grouse and sagebrush habitats was published in 1952 (Patterson 1952) at a time when the primary management objective was to eradicate as much sagebrush as possible. Public perception has progressed beyond the prediction that the much-maligned sagebrush will be regarded with increasing favor by land managers (McArthur and Plummer 1978:241) to concern about management of these systems (Braun et al. 1976, Dobkin 1995, Knick 1999, Knick et al. 2003). Recent legal actions that draw attention to Greater Sage-Grouse also emphasize the importance of the ecological health of their sagebrush habitats (United States Department of the Interior 2005b). Consequently, management actions often are designed to restore ecosystem integrity, emphasize native seed and vegetation, or benefit wildlife rather than directed solely at traditional commodity development. Strategies are being developed to focus limited conservation resources in regions that will have greatest benefit to sage-grouse

(Wisdom et al. 2005b,c; Meinke et al. 2009). Regional, state, and local working groups also are planning and implementing actions intended to benefit sage-grouse by protecting and improving ecosystems (Stiver et al. 2006).

Much of the information presented here and in the recent literature paint a bleak picture for the future of Greater Sage-Grouse. We hope that through better understanding, increased appreciation, and effective conservation, we might ensure the trajectory of long-term declines of Greater Sage-Grouse populations is not their destiny. As such, the Greater Sage-Grouse truly serves as an icon of the West.

ACKNOWLEDGMENTS

This introduction was improved by reviews kindly provided by D. S. Dobkin, D. E. Naugle, and J. T. Rotenberry. This volume represents the collective effort and commitment of many individuals. Support and encouragement from Carol Schuler, of the United States Geological Survey; Cal Groen, Virgil Moore, and Tom Hemker, of the Idaho Department of Fish and Game; J. David Brittell, of the Washington Department of Fish and Wildlife; and Roy Elicker, of the Oregon Department of Fish and Wildlife through the many phases of this project, along with their advocacy for science, have been instrumental in maintaining the quality of the information. We thank the authors who generously contributed their knowledge to benefit sage-grouse and sagebrush systems. The reviewers' expertise and ability to question and make significant comments improved each chapter. Clait E. Braun's keen editorial eye for detail and attention to facts provided an important check on the original chapters. Most important, we thank our families, who tolerated our absence even when we were present and, on rare occasions, our apparent lack of good humor. We sincerely thank all.

PART ONE

Management and Conservation Status

CHAPTER ONE

Historical Development, Principal Federal Legislation, and Current Management of Sagebrush Habitats

IMPLICATIONS FOR CONSERVATION

Steven T. Knick

Abstract. The historical disposition and development of sagebrush (*Artemisia* spp.) landscapes have resulted in land ownership mosaics and differences in environmental qualities among land managers that influence today's conservation planning. Early land-use policies following major land acquisitions from 1776 to 1867 in the western United States were designed to transfer the vast public resources to private ownership. Federal legislation enacted during the late 1800s and early 1900s encouraged development of arable regions, facilitated livestock grazing, created transportation corridors, and provided for access to minerals, coal, and petroleum. Productive lands characterized by deeper soils and access to water were transferred to private entities and converted from native habitats to agriculture. Privately owned lands are a major constituent of sagebrush landscapes in the Great Plains and Columbia Basin and are intermixed with public lands in other Sage-Grouse (*Centrocercus* spp.) Management Zones. The public still retains large areas and 70% of current sagebrush habitats. The Bureau of Land Management has responsibility for almost half of the sagebrush habitat in the United States; however, those lands are relatively unproductive and characterized by xeric environments and shallow soils. More recent legislation reflects changing public values to maintain or restore natural components, such as plants and wildlife, and minimize the impact of land uses in sagebrush landscapes. Multiple use dominates the management policy of most sagebrush habitat on public land; very little of the lands used by Greater Sage-Grouse (*Centrocercus urophasianus*) has protected status in national parks or reserves. Conserving sagebrush landscapes required by Greater Sage-Grouse and other wildlife will depend on engaging the mosaic of public agencies and private ownerships in management programs, understanding the broad diversity of habitat characteristics, and recognizing the limitations of environments supporting the majority of sagebrush habitat on public lands.

Key Words: federal government agencies, legislation, public land management, sagebrush.

Desarrollo Histórico, Legislación Federal Principal, y Manejo Actual de Hábitats de Artemisa: Implicaciones para la Conservación

Resumen. La disposición histórica y el desarrollo de territorios de sagebrush (*Artemisia* spp.) han resultado en mosaicos de propietarios y diferencias en calidades ambientales entre los administradores de tierras que influyen en los planes de conservación actuales. Las primeras políticas de uso de

Knick, S. T. 2011. Historical development, principal federal legislation, and current management of sagebrush habitats: implications for conservation. Pp. 13–31 *in* S. T. Knick and J. W. Connelly (editors). Greater Sage-Grouse: ecology and conservation of a landscape species and its habitats. Studies in Avian Biology (vol. 38), University of California Press, Berkeley, CA.

tierras que siguieron a grandes adquisiciones de tierras en 1776–1876 en el oeste de EE.UU., fueron diseñadas para transferir los vastos recursos públicos hacia una propiedad privada. La legislación federal redactada durante finales de 1800 y comienzos de 1900 estimuló el desarrollo de las regiones arables, facilitó el pastoreo de ganado, creó corredores de transporte, y subvencionó el acceso a minerales, carbón y petróleo. Las tierras productivas caracterizadas por suelos más profundos y acceso al agua fueron transferidas a las entidades privadas y transformadas de hábitats nativos a la agricultura. Tierras de propiedad privada son el principal constituyente de los territorios de artemisa en el Great Plains y en el Columbia Basin, y están mezcladas con terrenos públicos en otras zonas de manejo de sage-grouse (*Centrocercus* spp.). Siguen siendo de manejo público grandes áreas y el 70% de las áreas actuales de artemisa. El Bureau of Land Management es responsable por casi la mitad de los territorios de artemisa en los EE.UU., sin embargo, esas tierras son relativamente improductivas y caracterizadas por un medioambiente xérico (adaptado a las condiciones desérticas) y suelos poco profundos. La legislación más reciente refleja valores públicos cambiantes para mantener o restaurar los componentes naturales, tales como plantas y vida silvestre, y minimizar el impacto de uso de la tierra sobre el paisaje de artemisa. Un uso variado domina la política de manejo de la mayoría de los hábitat de artemisa en las tierras públicas; muy poco de la tierra empleada por el Greater Sage-Grouse (*Centrocercus urophasianus*) tiene el estado de especie protegida en los parques o reservas nacionales. La conservación de los paisajes de artemisa requeridos por el Greater Sage-Grouse y el resto de la vida silvestre dependerá de involucrar el mosaico de agencias públicas y propietarios privados en programas de manejo, que entiendan la amplia diversidad de características del hábitat, y que reconozcan las limitaciones que tienen los ambientes que sostienen la mayoría de los hábitats de artemisa en las tierras públicas.

Palabras Clave: agencias federales de gobierno, artemisa, gestión de tierras públicas, legislación

The mosaic of land ownerships and land uses in the western United States presents challenges for conserving large sagebrush (*Artemisia* spp.) landscapes required by Greater Sage-Grouse (*Centrocercus urophasianus*). Today's mix of ownership and administrative responsibilities is a direct result of the historical sequence of public land law policies that have guided disposition and use of sagebrush habitats since Euro-American settlement. These policies also left a landscape in which environmental characteristics and management objectives differ among ownership and management agencies, thus affecting conservation options and restoration potential.

Early settlers described an endless sea of sagebrush as they traveled westward (Frémont 1845, Young and Sparks 2002). The vast open landscapes acquired through purchases and cessions appeared to have unlimited forage for grazing livestock, minerals for mining, soils that could be converted to agriculture, and petroleum for an expanding nation's defense and infrastructure (National Research Council 1989, Flores 2001, Dombeck et al. 2003). Early policies by the federal government encouraged privatization, development, and use of these extensive lands (Bean and Rowland 1997). Resource availability (especially water for irrigation) dictated which lands were transferred to private entities, which were claimed for grazing or minerals, and which remained in the public trust (Talbert et al. 2007). But resources were not unlimited, and land-use policy increasingly recognized management and conservation beginning in the early 1900s (Clawson and Held 1957, Poling 1991). Most recently, greater demand for resources coupled with human population growth, urban development, and increasing intensity of land use are decreasing the amount of land available throughout much of the western United States for both consumptive and nonconsumptive uses (Holechek et al. 2006).

I first review the important federal legislation that has governed sagebrush habitats in the western United States through three primary periods: (1) initial acquisition, (2) disposition and transfer from public to private enterprise, and (3) management and conservation (Clawson and Held 1957, Poling 1991). These statutes guiding agricultural development, livestock grazing, and use of mineral and petroleum resources form the legal and policy foundation underlying current distribution of ownership and management of sagebrush habitats. I then describe the environmental characteristics of sagebrush habitats within the Sage-Grouse Conservation Area (SGCA), the region containing the pre–Euro-American settlement distribution of sage-grouse (Connelly et al. 2004) relative to public and private ownership. These characteristics, a direct result of the legislative disposition of sagebrush lands, are significant components affecting conservation strategies. I focus primarily on historical use of sagebrush habitats in the United States, but have added Canadian legislation where parallel developments are relevant.

PUBLIC LAND DEVELOPMENT IN THE UNITED STATES

Land Acquisition

Sagebrush lands were contained in the territories acquired by the United States in the Louisiana Purchase (1803) and the Oregon Compromise (1846) and ceded by Mexico (1848). Consequently, the federal government became responsible for large amounts of land and resources within a short period. In addition to western lands, the Alaska Purchase in 1867 added 1,500,000 km^2 to the federal land base. Acquisitions outpaced dispositions until 1850; the total net area in the public domain has subsequently declined following this peak (Clawson and Held 1957). Lands not specifically titled to individuals or corporations remained in public ownership as part of conditions for statehood (Dombeck et al. 2003).

Land and Resource Disposition

Disposition of land and resources to the private sector became a major priority of the federal government as an encouragement for settlement and as a means to fuel national economic growth (Clawson and Held 1957). Thus, initial legislation guiding use of public lands was directed toward granting lands to states for schools, developing irrigated agriculture, providing forage for livestock, creating or improving transportation, and mining for coal and mineral resources (Table 1.1).

Agricultural Development and Land Transfer

A succession of homestead acts beginning in the 1800s initiated a series of legislative actions by the federal government to transfer resources contained on public lands to the private sector for development and agriculture. The original Homestead Act, signed by President Lincoln in 1862, allowed a person older than 21 years of age to obtain title to 160 acres (65 ha) of undeveloped land, provided the individual built a home on it and developed the land for 5 years. Sixty-five hectares of land was too small for viable agriculture business in much of the arid regions of the western United States, and subsequent homesteading acts were passed to increase the acreage of lands that could be settled and to encourage dry-land farming (Table 1.1). Approximately 1,200,000 km^2 of public lands were transferred to private ownership under the homestead acts (Ross 1984).

The Desert Land Act (1877) allowed a settler to obtain 640 acres (259 ha) at $1.25/acre, provided the lands would be irrigated within 3 years. The Carey Act (1894) transferred an additional 404,700 ha of federal lands to the states (Colorado, Nevada, and Wyoming each were awarded a total of 809,371 ha, and Idaho received a total of 1,214,057 ha), on the condition that the lands were irrigated for agriculture. States then could pass 65-ha tracts to private ownership.

The Dominion Lands Act (1872) promoted a similar pattern of development in the prairies in western Canada. A male farmer who was a born or naturalized British subject could purchase 65 ha of land for $10 (female farmers could buy the same amount of land for up to $5,000), provided he agreed to cultivate at least 16.25 ha and build a permanent dwelling within 3 years. Unlike the American Homestead Act, a Canadian farmer could purchase the adjacent lot for the same price, thus doubling the size of his landholding. The Dominion Lands Act also established cadastral surveys similar to the United States Land Ordinance (1785) that gridded the land into townships consisting of 36 sections.

Railroads

Large amounts of land were also granted to private companies to build railroads connecting eastern and western states. The Pacific Railway Act of 1862 facilitated building a railroad and telegraph line connecting the Pacific Coast and Missouri. The Union Pacific Railroad and the Central Pacific Railroad were granted 25.9 km^2 under the 1862 act to be distributed in alternate sections on each side for every 1.6 km of completed track (Table 1.1). A subsequent amendment in 1864 increased the land area given to railroad companies to 51.8 km^2 for each mile of track completed (note the private railway corridor through predominantly public

TABLE 1.1

Principal federal legislation governing the management and use of public lands containing sagebrush in the United States.

Size of areas are given in acres to correspond with units of the original cadastral surveys (1 acre = 0.4047 ha). Laws are grouped in the United States Code (USC) under Titles 16 (Conservation), 30 (Mineral Lands and Mining) and 43 (Public Lands).

Year	Legislative act	Citation	Land use
1785	Land Ordinance Act	Continental Congress XXVIII, 375–381	Established the grid system of townships and sections for the Public Land Survey System that facilitated land sales. Section 16 of each township was to be given the states for public schools.
1862	Homestead Act	37th Congress, Chapter 75, 12, Stat. 392	Permitted entry on 160 acres (65 ha) provided the settler built a home, lived on the land, made improvements, and farmed it for 5 years.
1862	Pacific Railway Act	U.S. Statutes at Large, 12, 489 ff.	Granted the right of way through public lands for constructing a railroad connecting the Pacific Ocean and the Missouri River. The Union Pacific and Central Pacific railroad companies were to receive 10 mi^2 (26 km^2) of land distributed on each side for every 1 mi (1.6 km) of completed track.
1864	Pacific Railway Act	U.S. Statutes at Large, 13, 356 ff.	Increased the lands distributed to the railroad companies to total 20 mi^2 (52 km^2) for each mile of track completed.
1872	General Mining Act	30 USC 21–54	Declared that all valuable mineral deposits on lands belonging to the United States were free and open for purchase. Anyone could stake a claim at no cost. The holder of a claim has exclusive rights to mine and remove minerals. No royalty is paid to the U.S. on minerals. A mining claim conveys title to surface rights as well as minerals. Mineral lands containing oil, gas, coal, phosphate, potash, nitrate, oil shale, and asphaltic mineral can be homesteaded or sold but the federal government retains mineral rights.
1877	Desert Land Act	43 USC 321–339	Permitted entry on 640 acres (259 ha) at $1.25/acre provided the lands would be irrigated within 3 years.
1885	Unlawful Inclosures of Public Lands Act	43 USC 1061–1066	Made private fencing of public ranges illegal.
1891	Forest Reserve Act	26 Stat. 1103	Gave the president authority to withdraw public lands covered by timber as reservations. Established the first national forests.
1894	Carey Act	43 USC 641	Provided for 404,685 ha of federal lands in each state to be transferred to the state or to the actual settlers (limited to 65 ha each) provided that the lands could be irrigated. Colorado, Nevada, and Wyoming were each awarded an additional 404,685 ha and Idaho an additional 809,372 ha under subsequent amendments.

TABLE 1.1 *(continued)*

Year	Legislative act	Citation	Land use
1897	USDA Forest Service Organic Administration Act	16 USC 475	Restricted the president's authority to withdraw public lands for forest reserves. Identified the primary purposes for forest reserves as forest protection, secure water flows, and supplying timber. Established grazing management and gave the U.S. Department of the Interior's General Land Office the right to regulate occupancy and use on forest reserves.
1897	Oil Placer Act	29 Stat. 526	Extended the rights under the General Mining Act to include exploration and extraction of petroleum reserves.
1905	Transfer Act	33 Stat. 628	Transferred the national forest reserves from the Department of the Interior's General Land Office to the Department of Agriculture. Created the U.S. Forest Service from the Division of Forestry.
1906	Act for the Preservation of American Antiquities	16 USC 431–433	Permitted the president of the United States to restrict use on public lands owned by the federal government having historical significance without congressional approval.
1909	Enlarged Homestead Act	43 USC 218–221	Permitted entry to 320 acres (130 ha) for dry-land farming. Homesteaders could live within 20 miles (32 km) of the land if no suitable water supplies were available.
1912	Three-Year Homestead Act	37 Stat. 123–125	Reduced the occupancy period from 5 to 3 years.
1916	Stock Raising Homestead Act	43 USC 299	Permitted entry to 640 acres (259 ha) that had been designated for grazing. Federal government retained subsurface rights to minerals and coal.
1920	Mineral Leasing Act	41 Stat. 437	Directed management of the energy resources on federal lands to be developed by leasing exploration and development rights. Rather than transferring the public domain, the law retained public lands by the United States but allowed economic development of resources.
1929	Migratory Bird Conservation Act	16 USC 715	Established a commission to review and approve land and water area proposed by the Secretary of the Interior for purchase or rental for wildlife protection. Has been used primarily to establish waterfowl refuges.
1934	Taylor Grazing Act	43 USC 315–316	Established grazing fees and districts. Lands were classified as to their best use. Federal government recognized the need to care for the land and take into account the people who use it.
1945	Reorganization Act	28 USC 403	Section 403 of the Act merged the Grazing Service and the General Land Office to form the Bureau of Land Management within the U.S. Department of the Interior in 1946.

TABLE 1.1 (*continued*)

TABLE 1.1 (CONTINUED)

Year	Legislative act	Citation	Land use
1954	Recreation and Public Purposes Act	43 USC 869	Authorized the sale or lease of public lands to states, state agencies, other political subdivisions, or nonprofit organizations for recreational or public uses (campgrounds, parks, fairgrounds, landfills, and historic monuments).
1960	Multiple-Use Sustained Yield Act	16 USC 528–531	Directed that national forests be managed for multiple use and sustained yield. Multiple use meant the relative values of all resources should be considered, not necessarily the uses (or combination) that provide the greatest economic return or yield of production.
1960	Sikes Act	16 USC 670a–670o	Provided for cooperation by the U.S. Departments of the Interior and Defense with state agencies in planning, development, and maintenance of fish and wildlife resources on military reservations.
1964	Wilderness Act	16 USC 1131–1136	Recognized the need for protection and preservation of lands in their natural condition. A wilderness was defined as an area, generally >5000 acres (2,000 ha), of underdeveloped federal land retaining its primeval condition without permanent improvements, such as roads, or human habitation. These lands were to be protected and managed to preserve the natural character for future generations.
1964	Classification and Multiple Use Act	43 USC 1411–1418	Directed that natural resource lands under the authority of the U.S. Department of the Interior be managed under the principles of multiple use consistent with the Taylor Grazing Act.
1969	National Environmental Policy Act	42 USC 4321–4347	Federal agencies must consider the impact of their actions on the quality of the environment.
1971	Wild Free-Roaming Horses and Burros Act	26 USC 1331–1340	Stated that wild free-roaming horses and burros were a symbol of the western landscape. Gave the Secretary of the Interior the authority to control the proliferation of free-roaming horses and burros.
1972	Executive Order 11644	37 Fed. Reg. 2877	Directed federal agencies to identify areas where off-road vehicles can be used (or prohibited) and required that roads or trails be located to minimize harassment of wildlife or damage to habitat.

TABLE 1.1 *(continued)*

Year	Legislative act	Citation	Land use
1973	Endangered Species Act	16 USC 1531–1543	Provided for protection of endangered (any species in danger of extinction across all or a significant portion of its range) and threatened species (any species likely to become endangered) as well as for critical habitats. Recognized that species ranges included distinct geographic populations. Section 4 delineated the criteria and procedures for listing a species. Section 7 required federal agencies to consult the U.S. Fish and Wildlife Service to insure that any action authorized, funded or carried out by them is not likely to jeopardize the continued existence of listed species or modify their critical habitat. Section 9 prohibited any person under the jurisdiction of the United States to take (harm, harass, hunt, wound, kill, trap, capture, etc.) an endangered species.
1974	Federal Noxious Weed Act	7 USC 2801	Gave the Secretary of Agriculture the authority over entry into and movement within the United States by noxious weeds. Defined noxious weed as a living plant of foreign origin that can (in)directly injure not only agricultural interests but also the fish and wildlife resources of the United States.
1974	Sikes Act Extension	16 USC 670g–670o	Required Departments of the Interior and Agriculture to manage lands to help protect state-listed species of concern.
1975	Energy Policy and Conservation Act	42 USC 6201–6202	Developed the provisions to stabilize the energy supply through creation of the Strategic Petroleum Reserve, establish energy conservation programs and regulatory mechanisms, increase the supply of fossil fuels in the United States through price incentives and production requirements, reduce the demand for petroleum products and natural gas by making coal a more feasible alternative, assure the reliability of energy data, and conserve water by improving the water efficiency of certain plumbing products and appliances.
1976	Federal Land Policy and Management Act	43 USC 1701–1784	Public lands must be managed for multiple use and sustained yield and maintain quality of land. Directed that a portion of grazing fees should be returned for range improvements. The United States must receive fair market values for the use of public lands and resources unless otherwise provided for by statute.
1976	Executive Order 11987	Exec. Order No. 11987 2(a)	Directed federal agencies to restrict introduction of exotic species (those not naturally occurring within the United States) into lands under their administration.

TABLE 1.1 *(continued)*

TABLE 1.1 (CONTINUED)

Year	Legislative act	Citation	Land use
1977	Surface Mining and Reclamation Act	30 USC 1201–1328	Required that adverse impacts on fish, wildlife, and related environmental values be minimized. Recognized the need for reclamation of coal and other surface mining areas.
1978	Public Rangelands Improvement Act	43 USC 1901–1908	Provided for restoration of damaged rangelands, and recognized the need for a policy of inventory and monitoring. Established a formula for calculating grazing fees.
1985	Food Security Act	16 USC 3831–3836	Established the Conservation Reserve Program. Farmers enrolled in the program receive annual payments to set aside lands for at least 10 years and plant permanent cover, such as grasses, perennial forbs, and trees. The Secretary of Agriculture, through the National Resources Conservation Service, administers the program.
1990	Food, Agriculture, Conservation, and Trade Act	16 USC 3831	Expanded the lands that could be included in the Conservation Reserve Program to highly erodible or marginal croplands on which cultivation would damage water or environmental qualities.
1990	Federal Noxious Weed Act (amendment)	7 USC 2801	Extended the 1974 provisions to Department of the Interior and required each federal land management agency to fund programs in cooperation with states to control undesirable plants. Defined "undesirable" plants as to include noxious, harmful, injurious, or poisonous species.
1999	Executive Order 13112	CFR 64:6183–6186	Federal agencies were required to prevent introduction, provide for control, and minimize the impacts of invasive species. Defined invasive species as an "alien" (not native to a given ecosystem) whose introduction or presence is capable of causing harm to the economy, environment, or human health.
2000	Energy Policy and Conservation Act (reauthorization)	PL 106–469	Called for an inventory of all onshore federal lands to identify and estimate oil and gas reserves and the extent or nature of any restrictions or impediments to the development of such resources.
2005	Energy Policy Act	PL 109–058	Broad-based legislation designed to promote alternative fuel technologies, such as wind and solar energy, increase coal and biofuels production, and develop nuclear energy.
2007	Energy Security and Independence Act	PL 110–140	Provided for increased use of biofuels and required a 40% increase in fuel economy by automobiles by 2020.

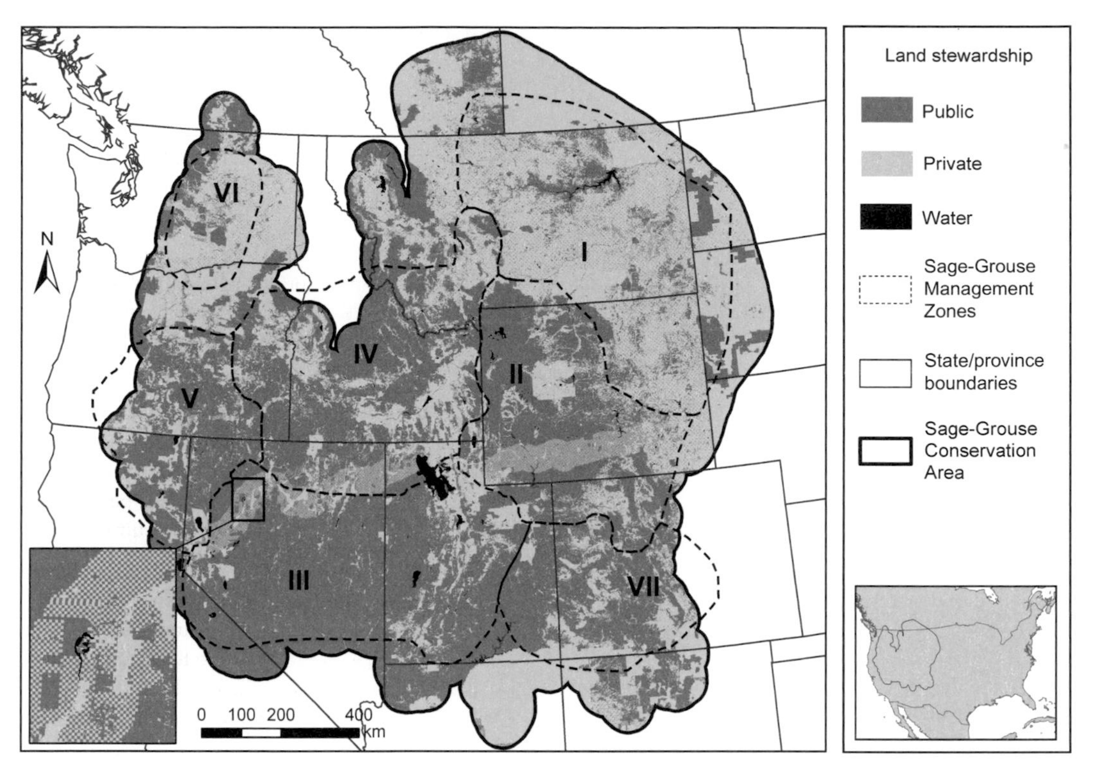

Figure 1.1. Distribution of public and private lands within the Sage-Grouse Conservation Area (Connelly et al. 2004). Land ownership information was compiled from state GAP analysis programs, the U.S. Department of the Interior (USDI) Geological Survey National Land Cover Database, USDI Bureau of Land Management, and individual state sources. Sage-Grouse Management Zones are listed in Table 1.2.

lands and checkerboard ownership in southern Wyoming, northern Utah, and Nevada; Fig. 1.1). Similarly, the Canadian Pacific Railway Company was granted 100,000 km^2 in alternating sections (1.6 km) to build a railroad across Canada. The Canadian Pacific Railway Company could replace land not suited for settlement by selecting other lands vested in the Canadian government (Innis 1923).

Livestock Grazing

The livestock industry expanded rapidly during the mid- to late 1800s, facilitated by establishment of railroads that connected open ranges with markets and transported people seeking opportunities in the western United States (Holechek 1981). Little effort was directed toward grazing restrictions or land management because the federal government was attempting to dispose of public lands at this time (Lieurance 1979, Ross 2006). Large ranching enterprises capitalized on free use of vast expanses of unfenced rangelands and forage (Box 1990, Flores 2001, Young and Sparks 2002, West 2003a).

Mineral and Energy Resources

The primary legislation governing mining and minerals has remained unchanged since passage of the General Mining Act of 1872 (Table 1.1). Discoverers had the right to extract gold, silver, and other valuable deposits under the General Mining Act by staking a claim to lands in the federal public domain. The federal government did not require royalties to be paid on minerals mined and sold from claims because miners and industry were important to settling western lands. The Oil Placer Act, passed in 1897, further extended the General Mining Act to include disposition of petroleum resources on public lands, which were declared free and open to exploration and purchase.

Management and Conservation

Legislation that guided separate development of land uses was enacted concurrently but not systematically, often creating conflicts over appropriate purposes for public lands that have resulted in today's multiple-use policies (National Research Council 1989, Bean and Rowland 1997). The prevailing attitude that public lands were best transferred to private citizens during the early period of territory acquisition and settlement was also changing by the mid-1900s, concurrent with an increasing awareness of conservation, preservation, and later environmental issues (Dombeck et al. 2003). The importance of nonconsumptive purposes began to dominate legislation beginning in the 1960s. Conflicts with traditional uses of public lands were to be resolved by land management agencies through land-use planning (Holechek 1981, Ross 2006).

Livestock Grazing

Unrestricted grazing coupled with increasing and unsustainable numbers of livestock, particularly on higher-elevation rangelands used during summer, resulted in conflicts among cattle and sheep grazers and calls for regulation (Carpenter 1981, Poling 1991, Donahue 1999). A series of legislative actions beginning in 1891 created forest reserves, regulated grazing on public lands, developed permit and fee systems, and delegated responsibility for administrating public land grazing to the U.S. Forest Service (USFS) in the Department of Agriculture and to the Grazing Service in the Department of the Interior (Donahue 1999). The Organic Act of 1897 (Table 1.1) gave the USFS the right to manage grazing on forest reserves. Subsequently, permits were established to graze a specified number of animals exclusively on tracts within the forest system.

The Taylor Grazing Act of 1934 (Table 1.1), the first statute to govern management of unreserved and unappropriated lands in the public domain, was enacted to achieve orderly use of rangelands and allow for their regeneration (Lieurance 1979, Poling 1991, Bean and Rowland 1997). The Taylor Grazing Act of 1934 terminated the open range policies that still existed on non-USFS lands and authorized the Secretary of the Interior to establish grazing districts of vacant, unappropriated, and unreserved land from any parts of the public domain, excluding Alaska, which were not national forests, parks, and monuments; Indian reservations; railroad grant lands; or revested Coos Bay Wagon Road grant lands, and which were valuable chiefly for grazing and raising forage crops (Taylor Grazing Act: 43 USC 315–316). The Secretary of the Interior was also authorized to issue permits to graze livestock upon annual payment of fees, of which a portion was returned to individual states. Public lands not reserved or withdrawn as refuges were designated as national resource lands and placed under the jurisdiction of the federal Grazing Service. The Grazing Service was merged with the General Land Office to form the Bureau of Land Management (BLM) in 1946.

Mineral and Energy Resources

The rate at which lands were being claimed by private interests within 10 years after passing the General Mining Act and Oil Placer Act threatened to remove petroleum reserves from the federal government at the same time that national dependence on oil was increasing (Costigan 1912). The Pickett Act was passed in 1910 to allow the president to withdraw any public lands of the United States from settlement or exploration and prevent loss of federal control over petroleum reserves.

The Mineral Leasing Act, passed in 1920, retained public lands in the federal trust but authorized leasing for coal, phosphates, oil, and natural gas. Thus, Congress now managed these resources on federal lands rather than transferring ownership to private enterprise (National Research Council 1989). The BLM is the principal administrator of the Mineral Leasing Act. Royalties were to be paid to the United States on gross revenues of extracted resources; half of the total went into the Reclamation Fund, and almost 40% was to be passed to individual states. Revenues derived from the Mineral Leasing Act may be an important source of revenue in some state budgets (Poling 1991).

The Energy Policy and Conservation Act of 1975 (Table 1.1) emphasized the need to stabilize the supply of energy and develop fossil fuels on federal public lands. The reauthorization of the Energy Policy and Conservation Act in 2000 directed the United States Departments of Agriculture, Energy, and the Interior to inventory all onshore oil and gas reserves and identify impediments to the development of those resources. Most of the onshore natural gas and much of the oil under federal ownership within the 48 contiguous states is contained in five geologic basins: (1) Uinta-Piceance of Colorado and Utah, (2) southwestern Wyoming (Greater Green River Basin), (3) San Juan Basin of New Mexico and Colorado,

(4) Montana Thrust Belt, and (5) Powder River Basin of Wyoming and Montana (United States Departments of the Interior, Agriculture, and Energy 2003). These geological basins span a large portion of sagebrush habitats in the eastern regions of the SGCA.

Multiple Use and Conservation

The Multiple Use and Sustained Yield Act passed in 1960 (Table 1.1) directed that national forests were to be managed for fish, wildlife, and outdoor recreation in addition to traditional uses such as timber or grazing. Furthermore, relative values of all uses or commodities were to be considered, not only those that provided the greatest dollar or unit of return. Sustained yield meant that annual production of renewable resources should be maintained in perpetuity without impairment of the productivity of the land (Multiple Use and Sustained Yield Act: Sec. 4 [16 USC 531(b)]). The Classification and Multiple Use Act of 1964 extended this policy to lands managed by the United States Department of the Interior. The Secretary of the Interior also had the authority to classify lands according to the best purpose for retention in the federal trust or disposition to private enterprise. Management now included fish, wildlife, and recreation interests, although both multiple-use acts were to be consistent with and not supersede the Organic Act (1897) governing lands managed by the USFS and the Taylor Grazing Act (1934) for lands under BLM management (Bean and Rowland 1997).

The Federal Land Policy and Management Act in 1976 (Table 1.1) directed that public lands were no longer to be disposed of but were now to be retained under control of the federal government and managed for multiple uses and sustained yield, and with fair market value received for their resources (Ross 2006). Land-use planning, involving public participation and based on an inventory of resources within a tract or area, was to be implemented to guide management. Homesteading was no longer allowed after passage of the Federal Land Policy and Management Act in 1976. Provisions in the Federal Land Policy and Management Act also halted unnecessary and undue degradation of public lands during mining activities that had been allowed under the 1872 Mining Law. The Federal Land Policy and Management Act further established that wildlife, fish, and natural scenic, scientific, and historical values be considered in long-term needs for future generations for renewable and nonrenewable resources (Federal Land Policy and Management Act: Sec. 103 [43 USC. 1702]). Half of the grazing fees were to be used for rangeland improvements. Land management agencies now could legally reduce consumptive uses, such as grazing, to meet long-term needs of wildlife and fish. The secretaries of the Interior and Agriculture were given authority to withdraw federal lands for wildlife protection, as well as for other public values. Further, the secretaries had the authority to prohibit hunting or fishing on public lands under their stewardship to protect declining populations (Bean and Rowland 1997).

The Public Rangelands Improvement Act of 1978 provided for restoration of damaged lands, established a policy of inventory and monitoring, and required periodic reporting of conditions and trends of lands to the secretaries of the Interior and Agriculture (Table 1.1). At least 80% of the appropriated funding for the act was to be used for on-ground range rehabilitation and improvements. The act also established fees tied to forage value and costs of production rather than to market value of the land that would be paid by users to the secretaries of the Interior or Agriculture.

The importance of minimizing damage from nonconsumptive and traditional uses, protecting wilderness or areas of significant historical or natural value, and ensuring survival of plant and wildlife was legally mandated in three principal conservation measures. The Wilderness Act (1964) explicitly recognized the need to protect areas from permanent human encroachment and use, and the need to preserve those places for future generations. The National Environmental Policy Act (1969) required an assessment of potential impacts of any activity or land use that could affect the environment prior to approval. The Endangered Species Act (1973) required an assessment, review, or consultation on the potential for management actions to adversely impact species, their habitats, or the quality of their environments; the intent of the Endangered Species Act is to prevent a species' extinction without a mandated balancing of other goals.

Species and their habitats in Canada are protected primarily under the Canada Wildlife Act (1972) and the Species at Risk Act (2002) as part of a broader program to protect biodiversity. The Canada Wildlife Act (Statutes of Canada 1985, c. W-9, s. 1) allows wildlife areas to be created or managed to protect or conserve wildlife, particularly species at risk. The Species at Risk Act (Statutes of Canada 2002, c. 29) established an independent committee of experts to assess and identify species at risk, provided for action plans to identify specific conservation actions to benefit endangered and threatened species, and provided for protection of listed species and their habitats (Canada Gazette 2003).

The Sikes Act (1960) and its extension (1974) provided the foundation for cooperative effort between federal land management agencies and state wildlife agencies (Table 1.1). Under the Sikes Acts, federal agencies, including BLM, could manage habitat to protect wildlife species, such as Greater Sage-Grouse, listed by states as sensitive or of conservation concern.

The increased recreational use of public lands has resulted in efforts to control its effects, particularly from off-road vehicles (Wilson 2008). Cooperative agreements established between federal and state agencies to protect fish and wildlife resources under the Sikes Act also required control of off-road vehicle traffic (Sikes Act: Sec. 670h.3.e). Executive Order 11644 (Nixon 1972) directed federal agencies to identify areas where off-road vehicles could be used or prohibited, and required that areas and trails be located to minimize the effects on wildlife or their habitat.

Fifteen national monuments administered within BLM's National Landscape Conservation System were established by President Clinton beginning in 1996 under the Antiquities Act (Dombeck et al. 2003). The Antiquities Act of 1906 (Table 1.1) gave the president authority to restrict use on public lands having historical, archeological, and cultural significance without congressional approval. Traditional uses, including livestock grazing, were allowed to continue on these national monuments (Ross 2006). Congress can set aside other lands designated as wilderness areas under the Wilderness Act (1964) to restrict commercial uses, motorized and mechanical transport, and permanent roads and structures in order to preserve their wilderness character.

CHARACTERISTICS OF SAGEBRUSH HABITATS BY MANAGEMENT AND OWNERSHIP

Sagebrush land cover within the SGCA covers 486,770 km^2 and is distributed across 13 states and three provinces. Almost all of the current distribution of Greater Sage-Grouse and sagebrush is within the western United States; <2% extends into Canada. Greater Sage-Grouse currently occupy about half of their historical distribution range-wide (Schroeder et al. 2004) and 10% of their historical range in Canada (Aldridge and Brigham 2003).

Seven management zones have been delineated for sage-grouse (Fig. 1.1) (Stiver et al. 2006) based on similarities in sagebrush environments within zones (Miller et al., this volume, chapter 10). The largest total area of sagebrush is within the Wyoming Basin and Snake River Plain (Table 1.2); these two management zones plus the Northern Great Basin also have the largest proportion of their total area dominated by sagebrush habitat.

Approximately 30% (150,186 km^2) of all sagebrush habitat in the SGCA is privately owned (Table 1.3). The greatest proportion of privately owned sagebrush habitat is in the Great Plains and Columbia Basin management zones (Fig. 1.1). Privately owned lands were consistently characterized by deeper soils and greater available water capacity—the ability of soils to store water that is available to plants—within soils in each management zone in comparison to public lands (Table 1.4). Private landholdings generally are located in valley bottoms, which have greater access to water (Fig. 1.2).

Federal agencies in the United States are responsible for almost two-thirds of the sagebrush landscape (Table 1.3, Fig. 1.3). The BLM is the principal public management agency and is responsible for 51% of sagebrush habitat in the United States (Table 1.3). Lands managed by the BLM contain higher amounts of sagebrush within the landscape (represented by the proportion of sagebrush within 5- and 18-km radii) compared to private lands. However, these lands are characterized by shallow soils, lower soil water capacity, and lower annual precipitation within each Sage-Grouse Management Zone (Table 1.4).

The USFS manages 8% of the sagebrush habitats in the United States, most of which are

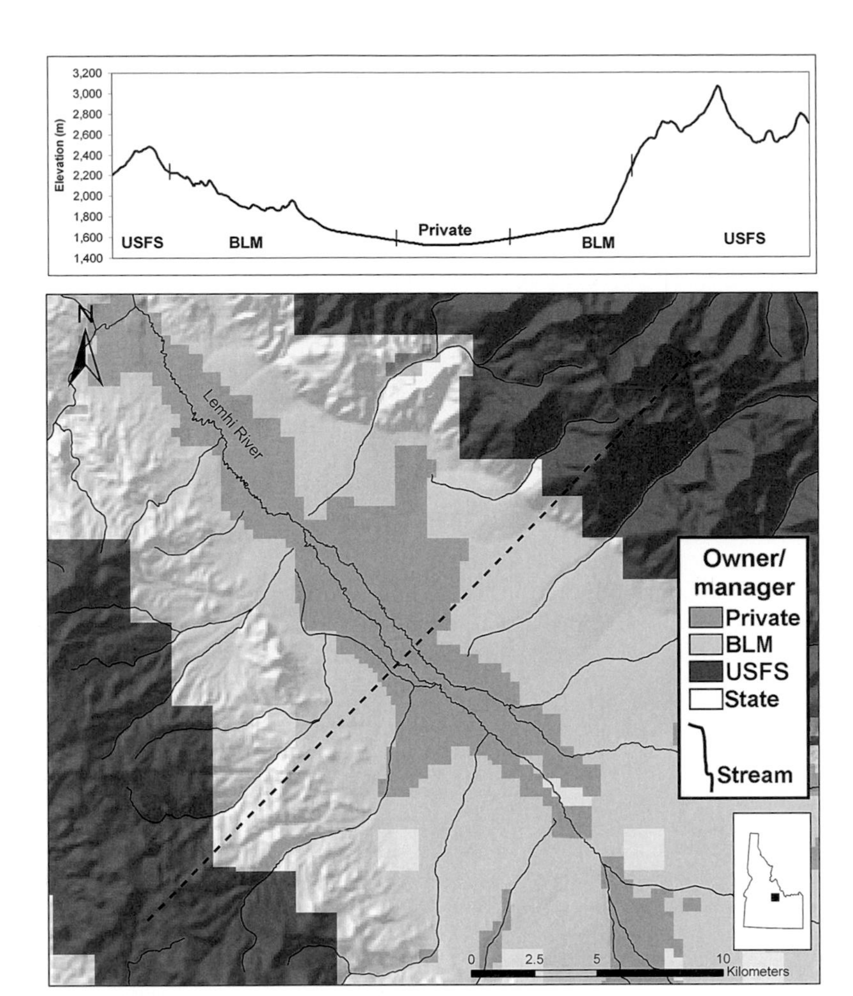

Figure 1.2. Elevational gradient (top) among land ownerships (bottom) in southeastern Idaho.

TABLE 1.2

Area statistics (km²) for Sage-Grouse Management Zones.

Sage-Grouse Management Zone	Area	Sage-Grouse Conservation Area		
		Area	Sagebrush[a]	(%)
Great Plains SMZ I	352,343	352,257	50,927	(14)
Wyoming Basin SMZ II	241,986	241,943	109,013	(45)
Southern Great Basin SMZ III	319,221	319,189	93,788	(29)
Snake River Plain SMZ IV	316,833	302,389	136,574	(43)
Northern Great Basin SMZ V	159,918	144,654	65,593	(41)
Columbia Basin SMZ VI	64,341	64,337	13,271	(21)
Colorado Plateau SMZ VII[b]	157,927	143,535	17,604	(11)
Totals	1,612,569	1,568,304	486,770	(100)

[a] Sagebrush communities include: intermountain basins big sagebrush shrubland, intermountain basins big sagebrush steppe, *Artemisia tridentata* ssp. *wyomingensis* shrub herbaceous alliance, Colorado Plateau mixed low sagebrush shrubland, Columbia Plateau scabland shrubland, Wyoming basin low sagebrush shrubland, Great Basin xeric mixed sagebrush shrubland, Columbia Plateau low sagebrush steppe, Columbia Plateau low sagebrush steppe, inter-mountain basins montane sagebrush steppe, *Artemisia tridentata* ssp. *vaseyana* shrubland alliance.

[b] The Colorado Plateau Management Zone includes Greater Sage-Grouse and Gunnison Sage-Grouse (*Centrocercus minimus*).

TABLE 1.3

Sagebrush area (km^2 and %) by management authority and ownership of sagebrush lands within Sage-Grouse Management Zones.

Totals for sagebrush area by management zone are in Table 1.2. Specific agencies include the USDI Bureau of Land Management (BLM), USDA Forest Service (FS), USDI Bureau of Indian Affairs (BIA), USDI Fish and Wildlife Service (FWS), and USDI National Park Service (NPS).

Sage-Grouse Management Zone	Sagebrush management and ownership[a]															
	Private		BLM		FS		BIA		FWS		NPS		Other federal agencies[b]		State	
	km^2	(%)	km^2	(%)	km^2	(%)	km^2	(%)	km^2	(%)	km^2	(%)	km^2	%	km^2	(%)
Great Plains	33,365	(66)	8,682	(17)	1,037	(2)	1,979	(4)	916	(2)	22	(0)	601	(0)	3,662	(4)
Wyoming Basin	37,962	(35)	53,384	(49)	4,315	(4)	4,394	(4)	201	(0)	1,265	(1)	0	(0)	7,250	(1)
Southern Great Basin	12,102	(13)	66,890	(71)	9,522	(10)	931	(1)	37	(0)	134	(0)	62	(0)	2,495	(6)
Snake River Plain	39,211	(29)	71,477	(52)	14,229	(10)	1,640	(1)	114	(0)	119	(0)	7	(0)	7,390	(6)
Northern Great Basin	13,504	(21)	40,550	(62)	6,356	(10)	614	(1)	3,161	(5)	137	(0)	314	(0)	900	(8)
Columbia Basin	7,774	(59)	716	(5)	285	(2)	1,521	(11)	306	(2)	21	(0)	10	(0)	1,472	(0)
Colorado Plateau	6,268	(36)	7,427	(42)	1,078	(6)	1,535	(9)	1	(0)	137	(1)	0	(0)	1,088	(7)
Totals	150,186	(31)	249,127	(51)	36,823	(8)	12,613	(3)	4,735	(1)	1,835	(0)	491	(0)	24,257	(5)

[a] Summary statistics derived in a Geographic Information System from individual state or province coverages of land ownership and management authority.

[b] Other federal agencies included U.S. Department of Defense, U.S. Department of Energy, and USDI Bureau of Reclamation.

TABLE 1.4

Environmental characteristics of lands within the Sage-Grouse Conversation Area under private and public ownership and by management authority for the USDI Bureau of Land Management (BLM) and USDA Forest Service (USFS) within Sage-Grouse Management Zones.

Sage-Grouse Management Zone and owner/manager	Sagebrush cover (%)[a]		Annual precipitation[b]	Elevation[c]	Soil		Available water capacity	Soil depth
	5-km	18-km	cm	m	Salinity[d]	pH[d]	cm[d]	cm[d]
Great Plains								
BLM	24	22	34	1,028	3.56	6.8	13	90
Private	14	14	37	1,089	2.82	7.0	16	107
USFS	5	6	52	1,553	1.81	5.9	12	94
Wyoming Basin								
BLM	64	62	31	1,991	2.42	6.8	13	98
Private	53	52	38	2,025	1.76	6.8	15	113
USFS	9	14	79	2,716	0.34	5.2	10	100
Southern Great Basin								
BLM	36	34	27	1,772	2.91	6.7	10	108
Private	25	26	36	1,729	3.24	7.1	15	126
USFS	21	25	56	2,467	0.70	5.8	9	90
Snake River Plain								
BLM	68	65	34	1,566	1.36	6.1	10	100
Private	41	41	40	1,505	1.51	6.8	16	123
USFS	20	26	76	2,124	0.33	5.2	10	102
Northern Great Basin								
BLM	69	66	32	1,536	1.75	5.8	13	92
Private	36	36	46	1,397	1.26	6.0	16	112
USFS	18	21	72	1,626	0.16	5.2	12	112
Columbia Basin								
BLM	40	28	27	520	0.35	6.3	13	101
Private	21	22	31	507	0.44	6.6	18	120
USFS	5	7	91	1,287	0.04	5.4	11	119
Colorado Plateau								
BLM	17	16	33	1,941	1.36	5.9	10	81
Private	16	15	41	2,316	1.53	6.6	14	116
USFS	3	6	71	3,017	0.25	5.3	12	112

[a] Calculated from a moving window analysis of a Geographic Information System coverage summing percent land cover of sagebrush within 5- and 18-km radius of each grid cell (Knick et al., this volume, chapter 12).

[b] Elevation (m) was ascertained from digital elevation models.

[c] Annual precipitation (cm) was estimated from PRISM models (Daly et al. 1994).

[d] Soil properties were obtained from the STATSGO soils database (United States Department of Agriculture 1995). Available water capacity was the total depth (cm) of available water in the soil profile. Soil pH represented the maximum value for soil reaction of the surface soil layer. Salinity (millimhos per centimeter) was measured as electrical conductivity of the soil in a saturated paste.

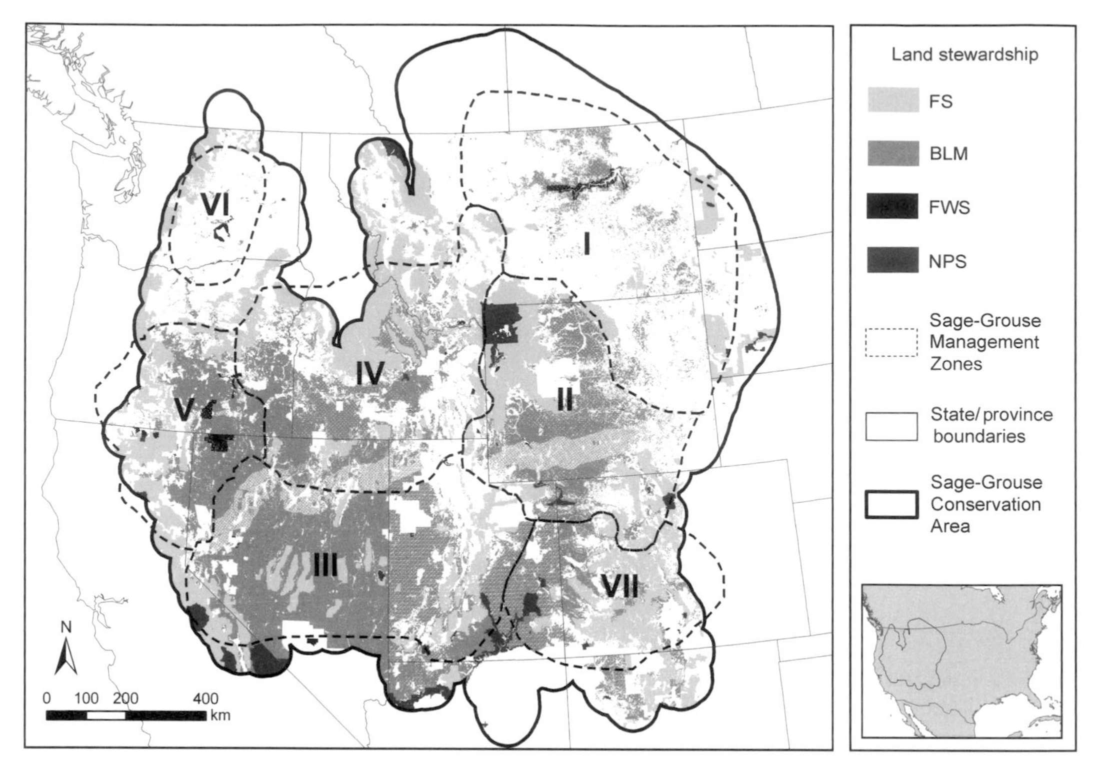

Figure 1.3. Distribution of sagebrush lands by public agency within the Sage Grouse Conservation Area in the United States and Canada. Sage-Grouse Management Zones are listed in Table 1.2.

located on boundaries of the dominant sagebrush regions (Figs. 1.2, 1.3). Lands managed by the USFS are usually at the highest elevations (Fig. 1.2) and receive the most precipitation (Table 1.4). Lands managed by the USFS also tend to be steeper and rockier than BLM or private lands.

Other federal agencies, including the United States Department of Defense, the United States Department of Energy, and the United States Department of the Interior (including the Bureau of Indian Affairs, the Fish and Wildlife Service, and the National Park Service), manage <1% of the sagebrush lands within the United States (Table 1.3). State agencies manage 5% of the total landscape dominated by sagebrush in the United States, with the greatest proportion in the Northern Great Basin Management Zone (Table 1.3).

A small portion (<1%) of the area currently occupied by Greater Sage-Grouse lies within wilderness or protected areas (Fig. 1.4). Similarly, little sagebrush habitat is legally protected from conversion of land cover (Fig. 1.5) (Stoms et al. 1998, Scott et al. 2001, Wright et al. 2001, Knick et al. 2003). Only the National Wildlife Refuge System, developed under the Migratory Bird Conservation Act (1929), was established for the purpose of wildlife conservation (Bean and Rowland 1997); little of the current sagebrush distribution is contained on refuges, which were located in areas that benefited waterfowl and other migratory birds. National parks were established primarily to preserve unique scenic or wilderness values (Clawson and Held 1957).

CONSERVATION IMPLICATIONS

Our ability to manage and conserve sagebrush habitats today has been strongly influenced by actions taken more than 100 years ago that encouraged settlement, transferred public land to private entities for agricultural development, governed livestock use, facilitated resource extraction, and shaped federal land-use policy. The system of land surveys established under the Land Ordinance of 1785 resulted in rectangular grids to map landholdings. The checkerboard mosaic of landholdings that exists today resulted from

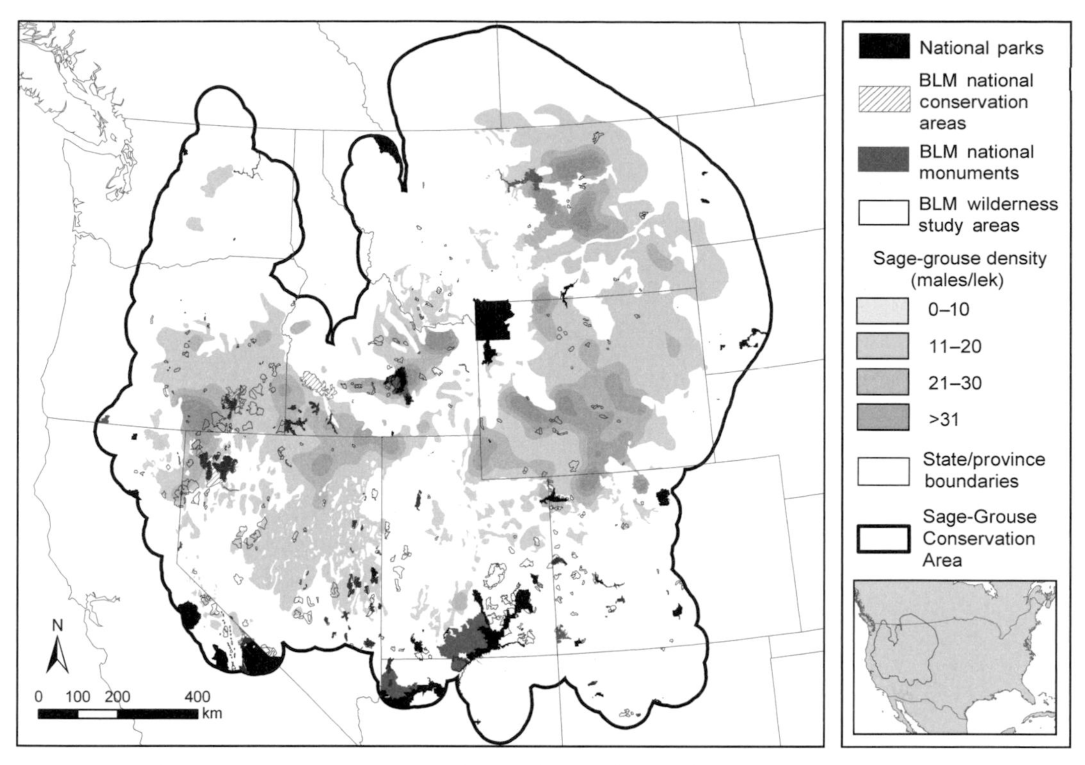

Figure 1.4. Distribution of Greater Sage-Grouse populations relative to federal wilderness and protected areas within the Sage-Grouse Conservation Area. Only sagebrush lands within United States national parks are legally protected from conversion of natural land cover and maintained in a natural state (Categories S1, S2 of Scott et al. 1993). Historical land uses, such as livestock grazing, are permitted in other wilderness or conservation areas. Contoured estimates of Greater Sage-Grouse numbers were developed by contouring counts from individual lek surveys conducted in 2003 (Fig. 13.1 of Connelly et al. 2004).

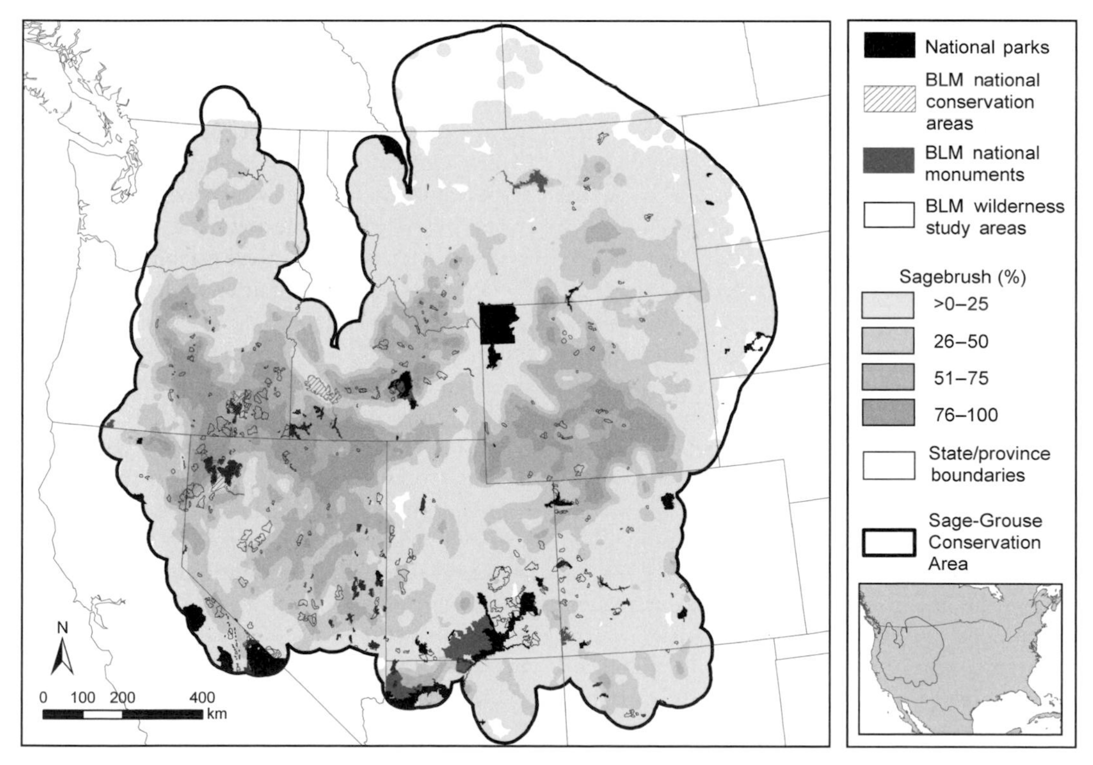

Figure 1.5. Distribution of sagebrush habitats relative to federal wilderness and protected areas within the Sage-Grouse Conservation Area. Only sagebrush lands within United States national parks are legally protected from conversion of natural land cover and maintained in a natural state (Categories S1, S2 of Scott et al. 1993). Historical land uses, such as livestock grazing, are permitted in other wilderness or conservation areas.

deeding land to private enterprises and giving incentives to states and railroad companies to develop the western United States. Ownership boundaries that may not follow geographic or topographical features can complicate management of landscape processes (Clawson and Held 1957).

Strategies among public agencies are also potential sources of conflict. Federal agencies primarily concerned with habitat management planning for multiple uses and sustained yield may pursue different goals than state agencies mandated with managing wildlife. Thus, current land-use planning will need to include multiple partners and objectives to manage the large landscapes used by Greater Sage-Grouse (Forbis et al. 2006). Establishment of local working groups across much of the SGCA represents a primary effort to link public and private interests (Hemker and Braun 2001, Stiver et al. 2006).

Differences in sagebrush habitats relative to public or private ownership largely influence the current and future status of these landscapes (Talbert et al. 2007). Legislation passed to encourage settlement during the late 1800s and early 1900s resulted in highly productive regions deeded to private enterprise. Lands converted to agriculture were concentrated in regions having deep, fertile soils and water for irrigation (Scott et al. 2001). An estimated 420,000 km^2 of shrub steppe existed in Washington state prior to settlement in the 19th century (Dobler et al. 1996). Summer fallowing had started by 1879, and by 1920, 80% of southeastern Washington was under cultivation (Buss and Dziedic 1955). Only 170,000 km^2 of shrub steppe across the state was present in 1986 (Dobler et al. 1996). Agriculture has replaced 75% of the shrub steppe in deep soils in eastern Washington but only 15% in shallow soils (Vander Haegen et al. 2000). In mountainous regions, valley bottoms with access to water often were claimed by private enterprise, leaving steeper, rockier hills in federal ownership (Fig. 1.2) (Clawson and Held 1957, Leu et al. 2008). Private grazing lands in southern Wyoming, Colorado, and northern New Mexico had more productive soils, shallower slopes, and greater water availability compared to adjacent public lands managed by BLM or the USFS (Talbert et al. 2007). The human footprint, a collective measure of anthropogenic use, was greatest in high-productivity regions, defined by deep soils, high precipitation, and shallow topographic terrain (Leu et al. 2008). This disproportionate loss of more-productive regions to agriculture or from diversion of water for irrigation or other consumptive purposes carries disproportionate impact to sagebrush landscapes and their capacity to maintain themselves by leaving regions that are more sensitive to disturbance and less able to recover.

Almost two-thirds of the total sagebrush distribution in the United States still remains within the public ownership. However, lands managed by BLM, which represent half of the total distribution of sagebrush, are relatively unproductive and are characterized by shallow soils in xeric environments. Thus, because BLM has responsibility for a large percentage of sagebrush habitats, management and restoration actions on large proportions of sagebrush-dominated landscapes in the SGCA will be conducted on lands that are less resistant to disturbance, have lower resilience in recovering from an increasing amount of disturbance, and are highly susceptible to invasion by exotic plant species (Miller et al., this volume, chapter 10).

Potential challenges and solutions to long-term conservation and management actions also differ because environmental qualities of sagebrush habitats vary among land management agencies (McIver and Starr 2001, Wisdom et al. 2005b). Treatments by the USFS to control the expansion of pinyon (*Pinus* spp.) and juniper (*Juniperus* spp.) will be conducted on sites at higher elevations receiving greater precipitation and having relatively short recovery periods. In contrast, management actions by BLM will occur primarily in lower elevation regions with little precipitation, long recovery periods, and threat of invasion by exotic plants. Recovery to a sagebrush-dominated landscape may require more than 100 years in these areas (Hemstrom et al. 2002). Because thresholds may have been crossed on these lands, recovery to a previous ecological state may not occur at all or may be complicated by significant potential for invasion by exotic plants (West 1996, Hemstrom et al. 2002, Meinke et al. 2009, Pyke, this volume, chapter 23).

Almost all sagebrush habitat in primary regions for Greater Sage-Grouse is undergoing use and resource development (Knick et al. 2003, Wisdom et al. 2005a, Holechek 2007, Knick et al., this volume, chapter 12). Wildlife conservation is not the exclusive or dominant objective on any major federal lands, except for the National Wildlife

Refuge System (Bean and Rowland 1997). Consequently, conservation objectives often compete with commodity production and nonconsumptive uses, such as recreation involving off-road vehicles, under the multiple-use mandate. Challenges to land uses increasingly are brought under the National Environmental Policy Act (1969) or to protect plants and animals through the Endangered Species Act (1973) (Bean and Rowland 1997, Quigley 2005). Petitions to list Greater Sage-Grouse (United States Department of the Interior 2005b), restrictions on land use, and wilderness designations across sagebrush lands have significant implications for energy, national security, grazing, and recreation interests (Wambolt et al. 2002, Holechek 2007).

Conservation of grass and shrublands is a concern beyond sagebrush habitats in the SGCA. Temperate grasslands, savannahs, and shrublands represent the least protected biomes: 5% of their total global area receives protection (Brooks et al. 2004b). Land use would need to be restricted on 50,000 km^2 of sagebrush habitats if 10% of the total geographic area, a minimum target to conserve species distributions (Svancara et al. 2005), is to be protected across the SGCA or set aside in a reserve system (Bock et al. 1993). Much larger areas, ranging from 33 to 75% of the range-wide distribution, may be necessary to conserve biodiversity and ecosystem integrity (Soulé and Sanjayan 1998). Extensive restrictions are unlikely because of the resource value of these lands for nonconsumptive and traditional uses. Thus, a large proportion of sagebrush habitat will continue to be managed for multiple purposes. Ultimately, our ability to develop long-term conservation strategies that maintain or increase populations of Greater Sage-Grouse will depend on involving a wide array of interests and perspectives in managing a broad diversity of uses for sagebrush habitats.

ACKNOWLEDGMENTS

K. A. Fesenmyer and K. M. Ross developed data sets and conducted geographic information system analyses. I benefited from many discussions with M. A. Hilliard, who wove agency, legislative, and ecological realities into direction for conserving sagebrush ecosystems. S. M. McAdam provided information on Canadian legislation. F. E. Busby, J. W. Connelly, J. C. Freemuth, M. A. Hilliard, and anonymous reviewers commented on earlier versions of the manuscript. The United States Geological Survey Forest and Rangeland Ecosystem Science Center supported writing the manuscript.

CHAPTER TWO

The Legal Status of Greater Sage-Grouse

ORGANIZATIONAL STRUCTURE OF PLANNING EFFORTS

San J. Stiver

Abstract. Range-wide conservation efforts to benefit Greater Sage-Grouse (*Centrocercus urophasianus*) began in 1954 with the Western Association of Fish and Wildlife Agencies' establishment of the Sage-Grouse Technical Committee. Contemporary conservation efforts expanded in the mid-1990s in response to increased concern about declining trends of sage-grouse populations and habitats. Seven petitions have been filed with the United States Fish and Wildlife Service (USFWS) to protect Greater Sage-Grouse under provisions of the Endangered Species Act (1973). Endangered species protection for Greater Sage-Grouse in the state of Washington was warranted but precluded. The 90-day finding determined that endangered species status was not warranted for the three petitions to protect Greater Sage-Grouse in Mono Basin in California and Nevada; the western subspecies of sage-grouse; and the eastern subspecies of sage-grouse. The remaining three petitions requesting range-wide protection for Greater Sage-Grouse were combined into one 12-month finding. The USFWS completed a finding in 2005 and determined that listing was not warranted. In 2010, the USFWS found that Greater Sage-Grouse warranted listing, but action was precluded by higher-priority listings. All western states and both Canadian provinces in the range of Greater Sage-Grouse have completed state or provincial strategic plans to manage Greater Sage-Grouse. Some conservation planning and conservation actions are being accomplished by local sage-grouse working groups. These groups are locally based, with membership composed of agency representatives and stakeholders in sagebrush (*Artemisia* spp.) ecosystems. More than 60 community-based sage-grouse conservation groups are active in the western United States and Canada. Conservation actions are planned, coordinated, funded and accomplished by a partnership of state and federal agencies, landowners, industry, nongovernmental organizations, and the public.

Key Words: Centrocercus urophasianus, community-based conservation, conservation efforts, conservation plans, Endangered Species Act, Greater Sage-Grouse, local working groups, sagebrush.

La Situación Jurídica del Greater Sage-Grouse: Estructura Organizacional de Los Esfuerzos de Planeamiento

Resumen. Los esfuerzos de conservación para beneficiar al Greater Sage-Grouse (*Centrocercus urophasianus*) en toda la extensión de su territorio comenzaron en 1954 con el establecimiento del Sage-Grouse Technical Committee (Comité técnico del Sage-Grouse) de la Western Association of Fish and Wildlife Agencies (Asociación occidental

Stiver, San J. 2011. The legal status of Greater Sage-Grouse: organizational structure of planning efforts. Pp. 33–49 *in* S. T. Knick and J. W. Connelly (editors). Greater Sage-Grouse: ecology and conservation of a landscape species and its habitats. Studies in Avian Biology (vol. 38), University of California Press, Berkeley, CA.

de las agencias de pesca y fauna). Esfuerzos contemporáneos de conservación fueron ampliados a mediados de 1990 en respuesta a la creciente preocupación por tendencias de descenso de las poblaciones y de los hábitats del sage-grouse. Se presentaron siete peticiones con el United States Fish and Wildlife Service (servicio de pesca y fauna de EE.UU.) para proteger al Greater Sage-Grouse bajo disposiciones del Endangered Species Act (ley que protege a las especies en peligro) (1973). En el estado de Washington, la protección del Greater Sage-Grouse como especie en peligro fue autorizada pero impedida. Los resultados de una investigación de 90 días determinaron que la categorización de especie en peligro no fuera autorizado para las tres peticiones para proteger al Greater Sage-Grouse en Mono Basin, California y Nevada; a la subespecie occidental del sage-grouse; y a la subespecie oriental del sage-grouse. Las tres peticiones restantes que solicitaban la protección del Greater Sage-Grouse en todo su territorio fueron combinadas en una investigación de 12 meses. El United States Fish and Wildlife Service concluyó una investigación en 2005 mediante la cual se determinó que el listado de la especie no fuera autorizado. Esta decisión fue litigada y remitida al United States Fish and Wildlife Service en diciembre de 2007 y actualmente está siendo reevaluada. Todos los estados occidentales de EE.UU. y ambas provincias canadienses dentro del territorio del Greater Sage-Grouse han completado planes estratégicos estatales o provinciales para el manejo de esta especie. Algunos planes y acciones de conservación del sage-grouse están siendo logrados por grupos de trabajo locales. La membresía de estos grupos con bases locales está compuesta por representantes de agencias federales y estatales, y tenedores de acciones en ecosistemas de sagebrush (*Artemisia* spp.). Más de 60 grupos comunitarios centrados en la conservación del sage-grouse se encuentran activos en el oeste de los EE.UU. y Canadá. Las acciones de conservación son planeadas, coordinadas, financiadas y logradas a través de la colaboración entre agencias estatales y federales, terratenientes, industrias, organizaciones no gubernamentales, y el público.

Palabras Clave: artemisa (sagebrush), *Centrocercus urophasianus*, conservación a nivel comunidad, esfuerzos de conservación, Greater Sage-Grouse (Urogallo), grupos de trabajo locales, Ley de Especies Amenazadas, planes de conservación.

Declining Greater Sage-Grouse (*Centrocercus urophasianus*) populations and habitat trends warranted concern for their long-term viability by state and provincial wildlife management, land management agencies, and conservationists by the mid-1990s (Connelly and Braun 1997, Braun 1998). Management authorities reacted to these trends by enhancing conservation efforts, adjusting hunting seasons, redirecting funding, and entering into cooperative agreements for coordinated management. Additionally, an assessment of sage-grouse population trends and habitat was prepared, as was a range-wide conservation strategy (Connelly et al. 2004, Stiver et al. 2006). Concern about sage-grouse and sagebrush declines also caused citizens, nongovernmental organizations, and industry to respond significantly by beginning or cooperating in conservation activities that benefit sage-grouse and sagebrush habitats. This chapter summarizes existing petitions filed to list Greater Sage-Grouse under the United States Endangered Species Act (ESA) (16 U.S.C 1531 et seq., wildlife laws, and ongoing conservation planning). The purpose is to document laws, regulations, policies, executive orders, judgments, findings, conservation planning and efforts, organizations, industry, and governments that frame the conservation status of Greater Sage-Grouse.

HISTORICAL BACKGROUND

Species Status

Greater Sage-Grouse historically occurred in parts of 15 states within the United States and three Canadian provinces prior to Euro-American settlement (Schroeder et al. 2004). Greater Sage-Grouse have declined throughout much of their former range and have been extirpated from 4 states and British Columbia (Connelly and Braun 1997, Schroeder et al. 1999, Schroeder et al. 2004). Greater Sage-Grouse currently occupy ~670,000 km^2, or 56% of their potential pre-settlement distribution (Schroeder et al. 2004). Existing populations of Greater Sage-Grouse increasingly are separated from adjacent populations, increasing conservation concerns (Schroeder et al. 1999).

Conservation and Management

The Western Association of Fish and Wildlife Agencies (WAFWA) formed the Western States Sage-Grouse Technical Committee in 1954 in part to develop strategies to monitor and manage sage-grouse. This committee had its first official meeting in 1959 and eventually expanded to include Columbian Sharp-tailed Grouse (*Tympanuchus phasianellus columbianus*). Contemporary sage-grouse conservation efforts began in 1995, when the Technical Committee determined, through harvest estimates and lek counts, that sage-grouse across their range had a sustained downward trend. The Technical Committee evaluated trends in number and distribution and recommended that WAFWA develop conservation measures to protect sage-grouse and sagebrush habitats. The Technical Committee also recognized the need to manage sage-grouse without regard to jurisdictional boundaries while respecting governmental authority of state and provincial management and federal land management. This approach allowed sage-grouse and their habitats to be managed according to populations and ecoregions.

WAFWA developed a memorandum of understanding (MOU) between member states and provinces regarding sage-grouse conservation in 1995. That MOU suggested that members begin cooperative partnerships with private, state, and federal land managers to address sage-grouse conservation issues at population and local levels. An update of this MOU, along with a separate MOU between WAFWA, the U.S. Bureau of Land Management (BLM), U.S. Forest Service (FS), and U.S. Fish and Wildlife Service (FWS) was executed in 1999 and 2000 (Appendix 2.1).

Petitions to List Greater Sage-Grouse under the United States Endangered Species Act

Concern about the status of sage-grouse generated seven petitions to protect Greater Sage-Grouse under provisions of the Endangered Species Act. The ESA evaluates species viability based upon five listing criteria (Section 4. (a)(1)). A species may be listed if it meets any of these:

1. The present or threatened destruction, modification, or curtailment of its habitat or range
2. Overutilization for commercial, recreational, scientific, or educational purposes
3. Disease or predation
4. The inadequacy of existing regulatory mechanisms
5. Other natural or artificial factors affecting its continued existence

Petitions to list sage-grouse have been submitted by multiple petitioners appearing in several combinations on individual petitions. Findings have been generated for all petitions or derivatives and most have been litigated or are continuing to be litigated (Table 2.1).

The first petition, dated 14 May 1999, was directed at the Washington state population of sage-grouse, asserting it was a distinct population segment (United States Department of the Interior 2000a). The USFWS made a positive 90-day finding on 24 August 2000 (United States Department of the Interior 2000a) for the Washington state petition (Table 2.1). A positive 90-day finding suggests that adequate information exists to begin a more in-depth analysis of the status of a species. This review should be completed in 270 days following the positive 90-day finding. The 12-month finding was published on 7 May 2001 (United States Department of the Interior 2001a) for the petition directed at the Washington state population. The 12-month finding was that the petition presented substantial information and that listing was warranted but precluded because of higher-priority listing actions. The USFWS classified the Columbia Basin Distinct Population Segment as a candidate species priority number 9 under USFWS policies and Section 4 of the ESA. A species listed as a candidate receives no statutory protection under ESA, but its status is reviewed annually to determine if any changes have occurred. The priority listing number is on a scale from 1 to 12 based upon the magnitude of threats faced by the species, the immediacy of the threats, and their uniqueness. The urgency to list increases with lower numbers.

Two petitions for the Mono Basin area (California–Nevada) were filed in 2001 and 2005. The USFWS rejected the 2001 petition in 2002 as not presenting substantial information indicating that listing was warranted (United States Department of the Interior 2002a). The USFWS considered both petitions in response to the 2005 petition and ongoing litigation and found the petitions did not present substantial information indicating that listing was warranted (United States Department of the Interior 2006b). This

TABLE 2.1
Sage-Grouse listing petitions, August 2008.

Petition	Date	Determination
Washington population of the western Sage-Grouse	14 May 1999	Warranted but precluded.
Mono Basin region Greater Sage-Grouse	28 Dec 2001	Petition not substantial.
Western subspecies Greater Sage-Grouse	24 Jan 2002	Petition not substantial.
Greater Sage-Grouse	8 Jun 2002	Positive 90-d finding. Not warranted. 12-mo finding. Litigated. Warranted but precluded.
Eastern subspecies Greater Sage-Grouse	3 Jul 2002	Petition not substantial.
Greater Sage-Grouse	19 Mar 2003	Positive 90-d finding. Not warranted. 12-mo finding. Litigated. Warranted but precluded.
Greater Sage-Grouse	22 Dec 2003	Positive 90-d finding. Not warranted. 12-mo finding. Litigated. Warranted but precluded.
Mono Basin region Greater Sage-Grouse	10 Nov 2005	Petition not substantial. In litigation.

decision was challenged and the USFWS began a 12-month finding in April 2008 (United States Department of the Interior 2008b).

The USFWS combined one petition for the western subspecies (*Centrocercus urophasianus phaios*), one for the eastern subspecies (*C. u. urophasianus*), and two petitions to list sage-grouse range-wide into one in 2004 (United States Department of the Interior 2000a, 2001a, 2003b, 2004c, 2004d, 2005b, 2008a). The USFWS made a positive 90-day finding on 5 April 2004 (United States Department of the Interior 2004c,d) on combined and separate petitions to list Greater Sage-Grouse range-wide. However, the USFWS determined that Greater Sage-Grouse did not warrant listing in their 12-month finding (United States Department of the Interior 2005b). The USFWS made the not-warranted finding based in part on a structured decision process using a panel of experts to evaluate information and project future scenarios. The panel projected that Greater Sage-Grouse would not become extinct in the foreseeable future (30–100 years). However, the panel did have concerns about risks faced by Greater Sage-Grouse and threats to their habitat (United States Department of the Interior 2005b). A notice of intent was filed on 7 January 2005 objecting to the not-warranted finding, although no additional action was taken. A complaint filed on 14 July 2006 alleged the 12-month finding was incorrect, arbitrary, and unwarranted. The United States District Court for Idaho issued a memorandum finding on 4 December 2007 deciding the USFWS determination was arbitrary and capricious. The United States District Court remanded the 12-month finding to the USFWS (court case CV-06-277-E-BLW). The USFWS published a finding of warranted, but precluded it because of higher-priority listing actions on 5 March 2010 (United States Department of the Interior 2010).

The remaining petitions to list Greater Sage-Grouse by subspecies or geographic area provided insufficient or unsubstantial information for a positive 90-day finding (Table 2.1). All but one petition has been litigated in federal court, with one extending to the United States Court of Appeals for the 9th Circuit.

LEGAL STATUS

Greater Sage-Grouse are classified as resident wildlife by all states and provinces. Nearly 75% of Greater Sage-Grouse habitats are on federally owned lands and subject to federal land management laws and regulations (Connelly et al. 2004; Knick, this volume, chapter 1).

Federal, State, and Provincial Laws

Greater Sage-Grouse are managed by individual states as resident native game birds in the

TABLE 2.2
State and provincial laws related to Greater Sage-Grouse.

State or province	Legal references
California	Fish and Game Codes (Title 14). Division 2, Part 1, Chapter 8, Article 1, Section 1801, Sections 1802 and 1802; Part 2, Chapter 1, Section 3500
Colorado	Colorado Revised Statutes, Title 33 Article 1-101
Idaho	Title 36 of Idaho State Code. Chapter 1, 36-102 and 36-103
Montana	Title 87 of Montana Code Annotated 2003 MCA 87-1-201 MCA 87-1-301, 87-2-101 and 87-1-102, chapter 3, (15)
Nevada	Chapter 46, NRS 501.110, NRS 501.181 and NAC 503.040, 1
North Dakota	21.1 of State Statutes, 20.1-01-03, 20.1-01-02, 16 and 20.1-04-02
Oregon	Oregon Revised Statutes (ORS) Chapter 496, 496.138, 496.012, 496.118 and 240, and 496.007
South Dakota	41-1-1(24), 41-1-2, 41-2-18, 41-3-1, 41-3-2, and 41-2-23
Utah	Title 23, Sections 23-13-3, 23-14-18, and 23-14-19
Washington	RCW 77.04.012, WAC 232-12-297(1.1) and WAC 232-12-011
Wyoming	Title 23, 23-1-302(a)(i)(B)
Alberta	Wildlife Act, ESCC
Saskatchewan	Wildlife Act, Chapter W-13.12

United States. No federal laws provide Greater Sage-Grouse legal status beyond state protections. Greater Sage-Grouse are cooperatively managed by provincial and federal governments in Canada. The Greater Sage-Grouse is afforded legal protection under Schedule 1 of the Species at Risk Act (SARA) (http://www.sararegistry.gc.ca/species/schedules_e.cfm?id=1). The purpose of SARA is to prevent extinction or extirpation of any indigenous Canadian wildlife species, subspecies, or distinct population segments. The Species at Risk Act also provides for recovery of endangered or threatened wildlife and encourages management of other species to prevent them from becoming species at risk. Greater Sage-Grouse is classified as endangered in Alberta and Saskatchewan and is protected under the provisions of SARA.

State statutes generally classify Greater Sage-Grouse as resident wildlife and game birds. These statues provide for the protection, conservation, propagation, management, and use of the species. Regulations, administrative codes, and commission regulations provide for the implementation of state statutes (Table 2.2).

Conservation Plans and Efforts

Sage-grouse conservation plans have been developed at several jurisdictional scales and among agencies. Local working groups have developed conservation plans or efforts at local scales. States have developed conservation plans to facilitate local working group efforts, state-level conservation, and integration with regional conservation efforts. The BLM has developed a national sage-grouse habitat conservation plan to guide management of sagebrush habitats on lands under their management (United States Department of the Interior 2004a). The Western Association of Fish and Wildlife Agencies, in cooperation with the USFS, BLM, USFWS, and the U.S. Geological Survey (USGS), developed a Greater Sage-Grouse conservation framework. The framework was composed of two distinct but complementary phases: Conservation Assessment of Greater Sage-Grouse and Sagebrush Habitats (Connelly et al. 2004) and the Greater Sage-Grouse Comprehensive Conservation Strategy (Stiver et al. 2006). The Assessment evaluated the status, trend, and trajectory of Greater Sage-Grouse and sagebrush habitats, while the Strategy proposed range-wide and administrative actions to conserve habitat and species in concert with existing conservation efforts.

Coordination and Standards

The distributions of many Greater Sage-Grouse populations and their habitats span one or more jurisdictional boundaries. Effective management requires coordination among multiple landowners, wildlife managers, and the public (Hemker and Braun 2001). Greater Sage-Grouse conservation efforts have been developed through the formation of partnerships between wildlife agencies, private landowners, land management agencies, and Native American tribes. Stakeholders engage in both species and habitat management as partners. These conservation efforts have occurred at the range-wide, country, state, provincial, local area, and project scales. Conservation efforts are directed at evaluating populations and their habitats, identifying risks or vulnerabilities, selecting actions to meet objectives, and implementing those actions.

The partnership formed by the 2000 interagency MOU (Appendix 2.1) recognized that conservation plans and conservation actions should meet specified standards for evaluation. The USFWS and the National Oceanic and Atmospheric Administration developed the Policy for the Evaluation of Conservation Efforts (PECE) in 2003 to provide a framework for determining the adequacy and standards of conservation efforts (United States Departments of Commerce and the Interior 2003). The fundamental requirements of PECE are that conservation efforts have a predictable likelihood of implementation and effectiveness. Many planning groups are aware of these evaluation criteria and have developed plans incorporating components of PECE. Conservation efforts must be considered by the USFWS when making a positive ESA determination for a species, subspecies, or distinct population segment.

Local Working Groups

The earliest sage-grouse local working group (LWG) was organized in Idaho in 1994 (Connelly et al. 2004). Twelve LWGs had been organized by 2000, and more than 60 were in place by 2008. Generally, LWGs have frameworks for conservation planning and project identification that are suggested by statewide plans. These plans provide guidance in identifying threats and prescriptions for addressing threats, and outline the need for monitoring the effectiveness of conservation actions and the need for adaptive management. LWGs and their progress are tracked on the local working group locator project as part of the National Biological Information Infrastructure (http://greatbasin.wr.usgs.gov/LWG/). Sixty LWGs have developed conservation plans and have implemented or begun to implement conservation efforts (http://greatbasin.wr.usgs.gov/LWG/conservationplans.asp). The USFWS is conducting a comprehensive analysis of conservation efforts and the efficacy of those efforts as part of the 2007 remanded finding.

State and Provincial Conservation Plans

All western states have developed statewide conservation plans to address conservation actions at population scales and to facilitate conservation needs at community scales. Most states meet periodically to assess the status of conservation efforts within their state and identify what should be done at that scale as well as how to help facilitate conservation efforts at the LWG scale and between states (Tirhi 1995, Utah Department of Wildlife Resources 2002, Montana Sage-Grouse Work Group 2003, Wyoming Sage-Grouse Working Group 2003, Nevada Governor's Sage-Grouse Conservation Team 2004, Stinson et al. 2004, Hagen 2005, McCarthy and Kobriger 2005, Idaho Sage-Grouse Advisory Committee 2006, Colorado Greater Sage-Grouse Steering Committee 2008, South Dakota Game, Fish and Parks 2008).

The Committee on the Status of Endangered Wildlife in Canada listed sage-grouse in Canada as threatened in 1997 and, after further review, changed the status to endangered in 1998 (Canadian Sage-Grouse Recovery Team 2001, Species at Risk Act 2002). The species was listed as threatened in Saskatchewan in 1987 and as endangered in 1999 under Saskatchewan's Species at Risk revisions of the Wildlife Act.

The provinces of Saskatchewan and Alberta jointly formed a sage-grouse recovery team outside of the Recovery of Nationally Endangered Wildlife Committee process in November 1997. The sage-grouse recovery team is composed of representatives from government (provincial and federal), land managers, landowners, conservation organizations, and industry from Saskatchewan and Alberta. The team developed the Canadian Sage-Grouse

Recovery Strategy in June 2001 (Canadian Sage-Grouse Recovery Team 2001). This strategy calls for the formation of local working groups to implement actions directed in the plan. In practice, sage-grouse conservation efforts have been undertaken by partnerships (S. M. McAdams, pers. comm.).

State Governors' Actions

Concern for Greater Sage-Grouse and the potential of significant adverse economic and societal consequences of an ESA listing prompted action from several state governors and the Western Governors' Association. Governor Dave Freudenthal of Wyoming hosted a sage-grouse summit in 2007 and formed a statewide sage-grouse implementation team. Wyoming identified 21 priority strategies, assigned agencies, and provided completion dates and funding to accomplish the tasks (http://governor.wy.gov/Media.aspx?MediaId=271). Governor Freudenthal also issued Executive Order 2008-02 on 1 August 2008 to facilitate sage-grouse conservation and updated it in 2010 with Executive Order 2010-4 (http://wyld.state.wy.us/uhtbin/cgisirsi/20101208225655/SIRSI/0/520/WS-GOV-EO-2010-04.pdf). Governor Jim Gibbons of Nevada issued an executive order that presented a state policy to preserve and protect sage-grouse habitat whenever possible. The order further directs full implementation of the 2004 Governor's Sage-Grouse Plan (http://gov.state.nv.us/EO/2008/EO-2008-09-26_SageGrouse.pdf).

The Western Governors' Association passed a resolution in June 2008 that identified a series of policies and management directives. These policies include the need for continued habitat and species assessments and plans, conservation efforts to meet PECE criteria, funding, passage of the North American Sagebrush Ecosystem Conservation Act (NASECA), support for LWGs, and signing the sage-grouse implementation MOU (http://www.westgov.org/wga/policy/08/sagegrouse8-12.pdf).

Nongovernmental and Industry Conservation

Traditional nongovernmental organizations (NGOs) and newly formed organizations have contributed and continue to contribute to sage-grouse and sagebrush conservation efforts. These groups have supported sage-grouse conservation in a variety of ways, including fee title purchase of habitat, easements, facilitation of conservation planning, funding of conservation actions, monitoring, and development of nontraditional conservation incentives. NGOs including but not limited to the National Wildlife Federation, the National Audubon Society, the North American Grouse Partnership, The Nature Conservancy, Environmental Defense Fund, and the Cooperative Sagebrush Initiative are working on sage-grouse and sagebrush conservation issues.

Industries such as mineral and coal mining, oil and gas exploration and production, renewable energy, energy transmission, ranching, and farming have a vested interest in the sagebrush steppe. These industries have supported and continue to support sage-grouse conservation within the constraints of conducting their business. Industries have provided funding and support for conservation actions, monitoring, and planning.

Federal Agency Conservation Plans

The BLM director issued a memorandum in July 2003 (Director's Office Instruction Memorandum No. 2003-003) regarding development of a BLM sage-grouse habitat conservation strategy and interim program guidance on sage-grouse habitat conservation. The memorandum directs that each national program immediately review and evaluate program policies that potentially impact or threaten long-term health of sage-grouse populations and their habitat on BLM lands. The memorandum also states that the BLM would issue interim guidance focusing on actions that can be taken immediately to minimize or eliminate threats to sage-grouse and their habitat and that do not require National Environmental Policy Act review before implementation.

The BLM developed a National Sage-Grouse Habitat Conservation Strategy (United States Department of the Interior 2004a) using an interdisciplinary team composed of senior agency staff and managers representing all affected program areas and administrative levels. The stated intent of the BLM Strategy is to provide a framework to address the conservation of sage-grouse habitats on BLM-managed land (http://www.blm.gov/wo/st/en/prog/more/fish__wildlife_and/wildlife/national_sage-grouse.html). The BLM Strategy presents four goals with 48 action items. The four goals are as follows:

1. Set forth the management framework for addressing conservation of sage-grouse on lands administered by the BLM.

2. Enhance knowledge of resource conditions and priorities to support habitat maintenance and restoration efforts.
3. Expand partnerships, available research, and information that support effective management of sage-grouse and sagebrush habitats.
4. Ensure leadership and resources are adequate to implement national and state-level sage-grouse and sagebrush habitat conservation strategies and/or plans.

The BLM also issued formal directives (U.S. Bureau of Land Management Instruction Memorandum 2004-136) related to sage-grouse habitat mapping and participation in state-led sage-grouse conservation planning, and gathering habitat data on BLM lands.

Range-wide Conservation Assessments and Strategy

The agencies' partnership formed by the 2000 MOU (Appendix 2.1) decided in 2003 that a range-wide assessment of the status of Greater Sage-Grouse and sagebrush habitats would be important to the ESA listing determination and to identifying threats that should be addressed in conservation efforts at all scales. The partnership also decided the conservation strategy should incorporate all other plans and a treatment of issues at the ecoregional and range-wide scales.

Conservation Assessment of Greater Sage-Grouse and Sagebrush Habitats

The conservation assessment of Greater Sage-Grouse and sagebrush habitats provided a baseline assessment of status and trends of components of the sagebrush steppe (Connelly et al. 2004). The report presented information important to understanding sage-grouse populations and habitat, as well as human and nonhuman influences, for an area in western North America that exceeds >2,000,000 km^2.

Greater Sage-Grouse Comprehensive Conservation Strategy

The strategy addressed conservation issues identified in the assessment and a range-wide issues forum (Stiver et al. 2006). The strategy is composed of seven substrategies that address the following: (1) conservation activities at all scales, (2) monitoring of the effectiveness of conservation actions, (3) implementation of conservation actions, (4) research and technology, (5) funding, (6) communication and outreach, and (7) adaptive management.

Significant direction included in the strategy was provided from sage-grouse local working groups; state, provincial, and federal conservation plans; and a range-wide issues forum (Stiver et al. 2006). The forum suggested that the guiding principle of sage-grouse and sagebrush conservation should be (1) protect what we have, (2) retain what we are losing, and (3) restore what has been lost. Furthermore, the forum identified three essential resources needed to facilitate the conservation effort: (1) funding; (2) leadership committed to organizing, supporting, and guiding a long-term effort; and (3) the appropriate organizational structure to sustain it. The strategy identifies short- and long-term objectives for accomplishing the elements of the strategy. Management agencies should develop a sage-grouse conservation implementation MOU. This partnership will provide guidance, organizational structure, and funding to implement the provisions of the strategy. This partnership will guide the conservation efforts in an interim period. The long-term objective is the passage of NASECA, which is envisioned to provide a range-wide perspective on sagebrush conservation with the governance, funding, and organization structure to fully implement conservation efforts in the sagebrush ecosystem (Stiver et al. 2006). The strategy provides a cost estimate of $426,000,000 for the implementation of all substrategies for the first five years.

Strategy Implementation Memorandum of Understanding

The interagency MOU executed in 2000 for the purpose of coordination and conservation planning expired in 2008 and was replaced with a conservation and management implementation MOU (Appendix 2.2). The 2008 interagency MOU expanded membership to include USGS, the Natural Resources Conservation Service, and the U.S. Department of Agriculture Farm Service Agency. The purpose of the MOU is to provide for cooperation among the various agencies in the conservation and management of sage-grouse, sagebrush habitat, and other sagebrush-dependent species. The MOU identified objectives including implementation of the conservation strategy, conservation of other sagebrush-dependent species of conservation concern, adaptive management of the ecosystem, and development of a partnership. The MOU established a technical team (the Range-wide Interagency Sage-Grouse Conservation Team) to lead operations of the MOU and a policy team (the Executive Oversight Committee) to review progress and provide guidance to conservation efforts. The MOU effective date was 18 September 2008.

CONSERVATION IMPLICATIONS

Greater Sage-Grouse conservation planning and conservation efforts are being conducted on three primary scales. Range-wide conservation issues, strategies, and actions are defined in the assessment, the strategy, and the BLM National Sage-Grouse Habitat Conservation Strategy. The assessment and strategy provide a broad view across jurisdictions; the BLM conservation strategy provides jurisdictional direction for >50% of the sage-grouse habitat managed by the BLM. Regional conservation efforts are defined in the state conservation plans. Conservation efforts appropriate for that scale are conducted by the state or province or combinations of LWGs. Fine-scale conservation efforts generally occur at the LWG level.

Few conservation efforts have been evaluated for likelihood of implementation or effectiveness. The USFWS is conducting a PECE analysis on all conservation efforts submitted for the current Greater Sage-Grouse ESA determination, and the strategy explicitly requires periodic evaluations.

The conservation risks and issues for sage-grouse and sagebrush habitats have been identified by a variety of planning processes at multiple scales. Prescriptions to address these risks and issues are generally known and available to management agencies. The interim framework for organizational structure and leadership that will allow conservation efforts across jurisdictional boundaries at multiple scales are in place. Funding and capacity to conduct limited conservation efforts at fine scales are generally lacking, even though conservation efforts are under way. Treatment prescriptions at a landscape scale are largely untested, and funding and capacity to conduct widespread treatments are lacking. The conservation challenges for Greater Sage-Grouse are significant, and efforts to abate declines in bird numbers and habitat degradation must be expanded.

ACKNOWLEDGMENTS

I would like to thank all of the members, past and present, of the Western Sage and Columbian Sharp-tailed Grouse Technical Committee for their call to action in 1995, and their tireless commitment to the conservation of sage-grouse. Directors of the Western Association of Fish and Wildlife Agencies led by W. A. Molini, Nevada Department of Wildlife, and P. J. Graham, Montana Fish, Wildlife, and Parks, provided leadership, direction, and support to begin conservation planning and conservation actions for sage-grouse. T. P. Hemker and B. D. Deeble provided thoughtful comments that greatly improved this manuscript. I would particularly like to acknowledge S. T. Knick for his encouragement, collaboration, and editing of this manuscript. I appreciate his efforts.

APPENDIX 2.1

Memorandums of Understanding, 1999 and 2000

1999 Memorandum of Understanding among Members of Western Association of Fish and Wildlife Agencies

CONSERVATION AND MANAGEMENT OF SAGE-GROUSE IN NORTH AMERICA

I. Purpose

The purpose of this memorandum of understanding (MOU) is to provide guidance for conservation and management of sage-grouse (*Centrocercus* spp.) and sagebrush (*Artemisia* spp., primarily *A. tridentata tridentata, A. t. vaseyana, A. t. wyomingensis, A. tripartita*) shrub steppe habitats upon which the species depends. Sage-grouse historically occurred in at least 15 states and three provinces. This species has become extirpated in five states (Arizona, Kansas, Nebraska, New Mexico, and Oklahoma) and one province (British Columbia). The current distribution of sage-grouse is reduced throughout the species' historic range. Reasons for the reduction in area occupied from pre-settlement periods relate to habitat loss, habitat degradation, and habitat fragmentation. The long-term trend in sage-grouse abundance is downward. The members of the Western Association of Fish and Wildlife Agencies agree that cooperative efforts are necessary to collect and analyze data on sage-grouse and their habitats so that cooperative plans may be formulated and initiated to maintain the broadest distribution and greatest abundance possible within the fiscal realities of the member agencies and cooperating partners.

II. Objectives

All member affected agencies agree that sage-grouse are an important natural component of the sagebrush shrub steppe ecosystem. As such, sage-grouse serve as an indicator of the overall health of this important habitat type in western North America. Further, the presence and abundance of sage-grouse reflects humankind's commitment to maintaining all natural components of the sagebrush shrub steppe ecosystem so that all uses of this type are sustainable over time. Specific objectives are:

1. Maintain and increase where possible the present distribution of sage-grouse.
2. Maintain and increase where possible the present abundance of sage-grouse.
3. Develop strategies using cooperative partnerships to maintain and enhance the specific habitats used by sage-grouse throughout their annual cycle.
4. Conduct management experiments on a sufficient scale to demonstrate that management of habitats can stabilize and enhance sage-grouse distribution and abundance.
5. Collect and analyze population and habitat data throughout the range of sage-grouse for use in preparation of conservation plans.

III. Actions

It is the intent of the members of the Western Association of Fish and Wildlife Agencies to sustain and enhance the distribution and abundance of sage-grouse through responsible collective management programs. These programs will include:

1. Identification of the present distribution of sage-grouse in each member state/province.
2. Collection of sage-grouse population data following standardized protocols throughout the range of the species.
3. Continuation of development of conservation plans based on the local working group concept.
4. Validation of habitat evaluation models.
5. Completion of genetic analyses across the range of sage-grouse to more effectively define and manage individual populations.
6. Development of cooperative partnerships with interested individuals and private, state, and federal land managers.
7. Support and implement the revised sage-grouse population and habitat management guidelines.

IV. Responsibilities

1. Each state/province will collect data as recommended by the Western States Sage-Grouse and Columbian Sharp-tailed Grouse Technical Committee within the constraints of their budgetary process.
2. All member states/provinces will work cooperatively to maintain and enhance sage-grouse and their habitats.

V. Approval

We, the undersigned designated officials, do hereby approve this memorandum of understanding as recommended by resolution at the summer meeting of the Western Association of Fish and Wildlife Agencies in Durango, Colorado, on 14 July 1999.

2000 Memorandum of Understanding among Western Association of Fish and Wildlife Agencies and United States Forest Service and United States Department of the Interior, United States Bureau of Land Management, and United States Fish and Wildlife Service

I. Purpose

The purpose of this memorandum of understanding (MOU) is to provide for cooperation among the participating state and federal land and wildlife management agencies in the development of a range-wide strategy for the conservation and management of sage-grouse (*Centrocercus* spp.) and their sagebrush (*Artemisia*) habitats. The sage-grouse is an obligate sagebrush habitat species that requires large tracts of sagebrush habitat for its survival. Sage-grouse historically occurred in at least 16 states and three provinces. This species has been extirpated in five states (Arizona, Kansas, Nebraska, New Mexico, and Oklahoma) and one Canadian province (British Columbia). Its current range includes portions of California, Oregon, Washington, Nevada, Idaho, Utah, Montana, Wyoming, Colorado, North Dakota and South Dakota. The long-term trend in sage-grouse abundance is downward throughout its range.

Member state agencies (state agencies) of the Western Association of Fish and Wildlife Agencies (WAFWA) have signed a memorandum of understanding (MOU) among members of the Western Association of Fish and Wildlife Agencies for the Conservation and Management of sage-grouse in North America. That MOU, signed in July of 1999, outlines the purpose, objectives, actions and responsibilities for cooperation among WAFWA states.

The United States Bureau of Land Management (BLM), the United States Forest Service (USFS) and the United States Fish and Wildlife Service (USFWS), and WAFWA (collectively, the parties) herein agree that cooperative efforts among the parties, consistent with the applicable statutory requirements, are necessary to conserve and manage the nation's sagebrush ecosystems for the benefit of sage-grouse and all other sagebrush-dependent species.

II. Objectives

The parties agree that sage-grouse are an important natural component of the sagebrush ecosystem. Sage-grouse serve as an indicator of the overall health of this important ecosystem in Western North America. Providing for the presence and abundance of sage-grouse reflects the parties' commitment to maintaining all natural components and ecological processes within sagebrush ecosystems. Specific objectives are to:

- Maintain, and increase, where possible, the present distribution of sage-grouse.
- Maintain, and increase, where possible, the present abundance of sage-grouse.
- Identify the impacts of major land uses and hunting on sage-grouse, and determine the primary causes for declines in sage-grouse populations.
- Develop a range-wide conservation framework to provide for cooperation and integration in the development of conservation plans to address conservation needs across geographic scales as appropriate.
- Develop partnerships with agencies, organizations, tribes, communities, individuals and private landowners to cooperatively accomplish the preceding objectives.

III. Actions

The States will convene working groups to develop state or local conservation plans. Working groups will comprise representatives of local, state, federal and tribal governments, as appropriate. Participation will be open to all other interested parties. Federal participation in working groups will operate in a manner consistent with the Federal Advisory Committee Act. Working groups will be convened within 60 days of the effective date of this agreement.

The parties will establish a Conservation Planning Framework Team consisting of four (4) representatives from WAFWA and one (1) representative each from BLM, USFS and USFWS. The Framework Team will develop a range-wide conservation framework and provide recommendations and guidance to the working groups concerning the contents of state and local conservation plans.

The parties will collect, analyze and distribute sage-grouse population and habitat data to the working groups for conservation planning. These data include, at a minimum: data on fire history, habitat composition and trend, known wintering and nesting habitat, and lek locations. Population data will be collected as recommended by the Western States Sage-Grouse and Columbian Sharp-tailed Grouse Technical Committee.

Each state conservation plan will provide recommendations:

- To protect and improve important sage-grouse sagebrush habitats.
- To actively manage to improve degraded sagebrush ecosystems.
- To reduce the fragmentation and isolation of sagebrush habitats.
- To address non-habitat issues, such as hunting, if such issues are identified to limit sage-grouse populations in an area.
- For desired population levels, distribution and habitat conditions.

The BLM, USFS and USFWS will provide for habitat protection, conservation and restoration, as appropriate, consistent with the National Environmental Policy Act and other applicable laws, regulations, directives and policies. In doing so, the BLM, USFS, and USFWS will consider the WAFWA Guidelines for Management of Sage-Grouse Populations and Habitats, state and local conservation plans, and other appropriate information in their respective planning processes. Parties to this agreement will work together to identify research needs and strategies and conduct joint assessments, monitoring and research.

IV. Authorities

This MOU is among the USFWS, BLM, USFS, and WAFWA under the provisions of the following laws:

Federal Land and Policy Management Act of 1976 (43 U.S.C. 1701 et seq.)

Fish and Wildlife Act of 1956 (16 U.S.C. 742 et seq.)

Fish and Wildlife Coordination Act of 1934 (16 U.S.C. 661-667)

Multiple-Use Sustained-Yield Act of 1960 (16 U.S.C. 528-531)

Forest and Rangeland Renewable Resources Research Act of 1978 (16 U.S.C. 1641-48)

National Forest Management Act of 1976 (16 U.S.C. 1600 et seq.)

Endangered Species Act of 1973 (16 U.S.C. 1531 et seq.)

National Wildlife Refuge Administration Act of 1966, as amended by the National Wildlife Refuge System Improvement Act of 1997 (16 U.S.C 668dd et seq.)

V. Approval

It is mutually agreed and understood by and between the parties that:

1. This MOU is neither a fiscal nor a funds obligation document. Nothing in this agreement may be construed to obligate federal agencies or the United States to any current or future expenditure of resources in advance of the availability of appropriations from Congress. Any endeavor involving reimbursement or contribution of funds between the parties to this MOU will be handled in accordance to applicable regulations, and procedures including those for federal government procurement and printing. Such endeavor will be outlined in separate agreements that shall be made in writing by representatives of the parties and shall be independently authorized in accordance with appropriate statutory authority. This MOU does not provide such authority.
2. This MOU in no way restricts the parties from participating in similar activities with other public or private agencies, organizations and individuals.
3. This MOU is executed as of the last date shown below and expires five years from the execution date, at which time it will be subject to review, renewal or expiration.
4. Modifications within the scope of this MOU shall be made by the issuance of a mutually executed modification prior to any changes being performed.
5. Any party to this MOU may withdraw with a 60-day written notice.
6. Any press releases with reference to this MOU, the parties, or the relationship established between the parties of this MOU, shall be reviewed and agreed upon by all of the parties.
7. In any advertising done by any of the parties, this MOU should not be referred to in a manner that states or implies that any party approves of or endorses unrelated activities of any other.
8. During the performance of the MOU the participants agree to abide by the terms of Executive Order 11246 on nondiscrimination and will not discriminate against any person because of race, age, color, religion, gender, national origin or disability.
9. No member of, or delegate to Congress, or resident commissioner, shall be admitted to any share or part of this agreement, or to any benefit that may arise from, but these provisions shall not be construed to extend to this agreement if made with a corporation for its general benefits.
10. The parties agree to implement the provisions of this MOU to the extent personnel and budgets allow. In addition, nothing in the MOU is intended to supersede any laws, regulations or directives by which the parties must legally abide.

IN WITNESS THEREOF, the parties hereto have executed this memorandum of understanding as of the last written date below.

APPENDIX 2.2

2008 Memorandum of Understanding among Western Association of Fish and Wildlife Agencies and U.S. Department of Agriculture, Forest Service and U.S. Department of the Interior, Bureau of Land Management and U.S. Department of the Interior, Fish and Wildlife Service and U.S. Department of the Interior, Geological Survey and U.S. Department of Agriculture, Natural Resources Conservation Service and U.S. Department of Agriculture, Farm Service Agency

I. Purpose

The purpose of this memorandum of understanding (MOU) is to provide for cooperation among the participating state and federal land, wildlife management, and science agencies in the conservation and management of Greater Sage-Grouse (*Centrocercus urophasianus*) sagebrush (*Artemisia* spp.) habitats and other sagebrush-dependent wildlife throughout the western United States and Canada.

The sagebrush biome has experienced long-term downward trends in both the abundance and distribution of sagebrush plant communities and the wildlife that depend on them. Successful long-term conservation, recovery, and restoration of these habitats and wildlife will require sustained, concerted, and well-coordinated efforts among a spectrum of landowners, land managers, resource specialists, scientists, and land users.

II. Background

In July 1999, responding to continuing range-wide declines in sage-grouse populations, member agencies of the Western Association of Fish and Wildlife Agencies (WAFWA) signed the MOU among members of the WAFWA for the conservation and management of sage-grouse in North America. The 1999 MOU outlines the purpose, objectives, actions, and responsibilities for cooperation among WAFWA members in further actions to conserve sage-grouse (Appendix 2.1).

In 2000, interagency cooperation was extended further through a MOU among the WAFWA, U.S. Bureau of Land Management (USBLM), U.S. Fish and Wildlife Service (USFWS), and U.S. Forest Service (USFS) (Appendix 2.1). The major focus of the 2000 MOU, described in Section III (Actions), was on conservation planning for sage-grouse and sagebrush habitats. Although early in 2007 some local and state conservation planning remained incomplete, the December 2006 delivery by WAFWA to USFWS of the Greater Sage-Grouse Comprehensive Conservation Strategy (comprehensive strategy) marked the need to shift emphasis from conservation planning to conservation action implementation incorporating adaptive management principles to inform and guide future management practices.

III. Objectives

The U. S. Department of the Interior—USBLM, USFWS, Geological Survey (USGS); and the U. S. Department of Agriculture—USFS, Natural Resources Conservation Service (NRCS), and Farm Service Agency (FSA); and the WAFWA; hereafter referred to collectively as the parties, herein acknowledge and agree that:

- sage-grouse are an important component of sagebrush ecosystems, and serve as an important indicator of the overall health of this important western North American biome, and
- cooperative efforts among the parties, consistent with applicable statutory and regulatory requirements, are necessary to conserve and manage North America's sagebrush biome ecosystems for the benefit of sage-grouse and all other sagebrush-dependent species and to maintain the many other values sagebrush systems provide.

Providing for the long-term presence and abundance of sage-grouse and other sagebrush-dependent species reflects the parties' commitment to understand and maintain all natural components and ecological processes and systems within the sagebrush biome. Specific objectives of this MOU are to:

- Implement the comprehensive strategy and provide for cooperation and integration in the development, implementation, and evaluation of actions, premised upon the best available science, and designed to address conservation needs across geographic scales, to maintain, enhance, and restore sagebrush habitats where possible.
- Implement conservation actions for other sagebrush-dependent species identified by the parties as being of conservation concern and provide for cooperation and integration in the development, implementation, and evaluation of actions designed to address conservation needs across geographic scales, as appropriate, to maintain and increase, where possible, their respective distribution and abundance;
- Adopt an adaptive management approach to the implementation of the conservation strategy that acknowledges that in the face of uncertainties as outcomes from management actions and other events become better understood through monitoring, evaluation of actions, incorporation of new scientific understanding, and the sharing of data and information, we produce better understanding and improve the management and conservation of the sagebrush biome, sage-grouse, and all other sagebrush-dependent species; and,
- Develop partnerships with agencies, organizations, tribes, communities, individuals, and private landowners to cooperatively accomplish the preceding objectives.

IV. Actions

Primary, but not exclusive, emphasis under this MOU will focus on conserving both Greater Sage-Grouse and Gunnison Sage-Grouse (*C. minimus*) through the implementation of range-wide, state and local conservation strategies, and/or plans for these species, including the comprehensive strategy. Management for the conservation or recovery of other sagebrush-dependent species of conservation concern shall be similarly guided by existing plans, premised upon the best available science, and approved by appropriate state, provincial, and/or federal agencies.

Sage-Grouse Working Groups

The states and provinces will continue support for working groups to develop and implement state, provincial, management zone, agency, and local conservation plans. Participation will be open to all interested parties including, but not limited to, landowners, land users, industry, other interested publics, and representatives of local, state, federal, and tribal governments, as appropriate. U.S. federal agency participation in working groups will be in a manner consistent with the Federal Advisory Committee Act.

Range-Wide Interagency Sage-Grouse Conservation Team

The parties will establish a range-wide interagency sage-grouse conservation team (RISCT or team) to be composed of the voting members of the Sage and Sharp-Tailed Grouse Technical

Committee, and one (1) technical expert each from the USBLM, FWS, USFS, USGS, FSA, and NRCS. The RISCT will provide technical expertise to the Executive Oversight Committee (EOC) in facilitating implementation of the comprehensive strategy, where consistent with applicable statutory authorities, and otherwise assisting with its implementation, evaluation, and long-term success using adaptive management principles. Internal team operational procedures will be determined by the RISCT. The RISCT will develop an initial plan of action for the implementation of the strategy to the EOC six (6) months from the effective date of the MOU and report annually to the EOC for review, redirection, and revision.

Executive Oversight Committee

The parties will establish an EOC to be composed of the director of each WAFWA member agency, or their designee, from each state and province within the range of the Greater Sage-Grouse, and one (1) management representative from each of the signatory federal agencies to this agreement, to periodically review overall progress in implementing the comprehensive strategy and conservation measures for other species of conservation concern in the sagebrush biome. Based on such review, the EOC will meet with the RISCT at least annually to provide general guidance, as needed, for continuing implementation of the comprehensive strategy and conservation measures for other species of conservation concern.

WAFWA Member Agencies

The member state and provincial agencies will, as appropriate and consistent with each state and provincial mission and authority, provide for species management, population monitoring, and evaluation consistent with adaptive management principles and guided by the best available science. Member agencies will consider the comprehensive strategy, state, provincial, and local working group plans, and the most current sage-grouse guidelines to manage sage-grouse populations. Member agencies will work collaboratively to facilitate data and information management and access, to the extent possible; provide technical, management, and scientific information in support of understanding the sagebrush biome and sage-grouse populations; and where appropriate ensure that all products resultant from this MOU reflect the best available science and have received independent, scientific peer review where appropriate and applicable.

U.S. Federal Agencies

The USBLM, USFWS, USFS, USGS, FSA, and NRCS will as appropriate and consistent with each agency's mission and authorities, provide for habitat protection, conservation, habitat monitoring, restoration, and evaluation consistent with adaptive management principles and guided by the best available science of the sagebrush biome, for sage-grouse and other sagebrush-dependent species of conservation concern, and consistent with the National Environmental Policy Act and other applicable laws, regulations, directives, and policies. In doing so, these agencies will consider the WAFWA Greater Sage-Grouse Comprehensive Conservation Strategy, existing guidelines to manage sage-grouse populations and their habitats (Connelly et al., 2000) and subsequent revisions thereof, state and local conservation plans, and other appropriate information in their respective planning and implementation processes. Parties will work collaboratively to facilitate data and information management and access, to the extent possible; provide technical, management, and scientific information in support of understanding the sagebrush biome; and where appropriate ensure that all products resultant from this MOU reflect the best available science and have received independent, scientific peer review where appropriate and applicable.

V. Authorities

This MOU is among the USBLM, USFWS, USFS, USGS, FSA, NRCS, and WAFWA under the provisions of the following laws:

Endangered Species Act of 1973 (16 U.S.C. 1531 et seq.);

Federal Advisory Committee Act of 1972 (5 U.S.C. Public Law 92-463, App);

Federal Land Policy and Management Act of 1976 (43 U.S.C. 1701 et seq.);

Fish and Wildlife Act of 1956 (16 U.S.C. 742 et seq.);

Fish and Wildlife Coordination Act of 1934 (16 U.S.C. 661-667);

Fish and Wildlife Improvement Act of 1978;

Forest and Rangeland Renewable Resources Research Act of 1978 (16 U.S.C. 1641-48);

Multiple-Use Sustained-Yield Act of 1960 (16 U.S.C. 528-531);

National Forest Management Act of 1976 (16 U.S.C. 1600 et seq.);

National Wildlife Refuge Administration Act of 1966, as amended by the National Wildlife Nonindigenous Aquatic Nuisance Prevention and Control Act, 1990;

Office of Management and Budget, Final Information Quality Bulletin for Peer Review, 2004;

Organic Act of 1879 (43 U.S.C. 31 et seq., 1879);

Refuge System Improvement Act of 1997 (16 U.S.C 668dd et seq.);

Section 1231 of the Food Security Act of 1985, as amended (16 U.S.C. 3831); and

Water Resources Development Act of 1990.

VI. Approval

It is mutually agreed and understood by and between the parties that:

1. This MOU is neither a fiscal nor a funds obligation document. Nothing in this agreement may be construed to obligate federal agencies or the United States to any current or future expenditure of resources in advance of the availability of appropriations from Congress. Any endeavor involving reimbursement or contribution of funds between the parties to this MOU will be handled in accordance with applicable regulations, and procedures including those for federal government procurement and printing. Such endeavor will be outlined in separate agreements that shall be made in writing by representatives of the parties and shall be independently authorized in accordance with appropriate statutory authority. This MOU does not provide such authority.
2. This MOU in no way restricts the parties from working together or participating in similar activities with other public or private agencies, organizations, and individuals.
3. This MOU is executed as of the date of the final signatory and expires five years from that date, at which time it will be subject to review, renewal, or expiration.
4. Modifications, including but not limited to adding new partners to the agreement, within the scope of this MOU shall be made by the issuance of a mutually executed written modification prior to any changes being performed.
5. Any party to this MOU may withdraw with a 60-day written notice. Such withdrawal shall be effective 60 days from the date such written notice is provided to the other parties.
6. Any advertising done by any of the parties with respect to this MOU or any related activities shall be subject to review and approval, in advance, by the RISCT.
7. During the performance of the MOU the participants agree to abide by the terms of Executive Order 11246 on nondiscrimination and will not discriminate against any person because of race, age, color, religion, gender, national origin, or disability.
8. No member of, or delegate to Congress, or resident commissioner, shall be admitted to any share or part of this agreement, or to any benefit that may arise from, but these provisions shall not be construed to extend to this agreement if made with a corporation for its general benefits.
9. The parties agree to implement the provisions of this MOU to the extent personnel and budgets allow. In addition, nothing in the MOU is intended to supersede any laws, regulations, or directives by which the parties must legally abide.

IN WITNESS THEREOF, the parties hereto have executed this memorandum of understanding as of the last written date below.

PART TWO

Ecology of Greater Sage-Grouse

CHAPTER THREE

Characteristics and Dynamics of Greater Sage-Grouse Populations

John W. Connelly, Christian A. Hagen, and Michael A. Schroeder

Abstract. Early investigations supported the view that Greater Sage-Grouse (*Centrocercus urophasianus*) population dynamics were typical of other upland game birds. More recently, greater insights into the demographics of Greater Sage-Grouse revealed this species was relatively unique because populations tended to have low winter mortality, relatively high annual survival, and some populations were migratory. We describe the population characteristics of Greater Sage-Grouse and summarize traits that make this grouse one of North America's most unique bird species. Data on movements, lek attendance, and nests were obtained from available literature, and we summarized female demographic data during the breeding season for the eastern and western portions of the species' range. Lengthy migrations between distinct seasonal ranges are one of the more distinctive characteristics of Greater Sage-Grouse. These migratory movements (often >20 km) and large annual home ranges (>600 km^2) help integrate Greater Sage-Grouse populations across vast landscapes of sagebrush (*Artemisia* spp.)–dominated habitats. Clutch size of Greater Sage-Grouse averages seven to eight eggs and nest success rates average 51% in relatively nonaltered habitats while those in altered habitats average 37%. Adult female Greater Sage-Grouse survival is greater than adult male survival and adults have lower survival than yearlings, but not all estimates of survival rates are directly comparable. The sex ratio of adult Greater Sage-Grouse favors females but reported rates vary considerably. Long-term age ratios (productivity) in the fall have varied from 1.4 to 3.0 juveniles/adult female.

Key Words: Artemisia, Centrocercus urophasianus, demographics, Greater Sage-Grouse, movements, nesting, populations, reproduction, sagebrush, survival.

Características y Dinámicas de Poblaciones del Greater Sage-Grouse

Resumen. Las investigaciones tempranas del Greater Sage-Grouse (*Centrocercus urophasianus*) apoyaron la visión que la dinámica de población de esta especie era típica de otras aves de caza de la altiplanicie. Más recientemente, mayores discernimientos en los datos demográficos del Greater Sage-Grouse revelaron que esta especie es relativamente única porque las poblaciones tendieron a tener mortalidad baja en el invierno, supervivencia anual relativamente alta, y algunas poblaciones eran migratorias. Describimos las características de la población del Greater Sage-Grouse y resumimos los rasgos que hacen a este grouse una de las especies de aves más

Connelly, J. W., C. A. Hagen, and M. A. Schroeder. 2011. Characteristics and dynamics of Greater Sage-Grouse populations. Pp. 53–67 *in* S. T. Knick and J. W. Connelly (editors). Greater Sage-Grouse: ecology and conservation of a landscape species and its habitats. Studies in Avian Biology (vol. 38), University of California Press, Berkeley, CA.

únicas de Norteamérica. Datos sobre los movimientos, concurrencia a los leks (asambleas de cortejo), y nidos fueron obtenidos de la literatura disponible, y resumimos datos demográficos de las hembras durante la temporada de cría para las partes orientales y occidentales del territorio de esta especie. Las largas migraciones entre los distintos territorios estacionales son una las características más distintivas del Greater Sage-Grouse. Estos movimientos migratorios (a menudo >20 km) y gran extensión del territorio anual habitado (>600 km^2) ayudan a integrar poblaciones del Greater Sage-Grouse a través de vastos paisajes de habitats dominados por artemisa (*Artemisia* spp.). El tamaño de puesta promedio del Greater Sage-Grouse es de entre siete y ocho huevos, y el éxito de la anidada promedia en un 51% en habitats relativamente no alterados mientras que en habitats alterados el promedio de éxito es de un 37%. La supervivencia de hembras adultas de Greater Sage-Grouse es mayor que la supervivencia de machos adultos, y los adultos tienen menor tasa de supervivencia que los juveniles, pero no todas las estimaciones de las tasas de supervivencia son directamente comparables. La proporción de sexos en adultos del Greater Sage-Grouse favorece a las hembras, pero las tasas divulgadas varían considerablemente. Las tasas de productividad (relación entre grupos de distinta edad) a largo plazo en el otoño han variado de 1.4–3.0 juveniles/hembra adulta.

Palabras Clave: anidación, artemisa (sagebrush), *Artemisia, Centrocercus urophasianus,* datos demográficos, Greater Sage-Grouse, movimientos, poblaciones, reproducción, supervivencia.

A population has been defined as a group of individuals of the same species that occupy an area of sufficient size to permit normal dispersal and/or migration behavior and in which numerical changes are largely determined by birth and death processes (Berryman 2002). For many years it was assumed that the demographics of populations (e.g., reproductive rates, survival, and effects of exploitation) were the same for all species of upland game (Allen 1962, Strickland et al. 1994). Allen (1962) summarized this paradigm well when he reported that small animal populations operate under a 1-year plan of decimation and replacement, nature habitually maintains a wide margin of overproduction, and a huge surplus of animals dies whether or not they are harvested. Early management investigations reinforced the view that the Greater Sage-Grouse (hereafter sage-grouse) was a typical upland game bird, similar in demographics and movements to Ring-necked Pheasant (*Phasianus colchicus*), Northern Bobwhite (*Colinus virginianus*), and Ruffed Grouse (*Bonasa umbellus*). For example, Wallestad (1975a) reported that sage-grouse populations in Montana during the breeding season, and possibly for the entire year, were centered and localized around strutting grounds (leks). This view of localized sage-grouse populations was prevalent through the 1970s, despite earlier work demonstrating that sage-grouse had relatively large annual ranges (Dalke et al. 1963). Early research also suggested these localized sage-grouse populations had relatively high annual turnover with high overwinter mortality (Wallestad 1975a).

With the advent of improved telemetry techniques in the 1980s and more recently the use of genetics, biologists gained greater insight into characteristics of sage-grouse populations. It slowly became evident that sage-grouse did not fit the typical paradigm for upland game birds because populations tended to have low winter mortality and relatively high annual survival, and many populations were migratory (Schroeder et al. 1999; Connelly et al. 1988, 2000c). Little evidence suggests that populations of sage-grouse produce a large annual surplus (Connelly et al. 2000a,c; Hausleitner 2003; Holloran et al. 2005). Recognition of these characteristics has influenced conservation and management of sage-grouse populations. Our objectives were to describe population characteristics of sage-grouse and summarize the traits that make this grouse one of North America's most interesting and unique bird species.

DATA SYNTHESIS

We summarized data on movement, fidelity, home range, lek attendance and timing, and nest locations from available literature. We used descriptive statistics for comparisons to avoid quantitative ritual (Guthery 2008) and because of differences in field and analytical techniques among studies. Stiver et al. (2006) established seven Sage-Grouse Management Zones (SMZs); they suggested that stressors to sage-grouse populations differ across the range of the species and identified different management goals for sage-grouse in the eastern and western portions of its range. Therefore, we summarized demographic data for female sage-grouse during the breeding season for the eastern (Great Plains, Wyoming Basin, and Colorado Plateau SMZs) and western (Southern Great Basin, Northern Great Basin, Snake River Plain, and Columbia Basin SMZs) portions of the species range (Fig. 3.1). We synthesized nest success for radio-marked sage-grouse by relative quality of the habitat within study areas, because of the documented importance of habitat quality to nest success. We assessed habitat qualitatively based on descriptions of habitat provided by authors of the reports and classified data as being from nonaltered or altered habitats. For example, if an author reported his or her study area was recently burned or highly fragmented, data would be presented under altered habitat. Virtually all sagebrush habitats have sustained anthropogenic alterations (Connelly et al. 2004); thus, we refer to these terms exclusive of anthropogenic landscape features such as roads and power lines. We then compared nest success by age group among these habitats and compared nest success rates to those of other prairie and steppe-nesting grouse.

Several studies have urged caution to minimize observer-induced nest desertion or loss by sage-grouse (Patterson 1952, Wakkinen 1990, Sveum et al. 1998b, Wik 2002, Slater 2003). Holloran (2005) wore rubber boots to reduce human scent while confirming nest locations and subsequently monitored nests from ≥60 m to minimize human-induced nest abandonment or predation. Therefore, although these data were retained in tables presenting demographics (Tables 3.1, 3.2), we excluded data from our summary statistics of nest success (Table 3.3) if investigators indicated they used intrusive methods of monitoring—for example, flushing females from nests, candling or floating eggs, or using flagging or other highly visible markers to locate nests.

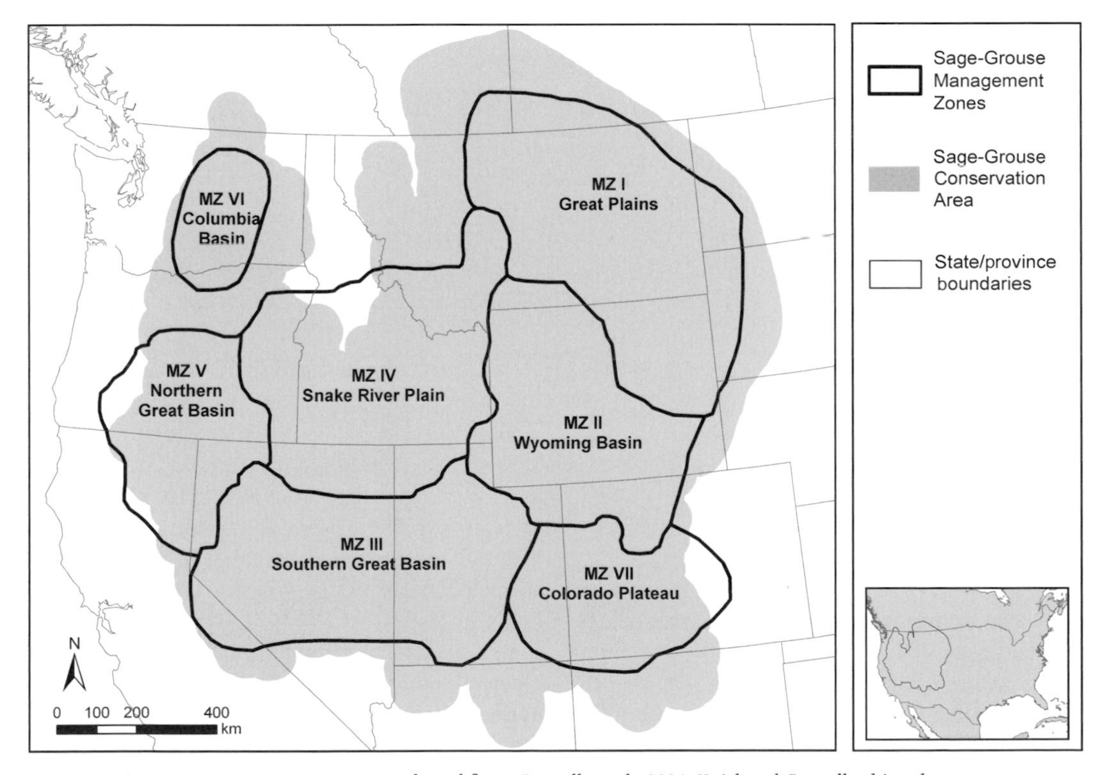

Figure 3.1. Sage-Grouse Conservation Area (adapted from Connelly et al., 2004; Knick and Connelly, this volume, Introduction) and Sage-Grouse Management Zones (Stiver et al. 2006).

TABLE 3.1

Demographic data for female Greater Sage-Grouse during the breeding season in the eastern portion of the species' range (after Schroeder et al. 1999).

All data represent average values for a given study.

State or province	Clutch size (N)	Percent likelihood of nesting (N)	Percent likelihood of renesting (N)	Percent nest success (N)	Percent population breeding success (N)	Source
Alberta	8 (28)	100 (28)	36 (14)	46 (26)	55 (22)	Aldridge and Brigham (2001)
				30 (93)		Aldridge (2005)
Colorado	7 (29)		38 (16)	45 (31)	47 (30)	Petersen (1980)
	7 (81)	85 (119)	12 (34)	55 (107)	57 (130)	Hausleitner (2003)
Montana	8 (22)	71 (31)		70 (20)	70 (20)	Wallestad and Pyrah (1974)
	8 (119)	78 (unknown) 93 (unknown)	19 (36) 43 (83)	46 (258)		Moynahan (2004)
North Dakota	8.3 (56)	96 (73)	28.6 (35)	47 (73)		Kaczor (2008)
South Dakota	8	92 (36)	10 (36)	31 (36)		Herman-Brunson (2007)
Wyoming	7 (154)			38 (216)[a]		Patterson (1952)
		77 (70)		50 (54)		Lyon and Anderson (2003)
		78 (111)	14 (73)	25 (97)		Slater (2003)
				60 (82)		Holloran (1999)
	7 (66)					Holloran (2005)
		81 (597)	9 (597)	49 (597)		Holloran et al. (2005)
		82 (73)		57 (14)	64 (42)	Kaiser (2006)
		67 (52)		66 (52)		Jensen (2006)
Range	7–8	67–100	9–38	25–70	47–70	
Average (SD)	7.5 (0.5)	82 (10.2)	19.7 (12.3)	48 (12.9)	58.6 (8.8)	

[a] Studies without radio telemetry.

MOVEMENTS, FIDELITY, AND HOME RANGE

Extensive movements between seasonal ranges and large annual home ranges are two of the more unique characteristics of sage-grouse life history (Dalke et al. 1960, Gill and Glover 1965, Berry and Eng 1985, Connelly et al. 1988, Bradbury et al. 1989). Movements of sage-grouse can be categorized into different types including: (1) dispersal from place of hatching to place of breeding or attempted breeding, (2) movements of individuals within a season, (3) migration between distinct and spatially separated seasonal ranges, and (4) home ranges that sum all movement types seasonally or annually. These movement categories share considerable overlap, especially in the context of annual home or seasonal ranges that include daily movements to obtain food, visit loafing and roosting sites, and engage in breeding

TABLE 3.2

Demographic data for female Greater Sage-Grouse during the breeding season in the western portion of the species' range (after Schroeder et al. 1999).

All data represent average values for a given study.

State or province	Clutch size (N)	Percent likelihood of nesting (N)	Percent likelihood of renesting (N)	Percent nest success (N)	Percent population breeding success (N)	Source
California				40 (88)		Popham and Gutiérrez (2003)
				43 (95)		Kolada (2007)
Idaho	8 (47)			77 (47)[a]		Bean (1941)
	7 (30)	69 (242)	15 (79)	52 (166)	48 (52)	Connelly et al. (1993)
				51 (41)		Apa (1998)
	6 (25)	92 (38)	17 (18)	45 (38)		Wik (2002)
				45 (47)		Lowe (2006)
Oregon		78 (119)	9 (75)	15 (124)	15 (119)	Gregg (1991), Gregg et al. (1994)
			22 (19)	37 (76)		Coggins (1998)
			34 (143)			Gregg et al. (2006)
Nevada	7.3 (35)	90 (84)	38 (50)	40 (87)		Rebholz (2007)
Utah	7 (147)			60 (161)[a]		Rasmussen and Griner (1938)
				86 (36)[a]		Trueblood (1954)
		63 (19)		66 (19)		Bunnell (2000)
		63 (103)		71 (84)		Chi (2004)
	6 (unknown)	82 (28)		59 (22)		Dahlgren (2006)
Washington	9 (55)	100 (129)	87 (69)	37 (188)	61 (111)	Schroeder (1997)
	7 (38)	80 (95)	25 (44)	41 (93)	40 (95)	Sveum et al. (1998b)
Range	6–9	63–100	9–87	15–86	15–61	
Average (SD)	7.1 (1.1)	78.4 (13.3)	29.9 (26.4)	52.1 (18.2)	41.0 (19.4)	

[a] Studies without radio telemetry

TABLE 3.3

Nest success rates reported for radio-marked Greater Sage-Grouse throughout the species' range.

State or province	Percent nest success (N)			Source
	Yearlings	Adults	Both	
Nonaltered habitats				
California			40 (88)[b]	Popham and Gutiérrez (2003)
Colorado	46 (21)[a]	58 (81)	55 (107)	Hausleitner (2003)
Idaho	73 (15)	52 (23)	63 (40)	Wakkinen (1990)
			52 (75)	Connelly et al. (1991)
	56 (16)	55 (29)	54 (82)	Fischer (1994)
	42 (12)	52 (23)	45 (38)	Wik (2002)
Montana	44 (9)	77 (13)	64 (22)	Wallestad and Pyrah (1974)
Oregon			24 (49)	Gregg (1991)
			37 (63)	Coggins (1998)
Utah	66 (24)	70 (24)	69 (48)	Chi (2004)
			59 (22)	Dahlgren (2006)
Wyoming			35 (78)	Heath et al. (1997)
	67 (21)	76 (21)	71 (42)	Heath et al. (1998)
	57 (34)	64 (48)	61 (82)	Holloran (1999)
			50 (54)	Lyon (2000)
			29 (21)	Slater (2003)
	47 (unknown)	49 (unknown)	49 (484)[b]	Holloran et al. (2005)
	43 (21)	74 (31)	66 (52)	Jensen (2006)
Mean	54 (173+)	63 (293+)	51 (1447)	
Altered habitats				
Alberta	25 (4)	50 (22)	46 (26)	Aldridge and Brigham (2001)
			30 (93)	Aldridge (2005)
Idaho	33 (15)	50 (26)	44 (41)	Apa (1998)
	31 (16)	50 (30)	45 (47)	Lowe (2006)
Oregon			12 (51)	Gregg (1991)
			30 (20)	Hanf et al. (1994)
Washington			41 (93)	Sveum (1995)
			37 (188)	Schroeder (1997)
Wyoming			24 (76)	Slater (2003)
	42 (53)	44 (211)	NA	Holloran (2005)
	57 (14)	64 (42)	62 (56)[b]	Kaiser (2006)
Mean	38 (102)	52 (331)	37 (691)	
Overall mean	49 (275+)	59 (624+)	46 (2,138)	

[a] Sample size in parentheses.

[b] Estimated from data provided in publication.

behavior, as well as migrations. Migration for sage-grouse has been defined as occurring when grouse move >10 km between seasonal ranges (Connelly et al. 2000c). Dispersal and migratory movements help quantify spatial attributes of a population and are fundamental components of the definition of a population (Berryman 2002).

Dispersal

Dispersal is paramount for integrating populations, recolonizing habitats, and maintaining genetic flow (Greenwood and Harvey 1982, Linberg et al. 1998, Barrowclough et al. 2004). Unfortunately, few data are available regarding dispersal by sage-grouse. In Colorado, median dispersal of females ($N = 12$) and males ($N = 12$) was 9 and 7 km, respectively, from their approximate places of hatching to their approximate places of breeding or attempted breeding (Dunn and Braun 1985). Dispersal appears to be discrete from brood breakup (Browers and Flake 1985), and the relatively few movements described seem to be rather gradual and sporadic (Dunn and Braun 1986a).

Seasonal Movements and Migration

Seasonal movement patterns are highly variable both within and among populations (Dalke et al. 1960, Connelly et al. 1988). Connelly et al. (2000c) reported that sage-grouse may have distinct winter, breeding, and/or summer use areas, or the seasonal-use areas may be integrated. For example, winter and breeding areas may be in the same general location or breeding and summer areas may be in the same location. Hence, it is possible for sage-grouse to migrate between two or three distinct seasonal home ranges, or not migrate. Regardless of their migratory status, sage-grouse tend to have large movements within a season when compared with other game birds, including Gray Partridge (*Perdix perdix*) (Weigand 1980, Church and Porter 1990), Ring-necked Pheasant (Hill and Robertson 1988), and Northern Bobwhite (Rosene 1969), which all exhibit relatively short (generally <10 km) movements. Grouse in nonmigratory populations tend to be relatively sedentary with seasonal movements <10 km, while birds in migratory populations may travel well over 100 km (Patterson 1952, Hulet 1983, Hagen 1999). Because of this variation, Connelly et al. (2000c) suggested that three sage-grouse movement patterns can be defined: (1) nonmigratory—sage-grouse make one-way movements <10 km between or among seasonal ranges; (2) one-stage migration—grouse move ≥10 km between two distinct seasonal ranges; and (3) two-stage migration—grouse move ≥10 km among three distinct seasonal ranges.

The close configuration of winter and breeding habitat in some areas may result in comparatively short or nonexistent movements between winter and breeding areas, whereas long distances between breeding and summer habitat result in extensive movements (Connelly et al. 1988, Hagen 1999). Breeding habitat in some areas may be positioned between winter and summer range, such as Idaho (Dalke et al. 1960) and California (Bradbury et al. 1989). In contrast, a study in Wyoming indicated that breeding and summer ranges tended to be relatively close together and winter range was more distant (Berry and Eng 1985). In eastern Idaho, male ($N = 47$) and female sage-grouse ($N = 27$) moved an average of 32 km and 18 km, respectively, between seasonal areas (Connelly et al. 1988). Similarly, male ($N = 27$) and female sage-grouse ($N = 22$) moved an average of 24 km and 17 km, respectively, from breeding to summer range in southwestern Idaho (Wik 2002). In Colorado, female sage-grouse ($N = 76$) moved an average distance of 10 km between winter range and breeding areas (Hausleitner 2003). Numerous ideas have been proposed to explain sage-grouse movement patterns, including differences in seasonal habitat selection (Connelly et al. 1988), desiccation of succulent forbs during summer (Wallestad 1971, Fischer et al. 1996a), harsh winter weather (Dalke et al. 1960, Gill and Glover 1965, Connelly et al. 1988), seasonal site fidelity (Berry and Eng 1985, Connelly et al. 1988, Fischer et al. 1997), and sex class (Beck et al. 2006).

Peak autumn migration is mid-October through late November, spring migration is mid-February through mid-March, and summer migration occurs from late May through early August (Schroeder et al. 1999). Onset of migration may be associated with weather conditions (Berry and Eng 1985). Autumn movements by sage-grouse in Idaho are generally slow and meandering, with a travel rate of 0.3 km/day, and summer movements are more direct and rapid, with a rate of 0.9 km/day; males moved at a faster rate than females (Connelly et al. 1988). Robertson (1991) reported that migratory sage-grouse in southeastern Idaho moved an average of 0.8 km/day during winter.

Weather and habitat distribution influence migration patterns but are not always sufficient to explain relatively long migration distances in relation to short distances between seasonal habitat types. Sage-grouse often migrate farther than would be necessary to reach suitable habitat (Connelly et al. 1988, Jensen 2006). One possible explanation for this discrepancy is that in subsequent years birds may display fidelity to their first seasonal ranges (Berry and Eng 1985, Connelly et al. 1988, Schroeder and Robb 2003) and thus bypass other suitable habitat to reach these areas.

Site Fidelity

Understanding site fidelity is paramount to understanding seasonal movements. Fidelity to display sites (leks) has been well documented in sage-grouse populations (Dalke et al. 1963, Wallestad and Schladweiler 1974, Emmons and Braun 1984, Dunn and Braun 1985), a trait they share with other species of grouse that breed on leks (Schroeder and Robb 2003). In addition, visits to multiple leks tend to be less frequent for adult males than yearlings, suggesting an age-related period of establishment (Emmons and Braun 1984, Schroeder and Robb 2003).

Researchers have also examined fidelity of females to nesting areas. The distance between a female's nests in consecutive years was a median of 0.7 km (range = 0.0–2.6 km) in Idaho (Fischer et al. 1993) and averaged 3.0 km (SD = 6.8 km) in Washington (Schroeder and Robb 2003), 2.0 km (SD = 5.5 km) in Montana (Moynahan et al. 2007), 0.7 km in Wyoming (Holloran and Anderson 2005), 2.4 km (SD = 0.1) in North Dakota (Herman-Brunson 2007), and 1.1 km (SD = 0.4 km) in South Dakota (Kaczor 2008). Studies in Colorado, Washington, and Wyoming indicated that unsuccessful females moved farther between consecutive nests than successful females (Hausleitner 2003, Schroeder and Robb 2003, Holloran and Anderson 2005), but not in the Dakotas (Herman-Brunson 2007, Kaczor 2008).

Greater distance between nests did not increase the likelihood of nesting success. The average distance between first nests and renests was 2.3 km (SD = 5.2 km) in Montana (Moynahan et al. 2007), 2.6 km (SD = 4.5 km) in Washington (Schroeder and Robb 2003), and 1.9 km (SD = 0.4 km) in South Dakota (Kaczor 2008); within years, consecutive nests were closer together for adult than for yearling females in Washington. This behavior of yearling females suggests a period of establishment similar to that of yearling males. The relatively large distances in Washington may be explained by extensive habitat fragmentation; one exceptional female had consecutive nests 32 km apart (Schroeder and Robb 2003). In contrast, Hulet (1983) described a female sage-grouse in southeastern Idaho that moved 170 km among seasonal ranges and returned to nest within 25 m of her previous year's nest. Strong fidelity between seasonal ranges for radio-marked ($N = 5$) sage-grouse was also reported for sage-grouse in northwestern Colorado (Hagen 1999). Birds returned within 1.2 km (median = 0.7, range = 0.2–4.0 km) of the geometric center of seasonal use areas between years.

Home Range

Relatively large seasonal movements and highly clustered distributions of sage-grouse have made estimating home range size difficult (Hagen et al. 2001), and emphasize the wide range of natural variation in home ranges for the species. Some variation is associated with seasonal behavior, habitat requirements, and juxtaposition of habitats (Connelly and Markham 1983, Holloran 1999, Hausleitner 2003). On an annual basis, individuals may occupy areas 4–615 km^2 (Hagen 1999, Connelly et al. 2000c, Hausleitner 2003). Home ranges within a season can vary from <1 to 29 km^2 during the breeding season (Schroeder et al. 1999), <1 to 26 km^2 during summer (Wallestad 1975a, Connelly and Markham 1983, Hagen 1999), 23 to 44 km^2 during autumn (Schroeder et al. 1999), and <1 to 195 km^2 during winter (Wallestad 1975a, Robertson 1991, Hagen 1999). Migratory sage-grouse in southeastern Idaho occupied mean areas of 140 km^2 in winter (Robertson 1991), but a nonmigratory population in central Montana occupied winter home ranges of 11–31 km^2 (Wallestad 1975a). Relatively large seasonal and annual home ranges coupled with extensive movements make Greater Sage-Grouse a true landscape species (Connelly et al. 2004).

Relevance of Movement to Identification of Populations

Dispersal and migratory movements have been studied in relatively small portions of the range of sage-grouse and on rather small numbers of individuals (Connelly et al. 1988). In some cases, sufficient individuals may have been marked to allow inferences to be drawn about the demographic importance of movements. Unfortunately, it is often impractical to use these movements to define populations of sage-grouse in North America using the approach recommended by Berryman (2002). Nevertheless, knowledge of movements in some portions of the range can be used to help spatially define some populations and extrapolated for assessments of the overall distribution of sage-grouse. Populations of sage-grouse have been considered distinct when they are separated from adjacent populations by at least 20 km of inhospitable and unoccupied habitat (Connelly et al. 2004). This distance is largely based on direct observations of grouse movements within occupied range and absence of movements through or over nonhabitat of sufficient size to apparently act as a barrier (Connelly et al. 1988, 2004).

BREEDING BIOLOGY

The Breeding Period

Male sage-grouse congregate on leks—arenas with relatively sparse cover—to display to and breed with females (Scott 1942, Patterson 1952, Giezentanner and Clark 1974, Connelly et al. 1981, Bergerud 1988a). An important factor affecting lek location appears to be proximity to, and configuration and abundance of, nesting habitat (Connelly et al. 1981, Connelly et al. 2000c). In fact, males may form and visit leks that are in areas of high female traffic (Bradbury and Gibson 1983, Gibson 1996).

Because up to 400 males may attend a lek, and because males form territories on leks, a lek can cover up to 20 ha (Scott 1942). Leks normally occur in the same location each year (Scott 1942, Patterson 1952). Some leks studied by early investigators (Scott 1942, Batterson and Morse 1948) have persisted for 28–67 years since first counted (Wiley 1973b, Hagen 2005). Dalke et al. (1963) reported the presence of broken bird-point arrowheads on one of the leks in their study area, suggesting that sage-grouse had used the site for at least 85 years. Leks and the number of attending males are regularly used to monitor the long-term status of populations because of their traditional locations (Connelly et al. 2003b, Connelly and Schroeder 2007).

Despite the traditional nature of leks, shifts in location may occur for numerous reasons. Gibson and Bradbury (1987) observed a shift in lek location following a severe winter, when traditional lek sites were covered in snow until May. Males may continue to use these new locations even when the snow is gone or in subsequent years. Sage-grouse may shift or abandon lek locations because of persistent disturbance and/or alteration of vegetative cover (Commons et al. 1999, Holloran 2005, Walker et al. 2007a). Intra- and intersexual behavioral interactions may also result in annual variation in lek location, but mechanisms are poorly understood. For example, female selection of specific males may encourage unselected males to alter location of lek territories. Over time female selection could result in the gradual shifting of a lek's location (Beehler and Foster 1988). This effect can be further exacerbated due to the formation of satellite leks during periods of relative abundance (Dalke et al. 1963). To consistently monitor leks spatially and temporally, researchers often consider multiple locations for a lek (including annual shifts and satellites) together as a lek complex (Connelly and Schroeder 2007).

Depending on snow depth, elevation, weather, and region, male sage-grouse begin the display season between the end of February and early April and end the display season in late May or early June (Eng 1963, Schroeder et al. 1999, Aldridge 2000, Hausleitner 2003). Adult males arrive at leks earliest in the season, followed by females and yearling males (Dalke et al. 1960, Eng 1963, Jenni and Hartzler 1978, Emmons and Braun 1984). Female attendance tends to be relatively synchronous, peaking in mid- to late March in Washington (Schroeder 1997), late March to early April in California (Bradbury et al. 1989) and Oregon (Hanf et al. 1994), and early to mid-April in Alberta (Aldridge and Brigham 2001), Colorado (Petersen 1980, Walsh 2002, Hausleitner 2003), Montana (Jenni and Hartzler 1978), and Wyoming (Patterson 1952). Females may also irregularly visit leks later in the breeding season due to renesting efforts (Eng 1963). Weather conditions may cause female attendance patterns to vary by up to 2 weeks (Jenni and Hartzler 1978, Schroeder 1997).

Male sage-grouse usually begin displaying on leks just prior to sunrise and depart shortly after sunrise as the display season begins (Jenni and Hartzler 1978). Males arrive on leks earlier and remain later as the season progresses, especially when females are present (Jenni and Hartzler 1978). During peak female attendance, males may display for up to 3–4 hours each morning and often during late evening and night (Scott 1942, Patterson 1952, Hjorth 1970, Walsh 2002). Because of the delay in attendance by yearling males, peak male attendance typically occurs about 3 weeks after peak female attendance (Patterson 1952, Eng 1963, Jenni and Hartzler 1978, Emmons and Braun 1984, Walsh et al. 2004). Emmons and Braun (1984) reported an average attendance rate over 5-day observation periods of 86% for radio-marked yearling males and 92% for adult males, and that 90% and 94% of yearling and adult male sage-grouse, respectively, attended leks during the period of high male counts (about 3 weeks after peak female attendance). These rates were pooled over 5-day periods and likely overestimated attendance. In contrast, Walsh et al. (2004) reported average daily male attendance rates of 42% and 19% for adult and yearlings, respectively. These authors indicated that on 58% of days in which seven radio-marked adult males were observed, they did not attend a lek. These rates were not adjusted for detection rate and are likely biased low. These studies are not directly comparable because investigators used different methodologies to measure attendance rates, but each study indicates that counts of males on leks represent minimum counts.

Nesting

Nest Location

An early synthesis of sage-grouse biology and management guidelines indicated that most females nest within 3.2 km of a lek (Braun et al. 1977), but recent literature suggests that many females nest farther from leks than previously suspected. The average distance between a female's nest and the nearest lek was 1.3–1.5 km in Idaho (Wakkinen et al. 1992, Fischer 1994), 2.7 km in North Dakota (Herman-Brunson 2007), 2.8 km in Colorado (Petersen 1980), 4.9 km in Alberta (Aldridge 2005), and 5.1 km in Washington (Schroeder et al. 1999). Similar nest-to-lek distances have also been recorded in Nevada (J. S. Sedinger, pers. comm.). Distances between a female's nest and her lek of capture were substantially larger than distances to the nearest lek (1.2–3.1 km larger) because females may not be captured or first observed at the lek nearest their nest (Petersen 1980, Wakkinen et al. 1992, Fischer 1994, Schroeder et al. 1999, Herman-Brunson 2007). Other studies have illustrated similar variation in nest-to-lek distances (Berry and Eng 1985, Hanf et al. 1994, Holloran 1999, Lyon and Anderson 2003, Slater 2003, Moynahan et al. 2007). Juxtaposition of habitats, disturbance, and extent of habitat fragmentation may influence location of nests with respect to leks (Lyon and Anderson 2003, Connelly et al. 2004, Schroeder and Robb 2003). Females in highly fragmented habitats of Washington moved almost twice as far to nest (Schroeder et al. 1999) as females in relatively intact habitats of southeastern Idaho (Wakkinen et al. 1992, Fischer 1994). Similarly, females from undisturbed leks in southwestern Wyoming moved an average of 2.1 km to nests, while females from disturbed leks moved 4.1 km (Lyon and Anderson 2003).

Timing of Nesting

Peak egg laying and incubation periods vary from late March through mid-June, with renesting stretching into early July (Schroeder et al. 1999, Gregg 2006). The typical date for initiation of incubation appears to be about 3–4 weeks following peak female attendance on leks (Schroeder 1997, Aldridge and Brigham 2001, Hausleitner 2003). Adults initiated incubation on average nine days earlier than yearlings in north-central Washington (Schroeder 1997). Following an approximate incubation period of 27 days (Schroeder et al. 1999), the precocial chicks leave the nest soon after hatching, are capable of weak flight by ten days of age, and are capable of strong flight by five weeks of age (Schroeder et al. 1999).

Clutch Size

The clutch size of sage-grouse is variable but relatively low compared to other game birds (Edminster 1954, Schroeder 1997). The average clutch size is seven eggs for sage-grouse, but varies from six to nine eggs (Tables 3.1, 3.2). In contrast, clutch sizes for Ring-necked Pheasant and Sharp-tailed Grouse (*Tympanuchus phasianellus*) range from 10 to 12 and 11 to 12 eggs, respectively (Hill and Robertson 1988, Connelly et al. 1998). Variation in clutch size has been attributed to age (Wallestad and Pyrah 1974, Petersen 1980, Hausleitner 2003); nesting attempt, where first nest clutches were larger than those of renests (Kaczor 2008); and annual variation in conditions

(Schroeder 1997, Walker 2008). Nevertheless, Wik (2002) did not detect measurable variation in clutch size either annually or by nesting attempt. Sage-grouse clutch size estimates may be biased low if these estimates have been based on post-hatching nest examinations and partial clutch loss occurs prior to hatch.

Nest Likelihood

The average likelihood of a female nesting in a given year varies from 63% to 100% and averages 82% in the eastern part of the species' range (Table 3.1) and 78% in the western portion of the range (Table 3.2). Nest initiation rates tended to be higher for adults (78–100%) than yearlings (55–79%) in three separate studies in Idaho (Connelly et al. 1993; Wik 2002; N. A. Burkepile, unpubl. data). Similarly, Holloran (2005) reported that nesting likelihood in Wyoming was greater for adult (85%) than yearling (67%) sage-grouse. Nest initiation rate was also higher for females captured on undisturbed leks in western Wyoming than for females captured on disturbed leks (Lyon and Anderson 2003).

Direct evidence from radiotelemetry studies has illustrated dramatic variation in renesting likelihood (Tables 3.1, 3.2). The average likelihood of renesting is 25% in the eastern portion of the range (Table 3.1) and 30% in the western portion (Table 3.2). Females were observed nesting two times after loss of first nest in Washington (Sveum 1995, Schroeder 1997), Nevada (Rebholz 2007), and Idaho (J. W. Connelly, pers. obs.). Adults were more likely to renest than yearlings in Washington (Schroeder 1997). The lower likelihood of renesting by yearlings may be due to later initiation of first nests and their shorter nesting season (Schroeder 1997), and drier conditions later in the nesting season (Hulet 1983). Higher renesting rates in southern Oregon were associated with improved habitat conditions; renesting rates increased from 14% to 30% as habitat improved (Coggins 1998). Gregg et al. (2006) reported that renesting was related to age of hen, nest initiation period, nest-loss period, and total plasma protein. Hens that renested had greater total plasma protein levels than nonrenesting hens (Gregg et al. 2006). Renesting rates in other game bird species, including Northern Bobwhite (Rosene 1969), Ring-necked Pheasant (Hill and Robertson 1988), Gray Partridge (Carroll 1993), and Greater Prairie-Chicken (*Tympanuchus cupido*; Norton 2005), tend to be higher, often exceeding 75%.

Nest Success

Reported nest success rates of sage-grouse vary between 15% and 86%, depending on habitat condition, methodology, and female age (Table 3.3). Nest success for sage-grouse appears comparable to that of other shrub and grassland grouse species, including 50–72% for Sharp-tailed Grouse (Connelly et al. 1998), 22–65% for Greater Prairie-Chicken (Schroeder and Robb 1993), 14–41% for Lesser Prairie-Chicken (*T. pallidicinctus*; Hagen and Giesen 2005), and 36–57% for White-tailed Ptarmigan (*Lagopus leucura*; Braun et al. 1993).

Baxter et al. (2008) stated that sage-grouse population declines are often related to poor nest success, but most reported nest success rates for sage-grouse are relatively high. The average nest success for 29 studies using radiotelemetry is 46%; 12 (37%) reported nest success rates are ≥50%, and 9 (30%) are <40% (Table 3.3). Moreover, when one compares relatively altered and unaltered habitats, success rates in unaltered habitats appear even higher. In relatively unaltered habitats, 11 of 18 (61%) studies report overall nest success rates ≥50%, and only four (22%) studies report nest success rates <40% (Table 3.3). Two of 12 (17%) studies in altered habitats report nest success rates ≥50%, and five (42%) studies report nest success <40% (Table 3.3).

Sage-grouse nest success rates were reported in central Montana using both apparent and maximum likelihood estimates (Moynahan et al. 2007). However, Moynahan et al. (2007) used repeated nest visits and flushed females on ≥2 visits to the nest site as part of their study methodology, with at least 24 (8%) nests abandoned due to observer influence. Similarly, Herman-Brunson (2007) reported low nest survival (31%, $N = 29$) but used multiple nest visits, flushed females from nests, and marked nests by placing flagging 20–40 m from the nest. Nest desertion by sage-grouse is relatively common during laying and early incubation, and there are a variety of causes of desertion, including field investigators (Patterson 1952). Most researchers avoid disturbing nesting sage-grouse because of concern that this species readily abandons nests following disturbance (Fischer et al. 1993, Sveum et al. 1998b, Wik 2002, Chi 2004, Holloran et al. 2005, Kaiser 2006,

Baxter et al. 2008). Thus, nest success rates reported by Moynahan et al. (2007) and Herman-Brunson (2007) may be related, in part, to field techniques rather than representative samples of nest survival. Recognizing this potential bias, Moynahan et al. (2007) attempted to identify nests abandoned due to observer influence and removed them from their analyses.

Apparent nest success rates derived from radiotelemetry studies may be biased high and unreliable (Kolada 2007, Moynahan et al. 2007). Apparent success likely overestimates nest success when nest visitation does not include the laying period (7–10 days) or sampling is inadequate for detecting a nesting attempt. However, Walker (2008) provided both apparent and maximum likelihood estimates for different areas and years in southeastern Montana and northeastern Wyoming. Most of these estimates were quite similar, and many of the maximum likelihood estimates were actually greater than apparent estimates. Clearly some caution should be used when interpreting levels of nest success from most studies until there is a better understanding of bias associated with different methods of estimating nest success.

Most investigators have not detected statistically different age-specific rates of nest success, except in central Montana, where adults had greater nest success than yearlings (Wallestad and Pyrah 1974). Nevertheless, 13 of 15 radiotelemetry studies indicated that nest success of adult hens was numerically greater than that of yearlings (Table 3.3).

Annual Reproductive Success

Annual reproductive success (probability of a female hatching ≥1 egg in a season) is more complex than nest success because it includes the likelihood of nesting and renesting. For example, if eastern range values are used (Table 3.1), approximately 39% of females successfully hatch one or more eggs in their first nest attempt (average nest likelihood = 82% × average nest success of 48%). When renest attempts are considered (average renesting likelihood = 20%), the average annual reproductive success is increased to approximately 43%, with about 9% of the average productivity due to renesting. Despite the averages, the high rate of renesting in north-central Washington was atypical for sage-grouse, resulting in 61% annual reproductive success, compared with the modest 37% nest success rate (Schroeder 1997; Table 3.2).

SURVIVAL AND POPULATION DYNAMICS

The definition of a population proposed by Berryman (2002) is somewhat unusual because it includes dispersal and migration behavior while ensuring that numerical changes within a population are largely driven by natality and mortality. In the accepted paradigm of upland game productivity, sage-grouse should regularly overproduce young or at least produce young in sufficient numbers to replace those lost through annual mortality (Allen 1962). Consequently, a strong relationship should exist among movement, productivity, and survival within a population, but little interaction should exist among populations. Information is now available on sage-grouse reproduction that allows us to explore this concept of overproduction and relationships among movement, productivity, and survival.

Survival

Survival in a sage-grouse population can be partitioned into three basic stages: (1) survival of chicks from hatching to brood breakup, usually early September; (2) survival of juveniles from early September to their recruitment to the breeding population, usually March; and (3) annual survival of breeding-age males and females. Recently, results from several studies were averaged to obtain an estimate of 10% survival for juveniles from hatch to breeding age, basically a combination of the first two stages (Crawford et al. 2004). This estimate was based in part on estimates of early juvenile survival, including 33% for Washington (Schroeder 1997), 60% for Wyoming (Holloran 1999), 7% for Utah (Bunnell 2000), and 19% for Alberta (Aldridge and Brigham 2001). At least two of these studies were from areas with fragmented or otherwise marginal sage-grouse habitats; thus, this estimate is likely biased low.

Direct information on survival of radio-marked chicks and juveniles has recently become available (Beck et al. 2006, Gregg 2006, Aldridge and Boyce 2007, Herman-Brunson 2007, Rebholz 2007, Kaczor 2008, Walker 2008). However, the time periods (18–51 days) and estimators (chick or brood survival) used make direct comparisons challenging. Gregg (2006) estimated that chick survival at 28 days posthatching was 39%; survival was higher for chicks from adult females than for chicks from yearling females. Rebholz (2007) estimated chick survival to 18 days posthatching at 44% in Nevada using methods similar to those of Gregg (2006). Apparent chick survival to 21 days posthatching was estimated between 34% and 42% in North Dakota and 32% and 50% in South Dakota, respectively (Herman-Brunson 2007, Kaczor 2008). In contrast, Aldridge and Boyce (2007) reported 12% survival of chicks to 51 days of age. Recruitment estimates from the Dakotas suggest that 6–17% of chicks were recruited into the spring breeding population (Herman-Brunson 2007, Kaczor 2008). Beck et al. (2006) reported relatively high survival of radio-marked juveniles (64–86%) from September until recruitment to the breeding population. Battazzo (2007) also found high winter survival of juveniles in Montana. Gender (Swenson 1986), food availability (Swenson 1986, Pyle and Crawford 1996, Holloran 1999, Huwer 2004), habitat quality (Pyle and Crawford 1996, Sveum et al. 1998a, Aldridge 2000, Huwer 2004, Gregg 2006), harvest rates (Wik 2002), age of brood female (Gregg 2006), and weather (Rich 1985) may impact juvenile survival, but lack of adequate survival estimates has made these potential relationships difficult to evaluate.

More information is available on adult survival than survival of other age groups, but a variety of field techniques (bands, radio transmitters, poncho-tags) reduce the comparability among studies. Survival for sage-grouse (N = 6,021) in Colorado was estimated using bands recovered from hunters (Zablan et al. 2003). These authors estimated survival to be 59% (95%CI, 57–61%) for adult females, 78% (95%CI, 72–75%) for yearling females, 37% (95%CI, 35–45%) for adult males, and 64% (95%CI, 57–65%) for yearling males. They recovered several sage-grouse ≥7 years of age. Yearling male prairie grouse may improve their survival to adulthood because they remain inconspicuous during their first year (Wittenberger 1978, Bergerud 1988b, Hagen et al. 2005). Annual survival of male sage-grouse was estimated to be 59% in Wyoming (June 1963), 58–60% in Idaho (Connelly et al. 1994, Wik 2002), and 30% in Utah (Bunnell 2000). Female survival was estimated to be 48–78% in Wyoming (June 1963, Holloran 1999, 2005), 48–75% in Idaho (Connelly et al. 1994, Wik 2002), 57% in Alberta (Aldridge and Brigham 2001), 61% in Colorado (Hausleitner 2003), and 37% in Utah (Bunnell 2000). Adult sage-grouse were similar to juvenile grouse in having high winter survival. Winter survival rates ranged from 82% to 100% (Hausleitner 2003) in Colorado and from 85% to 100% in southwestern Idaho (Wik 2002). All estimates except June (1963) were based on known-fate analyses from telemetry data.

Mortality Patterns

Little range-wide effort has been expended to examine seasonal patterns of mortality for sage-grouse. The greatest seasonal mortality for adult male and female sage-grouse appears to occur in spring, summer, and fall (Connelly et al. 2000a, Wik 2002). In Idaho, 43% of all documented deaths of radio-marked sage-grouse occurred from March through June with no difference by gender (Connelly et al. 2000a). In contrast, only 2% of radio-marked sage-grouse deaths occurred from November through February. Similarly, Wik (2002) reported that in southwestern Idaho, overwinter survival of adult males ranged from 85–90%, and overwinter survival of adult females ranged from 88% to 100%. Generally, most research suggests that overwinter mortality of sage-grouse is low (Robertson 1991, Connelly et al. 2000a, Wik 2002, Hausleitner 2003, Beck et al. 2006), but Moynahan et al. (2006) reported relatively high mortality during one of three winters that spanned their study period in central Montana.

Causes of Mortality

Like all species of upland game birds, sage-grouse have a variety of predators. Nest predators include Common Raven (*Corvus corax*), gulls (*Larus* spp.), red fox (*Vulpes vulpes*), coyote (*Canis latrans*), American badger (*Taxidea taxus*), and ground squirrels (*Spermophilus* spp.)(Patterson 1952; Hulet 1983; DeLong et al. 1995; Johnson and Braun 1999; Schroeder and Baydack 2001; Moynahan 2004; Coates 2007; Hagen, this volume, chapter 6). All of these predators except ground squirrels will also likely prey on young chicks. Investigators have recorded a western rattlesnake (*Crotalus viridus*) feeding on a sage-grouse chick and an adult female sage-grouse attacking a bull snake (*Pituophis catenifer*) adjacent to her nest (N. A. Burkepile, pers. comm.). Predators of older juvenile and adult sage-grouse include red fox, coyote, badger, bobcat (*Lynx rufus*), domestic cat, weasels (*Mustela* spp.), and a variety of raptor species (Dunkle 1977; Schroeder et al. 1999; Schroeder and Baydack 2001; Hagen, this volume, chapter 6).

Other causes of mortality for sage-grouse include collisions with vehicle or agricultural machinery; disease; flying into fences, power lines, and other obstacles; and pesticide application (Blus et al. 1989, Connelly et al. 2000a, Connelly et al. 2004, Walker 2008). Five percent of radio-marked sage-grouse deaths in southeastern Idaho were attributed to causes other than predation or hunting (Connelly et al. 2000a). Additionally, 33% of mortalities of juvenile sage-grouse in a southeastern Idaho study area were caused by collision with power lines (Beck et al. 2006).

Although most mortality of sage-grouse is due to predation (Connelly et al. 2000a), a substantial amount of mortality in some areas may be associated with hunting (Johnson and Braun 1999; Wik 2002; Connelly et al. 2003a; Zablan et al. 2003; Reese and Connelly, this volume, chapter 7). Reporting rates of 14.0–18.7% were estimated for a Colorado population of sage-grouse (Zablan et al. 2003). In contrast, average harvest rates were relatively low (3.3%, SE = 1.6%) in Oregon from 1993 to 2006 (Broms 2007). In southern Idaho, 15% of documented male sage-grouse deaths and 42% of female deaths were attributed to hunting (Connelly et al. 2000a), but <1% of female mortalities ($N = 217$) in northeast Wyoming and southeast Montana (2003–2006) were related to legal harvest (Walker 2008).

Sex and Age Ratios

Data on sex ratios are available from numerous states, but much of this information was obtained from hunter-harvested birds. The ratio of birds harvested in late summer in Wyoming ranged from 2.2 to 2.7 females per male (Patterson 1952), while the sex ratio for sage-grouse harvested in Colorado was 1.9 females per male (Rogers 1964). Autenrieth (1981) found highly variable sex ratio data among areas and years for adult sage-grouse shot throughout southern Idaho. From 1993 to 2006, sex ratios varied from 1.2 to 2.4 females per male in Oregon (Broms 2007). Sex ratio data obtained from harvested samples should be viewed with caution because there is likely a differential vulnerability by sex to hunting (Connelly et al. 2000a, Wik 2002).

Some information on sex ratios has been obtained by monitoring winter and spring populations. The sex ratio of sage-grouse spring breeding populations varied from 2.3 to 3 females per male in Colorado (Walsh et al. 2004) and Wyoming (Patterson 1952: 140), respectively. Primary sex ratios in northeast Nevada were approximately 1:1 over a 2-year period (Atamian 2007). During winter, the sex ratio for sage-grouse in northern Colorado was 1.6 females per male (Beck 1977).

Evidently the sex ratio of adult sage-grouse favors females, but reported rates vary considerably. Lower survival of males, possibly associated with breeding activities, is the primary reason the female-to-male ratio appears to increase for birds in older cohorts (Patterson 1952, Braun 1984, Swenson 1986, Zablan et al. 2003). However, Swenson (1986) suggested that lower male survival was due to greater physiological demands of male growth rates as chicks, thus indicating that disparate sex ratios begin to occur relatively soon after hatch. These survival rates are likely related to habitat quality (Swenson 1986, Barnett and Crawford 1994, Johnson and Braun 1999), which varies geographically and temporally; consequently, variation in sex ratios for sage-grouse should be expected.

Fish and wildlife agencies often use age ratios as an index to sage-grouse production (Beck et al. 1975, Hagen and Loughin 2008). The average age ratio in southern Idaho indicated by wing samples collected from 1961 to 1980 was 2.4 juveniles per adult female (Autenrieth 1981). These ratios appeared to vary substantially among areas and ranged from 2.0 to 2.8 juveniles per adult female. Overall, long-term age ratios in the fall have ranged from 1.4 to 3.0 juveniles per adult female. Since 1985, these ratios have been lower than long-term averages and generally ranged from 1.2 to 2.2 juveniles per adult female (Connelly and Braun 1997). Similarly, productivity (1993–2005) has varied from 0.8 to 2.3 juveniles per female in Oregon (Hagen and Loughin 2008).

CONSERVATION IMPLICATIONS

Sage-grouse do not fit the commonly accepted paradigm of upland game bird demographics (Allen 1962). Sage-grouse exhibit relatively high survival of breeding-age birds, especially in winter, and comparatively low productivity. Although average nest success is moderate (range-wide average of 46%), the large number of nonnesting females (18–22%), low rate of renesting (20–30%), inability to produce more than one brood in one season, and possibly low chick survival combine to ensure that sage-grouse are unlikely to produce rapidly increasing populations, even under the best of circumstances. This is contrary to most other game bird species that yield a wide margin of overproduction (Allen 1962), and lends strong support to the contention that management decisions should not be based on dogmatic beliefs and findings from early wildlife studies (Williams et al. 2004a).

Sage-grouse populations occupy vast landscapes on an annual basis as a result of large home ranges and substantial dispersal/migratory movements of individual birds (Robertson 1991, Connelly et al. 2000c, Wik 2002, Hausleitner 2003, Zablan et al. 2003, Beck et al. 2006). High survival is an advantage, but low productivity and vast areas of occupation substantially increase the difficulty of managing sage-grouse populations. Our characterization of a population that incorporates movement has important ramifications (Berryman 2002). Juvenile sage-grouse that moved farther distances to seasonal ranges had lower overall survival than did juveniles that moved relatively short distances (Beck et al. 2006). Large movements of sage-grouse help to integrate vast areas of occupied landscape, but they may have a cost in terms of increased mortality and may also place populations at risk of disturbance from factors that would seem superficially to be insignificant. For example, large linear disturbances such as a highway and/or power line may effectively divide a population (Connelly et al. 2004, Jensen 2006, Doherty et al. 2008), and, even if the direct loss of habitat is small, effects of fragmentation may be dramatic. Clearly, assessment of population effects due to habitat loss or fragmentation must include an understanding of demographic attributes of sage-grouse populations. Even though much is known about demographic characteristics of sage-grouse, more information is needed. Moreover, our findings clearly underscore the need to standardize approaches to assessing demographics.

Estimates of breeding propensity could be biased high if most hens are captured on or near leks. Females that are relatively far from leks may have a lower likelihood of nesting. Monitoring females radio-marked during summer or fall or following females over multiple breeding seasons would provide some insight into this issue.

An assessment of density dependence in sage-grouse populations would provide a framework for improving our understanding of population dynamics. Sage-grouse populations may not typically be regulated by density-dependent mechanisms, but a better understanding of density dependence can help guide research directed at examining causal linkages between habitat and population dynamics (Garton et al., this volume, chapter 15). In addition, research that can relate variation in annual rates of change to changes in habitat and environmental conditions would allow development of powerful predictive models useful in guiding management decisions and evaluating proposed projects.

Substantial heterogeneity in fitness seems to occur among nesting females; females that successfully nested had higher annual survival than those that failed (Moynahan et al. 2006). This finding suggests that if there is a cost of breeding, it is masked by marked variation in quality among females and that populations may be maintained by a small core of highly fit individuals. Individual heterogeneity in fitness needs further investigation and will hopefully be addressed in the near future.

ACKNOWLEDGMENTS

We thank the Idaho Department of Fish and Game, Oregon Department of Fish and Wildlife, and Washington Department of Fish and Wildlife for the support necessary to conduct research on Greater Sage-Grouse and to write this document. Much of this research has continued for many years, and long-term support is particularly critical in these situations. Reviews by M. J. Holloran, L. A. Robb, and J. S. Sedinger improved an early draft of this chapter.

CHAPTER FOUR

Characteristics of Greater Sage-Grouse Habitats

A LANDSCAPE SPECIES AT MICRO- AND MACROSCALES

John W. Connelly, E. Thomas Rinkes, and Clait E. Braun

Abstract. Greater Sage-Grouse (*Centrocercus urophasianus*) depend on sagebrush (*Artemisia* spp.) for much of their annual food and cover. This close relationship is reflected in the North American distribution of sage-grouse, which is closely aligned with sagebrush, and in particular big sagebrush (*Artemisia tridentata*) and silver sagebrush (*A. cana*). This association is most pronounced in late autumn, winter, and early spring when sage-grouse are dependent on sagebrush for both food and cover. However, sage-grouse also rely on sagebrush at other times of year, especially for nesting cover during the breeding season. Other habitat characteristics may not be as obviously important as sagebrush, but may be nearly as essential. For example, herbaceous vegetation provides important food and cover during nesting and early brood-rearing seasons, and thus has a major role in the population dynamics of sage-grouse. Available evidence clearly supports the conclusion that conserving large landscapes with suitable habitat is important for conservation of sage-grouse. Moreover, natural variation in vegetation and the dynamic nature of mature sagebrush stands should be considered for all habitat descriptions and prior to any management action. Sagebrush habitats have been lost, fragmented, and degraded as a result of many different anthropogenic disturbances. Complicating matters, the traditional nature of seasonal movements by Greater Sage-Grouse suggests this species has little ability to adapt to habitat change. Therefore, land management agencies must establish sagebrush conservation as one of their highest priorities if remaining habitats are to be maintained. Additionally, these agencies must develop and implement effective habitat reclamation measures to offset unavoidable losses. Given the strong dependence of Greater Sage-Grouse on sagebrush habitats, failure to protect what is left and fix what is broken will likely result in extirpation of many populations of Greater Sage-Grouse.

Key Words: Artemisia, Centrocercus urophasianus, Greater Sage-Grouse, habitat selection, herbaceous vegetation, sagebrush, shrub steppe.

Caracteristicas de los Habitats del Greater Sage-Grouse: Una Especie Que Depende del Paisaje en Micro y Macro Escalas

Resumen. El Greater Sage-Grouse (*Centrocercus urophasianus*) depende del sagebrush (*Artemisia* spp.) para la mayoría de su alimento y albergue anual. Esta estrecha relación se refleja en la distribución del sage-grouse en Norte América, la cual está estrechamente asociada con la de artemisa, especialmente el big sagebrush (*Artemisia tridentata*) y silver sagebrush

Connelly, J. W., E. T. Rinkes, and C. E. Braun. 2011. Characteristics of Greater Sage-Grouse habitats: a landscape species at micro- and macroscales. Pp. 69–83 *in* S. T. Knick and J. W. Connelly (editors). Greater Sage-Grouse: ecology and conservation of a landscape species and its habitats. Studies in Avian Biology (vol. 38), University of California Press, Berkeley, CA.

(*A. cana*). Esta asociación se acentúa a finales de otoño, invierno y comienzos de la primavera cuando el sage-grouse depende de artemisa tanto para su alimentación como para su alojamiento. Sin embargo, el sage-grouse también depende de artemisa en otras épocas del año, especialmente para cubrir los nidos durante la estación de reproducción o cría. Otras características del hábitat pueden no ser tan obviamente importantes como la artemisa, pero pueden ser casi tan esenciales. Por ejemplo, la vegetación herbácea proporciona importante alimento y cobertura durante la anidación y el comienzo de la estación de crecimiento de la crías, y de este modo tiene un rol principal en la dinámica de población del sage-grouse. La evidencia disponible claramente respalda la conclusión de que conservar amplios territorios con hábitat apropiados es importante para la conservación del sage-grouse. Más aún, las variaciones naturales de la vegetación y la naturaleza dinámica de la artemisa madura deberían de considerarse para todas las descripciones de hábitat, y antes de tomar cualquier acción de manejo. Los hábitats de artemisa se han perdido, fragmentado y degradado como consecuencia de diversas perturbaciones antropogénicas. Complicando las cosas, la naturaleza de los movimientos estacionales realizados por el Greater Sage-Grouse sugiere que esta especie tiene poca capacidad para adaptarse a cambios en el hábitat. Por lo tanto, las agencias de manejo de tierra deben establecer la conservación de la artemisa como una de sus más altas prioridades si se pretende mantener los hábitats que quedan. Adicionalmente, estas agencias deben desarrollar e implementar medidas efectivas de reclamo de hábitat para compensar las pérdidas inevitables. Dada la fuerte dependencia del Greater Sage-Grouse sobre los hábitats de artemisa, el fracaso en la protección de lo que queda y en arreglar lo dañado probablemente terminará en la extirpación de muchas poblaciones de Greater Sage-Grouse.

Palabras Clave: artemisa (sagebrush), *Artemisia, Centrocercus urophasianus,* estepa arbustiva, Greater Sage-Grouse, selección de hábitat, vegetación herbácea.

Studies of habitat requirements of Greater Sage-Grouse (*Centrocercus urophasianus*) (hereafter sage-grouse) date to at least the mid-1930s (Griner 1939), and numerous papers, reports, books, and management guidelines have been published on this aspect of sage-grouse biology since that time (reviews in Schroeder et al. 1999; Connelly et al. 2000c, 2004; Braun et al. 2005). Thus, it is likely that we know more about habitats used by this game bird than we know about habitats of any other native game bird in western North America.

Virtually all studies of sage-grouse habitats have described the importance of large, woody sagebrushes (*Artemisia* spp.) of western North America and the bird's dependence on these plants for food and cover during all periods of the year (Patterson 1952; Dalke et al. 1963; Connelly et al. 2000c, 2004). Because of this relationship, sage-grouse are considered a sagebrush obligate (Braun et al. 1976). Most early studies focused on the sagebrush component of sage-grouse habitat (Patterson 1952; Rogers 1964; Wallestad 1971, 1975b), and little attention was placed on the herbaceous vegetation component until more recently (Connelly et al. 1991, Gregg 1991, Gregg et al. 1994, Holloran et al. 2005). Thus, earlier papers addressing habitat management (Wallestad 1975a,b; Braun et al. 1977; Autenreith 1981) emphasized shrub overstory because of its documented importance, leaving a somewhat incomplete picture of the bird's habitat needs.

Sagebrush-dominated plant communities vary considerably across the range of sage-grouse (Tisdale and Hironaka 1981; West 1983a,b; West and Young 2000), and the specific habitat components used by grouse can vary due to biotic and abiotic factors. Generally, tall, woody species of sagebrush including big sagebrush (*Artemisia tridentata* ssp.), silver sagebrush (*A. cana*), and threetip sagebrush (*A. tripartita*) are used by sage-grouse throughout the year in all seasonal habitats (Griner 1939, Patterson 1952, Dalke et al. 1963). Other species of short sagebrush, such as little sagebrush (*A. arbuscula*) and black sagebrush (*A. nova*), may provide important seasonal habitat components during spring and winter (Griner 1939, Patterson 1952, Dalke et al. 1963). Shrub species such as rabbitbrush (*Chrysothamnus* spp.), antelope bitterbrush (*Purshia tridentata*), and spineless horsebrush

(*Tetradymia canescans*) are at times used for nesting and hiding cover by grouse (Patterson 1952, Dalke et al. 1963, Connelly et al. 1991), but they are not critical for persistence of sage-grouse populations.

Sage-grouse habitat use varies throughout the year; consequently, the importance of some vegetation characteristics changes seasonally. Moreover, assessments of sage-grouse habitat use are often made at varying scales. Much information is available on use areas (e.g., nest habitat and winter foraging areas); and more recently, data on habitat characteristics at landscape scales have been published. We summarize seasonal habitat needs (microscale), compare habitat characteristics in different segments of the species' range, and assess characteristics of landscapes (macroscale) occupied by sage-grouse. Unfortunately, less is known about sage-grouse use of habitats dominated by silver sagebrush compared to those of big sagebrush, and care should be taken when extrapolating the reported values for range-wide considerations. We attempt to provide comparable measures of seasonal habitats in the following examination, but estimated habitat values depend on field techniques used (Connelly et al. 2003b). Thus, not all studies are directly comparable, and values should be viewed cautiously.

BREEDING HABITAT

Sage-grouse breeding habitats are defined as those where lek attendance, nesting (including prenesting activity by hens), and early brood-rearing occur (Connelly et al. 2000c; 2003b). These habitats are sagebrush-dominated shrub steppe, typically consisting of large, relatively contiguous sagebrush stands, and are critical for survival of sage-grouse populations (Connelly et al. 2000c, Leonard et al. 2000, Holloran et al. 2005). Our discussion includes information on habitat selection and functions of the four components of breeding habitat: lekking, prenesting, nesting, and early brood-rearing.

Leks

Sage-grouse are polygamous and exhibit consistent breeding behavior each year on traditionally occupied leks (Patterson 1952, Wiley 1978). Connelly et al. (2003b) defined a *lek* as a traditional display area where two or more male sage-grouse have attended in two or more of the previous 5 years. The area is normally located in an open site in or adjacent to sagebrush-dominated habitats. Other investigators have also reported sage-grouse leks in or adjacent to sagebrush-dominated nesting habitat (Patterson 1952, Wakkinen et al. 1992). Lek locations may persist for many years. Scott (1942) reported a lek as active in 1940; this lek was still active 28 years later (Wiley 1973b). Moreover, Dalke et al. (1963) described the presence of small bird-point arrowheads on and near leks in eastern Idaho and suggested their occurrence indicated sage-grouse use for at least 80 years.

Leks are normally in relatively open areas, with less herbaceous and shrub cover than surrounding areas (Dingman 1980, Klott and Lindzey 1990). Leks commonly occur on comparatively gentle terrain (Rogers 1964), but they may also occur in valley bottoms or draws (Patterson 1952, Rogers 1964). Nisbet et al. (1983) developed a lek preference model that included slope (<10%), precipitation (>25 cm), distance to nearest water source (<2 km), and predicted encroachment by pinyon (*Pinus* spp.)-juniper (*Juniperus* spp.) woodlands. Rogers (1964) reviewed characteristics of 120 leks throughout Colorado during 1953–1961 and reported 50% were in sagebrush, 54% on gentle slopes, and 55% in bottoms; only 5% were within 200 m of a building; and, although 42% were >1.6 km from an improved county road, and 26% were within 100 m of a county or state highway. These estimates are likely biased because leks relatively close to roads would be more visible to observers than those farther away. In Nevada and Utah, 17 of 41 leks were in black sagebrush habitats, and 9 of 41 were in big sagebrush habitat; the other leks were in areas dominated by other shrubs and grass (Nisbet et al. 1983). Sagebrush within 1.5 km of active leks in North and South Dakota was taller than sagebrush around inactive leks (Smith et al. 2006). In addition, sagebrush density, forb cover, and bare ground were greater within 1.5 km of active leks in North Dakota than around inactive leks (Smith et al. 2006). Mating areas (arenas) within leks in North Park, Colorado, had an average vegetation canopy cover of only 7.3% and a mean height of 5.3 cm; sagebrush species present included big sagebrush, alkali sagebrush (*Artemisia arbuscula* ssp. *longiloba*), and black sagebrush (Petersen 1980).

Leks may be natural openings within sagebrush communities (e.g., dry stream channels, ridges, and grassy meadows) or openings created by human disturbances (e.g., edges of stock ponds, burned areas, gravel pits, sheep bed grounds, remote landing strips, plowed fields, and roads) (Patterson 1952, Dalke et al. 1963, Rogers 1964, Connelly et al. 1981, Hofmann 1991). Leks are typically adjacent to relatively dense sagebrush stands (Wakkinen et al. 1992) used for escape, nesting, and feeding cover (Patterson 1952, Gill 1965).

Wiley (1973b) reported that selection of leks by sage-grouse occurs at coarse and fine resolutions. At coarse resolutions, leks appear to be in sparse shrubby vegetation (Wiley 1973b). At fine resolutions, male sage-grouse choose sod-forming grasses or bare ground for display. Lek selection at both resolutions increased the conspicuousness and freedom of movement of displaying males (Wiley 1973b). The most important characteristic for leks may be their proximity to nesting habitat, which would be consistent with theories of lek evolution and mating behavior (Gibson 1996). Lek habitat is not considered limiting to sage-grouse populations (Schroeder et al. 1999).

Leks in Wyoming generally ranged from 0.25 to 16 ha in size, but one lek was 20 ha, with 400 strutting males (Scott 1942). Leks on open or cleared areas in southeastern Idaho ranged from 0.04 to 4 ha in size (Dalke et al. 1963, Klebenow 1973), while in central Washington, the four largest leks in one study had a mean size of 36 ha (Hofmann 1991).

Sage-grouse populations may be nonmigratory or migratory, moving >10 km between or among seasonal habitats (Connelly et al. 2000c). Leks for nonmigratory populations may occur near the center of the annual range (Eng and Schladweiler 1972, Wallestad and Pyrah 1974, Wallestad and Schladweiler 1974). Migratory populations typically do not exhibit this pattern (Dalke et al. 1963, Wakkinen et al. 1992). Travel by females dispersing between wintering and nesting areas, rather than vegetation type, may influence lek locations (Bradbury et al. 1989, Gibson 1996).

Leks often occur in complexes composed of one to two primary or large leks (>50 males), one or more smaller leks, and at times, satellite leks. Smaller or declining populations may simply consist of a few smaller leks (Connelly et al. 2004). A satellite lek is defined as a relatively small lek (usually <15 males) that develops near a large lek during years with relatively high grouse populations (Connelly et al. 2003b). Occurrence of satellite leks fluctuates depending upon population size, and satellite leks may not be used in years when populations are low (Dalke et al. 1963). In a study of 31 leks in Idaho, mean interlek distance was about 1.6 km (Wakkinen et al. 1992). Of 13 leks examined in the Upper Snake River Plain in Idaho (Klebenow 1969), 2 had an interlek distance of 0.8 km, but the distance was 2.4 km for 8 others. In Wyoming, lek density averaged 6.8 leks per 100 km^2 within a water-reclamation project area, and 8.4 leks per 100 km^2 in nearby, undeveloped sagebrush habitats (Patterson 1952). Willis et al. (1993) reported similar lek densities in Oregon.

During the breeding season, males display in early morning and evening hours, traveling up to 2.1 km (Ellis et al. 1987) from the lek to day-use feeding and resting areas. Male day roost locations in northeastern Utah were generally 0.5–0.8 km from the lek (Ellis et el. 1989), and 82% of male day roost locations in central Montana were between 0.3 and 1.8 km from the lek (Wallestad and Schladweiler 1974). In central Montana, sagebrush canopy cover at 51% of male day roost locations was between 20% and 40%, and no day roost locations were recorded in areas with ≤10% sagebrush canopy cover (Wallestad and Schladweiler 1974).

After attending leks, male sage-grouse in northeastern Utah used areas near leks that had comparatively greater canopy cover ($\bar{x}$ = 31%) and taller shrubs ($\bar{x}$ = 53 cm) than did nearby nonuse areas (Ellis et al. 1989). Minimum core day-use areas of males were 0.25 km^2, and grouse often walked to such sites from leks for feeding and loafing (Ellis et al. 1989).

Prenesting Habitat

Sage-grouse females move to the vicinity of their nest location within a few days of mating and remain relatively sedentary until they nest (Patterson 1952). Sage-grouse diets change from sagebrush to forbs during spring, as forbs become available (Barnett and Crawford 1994). Forbs provide increased levels of calcium, phosphorus, and protein that may affect nest initiation rate, clutch size, and reproductive rates (Barnett and Crawford 1994, Coggins 1998). Little information is available on prenesting habitat selection.

Nesting Habitat

Sage-grouse nesting habitat usually includes a broad area within or adjacent to winter range or between winter and summer range (Klebenow 1969, Wakkinen 1990, Fischer 1994) dominated by sagebrush with horizontal and vertical structural diversity (Wakkinen 1990, Gregg 1991, Schroeder et al. 1999, Connelly et al. 2000c). The understory of nesting habitat is composed of native grasses and forbs that provide herbaceous forage for prelaying and nesting hens, concealment of the nest and hen, and a food source of insects (Gregg 1991, Schroeder et al. 1999, Connelly et al. 2000c, Holloran et al. 2005).

Sage-grouse nest in many different sagebrush-dominated cover types; most nests are under sagebrush plants (Patterson 1952, Gill 1965, Wallestad and Pyrah 1974, Petersen 1980, Drut et al. 1994a, Gregg et al. 1994, Sveum et al. 1998b). Throughout Wyoming, 92–100% of nests were under sagebrush (Patterson 1952, Rothenmaier 1979, Holloran 1999); 90% of nests were under silver sagebrush plants in southern Canada (Aldridge and Brigham 2002); and 94% of nests were under big sagebrush plants in northern Colorado (Petersen 1980). In Utah, 70% of nests were under big sagebrush, 17% under black sagebrush, and 13% under other shrubs or grass (Dahlgren 2006). In southeastern Idaho, 21% of sage-grouse hens nested under shrub species other than sagebrush (Connelly et al. 1991). Similarly, Popham and Gutiérrez (2003) reported that 41% of sage-grouse nests in California were under shrubs other than big sagebrush. Other plants used by sage-grouse for nest cover include greasewood (*Sarcobatus vermiculatus*), bitterbrush, rabbitbrush, horsebrush, snowberry (*Symphoricarpos* spp.), shadscale saltbush (*Atriplex confertifolia*), mountain mahogany (*Cercocarpus* spp.), and basin wildrye (*Leymus cinereus*) (Patterson 1952, Klebenow 1969, Wakkinen 1990, Connelly et al. 1991, Popham and Gutiérrez 2003).

Sage-grouse nests have been consistently described in the literature as being under larger bushes (Wakkinen 1990, Gregg 1991, Fischer 1994, DeLong et al. 1995, Holloran 1999) with more obstructing cover (Wakkinen 1990, Fischer 1994, Popham and Gutiérrez 2003) within shrub patches when compared to random sites. In addition, nesting habitat had more sagebrush canopy cover (Klebenow 1969, Fischer 1994, Sveum et al. 1998b, Holloran 1999, Aldridge and Brigham 2002) and taller sagebrush (Wallestad and Pyrah 1974, Sveum et al. 1998b, Holloran 1999, Lyon 2000, Slater 2003, Holloran et al. 2005) compared to available habitats.

Selection of specific habitat features, such as sagebrush height and canopy cover within a landscape, by nesting sage-grouse has been extensively documented. Connelly et al. (2000b) suggested that nesting habitat within sagebrush stands should contain between 15% and 25% canopy cover. In central Montana, all nests located were in areas with >15% sagebrush canopy cover (Wallestad and Pyrah 1974). Females preferentially selected areas with sagebrush canopy cover of 15–50% for nesting in Utah (Rasmussen and Griner 1938). Silver sagebrush canopy cover was dominant at nest sites (31.9 ± 4.1%) in southern Canada and greater than at random sites (15.7 ± 2.4%) (Aldridge and Brigham 2002). Moreover, canopy cover of silver sagebrush was the only variable that discriminated between nests and random sites (Aldridge and Brigham 2002). Female sage-grouse in South Dakota selected areas with greater canopy cover of sagebrush and nest bowl visual obstruction compared to random sites (Kaczor 2008). Moreover, taller grass structure was associated with increased nest success in South Dakota (Kaczor 2008). In southeastern Wyoming, mean sagebrush canopy cover was 21.6% (range = 12–29%) and average sagebrush height was 30.6 cm (range = 22.6–38.1 cm) within a 37.2-m^2 plot surrounding nests (Rothenmaier 1979), while females selected areas with 36- to 63.5-cm tall sagebrush for nesting in Utah (Rasmussen and Griner 1938). In western Wyoming, 83% of nests were under bushes between 25 and 51 cm in height ($\bar{x}$ nest bush height = 35.6 cm; Patterson 1952). Petersen (1980) reported sagebrush height and canopy cover was 32 cm and 24% within 15 m of nests in northern Colorado. In Wyoming, higher total shrub canopy cover and taller live sagebrush occurred in the nest area than at random sites (Holloran 1999, Lyon 2000, Slater 2003), and mean height of nest bushes (46.4 cm) was greater than mean height of shrubs in the surrounding area (Holloran 1999). Taller average sagebrush heights (40.4 vs. 23.4 cm) occurred near nests compared to random locations in central Montana (Wallestad and Pyrah 1974). Sage-grouse hens in southeastern Idaho nested under taller bushes with a larger area and greater lateral obstructing cover compared to random sites within the same shrub patch (Wakkinen 1990). Fischer (1994) continued Wakkinen's study for an additional 3 years and indicated that nests were in areas with increased nest bush total area and height, ground-obstructing cover, lateral obstructing cover, sagebrush density of shrubs ≥40 cm tall, and total shrub canopy cover compared to dependent random sites between 40 and 200 m from the nest (Fischer 1994). Additionally, in southeastern Idaho, nests within habitat dominated by threetip sagebrush were in areas with higher big sagebrush density (13.3 vs. 1.6 plants/122 m^2), threetip sagebrush canopy cover (14.1% vs. 12.5%), and basal area of grasses (3.7% vs. 2.9%) compared to random plots within the same habitat type (Klebenow 1969). Overall, total shrub canopy cover was greater at nests compared to random locations (18.4% vs. 14.4%; Klebenow 1969). In south-central Washington, nests were consistently in areas with more shrub cover at or within 5 m of the nest compared to randomly selected sites (Sveum et al. 1998b).

Early investigations tended to focus on sagebrush overstory, but more recent research has also addressed the importance of herbaceous characteristics of nesting habitat. Numerous investigators reported nest locations had taller live and residual grasses, more residual grass cover, and less bare ground (Klebenow 1969; Heath et al. 1997, 1998; Wakkinen 1990; Sveum et al. 1998b; Holloran 1999; Lyon 2000; Slater 2003) compared to random plots. Similarly, residual grass cover and residual grass height in Wyoming were important determinants of Greater Sage-Grouse nesting habitat (Holloran et al. 2005). Nests in southeastern Idaho had higher ground obstructing cover and lateral obstructing cover compared to random sites (Fischer 1994). Additionally, increased forb cover (14.5–18.2% vs. 6.8–12.8%), food forb (within the study area, forbs known to be used as food by sage-grouse) cover (3.1–5.6% vs. 0.5–1.9%), and tall (>18 cm) grass cover (4.7–17.2% vs. 0.3–4.7%) was correlated with increased overall nest initiation rates (99% vs. 65%), renesting rates (30% vs. 14%), and nesting success rates (37% vs. 22%) in southeastern Oregon (Coggins 1998). Herbaceous differences within 2.5 m of nests compared to random plots in Wyoming included more total herbaceous (Lyon 2000), nonfood forb (forbs not known to be used for food by sage-grouse) (Holloran 1999), and total forb cover (Lyon 2000).

A meta-analysis of sage-grouse nesting habitat data indicated sagebrush cover and grass height were greater at nest sites than at random sites (Hagen et al. 2007). This analysis concluded that overall estimates of nest vegetation variables were consistent with those published in guidelines for managing sage-grouse habitats (Connelly et al. 2000c). Undoubtedly, a great deal of evidence now indicates that live sagebrush overstory and herbaceous understory are critical components of sage-grouse nest habitat.

Mean distance from lek of capture to nest sites was 3.4–4.6 km in southeastern Idaho (Wakkinen et al. 1992, Fischer 1994), 8.6 km (range = 0.4–63.8 km) in west-central Wyoming (Lyon 2000), 2.7 km in Montana (Wallestad and Pyrah 1974), 4.0 km in Colorado (Hausleitner 2003), and 7.8 km in Washington (Schroeder et al. 1999). Other studies have illustrated similar variation (Berry and Eng 1985, Hanf et al. 1994, Holloran 1999, Lyon and Anderson 2003, Slater 2003). Although Braun et al. (1977) indicated that most hens nest within 3.2 km of a lek, recent literature suggests many hens nest farther from their lek of capture than previously documented. In southwestern Wyoming, between 75% and 87% of nests found were within 5 km of the lek of capture (Slater 2003). In a nonmigratory population in central Montana, 68% of hens nested within 2.5 km of the lek of capture (Wallestad and Pyrah 1974). No differences were found among nest-to-lek versus random point–to-lek distances in southeastern Idaho, suggesting hens chose nests without regard to lek locations (Wakkinen et al. 1992).

Distances between consecutive-year nests (females followed through consecutive nesting seasons) suggests female fidelity to nesting areas. Mean distance between consecutive-year nests averaged 552 m in southwestern Wyoming (Berry and Eng 1985), 710 m in central Wyoming (Holloran 1999), and 683 m in west-central Wyoming (Lyon 2000). Median distance between consecutive-year nests was 740 m in southeastern Idaho (Fischer et al. 1993), and 67% of consecutive-year nests were <600 m apart in southwestern Wyoming (Slater 2003).

Researchers reported somewhat conflicting results regarding the relationship between nest success and vegetation. Successful nests in Montana were in areas of higher sagebrush density than unsuccessful nests, and canopy cover of sagebrush was greater (27%) in stands with successful nests compared to unsuccessful nests (20%) (Wallestad and Pyrah 1974). In southeastern Oregon, the canopy cover of medium-height shrubs (40–80 cm) and tall grasses (>18 cm) was higher at successful nests than unsuccessful nests or random sites (Gregg et al. 1994). Similarly, increasing grass cover was related to improved nest success in northwestern Nevada (Rebholz 2007). Connelly et al. (1991) reported that sage-grouse nesting under sagebrush experienced greater nest success (53%) than those nesting under nonsagebrush (22%). In contrast to many studies, vegetation characteristics had no relationship to nesting success in southern Idaho (Wakkinen 1990), nor were there differences in success for nests beneath big sagebrush compared to other shrub species in Washington (Sveum et al. 1998b). Moreover, nesting success for nonsagebrush nests (42%) was higher than sagebrush nests (31%) in California (Popham and Gutiérrez 2003). Use of nonsagebrush shrubs may occur because grouse are seeking an appropriate nest shrub associated with herbaceous cover (Connelly et al. 1991). If grouse cannot find this cover associated with a sagebrush plant, they may seek another species of shrub. Thus, differences in herbaceous cover may be as important or more important than shrub species in affecting nest success and may explain the different findings of various investigators.

Herbaceous vegetation characteristics that were consistently higher at successful versus unsuccessful sage-grouse nests throughout the range of studied populations included live and residual grass height and cover (Wakkinen 1990, Gregg et al. 1994, Sveum et al. 1998b, Aldridge and Brigham 2002, Hausleitner 2003, Holloran et al. 2005, Woodward 2006, Moynahan et al. 2007), forb cover (Holloran 1999, Hausleitner 2003), and visual obstruction (Wakkinen 1990, Popham 2000, Slater 2003). Moreover, in California, percent rock cover (rocks >10 cm in diameter; 27.7% vs. 14.5%), total shrub height (65.5 vs. 49.2 cm), and visual obstruction (40.2 vs. 32.5 cm) were greater at successful than unsuccessful nest sites (Popham and Gutiérrez 2003).

No relationships between lek of capture–to-nest distances and nesting success were reported in southeastern Idaho (Wakkinen et al. 1992), but Popham and Gutiérrez (2003) found that mean distance from lek to nest site was greater for successful than for unsuccessful nests in California (3.6 vs. 2.0 km, respectively). High nest densities surrounding a lek may result in increased predation through destruction of multiple nests in a given area by predators (Niemuth 1992, Popham 2000).

Early Brood-Rearing Habitat

Some investigators have grouped brood-rearing habitat data for spring and summer, but hens rear their broods for at least the first 2–3 weeks in the vicinity of their nest (Berry and Eng 1985) before moving to summer range (Connelly et al. 1988). Thus, early brood-rearing habitat is defined as sagebrush-dominated habitat within the vicinity of the nest used by sage-grouse hens with chicks up to three weeks following hatching (Connelly et al. 2000c). These areas are characterized by a sagebrush overstory and a healthy herbaceous understory containing an abundance of insects critical for survival of young chicks (Connelly et al. 2000c, Johnson and Boyce 1990). Early brood-rearing areas were 0.2–5.0 km ($\bar{x}$ = 1.1 km) from the nest in west-central Wyoming (Lyon 2000). Slater (2003) found 80% of early brood locations were within 1.5 km of the nest in southwest Wyoming. Movements from nest to early brood-ing areas in northern Colorado were between 0.3 and 2.3 km ($\bar{x}$ = 0.8 km; Petersen 1980).

In central Montana, 100% of brood observations during June were in sagebrush-grassland habitats (Peterson 1970); during June and July, size of brood-use areas averaged 86 ha (Wallestad 1971). Between 75% and 80% of brood locations from 1 through 15 June were in areas with 1–25% sagebrush canopy cover (Wallestad 1971). In south-central Wyoming, 68% of sage-grouse brood locations were in sagebrush-grass or sagebrush-bitterbrush habitats (Klott and Lindzey 1990). During early brood-rearing in south-central Washington, brooding females selected for big sagebrush–bunchgrass habitats and against grassland habitats (areas devoid of sagebrush); 70% of locations were within big sagebrush–bunchgrass habitats (Sveum et al. 1998a).

Compared to nesting habitat, early brood-rearing locations in central Wyoming had less live sagebrush (15.8% vs. 25.4%) and total shrub (19.3% vs. 30.5%) canopy cover, shorter average sagebrush heights (25.5 vs. 31.4 cm), and more total herbaceous (37.3% vs. 29.6%) cover (Holloran 1999). Additionally, total forb (9.3% vs. 7.3%), food-forb (3.6 vs. 1.8%), and bare ground (7.3% vs. 5.0%) cover tended to be higher in early brood-rearing habitats than in nesting habitats (Holloran 1999). Similarly, brood-use sites within big sagebrush–dominated habitats in southeastern Idaho had lower big sagebrush density (64 vs.

104 plants/37 m^2 plot) and canopy cover (8.5% vs. 14.3%), and higher percent frequency of common yarrow (*Achillea millefolium*; 23.5% vs. 9.4%), tailcup lupine (*Lupinus caudatus*; 18.3% vs. 7.5%), common dandelion (*Taraxacum officinale*; 12.0% vs. 3.1%) and common salsify (*Tragopogon dubius*; 2.2% vs. 0.3%) compared to random locations within the same vegetation type (Klebenow 1969). A combination of more residual grass and total forb cover, and shorter effective vegetation height was the best predictor of early brood-rearing-use areas compared to available habitats in central Wyoming (Holloran 1999). Early brood-rearing locations had less live sagebrush (15.8% vs. 20.2%) and total shrub (19.3% vs. 24.1%) canopy cover, more residual grass (2.9% vs. 2.0%), total forb (9.3% vs. 6.6%), and total herbaceous (37.3% vs. 29.4%) cover relative to available habitats (Holloran 1999). In west-central Wyoming, early brood-rearing locations had less live sagebrush density (1.9 vs. 2.3 plants/m^2), live sagebrush cover (21.5% vs. 27.0%), total shrub canopy cover (30.0% vs. 35.0%), and bare ground (23.5% vs. 39.6%) compared to available habitat (Lyon 2000). Lyon (2000) also found more total herbaceous (24.8% vs. 9.1%) cover at early brood-rearing locations compared to available habitat. In comparison, early brood-rearing locations had more sagebrush cover compared to random locations (8.7% vs. 4.5%) in southern Canada (Aldridge and Brigham 2002), where overall sagebrush cover values are much lower than in other portions of the species' range. Total forb (25% vs. 8%) and food forb (8% vs. 2%) cover were higher, and residual herbaceous cover (1% vs. 3%) and height (1 vs. 3 cm) were lower within 10 m of early brooding areas compared to random locations in south-central Washington (Sveum et al. 1998a).

When sage-grouse broods were found in grass-forb open areas in south-central Wyoming, use sites had more shrub cover relative to random openings, and dandelion, knotweed (*Polygonum* spp.), yarrow, and common salsify were more abundant at brooding sites than at random sites (Klott and Lindzey 1990). In southeastern Oregon, key forbs (those occurring in the crops of at least 10% of collected chicks or having aggregate mass ≥1%) were higher in habitats preferentially selected by broods relative to other habitats (Drut et al. 1994b). In northwestern Nevada, increasing grass cover was negatively associated with chick survival, while increasing total forb cover was associated with increased chick survival (Rebholz 2007). Rebholz (2007) also indicated that grouse generally did not forage in areas with dense grass cover and speculated that forb cover in these areas may provide fewer foraging opportunities compared to areas with higher forb cover. In southeastern Idaho, Fischer (1994) reported higher Hymenoptera abundance and higher Orthoptera frequency (no difference in abundance), but no difference in abundance of Coleoptera at brood-use versus random sites. In contrast, vegetation and insect quality differences between selected early brood-rearing and random locations were not detected in a Wyoming study (Slater 2003). During the early stages of life, sage-grouse broods consistently selected areas with more forb (Klebenow 1969, Klott and Lindzey 1990, Sveum et al. 1998a, Holloran 1999) and total herbaceous (Holloran 1999, Lyon 2000) cover, and less shrub canopy cover (Klebenow 1969, Holloran 1999, Lyon 2000), than at randomly selected areas. Clearly, availability of forb-rich habitats in close proximity to adequate protective cover is an important characteristic of early brood habitat (Klebenow 1969, Klott and Lindzey 1990, Sveum et al. 1998a, Holloran 1999, Lyon 2000). These habitat features result in an abundance of food items for chicks, particularly forbs (Drut et al. 1994b) and insects (Johnson and Boyce 1990, Fischer 1994).

Meta-analysis of data on sage-grouse brood-rearing habitat provided evidence that brood-use areas had less sagebrush cover, taller grass, greater forb cover, and greater grass cover than random locations (Hagen et al. 2007); these patterns were especially evident when researchers analyzed early and late brood-rearing habitats separately. Hagen et al. (2007) concluded that overall estimates of brood-area vegetation variables were consistent with those published in guidelines for managing sage-grouse habitats (Connelly et al. 2000c).

SUMMER AND LATE BROOD-REARING HABITAT

Summer or late brood-rearing habitats are those areas used by sage-grouse following desiccation of herbaceous vegetation in sagebrush uplands (Klebenow and Gray 1968, Savage 1969, Fischer et al. 1996a). The beginning of late brood-rearing coincides with the change in diets of sage-grouse chicks from predominantly insects to forbs

(Patterson 1952, Klebenow and Gray 1968, Klebenow 1969, Peterson 1970, Drut et al. 1994b). Sage-grouse exploit a variety of habitats—riparian, wet meadows, and alfalfa (*Medicago* spp.) fields—during summer. These habitats are generally used from July to early September, but use varies annually due to yearly weather conditions (Patterson 1952, Dalke et al. 1963, Gill and Glover 1965, Savage 1969, Wallestad 1971, Connelly et al. 1988, Gregg et al. 1993). Sage-grouse often use sagebrush-dominated habitats throughout the summer but select habitats based on availability of forbs. This is frequently accomplished by moving up in elevation or selecting sites where moisture collects and maintains forbs throughout the summer (Martin 1970, Wallestad 1971, Fischer et al. 1996a, Hausleitner 2003). These movements vary in response to plant moisture, vegetal cover, and elevation (Dalke et al. 1960, Wallestad et al. 1975, Connelly 1982). Broods in Idaho typically move up in elevation, following vegetation phenology to feed on succulent forbs (Klebenow 1969). Sage-grouse in southeastern Idaho moved as far as 82 km from breeding and nesting to summer ranges (Connelly et al. 1988, Fischer et al. 1997). Klebenow and Gray (1968) observed grouse migrating as far as 8–24 km to summer ranges at higher elevations ranging from 1,600 to >2,150 m. In comparison, broods in Montana only traveled relatively short distances to summer habitats, moving up to 5 km; some broods also remained in sagebrush by seeking out microhabitats such as small swales or ditches where forbs were still available (Wallestad 1971). A lack of shift in habitat selection between early and late brood-rearing may also suggest no difference in availability of forbs in the area (Aldridge and Brigham 2002). Aldridge (2000) suggested that broods do not move from sagebrush uplands to more mesic sites during wet years, and that wetland complexes may be limiting in dry years because of low food availability, ultimately causing low recruitment. Sage-grouse may also associate with agricultural lands and irrigated lawns during summer (Gates 1983, Connelly et al. 1988). Sage-grouse in Idaho moved to late brood-rearing habitats when moisture amounts in vegetation declined to ≤60% (Fischer et al. 1996a). Sage-grouse movements to breeding, nesting, and summer ranges may also be influenced by tradition. For instance, in Idaho, significantly more radio-marked sage-grouse than expected moved to traditional summer areas, rather than to closer (15–20 km) irrigated agricultural fields (Fischer et al. 1997). Wallestad (1971) also observed hens moving 5 km to summer habitat, bypassing a comparable area that was 3.2 km closer.

Adult and juvenile sage-grouse in Nevada used sagebrush adjacent to mesic areas during summer as loafing sites and for cover (Savage 1969), but midday locations in Washington had greater shrub cover and height compared to morning and afternoon loafing locations (Sveum et al. 1998a). Sage-grouse selected feeding habitat near edges of cover types with more horizontal and vertical cover and less variation in shrub density and size compared to random sites (Dunn and Braun 1986b). Hens with broods also used sites with more horizontal cover and greater variation in sagebrush canopy cover than random sites to roost, but they fed in open, relatively homogeneous areas during the morning and afternoon periods (Dunn and Braun 1986b). Similarly, broods in Wyoming used large openings and meadows, foraged on edges, and avoided the centers (Klott and Lindzey 1990). In Colorado, female night-roost sites had less bare ground and visual obstruction, but greater forb cover than at random sites (Hausleitner 2003). These areas may increase foraging opportunity by hens with broods with high energetic demands and provide greater escape potential from predators (Hausleitner 2003).

Forb canopy cover averaged 33% at brooding sites in Montana over two distinctly different (in terms of precipitation) summers (Peterson 1970). In north-central Colorado, young broods used areas with low forb canopy cover ($\bar{x}$ = 6.9%) after hatching and then quickly moved to meadows with far greater ($\bar{x}$ = 41.3%) forb canopy cover (Schoenberg 1982). Female grouse selected brood-rearing sites with higher average forb canopy cover (8% vs. 4%) and less bare ground than random sites (Hausleitner 2003). In Washington, hens selected areas with 19–27% forb canopy cover for late brood rearing (Sveum et al. 1998a). In Idaho, sites used by sage-grouse broods had twice as much forb cover as did independent sites (Apa 1998). Researchers in southeast Alberta recorded average forb cover in late brood-rearing habitat of 12.6% and suggested that 12–14% forb canopy cover might represent the minimum needed for brood-rearing habitat (Aldridge and Brigham 2002). Characteristics of brood sites were measured from mid-June through August in

South Dakota; these areas had greater total vegetation cover, grass cover, sagebrush cover, Japanese brome (*Bromus japonicus*), and bluegrass (*Poa* spp.) cover than did random sites (Kaczor 2008). In addition, brood sites in South Dakota had greater visual obstruction and taller grass heights than random sites (Kaczor 2008).

Water from sources such as stock ponds, springs, or intermittent streams may be important to sage-grouse during hot, dry periods and could affect summer distribution (Patterson 1952). Nevertheless, movements to irrigated agricultural lands or high-elevation summer ranges are probably in response to lack of succulent forbs in an area rather than a lack of free water (Connelly and Doughty 1989). Connelly (1982) also suggested that grouse do not commonly use water developments even during relatively dry years but instead obtain moisture from consuming succulent vegetation. Water developments tend to attract other animals and may serve as a predator sink for sage-grouse (Connelly and Doughty 1989). Small reservoirs, streams, and other water bodies can, however, provide islands of succulent vegetation (Wallestad 1971).

AUTUMN HABITAT

Autumn is a transitional period for sage-grouse (Wambolt et al. 2002), when their diets change from a variety of forbs, insects, and sagebrush to predominantly sagebrush (Rasmussen and Griner 1938, Patterson 1952, Leach and Hensley 1954, Gill 1965, Wallestad et al. 1975). Autumn habitats used by sage-grouse can vary widely based on availability, elevation, topography, water, distance between summer and winter habitats, and weather conditions. These habitats are generally used from as early as late August to as late as mid-December (Patterson 1952, Dalke et al. 1963, Gill and Glover 1965, Savage 1969, Wallestad 1971, Connelly 1982, Connelly et al. 1988).

During early autumn, in addition to sagebrush, habitats used may include upland meadows, riparian areas, greasewood bottoms, alfalfa fields, and irrigated native hay pastures (Patterson 1952, Gill 1965, Savage 1969, Wallestad 1971, Connelly 1982). As vegetation in these habitats desiccates or is killed by frost, sage-grouse begin using sagebrush habitats more often and form larger flocks (Patterson 1952, Savage 1969). During early autumn in north-central Colorado, sage-grouse abandoned irrigated native hay meadows in response to the cessation of irrigation, mowing of hay, and killing frosts (Gill and Glover 1965). During a 7-year study in eastern Idaho, sage-grouse gathered in large flocks near water during autumn migration, watering from 10 to 30 minutes daily (Dalke et al. 1963). Autumn habitats used by sage-grouse in northeastern Wyoming had higher densities of sagebrush (3.1–7.4 plants/m^2) than average for the entire study area (Postovit 1981). Sage-grouse in Montana used habitats with greater sagebrush cover in autumn than during the late brood-rearing period (Wallestad 1971). This shift coincided with the transition to a diet of sagebrush, as sage-grouse broods that had occupied bottomland vegetation types (greasewood and alfalfa fields) shifted into sagebrush in late August and September (Wallestad 1971). However, some sage-grouse in southeastern Idaho did not return to sagebrush habitats until October or November (Connelly and Markham 1983). During autumn in Colorado, sage-grouse occupied the same upland sagebrush habitats used for breeding; their use in the autumn appeared random and not tied to lek location, as it was during the breeding season (Gill 1965). In Idaho, movements from autumn sagebrush habitats to winter range were generally slow and meandering, beginning in late August and continuing into December (Connelly et al. 1988). During periods of early, severe winter snowstorms, sage-grouse may begin migrations to winter habitats but, at the onset of milder weather later in the autumn, may return to sites adjoining late brood-rearing habitat (Patterson 1952). Sage-grouse in Utah typically moved to winter range in mid-November; this movement appeared to be independent of snow depth (Welch et al. 1990).

WINTER HABITAT

Winter habitats of sage-grouse are dominated by sagebrush, and during this period sage-grouse rely almost exclusively on sagebrush exposed above snow for forage and shelter (Rasmussen and Griner 1938, Patterson 1952, Remington and Braun 1985, Robertson 1991, Schroeder et al. 1999, Connelly et al. 2000c, Crawford et al. 2004). Sage-grouse habitat selection during winter is influenced by snow depth and hardness, topography (elevation, slope, and aspect), and vegetation height and cover (Gill 1965, Beck 1977,

Connelly 1982, Schoenberg 1982, Robertson 1991, Schroeder et al. 1999). In North Park, Colorado sage-grouse selected either relatively exposed, windswept ridges or draws and swales (Beck 1977, Schoenberg 1982). Both windswept ridges and draws provided access to sagebrush above snow for food and cover (Beck 1977, Schoenberg 1982). Winter habitats of sage-grouse generally are dominated by big sagebrush; however, little sagebrush and silver sagebrush communities are also used during winter (Schroeder et al. 1999, Crawford et al. 2004). Sage-grouse in Idaho and Nevada often use little sagebrush habitats, while other sagebrush-dominated habitats are used in proportion to their availability (Connelly 1982, Klebenow 1985). In contrast, 38% of winter observations occurred in little sagebrush areas in 1 year of a 2-year Oregon study, while 98% of winter observations were in mountain big sagebrush (*Artemisia tridentata* ssp. *vaseyana*) during the second year (Hanf et al. 1994).

The spatial distribution of sage-grouse in winter often is related to snow depth (Patterson 1952; Dalke et al. 1963; Gill 1965; Klebenow 1973, 1985; Beck 1975, 1977; Welch et al. 1990). During relatively severe winters, a large proportion of sagebrush may be snow covered and unavailable for roosting or foraging. At the onset of winter, sage-grouse typically move to lower elevations with greater exposure of sagebrush above snow (Patterson 1952) and taller sagebrush; in migratory populations, this movement may extend up to 160 km (Patterson 1952). Shrub density and structure, including height and canopy cover, also influence habitat selection by sage-grouse during winter. Connelly et al. (2000c) recommended that canopy cover of sagebrush in both arid and mesic sites should be maintained at 10–30% in wintering habitat and further reported that grouse used shrub heights of 25–35 cm above snow. In Colorado, female sage-grouse used more dense (68 plants/0.004 ha) stands of mountain big sagebrush than did males (46 plants/0.004 ha; Beck 1977). Sage-grouse selected wintering areas in Colorado having greater sagebrush cover and sagebrush heights (two to three times taller) at use versus random sites (Schoenberg 1982). Height of sagebrush at sites used by wintering sage-grouse was greater than at random sites; evidence also suggested that sage-grouse moved to taller sagebrush as snow depth increased (Connelly 1982).

In central Montana, sage-grouse foraged during winter in big sagebrush with a mean canopy cover of 28%; observations of grouse in relatively high (>20%) cover were more common than in areas with lower sagebrush cover (Eng and Schladweiler 1972). Sage-grouse may roost in snow burrows or snow forms, apparently for energy conservation (Beck 1977, Back et al. 1987). In Montana, winter roost sites were in sagebrush with a mean canopy cover of 26%, and usually on flat terrain (Eng and Schladweiler 1972). In Colorado, shrub height and percent slope of winter feeding-loafing sites did not differ from roosting sites (Beck 1977).

Shrub canopy cover on winter ranges generally varies from 6% to 43% (Schroeder et al. 1999). Investigators in central Oregon found that sagebrush canopy cover was typically >20% at winter-use sites (Hanf et al. 1994). Within these sites, however, grouse tended to use patches with lower canopy cover (12–16%); in this study, most of the winter-use sites were in mountain big sagebrush (Hanf et al. 1994). In central Montana, sage-grouse selected dense (>20% canopy cover) stands of big sagebrush during winter (Eng and Schladweiler 1972), whereas in central Idaho they preferred black sagebrush when these shrubs were available above the snow (Dalke et al. 1963). Robertson (1991) reported Wyoming big sagebrush (*Artemisia tridentata* ssp. *wyomingensis*) canopy cover and height were consistently greater at use sites when compared to random sites. In Utah, satellite imagery was used to classify winter habitat of sage-grouse into seven shrub categories (Homer et al. 1993). Wintering grouse preferred shrub habitats with medium to tall (40–60 cm) shrubs and moderate shrub canopy cover (20–30%; Homer et al. 1993). Sage-grouse avoided winter habitats characterized by medium (40–49 cm) shrub height with sparse (<14%) sagebrush canopy cover. Cover of grasses and forbs for wintering habitats generally is irrelevant, because of the nearly complete reliance of sage-grouse upon sagebrush during this period (Homer et al. 1993).

Topography also influences use of winter habitats by sage-grouse. Flocks typically occur on south- or southwest-facing aspects (Beck 1977, Crawford et al. 2004) and on gentle slopes (<5%; Beck 1977). Sheltered microsites ameliorate effects of wind, especially at low temperatures (Sherfy and Pekins 1995), and contribute to maintaining energy balance. In eastern Idaho, movements between summer and winter ranges involved a decrease in mean elevation of 446 m (Hulet 1983).

Fidelity to winter areas has not been well studied, although some evidence of fidelity to winter areas among years has been demonstrated in Washington (Schroeder et al. 1999) and Wyoming (Berry and Eng 1985). In Utah, sage-grouse showed less fidelity to winter range than to other seasonal ranges (Welch et al. 1990).

LANDSCAPE COMPONENTS OF SAGE-GROUSE HABITAT

Sage-grouse populations typically inhabit large, interconnected expanses of sagebrush and, thus, have been characterized as a landscape-scale species (Connelly et al. 2004). As an example, sage-grouse used an annual range of at least 2,764 km^2 in eastern Idaho (Leonard et al. 2000). Historically, the distribution of sage-grouse was closely tied to the distribution of the sagebrush ecosystem (Wambolt et al. 2002, Schroeder et al. 2004). However, populations of sage-grouse have been extirpated from areas throughout their former range (Wambolt et al. 2002, Schroeder et al. 2004, Aldridge et al. 2008), concomitant with habitat loss and degradation, and the species' current distribution is more fragmented than it was prior to settlement by Europeans and, perhaps, less closely aligned with that of sagebrush—some areas support sagebrush but not sage-grouse.

Causes of habitat loss, fragmentation, and degradation in sagebrush habitats vary and include brush control by fire, chemical, and mechanical means (Klebenow 1970, Martin 1970, Wallestad 1975b); inappropriate livestock management; energy development; urbanization; and infrastructure necessary to maintain these activities (Hulet 1983, Evans 1986, Braun 1986, Beck and Mitchell 2000, Braun et al. 2002, Bunting et al. 2002, Lyon and Anderson 2003). Increased fire frequency in lower-elevation sagebrush habitats, often closely tied to invasion of annual grasses such as cheatgrass (*Bromus tectorum*), has resulted in losses of sagebrush over large expanses in the Intermountain West and Great Basin (Mack 1981, Miller et al. 1994, Crawford et al. 2004). In addition, decreased fire frequency in higher-elevation sagebrush habitats and impacts from inappropriate livestock grazing and other factors have resulted in conifer encroachment and subsequent reduction of the herbaceous understory and sagebrush canopy cover over large areas (Miller and Rose 1995, Miller and Eddleman 2001, Crawford et al. 2004).

Few studies have examined landscape-level issues regarding sage-grouse populations and habitats. A negative relationship was demonstrated between mean numbers of males/lek and agricultural development during a 17-year period on the upper Snake River Plain in Idaho; nearly 30,000 ha of sagebrush in the study area were converted to cropland from 1975 to 1992 (Leonard et al. 2000). In North Park, Colorado, Braun and Beck (1996) examined lek counts in relation to habitat loss from both plowing and spraying with 2,4-D of >28% of the study area. Initial spraying of >1,600 ha occurred in 1965, with an additional 500 ha sprayed and 1,460 ha plowed and seeded during the following 5 years (Braun and Beck 1996). The 5-year mean of males on active leks declined 25%, from 765 (1961–1965) to 575 (1971–1975; Braun and Beck 1996). Numbers rebounded by 1976–1980, however, and even exceeded the pretreatment levels (5-year $\bar{x}$ = 1,109 males). A recent study comparing percentage of tilled versus nontilled land surrounding sage-grouse leks in North Dakota revealed that abandoned leks had a higher percentage of tilled lands within a 4-km buffer of leks than did active leks (Smith et al. 2005). Nevertheless, there was no increase in percentage of tilled land from the 1970s to the late 1990s, suggesting that if the amount of tilled land was a factor in lek abandonment, this effect had occurred prior to the 1970s (Smith et al. 2005).

Variables other than vegetation may influence nest site selection at a broader scale (Jensen 2006). Landscape characteristics in central Wyoming potentially influencing nest site selection included surface roughness (actual surface area/flat land area), surface ruggedness (standard deviation of the landscape), elevation, and slope (Jensen 2006).

Characteristics of sage-grouse winter habitat are relatively similar throughout most of the species' range and, on a landscape scale, winter habitats should allow grouse access to sagebrush under all snow conditions (Connelly et al. 2000c). Doherty et al. (2008) reported that percent sagebrush cover on the landscape was an important predictor of winter use by sage-grouse and that strength of sagebrush habitat selection by sage-grouse was strongest at the 4-km^2 scale. Sage-grouse in North Park, Colorado, concentrated during winter in seven small areas that totaled 85 km^2; these areas made up only 7% of the sagebrush in the entire study area (Beck 1977). Marked declines occurred in sage-grouse abundance in Montana when a large (30%) portion of the winter habitat was plowed, primarily for grain production (Swenson et al. 1987). Sagebrush removal in winter habitats may be especially detrimental because of the relatively long periods that winter habitat may be occupied by sage-grouse annually (Eng and Schladweiler 1972). Maintaining intact winter habitat for sage-grouse may also be an issue in areas of energy development, such as natural gas fields and coalbed methane, especially if several populations converge in a common wintering area (Lyon 2000, Doherty et al. 2008). Fragmenting sagebrush habitats may also change coarse-resolution distribution patterns of sage-grouse. During a study in Colorado, in which >120 flocks (>3,000 birds total) were observed during two winters, only four flocks were found in altered (by spraying with 2,4-D, plowing, burning, or seeding) sagebrush habitats, although >30% of the study area had been treated (Beck 1977).

Sage-grouse are considered a landscape species, but conclusive data are unavailable on minimum patch sizes of sagebrush necessary to support viable populations of sage-grouse. In Wyoming, sage-grouse flocks could range as widely as several thousand square kilometers (Patterson 1952). Migratory populations of sage-grouse may use areas exceeding 2,700 km^2 (Connelly et al. 2000c, Leonard et al. 2000). Sagebrush patches used by broods averaged 86 ha in early summer (June and July) in central Montana but diminished to 52 ha later in summer (August and September; Wallestad 1971). However, brood-use areas are relatively small compared to areas used on a year-round basis and thus only partially representative of the broader landscape needs of the species during the year.

Edge density (amount of edge per unit area) for sagebrush and grass-forb–dominated habitats could be combined with cover type information to assess the potential of areas that could be used by sage-grouse (Shepherd 2006). Moreover, amount of sagebrush cover may not necessarily predict sage-grouse use patterns without also considering amount of open areas and interspersion levels (Shepherd 2006).

Diversity, Mosaics, and Juxtaposition

Sagebrush overstory essential to sage-grouse is both spatially and temporally diverse due to the extensive geographic region occupied by the sagebrush ecosystem in North America (Schroeder et al. 1999, Miller and Eddleman 2001). Sage-grouse use different heights and canopy cover of sagebrush seasonally, and overall, these values range from 25 to 80 cm in height and from 12% to 43% in canopy cover (Eng and Schladweiler 1972, Schoenberg 1982, Gregg 1991, Heath et al. 1997, Apa 1998, Holloran 1999, Aldridge and Brigham 2002). Reviews of available literature (Connelly et al. 2000c, Hagen et al. 2007) indicate that sage-grouse use sagebrush habitats characterized by a range of values. One reason for variation in sagebrush canopy cover within seasons is likely due to the dynamic nature of sagebrush stands. Sagebrush canopy cover is not static but changes both before and after the stand matures. In southwestern Montana, canopy cover of Wyoming big sagebrush varied from 10.6% to 16.1% over an 18-year period (Wambolt and Payne 1986). Moreover, browsing by black-tailed jackrabbits (*Lepus californicus*) can affect shrub cover (including big sagebrush) in sagebrush steppe habitat (Anderson and Shumar 1986).

Within the sagebrush ecosystem, a wide range of understory vegetation is used by sage-grouse during the breeding and brood-rearing periods (Wakkinen 1990, Gregg 1991, Fischer 1994, Holloran 1999, Aldridge and Brigham 2002). Sage-grouse selected nest sites with grass heights averaging 18 cm and grass canopy cover of 3–10% in southeastern Idaho compared to grass heights averaging 16 cm and canopy cover averaging 32% in more mesic southern Alberta (Wakkinen 1990, Aldridge and Brigham 2002). Early brood-rearing habitats reported by Holloran (1999) in Wyoming averaged 19 cm in height and 5% canopy cover for grasses, while Aldridge and Brigham (2002) reported findings of 45 cm and 34%, respectively. However, the greater height and canopy cover for grasses in brood habitat in the more mesic sagebrush steppe–grassland transition in southern Alberta compared to drier intermountain shrub steppe is not unexpected.

Sage-grouse typically occupy habitats with a diversity of species and subspecies of sagebrush but may also use a variety of other habitats, including riparian meadows, agricultural lands, and steppe dominated by native grasses and forbs, shrub willow (*Salix* spp.), and sagebrush habitats with some conifer or quaking aspen (*Populus tremuloides*) (Patterson 1952, Dalke et al. 1963). These habitats are usually intermixed in a sagebrush-dominated landscape (Griner 1939, Patterson 1952, Dalke et al. 1963, Savage 1969). Sage-grouse have used habitats altered by man throughout the species' range, including crested wheatgrass (*Agropyron cristatum*) seedings and different agricultural crops (Patterson 1952, Gates 1983, Connelly et al. 1988, Blus et al. 1989, Sime 1991). Leks are often in altered areas, including dirt roads and areas seeded with crested wheatgrass; however, these areas are adjacent to sagebrush stands that provide nesting and early brood-rearing habitat. By itself, evidence of use does not imply importance. The value of these habitats to sage-grouse in meeting their seasonal habitat requirements is dependent on the juxtaposition of these habitats in relation to sagebrush and the hazards (Connelly et al. 2000c, Beck et al. 2006) to grouse using these areas.

Remotely sensed imagery and landscape metrics were used to assess patterns of habitat use by sage-grouse in areas with different levels of fragmentation in southwestern Idaho (Shepherd 2006). Results of this landscape-scale study indicated that sagebrush patches adjacent to large, abrupt patches of grass-forb–dominated habitats (usually burned areas or crested wheatgrass seedings) may be effectively smaller than more interspersed sagebrush patches because these large patches received much less sage-grouse use on their periphery (Shepherd 2006). Shepherd (2006) concluded that a large proportion of the landscape with 60–70% sagebrush land cover may be used by sage-grouse if it is well interspersed with grass-forb cover. Thus, interspersion and juxtaposition of cover types likely have a great influence on the effectiveness of a given part of the landscape to provide sage-grouse with usable habitat.

Migratory Corridors

Sage-grouse may move long distances between and among seasonal ranges (Connelly et al. 1988, Connelly et al. 2000c). Movements of 40–160 km by sage-grouse along established routes from leks to summer range and winter habitats to leks have been described (Dalke et al. 1963, Connelly 1982, Leonard et al. 2000). In eastern Idaho, the mean distance moved between summer and winter ranges was 48.2 km for 28 hens (Hulet 1983). Travel of 35 km from a lek to a winter area was recorded in southwestern Wyoming (Berry and Eng 1985). Movement corridors, distances moved by grouse, and landscape relationships of spring and summer seasonal ranges were described for four sage-grouse breeding populations on a 231,600-ha study area in southern Idaho (Connelly et al. 1988). Unfortunately, the distribution, configuration, and characteristics of migration corridors are often unknown throughout much of the range of sage-grouse. Moreover, differences in techniques used to measure movements and average seasonal ranges of sage-grouse make comparisons among studies difficult (Schroeder et al. 1999).

Migratory corridors are generally defined by the relationship between habitat configuration and seasonal movements and habitat requirements of sage-grouse (Dalke et al. 1963, Connelly et al. 1988, Beck et al. 2006). Seasonal movements by sage-grouse are traditional and may occur between two or among three seasonal ranges (Dalke et al. 1963, Beck 1977, Schoenberg 1982, Connelly et al. 1988, Wakkinen 1990, Robertson 1991, Fischer 1994, Connelly et al. 2000c, Jensen 2006). Sage-grouse in southeastern Idaho did not readily change traditional movements (Wakkinen 1990, Fischer et al. 1997) following habitat disturbance. Similarly, Leonard et al. (2000) re-examined migration routes first described in the 1950s (Dalke et al. 1963) and reported that these routes had not changed over that period, despite significant changes in seasonal habitat.

CONSERVATION IMPLICATIONS

Sage-grouse may use agricultural areas and other altered habitats during summer, but relatively large blocks of sagebrush habitat (>4,000 ha) are critical to successful reproduction and overwinter survival (Leonard et al. 2000, Walker et al. 2007a). Available evidence clearly supports the conclusion that conserving large landscapes with suitable habitat is important for conservation of sage-grouse (Eng and Schladweiler 1972, Connelly et al. 2004, Doherty et al. 2008). At times, some habitat loss may be unavoidable, especially if it is associated with habitat improvement programs, e.g., improving understory herbaceous cover. In those cases, Connelly et al.

(2000c) recommended retaining at least 80% of winter and breeding habitats if habitat alteration was necessary. Woodward (2006) argued that some portions of sage-grouse habitat may benefit from management actions that resulted in more herbaceous cover, but not if it results in loss of sagebrush. In central Montana, differences in sagebrush cover among nesting, brood, and winter habitats of sage-grouse were relatively small, and any manipulation that attempted to improve one seasonal habitat would impact others for this nonmigratory sage-grouse population (Woodward 2006). Pedersen et al. (2003) provided evidence that relatively large and persistent habitat disturbance may cause extirpation of sage-grouse populations. Given the strong reliance of sage-grouse on sagebrush habitats, the persistence of this grouse is obviously linked to the conservation of relatively large blocks of sagebrush-dominated habitat in good ecological condition. A broad-scale model recently related persistence of sage-grouse populations to human population density, amount of cultivated land, drought, distance from periphery of sage-grouse range, and sagebrush cover (Aldridge et al. 2008). These authors concluded that sage-grouse conservation efforts should begin by maintaining large expanses of sagebrush habitat and enhancing the quality and connectivity of those patches. Variation in cover and height of shrub overstory and herbaceous understory has been documented by numerous investigators. Thus, characterizing habitats using a single value or narrow range of values (e.g., 15% or 20–25% sagebrush canopy cover) is inappropriate (Connelly et al. 2000c). Natural variation in vegetation and the dynamic nature of mature sagebrush stands should be considered for all habitat descriptions and prior to any management action.

Sagebrush habitats have been lost, fragmented, and degraded as a result of many different anthropogenic disturbances (Connelly et al. 2004). Moreover, energy development activities have recently been linked to facilitating establishment of non-native plants (Bergquist et al. 2007). Complicating matters, the traditional nature of seasonal movements by sage-grouse suggests this species has little ability to adapt to habitat change by moving to other areas when a significant portion of its seasonal habitat is altered. Clearly, land management agencies must establish sagebrush conservation as one of their highest priorities if remaining habitats are to be maintained. Moreover, these agencies will have to develop and implement effective habitat reclamation measures to offset unavoidable losses. Failure to protect what is left and fix what is broken will likely result in extirpation of many, if not most, populations of Greater Sage-Grouse.

ACKNOWLEDGMENTS

We thank our many colleagues for stimulating discussions on these topics as they enhanced our awareness of issues across the range of sage-grouse. In particular, we are indebted to members of the Western Sage and Columbian Sharp-tailed Grouse Technical Committee for their support of sage-grouse research and willingness to share ideas and concerns. Reviews by L. D. Flake and J. L. Beck greatly improved an earlier draft of this chapter, and we sincerely appreciate their efforts. Much of our individual and collaborative research was funded through Federal Aid in Wildlife Restoration projects, and we express our thanks to our individual agencies. This essay is a contribution of Idaho Federal Aid in Wildlife Restoration Project W-160-R.

CHAPTER FIVE

Molecular Insights into the Biology of Greater Sage-Grouse

Sara J. Oyler-McCance and Thomas W. Quinn

Abstract. Recent research on Greater Sage-Grouse (*Centrocercus urophasianus*) genetics has revealed some important findings. First, multiple paternity in broods is more prevalent than previously thought, and leks do not comprise kin groups. Second, the Greater Sage-Grouse is genetically distinct from the congeneric Gunnison Sage-Grouse (*C. minimus*). Third, the Lyon–Mono population in the Mono Basin, spanning the border between Nevada and California, has unique genetic characteristics. Fourth, the previous delineation of western (*C. u. phaios*) and eastern Greater Sage-Grouse (*C. u. urophasianus*) is not supported genetically. Fifth, two isolated populations in Washington show indications that genetic diversity has been lost due to population declines and isolation.

Key Words: Centrocercus urophasianus, genetics, lek evolution, mating system, population genetics, taxonomic boundaries.

Nuevos Conocimientos Sobre la Biología Molecular del Greater Sage-Grouse

Resumen. Estudios recientes sobre la genética del Greater Sage-Grouse (*Centrocercus urophasianus*) han revelado algunos hallazgos importantes. Primero, la paternidad múltiple en las crías es mas prevalente que lo que se creía anteriormente, y los leks (asambleas de cortejo) no están compuestos por grupos de parentesco. Segundo, el Greater Sage-Grouse es genéticamente distinto del Gunnison Sage-Grouse (*C. minimus*). Tercero, la población de Lyon–Mono en el Mono Basin, en los límites entre Nevada y California, tiene unas características genéticas únicas. Cuarto, la delineación previa entre el Western Greater Sage-Grouse (*C. u. phaios*) y el Eastern Greater Sage-Grouse (*C. u. urophasianus*) no tiene soporte genético. Quinto, dos poblaciones aisladas en Washington muestran indicios de que la diversidad genética ha sido perdida debido a la disminución y al aislamiento de las poblaciones.

Palabras Clave: Centrocercus urophasianus, evolución de los leks, fronteras taxonómicas, genética, genética de poblaciones, sistema reproductivo.

Oyler-McCance, S. J., and T. W. Quinn. 2011. Molecular insights into the biology of Greater Sage-Grouse. Pp. 85–94 *in* S. T. Knick and J. W. Connelly (editors). Greater Sage-Grouse: ecology and conservation of a landscape species and its habitats. Studies in Avian Biology (vol. 38), University of California Press, Berkeley, CA.

Molecular genetic approaches to the study of wildlife began in the late 1970s and early 1980s, but their initial use was very limited, in part because of the amount of expertise in molecular genetic techniques that was then both necessary and rare. In the 1980s, the development of the polymerase chain reaction (PCR) and concurrent improvements in DNA sequencing technology led to a major increase in efficiency and decrease in cost. As a result, it is now commonplace to gather DNA sequence information from targeted regions of the mitochondrial or nuclear genomes. Such techniques are now being widely applied to investigate relationships among species, populations, family groups, and individuals, even in species for which little was previously known at the genetic level (Haig 1998, DeYoung and Honeycutt 2005, Oyler-McCance and Leberg 2005).

In the nuclear genome, tandem repetitive DNA sequences, such as microsatellites, tend to undergo rapid change in the number of tandem repeats (Li et al. 2002). Within populations, such regions thereby generate numerous alleles that can be used in various ways to estimate different aspects of the individual population, such as past effective population size; to estimate levels of historical gene flow between populations; and even to determine parentage of offspring. While such repetitive regions are typically but not always absent in most vertebrate mitochondrial genomes, the rate of base substitution in parts of the mitochondrial genome typically exceed those of comparable regions in the nuclear genome (Pesole et al. 1999). One of the most widely targeted hypervariable mitochondrial regions is found at both ends of the major noncoding control region (Baker and Marshall 1997). It is these and other rapidly evolving segments of DNA that have provided a great deal of new insight into relationships among vertebrate populations and closely related species in recent decades (Avise 1994).

Sage-grouse (*Centrocercus* spp.) are ground-dwelling galliforms found exclusively in North America. Historically, sage-grouse occupied virtually any habitat dominated by sagebrush (*Artemisia spp.*), including much of the western United States (Johnsgard 1983). Until recently, all sage-grouse were considered to be one species. In 2000, sage-grouse were split into two species (Fig. 5.1): Greater Sage-Grouse (*C. urophasianus*) and Gunnison Sage-Grouse (*C. minimus*), following Young et al. (2000). The range of Greater Sage-Grouse historically spanned 15 western states and three Canadian provinces (Schroeder et al. 2004), yet they currently occupy only 56% of the range

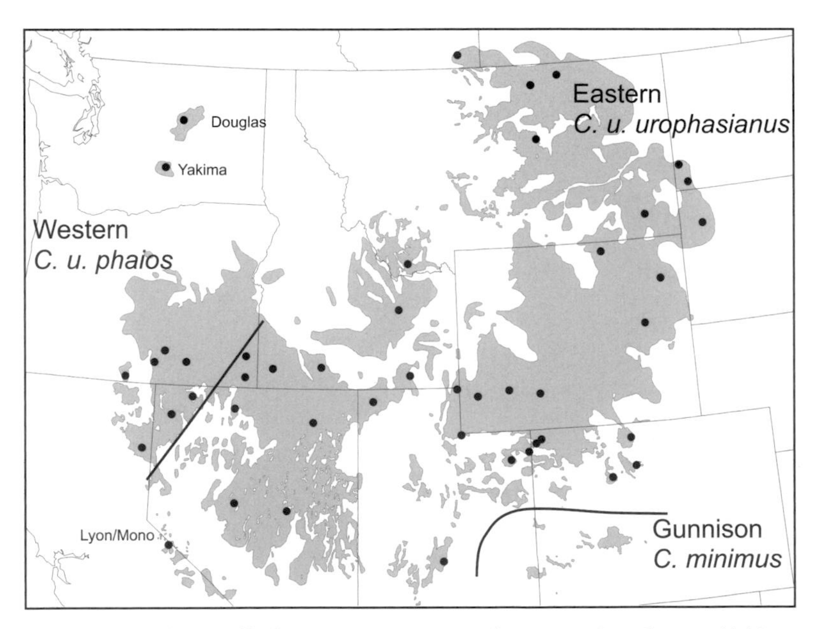

Figure 5.1. Distribution of both Greater Sage-Grouse and Gunnison Sage-Grouse with historical subspecies noted. Dots represent sampling sites for the population genetic studies. The two populations in Washington and the population along the border of Nevada and California are noted, as they are discussed in detail.

occupied before settlement by Europeans (Fig. 5.2) and were extirpated in at least four states and one province (Connelly and Braun 1997, Schroeder et al. 2004). Regional population declines have been dramatic, ranging from 17% to 47% (Connelly and Braun 1997). These declines are likely linked to the loss, fragmentation, and degradation of sagebrush habitat (Braun 1998), resulting in the isolation of small populations that formerly existed in more contiguous habitat (Fig. 5.2).

As sagebrush obligates, Greater Sage-Grouse rely on sagebrush for virtually all aspects of their life cycle (Patterson 1952, Braun et al. 1977, Connelly et al. 2000c). Seasonal movements of Greater Sage-Grouse have been reported to be quite variable, depending on the population being studied. Some populations are considered to be resident, with little movement throughout the year, while others have been documented to move up to 161 km (Patterson 1952). Still other populations have been described as migratory, with consistent seasonal movements up to 82 km (Connelly et al. 1988). With that stated, the amount of movement among populations and connectedness of populations remain largely unknown.

Greater Sage-Grouse have a polygynous mating system that has been the focus of much research (Wiley 1974, Wittenberger 1978, Gibson and Bradbury 1986, Bergerud 1988a, Gibson et al. 1991). In the spring, males gather together on leks, where they engage in an elaborate strutting display for females. The consistency of the behavioral components of these strut displays, even when compared between widely separated populations, led Wiley (1973a) to use them as an example of a fixed action pattern. Males establish territories on leks and defend them throughout the breeding season (Gibson and Bradbury 1986). Females arrive on leks, usually in groups, and are attracted to dominant males (Wiley 1978). While observational studies have provided some guidance on what proportion of males actually breed (Patterson 1952, Wiley 1973b), until recently this information had not been complemented with studies at the genetic level.

Because of declining habitat and population numbers, Greater Sage-Grouse have become the focus of conservation concern. As a result, nearly all aspects of their ecology are being examined in an attempt to better understand and manage this species. Several molecular genetic approaches have

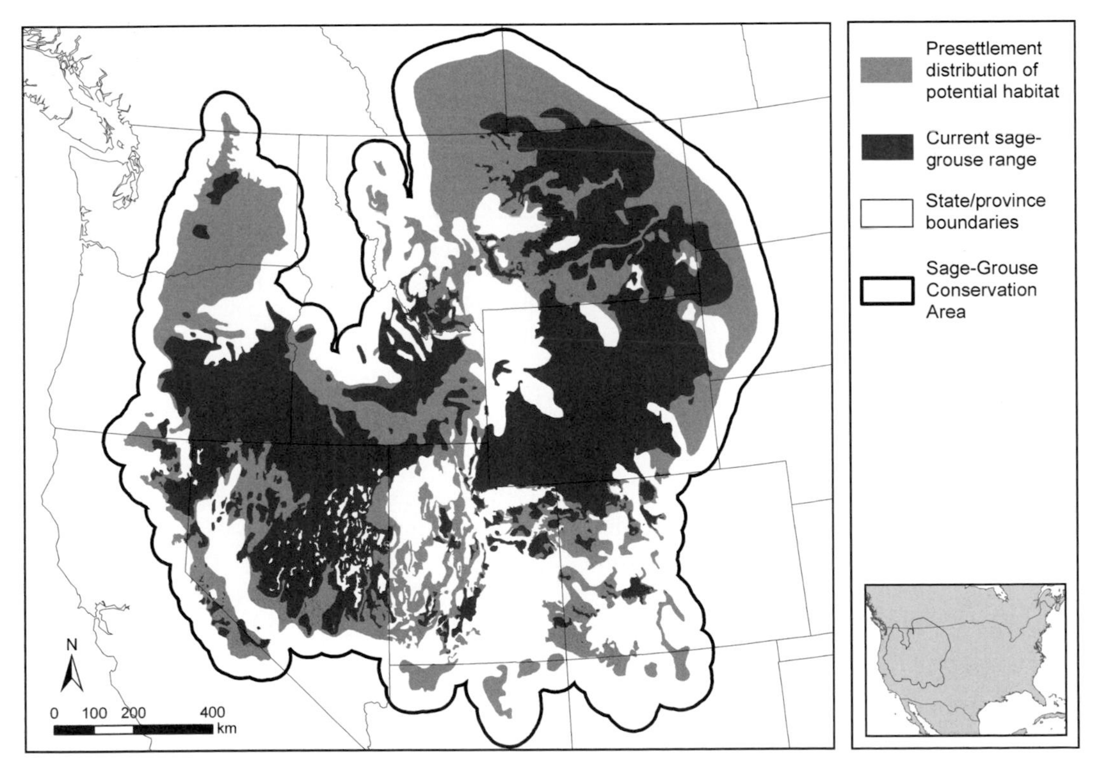

Figure 5.2. Historical and current distribution of the Greater Sage-Grouse (from Connelly et al., 2004, adapted from Schroeder et al. 2004).

been employed to aid in this effort. Because molecular methods can be used to more definitively assess parentage, such techniques have been used to augment observational studies of the mating system of Greater Sage-Grouse and to investigate rates of multiple paternity. Such information is needed for management to better estimate effective population sizes. Additionally, molecular methods can be used to reassess taxonomic boundaries defined using morphological and behavioral characters. Finally, such methods can also be used to examine population structure and gene flow, i.e., connectivity of populations, as well as to document levels of genetic diversity. All those metrics are paramount for conservation efforts. This chapter reviews the current published literature using molecular genetic techniques to obtain information useful for the conservation and management of Greater Sage-Grouse and put it into the context of avian ecology using selected molecular studies.

Molecular genetic studies of sage-grouse have focused on several questions. One study investigated the mating system of Greater Sage-Grouse by comparing behavioral data and genetic data (Semple et al. 2001). Another study examined the evolution of lek formation in Greater Sage-Grouse by determining whether leks consist of kin groups (Gibson et al. 2005). Four other studies have looked at taxonomic boundaries and population genetics (Kahn et al. 1999, Oyler-McCance et al. 1999, Benedict et al. 2003, Oyler-McCance et al. 2005b). We provide a review of these studies and discuss how they can be used in the management and conservation of Greater Sage-Grouse and how they fit into a larger ecological perspective.

INVESTIGATING THE LEK MATING SYSTEM

Behavioral observations of the lek mating system of Greater Sage-Grouse have led to the suggestion that mating distributions are strongly skewed and that females typically mate with only one dominant male (Wiley 1973b, Gibson and Bradbury 1986, Gibson et al. 1991). Evidence suggests that a component to the mating system exists that was unobserved by previous studies, whereby females may be mating off the lek with purportedly non-dominant males. This can be inferred in part by the study of Semple et al. (2001), who assessed the parentage of 10 broods of Greater Sage-Grouse in Long Valley, California, near Mono Lake (Fig. 5.1). In this study, behavioral observations on attendance, territoriality, and mating behavior of males on one lek were made. Each male was uniquely identified using colored leg bands. Both males and females associated with this lek were trapped, and blood samples were collected. Females were radio-collared and followed throughout the breeding process. After 7 to 9 days of incubation, nests were located, all eggs were collected, and embryos were sacrificed for genetic analysis.

Six variable microsatellite loci were used to identify the parentage of all embryos. The genotypes of the embryos were compared to the associated mother to identify which alleles were maternal and which were paternal. The number of paternal alleles was then used to determine whether the brood had been sired by more than one male. An attempt was made to exclude potential males as fathers based on their microsatellite genotypes.

Semple et al. (2001) found that all broods had genotypes consistent with their putative mother and that eight of 10 broods showed results consistent with only one father. Two broods did show results suggesting that at least two males had fathered the brood. Of the 10 females with broods, only four mated on the lek from which behavioral observations were made and thus were seen copulating. Three of those four females mated with males that had been genotyped, and in each case, the putative father could not be excluded as the father based on genotype data. The fourth female mated with a male who was not banded or genotyped. Semple et al. (2001) conclude that in all cases for which they had data, the genetic data were consistent with behavioral data. Although previous studies suggested very low instances of multiple paternity, this molecular genetic study showed that it may be more prevalent than was once thought.

Such comparisons between observation-based studies and molecular ones have been conducted in other species (Kellogg et al. 1995, Coltman et al. 1999, Heckel et al. 1999). The addition of molecular data in this case can add another layer of information, as mating behavior does not always reflect parentage (Coltman et al. 1999), and can provide better estimates of effective population size, which is important for the management of species (Lande and Barrowclaugh 1987).

Among avian lekking species, many have been shown to exhibit the behavior of polyandry (Trail 1985, Petrie et al. 1992, Fiske and Kålås 1995), although fewer studies have actually been able to document parentage (Lanctot et al. 1997, Lank et al.

2002). Only one other lekking grouse has been examined in this way. The mating behavior and genetic paternity of Black Grouse (*Tetrao tetrix*) have been documented in several studies (Alatalo et al. 1996, Kokko et al. 1999, Lebigre et al. 2007). Unlike the results found by Semple et al. (2001) in Greater Sage-Grouse, multiple mating and multiple paternity in Black Grouse were found to be rare (Alatalo et al. 1996, Lebigre et al. 2007). The studies of Black Grouse were larger, examining 135 broods as opposed to 10 in Greater Sage-Grouse, and included data from multiple leks (Lebigre et al. 2007). Given this difference in observed trends between related species, larger-scale studies of Greater Sage-Grouse are suggested in order to better estimate rates of multiple paternity. Although the study by Semple et al. (2001) provides an indication that some females are mating with more than just the dominant males on leks, it would be prudent to extend such studies to larger sample sizes and additional leks in order to better understand the impact that such behaviors have on the effective population sizes of Greater Sage-Grouse in general.

In another study that included molecular genetic data, Gibson et al. (2005) tested the hypothesis that lekking behavior evolved in Greater Sage-Grouse because kin selection favored subordinate males joining the leks of their close relatives in order to increase their inclusive fitness. This might occur if females were more attracted to leks with larger numbers of attending males. The authors collected blood samples and made behavioral observations at three major leks in one population in Long Valley, California, near Mono Lake (Fig. 5.1). Their data included two time periods (1984–1988 and 1997–2001, except 1999). Males were categorized as territorial or nonterritorial based on daily observations as to whether males aggressively excluded other males from territories on leks. A measure of spatial association of males off leks was also estimated using the degree to which males with overlapping ranges moved collectively from day to day as an index of association. These behavioral observations were coupled with inferences that could be made from their molecular data set. The mean relatedness within leks was estimated using 11 highly variable microsatellite loci, and it was then compared to mean relatedness estimates from within known family groups. They determined that the mean relatedness values on leks were statistically indistinguishable from zero, suggesting that males on leks are not closely related. Additionally, those mean relatedness values were significantly lower than relatedness values calculated among known family members (Gibson et al. 2005). They found no evidence for local clustering of related individuals while on leks, yet they did find that related individuals tended to cluster together when off the lek. Their data reveal that Greater Sage-Grouse leks consist of groups of largely unrelated males with little if any spatial association among kin either within or between leks.

Several other avian lekking species have been examined in a similar way, testing whether or not males on leks are more related than random. Similar to the findings of Gibson et al. (2005), males on leks were not found to be related in both the Long-tailed Manakin (*Chiroxiphia linearis*; McDonald and Potts 1994) and the Great Bustard (*Otis tarda*; Martin et al. 2001). Studies of the Peacock (*Pavo cristatus*) and White-bearded Manakin (*Manacus manacus*), however, have documented that males on leks are related (Petrie et al. 1999, Shorey et al. 2000, Höglund and Shorey 2003). Among lekking grouse, both Black Grouse and Lesser Prairie-Chicken (*Tympanuchus pallidicinctus*), have been examined in this manner, and both have shown males on leks to be related (Höglund et al. 1999, Bouzat and Johnson 2003), contrary to what has been found for Greater Sage-Grouse (Gibson et al. 2005). Again, it would be informative for future studies to assess additional populations in this regard to determine whether the population studied in Long Valley, California, is typical of the species. Ultimately such knowledge impacts management in the sense that genetic diversity contained within single leks may (or may not be) reduced relative to the species as a whole.

REASSESSING TAXONOMIC BOUNDARIES

Investigating a Morphologically and Behaviorally Different Group

Because morphological (Hupp and Braun 1991) and behavioral (Young et al. 1994) evidence suggested the sage-grouse in southwestern Colorado and southeastern Utah were distinct from sage-grouse elsewhere, two molecular studies were initiated examining molecular genetic variation across the boundary of these two groups of sage-grouse in Colorado (Kahn et al. 1999,

Oyler-McCance et al. 1999). The goal of the two genetic studies was to determine whether there were genetic differences between sage-grouse from southwestern Colorado and those from northwestern Colorado.

Kahn et al. (1999) sequenced 141 base pairs from a rapidly evolving part of the control region of mitochondrial DNA (mtDNA) from approximately 20 individuals each from seven populations. Six of those populations were within the range of the larger-bodied Greater Sage-Grouse, sampled in northern Colorado and Utah. The remaining population was within the range of the smaller-bodied grouse, sampled in the Gunnison Basin of Colorado. That population, along with others nearby, was later assigned species status as the Gunnison Sage-Grouse (Young et al. 2000).

This study revealed that the Greater Sage-Grouse populations were more diverse than the Gunnison population, with many mtDNA haplotypes present per population (average of 7.3), while the Gunnison Sage-Grouse population had only two haplotypes present. The Greater Sage-Grouse populations had four ubiquitous haplotypes, and the Gunnison Sage-Grouse population had only one of those four haplotypes and also included a haplotype that was unique. Using maximum parsimony analysis, Kahn et al. (1999) revealed that all haplotypes fell into one of two deep clades. All populations of Greater Sage-Grouse had individuals from both clades. From these observations, they hypothesized that the two deep clades might reflect what were historically two distinct populations or metapopulations that began diverging at least 850,000 years ago during the Pleistocene. More recently, the two groups intermingled, leading to the modern pattern.

Oyler-McCance et al. (1999) expanded on the mitochondrial study of Kahn et al. (1999) by including data from four nuclear microsatellites and by adding both mitochondrial and nuclear data from three additional populations of Gunnison Sage-Grouse. They, too, showed that the Greater Sage-Grouse had more genetic diversity in mitochondrial haplotypes, and their microsatellite data showed concordant patterns in the nuclear genome, with higher levels of allelic diversity among the Greater Sage-Grouse (average of 5.7 alleles per locus) than among the Gunnison Sage-Grouse (average of 2.6 alleles per locus). Population genetic analysis revealed a lack of evidence of gene flow among the two groups of sage-grouse, which was consistent with the idea that the Gunnison Sage-Grouse should be recognized as a new species (Young et al. 2000). Furthermore, Oyler-McCance et al. (1999) showed that levels of gene flow were much higher among populations of Greater Sage-Grouse ($F_{ST} = 0.03$) than among populations of Gunnison Sage-Grouse ($F_{ST} = 0.21$).

These two genetic studies provided evidence of a lack of gene flow between the two morphologically and behaviorally distinct groups of sage-grouse in Colorado. This was consistent with the previous morphological (Hupp and Braun 1991) and behavioral (Young et al. 1994) evidence suggesting that the small-bodied sage-grouse in southwestern Colorado were, in fact, a distinct species. This newly recognized species was subsequently named and described (Young et al. 2000).

Examining a Subspecific Boundary

The establishment of the Gunnison Sage-Grouse as a separate species was a recent event, and because its range is comparatively small, the Greater Sage-Grouse range was affected only slightly when this distinction was made. Within the remaining larger range of the Greater Sage-Grouse, the species has been divided into eastern (*C. u. urophasianus*) and western (*C. u. phaios*) subspecies (Aldrich 1946). This delineation was based on plumage and coloration differences in just eight individual Greater Sage-Grouse collected from Washington, Oregon, and California (Aldrich 1946). The western subspecies presumably occurred in southern British Columbia, central Washington, east-central Oregon, and northeastern California (Aldrich 1946). Populations in other areas of the range were considered to be the eastern subspecies (Fig. 5.1). The validity of the subspecies distinction was later questioned, as many biologists were unable to recognize this boundary in the field (Johnsgard 1983). Benedict et al. (2003) published a molecular investigation that covered the western part of the range of the Greater Sage-Grouse, spanning the boundary between the two subspecies, as described by Aldrich (1946, 1963).

In that study, tissue samples from 16 populations crossing the boundary separating the eastern and western subspecies, as described by Aldrich (1946, 1963), were collected. These were used to sequence part of the mitochondrial DNA control region (Kahn et al. 1999). Among the 332 individuals sampled across the range, Benedict et al. (2003) identified a total of 38 haplotypes, all falling into either one of the two monophyletic clades described by Kahn et al. (1999). Five of these were common, widespread haplotypes, and upon examination of their distribution, no obvious genetic subdivision between the eastern and western subspecies existed. On this basis, the authors concluded that no genetic evidence supported the original subspecies distinction, although they did point out that further behavioral and morphological measurement across the boundary should be considered.

In the course of addressing the subspecies question, the authors found that several individual sampled populations were notable for other reasons. In their broad survey, they noted that novel haplotypes—those found in only one population—were common, but also that they usually occurred at low frequency, typically <10% of the individuals sampled within each population (Benedict et al. 2003). Along the border between Nevada and California, however, they found an unusual population (Lyon–Mono, Fig. 5.1) in which 87.5% of the haplotypes found were unique, constituting 97.7% of the birds sampled. This led the authors to suggest that the Lyon–Mono population has been isolated from neighboring populations for a considerable amount of time. Because this population has closely related but novel haplotypes belonging to each clade, they believe that it is likely that this isolation occurred after the intermixing of populations representing the two major haplotype clades, as originally hypothesized by Kahn et al. (1999). Presumably, over tens of thousands of years of isolation, factors such as mutation and genetic drift resulted in the significant divergence of the Lyon–Mono population from other Greater Sage-Grouse populations. Benedict et al. (2003) indicate that the preservation of genetic diversity represented by the unique allelic composition of the Lyon–Mono population is of particular importance for conservation. Because the likelihood that the distinctiveness of neutral genetic markers extends to genes under adaptive selection, the authors believe that this population should be managed independently to avoid the translocation of other Greater Sage-Grouse into this area. They also maintain that it will be critical that additional morphological and behavioral studies of the Lyon–Mono population be undertaken to address taxonomic questions.

In addition, Benedict et al. (2003) found that the two populations in Washington (Fig. 5.1) contained the lowest level of haplotype diversity observed, perhaps resulting from a recent genetic bottleneck given that these populations now occupy just 8–10% of their original range (Friedman and Carlton 1999) and have shown significant declines in population size (Schroeder et al. 2000). Benedict et al. (2003) suggest that the probable loss of genetic variation caused by this bottleneck and its potentially long-term adverse impact (Bouzat et al. 1998, Le Page et al. 2000) should be addressed as management strategies are developed for these populations. They advocate that active management, such as translocation of birds, may be justified to ensure their continued persistence.

Examining Taxonomy and Population Genetics Across the Range

In order to investigate taxonomic questions and examine levels of gene flow and connectedness among populations, Oyler-McCance et al. (2005b) greatly extended the sampling range and density of previous studies and provided a comprehensive examination of the distribution of genetic variation across the entire range of Greater Sage-Grouse using both mtDNA sequence data and data from nuclear microsatellites. They collected 1,080 samples from 46 populations from all U.S. states with populations of Greater Sage-Grouse (11) and one Canadian province (Alberta), spanning the entire range of the species.

Overall, Oyler-McCance et al. (2005b) found that the distribution of genetic variation showed a gradual shift across the range in both mitochondrial and nuclear data sets. They indicate that this pattern suggests localized gene flow with isolation by distance—that is, movement among neighboring populations yet not likely across the range. Results from their nested clade analysis of the mitochondrial data confirm this finding, with seven clades characterized by restricted gene flow with isolation by distance. Analysis of their microsatellite data showed similar results, with a positive correlation between genetic distance and geographic distance. Furthermore, a genetic clustering analysis (STRUCTURE; Pritchard et al. 2000) revealed that unique genetic clusters were made up of populations geographically adjacent to one another, and while most genetic clusters consisted of many populations, the smaller, more fragmented populations on the periphery of the range in Colorado, Utah, Lyon–Mono in Nevada and California, and Washington made up their own clusters, suggesting lower amounts of gene flow in these areas (peripheral isolates). These data are consistent with previous research on dispersal (Dunn and Braun 1985) suggesting that gene flow is likely limited to the movement of individuals between neighboring populations and not likely the result of long-distance movements of individuals across large portions of the range. Oyler-McCance et al. (2005b) believe that this information is important because conservation efforts often consider translocations and augmentation of existing populations using animals from outside populations. Their data suggest linkages among neighboring populations and differences among distant populations, raising the possibility that local adaptations may exist and that translocations should involve neighboring populations rather than geographically distant populations.

Isolation by distance has been reported in many other grouse species, including Red Grouse (*Lagopus lagopus scoticus*; Piertney et al. 1998), Black Grouse (Caizergues et al. 2003b), Capercaillie (*Tetrao urogallus*; Segelbacher et al. 2003), Greater Prairie-Chickens (*Tympanuchus cupido*; Johnson et al. 2003), White-tailed Ptarmigan (*Lagopus leucura*; Fedy et al. 2008), and Rock Ptarmigan (*Lagopus mutus*; Caizergues et al. 2003a). Thus, it is not surprising that, in general, grouse exhibit this pattern, being ground-dwelling birds with short-distance seasonal movements compared to other avian migratory species.

Because of the comprehensive sampling regime and addition of nuclear markers, Oyler-McCance et al. (2005b) were also able to address some of the issues raised by Benedict et al. (2003). Similar to the findings of Benedict et al. (2003), Oyler-McCance et al. (2005b) discovered that in both mtDNA and microsatellite data sets, the least amount of genetic diversity was in the two Washington populations, likely due to habitat loss and subsequent population decline. The authors also reinforced the observation that the Lyon–Mono population was genetically unique compared to all other populations and that the difference was striking. While an additional 24 populations were added in their study, the fact remained that 93% of individuals in the Lyon–Mono population contained novel haplotypes not found elsewhere across the range. The genetic diversity present in Lyon–Mono was comparable to, if not higher than, most other populations, suggesting that the differences were not due to a genetic bottleneck or founder event. Their nuclear data corroborated these facts, because Lyon–Mono was significantly different from almost all other populations; and in their STRUCTURE analysis, the Lyon–Mono population was the only population forming its own cluster.

Taxonomic classifications have traditionally been defined using morphological and behavioral characteristics. With the advent of PCR and with advances in molecular techniques, genetic information is now being gathered and used to address such issues. Taxonomic delineations derived only from morphological characteristics can be erroneous (Avise 1989), as they can either fail to recognize distant forms (Avise and Nelson 1989) or recognize forms that exhibit little evolutionary differentiation (Laerm et al. 1982). While classifications based on morphology and behavior have been acceptable, use of molecular techniques can often resolve discrepancies and help to augment or refine taxonomic definitions. This is particularly important as it applies to listing under the Endangered Species Act (ESA), for defining management units, and for recovery of the species/subspecies, because funding priorities generally are based on taxonomic status (O'Brien and Mayr 1991).

The taxonomic status of many different species has recently been reevaluated using genetic data. In addition to the Gunnison Sage-Grouse, several new species have been recognized, including the Kemp's ridley sea turtle (*Lepidochelys kempii*), the north Pacific right whale (*Eubalaena japonica*), and Perrin's beaked whale (*Mesoplodon pewini*), all the result of molecular genetic analyses (Bowen et al. 1991, Rosenbaum et al. 2000, Dalebout et al. 2002). More common, however, is the use of molecular genetic data to investigate and often redefine subspecies. As difficult as it is for scientists to concur on a definition of species (Avise 1994), it is even more difficult to come to an agreement on the appropriate way to identify subspecies (Zink 2004, Cronin 2006, Haig et al. 2006). This is perhaps why the assessment of subspecies, fueled by the potential economic impacts of their protection, can become so controversial and divisive (Ramey et al. 2005, King et al. 2006, Vignieri et al. 2006). Examples of such studies include examination of the Preble's meadow jumping mouse (*Zapus hudsonius preblei*) and the Northern Spotted Owl (*Strix occidentalis*), both of which have been listed under the ESA and have been studied extensively by more than one genetics lab (Barrowclaugh and Gutiérrez 1990; Haig et al. 2001, 2004; Barrowclaugh et al. 2005; Ramey et al. 2005; King et al. 2006; Vignieri et al. 2006).

Several authors have investigated phylogenetic relationships among grouse, including the Greater Sage-Grouse (Ellsworth et al. 1996, Gutiérrez et al. 2000, Drovetski 2002). The focus of these studies was to elucidate the evolutionary relationships among grouse. At a finer scale, genetic data (Barrowclaugh et al. 2004) have recently been used to raise two subspecies of Blue Grouse (*Dendragapus obscurus*) to species level, now recognizing the Dusky Grouse (*D. obscurus*) and the Sooty Grouse (*D. fuliginosus*) as separate species. Additionally, the boundaries among several subspecies of Sharp-tailed Grouse (*Tympanuchus phasianellus*) have recently been questioned because of analysis of genetic data (Spaulding et al. 2006). It is clear that this type of information will continue to aid in the refinement of taxonomic classifications in the future.

Given the anomalous nature of the Lyon–Mono population relative to the numerous other populations of the species that were sampled, it becomes interesting to compare it to the newly recognized and closely related Gunnison Sage-Grouse. Oyler-McCance et al. (2005b) noted that the Lyon–Mono population is at least as divergent from other populations of Greater Sage-Grouse as Gunnison Sage-Grouse are from Greater Sage-Grouse, by virtue of the large number of new haplotypes unique to that population (novel haplotypes). Gunnison Sage-Grouse were recognized as a new species based on morphological, behavioral, and genetic data (Young et al. 2000), and those authors felt it important to consider these features in concert, as it can be difficult to define species based on molecular data alone, at least if one is applying the biological species concept. Such considerations led Taylor and Young (2006) to initiate a behavioral study of the Lyon–Mono population whereby they compared the strutting behavior of males from that population with several other populations of Greater Sage-Grouse. Their comparisons of behavior revealed few or no differences (Taylor and Young 2006). This suggests that while Lyon–Mono may have been isolated for an amount of time similar to the isolation of Gunnison Sage-Grouse, they have not experienced a significant divergence in behavioral characteristics as has been documented in Gunnison Sage-Grouse (Young et al. 2000), which ultimately led to their reproductive isolation (Oyler-McCance et al. 2005a). Precisely because of findings such as this one, some argue that morphological and behavioral data as well as genetic data are essential in the description and revision of taxonomic boundaries (Haig et al. 2006).

The study of Oyler-McCance et al. (2005b) reinforced the findings of Benedict et al. (2003) that the Lyon–Mono population is sufficiently genetically different to warrant special attention. They appear similar to other populations in behavior (Taylor and Young 2006), but Oyler-McCance et al. (2005b) believe that more comprehensive morphological comparisons should be performed before a change in taxonomic status should be considered. Regardless of the label placed on this population, Oyler-McCance et al. (2005b) believe that if the population could be managed separately and protected due to its genetic distinctiveness, it may become important to the survival of the species over large timescales.

The molecular examination of the mating system of Greater Sage-Grouse by Semple et al. (2001) and Gibson et al. (2005) focused on leks in Long Valley, California, near Mono Lake (Fig. 5.1), which is the same general area (Lyon–Mono) described to be genetically unique by Benedict et al. (2003) and Oyler-McCance et al. (2005b). Therefore, it would be very interesting to conduct studies similar to that of Semple et al. (2001) and Gibson et al. (2005) in areas outside of the Mono Basin to determine whether the patterns documented in the Mono Basin (Lyon–Mono) hold true outside of the genetically unique population. It is possible that while the isolation of the Mono Basin birds did not lead to the gross morphological and behavioral changes that have been documented in Gunnison Sage-Grouse, there may be more subtle differences in territoriality of males or in overall lek behavior between the Mono Basin group and Greater Sage-Grouse elsewhere.

CONSERVATION IMPLICATIONS

Molecular genetic techniques have only recently been used to help achieve a better understanding of avian biology and the interaction between species and their changing environments. Here we have attempted to discuss the contributions to Greater Sage-Grouse biology that molecular genetic techniques have afforded. Advances toward a better understanding of the mating system have already been made, and increased availability of microsatellite markers and technological improvements will undoubtedly lead to more broad and comprehensive studies in the future. Similarly, more fine-scale population-level analyses of Greater Sage-Grouse will ensue as localized questions arise. Finally, these studies have provided an interesting case study for examining the implications of sexual selection generated by the different evolutionary pathways of the Gunnison Sage-Grouse and the Lyon–Mono population of Greater Sage-Grouse.

ACKNOWLEDGMENTS

This chapter represents a summary of genetic studies on Greater Sage-Grouse from several papers. We thank the authors of those papers both for their published contributions and for various enlightening discussions through the years. Both authors particularly thank C. E. Braun for providing the initial impetus and support for that work via the Colorado Division of Wildlife, which launched several early studies and which eventually led to the wider survey of Greater Sage-Grouse.

CHAPTER SIX

Predation on Greater Sage-Grouse

FACTS, PROCESS, AND EFFECTS

Christian A. Hagen

Abstract. Although Greater Sage-Grouse (*Centrocercus urophasianus*) face a suite of predators in sagebrush (*Artemisia* spp.) communities across the species' range, none of these predators specializes on sage-grouse. Greater Sage-Grouse are susceptible to predation from egg to adult, leading to the hypothesis that predator control would be an effective conservation tool for sage-grouse populations. Therefore, I reviewed the literature pertaining to predator communities across the range of Greater Sage-Grouse and assessed the effects of predation on sage-grouse life history. I then provided a framework for evaluating when predator management may be warranted. Generally, nest-success rates and adult survival are high, suggesting that on average predation is not limiting. However, in fragmented landscapes or in areas with subsidized predator populations, predation may limit population growth. Few studies linked habitat quality to mortality rates, and fewer still linked these rates to predation. Predator management studies have not provided sufficient evidence to support implementation over broad geographic or temporal scales, but limited information suggests predator management may provide short-term relief for a population sink. Evaluating the need for predator management will require linking reduced demographic rates to habitat quality (fragmentation or degradation) or predator populations out of the natural range of variability (exotic species of subsidized populations). Alternatively, managers might consider predator management in translocation efforts to buffer recently released individuals from potentially elevated predation rates. Future work should quantify predator and alternate prey communities in habitats used by Greater Sage-Grouse.

Key Words: Artemisia, Centrocercus urophasianus, Greater Sage-Grouse, habitat quality, predator control, predator-cover complex, sagebrush.

Depredación del Greater Sage-Grouse: Hechos, Procesos, y Efectos

Resumen. Aunque el Greater Sage-Grouse (*Centrocercus urophasianus*) enfrenta una serie de depredadores en las comunidades de sagebrush (*Artemisia* spp.) a lo largo del rango de distribución de la especie, ninguno de los depredadores se especializa en sage-grouse. Los Greater Sage-Grouse son susceptibles a la depredación desde el huevo hasta la adultez, llevando a la hipótesis de que el control de depredadores sería una herramienta de conservación efectiva para las poblaciones de sage-grouse. Por esto, revisé la literatura relacionada con comunidades de depredadores a lo largo del

Hagen, C. A. 2011. Predation on Greater Sage-Grouse: facts, process, and effects. Pp. 95–100 *in* S. T. Knick and J. W. Connelly (editors). Greater Sage-Grouse: ecology and conservation of a landscape species and its habitats. Studies in Avian Biology (vol. 38), University of California Press, Berkeley, CA.

territorio de Greater Sage-Grouse y determiné los efectos de la depredación sobre la historia de vida del sage-grouse. Luego proporcioné un sistema para evaluar cuándo está justificado el manejo de depredadores. Generalmente, las tasas de éxito de anidación y de supervivencia de adultos son altas, sugiriendo que, en promedio, la depredación no es limitante. Sin embargo, en paisajes fragmentados o en áreas con poblaciones de depredadores subsidiados, la depredación puede limitar el crecimiento de la población. Pocos estudios asocian calidad de hábitat con tasas de mortalidad, y aún menos asocian estas tasas con depredación. Los estudios de manejo de depredadores no han aportado evidencia suficiente para justificar la implementación a gran escala geográfica o temporal, pero alguna información escasa sugiere que el manejo de depredadores puede proporcionar un alivio a corto plazo para una población en descenso. La evaluación de la necesidad de manejo de depredadores requerirá asociar tasas demográficas reducidas con calidad de hábitat (fragmentación o degradación) poblaciones de depredadores fuera del rango natural de variabilidad (especies exóticas de poblaciones subsidiadas). Alternativamente, los administradores pueden considerar el manejo de depredadores en los esfuerzos de translocación para proteger de tasas potencialmente elevadas de depredación a individuos recientemente liberados. El trabajo futuro debe cuantificar depredadores y comunidades de presa alternativas en los hábitats usados por el Greater Sage-Grouse.

Palabras Clave: artemisa (sagebrush), *Artemisia,* calidad del hábitat, *Centrocercus urophasianus,* complejo depredador-cubierta vegetal, control de depredadores, Greater Sage-Grouse.

Greater Sage-Grouse (*Centrocercus urophasianus*) coevolved with a suite of predators from egg to adult life-history stages (Schroeder et al. 1999). Most game birds are short-lived with high reproductive effort (Schroeder et al. 1999). On the contrary, Greater Sage-Grouse are longer-lived than most game birds (>3 years) and have smaller clutch sizes. Most galliform mortality, both egg and individual, is due to predation (Johnsgard 1973). Game bird species have adapted and persisted despite this high mortality. Thus, predation is a fact of life for game birds, and the debate should focus on the process of predation and its seasonal effects on populations. Given an adequate amount of quality habitat, game birds should coexist and proliferate with their endemic predator community (Bergerud 1988b). However, in a landscape context, when quantity and/or quality of habitat is diminished, predators may gain an advantage over their prey and affect populations. Similarly, prey species may be more vulnerable to exotic predators and slow to adapt antipredator strategies in the short term, with considerable losses to populations (Côté and Sutherland 1997).

Cryptic plumage, behavioral adaptations (e.g., flocking and habitat selection) have enabled Greater Sage-Grouse (hereafter sage-grouse) to occupy and persist in vast regions dominated by sagebrush (*Artemisia* spp.) cover (Schroeder et al. 1999). Presumably, sage-grouse select habitats that minimize risk of predation relative to a life-history stage, in a trade-off that is referred to as the predator-cover complex (Bergerud 1988b). Males select areas with sparse vegetation to conduct breeding displays; these areas provide opportunities for early detection of predators and greater visibility for prenesting female sage-grouse (Gibson et al. 2002, Boyko et al. 2004). Nesting females select areas with greater sagebrush cover and taller grass than females rearing broods (Hagen et al. 2007). The selection of nesting cover focuses on concealment from predation and protection from weather, while early brood-rearing habitats are more sparsely vegetated. Brood-rearing habitats may pose greater predation risk to young chicks, but these risks may be outweighed by potentially increased growth rates and greater chick survival (Bergerud 1988b, Drut et al. 1994a, Gregg 2006). Sage-grouse winter range is largely defined by sagebrush availability and may include some of the densest stands of shrub cover (Hupp and Braun 1989, Moynahan et al. 2006). Sage-grouse in some regions use windswept ridges for foraging and roosting and use snow roosts in other regions (Beck 1977, Back et al. 1987, Hagen 2005). The common thread of seasonal habitat selection is a balance of concealment to avoid predation while meeting biological demands. This chapter is organized to review the facts of predation as they relate to predator communities and life-history stages of sage-grouse, describe effects of predation on populations, summarize studies that assessed the effectiveness of predator management programs, and provide a context as to when predator management might be considered as a conservation tool.

COMPONENTS OF PREDATION

Predator Community

Coyote (*Canis latrans*), American badger (*Taxidea taxus*), bobcat (*Lynx rufus*), and several species of raptors are common predators of juvenile and adult sage-grouse through most of the range of the species (Patterson 1952, Schroeder et al. 1999, Schroeder and Baydack 2001). Additionally, coyote, American badger, Common Raven (*Corvus corax*), and Black-billed Magpie (*Pica hudsonia*) commonly prey on sage-grouse eggs (Patterson 1952, Schroeder et al. 1999, Schroeder and Baydack 2001). Many additional predators can kill and consume younger birds, including Common Raven, Northern Harrier (*Circus cyaneus*), ground squirrels (*Spermophilus* spp.), and weasels (*Mustela* spp.)(Schroeder et al. 1999, Michener 2005). The role of snakes in predation of sage-grouse nests has not been well documented (Coates 2007). The abundance of red fox (*Vulpes vulpes*) and common raccoon (*Procyon lotor*) may have substantially increased in sage-grouse range in association with landscape-level changes (Fichter and Williams 1967, Bunnell 2000, Connelly et al. 2000a, Baxter et al., 2007). However, the predator communities to which sage-grouse are susceptible are made up of generalists. No one predator focuses solely on sage-grouse as their primary food source; most predators are dependent upon rodents and lagomorphs as primary prey (Schroeder et al. 1999).

Communal Display Grounds

Male sage-grouse are predisposed to predation during conspicuous breeding displays on communal display grounds called leks (Hartzler 1974, Wiley 1974, Boyko et al. 2004). Displaying males have been hunted by Golden Eagle (*Aquila chrysaetos*), Prairie Falcon (*Falco mexicanus*), and Northern Harrier, although mortalities are rarely observed for the latter. Hartzler (1974) observed a predawn kill of a male sage-grouse by a bobcat and found evidence suggesting that two others had met a similar fate. The timing of early morning displays and aggregation of males are likely adaptations to reduce predation during the breeding season (Gibson et al. 2002, Boyko et al. 2004).

Nesting and Brood Rearing

Mortality of incubating females during nesting occurs but at low rates (Schroeder et al. 1999, Gregg 2006), likely because of their tendency to abandon a nest when disturbed by predators (Patterson 1952). However, partial clutch loss occurs in ~50% of nests and nest success rangewide averages 49% (range 15–86%; Connelly et al., this volume, chapter 3). Corvids are effective nest predators, taking eggs and possibly recently hatched chicks (Autenrieth 1981, Coates 2007). Coyote and American badger are also known to depredate sage-grouse nests; the latter do so somewhat more commonly (Schroeder et al. 1999). Red fox have become more common in limited portions of sage-grouse range and are effective nest predators (Baxter et al. 2007). Ground squirrels are assumed to be egg predators; recent evidence indicates that the mandibles of some ground squirrel species are physically incapable of puncturing sage-grouse eggs (Holloran and Anderson 2003, Michener 2005, Coates 2007). Ground squirrels potentially can kill altricial hatchlings (Pietz and Granfors 2000), but the extent that this occurs to precocial young is unknown. Young chicks are susceptible to both exposure and predation, but documentation of predators for this life stage is largely speculative—likely predators include Common Raven, Northern Harrier, coyote, red fox, snakes, and weasels (Gregg 2006).

Adult Survival

Predation of adult sage-grouse is at its lowest frequency outside of the lekking, nesting, and brood-rearing season (Connelly et al. 2000a, Naugle et al. 2004, Moynahan et al. 2006). Late summer is a period of high survival (90–100%) for adults, except for events outside the natural range of variability such as exposure to West Nile virus (Naugle et al. 2004). Losses of sage-grouse to predation in winter have not been well documented, but the growing evidence suggests that survivorship is high (70–95%), and significant losses only occur as a result of extreme weather events (Robertson 1991, Beck et al. 2006, Moynahan et al. 2006, Battazzo 2007). Annual survival of adult males (40–70%) and females (60–80%) is higher than most other game birds in North America (Connelly et al., this volume, chapter 3).

PROCESS AND EFFECTS OF PREDATION

Predator-Prey Dynamics

The dynamics between predators and prey in the sagebrush biome are complex (Mezquida et al. 2006). Coyote and Golden Eagle, the top predators in this system, can be considered generalists, because rodents and lagomorphs make up most of their diets. Primary-prey population oscillations have been linked to fluctuations in predator abundances (Norrdahl and Korpimaki 2000, Valkama et al. 2005). Two mechanisms explain how alternate prey, including sage-grouse, may have higher predation rates from generalist predators: (1) alternate-prey hypothesis, which states that when primary-prey populations decrease, predators switch their diets from declining prey to alternate prey, or (2) shared-predation hypothesis, which states that if predators kill indiscriminately as predator abundances increase (with more primary prey available), more alternate-prey species are susceptible to higher predation rates (Norrdahl and Korpimaki 2000). Which of these two hypotheses best fits the sagebrush biome is outside the scope of this chapter, but it is important to recognize these as possible mechanisms for limiting prey and predator populations.

Predation may influence the population dynamics of grouse by reducing nest success, survival of juveniles especially during the first few weeks after hatching, and annual survival of breeding-age birds (Valkama et al. 2005). The low survival of sage-grouse in Strawberry Valley, Utah, has been attributed to an unusually high density of red fox (Bunnell 2000, Baxter et al. 2007). Nest success is extremely variable, and differences in success have been attributed to variation in habitat and management strategy (Connelly et al. 1991, Gregg et al. 1994, Connelly et al. 2000a, Holloran et al. 2005, Moynahan et al. 2007). Sage-grouse may partly compensate for predation pressure on nests by renesting (Schroeder 1997). Habitat in sufficient quality and quantity often has been identified as an important goal for reducing effects of predation (Connelly et al. 1991, 2000a; Schroeder and Baydack 2001; Coates 2007). Survival of juvenile sage-grouse is presumably low, but it is also difficult to accurately assess because few studies have examined this life stage (Gregg 2006, Aldridge and Boyce 2007). Unlike nesting habitat, management of brood-rearing habitat has focused on increasing density and diversity of forbs (Klott and Lindzey 1990, Pyle and Crawford 1996, Sveum et al. 1998a), rather than improving vegetation to reduce predation (Edelmann et al. 1998). Many observations and recommendations have been made concerning the importance of suitable habitat for reducing predation pressure on adults, but detailed data have been difficult to obtain (Schroeder and Baydack 2001).

PREDATOR CONTROL AS A CONSERVATION TOOL

Management of predator populations to benefit avian species can include both nonlethal and lethal techniques; the latter is often highly controversial. Nevertheless, in the case of threatened and endangered species, lethal control techniques have been used with some short-term successes (Witmer et al. 1996). Predator control programs for upland game birds are common in Europe, where wildlife is considered private property and habitat is often limited (Gottschalk 1972). Conversely, predator control programs to benefit game populations are less common in North America, where wildlife is held in the public trust and habitat can often be extensive (Witmer et al. 1996). Historically, most predator control programs were designed to protect domestic livestock, not wildlife. The question remains as to whether or not predator control programs are effective conservation tools, and if so, under what conditions it may be appropriate to use them.

Evidence for Effectiveness of Predator Removal

Côté and Sutherland (1997), using meta-analytic techniques, found positive effects in postbreeding population size and hatching success for several species of game birds for which predator removal programs had been instituted. Breeding population sizes were not measurably affected by such programs, and Côté and Sutherland (1997) concluded that these programs have short-term gains that may benefit harvest management but do not meet the goals of conservation. Habitat fragmentation was acknowledged as the underlying factor leading to predator removal programs, and neither habitat quality nor extent was controlled in the meta-analysis. Recent research on Northern Bobwhite (*Colinus virginianus*) at Tall Timbers Research Station strongly suggests that predator removal has little effect on nest success and survival in areas with high-quality habitat, but it has greater impacts in areas with lower-quality habitat (W. E. Palmer, pers. comm.). Thus, the concept that game birds can proliferate given adequate-quality habitat in spite of predation is further supported. However, variation in annual fluctuations in demographic rates was reduced during removal efforts. It was also demonstrated that predator removal had to be an operational effort, because predator numbers quickly rebounded once control efforts ceased (W. E. Palmer, pers. comm.).

Predator management specifically to benefit grouse comes primarily from Europe, with a few examples from North America. Predator control experiments in Europe have demonstrated short-term benefits for fall populations of Capercaille (*Tetrao urogallus*), Black Grouse (*T. tetrix*), Hazel Grouse (*Bonasa bonasia*), and Willow Ptarmigan (*Lagopus lagopus*), but with little evidence that adult survival was improved (Marcström et al. 1988, Baines 1996, Kauhala et al. 2000, Baines et al. 2004). These studies also provided some evidence that grouse were more susceptible to predation when small mammal populations were low.

The endangered Attwater's Prairie-Chicken (*Tympanuchus cupido attwaterii*), of the Texas coast, has an extremely limited range, and population growth has been limited by poor recruitment. Lawrence and Silvy (1995) found improvements in nest success after mammalian predator reductions, but adult prairie-chicken survival was lower at treatment areas than at nonreduction areas. Thus, it appears that predator control may have short-term positive benefits for populations that require increases in nest success and chick survival. These populations are exposed to large numbers of raptors during winter migration that were not part of the predator management experiment.

Few studies have specifically evaluated predator removal as a method to improve demographic rates in sage-grouse, and only one of these studies measured both vegetation and predator communities relative to treatment and control (Slater 2003). Batterson and Morse (1948) evaluated sage-grouse nest success in predator removal and nonremoval areas. Methods for nest searching were not described, but one can only assume that detection probabilities were unequal among different cover types. Most nests found were under isolated shrubs away from other associated cover (Batterson and Morse 1948). It was this type of nest for which raven removal was evaluated. Nest success was <6% in nonremoval areas and 13–51% in the treatment area. However, habitat quality was not quantified in either of these study areas. More recently, the effects of coyote and red fox control programs on sage-grouse populations were evaluated in Wyoming and Utah (Slater 2003, Baxter et al. 2007). Coyote control annually from June to February had no measurable effect on nest success or chick survival (Slater 2003). Generally, habitat conditions were similar between the two sites, except the treatment area had significantly more herbaceous cover (Slater 2003). Conversely, year-round red fox removal appeared to increase adult survival and productivity in Utah (Baxter et al. 2007); unfortunately, this study did not have a nonremoval site to compare these increases in demographic rates. In Utah and elsewhere (Garton et al. this volume, chapter 15), sage-grouse population trends were increasing during a 1998–2005 study by Baxter et al. (2007). Thus, the inference on red fox removal as it relates to sage-grouse demography is limited.

Conceptual Framework for Evaluating Effects of Predation

Quantity, quality, and configuration of habitat have the potential to impact predator behavior and dynamics. These considerations include, but are not limited to, hiding cover at nests (Connelly et al. 1991, Gregg et al. 1994, Holloran et al. 2005, Coates 2007) and visibility at leks (Hartzler 1974). Additionally, feeding areas adequate in quantity and size may minimize risks associated with increased travel and time spent in riskier habitats (Gregg et al. 1994, Fischer et al. 1996a, Pyle and Crawford 1996, Beck et al. 2006). However, changes in the species composition and abundance of predator communities may affect nest success and survival of ground-nesting birds

(Greenwood and Sovada 1996, Evans 2004). These changes may be induced by anthropogenic developments in native habitats. Specifically, predator populations may respond positively to anthropogenic food sources such as road kills and dump sites, shelter, and nest substrate (Knight and Kawashima 1993). Increased predation pressure in fragmented habitats has been well documented for grouse in Europe (Andrén et al. 1985, Andrén and Angelstam 1988, Bernard-Laurent and Magnani 1994, Kurki et al. 1997). Poor concealment cover at the microscale can increase nest predation rates by Common Raven (Coates 2007); and, if raven populations are artificially high because of other human-induced conditions, prey species may have elevated predation rates (shared-predation hypothesis). Corvid abundances have been positively correlated with higher nest-predation rates of many bird species, including grouse (Angelstam 1986, Andrén 1992, Manzer and Hannon 2005).

Landscape fragmentation, habitat degradation, and human populations have the potential to increase predator populations through increasing kill efficiency and subsidizing food sources and nest or den substrate. Thus, otherwise suitable habitat may change into a habitat sink for grouse populations (Aldridge and Boyce 2007). Temporary measures to buffer a sink population or increase low vital rates (i.e., apparent nest success <30%, female survival <45%) resulting from above-average predation rates using short-term predator reduction programs may be warranted in instances when habitat restoration cannot be achieved in a timely manner (Connelly et al. 2000a).

Reintroductions

Individuals captured and translocated for population augmentation or reintroduction may suffer higher mortality rates than individuals native to a particular area (Musil et al. 1993; Baxter et al. 2008; M. F. Livingston, pers. comm.). Because of the stress of transporting and the unfamiliarity of the habitat and hiding cover, translocated individuals may suffer from elevated rates of predation. Predator reductions in these instances may be warranted in the short term to buffer a translocated population from elevated levels of predation (Connelly et al. 2000a, Hagen et al. 2004). Conversely, if an exotic predator has become established in native habitat and is negatively affecting local sage-grouse populations, predator management methods may be warranted (Connelly et al. 2000a).

CONSERVATION IMPLICATIONS

Predator management has been tried within the range of sage-grouse, but sufficient evidence has not been provided to support implementing control programs over broad geographic or temporal scales. Predator management can include both lethal and nonlethal methods to buffer population sinks in the short term (2–3 years) from the adverse effects of predation rates outside the range of natural variability. A more recent recognition is that the broader financial and political cost to removing predators at a scale and extent that may be effective is no longer socially or ecologically viable (Messmer et al. 1999). The most effective long-term predator management for sage-grouse population may be through maintaining connectivity of suitable habitats (Schroeder and Baydack 2001). However, most sage-grouse research has failed to quantify predator community structure or predation rates in relation to habitat variables, let alone within the landscape context. Thus, it is not currently possible to understand relationships among habitat structure, demographic rates of sage-grouse, and the predator community of an area and to incorporate these into a broad-based predator management program for sage-grouse. It is critical for future sage-grouse conservation efforts to quantify predator communities as they relate to demographic rates and habitat variables so we can better understand the predator-cover complex as it pertains to Greater Sage-Grouse life history.

ACKNOWLEDGMENTS

I thank J. W. Connelly and S. T. Knick for providing the opportunity to participate in this project. M. A. Gregg, J. C. Pitman, and an anonymous reviewer provided comments on an earlier draft of the chapter. W. E. Palmer of Tall Timbers Research Station provided information and insights to unpublished data on Northern Bobwhite and predator removal. This work was supported in part by the Oregon Department of Fish and Wildlife.

CHAPTER SEVEN

Harvest Management for Greater Sage-Grouse

A CHANGING PARADIGM FOR GAME BIRD MANAGEMENT

Kerry Paul Reese and John W. Connelly

Abstract. Harvest of Greater Sage-Grouse (*Centrocercus urophasianus*) has occurred throughout recorded history, but relatively few studies address the impact of harvest on sage-grouse numbers. Harvest of Greater Sage-Grouse occurs in 10 of 11 western states in which they reside. Hunting seasons and bag and possession limits have often become more conservative over the species' range during the past decade, as states responded to changing population numbers and perceived threats to the birds and then acted to reduce harvest opportunities. By 2007, hunting season lengths ranged from 2 to 62 days with a mean length of 10 days. Annual harvest estimates range from 10 birds in South Dakota to 10,378 in Wyoming. Total estimated annual harvest of Greater Sage-Grouse in the 10 states in 2007 was 28,180 birds. The effects of hunting on sage-grouse populations remains equivocal based on published literature, but the paradigm of harvest as compensatory may be shifting as evidence accumulates that populations of Greater Sage-Grouse require more conservative hunting regulations to reduce the potential for excessive harvest. Recent research suggests that because Greater Sage-Grouse normally experience low mortality over winter, mortality from hunter harvest in September and October may not be compensatory. Harvest mortality on most populations of Greater Sage-Grouse appears to be low, but both harvest levels and population abundance must be closely monitored in every population to improve management regulations for the harvest of the species. Biological data obtained from harvested birds are vital for continued management of sage-grouse populations. No studies have demonstrated that hunting is a primary cause of reduced numbers of Greater Sage-Grouse, and cessation of harvest in Washington 20 years ago has not resulted in increasing population levels. Continued concern over general population declines in Greater Sage-Grouse populations from known (disease, habitat loss, and habitat fragmentation) and unknown origins requires new research and continued routine collection of biological data for each population to optimize future harvest strategies.

Key Words: additive, *Centrocercus urophasianus*, compensatory, exploitation, Greater Sage-Grouse, harvest, hunting, mortality.

Gestion de la Cosecha de Greater Sage-Grouse: Un Paradigma Cambiante Para la Gestion de Aves de Caza

Resumen. La cosecha del Greater Sage-Grouse (*Centrocercus urophasianus*) ha sido registrada a lo largo de la historia, pero relativamente pocos

Reese, K. P., and J. W. Connelly. 2011. Harvest management for Greater Sage-Grouse: a changing paradigm for game bird management. Pp. 101–111 *in* S. T. Knick and J. W. Connelly (editors). Greater Sage-Grouse: ecology and conservation of a landscape species and its habitats. Studies in Avian Biology (vol. 38), University of California Press, Berkeley, CA.

estudios evaluaron el impacto de la caza en los números del sage-grouse. La cosecha del Greater Sage-Grouse ocurre en 10 de 11 Estados occidentales en donde reside. Las estaciones de caza, y los límites de caza y de posesión han llegado a ser a menudo más conservadores dentro del territorio de esta especie durante la última década, a medida que los estados respondieron a los números de población cambiantes y a las amenazas percibidas para las aves, y luego actuaron para reducir oportunidades de la cosecha o caza. Antes del 2007, la duración de la estación de caza se encontraba en el rango de 2 a 62 días con una duración promedio de 10 días. Estimaciones anuales de la cosecha varían entre 10 aves en Dakota del Sur a 10.378 en Wyoming. El total anual estimado de cosecha del Greater Sage-Grouse en los 10 estados en 2007 fue de 28.180 aves. Los efectos de la caza sobre poblaciones del sage-grouse siguen siendo ambiguos basados en la literatura publicada, pero el paradigma de la cosecha como compensatorio puede estar cambiando a medida que se acumula evidencia de que las poblaciones de Greater Sage-Grouse requieren regulaciones más conservadoras de la caza para reducir el potencial de una cosecha excesiva. Investigaciones recientes sugieren que, debido a que el Greater Sage-Grouse normalmente percibe una baja mortalidad durante el invierno, la mortalidad por la cosecha de los cazadores en septiembre y octubre puede no ser compensatoria. La mortalidad por cosecha en la mayoría de las poblaciones de Greater Sage-Grouse aparenta ser baja, pero tanto los niveles de cosecha como la abundancia de la población deben supervisarse de cerca en cada población para mejorar las regulaciones que gestionan la cosecha de la especie. Los datos biológicos obtenidos de aves cosechadas son vitales para el manejo continuo de las poblaciones de sage-grouse. Ningún estudio ha demostrado que la caza es una de las principales causas de números reducidos de Greater Sage-Grouse, y la suspensión de la cosecha en Washington hace 20 años no ha dado lugar a poblaciones más numerosas. La preocupación continua por el descenso de la población en general de Greater Sage-Grouse ya sea por causas conocidas (enfermedad, pérdida del hábitat, y fragmentación del hábitat) o de origen desconocido, requieren de nuevas investigaciones y la colección continua y rutinaria de datos biológicos para cada población para optimizar las estrategias futuras de la cosecha.

Palabras Clave: aditivo, caza, *Centrocercus urophasianus*, compensatoria, explotación, Greater Sage-Grouse, mortalidad, recolección.

The impact of harvest on populations of many game bird species remains uncertain (Gutiérrez 1994, Roy and Woolf 2001, Otis 2002, Williams et al. 2004a). Numerous examples exist of sustainable harvest within high-quality habitats (Potts 1986, Hudson and Dobson 2001, Sutherland 2001, Willebrand and Hornell 2001), but excessive harvest can reduce spring breeding population size of game birds (Anderson and Burnham 1976, Small et al. 1991, Williams et al. 2004b). The general approach to harvest management of upland game was developed during the 1930s and 1940s (Wing 1951, Allen 1962, Dasmann 1964). This approach assumed small game populations produce a large number of young each year, most of which are available for harvest because they would not survive the winter and add to the next season's breeding population. These ideas suggested hunting was a compensatory form of mortality and implied that a large portion of a small game population could be harvested each fall because, if not taken by hunters, they would likely die from other causes prior to the next breeding season. Early researchers provided evidence to support these ideas from studies of small game (Leopold et al. 1943; Errington 1945, 1956; Bump et al. 1947; Allen 1962). Reproductive characteristics and effects of exploitation were believed to be the same for all species of upland game (Allen 1962, Strickland et al. 1994). Allen (1962) summarized this situation well when he wrote that populations of small animals operate under a 1-year plan of decimation and replacement, that nature maintains overproduction, and that a huge surplus of animals die whether harvested or not. These early studies provided little empirical evidence to assess the impacts of exploitation.

Our objectives are (1) to demonstrate how wildlife managers embraced this paradigm for harvest management of Greater Sage-Grouse (*Centrocercus urophasianus*) and (2) to show how this paradigm shifted with the addition of reliable data on vital rates and effects of exploitation. We also compare the current approach to harvest management of Greater Sage-Grouse to that of other species of upland game birds and summarize available information on sage-grouse harvest across the species' range.

HARVEST MANAGEMENT OF GREATER SAGE-GROUSE

Harvest of Greater Sage-Grouse has occurred throughout recorded history (Patterson 1952, Autenreith 1981), but season length and bag and possession limits have varied greatly over time and among areas. Exploitation rates may have changed markedly, but relatively few studies have addressed the impact of harvest on sage-grouse numbers. Presently, harvest of Greater Sage-Grouse occurs in 10 of the 11 western states in which they reside; only Washington prohibits harvest. Neither Alberta nor Saskatchewan allow harvest. Gunnison Sage-Grouse (*Centrocercus minimus*) are not presently harvested in Colorado or Utah, the extent of their distribution.

Early Harvest Management (Pre-1990)

Excessive harvests in the 1800s, possibly coupled with intense livestock grazing, decimated populations of Greater Sage-Grouse, forcing most states to then prohibit harvest (Patterson 1952). Populations increased after a prohibition on hunting, and hunting was resumed in many areas (Patterson 1952). Harvest regulations over the past 100 years have changed as populations were perceived to increase or decline (Autenreith 1981, Wambolt et al. 2002). For example, season length for Greater Sage-Grouse in Idaho varied from 137 to 15 days from 1903 to 1933 and from 1.5 to 30 days from 1956 to 1990 (Connelly et al. 2005). By the mid-1970s to early 1980s, many western states attempted to standardize and often liberalize sage-grouse hunting seasons, at least compared to the seasons held in the 1950s and 1960s. Wallestad (1975a:27) supported this approach to sage-grouse harvest in Montana by providing the statements quoted from Allen (1962) and reported that Montana maintained liberal sage-grouse seasons because of high annual turnover, law of diminishing returns, and opening day phenomena. Moreover, this approach was also supported by studies suggesting that hunting had minimal impact on sage-grouse populations (June 1963, Crawford 1982, Braun and Beck 1985, Braun 1987).

Current Harvest Management (Post-1990)

Evidence began to accumulate during the 1980s and 1990s suggesting that, under some circumstances, harvesting of game birds may have an additive effect on mortality for a number of species (Gregg 1990, Robinette and Doerr 1993, Dixon et al. 1996). Publications began to reflect new information suggesting that earlier views of harvest management were not always correct (Peek 1986, Caughley and Sinclair 1994). In referring to upland game management, Brennan (1994) observed that wildlife professionals have been missing the mark with respect to operating as responsible stewards for vertebrate resources. Gutiérrez (1994) reinforced this idea by also concluding that research and management have not met the challenge of upland game management. The Policy Analysis Center for Western Public Lands (Wambolt et al. 2002) concluded that, in some instances, hunting might slow population recovery of sage-grouse or stabilize populations at lower than desirable levels. The Center also concluded that agencies should not assume that hunting was a totally compensatory form of mortality, nor should they base hunting seasons on the concept that seasons and bag limits can be liberal because of high annual turnover in the population (Allen 1962, Wallestad 1975a). In the mid-1990s, after obtaining new information on Greater Sage-Grouse vital rates suggesting that this species was relatively long-lived and had a low annual mortality and high overwinter survival (Wakkinen 1990, Robertson 1991, Connelly et al. 1993, Zablan 1993), Idaho and Wyoming reduced harvest on sage-grouse. Wyoming also changed the opening date for Greater Sage-Grouse from 1 September to mid-September and reported these changes resulted in lower harvest (Heath et al. 1997).

By 2007, opening dates for hunting Greater Sage-Grouse ranged from 1 September to 5 October, but most states began their seasons in mid-September (Table 7.1). Season lengths ranged from 2 to 62 days, with a mean length of 10 days. Only four states (Montana, Idaho, Wyoming, and Nevada) retained seasons of >10 days, while 4 of the 10 states had open seasons of seven days or less. One state in 2007 had a season longer than four weeks.

TABLE 7.1
Calendar year 2007 hunting season structure for Greater Sage-Grouse by state.
Numbers in parentheses indicate different hunt units or seasons.

State	Opening date for state or area	Season length (days)	Bag/possession limit
California			
(1)	8 Sep	2	2/2 season limit, 35 permits
(2)	8 Sep	2	1/1 season limit, 60 permits
Colorado	13 Sep	7	2/4
Idaho			
(1)	15 Sep	23	2/4
(2)	15 Sep	7	1/2
Montana	1 Sep	62	2/4
Nevada			
(1)	5 Oct	10	2/4
(2)	25 Sep	15	2/4
(3)	15 Sep	2	3/6 with 75 permits
(4)	22 Sep	2	3/6 with 75 permits
North Dakota	24 Sep	3	1/1 season limit
Oregon	8 Sep	9	2/2 season limit, 13 areas with 10–225 permits, total of 1,175 permits
South Dakota	26 Sep	2	1/1 season limit
Utah			
(1)	15 Sep	9	2/2 season limit, 395 permits
(2)	15 Sep	9	2/2 season limit, 283 permits
(3)	15 Sep	9	2/2 season limit, 195 permits
(4)	5 Sep	9	2/2 season limit, 479 permits
Wyoming	22 Sep	11	2/4

Hunting seasons and bag and possession limits for Greater Sage-Grouse have fluctuated and often became more conservative over the species' range during the past decade as states responded to changing population numbers and perceived threats to the birds, and then acted to reduce harvest opportunities. For example, between the hunting seasons of 1992 and 2007, Idaho reduced season length from 30 to 23 days, and reduced bag and possession limits from three and six to two and four, respectively (Table 7.2). Harvest in Idaho declined from 29,900 birds in 1992 to 4,935 in 2007 (Table 7.2). In addition to shifting the opening date of the hunting season from 1 September to 22 September, Wyoming shortened the season from 30 to 11 days and reduced the bag and possession limits from three and six to two and four birds, respectively (Table 7.2). More than 34,000 sage-grouse were harvested in Wyoming in 1992, and 10,378 were harvested in 2007. Utah changed the regulations for Greater Sage-Grouse from a general hunt with a three- and six-bird bag and possession limit to permit-only hunts. Total number of permits in 2007 was 1,352 for a 9-day season, with a two-bird season limit. Harvest in 2007 was 10% of the level in 1992 (Table 7.2). California reduced the number of permits to 95 in 2007, and harvest declined

TABLE 7.2
Hunting season structure and estimated number of Greater Sage-Grouse harvested by Idaho, Utah, and Wyoming from 1992–2007.

State	Year	Season length (days)	Bag/possession limits	Estimated harvest[a]
Idaho	1992	30	3/6	29,900
	1997	23	2/4	16,000
	2002	23	2/4	7,576
	2007	23	2/4	4,935
Utah	1992	9	3/6	12,156
	1997	9	1/2	4,489
	2002	9	2/2 season limit, 640 permits	511
	2007	9	2/2 season limit, 1,352 permits	1,255
Wyoming	1992	30	3/6	34,388
	1997	16	3/6	11,551
	2002	9	2/4	4,835
	2007	11	2/4	10,378

[a] Harvest estimated through mail or telephone surveys conducted independently by each state. For details, contact state of interest.

from 493 in 1999 to 47 in 2007. California maintained the same basic season structure over the past few years, but concern over drastic declines in lek counts in Lassen County led to a reduction from 175 two-bird permits in 2006 to 35 two-bird permits in 2007. Montana retained a 1 September opening date with a 62-day season, but reduced bag and possession from three and six to two and four birds. South Dakota retained a 2-day season, but reduced the season limit from two birds to one.

In contrast, Oregon slightly increased hunter opportunity from 2001 to 2003 through increasing the number of Greater Sage-Grouse permits to 1,275, only 10 more than in 2001. This was a <1% increase in permit numbers. Oregon reduced the number of permits to 1,175 in 2007. Regulations in Colorado, Idaho, Montana, Nevada, North Dakota, and South Dakota remained relatively constant since 2001.

In 2007, Colorado offered a 7-day season with a two-bird bag limit and four-bird possession limit. North Dakota had a 3-day season and South Dakota had a 2-day season, with each state having a one-bird bag and possession/season limit. North Dakota closed the sage-grouse hunting season for 2008 for the first time in almost 50 years as a result of a significant decline in the breeding population from 2007 to 2008. Each of these states may have opened or closed specific hunting areas not reported here. For example, an area in Idaho closed to harvest since 1998 was opened in 2002. Idaho also reduced the season in 2007 in two subunits from 23 days, two-bird bag and four in possession, to a 7-day season, with one bird bag and two in possession, due to drought and poor production of young. Concern over impacts of West Nile virus prompted reduced hunting opportunities in some areas of Oregon and Idaho in 2006 and 2007. West Nile virus is recognized as a severe threat to sage-grouse (Naugle et al. 2004, Walker et al. 2007b) that may impact populations in late summer, when adult survival is typically high (Connelly et al. 2000a). Mortality from West Nile virus may occur in sage-grouse within just weeks of the hunting season (Walker and Naugle, this volume, chapter 9), and managers may be forced to respond with emergency hunting closures to protect populations reduced by the disease.

Two states increased hunting opportunities for Greater Sage-Grouse from 2002 to 2007. Wyoming extended season length from 9 to 11 days, and Nevada increased season length in some areas from 9 to 10 or 15 days. Nevada also increased

TABLE 7.3

Annual harvest estimates for Greater Sage-Grouse by state, 2001–2007[a].

State	Year						
	2001	2002	2003	2004	2005	2006	2007
California	234	140	170	180	152	191	47
Colorado	782	307	433	1,748	1,197	745	1,197
Idaho	7,000	7,576	—[b]	8,100	10,500	12,509	4,935
Montana	4,525	5,475	7,843	3,056	3,515	4,927	3,116
Nevada	2,692	3,940	4,557	5,244	3,176	3,701	6,355
North Dakota	37	45	15	12	20	9	120
Oregon	1,015	973	979	1,228	1,277	1,092	767
South Dakota	12	16	12	25	26	15	10
Utah	1,182	511	1,049	1,361	1,744	1,498	1,255
Wyoming	12,742	4,835	5,666	11,783	13,176	12,920	10,378
Totals	30,221	23,181	20,724	32,737	34,783	37,607	28,180

[a] Harvest estimated through mail or telephone surveys conducted independently by each state. For details, contact state of interest.

[b] No estimate available.

general hunting opportunity with fewer permit-only areas.

Harvest regulations for Greater Sage-Grouse should minimize the possibility of negative effects on populations. Thus, geographic units with dissimilar vegetative, physical, or ecological attributes may differ in appropriate hunting season length and bag limits. However, all hunting seasons should be established with relatively low rates of harvest (Schroeder et al. 1999, Connelly et al. 2000a, Wambolt et al. 2002), which should allow populations to increase if habitat quality is not limiting population numbers (Connelly et al. 2003b). State wildlife management agencies have recently reduced harvest opportunities to minimize the possibilities that hunting may have a negative impact on Greater Sage-Grouse populations. Based on a review of the literature, Connelly et al. (2000c) suggested that no more than 10% of the fall population be removed through harvest. Harvest regulations as currently structured (Table 7.1) tend to be conservative and may keep harvest (Table 7.3) to <10% of fall population size. However, states do not presently measure fall population size of Greater Sage-Grouse and no recognized protocol has been established to do so.

HARVEST EFFECTS

Currently available data indicate that Wyoming, Nevada, Idaho, and Montana, with general (non-permit) Greater Sage-Grouse hunting seasons, lead in estimated numbers of harvested birds (Table 7.3). Annual harvest estimates range from 10 in South Dakota to 10,378 in Wyoming. Total estimated annual harvest in the 10 states is 28,180 Greater Sage-Grouse. This is a reduction from the 2006 harvest estimate of 37,607 sage-grouse, the largest harvest over the past 7 years (Table 7.3). Many state wildlife agencies appear to monitor Greater Sage-Grouse population changes closely, and recently there have been corresponding changes in season regulations as warranted to provide harvest opportunities but avoid overharvest of birds.

During the 1950s, 1960s, and 1970s, state wildlife agencies tended to use a one-size-fits-all approach to harvest management for game birds (Allen 1962, Wallestad 1975a, Connelly et al. 2005). Wildlife managers and scientists over the past 20 years have increasingly recognized that sport hunting can reduce populations of some upland game species (Bergerud 1985, Ellison 1991,

Dixon et al. 1996, Williams et al. 2004b) unless prevented through appropriate harvest regulations tailored to the species' biology and habitat requirements. Harvest mortality that reduces the population the subsequent spring breeding season is termed additive (Anderson and Burnham 1976). Additive hunting mortality will result in a spring breeding population lower than if harvest had not occurred in the previous fall, because each bird harvested is in addition to those that die naturally through disease, starvation, accidents, or predation. In contrast, mortality from hunters could act to reduce natural mortality through a number of density-dependent mechanisms, including reduced depredation or decreasing competition for food or shelter, so that total mortality is no higher than without harvest. Compensatory mortality does not reduce subsequent spring breeding population size below what it would have been due to natural mortality (Anderson and Burnham 1976). Partial compensation, or partial additivity, could also occur in exploited populations (Anderson and Burnham 1976).

Life-history characteristics of Greater Sage-Grouse differ from most other upland game birds. Many game bird species have an r-selected strategy (Anderson 2002:54)—high fecundity with large clutch sizes of 10–17 eggs; high annual rates of natural mortality, especially over winter (40–70%); and short life spans of 1–2 years (Gullion 1984, Potts 1986, Petersen et al. 1988, Christensen 1996, Giudice and Ratti 2001). Removal of individuals through hunting likely compensates for the many birds that would die naturally during their first or second winter (Kokko 2001, Sutherland 2001). Greater Sage-Grouse, however, tend toward a K-selected life-history strategy, with relatively low productivity through clutch sizes of six to nine eggs, low overwinter mortality rates of 2–20%, and long life spans of 3–6 years (Schroeder et al. 1999). Hunting will not have as large an impact on a population exhibiting an r-selected strategy as it will on one having a K-selected strategy (Anderson 2002).

The appropriate harvest rate, expressed as percentage of fall population size, remains elusive for Greater Sage-Grouse, although several studies have addressed this issue during recent decades (Autenreith 1981, Crawford 1982, Braun and Beck 1985, Connelly et al. 2000a). A harvest rate of 30% for Greater Sage-Grouse in Idaho was deemed allowable by Autenreith (1981), but he believed this high rate was not reached in any area. He emphasized that harvest should be more conservative in xeric areas close to urban centers than in more mesic areas. Forbs are readily available to grouse throughout mesic ranges, and grouse do not congregate in restricted feeding areas in August and September as they do in xeric ranges with limited mesic sites. As a result, dispersed birds in more mesic ranges are not as vulnerable to harvest as aggregated birds in xeric ranges nearer urban centers (Autenreith 1981).

Crawford (1982), Crawford and Lutz (1985), and Braun and Beck (1985) all examined impacts of harvest on Greater Sage-Grouse through use of harvest estimates, lek-count trends, brood counts, or band recoveries. In Oregon, Crawford (1982: 376) analyzed 20 years of data and reported that the mortality from harvest may have been compensatory. Crawford and Lutz (1985) concluded that while harvest may have short-term effects on Greater Sage-Grouse populations through increased mortality, sport hunting was not responsible for the long-term decline in sage-grouse numbers in Oregon. Braun and Beck (1985) reported that 7–11% of the fall population was harvested in an area of Colorado and that harvest had no measurable effect on sage-grouse densities in the spring. They concluded that hunting mortality could remove 20–25% of the fall population without being additive.

Van Kooten et al. (2007) suggested that hunting mortality for a population of Greater Sage-Grouse in Nevada is likely compensatory to other causes of mortality, but based their conclusion on analysis using an incomplete data set of historical lek counts coupled with observational data. It is not possible to tell from their paper if the observational data, which they referred to as trend data, resulted from casual observations made by field biologists, systematic counts, aerial surveys, or some combination. The time of year these observations were made was not provided, nor did the authors indicate what was counted (e.g., hens with broods, all birds, adult males, or some combination). They apparently did not consider how lek data were obtained, when counts were conducted, or whether data were collected from the air or ground. These problems were confounded by the author's lack of direct knowledge of sage-grouse biology and inappropriate use of supporting data.

Greater Sage-Grouse populations declined further in the 1990s, and some states began to reduce harvest opportunities, but the influence of harvest received little attention in most states until late in the decade. R. M. Gibson (pers. comm.) examined dynamics of two populations of Greater Sage-Grouse in Mono County, California, using data over a 45-year span. One population was isolated, and the other was contiguous with populations in Nevada. Data used consisted of lek counts, numbers of birds shot per hunter in fall, juveniles per hen in brood counts and in the fall bag, and number of birds inspected at check stations. He reported that the population contiguous with Nevada fluctuated independently of hunting mortality. However, changes in the isolated population were correlated with number of birds examined at check stations the previous fall. Gibson (1998) concluded that hunting mortality could depress and hold population levels of sage-grouse well below carrying capacity and that this should be of widespread concern in the light of long-term population declines and range fragmentation in this species.

Johnson and Braun (1999) used 23 years of hunter harvest and lek-count data in a population viability analysis (PVA) for Greater Sage-Grouse in North Park, Colorado. A PVA can assess the risk of a species population going extinct based on its survival and production values (Boyce 1992). Johnson and Braun (1999) reported that hunting mortality for Greater Sage-Grouse was compensatory up to a threshold level, and at or beyond that threshold, harvest may be additive. The level where harvest becomes additive was not reported.

Connelly et al. (2000a) assessed mortality of radio-marked adult Greater Sage-Grouse in Idaho in another 20-year study. Adult females were affected by harvest more than adult males. Fall harvest caused 15% of known male mortality and 42% of known female mortality. Forty-six percent of all female mortality occurred during the hunting season (September–October), and harvest accounted for 91% of female deaths. In contrast, only 2% of the deaths of either sex occurred during the four post-hunting-season months of November through February. The low overwinter mortality rate supports the contention that winter is not typically a difficult season for Greater Sage-Grouse (Beck and Braun 1978, Remington and Braun 1985, Sherfy 1992, Sika 2006).

Recognizing the typically low overwinter mortality of Greater Sage-Grouse is vital to understanding impacts of harvest. Robertson (1991) monitored radio-marked sage-grouse over three winters in southeastern Idaho. Sample sizes were small (seven, seven, and nine birds, respectively), but only one death occurred over three winters, and the average survival over three winters was 96%. A recent analysis of 20 years of band-recovery data from North Park, Colorado, provides more evidence that winter is not a period of great loss for sage-grouse. Zablan et al. (2003) found no evidence that variation in winter precipitation affected annual survival rates of sage-grouse, adding further support for winter as a time of minimal mortality. Wik (2002) measured seasonal survival rates of radio-marked Greater Sage-Grouse in southwestern Idaho from spring 1999 through fall 2001. Six estimates of winter survival rates of different age and gender classes of grouse produced point estimates of 0.85, 0.87, 0.88, 0.90, 1.00, and 1.00, all exceeding the 0.80 minimum overwinter survival reported by Schroeder et al. (1999). Sika (2006) reported the overwinter survival rate of 94 female Greater Sage-Grouse in central Montana as 0.98 during winter 2004–2005 and as 0.97 in 2005–2006. Similarly, Beck et al. (2006) found high overwinter survival of juvenile sage-grouse as they recruited into the breeding population during two winters in Idaho. Of 48 radio-marked juveniles on 1 December, only three birds died over both 4-month winters; two died in December, no birds died during January or February, and one bird died in March. In contrast, Moynahan et al. (2006) reported relatively low overwinter survival for a sage-grouse population in Montana and suggested that this mortality was related to unusually severe winter conditions.

Greater Sage-Grouse normally experience low mortality over winter, and mortality from hunter harvest in September and October may not be compensated after October. What is the threshold point where harvest becomes additive? This value has not been conclusively identified, but Connelly et al. (2000c) suggested that 10% of the fall population could be removed through harvest.

What are current harvest rates of fall populations of Greater Sage-Grouse? In Idaho, with harvest regulations of a 30-day season, three-bird limit, and six birds in possession, the harvest rate of adult females over 20 years averaged 8%, but in six of those years exceeded 10%. Connelly et al. (2000a) concluded that for adult females, hunting losses are likely additive to winter mortality and may result in lower breeding populations. Wik (2002) reported harvest rates in an area of southwestern Idaho with a 23-day season and two- and four-bird bag and possession limits, respectively, to be 5% for males, 6% for adult females and 18% for subadult females and suggested limiting harvest, especially of females. In a study of 93 radio-marked female Greater Sage-Grouse equally distributed on a hunted and a nonhunted area in central Montana, Sika (2006) reported no deaths from hunters over two hunting seasons (90 days with three- and six-bird bag and possession limits, respectively). However, apparently 30% of her hunted study area was closed to hunting by the landowner, and it was not clear how this may have affected vulnerability of the radio-marked birds to exploitation. Some birds were reportedly harvested by hunters during this study, but none of the radio-marked birds was shot. Wik (2002) concluded that sage-grouse use of private and tribal lands closed to hunting in his study area may have confounded his harvest estimates. Similarly, Browers (1983) used both reward bands and radiotelemetry to assess harvest of sage-grouse summering near a nuclear facility in southeastern Idaho. He reported that, during the 3 years of his study, only one of 43 marked birds was harvested and indicated the low harvest rate was due to most sage-grouse remaining in habitats on the Idaho National Engineering Laboratory, which was closed to hunting.

Zablan et al. (2003) considered recovery rates to reflect harvest rates, since most recovered bands from Greater Sage-Grouse were from hunter-killed birds. In a 20-year study, where bag and possession limits varied from two and four birds, respectively, in a 3-day season to three and six birds in a 30-day season, they reported that recovery rates varied from 14.0–18.7% in North Park, Colorado, but found no correlation between population fluctuations, as measured by lek-count data, and recovery data over the period of study. Zablan et al. (2003) suggested that lek counts are inadequate to index population changes or that sample sizes of recovered banded birds from hunters were low. Wildlife management agencies have reduced hunting pressure and harvest rates through regulatory changes but have not universally accepted a specific harvest rate. One harvest rate may not be appropriate for all populations of Greater Sage-Grouse.

What is known about the effects of exploitation on populations of Greater Sage-Grouse? Zunino (1987) conducted a nonreplicated experiment on two areas of Nevada to examine density of Greater Sage-Grouse in relation to harvest. He used fall helicopter surveys that were assumed to count a constant proportion of the population and estimated fall grouse densities in 1984 and 1985. Fall sage-grouse populations on both the control (nonhunted) and treatment (hunted) areas increased between years. However, the increase in fall population density was four times greater on the control area than on the treatment area. These differences were attributed to the treatments no harvest and harvest (Zunino 1987). In contrast, Stigar (1989) conducted a similar study in Nevada but did not identify any effects of harvest on sage-grouse populations.

Connelly et al. (2003a) conducted the only other experimental study of Greater Sage-Grouse response to exploitation. They used lek counts from 1996 to 2001 on 19 lek routes to assess response to three levels of exploitation. All lek routes were in areas with the same harvest regulations in 1996 (30-day season, three-bird bag, six in possession). In 1997 and continuing through 2001, regulations included 1 of 3 policies: no hunting; a restrictive 7-day season with one-bird bag, two in possession; or a moderate 23-day season with two-bird bag and four in possession. Treatments (no hunting, restrictive, or moderate seasons) consisted of five, seven, and seven lek routes, respectively. Lek routes were also categorized as being in lowland areas close (≤1.5 h drive) to major cities and towns or in high-elevation mountain valleys farther from urban centers. Areas that remained open to hunting after reducing harvest opportunities had lower rates of spring population increase than did areas with no hunting (Connelly et al. 2003a). Both the moderate and restrictive hunting seasons produced harvests that apparently slowed population growth (Connelly et al. 2003a). Populations in low-elevation habitats, close to urban centers, and isolated because of habitat fragmentation may be less able to withstand a harvest rate that has little or no effect on populations in more extensive, contiguous, remote, or mesic areas (Autenrieth 1981, Gibson 1998, Connelly et al. 2003a).

Sedinger and Rotella (2005) critiqued the work by Connelly et al. (2003a) as being inconclusive. However, Reese et al. (2005) explained that they misinterpreted several aspects of the study due, in part, to errors in the subheadings of two tables in Connelly et al. (2003a). Sedinger and Rotella (2005) did not espouse compensatory or additive mortality from harvest of Greater Sage-Grouse, but they presented several important topics that require more research to further refine understanding of the relationships of density dependence, life-history stages, and population size and growth rates to impacts of harvest on population dynamics of sage-grouse. Both Sedinger and Rotella (2005) and Reese et al. (2005) indicated that more research is necessary into sage-grouse population dynamics, especially on the relative influences of density dependence as depicted by Williams et al. (2003) and influences of West Nile virus (Walker and Naugle, this volume, chapter 9), to improve management regulations for the harvest of Greater Sage-Grouse. Different populations of sage-grouse may respond similarly to populations of Northern Bobwhite (*Colinus virginianus*) and Ring-necked Pheasant (*Phasianus colchicus*)—at low abundance on the periphery of their ranges, populations of these species differed in density-dependent and density-independent processes (Williams et al. 2003). Obtaining more detailed population-specific data for future harvest management strategy decisions for Greater Sage-Grouse appears to be essential.

Sika (2006) reported results on survival of 93 female Greater Sage-Grouse on two areas of central Montana. No radio-marked females were killed by hunters during the 62-day season on the hunted area, but female survival on the hunted area (2004 hunting season survival was 0.75 and in 2005 was 0.90) was lower compared to survival on the nonhunted area (2004 survival was 0.95 and 2005 was 0.98). These values might be interpreted as a harvest response (Sika 2006), but summer survival was also lower for birds on the hunted area than on the nonhunted area. These multiseasonal differences in survival may have been caused by site-level differences such as habitat characteristics, land use, or predator communities rather than hunting (Sika 2006).

No studies have demonstrated that hunting is a primary cause of reduced numbers of Greater Sage-Grouse. Many studies support habitat-based reasons for sage-grouse population declines (Swenson et al. 1987; Dobkin 1995; Connelly and Braun 1997; Connelly et al. 2000a,b; Leonard et al. 2000; Aldridge and Brigham 2002; Pedersen et al. 2003; Walker et al. 2007a; Doherty et al. 2008). Game managers must be cautious with harvest given the current status of most populations. Eliminating harvest as a source of mortality did not produce any detectable increase in Greater Sage-Grouse numbers in Washington, probably because of large-scale habitat issues (Schroeder et al. 2000); however, reducing harvest may aid in population recovery in specific cases (Connelly et al. 2003b).

MONITORING HARVEST

Monitoring harvest of Greater Sage-Grouse provides population data not easily obtained except through costly radiotelemetry studies of specific populations. Wings from hunter-harvested birds allow calculation of sex and age ratios, young per hen, distribution of hatch, and percent of successful hens at relatively large scales (Beck et al. 1975, Autenrieth et al. 1982). These data, in conjunction with population-trend counts, contribute to understanding the dynamics of sage-grouse at landscape scales. Harvest of Greater Sage-Grouse also creates a constituency of advocates in each state interested in maintaining healthy populations of the species (Christiansen 2008). For example, >5,400 people pursued Greater Sage-Grouse in Wyoming in 2006 (Christiansen 2008), and 15,753 nonrestricted permits for Greater Sage-Grouse and Sharp-tailed Grouse (*Tympanuchus phasianellus*) were purchased in Idaho in 2007 (B. B. Ackerman, unpubl. data). A total of 2,772 permits was available in 2007 in those states that restrict hunting through a permit-only season. These hunters and the biological data they might provide are vital for continued management of sage-grouse populations.

An appropriate harvest rate has not been determined for Greater Sage-Grouse populations. Harvest equal to 5–10% of the fall population may be appropriate but assumes detailed and specific knowledge of population size in September or October. Given the uncertainty in abundance estimates for breeding season populations, expecting any state to adequately determine the size of any population of Greater Sage-Grouse in fall is not realistic. Populations of Greater Sage-Grouse, except those under extremely adverse conditions, should increase from the lek-count period in spring through production of chicks over summer to the fall hunting period. Thus, harvest management should be based on spring population size. Managers could propose harvest of a percentage of the spring breeding population estimate, perhaps 5%; devise and implement survey protocols to obtain breeding season population size (Reese and Bowyer 2007); and subsequently be assured that hunter harvest would not likely exceed the threshold to become additive. A conservative, yet data-centered, approach to harvest may be warranted given the results of recent studies and the continuing concerns over population and habitat trends of Greater Sage-Grouse.

CONSERVATION IMPLICATIONS

Harvest of Greater Sage-Grouse continues to be a cultural, recreational, and economic activity across 10 western states. Sage-grouse are viewed by many as a trophy species with increasing intrinsic value. In California, where harvest of Greater Sage-Grouse was limited to 95 permits in 2007, numerous hunters reported harvesting a bird but did not send in a wing because they stated that they will have the bird mounted as a trophy (K. P. Reese, unpubl. data). As Greater Sage-Grouse become more valued as trophies, management agencies might consider moving the hunting season dates to later in the fall to provide birds in full plumage. A later season date could also reduce harvest on young birds (Connelly et al. 2000a) and shift pressure to adult males as the preferred trophy. A spring season on males is theoretically possible if this would not interfere with breeding of females. Given the declines in sage-grouse populations across the West (Connelly and Braun 1997, Connelly et al. 2004), many hunters, conservationists, and biologists have expressed concern over the possible impacts of continued sport hunting of the species.

Clearly, the paradigm governing harvest management of Greater Sage-Grouse is changing, as it is for other species of upland game birds. Hunting season timing, length, bag and possession limits, and season limit need to be examined and established for each population of Greater Sage-Grouse to allow development of biologically relevant regulations. Monitoring and critical examination of numbers harvested as a proportion of population size, sex and age ratios, and population fluctuation are vital to proper management on a population-specific basis. Williams et al. (2004b) suggested that management efforts for Northern Bobwhite should shift from fine-scale small-area habitat management to larger-scale regional management of usable space. In essence, they espouse population-based management of the species. All states could enhance their data on attributes of specific populations of Greater Sage-Grouse through more concentrated efforts to obtain information from harvested birds through wing barrels, check stations, and/or mandatory wing returns from hunters. Some states might consider adopting a permit-only hunting season structure to more fully control the number of Greater Sage-Grouse harvested each year by population.

Hunting opportunity for Greater Sage-Grouse has been reduced where data suggested a possible negative impact from hunting, and it has also been reduced in response to general population declines of known (e.g., disease and habitat loss) and unknown origin. Hunting has not been demonstrated as the primary cause of decline in any Greater Sage-Grouse population (Connelly et al. 2000a, Reese et al. 2005), but the cautionary recommendations outlined in Connelly et al. (2000c) remain appropriate. Monitoring to produce the data necessary to set harvest regulations must be tailored to each population of Greater Sage-Grouse to optimize harvest strategies. States must continue to adjust seasons statewide and for specific populations. A risk-sensitive harvest strategy (Williams et al. 2004a) that avoids reducing individual populations of sage-grouse will require new research and continued routine collection of data for population monitoring. Harvest should be controlled and the social aspects, as well as biological implications of changes to harvest seasons, should be included in all future alterations to the harvest of Greater Sage-Grouse.

ACKNOWLEDGMENTS

We thank state agency wildlife biologists for providing information on Greater Sage-Grouse hunting season structure and harvest numbers. This chapter was improved by reviews from K. C. Jensen, D. L. Mitchell, and T. R. Thompson. Funding was provided by the Idaho Department of Fish and Game and the University of Idaho. This is contribution 1028 of the University of Idaho Forest, Wildlife and Range Experiment Station and Idaho Federal Aid in Wildlife Restoration Project W-160-R.

CHAPTER EIGHT

Parasites and Infectious Diseases of Greater Sage-Grouse

Thomas J. Christiansen and Cynthia M. Tate

Abstract. We report the parasites, infectious diseases, and noninfectious diseases related to toxicants found in Greater Sage-Grouse (*Centrocercus urophasianus*) across its range. Documentation of population-level effects is rare, although researchers have responded to the recent emergence of West Nile virus with rigorous efforts. West Nile virus shows greater virulence and potential population-level effects than any infectious agent detected in Greater Sage-Grouse to date. Research has demonstrated that (1) parasites and diseases can have population-level effects on grouse species; (2) new infectious diseases are emerging; and (3) habitat fragmentation is increasing the number of small, isolated populations of Greater Sage-Grouse. Natural resource management agencies need to develop additional research and systematic monitoring programs for evaluating the role of micro- and macroparasites, especially West Nile virus, infectious bronchitis and other corona viruses, avian retroviruses, *Mycoplasma* spp., and *Eimeria* spp. and associated enteric bacteria affecting sage-grouse populations.

Key Words: Centrocercus urophasianus, disease, Greater Sage-Grouse, parasite, pathogen.

Parásitos y Enfermedades Infecciosas del Greater Sage-Grouse

Resumen. Divulgamos los parásitos, las enfermedades infecciosas, y las enfermedades no contagiosas relacionadas con tóxicos encontrados en el Greater Sage-Grouse (*Centrocercus urophasianus*) a lo largo de su distribución. La documentación de los efectos a nivel poblacional es poco común aunque los investigadores han respondido con esfuerzos rigurosos a la reciente aparición del virus del Nilo occidental. El virus del Nilo occidental ha mostrado la mayor virulencia y potenciales efectos a nivel de población que cualquier otro agente infeccioso detectado en el Greater Sage-Grouse hasta la fecha. Investigaciones han demostrado que: (1) los parásitos y las enfermedades pueden tener efectos a nivel poblacional sobre especies de sage-grouse, (2) nuevas enfermedades infecciosas están emergiendo, y (3) la fragmentación del hábitat está aumentando el número de poblaciones pequeñas y aisladas de Greater Sage-Grouse. Las agencias encargadas del manejo de recursos naturales necesitan desarrollar investigaciones adicionales y programas de monitoreo sistemáticos para evaluar el rol de los micro y macro parásitos, especialmente el virus del Nilo occidental, bronquitis

Christiansen, T. J., and C. M. Tate. 2011. Parasites and infectious diseases of Greater Sage-Grouse. Pp. 113–126 *in* S. T. Knick and J. W. Connelly (editors). Greater Sage-Grouse: ecology and conservation of a landscape species and its habitats. Studies in Avian Biology (vol. 38), University of California Press, Berkeley, CA.

infecciosa y otros virus "corona", retrovirus aviares, *Mycoplasma* spp., y *Eimeria* spp. y bacterias entéricas asociadas que afectan a las poblaciones del sage-grouse.

Palabras Clave: Centrocercus urophasianus, enfermedad, Greater Sage-Grouse, parásito, patógeno.

Greater Sage-Grouse (*Centrocercus urophasianus*) host a variety of potentially pathogenic organisms, including macroparasitic arthropods (e.g., lice and ticks), helminths (e.g., nematodes, cestodes, and trematodes), and microparasites (protozoa, bacteria, fungi, and viruses). Few reports exist of fatal disease outbreaks in sage-grouse populations, with the exception of those caused by *Eimeria* spp. (coccidiosis; Honess and Post 1968) and West Nile virus (Naugle et al. 2004, 2005; Walker et al. 2007b; Walker and Naugle, this volume, chapter 9).

The presence of pathogenic organisms within a host species does not necessarily result in disease of the individual or indicate a population-level effect. Some level of parasites and disease-producing organisms is normal and does not typically cause significant alteration in structure or function of the population of the host species (Patterson 1952, Herman 1963). However, during the 1940s and 1950s, both managers and laypersons believed that infectious disease caused heavy annual losses of game birds (Herman 1963), including Greater Sage-Grouse (Patterson 1952). Herman (1963) reported frequent references to the grouse disease, as though investigators were convinced that one specific agent could be the universal root of all losses. Under certain circumstances, effects of a disease-causing agent may increase to a level that impacts local populations (Herman 1963). Such circumstances may include extremes in precipitation, poor nutrition, or introduction of an exotic disease agent.

Researchers in the United Kingdom demonstrated that parasites regulated Red Grouse (*Lagopus lagopus* ssp. *scoticus*) population cycles (Hudson 1986, Hudson et al. 1998), increased grouse vulnerability to predators (Hudson et al. 1992), and affected grouse behavior (Fox and Hudson 2001). Most macro- and microparasites and infectious diseases (Table 8.1) have not been documented to result in widespread population-level impacts to Greater Sage-Grouse. Noteworthy morbidity and mortality were typically discovered when dry conditions caused grouse to concentrate near water sources, likely contaminating water and soil with fecal material. (Scott 1940, Honess and Post 1968). Even in these circumstances, evidence was not sufficient to conclude that disease was responsible for major population declines across any extensive area of the bird's range (Patterson 1952).

Researchers have conducted few systematic surveys for parasites or pathogens in Greater Sage-Grouse. This problem has been exacerbated by the ineffectiveness of some techniques for detecting pathogens that are present, or by mistakes in identifying pathogens that were present. Thus, the role parasites have, if any, in population declines of sage-grouse across the species' range is essentially unknown. The recent emergence of West Nile virus (WNV) and its relative virulence in Greater Sage-Grouse underscore the importance of understanding the role of diseases in host population dynamics (Naugle et al. 2004; Clark et al. 2006; Walker et al. 2007b; Walker and Naugle, this volume, chapter 9). This information—coupled with (1) existing data that demonstrate population-level effects of parasites and disease on grouse species, (2) emergence of new infectious diseases, and (3) habitat fragmentation that increases the number of small, isolated populations of Greater Sage-Grouse—suggests diseases and parasites deserve further study and greater attention by managers and policy makers as suggested for prairie grouse (Peterson 2004).

Traditionally, the term *parasite* was interpreted to mean a set of organisms from one of three taxonomic groups—protozoa, helminths, and arthropods—that during at least one life stage take nutrients from a host. Here, we use the contemporary, ecologically based terms *macroparasite* (arthropods and helminths) and *microparasite* (protozoa, viruses, bacteria, and fungi; Anderson and May 1979, 1981). We describe known macro- and microparasites of Greater Sage-Grouse and their implications. Many sources of information were not published in the refereed literature but appeared in agency reports and proceedings. We included these studies in addition to those in the published literature. We

TABLE 8.1

Known parasites and infectious diseases of Greater Sage-Grouse (Centrocercus urophasianus) *by state or province.*

Parasite/disease	State/province	Prevalence[a]	Reference
	Macroparasites		
	Ectoparasites		
Mallophaga			
Gonoides centrocerci and/ or *Lagopoecus gibsoni*	UT		Griner (1939)
	WY		Simon (1940)
	WY		Honess and Winter (1956)
			Malcomson (1960)
	WY		Boyce (1990)
	WY	39.1 (N = 248)	Johnson and Boyce (1991)
	WY	22.1 (N = 730)	Deibert (1995)
L. perplexus	MT, OR, WY		Simon (1940)
	CO	27.0 (N = 84)	Keller et al. (1941)
	CO	18.6 (N = 59)	Dargan et al. (1942)
	WY		Honess and Winter (1956)
Acarina			
Haemaphysalis chordeilis (*cinnabarina*)	MT	92.3 (N = 26)	Parker et al. (1932)
			Simon (1940)
			Hopla (1974)
H. leporis-palustris	MT	3.9 (N = 26)	Parker et al. (1932)
			Simon (1940)
Diptera (true flies, mosquitoes, midges, keds)			
Ornithomyia anchineuria (ked)	NV	~25.0 (N = ~200)	M. R. Dunbar (pers. comm.)
	Endoparasites		
Platyhelminthes (flukes and tapeworms)			
Brachylaema fuscata[b]			Babero (1953)
Raillietina centrocerci	WY		Simon (1937)
	UT		Griner (1939)
	ID, MT, UT, WY		Simon (1940)
R. centrocerci	CO	34.6 (N = 52)	Keller et al. (1941)
R. cesticillus	CO	57.6 (N = 59)	Dargan et al. (1942)
	CO		Rogers (1964)
	CO, UT, WY		Honess (1982a)

TABLE 8.1 (*continued*)

TABLE 8.1 (CONTINUED)

Parasite/disease	State/province	Prevalence[a]	Reference
Rhabdometra nullicollis	CO		Ransom (1909)
			Boughton (1937)
	WY		Simon (1937)
	WY		Simon (1940)
			Babero (1953)
	WY		Honess (1982a)
Nematoda			
Acuaria (Cheilospirura) centrocerci	MT, WY		Simon (1939a)
			Simon (1940)
	WY		Bergstrom (1982)
A. (Cheilospirura) spinosa	MT, WY		Wehr (1933)
			Shillinger and Morley (1937)
Habronema urophasiana	MT		Wehr (1931)
	WY		Simon (1939a)
	WY		Simon (1940)
			Bergstrom (1982)
Heterakis gallinarum (syn. *gallinae)*			Schillinger and Morley (1937)
			Simon (1939a)
			Simon (1940)
			Bergstrom (1982)
Ornithofilaria tuvensis	WY	P (N = 3)	Hepworth (1962)
	WY		Thorne (1969)
Oxyspirura lumsdeni[b]	SK	P (N = 20)	Addison and Anderson (1969)
Microfilaria	CO	13.3 (N = 30)	Stabler and Kitzmiller (1974)
	CO	3.0 (N = 361)	Stabler et al. (1977)
	NV, OR	1.0 (N = 196)	Dunbar et al. (2003)
Microparasites			
Protozoa: Hematozoa			
Haemoproteus canachites	CO	20.0 (N = 30)	Stabler and Kitzmiller (1974)
	CO	9.1 (N = 361)	Stabler, Braun, and Beck (1977)
	CA	37.5 (N = 184)	Gibson (1990)
	WY	11.8 (N = 296)	Johnson and Boyce (1991)
Leukocytozoon lovati (bonasae)	CO	90.0 (N = 30)	Stabler and Kitzmiller (1974)
	CO	44.9 (N = 361)	Stabler, Braun, and Beck (1977)
	NV, OR	9.7 (N = 196)	Dunbar et al. (2003)
	WY	18.5 (N = 296)	Johnson and Boyce (1991)

TABLE 8.1 *(continued)*

Parasite/disease	State/province	Prevalence[a]	Reference
Plasmodium pediocetti	CO	0.0 (N = 30)	Stabler and Kitzmiller (1974)
	CO	0.6 (N = 361)	Stabler et al. (1977)
	WY		Boyce (1990)
	WY	33.8 (N = 296)	Johnson and Boyce (1991)
	WY	24.5 (N = 392)	Deibert (1995)
	NV, OR	0.5 (N = 196)	Dunbar et al. (2003)
Trypanosoma avium	CO	A (N = 3)	Stabler et al. (1966)
	CO	26.7 (N = 30)	Stabler and Kitzmiller (1974)
	CO	8.0 (N = 361)	Stabler et al. (1977)
Protozoa: Other			
Sarcocystis releyi	AL	P (N = 2)	Salt (1958)
Trichomonas simoni	WY		Simon (1940)
	WY		Honess (1955)
Trichomanas sp.	CO		Keller et al. (1941)
	CO		Dargan et al. (1942)
	CO		Rogers (1964)
Protozoa: Intestinal coccidia			
Eimeria angusta, E. centrocerci and/or *E. pattersoni*	WY		Scott and Honess (1933)
	WY		Scott and Honess (1937)
	WY		Scott (1940)
	WY		Simon (1939b)
	WY		Simon (1940)
	CO		Keller et al. (1941)
	CO		Dargan et al. (1942)
	WY		Honess (1942)
	CO		Carhart (1943)
	CO		Grover (1944)
	WY		Patterson (1952)
			Levine (1953)
	WY		Honess and Post (1955)
	WY		Honess and Winter (1956)
	CO		Rogers (1964)
	WY		Honess (1968)
	WY	~12.2 (N = ~1,888)[c]	Honess and Post (1968)

TABLE 8.1 (*continued*)

TABLE 8.1 (CONTINUED)

Parasite/disease	State/province	Prevalence[a]	Reference
Bacteria			
Clostridium perfringens	OR	P (N = 1)	Hagen and Bildfell (2007)
Escherichia coli (colibacillosis)	WY	P	Honess and Post (1968)
Salmonella spp. (salmonellosis)	WY	P	Post (1960)
	WY		Thorne (1969)
Francisella (pasturella) tularensis (tularemia)	MT	P (N = 28)	Parker et al. (1932)
	MT		Hopla (1974)
Mycoplasma spp. (mycoplasmosis)	CO	I (N = 115)	Hausleitner (2003)
	OR	A (N = 40)	Dunbar and Gregg (2003)
Fungi			
Aspergillus fumigatus (aspergillosis)	WY	P (N = 1)	Patterson (1952)
	WY		Honess and Winter (1956)
	WY		Thorne (1969)
Viruses			
Flavivirus (West Nile virus)[d]	AB, MT, WY		Naugle et al. (2004)
	MT, WY		Walker et al. (2004)
	AB, CA, CO, MT, WY		Naugle et al. (2005)
			Clark et al. (2006)
	MT, WY		Walker et al. (2007)
	AB, CA, CO, ID, MT, ND, NV, OR, SD, UT, WY		Walker and Naugle (this volume, chapter 9)
Avian retroviruses			
Leukosis/sarcoma viruses	CO	M.E.	Dimcheff et al. (2000)
Coronavirus (avian infectious bronchitis)	NV, OR	A.B. (N = 33–78)	Dunbar and Gregg (2004)
Avipoxvirus	TX (bird farm)	P (N = 1)	DuBose (1965)

[a] Prevalence (%; number sampled) was included where these data were available. P = present, A = absent, I = inconclusive, M.E. = molecular evidence, A.B. = antibodies only.

[b] These parasites have little documentation other than a brief mention in a published document. As a result, they are not discussed in the text.

[c] Fecal samples; sample sizes were not precisely provided.

[d] The references for West Nile virus provide more detail on survival, infection rates, and serology than can be adequately summarized in this table.

organized information to report (1) relevant details of host infection with the parasite; (2) description of the disease that may result from host infection with the parasite; (3) parasite prevalence data, if provided by the original authors, as well as individual case reports and documentation of mortality events associated with the disease in Greater Sage-Grouse or closely related species; and (4) examples of studies designed to evaluate effects of the parasite on populations of Greater Sage-Grouse or closely related species. The Greater Sage-Grouse is the focal species of our review, but information may also be relevant to Gunnison Sage-Grouse (*Centrocercus minimus*).

GENERAL DISEASE ECOLOGY

All organisms discussed in this review are potential pathogens, or disease-causing agents. The impact of a disease upon wildlife populations was, and for the most part still is, measured in terms of visible morbidity (clinical disease) or mortality (death).

The seminal work of Anderson and May (1978, 1979) and May and Anderson (1978, 1979) emphasized measuring less overt impacts of parasites on host populations—how parasites may reduce reproductive success and other host fitness parameters, and may therefore have a role in regulating or even, in some cases, extirpating host populations, without necessarily causing obvious signs of disease.

Parasites are organisms that use their hosts as habitat, are nutritionally dependent on the host, and cause harm to the host at some point in their life cycle (Anderson and May 1978). Parasites complete their life cycle either directly, via contact between hosts, ingestion, inhalation, or skin penetration, or indirectly, via an intermediate host or biting vector. Macroparasites generally occur as endemic host infections that are more likely to cause morbidity or epizootics than mortality. Conversely, microparasitic diseases often cause periodic epizootics where the host either dies or develops immunity. The periodicity of epizootics is dependent on densities of susceptible hosts.

Parasite communities and disease pathology also differ due to varying environmental conditions, especially climate. A pathogen that regulates a host population in one location may not even occur in another area, and generalizations should be viewed with caution (Peterson 2004). Many pathogens are sensitive to temperature, rainfall, and humidity (Harvell et al. 2002). Climate warming may increase pathogen development and survival, disease transmission, and host susceptibility, and scientists predict most host-parasite systems will experience more frequent or severe disease impacts with global warming (Harvell et al. 2002). Deem et al. (2001) maintained the amplified role of diseases as a factor limiting species' survival can be traced to anthropogenic changes on a global scale that have direct and indirect influences on the health of wildlife species. These changes include human population growth, habitat fragmentation and degradation, isolation of populations, and increased proximity of humans and their domestic animals to wildlife (Deem et al. 2001). Ample documentation shows these factors have influenced sage-grouse populations (Braun 1998, Schroeder et al. 1999, Connelly et al. 2004, Schroeder et al. 2004, Stiver et al. 2006).

PARASITES AND DISEASES

Macroparasites

Mallophaga

Bird lice are small, wingless, and dorsoventrally flattened insects that feed on fragments of skin and feathers. Investigators have identified three species of chewing lice on Greater Sage-Grouse: *Gonoides centrocerci*, *Lagopoecus gibsoni*, and *L. perplexus* (Table 8.1). Honess (1982b) suggested that presence of lice was little more than an irritant to the host. However, Clayton (1990) reported that chewing lice destroyed the insulating capacity of feathers of Rock Pigeons (*Columba livia*) and, as a consequence, infested males had increased thermoregulatory costs and were in poor condition in spring. Boyce (1990), Johnson and Boyce (1991), Spurrier et al. (1991), and Deibert (1995) suggested presence and prevalence of lice and other parasites can affect mate selection and breeding success by birds, including Greater Sage-Grouse. Breeding male Greater Sage-Grouse had lower lice prevalence (25.6%, $N = 39$) than nonbreeding males (34.5%, $N = 293$; Deibert 1995). Breeding success of females in relation to lice prevalence has not been investigated, although Deibert (1995) reported that females were less likely than males to have lice. Louse prevalence was 5.8% ($N = 173$) on females and 33.4% on males ($N = 332$).

Acarina

Two species of ixodid ticks have been identified on Greater Sage-Grouse: *Haemaphysalis chordeilis* and *Haemaphysalis leporis-palustris*. Early authors reported *Haemaphysalis cinnabarina* (Parker et al. 1932, Simon 1940). However, Kingston and Honess (1982) reported that *H. cinnabarina* does not occur in North America, and that all references to this species should be referred to as *H. chordeilis*. Parker et al. (1932) documented high numbers of ticks on Greater Sage-Grouse in August and September 1931 in a mortality event associated with tularemia in Montana and suggested that pathologic effects from the feeding activity of so many ticks may have partially contributed to the mortalities. During the investigation, 488 individual ticks (*H. cinnabarina = chordeilis*), including 30 partially to fully engorged females, were counted on a single recently dead sage-grouse (Parker et al. 1932). Of 26 sage-grouse observed both within and outside the epizootic area, 92.3% had ticks (Parker et al. 1932). The average number of ticks from 8 sage-grouse collected within the tularemia epizootic area and 18 sage-grouse outside of the epizootic area was 154 and 12, respectively (Parker et al. 1932). One rabbit tick (*H. leporis-palustris*) was recorded on one sage-grouse ($N = 26$), but it was the most common tick species found on cottontail rabbits (*Sylvilagus* spp.) and jackrabbits (*Lepus* spp.) in the same study area (Parker et al. 1932).

Hudson (1992) reported that louping ill virus, which the tick *Ixodes ricinus* transmits from sheep to Red Grouse in the United Kingdom, was associated with higher grouse mortality, lower densities, and reduced harvest. Some species of ticks cause fatal tick paralysis in some hosts, including birds (Luttrell et al. 1996), when adult female ticks release a neurotoxin into the host's bloodstream (Honess and Bergstrom 1982). We did not find a reference to tick paralysis affecting any bird species occupying Greater Sage-Grouse range.

Diptera

Some Diptera are important vectors of animal diseases, including avian malaria, avian pox, and West Nile virus. One veterinarian detected high, but unquantified, numbers of keds (louse flies [*Ornithomyia anchineuria*]) on about 25% of approximately 200 Greater Sage-Grouse captured in the Montana Mountains of northern Nevada in 2002 (M. R. Dunbar, pers. comm.). Keds have not been documented on Greater Sage-Grouse elsewhere.

Helminths

Greater Sage-Grouse are the only known host of the tapeworm *Raillietina centrocerci* (Honess 1982a). Two additional tapeworms, *Raillietina cesticillus* and *Rhabdometra nullicollis*, have been reported in Greater Sage-Grouse (Simon 1940, Keller et al. 1941). In Colorado, 46.8% of 111 sage-grouse sampled over 2 years were infected by *R. cesticillus* or *R. centrocerci* (Keller et al. 1941, Dargan et al. 1942). Leidy (1887) reported *Hymenolepis microps* in Greater Sage-Grouse from Wyoming, but Simon (1940) believed this to be a mistaken identification. Greater Sage-Grouse show no apparent clinical signs of infection, even though tapeworms may distend the intestine or protrude from the vent, and may reflect an almost perfect adjustment between the host and its parasite (Honess 1982a). Even heavily parasitized sage-grouse appear to be in good physical condition (Thorne 1969). Empirical studies of potential fitness-reducing effects of cestodes on Greater Sage-Grouse have not been conducted. Cole and Friend (1999), however, suggested heavy burdens of tapeworms could have direct and/or indirect adverse impacts on individual birds, such as intestinal occlusion, reduction in vigor, and increased susceptibility to other parasites.

Nematode worms of upland game birds belong to *Habronema* and *Acuaria* (*Cheilospirura*) and have indirect life cycles involving an arthropod intermediate host. Species documented in Greater Sage-Grouse (Table 8.1) include *Acuaria centrocerci*, *A. spinosa*, and *Habronema urophasiana* (Braun and Willers 1967). Sage-grouse infection with *Acuaria* (*Cheilospirura*) *spinosa* resulted in necrosis, at times accompanied by slight hemorrhage of the horny layer of the gizzard and underlying muscles, but in no case did the health of infected birds seem seriously compromised (Simon 1940). Bergstrom (1982:224) indicated severe infections of *A. centrocerci* could threaten Greater Sage-Grouse in local areas, but gizzard worms have only been seen occasionally and likely do not present an important threat. No prevalence data have been reported.

Heterakis gallinarum is a cosmopolitan nematode that lives in the cecae of domestic chickens and related birds. This nematode likely arrived in the United States with imported Ring-necked Pheasants (*Phasianus colchicus*; Schmidt and Roberts

1996). The life cycle of *H. gallinarum* is direct, and the eggs can remain infective for up to 4 years in soil (Schmidt and Roberts 1996). Heavy infections of *H. gallinarum* may result in thickening and bleeding of the cecal mucosa, but the cecal worm is not highly pathogenic by itself (Schmidt and Roberts 1996). *H. gallinarum* is considered an important avian macroparasite because its eggs can harbor *Histomonas meleagridis*, the flagellated protozoan microparasite that causes avian blackhead. Avian blackhead is economically important in domestic fowl and game bird farms (Bergstrom 1982). Peterson (2004) reported relatively high occurrence and prevalence of *H. gallinarum* in Plains Sharp-tailed Grouse (*Tympanuchus phasianellus jamesii*), Greater Prairie-Chicken (*T. cupido pinnatus*), and Lesser Prairie-Chicken (*T. pallidicinctus*), but the sole documentation of this nematode in sage-grouse was provided by Simon (1940), who reported a personal communication with J. E. Schillinger who identified *H. gallinae* (= *gallinarum*) in a single Greater Sage-Grouse specimen (location not reported). The single documentation of cecal worms in Greater Sage-Grouse suggests the possibility of exposure to *H. meleagridis*, but we found no records of avian blackhead in sage-grouse. Temperature influences development (embryonation) of *H. gallinarum*, so fluctuations in macroparasite abundance could be linked with weather conditions (Saunders et al. 2002). The relatively cold, dry climate of the sage steppe environment inhabited by Greater Sage-Grouse may be unfavorable, at least historically, to embryonation of *H. gallinarum*.

Parasites shared between sympatric host species may be more detrimental to the fitness of one of the host species than the other (Tompkins et al. 2000a). For example, *H. gallinarum* can parasitize Ring-necked Pheasants with few effects. However, impacts are much more severe if Gray Partridge (*Perdix perdix*) become infected (Tompkins et al. 2000a,b). Shared parasite relationships have been neither documented nor explored in Greater Sage-Grouse, but pheasants and Greater Sage-Grouse share habitat in localized areas. Peterson (2004) questioned the wisdom of perpetuating pheasants in areas inhabited by at-risk populations of prairie grouse, due to concern for the potential transmission of macro- and microparasites.

Filarids are tissue-dwelling nematodes that use arthropods as intermediate hosts. The first-stage juvenile forms of some filarid worms circulate in the bloodstream of the definitive host and are known as microfilaria (Schmidt and Roberts 1996). Microfilaria have been reported in sage-grouse (Table 8.1), but no reports of disease are associated with their occurrence. Adults of one species of filarial nematode, *Ornithofilaria tuvensis*, were identified in three Greater Sage-Grouse in southwest Wyoming on one occasion (Hepworth 1962). These white, hairlike worms occurred in connective tissue between the skin and breast muscles of apparently healthy hunter-harvested birds (Hepworth 1962). Hunters in the area reported additional sage-grouse with similar infections in the breast; some infections were apparently severe enough to render the birds unable to fly (Hepworth 1962). However, none of these affected birds were collected for analysis, and effects of *O. tuvensis* infection in Greater Sage-Grouse are unknown.

Microparasites

Protozoa

Hematozoa are arthropod-borne protozoan microparasites found in the bloodstream of vertebrate hosts (Jolley 1982, Atkinson 1999). Avian hematozoa include but are not limited to the group of intracellular microparasites responsible for avian malaria. Avian malaria microparasites, also known as hemosporidia, live inside blood cells at some point in their life cycle and belong to three closely related genera: *Plasmodium*, *Haemoproteus*, and *Leucocytozoon*. A distantly related flagellated hematozoan, *Trypanosoma avium*, swims free in blood plasma during all in-host life stages. Hemosporidians and other hematozoans, such as *T. avium*, require biological vectors such as mosquitoes, black flies, and biting midges for efficient transmission. Transmission and incidence peaks of hematozoans in temperate areas are seasonal. However, infections may be chronic and persist through winter, recrudescing in response to the physiologic stress of reproduction in the host in spring prior to vector emergence (Nordling et al. 1998).

Hematozoans are commonly found in wild birds (Atkinson 1999) including Greater Sage-Grouse (Table 8.1; Stabler and Kitzmiller 1974, Stabler et al. 1977, Boyce 1990, Gibson 1990, Johnson and Boyce 1991, Deibert 1995, Dunbar et al. 2003). Overt disease of Greater Sage-Grouse due to hematozoan infection has not been reported. Infections with hemosporidia may result in death in highly susceptible bird species and age classes (Bennett et al. 1993, Atkinson 1999). Avian malaria tends to be most pathogenic in domesticated birds and in wild avian species native to areas where avian malaria is not endemic when they are translocated to areas where malaria is endemic, or where hemosporidia and vectors are exotic invaders in native habitat.

Researchers have investigated the potential fitness-reducing effects of hemosporidiosis in grouse species including Greater Sage-Grouse. Male Greater Sage-Grouse in Wyoming infected with *Plasmodium pediocetii* had significantly lower reproductive success than noninfected males (Boyce 1990). However, there was no difference in reproductive success between *Haemoproteus*-infected and noninfected Greater Sage-Grouse males in a California population (Gibson 1990). Greater Sage-Grouse from Nevada and Oregon with *Leucocytozooan lovati*–infected hens ($N = 7$) reportedly experienced lower chick production (0.7 chicks/hen) than that of uninfected hens ($N = 10$; 1.6 chicks/hen), but the small sample size makes interpretation difficult (Dunbar et al. 2003). Stabler and Kitzmiller (1974) and Stabler et al. (1977) documented *Trypanosoma avium* in Greater Sage-Grouse (Table 8.1). The authors did not describe clinical signs or suggest implications to individual hosts or populations.

Sarcocystis spp. are tissue-dwelling protozoan microparasites that cycle through predator-prey systems. Birds are some of the numerous intermediate hosts of sarcocyst species. Avian sarcocystosis is often a highly noticeable condition, because the sarcocysts localize in the pectoral muscle of the intermediate host, resulting in a condition commonly referred to as rice breast. Sarcocysts are ubiquitous in wild species and generally cause little to no pathology. Salt (1958) described *Sarcocystis rileyi* in two apparently healthy Greater Sage-Grouse in Alberta. However, Jolley (1982) maintained that reports of *S. rileyi* in gallinaceous and other birds were probably incorrect and involved other species of *Sarcocystis*, citing the general tendency of sarcocysts to be confined to one taxonomic order of intermediate hosts and the specific documentation of *S. rileyi* as a commonly reported species in several genera of ducks.

Trichomonads are flagellated protozoan microparasites that have a direct lifecycle and no cyst or larval stage to confer environmental hardiness (Jolley 1982). Many trichomonad species likely remain unknown because no one has looked for them, and little attention is given to them because most are nonpathogenic (Jolley 1982). An exception to this generality is *Trichomonas gallinae*. Some virulent *T. gallinae* strains have caused severe epizootics of oropharyngeal disease in columbids, raptors, domestic turkeys, and chickens (Jolley 1982, Cole 1999). *Tritrichomonas simoni* was found in large unquantified numbers in the cecae of all healthy Greater Sage-Grouse examined (Honess 1955). This species is not known to be pathogenic (Jolley 1982).

Three species of enteric coccidia are known to infect Greater Sage-Grouse: *Eimeria angusta, E. centrocerci*, and *E. pattersoni* (Jolley 1982). These microparasites develop within cells of the intestinal lining. Birds infected with *Eimeria* spp. intermittently shed oocysts in feces. Once deposited in the external environment, oocysts rapidly develop into an infectious sporulated form in optimally warm temperatures. Transmission occurs when susceptible birds ingest contaminated feed or water (Jolley 1982). Greater Sage-Grouse chicks affected with *Eimeria* displayed extreme weakness (inability to sustain flight or running), watery diarrhea, and nasal discharge (Honess and Post 1968).

Prior to the emergence of West Nile virus, coccidiosis was the most important known disease of Greater Sage-Grouse. Losses of young sage-grouse were documented in Wyoming, Colorado, and Idaho from 1932 to 1953, typically in areas where large numbers of birds congregated, resulting in fecal contamination of soil and water (Honess and Post 1968). One report from 1932 cites a local game warden's estimate that about 400 young birds died out of a total sage-grouse population of 2,000 in an area covering about 1.6 km^2 (Honess and Post 1968: 6). Of 1,888 sage-grouse fecal samples collected from 1935 to 1957 across Wyoming, 231 (12.2%) contained *Eimeria* spp. oocysts (Honess 1968, Honess and Post 1968). Prevalence of infection varied seasonally from 25.0%, based on presence/absence of oocysts in fecal samples collected between 15 June and 15 September ($N = 788$), to 5.8% in samples collected outside those dates ($N = 1{,}100$).

Johnson and Boyce (1991) reported no evidence of coccidia in fecal samples they collected in southeast Wyoming in 1987 and 1988. Sporadic occurrence of coccidiosis-associated morbidity and mortality in individual birds is reported; however, notable mortality events attributed to coccidiosis in Greater Sage-Grouse have not been documented since the early 1960s. This change in disease dynamic may be the result of decreased sage-grouse densities. In 1932, when the first mortality event due to coccidiosis was reported, the estimated Greater Sage-Grouse density was 2,000 birds/2.59 km^2 at the site of the epizootic. Sage-grouse have not recently been reported approaching a density that high, although the Wyoming Game and Fish Department (unpubl. data) reports observations of several hundred sage-grouse feeding in irrigated meadows of 2.59 km^2 or less near Farson and Boulder over the last decade.

Bacteria

Clostridium perfringens is an anaerobic bacterium that occurs in the intestinal tract of humans and other animals as well as in most soils, where it persists in an extremely resistant spore form. *C. perfringens* type A is associated with a variety of diseases, including necrotic enteritis, a disease of worldwide economic importance in poultry. Hagen and Bildfell (2007) recently described fatal disease due to infection with *C. perfringens* type A in a free-ranging adult male sage-grouse in Oregon. Histopathologic examination of the small intestine revealed necrotizing and hemorrhagic mucosal lesions and large numbers of gram-positive rods consistent with *C. perfringens*, which polymerase chain reaction analysis confirmed to be *C. perfringens*. Considering the ubiquity and environmental hardiness of *C. perfringens* and the fact that sage-grouse are at times infected with coccidia, necrotic enteritis may be a disease concern, particularly in small, isolated sage-grouse populations.

Sick Greater Sage-Grouse associated with an outbreak of coccidiosis were also bacteremic with *Escherichia coli* (Honess and Post 1968). Presence of *E. coli* is probably underreported, but we do not believe it is a threat to wild populations of Greater Sage-Grouse because it has only been shown to cause acute mortality in captive birds kept in unsanitary conditions (Friend 1999b).

Bacteria of the genus *Salmonella* are responsible for several diseases of birds, such as pullorum and fowl typhoid (Thorne 1969). Post (1960) documented the single instance of Greater Sage-Grouse mortality associated with *Salmonella* sp. and attributed its occurrence to stagnant and contaminated water supplies. Dunbar and Gregg (2004) found no antibodies to *S. typhimurium* and *S. pullorum* when they tested sera from 40 Greater Sage-Grouse from southeastern Oregon. The much lower infection rate in free-ranging birds than in captive birds caused Friend (1999c) to conclude that, in general, salmonellosis was not an important disease of free-ranging wild birds, except for passerines using bird feeders. Diseases associated with *Salmonella* spp. pose little risk to sage-grouse unless environmental conditions concentrate birds and allow fecal material to contaminate limited water supplies.

Tularemia is primarily a tick-borne disease of mammals, but natural infections by the bacteria *Francisella tularensis* have been documented in die-offs of Ruffed Grouse (*Bonasa umbellus*) and other grouse species (Friend 1999b). Parker et al. (1932) reported a die-off of Greater Sage-Grouse in Montana in which birds were heavily infested with bird ticks (*Haemaphysalis chordeilis*) and infected with *Francisella tularensis*. Hopla (1974) concluded that it was likely that tularemia, ticks, or both factored in the mortalities of these grouse. There are no additional reports of tularemia in Greater Sage-Grouse.

Mycoplasma spp. are one of the most common causes of respiratory disease in game birds; in addition to significant morbidity and mortality, common species such as *M. gallisepticum*, *M. meleagridis*, and *M. synoviae* can decrease fitness parameters such as reproductive performance and recruitment (Nettles and Thorne 1982, Nettles 1984, Hagen et al. 2002). However, only a single tentative documentation exists of Greater Sage-Grouse exposure to *Mycoplasma* spp. Using the serum plate agglutination test for detection of antibodies to *M. synoviae*, Hausleitner (2003) tested paired sera collected in consecutive years from 115 Colorado Greater Sage-Grouse. Results indicated a large variation in seroprevalence between years and between two laboratories, leading the author to consider potentially confounding factors, including compromise of sample quality due to inconsistent handling and inherent variation of the assay due to the small size of *M. synoviae*. Future attempts to evaluate Greater Sage-Grouse exposure to, and potential natural infection with, *Mycoplasma* spp. should use the hemoagglutination inhibition test—HI in conjunction with attempted isolation of the organism from seropositive individuals (Hausleitner 2003). Dunbar and Gregg (2004) found no antibodies to *M. gallisepticum* and *M. synoviae* (unspecified test) in 40 Greater Sage-Grouse from southeastern Oregon.

Fungi

The saprophytic mold, *Aspergillus fumigatus*, causes the disease aspergillosis, which is highly pathogenic in many birds (Thorne 1969). The fungus is closely associated with agriculture, thriving in damp, decaying grain and vegetation (Friend 1999a). Physiological stress may enhance clinical manifestation of aspergillosis (Friend 1999a). One case of aspergillosis has been reported in Greater Sage-Grouse (Honess and Winter 1956), but no evidence suggests aspergillosis has a significant role in sage-grouse ecology. Sage-grouse habitats and feeding habits are not generally compatible with the ecology of aspergillosis.

Viruses

More than 300 species of birds, including Greater Sage-Grouse, have been infected with West Nile virus since its introduction to North America in 1999 (Centers for Disease Control and Prevention 2007). The dominant vector of WNV in sagebrush habitats is the mosquito, *Culex tarsalis* (Naugle et al. 2004, Doherty 2007). Many birds act as reservoirs and amplifying hosts for the virus, infecting mosquitoes that may transmit the virus to more birds or to other hosts. Host species differ in the extent to which the virus replicates and circulates in blood at levels high enough to infect mosquitoes. Greater Sage-Grouse are competent amplifying hosts for WNV (Clark et al. 2006, Walker et al. 2007b).

While conducting studies unrelated to disease, researchers documented elevated late-summer mortality of Greater Sage-Grouse across southeastern Alberta, eastern Montana, and northeastern Wyoming during an outbreak of WNV in 2003 (Naugle et al. 2004). WNV mortalities in Greater Sage-Grouse were confirmed in 10 of 11 states and one of two Canadian provinces in the current Greater Sage-Grouse range by the end of 2008 (Walker and Naugle, this volume, chapter 9; Table 8.1). Survival of sage-grouse in three study locations declined an average of 25% between pre-WNV years and the first year WNV was detected (2003), whereas survival did not decline in a study area where WNV was not detected (Naugle et al. 2004). Additionally, survival in four study areas with WNV-induced mortality in 2003 was, on average, 26% lower than the study site where WNV was not detected (Naugle et al. 2004). Overall, individuals in populations exposed to the virus were 3.3 times more likely to die during the late summer WNV period than birds in uninfected populations (Naugle et al. 2004).

Initially, no live sage-grouse tested seropositive for WNV antibodies, suggesting sage-grouse rarely survived infection (Naugle et al. 2004, Walker et al. 2007b). The extreme susceptibility of sage-grouse was confirmed in 2004 when, in separate trials, all nonvaccinated birds experimentally infected with WNV died within 6–8 days, regardless of dosage (Clark et al. 2006; T. E. Cornish, unpubl. data). Six of 58 (10.3%) females captured in Wyoming in spring 2005 tested seropositive, and in spring 2006, 2 of 109 (1.8%) tested seropositive, suggesting some grouse likely survived because they successfully mounted an immune response to infection (Walker et al. 2007b). However, data also demonstrated that while susceptibility to WNV was high (up to 100%), the rate of infection was relatively low (2.4–28.9%), and resistance was neither widespread nor common (1.8–10.3%)(Walker et al. 2007b).

Summer temperatures may affect WNV viremia (Naugle et al. 2005). Sage-grouse mortalities attributable to WNV in 2003 coincided with mean daily July and August temperatures that were 3°C above long-term averages. Lower incidence of WNV in 2004 corresponded with mean daily temperatures that remained well below 21°C (Naugle et al. 2005). Brust (1991) suggested a mean daily temperature below 21°C reduces autogeny in *C. tarsalis*. The findings of Naugle et al. (2005) and Brust (1991) suggest increasing temperatures associated with climate change may increase WNV risk for sage-grouse. We recommend further investigation into how weather, climate, and elevation interact and affect WNV viremia.

Availability of surface water in sagebrush habitats may directly influence exposure of sage-grouse to WNV (Schmidtmann et al. 2005, Zou et al. 2006b, Doherty 2007). Throughout their range, Greater Sage-Grouse hens and broods congregate in mesic habitats in mid- to late summer (Connelly et al. 2000c). *C. tarsalis* exploits mesic habitats as breeding sites (Goddard et al. 2002). Risk of sage-grouse exposure to WNV may be particularly acute when WNV outbreaks coincide with environmental factors, such as drought, that aggregate birds around remaining water sources (Naugle et al. 2004).

Avian retroviruses cause various forms of transmissible neoplasia in birds. They are divided into two major groups—the leukosis/sarcoma (L/S) viruses and reticuloendotheliosis (RE) viruses (Drew 2007). Both L/S and RE viruses can be transmitted vertically through the egg and via direct or indirect horizontal transmission. Additionally, mechanical transmission via biting insects may be important in transmission of RE viruses (Motha et al. 1984, Drew et al. 1998, Drew 2007). Molecular evidence shows endogenous avian L/S virus infection of a Greater Sage-Grouse from Colorado (Dimcheff et al. 2000). RE virus infection has also been associated with disease and/or antibodies in prairie-chickens (*Tympanuchus* spp.; Drew et al. 1998, Wiedenfeld et al. 2002, Bohls et al. 2006, Zavala et al. 2006). Peterson (2004) concluded that RE virus in wild galliforms is of considerable management importance for captive breeding and translocation projects, and wild bird populations of unknown or confirmed-negative exposure status should be protected from unintentional introduction of RE virus.

Positive antibody titers to avian infectious bronchitis (AIB) virus were detected in Greater Sage-Grouse in southeastern Oregon and northwestern Nevada in 2003; however, subsequent attempts to isolate AIB virus from sage-grouse in Nevada using tracheal and cloacal swabs ($N = 21$) were unsuccessful (Dunbar and Gregg 2004). Positive antibody titers to AIB were also found in Lesser Prairie-Chickens, but no clinical signs of disease were observed (Peterson et al. 2002).

Avian pox virus has not been documented in wild Greater Sage-Grouse in their natural habitat, and only a single historical report of avian pox in a captive sage-grouse exists. Natural infection was documented in one captive Greater Sage-Grouse from a Texas farm cohousing at least four other species of the grouse family (DuBose 1965). This sage-grouse had a small lesion on one leg and extensive lesions complicated by secondary bacterial and fungal infection on the head. Experimental infection and cross-protection studies using domestic chickens indicated the virus isolated from Greater Sage-Grouse was identical or similar to fowlpox (Dubose 1965).

Nonparasitic Disease

Various noninfectious diseases or disease-like conditions such as neoplasia (tumors), starvation/malnutrition, genetic disorders, trauma, and toxicosis (poisoning) can result in morbidity and mortality for wildlife populations, some with population-level effects.

During 1949–1950, 1,700,000 ha of Wyoming rangeland (~15% of the state) were aerially treated with toxaphene and chlordane bran bait to control grasshoppers (family: *Acrididae*; Post 1951). Post (1951) monitored treatment and control plots via unspecified observational mortality surveys and determined game bird (Greater Sage-Grouse, Ring-necked Pheasant, Sharp-tailed Grouse, and Gray Partridge) mortality was 23.4% on treated areas and 10.1% on control plots. Of the 45 sage-grouse mortalities recorded on 16 40-ha treatment plots, pesticide poisoning was suspected in 11. Additional deaths from automobiles and mowing machines were indirectly related, as all showed symptoms of toxemia (Post 1951). The scale and toxicity of grasshopper control efforts at that time suggest widespread but unquantified negative impacts to sage-grouse populations. We were unable to find any research efforts that retrospectively estimated the effect of insecticide toxicosis and its proportional contribution, if any, to population declines in Greater Sage-Grouse reported by Braun (1998) and Connelly et al. (2004). Neither chlordane nor toxaphene have been registered for grasshopper control since the early 1980s (United States Department of Health and Human Services 1995, 1997).

Blus et al. (1989) documented effects of pesticides on Greater Sage-Grouse following a 1981 observation of sage-grouse mortality associated with an Idaho potato field sprayed with the organophosphorus insecticide methamidophos. Cholinesterase assays of brains and residue analysis of crop contents indicated that 5% and 16% of the radio-marked sample ($N = 82$) died from organophosphorus insecticides in 1985 and 1986, respectively (Blus et al. 1989). Approximately 200 sage-grouse were present in a block of alfalfa sprayed with dimethoate; 63 of these were later found dead, and cholinesterase activity in 43 brains suitable for assay were depressed >50% (Blus et al. 1989). Of 31 intoxicated grouse radio-tagged after being found in dimethoate-sprayed alfalfa, 20 died (Blus et al. 1989). Both methamidophos and dimethoate remain registered for use in the United States.

CONSERVATION IMPLICATIONS

Parasites can regulate host populations (Tompkins et al. 2002, Peterson 2004). This is a concern for small, isolated populations, because if the infection suppresses the population or decreases reproductive fitness, that population becomes more vulnerable to effects of other stressors that could cause extirpation or extinction.

Microparasitic epizootics characterized by high host mortality can also lead directly to extinction of small populations (Peterson 2004). Widespread, robust host populations that have high mortality due to an epizootic are likely to recover due to reproduction and immigration, but small, isolated host populations may be much less likely to recover, since such populations are typically characterized by limited reproduction and immigration is not possible.

Prior to the emergence of West Nile virus in sage-grouse in recent years, monitoring and research on sage-grouse disease was minimal. Consequently there was little evidence to suggest whether or not parasites and infectious disease were major threats to sage-grouse. Parasites and disease are not mentioned in the 1997 Idaho Sage Grouse Management Plan (Idaho Department of Fish and Game 1997), nor were they addressed in the most recent sage-grouse management guidelines (Connelly et al. 2000c). Early versions of local, state, and national conservation plans/strategies prepared prior to 2004 only briefly discussed disease issues, and these documents largely concluded diseases were not a priority. Most, if not all, state and local sage-grouse conservation plans prepared or updated since 2003 identify WNV as an issue to monitor and consider in management. A panel of sage-grouse experts convened to assist the United States Fish and Wildlife Service with its status review of the Greater Sage-Grouse did not identify disease as a primary extinction risk factor, although the experts expressed concerns about the potential effects of future WNV outbreaks (United States Department of the Interior 2005b).

The focus on WNV is appropriate, but the long-term impacts of most macro- and microparasites and their associated infectious diseases on Greater Sage-Grouse populations remain largely unknown. We recommend that avian infectious bronchitis virus and other avian corona viruses, avian retroviruses, *Mycoplasma* spp., and the *Eimeria* coccidians and associated enteric bacteria be evaluated or at least monitored more closely in addition to WNV. These parasites and diseases may be subject to amplification by climate change or anthropogenic disturbance, or have a history of impacting sage-grouse.

We also recommend systematic, or at least opportunistic, range-wide infectious disease surveillance for sage-grouse. The level of interest in Greater Sage-Grouse will continue to generate many research efforts across the species' range, during which hundreds of birds will be captured and handled. Adding disease surveillance to these efforts by collecting, processing, and analyzing blood and fecal samples would be an efficient and effective method of securing basic parasite and disease occurrence and prevalence data, as well as basic blood chemistry levels. Better still would be an integration of disease results into overall research efforts (Hausleitner 2003). These collections and analyses will require investigators trained in venipuncture techniques as well as proper storage and shipment of samples. We recommend that a standard protocol for techniques such as venipuncture and use of field centrifuges be developed and distributed. We also recommend developing a process for centralized archiving of samples for future use.

Determining ecological implications where parasites and diseases are documented requires relatively rigorous methods. Detailed studies on Red Grouse (Hudson 1986, 1992; Hudson et al. 1992, 1998, 1999; Fox and Hudson 2001) have demonstrated significant population-level effects of parasites to a tetraonid game bird and provide a model on which similar research could be conducted on sage-grouse, as recommended by Peterson (2004) for prairie grouse and Tompkins et al. (2002) for wildlife in general. The potential implications of climate change, anthropogenic disturbance, and an apparent increase in emerging infectious diseases amplify the need to effectively monitor disease impacts to Greater Sage-Grouse. Hoberg et al. (2008) argued for and provided a model for an integrated, multidisciplinary approach to detect, understand, and forecast effects of emerging infectious diseases in wildlife and humans in an environment affected by rapid climate change and anthropogenic disturbance.

Finally, we recommend any attempt to hold or propagate sage-grouse in captivity address the enhanced susceptibility to infection and transmission of parasites and infectious disease inherent to captive birds and prevent transmission to wild, free-ranging populations. Conservation of Greater Sage-Grouse is receiving increased attention and funding. The opportunity exists for well-designed research and monitoring efforts to examine if and how parasites affect sage-grouse populations.

ACKNOWLEDGMENTS

We thank E. Tom Thorne (deceased), Walt Cook, and several anonymous referees who reviewed earlier versions of the chapter and provided detailed comments that greatly improved the final product. We also thank the Wyoming Game and Fish Department for supporting our preparation of the chapter. Most important, we thank the researchers cited in this review for their valuable work.

CHAPTER NINE

West Nile Virus Ecology in Sagebrush Habitat and Impacts on Greater Sage-Grouse Populations

Brett L. Walker and David E. Naugle

Abstract. Emerging infectious diseases can act as important new sources of mortality for wildlife. West Nile virus (Flaviviridae, *Flavivirus*) has emerged as a potential threat to Greater Sage-Grouse (*Centrocercus urophasianus*) populations since 2002. We review the ecology of West Nile virus in sagebrush (*Artemisia* spp.) ecosystems of western North America, summarize the influence of the virus on Greater Sage-Grouse mortality and survival, use demographic models to explore potential impacts on population growth, and recommend strategies for managing and monitoring such impacts. The virus was an important new source of mortality in low- and mid-elevation Greater Sage-Grouse populations range-wide from 2003 to 2007. West Nile virus can simultaneously reduce juvenile, yearling, and adult survival—three vital rates important for population growth in this species—and persistent low-level West Nile virus mortality and severe outbreaks may lead to local and regional population declines. West Nile virus mortality in simulations was projected to reduce population growth (i.e., finite rate of increase, λ) of susceptible populations by an average of 0.06–0.09/yr. However, marked spatial and annual fluctuations in nest success, chick survival, and other sources of adult mortality are likely to mask population-level impacts in most years. Impacts of severe outbreaks may be detectable from lek-count data, but documenting effects of low to moderate mortality will require intensive monitoring of radio-marked birds. Resistance to West Nile virus–related disease appears to be low and is expected to increase slowly over time. Eliminating mosquito breeding habitat from anthropogenic water sources is crucial for reducing impacts. Better data are needed on geographic and temporal variation in infection rates, mortality, and seroprevalence range-wide. Small, isolated, and peripheral populations, particularly those at lower elevations, and those experiencing large-scale increases in distribution of surface water may be at higher risk.

Key Words: Centrocercus urophasianus, Culex tarsalis, emerging infectious disease, flavivirus, Greater Sage-Grouse, resistance, sagebrush, survival, West Nile virus.

Ecología del Virus de West Nile en Hábitat de Sagebrush e Impactos en Poblaciones de Greater Sage-Grouse

Resumen. Las enfermedades infecciosas emergentes pueden actuar como nuevas fuentes importantes de mortalidad para la vida silvestre.

Walker, B. L., and D. E. Naugle. 2011. West Nile virus ecology in sagebrush habitat and impacts on Greater Sage-Grouse populations. Pp. 127–142 *in* S. T. Knick and J. W. Connelly (editors). Greater Sage-Grouse: ecology and conservation of a landscape species and its habitats. Studies in Avian Biology (vol. 38), University of California Press, Berkeley, CA.

El virus del Nilo occidental (Flaviviridae, *Flavivirus*) ha emergido como una amenaza potencial para poblaciones del Greater Sage-Grouse (*Centrocercus urophasianus*) desde 2002. Revisamos la ecología del virus del Nilo occidental en ecosistemas de artemisa (*Artemisia* spp.) en el oeste de Norte América, resumimos la influencia del virus sobre la mortalidad y la supervivencia del Greater Sage-Grouse, usamos modelos demográficos para explorar impactos potenciales sobre el crecimiento de la población, y recomendamos estrategias para monitorear y manejar tales impactos. El virus fue una importante fuente nueva de mortalidad en poblaciones de Greater Sage-Grouse en altitudes medias y bajas en el lapso de 2003–2007. El virus Nilo occidental puede reducir simultáneamente la supervivencia del Greater Sage-Grouse juvenil, de individuos de un año y de adultos-tres componentes vitales importantes para el crecimiento de la población en éstas especies-y una mortalidad baja persistente, por virus del Nilo occidental, y unos brotes severos pueden llevar a una disminución de la población local y regional. En simulaciones se proyectó que la mortalidad por virus del Nilo occidental reduce el crecimiento de la población (es decir, la tasa finita de incremento, λ), de poblaciones susceptibles, en un promedio de 0.06–0.09 por año. Sin embargo, fluctuaciones marcadas, de tipo anual y de espacio, en éxito de a nidación, supervivencia de polluelos, y otras fuentes de mortalidad de adultos es probable que enmascaren los impactos, a nivel poblacional, la mayoría de los años. Los impactos de brotes severos se pueden detectar en los datos de los lek-count, pero documentar los efectos de la mortalidades baja a moderada requerirá de un intenso monitoreo de pájaros radio-marcados. La resistencia a enfermedades relacionadas con el virus del Nilo occidental parece ser baja y se espera que se incremente lentamente en el tiempo. Eliminar los hábitats de reproducción del mosquito de fuentes antropogénicas de agua es crucial para reducir impactos. Se necesitan mejores datos acerca de la variación geográfica y temporal en las tasas de infección, mortalidad, y de seroprevalencia en su distribución. Las poblaciones pequeñas, aisladas, y perisféricas, especialmente aquellas en altitudes bajas y aquellas que están experimentando incrementos a gran escala en la distribución de superficies de agua, pueden estar en mayor riesgo.

Palabras Clave: artemisa (sagebrush), *Centrocercus urophasianus, Culex tarsalis,* enfermedad infecciosa emergente, flavivirus, Greater Sage-Grouse, resistencia, supervivencia, virus Oeste del Nilo occidental.

Infectious diseases are now widely recognized as important sources of mortality in wild bird populations and have emerged as a major issue in avian conservation, particularly for sensitive, threatened, and declining species (Daszak et al. 2000, Dobson and Foufopoulos 2001, Friend et al. 2001, Chomel et al. 2007). Timely and appropriate management and mitigation of disease impacts requires detailed information on ecological interactions between the pathogen and its hosts, vectors, and environment. Assessing the importance of disease for prioritizing conservation efforts requires data on disease spread, distribution, and impacts on population demographics and growth.

A major new concern for conservation of wild bird populations in North America is the recent arrival and rapid spread of West Nile virus (WNV; Flaviviridae, *Flavivirus*). West Nile virus is a mosquito-borne flavivirus that can cause debilitating or fatal neuroinvasive disease in wild birds (Marra et al. 2004, Hayes et al. 2005b, McLean 2006). The virus persists largely within a mosquito-bird-mosquito infection cycle (Campbell et al. 2002). West Nile virus has expanded across the continent at an unprecedented rate since 1999 (Marra et al. 2004, McLean 2006, Kilpatrick et al. 2007) and is now considered the predominant arthropod-borne disease in the United States (Kilpatrick et al. 2007, Kramer et al. 2008). The virus is known from at least 317 wild, captive, and domestic bird species in North America, of which 254 are native (Centers for Disease Control and Prevention 2008). Over 48,000 infected dead birds had been reported as of 2005 (McLean 2006), but because most WNV mortality in wild populations goes unnoticed or unreported (Ward et al. 2006), the virus is thought to have caused the deaths of millions of wild birds since 1999 (McLean 2006, Gubler 2007). Although confirmed as a new source of mortality, population-level effects of the virus on wild bird populations remain largely unknown (Marra et al. 2004, McLean 2006). Only recently have studies documented local and regional population declines in common and widespread birds following the arrival of WNV, e.g., American Crow (*Corvus brachyrhynchos*), Blue Jay (*Cyanocitta cristata*), Yellow-billed Magpie (*Pica nuttalli*), Western Scrub-Jay (*Aphelocoma californica*), Steller's Jay (*Cyanocitta stelleri*), American Robin (*Turdus migratorius*), Eastern Bluebird (*Sialia sialis*), Black-capped Chickadee (*Poecile atricapillus*), Carolina Chickadee (*Poecile carolinensis*), Tufted Titmouse (*Baeolophus bicolor*), and House Wren (*Troglodytes aedon*) (Koenig et al. 2007, LaDeau et al. 2007). WNV-related mortality rates in the American Crow can reach 40–68% (Caffrey et al. 2003, 2005; Yaremych et al. 2004; Koenig et al. 2007; LaDeau et al. 2007). West Nile virus reduced Yellow-billed Magpie populations in California by as much as 49% from 2003 to 2006 (Crosbie et al. 2008). WNV-related mortality resulted in a tenfold reduction in survival, from 0.44 to 0.04, in American White Pelican (*Pelecanus erythrorhynchos*) chicks (Sovada et al. 2008).

West Nile virus has also recently emerged as a potential threat to sage-grouse (*Centrocercus* spp.) populations (Naugle et al. 2004). Greater Sage-Grouse (*C. urophasianus*) and Gunnison Sage-Grouse (*C. minimus*) are gallinaceous birds native to western sagebrush (*Artemisia* spp.) habitats (Schroeder et al. 1999). Previously widespread, both species have been extirpated from much of their original range (Schroeder et al. 2004) and have experienced long-term population declines due to loss, fragmentation, and degradation of sagebrush habitat (Schroeder et al. 1999, Connelly et al. 2004). Historical population declines and continued loss and degradation of sagebrush habitat have led to concern over the conservation status of sage-grouse (Connelly et al. 2004, Stiver et al. 2006), repeated attempts to list both species under the Endangered Species Act, and range-wide efforts to assess risks to populations (Connelly et al. 2004, Stiver et al. 2006, Aldridge et al. 2008).

A series of studies on Greater Sage-Grouse starting in 2003 documented reduced survival due to WNV (Naugle et al. 2004, 2005; Walker et al. 2004, 2007b; Aldridge 2005; Kaczor 2008; Walker 2008), near-extirpation of a local population following a WNV outbreak (Walker et al. 2004), high rates of mortality following infection (Clark et al. 2006), WNV-related mortality events among unmarked birds (United States Geological Survey 2006), and links between WNV mortality, mosquito abundance, and changes in land use (Zou et al. 2006b, Doherty 2007, Walker 2008). Understanding the impact of WNV on Greater Sage-Grouse populations is important for assessing this species' conservation status, but requires an updated synthesis of recent scientific data. The objectives of this chapter are to: (1) review the ecology of WNV in sagebrush ecosystems of western North America; (2) summarize recent data on distribution of WNV mortality events, impacts on mortality and survival rates, and resistance to WNV disease; (3) use demographic models to explore potential impacts of WNV-related mortality on population growth; and (4) recommend strategies for monitoring and mitigating impacts of the virus on sage-grouse populations.

ECOLOGY OF WEST NILE VIRUS IN SAGEBRUSH HABITATS

The transmission cycle of WNV in sagebrush habitats involves complex interactions among vectors, reservoirs, amplifying hosts, and environmental factors, including temperature and the distribution of surface water. The main vectors for WNV worldwide are mosquitoes, particularly those in the genus *Culex* (Goddard et al. 2002; Turell et al. 2001, 2005). Other ectoparasites, including ticks (Hutcheson et al. 2005, Dawson et al. 2008), hippoboscid flies (Farajollahi et al. 2005), and biting midges (Naugle et al. 2004) may also be involved in WNV transmission, but few data are available on their role as WNV reservoirs or vectors (van der Meulen et al. 2005). WNV infection has been documented in several genera of mosquitoes (*Culex, Aedes, Ochlerotatus, Culiseta*; Goddard et al. 2002, Doherty 2007) and at least one other arthropod family, biting midges (Naugle et al. 2004), in sagebrush habitats of western North America. The dominant vector of WNV in sagebrush habitats is the mosquito *Culex tarsalis* Coquillett (Goddard et al. 2002, Naugle et al. 2004, Turell et al. 2005, Doherty 2007). *Culex tarsalis* is a highly competent vector (Goddard et al. 2002, Turell et al. 2005), in part because it can inoculate hosts with high doses of virus ($10^{4.3}$–$10^{5.0}$ plaque-forming units [PFU]) directly into the bloodstream while feeding (Reisen et al. 2007; Styer et al. 2007a,b). The species is abundant and widely distributed in arid sagebrush habitats (DiMenna et al. 2006, Doherty 2007), and individuals may disperse as much as 18 km to colonize newly available surface water (Bailey et al. 1965, Beehler and Mulla 1995, Reisen et al. 2003). The species prefers sites with submerged vegetation on which to oviposit and warm, standing water that promotes rapid larval development, including ephemeral puddles, vegetated pond edges, and hoofprints (Milby and Meyer 1986, Buth et al. 1990, Doherty 2007). *Culex tarsalis* feeds primarily on birds in spring and early summer, then shifts its feeding patterns to include mammals in late summer (Lee et al. 2002). The important role of *Culex* mosquitoes in WNV epidemics may be due to their broad range of hosts and seasonal shifts in host preferences (Kilpatrick et al. 2006a,b). *Aedes vexans*, a floodwater mosquito common in western sagebrush habitats that primarily feeds on mammals, has recently been demonstrated capable of transmitting WNV from infected chickens (*Gallus gallus domesticus*; Tiawsirisup et al. 2008).

Much is known about WNV vectors in sagebrush habitat, but reservoirs for WNV are poorly understood. Reservoirs are those species that harbor the virus and serve as sources for naïve host-feeding mosquitoes that initiate the WNV transmission cycle each year. Both resident and migratory birds can be competent hosts and may act as a source of virus in spring or early summer due to reactivation of a chronic infection (McLean 2006). Infected birds are known to exhibit migratory behavior and may be able to carry the virus long distances (Owen et al. 2006). Migratory birds are widely thought to be responsible for the spread of WNV across North America, but direct evidence is lacking (Peterson et al. 2003, Rappole and Hubálek 2003, Reed et al. 2003.). Most migratory breeding passerines in sagebrush habitats, e.g., Brewer's Sparrow (*Spizella breweri*), Vesper Sparrow (*Pooecetes gramineus*), Sage Sparrow (*Amphispiza belli*), Horned Lark (*Eremophilus alpestris*), and Western Meadowlark (*Sturnella neglecta*), arrive in early spring prior to the emergence of host-feeding mosquitoes, so it is unclear whether they are involved in initiating WNV transmission in sagebrush habitat. Migratory birds passing through in late spring or early summer or those returning south in mid- to late summer that congregate on or near water sources in sagebrush habitat—songbirds, waterfowl, shorebirds—may also be a source of the virus. Exotic species commercially raised and released into sage-grouse habitat that carry the virus but are largely resistant to WNV disease—for example, Ring-necked Pheasant (*Phasianus colchicus*), Chukar (*Alectoris chukar*), and Gray Partridge (*Perdix perdix*)—may also serve as WNV reservoirs (Meece et al. 2006, Wünschmann and Ziegler 2006). WNV in some regions is known to overwinter in infected diapausing mosquitoes, including *Culex tarsalis* (Nasci et al. 2001, Goddard et al. 2003, Reisen et al. 2006a), and it is possible that infected mosquitoes emerge in spring to begin WNV transmission anew. Offspring of *C. tarsalis* infected via vertical transmission from mother to offspring via eggs may also overwinter as eggs or larvae and emerge as infected adults the following spring (Goddard et al. 2003).

Wild birds are clearly the most important amplifying hosts for WNV (Marra et al. 2004, McLean 2006, Kramer et al. 2008), but identifying and targeting specific species for management is extremely difficult (Lord and Day 2001, Kilpatrick et al. 2006b). Sagebrush habitats typically support lower avian diversity than other western ecosystems, e.g., riparian areas, but numerous avian hosts, mammals, reptiles, and amphibians could be involved in either maintaining or attenuating transmission (Marra et al. 2004, van der Meulen et al. 2005, Lord et al. 2006, McLean 2006), including sparrows, ducks, Wilson's Snipe (*Gallinago delicata*), Sora (*Porzana carolina*), Short-eared Owl (*Asio flammeus*), Red-tailed Hawk (*Buteo jamaicensis*), Ring-necked Pheasant, Greater Sage-Grouse, House Wren, American Robin, Common Yellowthroat (*Geothlypis trichas*), Western Meadowlark, and Bullock's Oriole (*Icterus bullockii*; C. Y. Kato, unpublished data). Potential mammalian hosts were also detected—cows (*Bos taurus*), sheep (*Ovis aries*), horses (*Equus caballus*), deer (*Odocoileus* spp.), pronghorn (*Antilocapra americana*), moose (*Alces alces*), rabbits, felines, and skunks (C. Y. Kato, unpubl. data). Viremia in mammals less commonly reaches levels required to infect host-feeding mosquitoes (Turell et al. 2000, Sardelis et al. 2001, van der Meulen et al. 2005), but recent studies have documented several potential mammalian hosts for WNV (Tiawsirisup et al. 2005; Platt et al. 2007, 2008), and mammals may be involved in nonviremic transmission (Higgs et al. 2005, Reisen et al. 2007).

Numerous studies purport to have identified key amplifying hosts or species- and habitat-specific exposure or infection rates of WNV based on seroprevalence—the proportion of live individuals with neutralizing antibodies to WNV (Komar et al. 2005, Beveroth et al. 2006). However, species with low seroprevalence do not necessarily experience low infection rates nor are they precluded from transmitting WNV (Walker et al. 2007b). Species that are immune to the virus and highly susceptible species that die quickly prior to infecting additional vectors may serve as dead-end hosts that attenuate transmission (Lord and Day 2001, Reisen et al. 2006c). The relative abundance of different reservoir and amplifying host species can vary by season, among years, and among locations. Levels of viremia in infected Greater Sage-Grouse exceed the host-to-vector transmission threshold of $10^{5.0}$ PFU/ml and the birds live sufficiently long to infect new mosquitoes; thus, despite their susceptibility, sage-grouse are considered competent amplifying hosts (Clark et al. 2006, but see van der Meulen et al. 2005). In mid-summer, sage-grouse often congregate in flocks near both natural and artificial water sources (Schroeder et al. 1999, Connelly et al. 2000c, Walker et al. 2004). These habitats often support populations of breeding mosquitoes (Doherty 2007) and, because sage-grouse are competent hosts, congregations of sage-grouse around water sources may lead to rapid spread of the virus within sage-grouse flocks and severe local mortality events (Walker et al. 2004, 2007b). Host competency of other avian species using sagebrush habitats in late summer has not been studied. The difficulty of identifying both reservoir and amplifying hosts severely limits management options for WNV, with most strategies focusing on water management and vector control.

West Nile virus transmission is also regulated by environmental factors, including temperature, precipitation, and distribution of anthropogenic water sources that support breeding mosquito vectors (Brust 1991; Dohm et al. 2002; Reisen et al. 2006b; Zou et al. 2006a,b). Sagebrush habitats are characterized by cold winters; cool, wet springs; and hot, dry summers. Extremely cold temperatures largely preclude mosquito activity and virus amplification in sagebrush habitats in winter, and it is unlikely that enzootic transmission occurs outside the known summer transmission period. Spring temperatures may allow WNV transmission as early as mid-May (Zou et al. 2006a) and in fall, as late as mid-September. All documented WNV-related mortality in sage-grouse has occurred from mid-May through mid-September, with a peak in July and August (Walker et al. 2007b; Walker 2008; D. E. Naugle, unpubl. data).

Temperature and precipitation both directly influence potential for WNV transmission. The specific annual or seasonal temperature and precipitation profiles that promote outbreaks in sagebrush habitats have not been identified, but some general patterns are evident. In years with lower summer temperatures, reduced and delayed WNV transmission has been documented in sage-grouse (Naugle et al. 2005, Walker et al. 2007b) and migratory passerines (Bell et al. 2006). It has been suggested in other ecosystems that high temperatures associated with drought conditions increases WNV transmission (Epstein and Defilippo 2001, Shaman et al. 2005). Higher temperatures facilitate greater nocturnal host-seeking activity by mosquitoes, more rapid larval development, and shorter extrinsic incubation periods for the virus—the time it takes for the virus to replicate inside the mosquito and invade its salivary glands (Reisen et al. 2006b). Summer drought is an annual occurrence in sage-grouse habitats range-wide. Temperature can also influence exposure of Greater Sage-Grouse to WNV by influencing habitat use. Throughout their range, Greater Sage-Grouse congregate in mesic habitats in mid- to late summer (Connelly et al. 2000c) and often use ponds, springs, and other standing water sources during hot weather (Dalke et al. 1963, Connelly and Doughty 1989). *Culex tarsalis* exploits such habitats for breeding (Goddard et al. 2002, Doherty 2007), and risk of exposure to WNV for Greater Sage-Grouse may be elevated if WNV outbreaks coincide with drought conditions that aggregate birds in mesic areas or near remaining water sources (Naugle et al. 2004). Temperature, mosquito activity, and *Culex tarsalis* abundance decrease with elevation, and Greater Sage-Grouse inhabiting high-elevation sites in summer are generally thought to be less vulnerable than low-elevation populations (Naugle et al. 2004, Kaczor 2008). Similarly, populations farther north may be relatively less susceptible than those at similar elevations farther south, because summer temperatures are generally lower at higher latitudes (Naugle et al. 2005). The highest confirmed elevation at which Greater Sage-Grouse have been infected with WNV is ~2,300 m, in the Lyon–Mono population of eastern California (Naugle et al. 2005). Increasing temperatures associated with changing climate may exacerbate WNV risk for sage-grouse (Epstein 2001), but risk also depends on complex interactions with other environmental factors, including precipitation and distribution of water.

Artificial water sources may also facilitate the spread of WNV within sage-grouse habitats (Zou et al. 2006b, Doherty 2007, Walker et al. 2007b). For example, construction of ponds for water produced during coal-bed natural gas extraction increased larval mosquito habitat around pond edges by 75%, from 619 to 1,085 ha, during a 5-year period of development (1999–2004) across a 21,000-km^2 area of northeastern Wyoming (Zou et al. 2006b). These ponds support abundant *Culex tarsalis*, and they support them longer than natural, ephemeral water sources (Doherty 2007). West Nile virus mortality associated with coal-bed natural gas ponds is thought to have contributed to extirpation of at least one local sage-grouse population in northeastern Wyoming (Walker et al. 2004, Walker 2008). Projects that create mesic zones around stock tanks or ponds as habitat improvements for sage-grouse may inadvertently contribute to the WNV problem, because *Culex tarsalis* readily take advantage of water-filled hoofprints around tanks and ponds for breeding (Doherty 2007). Sage-grouse may use standing water in summer and fall when it is available, but they do not require standing water (Dalke et al. 1963, Schroeder et al. 1999, Connelly et al. 2004). Estimated WNV infection rates were relatively low from 2003 to 2005 in undeveloped sagebrush habitats of the Powder River Basin (Walker et al. 2007b, Walker 2008), due in part to lack of available surface water in late summer, but were higher in areas with surface water provided by coal-bed natural gas ponds (Walker 2008).

The major ecological factors that regulate WNV transmission are known, but local outbreaks remain difficult to predict. Specific environmental conditions, e.g., temperature-precipitation profiles and water sources, must coincide with biotic factors, including infected reservoirs, competent host-feeding vectors, suitable amplifying hosts, and susceptible naïve individuals, for an outbreak to occur. Recent attempts to model WNV transmission events based on degree-day models appear promising (Zou et al. 2006a), but need to incorporate changes in the distribution of larval breeding sites over time (Zou et al. 2006b) and spatial variation in temperature-precipitation profiles to improve predictive ability (Walker 2008).

Several recent discoveries further complicate our understanding of WNV transmission and may have important implications for how WNV might affect sage-grouse populations, including: (1) acquired temporary immunity in juveniles, (2) vertical (mother-to-offspring) or horizontal (bird-to-bird) virus transmission, (3) changes in virulence, (4) impacts of the virus on mosquito demographics and behavior, and (5) nonviremic or nonpropagative virus transmission among cofeeding mosquitoes. First, in diurnal raptors, owls, and domestic chickens, young can acquire temporary immunity for up to 33 days via maternal transmission of antibodies (Gibbs et al. 2005, Hahn et al. 2006, Nemeth and Bowen 2007). Chicks of infected females may be temporarily buffered from impacts of the virus if this phenomenon occurs in sage-grouse. Second, vertical transmission of WNV from mother to offspring has not been documented and is considered unlikely, but horizontal transmission between adult sage-grouse has been demonstrated in captivity (Clark et al. 2006). Whether horizontal transmission occurs in free-ranging populations remains unknown. Third, birds can contract arthropod-borne viruses by consuming infected vectors (Gilbert et al. 2004), such as in Red Grouse (*Lagopus lagopus*), but sage-grouse have not been reported to actively feed on adult or larval mosquitoes or ticks (Schroeder et al. 1999). Fourth, infection with WNV increases blood-feeding rates in female *Culex tarsalis* but may also decrease fecundity (Styer et al. 2007b), so it is unclear whether these effects together result in acceleration or attenuation of WNV transmission. Fifth, studies have documented multiple strains of WNV and competitive displacement of the NY99 strain by the WN02 strain since 1999 (Davis et al. 2005); implications of these discoveries are unclear. One study documented a virus strain with a shorter extrinsic incubation period that could lead to shorter intervals between transmission events (Moudy et al. 2007), while other studies have reported decreased replication rates and reduced neuroinvasiveness (Davis et al. 2004). Most disturbing, however, are reports of transmission of WNV between infected and uninfected *Culex* mosquitoes cofeeding on uninfected vertebrate hosts in a laboratory setting (Higgs et al. 2005, Reisen et al. 2007). Amplifying hosts may not be required for transmission if nonviremic transmission or transmission via nonpropagative viremia occur in the wild, and transmission among vectors could occur much more rapidly.

SAGE-GROUSE AND WEST NILE VIRUS

Demographic impacts of WNV on Greater Sage-Grouse are relatively well-known compared with other North American species. Recent studies of radio-marked sage-grouse have allowed testing for neutralizing antibodies to WNV at capture and for WNV infection following mortality (Naugle et al. 2004, 2005; Walker et al. 2004, 2007b; Aldridge 2005; Kaczor 2008). The most reliable data on WNV mortality and infection rates come from research studies using marked individuals. However, WNV mortality rates using data from radio-marked birds may be underestimated because many carcasses cannot be recovered and tested (Walker et al. 2004).

Distribution and Spread

West Nile virus was first detected within Greater Sage-Grouse range in 2002 (Kilpatrick et al. 2007), and a WNV-positive Greater Sage-Grouse mortality was first documented in Wyoming that same year (Naugle et al. 2004). WNV infections in humans, horses, and sentinel species (mosquitoes, chickens) had been documented in all 11 states and two Canadian provinces within current sage-grouse range as of December 2007 (Kilpatrick et al. 2007), and WNV-positive mortalities in Greater Sage-Grouse had been confirmed in 10 states and one province (Table 9.1, Fig. 9.1). No WNV-positive Greater Sage-Grouse have been reported from Washington or Saskatchewan (Fig. 9.1). However, the combination of WNV-positive mortalities in extreme northeastern Montana in 2007, regular cross-border and long-distance movements between Montana and Saskatchewan (J. D. Tack, pers. comm.), and previously documented mortalities in southeastern Alberta in 2003–2005 (Naugle et al. 2004, 2005; Walker 2006) suggest that Saskatchewan populations have also been affected.

WNV Mortality and Survival

Impacts of WNV on Greater Sage-Grouse have been reported in the literature in different ways: the

TABLE 9.1

U.S. states and Canadian provinces with confirmed (+) West Nile virus–positive Greater Sage-Grouse mortalities, 2002–2007.

West Nile virus was detected in other species (horses, humans, mosquitoes, or sentinel species) in all states and provinces within Sage-Grouse range by 2002 except Alberta (2003), Nevada (2003), Utah (2003), and Oregon (2005).

State or province	2002	2003	2004	2005	2006	2007
California			+	+		
Colorado			+			
Idaho				+		
Montana		+	+	+	+	+
Nevada				+		
North Dakota					+	
Oregon				+		
South Dakota					+	+
Utah			+			
Washington						
Wyoming	+	+	+	+	+	+
Alberta	+	+	+			
Saskatchewan						

SOURCE: Kilpatrick et al. 2007, Centers for Disease Control and Prevention 2008.

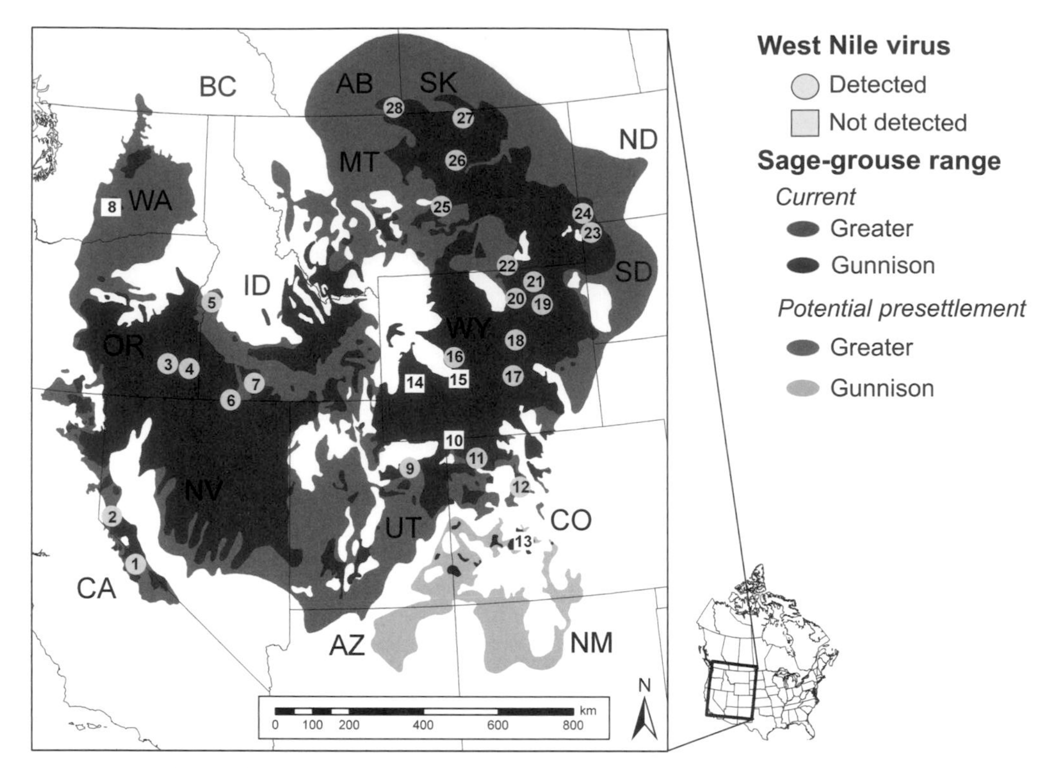

Figure 9.1. Distribution of Greater Sage-Grouse and Gunnison Sage-Grouse and locations where birds were monitored or tested for West Nile virus from 2002 to 2007. Numbered sites include (1) Mono Co., CA; (2) Douglas Co., NV; (3) Harney Co., OR; (4) Malheur Co., OR; (5) Washington Co., ID; (6) Owyhee Co., ID, and Elko Co., NV; (7) Twin Falls Co., ID; (8) Yakima Co. and Kittitas Co., WA; (9) Duchesne Co., UT; (10) Moffat Co. (Hiawatha/Cold Springs Mtn.), CO; (11) Moffat Co. (Axial Basin), CO; (12) Routt Co., CO; (13) Gunnison Co., CO; (14) Sublette Co. (Pinedale), WY; (15) Fremont Co. (Lander), WY; (16) Fremont Co. (Wind River Indian Reservation), WY; (17) Carbon Co., WY; (18) Natrona Co., WY; (19) Campbell Co. (Wright), WY; (20) Johnson Co., WY; (21) Campbell Co. (Spotted Horse), WY; (22) Bighorn Co., MT, and Sheridan Co., WY; (23) Butte Co. and Harding Co., SD, Crook Co., WY, and Carter Co., MT; (24) Bowman Co., ND; (25) Golden Valley Co. and Musselshell Co., MT; (26) Phillips Co., MT; (27) Valley Co., MT; and (28) southeastern AB. Map is based on data reported in Naugle et al. (2004, 2005), Walker et al. (2004), Walker (2006), United States Geological Survey (2006), Kaczor (2008), Walker (2008), and unpublished data provided by state agencies and researchers.

number of confirmed WNV-positive mortalities (United States Geological Survey 2006), minimum and maximum possible WNV-related mortality rates (Walker et al. 2007b, Kaczor 2008), and differences in survival between areas with and without WNV mortality (Naugle et al. 2004, 2005). Most published data are from the eastern half of the species' range.

WNV-related mortality reduced late-summer survival of adult females across much of the eastern edge of the species' range in 2003, a year with persistent high summer temperatures and extreme drought (Naugle et al. 2004). Late-summer survival (15 July–31 August) at four study locations with confirmed WNV mortality in Wyoming, Montana, and Alberta declined an average of 0.25 between pre-WNV years (1998–2002; 0.89 ± 0.01) and the first year that WNV was detected (2003; 0.64 ± 0.07), whereas survival remained high (0.90 pre-WNV vs. 0.85 in 2003) at a study site in western Wyoming where WNV was not detected. Late-summer survival across the four study areas with WNV mortality averaged 0.26 lower (0.64) than at the one study area where WNV was not detected (0.90). Individuals in populations exposed to the virus during July–August 2003 were 3.3 times more likely to die than birds in uninfected populations (Naugle et al. 2004). Female survival in the Powder River Basin of northeastern Wyoming and southeastern Montana during the July–September WNV transmission season was 0.20 (95% CI 0.01–0.44; N = 10) in areas with confirmed WNV mortality and 0.76 (95% CI 0.63–0.91; N = 34) in areas without WNV mortality (Walker et al. 2004). The 2003 outbreak near Spotted Horse, Wyoming, was associated with extirpation of the local breeding population. The five leks in that region showed 76%, 95%, and 91% declines in maximum, median, and mean male counts, respectively, from spring 2003 to spring 2004 (Walker et al. 2004). Mean number of males per count declined from 5.1 ± 0.5 SE in 2003 to 0.5 ± 0.2 SE in 2004, whereas counts at nearby unaffected leks did not change (10.2 ± 1.5 SE in 2003 vs. 10.4 ± 1.4 SE in 2004). Females also largely disappeared. At the five affected leks, 36 females were counted on 19 lek visits in spring 2003, whereas only 1 female was counted on 21 visits in spring 2004 (Walker et al. 2004). All five affected leks were inactive by 2005 and remained inactive through 2007 (Wyoming Game and Fish Department, unpubl. data).

Later timing of mortalities and dramatically fewer case rates of WNV in humans, horses, and other wild birds in 2004 in the eastern portion of the species' range suggested that below-average spring precipitation and summer temperatures limited mosquito production and reduced WNV transmission compared with 2003 (Naugle et al. 2005, Bell et al. 2006, McLean 2006). July–September survival in 2004 was consistently lower (0.86, range 0.83–0.92) at four sites across the species' range with confirmed WNV-positive mortalities than at eight sites without (0.96, range 0.92–0.100)(Naugle et al. 2005). WNV-related mortality among radio-marked females from 1 July–15 September 2004 in the Powder River Basin was between 3.7% and 9.4% ($N = 118$) (Walker et al. 2007b, Walker 2008).

Moderate summer temperatures may have again attenuated mosquito production, virus amplification, or transmission in the eastern half of the species' range in 2005 (Walker 2006). WNV-related mortality rates in northeastern Wyoming and southeastern Montana from 1 July to 15 September 2005 were between 2.4% and 8.2% ($N = 123$)(Walker et al. 2007b, Walker 2008). California, Nevada, Utah, and Alberta reported WNV-positive mortalities in 2005, but they did not report mortality or survival rates.

The first confirmed WNV-positive mortality in 2006 was documented on 14 June in Bighorn County in southeastern Montana, almost a month earlier than in previous years (Walker 2008). Elevated late-summer mortality was also reported on the Charles M. Russell National Wildlife Refuge in Montana (M. R. Matchett, pers. comm.). WNV-related mortality from 15 June to 15 September 2006 in southeastern Montana and northeastern Wyoming was between 5% and 15% of radio-marked females ($N = 123$)(D. E. Naugle, unpubl. data). Kaczor (2008), working in northwestern South Dakota, reported minimum and maximum possible WNV-related mortality rates among radio-marked juvenile sage-grouse as 6.5–71.0% ($N = 31$) from 12 July to 31 September 2006.

A confirmed outbreak of WNV in South Dakota in 2007 contributed to a 44% mortality rate ($N = 80$) among radio-marked females from mid-July to mid-September (K. C. Jensen, pers. comm.). Kaczor (2008) reported minimum and maximum possible WNV-related mortality rates among juveniles as 20.8–62.5% ($N = 24$) from 12 July to 31 September 2007 in northwestern South Dakota. In northeastern Montana (Valley Co.), 26% of radio-marked females ($N = 30$) died during a 2-week period in early August immediately following the first detection of WNV in mosquito pools, with confirmation of two WNV-positive mortalities (J. D. Tack, pers. comm.). WNV-related mortality among radio-marked females from 15 June to 6 September in the Powder River Basin was between 8% and 21% ($N = 85$)(D. E. Naugle, unpubl. data), with one WNV-positive mortality collected May 17.

Reports of WNV-related mortality events among unmarked birds provide additional evidence that sage-grouse populations are impacted by WNV. For example, mortalities reported by landowners near the town of Burns, Oregon, in August 2006 resulted in recovery of several freshly dead sage-grouse that tested positive for WNV and discovery of >60 other decomposed sage-grouse carcasses and a sick, WNV-positive Northern Harrier (*Circus cyaneus*)(United States Geological Survey 2006). Summer mortality events also occurred in several areas of Idaho and along the Idaho-Nevada border in 2006; at least 55 carcasses were discovered, and although not all were testable, 11 tested positive for WNV infection (United States Geological Survey 2006). Unusually large mortality events reported by hunters and landowners in Owyhee County, Idaho, led to closure of the hunting season in that area in 2006 (United States Geological Survey 2006). Another large but unexplained sage-grouse mortality event was reported near Jordan Valley, Oregon, in 2006, but remains were either not available or not testable (United States Geological Survey 2006). Severe declines in North Dakota populations between 2007 and 2008 were associated with high WNV mortality in summer 2007 (A. C. Robinson, pers. comm.).

Resistance to WNV

The prevalence, geographic distribution, and spread of resistance to WNV disease among sage-grouse populations will have important implications for both short- and long-term effects of the virus. Here we define resistance as the ability to survive WNV exposure, WNV infection, or both, and we assume that individuals with neutralizing antibodies to WNV were at minimum exposed to the virus. Under this definition, resistant individuals may still experience sublethal or residual effects of WNV infection.

The extent and distribution of resistance to WNV in wild populations remains unknown, but high mortality rates during severe WNV outbreaks and following experimental infection suggest that resistance is extremely low (Naugle et al. 2004, Clark et al. 2006, Walker et al. 2007b). Serum and tissues from 363 live and hunter-killed birds were tested for WNV in late 2003 and early 2004 following the 2003 outbreak, but no evidence of resistance to WNV was found—no birds tested seropositive for neutralizing antibodies to WNV (Naugle et al. 2004, 2005). The susceptibility of Greater Sage-Grouse to WNV was confirmed in 2004 when, in separate laboratory trials, all unvaccinated birds ($N = 44$) experimentally infected with WNV died within 6–8 days, regardless of dosage (Clark et al. 2006; T. E. Cornish, pers. comm.). Infected birds exhibited copious oral and nasal discharge, loss of mobility, shivering and piloerection of feathers, weakness, drooped wings, tilted heads, ataxia, labored breathing, and shedding of the virus from the cloaca (Clark et al. 2006). The first report of Greater Sage-Grouse surviving exposure to WNV was in the Powder River Basin of northeastern Wyoming and southeastern Montana in 2005, when 10.3% of 58 individuals captured in spring tested seropositive. However, in spring 2006, only 1.8% of 109 birds tested seropositive (Walker et al. 2007b). Seropositive live birds have not yet been reported from other parts of the species' range, but because sage-grouse are capable of dispersing long distances and demonstrate a genetic pattern of isolation by distance (Oyler-McCance et al. 2005b), other populations may also contain resistant individuals. The duration of immunity among birds that survive WNV infection is unknown (Marra et al. 2004). As in other flaviviruses, immunity is suspected to confer lifelong resistance to WNV, but it may or may not cross-protect seropositive individuals from other flaviviruses (Fang and Reisen 2006).

Carryover Effects of WNV Infection

It remains unclear whether sage-grouse experience sublethal or residual effects of WNV infection on productivity or overwinter survival, in part because high mortality during outbreaks has left few infected survivors for observation (Walker et al. 2004, 2007b). However, as in other birds (e.g., diurnal raptors and owls; Nemeth et al. 2006a,b; Saito et al. 2007), sage-grouse infected with WNV may suffer persistent symptoms that reduce subsequent survival, reproduction, or both. Nonlethal cases of WNV infection often result in chronic symptoms and lengthy recovery periods in other species (Marra et al. 2004; Hayes et al. 2005b; Nemeth et al. 2006a,b). The nature and severity of carryover effects of WNV on Greater Sage-Grouse deserve further study.

IMPACTS OF WNV ON POPULATION GROWTH

Matrix population models are valuable for understanding how impacts of potential stressors on vital rates translate into consequences for population growth. Life-stage simulation analyses (LSA), in particular, allow consideration of changes in both the mean and variance of specific vital rates on changes in population growth (Wisdom et al. 2000b, Reed et al. 2002). However, assumptions associated with matrix models suggest these models are best used to identify changes in population growth rate under different scenarios, rather than absolute estimates of growth rate (Reed et al. 2002). To better understand population-level impacts of WNV on sage-grouse, we estimated differences in population growth under different scenarios of WNV impacts using a life-stage simulation analysis model (Wisdom et al. 2000b). We parameterized the model with vital rate means and variances from across the species' range to adequately capture the full background range of spatial and temporal variation in demographics.

ANALYSES

We conducted life-stage simulation analysis in MATLAB version R2007a (MathWorks, Inc. 2007) to test the importance of mean vital rate values and their variability in predicting population growth (finite rate of increase, λ) for each of three WNV impact scenarios. We then generated and compared means for λ for each scenario based on 1,000 LSA simulations. Variance of demographic rates can strongly influence model results and interpretation (Wisdom et al. 2000b). We used the variance discounting method of White (2000) to remove sampling variance from total variance estimates and obtain an estimate of actual spatial and temporal variance for each vital rate. We used a two-stage, female-based life-cycle model to summarize stage-specific rates of fertility and survival. We then used vital rates for each stage and associated estimates of process variance based on range-wide data (Appendix 9.1) to parameterize a corresponding 2 × 2 stage-specific population projection model based on a prebreeding, birth-pulse census and a 1-year projection interval with birds censused on ~1 April, just prior to the initiation of nesting. The two stages were yearling and adult. Chick (<35 days of age) and juvenile (>35 days of age) survival were not considered separate stages but were incorporated into fertility rates. Vital rates for each simulation were randomly selected from either a beta or stretched beta distribution (Morris and Doak 2002). We conducted analyses both with and without correlations among vital rates to see how correlation structure influenced estimates of λ (Morris and Doak 2002). Complete details regarding model structure, vital rate estimation, variance discounting, and correlations among vital rates are summarized in Walker (2008).

WNV IMPACT SCENARIOS

Scenarios included models (1) without WNV-related mortality (i.e., based on vital rate data prior to 2003, or data excluding WNV-related mortalities); (2) with WNV-related mortality based on observed infection and mortality rate data reported from 2003 to 2007 (Walker et al. 2007b); and (3) with WNV-related mortality, but with increasing resistance to WNV over time. Scenarios for WNV impacts that model the effects of increasing temperature due to climate change and of increasing anthropogenic water sources due to energy development would also be valuable, but they were beyond the scope of the current analysis. We estimated means and variances of survival for juveniles >35 days of age, yearlings, and adults from range-wide data collected prior to 2003 or from data that excluded WNV-related mortalities (Walker 2008). We randomly selected infection rates for each simulation replicate in scenarios 2 and 3 from a stretched beta distribution with $\bar{x} = 0.07$, SD = 0.05, minimum = 0.005, maximum = 1.0 (Morris and Doak 2002: Box 8.3). This resulted in a distribution of infection rates (0–50%) and mortality rates (0–38%) consistent with published estimates (Walker et al. 2007b), allowed most simulated years to have low rates of WNV infection (median = 0.05) and mortality (median = 0.05), and produced some years with extreme values for infection rate (~50%)(Walker 2008). We calculated mortality due to WNV (M) for each simulation replicate using infection rate (I) and resistance to WNV-related disease (R) as $M = I - (I \times R)$, i.e., proportion infected minus proportion infected but resistant to disease following exposure or infection. We used mortality rates to appropriately reduce juvenile, yearling, and adult survival by increasing mortality during the typical 2.5-month WNV period (1 July–15 September) for each replicate. We assumed that resistance was constant in scenario 2 and used a value of 0.04, the mean spring seroprevalence value reported by Walker et al. (2007b). We assessed in scenario 3 how an increase in resistance to WNV might change population growth rate by calculating changes in the proportion of resistant individuals in the population under simulated rates of WNV infection and WNV mortality using 0.04 as the starting value for resistance. We assumed in this scenario that all resistance to WNV infection and disease was heritable and that all female offspring of a resistant female inherited traits that conferred resistance (i.e., heritability of resistance = 1). We conducted each simulation with 20 replicates to simulate responses within a 20-year management time frame, and conducted the entire simulation 1,000 times to generate means and standard deviations for λ for each year during the 20-year period.

FINDINGS

The addition of WNV mortality resulted in a projected average estimated reduction in λ of −0.059 to −0.086, depending on the scenario and whether vital rates were correlated or uncorrelated (Table 9.2). However, substantial annual variation in vital rates that influence λ resulted in wide variation in simulated values for λ in all scenarios (Fig. 9.2). Results of LSA indicated that several different groups of vital rates were important for population growth; vital rates most highly correlated with population growth in LSA included nest success, chick survival, juvenile survival, and adult and yearling survival (Fig. 9.3). The proportion of resistant

TABLE 9.2

Estimated average reduction in annual population growth rate (finite rate of increase, λ) under different West Nile virus (WNV) impact scenarios relative to no WNV mortality.

Data are based on life-stage simulation analyses using vital rates for female Greater Sage-Grouse from range-wide data. Results are based on 1,000 LSA simulation replicates. Reductions in λ due to WNV mortality may be masked in any given year by annual fluctuations in vital rates influential for population growth (nest success, chick survival, juvenile survival, survival of breeding-age females).

Scenario	Correlated[a] Δλ	Uncorrelated[b] Δλ
No WNV	0.000	0.000
Current WNV	−0.086	−0.060
Current WNV with increasing resistance	−0.081	−0.059

[a] Simulated vital rates for each replicate accounted for correlations among vital rates.

[b] Simulated vital rates for each replicate were uncorrelated.

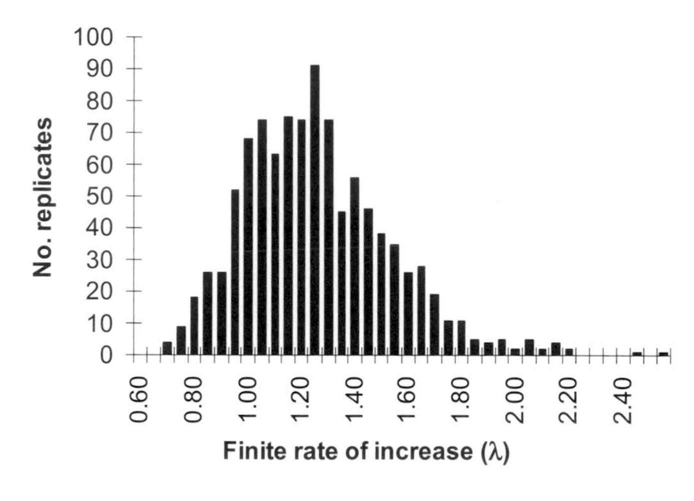

Figure 9.2. Distribution of simulated annual population growth rates (finite rate of increase, λ) for female Greater Sage-Grouse based on life-stage simulation analysis using range-wide data, assuming no WNV impacts. Absolute values of range-wide population growth based on simulated data from population models cannot be used to infer range-wide population trends.

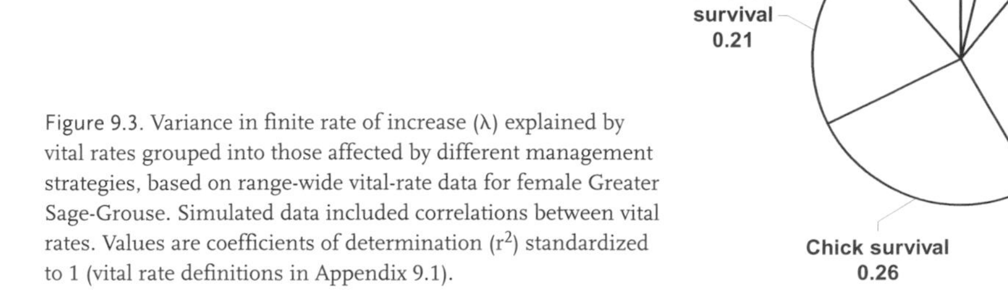

Figure 9.3. Variance in finite rate of increase (λ) explained by vital rates grouped into those affected by different management strategies, based on range-wide vital-rate data for female Greater Sage-Grouse. Simulated data included correlations between vital rates. Values are coefficients of determination (r^2) standardized to 1 (vital rate definitions in Appendix 9.1).

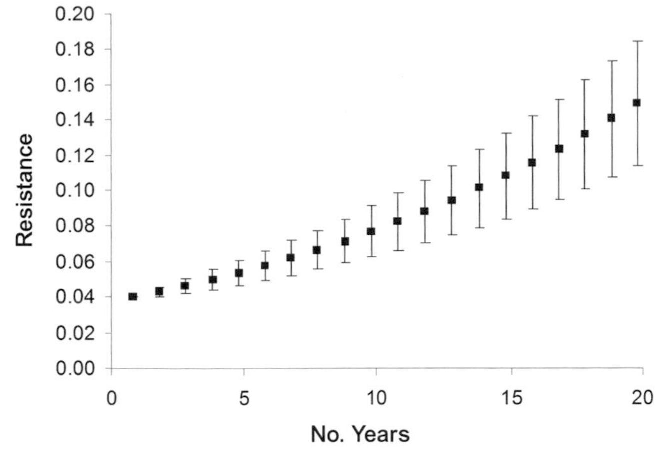

Figure 9.4. Projected change in resistance to WNV disease of female Greater Sage-Grouse at the start of the breeding season over a 20-year period based on simulated vital rates in life-stage simulation analyses. Error bars represent 1 standard deviation. The initial value for resistance was set at 0.04 in year 1 (i.e., 4% of the population resistant to WNV).

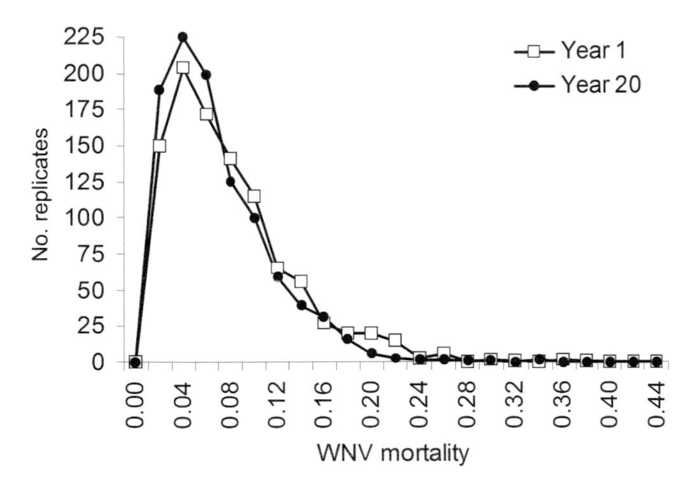

Figure 9.5. Distribution of WNV mortality among female Greater Sage-Grouse in year 1 (open squares) and year 20 (black circles) of the simulation, with increasing resistance over time based on simulated infection rates from range-wide data, assuming no carryover effects of WNV infection.

individuals in the population was projected to increase only marginally over a 20-year time period, from 0.04 to 0.15 using current estimates of infection rates (Fig. 9.4). The increase in resistance was projected to shift the distribution of WNV-related mortality rates lower over time (Fig. 9.5).

DISCUSSION

West Nile virus is a new source of mortality that complicates efforts to conserve Greater Sage-Grouse. Growth is expected to decline in susceptible populations, birds appear to show little resistance to WNV, and management options for controlling the spread of WNV in sage-grouse habitat are limited. Prior to emergence of WNV, little evidence implicated disease, exotic or otherwise, as a major threat to Greater Sage-Grouse (Schroeder et al. 1999; Connelly et al. 2000c, 2004). Several lines of evidence suggest that WNV represents a new risk to sage-grouse populations, including: (1) documented presence of the virus throughout the species' range; (2) persistent, and sometimes substantial, declines in late-summer survival due to WNV mortality; (3) large mortality events attributed to WNV in both marked and unmarked populations; (4) local extirpation of a breeding population following a severe WNV outbreak; (5) projected declines in simulated population growth in susceptible populations based on empirical infection and mortality rate data; (6) documented low levels of resistance to WNV disease in captive populations; (7) low potential for increasing resistance over time; and (8) potential for large-scale increases in mosquito breeding habitat, and consequently WNV risk, due to increases in surface water associated with energy development.

These conclusions may be conservative regarding the impact of WNV. First, limited evidence suggests that in some years, early-season WNV mortality may also reduce survival of chicks <35 days of age either directly or indirectly by affecting survival of hens with dependent broods (Aldridge 2005, Walker 2008). Second, the distribution of mortality rates used in simulations may underestimate mortality in wild populations. Third, the presence of neutralizing antibodies in seropositive live birds does not always reliably indicate resistance. Fourth, as in other species, it is entirely plausible that birds that survive WNV infection may later experience reduced survival or reproduction. Finally, WNV mortality usually comes at a time of year (July–September) when survival is typically high (Schroeder et al. 1999), suggesting it is additive to other sources of mortality.

CONSERVATION IMPLICATIONS

The long-term response of different sage-grouse populations to WNV is expected to vary markedly depending on factors that influence susceptibility, including: (1) annual and seasonal temperature-precipitation profiles, (2) land uses that influence the distribution of surface water, (3) population size, (4) genetic diversity, and (5) connectivity with other populations. Small, isolated, or genetically depauperate populations and those on the fringe of the species' range, as in eastern California, Washington, North and South Dakota, Alberta, and Saskatchewan, are likely at higher risk. WNV outbreaks in small populations are more likely to reduce population size below a threshold from which recovery is unlikely and the likelihood of demographic or genetic rescue by adjacent populations is low (Morris and Doak 2002). Large, intact, low- to mid-elevation populations affected annually by WNV in northern Nevada, southeastern Idaho, and central Montana may absorb impacts of WNV if the quality and extent of available habitat still supports positive population growth. Impacts from WNV in some populations, such as northeastern Wyoming, may act synergistically with other stressors like energy development and tillage agriculture to substantially reduce population size, distribution, or persistence (Walker et al. 2007a). Conservation of large, high-elevation populations, such as those in northwestern Colorado and western Wyoming, where WNV risk is relatively lower, will be important for offsetting impacts of WNV at a range-wide scale. Changes in virulence or epizootiology as the virus adapts to new environments and new hosts and vectors in North America will also influence long-term impacts of WNV on sage-grouse populations, but whether such changes will ameliorate or exacerbate current impacts is unknown.

Understanding long-term impacts of WNV will require intensive monitoring of radio-marked populations. Population models suggest that, except during severe outbreaks (Walker et al. 2004), natural geographic and temporal fluctuation in vital rates that drive population growth can mask impacts of WNV in any given year. Impacts of WNV mortality, and even severe WNV outbreaks, may go undetected without radio-marked individuals (Walker et al. 2004, United States Geological Survey 2006) and lead to the misperception among managers and policy makers that WNV is no longer an issue for Greater Sage-Grouse. Moreover, in the absence of radio-marked birds, population declines due to severe or persistent WNV mortality may be incorrectly attributed to other potential stressors and lead to inappropriate policy decisions and management or mitigation actions. Radio-marking is known to reduce reproductive effort and survival in other species (Withey et al. 2001), but mass mortality events of unmarked individuals and high late-summer survival among unaffected, radio-marked birds range-wide (Naugle et al. 2005) suggest that radio marking itself does not predispose individuals to greater risk of mortality following WNV infection. We strongly recommend continued range-wide monitoring and testing of radio-marked populations to estimate WNV-related mortality and testing of serum samples from live birds to document the extent and distribution of resistance to WNV. Failure to do so will hinder our understanding of how this emerging disease influences sage-grouse populations and may diminish our ability to maintain the species' distribution and abundance (Friend et al. 2001).

Vaccines have been used to guard against catastrophic mortality in captive populations but are unlikely to be available or effective for protecting wild sage-grouse populations from impacts of WNV (Clark et al. 2006, Kilpatrick et al. 2007). Equine and fowl WNV vaccines administered via intramuscular injection have reduced mortality in captive birds (Bertelsen et al. 2004, McLean 2006, Bunning et al. 2007). However, fowl vaccine used on captive sage-grouse was only marginally effective; the vaccine reduced mortality rates from 100% to 80% ($N = 5$), increased average time to mortality from 3.7 to 6.7 days, and reduced average peak viremia from $10^{6.4}$ PFU to $10^{2.4}$ PFU (Clark et al. 2006). The lack of market incentives to produce a vaccine specifically for sage-grouse and lack of an effective delivery mechanism to large numbers of wild birds are major barriers to implementation (Clark et al. 2006, McLean 2006). Moreover, vaccinations only benefit treated individuals, rather than conferring long-term immunity to offspring, so any vaccination program would require annual treatments (Kilpatrick et al. 2007).

Managing production of mosquito vectors from human-made water sources, reducing the distribution of human-made mosquito breeding habitats in sage-grouse habitat, or both are potential options for reducing impacts of WNV. Previous studies published prior to the arrival of WNV have recommended use of water developments in arid sagebrush landscapes to benefit sage-grouse, but they also cautioned this should be done only if such actions clearly benefit the birds (Paige and Ritter 1999, Connelly et al. 2000c). Sage-grouse use standing water when it is available (Dalke et al. 1963), but they do not require it (Connelly and Doughty 1989, Schroeder et al. 1999, Connelly et al. 2004). Addition of artificial water sources that increase the distribution or abundance of *Culex tarsalis* in sage-grouse habitat may be particularly detrimental. Artificial water sources known to support breeding *Culex tarsalis* in sage-grouse habitat include overflowing stock tanks, stock ponds, seep and overflow areas below earthen dams, irrigated agricultural fields, and ponds constructed for coal-bed natural gas development (Zou et al. 2006b, Doherty 2007). Several strategies are recommended to reduce mosquito production from artificial water sources without eliminating the water source. First, ponds and tanks can be constructed, modified, or managed in ways that discourage breeding mosquitoes (Doherty 2007). It may also be possible to control mosquitoes with mosquitofish (*Gambusia* spp.) or native fish species that eat mosquito larvae, biological or chemical larvicides (BTI, *Bacillus thuringiensis v. israelensis*), or spraying for adults (Doherty 2007). Mosquito control programs appear effective for reducing WNV risk, but only if applied appropriately and consistently by qualified mosquito control personnel (Gubler et al. 2000, Reisen and Brault 2007). The costs and benefits of control need to be weighed against potential detrimental or cascading ecological effects of widespread spraying (Marra et al. 2004). Requiring infectious disease impact statements as part of planned large-scale changes in land use for energy development (McSweegan 1996) may also improve coordinated management of WNV risk in sage-grouse habitat.

ACKNOWLEDGMENTS

We thank J. C. Owen, V. O. Ezenwa, and M. W. Miller for discussion and comments on earlier drafts of this chapter; L. S. Mills and C. R. Hartway for assistance with population modeling in MATLAB; and researchers and wildlife managers throughout the West for sharing unpublished data on WNV mortality. Comments by C. E. Braun and S. T. Knick improved the chapter.

APPENDIX 9.1

Range-wide vital rate summary.

Vital Rate[a]	Mean	Variance[b]
$INIT_{YR1}$	0.829	0.0166
$INIT_{AD1}$	0.930	0.0038
$INIT_{YR2}$	0.148	0.0368
$INIT_{AD2}$	0.395	0.0599
$INIT_{AD3}$	0.074	0.0051
$FCLUTCH_{YR1}$	3.81	0.118
$FCLUTCH_{YR2}$	3.29	0.316
$FCLUTCH_{AD1}$	4.16	0.040
$FCLUTCH_{AD2}$	3.52	0.200
$FCLUTCH_{AD3}$	3.02	0.200[c]
$SUCC_{YR1}$	0.481	0.0268
$SUCC_{AD1}$	0.569	0.0183
$SUCC_{YR2}$	0.540	0.1309
$SUCC_{AD2}$	0.553	0.0623
HATCH	0.921	0.0018
$CHSURV_{YR}$	0.391[d]	0.0084[d]
$CHSURV_{AD}$	0.391[d]	0.0084[d]
$JUVSURV_{83}$	0.799	0.0154
$JUVSURV_{91}$	0.782	0.0177
$SURV_{YR}$	0.684	0.0182
$SURV_{AD}$	0.582	0.0050

[a] Variables defined as: $INIT_{YR1}$ = nest initiation rate of yearlings; $INIT_{AD1}$ = nest initiation rate of adults; $INIT_{YR2}$ = renesting rate of yearlings; $INIT_{AD2}$ = renesting rate of adults; $INIT_{AD3}$ = second renesting rate of adults; $FCLUTCH_{YR1}$ = clutch size (female eggs only) of yearling first nests; $FCLUTCH_{YR2}$ = clutch size (female eggs only) of yearling renests; $FCLUTCH_{AD1}$ = clutch size (female eggs only) of adult first nests; $FCLUTCH_{AD2}$ = clutch size (female eggs only) of adult renests; $FCLUTCH_{AD3}$ = clutch size (female eggs only) of adult second renests; $SUCC_{YR1}$ = nest success of yearling first nests; $SUCC_{AD1}$ = nest success of adult first nests; $SUCC_{YR2}$ = nest success of yearling renests; $SUCC_{AD2}$ = nest success of adult renests (and second renests); HATCH = hatching success; $CHSURV_{YR}$ = survival of chicks from yearling females from hatch to 35 d; $CHSURV_{AD}$ = survival of chicks from adult females from hatch to 35 d of age; $JUVSURV_{83}$ = survival of juveniles from 35 d of age to 10 September for renests; $JUVSURV_{91}$ = survival of juveniles from 35 d of age to 10 September for first nests; $SURV_{YR}$ = annual survival of yearlings; and $SURV_{AD}$ = annual survival of adults.

[b] Process variance estimated following White (2000).

[c] Process variance for clutch size of adult second renests could not be estimated from range-wide data; the value for clutch size of adult renests was used.

[d] Mean and process variance for chick survival of yearling and adult females were the same in range-wide data because most previous publications did not present chick survival estimates separately for each stage.

[e] Process variance for juvenile survival could not be estimated; values represent raw variance estimates from range-wide data.

PART THREE

Ecology of Sagebrush

CHAPTER TEN

Characteristics of Sagebrush Habitats and Limitations to Long-Term Conservation

Richard F. Miller, Steven T. Knick, David A. Pyke, Cara W. Meinke, Steven E. Hanser, Michael J. Wisdom, and Ann L. Hild

Abstract. The distribution of sagebrush (*Artemisia* spp.) within the Sage-Grouse Conservation Area (SGCA, the historical distribution of sage-grouse buffered by 50 km) stretches from British Columbia and Saskatchewan in the north, to northern Arizona and New Mexico in the south, and from the eastern slopes of the Sierra Nevada and Cascade mountains to western South Dakota. The dominant sagebrush (sub)species as well as the composition and proportion of shrubs, grasses, and forbs varies across different ecological sites as a function of precipitation, temperature, soils, topographic position, elevation, and disturbance history. Most important to Greater Sage-Grouse (*Centrocercus urophasianus*) are three subspecies of big sagebrush (*Artemisia tridentata*)(basin big sagebrush [*A. t.* ssp. *tridentata*], Wyoming big sagebrush [*A. t.* ssp. *wyomingensis*], and mountain big sagebrush [*A. t.* ssp. *vaseyana*]); two low or dwarf forms (little sagebrush [*A. arbuscula*] and black sagebrush [*A. nova*]); and silver sagebrush (*A. cana*), which occurs primarily in the northeast portion of the sage-grouse range. Invasive plant species, wildfires, and weather and climate change are major influences on sagebrush habitats and present significant challenges to their long-term conservation. Each factor is spatially pervasive across the Greater Sage-Grouse Conservation Area and has significant potential to influence processes within sagebrush communities. Cheatgrass (*Bromus tectorum*), the most widespread exotic annual grass, has invaded much of the lower-elevation, more xeric sagebrush landscapes across the western portion of the Greater Sage-Grouse Conservation Area. A large proportion of existing sagebrush communities are at moderate to high risk of invasion by cheatgrass. Juniper (*Juniperus* spp.) and pinyon (*Pinus* spp.) woodlands have expanded into sagebrush habitats at higher elevations creating an elevational squeeze on the sagebrush ecosystem from both extremes. Number of fires and total area burned have increased since 1980 throughout the SGCA except in the Snake River Plain, which has a long-term history of high fire disturbance. Climate change scenarios for the sagebrush region predict increasing trends in temperature, atmospheric carbon dioxide, and frequency of severe weather events that favor cheatgrass expansion and increased fire disturbance resulting in a decline in sagebrush. Approximately 12% of the current distribution of sagebrush is predicted to be replaced by expansion of other woody vegetation for each 1°C increase in temperature. Periodic drought regularly influences sagebrush ecosystems; drought duration and severity have increased throughout the 20th century in much of the interior western

Miller, R. F., S. T. Knick, D. A. Pyke, C. W. Meinke, S. E. Hanser, M. J. Wisdom, and A. L. Hild. 2011. Characteristics of sagebrush habitats and limitations to long-term conservation. Pp. 145–184 *in* S. T. Knick and J. W. Connelly (editors). Greater Sage-Grouse: ecology and conservation of a landscape species and its habitats. Studies in Avian Biology (vol. 38), University of California Press, Berkeley, CA.

United States. Synergistic feedbacks among invasive plant species, fire, and climate change, coupled with current trajectories of habitat changes and rates of disturbance (natural and human-caused), will continue to change sagebrush communities and create challenges for future conservation and management.

Key Words: Artemisia, Bromus tectorum, climate change, community dynamics, drought, exotic plant species, juniper, pinyon, sagebrush, weather, wildfire.

Características de los Hábitats de Artemisa y Limitaciones para la Conservacion a Largo Plazo

Resumen. La distribución de artemisa (*Artemisia* spp.) dentro del área de conservación del Greater Sage-Grouse (SGCA; la distribución histórica del sage-grouse amortiguada por 50 km) se extiende desde la British Columbia y Saskatchewan, en el norte, al norte de Arizona y New Mexico en el sur, y desde las laderas orientales de la Sierra Nevada y Montañas Cascade, al oeste de Dakota del Sur. Las (sub)especies de artemisa dominantes, así como la composición y la proporción de arbustos, gramíneas, y malezas varían a lo largo de los diferentes sitios ecológicos en función de las precipitaciones, la temperatura, los suelos, la posición topográfica, la altitud, y el historial de disturbios en el área. Lo más importante para el Greater Sage-Grouse (*Centrocercus urophasianus*) son tres subespecies de big sagebrush (*Artemisia tridentata*)(basin big sagebrush [*A. t.* spp. *tridentata*], Wyoming big sagebrush [*A. t.* ssp. *wyomingensis*], y mountain big sagebrush [*A. t.*, ssp. *vaseyana*]); dos formas bajas (o enanas, la little sagebrush [*A. arbuscula*] y black sagebrush [*A. nova*]); y silver sagebrush (*A. cana*) que ocurre principalmente en la porción nordeste de la zona de Greater Sage-Grouse. Especies de plantas invasoras, incendios forestales, el tiempo, y el cambio climático son las principales influencias sobre los hábitats de la artemisa y también presentan desafíos importantes para su conservación a largo plazo. Cada uno de ellos está presente a través de la zona de SGCA y tiene un importante potencial para influir en los procesos dentro de las comunidades de artemisa. Cheatgrass (*Bromus tectorum*), la más extendida hierba anual exótica, ha invadido gran parte de la altitudes más bajas, los paisajes de artemisa más adaptados a las condiciones desérticas a través de la parte occidental de la SGCA. Una gran proporción de las comunidades existentes de artemisa están entre un moderado a alto riesgo de invasión por cheatgrass. Bosques de enebro (*Juniperus* spp.) y piñón (*Pinus* spp.) se han expandido a los hábitats de artemisa en elevaciones más altas creando así presión sobre el ecosistema de artemisa en ambos extremos. El número de incendios y el área total incendiada se han incrementado desde 1980, en toda la SGCA, excepto en el área del Snake River Plain, el cual tiene una largo historial de alteraciones por fuegos grandes. Escenarios de cambios climáticos en la región de la artemisa predicen tendencias de incremento en la temperatura, dioxodo de carbono atmosférico, y frecuencia de eventos severos de clima que favorecen la expansión del cheatgrass e incrementan la alteración por fuego que produce una disminución de artemisa. Se predice que aproximadamente el 12% de la actual distribución de artemisa será remplazada, por la expansión de otra vegetación leñosa, por cada grado centígrado de incremento en temperatura. Sequías periódicas influencian regularmente el ecosistema de la artemisa; la duración y severidad de las sequías se han incrementado a lo largo del siglo XX en la mayoría del interior del oeste de EE.UU. Las retroalimentaciones climáticas sinérgicas entre especies de plantas invasivas, fuego y cambios climáticos, sumados a las actuales trayectorias en cambios de hábitat y los grados de alteración (naturales y causadas por el hombre) continuarán cambiando las comunidades de artemisa y creando desafíos para la futura conservación y manejo.

Palabras Clave: artemisa, *Artemisia, Bromus tectorum,* cambio climático, clima, dinámica de la comunidad, especies de plantas exóticas, enebro, incendios forestales, pinos, sequía.

Ecological sites supporting sagebrush (*Artemisia* spp.) within the Sage-Grouse Conservation Area (SGCA, the historical distribution of sage-grouse buffered by 50 km [Connelly et al. 2004, Schroeder et al. 2004]) represent some of the largest and most imperiled ecosystems in North America (Noss et al. 1995; Center for Science, Economics, and Environment 2002). The primary patterns, processes, and components of sagebrush ecosystems have been altered significantly since Euro-American settlement in the late 1800s (West and Young 2000, Bunting et al. 2003). Few, if any, landscapes remain intact and unchanged throughout the SGCA (Miller et al. 1994, West 1996, Miller and Eddleman 2001). It is unlikely that we can return to pre-settlement conditions because size of the area and magnitude of changes far exceed any financial or logistical resources available (Hemstrom et al. 2002, Wisdom et al. 2005c, Meinke et al. 2009). In addition, loss of available parts from native systems (West 1996, Longland and Bateman 2002) coupled with continual short- and long-term changes in climate further complicate our ability to recreate previous sagebrush communities. However, learning how sagebrush communities function and the potential effects of primary disrupters can help increase understanding how these systems respond to land use (Knick et al., this volume, chapter 12) or management actions, such as restoration (Pyke, this volume, chapter 23) and may help maintain at least a portion of these ecosystems.

We describe the distribution of sagebrush within the SGCA followed by a general description of the characteristics of sagebrush alliances (a physiognomically uniform group of plant associations sharing one or more dominant or diagnostic species, which, as a rule, are found in the uppermost stratum of the vegetation; Grossman et al. 1998) and plant associations (a plant community type of definite floristic composition, uniform habitat conditions, and uniform physiognomy; Daubenmire 1978, Grossman et al. 1998). We then describe the extent and potential for invasive plant species, wildfire, and weather and climate change to alter the character of sagebrush systems and further affect our ability to conserve and manage these habitats. Sagebrush ecosystems are influenced by numerous stressors: Wisdom et al. (2005b:30–35) list more than 25 potential disrupters. Of these, we focused on exotic plants, wildfire, and climate because they are spatially pervasive agents having a significant potential to influence long-term changes in patterns and processes across the SGCA. Other disrupters that more directly originate from human actions are considered elsewhere (Knick et al., this volume, chapter 12; Leu et al. 2008; Leu and Hanser, this volume, chapter 13).

The SGCA includes Küchler's (1970) three sagebrush vegetation types in addition to a portion of the northern Great Plains, which also supports stands of upright woody sagebrush. Sagebrush is a dominant land cover across much of the unforested parts of this region. However, many areas now contain only islands of sagebrush habitats embedded within larger expanses of highly altered landscapes. Sage-grouse have been extirpated from many of these islands (Schroeder et al. 2004; Aldridge et al. 2008; Wisdom et al., this volume, chapter 18); other sagebrush-dependent wildlife, such as Sage Sparrows (*Amphispiza belli*), Brewer's Sparrows (*Spizella breweri*), and Sage Thrashers (*Oreoscoptes montanus*), continue to use these sagebrush habitats (Wisdom et al. 2000b, Knick and Rotenberry 2002, Dobkin and Sauder 2004, Wisdom et al. 2005b). These remnants also may retain critical plant and wildlife components that could be valuable in restoration of adjacent potential or at-risk sagebrush communities (West 1996, West and Young 2000, Longland and Bateman 2002, Bunting et al. 2003). These remaining sagebrush islands continue to interact with other habitats in the landscape matrix by providing seed sources and habitat for resident and transient wildlife.

GEOGRAPHIC DISTRIBUTION OF SAGEBRUSH WITHIN THE GREATER SAGE-GROUSE CONSERVATION AREA

Sagebrush habitats are distributed throughout the SGCA (Fig. 10.1). The SGCA extends from British Columbia and Saskatchewan in the north to northern Arizona and New Mexico in the south, and from the eastern slopes of the Sierra Nevada and Cascade Mountains in the west to western North and South Dakota in the east (Fig. 10.1). Küchler (1964, 1970)(Fig. 10.2) separated sagebrush in this region into two potential natural vegetation types: (1) sagebrush steppe (type 55; Küchler 1970), where sagebrush is frequently a codominant with perennial bunchgrasses under potential natural conditions; and (2) Great Basin sagebrush (type 38), where sagebrush can often be the dominant plant

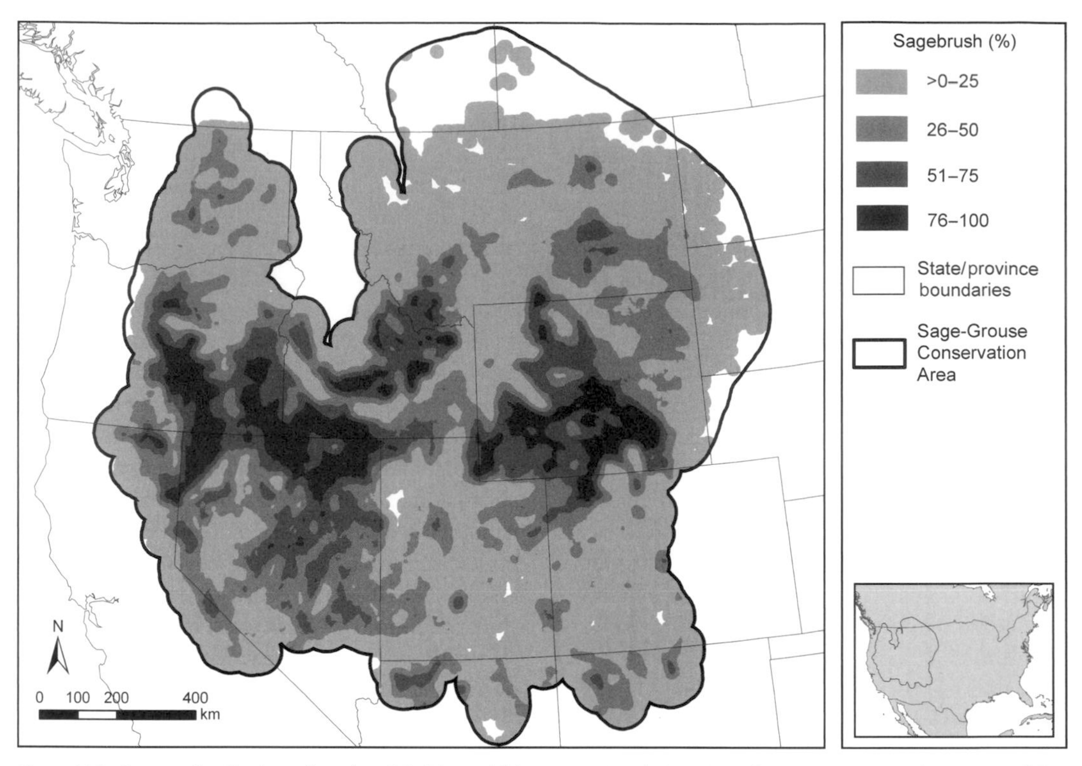

Figure 10.1. Current distribution of sagebrush habitats within western North America. The map represents the percent of the landscape dominated by sagebrush habitats and not site-specific values of ground cover. As such, it is intended to be a general representation of sagebrush distribution.

layer with a sparse understory. These two vegetation types, which exclude the Great Plains region, occupy >500,000 km^2 (Table 10.1)(Küchler 1970). West (1983a), naming Küchler's Great Basin sagebrush type the Great Basin–Colorado Plateau sagebrush semidesert, addressed some of the differences between these two subregions. Sagebrush steppe occupies parts of British Columbia, the Columbia Basin, northern Great Basin, Snake River Plain, Montana, Wyoming Basin, and northern Colorado (Fig. 10.2). The Great Basin sagebrush type lies to the south below the polar front gradient (Miller and Eddleman 2001) where temperatures are warmer, summer precipitation increases, and winter precipitation decreases (Mitchell 1976). The Great Basin sagebrush type includes portions of the Colorado Plateau and extends across Nevada, Utah, southeastern Colorado, northern Arizona, northern New Mexico, and central-eastern California. A third vegetation type, the mixed desert shrubland (Knight 1994; Küchler's type 56) occurs in the Bighorn Basin in north-central Wyoming (Fig. 10.2). Portions of the northern Great Plains that Küchler (1964, 1970) mapped as *Grama*-needlegrass-wheatgrass (type 64) in eastern Montana and eastern Wyoming (Fig. 10.2) support plains silver sagebrush (*Artemisia cana* ssp. *cana*) and sand sagebrush (*A. filifolia*).

Geographic subdivisions across the SGCA have been delineated into Sage-Grouse management zones based on general similarities in climate, elevation, topography, geology, soils, and floristics (West 1983b, Miller and Eddleman 2001)(Fig. 10.3). Sagebrush habitats in the Columbia Basin, Northern Great Basin, Snake River Plain, Wyoming Basin, Southern Great Basin, and Silver Sagebrush floristic provinces (Fig. 10.3) are of primary importance to Greater Sage-Grouse (*Centrocercus urophasianus*). Boundaries of floristic provinces were used as general guides for delineating the current Sage-Grouse Management Zones (Stiver et al. 2006).

SAGEBRUSH TAXA

Intermountain Region

Shultz (2009) recognized 13 species and 12 subspecies in the genus *Artemisia* subgenus *Tridentate*. The most predominant sagebrush taxa, which provide important annual and seasonal

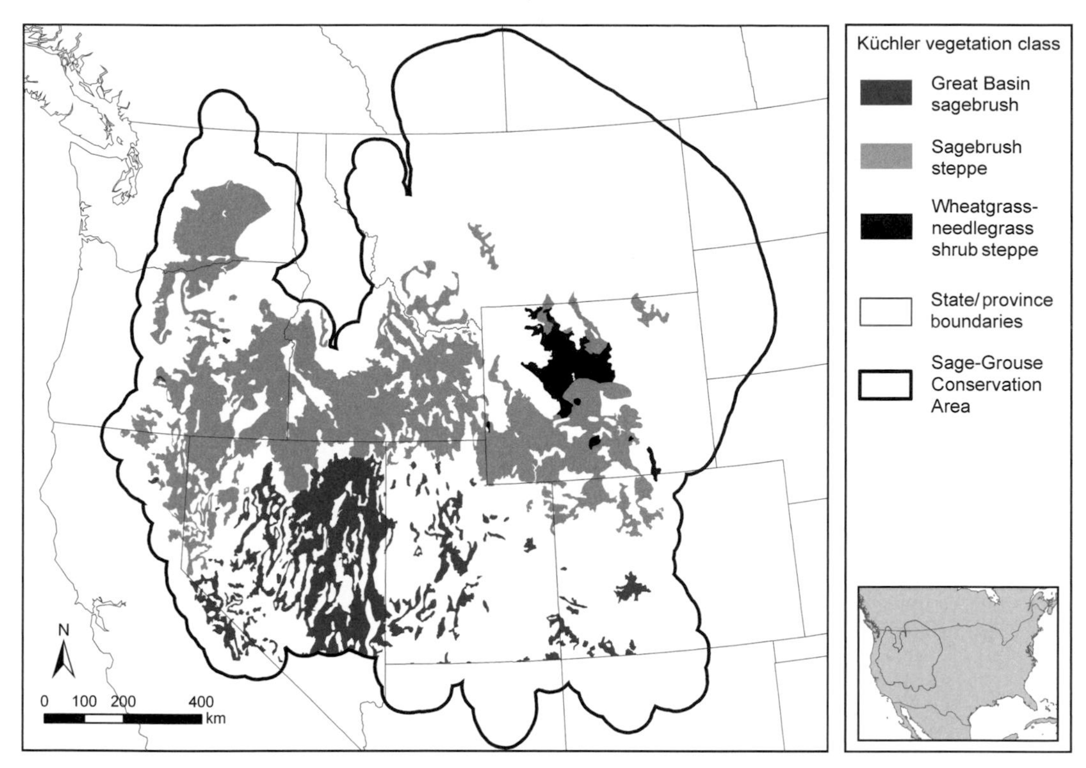

Figure 10.2. Küchler map (1970) of potential sagebrush distribution. The map illustrates the potential distribution of Great Basin sagebrush (type 38), sagebrush steppe (type 55), and wheatgrass-needlegrass mixed desert shrubland (type 56) vegetation types within the Sage-Grouse Conservation Area. The vegetation classes represent the potential vegetation that would be present in the absence of natural or human disturbance.

habitats for sage-grouse across the SGCA, are three subspecies of *Artemisia tridentata*, represented by Wyoming big sagebrush (*A. t.* ssp. *wyomingensis*), basin big sagebrush (*A. t.* ssp. *tridentata*), and mountain big sagebrush (*A. t.* ssp. *vaseyana*); two low (or dwarf) forms of sagebrush, little sagebrush (*A. arbuscula*) and black sagebrush (*A. nova*); and silver sagebrush (Connelly et al. 2000c, Crawford et al. 2004). The abundance and frequency of occurrence of sagebrush taxa characterizing different ecological sites is influenced primarily by soils, climate, topographic position, and disturbance history (West 1983b)(Fig. 10.4). An ecological site is defined as having specific physical characteristics that differ from other sites in the ability to produce distinctive compositions and amounts of vegetation and in response to management.

The three subspecies of big sagebrush usually occur on well-drained, moderately deep, sandy to clay loam soils. Wyoming big sagebrush typically occurs on warmer, drier sites at elevations between 150 and 2,140 m in valleys and foothills (Mahalovich and McArthur 2004). Soils are often underlain by an argillic, caliche, or silica layer. Basin big sagebrush also occurs in the valleys and foothills on deep (often ≥1 m), dry, fertile soils. Mountain big sagebrush occurs on relatively cooler sites in foothills and mountains on moderate to deep well-drained soils where summer moisture is available. It typically occurs at elevations varying from 1,200 to 3,100 m.

The low (or dwarf) forms of sagebrush, including little sagebrush, black sagebrush, and rigid (stiff or scabland) sagebrush (*A. rigida*), generally occur on shallow or poorly drained soils (Eckert 1957, Fosberg and Hironaka 1964). A strong argillic horizon, duripan, or bedrock in these areas that keeps water from draining generally is present <33 cm from the surface or <50 cm in wet areas. Depth to the wetting horizon is usually limited when black or little sagebrush occur on deeper soils, and soils are coarse textured (Fosberg and Hironaka 1964, Sabinske and Knight 1978, Tisdale 1994). Three subspecies of little sagebrush grow in the western portion of the sagebrush region extending east into western Wyoming (Cronquist et. al. 1994, Mahalovich and McArthur 2004). Little

TABLE 10.1

Area occupied by sagebrush in the Intermountain region estimated from Küchler's map of potential vegetation (Küchler 1970, West 1983b).

The regional boundaries used for these area estimates were delineated in Küchler's map (1970) of potential vegetation (Fig. 10.2) and did not include eastern parts of the Sage-Grouse Conservation Area.

Type/state	Area (km²)	% of type total
Sagebrush steppe		
Wyoming	99,642	25.5
Idaho	96,162	24.6
Oregon	75,728	19.3
Nevada	38,498	9.8
Washington	34,478	8.8
Colorado	14,399	3.7
California	12,558	3.2
Montana	11,089	2.8
Utah	8,936	2.3
Total	391,490	100.0
Great Basin sagebrush		
Nevada	96,908	70.8
Utah	26,040	19.0
Colorado	7,606	5.6
California	6,373	4.7
Total	136,927	100.0

sagebrush (*Artemisia arbuscula* spp. *arbuscula*) occurs from western Wyoming and Colorado to south-central Washington to northern California between elevations of 700 and 3,780 m. Lahontan little sagebrush (*A. a.* spp. *longicaulis*) occupies elevations between 1,050 and 2,000 m in northwest Nevada and neighboring Oregon and California (in the vicinity of old Lake Lahontan). Hot springs little sagebrush (*A. a.* spp. *thermophila*) occurs across western Wyoming, northern Utah, and eastern Idaho between 1,800 and 2,500 m. Black sagebrush extends farther south and east than little sagebrush and commonly occurs on calcareous soils between 625 and 2,990 m elevation. Early sagebrush (*A. longiloba*) was recently recognized as a fourth subspecies of little sagebrush (Shultz 2009). Early sagebrush flowers earlier than other low-statured sagebrush and is an important taxa for sage-grouse. It also is one of the most palatable sagebrush species and often is heavily browsed (Winward 2004, Rosentreter 2005). Early sagebrush occurs from North Park, Colorado, to central Oregon and central California. Sagebrush taxa in Nevada follow an increasing gradient of soil fertility from black sagebrush, little sagebrush, early sagebrush, Wyoming big sagebrush, basin big sagebrush, and mountain big sagebrush (Jensen 1989a).

Northern Great Plains

The primary *Artemisia* species in the northeastern range of Greater Sage-Grouse, including northeastern Wyoming, eastern Montana, southeastern Alberta, southwestern Saskatchewan, and the extreme western portions of South Dakota, southwest North Dakota, and northwest Nebraska are Wyoming and basin big sagebrush, prairie sagewort (or fringed sagebrush [*A. frigida*]), plains

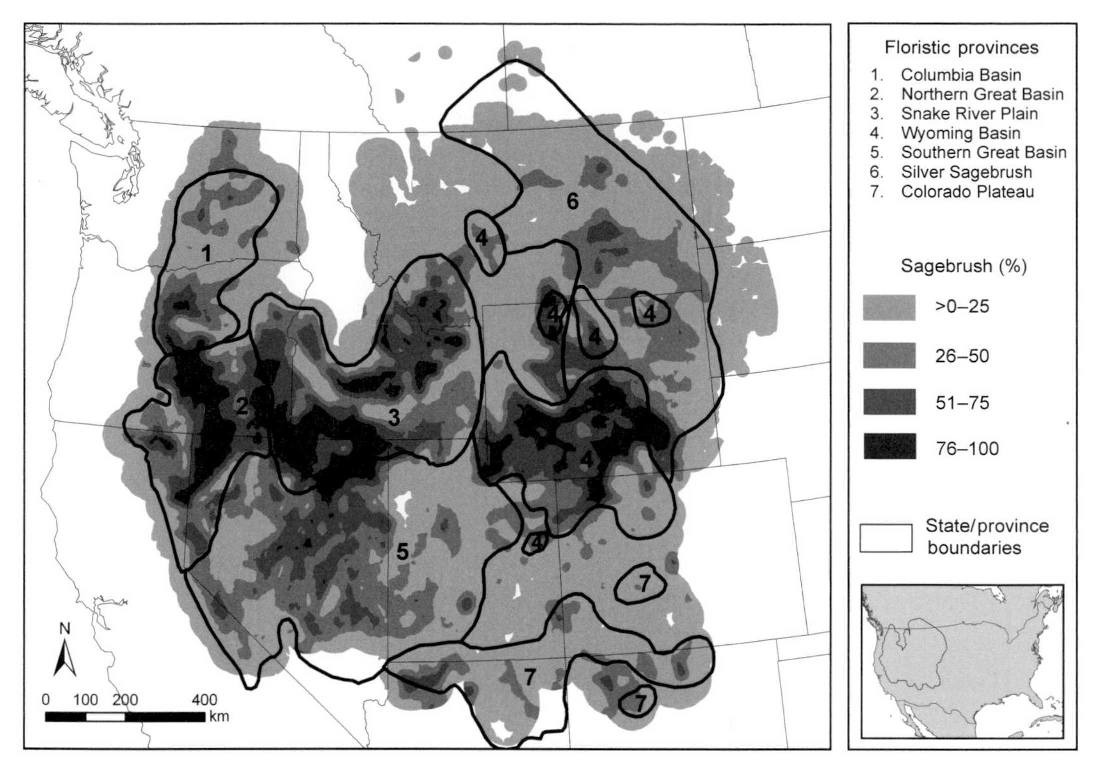

Figure 10.3. Geographic subdivisions within the sagebrush steppe are (1) Columbia Basin, (2) Northern Great Basin, (3) Snake River Plain, and (4) Wyoming Basin. The Great Basin includes (5) Southern Great Basin and (7) Colorado Plateau. The Northern Great Plains grasslands overlap the Silver Sagebrush (6) subdivision (derived from West 1983b, Küchler 1970, Miller and Eddleman 2001, and this study). Percent sagebrush habitat is the general landscape distribution of sagebrush.

silver sagebrush, and sand sagebrush. Wyoming big sagebrush is the most common subspecies of the *tridentata* group in this region, and typically occurs in marine shales and upland soils (Morris et al. 1976, Wambolt and Frisina 2002).

Prairie sagewort, a low-growing subshrub, is widely distributed and characteristic of the high plains of central North America extending west into south-central Idaho, eastern Washington, and central Utah and east throughout eastern Montana (Morris et al. 1976, Wambolt and Frisina 2002). Prairie sagewort grows in dry, open sites from plains and foothills to middle and upper elevations in the mountains up to 3,400 m and is common on disturbed sites (Bai and Romo 1996).

Plains silver sagebrush is widespread throughout the northern Great Plains at 1,200- to 2,100-m elevation and occupies well-drained, coarse-textured soils on alluvial flats, terraces, valley bottoms, and drainage ways. In Montana, plains silver sagebrush is distributed primarily throughout the central and eastern portions of the state and is the most common upright shrubby *Artemisia* species in the north and northeastern plains (Morris et al. 1976). The silver sagebrush–western wheatgrass (*Pascopyrum smithii*) type is of major importance throughout this region.

Sand sagebrush is a widespread but low-abundance species that commonly grows in dunes and coarse soils. The species has an extensive distribution across the Great Plains, southward from the Black Hills to the Texas panhandle, and west through New Mexico and Utah to Arizona and Nevada (McKean 1976). Sand sagebrush is associated with Lesser Prairie-Chicken (*Tympanuchus pallidicinctus*) habitat in Colorado, Kansas, and Oklahoma (Cannon and Knopf 1981, Pitman et al. 2006). It is important to note that, unlike the big sagebrush subspecies, both sand and silver sagebrush are capable of resprouting following fire.

CLASSIFICATION OF ECOLOGICAL COMMUNITIES WITHIN SAGEBRUSH SYSTEMS

Sagebrush habitats within the SGCA have been described and mapped using the International Classification of Ecological Communities, which

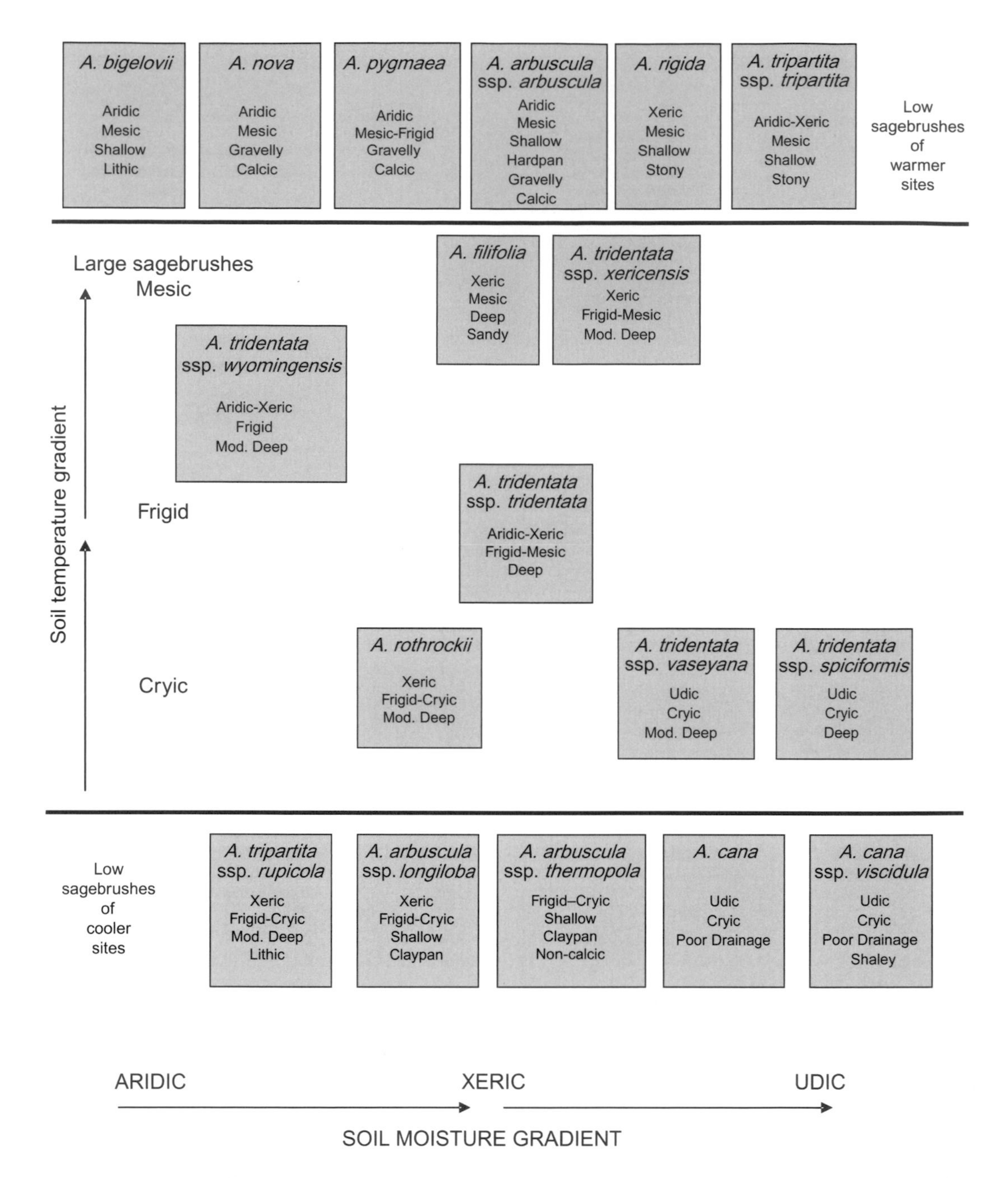

Figure 10.4. Ordination of major sagebrush taxa in the Intermountain Region against gradients of soil temperature and soil moisture (adapted from West and Young 2000; with additions from Robertson et al. 1966, McArthur 1983, and this study). Sagebrush species not shown were prairie sagewort (*Artemisia frigida*), Owyhee sage (*A. paposa*), birdfoot sagebrush (*A. pedatifida*), and bud sagebrush (*Picrothamnus desertorum*).

separates sagebrush communities based on floristics into alliances and plant associations (Reid et al. 2002). Sagebrush alliances are delineated by a species or subspecies of sagebrush and, at times, a second diagnostic shrub that may include antelope bitterbrush (*Purshia tridentata*), snowberry (*Symphoricarpos* spp.), and saltbush (*Atriplex* spp.). Rabbitbrush (*Chrysothamnus* spp. and *Ericameria* spp.), which often increase with disturbance, are typically not used as a diagnostic species distinguishing plant alliances.

Plant associations, which further separate sagebrush alliances, are usually delineated by understory perennial grass. Tussock-forming grasses typically dominate the undergrowth in the intermountain region. Common examples include Idaho fescue (*Festuca idahoensis*), bluebunch wheatgrass (*Pseudoroegneria spicata*), Thurber's needlegrass (*Acnatherum thurberianum*), needle and thread (*Hesperostipa comata*), Columbia needlegrass (*Acnatherum nelsonii*), western needlegrass (*A. occidentalis*), California brome (*Bromus carinatus*), squirreltail (*Elymus elymoides*), and Sandberg bluegrass (*Poa secunda*).

Sagebrush–grass type alliances in the northern Great Plains are composed of big sagebrush and silver sagebrush often mixed with rabbitbrush, saltbush, or winterfat (*Krascheninnikovia lanata*). The understory is dominated by rhizomatous grasses. Common understory grasses associated with sagebrush-shrub mixes are western or bluebunch wheatgrass, needle and thread, prairie Junegrass (*Koeleria macrantha*), and blue grama (*Bouteloua gracilis*). Prairie sandreed (*Calamovilfa longifolia*), sand dropseed (*Sporobolus cryptandrus*), and sand bluestem (*Andropogon hallii*) are commonly present in coarse soils, whereas alkali sacaton (*S. airoides*) is more abundant on moist bottomlands and more alkaline sites in association with greasewood (*Sarcobatus vermiculatus*). Sagebrush and other shrubs in these associations generally are more scattered and reduced in stature on the plains where wheatgrasses dominate the visual aspect. Sand sagebrush–bluestem associations commonly occur on sandy soils within the northern Great Plains. Sand sagebrush may be associated with mid and tall grasses, including sand bluestem, little bluestem (*Schizachyrium scoparium*), and prairie sandreed.

Ecological site descriptions and state-and-transitional models are currently being developed across shrublands in the western United States. These descriptions use soil physical characteristics in combination with environmental factors, hydrology, and vegetation as interrelated components to describe land units with particular kinds and amount of vegetation and their disturbances and potential successional transitions and pathways (United States Department of Agriculture 2003). In addition, ecological site descriptions incorporate temporal changes associated with management, disturbance, and subsequent plant community development on the site. Ecological site descriptions are based on physical site characteristics and recognize the transitions between vegetative physiognomic states; thus, they are particularly well suited to sagebrush-dominated landscapes. Such site descriptions allow for the variety of plant dominance possible in shrubland systems following fire, invasive species, and site disturbances. One or more sagebrush plant associations may occupy an ecological site at different times.

The structure of the sagebrush steppe and Great Basin sagebrush vegetation types is typically characterized by four layers: (1) shrubs 0.3–1.0 m tall, (2) forbs and caespitose grasses 0.2–0.6 m tall, (3) low-growing grasses and forbs <0.2 m tall, and (4) biological soil crust. Potential plant cover varies widely with soils and moisture availability ranging from considerable exposed bare ground, but it can approach 100% in wet sagebrush communities. Biological soil crusts typically increase on more arid sites where vascular plant cover is low (Ponzetti et al. 2007). These crusts are less common in the eastern plains because vascular plant cover often increases.

Annual herbaceous production varies widely within and among communities dominated by different sagebrush species and subspecies, ranging from 120 to 2,350 kg/ha (Table 10.2). This is due primarily to moisture availability during the growing season, which in turn is related to climate, soils, and topographic position. The average number of days in Nevada when soil temperature and moisture were not limiting to herbaceous plant growth across 372 range sites varied from 28–32 days on little sagebrush sites, to 50–56 days on mountain big sagebrush sites, to 130 days on meadows (Jensen 1989b).

Species composition of vascular plants in sagebrush communities is strongly influenced by moisture availability and edaphic characteristics (Passey et al. 1982, Barker and McKell 1983, Jensen et al. 1990), particularly soil texture, nutrients, and depth to Bt horizon (Davies et al. 2007a). Floristic diversity is usually considered as moderate (West 1983b). Species numbers on sites with minimal disturbance ranged from 20 in the Columbia Basin in central Washington (Daubenmire 1975a), to 13–24 in the Snake River Plain (Tisdale et al. 1965), to 54 across several sagebrush communities in Nevada (Zamora and Tueller 1973), to 24–56 in mountain big sagebrush communities in the northern Great Basin (Miller et al. 2000). Forb species usually outnumber grass species, but forbs often constitute a much smaller portion of the biomass and ground

TABLE 10.2

Ranges of annual precipitation, elevation, soil depth, and aboveground annual herbaceous production for communities with a dominant overstory of common Artemisia *species and subspecies.*

Species/subspecies	Annual precipitation (mm)	Elevation (m)	Site adaptation and soils	Annual productivity (kg/ha)
Artemisia tridentata spp. *spiciformis*	>400	2,300–3,200	High mountain areas	>1,850
A. t. spp. *vaseyana*	350–450	1,200–3,200	Moderate to deep summer moisture	1,120–3,080
A. t. spp. *tridentata*	200–400	610–2,140	Deep, dry fertile soils	868–2,350
A. t. spp. *wyomingensis*	180–300	150–2,150	Shallower to moderately deep; hotter than *A.t.t, A.t.v*	490–990
A. tripartita	300–400	1,100–2,300	Rocky knolls to moderately deep	560–1,370
A. arbuscula	200–400	700–3,780	Shallow, often alkaline, or if deep, usually an abrupt textural change between the A and B horizons	370–1,000
A. nova	200–300(400)	625–2,990	Shallow, often calcareous	440–620
A. rigida	200–400	230–2,130	Shallow, rocky scablands	120–250
A. cana spp. *cana*		up to 3,300	Well-watered, deep soils along stream bottoms and drainages	NA

SOURCE: Adapted from Miller and Eddleman 2001; and derived from Passey et al. 1982, Cronquist et al. 1994, Shiflet 1994, and Mahalovich and McArthur 2004.

cover. Mean perennial forb cover in five minimally disturbed Wyoming big sagebrush plant associations across southeast Oregon was usually <5%, compared to 10–20% cover of perennial grasses (Davies et al. 2007a). Forb abundance can be highly variable from year to year and is largely affected by amount and timing of precipitation.

TEMPORAL SCALES OF SAGEBRUSH DYNAMICS

Long-Term Dynamics

Long-term dynamics of sagebrush ecosystems extend over centuries or millennia. Pre-settlement shifts in potential natural vegetation were caused primarily by long-term changes in climate and severe disturbances (e.g., volcanic eruptions and floods), resulting in a change in plant associations, alliances, and disturbance regimes. Climate has fluctuated since the end of the Pleistocene 10,000 years before present (BP), with cooler and wetter, warmer and drier, and warmer and more humid periods (Antevs 1938, 1948; Bright and Davis 1982; Davis 1982). The duration of these periods extended from centuries to several millennia and resulted in changes in abundance between sagebrush and graminoids, and the distribution of pinyon (*Pinus* spp.), juniper (*Juniperus* spp.), sagebrush, grassland, and salt desert communities (Mehringer 1985, Mehringer and Wigand 1987, Wigand 1987).

Severe drought and major fires followed the Neoglacial during the beginning of the late Holocene (2,500 BP) in the northern Great Basin and resulted in rapid regional declines in juniper and perennial grasses, and expansion of sagebrush at the upper elevations and salt-desert shrub at lower elevations (Mehringer 1985, Mehringer and Wigand 1987, Wigand 1987). Examination of charcoal layers, pollen cores, and sediments indicate that frequent large fires in combination with climate were primary drivers of pinyon and juniper abundance and distribution (Miller et al. 2001). The Little Ice Age (700–150 years BP), which ended just prior to Eurasian settlement, was the wettest and coolest period during the last half of the Holocene. A general warming trend has occurred since the end of the Little Ice Age (ca. 1850) similar to post-Neoglacial conditions (Ghil and Vautgard 1991). However, major fires, which immediately followed the post-Neoglacial period, are in contrast to region-wide declines in fire events in the late 1800s and early 1900s that were synchronous with conifer expansion rates that exceed any that have occurred during a similar length of time in the southern and northern Great Basin (Miller and Wigand 1994, Miller and Tausch 2001, Miller et al. 2008).

Short-Term Dynamics

Short-term changes, usually calculated in years or decades, are a function of weather and disturbance (e.g., fire, diseases, molds, insects, and changes in herbivory) resulting in fluctuation or permanent change in relative abundance of species and structure of plant communities. Short-term climatic cycles measured in years can affect plant community dynamics, particularly in combination with disturbance, through influencing plant succession, annual abundance and diversity of plant species, and length of the growing season. Vegetation composition and structure that is most persistent through time within a defined climate regime or ecological site is often affected by severity and frequency of disturbance events (Miller and Heyerdahl 2008). There are two potential outcomes resulting from disturbance or climatic change: (1) plant communities shift within their range of natural variability (e.g., succession from one phase to another within a steady state), or (2) they cross a threshold and shift to a new steady state. A phase is defined as a plant community within a state that is hypothesized to replace other communities along traditional succession-retrogression pathways; succession from one community to the next is readily reversible over short time periods (years to decades) without management intervention because they are not separated by thresholds. However, an at-risk community phase may not progress directly to the most resilient community phase without passing through an intermediate phase. A state is a suite of plant community successional phases occurring on similar soils that interact with the environment to produce resistant functional and structural attributes with a characteristic range of variability maintained through autogenic repair mechanisms. Shifts between multiple stable states represent a transition across a threshold that requires large changes to return the site to the previous state (Westoby et al. 1989b, Bestelmeyer et al. 2003, Briske et al. 2003).

CURRENT AND POTENTIAL DISTRIBUTION OF SAGEBRUSH HABITATS

Accurate estimates of the amount of sagebrush habitat that has been lost from what was present during pre-settlement are not possible because of our inability to map the historical distribution with an accuracy or resolution comparable to that in modern satellite image data. Therefore, we estimated the difference between the area that could be dominated by sagebrush in Küchler's (1970) potential vegetation map for Great Basin, sagebrush steppe, and wheatgrass-needlegrass shrub steppe (Fig. 10.2) to the current distribution of sagebrush habitats (Fig. 10.1). Küchler's map depicts the vegetation that would occur if there were no disturbances from humans or nature (Küchler 1964). Therefore, our analysis compared the difference between the vegetation type that could potentially occur and what currently was present (LANDFIRE 2006). We used only regions included in Great Basin sagebrush, sagebrush steppe, and wheatgrass-needlegrass shrub steppe types. We recognize that some of the difference is a function of the coarse resolution in Küchler's map compared to the finer resolution in the sagebrush map. We subtracted forested, water, marsh, and wetland habitats delineated in the map of current habitats (LANDFIRE 2006) from the total area for each sagebrush type in Küchler's (1970) map to partially correct for differences in thematic and spatial resolution. We emphasize that the analysis only identified broad-scale differences between current and potential distribution in sagebrush and was not intended to identify specific locations where sage-grouse habitat had been lost. Additionally, because of scale differences between Küchler's (1970) map and our habitat map, we cannot clearly distinguish where Küchler's broad delineations may have masked smaller parcels of nonhabitat.

Fifty-five percent of the area delineated on Küchler's maps as potentially dominated by sagebrush across Washington, Montana, Wyoming (sagebrush habitats in eastern portions of Montana and Wyoming were not included), Idaho, Oregon, Nevada, Utah, California, and Colorado are currently occupied by sagebrush (Fig. 10.5, Table 10.3). Wyoming (66%) and Oregon (65%) have the highest portion of potential area that is currently mapped as sagebrush, whereas Utah (38%) and Washington (24%) had the lowest. Within the areas not currently mapped as sagebrush, agriculture made up the largest category of land cover (10% of the potential area) and was the dominant land cover within potential sagebrush areas in Washington (42%) and Idaho (19%). Urban areas covered 1% of the potential sagebrush areas. The remaining 31% of potential sagebrush vegetation was mapped as barren, grassland, burn, exotic grassland, shrubland, and juniper woodland.

Previous estimates of potential sagebrush vegetation currently designated as urban, agriculture, or converted to land-cover types that no longer can support sagebrush vegetation were 3% for the Great Basin sagebrush type, 5% for wheatgrass-needlegrass-shrub steppe, and 15% for sagebrush steppe (Klopatek et al. 1979). Based on updated maps of urban and agriculture areas (and corrected for other nonsagebrush habitats), we estimated that 5% (6,293 km^2) of the area occupied by Great Basin sagebrush now was in agriculture, urban, or industrial areas; 46% (63,379 km^2) still supported sagebrush; and 49% (67,635 km^2) was dominated by other vegetation types. In the wheatgrass-needlegrass-shrub steppe, 5% (1,339 km^2) of the potential area that could support sagebrush

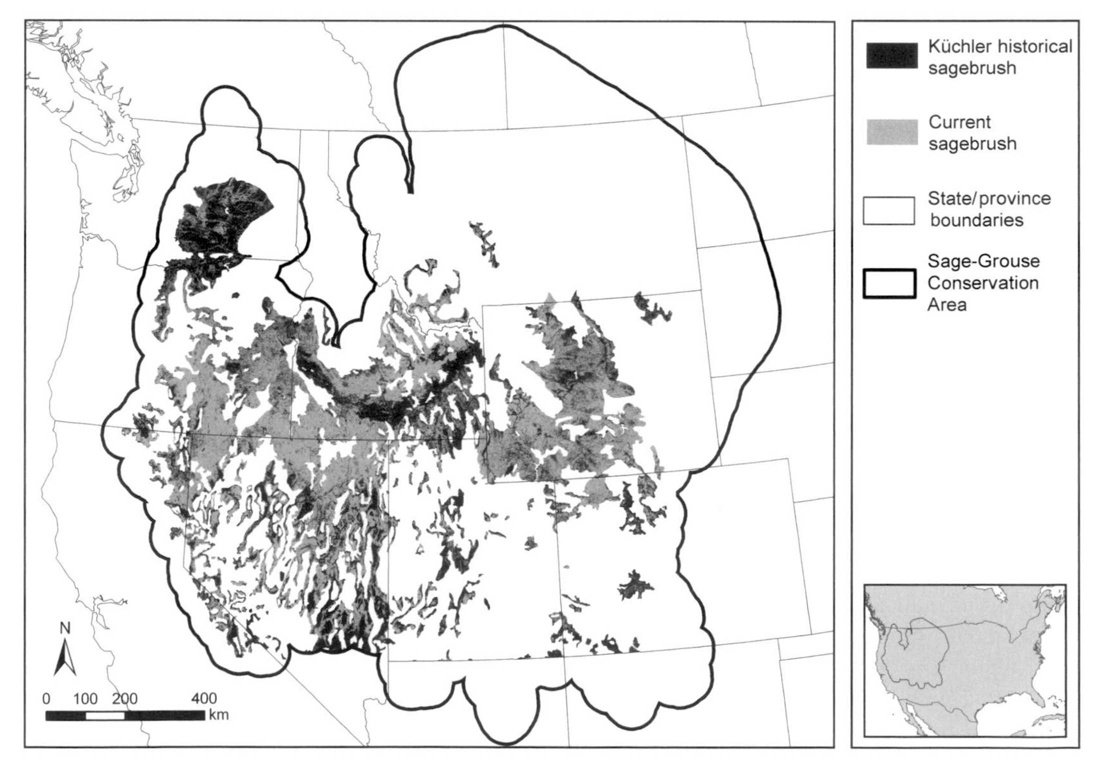

Figure 10.5. Difference between Küchler's map (1970) of potential sagebrush distribution (Fig. 10.2) and current distribution of sagebrush (Fig. 10.1). Only the distribution of Küchler's categories for Great Basin sagebrush, sagebrush steppe, and wheatgrass-needlegrass shrub steppe are used in this analysis. Sagebrush habitats also exist outside of the distribution of these habitat types.

TABLE 10.3

Percent of sagebrush in Küchler's (1970) map of potential vegetation (Fig. 10.2) that is currently in sagebrush habitat (Fig. 10.1).

Only comparisons for Küchler's categories of Great Basin sagebrush (type 38), sagebrush steppe (type 55), and wheatgrass-needlegrass shrub steppe (type 56) were used in this analysis. The remaining percentage consisted of habitat categories describing barren, burn, grassland, exotic grassland, non-sagebrush shrublands, and juniper woodland habitats (derived from the LANDFIRE 2006 Existing Vegetation Map).

		Current condition	
State	Potential area (km^2)	Sagebrush km^2 (%)	Agriculture km^2 (%)
Washington	34,478	8,158 (23.7)	14,618 (42.4)
Montana[a]	11,089	6,229 (56.2)	834 (7.5)
Wyoming[a]	99,642	65,809 (66.0)	3,391 (3.4)
Idaho	96,162	52,853 (55.0)	17,896 (18.6)
Oregon	76,107	49,062 (64.5)	6,582 (8.6)
Nevada	135,406	79,540 (58.7)	1,797 (1.3)
Utah	34,976	13,147 (37.6)	3,399 (9.7)
California	18,931	9,419 (49.8)	1,513 (8.0)
Colorado[a]	22,005	8,932 (40.6)	2,560 (11.8)
Total	528,796	293,194 (55.4)	52,619 (10.0)

[a] Does not include sagebrush lands in eastern portions of the state outside the boundaries of the Küchler's vegetation types used in this analysis.

vegetation has been converted to agriculture, urban, or industrial habitats; 59% (15,864 km^2) currently was mapped as sagebrush, and 36% (9,601 km^2) was occupied by other vegetation types. In the sagebrush steppe, 14% (52,865 km^2) of the area that could potentially support sagebrush has been converted to agriculture, urban, or industrial categories; 59% (213,908 km^2) was occupied by sagebrush habitat; and 27% (97,913 km^2) has been converted to other vegetation types. These analyses are based only on dominant land cover across large regions. Information about understory, soil, and other characteristics not mapped in satellite imagery or captured in coarse-resolution maps is not implied.

PRIMARY DISRUPTERS OF SAGEBRUSH HABITAT DYNAMICS

Invasive Species

A broad array of invasive plants influences the structure and function of habitats used by Greater Sage-Grouse (Table 10.4). An invasive species is defined as an exotic or native species that is non-native to the specific ecosystem under consideration and whose introduction causes or is likely to cause economic or environmental harm or harm to human health (Clinton 1999); this definition also includes species native to other parts of North America; however, species that occur within the region are classified as increasers. Sheley and Petroff (1999) list 29 species of rangeland weeds. The Interior Columbia Basin Ecosystem Management Project compiled a similar list of 25 species (United States Departments of Agriculture and the Interior 1997a,b). With the exception of the snakeweeds (broom snakeweed [*Gutierrezia sarothrae*], threadleaf snakeweed [*G. microcephala*], and poison hemlock [*Conium maculatum*]), which are native to the western United States (Sterling et al. 1999) and considered increasers, the remaining species on these lists are defined as nonnative invasive in one or more of the sagebrush communities important to Greater Sage-Grouse. No scientific reports, models, or maps currently exist to provide a comprehensive list of the susceptibility of habitats within the SGCA to invasion by these weeds. Estimates of susceptibility (Table 10.4) to

TABLE 10.4

Major vegetation cover types within the Sage-Grouse Conservation Area and their susceptibility to invasion by nonindigenous plant species.

Susceptibility to invasion is defined by four categories: (H) high[a], (M) moderate[b], (L) low[c], and (U) unknown[d].

	Upland communities				
Invasive species	Basin big sagebrush, Wyoming big sagebrush, three-tip sagebrush	Mountain big sagebrush	Low sagebrush, black sagebrush	Salt-desert shrub	Wheatgrass, bunchgrass
Cheatgrass (*Bromus tectorum*)	H	M	M	M	H
Musk thistle (*Carduus nutans*)	U	M	U	M	M
Whitetop (*Cardaria* spp.)	M	M	M	M	M
Diffuse knapweed (*Centaurea diffusa*)	M	M	M	L	H
Spotted knapweed (*C. maculosa*)	M	M	U	L	H
Russian knapweed (*C. repens*)	M	M	U	M	M
Yellow starthistle (*C. solstitialis*)	M	M	M	L	H
Squarrose knapweed (*C. virgata*)	M	M	M	M	M
Rush skeletonweed (*Chondrilla juncea*)	M	M	U	L	M
Oxeye daisy (*Chrysanthemum leucanthemum*)	U	U	U	L	M
Canada thistle (*Cirsium arvense*)	M	M	M	M	H
Bull thistle (*C. vulgare*)	M	M	M	M	M
Poison hemlock (*Conium maculatum*)	L	L	L	L	L
Common crupina (*Crupina vulgaris*)	L	M	L	L	M
Leafy spurge (*Euphorbia esula*)	M	L	M	M	M
Halogeton (*Halogeton glomeratus*)	M	M	M	H	M
Orange hawkweed (*Hieracium aurantiacum*)	L	M	L	L	L
Meadow hawkweed (*H. pratensis*)	L	L	L	L	L

Dyer's wood (*Isatic tinctoria*)	H	L	H	L	H
Perennial pepperweed (*Lepidium latifolium*)	L	L	L	L	L
Dalmation toadflax (*Linaria dalmatica*)	M	H	M	L	H
Yellow toadflax (*L. vulgaris*)	M	M	U	L	M
Purple loosestrife (*Lythrum salicaria*)	L	M	L	L	L
Scotch thistle (*Onopordum acanthium*)	M	L	U	L	M
Sulphur cinquefoil (*Potentilla recta*)	U	M	U	L	H
Mediterranean sage (*Salvia aethiopis*)	H	M	U	L	H
Russian thistle (*Salsola kali*)	M	M	L	M	M
Tansy ragwort (*Senecio jacobaea*)	U	U	U	U	U
Sowthistles (*Sonchus* spp.)	M	M	M	M	M
Medusahead (*Taeniatherum caput-medusae*)	M	M	L	M	M

SOURCE: Compiled from United States Departments of Agriculture and the Interior 1997b, Sheley and Petroff 1999.
[a] Invades the cover type successfully and becomes dominant or codominant even in the absence of intense or frequent disturbances.
[b] Invades the cover type successfully because high intensity or frequency of disturbance alters the soil surface or removes the normal canopy cover.
[c] The species typically does not invade the cover type because the cover type does not provide suitable habitat for the species.
[d] Distribution records are limited and interpretation of the susceptibility would be difficult.

invasion are based upon the knowledge of experts and written descriptions of the types of vegetation communities where infestations or colonization populations currently exist.

Estimates of the size of infestations of any of these species are subjective because of the lack of a definition of what constitutes an infestation. For example, cited estimates of diffuse (white) knapweed (*Centaurea diffusa*) infestation in Idaho ranged from 410 to 5,670 km^2 (Roché and Roché 1999). Thus, it is extremely difficult to ascertain a reasonable estimate of the area of lands currently occupied or level of dominance by invasive plants within any area across the range-wide distribution of Greater Sage-Grouse. Sheley and Petroff (1999) and United States Departments of Agriculture and the Interior (1997a,b) relied heavily on distribution maps of counties in the five-state area (Idaho, Montana, Oregon, Washington, and Wyoming) covered by the Invaders Database (Rice 2004), where counties are considered occupied by the plant if at least one occurrence of a species has been recorded and verified through herbarium collections or reports. Sheley and Petroff (1999) extended these maps into surrounding states, but we were unsuccessful in obtaining and verifying their data.

Many of the species listed (Table 10.4) may be widely distributed across the SGCA, but infestations are localized because of the narrow environmental needs of the invasive species. For example, diffuse knapweed is estimated to reach its greatest competitiveness within shrub-grassland communities where antelope bitterbrush may dominate or codominate along the eastern side of the Cascade Range in Washington (Roché and Roché 1999).

Invasions into native plant communities also may be sequential, as the initial invaders are replaced by a series of new exotics or by species adapting to new habitats within their range (Young and Longland 1996). Areas that once were dominated by cheatgrass (*Bromus tectorum*) in some locations along the Snake River Plain and the Boise Front Range in Idaho have been replaced by medusahead (*Taeniatherum caput-medusae*). Rush skeletonweed (*Chondrilla juncea*), which originally was localized to disturbed areas in the drier sagebrush-grassland communities, now is invading areas dominated by medusahead (Sheley et al. 1999) and following wildfire (Kinter et al. 2007).

Exotic Annual Grasses

Cheatgrass and medusahead have become the most problematic of the exotic annual grasses within the SGCA. These Eurasian annual grasses were introduced in the 1890s and have continued to expand their range (Mack 1981). Both grasses are winter annuals that rely on winter precipitation to invade and dominate lands. They tend to be more dominant in the Intermountain West (Washington, Oregon, Idaho, Nevada, and Utah) than in the Rocky Mountain states that receive more summer precipitation (parts of Montana, Wyoming, and Colorado can have local infestations of cheatgrass or other *Bromus* species). In particular, cheatgrass invasion can result in a dominant near-monoculture in the more arid, lower-elevation, Wyoming big sagebrush communities (Chambers et al. 2007). Annual-dominated communities can be considered a new steady state (Laycock 1991) over much of eastern Washington, eastern Oregon, southern Idaho, Nevada, and Utah.

Cheatgrass likely was first introduced within the intermountain region of the United States via contaminated imports of grains and expanded along transportation routes and in locations of documented severe livestock grazing and reductions in native perennial grasses (Young and Evans 1973, Mack 1981). Cheatgrass reached most of its current range expansion during the 1930s (Billings 1990) but has continued to expand southward into the Mojave Desert (Hunter 1991) and eastward.

Cheatgrass has been a major factor in loss of Wyoming big sagebrush communities (Chambers et al. 2007). Medusahead is filling a similar niche in more mesic communities with heavier clay soils (Dahl and Tisdale 1975). Since the initial occurrences in the late 1800s and early 1900s, medusahead has continued to spread and occupy new locations (Miller et al. 1999). These communities now include little sagebrush and mountain big sagebrush communities, as well as some Wyoming big sagebrush communities at lower elevations.

Cheatgrass Distribution in the Intermountain Western United States

One estimate of land area in sagebrush ecosystems dominated by introduced annual grasses comes from a qualitative survey conducted by the Bureau of Land Management (BLM) in 1991 (Pellant and

Hall 1994). This survey covered 400,000 km^2 of BLM-managed lands in Washington, Oregon, Idaho, Nevada, and Utah. Cheatgrass and medusahead now either dominate or have a significant presence (estimated >10% composition based on biomass) on 70,000 km^2 of public land within these five states.

Recent surveys using combinations of field data and remote imagery estimated that cheatgrass now dominates >20,000 km^2, or 7% of land cover, within a portion of the northern Great Basin (Bradley and Mustard 2005, 2006; Peterson 2005). Whisenant (1990) indicated that cheatgrass has become a major herbaceous species in the West, dominating over 400,000 km^2, but his estimate was actually a major overestimate and a misinterpretation of the original citation that indicates cheatgrass now dominates on many rangelands within 410,000 km^2 of potential steppe vegetation in the intermountain western United States (Mack 1981). This incorrect figure has been repeated in other prominent review papers on the topic (d'Antonio and Vitousek 1992).

We mapped the probability of presence of cheatgrass within five floristic provinces in the Intermountain West. Using environmental variables from field surveys in Washington, Oregon, Idaho, Utah, and Nevada (Connelly et al. 2004), we developed a statistical function predicting the probability of presence by cheatgrass using logistic regression and Akaike's Information Criterion to select the best model (Burnham and Anderson 2002). Initial predictor variables in the model were elevation (m), slope (%), aspect (degree), annual precipitation (cm), depth to rock (cm), soil pH, salinity (mmhos/cm), and available water capacity (cm) (Meinke et al. 2009). We mapped the probability of cheatgrass presence across each floristic province based on the values of predictor variables within each 2-km grid cell (Fig. 10.6). A moderate to high probability of presence by cheatgrass was predicted for 281,000 km^2, or almost half of the area within the mapped region of the Intermountain West (Fig. 10.6) (Meinke et al. 2009). Approximately 65% the Great Basin ecoregion (1,500,000 km^2) has environmental conditions suitable for moderate to high risk of cheatgrass invasion; 38% of the existing sagebrush was at moderate risk and 20% at high risk (Suring et al. 2005b).

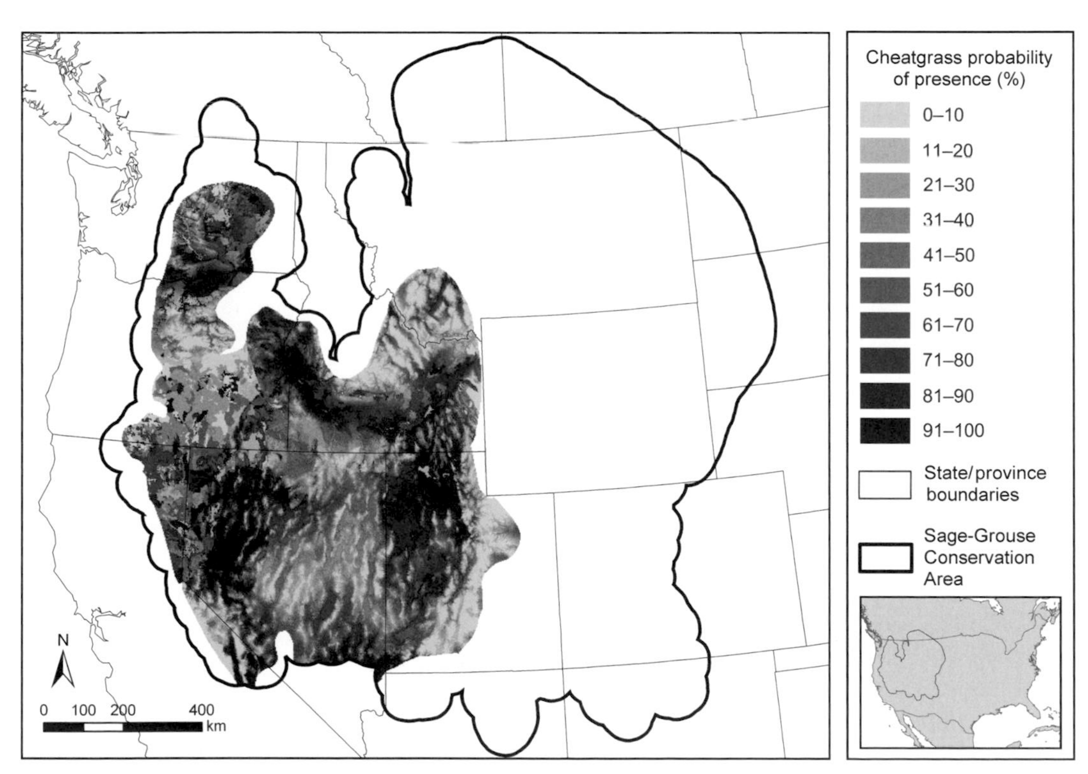

Figure 10.6. Predicted distribution of cheatgrass modeled from logistic regressions of presence/absence of cheatgrass and environmental factors measured at survey points (Meinke et al. 2009).

Postsettlement Woodland Expansion

Utah juniper (*Juniperus osteosperma*), western juniper (*J. occidentalis*), single-leaf pinyon (*Pinus monophylla*), and two-needle pinyon (*P. edulis*) are the primary conifer species occurring in the SGCA and are encroaching and infilling into large portions of sagebrush steppe–dominated communities at higher elevations (Cottam and Stewart 1940, Tausch et al. 1981, Knapp and Soulé 1998, Tausch and Nowak 1999, Miller and Tausch 2001, Weisberg et al. 2007, Miller et al. 2008). To a lesser extent, Rocky Mountain juniper (*J. scopulorum*) also is expanding into sagebrush communities in portions of its range. The increase in woodlands after Euro-American settlement, which began in the late 1800s, is occurring at rates exceeding that of any expansions during the Holocene (Miller and Wigand 1994). The expansion of pinyon and juniper co-occurred with introduction of livestock and surface fire exclusion (Tausch et al. 1981, Miller and Rose 1999, Miller and Tausch 2001, Swetnam et al. 2001).

Cover of sagebrush and other associated shrubs decline with increasing tree dominance (Adams 1975, Miller et al. 2000, Roberts and Jones 2000, Schaefer et al. 2003)(Fig. 10.7). Juniper and pinyon woodlands occupy approximately 189,000 km^2 in the Intermountain West (Miller and Tausch 2001). As much as 90% of the areas currently dominated by pinyon and juniper in the sagebrush steppe and Great Basin sagebrush vegetation types were predominantly persistent sagebrush vegetation types prior to the late 1800s (Tausch et al. 1981, Johnson and Miller 2006, Miller et al. 2008). However, the proportion varies spatially across the SGCA. A greater proportion of extensive old woodlands occupy the Colorado Plateau (Floyd et al. 2000, 2008; Eisenhart 2004) and the Mazama ecological province in central Oregon (Waichler et al. 2001, Miller et al. 2005). The greatest proportion of woodland encroachment has occurred at higher elevations in mountain big sagebrush plant associations and little and black sagebrush plant associations that occur on moderate to deep soils. Expansion and infill have also occurred in little and black sagebrush plant associations. Millions of hectares of potential sagebrush vegetation types are considered at high risk of displacement by juniper and pinyon woodlands (Suring et al. 2005b) by both encroachment and infill (Weisberg et al. 2007). We have limited documentation, but other conifer species such as Douglas fir (*Pseudotsuga menziesii*) have been expanding into mountain big sagebrush in Montana (Sindelar 1971, Dando and Hansen 1990, Hansen et al. 1995, Heyerdahl et al. 2006). Pinyon and juniper currently occupy far less land

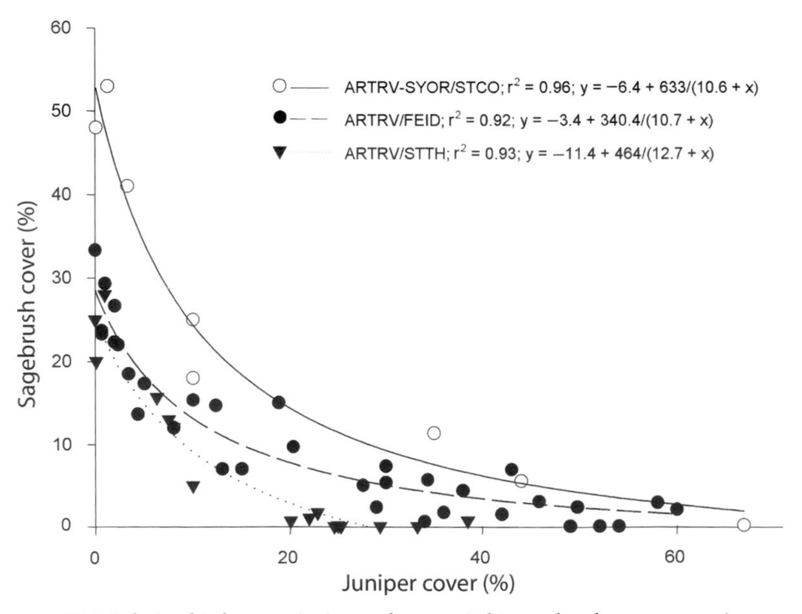

Figure 10.7. Relationship between juniper and mountain big sagebrush canopy cover in three plant associations: *Artemisia tridentata* ssp. *vaseyana-Symphoricarpos oreophilus/Stipa columbiana* (ARTRV-SYOR/STCO), *A. tridentata* ssp. *vaseyana/Festuca idahoensis* (ARTRV/FEID), and *A. tridentata* ssp. *vaseyana/Stipa thurberiana* (ARTRV/STTH) (Miller et al. 2000).

than their potential under current climatic conditions (Betancourt 1987, West and Van Pelt 1987, Miller et al. 2001). In addition, many of these woodlands are in different phases of development, where tree densities and cover are continuing to increase, resulting in the continued loss of sagebrush communities (Miller et al. 2000, 2008).

Mechanisms of Ecosystem Disruption

The effect of invasive species may be evaluated in how they alter the community structure (type, number, and relative abundance of species) and ecological function (nutrient, energy, and water cycles). In addition, invasive plant species can influence or disrupt the functional relationships among organisms in a community.

The altered structure of plant communities that results when invasive species replace native plant species can have significant consequences on the dynamics of sagebrush ecosystems. For example, the invasion of cheatgrass into the Wyoming big sagebrush alliance changes the structure of the understory to provide more complete and continuous ground cover in comparison to sparse, discontinuous cover of native perennial grasses that often occur in this alliance (Klemmedson and Smith 1964, Billings 1990). Consequently, these continuous and extremely flammable fuels result in fires in cheatgrass-dominated systems that are more frequent and often less complex, with few unburned patches. Sagebrush species are intolerant of fire and easily killed, with the exception of threetip (*Artemisia tripartita*), silver, and sand sagebrush, which are sprouters (Billings 1990, West and Young 2000). Reestablishment of most sagebrush species following fire depends on seed from the soil seed bank and dissemination from unburned areas in addition to appropriate conditions for germination and establishment (Hemstrom et al. 2002, Lesica et al. 2007). Seeds of mountain big sagebrush can remain in the seed bank over multiple years, compared to the seeds of Wyoming big sagebrush, which are not viable for more than 1 year unless covered by soil (Wijayratne and Pyke 2009). Dissemination of sagebrush seed is primarily from wind, the majority falling within 9–12 m (and under rare conditions, up to 33 m) of the parent plant (Blaisdell 1953, Mueggler 1956, Johnson and Payne 1968, Daubenmire 1975b, Frischknecht 1978). More complete fires in cheatgrass-dominated communities result in fewer, more widely dispersed seed sources in remaining unburned islands. Cheatgrass is highly competitive for resources in the community, making it difficult for new perennial grass and shrub seedlings to establish (Harris 1967, Francis and Pyke 1996, Beckstead and Augspurger 2004, Chambers et al. 2007). Consequently, the primary mechanisms to reestablish native plants are severely compromised or eliminated. A sequence is started in which native species die, are not replaced and possibly eventually eliminated from the species pool.

A second structural change that may occur with infestations of invasive plants is a change in life-forms represented in the community. Elimination of woody plants, such as sagebrush, may be more permanent in communities dominated by annual grasses than if the herbaceous understory consisted of native perennial bunchgrasses and forbs. This structural alteration becomes extremely important for sagebrush-obligate animals that require sagebrush for nesting or sparse ground cover for foraging (Rotenberry 1998, Dobkin and Sauder 2004, Knick et al. 2005, Suring et al. 2005a, Beck et al. 2009).

A third form of structural change can occur belowground in the form of the composition and distribution of roots and mycorrhiza, which may affect nutrient cycling and organic matter content in the soil. Shrub grassland communities that are dominated by perennial plants have a mixture of shallow and deep roots, which are associated with mycorrhiza, and have varying forms of carbon that decompose at different rates. As a native shrub grassland community is converted to a community dominated by cheatgrass with few mycorrhiza associations, which has roots that contain fewer structural cells and decompose more quickly than woody plant roots, the distribution of amount and type of organic matter in the soil is changed. Surface soils of cheatgrass communities tend to contain more of the easily decomposed organic matter, while little of the recalcitrant forms of organic matter are found at any depth. In native shrub grasslands, less of the easily decomposed form exists near the surface and more of the recalcitrant form occurs throughout the soil depth (Norton et al. 2004).

Functional relationships may also change with conversion of a community from a diverse native plant system to dominance by an invasive plant. Postfire communities dominated by cheatgrass become net sources of carbon to the atmosphere because of decreased net carbon exchange compared to native sagebrush communities, which are net carbon sinks (Prater et al. 2006). Communities dominated by cheatgrass also have lower evapotranspiration compared to native sagebrush systems; lower soil surface moisture and increased surface temperatures can increase the potential severity of summer droughts and stress to native plants (Prater et al. 2006). Shifts from fibrous to taprooted forb species can result in reduced water infiltration in some soils (Tisdall and Oades 1982).

The native sagebrush steppe and Great Basin sagebrush mosaics important for Greater Sage-Grouse often are characterized by a discontinuous spatial arrangement of perennial plants within the community. The interspaces among perennial plants may be partially filled with biological soil crusts (lichens, mosses, and cyanobacteria; Johansen 1993, Belnap and Lange 2001, Belnap et al. 2006, Muscha and Hild 2006), but percent cover can be highly variable across the SGCA. The cover of crusts in ungrazed Wyoming big sagebrush communities was <12% in Wyoming (Muscha and Hild 2006); crusts were more prominent on finer-textured soils and in more arid environments, where perennial plants tend to be widely spaced. Communities with wide spatial arrangements of perennial plants or with woody plants tend to concentrate soil nutrients around these plants. This creates a heterogeneous distribution of nutrients, with resource-rich patches surrounding perennial plants and resource-poor interspaces (Charley and West 1977, Doescher et al. 1984, Bolton et al. 1993, Jackson and Caldwell 1993, Halvorson et al. 1995, Ryel et al. 1996). In addition, soil surface temperatures are moderated and soil water availability is greater beneath sagebrush canopies compared to the interspaces (Davies et al. 2007b).

The shift from a native shrub grassland community to a near monoculture of annual invasive grasses changes the temporal availability of water. Stands of cheatgrass reduced growth of native perennial plants, which was a function of significant reduction in water availability and reduced native plant water content (Melgoza et al. 1990, Booth et al. 2003). Paradoxically, conversion of sagebrush habitats to a cheatgrass monoculture can close the community to reestablishment of native plants (Robertson and Pearse 1945), but open the community to subsequent invasions by other exotics, such as medusahead, Russian thistle (*Salsola kali*), and rush skeletonweed (Kinter et al. 2007), and can lead to dominance by perennial weed species (Young and Longland 1996, Shaw and Hild 2007).

There is considerable speculation regarding the effect of cheatgrass on nitrogen cycles and how nitrogen levels relate to native plant maintenance and establishment in communities. Evans et al. (2001) speculated that cheatgrass monocultures would lead to reduced nitrogen availability at a site, but others have found no evidence for this relationship (Bolton et al. 1990, Svejcar and Sheley 2001). Increased native perennial establishment with reduced nitrogen availability has occurred in sagebrush ecosystems in northwestern Colorado and northwestern Nevada (McLendon and Redente 1990, 1992; Young et al. 1997), but whether this is the primary functional driver of succession or a secondary driver associated with water uptake (Booth et al. 2003) is unclear.

Maintaining resistance to cheatgrass invasion may be a function of containing nitrogen and other resources within the native plant community (Beckstead and Augspurger 2004, Humphrey and Schupp 2004). Cheatgrass and other invasive species are characterized as benefiting from disturbance, but a disturbance event is not required for cheatgrass germination and dominance to occur (Roundy et al. 2007). Thus, the primary effect of disturbance may be expressed primarily through reductions in native plant communities that release resources better exploited by invasive plants (Chambers et al. 2007) and allow invasion into more resistant plant communities. Cheatgrass then can dominate the site until resources become limiting and perennial species can reestablish dominance (Mata-González et al. 2008).

Wildfire

Characteristics of Fire and Fire Regimes

Characteristics of individual fire events as well as the collective fire regime are important drivers of structure, composition, and abundance of vegetation within sagebrush communities. At broader spatial scales, fire events and regime are dominant determinants of habitat configuration within the landscape. Individual fires are described by severity (the level of biological and physical effect of fire on all plant layers, soils, and animals), intensity (the amount of energy released during a fire), season, extent or size, and complexity (patchiness of burned and unburned areas within the fire boundary).

Severity in forestry terms is defined as the percent mortality on the overstory vegetation layer and does not consider the influence on understory vegetation layers, soils, and other fauna. Fire regime is a function of the mean and range of the interval (usually in years) between fire events for a defined area. The fire regime for a specific area is influenced by climate, regional location, fuel characteristics (biomass and structure), and recovery time following disturbance, topography, season and frequency of ignition, and vegetation composition.

Relative frequency of fire in sagebrush systems has been estimated using composite fire intervals (CFI) and natural fire rotation (NFR)(Heinselman 1973; Baker, this volume, chapter 11). Both estimation methods are critically dependent on the spatial and temporal period over which they are computed. Each provides a perspective on the role of fire in sagebrush systems that must be interpreted from the appropriate scale.

Composite fire intervals are derived by documenting burns occurring at individual sites that are typically several hectares or less in size. Composite fire intervals can reveal fine-scale variation in fire frequency in both time and space within a specific landscape. Heterogeneity in fire occurrence at fine scales can have important ecological consequences related to seed dissemination, succession, rate of recovery, diversity, landscape complexity, and habitat suitability. Each of these small areas is often within a single plant association or ecological site, allowing evaluation of fire occurrence within these ecological units. Historical fire frequencies can vary at relatively fine scales (tens to hundreds to thousands of hectares) across a landscape in at least some sagebrush systems, affecting the spatial distribution of vegetation (Heyerdahl et al. 2006, Miller and Heyerdahl 2008). Small fires or fires with moderate to high complexity can have important ecological consequences for vegetation. In addition, site-specific estimates can place local fire intervals within a historic range of variation specific to a location. Computer simulations indicate that CFI can be an accurate estimator of mean fire interval (Van Horne and Fulé 2006, Parsons et al. 2007).

The probability of fire occurrence estimated from NFR substitutes space for time by incorporating multiple burns within landscapes or regions. Natural fire rotation does not directly consider variation across space or time (Heinselman 1973, Agee 1993, Reed 2006) but instead estimates the probability that, on average, a location will burn within a year (e.g., a fire rotation of 100 years means that fire will burn the entire landscape over a 100-year period and that each point in the landscape will burn, on average, once during that period; Baker 2006b). Large fires dominate the computation of NFR and are best computed for an area that exceeds the largest expected fire in one rotation. Broad estimates at regional scales present differences in dominant disturbances (e.g., fire is a dominant and frequent disturbance in the Snake River Plain compared to the Wyoming Basin) or relative changes within regions between periods used for estimating NFR (Knick and Rotenberry 2000; Baker, this volume, chapter 11).

Pre–Euro-American Dynamics

A clear picture of the complex spatial and temporal patterns of historic fire regimes in many sagebrush communities prior to Euro-American settlement is unlikely. The vast variation in fuel composition and structure, landscape heterogeneity, ignition from aboriginal and lightning sources, weather, and topography contained within the SGCA makes this challenge even more daunting. We can only estimate the potential of different vegetation phases, plant associations, and ecological sites, each on unique landscapes to burn based on proxy data, which include variables that drive fire regimes. Even then, pre–Euro-American settlement fire regimes reconstructed from cross-dated fire scars across the SGCA are few and spatially limited.

The magnitude of aboriginal burning and its impacts on western landscapes prior to the effects of Euro-American diseases and settlement is highly controversial; literature accounts range from limited (Baker 2002, Vale 2002) to significant (Keeley 2002, Kay 2007). Lightning, a primary ignition source, varied greatly in the Intermountain West and is influenced by regional location, topographic variability, and moisture availability (Knapp 1997). Ignitions from lightning increased moving from east to west toward the eastern slopes of the Sierra and Cascade Mountains, with increased variation in elevation (Knapp 1997) and increasing elevation (Gruell 1985). Early explorers observed fires in higher elevations but seldom reported fires in the sagebrush valleys at lower elevations (Gruell 1985).

The temporal dynamics of regional fire occurrences and shifts in vegetation over long time periods (5,000 years) in the SGCA have been based on abundance of charcoal and ash collected in pond and lake sediment cores in the Great Basin (Mehringer 1985, Mehringer and Wigand 1987, Wigand 1987). These data imply that fire occurrence in the surrounding area increased during relatively wet periods, which increased fuel abundance and resulted in distinct long-term patterns. Recent macroscopic charcoal work in central Nevada also correlated an increase in fire occurrences in Wyoming big sagebrush during periods of wetter climate (Mensing et al. 2006).

Most estimates of fire-return intervals in drier Wyoming big sagebrush communities were based on opinion and circumstantial evidence (e.g., estimated time for sagebrush to reestablish), rather than on experimental data. Fire-return intervals of up to a century were recorded in Wyoming big sagebrush communities located in valley bottoms in central Nevada and varied with climate and fuel accumulation (Mensing et al. 2006). More arid sites in the Wyoming big sagebrush alliances characterized by long fire-return intervals also may have had an extremely wide historic range of variation in years between fires, possibly resulting in a range of several burns in a single century to not burning at all in more than 100 years (Fig. 10.8). High variability in fire occurrence in both time and space illustrates the difficulty in describing fire regimes as well as the complexity of this disturbance as a mechanism in structuring sagebrush communities.

Fire generally was more common in mountain big sagebrush at higher elevation sites. Mean composite fire intervals in southwestern Montana were estimated to be 32 years in mountain big sagebrush communities (Lesica et al. 2007). Composite fire intervals for ecological sites supporting mountain big sagebrush and an Idaho fescue understory that were adjacent to large stands or surrounding small patches of ponderosa pine (*Pinus ponderosa*) or Douglas fir were as frequent as 10–35 years (estimated from 1- to 10-ha plots) (Miller and Rose 1999, Heyerdahl et al. 2006, Miller and Heyerdahl 2008). Fire-return intervals within small sample plots (1 to several hectares) systematically located across a 1,030-ha study area experienced fires prior to livestock grazing every 2–84 years (Heyerdahl et al. 2006). Communities with relatively short intervals (<20 years) likely would have been predominantly grasslands with scattered patches of shrubs (Fig. 10.8). These areas are typically in moist (35- to 40-cm precipitation zone) habitats on deep to moderately deep soils with strong, well-developed mollic horizons (Heyerdahl et al. 2006, Miller and Heyerdahl 2008). Fine-scale variation in CFI in northeastern California ranged from 10 to more than 100 years among plots located systematically within plant associations containing mountain big sagebrush in at least one successional state but differing in soil depth and texture, aspect, slope, and dominant diagnostic grass species (Miller and Heyerdahl 2008).

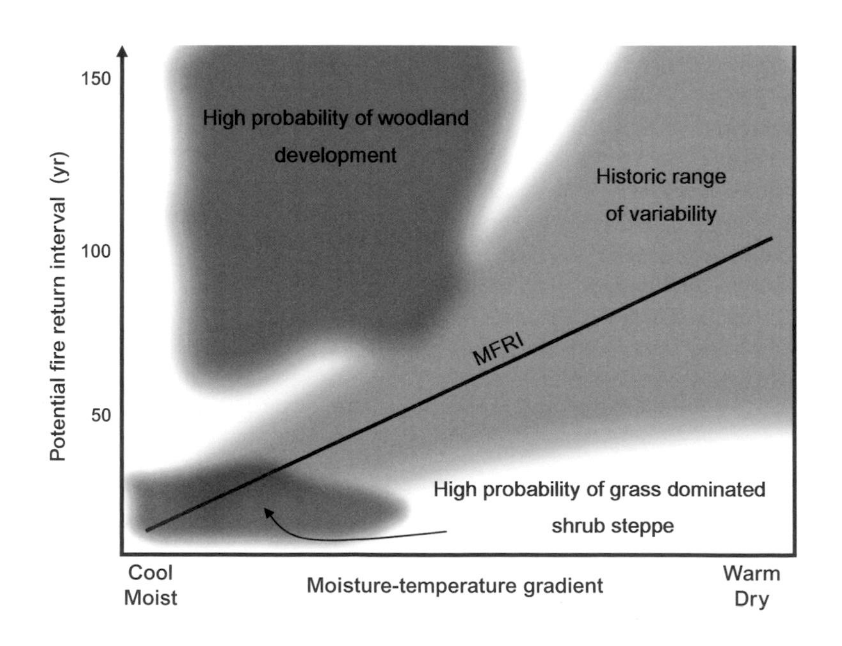

Figure 10.8. Conceptual model illustrating the historic potential mean fire return interval (MFRI) and historic range of variation (light gray area) in sagebrush steppe as it relates to temperature and moisture, resulting in a change in structure, composition, and abundance of fuels. Persistent vegetation that occupies the light gray area would likely be a sagebrush herbaceous mix, although herbaceous vegetation would occupy the site immediately following fire until the sagebrush stand redeveloped (derived by S. C. Bunting and R. F. Miller).

Post–Euro-American Settlement Changes at Low Elevations—Exotic Grasses

Wyoming big sagebrush communities in the Intermountain West prior to Euro-American settlement consisted of sagebrush and perennial grasses that were spatially discontinuous and generally did not carry fires except under extreme weather conditions (e.g., low humidity and high winds) or several wet years that allowed fine fuels to accumulate (Hull and Hull 1974, Vale 1975, Mensing et al. 2006). Invasion by exotic annuals into this sagebrush alliance has resulted in dramatic increases in number and frequency of fire occurrences (Young and Evans 1973, West 2000). Cheatgrass readily invades sagebrush communities, especially sites where native perennials have been depleted. Cheatgrass fills the interspaces between the shrubs and grasses, providing a continuous fuel source that facilitates fire spread. Fires kill sagebrush, which needs to regenerate from seed, while cheatgrass recovers within the first 2 years after fire (Young and Evans 1978). Observations of increased fire frequencies were reported in the early to mid-1900s after annual grasses had invaded much of the Intermountain West (Pickford 1932, Piemeisel 1951, Robertson 1954). Recent fire-return intervals in Wyoming big sagebrush were as low as 5 years in portions of the Snake River Plain where cheatgrass now dominates (Whisenant 1990), although how these estimates were derived was not documented.

Post–Euro-American Settlement Changes at High Elevations–Woodland Expansion

The initial increase in postsettlement conifer expansion beginning in the mid- to late 1800s influenced large portions of today's sagebrush habitats at higher elevations. Tree establishment that created lasting woodlands in sagebrush steppe communities was not synchronous but varied spatially and temporally within six different mountain ranges in Utah, Nevada, Idaho, and Oregon, and it continues to be an ongoing process today (Fig. 10.9) (Miller et al. 2008). Current stages of woodland development at sites that historically were sagebrush steppe communities range from initial phases of encroachment to tree-dominated stands that have been closed for >50 years. The rate of woodland expansion across these six mountain ranges has declined since the 1960s because few sagebrush steppe communities that do not contain juniper or pinyon pine remain within the woodland belt.

Woodlands can encroach into sagebrush communities when the interval between fires becomes long enough for seedlings to establish and trees to mature and dominate a site. Juniper and pinyon trees are killed by fire. However, the probability of western juniper being killed by fire decreases for trees >50 years of age on productive sites and >90 years on low productive sites (Burkhardt and Tisdale 1976, Bunting 1984, Miller and Rose 1999). Trees that have escaped fire can attain >100–200 years of age. Old trees in the Great Basin most frequently occupy sites characterized by shallow rocky soils supporting limited fuels that often are intermixed with ecological sites or plant associations on deeper soils.

The probability of woodlands displacing sagebrush communities increases where seed sources are nearby and fire intervals in more productive sagebrush steppe sites increase to >50 years (Fig. 10.10). Pinyon or juniper trees can reestablish within the first decade following fire if seed is available. Most seeds are dispersed by birds; the majority of seeds are deposited <100 m from the seed tree (Schupp 1993; Chavez-Ramirez and Slack 1994; Schupp et al. 1997; Chambers et al. 1999a,b).

The initial increase in rates of establishment and expansion of conifers into sagebrush steppe communities in the 1800s did not have an immediate effect until density and size allowed trees to dominate the vegetation. Stands that established with sufficient densities in the 1860s and 1870s in southeastern Idaho began to close in the 1950s (Johnson and Miller 2006). The majority of these trees would not have reached maturity under the historic fire regime.

An immediate widespread decline in fires coincided with the introduction of large numbers of livestock in the late 1800s (Miller and Rose 1999; Heyerdahl et al. 2006; Swetnam et al. 2001). Fire declines in the 17th–19th centuries in some areas in the Southwest coincided with the early introduction of sheep, goats, and cattle by Navajos and Hispanic settlers (Savage and Swetnam 1990, Touchan et al. 1995, Baisan and Swetnam 1997). The majority of these sudden declines in fire occurred prior to fire suppression efforts that began in the early 1900s (Miller and Rose 1999; Swetnam et al. 2001). Livestock grazing can

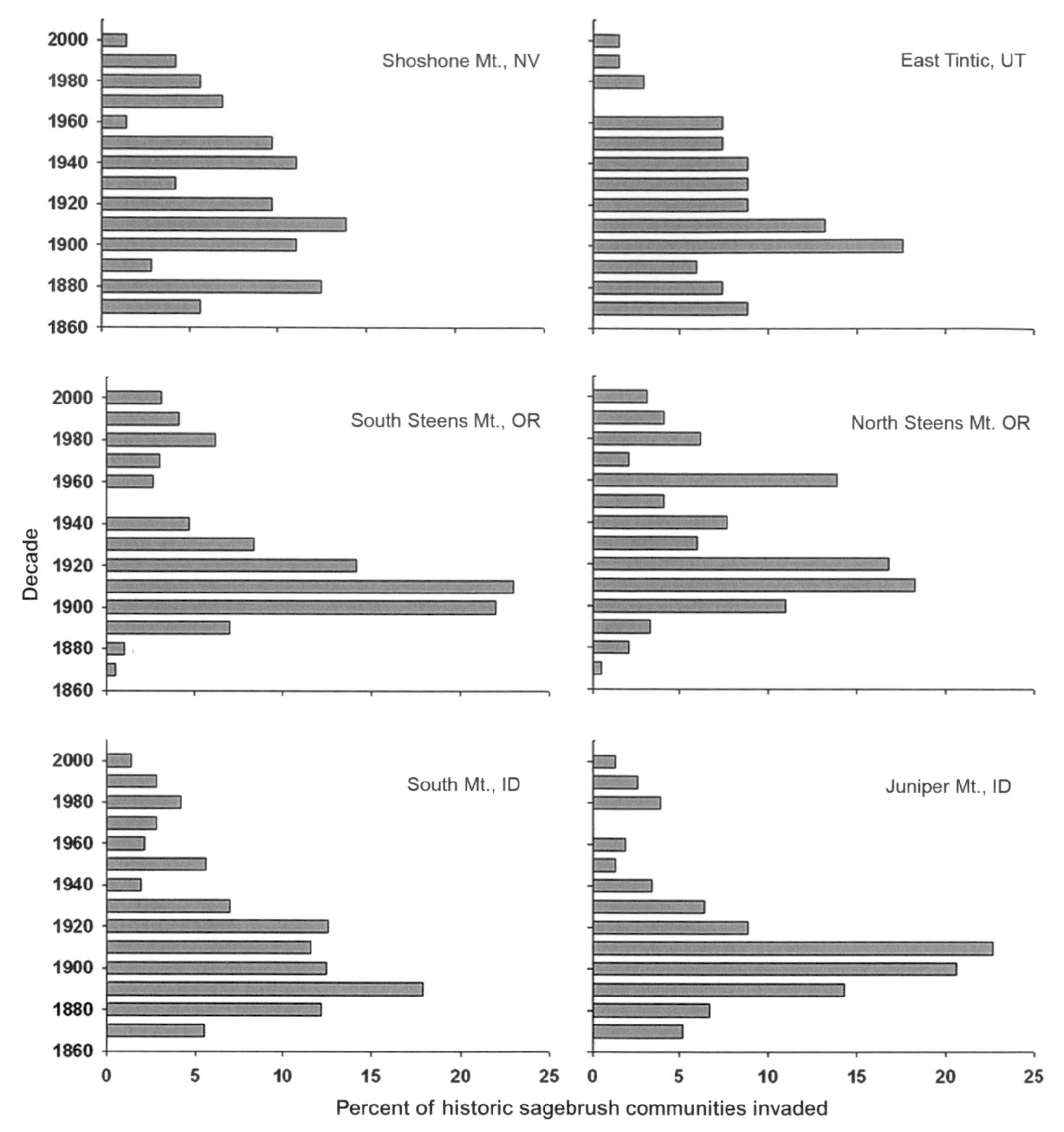

Figure 10.9. The proportion of decadal encroachment across six mountain ranges of pinyon and/or juniper between 1860 and 2000 into historic sagebrush steppe stands with no evidence of pre-settlement trees. Data were collected from transects established along an elevation gradient (14–24 km in length), which extended from the lower to upper boundaries of each woodland (adapted from Miller et al. 2008).

influence fire by reducing the current year's fuel loads and changing the structure, composition, and abundance of vegetation. A reduction in fine fuels through livestock's heavy use of herbaceous plants during the late 1800s then could have reduced fire occurrence across western landscapes and helped drive postsettlement increase and expansion of pinyon and juniper woodlands. The establishment of tree seedlings in mountain big sagebrush plant associations can occur within a decade following fire where a seed source is available (Miller and Rose 1999; Johnson and Miller 2006, 2008; Miller and Heyerdahl 2008). However, the reduction of fire in the late 1800s allowed these trees to reach maturity in historic sagebrush steppe communities, forming woodlands with different fuel characteristics and resulting in a different fire regime.

Evidence for a direct relationship between livestock grazing and woodland encroachment through a reduction of competition from native grasses, forbs, and shrubs is difficult to document.

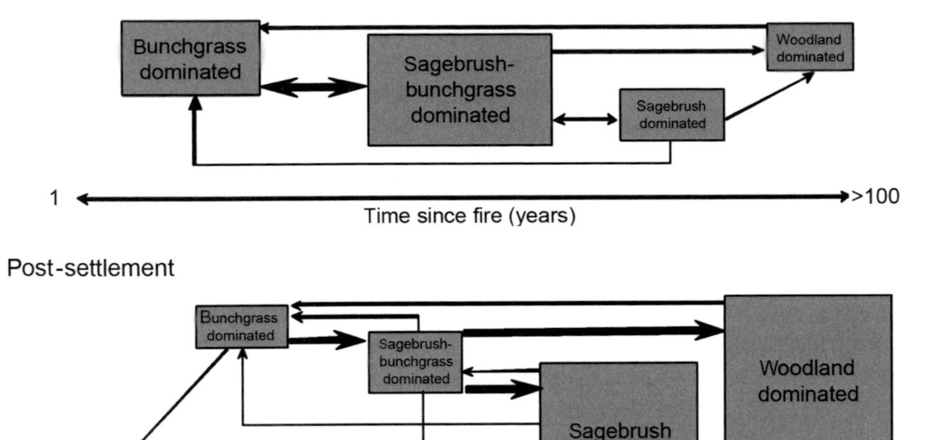

Figure 10.10. Conceptual model of pre- and post-settlement dynamics for plant communities in the Sage-Grouse Conservation Area. Differences in box and arrow sizes imply a difference in proportion of phases and transition from one phase to another within and across steady states.

Both the extent (size) and complexity of fires prior to the introduction of livestock are unknown. Conifer encroachment patterns in sagebrush steppe have not been related to fence line contrasts, distance to water, or ecological conditions that are often observed with other vegetation layers. Western juniper also has increased in the absence of grazing on a relict area (Soulé and Knapp 1999, 2000). Elevated carbon dioxide levels also may have accelerated woodland expansion (Knapp and Soulé 1999, Knapp et al. 2001) but were unlikely to be a factor during the late 1800s and early 1900s.

Change in Fire Regimes

We developed a database of fire statistics from records assembled across the SGCA. Although the fire records include forested areas, we attempted to eliminate those portions of fire polygons that did not burn sagebrush habitats by masking forested areas identified in the vegetation coverage (LANDFIRE 2006). We plotted frequency of fires for all years since 1900 for which fires were documented (Figs. 10.11–10.15); records of fires in some regions were present from 1870. We recognize, however, that analyses of these data are confounded by: (1) an increase in reporting effort by the agencies, (2) differential reporting across regions, (3) lack of recordkeeping in some districts until the 1980s, (4) lack of records of geographic information system–based polygons of fire data until the 1990s, and (5) the fact that the most recent source for fire polygons, the United States Geological Survey Geospatial Multi-Agency Coordination Group, began maintaining fire records in 2004. Therefore, we mapped previous fires only from 1960 through 2007 (Fig. 10.16) for descriptive purposes and conducted statistical analysis on fire size, number of fires, total area burned, and within-year variation in fire sizes recorded from 1980 through 2007 (Table 10.5).

Number of fires and total area burned across the SGCA increased in each of the geographic subdivisions except the Snake River Plain from 1980 through 2007 (Table 10.5). Average fire size increased during this period only in the Southern Great Basin. Within-year variation in fire size decreased in all geographic regions except the Snake River Plain (Table 10.5). The decrease in variation within years is probably because of

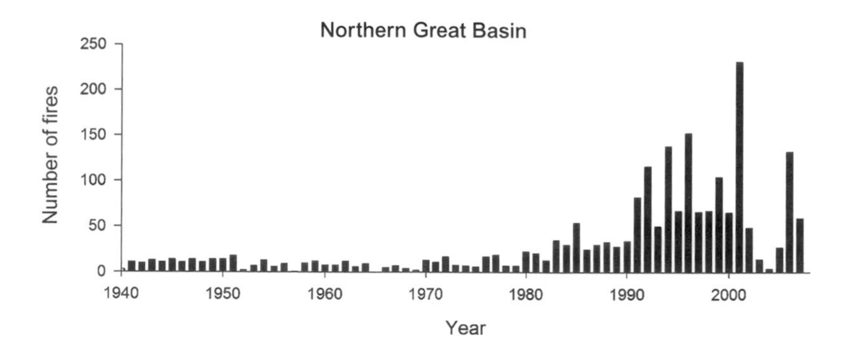

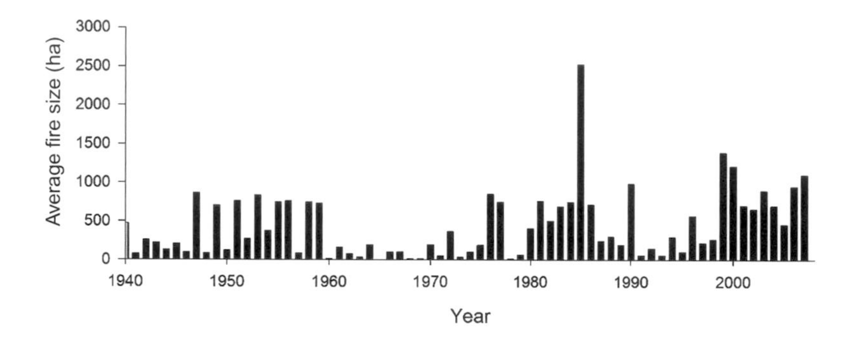

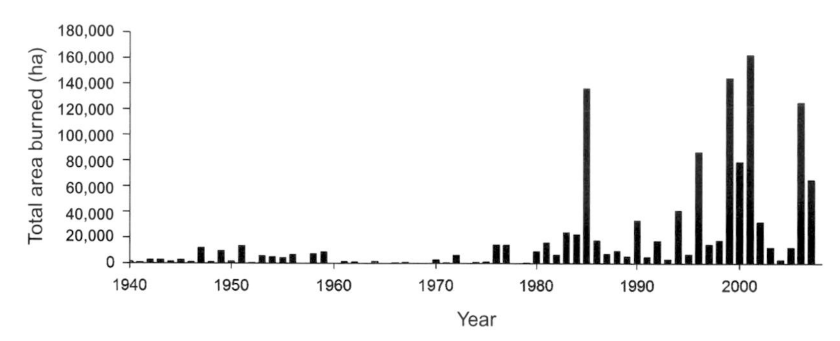

Figure 10.11. Number of fires, average fire size (ha), and total area burned (ha) within the Northern Great Basin (Fig. 10.3). Regression models of changes in fire statistics from 1980 through 2007 are presented in Table 10.5.

greater suppression capabilities. The increased number of fires resulted in a significant increase in total area burned since 1980.

Location of fires mapped since 1960 was related to the distribution of cheatgrass, particularly within the Snake River Plain and Northern Great Basin (Figs. 10.6, 10.16). Cheatgrass was established throughout the area by the 1920s and 1930s (Klemmedson and Smith 1964, Mack 1981, Billings 1990). Consequently, much of the Snake River Plain occupied by cheatgrass in southern Idaho has been well-defined by fires. Fires in northern Nevada and eastern Oregon, also within the cheatgrass region, were more pronounced since 1980. Fires in the eastern section of the SGCA have been recorded only in more recent years.

The total area burned each year on or adjacent to lands managed by the Bureau of Land Management (BLM) was highly variable from 1997 through 2005, illustrating the difficulty in planning for an average year. Area burned per year varied almost sixfold, from 1,455 km^2 (sum of force and contract accounts, BLM and non-BLM lands) in 1998 to 8,142 km^2 in 1999, and tenfold to 14,365 km^2 in 2006 (Tables 10.6, 10.7). The

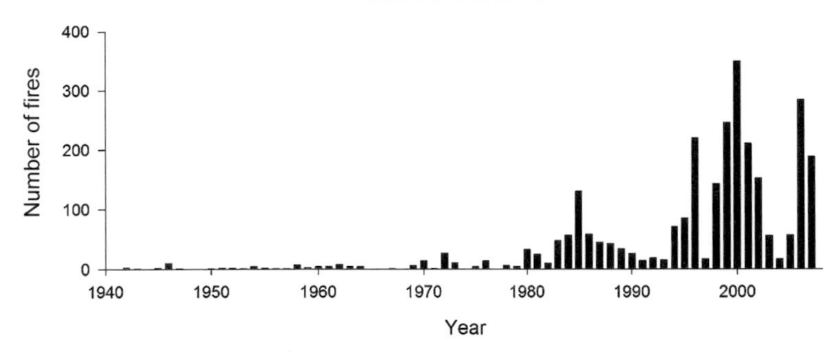

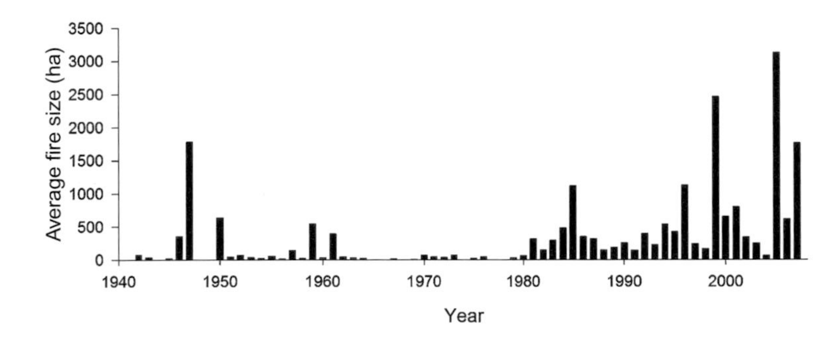

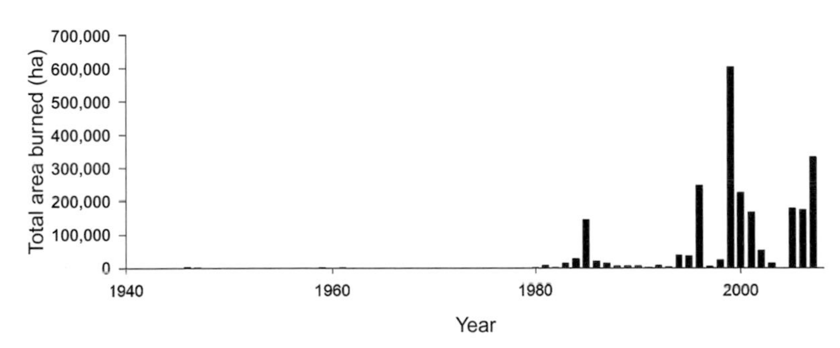

Figure 10.12. Number of fires, average fire size (ha), and total area burned (ha) within the Southern Great Basin (Fig. 10.3). Regression models of changes in fire statistics from 1980 through 2007 are presented in Table 10.5.

most total area burned was consistently found in cheatgrass regions throughout Idaho, Nevada, and Oregon.

Human-caused fires within the SGCA were related to the network of roads (Fig. 10.17). Fire ignitions are an additional consequence of roads and access by humans, in addition to road influences on habitat fragmentation and spread of exotic plant species (Trombulak and Frissell 2000). Of 3,465 fires ignited during 2006 on or adjacent to lands managed by the BLM in Colorado, Idaho, Montana, Oregon, Utah, Washington, and Wyoming, 822 (24%) were caused by humans (United States Department of Interior 2006a:Public Land Statistics Table 6-1).

Global Climate Change

Seasonal and Annual Patterns

Areas dominated by sagebrush habitats are characterized as well-vegetated semideserts in semiarid climates (West and Young 2000, Miller and Eddleman 2001). Timing and abundance of water availability, which varies seasonally and annually, are the major factors that affect the structure, composition,

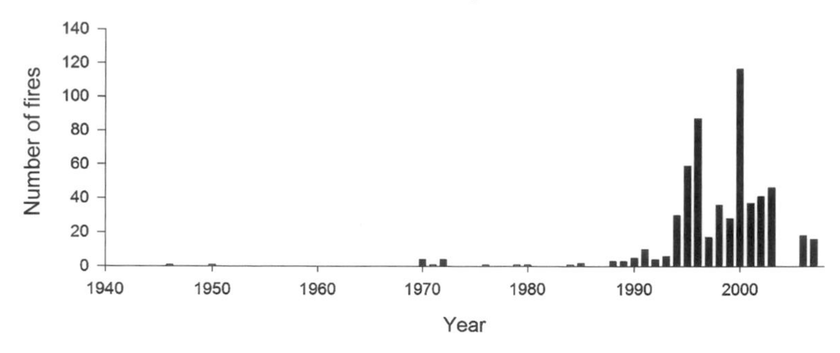

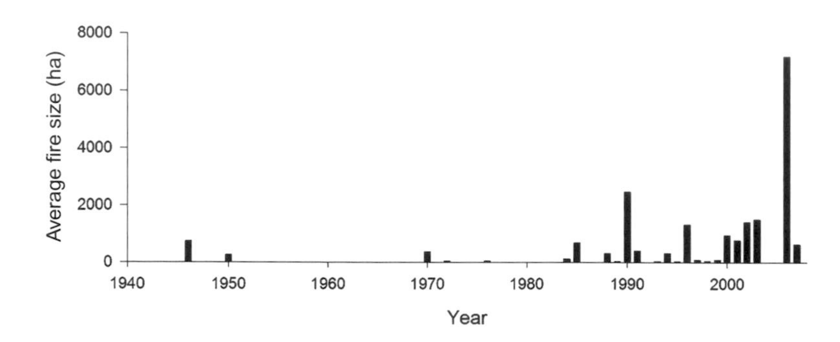

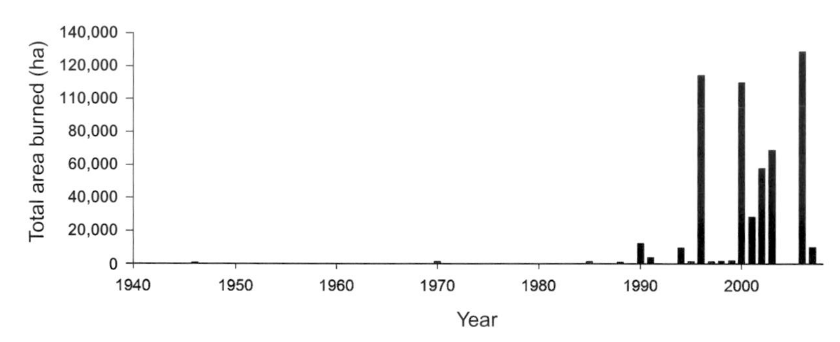

Figure 10.13. Number of fires, average fire size (ha), and total area burned (ha) within the Silver Sagebrush (Fig. 10.3). Regression models of changes in fire statistics from 1980 through 2007 are presented in Table 10.5.

and abundance of vegetation (Toft et al. 1989, West 1996, Anderson and Inouye 2001). Sagebrush systems maximize net annual production in late spring and early summer, when available soil moisture and warm temperatures coincide (West 1983b). Most of the water available to plants in the surface soil layers is depleted by midsummer (Anderson et al. 1987, Jensen 1989b, Obrist et al. 2004). Plants in sagebrush communities have evolved mechanisms, such as deep rooting systems, shedding of leaves (Miller and Shultz 1987), or becoming dormant, to survive periods of high water stress.

Most moisture available to sagebrush communities occurs seasonally as precipitation during winter and spring. The short-term, high-intensity rainfall typical of isolated convective storms during summer generally is of minor importance, because little moisture infiltrates the fine-textured soils characteristic of sagebrush habitats and annual occurrence of such rainfall in any one area is undependable (West and Young 2000). Almost all water recharge in the soils was through roots by hydraulic redistribution (Ryel et al. 2003). Loss of deep-rooted sagebrush plants and conversion to systems dominated by annual grasses thus reduce the potential for deep soil water recharge.

The amount of summer precipitation increases across the SGCA from west to east and north to

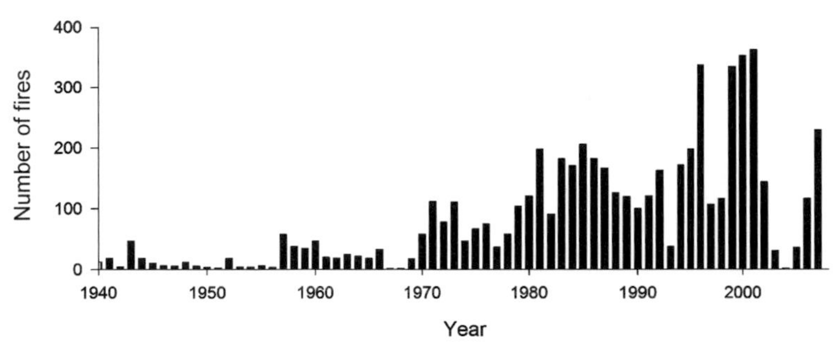

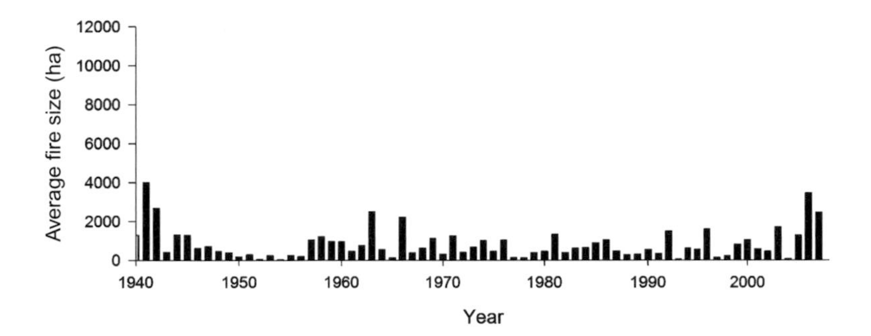

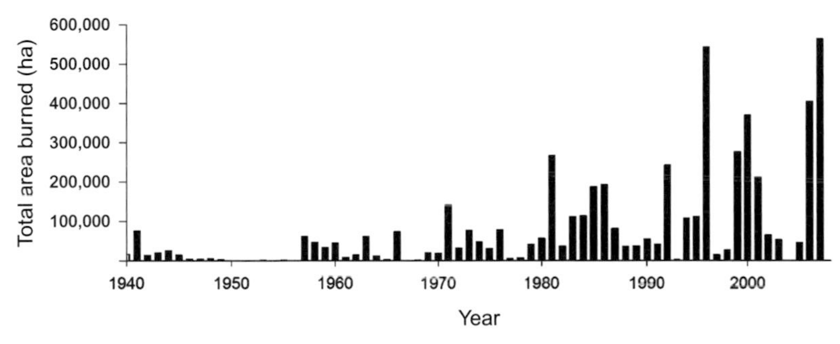

Figure 10.14. Number of fires, average fire size (ha), and total area burned (ha) within the Snake River Plain (Fig. 10.3). Regression models of changes in fire statistics from 1980 through 2007 are presented in Table 10.5.

south due to mixing of air from the Pacific Ocean with summer monsoons from the Gulf of Mexico (Houghton 1969, West 1983a). Elevation, aspect, and soil depth, structure, and texture are also important factors influencing the amount of available moisture (Jensen 1989a,b) and the potential sagebrush community it can support (Jensen et al. 1990).

Interannual variation of precipitation varies greatly across the SGCA, which is subject to periodic drought (Patterson 1952, Thurow and Taylor 1999, Seager et al. 2005). An operational definition of drought is based on the departure from an average amount of precipitation or other climate variables that have been derived from a historical (usually 30-year) average (United States Department of Commerce 2004). Thus, drought defined relative to an average set of conditions has occurred periodically but not regularly in sagebrush habitats. Drought affected sagebrush landscapes during the periods approximated by the late 1890s to 1905, mid-1920s to 1940, early 1950s to mid-1960s, mid-1970s, mid-1980s to mid-1990s, and 1999–2004 (Fig. 10.18). Local scale patterns of drought also have occurred with varying length and severity; water-year precipitation was above average in only 4 years between 1933 and 1956 in the upper Snake River Plain (Anderson and Inouye

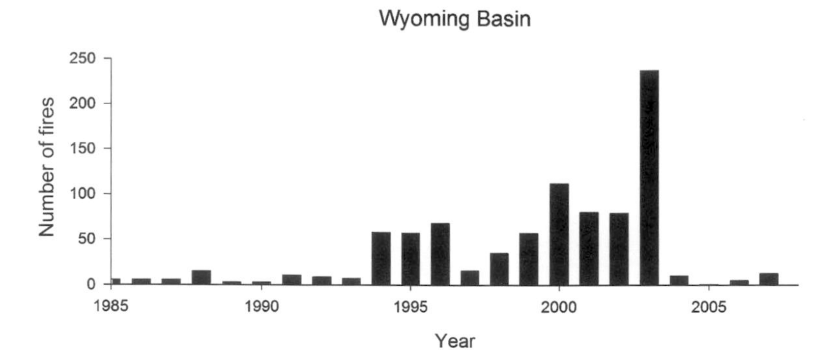

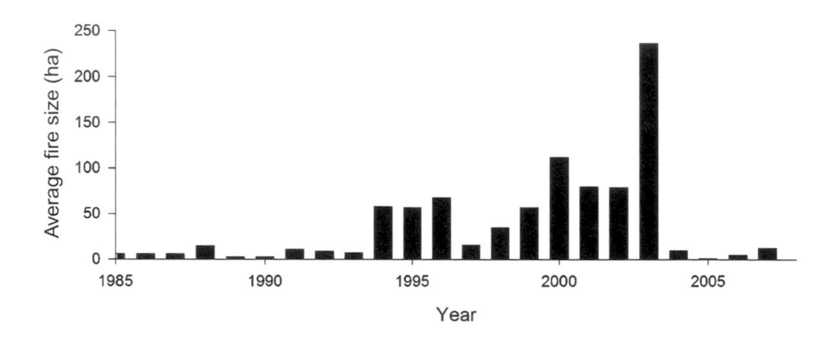

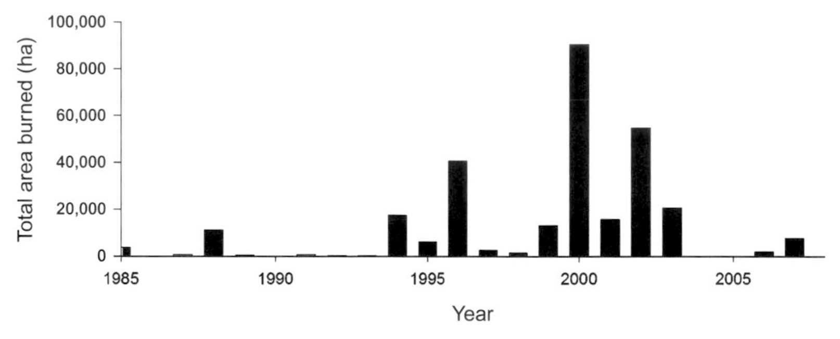

Figure 10.15. Number of fires, average fire size (ha), and total area burned (ha) within the Wyoming Basin (Fig. 10.3). Regression models of changes in fire statistics from 1980 through 2007 are presented in Table 10.5.

2001). In contrast to much of the western United States, drought duration and severity have increased in interior regions and across much of the SGCA during the 20th century (Andreadis and Lettenmaier 2006; Seager et al. 2005, 2007).

Soil erosion is considered the greatest threat to long-term sustainability of shrublands (Society for Range Mangement 1995). Semiarid shrublands are subject to soil erosion during drought because precipitation is insufficient to maintain vegetative cover (Morrison 1964, Thurow and Taylor 1999). Reduced vegetation cover and increased soil erosion result in long-term changes characterized by reduced soil depth, decreased water infiltration, and reduced water storage capacity (Milton et al. 1994, Thurow and Taylor 1999) possibly resulting in a shift of vegetation to a new steady state.

Climate Change

Climate change is a complex process in which interactions among natural and anthropogenic sources affect long-term trends in temperature, precipitation, and atmospheric characteristics (Notaro et al. 2006). The Intergovernmental Panel on Climate Change defined climate change as "a change in the

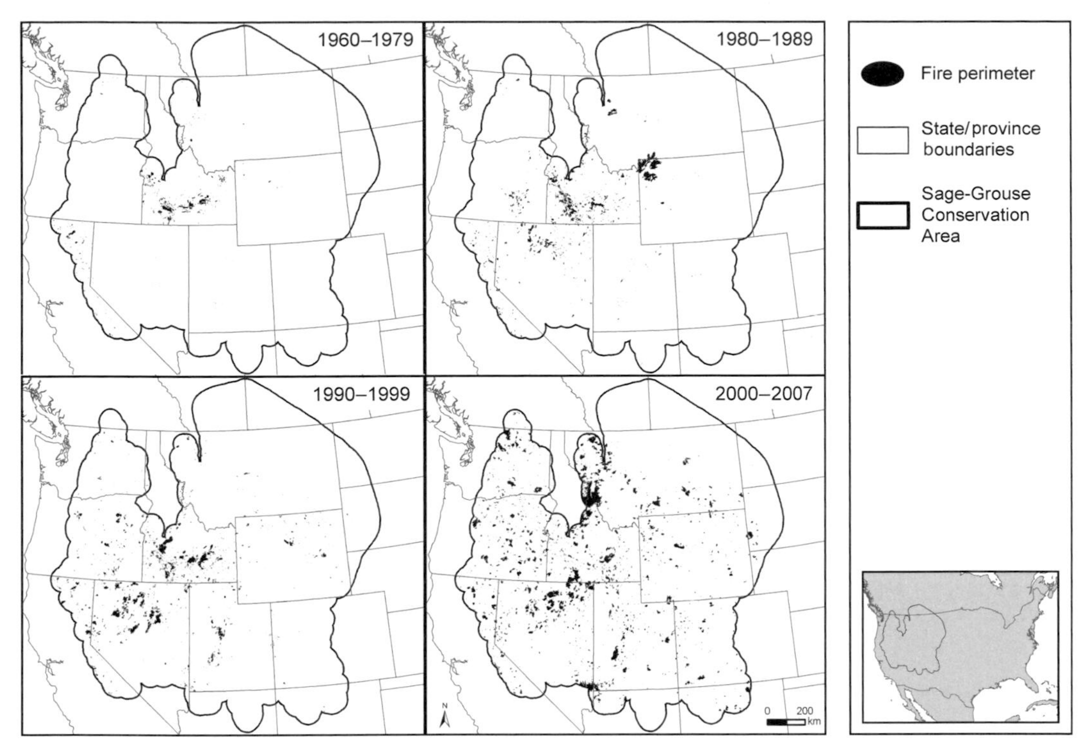

Figure 10.16. Fires mapped in the western United States from 1960 to 2007. Fire information was obtained from >500 source data sets obtained from the U.S. Bureau of Land Management, U.S. Forest Service, U.S. National Park Service, and other state and federal agencies.

state of the climate that can be identified, using statistical tests, by changes in the mean and/or the variability of its properties, and that persists for an extended period, typically decades or longer" (Intergovernmental Panel on Climate Change 2007:30). The definition includes any change and does not distinguish between human-induced or natural causes. Analysis of historical patterns presents dominant trends that may (or may not) continue into the future because of feedbacks among drivers and changes in their dynamics at different equilibria (Intergovernmental Panel on Climate Change 2007).

Global climate-change models predict that more variable and severe weather events (drought, storms), higher temperatures, drier summer soil conditions, and wetter winter seasons will dominate future weather patterns at mid-latitude, semiarid regions (Schlesinger et al. 1990, Schneider 1993, Christensen et al. 2007, Intergovernmental Panel on Climate Change 2007). Potential changes for locations or vegetation communities across arid and semiarid landscapes in specific regions are difficult to predict, because most climate change models project scenarios over extremely broad geographic or continental regions (Reynolds et al. 1997). Average temperatures may warm by as much as 2.8°C to >6°C for the Great Basin and much of the adjacent areas that support sagebrush (Bachelet et al. 2001, Neilson et al. 2005, Christensen et al. 2007). Atmospheric carbon dioxide, methane, nitrous oxide, and halocarbons including fluorine, chlorine, and bromine also are predicted to increase under most global climate change scenarios (Christensen et al. 2007, Intergovernmental Panel on Climate Change 2007).

Long-term changes in global climate and atmospheric conditions, particularly increased temperature and levels of carbon dioxide, will shift competitive advantage among individual plant species. Increased temperatures could exert additional stresses on sagebrush if increased transpiration rates during winter months coupled with changes in precipitation patterns result in increased soil moisture loss throughout the year (Shafer et al. 2001). In addition, the trend for decreased snowpack, earlier onset and warmer spring periods, and reduced summer water flows in the western United States further extends the period of summer stress (Cayan et al. 2001, Intergovernmental Panel on Climate Change 2007). A substantial increase in

TABLE 10.5

Changes in number and average size of fires, total area burned, and variation in fire size by year within floristic divisions from 1980–2007.

We used a time-series linear model regressing year by fire variables with dependent variables log-transformed. We conducted our analysis only for fires recorded after 1980 because of more consistent reporting, even though fire statistics were available prior to 1980 (Figs. 10.11–10.15). Data were incomplete for 2004 and 2005, and were not included in the analysis.

Floristic Provinces (no. of fire years 1980–2007)	Dependent variable	F	P	r^2	Coefficient[a]
Northern Great Basin (25)	Fires (N)	8.95	<0.01	0.25	0.05
	Mean size (ha)	0.29	0.59	0.01	0.01
	Total area (ha)	5.31	0.03	0.19	0.07
	CV[b]	8.28	<0.01	0.26	−0.01
Southern Great Basin (25)	Fires (N)	13.71	<0.01	0.37	0.08
	Mean size (ha)	4.57	0.04	0.17	0.04
	Total area (ha)	10.71	<0.01	0.29	0.13
	CV	9.20	<0.01	0.25	−0.01
Silver Sagebrush (19)	Fires (N)	14.87	<0.01	0.44	0.14
	Mean size (ha)	2.77	0.11	0.09	0.11
	Total area (ha)	10.70	<0.01	0.35	0.24
	CV	5.29	0.03	0.19	−0.04
Snake River Plain (25)	Fires (N)	0.02	0.90	0.00	0.00
	Mean size (ha)	1.64	0.21	0.03	0.03
	Total area (ha)	0.92	0.35	0.00	0.03
	CV	1.91	0.18	0.04	0.00
Wyoming Basin (21)	Fires (N)	8.31	0.01	0.27	0.11
	Mean size (ha)	2.59	0.12	0.07	0.06
	Total area (ha)	8.66	0.01	0.28	0.17
	CV	5.79	0.03	0.19	−0.02

[a] Log-transformed coefficient.

[b] Coefficient of variation (CV) used to standardize estimates of standard deviations across years.

temperature also confers a competitive advantage to frost-sensitive woodland vegetation that currently dominates the Chihuahuan and Sonoran Deserts but is temperature limited in its northern distribution. These woodlands may expand substantially northward into and displace large areas of sagebrush and other shrublands in the western portions of the sagebrush ecosystem, based on variety of projected conditions of climate change (Neilson et al. 2005). However, a large portion of these woody species also may be replaced by exotic grasses and forbs.

Atmospheric carbon dioxide has increased from preindustrial levels of 280 μmol mol^{-1} to current levels of 360 μmol mol^{-1} and are projected to be 420 μmol mol^{-1} in 2020 and may exceed 600 μmol mol^{-1} by 2100 (Bazzaz et al. 1996, Ziska et al. 2005). Over the last century, atmospheric carbon dioxide has increased by >20% (Polley 1997). Ecosystem responses to enhanced carbon dioxide involve interactions of biogeochemical cycles, water and energy fluxes, and vegetation dynamics dependent on the temporal scale over which

TABLE 10.6

Fire area (km²) on or adjacent to lands managed by the Bureau of Land Management from 1998–2006.

Fire protection given public land administered by the BLM using protection forces and facilities supervised and operated by the BLM.

State	2006		2005		2004		2003		2002		2001		2000		1999		1998	
	BLM	Non-BLM	BLM	Non-BLM	BLM	Non-BLM	BLM	Non-BLM	BLM	Non-BLM	BLM	Non-BLM	BLM	Non-BLM	BLM	Non-BLM	BLM	Non-BLM
Arizona	185	36	1,158	132	29	12	20	0	33	0	8	2	4	0	92	6	8	0
California	38	43	95	95	25	97	45	8	88	97	30	7	34	11	81	96	17	30
Colorado	31	25	12	11	15	20	28	1	26	24	1	0	1	4	34	7	9	1
Idaho	1,844	433	1,170	346	21	19	104	69	127	85	325	178	1,146	460	1,265	127	87	70
Montana	221	720	10	22	7	27	281	312	3	6	0	0	0	24	3	0	4	0
Nevada	3,597	806	5,682	290	86	4	29	6	84	30	1,456	168	1,951	229	3,995	1,254	160	69
New Mexico	25	55	2	6	1	0	4	0	82	82	10	5	250	33	16	20	1	0
Oregon	1,242	375	61	10	4	2	24	0	415	23	600	50	600	143	46	6	130	107
Utah	728	281	526	291	65	20	233	96	16	16	57	62	315	99	249	119	136	79
Washington	10	29	2	30	0	0	38	16	4	16	0	0	3	25	4	12	0	0
Wyoming	95	520	9	18	4	5	20	6	1	1	14	33	101	200	3	3	2	1
Total km²	8,017	3,324	8,726	1,249	256	206	824	513	878	380	2,502	504	4,405	1,228	5,787	1,650	554	356

SOURCE: United States Department of the Interior 1998, 1999, 2000d, 2001b, 2002b, 2003f, 2004e, 2005d, 2006c.

TABLE 10.7

Fire area (km^2) on or adjacent to lands managed by the BLM from 1998–2006.

Fire protection given public land administered by the BLM using contracted protection forces and facilities.

	2006		2005		2004		2003		2002		2001		2000		1999		1998	
State	BLM	Non-BLM	BLM	Non-BLM	BLM	Non-BLM	BLM	Non-BLM	BLM	Non-BLM	BLM	Non-BLM	BLM	Non-BLM	BLM	Non-BLM	BLM	Non-BLM
Arizona	0	0	0	313	0	0	0	0	0	0	0	0	0	0	0	2	0	1
California	179	651	71	20	206	1,603	21	122	68	106	4	50	99	181	37	73	29	71
Colorado	3	9	0	0	0	3	0	1	0	0	0	0	0	0	0	2	0	0
Idaho	1	17	0	5	0	7	4	24	9	7	7	13	61	233	20	45	2	227
Montana	35	1,099	19	64	0	8	0	36	4	21	0	0	2	2	2	2	4	2
Nevada	277	478	194	89	0	2	8	9	0	10	1	190	24	84	146	202	7	56
New Mexico	9	26	0	2	0	0	0	1	0	1	1	2	7	28	0	2	4	7
Oregon	2	76	13	9	8	13	1	1	1	1	0	1	3	8	0	0	29	100
Utah	2	60	1	26	0	0	1	0	5	15	0	0	43	26	47	124	5	1
Washington	45	6	1	10	7	80	3	6	8	3	3	5	0	0	0	0	0	0
Wyoming	9	42	1	4	1	1	10	1	1	2	1	22	1	48	0	0	0	0
Total km^2	565	2,463	299	540	222	1,718	49	201	95	165	16	283	239	609	253	453	80	465

SOURCE: United States Department of the Interior 1998, 1999, 2000d, 2001b, 2002b, 2003f, 2004e, 2005d, 2006c.

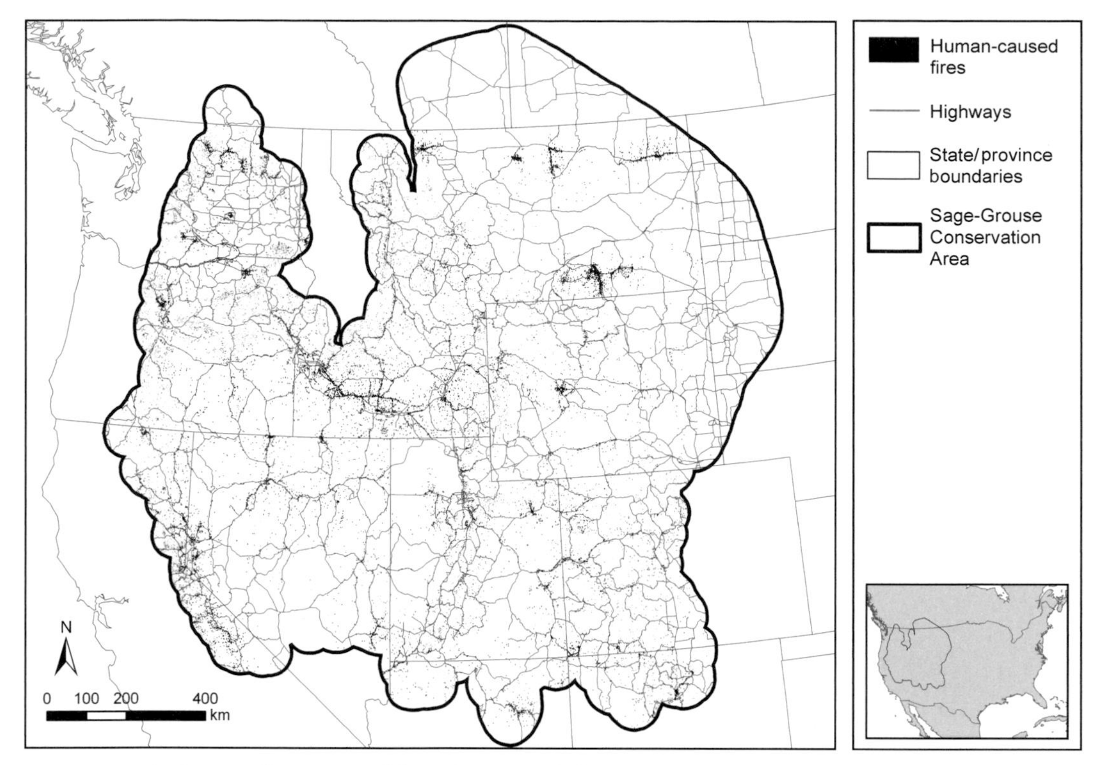

Figure 10.17. Distribution of human-caused fires relative to roads from 1960 to 2007. Fire information was obtained from the National Fire Occurrence database.

carbon dioxide increases (Körner 1996, Walker and Steffen 1996). The trend of increasing atmospheric carbon dioxide is likely to continue, and natural feedbacks or anthropogenic changes on the landscape that influence ecosystem processes may either ameliorate or intensify those effects (Noble 1996).

Increased levels of carbon dioxide favor exotic annual grasses at the expense of native vegetation in arid shrubland ecosystems (Mooney and Hobbs 2000, Smith et al. 2000). Cheatgrass responded positively to elevated carbon dioxide when compared to native grasses (Smith et al. 1987). In controlled laboratory tests, reproductive biomass of cheatgrass doubled and time to maturation decreased by 10 days at the current atmospheric carbon dioxide levels compared to lower levels recorded in preindustrial time (Ziska et al. 2005). Cheatgrass already competes successfully against native grasses because of earlier maturation, shallow root systems to collect water in soils, greater seed production, and the ability to respond quickly to resources released during disturbance (Klemmedson and Smith 1964). Thus, the ability of cheatgrass to compete in sagebrush ecosystems created by enhanced carbon dioxide or changes in annual precipitation, temperature, or severe storms will facilitate its spread and exacerbate the cycle of fire and cheatgrass dominance (d'Antonio and Vitousek 1992, d'Antonio 2000, Ziska et al. 2005).

Each 1°C increase in temperature was predicted to result in a loss of 87,000 km^2 of existing sagebrush habitat, primarily to increasing distribution of other woody vegetation (Neilson et al. 2005). Only 20% of the current sagebrush distribution would remain under the most extreme scenario of an increase of 6.6°C (Neilson et al. 2005). Moreover, these scenarios have largely overlooked the potential response of exotic plant species and changes in fire cycles to further reduce sagebrush from its current distribution. These models predict that future remaining sagebrush habitats will primarily be in the more northerly latitudes and higher elevations (Fig. 10.19) (Shafer et al. 2001, Neilson et al. 2005).

CONSERVATION IMPLICATIONS

Altered disturbance regimes in many regions of the SGCA are shifting the potential natural community outside of the range of historic variation

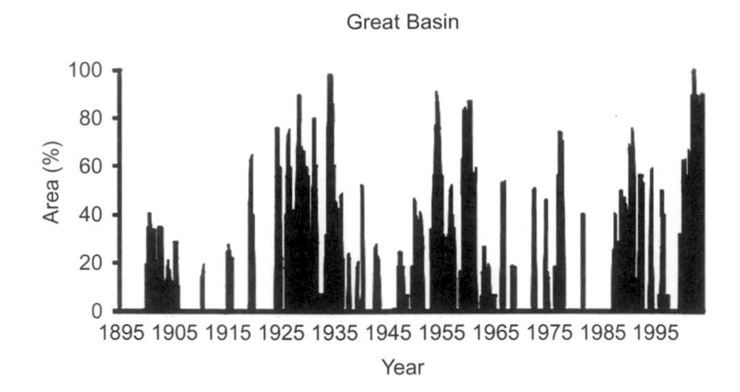

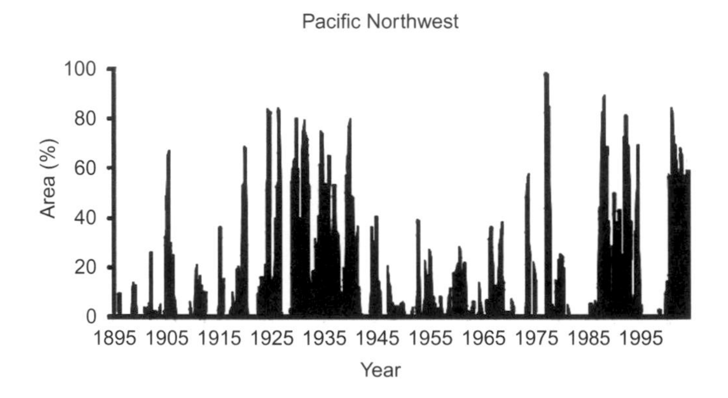

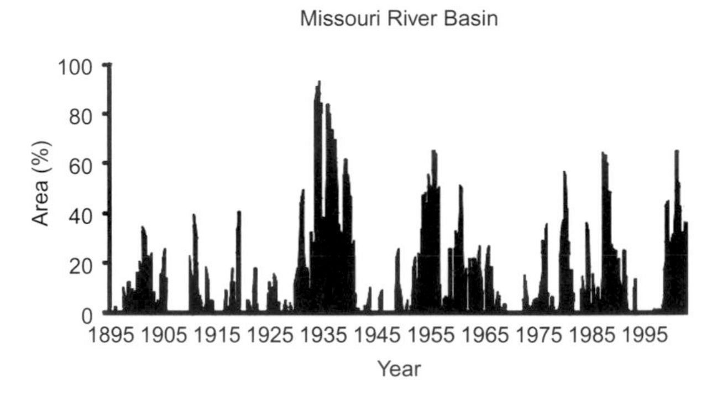

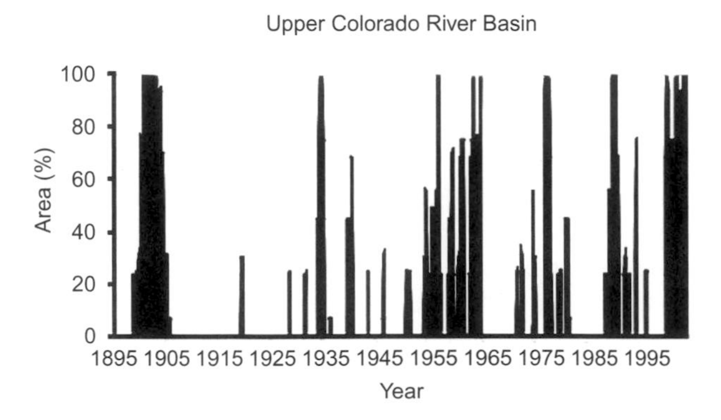

Figure 10.18. Percent of major river basins experiencing drought conditions from 1895 to 2004 (United States Department of Commerce 2004). The graphs represent the Palmer (1965) Drought Severity Index, which measures the extent of departure from the long-term average based on precipitation, temperature, and available water capacity.

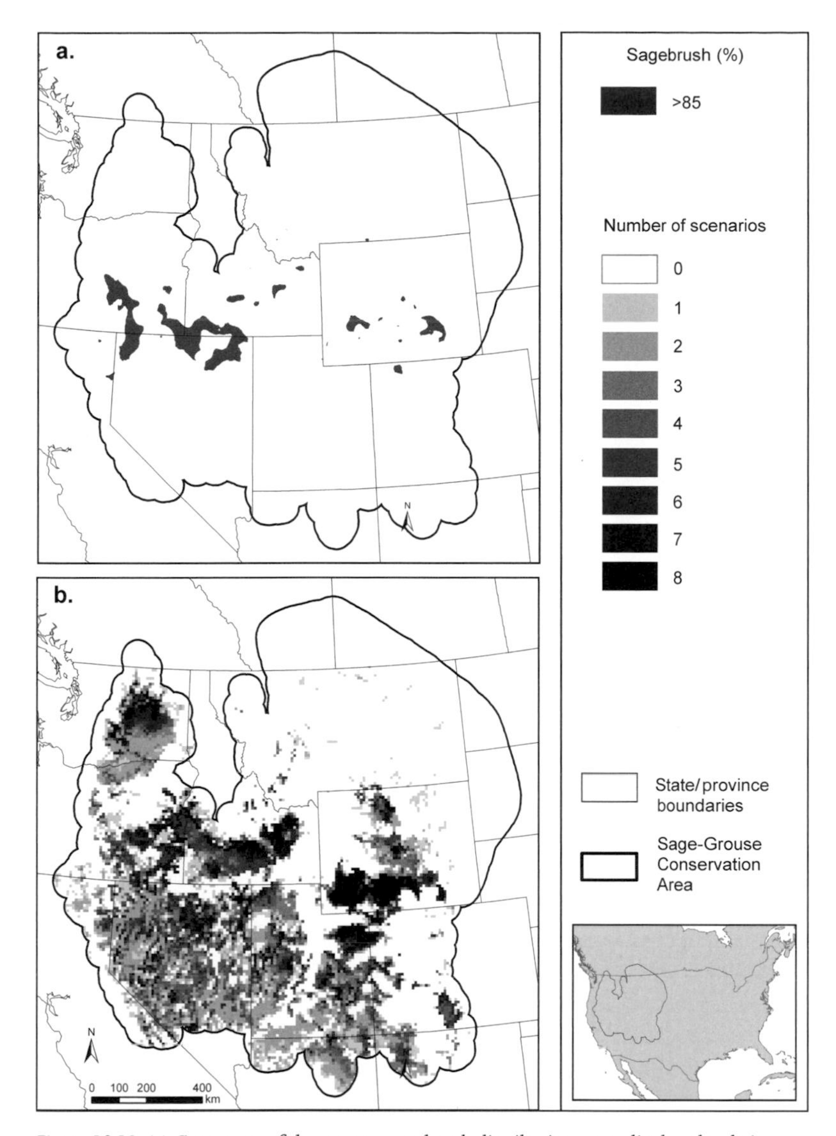

Figure 10.19. (a) Core areas of the current sagebrush distribution were displayed only in those regions in which >85% of the landscape (Fig. 10.1) was dominated by sagebrush. (b) Distribution of sagebrush predicted under current climate and seven models of future scenarios (Neilson et al. 2005). Each cell in the map is the sum of alternate models for future climate scenarios predicting that sagebrush will remain in that location.

to new steady states. Changes in plant composition and structure of sagebrush habitats as a result of disturbance events or successional trends may occur over a decade, over multiple decades, or within a single unique combination of weather and disturbance events over a few weeks. Resistance and resilience to change resulting from disturbance generally increase with increasing moisture and decreasing temperatures, and vary widely across different sagebrush plant associations and ecological sites. Sagebrush communities that have persisted may change if disturbances become chronic or severe, replaced by exotic plants that are better adapted to more frequent disturbance or changing climate, and/or are better able to exploit new resources.

Cheatgrass currently is present throughout much of the western United States (Wisdom et al. 2005b, Meinke et al. 2009). Once established in a sagebrush community, the effects cascade in synergistic feedbacks toward increasing cheatgrass dominance resulting from increased fire disturbance, loss of

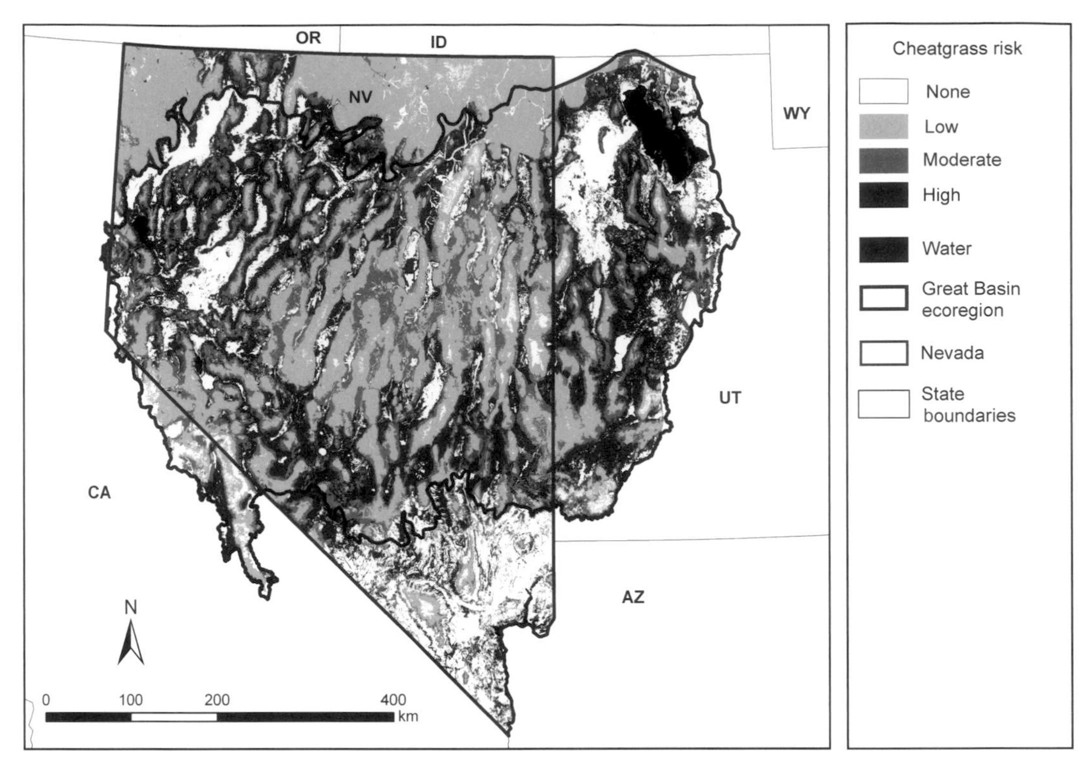

Figure 10.20. Estimated risk of cheatgrass displacement of sagebrush and other susceptible land cover types in the Great Basin ecoregion during the next 30 years (from Suring et al. 2005b, with permission from Alliance Communications).

perennial species and their seed banks, and decreased stability and resilience to changes in interannual weather and long-term climate patterns (d'Antonio and Vitousek 1992, d'Antonio 2000, Brooks et al. 2004a, Chambers et al. 2007).

The extensive distribution of cheatgrass, combined with its aggressiveness in replacing sagebrush, poses substantial risk of increased fire and loss of existing sagebrush communities. The risk of fire was <60% when ground cover of cheatgrass was <20%; areas with >45% ground cover of cheatgrass had a 100% risk of fire (Link et al. 2006). Of the 83,000 km^2 of existing sagebrush cover types in the Great Basin ecoregion of Nevada, Utah, and California, nearly 58% (48,000 km^2) is estimated to be at moderate or high risk of elimination from continued invasion of cheatgrass during the next 30 years (Fig. 10.20) (Suring et al. 2005b). Similar estimates of risk posed by cheatgrass are projected for a vast area of 200,000 km^2 of existing sagebrush that encompasses southeast Oregon, southern Idaho, northeast California, Nevada, and western Utah (Suring et al. 2005b). Approximately 56% (113,000 km^2) of existing sagebrush among these provinces was estimated at moderate or high risk of loss from cheatgrass during the next 30 years (M. J. Wisdom, unpubl. data). Sagebrush sites considered to be at higher risk were those at lower elevations and on south- and west-facing slopes; these conditions generally represent the warmer, drier sagebrush sites on which cheatgrass has most successfully displaced big sagebrush communities in the past (Suring et al. 2005b). Sites considered at lower risk were those at higher elevations and on north- and east-facing slopes, which represent colder, wetter sagebrush communities. Cheatgrass is a self-pollinator and evolves slowly, but the species is comprised of many different populations that have different adaptive strategies (Kinter and Mack 2004). This results in uncertainty of its ability to invade plant associations currently thought to be resistant to encroachment. Elevated temperatures due to climate change may also increase the competitive ability of cheatgrass at higher elevations, thus expanding the range of cheatgrass into regions where it currently is limited.

The past and present role of wildfire in the SGCA is temporally and spatially variable. The increase in fine fuel continuity as a result of cheatgrass invasion (Whisenant 1990, Knapp 1995) and encroachment of juniper and pinyon into some sagebrush plant associations and ecological sites (Tausch et al. 1981, Miller et al. 2008) suggest that fire return intervals have changed significantly across portions of the SGCA. Present-day fire regimes across the SGCA also have shifted spatially; much of the increased fire has occurred in the more arid Wyoming big sagebrush communities and decreased across many mountain big sagebrush communities. Both scenarios of altered fire regimes, resulting in conifer expansion at high-elevation interfaces and exotic weed encroachment at lower elevations, have caused a significant loss in Greater Sage-Grouse habitat (Knick et al. 2005, Wisdom et al. 2005b).

Shifts in fire regimes have significant conservation implications through changing the proportion of phases and new steady states in both time and space (Fig. 10.10). Repeat fires that eliminate or reduce shrubs, native grasses, and forbs; disturb soils and biological crusts; and release nutrients have allowed cheatgrass and other introduced annuals to replace the native shrub and herb layers. The converted landscape, which is largely composed of introduced annuals, is more susceptible to annual weather patterns and varies greatly from year to year, depending on moisture availability (Knapp 1995). In addition, shrub cover is absent from these landscapes, the season of available green plant material is shortened, high-quality perennial forbs are scarce, food for wildlife is limited or absent in late summer through winter, and the fire season is lengthened (Miller and Eddleman 2001).

Fires are an increasingly significant disturbance throughout much of the SGCA. Part of our recorded increases may be a function of differences in reporting fires and better technology to map fire polygons. However, the increases also suggest a significant spatial shift of more frequent fires occurring in the drier sagebrush plant associations than occurred prior to Euro-American settlement. The increase in areas burned each year coupled with decreases in total area of sagebrush habitats will further accelerate the trajectory of habitat loss for Greater Sage-Grouse.

Future effects of global climate change for sagebrush-dominated ecosystems must be considered in the context of the current short-term, large-scale habitat changes. Exotic annuals, especially cheatgrass, have increased the frequency and intensity of wildfires from the historical disturbance regime and facilitated large-scale conversion of shrublands into exotic annual grassland (Brooks et al. 2004a). Long-term changes in climate that facilitate or enhance invasion and establishment by invasive annual grasses further exacerbate the fire regime and accelerate loss of sagebrush habitats (d'Antonio and Vitousek 1992). Similarly, increases in woody vegetation will also increase fire potential (Neilson et al. 2005). Thus, effects of disturbances will be amplified by greater susceptibility for habitats to burn as well as the decreased likelihood for recovery of shrublands. The increasing amount of land-use activities also will have a significant influence on the soils, biological crusts, and vegetation of these systems and their ability to recover from the cascading effects created by invasive plants, fire, and climate change (Foley et al. 2005, Belnap et al. 2006).

The effects of climate change also must be factored into long-term conservation actions. The U.S. Department of the Interior, in considering potential listing for Greater Sage-Grouse under the Endangered Species Act, projected scenarios for extinction for 100–200 years into the future (United States Department of the Interior 2005b). Over these periods, the predicted changes in atmospheric carbon dioxide and temperature become increasingly important and are sufficiently large to overwhelm any current trajectories of habitat loss and alteration. We caution that projections of the effects of global warming become less reliable as they extend further into the future because of the complexity of interactions among the primary variables driving changes.

Land management agencies have placed major emphases on restoration programs, which are difficult and expensive, and may take centuries for complete restoration of a functioning system of sagebrush habitats within a landscape mosaic (United States Department of the Interior 1996, Hemstrom et al. 2002). The primary disturbance variables presented in this chapter—introduction of exotic plants, changes in wildfire, and climate change—will further complicate our ability to conserve and manage sagebrush communities as they now exist. They also call into question our ability to restore these ecosystems in the future. For example, sagebrush may be particularly susceptible to climatic influences on seedling recruitment (Maier et al. 2001, Perryman et al. 2001). The addition of new disturbance factors since Euro-American settlement of the region has created new steady states, including plant associations that did not exist in the past. Sagebrush communities that currently are the dominant land cover across large areas are at risk of shifting to new states from which returns to previous states following land use or disturbance are unlikely (Laycock 1991, West and Young 2000).

In summary, sagebrush habitats are severely stressed across much of the SGCA, and their total area likely will decline in the relatively near future as a result of invasive species, fire, and climate change. Restoration programs are in relative infancy, and, together with the long-term dynamics of sagebrush systems, will require many years to have a substantial impact in slowing or stabilizing this loss. The end result is likely to be continuing and perhaps accelerated declines of many sagebrush-dependent species, including Greater Sage-Grouse.

ACKNOWLEDGMENTS

This chapter benefited from comments by nine anonymous reviewers from the Ecological Society of America, who reviewed previous material in the Conservation Assessment for Greater Sage-Grouse and Sagebrush Habitats (Connelly et al. 2004); and J. W. Connelly; and two reviewers of the revised manuscript. The U.S. Geological Survey Forest and Rangeland Ecosystem Science Center supported STK during development and writing. K. A. Fesenmyer, S. E. Hanser, and K. M. Ross updated the GIS figures and tables. We thank R. P. Neilson and R. J. Drapek for the global climate change data used in Figure 10.19.

CHAPTER ELEVEN

Pre–Euro-American and Recent Fire in Sagebrush Ecosystems

William L. Baker

Abstract. Sagebrush (*Artemisia* spp.) ecosystems are under threat from a variety of land uses, disturbance, and invasive species, and are also thought by some to have been affected by fire exclusion and require burning as a part of restoration. To better understand the historical range of variation (HRV) of fire in sagebrush ecosystems and whether sagebrush fire regimes today have too much or too little fire, I estimated fire rotation (expected time to burn the area of a landscape) in sagebrush ecosystems under the HRV. Estimates derived from five sources are >200 years in little sagebrush (*A. arbuscula*), 200–350 years in Wyoming big sagebrush (*A. tridentata* ssp. *wyomingensis*), 150–300 years in mountain big sagebrush (*A. t.* ssp. *vaseyana*), and 40–230 years in mountain grasslands containing patches of mountain big sagebrush with longer rotations in areas where sagebrush intermixes with forests. Landscape dynamics under the HRV were likely dominated in all sagebrush areas by infrequent episodes of large, high-severity fires followed by long interludes with smaller, patchier fires, allowing mature sagebrush to dominate for extended periods. Fire rotation, estimated from recent fire records, suggests fire exclusion had little effect on fire in sagebrush ecosystems. Instead, cheatgrass (*Bromus tectorum*), human-set fires, and global warming may have led to too much fire relative to the HRV in four floristic provinces within the range of sagebrush in the western United States. Sagebrush ecosystems would generally benefit from rest from disturbance. Global warming is likely to increase fire, and widespread prescribed burning of sagebrush is unnecessary. Where cheatgrass occurs, fire suppression is sensible. In areas of depleted understories, restoration to reestablish native plants is needed if sagebrush ecosystems are to effectively recover from future disturbance.

Key Words: Artemisia, fire, fire exclusion, fire rotation, landscape dynamics, mean fire interval.

Incendios Pre–Euro-Americanos y Recientes en Ecosistemas de Artemisa

Resumen. Los ecosistemas de sagebrush (*Artemisia* spp.) están bajo amenaza por una variedad de usos del suelo, de alteraciones, de especies invasivas, y también hay personas que consideran que han sido afectados por la exclusión de incendios y

Baker, W. L. 2011. Pre–Euro-American and recent fire in sagebrush ecosystems. Pp. 185–201 *in* S. T. Knick and J. W. Connelly (editors). Greater Sage-Grouse: ecology and conservation of a landscape species and its habitats. Studies in Avian Biology (vol. 38), University of California Press, Berkeley, CA.

que requieren de los mismos como parte de la restauración. Para entender mejor el rango de variación histórico (HRV) de los incendios en los ecosistemas de artemisa y si los regímenes actuales de incendios de la artemisa tienen mucho o muy poco fuego, estimé la rotación de incendio (tiempo estimado para quemar el área de un territorio) en ecosistemas de artemisa bajo el HRV. Las estimaciones, derivadas de cinco fuentes, son >200 años en little sagebrush (*A. arbuscula*), 200–350 años en Wyoming big sagebrush (*A. tridentata* ssp. *wyomingensis*), 150–300 años en mountain big sagebrush (*A. t.* ssp. *vaseyana*), y 40–230 años en terrenos de pastos de montaña que contienen parches de mountain big sagebrush, con rotaciones más largas en áreas donde la artemisa se mezcla con bosques. La dinámica del territorio bajo HRV probablemente fue dominada, en todas las areas de artemisa, por episodios poco frecuentes de incendios grandes de alta severidad, seguidos por largos intervalos con fuegos mas pequeños y por parches, permitiendo a la artemisa madura dominar por extensos períodos. La rotación de incendios, estimada a partir de registros recientes de fuego o incendios, sugiere que la exclusión de incendios tiene poco efecto sobre los incendios en ecosistemas de artemisa. En cambio el cheatgrass (*Bromus tectorum*), los fuegos iniciados por humanos, y el calentamiento global pueden haber llevado a demasiado fuego con relación al HRV en cuatro provincias florales localizadas dentro del rango de distribución la artemisa en el del oeste de EEUU. Los ecosistemas de artemisa generalmente se beneficiarían si tuviesen un periodo de reposo luego de un disturbio. El calentamiento global es probable que incremente los incendios, y el quemado diseminado prescrito de artemisa no es necesario. Donde hay cheatgrass la supresión del fuego es sensible. En áreas donde la vegetación debajo de los árboles o arbustos está agotada, la restauración para reestablecer plantas nativas es necesaria si se pretende que los ecosistemas de artemisa se recuperen efectivamente de alteraciones o disturbios futuros.

Palabras Clave: Artemisia, dinámica del paisaje, exclusión del fuego, incendio, intervalo medio de incendios rotación del fuego.

Sagebrush (*Artemisia* spp.)–dominated ecosystems are threatened by energy development, housing, roads, domestic livestock grazing, invasive species, global warming (Knick et al. 2003, this volume, chapter 12; Miller et al., this volume, chapter 10), and by fire. Large wildfires are occurring and managers also burn sagebrush for a variety of purposes. However, understanding of the historical role of fire in sagebrush ecosystems is under revision. Scientists commonly suggested fire was frequent in sagebrush ecosystems prior to Euro-American settlement, and subsequent exclusion of fire allowed tree invasion, particularly into high-elevation sagebrush communities (Miller and Rose 1999; Miller et al., this volume, chapter 10) or an increase in sagebrush density (Winward 1991). However, new evidence suggests fire was not historically frequent in sagebrush (Baker 2006a), and recent scientific consensus is that data on historical fire regimes are insufficient to ascertain how important fire exclusion has been for tree invasion into sagebrush ecosystems (Romme et al. 2008). Better understanding of the historical range of variability (HRV; Landres et al. 1999) of fire is needed to determine if sagebrush ecosystems today are deficient in fire or have a surplus, and whether fire maintained sagebrush and prevented tree invasions. The few centuries before Euro-American settlement have the most relevant information for the HRV.

Part of the current revision regarding the historical role of fire is because past methods used to study fire under the HRV have been shown to be inaccurate (Kou and Baker 2006a,b; Baker 2006b), and replacements have been developed (Kou and Baker 2006a). Past studies used composite fire intervals (CFIs) in adjoining forests to conclude that sagebrush burned frequently under the HRV (Crawford et al. 2004; Miller et al., this volume, chapter 10). This idea was mirrored in a prototype national assessment by the LANDFIRE program (Rollins and Frame 2006). However, first estimates of fire rotation and new estimates of mean fire interval were presented in a recent critical review, which concluded that fire in sagebrush ecosystems had long rotations and long mean fire intervals under the HRV (Baker 2006a).

New research has since appeared, and the purpose of this chapter is to update what is known about HRV for fire regimes in sagebrush ecosystems building on parts of Baker (2006a), but with added evidence and an emphasis on landscape dynamics. Baker (2006a) estimated only low values for fire rotation and mean fire interval, but here I estimate the full range. I also include the first analysis of recent fire rotation in sagebrush ecosystems relative to fire rotation under the HRV. I focus on fire and sagebrush taxa, but emphasize that these ecosystems include a diversity of other shrubs, grasses, and forbs (Knick et al. 2005; Miller et al., this volume, chapter 10). Fire has largely negative effects on Greater Sage-Grouse (*Centrocercus urophasianus*; Nelle et al. 2000, Knick et al. 2005, Beck et al. 2009). In this chapter, I focus on the historical role of fire in sagebrush ecosystems rather than indirect influence on sage-grouse.

FIRE CONTROLS THE LANDSCAPE MOSAIC

The intensity or energy release of fires in sagebrush can vary over a sevenfold range because of variation in fuel loads, fuel moisture, and wind speed (Sapsis and Kauffman 1991), but sagebrush fires are nearly all high-severity or stand-replacing, not low- or mixed-severity (Baker 2006a). A low-severity fire would burn beneath the shrubs but not kill aboveground stems. A mixed-severity fire would burn in places beneath shrubs without top killing them, while killing them in other places. High-severity fire implies that the sagebrush or other shrub that burns is top killed; in most sagebrush taxa, the plant is also killed, because most taxa do not resprout after fire. The evidence for high-severity fires in sagebrush was summarized by Britton and Clark (1985:23): "It is relatively unimportant how fast the fire moves, how hot the fire is, or what the fire intensity is . . . if a fire front passes through an area, the sagebrush will be killed." Since fire does not generally burn through sagebrush stands at low severity, low-severity fire does not thin sagebrush stands, and exclusion of low-severity fire cannot lead to increases in sagebrush density within a stand, as implied by Winward (1991).

Fire or fire exclusion instead shapes the landscape mosaic, which consists of patches of burned, recovering, and long-unburned sagebrush varying in density, height, cover, and other attributes (Figs. 11.1a,b). Characteristics of the mosaic are shaped by several aspects of the fire regime, including fire rotation and mean fire interval, fire sizes, and pattern of unburned area.

TRANSITION IN FIRE HISTORY AND THE ROLE OF FIRE IN SAGEBRUSH

Fire-history terms and methods must be discussed, because revision of methods is a primary reason that understanding of the role of fire in sagebrush is in transition. Fire rotation is the expected time to burn once through a land area equal to that of a landscape of interest (Baker and Ehle 2001, Reed 2006). Fire rotation is the only fire-regime parameter that quantifies how long it takes, on average, for fire to burn across a particular landscape, and it is the key parameter to know and understand in managing fire. Historical fire rotation under the HRV must be known to determine whether a particular landscape today has a deficiency or surplus of fire.

The population-mean fire interval, which is the average point-mean fire interval (mean interval between fires at a point in the landscape) across a landscape, is equal to the fire rotation (Baker and Ehle 2001), and this equivalency is a central concept. Reed (2006) suggested this equivalency did not hold, but this is incorrect because of an error in Reed's analysis (Baker 2009). This equivalency means that each individual estimate of point-mean fire interval also estimates the fire rotation near that point in the landscape. Fire rotation across a landscape can be estimated from a statistical sample of point mean fire intervals across a landscape. This can also be reversed. Fire rotation provides expected mean fire interval at any point in the landscape. If mean fire interval across sample points is 100 years, fire rotation is an estimated 100 years. If fire rotation is 100 years, then we expect to find mean fire intervals of about 100 years at points.

Fire rotation can be calculated at any spatial extent by summing areas of individual fires over a period of observation and dividing the period by the fraction of the landscape burned:

$$FR = \frac{t}{\sum_{i=1}^{n} a_i/A} \qquad \text{(Eqn. 11.1)}$$

where FR is fire rotation in years, t is the period of observation in years, a_i is the area of fire i of n total fires observed over period t, and A is the areal extent for which fire rotation is being estimated. For example, in a period of 50 years, if 7,500 ha of a 10,000-ha study area burns, the fire rotation is 50/(7,500/10,000) = 66.67 years. Fire rotation can

Figure 11.1. Fires in sagebrush in the last few years illustrating spatial heterogeneity and human effects: (a) 2007 fire near Burns, Oregon, illustrating a mosaic of burned and unburned areas, including topographic effects (background on hillside) and either shifts in wind or variation in fuel loading or fuel moisture (foreground); (b) Murphy complex fires of 2007 in southern Idaho and northern Nevada, illustrating a large, high-severity fire with few unburned areas; (c) cheatgrass and dead sagebrush on the Snake River Birds of Prey National Conservation Area, Idaho; (d) bulldozer fire line up a hill in a 2007 fire in northern Nevada.

also be estimated by the mean of a set of estimates of point-mean fire interval.

Spatial heterogeneity in fire regimes can be analyzed at multiple spatial scales by estimating either fire rotation in a set of subareas or mean fire interval in a set of points across a landscape (Baker 1989). Point estimates are smaller samples that are inherently less accurate. Fire rotation is not just a coarse-scale measure, and mean fire interval is not just a fine-scale measure of fire, as suggested (Miller and Heyerdahl 2008; Miller et al., this volume, chapter 10). Both measures can be calculated at any spatial scale, and the two measures are directly related by the equivalency of fire rotation and average mean fire interval across a sample of points.

Revision of fire-history methods has arisen because a commonly calculated measure has been shown not to accurately estimate the mean fire interval at a point. The common measure derives from composite fire intervals (CFIs), which are obtained by making a complete list (composite) of fires that burned a particular number or fraction of sample trees in a study area. Intervals are calculated between the fires in the list, along with summary parameters such as mean CFI.

Mean CFI nearly always underestimates the length of the actual mean fire interval at a point (Baker and Ehle 2001; Baker 2006b; Kou and Baker 2006a,b). These studies explain the reasons: (1) fires are commonly included in the composite list that did not actually burn the point because sampling areas are too large; (2) most fires are small and do not burn the whole study area, but CFI does not adjust for fire size; (3) mean CFI declines as sample size increases, an undesirable property that means its value may be more related to sample size than a property of a fire regime; (4) intentional targeting of particular sample areas and particular sample trees has been common and biases CFI estimates toward shorter intervals; and (5) the longest fire intervals, which are often incomplete, are commonly omitted, biasing CFI toward shorter intervals.

An empirical study (Van Horne and Fulé 2006) and a simulation study (Parsons et al. 2007) suggested CFI was accurate, but only compared among different methods of estimating CFI, rather than to a known mean fire interval or fire rotation. Computer simulation that did compare CFI to known mean fire interval and fire rotation (Kou and Baker 2006b) and empirical analysis that compared CFI to fire rotation known from mapped fires (Baker 2006b) both confirm CFI is inaccurate. Modifications of CFI also fail and CFI likely cannot be fixed (Kou and Baker 2006a,b).

A new estimator of point-mean fire interval has been derived and shown by simulation to be accurate and unbiased (Kou and Baker 2006a), but has not yet been applied near sagebrush landscapes. In the meantime, calibration shows that point-mean fire interval, as well as fire rotation and population-mean fire interval, can be estimated as a multiple of mean CFI (Baker and Ehle 2001, Kou and Baker 2006b). For example, in a Grand Canyon calibration, mean CFIs of about 5–20 years were found where fire rotations and average mean fire intervals actually were about 45–292 years (Baker 2006b). The best current estimate is that mean CFI can be multiplied by 3.6–16.0 to estimate point-mean fire interval or average mean fire interval and fire rotation (Baker and Ehle 2001, Baker 2006b). I will use this estimate later in this analysis. Review of sagebrush fire regimes in Miller et al. (this volume, chapter 10) is based on uncorrected CFIs.

ESTIMATING HISTORICAL FIRE ROTATION USING MULTIPLE LINES OF EVIDENCE

Fire rotation and mean fire interval are difficult to reconstruct in large expanses of sagebrush. Sagebrush plants do not record fire scars that can be dated, older stand origins from fire cannot be dated, and few locations are likely to yield adequate pollen and charcoal evidence. Thus, sources of information about fire rotation and mean fire interval in sagebrush landscapes are limited to: (1) fire-scar records from adjoining forests, (2) fire-rotation estimates in adjoining forests, (3) fire-frequency estimates from macroscopic charcoal records in sediments near sagebrush, and (4) sagebrush recovery time. All of these lines of evidence were used to estimate ranges of fire rotation and mean fire interval in sagebrush.

A New Estimate of Adjacency Correction

The first two sources, both from forests, require correction to estimate fire rotation and mean fire interval in adjacent sagebrush. These two are valuable sources that can provide annual-resolution fire history, but the need for correcting them is unavoidable, since they are from forests that may have different fire regimes. Miller and Heyerdahl (2008) used inference based on fuels, soils, and succession to qualitatively suggest fire regimes in sagebrush near forests, but a repeatable quantitative estimation method is needed, which I call adjacency correction.

To estimate fire rotation in sagebrush, I proposed an adjacency correction of 2.0 (Baker 2006a), so if fire rotation was 400 years in woodlands, it would be estimated as 800 years in adjacent sagebrush. The basis for this correction was the number of lightning strikes per fire start in sagebrush ($N = 144$), in Douglas-fir (*Pseudotsuga menziesii*; $N = 42$), and in ponderosa pine (*Pinus ponderosa*; $N = 24$) in Idaho (Meisner et al. 1994); thus, fire was three to six times less likely to ignite in sagebrush than in forests. Seven to 23 times fewer fires per unit area occurred in sagebrush than in forests in Colorado (Fechner and Barrows 1976). I chose 2.0 as a conservative estimate of the needed correction. However, ignition rates and fire-density estimates may have little relationship with fire rotation.

To better refine needed adjacency correction, I estimated recent fire rotations in sagebrush and pinyon-juniper (*Pinus-Juniperus*) woodlands throughout the western United States; the ratio of fire rotations likely provides a better correction. I used a point map of fires that gives area burned by each fire (but not its boundary) from 1980 to 2003 (United States Bureau of Land Management 2004). The Westgap map of the national GAP program, a mapping program using satellite imagery (United States Geological Survey 2005b), was used to select fires in sagebrush (codes 74, 75) and pinyon-juniper (codes 67–69). Areas of individual fires were used to estimate fire rotation for the 24-year period (Equation 11.1). Fire rotation was 235 years for sagebrush and 409 years for pinyon-juniper, yielding a ratio of 0.57. This suggests Baker's (2006a) adjacency correction was incorrect.

Fires may ignite at a lower rate and occur at lower density in sagebrush than woodlands and forests (Baker 2006a), but must burn more land area per fire in sagebrush. Perhaps this is because of the more open and windy environment of sagebrush and relatively few barriers to fire spread.

The maps (United States Bureau of Land Management 2004, United States Geological Survey 2005) have limitations, and some assumptions must be made. The map of fires is a point map, but one with area burned for each fire, not a polygon map with explicit fire boundaries; some incorrect selections could have occurred because of where the point was mapped. Some duplicate records were found and some areas of sagebrush appear lacking in fire records. A period of 24 years is insufficient to accurately estimate long rotations. Polygon records would overcome some of these problems. Fire rotation for 1980–2007 in sagebrush from polygon records was 169 years, based on a weighted mean of province estimates using province area as weights (Table 11.1), compared to 235 years for the point map. This suggests the point map underestimates fire rotation, but there is no comparable polygon fire map for pinyon-juniper woodlands. Finally, these recent fire rotations may differ from fire rotations under the HRV, but the ratio of rotations may be less affected by land uses, particularly if land uses had similar effects in these ecosystems.

Thus, as an improved but still approximate adjacency correction, I assume fire rotation in sagebrush under the HRV was 0.57 times (235 / 409) fire rotation in pinyon-juniper woodlands, where correction is needed. Two situations occur. Where woodlands or forests surround, or are in close proximity to, small areas of sagebrush, it is more likely the fire regime is dominated by the woodland or forest fire regime, and no adjacency correction is applied. The adjacency correction is applied only where woodlands or forests adjoin large areas (hundreds of hectares) of sagebrush. Correction is not available by sagebrush taxon.

This adjacency correction has some calibration, but could be improved by further research. It would be desirable to have a full modern calibration (sensu Baker and Ehle 2001) that analyzes how to accurately estimate fire rotation in sagebrush from data in nearby forests, using a modern sagebrush landscape in which data from mapped fires are available. Modern calibration has been completed for few fire-history methods (Baker and Ehle 2001, Baker 2006b).

Estimating Sagebrush Fire Rotation from Fire History in Adjoining Forests

Adjacency correction is needed for both fire-scar and fire-rotation estimates from adjoining forests. Fire-scar records on trees near sagebrush stands actually require two corrections to estimate fire rotation in sagebrush landscapes. I first used the known range of multipliers for CFI, 3.6–16.0, and the 0.57 adjacency correction to estimate mean fire interval and fire rotation. Adjacency correction is likely not needed if sagebrush area is small and surrounded by forest, but otherwise the 0.57 correction is applied. Estimated fire rotations, after corrections, are from all available fire-scar studies of sagebrush near forests (Table 11.2). These estimates have a large range because of large correction ranges and are of less value than other sources.

Fire-rotation estimates are inherently better estimates because they do not require correction using CFI multipliers, as do CFI estimates. New studies since Baker (2006a) are included (Table 11.1). Pinyon-juniper woodlands adjoin much sagebrush, particularly Wyoming big sagebrush (*Artemisia tridentata* ssp. *wyomingensis*), little sagebrush (*A. arbuscula*), and mountain big sagebrush (*A. tridentata* ssp. *vaseyana*). No adjacency correction is likely needed or used where woodlands surround smaller areas of sagebrush.

In one case, researchers estimated the actual areas of each reconstructed fire, and fire rotation can be estimated using Equation 11.1. This was a study of a mosaic of mountain big sagebrush and Douglas-fir forest in Montana (Heyerdahl et al. 2006). I interpolated the area of 11 fires from 1700–1860, representing the HRV, from the study's Fig. 11.3b to be 10, 175, 75, 40, 110, 210, 30, 50, 210, 110, and 300 ha, for a total of 1,320 ha burned in the 1,030-ha study area in a 160-year period. This is a fire rotation of 125 years, based on Equation 11.1. No correction is needed for adjacency, since the mosaic is an intermixture of sagebrush and Douglas-fir forests. The fire-size estimates of Heyerdahl et al. (2006) are based on a convex hull around locations with evidence of a fire, but unburned areas likely occurred in the sagebrush. Using the best available estimate of unburned area in mountain big sagebrush (21%; Baker 2006a), corrected fire rotation and average

TABLE 11.1

Estimates in years for pre-Euro-American fire rotations and mean fire interval in sagebrush.

Sources are fire scars and fire rotation in adjoining forests, fire frequency in paleo-charcoal records, and time for sagebrush to recover fully after fire.

		Original sources			Corrected estimates			
Taxon		Source	Setting	Estimate	After 3.6 mult. corr.	After 16.0 mult. corr.	Small sagebrush areas after no adj. corr.	Large sagebrush areas after 0.57 adj. corr., if needed
Little sagebrush	Young and Evans (1981)	Scars	Adjacent	95	342	1,520	—	195–866
	Miller and Rose (1999)	Scars	Intermix	138	497	2,208	497–2,208	—
	Bauer (2006)[a]	Rotation	Intermix	427	—	—	427	—
	Summary						>425	>200
Wyoming big sagebrush	Young and Evans (1981)	Scars	Adjacent	95	342	1,520	—	195–866
	Floyd et al. (2004)[b]	Rotation	Intermix	~400	—	—	~400	—
	Bauer (2006)[a]	Rotation	Both	427	—	—	427	243
	Shinneman (2006)	Rotation	Both	400–600	—	—	400–600	228–342
	Mensing et al. (2006)	Charcoal	Expanses	200–500[c]	—	—	—	200–500
	This chapter	Recovery	Expanses	Uncertain	—	—	—	Uncertain
	Summary						400–600	200–350
Mountain big sagebrush								
Fast track	This chapter	Recovery	Expanses	>50–70	—	—		>50–70
Slow track	This chapter	Recovery	Expanses	>150–200	—	—		>150–200
	Jacobs and Whitlock (2008)	Charcoal	Expanses	150–200	—	—	—	150–200
	Nelson and Pierce (2010)	Charcoal	Expanses	183	—	—	—	183

TABLE 11.1 *(continued)*

TABLE 11.1 (CONTINUED)

Taxon		Original sources			Corrected estimates			
		Source	Setting	Estimate	After 3.6 mult. corr.	After 16.0 mult. corr.	Small sagebrush areas after no adj. corr.	Large sagebrush areas after 0.57 adj. corr., if needed
Mountain big sagebrush (cont.)								
Near piñon-juniper	Burkhardt and Tisdale (1976)	Scars	Adjacent	>30–40	>108–144	>480–2,304	—	>62–1,313
	Wangler and Minnich (1996)[a]	Rotation	Intermix	480	—	—	480	—
	Floyd et al. (2008)[b]	Rotation	Intermix	400–600	—	—	400–600	—
	Bauer (2006)[a]	Rotation	Both	427	—	—	427	243
	Shinneman (2006)	Rotation	Both	400–600	—	—	400–600	228–342
Near Douglas fir	Heyerdahl et al. (2006)	Rotation	Intermix	160[d]	—	—	160	—
	Summary						160, 400–600	150–300
Mountain grasslands/patchy sagebrush	Houston (1973)[b]	Scars	Both	20–25	72–90	320–400	72–400	41–228
	Arno and Gruell (1983)	Scars	Both	<35–40	<126–144	<560–2304	<126–2,304	<72–1,313
	Miller and Rose (1999)	Scars	Intermix	12–15	43–54	192–864	43–864	—
	Summary						Uncertain	40–230

[a] Bauer lists little sagebrush and Wyoming big sagebrush, but the correct taxa are mountain big sagebrush and Wyoming big sagebrush in the valley bottom, mountain big sagebrush higher in the watershed, and little sagebrush on ridgetops (P. J. Weisberg, pers. comm.).

[b] Authors do not identify the sagebrush taxon nearby; I assigned this tentatively based on elevation or other aspects of the environmental setting.

[c] This estimate is related to fire frequency, and may require correction to estimate fire rotation and mean fire interval, but the needed correction is unknown.

[d] Estimated from data in Heyerdahl, et al. (2006).

TABLE 11.2

Sagebrush area, sagebrush area burned, and estimated recent fire rotation and mean fire interval, using equation 11.1 by floristic province (Fig. 11.4).

Estimates are based on total area burned over the period 1980–2007 (data from Miller et al., this volume, chapter 10). Years of record are less than the full 28 yr between 1980–2007 because of missing or incomplete data. Sagebrush area by floristic province is from GIS analysis of a map of sagebrush from the U.S. LANDFIRE program (S.P. Finn, pers. comm.).

Floristic province	Sagebrush area (ha) = a	Total sagebrush area burned (ha) = b	Fraction of sagebrush burned = b/a	Years between 1980–2007 with usable data	No. of usable years (N)	Estimated fire rotation (yr) = N/(b/a)
Columbia Basin	2,677,204	327,814	0.12	1994, 1996, 1999–2003, 2006, 2007	9	74
Northern Great Basin	6,749,895	1,104,730	0.16	1980–2003, 2006, 2007	26	158
Snake River Plain	11,285,028	4,208,448	0.37	1980–2003, 2006, 2007	26	70
Wyoming Basin	9,113,938	263,983	0.03	1994–2003	10	345
Southern Great Basin	10,212,674	2,180,419	0.21	1980–2003, 2006, 2007	26	122
Colorado Plateau	2,031,744	89,787	0.04	1994–1996, 1998–2003, 2006, 2007	11	249
Silver Sagebrush	5,474,227	555,417	0.10	1984, 1985, 1988–2003, 2006, 2007	20	197

mean fire interval in mountain big sagebrush would be ~160 years, based on Equation 11.1. This is the only available estimate of fire rotation in mountain big sagebrush near Douglas-fir forests. The mean CFI across the study area would be 16 years (160/(11 − 1)), reinforcing the conclusion that fire rotation and mean fire interval are 3.6–16 times the mean CFI (10.0 times in this case).

Estimating Sagebrush Fire Rotation from Macroscopic Charcoal in Sediments

Macroscopic charcoal fragments and pollen from a permanent spring show the relationship between fire, drought, and sagebrush abundance over the last 5,500 years in central Nevada (Mensing et al. 2006). Macroscopic charcoal generally reflects larger fires within watersheds (Mensing et al. 2006), and this was supported by a charcoal peak associated with a >1,400-ha fire in 1986 near the spring. Upland vegetation was Wyoming big sagebrush with some basin big sagebrush (*A. tridentata* ssp. *tridentata*), salt desert shrubs, and wetland vegetation near the spring. Inferred fire frequency is a count of the number of fire events per 1,000-year period in the watershed based on charcoal peaks identified by particular thresholds above a background level (Fig. 11.2). The A/C ratio represents the ratio of *Artemisia* pollen to pollen of salt desert plants (Chenopodiaceae and Sarcobataceae families); positive values represent wetter climate with more *Artemisia*, and negative values represent drier climate with more salt desert plants.

The A/C ratio suggests that as the climate became wetter and sagebrush increased, so did the fire frequency (Fig. 11.2). The authors suggest that fire frequency could not be calculated, because individual fires could not be resolved. However, macroscopic charcoal likely identifies periods of large total burned area that most

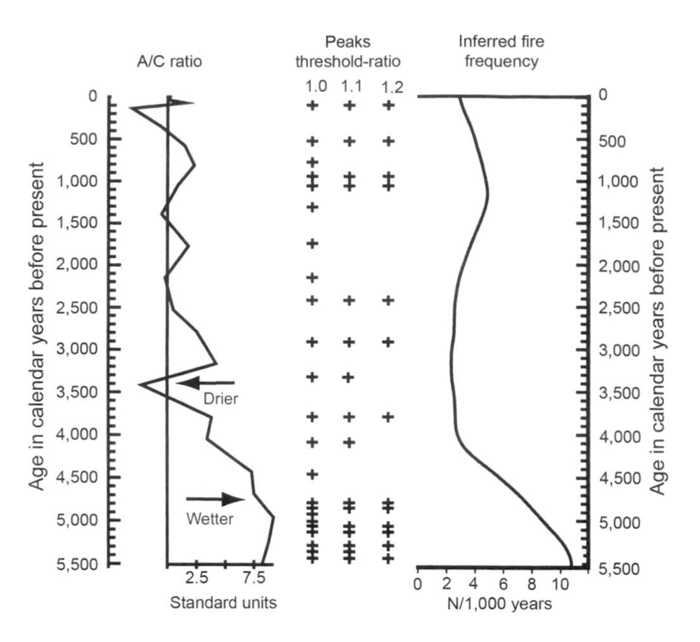

Figure 11.2. A paleo-indicator of climate, the A/C ratio, versus inferred fire frequency for Wyoming big sagebrush in central Nevada. Pluses represent charcoal peaks (fires) at varying threshold levels, and inferred fire frequency is a running count of number of peaks per 1,000 years. The A/C ratio is the ratio of pollen of *Artemisia* to Chenopodiaceae + Sarcobataceae, with low values representing drier climate and high values representing wetter climate. Reproduced from Fig. 6 of Mensing et al. (2006:74) with permission of *Western North American Naturalist*.

contribute to fire rotation, which does not require individual fire years, only accumulated burned area (Equation 11.1). It is not likely each charcoal peak represents a fire that burned all of a study area; it may require >1 detected peak to accumulate burned area equaling a study area. The inverse of charcoal-derived fire frequency (Fig. 11.2) likely estimates fire rotation, but the needed correction is unknown. Intervals between peaks varied from about 200–500 years over the last 1,000 years, corresponding to five to two peaks per 1,000 years (Fig. 11.2). Jacobs and Whitlock (2008) and Nelson and Pierce (2010) provide similar charcoal-based evidence of fire at 150-200 year intervals in mountain big sagebrush (Table 11.1).

Estimating Sagebrush Fire Rotation Using Sagebrush Recovery Time

Another method to estimate fire rotation in sagebrush is from the time required for sagebrush to regain full coverage and maturity after fire. The premise is that fires likely did not burn, on average, more often than the time required for sagebrush to recover (Wright and Bailey 1982). Data on sagebrush cover and frequency, from chronosequences of sites varying in time since fire, are compared to similar data in an unburned control site (Fig. 11.3). Cover is the essential measure, as frequency indicates only plant presence, not recovery to mature size and cover. This definition of recovery is a simplification, as recovery of sagebrush plants to a mature state could occur before pre-burn cover is reached, and pre-burn cover could have been outside the HRV because of past domestic livestock grazing or other land-use effects. Individual sagebrush plants are able to grow from seed to full maturity in a shorter period, but full coverage of mature plants across a burn best measures recovery. Recovery across a burn is hampered by slow sagebrush seed dispersal (Young and Evans 1989) and infrequent years favoring germination (Maier et al. 2001).

The available data suggest mountain big sagebrush recovers faster than does Wyoming big sagebrush (Fig. 11.3). New data since Baker (2006a) suggest two possible recovery tracks for mountain big sagebrush: a fast track represented by the 16 upper points, with nearly full recovery by about 25–35 years after fire (Fig. 11.3a), and a slower track represented by >40 points with 75 or more years for full recovery (Fig. 11.3a). The slow track could occur in larger fires, particularly if seed survival is low and seed must disperse into the fire from distant unburned areas. Welch and Criddle (2003) estimated 70 years for mountain big sagebrush to reach the middle of a large burned area and a few decades more for plants to mature. Thus, full recovery on the slow track may require up to 100 years (range = ~75–100 years). The fast track may be favored by more precipitation or otherwise favorable environment for sagebrush regeneration, smaller fires, or more survival of seed on the surface or in the seed bank. However, there may be a

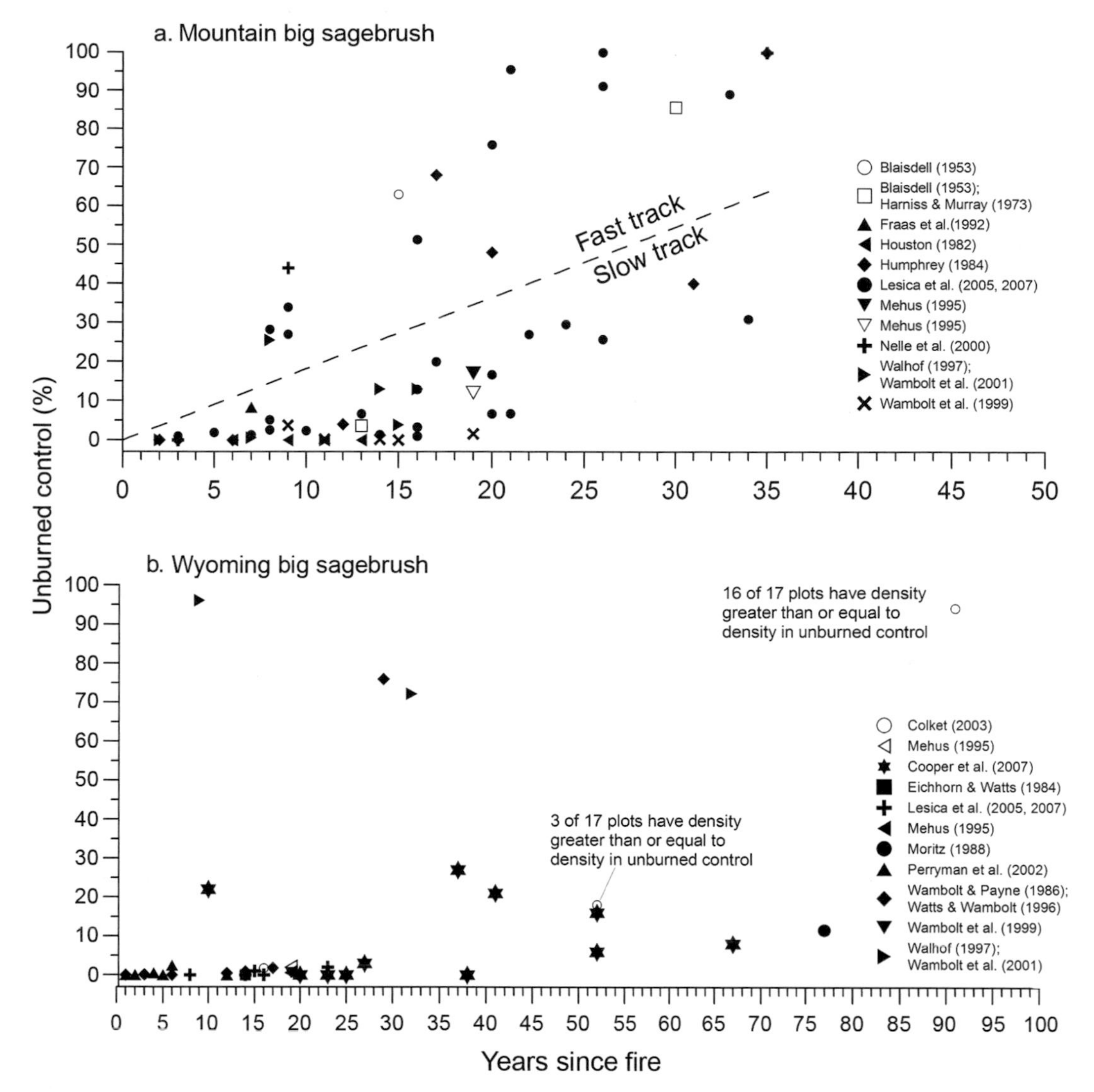

Figure 11.3. Sagebrush cover (closed symbols) and density (open symbols) versus time since fire as a percentage of these values in unburned control areas, for (a) mountain big sagebrush and (b) Wyoming big sagebrush. Each point represents a single sample plot, and similar symbols indicate a single study. In (a), apparent fast-track and slow-track recovery trajectories are separated by a dashed line. Only data for >20 years since fire, except one point, are presented from Cooper et al. (2007). I could not estimate the years for points <20 years since fire from their graph, but all missing points <20 years since fire have 0% recovery. From *Fire Ecology in Rocky Mountain Landscapes* by William L. Baker. Copyright © 2009 Island Press. Reproduced by permission of Island Press, Washington, DC.

continuum of rates of recovery rather than just two tracks. More research is needed on recovery rates, as neither fire severity nor differences in mean annual precipitation, heat load, or soil texture explained rate of recovery after fire in mountain big sagebrush in Montana (Lesica et al. 2007).

Wyoming big sagebrush recovery after fire is highly variable and often slow (Fig. 11.3b). Baker (2006a) estimated 50–120 years for full recovery, but recovery can occur more quickly (Fig. 11.3b), suggesting the possibility of a slow and fast track as well. My estimate was based on only two points that were close to full recovery, and only one was for cover (Fig. 11.3b). This evidence is too limited to accurately estimate the time for full recovery of Wyoming big sagebrush after fire. What is known is that by 25 years after fire, Wyoming big sagebrush typically has <5% of pre-fire cover (Fig. 11.3b). Further research is needed on effects of fire size and environmental setting on rate of recovery, and more data are needed from fires >50 years old.

Fire rotation in most ecosystems appears to be commonly much longer than the period to regain pre-fire cover of mature dominant plants. Perhaps this occurs because communities tend to become dominated by plants that, among other attributes, also can regrow sufficiently fast to have a reasonable period of maturity and seed production before suffering widespread mortality. Fire rotation appears to be commonly at least two to three times the period to regain pre-fire cover of mature plants. For example, mature pinyon-juniper woodlands recover within ~200 years where fire rotation is 400–600 years (Baker and Shinneman 2004, Floyd et al. 2004). In lodgepole pine forests, mature trees dominate after ~150 years where fire rotation is ~300 years (Buechling and Baker 2004). Similarly, it requires 20–30 years after fire in chaparral in California for shrubs to fully recover where fire rotation was ~80 years (Keeley et al. 1999). Thus, I conservatively estimate fire rotation and mean fire interval for sagebrush are at least twice the recovery period: >50–70 years for fast-track and >150–200 years for slow-track mountain big sagebrush.

Summary Estimates of Historical Fire Rotation in Sagebrush Under the HRV

Combining fire-scar, fire-rotation, charcoal, and recovery data leads to summary estimates of fire rotation and mean fire interval in sagebrush under the HRV (Table 11.1). More extreme values and imprecise estimates are passed over in the pooled estimates (e.g., fast-track recovery of mountain big sagebrush is uncommon; Burkhardt and Tisdale's CFI estimates have a large range). Fire rotation in little sagebrush is estimated to be >425 years in intermix with pinyon-juniper and >200 years in larger areas; in Wyoming big sagebrush it is 400–600 years in intermix with pinyon-juniper and 200–350 years in larger expanses; in mountain big sagebrush it is 160 years in intermix with Douglas-fir, 400–600 years in intermix with pinyon-juniper, and 150–300 years in larger expanses; finally, in mountain grasslands with patchy mountain big sagebrush, rotation is uncertain in intermix and 40–230 years in larger areas (Table 11.1).

Estimates of fire rotation and mean fire interval are likely to be refined further as new data are collected. Most estimates have a large range. Nonetheless, these estimates improve upon Baker (2006a) and are the best available, as they provide a full range of estimates using new data, improved corrections, and multiple lines of evidence (Table 11.1). These estimates suggest sagebrush did not burn frequently under the HRV, but instead at multi-century intervals.

In chaparral, another shrub-dominated ecosystem that had a much shorter fire rotation of about 80 years under the HRV (Keeley et al. 1999), dominant shrubs have prominent fire adaptations, such as resprouting and heat-stimulated seed germination (Keeley et al. 1999). In contrast, most sagebrush taxa do not survive fire and do not resprout, appear to lack heat-stimulated seed, are slow dispersers, and are slow to recover in burned areas (Young and Evans 1989, Baker 2006a). Little adaptation to fire in sagebrush taxa is consistent with evidence of long fire rotations and mean fire intervals under the HRV (Table 11.1). Fire changed the distribution of resources in sagebrush ecosystems so rarely under the HRV that sagebrush ecosystems likely have little or no fire-dependence, although they can recover after fire.

LARGE FIRES LEAD TO MATURE BUT FLUCTUATING SAGEBRUSH LANDSCAPES

Little is known about the properties of individual fires in sagebrush under the HRV, because it is difficult to reconstruct the shapes, sizes, amount of unburned area, and other attributes of individual fires that occurred in past centuries in sagebrush. However, nearly every fire regime studied throughout the world shares some properties, and these basic theoretical relationships likely also apply to sagebrush landscapes.

Consistent properties of the distribution of fire sizes mean that large fires are the key fires that likely shaped sagebrush landscapes under the HRV. A nearly linear negative trend is common between number of fires and fire size on a log-log plot, thus there are exponentially more small than large fires (Cumming 2001, Weiguo et al. 2006). Yet a large fraction (usually >90%) of total burned area in fire regimes comes from the largest few percent of total fires, and small fires do not total to much burned area (Strauss et al. 1989). In the Rocky Mountains, for example, ~96% of total burned area between 1980 and 2003 was from the 2% of total fires that were >200 ha, whereas only 0.1% of the total fires (those >15,000 ha) accounted for about half the total burned area (Baker 2009, an analysis of United States Bureau of Land Management 2004). These fire-size relationships mean that the few largest fires that occur in particular places, which occur at intervals approaching the fire rotation (Table 11.1), are the fires that control nearly all of the reshaping of landscape mosaics that is done by fire.

The large fires that most shape sagebrush landscapes become large in part because they are relatively high intensity from consuming most of the sagebrush and other fuels, which increases their spread rate. Large fires are also promoted where fuels are continuous, winds are strong, topography is level, and natural firebreaks are rare or lacking. Natural firebreaks potentially include rivers and smaller streams, canyons, rock outcrops and talus, sand dunes, wetlands, or other areas with limited or moist fuels. Recent sagebrush fires have become largest under these conditions (Knapp 1998), as on the Snake River Birds of Prey National Conservation Area in Idaho (Fig. 11.1c). These large fires, both today and under the HRV, were promoted the year after cool, wet years, likely because cool, wet years increase fine-fuel production; weather conditions in the fire year itself are less important (Knapp 1998, Miller and Rose 1999, Westerling et al. 2003).

Large fires can burn much of the total sagebrush cover, leaving few unburned islands of surviving shrubs. Unburned area within a fire perimeter is likely higher when sagebrush cover or fine fuels are lower or fuel moisture is higher, or because shifting winds and variable topography and fuels (Baker 2006a, Wright and Prichard 2006) may leave a complex mosaic of burned and unburned area (Fig. 11.1a). However, large areas of sagebrush can and do burn, leaving little or no unburned area within a fire perimeter (Fig. 11.1b). For example, 8,300 ha of Wyoming big sagebrush in Idaho burned in 6.5 hr in 1994, leaving little unburned area within the fire perimeter (Butler and Reynolds 1997). Pre–Euro-American sagebrush fires may have had less unburned area than is typical in modern prescribed fires (Wrobleski and Kauffman 2003).

The interludes between large fires are nearly as long, on average, as the fire rotation (Table 11.1). During these long interludes, sagebrush could fully recover and dominate in spite of poor dispersal capability (Young and Evans 1989) and slow recovery (Fig. 11.3). Thus, sagebrush landscapes would have been dominated most of the time by large areas of mature sagebrush, as documented by early historical accounts of explorers (Vale 1975).

However, during these long interludes, the density, cover, and condition of sagebrush likely fluctuated naturally under the influence of a variety of other mortality and defoliation agents, including insects, disease, and drought (Anderson and Inouye 2001, Beck et al. 2009). Sagebrush cover may also decline if there is adequate cover of understory native plants to provide competition for regenerating sagebrush (Lommasson 1948, Anderson and Inouye 2001) and increase again during favorable climatic episodes (Maier et al. 2001). Mature sagebrush may at times have dead branches and relatively slow growth, but decadent is an inappropriate term for the mature but fluctuating condition of sagebrush plants during the long interludes between large fires.

The long-interlude mature landscape would have been sparsely peppered over time by small fires, each small fire probably leaving more unburned islands. Long interludes likely had higher landscape diversity, because of small burns and mosaics of burned/unburned area within large expanses of mature sagebrush. Fire did not maintain vegetation in an early- or mid-seral condition as suggested based on comparison with uncorrected CFI estimates of fire (Lesica et al. 2007). Fire rotations were instead long, and the amount of early-successional post-fire vegetation was likely low much of the time, because small interlude fires account for little total burned area. Early accounts of explorers document little area of grassland within large expanses of sagebrush (Vale 1975). Infrequent large fires ending the interludes could initiate multi-decadal periods of reduced landscape diversity and more prominent early-successional vegetation while sagebrush recovers.

Sagebrush landscapes likely fluctuated, as do all landscapes subject to fire, because all known fire regimes have episodes of extensive fire, accounting for most of the total burned area, followed by long interludes of small fires, accounting for little burned area (Baker 2009).

RECENT FIRE ROTATION AND LAND-USE EFFECTS IN SAGEBRUSH

Recent Fire Rotations Relative to Fire Rotations Under the HRV

To analyze whether decreased or increased fire might be evident recently, I used data on total area burned by year from 1980 to 2007 from polygon fire maps by Miller et al. (this volume, chapter 10: Figs. 10.11–10.15) to estimate recent fire

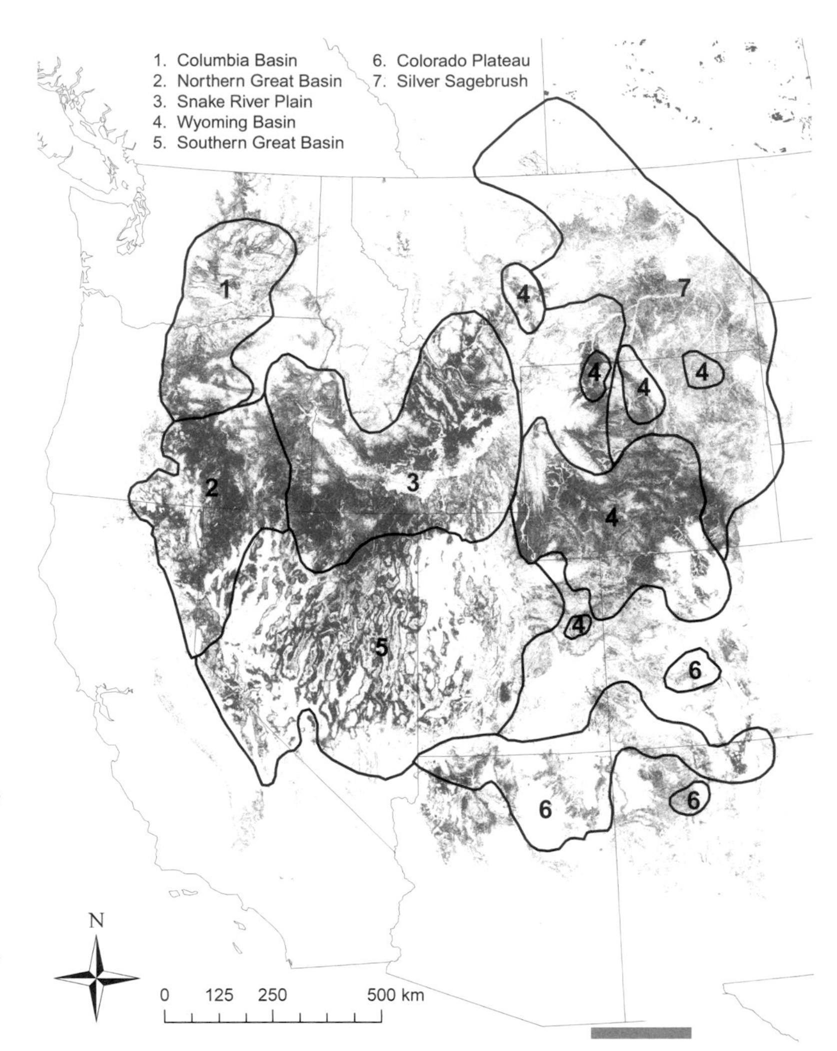

Figure 11.4. Sagebrush in the western United States (gray shading) and seven floristic provinces (Table 11.2) used in the analysis of recent fire rotations. Sage-Grouse Management Zones initially were derived from boundaries of floristic provinces (Stiver et al. 2006).

rotation for each of seven floristic provinces (Fig. 11.4). The estimates are fire rotations that would ensue if recent rates of burning continued. However, 26-year or shorter periods (Table 11.2) are too short to accurately estimate fire rotations or mean fire intervals in the hundreds of years; thus the estimates are imprecise. Rotations estimated from short periods of data are good indicators of present rates of burning, but may fluctuate or change in the future. Past suggestions that fire exclusion was having an effect in sagebrush were based on data from only a few small tree-ring sites (Miller and Rose 1999), but we now have area estimates of fire across the geographical range of sagebrush ecosystems in the western United States.

Recent fire rotations (Table 11.2) can be compared to estimates for the HRV (Table 11.1). Recent estimates are not available by sagebrush taxon, which is how estimates are available for the HRV (Table 11.1), and the tables are not directly comparable. Nonetheless, fire rotation since ~1980 (Table 11.2) is almost certainly outside the HRV and far too short in floristic provinces where Wyoming big sagebrush is common and for which fire rotation was likely in the 200–350 years range under the HRV (Table 11.1). These include the Snake River Plain and Columbia Plateau floristic provinces, as well as both the Southern and Northern Great Basin (Table 11.2), all of which have extensive cheatgrass invasions. Similarly, in central Nevada (Southern Great Basin floristic province), charcoal from fire in Wyoming big sagebrush was an order of magnitude higher after Euro-American settlement than any time in the previous 5,500 years (Mensing et al. 2006).

Longer recent fire rotations (200–350 years) in the Silver Sagebrush, Colorado Plateau, and Wyoming Basin floristic provinces (Table 11.2) are not likely outside the HRV, given the 200–350 years estimate for Wyoming big sagebrush and 150–300 years estimate for mountain big sagebrush in large sagebrush areas, and even longer rotations in intermix under the HRV (Table 11.1). However, Wyoming Basin and Colorado Plateau estimates are based on only 10–11 years of data (Table 11.2).

These estimates, although based on limited data, suggest there is now likely similar fire relative to the HRV in three floristic provinces and too much fire relative to the HRV in four floristic provinces. Estimates of fire rotation for individual sagebrush taxa, not presently feasible, might be different.

Reduced Fire: Fire Exclusion and Potential Fraction of Affected Land

Fire exclusion clearly had little overall effect, since fire rotations are likely either similar or shorter today than under the HRV, but it is worth considering whether fire exclusion might still have reduced fire in some communities. Fire exclusion can arise for several reasons, including reductions in fine fuels that enhance fire spread, intentional suppression, and landscape fragmentation by anthropogenic firebreaks such as roads. Landscapes were fragmented by roads, agricultural developments, and other human infrastructure early in Euro-American settlement, but fragmentation expanded over the 20th century (Knick et al., this volume, chapter 12). Intentional fire control began with Euro-American settlers (Baker 2009), but was relatively ineffective until the late 1950s, when aerial fire suppression expanded (Ely et al. 1957). Anthropogenic firebreaks expanded spatially and temporally in heterogeneous patterns. Maps of some of the potential agents (Leu et al. 2008) provide important clues, but local analysis of pattern, timing, and magnitude is needed.

The fire rotation under the HRV (Table 11.1) limits how fast a fire-exclusion effect is realized and how much of a landscape is affected. For example, if fire rotation under the HRV was 200 years, then 50 years of effective fire exclusion may have affected, on average, only about one-fourth (50/200) of the land area. Based on this approximation and estimated fire rotations (Table 11.1), fire exclusion had little expected effect in Wyoming big sagebrush or little sagebrush. In large areas of mountain big sagebrush intermixed with Douglas-fir, the expected affected area is only one-sixth to one-third of the land area (50/300 to 50/150), while an effect in intermixture with pinyon-juniper is likely minor (Table 11.1). In contrast, mountain grasslands with patchy mountain big sagebrush could have been largely affected if all fires had been excluded. Similarly, expectations are that tree invasions are likely not primarily due to fire exclusion in most sagebrush, except a part of the mountain big sagebrush ecosystem and in grasslands with patchy mountain big sagebrush. These are just theoretical expectations, not based on data showing that fire exclusion actually did occur.

However, spatial lags suggest fire exclusion cannot explain early, rapid tree invasions in any sagebrush ecosystems. Early reduction in fine fuels by livestock that likely occurred relatively simultaneously across landscapes is documented (Baker 2009), but a particular area of a landscape cannot be affected by fire exclusion until a fire ignites and its spread is restricted (Baker 1993). It requires half a fire rotation on average, for example, for a fire-exclusion effect to spread halfway across a landscape. Tree invasions that began immediately after Euro-American settlement (Miller and Rose 1999) must be primarily linked to causes that lack spatial lags. Tree invasions that accelerated after ~1960 (Soulé et al. 2003, Weisberg et al. 2007) match expanded aerial attack, but a multi-decadal lagged effect is still likely.

The net effect of land uses on sagebrush fire was similar or increased fire today relative to the HRV, not a decline in fire; thus, fire exclusion likely had little overall effect. Fire exclusion likely affected limited locations in parts of the mountain big sagebrush ecosystem and in grasslands with patchy sagebrush, but even in these locations effects are lagged and cannot explain invasions that began shortly after Euro-American settlement. More important, fire rotations and mean fire intervals were long in sagebrush under the HRV (Table 11.1)—certainly sufficiently long to allow trees to widely invade—yet millions of hectares of mature sagebrush without trees greeted early explorers (Vale 1975). This suggests that fire was not the primary factor preventing tree invasion into sagebrush under the HRV, and factors other than fire exclusion must be primary causes of tree invasions after Euro-American settlement.

Increased Fire—Cheatgrass, Human-Set Fires, and Global Warming

Where fires have increased in sagebrush ecosystems relative to the HRV, as in the four floristic provinces, this increase is significantly related to cheatgrass invasion, particularly in low elevation areas (Miller et al., this volume, chapter 10). Mechanisms causing increased fire in cheatgrass areas are explained in Miller et al. (this volume, chapter 10). Cheatgrass expansion after fire is particularly increased in vulnerable landscapes. In western Colorado, vulnerability to post-fire cheatgrass expansion was correlated with high pre-fire cheatgrass, low cover of biological soil crust, and low native forb and grass cover, all associated with degradation by domestic livestock grazing and with roads or other dispersal corridors into the fire area (Shinneman and Baker 2009). Cheatgrass expansion after fire is likely also increased by some fire-control practices, including bulldozer fire lines (Fig. 11.1d) and back burns set from roads, as these are areas that can have high cheatgrass populations (Gelbard and Belnap 2003).

Cheatgrass fires were evident by the 1920s or earlier (Pickford 1932). More than 50% of the Snake River Plain, Idaho, was dominated by cheatgrass by the 1990s, and fires burned ~36% of 290,000 ha in 18 years (Knick and Rotenberry 1997), a fire rotation of ~50 years. This is intermediate between the estimate of 27.5 years for 1979–1995 and 80.5 years for 1950–1979 in this area (Knick and Rotenberry 2000). This rotation is much shorter than in Wyoming big sagebrush under the HRV (Table 11.1) and is likely to prevent full sagebrush recovery (Fig. 11.3b).

Fires have also increased because of the expansion of the road network and increased ignitions by people (Baker 2009; Miller et al., this volume, chapter 10). Annual burned area has increased in general in the last few decades, relative to previous decades, because of global warming, identified for fires in forests near sagebrush (Westerling et al. 2006), but not specifically in sagebrush. However, climate projections are for more fire in sagebrush (Neilson et al. 2005).

Beyond Fire Exclusion as an Explanation of Recent Changes in Sagebrush Landscapes

Since fire is generally similar or has increased in sagebrush ecosystems since Euro-American settlement, a general fire-exclusion effect is lacking. Increased sagebrush density within sagebrush stands, where it has occurred, is likely caused by other land uses, such as domestic livestock grazing, or represents natural fluctuation (Anderson and Inouye 2001).

A complex of causes other than fire exclusion must largely explain tree invasions into Wyoming big sagebrush, little sagebrush, and even mountain big sagebrush, particularly invasions that began early. This complex includes (1) loss of competition from native grasses and forbs (Johnsen 1962), facilitation of tree regeneration by increased shrub cover, and enhanced seed dispersal, all related to domestic livestock grazing (Soulé et al. 2003); (2) climatic fluctuations favorable to tree regeneration (Soulé et al. 2003, 2004; Shinneman 2006); (3) enhanced tree growth because of increased water-use efficiency associated with a carbon dioxide fertilization effect (Soulé et al. 2004); and (4) natural recovery from past disturbance. Commonly overlooked is recovery from past human disturbances (e.g., deforestation from mining) and fires, droughts, and other natural disturbances (Romme et al. 2008). In New Mexico, for example, a rephotographic study found many cases in which tree-density increase and apparent invasion were recovery from earlier disturbances that had removed trees (Sallach 1986). Past focus on fire exclusion has left us with insufficient analysis of the relative importance of these more likely explanations of tree invasion into sagebrush.

CONSERVATION IMPLICATIONS

Sagebrush ecosystems, including all taxa reviewed here, do not require added disturbance today if the goal is restoring or maintaining sagebrush. Current fire rotations are likely short in Wyoming big sagebrush and short throughout four floristic provinces relative to the HRV, resulting in a surplus of fire. Moreover, large areas of sagebrush have been fragmented or converted to nonnative plants or agriculture (Knick et al., this volume, chapter 12); most natural disturbance agents—insects, drought, and fire—have not been reduced; and some, such as drought and fire, have increased in some areas and may increase further in the future under climate-change predictions. Most sagebrush landscapes are highly heterogeneous today because of land uses (Knick et al., this volume, chapter 12) and natural disturbances, and further increased heterogeneity is likely not needed.

If the goal is to mimic the disturbance regime in sagebrush under the HRV, these ecosystems need rest and recovery from past disturbances, particularly disturbances by land uses (Knick et al., this volume, chapter 12) and fire, not additional disturbance. Burning at rotations somewhat less than the fire rotation under the HRV (Table 11.1) might allow a mixture of plants, if the rotation were at least as long as the recovery period for sagebrush (Fig. 11.3). However, this would be atypical of sagebrush ecosystems under the HRV and could have deleterious effects on some plants and animals. Moreover, wildfire is expected to increase substantially in sagebrush because of global warming (Neilson et al. 2005), likely leading to fire rotations shorter than under the HRV, which is already evident in four floristic provinces. Prescribed burning is generally unnecessary and would not be restorative, given the present fire regimes in the floristic provinces and increased fire expected from global warming.

Where reversal of tree invasions or restoration of degraded sagebrush communities is the management goal, burning is also not essential or advisable, particularly if maintenance of the sagebrush canopy is needed. Degraded sagebrush communities deficient in native plants are unlikely to be restored by fire alone. Burning does not lower sagebrush density within a patch, since thinning fires are unknown, but instead reduces sagebrush cover across landscapes (Fig. 11.1a), and decades to centuries are required for sagebrush cover and density to fully recover (Fig. 11.3). Other means, including removal of livestock grazing (Anderson and Inouye 2001), allow native grasses and forbs to increase while also maintaining the sagebrush canopy. Similarly, where tree invasions were caused by livestock grazing or other land uses, invading trees are best controlled not using fire that would also kill sagebrush but by using the lowest-impact mechanical means that allows retention and survival of sagebrush, such as individual tree removal.

Restoration is likely to be ineffective if the specific causes of degradation or invasion are not identified and remedied. For example, if reduction in plant competition from livestock grazing is a primary cause of tree invasion, then burning or mechanically removing trees without restoring native plants and reforming management of livestock grazing is likely to lead to renewed tree invasion in a potentially endless cycle. Treatment of causes, not just symptoms, is essential for effective ecological restoration (Noss et al. 2006).

The most important ecological restoration needs in sagebrush are to control invasive species and restore the diversity and cover of native plants while retaining sagebrush cover, so the ecosystem has renewed capacity to resist fire and to recover effectively after fire and other disturbances (Baker 2006a, Link et al. 2006). This is an achievable goal (Link et al. 2006) that is especially important because of increased wildfire expected from global warming.

Protection of extant sagebrush ecosystems is increasingly a management goal, because of the decline of sage-grouse (Connelly and Braun 1997, Connelly et al. 2004) and other sagebrush obligates (Knick and Rotenberry 2002). Where the management goal is protection, active fire control is sensible wherever cheatgrass occurs. This includes much of the range of Wyoming big sagebrush and at least the lower elevations of the mountain big sagebrush zone. These sagebrush areas are vulnerable to potentially irreversible replacement by cheatgrass following fire, leading to sagebrush regeneration failure (Fig. 1c; Pellant 1990). Current fire rotations are likely too short in these areas to allow full recovery of Wyoming big sagebrush after fire. These areas warrant complete protection from fire until a solution is found to effectively control cheatgrass and until plant diversity can be sufficiently restored to allow natural recovery after fire.

ACKNOWLEDGMENTS

I appreciate funding from the United States Geological Survey to complete analyses for this article and from the Bureau of Land Management under agreement ESA020016, which supported earlier work in Baker (2006a) that is a foundation for this article. I am grateful for anonymous reviews, and particularly for thorough and helpful reviews by Peter Weisberg, Steven T. Knick, and John W. Connelly, which significantly improved the manuscript. I thank Sean P. Finn, Richard F. Miller, Steven T. Knick, David A. Pyke, Cara W. Meinke, Steven E. Hanser, Michael J. Wisdom, and Ann L. Hild for providing and sharing maps and data needed for the analyses.

CHAPTER TWELVE

Ecological Influence and Pathways of Land Use in Sagebrush

Steven T. Knick, Steven E. Hanser, Richard F. Miller, David A. Pyke, Michael J. Wisdom, Sean P. Finn, E. Thomas Rinkes, and Charles J. Henny

Abstract. Land use in sagebrush (*Artemisia* spp.) landscapes influences all sage-grouse (*Centrocercus* spp.) populations in western North America. Croplands and the network of irrigation canals cover 230,000 km^2 and indirectly influence up to 77% of the Sage-Grouse Conservation Area and 73% of sagebrush land cover by subsidizing synanthropic predators on sage-grouse. Urbanization and the demands of human population growth have created an extensive network of connecting infrastructure that is expanding its influence on sagebrush landscapes. Over 2,500 km^2 are now covered by interstate highways and paved roads; when secondary roads are included, 15% of the Sage-Grouse Conservation Area and 5% of existing sagebrush habitats are >2.5 km from roads. Density of secondary roads often exceeds 5 km/km^2, resulting in widespread motorized access for recreation, creating extensive travel corridors for management actions and resource development, subsidizing predators adapted to human presence, and facilitating spread of exotic or invasive plants. Sagebrush lands also are being used for their wilderness and recreation values, including off-highway vehicle use. Approximately 12,000,000 animal use months (AUM = amount of forage to support one livestock unit per month) are permitted for grazing livestock on public lands in the western states. Direct effects of grazing on sage-grouse populations or sagebrush landscapes are not possible to assess from current data. However, management of lands grazed by livestock has influenced sagebrush ecosystems by vegetation treatments to increase forage and reduce sagebrush and other plant species unpalatable to livestock. Fences (>2 km/km^2 in some regions), roads, and water developments to manage livestock movements further modify the landscape. Oil and gas development influences 8% of the sagebrush habitats with the highest intensities occurring in the eastern range of sage-grouse; >20% of the sagebrush distribution is indirectly influenced in the Great Plains, Wyoming Basin, and Colorado Plateau SMZs. Energy development physically removes habitat to construct well pads, roads, power lines, and pipelines; indirect effects include habitat fragmentation, soil disturbance, and facilitation of exotic plant and animal spread. More recent development of alternative energy, such as wind and geothermal, creates infrastructure in new regions of the sage-grouse distribution. Land use will continue to be a dominant stressor on sagebrush systems; its individual and cumulative

Knick, S. T., S. E. Hanser, R. F. Miller, D. A. Pyke, M. J. Wisdom, S. P. Finn, E. T. Rinkes, and C. J. Henny. 2011. Ecological influence and pathways of land use in sagebrush. Pp. 203–251 *in* S. T. Knick and J. W. Connelly (editors). Greater Sage-Grouse: ecology and conservation of a landscape species and its habitats. Studies in Avian Biology (vol. 38), University of California Press, Berkeley, CA.

effects will challenge long-term conservation of sage-grouse populations.

Key Words: agriculture, Conservation Reserve Program, energy development, land use, livestock grazing, off-highway vehicle, prescribed fire, sagebrush, sage-grouse, urbanization.

Influencia Ecológica y Circuitos del Uso del Suelo en Artemisa

Resumen. La utilización del suelo en paisajes de artemisa (*Artemisia* spp.) influencia a todas las poblaciones de sage-grouse (*Centrocercus* spp.) en Norteamérica occidental. Las tierras de cultivo y la red de canales de irrigación cubren 230,000 km^2 e influencian indirectamente hasta un 77% del Área de Conservación del Sage-Grouse (Sage-Grouse Conservation Area) y el 73% de la cubierta de suelo de artemisa, al subvencionar a depredadores sinántropos sobre el sage-grouse. La urbanización y las demandas del crecimiento demográfico humano han creado una extensa red de infraestructura de conexión la cual está ampliando su influencia en paisajes de artemisa. Hoy en día más de 2,500 km^2 se encuentran cubiertos por carreteras interestatales y caminos pavimentados; cuando se incluye a los caminos secundarios, el 15% del Área de Conservación del Sage-Grouse y el 5% de los hábitats existentes de sagebrush se encuentran a >2.5 km de los caminos. La densidad de caminos secundarios a menudo excede los 5 km/km^2, dando como resultado el acceso motorizado extensivo para fines recreativos, creando extensos corredores para acciones de manejo y desarrollo de recursos, subvencionando a depredadores adaptados a la presencia humana, y facilitando la expansión de plantas exóticas o invasoras. Las tierras de artemisa también están siendo utilizadas por sus valores naturales y recreativos, incluyendo el uso de vehículos fuera de las carreteras. Aproximadamente se permitieron 12,000,000 de meses de uso animal (animal use months o AUM = cantidad de forraje necesario para soportar una unidad de ganado por mes) para el ganado de pastoreo en tierras públicas en los estados occidentales. Los efectos directos del pastoreo sobre poblaciones del sage-grouse o paisajes de artemisa no son posibles de determinar utilizando datos actuales. Sin embargo, el manejo de tierras pastadas por el ganado ha influenciado ecosistemas de artemisa mediante el uso de tratamientos de la vegetación para aumentar el forraje y reducir el sagebrush y otras especies vegetales desagradables para el ganado. Las cercas (>2 km/km^2 en algunas regiones), los caminos, y las obras sobre los recursos de agua para manejar los movimientos del ganado han modificado el paisaje aun más. El desarrollo del petróleo y el gas influencian el 8% de los hábitats de artemisa con las intensidades más altas encontrándose en la extensión este del territorio de sage-grouse; >20% de la distribución del sagebrush se encuentra indirectamente influenciada en las áreas de manejo de los Great Plains, Wyoming Basin, and Colorado Plateau. El desarrollo de energía físicamente remueve el hábitat para construir las plataformas para los pozos, los caminos, las líneas eléctricas, y las tuberías; los efectos indirectos incluyen la fragmentación del hábitat, disturbio del suelo, y la facilitación de la expansión de plantas y animales exóticos. El reciente desarrollo de energías alternativas, tales como la eólica y geotérmica, crea la infraestructura en nuevas regiones de la distribución del sage-grouse. La utilización del suelo continuará siendo un factor de estrés dominante en sistemas de artemisa sus efectos individuales y acumulativos desafiarán la conservación a largo plazo de las poblaciones de sage-grouse.

Palabras Clave: agricultura, artemisa, desarrollo energético, fuego prescrito, pastoreo de ganado, Programa Reserva para la Conservación (Conservation Reserve Program), sage-grouse, urbanización, uso de la tierra, vehículos fuera de carretera.

Lands dominated by sagebrush (*Artemisia* spp.) provide a broad array of resources used by humans. Areas used primarily for traditional industries, such as mining, livestock grazing, and energy development, are juxtaposed with urban areas, crossed by infrastructure network to transport people and resources, or fragmented by agriculture and expanding exurban development. Recreation, wildlife conservation, and wilderness amenities also have intangible values, impose physical demands on sagebrush and surrounding landscapes, and often have legal designations or restrictions that can affect land management and other uses. Thus, resources contained in sagebrush systems range from consumptive commodities having a negotiated market value to aesthetic qualities that defy monetary currency despite efforts of natural resource economists.

Human land use has been a significant force shaping sagebrush systems, particularly since Euro-American settlement (West 1999, Griffin 2002). However, quantifying and understanding effects of land use across the range-wide distribution of Greater Sage-Grouse (*Centrocercus urophasianus*; hereafter, sage-grouse) is challenging because of the following conditions:

1. Demand for different resources and intensity of use have changed over time due to local, national, and global needs coupled with changes in technological capabilities to extract and use resources.
2. Social and economic importance of traditional use varies spatially and temporally relative to nonconsumptive use.
3. Effects of land use differ from distinct impacts occurring at delineated locations to more diffuse spatial and temporal influences.
4. Multiple land uses often combine in synergistic relationships with environmental factors to create a cumulative effect.
5. Sagebrush systems vary spatially along gradients of elevation, annual patterns of precipitation, and soils that influence their resistance and resilience to disturbance.
6. Complex dynamics of sagebrush systems can result in less predictable transitions to alternate states of vegetation communities rather than following well-defined trajectories of vegetation succession.

Adopting narrow management paradigms or espousing single solutions ignores these complexities (Crawford et al. 2004).

We described the dominant anthropogenic land uses in the Sage-Grouse Conservation Area (SGCA; the pre-settlement distribution of sage-grouse buffered by 50 km [Connelly et al. 2004, Schroeder et al. 2004]) and their influence on patterns and processes of sagebrush habitats and sage-grouse populations. We organized land uses into broad categories of agriculture, urbanization and infrastructure, livestock grazing, energy development (nonrenewable and renewable), and military training. The cumulative influence of anthropogenic land use is presented elsewhere as the human footprint (Leu et al. 2008; Johnson et al., this volume, chapter 17; Leu and Hanser, this volume, chapter 13).

Land uses have impacted sagebrush ecosystems at multiple temporal and spatial scales. Our first objective was to provide the historical development of each category of land use and the background against which it became a significant influence. We present temporal information on changes in frequency, intensity, or location of the land use when data were available. Our second objective was to describe the current status of land use. We mapped the distribution of land use and quantified, when possible, the potential area altered directly by physical habitat displacement or influenced indirectly through changes in plant or wildlife dynamics (Leu et al. 2008). Our last objective was to discuss the influence of each land use on sagebrush habitats or sage-grouse populations. We have presented a significant amount of background information, when available from published literature, because many of these interactions require a detailed understanding of the dynamics of sagebrush ecosystems.

We recognized but did not evaluate resource commodity needs or demand, nor did we assess public perceptions of land use (Kennedy et al. 1995, Donahue 1999). We did not determine societal benefits and costs associated with a land use. We also did not present strategies for mitigating use or recommend alternative levels of use. Rather, we attempted to answer the ecological questions about land uses (actions or choices), and how they influence pattern and functions of sagebrush systems (reactions or consequences). Thus, we presented the ecological context for management and land use discussions that also include political, economic, and social considerations (Mills and Clark 2001, Foley et al. 2005, Lackey 2007).

METHODS

Study Area

We considered the primary land uses within the range-wide distribution of Greater Sage-Grouse, which encompasses >2,000,000 km^2 (Connelly et al. 2004). Regional analyses were based on seven Sage-Grouse Management Zones (SMZs; Stiver et al. 2006), which were delineated from floristic regions that contain similar environmental influences on vegetation communities (Miller and Eddleman 2001). SMZs also reflected boundaries of broad regional populations of sage-grouse within the range-wide distribution. Estimated connectivity among sage-grouse leks primarily was contained within each SMZ rather than among zones (Garton et al., this volume, chapter 15; Knick and Hanser, this volume, chapter 16). State-wide summaries are presented when source data could not be delineated into SMZs.

Sagebrush is the dominant land cover on 530,000 km^2 within the sage-grouse range (Knick, this volume, chapter 1) and consists primarily of 20 taxa encompassing 11 major *Artemisia* species and subspecies groups (McArthur and Plummer 1978). Vegetation and wildlife communities vary greatly across the range covered by sagebrush as a function of differences in underlying soils, climate, topographic position, landform, and geographic location (West and Young 2000; Miller and Eddleman 2001; Davies et al. 2007a; Miller et al., this volume, chapter 10). Approximately 70% of the sagebrush land cover is on public land managed by state or federal agencies (Knick, this volume, chapter 1); reserves or landscapes managed to maintain a natural character represent <1% of the total land surface covered by sagebrush.

We assumed that land use at local scales can be aggregated across regional or range-wide extents (Allen and Starr 1982, Wiens 1989b, Peterson and Parker 1998) to assess broader influences on Greater Sage-Grouse populations or sagebrush landscapes. Dominant patterns then can form the basis for regional or range-wide conservation actions, such as prioritizing restoration (Wisdom et al. 2005c, Meinke et al. 2009), developing monitoring approaches (Connelly et al. 2003b, Washington-Allen et al. 2006), or relating environmental factors to sage-grouse population trends (Reese and Bowyer 2007; Garton et al., this volume, chapter 15).

Data Sources

Spatial Information

Most spatial information used in this study was developed initially for the range-wide Conservation Assessment for Greater Sage-Grouse and Sagebrush Habitats (Connelly et al. 2004) and was updated when new data became available. Analyses using land cover information first derived from the SageStitch map (Comer et al. 2002) now have been conducted using the more recent LANDFIRE Existing Vegetation Map (LANDFIRE 2006). We did not compare results to assess changes in land cover or other land-use estimates and accuracies when different methods were used.

Our primary challenge was obtaining comprehensive data spanning a study area that included all or part of 14 states and three provinces. Spatially explicit data (one or more values referenced to a map location) of important environmental and land-use factors often were limited to administrative boundaries such as counties, states, or Bureau of Land Management (BLM) field offices. Thus, we often merged spatial data from multiple sources to create seamless data layers in a geographic information system (GIS) for regional or range-wide analyses. We used the coarsest resolution when resolving GIS layers, thematic classifications, and terminology when compiling information from different sources.

We converted linear features to unit length/unit area to map relative density and create more meaningful measures of distribution across the landscape. We also converted point data to a contoured distribution of densities. Specific methods and values used for individual coverages are shown in figure or table captions, accompanying text, or included in online metadata.

Disturbance can influence an area beyond its immediate boundaries, although the function relating distance and magnitude of effect can be difficult to define (Turner and Dale 1991). For example, roads can provide dispersal corridors for invasive plants to spread into the surrounding region, influencing plant community dynamics beyond the physical road itself (Belcher and Wilson 1989, Gelbard and Belnap 2003). Other disturbances or threats (e.g., insecticide use on specific crops) may be confined locally, but long movements by sage-grouse among widely spaced seasonal ranges may result in exposure during a critical period in the annual life cycle or at a small location in the annual home range (Blus et al. 1989, Walker et al. 2007a, Doherty et al. 2008). We used an ecological rationale for estimating the area around points, lines, or polygons from which land use potentially influenced land cover or sage-grouse populations. Estimates for effect sizes into surrounding areas were based on foraging movements of human-subsidized predators, distance of exotic plant species spread, or on distribution data relative to land-use features (Gelbard and Belnap 2003, Connelly et al. 2004, Bradley and Mustard 2006, Leu et al. 2008).

Many of our primary analyses quantified overlap between land use or its influence and sagebrush cover types or other environmental variables. This approach can be appropriate for land uses or disturbances that have well-defined effects in space and time (Pickett and White 1985). Diffuse disturbances, such as livestock grazing, that alter plant composition or influence community processes may not result in altered landscape patterns readily mapped in a GIS and are more difficult to quantify at broad scales (Turner and Bratton 1987).

Nonspatial Information

Tabular and summaries of nonspatial data were created from online sources, field records or archives, or published literature (Connelly et al. 2004). We often relied on statistics available from or estimated for a single management entity. Public land statistics presented in annual reports by the BLM represent activities by the largest federal land management steward for sagebrush lands in the United States (Knick et al. 2003, Wisdom et al. 2005b). However, these statistics cannot be extrapolated to activities on private lands or to other agencies because environmental characteristics of lands differ among stewards (Knick, this volume, chapter 1). Thus, they represent only part of the collective scenario for sagebrush regions.

We also used public land statistics to assess land uses, management issues, habitat characteristics, and treatments. Public land statistics often compile summary numbers that may not adequately represent land use, may contain data that were not updated from previous years, or may fail to capture many complexities inherent in management and response of sagebrush ecosystems. We used that information to characterize management actions and use of sagebrush lands because these reports are used to document agency activities on public lands. Agency personnel reviewed our use of public land statistics, but interpretations and conclusions remain our own.

Documentation of Data and Sources

All nonproprietary and nonsensitive spatial data sets used in our analysis are available for download on the SAGEMAP website (United States Department of the Interior 2001c). Each data set is accompanied by a metadata record documenting original source and GIS procedures.

AGRICULTURE

Agriculture can include many types of production of food and goods through practices such as cultivating soils, producing crops, and raising livestock. The mapped category of agriculture used in our analysis was principally cropland although mapped pasturelands also were included in the land cover definition. We discuss agricultural land uses primarily related to croplands unless otherwise noted.

Historical Development and Current Status

Most lands dominated by sagebrush in the western United States were acquired by the federal government through purchases and cessions of large territories in the early and mid-1800s (Dombeck et al. 2003). The United States government followed policies to transfer these vast acquisitions to private entities by encouraging conversion and development of sagebrush and other arid lands (Clawson and Held 1957). Public lands were given to private entities under a succession of homestead acts (1862, 1877, 1909, 1912, and 1916) that required settlers to build homes and develop land for agriculture. The prime areas for growing crops, characterized by deep, fertile soils and available water for irrigation, were claimed early during Euro-American settlement. The initial establishment of settlements throughout the Great Basin occurred between the 1850s and 1860s (Oliphant 1968, Miller et al. 1994, Flores 2001, Young and Sparks 2002). Lands transferred to private entities and converted to cropland were also predominantly at lower elevations and on relatively flat terrain (Talbert et al. 2007; Knick, this volume, chapter 1).

Homesteaders arrived in Washington state by the late 1800s, and the proportion of virgin prairie and brush in southeastern Washington under cultivation increased from 25% in 1890 to >80% in 1920 (Buss and Dziedic 1955). Use of tractors for wheat farming further increased technological capabilities, and almost all available lands were cultivated by 1945 (Buss and Dziedic 1955). Grasslands, which once covered 25% of eastern Washington, were reduced to 2% (McDonald and Reese 1998). Approximately 170,000 km^2 of the pre-settlement distribution of 420,000 km^2 of shrub steppe in Washington was present in 1986 (Dobler et al. 1996).

The Carey Act, passed in 1894, transferred additional lands to states, which then could be privatized provided that irrigation was developed. Large irrigation projects too expensive for private enterprise were developed with federal funding through the Newlands Reclamation Act passed in 1902 (Clawson and Held 1957). The first private irrigation projects in southwestern Idaho were started in the 1840s and on the upper Snake River in Idaho near Wyoming in the 1870s (Nokkentved 2008). Almost all of the Snake River Plain in southern Idaho that contains deep loamy soils and once supported basin big sagebrush (*Artemisia tridentata* ssp. *tridentata*) now has been converted to cropland (Hironaka et al. 1983).

An estimated 10% of sagebrush steppe that existed prior to Euro-American settlement has been converted to agriculture; irrigation is not feasible on the remaining 90%, topography and soils are limiting, or temperatures are too extreme for certain crops (West 1996). However, economic advantages or technological improvements in irrigation methods now permit agriculture development on steeper terrains and in regions further from river floodplains (Brown et al. 2005, Vander Haegen 2007). For example, it was economically feasible for an additional 800 km^2 of public land in the Snake River Plain to be developed for irrigation in the late 1980s (Hamilton and Gardner 1986).

Agriculture, mostly mapped croplands, currently covers >230,000 km^2 (11%) of the Sage-Grouse Conservation Area (Table 12.1). The total area influenced by agriculture was >1,600,000 km^2 when a high effect buffer was used to include potential movements out to 6.9 km away from agricultural developments by human-subsidized predators (Boarman and Heinrich 1999, Leu et al. 2008). Almost three-fourths of all sagebrush within the range of Greater Sage-Grouse was influenced by agriculture at this level of effect (Table 12.1). Proportion of agricultural lands within SMZs varied due to topography, soil characteristics, temperature, water availability, and type of crop grown. Primary agricultural regions were within the Columbia Basin (32% of total area in agriculture) and Great Plains (19%) SMZs (Fig. 12.1). Less than 5% of the land area in the Wyoming Basin, Southern Great Basin, Northern Great Basin, and Colorado Plateau was mapped as agriculture land cover. Irrigation canals covered <0.1% of the land area within the SGCA or SMZ (Table 12.1).

Government Conservation Programs

The Conservation Reserve Program (CRP) is a voluntary program, authorized in 1985 under the Food Security Act, in which landowners receive annual payments in return for establishing permanent vegetation on idle or erodible lands that previously had been used for growing crops (Barbarika et al. 2004). The purpose of the program is to control soil erosion, improve water retention, and provide wildlife habitat. Lands placed into the program are to be set aside for 10 years and cannot be grazed except under emergency drought conditions. The amount of lands placed in CRP has markedly increased since the program inception in 1987 (Fig. 12.2). The increase was primarily in agricultural regions of the Columbia Basin and Great Plains SMZs. Other federal government programs, such as the Wildlife Habitat Incentives Program authorized in 1996, provide support to landowners to implement measures specific to conserving wildlife habitat (Riley 2004, Gray and Teels 2006).

The Permanent Cover Program was established in Canada in 1986 as part of the National Agricultural Strategy to reduce soil degradation by planting perennial vegetation. Lands placed in the Permanent Cover Program through 10- or 21-year agreements can be used for hay or pasture, unlike CRP under which lands set aside can be grazed only during emergency drought (McMaster and Davis 2001).

Ecological Influences and Pathways

The capability of agricultural regions within the sage-grouse range to support viable populations likely depends on the quantity and configuration of sagebrush remaining within the mosaic (Wisdom et al. 2002a,b; Aldridge et al. 2008; Johnson et al., this volume, chapter 17; Wisdom

TABLE 12.1

Area (km²) and percent of area influenced by agriculture (cropland and irrigation canals) in Sage-Grouse Management Zones and the Sage-Grouse Conservation Area (SGCA).

Effect area was the total area of land use and surrounding buffer.

	Area		Effect area Low		Effect area High		Sagebrush area[a] (%)	
Agriculture	km^2	(%)	km^2	(%)	km^2	(%)	Low	High
Cropland[b]								
Great Plains	66,052	(18.7)	240,021	(68.1)	319,483	(90.7)	47.4	83.7
Wyoming Basin	8,723	(3.6)	100,300	(41.5)	169,011	(69.8)	36.4	67.6
Southern Great Basin	6,943	(2.2)	87,437	(27.4)	196,668	(61.6)	24.4	61.1
Snake River Plain	28,657	(9.5)	156,729	(51.8)	254,539	(84.2)	48.1	83.1
Northern Great Basin	5,467	(3.8)	45,061	(31.1)	94,308	(65.1)	27.7	61.3
Columbia Basin	20,759	(32.3)	53,984	(83.9)	58,085	(90.3)	95.8	99.2
Colorado Plateau	6,804	(4.7)	73,353	(51.1)	115,933	(80.8)	58.2	85.4
SGCA	232,643	(11.2)	1,049,310	(50.4)	1,603,517	(77.0)	40.7	73.1
Irrigation canals[c]								
Great Plains	58	(<0.1)						
Wyoming Basin	158	(0.1)						
Southern Great Basin	101	(<0.1)						
Snake River Plain	430	(0.1)						
Northern Great Basin	119	(0.1)						
Columbia Basin	50	(0.1)						
Colorado Plateau	75	(0.1)						
SGCA	1,093	(<0.1)						

[a] Sagebrush area was delineated from the LANDFIRE (2006) map of land cover and covered 529,708 km^2 (25.4%) of the Sage-Grouse Conservation Area.

[b] Areas for low (2.5 km) and high (6.9 km) ecological effect were estimated around agriculture lands based on spread of exotic plant species (Bradley and Mustard 2006) and foraging distances of mammalian and corvid predators (Boarman and Heinrich 1999, Leu et al. 2008).

[c] Area of irrigation canals estimated by linear distance × width (12.8 m). We did not delineate areas for ecological effect associated with irrigation canals.

et al., this volume, chapter 18). Range-wide, sage-grouse were more likely to be extirpated from areas containing >25% cultivated cropland and in which <25% of the landscape was dominated by sagebrush (Aldridge et al. 2008). Sage-grouse populations declined by 73% on a study area in south-central Montana concurrent with a 16% decrease in sagebrush land cover, including 30% of the wintering habitat, which was plowed and converted to croplands (Swenson et al. 1987). Declining populations of sage-grouse in the upper Snake River Plain of southeastern Idaho were correlated with amount of cropland area (Leonard et al. 2000). Little residual sagebrush was left in areas converted to croplands. Cropland area increased 74% from 403 km^2 in 1975 to 635 km^2 in 1985 and to 701 km^2 in 1992. Cropland as a percentage of the 2,249 km^2 study area increased from 18% in 1975 to 28% in 1985 and was 31% of the land surface in 1992.

Loss of 20% of the sagebrush between 1958 and 1993 within the range of Gunnison Sage-Grouse

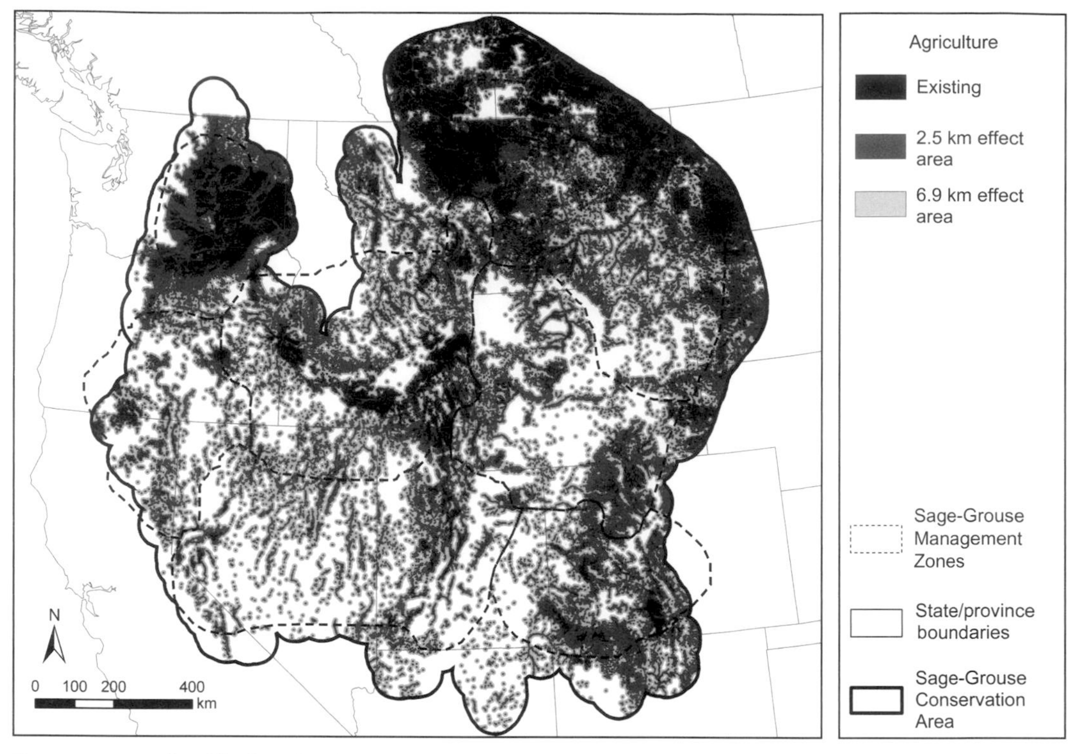

Figure 12.1. Agricultural lands within the Sage-Grouse Conservation Area. Mapped land cover primarily depicts croplands, although pasture was included in the agriculture category (LANDFIRE 2006).

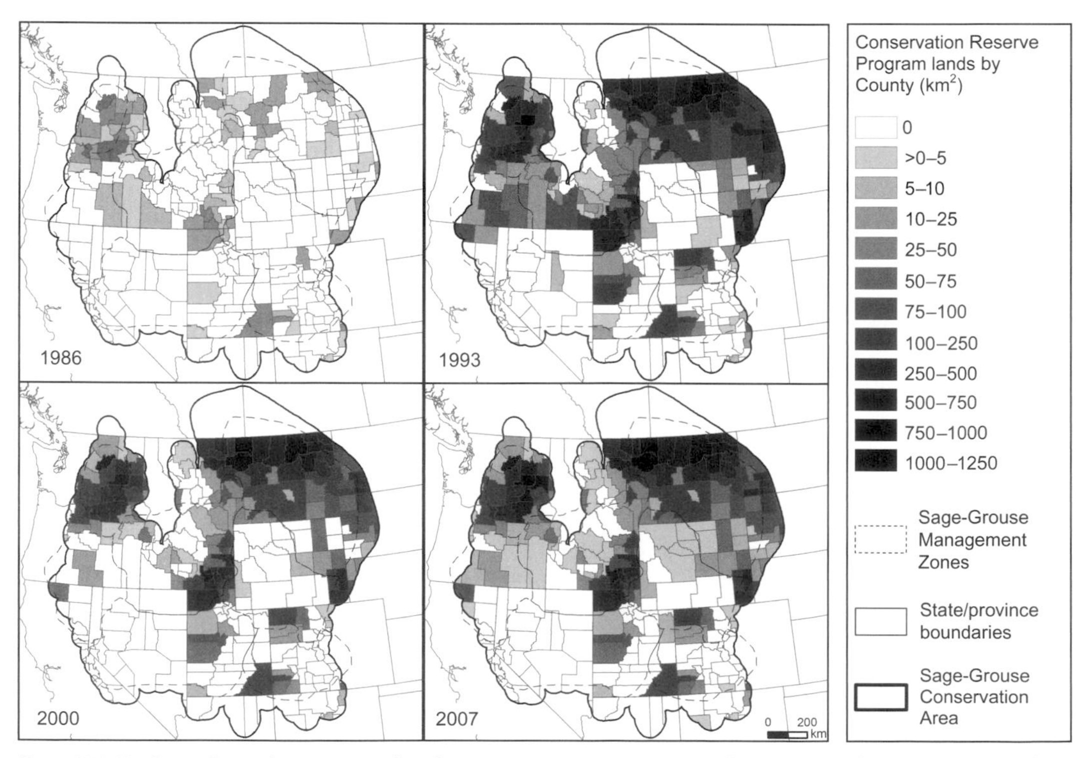

Figure 12.2. Total area (ha) under contract within the Conservation Reserve Program for counties in the United States by five-year intervals from 1987 through 2004 (United States Department of Agriculture 2009).

(*Centrocercus minimus*) in southwestern Colorado was coupled with increased landscape fragmentation by conversion to cropland (Oyler-McCance et al. 2001). This study was not directly linked to population trends, but Gunnison Sage-Grouse now occupy 10% of their historical range primarily due to habitat loss (Schroeder et al. 2004).

Numerous patches of sagebrush interspersed at small scales potentially can provide sufficient habitat if the landscape maintains sufficient sagebrush land cover at broader scales. Shrub steppe was increasingly isolated and mean size of remaining patches decreased in the Columbia Basin due to conversion to agriculture (McDonald and Reese 1998). Sage-grouse persist within this matrix (Schroeder and Vander Haegen, this volume, chapter 22), although populations are undergoing long-term declines (Garton et al., this volume, chapter 15) and are genetically isolated from populations in neighboring SMZs (Oyler-McCance et al. 2005b) by regions more intensively developed for agriculture.

Near-total conversion to cropland and other agriculture practices in southern Idaho has separated sagebrush-dominated landscapes into higher-elevation, less-productive, and less-arable regions north and south of the Snake River Plain (Knick et al. 2003). Sage-grouse populations that once were continuous across this region now are disconnected (Schroeder et al. 2004; Knick and Hanser, this volume, chapter 16). In comparison, agricultural regions of the Great Plains containing smaller-scale dispersion of sagebrush habitat of the Great Plains still support sage-grouse populations.

Cropland Insecticides

Cropland expansion into historic sagebrush habitat has resulted in some sage-grouse using irrigated lands in summer. Sage-grouse typically avoid cultivated croplands and areas dominated by human development (Aldridge and Boyce 2007). However, green vegetation, such as alfalfa, and insects associated with crops or livestock operations can be attractive foods during brood-rearing when forbs in native sagebrush communities become desiccated (Patterson 1952; Hagen et al. 2007; Connelly et al., this volume, chapter 4). At least 18% of 82 radio-marked birds in southern Idaho were attracted to lush croplands adjacent to sagebrush habitat that were sprayed with the insecticides dimethoate and methamidophos (Blus et al. 1989). Intoxicated or dead sage-grouse were found in or near six fields. Brain cholinesterase activities and residue analysis of crop contents indicated that 5% of the marked sage-grouse in 1985 and 16% in 1986 died from these organophosphorus insecticides. Another 63 of an estimated 200 sage-grouse using a block of alfalfa sprayed with dimethoate died of organophosphorus poisoning.

Croplands and insecticide spraying potentially influence sage-grouse populations over broad regions. Sage-grouse in southern Idaho are migratory; individuals can move >80 km from winter areas and leks to summer ranges and have annual ranges >2,500 km^2 (Dalke et al. 1963, Connelly et al. 1988). Distances traveled by females with broods from nesting areas to late brooding areas ranged from 3 to 21 km (Gates 1983). Total extent of the area of effect cannot be estimated because actual distances that sage-grouse move to irrigated and sprayed fields is unknown. Similarly, the impact on sage-grouse populations from exposure to organophosphorus insecticides has not been estimated relative to other causes of mortality or their timing within an annual cycle.

Government Conservation Programs

The CRP places vegetation cover on lands that otherwise would not be used by many species of wildlife. Large-area set-asides in CRP, however, are constrained by limits on total area that can be enrolled within a region. Instead, CRP set-asides typically focus on providing riparian or grassland buffer strips rather than large areas of upland vegetation (Riley 2004). Consequently, species that have benefited from CRP primarily include grassland birds with relatively small home ranges that use early successional vegetation or a mosaic of cropland bordered by grassland buffers (Haroldson et al. 2006).

Sage-grouse respond to vegetation and landscape structure, not program or landownership. Thus, the value of lands placed in the CRP program for sage-grouse likely is related to size and quality of individual CRP patches as well as the cumulative amount and distribution of CRP lands within a larger matrix of croplands and sagebrush. Sage-grouse in the Columbia Basin used CRP lands that had been set aside sufficiently long to permit perennial grasses and sagebrush to become dominant vegetation (Schroeder and Vander Haegen, this volume, chapter 22). Sage-grouse populations stabilized in the region containing 17% CRP, compared to a population that continued to decline in an adjacent region containing <2% CRP (Schroeder and Vander Haegen, this volume, chapter 22).

Capability of lands placed in CRP to benefit sage-grouse also should be considered relative to other alternative uses for those sites. Lands set aside in CRP provide some habitat components in contrast to cropland, development, or barren land susceptible to erosion. Gunnison Sage-Grouse used CRP lands in proportion to their availability in southeastern Utah (Lupis et al. 2006). Suitable, if not optimal, habitat was available through CRP on >30% of the study area that otherwise would have been used for crops and would not be likely to have supported sage-grouse.

URBANIZATION AND INFRASTRUCTURE

Historical Development and Current Status

Indigenous people have inhabited sagebrush regions for 13,000 years BP or earlier (Thomas 1973, Grayson 1993). Local densities of humans estimated for pre-American settlement periods were 0.3 persons/km^2 in the Columbia Plateau, 0.08 in the Great Basin, and 0.13 in the Great Plains (Vale 2002a). Low densities limited their impact on the biophysical landscape, although hunting, gathering, and burning may have had a locally significant influence (Flores 2001, Griffin 2002, Vale 2002b, Kay 2007).

Human populations have increased and expanded, primarily over the past century and in the western portion of the sagebrush distribution (Fig. 12.3). In 1900, 51% of the 325 counties within the historical distribution of sage-grouse had <1 person/km^2 and 4% of the counties had densities of >10 persons/km^2. By 1950, 39% of the counties had <1 person/km^2 and 9% had >10 persons/km^2. In the most recent census in 2000, 31% of the counties had <1 person/km^2 and 22% had >10 persons/km^2.

The Columbia Basin has the highest density of humans (Fig. 12.4) in contrast to the low densities in the Great Plains. Differences in population trends among SMZs reflect a general movement in the United States from midwestern plains to western states (Brown et al. 2005). Population densities in the Great Plains in the 2000 census had increased more slowly since 1990 (7%) and 1920 (49%) relative to other management zones (Fig. 12.4). In comparison, population densities have increased between 19% (Wyoming Basin) and 31% (Colorado Plateau) from 1990–2000, and between 166% (Wyoming Basin) and 666% (Northern Great Basin) from 1920–2000 (Fig. 12.4).

Availability of resources often limited areas that could be settled by early inhabitants of arid regions (West 1999). Early settlements by Euro-Americans were along transportation corridors, such as rivers or railroad lines, or in regions where minerals had been discovered (Young and Sparks 2002). The dominant urban areas in the sage-grouse range are in the Columbia River Valley of Washington, the Snake River Valley of southwestern Idaho, and the Bear River Valley of northern Utah. Most urban areas within the sage-grouse range are on the edge of regions dominated by sagebrush land cover with the exception of cities and towns in the Southern Great Basin SMZ. Area of influence of urban development ranged from 16% of the area within the Great Plains to 49% of the Columbia Basin (Table 12.2).

Rural areas also have been developed throughout the sagebrush region, particularly around urban centers and major highways. Much of this development has been in recent decades. The amount of uninhabited area (0 people/km^2) within the Great Basin ecoregion decreased from 90,000 km^2 in 1990 to <12,000 km^2 in 2004 (Torregrosa and Devoe 2008). Economic opportunities combined with availability of public lands for recreation and other natural or wilderness qualities are major reasons for this population expansion (Hansen et al. 2002, Brown et al. 2005). Increased affluence, change in social values, and ability to conduct business from remote locations through electronic commerce also has resulted in home development on large acreages surrounding cities. Many existing ranches that had been used primarily for livestock production are being sold for the rural or wildland amenities that they provide and subdivided into ranchettes (Riebsame et al. 1996, Holechek 2001, Gosnell and Travis 2005, Holechek 2006). This is best documented for the Colorado Front Range and the greater Yellowstone ecosystem, but ranchette and subdivisions are being developed throughout the sage-grouse range, particularly in the Great Basin (Torregrosa and Devoe 2008).

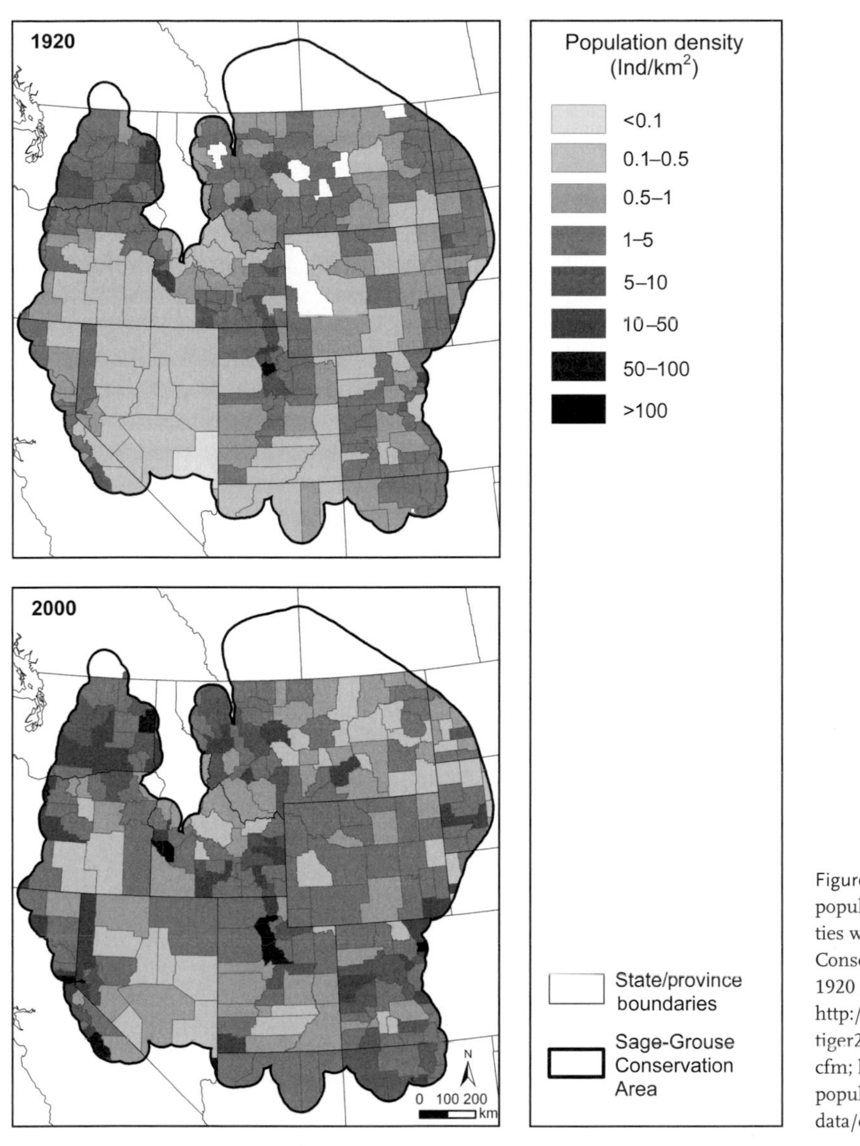

Figure 12.3. Change in population density for counties within the Sage-Grouse Conservation Area from 1920 (top) to 2000 (bottom). http://arcdata.esri.com/data/tiger2000/tiger_download.cfm; http://www.census.gov/population/www/censusdata/cencounts/index.html

Infrastructure

Interstate and other major paved highways covered an estimated 2,500 km^2 (0.1%) of the SGCA (Table 12.3). Interstates and highways influenced 851,044 km^2 (41%) of the total area and 38% of the sagebrush land cover (Table 12.3, Fig. 12.5). These major roads tend to follow river valleys and traverse lower elevations. In contrast, secondary paved roads exist throughout most sagebrush regions (Fig. 12.6) in densities up to >5 km/km^2. Less than 5% of the entire sage-grouse range was >2.5 km from a mapped road, and almost no area of sagebrush was >6.9 km (Table 12.3). Railroads covered 487 km^2 (<0.1%) of the landscape but influenced 10% of the SGCA and 7% of the sagebrush (Table 12.3).

Power lines covered a minimum of 1,089 km^2 (Fig. 12.7, Table 12.3) and had an ecological influence on almost 50% of all sagebrush within the SGCA. We were unable to map or estimate the density of smaller distribution lines in rural areas. Similar to roads, power lines also followed major river valleys and crossed lower elevations.

The Energy Policy Act of 2005 directed that corridors for transporting energy (oil, gas, hydrogen, electricity) be designated on federal land (United States Departments of Energy and the

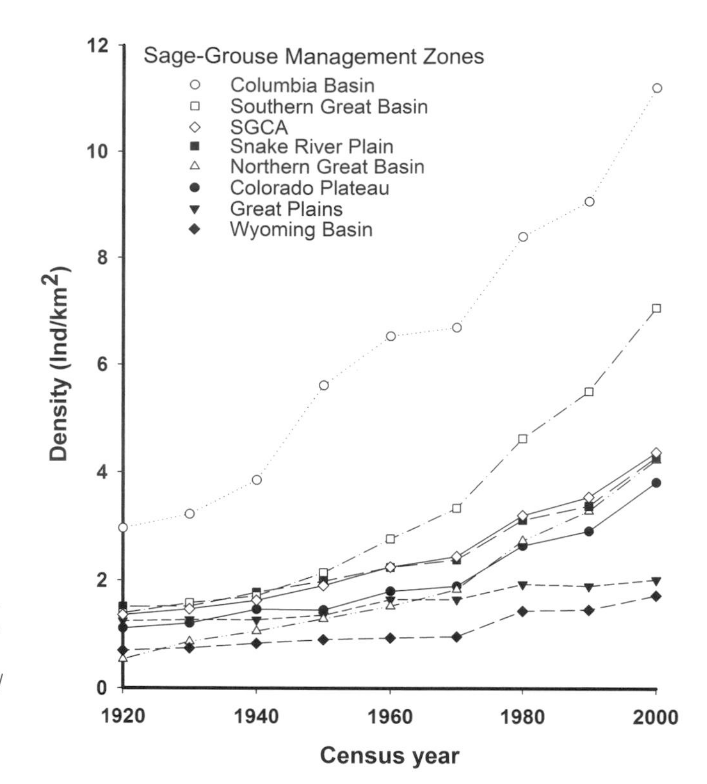

Figure 12.4. Population density (ind/km^2) within the Sage-Grouse Management Zones in the Sage-Grouse Conservation Area (United States Census Statistics 1900–2000). Densities were averaged from census results for individual counties within Sage-Grouse Management Zones. http://arcdata.esri.com/data/tiger2000/tiger_download.cfm; http://www.census.gov/population/www/censusdata/cencounts/index.html

TABLE 12.2

Area (km^2) and percent of area influenced by urban development in Sage-Grouse Management Zones and the Sage-Grouse Conservation Area (SGCA).

Effect area was the total area of land use and surrounding buffer.

	Area		Effect area[a]		
Urban development	km^2	(%)	km^2	(%)	Sagebrush area[b] (%)
Great Plains	653	(0.2)	57,652	(16.4)	16.0
Wyoming Basin	457	(0.2)	49,970	(20.7)	18.4
Southern Great Basin	1,824	(0.6)	54,587	(17.1)	13.5
Snake River Plain	1,131	(0.4)	78,159	(25.8)	18.8
Northern Great Basin	741	(0.5)	28,694	(19.8)	12.5
Columbia Basin	679	(1.1)	31,185	(48.5)	48.4
Colorado Plateau	780	(0.5)	42,288	(29.5)	29.2
SGCA	8,612	(0.4)	454,193	(21.8)	18.6

[a] Area for ecological effect (6.9 km) was estimated from foraging distances of mammalian and corvid predators (Boarman and Heinrich 1999, Leu et al. 2008).

[b] Sagebrush area delineated from the LANDFIRE (2006) map of land cover and included 529,708 km^2 (25.4%) of the Sage-Grouse Conservation Area.

TABLE 12.3

Area (km²) and percent of area influenced by infrastructure in Sage-Grouse Management Zones and the Sage-Grouse Conservation Area (SGCA).

Effect area was the total area of land use and surrounding buffer.

	Area		Effect area				Sagebrush area[a] (%)	
			Low		High			
Land use	km^2	(%)	km^2	(%)	km^2	(%)	Low	High
Interstates/highways[b]								
Great Plains	371	(0.1)	52,524	14.9	129,582	(36.8)	13.1	35.0
Wyoming Basin	283	(0.1)	42,431	17.5	103,479	(42.8)	17.9	45.5
Southern Great Basin	403	(0.1)	52,192	16.4	126,224	(39.5)	14.6	36.8
Snake River Plain	368	(0.1)	46,565	15.4	110,650	(36.6)	12.7	32.8
Northern Great Basin	125	(0.1)	20,314	14.0	50,160	(34.6)	11.5	29.5
Columbia Basin	168	(0.3)	17,904	27.8	39,338	(61.1)	28.3	61.6
Colorado Plateau	185	(0.1)	27,792	19.4	66,095	(46.0)	20.8	47.6
SGCA	2,552	(0.1)	353,803	17.0	851,044	(40.9)	15.3	38.3
All roads[b]								
Great Plains	2,738	(0.8)	302,754	(85.9)	323,551	(91.8)	91.3	98.7
Wyoming Basin	2,228	(0.9)	214,456	(88.6)	231,103	(95.5)	98.5	99.7
Southern Great Basin	2,495	(0.8)	285,839	(89.5)	313,396	(98.2)	94.3	99.8
Snake River Plain	2,575	(0.9)	272,400	(90.1)	297,782	(98.5)	94.7	99.7
Northern Great Basin	1,599	(1.1)	138,158	(95.4)	144,469	(99.7)	95.6	100.0
Columbia Basin	870	(1.4)	63,097	(98.1)	64,270	(99.9)	98.7	100.0
Colorado Plateau	1,210	(0.8)	127,848	(89.1)	142,121	(99.0)	95.3	99.3
SGCA	17,681	(0.8)	1,776,936	(85.3)	1,923,694	(92.3)	95.2	99.6
Railroads[c]								
Great Plains	65	(<0.1)	34,805	(9.9)			8.8	
Wyoming Basin	34	(0.0)	17,391	(7.2)			7.4	
Southern Great Basin	48	(<0.1)	20,646	(6.5)			3.8	
Snake River Plain	110	(<0.1)	28,668	(9.5)			7.0	
Northern Great Basin	57	(<0.1)	13,927	(9.6)			5.7	
Columbia Basin	34	(0.1)	15,013	(23.3)			26.2	
Colorado Plateau	20	(<0.1)	7,326	(5.1)			4.2	
SGCA	487	(<0.1)	200,109	(9.6)			7.1	
Power lines[d]								
Great Plains	159	<0.1	52,362	14.9	112,697	32.0	15.4	33.5
Wyoming Basin	162	0.1	53,545	22.1	107,774	44.5	27.7	54.2
Southern Great Basin	124	<0.1	43,799	13.7	98,506	30.9	11.8	27.8
Snake River Plain	158	0.1	53,217	17.6	112,855	37.3	17.1	36.8

TABLE 12.3 (*continued*)

TABLE 12.3 (CONTINUED)

	Area		Effect area				Sagebrush area[a] (%)	
			Low		High			
Land use	km^2	(%)	km^2	(%)	km^2	(%)	Low	High
Northern Great Basin	64	<0.1	22,275	15.4	49,634	34.3	13.8	31.4
Columbia Basin	85	0.1	25,280	39.3	44,116	68.6	43.8	76.1
Colorado Plateau	67	<0.1	24,364	17.0	55,635	38.8	21.4	46.9
SGCA	1,089	<0.1	366,298	17.6	777,983	57.3	18.7	39.0
Communication towers[e]								
Great Plains	9	<0.1			14,544	4.1		4.0
Wyoming Basin	6	<0.1			11,775	4.9		4.9
Southern Great Basin	12	<0.1			14,293	4.5		3.0
Snake River Plain	12	<0.1			16,615	5.5		3.6
Northern Great Basin	5	<0.1			6,556	4.5		2.7
Columbia Basin	6	<0.1			8,573	13.3		12.2
Colorado Plateau	5	<0.1			7,934	5.5		5.5
SGCA	74	<0.1			106,795	5.1		4.2

[a] Sagebrush area was delineated from the LANDFIRE (2006) map of land cover and included 529,708 km^2 (25.4%) of the Sage-Grouse Conservation Area.

[b] Effect area of 7 km estimated from distribution of Greater Sage-Grouse leks relative to Interstate 80 in Wyoming (Connelly et al. 2004). Surface area of roads was estimated from linear distance × width (interstate highways 73.2 m; federal and state highways 25.6 m; secondary roads 12.4 m).

[c] Surface area of railroads estimated from linear distance × width (9.4 m). Buffer size for ecological effect was 3 km to estimate spread of exotic plants. We did not estimate a high effect area for railroads.

[d] Low (2.5 km) and high (6.9 km) effect areas were estimated based on spread of exotic plant species (Bradley and Mustard 2006) and foraging distances of mammalian and corvid predators (Boarman and Heinrich 1999, Leu et al. 2008).

[e] We combined three categories of communications towers, which are based on height and location relative to glide paths around airports. Surface area was estimated at 1 ha/tower.

Interior 2008). New corridors, as currently proposed (Fig. 12.7), would affect an additional 2% (12,000 km^2) of the sagebrush across the SGCA currently not influenced by a mapped power line. Amount of additional sagebrush habitat that would be affected by new corridors ranged from <0.2% in the Columbia Basin (13 km^2) and Great Plains (85 km^2), which already have a large proportion influenced by power lines and infrastructure, to >5% in the Northern Great Basin (3,552 km^2).

A minimum of 10,182 communications towers >62 m in height were present in the SGCA (Fig. 12.5). The area potentially influenced included 4% of the current sagebrush distribution (Table 12.3).

Recreation and Off-highway Vehicle Use

Off-highway vehicle (OHV) use is defined as any motorized use by motorcycles, all-terrain vehicles, four-wheel-drive jeeps, and other vehicles capable of off-highway terrestrial travel, and which occurs predominantly on unpaved roads and single-track and two-track motorized trails (Ouren et al. 2007). A major part of OHV use on public lands is for recreation. Off-highway vehicle use and recreation, however, have not been well-documented for the SGCA or for sagebrush ecosystems. Over 8,400,000 people live within 5 km of sagebrush, and 7,600,000 live within 5 km of public lands. Many people live in these locations primarily because of access to public lands for recreation (Hansen et al. 2002, 2005).

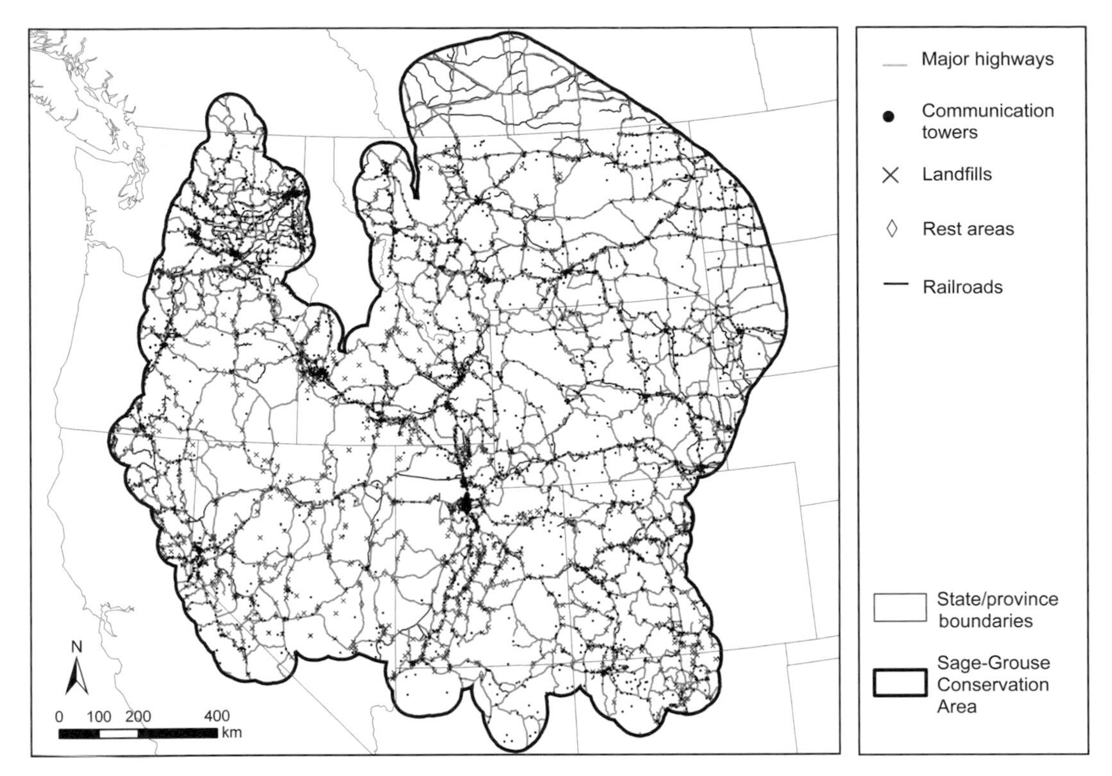

Figure 12.5. Communication towers, landfills, rest areas, major highways, and railroads within the western United States. Landfill locations obtained from waste management agencies of individual states. Location of rest areas obtained from transportation agencies of individual states. Communications tower locations obtained from Federal Communications Commission. Interstate and major highway (10-km buffer) locations obtained from U.S. Bureau of Transportation Statistics and Geo Gratis for Canada. Major railroad locations obtained from U.S. Bureau of Transportation Statistics and Geo Gratis for Canada.

Recreational use of OHVs is one of the fastest-growing outdoor activities, primarily in the western United States, where >27% of the population used OHVs for recreation during the survey period in 1999–2004 (Cordell et al. 2005). More than one in three persons in rural areas and one in four living in cities participated in OHV recreation an average of 24 days/year during 1999–2004.

Executive Order 11644 (Nixon 1972) required federal agencies to designate approved sites, or zones of use, for OHVs on public lands. Areas designated for OHV use were to minimize damage to soils, watersheds, vegetation, and wildlife as well as minimize conflict with other recreational uses. Agencies were required to monitor the effects of OHV use and adjust designated areas if required.

Ecological Influences and Pathways

Urban areas by themselves represent inhospitable environments for sage-grouse. However, most urban areas within the SGCA are on the edge of the current distribution of sagebrush and sage-grouse rather than within core regions. Nationally, urbanization is a major cause of species endangerment, particularly in association with recreation and agriculture, but had only a minor influence in the western states of Idaho, Utah, and Nevada because large amounts of public lands prevent urban development (Czech et al. 2000). However, the recent pace of development in these regions (Torregrosa and Devoe 2008) may increase the effect of urbanization on sagebrush and sage-grouse.

The expanding urbanization gradient through development of ranchettes or subdivisions influences wildlands through spatial and temporal changes in the physical environment that result in habitat loss and reduction of native species or changes in biodiversity (McKinney 2002, Hansen et al. 2005). Ranchettes and subdivisions continue to provide some sagebrush habitat, in contrast to total conversion in urban areas. Subdivided ranchettes had lower densities of shrub and grassland

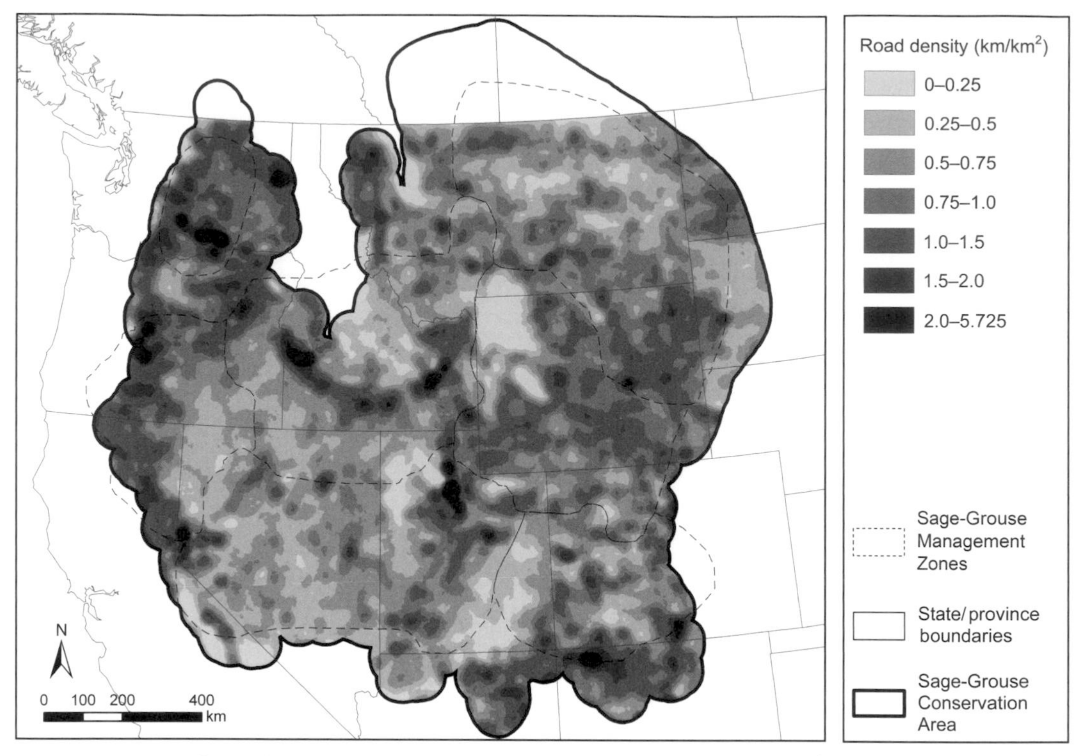

Figure 12.6. Contoured secondary roads in the Sage-Grouse Conservation Area (density [km/km^2] within an 18-km radius) (GIS coverages obtained from U.S. Census Bureau).

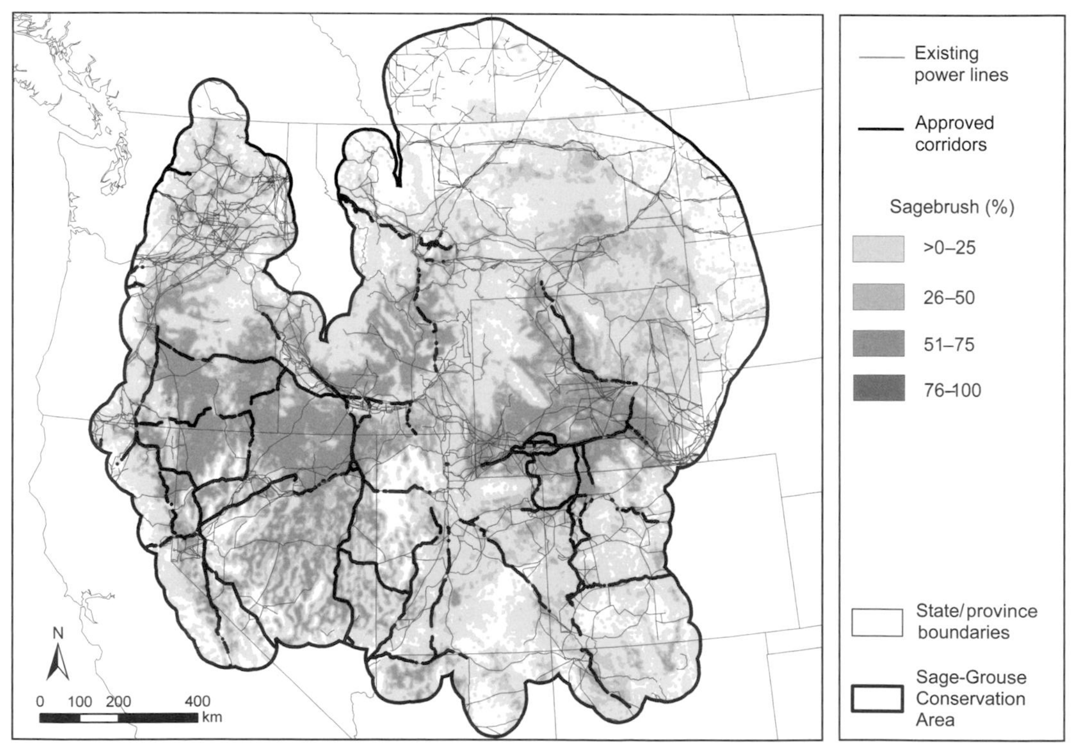

Figure 12.7. Power lines and proposed energy corridors within the Sage-Grouse Conservation Area (compiled from GIS coverages obtained from individual state, provincial, and federal sources).

birds, more domestic than native predators, and fewer native plant species when compared to full-scale ranches (Maestas et al. 2003, Hansen et al. 2005). Consequently, road fragmentation and disturbance from human activities in and around human dwellings (Mitchell et al. 2002) probably make many of these areas inhospitable to sage-grouse and other wildlife that depend on landscape arrangement and amount of sagebrush land cover (Aldridge et al. 2008; Hanser and Knick, this volume, chapter 19).

The physical boundaries of cities may be small relative to total area in the SGCA, but people living in cities require resources from immediately surrounding regions or that need to be transported from elsewhere. Often, the needs supporting urbanization extend well beyond the city (Theobald 2001). Increases in technological capabilities have reduced limitations to moving resources and increased connectivity between supply and demand. Proposed water development to support the growth of Las Vegas (located outside of the southern Great Basin SMZ) would potentially trigger declines in groundwater across at least 78 basins covering nearly 130,000 km^2, including areas within the Southern Great Basin (Deacon et al. 2007).

The connecting infrastructure of roads, motorized trails, railways, power lines, and communications corridors fragment or remove sagebrush land cover (Leu and Hanser, this volume, chapter 13). The ecological impact of roads and motorized trails include: (1) increased mortality of wildlife from collisions with vehicles, (2) modification of animal behavior because of habitat changes or noise disturbance, (3) alteration of physical environment, (4) alteration of chemical environment through leaching or erosion, (5) spread of exotic and invasive plant and wildlife, and (6) increased habitat alteration and use by humans (Forman and Alexander 1998, Forman 2000, Trombulak and Frissell 2000, Ouren et al. 2007). Unpaved roads fragment sagebrush landscapes as well as provide disturbed surfaces that facilitate spread of invasive plant species (Belcher and Wilson 1989, Gelbard and Belnap 2003).

Recreation, including hiking, hunting, fishing, and OHV use, was a major cause of species endangerment in the Great Basin and was a primary factor endangering 12 species in Utah and Nevada (Czech et al. 2000). Even activities that often are perceived as low impact or benign, such as hiking and mountain biking, have an influence on wildlife (Miller et al. 1998, Taylor and Knight 2003). Any human activity of high frequency along established roads or corridors, whether motorized or nonmotorized, can affect wildlife habitats and species negatively through habitat loss and fragmentation, facilitation of exotic plant spread, population displacement or avoidance, establishment of population barriers, or increased human-wildlife encounters that increase wildlife mortality (Gaines et al. 2003). These effects appear to be common across a variety of habitats and species that span the full range of forested to arid terrestrial environments (Gaines et al. 2003, Ouren et al. 2007). Nationally, recreation was listed in the Federal Register as a primary factor for 27% of the endangered, threatened, or proposed species (Wilcove et al. 1998).

Off-highway vehicle use can increase noise disturbance while also increasing soil compaction and erosion through surface disturbance and vegetation loss or alteration (Webb and Wilshire 1983, Lovich and Bainbridge 1999). OHVs were a primary factor given in the Federal Register for 13% of the species listed as endangered or threatened or proposed for listing (Wilcove et al. 1998). Direct effects of OHV use on wildlife populations have varied by species but appear similar to effects from motorized use of roads (Havlick 2002). Reduced nesting success of songbirds near OHV trails was offset by lower numbers of predators using these areas (Barton and Holmes 2007). Despite widespread and increasing use of OHVs (Cordell et al. 2005), potential effects on sagebrush habitats and most obligate wildlife species, including sage-grouse, have not been studied.

LIVESTOCK GRAZING

Historical Development and Current Status

Livestock grazing is the most widespread land use across the sage-grouse range (Brussard et al. 1994, Noss 1994, Crawford et al. 2004). Almost all sagebrush habitats have been grazed in the past century (Saab et al. 1995, West 1996, West and Young 2000, Hockett 2002). Areas ungrazed by livestock exist only in regions in which natural or human-developed water sources are not available, such as in inaccessible kipukas or tops of physically isolated buttes. Even lands on which grazing is currently restricted often have a previous history of livestock grazing (e.g., United States Department of Energy Idaho National Laboratory; Anderson and Holte 1981).

Livestock grazing in western shrublands became significant in the ecological and political landscape in the late 1800s (Oliphant 1968, Young and Sparks 2002). Completion of the first transcontinental railroad in the 1860s greatly expanded the livestock industry, because products could be shipped to markets on the East and West Coasts (Holechek 1981, Mitchell and Hart 1987, Flores 2001). Early grazing was largely unregulated by either fences or a legal system, and competition was intense among ranchers, homesteaders, and free-rangers as well as between cattle grazers and sheepherders (Carpenter 1981, Poling 1991, Donahue 1999).

Major increases in numbers of livestock and areas grazed from 1880 to 1905 altered the condition of western landscapes (Griffiths 1902, Mitchell and Hart 1987, Box 1990). Grazing on public lands was unregulated during this period without boundaries or controls on number or kind of livestock grazed (Carpenter 1981). Number of livestock increased from 4,100,000 cattle and 4,800,000 sheep in 1870 to 19,600,000 cattle and 25,100,000 sheep in 1900 (Donahue 1999). Native perennial bunchgrasses in the Intermountain West lacked seed production and morphological characteristics to sustain anything greater than low levels of grazing disturbance (Mack and Thompson 1982, Miller et al. 1994). Large declines of native grasses and winterfat (*Krascheninnikovia lanata*) occurred between 1870 and 1900 (Griffiths 1902, Cottam and Stewart 1940, Christiansen and Johnson 1964); reduction and loss of palatable forage species and increases in plant species of low palatability may have taken only 10–15 years at any given site under heavy uncontrolled grazing (Hull 1976). Forage production that would support livestock dropped between 60% and 90% of the site potential following depletion of the vegetation community by the early 1930s (Aldous and Shantz 1924, McArdle et al. 1936).

A drought in the 1920s followed the period of heavy grazing and became severe in the 1930s. Native grasses and forbs were depleted from the vegetation community and replaced in much of the Great Basin and surrounding region by exotic annual grasses (Robertson and Kennedy 1954, Young et al. 1972, Yensen 1981, Billings 1990, Miller et al. 1994). Loss of protective vegetation cover and trampling in some communities resulted in extensive soil disturbance and erosion (McArdle et al. 1936, Cottam and Stewart 1940). Shrub density also increased with heavy grazing and loss of understory grasses and forbs, although total distribution of shrubs across a landscape or region likely remained similar (Vale 1975).

The Taylor Grazing Act (1934) created grazing districts on public lands, established a permit and fee system to limit numbers of livestock, and developed administration of public land grazing under the United States Grazing Service, which became the BLM in 1946 (Poling 1991). A portion of the grazing fees was returned to the grazing districts for land improvements. Prescriptive management of grazing also was introduced to reduce grazing impacts through periodic rest and rotations rather than season-long grazing. Land improvements were implemented primarily to restore grasses in the herbaceous understory to maximize forage production for livestock. Other management objectives were to reduce plants poisonous to livestock, stabilize soils, and reduce shrub cover.

Herbicides, plowing, burning, and a diversity of other methods were used to remove the competitive woody overstory, particularly sagebrush, to maximize forage production for livestock by encouraging growth of grasses and forbs over large areas of the western states (Pechanec et al. 1965, Vale 1974, Young et al. 1981, Cluff et al. 1983). Estimates vary on how much sagebrush was treated or eradicated. Between 200,000 and 240,000 km^2 were estimated to have been treated over a 30-year period beginning in the 1940s (Schneegas 1967). Other studies estimated that 10–12% of the total sagebrush distribution (400,000–480,000 km^2) was treated by the 1970s (Vale 1974, Pechanec et al. 1965). Total area of sagebrush treated on lands managed by the BLM was 180,000 km^2 between 1940 and 1994; during the peak period of sagebrush removal in the 1960s, 11,000 km^2/year were treated (Miller and Eddleman 2001).

Replanting areas after sagebrush was removed with nonnative perennial grasses such as crested wheatgrass (*Agropyron cristatum*), desert wheatgrass (*A. desertorum*), and Siberian wheatgrass (*A. fragile*) reduced the necessity of a preexisting understory and was particularly successful at increasing forage that could be grazed by livestock (Shane et al. 1983). The first reseedings in southern Idaho were conducted in 1932 (Hull 1974).

Land treatments or burned areas currently are reseeded to reduce the spread of invasive plants, stabilize soils and reduce erosion, and create wildlife habitat (Monsen et al. 2004). With the exception of mineland reclamation, revegetation guidelines currently emphasize introduced species, although policies encourage the use of native seed mixes when available and affordable (Richards et al. 1998, Asay et al. 2001, Pyke et al. 2003). Total area seeded by the BLM to improve wildlife habitat averaged >400 km^2/yr and ranged from 55 km^2 in 1998 to 1,673 km^2 in 2000 (Table 12.4). However, seed mixtures used in these treatments may have been selected to benefit other wildlife and not targeted specifically to benefit sage-grouse.

Legislation enacted during the 1960s and 1970s directed management on public lands to maintain or improve ecosystem integrity and to provide sustainable uses of resources for multiple purposes, including recreation and wildlife, and not solely to produce livestock forage (Ross 1984, 2006; Poling 1991). Management paradigms also shifted because public lands were now to be retained, rather than disposed of to private entities. Consequently, sustaining resource productivity in perpetuity rather than as a temporary landholder placed importance on managing grazing impacts that were compatible with the land's productivity and in compliance with environmental laws.

Livestock Numbers on Public Lands

Numbers of livestock grazed on public lands are indexed by animal unit month (AUM), which is the amount of forage required to feed one 454-kg (1,000-lb) cow and her calf, one horse, five sheep, or five goats for 1 month. Permitted AUMs, as defined by BLM, represent potential maximum use based on land condition and trend. Actual use varies because of economics, nonuse due to forage or drought conditions, and unreported trespass. Number of AUMs increased following enactment of the Taylor Grazing Act in 1934 to >15,000,000 beginning in the 1940s (Clawson and Held 1957). Permitted AUMS have declined slowly since the late 1960s (Mitchell 2000). Approximately 12,700,000 AUMs were permitted annually on public lands in the western states from 2005 to 2007 (Table 12.5).

Condition of Public Lands

Land condition traditionally has been assessed by relative departure of the seral stage of the current vegetation from a climax community (West 2003a). Seral stage has been estimated for approximately half of the lands managed by the BLM in 13 states within the SGCA (Table 12.6). Less than 10% of the surveyed lands were potential natural communities at climax (excellent condition), 76% were in late or mid-seral stages (good to fair condition), and 15% were in early seral (poor) condition.

TABLE 12.4

Total area (km^2) seeded on lands managed by the BLM from 1997 to 2006 for wildlife habitat improvement.

Dash indicates that a state did not submit information for publication year.

State	1997	1998	1999	2000	2001	2002	2003	2004	2005	2006
Arizona	16	0	1	57	<1	1	0	1	1	1
California	4	5	<1	—	1	—	—	—	58	—
Colorado	<1	1	1	3	21	25	40	—	0	2
Idaho	283	—	4	—	—	—	101	—	—	116
Montana	3	1	2	—	8	<1	0	—	0	0
Nevada	23	45	90	1,582	202	265	300	26	<1	25
New Mexico	<1	—	6	2	1	<1	4	10	2	—
Oregon	6	3	1	20	93	—	0	40	40	32
Utah	74	—	67	7	—	14	21	220	46	83
Wyoming	0	—	<1	2	12	1	0	0	1	—
Total area	410	55	173	1,673	339	306	466	297	148	259

SOURCE: Public Land Statistics (United States Department of the Interior 1997, 1998, 1999, 2000d, 2001b, 2002b, 2003f, 2004e, 2005d, 2006c).

TABLE 12.5

Grazing permits and leases in force on public lands from 2005–2007.

AUMs = animal units/month.

State	2005 Permits	2005 AUMs	2006 Permits	2006 AUMs	2007 Permits	2007 AUMs
Arizona	758	660,528	757	660,007	765	660,086
California	555	361,430	548	355,726	541	345,164
Colorado	1,594	664,003	1,591	650,168	1,575	641,314
Idaho	1,889	1,351,806	1,890	1,348,526	1,889	1,358,417
Montana	3,743	1,283,126	3,755	1,281,144	3,765	1,281,748
Nebraska	17	578	17	578	17	578
Nevada	662	2,187,729	644	2,137,635	650	2,132,155
New Mexico	2,286	1,861,231	2,275	1,856,795	2,275	1,862,572
North Dakota	75	9,226	76	9,233	76	9,233
Oklahoma	4	138	4	132	4	132
Oregon	1,284	1,026,548	1,277	1,026,463	1,275	1,028,553
South Dakota	470	73,924	472	73,828	475	73,737
Utah	1,519	1,238,877	1,499	1,239,786	1,490	1,225,890
Washington	294	32,144	283	33,603	279	33,078
Wyoming	2,790	1,949,789	2,792	1,960,956	2,798	1,937,041
Total	17,940	12,701,077	17,880	12,634,580	17,874	12,589,698

SOURCE: Public Land Statistics (United States Department of the Interior 2005d, 2006c, 2007a).

NOTE: Number of grazing permits were combined for Section 15 (lands within grazing districts) and Section 3 (isolated tracts outside of grazing districts).

Current assessment of land condition is based on ecological criteria rather than on seral stages of plant communities, livestock preferences for a plant species, or a plant species resistance to livestock grazing (National Research Council 1994). Indicators of soil characteristics and erosion, hydrologic function, biotic integrity, and ecological processes are integrated into evaluations of the upland's functioning (Whitford et al. 1998, Pyke et al. 2002, West 2003b). Evaluation of land condition is not directly transferrable to assessments for seral stage. Similarly, riparian areas are evaluated for different criteria using techniques that reflect stream geomorphology, stream bank stability, and plant community potential (Pritchard et al. 1998).

Standards and guidelines for management of public grazing lands are established by local resource advisory councils and also must address habitats and conservation measures for endangered, threatened, proposed, candidate, or other at-risk or special status species (United States Department of the Interior 2003d,e). Sixty-three percent (>402,000 km^2) of the public lands managed by the BLM have been assessed according to standards and guidelines. Under established criteria, 40% of assessed lands (25% of all lands under management by the BLM, including nonsagebrush habitats) met standards or were making progress toward meeting those standards (category A, Table 12.7). Livestock were a factor in 19% of the assessed lands not meeting standards (categories B and C). Another 5% of the assessed lands were not meeting standards for causes other than livestock grazing (category D).

TABLE 12.6

Percent of lands inventoried and by seral condition on public lands managed by the BLM in 2007.

Similarity to the climax vegetation community at a site was 76–100% for potential natural communities, 51–75% for late seral, 26–50% for mid-seral, and 0–25% for early seral stages[a].

State	Inventoried	Potential natural community	Late seral	Mid seral	Early seral	Unclassified/not inventoried
Arizona	56	8	44	37	11	37
California	15	3	21	45	31	72
Colorado	47	7	27	41	25	50
Idaho	74	2	25	41	33	9
Montana[b]	74	9	66	24	1	25
Nevada	40	4	38	46	12	58
New Mexico	76	4	24	43	30	24
Oregon and Washington	57	1	28	59	12	36
Utah	61	12	30	44	13	33
Wyoming	59	27	38	30	5	41
Total	54	8	35	41	15	42

SOURCE: United States Department of the Interior (2007a). Statistics were reported for 2007, although surveys to classify sites may have been conducted in previous years or decades.

[a] Potential natural communities are considered by BLM to be in excellent condition, late seral stages to be good, mid-seral to be fair, and early seral to be poor, based on the relationship of current conditions relative to a perceived natural climax condition (West 2003a).

[b] Montana statistics also include North Dakota and South Dakota.

Management Actions on Public Lands

Large numbers of treatments are conducted on public lands, including those dominated by sagebrush, for different purposes (Table 12.8). The BLM addressed use of habitat treatments and concluded that treating vegetation was necessary to develop or restore a desired plant community, create biological diversity, increase forage or cover for animals, protect buildings and other facilities, manage fuels to reduce wildfire hazard, manage vegetation community structure, rejuvenate late successional (old-growth) vegetation, enhance forage/browse quality, or remove noxious weeds and poisonous plants (United States Department of the Interior 1991). Over 24,000 km^2 managed by the BLM, most of it in Nevada, Idaho, Oregon, and Wyoming, would be treated annually as the preferred strategy to reduce fire risk and to control unwanted vegetation (United States Department of the Interior 2007b,c).

Different combinations of herbicides and seasons of applications have been developed to remove sagebrush, other unwanted woody shrubs, and weedy annuals (Tueller and Evans 1969; Evans and Young 1975, 1977; McDaniel et al. 1991). More recent treatment objectives have emphasized thinning density of sagebrush and control of noxious weeds. Small irregular patchworks of habitat are created in some regions, in contrast to total removal of sagebrush from large areas. The most common herbicides used on lands managed by the BLM are 2,4-D, picloram, and tebuthiuron (United States Department of the Interior 2007c). Under the preferred alternative to reduce fire risk and control unwanted vegetation, tebuthiuron would be used on 25%, 2,4-D on 18%, and picloram on 15% of the 3,770 km^2 treated annually by the BLM (United States Department of the Interior 2007c). Herbicide use can result in short-term decreases in exotic plants. However, long-term benefits (4–16 years) may not be obtained because exotic plants are capable of recolonizing treated sites, particularly when native plant species have also been reduced by herbicide treatments (Rinella et al. 2009).

TABLE 12.7

Total area (km²) and percent of assessed lands[a] managed by the BLM meeting standards and guidelines established for rangeland health[b].

State	Category A[c]		Category B[d]		Category C[e]		Category D[f]		Assessed		Not assessed		Total area
	km²	%	km²	%	km²	%	km²	%	km²	%	km²	%	km²
Arizona	28,839	62	1,452	3	1,035	2	83	0	31,409	68	14,793	32	46,203
California	10,561	32	5,824	18	2,524	8	2,535	8	21,444	65	11,427	35	32,871
Colorado	19,013	60	5,418	17	272	1	4,284	14	28,987	92	2,621	8	31,609
Idaho	13,694	29	15,766	34	2,736	6	4,608	10	36,804	79	10,018	21	46,822
Montana[g]	25,740	77	3,997	12	420	1	1,598	5	31,755	96	1,478	4	33,233
Nevada	60,572	34	35,726	20	3,524	2	5,060	3	104,882	58	74,488	42	179,370
New Mexico	8,439	17	1,072	2	53	0	397	1	9,961	20	40,816	80	50,777
Oregon/ Washington	19,763	35	5,833	10	4,645	8	1,944	3	32,184	58	23,775	42	55,959
Utah	33,961	39	5,147	6	1,576	2	6,508	7	47,191	54	40,624	46	87,816
Wyoming	32,536	45	20,192	28	2,715	4	2,071	3	57,513	80	14,139	20	71,653
Total	253,117	40	100,427	16	19,499	3	29,088	5	402,131	63	234,181	37	636,312

[a] Percentages were calculated from total assessed area because data do not represent random or stratified sampling of lands managed by the BLM, and no inference could be assumed for lands that have not been assessed.

[b] http://www.blm.gov/nstc/rangeland/rangelandindex.html.

[c] Category A = Meeting standards or making significant progress toward meeting standards.

[d] Category B = Not meeting all standards or making significant progress, but appropriate action has been taken to ensure significant progress toward meeting standards (livestock is a significant factor).

[e] Category C = Not meeting standards or making significant progress toward meeting the standards, and no appropriate action has been taken to ensure significant progress (livestock is a significant factor).

[f] Category D = Not meeting standards or making significant progress toward meeting the standards, due to causes other than livestock grazing.

[g] Montana statistics include North Dakota and South Dakota.

Primary actions to facilitate prescriptive livestock grazing on lands managed by the BLM include construction of fences, development or control of water, and habitat modifications (Table 12.8). From 1962 to 1997, >51,000 km of fence were constructed on land administered by the BLM in states supporting sage-grouse populations (Connelly et al. 2000c). More than 1,000 km of fences were constructed each year from 1996 through 2002, and density of fences exceeds 2 km/km² in some regions; most fences were constructed in Montana, Nevada, Oregon, and Wyoming (Table 12.8, Fig. 12.8). Water developments were also widespread throughout public lands (Fig. 12.9).

Prescribed fire is one of the most common management tools for reducing density of sagebrush, facilitating growth of grasses and forbs, and controlling juniper (*Juniperus* spp.) and pinyon (*Pinus* spp.) woodland expansion into sagebrush habitats (Miller and Eddleman 2001, Baker and Shinneman 2004, Baker 2006a), and for assisting in controlling annual grasses (DiTomaso et al. 2006). Prescribed fires were conducted on >3,700 km² of public lands from 1997 through 2006 (Table 12.9). Most areas treated by prescribed fire were in Oregon and Idaho.

Area treated under the Emergency Stabilization and Rehabilitation program, designed to rehabilitate areas following fire, varied from 281 km² in 1997 to 16,135 km² in 2002 (Table 12.10). The majority of areas treated after burning were reseeded with differing mixes of shrubs, forbs, and grasses following extensive fires in Idaho, Nevada, and Oregon. The main purpose of this program is to stabilize soils and maintain site productivity, not to regain site suitability for wildlife.

TABLE 12.8

Number of habitat treatments (1929–2004) on lands managed by the BLM.

Habitat treatment	AZ	CA	CO	ID	MT	ND	NM	NV	OR	SD	UT	WA	WY	Totals
Unknown	9	20	15	23	30		12	40	18		91		25	283
Fences	1,489	669	3,051	3,717	5,790	48	1,445	3,489	3,451	357	3,265	239	5,278	32,288
Fence modification	1	6	9	149	17		7	23	89		17	4	23	345
Hazard reduction	1	8	28	47	9		1	5	10		2		4	115
Lake/wetland improvement	21		8	8	17		2		7	1	2	10	8	84
Land treatment[a]	30	15	253	158	82		78	77	54	4	141	11	73	976
Misc. facility improvement	468	173	1,038	1,254	1,420		368	1,480	1,263	10	1,300	41	644	9,459
Perch/nesting structure			2	6	7		2	3	3		6	11	44	84
Stream bank stabilization	1		18	4	13		1	2	11		1	3	2	56
Stream improvement		8	35	11	13		3	1	14		17	4	22	128
Timber stand improvement		8	23	9	1		3	1	15		1	1	12	74
Vegetation manipulation	220	207	880	1,585	445	9	681	895	1,089	14	824	66	481	7,396
Weed control[b]		2	138	67	320	3	31	12	126	15	14	294	42	1,064
Water development/control	2,180	1,180	9,189	5,017	10,587	56	2,247	5,163	7,555	561	5,292	130	7,813	56,970
Water facilities modified		15	49	65	15		3	3	126	1	20	2	38	337
Total habitat treatments	4,420	2,311	14,736	12,120	18,766	116	4,884	11,194	13,831	963	10,995	816	14,509	109,661

SOURCE: Bureau of Land Management Range Improvement Projects database.

[a] Includes contour furrowing, ripping and deep tillage, pitting, terracing, checks, and scalping.

[b] Includes biological and chemical treatments, blading, cutting or beating, chaining or cabling, chipping, log and scatter, plowing, prescribed fire, and wildfire.

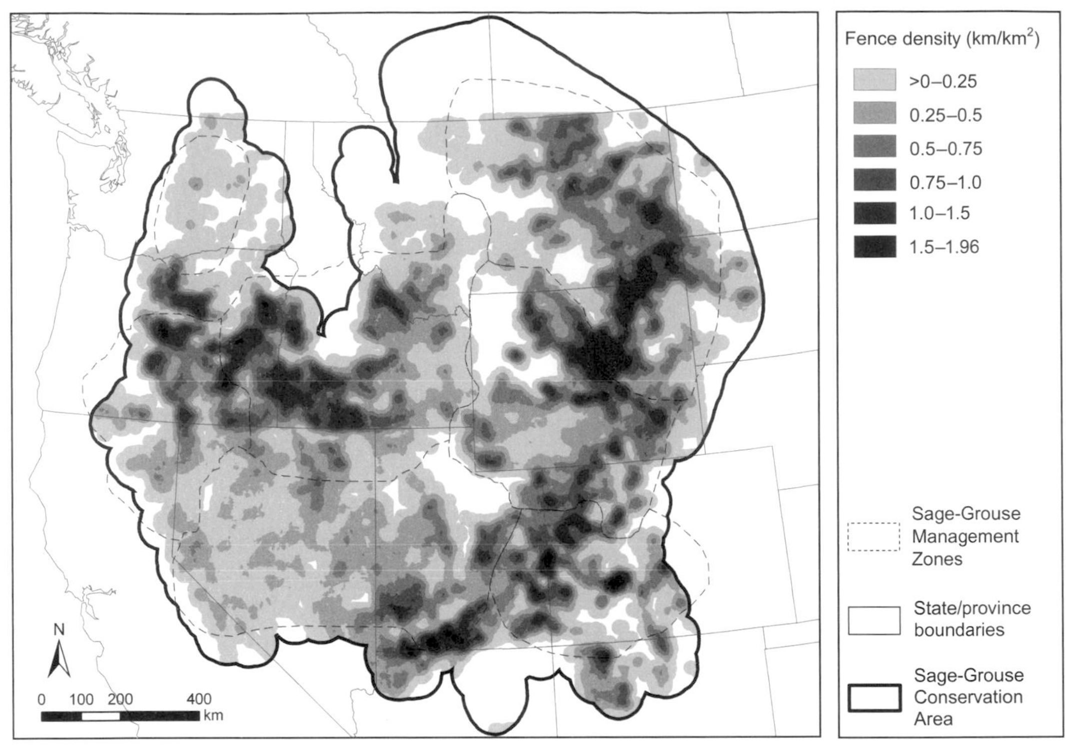

Figure 12.8. Linear density (km/km^2) of fences (estimated from allotment and pasture boundaries) on public lands in the Sage-Grouse Conservation Area (GIS coverages obtained from Bureau of Land Management Geocommunicator).

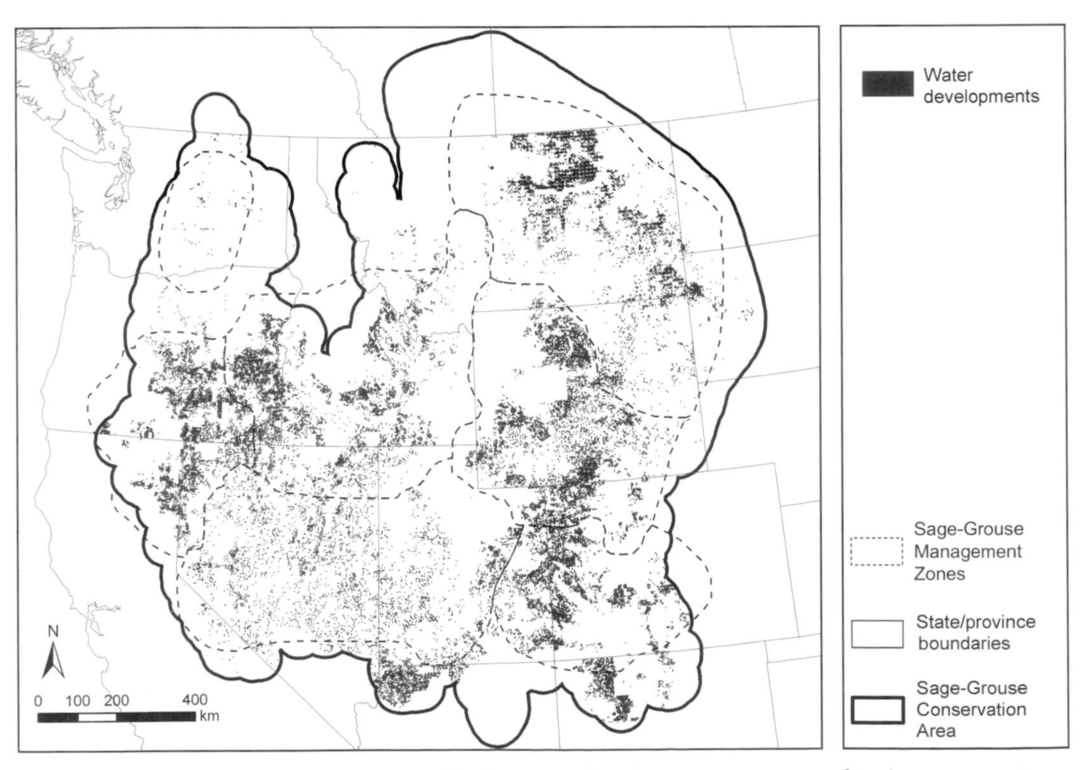

Figure 12.9. Water developments on lands managed by the Bureau of Land Management (Bureau of Land Management Range Improvement database). Locations of water development are recorded to the nearest 2.59 km^2.

TABLE 12.9

Total area (km^2) treated by prescribed fire on lands managed by the BLM from 1997–2006.

Dashes indicate states that did not submit information for publication year.

State	1997	1998	1999	2000	2001	2002	2003	2004	2005	2006
Arizona	8	95	94	40	48	52	15	3	57	0
California	16	45	21	22	3	7	—	—	4	—
Colorado	3	56	46	2	2	24	4	—	0	2
Idaho	11	64	102	43	100	59	8	—	—	6
Montana	3	18	31	15	34	12	10	—	0	20
Nevada	10	2	3	<1	10	28	13	7	24	2
New Mexico	60	131	71	15	34	53	95	3	8	—
Oregon	72	169	225	55	86	223	59	81	81	73
Utah	11	24	30	33	64	19	25	19	15	32
Wyoming	59	117	138	23	11	20	43	43	74	—
Total area	253	720	764	251	393	497	272	156	263	135

SOURCE: Public Land Statistics (United States Department of the Interior 1997, 1998, 1999, 2000d, 2001b, 2002b, 2003f, 2004e, 2005d, 2006c).

TABLE 12.10

Number of projects (N) and total area (km^2) treated for emergency fire rehabilitation on lands managed by the BLM from 2002–2007.

	2002		2003		2004		2005		2006		2007	
State	N	km^2	N	km^2	N	km^2	N	km^2	N	km^2	N	km^2
Arizona	0	0	2	1	4	9	15	4	17	138	15	6
California	19	34	11	278	37	467	27	432	27	396	21	476
Colorado	16	52	7	170	18	138	15	99	19	9	20	26
Idaho	85	2,900	51	1,524	53	1,345	52	926	92	3,611	115	5,943
Montana	3	8	4	82	6	601	6	311	11	335	4	211
Nevada	103	11,226	111	3,099	108	2,235	65	278	182	6,926	176	15,697
New Mexico		0	1	3	1	1	5	1	10	1	7	1
Oregon	50	1,392	46	2,858	39	2,318	24	493	58	593	63	1,689
South Dakota		0	1	8	1	4	1	<1	1	8	0	0
Utah	27	409	54	925	76	627	77	974	115	1,931	122	2,124
Washington		0	2	1	2	32	2	35	1	2	2	9
Wyoming	5	113	6	301	6	17	6	0	5	25	4	0
Total	308	16,134	296	9,250	351	7,794	295	3,553	538	13,975	549	26,182

SOURCE: Public Land Statistics (United States Department of the Interior 2002b, 2003f, 2004e, 2005d, 2006c, 2007a).

TABLE 12.11
Nonfire fuels treatment (km²) on lands managed by the BLM from 1999–2002.

State	1999	2000	2001	2002
Arizona	<1	10	32	46
California	7	5	4	14
Colorado	15	9	74	60
Idaho	20	277	324	228
Montana	0	<1	0	15
Nevada	0	123	6	40
New Mexico	4	5	4	150
North Dakota	0	0	0	0
Oregon	6	27	55	160
South Dakota	0	0	0	0
Utah	14	1	0	63
Washington	0	0	1	1
Wyoming	0	2	52	54
Total area	66	460	553	832

SOURCE: Public Land Statistics (United States Department of the Interior 1999, 2000d, 2001b, 2002b). Includes mechanical, hand, and chemical treatments.

Consequently, the primary grasses and forbs used are nonnative species; sagebrush seedings are limited to crucial areas for wildlife or to create small islands from which seed dispersal can colonize remaining unseeded parts of the landscape.

Replanting native vegetation in post-fire rehabilitation is increasingly being emphasized (Richards et al. 1998). More recent efforts have focused on techniques and environmental conditions needed to successfully reseed or promote specific forb species important to sage-grouse (Wirth and Pyke 2003, Wrobleski and Kauffman 2003). Use of nonnative species in post-fire rehabilitation will continue because of seed availability, ease of establishment, and desirable growth response to achieve short-term objectives that include stabilizing soils, controlling water erosion, and preventing permanent loss or damage of ecosystem functions (Asay et al. 2001, Beyers 2004, Booth and Vogel 2006). Moreover, nonnative grasses may be necessary in some closed communities to gain an advantage over cheatgrass (*Bromus tectorum*) (Robertson and Pearse 1945), after which native plants can become established (Cox and Anderson 2004).

Livestock grazing generally is not permitted for two growing seasons post-fire to allow for seeding or rehabilitation efforts to develop. Reintroduction of livestock grazing after two growing seasons but before the native or reseeded plant community has developed can result in increased levels of exotic grasses and failed rehabilitation efforts (Evans and Young 1978, Monsen et al. 2004, Eiswerth and Shonkwiler 2006).

The area on which nonfire fuels treatments, including mechanical (harrowing, chaining), hand (thinning, piling), and chemical (herbicide) methods, were conducted increased from 66 km^2 in 1999 to 832 km^2 in 2002 (Table 12.11). Mechanical treatments specifically to improve wildlife habitat were conducted on 1,242 km^2 from 1997 to 2006, primarily in Utah and Nevada (Table 12.12).

Ecological Influences and Pathways

Effects of livestock grazing are expressed as relative differences in the structure of sagebrush communities and their functioning, and not as disruptions that can be mapped as distinct categories over the larger landscape. Therefore, we present

TABLE 12.12

Total area (km^2) treated with disking and chaining for wildlife improvements on lands managed by the BLM from 1997–2006.

Dashes indicate states that did not submit information for publication year.

State	1997	1998	1999	2000	2001	2002	2003	2004	2005	2006
Arizona	1	<1	1	2	0	<1	0	0	0	0
California	<1	<1	0	—	0	—	—	—	0	—
Colorado	1	1	<1	<1	0	37	20	—	0	6
Idaho	0	0	0	—	—	—	0	—	—	<1
Montana	3	1	1	—	0	0	0	—	0	0
Nevada	0	3	0	8	0	13	0	51	417	9
New Mexico	<1	0	9	13	1	3	8	0	5	—
Oregon	3	4	<1	1	10	—	0	0	0	0
Utah	190	0	9	4	—	23	0	53	86	208
Wyoming	23	0	2	0	5	1	0	0	4	—
Total area	221	9	23	29	16	77	28	104	512	223

SOURCE: Public Land Statistics (United States Department of the Interior 1997, 1998, 1999, 2000d, 2001b, 2002b, 2003f, 2004e, 2005d, 2006c).

detailed background on the dynamics of sagebrush vegetation to better interpret the effects of past and current livestock grazing and management.

Livestock grazing is a press form of disturbance that exerts repeated pressure over time on a system (Bender et al. 1984, Turner and Bratton 1987). Point-sources of disturbance, such as fire, have measurable effects from a well-defined origin in space and time. Other land uses, such as croplands, have defined boundaries and influence sagebrush systems primarily by removing or fragmenting habitat and secondarily by altering system pathways. In contrast, the impact of livestock grazing is spread diffusely across the landscape. The exceptions are near locations of concentration, such as water points or mineral-nutrient blocks, where high amounts of disturbance concentrated around these locations can denude or radically change vegetation within 30 ha centered on this point of concentration (Washington-Allen et al. 2004, Brooks et al. 2006).

Management of livestock grazing and assessment of habitat condition is site-specific rather than at large scales or across landscapes (Mitchell 2000, Crawford et al. 2004). Reliable numbers of livestock, coupled with spatial and temporal information on grazing intensity and habitat characteristics, are not available to assess grazing effects for a large-scale analysis (Society for Range Management 1995, West 2003a). Most research on livestock grazing is short-term or small-scale; syntheses of results in meta-analyses have emphasized the difficulty of drawing broad generalizations about effects of grazing from site-based information (Milchunas and Lauenroth 1993, Jones 2000). Grazing potentially influences nest success of sage-grouse through loss of vegetation for cover (Connelly and Braun 1997, Beck and Mitchell 2000), but this local effect has not been translated to impacts at a population level.

Conceptual models of dynamics in grass and shrubland systems have developed along two basic, but not mutually exclusive, theories. Clementsian and state-and-transition models differ on assumptions about succession and the role of disturbance in shaping vegetation communities (Friedel 1991, Joyce 1993, Schacht 1993, Scarnecchia 1995, Society for Range Management 1995). Land managers use these models as tools to support decisions on grazing practices and to establish performance criteria. Expected vegetation response to grazing or land treatments is predicted from known and desired vegetative communities and ecological function. Evaluating effects or evaluating actions thus depends on how these conceptual models reflect actual dynamics of sagebrush systems.

Early models of shrubland dynamics were based on a Clementsian viewpoint of predictable, linear, and reversible succession to a climax community (Dyksterhuis 1949). Grasslands were colonized by shrubs, which ultimately dominated the community and reduced or excluded grasses and forbs. Disturbance was a minor but necessary component that removed shrubs and permitted grasses and forbs to reestablish. Net primary productivity of the vegetation was greatest at intermediate seral stages. Seral condition was assessed as a function of percent of vegetation that resembled the climax stage, and management actions were based on relationship to successional endpoints of the community (Vavra and Brown 2006). Grazing was treated as a disturbance that set communities back from climax to an earlier seral stage; release from grazing allowed the community to return to a climax stage. Old-growth or climax communities dominated by shrubs and having a minor component of grasses or forbs in the understory were considered to have low productivity. Treatments were applied to return these old-growth communities to and earlier successional stage.

Stocking rates for livestock are based on these conceptual relationships between the current vegetation community and an intended condition, seral stage, or climax community, and on estimates of an average amount of available forage plants and the area necessary to sustain livestock (Box 1990, Society for Range Management 1995, Holechek et al. 1998). Actual number grazed is based on season of use, distribution and use of available forage, an estimate of native herbivory (Beever and Aldridge, this volume, chapter 14), and class of livestock. Average conditions rarely exist, and western landscapes experience periodic droughts (Miller et al., this volume, chapter 10). Long-term effects of domestic livestock grazing also may be additive to native herbivory, thus increasing the total effect of changes in sagebrush community structure and function caused by herbaceous depletion (Manier and Hobbs 2007). Stocking rates are also based on livestock production and financial livelihood of grazing permittees, in addition to environmental considerations (Holechek et al. 1999). In the natural world, the number of herbivores is constrained by food supply. The interaction between domestic livestock grazing and food availability is less tightly coupled, because the relationship between food supply and numbers of livestock is buffered by administrative, management, or economic factors. Thus, time lags in changing numbers of livestock in response to changes in habitat conditions can magnify the effects of grazing (Thurow and Taylor 1999). Stocking rates and estimates of the amount of vegetation that can be removed based on differences between a current vegetation community and an ideal or climax seral stage also may not be sustainable over the long term (Holecheck et al. 1998).

Absence of information on vegetation responses coupled with grazing regime, type of livestock, or actual number of native or domestic animals grazed hinder our understanding of the effects of livestock grazing on long-term dynamics of sagebrush systems (Yorks et al. 1992). The subjectivity of the assessments of seral and climax categories, unstratified sampling, and percent of unsurveyed lands limit our effectiveness in assessing the relationship between livestock numbers and vegetation or for monitoring spatial or temporal trends (West 2003a,b). Also, surveys of land condition are not updated every year, even though annual statistics are reported, which has significant implications for proposed grazing regulations and the monitoring information and feedback required before taking action (United States Department of the Interior 2003e). Qualifiers, such as moderate or heavy utilization rates, are based on how much vegetation has been removed, and often are subjective and variable across sites (Frost et al. 1994).

Succession may follow predictable pathways in early phases of shrub establishment if nonnative plants were not part of the previous community (Seefeldt et al. 2007) and seed sources for shrubs are available from undisturbed islands (Longland and Bateman 2002). In relatively pristine conditions, mountain big sagebrush communities in southern Wyoming followed a long-term pathway in which total shrub cover increased over a 40-year period post-disturbance (Ewers and Pendall 2008). The length of time for sagebrush to recover varies based on amount and timing of precipitation; source, size, and intensity of disturbance; fire-return interval; sagebrush species or subspecies; native and domestic herbivory; and site-specific environmental factors that influence vegetation competition (Miller et al., this volume, chapter 10). However, the long-term dynamics of other sagebrush communities have been less predictable than expected under a Clementsian model (Anderson and Holte 1981, Anderson and Inouye 2001, West and Yorks 2002).

State-and-transition models, in which disturbance is a primary driver of change among multiple potential vegetative states, may better describe the dynamics of sagebrush systems than succession along a continuum to a climax (Bestlemeyer et al. 2003, Briske et al. 2005, Vavra and Brown 2006). Several community phases, analogous to successional or seral state, can potentially exist in one vegetative state for long periods until disturbance pushes them into a phase-at-risk, a plant community with lowered resilience, or an alternate steady state. Stability of a community phase within a vegetative state results from its resistance and resilience to disturbance or change (Stringham et al. 2003, Bestelmeyer et al. 2009). A wide range of shrub-grass compositions can be stable, length of time within seral stages may be nonlinear, and transitions among different states may be unpredictable (West et al. 1984, Westoby et al. 1989b, Laycock 1991). Community changes from among phases within a state are often reversible and can occur naturally or be brought about by management. However, a decrease in abundance of native species will potentially lower the resiliency of a community within a state, particularly when exotic or invasive plant species are present, which may result in a transition to an alternate state (Miller et al., this volume, chapter 10; Pyke, this volume, chapter 23). Transitions among vegetation states may be irreversible and return to a previous state no longer economically or ecologically possible once a threshold is crossed (Bestelmeyer et al. 2003). Consequently, plant species response to grazing may not be closely correlated with the level of grazing (Milchunas and Lauenroth 1993, West and Young 2000). Instead of setting the vegetation community back from climax, livestock grazing may alter vegetation, water and nutrient availability, and soils past thresholds from which the system cannot return (Society for Range Management 1995). Vegetation communities released from grazing may exhibit unpredicted results or not respond (Anderson and Holte 1981, West et al. 1984, Stohlgren et al. 1999, West and Yorks 2002), may not return to a previous vegetative state (Holechek and Stephenson 1983), or may remain opened to exotic plant invasions (Young and Allen 1997). Thus, removing disturbance associated with livestock grazing from sagebrush systems possibly will not have the intended effect (Curtin 2002).

Pattern and influence of livestock grazing in sagebrush habitats is fundamentally different from the disturbance regime in which plants have evolved over the past 10,000–12,000 years. Management using prescriptive grazing attempts to mimic previous natural grazing regimes and vegetation response but still maximize forage for livestock by using rest-rotation or other strategies that vary grazing intensities. However, plant communities are not given similar rest from this repetitive grazing, and nutrient recycling differs between nomadic herds and domestic grazing operations (Bock et al. 1993, Freilich et al. 2003, Wisdom et al. 2006). Grazing by American bison (*Bison bison*) was the primary disturbance in the eastern sagebrush steppe region (the *Bouteloua* region) and was locally intense but highly variable in space and time. The vegetation community in this portion of the sage-grouse range was adapted to withstand grazing disturbance even during frequent drought periods (Heitschmidt et al. 2005). Introduction of domestic livestock in these regions increased the site-specific frequency of grazing disturbance (Mack and Thompson 1982).

Much of the western sagebrush steppe (the *Agropyron* region) has had a long period in which large hoofed grazers were rare. Large herbivores became extinct at the end of the Pleistocene (10,000–12,000 years BP), and bison largely contracted their distribution (Mack and Thompson 1982, Billings 1990). Small numbers of bison still ranged in parts of the Great Basin region and western Montana and were relatively common in eastern Idaho, but disappeared from the landscape soon after Euro-American settlement.

Fire, not grazing by large herbivores, was the dominant historic disturbance in these western sagebrush systems. Fire kills most species of sagebrush (Pechanec et al. 1965), creates nonvegetated patches in the landscape, and releases nutrients. An uneven distribution of soils, micro- and macrotopography, environmental and moisture conditions, and biomass resulted in a mosaic of openings and uneven-aged distributions of sagebrush, both at local and landscape scales. The presence of perennial grass understory and intact biological crusts protecting soils were also primary components driving community recovery following burns. Introducing livestock grazing into these regions resulted in a series of synergistic interactions that fireproofed these systems. The intensity and season of grazing depleted grass and forb understories that previously spread fire, leading to increased shrub cover. Disrupted biological crusts and disturbed soils reduced the resiliency of these systems. Grazing intensities that depleted the grass and forb understory, altered soils, and destroyed protective biological crusts created new systems that not only were resistant to recolonization by the native flora but were vulnerable to invasion by exotic plants (Mack and Thompson 1982, Young and Sparks 2002).

New alternate stable states that were formerly sagebrush communities consist of a different suite of community phases differing in plant composition, structure, and function. These new states are often characterized at one endpoint by systems largely resistant to disturbance and by those systems maintained by frequent large-scale disturbance at another. These dominant systems present significant challenges for managing livestock grazing because new and additional disturbances are being applied onto a natural disturbance template to which the vegetation community previously has been adapted (West and Young 2000, Vavra and Brown 2006).

The percentage of public lands that are in improved range condition has increased since the early 1900s (Ross 1984, Society for Range Management 1989, Box 1990). In addition, permitted and authorized AUMs have decreased (Mitchell 2000). However, productivity of western shrublands has declined due to previous grazing history, changes in soils and vegetation, or drought and now is less than it was pre-settlement (Young et al. 1981, West 1983b, Holechek and Stephenson 1998, Holechek et al. 1999). Estimated area per month required to support a cow and calf (or equivalent) was 1.2 ha prior to settlement and unrestricted grazing, 3.7 ha in the 1930s, and 3.2 ha in the 1970s (West 1983b). The distribution of livestock has also changed, because water developments have increased the area that can be grazed (Freilich et al. 2003). We cannot conclude the effect of grazing across the sage-grouse range has been reduced, because fewer numbers of livestock may still exert a larger influence over a broader region on lands that now have lower inherent productivity.

Management of livestock grazing influences ecosystems by actions designed to control or protect livestock or to increase forage availability or improve foraging conditions (West 1996). Livestock movements are managed by establishing water sources, placing salt, and building fences and roads. Increased availability of water substantially influences movements and distribution of livestock in arid western habitats and has expanded the area that can be grazed (Valentine 1947, Freilich et al. 2003). Fences modify access and movements by humans and livestock, exerting a new mosaic of disturbance and use on the landscape (Freilich et al. 2003). Fences also increase potential mortality of sage-grouse directly due to collisions or indirectly by increasing predation rates by providing perches for raptors. Potential predators on livestock are controlled by lethal means or by fencing livestock to control their movements. Habitat manipulations to increase forage include prescribed fire, herbicides, and mechanical treatments to remove sagebrush followed by reseeding, generally with nonnative plants, primarily grasses. These habitat manipulations may alter the natural food web, influence fire and other disturbance regimes, change the nutrient dynamics, alter composition of predator guilds, and affect the vegetation structure used by wildlife (Freilich et al. 2003). Ultimately, livestock function as keystone species: grazing and management actions to manipulate habitats do not preclude wildlife and vegetation, but they influence ecological pathways and can affect which species persist (Bock et al. 1993).

The question of effects of livestock grazing at large spatial scales is difficult because we lack control areas sufficiently large to include landscape processes (Bock et al. 1993). Compounding site-specific results neither gives us a cumulative estimate of effect nor tells us what the landscape would be like in the absence of grazing (National Research Council 1994). We also lack an understanding of the way sagebrush ecosystems functioned before livestock grazing began in the 1800s (Freilich et al. 2003). Our inability to test for an effect does not demonstrate that livestock grazing has no effect or is a compensatory use of sagebrush habitats and therefore should be ignored. Concluding no effect when one exists (type II error) is as significant an error as concluding an effect when none exists (type I error) (Eberhardt and Thomas 1991, Wiens and Parker 1995).

Response to Fire

Effects of fire on sage-grouse habitat depend on site potential; site condition; functional plant groups; and burn pattern, intensity, and size (Miller and Eddleman 2001). Site-specific variables, including soil characteristics, climate, previous disturbance history, and presence of livestock grazing, influence the trajectory of vegetation recovery following burns (West and Yorks 2002, Eiswerth and Shonkwiler 2006, Seefeldt et al. 2007). Prescribed fires generally burn at lower intensities than wildfires because of management controls and season of burn. Therefore, plant responses following prescribed fires may differ from those after wildfires. Because of these complexities, the dynamics of recovery following burns in sagebrush-dominated landscapes may differ regionally, among sagebrush species and subspecies, and between natural and prescribed fires.

Grass and forb response following fire is a function of pre-burn site condition, fire intensity, and both pre- and post-fire patterns of precipitation. Herbaceous production typically decreases in the first year, followed by recovery or increase to pre-burn levels in the second or third year (Blaisdell 1953, Harniss and Murray 1973, Antos et al. 1983, Cook et al. 1994, Seefeldt et al. 2007). Short-term benefits to nesting concealment that result from increased herbaceous growth following fire (Klebenow 1973, Pyle and Crawford 1996, Wrobleski and Kauffman 2003) may not balance the loss of sagebrush canopy structure required by sage-grouse during the nesting season and winter (Fischer et al. 1996b, Connelly et al. 2000b, Nelle et al. 2000, Beck et al. 2009).

Sagebrush recovery from burns, particularly in more xeric sagebrush ecosystems, is long-term. Average time for big sagebrush subspecies to recover to pre-burn levels is 30–35 years following fire and ranges from 10 to >50 years (Harniss and Murray 1973, Barney and Frischknecht 1974, Watts and Wambolt 1996, Ziegenhagen and Miller 2009). Even within subspecies, a survey of more than 50 fires in mountain big sagebrush communities reported the variation in time for sagebrush canopies to approach 20–30% cover averaged 25–30 years but varied from 15 to >50 years (Ziegenhagen 2004). The rate of recovery for mountain big sagebrush within the perimeter of large fires was dependent upon establishment from seed in soil seed bank in the first two post-fire years (Ziegenhagen and Miller 2009). Short-lived viability of sagebrush seed in the soil further imposes limits on regeneration if unfavorable weather conditions follow a burn. Wyoming big sagebrush (*Artemisia tridentata* ssp. *wyomingensis*) seed is viable for one year in the soil; seeds of mountain big sagebrush (*Artemisia tridentata* ssp. *vaseyana*) may be viable for 1–3 years.

Landscape restoration may require centuries or longer in the absence of active restoration (Hemstrom et al. 2002; Baker 2006a, this volume, chapter 11; Miller et al., this volume, chapter 10). Size of the fire, fire frequency, and completeness of the burn influence the availability of viable seeds remaining in soils or in unburned islands of sagebrush (Longland and Bateman 2002; Baker, this volume, chapter 11). Seed dispersal and subsequent expansion of the boundary of a shrub patch from these islands can be slow. The majority of seeds fall <12 m from the parent plant (Mueggler 1956, Johnson and Payne 1968). In turn, from 3 to >30 years may be required for these seeds to grow to mature, seed-producing plants, depending on timing and amount of precipitation.

Sage-grouse are slow to recolonize disturbed areas, even though structural features of the shrub community may have recovered. Fire is one of the primary factors linked to population declines of Greater Sage-Grouse because of long-term loss of sagebrush and conversion to exotic grasses (Connelly and Braun 1997). Small increases in fire within the landscape surrounding leks significantly decreased the probability of persistence over a 30-year period (Knick and Hanser, this volume, chapter 16) and decreased trends in male sage-grouse counted at leks (Johnson et al., this volume, chapter 17). Adding more fire to these systems by prescribed burning has limited usefulness for managing sage-grouse habitats, particularly in more xeric Wyoming big sagebrush communities (Fischer et al. 1996b, Pedersen et al. 2003, Lesica et al. 2007, Beck et al. 2009), and the legitimacy of this management technique has been strongly criticized as a tool for improving sage-grouse habitat (Connelly and Braun 1997; Connelly et al. 2000b,c; Beck et al. 2009).

ENERGY DEVELOPMENT

Historical Development and Current Status

Oil and Gas Development

Oil and gas development and construction of accompanying power lines, roads, and pipelines began in the late 1800s with discovery of oil in the interior western region of the United States.

Development began in the 1880s across the primary oil and gas regions in Wyoming, during the 1920s in Colorado, and in the 1940s in Alberta (Braun et al. 2002). Oil was the primary resource developed until the 1970s, when natural gas production began to dominate the industry. The production of natural gas from the central Rocky Mountain area had first been constrained by lack of transportation capacity (Doelger and Barlow 1989). However, additional interstate pipelines have been constructed during the past 20 years to transport natural gas to California and midwestern markets.

The Energy Policy and Conservation Act, signed into law in 1975, was designed to stabilize domestic consumption with energy development on public lands, establish a Strategic Petroleum Reserve, and inventory onshore oil and natural gas reserves. The act's reauthorization and amendment in 2000 provided for re-inventory of federal oil and gas reserves and directed a study of the extent and nature of any restrictions or impediments to develop oil and gas resources. Seventeen geologic basins within the interior western United States currently contain oil and onshore natural gas under federal stewardship (Fig. 12.10) (United States Departments of the Interior, Agriculture, and Energy 2003). Of these, five geologic basins have most of the onshore petroleum resources and were included in phase I of the inventory and impediments study: (1) Uinta-Piceance area in Colorado and Utah; (2) southwestern Wyoming (Greater Green River Basin); (3) Paradox–San Juan Basin in Utah, Colorado, and New Mexico; (4) Montana thrust belt in Montana and Wyoming; and (5) the Powder River Basin in Wyoming, Montana, and South Dakota. Subsequently, phases II and III (United States Departments of the Interior, Agriculture, and Energy 2006, 2008) have redefined boundaries and increased the number of basins included in the inventory; 11 basins included within the SGCA now have been inventoried (Table 12.13). Individual oil and gas wells within all producing regions are primarily in sagebrush-dominated landscapes (Figs. 12.11–12.14).

Stipulations on 11.3% of the federal lands restrict the timing of construction activities to reduce disturbance during periods critical to sage-grouse or other wildlife. Leasing is not permitted because of administrative restrictions on 37% of

TABLE 12.13

Geologic basins where significant oil and gas resources have been inventoried within the Sage-Grouse Conservation Area (Fig. 12.10), total area (km²), and area (km²) of sagebrush included within the basin.

Basin	State	Total area (km²)	Sagebrush (km²)
Denver Basin	Colorado, Nebraska, South Dakota, Wyoming	37,329	1,726
Eastern Great Basin	Arizona, Nevada, Utah	244,638	93,779
Eastern Oregon–Washington	Oregon, Washington	92,081	24,792
Montana thrust belt	Montana	42,713	5,759
Paradox Basin	Arizona, Colorado, New Mexico, Utah	79,018	8,971
Powder River Basin	Montana, Nebraska, South Dakota, Wyoming	112,585	29,644
San Juan Basin	Colorado, New Mexico	32,821	7,343
Southwestern Wyoming	Colorado, Utah, Wyoming	114,137	59,694
Uinta-Piceance Basin	Colorado, Utah	76,555	17,142
Williston Basin	Montana, North and South Dakota	96,715	3,487
Wyoming thrust belt	Idaho, Utah, Wyoming	32,422	11,556

SOURCE: United States Departments of the Interior, Agriculture, and Energy (2008).

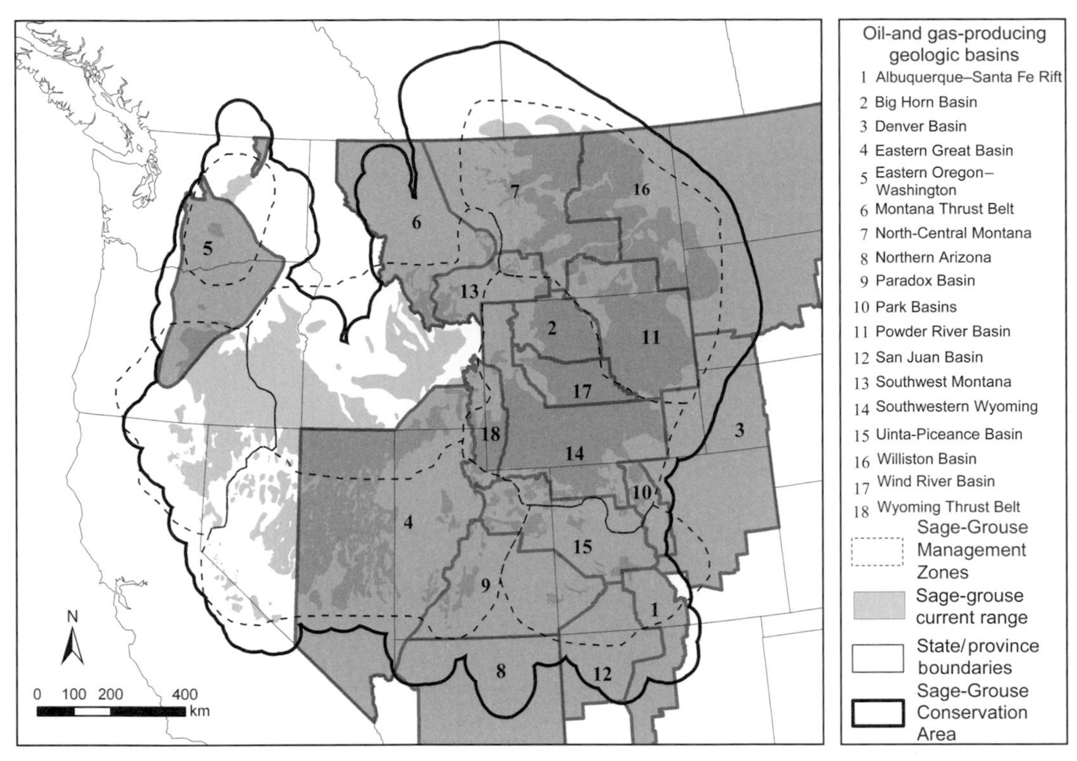

Figure 12.10. Geologic basins in which oil and gas resources have been developed within the Sage-Grouse Conservation Area (compiled from five state oil and gas conservation commissions, two state geological surveys, and Bureau of Land Management).

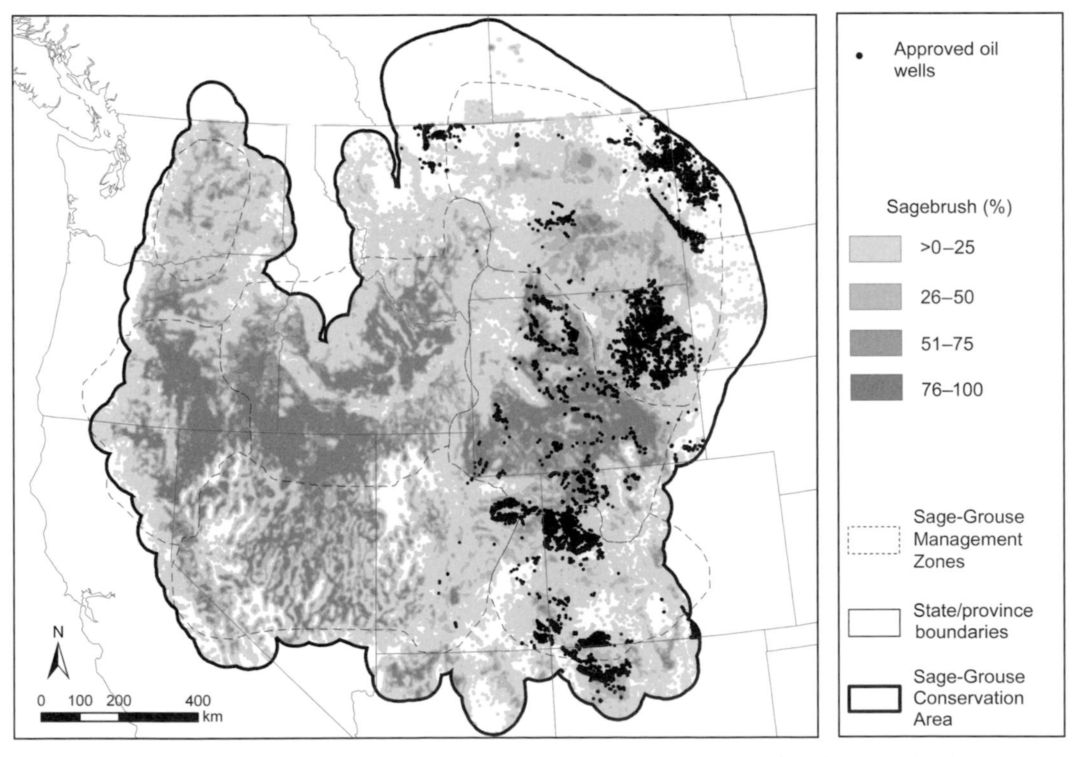

Figure 12.11. Current producing oil wells in the Sage-Grouse Conservation Area (compiled from five state oil and gas conservation commissions, two state geological surveys, and Bureau of Land Management).

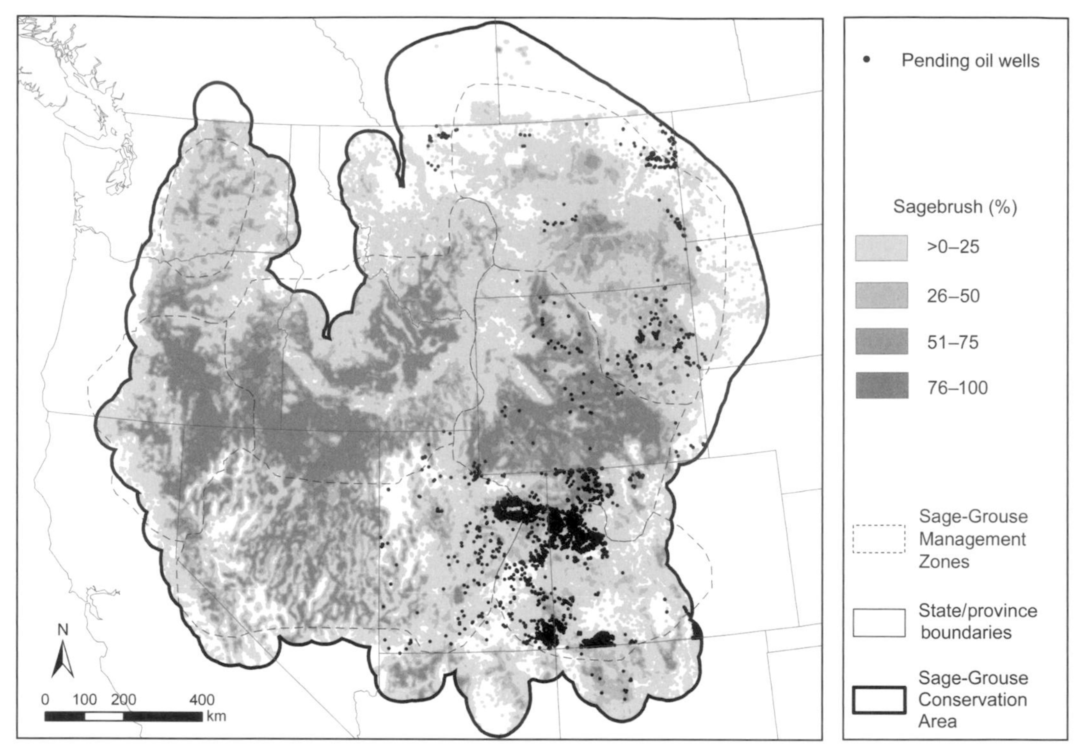

Figure 12.12. Permitted or pending oil wells in the Sage-Grouse Conservation Area (1929–2004) (compiled from five state oil and gas conservation commissions, two state geological surveys, and Bureau of Land Management).

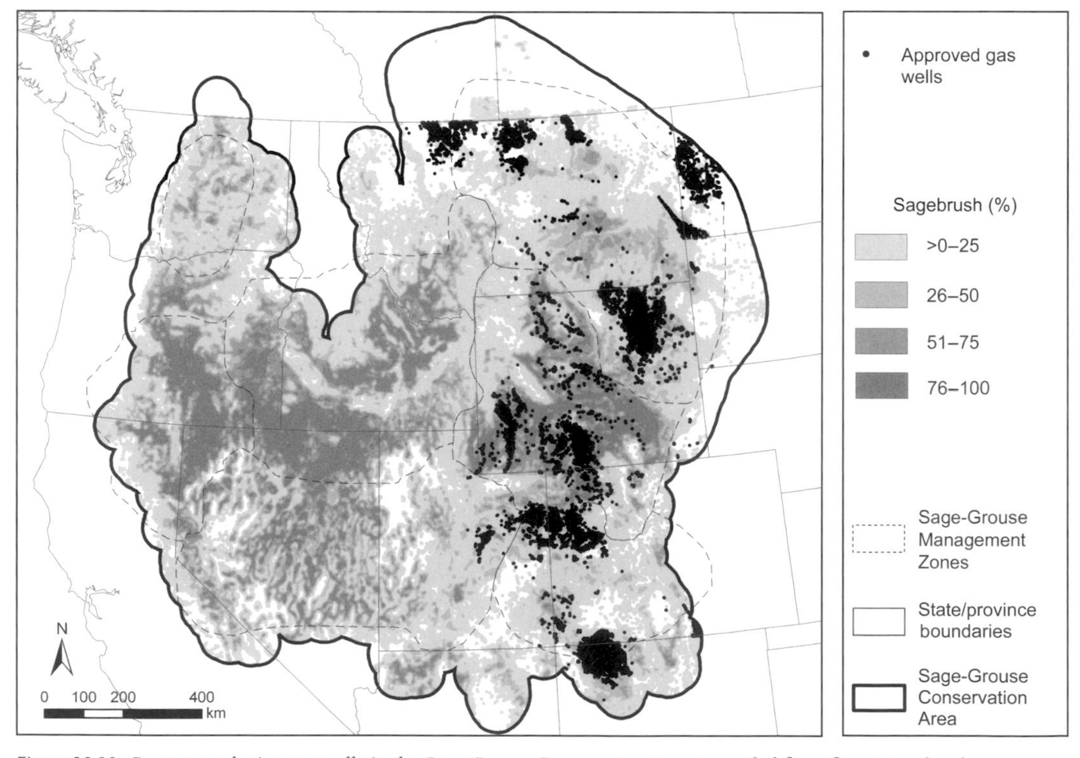

Figure 12.13. Current producing gas wells in the Sage-Grouse Conservation Area (compiled from five state oil and gas conservation commissions, two state geological surveys, and Bureau of Land Management).

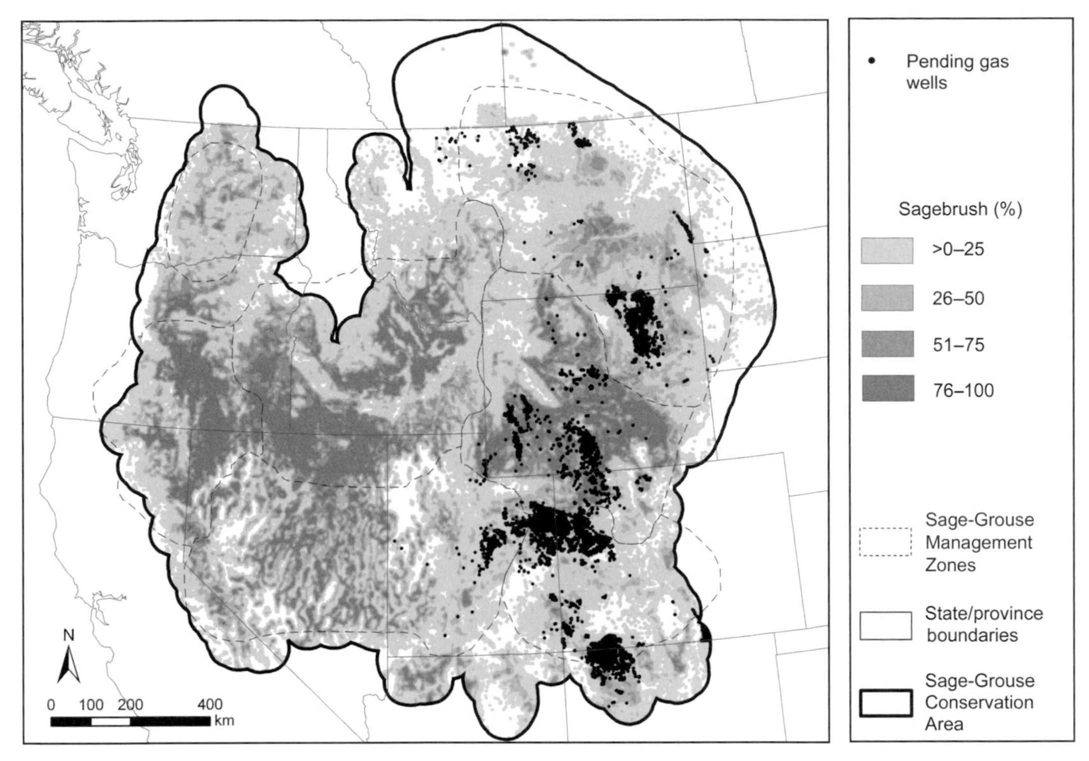

Figure 12.14. Permitted or pending natural gas wells in the Sage-Grouse Conservation Area (1929–2004) (compiled from five state oil and gas conservation commissions, two state geological surveys, and Bureau of Land Management).

the federal lands within the major geological basins of the SGCA; leasing under standard terms is permitted on 26% (Table 12.14).

A potential of >9,300 wells is being considered in the Greater Green River Basin, and at least 4,400 wells currently have been approved (Table 12.15). In the Powder River Basin, 27,522 wells have been drilled for coalbed methane natural gas; an additional 66,635 may potentially be drilled to develop the gas reservoirs (Table 12.15). An estimated 9,656 km of overhead power lines have been developed for coalbed methane natural gas production in the Powder River Basin (Braun et al. 2002). Over 28,000 km of new roads, 33,000 km of pipelines, and 8,400 km of overhead power lines would be developed as part of the infrastructure to construct an additional 50,000 wells in this 32,400-km^2 region (United States Department of the Interior 2003c). Similar increased development is planned for other oil- and gas-producing regions (Table 12.15).

Oil and gas development, including well pads, pipelines, and roads, influenced a minimum of 139,000 km^2 (7%) of the total area and 8% of the sagebrush within the SGCA (Table 12.16). The direct physical loss of habitat to well pads and pipelines was >4,500 km^2. The pipeline networks required to move oil and gas influence 12,000 km^2 and 1% of sagebrush habitats (Table 12.16, Fig. 12.15).

The BLM is responsible for leasing oil and gas rights on public lands. Total area and sagebrush area that has been leased increased rapidly during the late 1980s and early 1990s and continues to the present (Fig. 12.16). Total area leased is highest in the Wyoming Basin, Great Plains, and Southern Great Basin SMZs. The Wyoming Basin, Colorado Plateau, and Great Plains have the largest amount of sagebrush area leased. Lease holders of oil and gas rights on public lands file an application with the BLM for a permit to drill. Of 320,192 applications filed from 1929–2007, 86% were authorized, 1% are pending, and 13% were withdrawn or rejected (Table 12.17).

TABLE 12.14

Major category of stipulations on oil/gas development, total area (km²), and % of federal lands affected in 11 major and priority geological basins.

Access category	Area (km²)	Federal lands[a] (%)
No leasing (statutory/executive order)	65,865	12.1
No leasing (administrative) pending NEPA or LUP action.	93,959	7.2
No leasing (administrative), general	39,320	17.3
Leasing, no surface occupancy	51,416	9.4
Leasing, cumulative timing limitations on drilling >9 mo	787	0.1
Leasing, cumulative timing limitations on drilling 6–9 mo	9,126	1.7
Leasing, cumulative timing limitations on drilling 3–6 mo	51,677	9.5
Leasing, controlled surface use	88,984	16.3
Leasing, standard lease terms	143,488	26.3

SOURCE: Tables 3–5, 3–7, 3–8, 3–9, 3–10, 3–11, 3–12, 3–13, 3–14, 3–15, 3–16; United States Departments of the Interior, Agriculture, and Energy (2008).

[a] Federal lands influenced by each category within the geologic basin.

TABLE 12.15

Area (km²) of major oil- and gas-producing fields and number of wells and spacing for the Greater Green River Basin, Powder River Basin, and the Uinta-Piceance Basin (Fig. 12.10).

		Number of wells			Spacing (ha/well)[b]	
Field or area	Area (km²)[a]	Approved	Drilled	Potential	Range	Potential
			Greater Green River Basin			
Jonah	359		1,089	3,100		82 km² cap[c]
Pinedale Anticline	638	4,400	875	4,400		52 km² cap[c]
Big/Piney LaBarge per CAP	797	No limit		No limit	4–16	4–16
Moxa Arch[d]	1,930		1,839	1,861[e]		Additional 12 km² cap[d] from existing at any one time
Fontenelle	725		307			
Continental Divide-Creston[f]	4,451	2,100	1,721	8,950	16 downhole/ 32 surface	16 downhole/ 32 surface
Atlantic Rim	1,092	177	354	2,000 in EIS	2.6[g]	30 km² cap[c] at any one time
Little Snake Field Office, CO	497	50	731	3,327	16–30	8
			Powder River Basin			
Powder River Basin, WY[h]	32,376	15,300	26,422	51,000	32	32
Powder River Basin, MT[h]	11,028	1,200	1,100	15,635	32	32

TABLE 12.15 *(continued)*

Field or area	Area (km²)[a]	Number of wells: Approved	Number of wells: Drilled	Number of wells: Potential	Spacing (ha/well)[b]: Range	Spacing (ha/well)[b]: Potential
Uinta-Piceance Basin						
Piceance-Uintah Basin (BLM White River Field Office)	10,827	320	5,800	—[i]	2–16	2–16
Soldier Creek Field CBM[h]	20	private	16	56	65	32
Castlegate Field CBM[h]	103	154 in EIS	63	124	65	16–32
West Tavaputs Plateau[j]	558	—[k]	116	538 well pads (807 wells)	65	16–32
XTO's Little Canyon EA	134	—[l]	60	513	8–16	8–16
Gasco's EIS	836	—[l]	200	1,491	8–16	8–16
Enduring's Big Pack EA	139	—[l]	24	664	8–16	8–16
Kerr McGee's Greater Natural Buttes EIS	659	—[l]	600	3,496	8–16	8–16
Enduring Resources Southam Canyon EA	42	—[l]	6	152	8–16	8–16
Greater Chapita Wells EIS	170	—[l]	700	7,028	8–16	8–16
Castle Peak and Eightmile Flat oil and gas expansion project	471	—[m]	900	973	16	16
Greater Deadman Bench		—[m]	600	1,368	8–64	16–64
RDG EIS		—[m]	85	420	16–64	16–64
West Bonanza field development		—[m]	48	133	16–64	16–64

SOURCE: Compiled from U.S. Geological Survey National Oil and Gas Assessment and BLM state and field offices.

[a] Area was obtained from individual environmental impact statements or mapped estimates in a GIS.

[b] Spacing of oil and gas wells is the responsibility of each state's oil and gas agency or board, except on Native American lands, which are administered by the BLM. Well density on lands administered by BLM is based on land use plans and by accepting and applying the spacing decisions of individual state oil and gas agencies and boards. Oil and gas companies, along with mineral estate managers in the BLM, recommend spacing requirements to the state oil and gas agency or board during the development phase of oil or gas fields. In most cases, the estate developer submits a request to the state oil and gas agency or board to decrease well spacing. Each geologic basin has a standard set spacing unless exempted after petition to the state oil and gas agency or board. Well spacing generally defines the minimum number of wells per acre (5, 10, 20, 40, 80, 160, 320, 640 acres; or metric equivalent 2, 4, 8, 16, 32, 65, 130 ha). Exemptions to limits on spacing are possible in some fields. Where one oil and gas reservoir lies below another, multiple wells may be drilled within the same spacing unit.

[c] Disturbance cap or maximum allowable habitat loss.

[d] Proposed EIS.

[e] Proposed additional well pads in EIS.

[f] Natural Gas Project. EIS in progress. Combines CD/Wamsutter II and Creston/Blue Gap.

[g] Maximum eight well sites/2.5 km².

[h] Coalbed methane.

[i] New RMP Amendment in progress.

[j] Natural gas.

[k] New EIS in progress.

[l] New EA in progress.

[m] Completed EIS.

TABLE 12.16

Area (km²) and percent of area influenced by oil and gas development in Sage-Grouse Management Zones and the Sage-Grouse Conservation Area (SGCA).

Effect area was the total area of land use and surrounding buffer.

Energy development	Area km²	Area (%)	Effect area km²	Effect area (%)	Sagebrush area[a] (%)
Oil/gas wells[b, c]					
Great Plains	438	(0.1)	55,051	(15.6)	20.1
Wyoming Basin	210	(0.1)	34,786	(14.4)	20.3
Southern Great Basin	32	(<0.1)	4,967	(1.6)	1.6
Snake River Plain	<1	(<0.1)	28	(<0.1)	<0.1
Northern Great Basin	0	(0.0)	0	(0.0)	0.0
Columbia Basin	<1	(<0.1)	13	(<0.1)	<0.1
Colorado Plateau	279	(0.2)	23,512	(16.4)	29.4
SGCA	1,111	(0.1)	138,771	(6.7)	7.9
Pipelines[d]					
Great Plains	887	(0.3)	3,199	(0.9)	1.3
Wyoming Basin	1,142	(0.5)	2,982	(1.2)	1.6
Southern Great Basin	304	(0.1)	1,010	(0.3)	0.2
Snake River Plain	377	(0.1)	1,361	(0.5)	0.4
Northern Great Basin	82	(0.1)	280	(0.2)	0.2
Columbia Basin	66	(0.1)	257	(0.4)	0.3
Colorado Plateau	361	(0.3)	1,141	(0.8)	1.3
SGCA	3,924	(<0.1)	12,094	(0.6)	0.7

[a] Sagebrush area was delineated from the LANDFIRE (2006) map of land cover and covered 529,708 km² (25.4%) of the Sage-Grouse Conservation Area.

[b] Well pads vary from 0.1 ha for coalbed natural gas wells on flat ground in the Powder River Basin (United States Department of the Interior 2003c) used to develop reservoirs 600 m below ground. Well pads >7 ha are used to extract deep gas (7,000 m below ground) in the Madden Field of the Wind River Basin (United States Department of the Interior 2000c). Typical well pads are 0.8–1.9 ha for extracting oil and gas from reservoirs that are 4,300 m below ground and >1.9 ha for development of reservoirs >5,500 m below the ground surface. Pads for compressor stations along pipelines require 5–7 ha.

[c] Area for ecological effect (3 km) was estimated from influence of oil/gas construction activities on sage-grouse nesting (Lyon and Anderson 2003). Roads constructed to connect well pads were typically 4–7 m wide, not including drainage ditches (United States Department of the Interior 1989). Construction of new access roads averages 7 d of heavy equipment/1.6 km of road constructed.

[d] Pipelines to transport oil and gas range from 5–15 cm diameter for flow lines. Trunk lines (15–20 cm) and transmission lines (25–66 cm) are buried. Width of the surface area required to construct lines ranges from 0.4 m for smaller diameter pipes to >23 m for the larger transmission lines. Area for ecological effect (3 km) was estimated from spread of exotic plants (Gelbard and Belnap 2003, Bradley and Mustard 2006).

Alternative Energy Sources

The National Energy Policy, established in 2001, encouraged development of alternative or renewable energy sources (National Energy Policy Development Group 2001). Of these, wind and geothermal energy are the primary alternative sources that affect sagebrush and sage-grouse.

Federal lands in the western United States have significant potential to produce energy from windpower (Fig. 12.17). Wind power has been recognized for a long time (early explorers sailed around the world on wind power), but development of significant quantities of energy from wind power is relatively recent. Area leased/year on lands managed by the BLM more

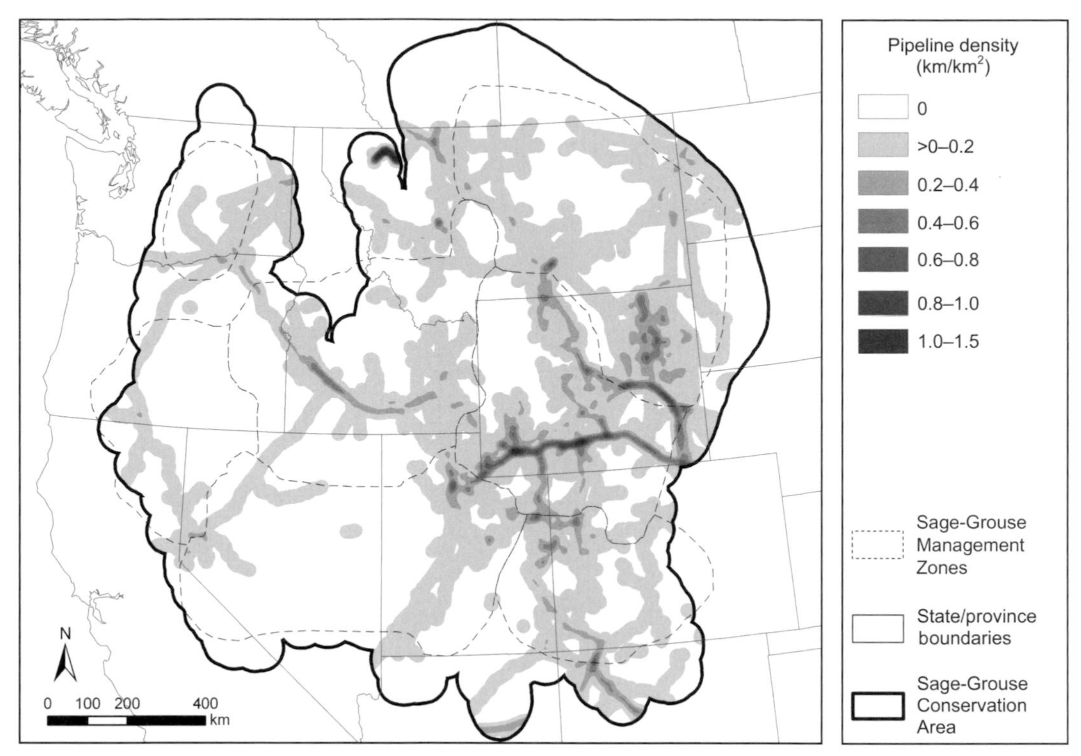

Figure 12.15. Density of primary natural gas and oil pipelines in the Sage-Grouse Conservation Area (developed from U.S. Department of Transportation and 17 other sources).

than doubled in 2007 and 2008 compared to leases in 2002–2006 (Table 12.18). The Wyoming Basin (1,387 km²), Southern Great Basin (1,176 km²), Northern Great Basin (1,088 km²), and Columbia Basin (920 km²) SMZs had the most total area leased for wind energy. Development of pending leases will more than triple the current area leased in Wyoming and significantly increase the area leased in the Snake River Plain and Northern Great Basin (Table 12.18). Total area leased for wind energy in the SGCA would double to 10,500 km² if all pending leases are developed.

Geothermal energy provides 17% of the renewable electricity generation in the United States, of which most production is in California (United States Department of the Interior 2005c). Geothermal production within the SGCA is primarily in the Southern Great Basin and Northern Great Basin SMZs (Fig. 12.18). Almost 2,000 km² has been leased for geothermal development in the SGCA, and leases on another 1,140 km² are pending (Table 12.19).

Ecological Influences and Pathways

Oil and Gas Development

Development of oil and gas resources includes direct loss of habitat for constructing well pads, roads, and pipelines connecting well locations and to transport oil and gas. A typical well has a well pad and an access road. Depending upon the individual oil and gas fields, wells may also have pump jacks, separators, storage tanks, electrical lines, and produced water ponds/pits or water discharge pipelines associated with development. In addition, ancillary development for flow lines, other roads, compressor stations, pumping stations, and electrical facilities are necessary to develop a field (Gerding 1986). Expected life of a well for economic production of coalbed methane natural gas is 12–18 years with advanced technology and 20–100 years for oil and gas wells (United States Department of the Interior 2003c). Oil and gas development fields may expand slightly as they mature, but development continues within the field. For example, a given field may have

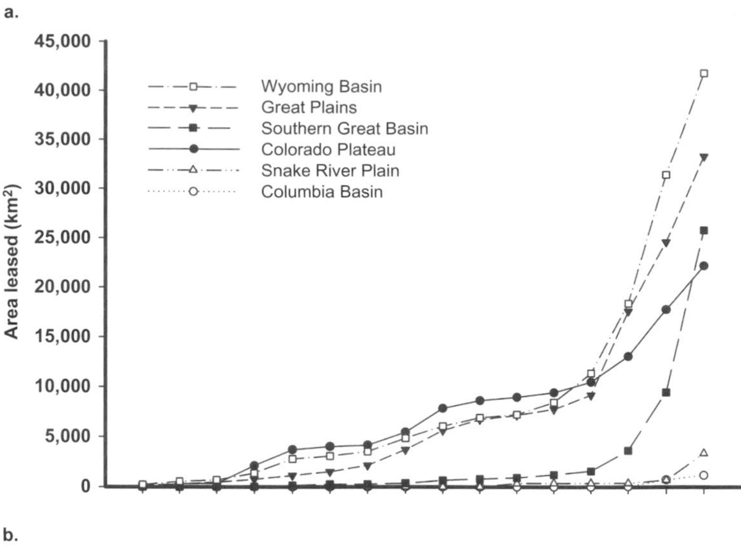

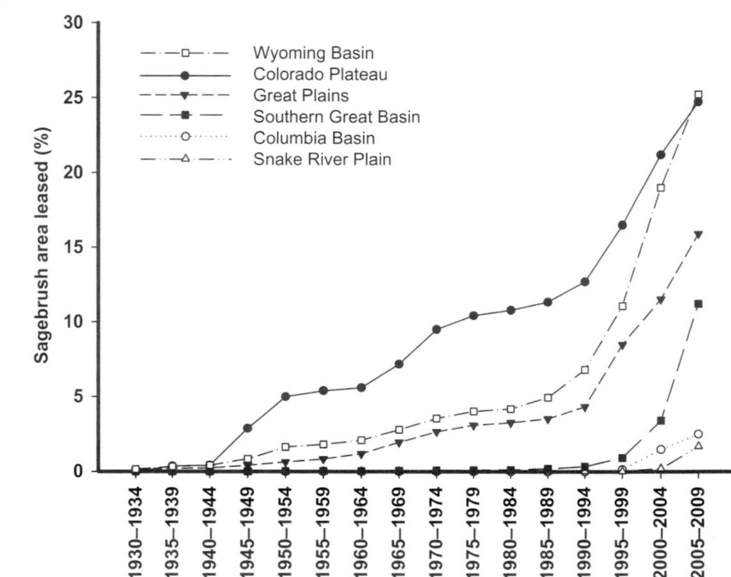

Figure 12.16. Cumulative total area (a) and sagebrush area (b) leased within Sage-Grouse Management Zones for oil and gas development (GIS coverages obtained from Bureau of Land Management Geocommunicator).

an initial spacing of 65 ha/well that decreases to 32 ha/well as the field matures, thus increasing the small-scale fragmentation of the landscape; habitat loss also is exponential because of areal reduction rather than linear with decreased spacing.

Construction of a well pad and drilling requires heavy equipment and can generate intense noise, exhaust, surface, and road disturbance. Vehicle traffic and noise disturbance on roads and at well sites is highest during drilling. Female sage-grouse moved greater distances from leks and had lower rates of nest initiation in areas disturbed by vehicle traffic (1–12 vehicles/day) (Lyon and Anderson 2003). Disturbances within 200 m of lek sites resulted in loss of attendance at sage-grouse leks (Braun et al. 2002).

Stipulations may regulate timing of construction activities and may be directed to avoid disturbance to sage-grouse near leks during breeding or nesting periods. However, direct and indirect effects on the surrounding landscape from roads, power lines, compressor stations, and pipelines remain following construction. Approximately 1% of the total area proposed for coalbed methane development in the Powder River Basin would experience surface disturbance due to well pads and construction, road access, and pipelines (United States Department of the Interior 2003c).

TABLE 12.17

Number of applications[a] for oil and gas leases received by federal agencies[b] from 1929–2007.

State	No. of leases								
	Authorized		Pending		Rejected		Withdrawn		Total
Wyoming	95,973		706		2,398		1,024		100,101
Arizona	4,450		10		234		1,073		5,767
California	7,015		83		1,222		768		9,088
Colorado	26,327		236		2,507		1,167		30,237
Idaho	4,223		0		430		849		5,502
Montana	26,880		1,467		3,384		2,783		34,523
New Mexico	41,786		29		4,177		3,465		49,457
Nevada	19,876		28		1,431		1,265		22,600
Oregon	4,655		6		834		2,842		8,337
Washington	1,722		5		1,323		4,154		7,204
Utah	31,354		417		2,248		1,833		35,852
South Dakota	3,247		93		368		181		3,889
North Dakota	6,782		180		470		203		7,635
Totals (%)	274,290	(85.7)	3,269	(1.0)	21,026	(6.6)	21,607	(6.7)	320,192

SOURCE: BLM LR2000 Lands and Records database.

[a] A lessee must file an application for permit with the BLM to drill that contains information on well specifics, emergency procedures, protection of groundwater and other resources, mitigation to wildlife and archeological resources, and weed control measures. The application is reviewed by the administering field office prior to approval to ascertain the application's adequacy for operational and environmental provisions. Specialists involved may include geologists, petroleum engineers, surface reclamation specialists, biologists, archeologists, and others. The reviews may recommend relocation of proposed wells or other modifications to mitigate impacts. Conditions of approval, including mitigation measures, are attached to every permit. On state or private lands, the lessee/owner must apply to the state's oil and gas agency or board for the development of oil and gas resources.

[b] The U.S. Geological Survey was the approval authority for applications for permit to drill prior to 1972. The BLM was authorized to provide recommendations into the decision process in 1972 and became the approving authority in 1982 following a merger with the Minerals Management Service. Lease applications that were closed prior to 1983 are not in the LR2000 database.

Stipulations require no surface infrastructure within 0.4 km of a lek in the Powder River Basin (United States Department of the Interior 2003c) but still permit development on large areas of surrounding habitat important for nesting and breeding populations. Development could still occur on 75% of the landscape within 0.8 km and 98% of the landscape within 3.2 km around leks. Development on 98% of the landscape was predicted to decrease the probability of lek persistence from 85% to 5% (Walker et al. 2007a). The cumulative effect of gas development, rather than any single source or infrastructure component, caused greater declines in number of males attending leks and lower lek persistence compared to regions not undergoing development (Walker et al. 2007a).

Construction of well pads and infrastructure create ground disturbance, fragment remaining landscapes dominated by sagebrush, subsidize avian predators by creating perches and refuse sites, and facilitate spread of exotic plants. Sites disturbed by coalbed methane development in the Powder River Basin had significantly greater exotic plant species richness, higher salinity and nutrient levels, and lower numbers and percent cover of native species when compared to sites not influenced by development activities (Bergquist et al. 2007).

Areas disturbed by oil and gas development are required to be reseeded in reclamation efforts to control soil erosion, establish desirable vegetation, and facilitate natural processes to restore the site. Sage-grouse continued to use highly

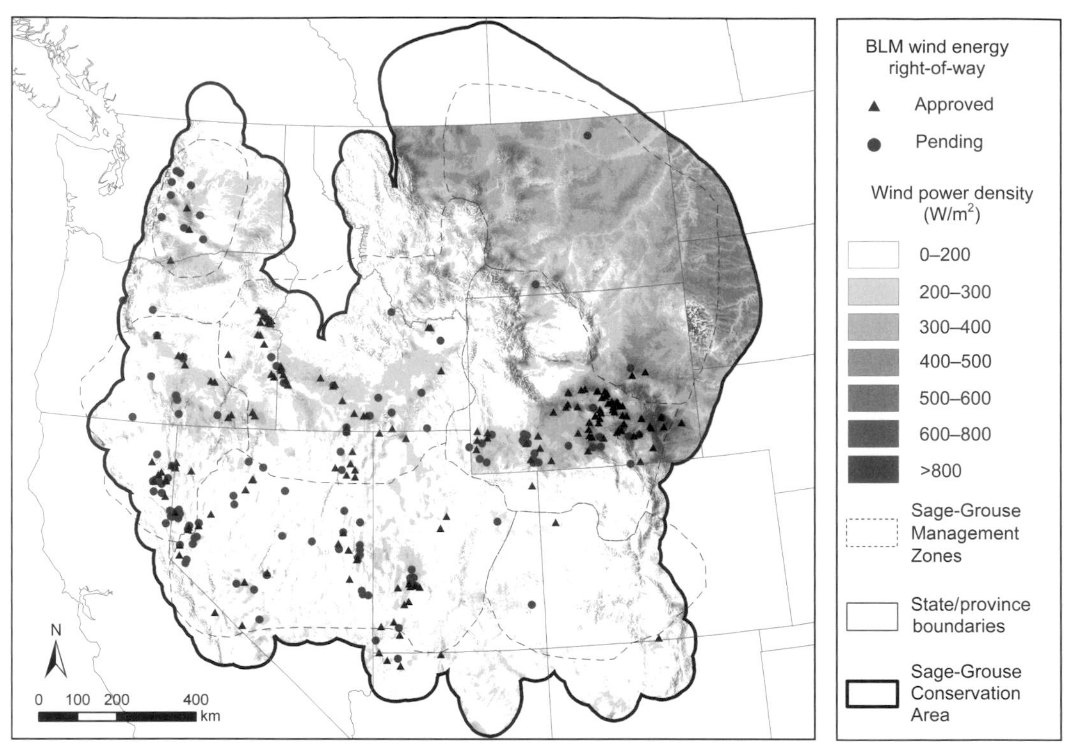

Figure 12.17. Approved and pending right of way for wind energy development on lands managed by the Bureau of Land Management (GIS coverages obtained from Bureau of Land Management Geocommunicator).

TABLE 12.18

Area (km^2) of active and pending wind power leases issued by the BLM from 2002–2008 (Fig. 12.17).

Sage-Grouse Management Zone	2002	2003	2004	2005	2006	2007	2008	Sum	Pending
Great Plains	0	0.0	0.0	8.2	0.0	0.0	147.2	155.4	44.1
Wyoming Basin	9.7	102.2	0.0	4.3	151.0	852.9	267.4	1,387.4	2,828.3
Southern Great Basin	0	0.0	15.1	149.2	169.9	143.8	698.4	1,176.4	713.8
Snake River Plain	0	8.2	0.0	54.8	24.2	118.0	20.1	225.3	744.5
Northern Great Basin	0	0.0	16.7	91.2	245.6	675.0	59.9	1,088.3	853.4
Columbia Basin	0	472.2	381.3	1.3	0.0	60.9	4.6	920.2	15.3
Colorado Plateau	0	0.0	0.0	0.0	0.0	0.0	3.9	3.9	0.6
SGCA Total	9.7	582.8	414.0	309.1	590.6	2,005.5	1,201.4	5,113.1	5,381.5
SGCA Cumulative	9.7	592.5	1,006.5	1,315.5	1,906.1	3,911.7	5,113.1		

SOURCE: BLM Geocommunicator database.

fragmented habitats in some oil fields and reclaimed areas, but population levels were below numbers prior to disturbance (Braun et al. 2002) or relative to similar regions without energy development (Doherty et al. 2008).

The proportion of wells on private lands (total number of producing, pending, abandoned, and unknown status) was 50% (36,329) in the Paradox–San Juan Basin, 42% (20,795) in the Uinta-Piceance Basin, 33% (19,034) in the Greater Green

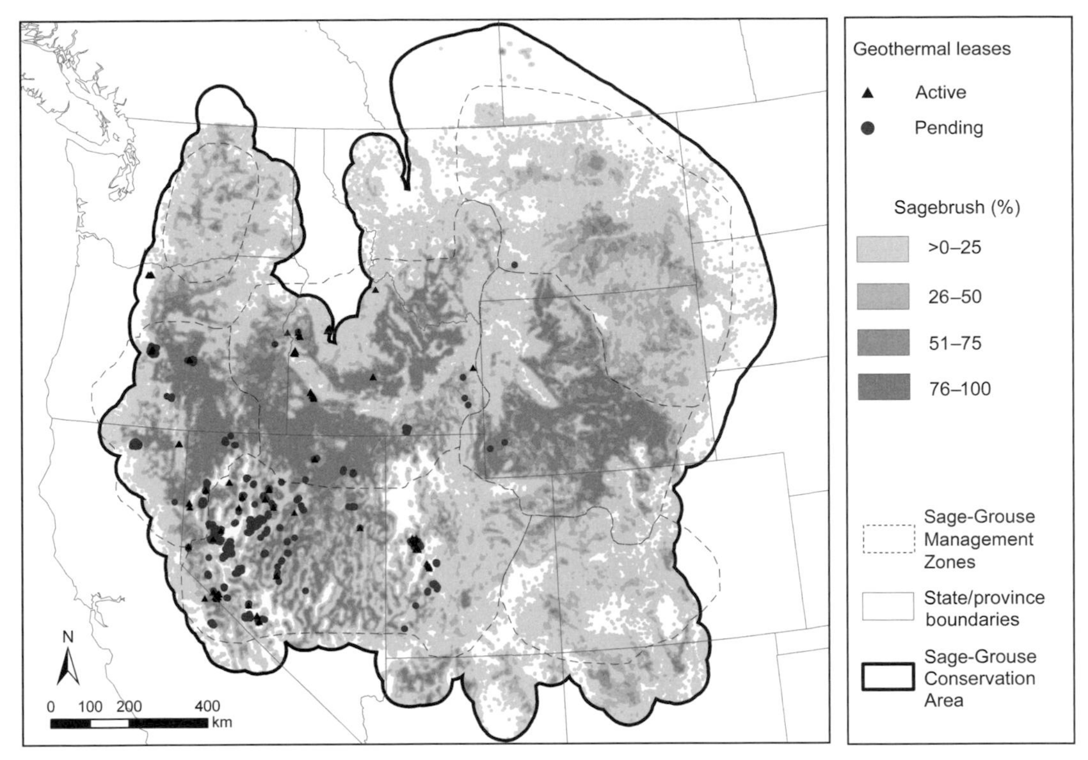

Figure 12.18. Active and pending geothermal developments in the western United States (GIS coverages obtained from Bureau of Land Management Geocommunicator).

TABLE 12.19

Area (km²) of active and pending geothermal leases (Fig. 12.18) issued by the BLM from 1970–2009.

Sage-Grouse Management Zone	1970–1979	1980–1989	1990–1999	2000–2009	Sum	Pending
Great Plains	0	0	0	0	0	0
Wyoming Basin	0	0	0	0	0	0
Southern Great Basin	116.1	77.9	364.4	494.7	1,053.1	714.0
Snake River Plain	0	0	0	32.3	32.3	326.3
Northern Great Basin	3.5	269.1	221.2	226.7	720.4	73.0
SGCA Total	119.6	347.0	585.6	753.7	1,805.9	1,137.5
SGCA Cumulative	119.6	466.6	1,052.3	1,805.9		2,943.4

SOURCE: BLM Geocommunicator database.

River Basin, 74% (67,668) in the Powder River Basin, and 77% (220) in the Montana thrust belt. Development on private lands does not require mitigation efforts. Increased development on private lands potentially shifts the ecological and conservation importance of remaining sagebrush habitats toward those in public stewardship.

Power line poles along transmission corridors provide nest and perching opportunities for Common Raven (*Corvus corax*), American Crow (*C. brachyrhynchos*), and raptors (Reinert 1984, Knight and Kawashima 1993, Steenhof et al. 1993, Lammers and Collopy 2007). Ravens are primary predators on sage-grouse and other prairie grouse nests (Manzer and Hannon 2005, Coates et al. 2008) and can travel >10 km from these locations (Boarman and Heinrich 1999). Collisions with power lines, in addition to increased predation risk, were a primary source of mortality for lowland populations of sage-grouse in Idaho (Beck et al. 2006).

Alternative Energy Sources

The effects of wind or geothermal energy development on sagebrush or sage-grouse are largely unknown because development has been too recent to identify immediate or lag effects. Specific environmental concerns of wind turbines were noise produced by the rotor blades, aesthetic (visual), and mortality to bats and birds flying into rotors (United States Government Accounting Office 2005). These considerations may be reduced through advanced technology, but the greater influence on ecosystems is likely to result from roads that are necessary to construct and maintain sites used for wind energy, and power lines that transfer electricity to users (Kuvlesky et al. 2007). Sage-grouse also may avoid areas with turbines because of the visual obstruction.

MILITARY TRAINING

Historical Development and Current Status

Many exercises conducted during military training are destructive. Environmental concerns began to be addressed in the 1980s through programs to monitor habitat and wildlife on lands used for military training and testing (Diersing et al. 1992). Consequently, meeting national security needs and conserving wildlife and habitat often compete, because lands on military bases must also be managed to minimize short- and long-term environmental impacts (Prosser et al. 2003). The Sikes Act, enacted in 1960 with subsequent amendments, provided for cooperation between the United States Department of Defense and Department of the Interior for planning, developing, and maintaining fish and wildlife resources on military lands.

The military trains on 87 installations within the SGCA (Fig. 12.19). Total area of sagebrush

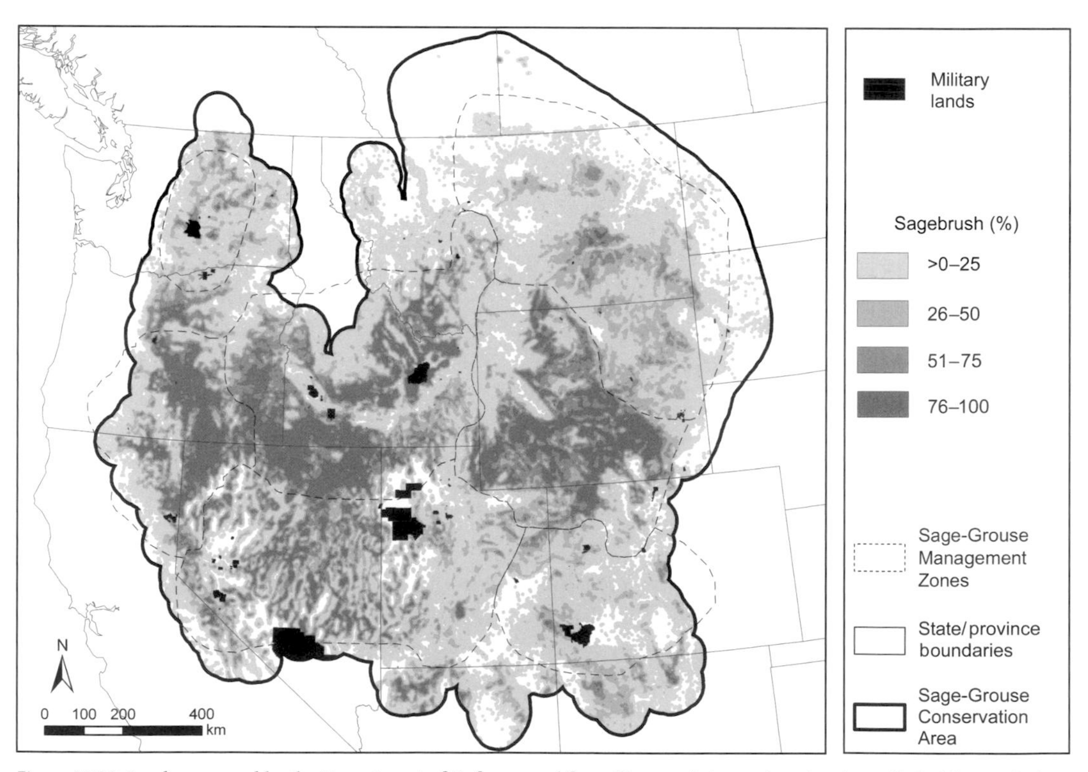

Figure 12.19. Lands managed by the Department of Defense used for military training and testing (compiled with permission from U.S. Department of Defense).

was 6,815 km^2 contained within 52 installations, or 26% of the total area used for military training.

All military facilities use the Range and Training Land Assessment (formerly the Land Condition Trend Analysis) to assess soil and vegetation characteristics and monitor their lands (Diersing et al. 1992, West 2003a). Thirty-six percent of Army training lands had moderately to highly erodible soils (Shaw and Kowalski 1996). Ground disturbance due to military activities was evident on 17% of lands surveyed; >60% of those areas exceeded soil-loss tolerance (Shaw and Kowalski 1996). However, some areas on military lands remain in good condition because of access restrictions to protect sensitive plant and wildlife, and concentration of high-impact maneuvers.

Ecological Impacts and Pathways

The direct influences of military training and testing on shrubland habitats result from maneuvers by tracked and wheeled vehicles, and from fires from ordnance impacts. Tracking by armored military vehicles created soil and vegetation disturbance at local scales that facilitated spread of exotic species, increased potential for soil erosion, and potentially reduced ecosystem productivity and stability (Belcher and Wilson 1989; Shaw and Diersing 1989, 1990; Watts 1998; Milchunas et al. 2000).

Military training exercises also affect landscape composition and pattern at larger scales. Frequent fires eliminated shrub cover within an ordnance impact area on a military training area in southeastern Idaho (United States Department of the Interior 1996). The area used for tracked vehicle maneuvers had smaller, more closely spaced shrubland patches compared to landscapes influenced by fire and livestock grazing (Knick and Rotenberry 1997).

DISCUSSION

Resource Demand and Intensity of Use

Total land use and demands on sagebrush lands for all purposes have increased since Euro-American settlement (West 1999), and their effects on sage-grouse are dynamic in time and space across the species range. Ultimately, Euro-American settlement has had a much greater effect on western landscapes through transforming or converting habitat, manipulating animal communities, and altering disturbance regimes than was exerted by indigenous peoples (West and Young 2000, Griffin 2002).

Past administrative policies and resource use continue to influence patterns and processes of sagebrush habitats and management options (Knick, this volume, chapter 1). Highly productive lands that could be converted to agriculture or irrigated were transferred to private entities during the era of disposal beginning in the late 1800s (Clawson and Held 1957). Less productive lands for growing crops remained in the public trust and have been used for livestock grazing, mineral extraction, energy development, and outdoor recreation. Conservation strategies for wide-ranging species like Greater Sage-Grouse thus need to encompass a broad range of landscapes, ownerships, and land uses (Forbis et al. 2006).

Unregulated livestock grazing more than a century ago changed soils and plant communities, lowered ecosystem productivity, and facilitated invasion by a suite of exotic plant species (Young et al. 1972, Young and Longland 1996) that influenced today's fire and disturbance regimes and limit management options (Vavra and Brown 2006). Widespread efforts beginning in the 1950s to eradicate sagebrush and increase forage for livestock resulted in population declines of Greater Sage-Grouse (Swenson et al. 1987, Connelly and Braun 1997, Beck and Mitchell 2000). Structural features of some low-elevation big sagebrush communities treated with prescribed burns have not recovered in >10 years (Nelle et al. 2000, Beck et al. 2009); Greater Sage-Grouse still had not recolonized other areas treated >20 years ago (Byrne 2002) or used only areas in close proximity (<30 m) to untreated sagebrush adjacent to treatment areas (Martin 1970, Dahlgren et al. 2006).

Relative demand for resources among traditional uses has also been dynamic since Euro-American settlement. Mining, livestock grazing, and ranching are decreasing as a percent of the economics in parts of the western United States (Hansen et al. 2002, 2005), but other extractive industries, such as oil and gas development, are increasing (Holechek 2006; Naugle et al., this volume, chapter 20). The United States National Energy Policy projected an increase in oil consumption by 33%, in natural gas consumption by >50%, and in electricity by 45% over the next

20 years (National Energy Policy Development Group 2001). Annual demand for petroleum is expected to increase annually by 1.4% and for natural gas by 2.3% between now and 2020 (National Petroleum Council 1999, 2003). Consumption of natural gas is projected to increase from 639,961,000 m^3 in 2002 to between 824,020,000 and 968,436,000 m^3 in 2025 (United States Department of Energy 2009). Consequently, efforts to extract existing energy resources have intensified and the spatial scale of exploration expanded across the entire sage-grouse range to develop new resources. Natural gas development is expected to increase for the next 15–20 years, and by >40% in the Greater Green River Basin by 2015. Leasing and development of wind and geothermal energy are also increasing rapidly in the sage-grouse range. These increases and changes in disturbance imposed on sagebrush landscapes by energy development will influence Greater Sage-Grouse populations and habitat dynamics far into the future and create challenges for conservation (Kiesecker et al. 2009; Doherty et al., this volume, chapter 21).

Changing Values for Use and Management

Debate over use of public lands is not a recent phenomenon and has often been contentious and litigious among competing interests (Donahue 1999, Dombeck et al. 2003). Ecological and botanical perspectives have often been at odds with commodity development beginning with early surveyors and scientists studying these habitats in the 1880s (Box 1990). Over 70% of the sagebrush habitat used by sage-grouse is public land managed by state or federal agencies for multiple use (Knick et al. 2003; Knick, this volume, chapter 1); consumptive and nonconsumptive use of public resources often overlap spatially and temporally.

Recreation and wilderness values of public lands are primary reasons for exurban development and increasing human densities in rural parts of the western United States. Recreation, including OHV use, is rapidly increasing (Cordell et al. 2005) although these effects have yet to be quantified for sagebrush landscapes. Society also considers wildland, wildlife, and aesthetic components as important objectives for managing land use or habitat treatments (Kennedy et al. 1995). Vegetation treatments that include hydrologic processes, soil stabilization, and wildlife habitat conservation are increasingly conducted in attempts to restore ecosystem functions rather than conducted solely to increase forage for livestock (Monsen et al. 2004). Treatment of large areas, use of herbicides, mechanical treatments, and planting nonnative plant species are used to control exotic plants, reduce fire hazards, and rehabilitate burned areas (Pyke, this volume, chapter 23). But other strategies, such as thinning sagebrush with tebuthiuron (Olson and Whitson 2002), reseeding and managing for specific forb species (Wirth and Pyke 2003, Wrobleski and Kauffman 2003), restoring herbaceous communities while retaining existing sagebrush densities (Huber-Sannwald and Pyke 2005), or improving our understanding of state-transition dynamics as decision-support tools (Bestelmeyer et al. 2003, 2009), reflect a larger emphasis on managing for sage-grouse or other sagebrush-obligate wildlife. Effects of these changes remain unknown because the landscape dynamics of sagebrush and sage-grouse population response is long-term and often includes lag effects to immediate actions. Monitoring vegetation and wildlife response to habitat treatments across appropriate spatial and temporal scales is important to inform future management (Connelly et al. 2003b, Crawford et al. 2004).

Cumulative Effect of Land Use

Almost all sagebrush-dominated lands are managed for multiple and often concurrent uses (Knick et al. 2003; Knick, this volume, chapter 1). Dominant land uses vary across the sage-grouse range relative to resource distribution. Crop production is the primary land use in the Columbia Basin and Great Plains, but also influences a large proportion of the sagebrush in the Colorado Plateau and Snake River Plain. Oil and gas development occurs primarily in the eastern portion of the Greater Sage-Grouse range but is expanding in the Southern Great Basin. Wind energy is being developed in the Columbia Basin, Northern Great Basin, Southern Great Basin, and Wyoming Basin; geothermal sources are primarily in the Northern Great Basin and Southern Great Basin SMZs. Livestock grazing and habitat treatments are widespread on almost all public and many private lands throughout the sage-grouse range. Similarly, recreation and OHV use is increasing throughout the sage-grouse range.

The collective influence rather than an individual local effect may translate into landscape change or influence population trends. An individual disturbance, such as a water impoundment or oil well pad, directly impacts or removes relatively small (<2 ha) areas of sagebrush or influences only a minor portion of the annual range of sage-grouse. However, these water impoundments, or standing water associated with irrigation projects, may become breeding areas for mosquitoes and result in sage-grouse deaths from West Nile virus that influence populations (Naugle et al. 2005; Walker and Naugle, this volume, chapter 9). Some land uses, such as military training, may have very intense effects on habitats (Shaw and Diersing 1990, Watts 1998) but are restricted to relatively small regions within the sage-grouse range. Other management actions, such as building fences, influence sagebrush ecosystems across the entire sage-grouse range but at lower intensities. Isolating individual sources that cause changes in population parameters may be difficult because the influence is exerted by the combined effect, rather than a single factor (Walker et al. 2007a).

Indirect effects of one land use also can amplify or facilitate other disturbances. For example, railways helped facilitate the initial spread of cheatgrass, which had invaded much of the Intermountain West by the 1930s after its initial establishment in the 1890s (Mack 1981, Young and Sparks 2002). Cheatgrass readily invaded disturbed soils adjacent to railroads, and fires ignited by trains also promoted expansion of cheatgrass. Cattle transported along the rail system depleted areas of native herbaceous vegetation and facilitated further spread of cheatgrass (Young and Longland 1996). Soil trampling further eliminated protective cover of biological crusts, lichens, and mosses between bunchgrasses and opened the system for cheatgrass invasion (Mack 1981, Young and Allen 1997). Landscapes in southwestern Idaho in which grazing was combined with other disturbances experienced the greatest rate of shrub loss and increase in cheatgrass compared to landscapes having a single source of disturbance (Knick and Rotenberry 1997). Ultimately, the cumulative impact of the individual disturbances and multiple land uses, rather than any single source, likely has the most significant influence on the trajectory of sagebrush ecosystems.

Spatial Variation in Sagebrush Systems

Sagebrush landscapes remain one of the most dominant features of western North America and cover approximately 500,000 km^2 within the Sage-Grouse Conservation Area (Connelly et al. 2004; Miller et al., this volume, chapter 10). Sagebrush communities vary widely across their distribution in response to gradients of elevation, amount and annual timing of precipitation, and dominant disturbance regime (Miller and Eddleman 2001). Consequently, the inherent resistance and resiliency of sagebrush communities to land use varies considerably (Washington-Allen et al. 2006). Sage-grouse use of these communities also varies seasonally and spatially (Connelly et al. 2000c). Their response to land use also varies across the species range and differs among land uses. Advocating a specific action, such as removing livestock grazing or using prescribed fire, ignores the variability of sagebrush systems across the sage-grouse range and differences among areas to sustain and respond to land use (Vavra and Brown 2006). Instead, holistic consideration of all human-associated disturbances is required to address the complexity of multiple, interacting, or additive factors if sagebrush areas are to be effectively maintained or restored (Wisdom et al. 2005a,c; Meinke et al. 2009).

Complex Dynamics in Sagebrush Systems

The relatively simple structure of sagebrush communities (West and Young 2000) masks complex multiscale dynamics of shrubland systems (Bestelmeyer et al. 2003, Briske et al. 2005) that often preclude identifying direct linkages between land use and system response. Conversion to cropland or pastures removes available habitat at one spatial scale. This habitat loss at one scale may be perceived as landscape fragmentation at broader scales. Landscape fragmentation and dispersed patches of available habitat lead to increasingly unsuitable conditions for sage-grouse (Schroeder et al. 2000, 2004; Aldridge et al. 2008). Long-term viability of sage-grouse populations is no longer possible when habitat within a landscape is reduced and connectivity among populations is decreased beyond some limiting threshold (Knick and Hanser, this volume, chapter 16). We have identified strong correlations between land uses and sage-grouse population responses (Swenson et al.

1987, Connelly and Braun 1997, Connelly et al. 2000b, Leonard et al. 2000, Walker et al. 2007a). However, threshold levels or underlying mechanisms often remain unclear.

Land use may also influence sage-grouse habitat and populations through multiple pathways. Noise and construction activities for oil and gas development directly influence sage-grouse during breeding and nesting periods (Lyon and Anderson 2003, Walker et al. 2007a). Habitat loss and fragmentation at smaller spatial scales created by construction of oil and gas well pads and infrastructure also decreased winter habitat use by Greater Sage-Grouse (Doherty et al. 2008). Changes in hydrology caused by extracting coalbed methane natural gas alter groundwater tables and increase the potential to spread of diseases such as West Nile virus (Zou et al. 2006b, Walker et al. 2007b) that may have significant local effects on sage-grouse populations (Walker and Naugle, this volume, chapter 9). Long-term indirect effects from facilitating spread of exotic plants along roads and pipelines ultimately may have the greatest influence on loss of sagebrush by increasing fire sizes and frequencies.

Conceptual models have yet to be applied to observed dynamics of sagebrush systems (Allen-Diaz and Bartolome 1998, Bestelmeyer 2009). The environmental conditions that promote stability or length of time over which communities can remain within a vegetative state are largely unknown. The multiple vegetative states that potentially emerge following disturbance also challenge to our ability to understand how sagebrush landscapes might differ with new disturbance regimes. Consequently, simple solutions proposed for some land uses are unlikely or not widely applicable.

CONSERVATION IMPLICATIONS

The primary stresses that human land use places on natural ecosystems include: (1) harvesting renewable resources, resulting in loss of productivity; (2) physical restructuring of the landscape; (3) introduction of exotic species; and (4) discharge of toxic substances to air, land, and water (Rapport and Whitford 1999). Ecosystems that are heavily stressed lack the capacity to maintain normal function, initiating a process of degradation and lowered resilience for further disturbance as well as reduced ability to continue to provide ecosystem services (functions that have some value to society) (Milton et al. 1994, Foley et al. 2005). Stresses at a population or local level are more acute than at an ecosystem level. Stresses observed at an ecosystem level may indicate that the system is breaking down (Odum 1985). All stresses are present in the sage-grouse range, and ecosystem services derived from sagebrush landscapes also have declined spatially and temporally. For example, number of AUMs for livestock grazing have been reduced, OHV access has been restricted, and the widespread conversion of sagebrush communities to exotic grasslands has increased costs for fire prevention and decreased their value to wildlife.

Disturbance in some form has been an integral component of sagebrush systems throughout their evolutionary history. Frequency and severity of disturbance both creates and responds to landscape pattern, quantity of habitat, and composition (Pickett and White 1985, Turner 1987). Climate, soils, precipitation, and characteristics of the previous community affect resistance (ability to withstand change) of sagebrush communities to disturbance (West and Young 2000, West 2003b). Variation in these environmental characteristics across the sagebrush distribution also influences their inherent resilience (the relative ability to return to a reference vegetative state or the amount of change required to transform a system to a different state) (West 2003b, Briske et al. 2005). Current levels of land use impose new multiscale effects on this background of natural disturbance. However, quantifying amount or level of disturbance provides little insight without associated information on response in sagebrush systems. To understand effects, livestock densities or AUMs across the sage-grouse range need to be interpreted in terms of vegetation, soils, productivity, and relationship to native herbivory. Similarly, number of oil and gas wells interpreted in the absence of measures of landscape heterogeneity and sage-grouse population response hinders developing cause-and-effect relationships. Thus, society's ability to use resources contained in sagebrush but also manage disturbance caused by land use is limited. The conundrum is how to manage this new disturbance regime to maintain or restore sagebrush ecosystems (Forbis et al. 2006, Vavra and Brown 2006, Provencher et al. 2007) without incurring deleterious effects of habitat loss and fragmentation, spread of exotic plants, and regional extirpation of wildlife (Raphael et al. 2001, Hemstrom et al. 2002, Knick et al. 2003, Wisdom et al. 2005b).

The rapidity with which humans can transform an entire landscape through land use is significantly greater than the natural disturbances that previously influenced dynamics in sagebrush ecosystems. In an ecological context, we are not far removed from the era of disposal in which suitable lands obtained through large purchases or cessions were transferred to private entities to settle and convert to agriculture (Clawson and Held 1957). The early history of livestock grazing following Euro-American settlement has influenced current soils and plant communities. Depletion of plant communities facilitated invasion by cheatgrass (Young et al. 1972, Young and Allen 1997), which increased frequency and size of fires (Klemmedson and Smith 1964, Billings 1990), a problem that today consumes vast amounts of personnel, financial, and logistic resources in fire control, suppression, and fuels management. It is an even briefer moment since an estimated 10–20% of the sagebrush range was treated to increase forage production for livestock (Schneegas 1967, Vale 1974). The large blocks of sagebrush that were eradicated from the 1950s through the 1970s and replanted with crested wheatgrass influence distributions of wildlife populations >20 years later (Reynolds and Trost 1981, Swenson et al. 1987). Recent national concerns have resulted in major efforts, such as through the Energy Policy Act passed in 2005, to develop renewable and nonrenewable energy resources. Exploration for new oil and gas resources (United States Departments of the Interior, Agriculture, and Energy 2003, 2008) and development of existing resources have been accelerated (United States Department of Energy 2009) and administrative processes expedited (Bush 2001). Development of wind and geothermal energy is being planned for most states in the Sage-Grouse Conservation Area. Sagebrush that is converted to concrete, asphalt, or cropland is effectively lost. Recent developments in rural areas also are unlikely to revert back from subdivided ranchettes to contiguous sagebrush landscapes.

Grasses and forbs may respond within 1 to 3 years if soils and seed sources permit recovery or restoration, but return to a shrub-dominated community often requires >20–30 years, and landscape restoration may require centuries or longer (Hemstrom et al. 2002). Even longer periods may be required for sage-grouse to use recovered or restored landscapes. Society is beginning to understand the local effects of some land uses and how legacies of past actions influence current ecosystem processes (Foley et al. 2005). However, we cannot fully appreciate how the cumulative effects of these unprecedented intensities and current trajectories of land use will influence long-term conservation of sagebrush landscapes and sage-grouse populations.

ACKNOWLEDGMENTS

Reviews by W. C. Gilgert and N. Devoe improved the clarity and quality of the manuscript. J. W. Connelly commented on portions of the draft manuscript relevant to sage-grouse. K. M. Ross and K. J. van Gunst conducted GIS analysis and data searches and compiled summary information. The land use chapter that appeared in the 2004 Conservation Assessment of Greater Sage-Grouse and Sagebrush Habitats was reviewed by M. A. Hilliard, M. S. Karl, N. E. West, and the Ecological Society of America.

CHAPTER THIRTEEN

Influences of the Human Footprint on Sagebrush Landscape Patterns

IMPLICATIONS FOR SAGE-GROUSE CONSERVATION

Matthias Leu and Steven E. Hanser

Abstract. Sagebrush (*Artemisia* spp.) ecosystems in the western United States have changed in quantity and configuration from a variety of causes including agriculture and human population growth since Euro-American settlement. Activities sustaining human society can decrease or fragment land cover and alter ecological processes within sagebrush systems. The extent of these activities, cumulatively called the human footprint, within the range of sage-grouse (Greater Sage-Grouse [*Centrocercus urophasianus*] and Gunnison Sage-Grouse [*C. minimus*]) has not been evaluated. Using a recent human-footprint model of the western United States, we evaluated human-footprint intensity: (1) across the sage-grouse range within seven Sage-Grouse Management Zones (SMZs), (2) across five sagebrush land-cover classes and a nonsagebrush land-cover class within SMZ, and (3) on landscape pattern of sagebrush land cover in relation to three scenarios differing in human-footprint effect area. Based on four criteria, we ranked SMZs from most to least human-footprint influence as follows: Columbia Basin, Colorado Plateau, Wyoming Basin, Great Plains, Snake River Plain, Southern Great Basin, and Northern Great Basin. Range-wide, black (*Artemisia nova*) and little (*A. arbuscula*) sagebrush land covers were least affected by the human footprint. Increasing human-footprint effect area decreased sagebrush land cover in the landscape between 33.5% and 97.0% and reduced mean patch size by 18.7% to 60.5%. A landscape-pattern analysis, using a lacunarity index, or measure of sagebrush patchiness, revealed sagebrush landscapes to be multiscaled, with dispersed sagebrush patches at small and clumped distributions at large scales, and organized at a scale between 4.5 and 9.0 km. This scale overlaps with published sage-grouse average dispersal and movement patterns. Our study supports growing evidence that sage-grouse respond to environmental factors at larger scales than those currently applied in management.

Key Words: Artemisia, Centrocercus spp., fragmentation, Greater Sage-Grouse, Gunnison Sage-Grouse, human footprint, lacunarity, sagebrush.

Influencias de la Huella Humana en Patrones del Paisaje de Artemisa: Implicaciones para la Conservación de Sage-Grouse

Resumen. Los ecosistemas de sagebrush (*Artemisia* spp.) en el oeste de los EE.UU. han cambiado en cantidad y configuración por una variedad de causas

Leu, M., and S. E. Hanser. 2011. Influences of the human footprint on sagebrush landscape patterns: implications for sage-grouse conservation. Pp. 253–271 *in* S. T. Knick and J. W. Connelly (editors). Greater Sage-Grouse: ecology and conservation of a landscape species and its habitats. Studies in Avian Biology (vol. 38), University of California Press, Berkeley, CA.

que incluyen a la agricultura y al crecimiento demográfico humano desde el establecimiento Euro-Americano. Las actividades que sostienen a la sociedad humana pueden hacer disminuir o fragmentar la cubierta vegetal del suelo y alterar procesos ecológicos dentro de los ecosistemas de artemisa. El grado de estas actividades, a las cuales acumulativamente se las denomina huella humana, dentro del territorio del sage-grouse (Greater Sage-Grouse [*Centrocercus urophasianus*] y Gunnison Sage-Grouse [*C. minimus*]) no se ha evaluado. Usando un modelo reciente de la huella humana en el oeste de los EE.UU., evaluamos la intensidad de la huella humana (1) a través del territorio de sage-grouse dentro de siete zonas de manejo del sage-grouse (SMZ o Sage-Grouse Management Zones), (2) a través de cinco clases de cubierta vegetal de sagebrush y de una clase de cubierta vegetal sin sagebrush dentro de SMZ, y (3) en el patrón del paisaje de la cubierta de suelo de artemisa en relación a tres escenarios que se diferencian en el área del efecto de la huella humana. Utilizamos cuatro criterios y organizamos las SMZ comenzando por las más influenciadas a las menos influenciadas por la huella humana, de la siguiente manera: Columbia Basin, Colorado Plateau, Wyoming Basin, Great Plains, Snake River Plain, Southern Great Basin, y Northern Great Basin. A nivel regional, black sagebrush (*Artemisia nova*) y little sagebrush (*A. arbuscula*) fueron los menos afectados por la huella humana. El área cada vez mayor del efecto de la huella humana hizo disminuir la cubierta de suelo de artemisa en el paisaje entre un 33.5% y 97.0%, redujo el tamaño promedio del parche de suelo entre un 18.7% y 60.5%, y creó paisajes en los cuales la cubierta de suelo de la artemisa estaba altamente agrupada. Un análisis del patrón del paisaje, usando un índice de lacunaridad, o una medida de la distribución de parches de artemisa, reveló que los paisajes de artemisa son multiescalares, con parches dispersos de artemisa en pequeña escala y distribuciones agrupadas a gran escala, y organizados en una escala entre 4.5 y 9.0 km. Esta escala se superpone con los patrones promedio publicados de la dispersión y del movimiento del sage-grouse. Nuestra investigación apoya la evidencia cada vez mayor que el sage-grouse responde a los factores ambientales en escalas más grandes a aquellas que son actualmente aplicadas en su manejo.

Palabras Clave: artemisa, *Artemisia*, especies de *Centrocercus*, fragmentación, Greater Sage-Grouse, Gunnison Sage-Grouse, huella humana, lagunaridad.

Historically, sagebrush ecosystems consisted of large contiguous patches of sagebrush with herbaceous understory, sagebrush patches interspersed with native grasslands, and/or native grassland patches interspersed with sagebrush patches. Composition and configuration are dependent on geographical location, climate, soils, elevation, and geographically differing disturbance regimes (Frémont 1845, Vale 1975, Mack and Thompson 1982, Young 1989, Miller and Eddleman 2001). Fire extent and frequency differed among sagebrush ecological systems of the sagebrush biome (Miller and Eddleman 2001; Miller et al., this volume, chapter 10). Grazing by large ungulates was relatively rare in the western portion of this biome, whereas sagebrush landscapes east of the Rocky Mountains were likely influenced by large-scale grazing by American bison (*Bison bison*) and other ungulates (Frémont 1845, Vale 1975, Mack and Thompson 1982, Knight 1994). Indigenous peoples sparsely populated sagebrush ecosystems, only about 230,000 people lived in the Intermountain West when Europeans first arrived in North America in 1492, and had only local effects on sagebrush ecosystems (Vale 2002a; Knick et al., this volume, chapter 12).

Few sagebrush landscapes remain intact since Euro-American settlement of the West; most have been fragmented, altered, or lost due to numerous factors including agriculture, improper livestock grazing, energy and natural resource development, rural sprawl, invasive plant and animal species, and fire (Noss et al. 1995; Hann et al. 1997; Miller and Eddleman 2001; Connelly et al. 2004; Brown et al. 2005; Miller et al., this volume, chapter 10; Knick et al., this volume, chapter 12). Presently, major disturbances influencing sagebrush landscape patterns differ geographically, with oil and gas development limited to the eastern portion of the sage-grouse range (Knick et al. 2003; this volume, chapter 12) and agriculture in the western portion of the range, particularly in the states of Idaho and Washington (Hironaka et al. 1983, Leonard et al. 2000, Vander Haegen 2000). Since 1950, the western United States experienced rapid human population growth with regional rates exceeding the United States average (Brown et al. 2005) and rural areas growing faster than urban areas in 60% of counties in the Rocky Mountain states (Odell et al. 2003). This increased influx of humans resulted in a rural sprawl away from urban centers because immigrants to this region tend to be outdoor enthusiasts who want to live in open spaces and near recreational opportunities (Brown et al. 2005, Hansen et al. 2005).

The cumulative effects of human actions on landscapes—the human footprint—can be delineated as the physical and/or the ecological human footprint. The physical footprint is the land surface occupied by anthropogenic features (Sanderson et al. 2002a, Leu et al. 2008) and in sagebrush ecosystems consists of large-scale conversion of sagebrush land cover to agricultural land (Leonard et al. 2000), rural development (Knight et al. 1995, Hansen et al. 2002, Odell et al. 2003), and small-scale, but widely distributed, energy developments (Walker et al. 2007a, Doherty et al. 2008). Agricultural and human-populated areas are the most common anthropogenic features, covering 10% of the western United States (Leu et al. 2008). The infrastructure—power lines and roads—needed to support rural sprawl and increased land use adds to the physical human footprint with roads covering 1.2% of the western United States (Leu et al. 2008). The ecological human footprint occurs where the physical human footprint influences ecological processes beyond its physical location. For example, population increases and range expansion of synanthropic nest predators, species benefiting from human resources (Restani et al. 2001, Kristan and Boarman 2003, DeLap and Knight 2004), influence Greater Sage-Grouse nest behavior (Coates and Delehanty 2008) and demography (Coates et al. 2008).

Relatively little information is available on the influence of human actions across large scales on sagebrush ecosystems (but see Wisdom et al. 2005b, Aldridge et al. 2008). We evaluated the extent of the human footprint in sagebrush ecosystems using a recently developed model for the western United States (Leu et al. 2008) and discuss potential effects in relation to sage-grouse conservation. Our objectives were: (1) to evaluate human-footprint intensity across the current sage-grouse range and within seven Sage-Grouse Management Zones (SMZs)—this comparison enables land managers to assess the cumulative influence of human actions and provides essential information for managers to direct potential management and conservation actions to restore or maintain sagebrush landscapes most cost-effectively (Wisdom et al. 2005c); (2) to evaluate human-footprint intensity specific to sagebrush land cover across the sage-grouse range within SMZs—this analysis informs sage-grouse management strategies related to habitat restoration and sage-grouse seasonal habitat use; and (3) to assess changes in sagebrush landscape pattern, using a lacunarity analysis, in relation to the ecological human footprint based on three scenarios differing in human-footprint-effect area extent—future sage-grouse conservation efforts can benefit from understanding the relationship between human-footprint intensity and fragmentation of sagebrush land cover and the influence of this relationship on daily, seasonal, and annual movements of sage-grouse.

METHODS

Study Area

We restricted our analyses to the current distribution of Greater Sage-Grouse and Gunnison Sage-Grouse (hereafter, sage-grouse) (Schroeder et al. 2004) within the western United States and each of seven SMZs (Stiver et al. 2006; Fig. 13.1). Sage-grouse management is likely to be similar within an SMZ (Stiver et al. 2006) given intra-SMZ similarities in climate, topography, soils, and elevation (Miller and Eddleman 2001).

The current sage-grouse range within our analysis region covered an area of 650,623 km^2, with the greatest proportion of range overlapping the Great Plains SMZ (26.8%), followed by the Wyoming Basin (22.5%), Snake River Plain (22.2%), Southern Great Basin (15.8%), Northern Great Basin (10.6%), Colorado Plateau (1.4%), and Columbia Basin (0.7%). Within the sage-grouse range, large expanses of unfragmented sagebrush patches are restricted to landscapes of south-central and southeast Oregon, northwest Nevada, southwest Idaho, and southwest and south-central Wyoming (Knick et al. 2003; Knick and Hanser, this volume, chapter 16).

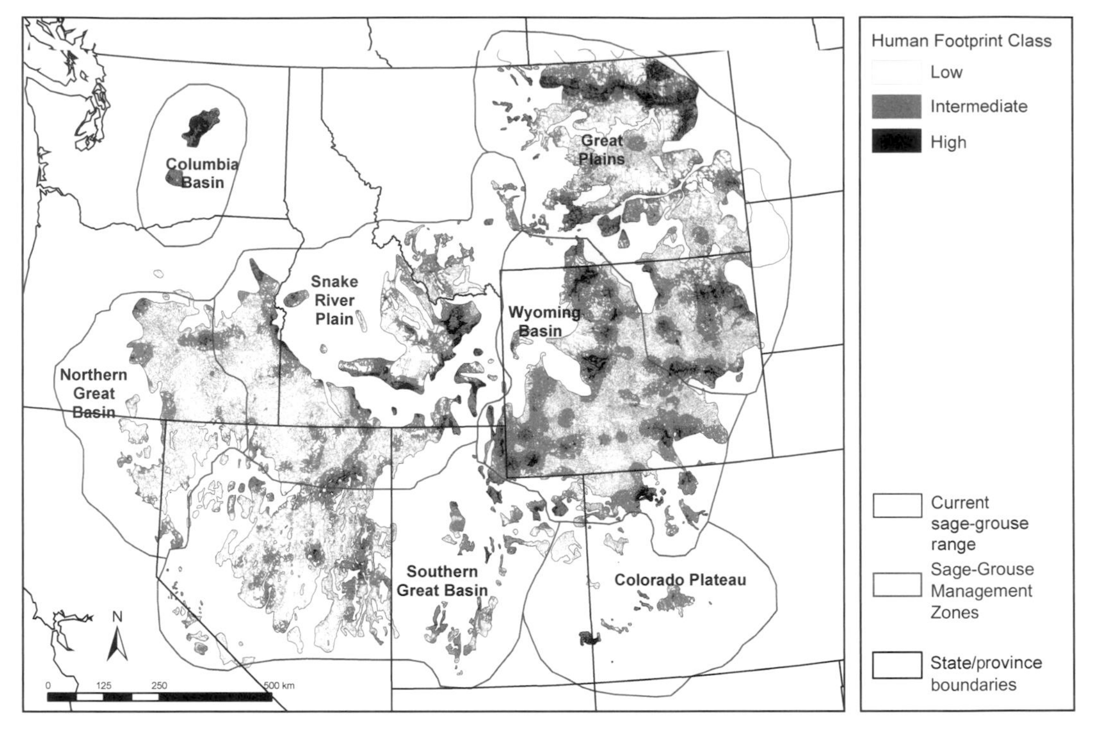

Figure 13.1. Spatial extent of three human-footprint-intensity classes for the conterminous United States within the current range of sage-grouse and Sage-Grouse Management Zones (Stiver 2006). Human-footprint-intensity classes are low (class 1–3; Leu et al. 2008), intermediate (class 4–6), and high (class 7–10).

Human-Footprint Intensity Within Sage-Grouse Management Zones and Sagebrush Land-Cover Classes

Within each SMZ, we first examined relative human-footprint intensity and its relationship to sagebrush communities and sagebrush landscape pattern. We converted SMZs vector to raster files of 0.18-km resolution. We evaluated human-footprint intensity, expressed as percent area within each of three intensity classes, among SMZs, five sagebrush land-cover classes, a non-sagebrush land-cover class, and range-wide. We performed all spatial analyses in ArcMap 9.2 (ESRI 2006) by intersecting each of seven SMZs and six land-cover classes with the three human-footprint intensity classes to derive percent area for each unique combination.

We used the recent human-footprint model developed for the western United States (Leu et al. 2008) as the input data set for SMZ and range-wide comparisons. The human-footprint model was based on seven input models—three models quantified top-down human influences of synanthropic predators (avian predators, domestic dogs and cats), and four models quantified bottom-up human influences on habitat (invasion of exotic plants, human-caused fires, energy extraction, and human wildland fragmentation). In this model, human-footprint intensity ranged from minimal effect (class 1) to a maximum effect (class 10). The human-footprint model is clearly not an exhaustive model of anthropogenic influences, but is based on spatial data available across the western United States. For example, cattle-stocking rates were not consistently available, precluding the development of a grazing model (Leu et al. 2008).

We merged the original 10 human-footprint classes into three human-footprint intensity classes (Fig. 13.2): low (classes 1–3), intermediate (classes 4–6), and high (classes 7–10). Binning of human-footprint classes was necessary to ease interpretation of human-footprint comparisons across SMZs and land-cover types. The three human-footprint intensity classes correspond to thresholds of human actions with the ecological human footprint of point (e.g., campgrounds, oil and gas wells), line (e.g., roads, railroads, and power lines), and polygon features (e.g., agriculture and urban areas), increasing with human-footprint intensity (Leu et al. 2008).

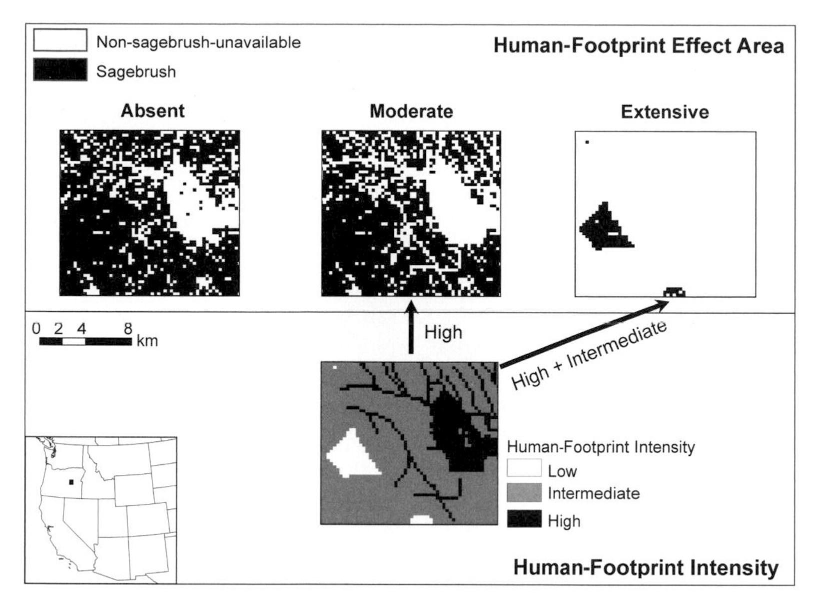

Figure 13.2. Relationship between human-footprint-intensity classes and human-footprint-effect-area scenarios. Human-footprint-effect-area classes are absent (human-footprint-intensity class 1–3; Leu et al. 2008), moderate (class 4–6), and extensive (class 7–10).

We used the LANDFIRE Existing Vegetation Map (LANDFIRE 2006) as our base land-cover map to evaluate intensity of human footprint on sagebrush land-cover types. We resampled LANDFIRE from 30-m to 180-m cell size to match the grain of the human-footprint model (Leu et al. 2008) and collapsed the eight LANDFIRE sagebrush land-cover classes into five functional sagebrush land-cover classes: (1) black sagebrush (*Artemisia nova*) land cover (Great Basin xeric mixed sagebrush shrubland); (2) little sagebrush (*A. arbuscula*) land cover (Colorado Plateau mixed low sagebrush shrubland, Columbia Plateau low sagebrush steppe, and Wyoming basin dwarf sagebrush shrubland and steppe); (3) montane big sagebrush (*A. tridentata* ssp. *vaseyana*) steppe land cover (Intermountain basins montane sagebrush steppe and *A. t.* ssp. *vaseyana* shrubland alliance); (4) big sagebrush (*A. t.* ssp. *wyomingensis* and *A. t.* ssp. *tridentata*) shrubland land cover (Intermountain basins big sagebrush shrubland); and (5) big sagebrush steppe land cover (Intermountain basins big sagebrush steppe). All land-cover classes, other than those included in the sagebrush land-cover classes, were combined into a nonsagebrush land-cover class, including anthropogenic features.

Ecological Human-Footprint Intensity and Land-Cover Fragmentation

We investigated landscape configuration of sagebrush land cover, a critical determinant of sagebrush land-cover connectivity, in relation to human-footprint intensity by examining landscape patterns and patch metrics across SMZs. For these analyses, we first buffered each SMZ by 18 km to avoid edge effects influencing landscape metrics, and then converted vector to raster files at 36-km extent (grid size = 0.18 km). Rasterization of SMZ shapefiles resulted in reduced SMZ extents for larger SMZs or slightly enlarged extents for smaller SMZs.

Because sagebrush landscape fragmentation is inherently complex and degree of fragmentation varies spatially throughout the sage-grouse range (Knick et al. 2003; Knick and Hanser, this volume, chapter 16), we chose a landscape metric that assessed complexity of sagebrush landscapes at various scales to mimic how sagebrush-obligate vertebrate species might perceive landscape patterns. We used the lacunarity landscape metric, which is a spatial statistic that simultaneously describes land cover contagion, that is, an index distinguishing clumped from dissected landscapes (Turner et al. 2001), and dispersion, an index measuring patch arrangement (Forman

1995), or in other words, landscape pattern along a gradient of patch dispersion from dispersed to clumped (Plotnick et al. 1993, Elkie and Rempel 2001). Lacunarity has several advantages over other, more common fixed-scale landscape metrics because it consists of a single metric evaluated at multiple scales, is not influenced by edge effects, and is not restricted to landscapes with high occurrence of habitat of interest (Plotnick et al. 1993). Lacunarity metrics can also be used to assess degree of relative fragmentation across diverse landscapes (Wu et al. 2000).

We evaluated lacunarity (Plotnick et al. 1993, Wu and Sui 2001) in relation to three scenarios differing in human-footprint-effect area: absent, moderate, and extensive (Fig. 13.2). The absent-effect area reflected baseline distributions of sagebrush land cover and the physical human footprint of dominant anthropogenic features such as agricultural land, urban areas, state and interstate highways. The absent-effect-area spatial data set was classified as sagebrush (value = 1) and nonsagebrush (value = 0) land cover. The sagebrush land cover, overlapping 51.9% of the sage-grouse range, combined all five sagebrush land-cover classes. The nonsagebrush land cover, overlapping 48.1% of the sage-grouse range, consisted mainly of woodland, shrubland, and grassland land-cover classes other than sagebrush (85.8%), and to lesser extent agricultural (12.7%) and developed land-cover classes (1.5%; urban area, interstate and state highways). The moderate-effect-area spatial data set (Fig. 13.2) consisted of nonsagebrush land cover, as in the absent-effect-area scenario, plus the ecological footprint of anthropogenic features mapped in LANDFIRE (LANDFIRE 2006) and additional anthropogenic features, such as power lines and secondary roads, not mapped in LANDFIRE. Overall, the moderate-effect area delineated the high-intensity human-footprint area (classes 7–10). All pixels classified as sagebrush within the moderate-effect area were reclassified as unavailable sagebrush land cover (value = 0) and combined with nonsagebrush land cover. We named this scenario the moderate-effect area because the human-footprint extent was intermediate between the absent- and extensive-effect areas. The extensive-effect area data set (Fig. 13.2) also consisted of nonsagebrush, plus the ecological footprint of both the intermediate- and high-intensity human footprint (classes 4–10). All pixels classified as sagebrush land cover within the extensive-effect area were reclassified as unavailable sagebrush (value = 0) and combined with nonsagebrush land cover.

We modeled lacunarity as:

$$\text{lacunarity} = (\sigma/\bar{x}^2) + 1 \qquad \text{(Eqn. 13.1)}$$

where σ = variance and $\bar{x}$ = mean (Plotnick et al. 1993, Wu and Sui 2001) are based on number of sagebrush pixels within a predetermined area. To calculate lacunarity, we first performed focal sum moving-window analyses on the three human-footprint-effect-area scenarios. A focal sum moving-window analysis computes an output data set where the value at each location is a function of the values in the input data set within a predetermined window size. For each location in the land-cover input data set, the analysis computes the sum of all sagebrush pixels within the window surrounding the processing location and yields this value to the output data set. In our analyses we used 10 window sizes (sampling scale) at progressively larger sampling scales: 0.18 km (1 cell), 0.36 km, 0.9 km, 1.8 km, 3.6 km, 4.5 km, 9 km, 18 km, 27 km, and 36 km (200 cells). Intermediate scales approximated average female sage-grouse dispersal distance (8.8 km)(Dunn and Braun 1985) and recommended conservation area around leks for nonmigratory (5 km) and migratory (18 km) Greater Sage-Grouse populations (Connelly et al. 2000c). We did not use sampling scales >36 km because increasingly more habitat outside of the current sage-grouse range would have been included in the analyses, leading to potential bias in SMZ estimates.

We derived means and variances of focal sum moving-window analysis outputs, within each SMZ at all sampling scales, using zonal summary statistics (ArcMap 9.2, ESRI 2006). Means and variances subsequently were exported to a spreadsheet application to calculate lacunarity (Eqn. 13.1). We plotted lacunarity versus sampling scale in natural-log graphs to normalize data (Plotnick et al. 1993). Lacunarity functions provide four pieces of information on landscape pattern. First, lacunarity values at a given scale reveal the relative proportion of land cover of interest (Plotnick et al. 1993, Wu and Sui 2001). Landscapes with a low proportion of sagebrush land cover have high lacunarity values because variances exceed means, whereas landscapes dominated by sagebrush land cover have low lacunarity values because means exceed variances. Second, shapes of lacunarity functions reveal landscape patterns (Plotnick et al. 1993). Convex functions represent landscapes in which sagebrush land cover is clumped and interspersed with large patches of nonsagebrush land cover; in this function, only at large scales are sagebrush patches included in the lacunarity calculation. In contrast, concave functions represent landscapes in which sagebrush land cover is dispersed, landscapes consisting of smaller patches of sagebrush and nonsagebrush land cover, resulting in rapidly declining functions. Multiconcavity functions represent landscapes in which sagebrush patch distribution is hierarchical, with nonsagebrush gaps varying over a range of sizes. Third, lacunarity functions reveal the scale where landscape patterns are repeating, or in other words, the scale at which interacting ecological processes in a given ecosystem produce landscapes of similar configuration (Holling 1992). The scale of landscape pattern or domain of scales (Wu et al. 2000) occurs where lacunarity asymptotically approaches zero. Variance of sagebrush land cover and lacunarity is high when sampling scales are less than the scale of landscape pattern. In contrast, when sampling scales exceed the scale of landscape pattern, variance of sagebrush cover and natural log (lacunarity) will approach zero. Lacunarity in natural landscapes rarely equals zero because the extent of the largest sampling scale does not equal the extent of land cover of interest, given the convoluted perimeter of patches in natural landscapes. The scale of landscape pattern is smaller in dispersed compared to clumped landscapes (Wu et al. 2000). Fourth, lacunarity functions reveal whether landscapes are hierarchically composed of smaller patches nested within larger patches and the scale(s) of this patch structure. The scale of landscape nestedness or the relevant scale (Elkie and Rempel 2001) can be detected by scanning lacunarity functions for inflection points at which slopes become more negative. This decrease in the slope occurs due to the same process as the scale of landscape pattern: sampling scale exceeds the scale of smaller patches, but lacunarity does not decrease to zero because sampling scale is smaller than larger-scale patch structure. To determine scale(s) of landscape nestedness in lacunarity functions, we calculated slopes for each sampling scale interval to identify inflection points.

Evaluation of Lacunarity in Artificial Landscapes

We analyzed artificial landscapes due to the lack of previous research evaluating lacunarity in natural landscapes demarcated by convoluted patch boundaries and to aid interpretation of lacunarity analyses from natural landscapes (Elkie and Rempel 2001). We developed artificial landscapes within the extent of the Columba Basin SMZ, which was large enough to encompass the range of conditions expected in the analysis of natural landscapes. We developed nine artificial single-scale landscapes in SIMMAP 2.0 (Saura and Martínez-Millán 2000, SIMMAP 2003) that differed in percent of land cover (20%, 35%, 55%), corresponding to the range of percentages of sagebrush land cover in SMZ landscapes, and three levels of fragmentation, the simulation parameter (SP)(SIMMAP 2003): dispersed (SP = 0.001); moderately dispersed (SP = 0.35), and clumped (SP = 0.59) (Fig. 13.3). Because sagebrush landscape patterns are complex, we developed multiscale landscapes within the three land-cover scenarios by merging single-scale landscapes across fragmentation levels; this resulted in four landscapes per land-cover proportion: dispersed–moderately dispersed landscapes (SP = 0.01 + SP = 0.35), dispersed–clumped landscapes (SP = 0.01 + SP = 0.59), all-pattern landscapes (all three fragmentation levels combined), and moderately dispersed–clumped landscapes (SP = 0.35 + SP = 0.59). We calculated lacunarity for each sampling scale in the 21 artificial landscapes (three single-scale and four multiscale landscapes for each of three percent-cover classes), developed natural-log graphs, and examined lacunarity curves for their shape, scale of landscape nestedness, and scale of landscape pattern.

Evaluation of Patch Metrics in Sagebrush Landscapes Within Sage-Grouse Management Zones

We calculated patch metrics for sagebrush and nonsagebrush-unavailable sagebrush land cover to compare results from the lacunarity analyses (FRAGSTATS ver. 3.2, McGarigal et al. 2002). In addition to number of patches, average patch size, and patch size variation, we also calculated a largest patch index (LPI), a measure of landscape dominance that relates area of the largest patch to total area of the landscape (LPI = max [area patch a_{ij}]/area of landscape × 100) (McGarigal et al. 2002). We used the LPI to differentiate patch dynamics in landscapes that differ in the degree of human-footprint-effect area because this metric describes

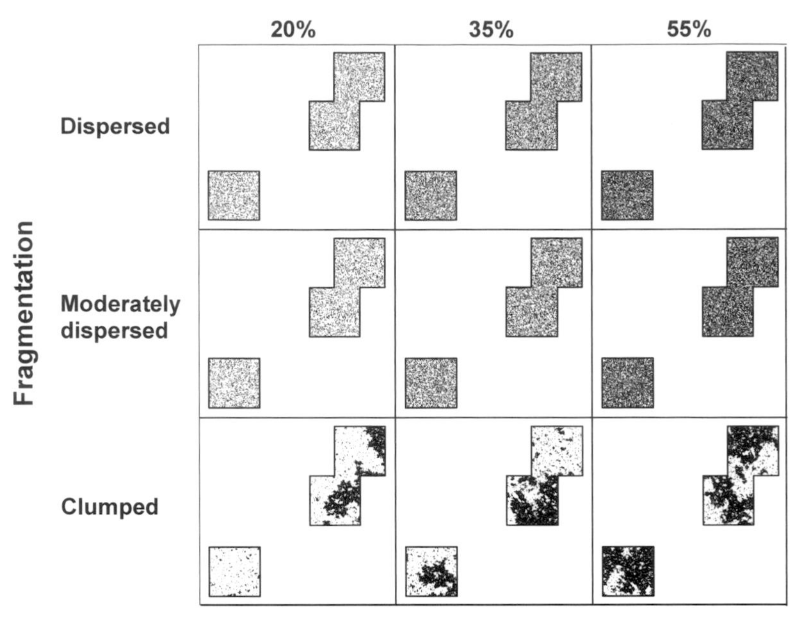

Figure 13.3. Single-scale artificial landscapes approximately delineating the Columbia Basin Management Zone (Stiver 2006). Landscapes differ in three levels of percent land cover (20%, 35%, and 55%) and three levels of fragmentation: dispersed (SP = 0.001; SIMMAP 2003); moderately dispersed (SP = 0.35), and clumped (SP = 0.59).

common features of landscapes from different physiographic regions (Cushman et al. 2008).

RESULTS

Human-Footprint Intensity Within Sage-Grouse Range and Management Zones

Human-footprint intensity varied spatially across the sage-grouse range (Fig. 13.1). High-intensity human-footprint classes were most prevalent in the northwestern, central, and eastern portion of the sage-grouse range, with the opposite geographical distribution for the low-intensity footprint. Overall, high-intensity human footprint covered 4.8% of current sage-grouse range, intermediate 47.0%, and low 48.2%.

Intensity of human footprint varied among SMZs (Fig. 13.4). The Northern Great Basin, Southern Great Basin, and Snake River Plain SMZs contained a larger proportion of low-intensity human-footprint area (range: 0.04–0.21) compared to the range-wide intensity. In contrast, the Colorado Plateau, Great Plains, and Columbia Basin SMZs had a higher proportion of high-intensity area (range: 0.001–0.47) compared to the range-wide intensity. In the Columbia Basin SMZ, the high-intensity human footprint (52.1%) covered more area than intermediate-intensity footprint (47.6%), and only 0.3% of this SMZ overlapped with the low-intensity footprint.

Human-Footprint Intensity and Sagebrush Land-Cover Classes

Human-footprint intensity varied among sagebrush land-cover types (Fig. 13.5). The least common land-cover classes across the sage-grouse range, black sagebrush covering 3.4% of current sage-grouse range, and little sagebrush covering 4.2% of current sage-grouse range, had the highest proportion of low-intensity and lowest proportion of high-intensity human-footprint area compared to range-wide values. Considering big sagebrush land-cover classes, the montane sagebrush steppe, covering 7.3% of current sage-grouse range, and big sagebrush steppe, covering 14.9% of current sage-grouse range, had higher proportions of low-intensity area compared to the big sagebrush shrubland, covering 23.2% of current sage-grouse

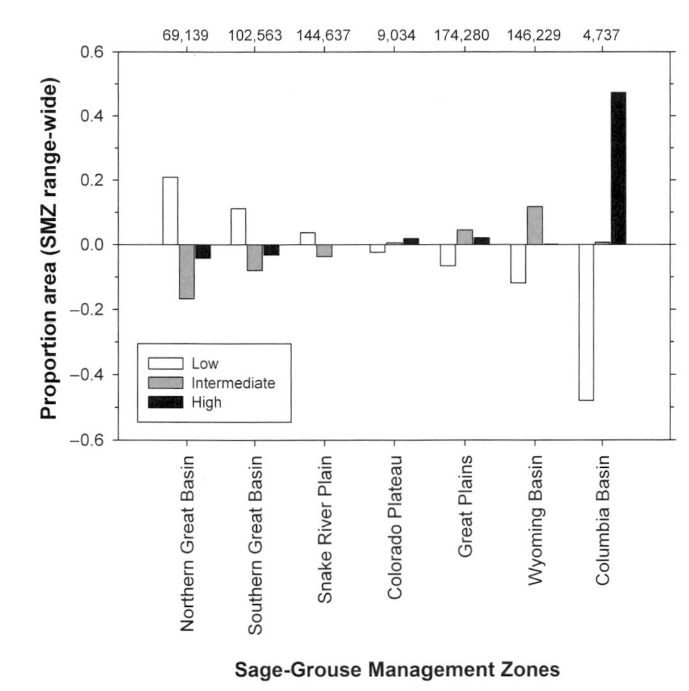

Figure 13.4. Difference in proportion area between seven Sage-Grouse Management Zones (SMZ; Stiver 2006) and sage-grouse range-wide values for three human-footprint-intensity classes across the sage-grouse range. Positive proportions indicate more area in human-footprint-intensity class compared to range-wide value. SMZs were ranked from highest to lowest proportion of area within the low-intensity human footprint. Total sage-grouse range (km^2) within each SMZ shown on top axis.

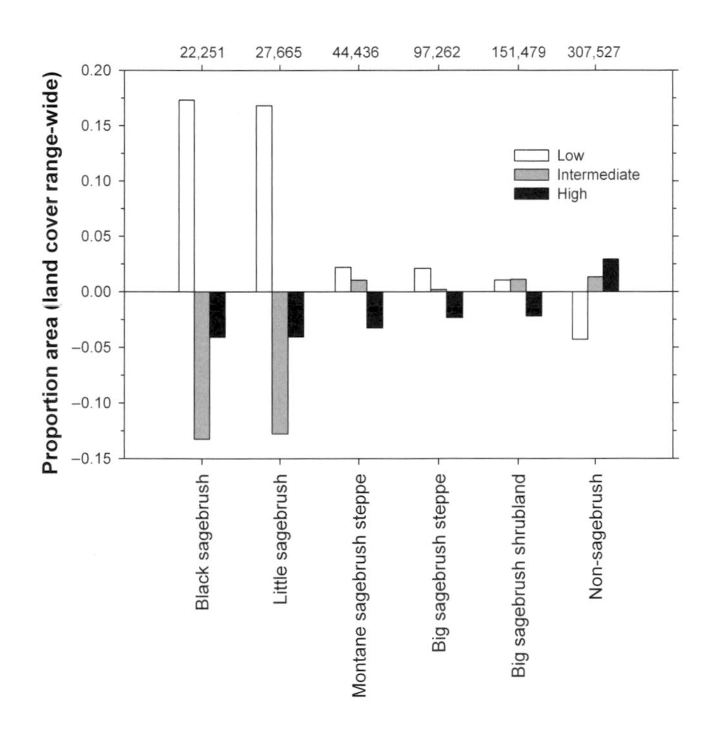

Figure 13.5. Difference in proportion area between six land cover classes and sage-grouse range-wide values for three human-footprint-intensity classes across the sage-grouse range. Positive proportions indicate more area in human-footprint-intensity class compared to range-wide values. Land cover types were ranked from highest to lowest proportion of area within the low-intensity human footprint. Total land cover area (km^2) within each land cover class shown on top axis.

range. In contrast, the montane sagebrush steppe had a lower proportion in high-intensity area compared to the big sagebrush shrubland and big sagebrush steppe land-cover types.

Ecological Human Footprint Intensity and Sagebrush Landscape Pattern

Artificial Landscapes

In single-scale landscapes, independent of proportion of land cover, clumped landscapes had convex lacunarity functions, dispersed landscapes had concave functions, and moderately dispersed landscapes had multiconcavity functions (Fig. 13.6). In multiscale landscapes, the shapes of lacunarity functions were both concave and multiconcave.

Scale of landscape pattern was related to degree of fragmentation (Fig. 13.6). For single-scale clumped landscapes, landscape pattern was not defined at the 36-km scale. Scale of landscape pattern increased with degree of fragmentation for dispersed and moderately dispersed landscapes and related inversely to percent land cover within a patch dispersion class. For dispersed landscapes, scale of landscape pattern was between 1.8 and 3.6 km for the 35% and 55% land-cover landscapes and between 3.6 and 4.5 km for the 20% land-cover landscapes. In contrast, the scale of landscape

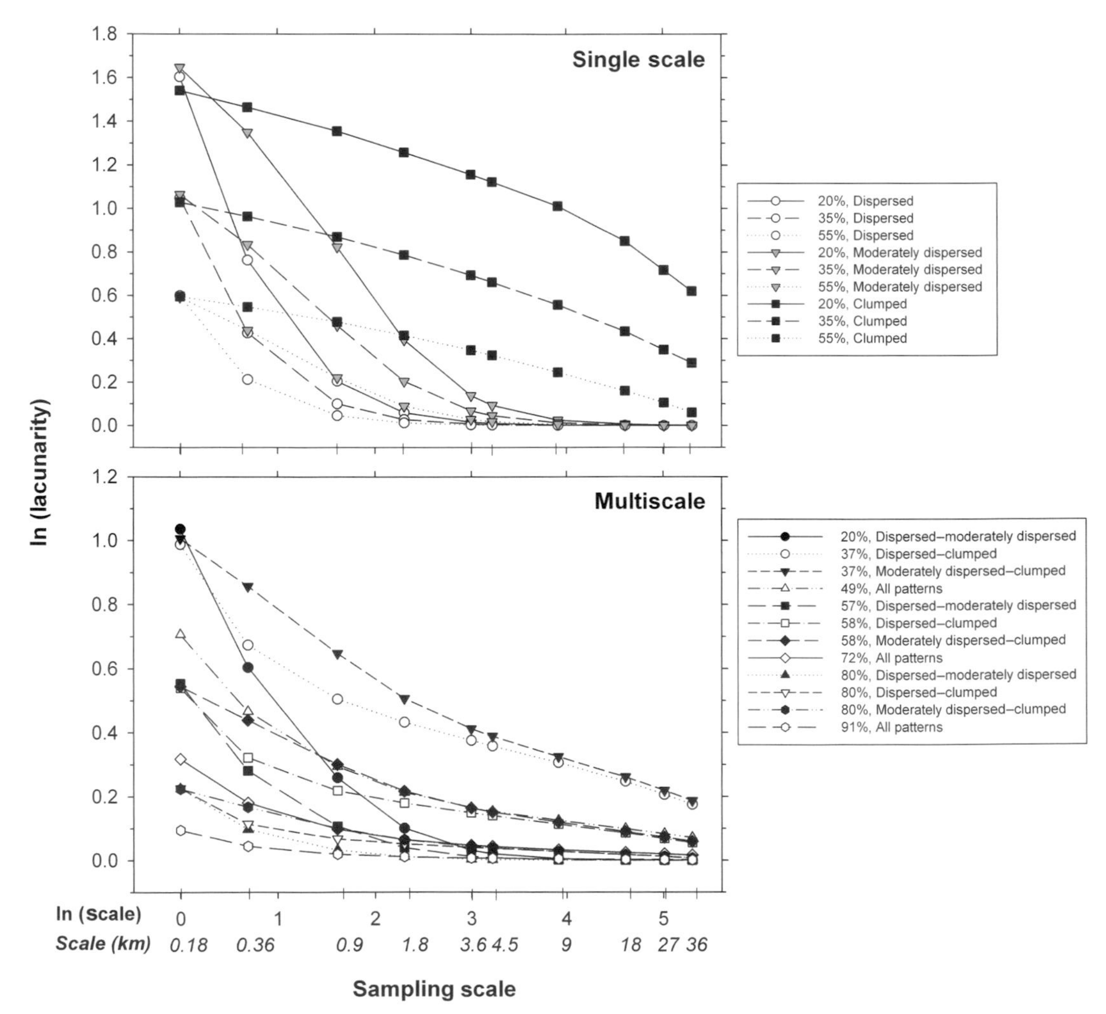

Figure 13.6. Lacunarity curves for simulated single-scale landscapes (N = 9; upper panel) and multi-scale landscapes (N = 12; lower panel) within the approximate delineation of the Columbia Basin Management Zone. Landscapes differ in three levels of percent land cover (20%, 35%, and 55%) and three levels of fragmentation for single-scale landscapes and four levels of fragmentation for multi-scale landscapes.

pattern for moderately dispersed landscapes was between 4.5 and 9.0 km for the 55% land-cover landscapes and between 9.0 and 18.0 km for 20% and 35% land-cover landscapes.

Scale of landscape pattern also was related to fragmentation in multiscale landscapes (Fig. 13.6). The scale of landscape pattern was not defined at 36 km for dispersed–clumped landscapes, moderately dispersed–clumped landscapes, and all-pattern landscapes. For dispersed–moderately dispersed landscapes, the scale of landscape pattern related inversely to percent land cover: 3.6–4.5 km for 55% land-cover landscapes, 4.5–9.0 km for 35% land-cover landscapes, and 9.0–18.0 km for 20% land-cover landscapes.

Single-scale dispersed and moderately dispersed landscapes were not nested at any scale, whereas single-scale clumped landscapes had a scale of landscape nestedness between 3.6 and 4.5 km that was independent of proportion land cover. For multiple-scale landscapes, the scale of landscape nestedness, regardless of percent land cover, was smaller for dispersed–clumped landscapes (between 9 and 18 km) compared to moderately dispersed–clumped and all-pattern landscapes (between 18 and 27 km) (Fig. 13.6).

Sagebrush Landscapes: Absent-Human-Effect Area

Lacunarity values at the 0.18-km sampling scale for absent-human-footprint-effect area, sagebrush landscape patterns in absence of small-effect-area anthropogenic features and the ecological human footprint, followed three patterns (Fig. 13.7). The group with highest lacunarity values, or lowest percent sagebrush land cover, included the Great Plains, Colorado Plateau, and Columbia Basin SMZs, in order of increasing proportion of sagebrush land cover. The Southern Great Basin SMZ had an intermediate lacunarity value, whereas the group with lowest lacunarity values, or highest percent sagebrush land cover, included the Wyoming Basin, Snake River Plain, and Northern Great Basin SMZs.

Lacunarity functions were multiconcave for all SMZs (Fig. 13.7) most closely resembling artificial landscapes that were multiscale dispersed–clumped (Fig. 13.6). At the 36-km scale, none of the sagebrush landscapes reached the scale of landscape pattern. All landscapes had a scale of nestedness between 4.5 and 9 km.

Sagebrush land-cover patch metrics corroborate the lacunarity analysis (Table 13.1). Mean patch size and largest patch index were highest for landscapes with extensive areas of sagebrush land cover (Northern Great Basin, Snake River Plain, and Wyoming Basin) and lowest for landscapes with small extents of sagebrush land cover (Colorado Plateau, Columbia Basin, and Great Plains).

Sagebrush Landscapes: Moderate and Extensive Human-Footprint-Effect Area

Increasing the human-footprint-effect area influenced SMZs differentially (Fig. 13.8). At the 0.18-km scale, the Columbia Basin SMZ, which had the third-highest lacunarity in the absent-effect-area scenario (Fig. 13.7), changed to the highest lacunarity value in the moderate- and extensive-effect-area scenarios (Fig. 13.8). The Great Plains, which had the highest lacunarity value in the absent-effect-area scenario (Fig. 13.7), changed to third-highest lacunarity value in the moderate- and extensive-effect-area scenarios (Fig. 13.8). Ranking of the Southern Great Basin, Wyoming Basin, Snake River Plain, and Northern Great Basin did not change between absent- and moderate-effect-area scenarios, but the Wyoming Basin had a higher lacunarity value compared to the Southern Great Basin in the extensive-effect-area scenario.

Increasing human-footprint-effect area decreased the proportion of available sagebrush (i.e., decreasing sagebrush patch size) but increased nonsagebrush land cover (i.e., gap size among sagebrush patches) at the 0.18-km scale; this change was small for moderate-effect-area scenarios but large for extensive-effect-area scenarios (Figs. 13.7, 13.8). Comparing absent- and moderate-effect-area scenarios, lacunarity increased by ~2% in the Great Plains, Northern Great Basin, and Southern Great Basin SMZs; ~6% in the Colorado Plateau, Snake River Plain, and Wyoming Basin SMZs; and 18% in the Columbia Basin SMZ. In contrast, comparing absent- and extensive-effect-area scenarios, lacunarity increased between 45% and 75% for the Colorado Plateau, Great Plains, Northern Great Basin, and Southern Great Basin SMZs, and >100% for the Snake River Plain (108%), Wyoming Basin (146%), and Columbia Basin (240%) SMZs.

Across all SMZ landscapes, increasing human-footprint-effect area was unrelated to changes in lacunarity functions; they remained multiconcave (Fig. 13.8), but related with an upward shift in functions. This shift occurred due to loss of sagebrush land cover and an increasingly

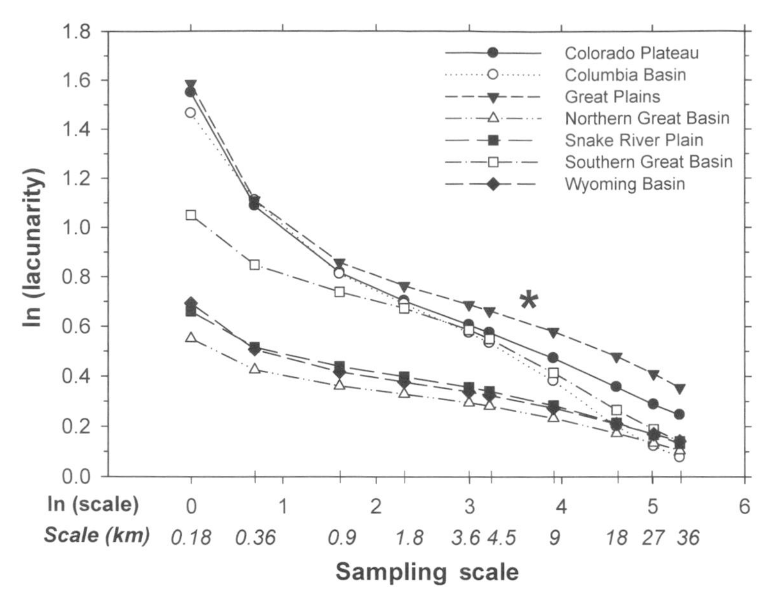

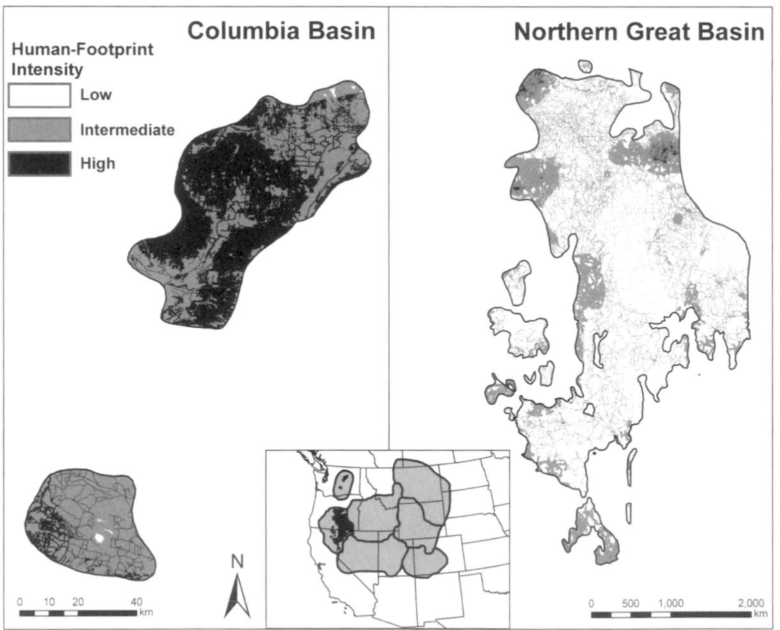

Figure 13.7. Lacunarity curves for the absent-human-footprint-effect area across seven Sage-Grouse Management Zones (SMZ, Stiver 2006) and spatial representation of a highly fragmented (Columbia Basin SMZ, upper panel) and a minimally fragmented (Northern Great Basin SMZ, lower panel) landscape. Sampling scale of lacunarity analyses ranged from 0.18–36 km. Asterisk in upper panel denotes scale of landscape nestedness (4.5–9.0 km).

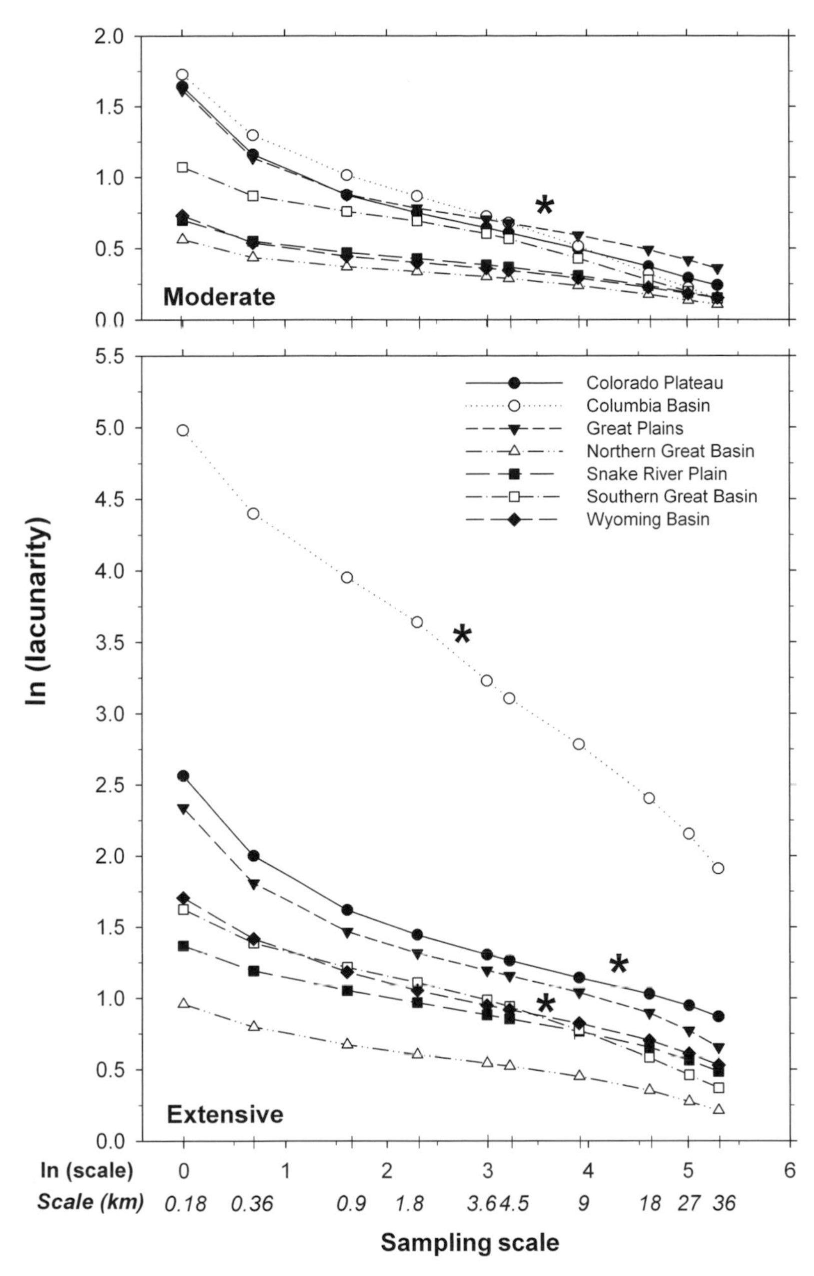

Figure 13.8. Lacunarity curves for the moderate-human-footprint-effect area scenario (upper panel) and the extensive-human-footprint-effect area scenario (lower panel). Sagebrush landscape pattern is more clumped with increasing-human-footprint-effect area. Note changes in scale between upper and lower panels. Asterisks denote scale of landscape nestedness: 4.5–9.0 km for all SMZs in the moderate-human-footprint-effect area, 4.5–9.0 km for all SMZs, other than the Colorado Plateau SMZ, in the extensive-human-footprint-effect area. Scale of landscape nestedness for the Colorado Plateau SMZ increased to 9.0–18.0 km, and the Columbia Basin SMZ had an additional scale between 1.8–3.6 km.

TABLE 13.1

Patch metrics for sagebrush and non-sagebrush land cover in relation to three scenarios differing in human-footprint effect area within seven Sage-Grouse Management Zones (SMZs)

SMZ	Human-footprint effect area	Sagebrush land cover (%)[a]	Number of patches		Mean patch size, ha (SE)		Largest patch index	
			Sagebrush	Non-sagebrush	Sagebrush	Non-sagebrush	Sagebrush	Non-sagebrush
Colorado Plateau	Absent	21.2	39,792	4,948	27.0 (4.8)	804.6 (772.2)	3.11	75.57
	Moderate	19.4	39,684	3,991	24.7 (4.3)	1,021.2 (982.7)	3.02	77.57
	Extensive	7.7	21,380	1,049	18.2 (1.5)	4,447.0 (4437.8)	0.44	92.15
Columbia Basin	Absent	23.1	9,090	2,043	36.2 (7.5)	536.5 (380.8)	3.24	46.29
	Moderate	17.8	7,662	1,270	33.1 (5.7)	922.8 (663.6)	2.44	50.44
	Extensive	0.7	686	26	14.3 (1.7)	54,454.3 (38405.9)	0.04	62.98
Great Plains	Absent	20.5	175,389	27,291	29.7 (4.4)	742.2 (730)	2.01	78.22
	Moderate	19.9	175,209	25,221	28.9 (4.2)	809.3 (797.4)	1.87	78.96
	Extensive	9.6	108,841	9065	22.6 (1.6)	2,539.1 (2530.1)	0.62	90.04
Northern Great Basin	Absent	57.6	26,196	57,787	247.9 (198.2)	82.8 (30.6)	45.89	10.70
	Moderate	57.0	26,079	55,731	246.4 (196.7)	87.0 (37.4)	45.37	12.53
	Extensive	38.3	27,090	26,290	159.5 (29.3)	264.5 (253.2)	6.18	59.04
Snake River Plain	Absent	51.6	75,708	109,255	175.9 (110.8)	114.2 (47)	29.45	14.08
	Moderate	49.8	72,524	101,407	177.0 (109.9)	127.7 (98.8)	28.18	38.79
	Extensive	25.4	52,535	38,679	124.9 (21.2)	497.1 (489.2)	3.70	73.37
Southern Great Basin	Absent	35.0	103,184	65,584	83.5 (44.3)	244.1 (107.7)	18.46	20.68
	Moderate	34.2	103,326	62,213	81.6 (42.9)	260.2 (115.7)	17.92	21.10
	Extensive	19.7	71,383	26,478	67.9 (5.4)	746.8 (554.5)	0.83	53.79
Wyoming Basin	Absent	50.0	78,225	124,778	139.9 (118.5)	87.8 (63.8)	42.31	36.20
	Moderate	48.2	77,768	115,237	135.7 (113.5)	98.5 (81.7)	40.30	43.00
	Extensive	18.2	61,685	24,888	64.5 (4)	720.2 (713.5)	0.81	81.07

[a]Cover (%) for non-sagebrush is complementary to sagebrush land cover, therefore not reported. For area of current sage-grouse range within each SMZ, see Fig. 13.4.

clumped patch structure. Compared to the absent-effect-area scenario at the 36-km scale, in the moderate-human-footprint-effect-area scenario, lacunarity increased <2% for the Colorado Plateau, Great Plains, and Northern Great Basin SMZs, 5% for the Southern Great Basin SMZ, 8% for the Wyoming Basin SMZ, 14% for the Snake River Plain SMZ, and 93% for the Columbia Basin SMZ. In the extensive-effect-area scenario, lacunarity increased 85% for the Great Plains SMZ, 103% for the Northern Great Basin SMZ, 163% for the Southern Great Basin SMZ, 253% for the Colorado Plateau SMZ, 266% for the Snake River Plain SMZ, 273% for the Wyoming Basin SMZ, and 2,385% for the Columbia Basin SMZ.

Scale of landscape nestedness changed in only two SMZs with increasing human-footprint-effect area (Fig. 13.8). Scale of landscape nestedness was between 4.5 and 9.0 km in the moderate- and extensive-effect-area scenarios, except for the Colorado Plateau SMZ, in which the scale increased to 9.0 and 18.0 km in the extensive-effect area. In the extensive-effect-area scenario, the Columbia Basin SMZ had an additional scale of landscape nestedness between 1.8 and 3.6 km.

Sagebrush land-cover patch metrics indicated a shift in landscape quantity and configuration with increasing human-footprint-effect area. Percent area of sagebrush land cover, number of patches, mean patch size, and largest patch index of sagebrush land cover decreased with increasing human-footprint-effect area (Table 13.1). The only exception was the Northern Great Basin SMZ, in which the number of patches slightly increased. For non-sagebrush land cover, number of patches also decreased, but mean patch size and largest patch indices increased with increasing human-footprint-effect area. These results corroborate the lacunarity analyses that with increasing human-footprint-effect area, the quantity of sagebrush land cover decreased and patches became more clumped.

Overall Human-Footprint Effect

We ranked SMZs from most to least human-footprint effect based on sums of ranking scores derived from (1) difference in proportion of area in low- and high-intensity human footprint compared to range-wide values, (2) mean sagebrush patch size in the absent-human-footprint-effect area, (3) percent decrease in all sagebrush land cover and mean patch size between the absent- and extensive-human-footprint-effect area, and (4) percent increase in lacunarity at the 36-km scale between absent- and extensive-human-footprint-effect areas. On a decreasing scale of human-footprint effect, our ranking was Columbia Basin, Colorado Plateau, Wyoming Basin, Great Plains, Snake River Plain, Southern Great Basin, and Northern Great Basin.

DISCUSSION

Parts of the sage-grouse range have undergone rapid changes during the past two to four decades with low-density human development creeping into sagebrush ecosystems. Within the Great Basin, an area overlapping the states of California, Idaho, Nevada, and Utah, human populations have increased by 69% and uninhabited areas decreased by 86% between 1990 and 2004 (Torregrosa and Devoe 2008). The majority of the sage-grouse range overlapped with human-footprint-intensity classes in which percent human population increase was above average between 1990 and 2000 (Leu et al. 2008). Sage-grouse were extirpated in areas with >4 people/km^2 (Aldridge et al. 2008) and only 5% of sage-grouse leks (N = 120) were within 200 m of a building in Colorado (Rogers 1964). Sage-grouse population persistence may not be influenced by a single anthropogenic line or point feature, but by a threshold of multiple human resources acting in synergy.

Human effects on sagebrush land cover varied among sagebrush ecological systems. In general, human-footprint intensity increases with proximity to valley floor, soil depth, and aboveground productivity, but decreases with increasing elevation (Leu et al. 2008). Therefore, regional differences in suitability of sagebrush lands to large-scale conversion to agriculture (McDonald and Reese 1998, Vander Haegen et al. 2000, Vander Haegen 2007), the human land use with the largest extent across the western United States (Leu et al. 2008), can be used to predict human-footprint intensity in sagebrush ecosystems. For example, black and little sagebrush, occurring in poorly drained and shallow soils (West and Young 2000; Miller et al., this volume, chapter 10), were least influenced by the human footprint range-wide, whereas big sagebrush shrubland and steppe, generally occurring in well-drained, deep sandy or clay-loam soils, were most affected by the human footprint. A large proportion of the more productive big sagebrush, particularly areas dominated by basin (*A. t.* ssp. *tridentata*) and mountain big sagebrush have been converted to agricultural land (Hironaka et al. 1983, Leonard et al. 2000). We observed a similar trend with elevation separating sagebrush ecological systems (Shumar and Anderson 1986; Jensen 1990; Miller et al., this volume, chapter 10), with the high-elevation montane sagebrush steppe land-cover class being least influenced, the mid-elevation big sagebrush steppe intermediate, and the low-elevation big sagebrush shrubland being most affected by the human footprint. Other large-scale human land uses, such as natural-gas extraction (Knick et al. 2003), often occur in big sagebrush systems dominated by Wyoming big sagebrush (*A. t.* ssp. *wyomingensis*). Wyoming big sagebrush typical of xeric sites (Miller et al., this volume, chapter 10) may have low resiliency to human disturbance, and therefore any disturbance in this ecosystem may have disproportionally adverse affects. Overall, the human footprint may have a disproportionately higher effect on sagebrush ecosystems in productive low-elevation regions.

The lacunarity landscape metric shows great promise for assessing complex sagebrush landscape patterns. Because lacunarity analyses are evaluated at multiple scales, this metric is more sensitive in identifying landscape patterns that influence animal movements (With and King 1999, McIntyre and Wiens 2000), and unlike conventional landscape metrics such as fractal geometry, patch dispersion, and percolation theory, lacunarity does not produce similar metric values for landscapes differing in patch arrangements (Plotnick et al. 1993). Despite their ease in calculation, lacunarity analyses have rarely been used to study patterns of natural landscapes (but see Wu et al. 2000, Derner and Wu 2001, Elkie and Rempel 2001), perhaps because interpretation of lacunarity curves can be difficult. However, we found that using lacunarity analyses of simulated landscapes, where degree of fragmentation and proportion of land cover reflect the range of values of landscapes studied, greatly aids in the interpretation of lacunarity functions of landscape patterns. For sage-grouse, we recommend comparing lacunarity curves of landscape pattern with those of sage-grouse landscape use (McIntyre and Wiens 2000). Such comparisons will reveal important information on sage-grouse landscape function, that is, how sage-grouse use a landscape in relation to landscape pattern.

We used lacunarity analyses to identify and evaluate changes in landscape pattern in relation to human-footprint-effect area. Sagebrush ecosystems most closely represented a multiscale dispersed–clumped landscape, typical of landscapes dominated by a mixture of small and large patches of sagebrush. Landscape pattern varied among SMZs because underlying mechanisms of fragmentation are shaped by natural processes, human disturbance, or both. For example, the Southern Great Basin SMZ consists of a basin and range geologic pattern (Fiero 1986) where sagebrush largely occurs in basins. The Columbia Basin SMZ, in contrast, consists of a repeating pattern of patches of sagebrush juxtaposed within larger patches of agricultural land (Vander Haegen et al. 2000). In both SMZs, landscape pattern was dispersed at small scales but changed to a more clumped pattern at larger scales. Although landscape patterns are similar between the two SMZs, the probability of Greater Sage-Grouse effective population size to decrease below 500 within the next 30 years is 0% in the Southern Great Basin SMZ (Garton et al., this volume, chapter 15), suggesting sage-grouse are well adapted to the natural landscape pattern. In contrast, that same probability is 76% for sage-grouse populations in the anthropogenic landscape pattern of the Columbia Basin SMZ.

Small-scale patterns were nested within larger-scale patterns in sagebrush landscapes at thresholds between 4.5 and 9.0 km. This threshold did not change with increasing ecological human-footprint intensity, with the exception of the Colorado Plateau SMZ, where the scale of landscape nestedness increased to 9.0–18.0 km in the extensive-human-footprint-effect area. The apparent universality of the 4.5–9.0 km scale of landscape nestedness persisted despite differences among SMZs in climatic, elevational, topographical, and geological regimes (Miller and Eddleman 2001; Miller et al., this volume, chapter 10). Thus, sagebrush landscapes may be organized at an endogenous scale, at which self-organizing processes shape landscapes (Holling 1992). The mechanisms for an endogenous scale remain unclear but could have been influenced by historical fire patterns in the western regions of sagebrush plant communities (Miller et al., this volume, chapter 10; Baker, this volume, chapter 11). If sagebrush habitats have such an endogenous scale, we would predict that sage-grouse adapted to this perceptive scale (Szabó and Meszéna 2006). Several studies provide evidence that sage-grouse adapted to an endogenous scale. Range-wide mean nearest-neighbor lek distance was 5.2 km ± 0.05 km (SE; $N = 4,838$; Knick and Hanser, this volume, chapter 16). The median dispersal distance for females was 8.8 km and 7.8 km for male Greater Sage-Grouse in Colorado (Dunn and Braun 1985). Mean movement distance between leks and nests was 5.1 km ± 0.8 (SE; $N = 7$ studies; Holloran and Anderson 2005; Connelly et al., this volume, chapter 3), and proportion of sagebrush within a radius of 6.4 km influenced Greater Sage-Grouse lek persistence in Wyoming (Walker et al. 2007a).

We also predicted that human disturbance at smaller scales than the endogenous sagebrush scale might negatively influence sage-grouse persistence. Human disturbances at scales smaller than the scale of landscape nestedness (4.5–9 km) have influenced sage-grouse habitat use and population trends. For example, coalbed natural-gas development lowered Greater Sage-Grouse lek persistence (Walker et al. 2007a) at 0.8- and 3.2-km scales and influenced wintering Greater Sage-Grouse sagebrush habitat use at the 1.1-km scale (Doherty et al. 2008). Moreover, from 1965 to 2007 the average rate of change was <1.0 in five of eight time intervals for Greater Sage-Grouse populations in the Columbia Basin SMZ (Garton et al., this volume, chapter 15), which had an additional scale of landscape nestedness between 1.8 and 3.6 km in the extensive-human-footprint-effect area scenario. The importance of scale of landscape nestedness in sagebrush ecosystems and its relation to sage-grouse population persistence should be further investigated given its potential to understand processes that organize sagebrush landscapes and influence sage-grouse. We suggest that scale of landscape nestedness should be determined by further partitioning the critical range of scales identified in this chapter into smaller intervals and evaluated outside of the SMZ framework in landscapes dominated by the low-intensity (e.g., Owyhee region in southwest Idaho) versus high-intensity (e.g., Columbia Basin SMZ) footprint to discern potential differences in scale of nestedness among regions.

Few studies have used lacunarity analyses to identify the scale of landscape pattern—the scale at which landscapes become self-organized or landscape patterns repeat themselves (Plotnick et al. 1993, Holling 1992). In tiger bush landscapes of Africa, an ecosystem characterized by alternating linear features of shrub (10–40 m wide) and bare ground (Wu et al. 2000), the scale of landscape pattern was well below 0.5 km and increased with degree of fragmentation. For sagebrush ecosystems, we found that the scale of landscape pattern was not reached even at a sampling scale of 36 km and that landscapes differed greatly in lacunarity at this scale. This suggests that sagebrush landscapes even at large scales (>36 km) are likely structured by small-scale patch dynamics and that processes influencing patch dynamics differ geographically. Clearly, our analysis needs to be repeated at scales exceeding 36 km to determine how sagebrush patch dynamics vary regionally. This is important because we do not understand how sage-grouse respond to patch dynamics and patch size (Connelly et al., this volume, chapter 4).

Influence of the human-footprint-effect area on sagebrush landscape patterns differed in relation to human-footprint intensity. Sagebrush landscapes were progressively more clumped with increasing human-footprint-effect area, that is, the ecological human footprint potentially converts suitable to unavailable sagebrush land cover. Our knowledge of how sagebrush landscape patterns influence ecological processes stems from studies that were based on land-cover spatial data sets, such as Sagestitch (Comer et al. 2002), LANDFIRE (LANDFIRE 2006), and Shrubmap (United States Geological Survey 2005), delineating only part of the physical human footprint limited to large-scale anthropogenic features and prominent linear features like interstate and state highways and point features like oil wells. Moreover, these studies did not incorporate the ecological human footprint. This is problematic given that recent sage-grouse studies show that during winter, Greater Sage-Grouse avoided suitable sagebrush habitat within 4 km^2 of natural-gas wells (Doherty et al. 2008), and during breeding, lek persistence was negatively influenced by the presence of gas wells within 0.8 and 3.2 km (Walker et al. 2007a). Additionally, power lines and fences cause direct mortality of sage-grouse (Beck et al. 2006; Connelly et al., this volume, chapter 3), but spatial data for these features are either underestimated in extent for power lines (M. Leu and S. E. Hanser, unpubl. data) or are not consistently mapped across the sage-grouse range. Therefore, studies that employ land-cover spatial data sets not adjusted for the physical and ecological human footprint are limited in identifying effects of the human footprint on sagebrush ecological processes and sage-grouse population, because any land cover mapped as sagebrush is assumed available.

CONSERVATION IMPLICATIONS

SMZs differed greatly in human-footprint intensity and ranked in descending order of influence as follows: Columbia Basin, Colorado Plateau, Wyoming Basin, Great Plains, Snake River Plain, Southern Great Basin, and Northern Great Basin. The ranking of SMZs will aid land managers in assigning importance to managing sage-grouse populations according to human-footprint influence. The Columbia Basin SMZ deserves special management consideration, given that Greater Sage-Grouse populations show low genetic diversity (Oyler-McCance et al. 2005b) and are declining (Schroeder et al. 2000; Garton et al., this volume, chapter 15). In this SMZ, the ecological human footprint is most influential, and therefore, the development of additional human features needs to be carefully planned to avoid further fragmentation of sage-grouse habitat. For example, when building additional power lines, which serve as perch sites from which raptors prey on sage-grouse (Connelly et al. 2000c), management agencies could restrict construction in sage-grouse habitats to areas already dominated by high-intensity human footprint. In addition, lek connectivity is lowest (Knick and Hanser, this volume, chapter 16) in the Columbia Basin SMZ, suggesting restoration of sagebrush habitat is needed to increase connectivity among leks (Schroeder et al. 2000). A significant proportion of agricultural land in the Columbia Basin SMZ is in Conservation Reserve Program land (Vander Haegen et al. 2001) and may benefit Greater Sage-Grouse (Schroeder et al. 2000). Similarly, the Mono Lake region of the Northern Great Basin SMZ also needs careful planning when adding additional human features to landscapes used by Greater Sage-Grouse. For the area overlapping with the Mono Lake Greater Sage-Grouse subpopulation, a population genetically distinct from the rest of Greater Sage-Grouse populations (Oyler-McCance et al. 2005b), proportion of low-intensity human footprint is lower, whereas proportion of intermediate human footprint is higher compared to the entire Northern Great Basin SMZ (Leu and Hanser, unpubl. data). The influence of the human footprint in the Wyoming Basin is important because this SMZ has one of the few remaining clusters of sagebrush landscapes, contains the most highly connected network of sage-grouse leks (Knick and Hanser, this volume, chapter 16), has the highest average lek size and largest average number of leks censused of all SMZs since 2000 (Connelly et al. 2004; Garton et al., this volume, chapter 15), and is predicted to be one of the last strongholds of sagebrush in the face of climate change (Neilson et al. 2005). We likely underestimated the ecological human-footprint intensity in the Wyoming Basin SMZ because the distribution of oil and gas wells used in our analyses represented a snapshot from 1994 (Leu et al. 2008); however, since 1994 the number of oil and gas wells in the Wyoming Basin has increased by 50.2% (S. E. Hanser, unpubl. data). Whether this increase occurred on sage-grouse habitat needs to be determined.

The human footprint can influence sage-grouse population regulation via top-down and/or bottom-up regulatory processes. Top-down human footprint effects increase the spread of synanthropic predators into areas in which they do not occur or are present only at low densities in the absence of human features (Restani et al. 2001). For sage-grouse, synanthropic predators include Common Ravens (*Corvus corax*). Common Ravens are sage-grouse nest predators (Schroeder et al. 1999, Coates et al. 2008) and removal of ravens may increase nest success (Batterson and Morse 1948), although the findings of this research are questionable (Coates 2008; Hagen, this volume, chapter 6). Regardless, culling Common Ravens is regularly implemented as a management strategy to increase Greater Sage-Grouse productivity in the state of Nevada (Coates et al. 2008). No recently published work has shown that raven removal results in increased sage-grouse populations, and several lines of evidence suggest that controlling Common Raven populations is at best a short-term fix. Although between 1966 and 2007 the western Common Raven population was increasing, populations in the Basin and Range, Columbia Plateau, and Great Basin ecoregions were stable during this period (Sauer et al. 2008). Therefore, in SMZs with stable Common Raven populations, culling of this species may have little effect on sage-grouse demography given that they have coevolved with this predator (Connelly 2000). Culling of territorial Common Ravens will result in reoccupation of vacant territories by floaters (Coates et al. 2007), whereas culling of individuals at resource subsidies, such as landfills and dairy operations, will kill individuals that only incidentally prey on wildlife (Webb et al. 2009). We argue that Common Raven predation effects occur locally and that culling programs are ineffective and not addressing the real problem: preventing further loss of sage-grouse habitat across the annual cycle of this species due to bottom-up human-footprint effects.

Human footprint bottom-up effects occur directly via habitat loss or indirectly via the conversion of suitable sagebrush habitat to habitat that is unsuitable or unavailable to sage-grouse. We argue that bottom-up effects may have large-scale implications on sage-grouse populations because Greater Sage-Grouse have low tolerance to human disturbance such as roads (Lyon and Anderson 2003, Holloran and Anderson 2005, Aldridge and Boyce 2007), oil and gas development (Braun et al. 2002, Lyon and Anderson 2003, Holloran and Anderson 2005, Aldridge and Boyce 2007, Walker et al. 2007a, Doherty et al. 2008), and exurban development (Aldridge et al. 2008) especially during the breeding season. The human footprint is most intense at low elevation near valley floors (Leu et al. 2008) and may have a disproportionate effect on sage-grouse populations that depend on low- to mid-elevation habitat. With projected increases in the human footprint along valley floors, careful planning is needed to avoid increasing the human footprint in sage-grouse habitat.

Future sage-grouse studies should focus on how potential synergistic effects of the human footprint affect sage-grouse demography and habitat use (Walker et al. 2007a, Doherty et al. 2008). We recommend that sage-grouse study sites should be selected a priori along the entire gradient of human disturbance, sagebrush cover, and ecological productivity (as measured by Normalized Difference Vegetation Index; Leu et al. 2008) and not focused primarily on the best habitat. Determining the effect area of various human features and actions is paramount to sage-grouse conservation and will significantly increase the predictive capabilities of sage-grouse human-footprint models. In turn, such models will benefit land managers when developing future management plans, identifying areas of conservation need, delineating areas for habitat restoration, and evaluating conservation success by comparing temporal changes in human actions in sage-grouse habitat (Haines et al. 2008).

Multiple studies have recommended that sage-grouse populations should be managed at large spatial scales (Connelly et al. 2000c, Connelly et al. 2004, Holloran and Anderson 2005, Walker et al. 2007a, Aldridge and Boyce 2007). Our study suggested an endogenous sagebrush landscape scale between 4.5 and 9.0 km to which sage-grouse dispersal and movement patterns have adapted. Growing evidence suggests that sage-grouse respond to large scales that exceed currently applied management scales of no disturbance within 0.5 km around leks and within 3.2 km around leks during the breeding season (15 March–15 June) in areas with coalbed natural gas extractions (Walker et al. 2007a, Doherty et al. 2008). Therefore, changes could be made to existing management strategies to better reflect sage-grouse movement and dispersal patterns.

A common strategy to assess human land use, land-cover composition and configuration, and management of sage-grouse populations is to use concentric zones of increasing size centered on sage-grouse leks (Holloran and Anderson 2005, Walker et al. 2007a) or sage-grouse sightings (Doherty et al. 2008). However, zones delineated around sage-grouse leks may not be relevant because proportionally more nonimportant habitat may be included in the managed area with increasing distance of radii (Connelly et al. 2000c). Spatially explicit models that delineate sage-grouse source and sink habitats (Aldridge and Boyce 2007) might ultimately be more appropriate and should include ecological human-footprint models delineating suitable sagebrush habitat avoided by sage-grouse. To develop human-footprint models specific to sage-grouse, more effort needs to be undertaken to accurately assess the distribution of human resources throughout the range of sage-grouse.

ACKNOWLEDGMENTS

We thank S. T. Knick, J. W. Connelly, C. D. Marti, and two anonymous reviewers for thoughtful inputs on earlier drafts of this paper. Any use of trade, product, or firm names in this publication does not imply endorsement by the United States government.

CHAPTER FOURTEEN

Influences of Free-Roaming Equids on Sagebrush Ecosystems, with a Focus on Greater Sage-Grouse

Erik A. Beever and Cameron L. Aldridge

Abstract. Free-roaming equids (horses [*Equus caballus*] and burros [*E. asinus*]) in the United States were introduced to North America at the end of the 15th century, and have unique management status among ungulates. Legislation demands that these animals are neither hunted nor actively managed with fences and rotation among pastures, but instead constitute an integral part of the natural system of the public lands. Past research has elaborated that free-roaming horses can exert notable direct influences in sagebrush (*Artemisia* spp.) communities on structure and composition of vegetation and soils, as well as indirect influences on numerous animal groups whose abundance collectively may indicate the ecological integrity of such communities. Alterations to vegetation attributes and invertebrates can most directly affect fitness of Greater Sage-Grouse (*Centrocercus urophasianus*) and other sagebrush-obligate species; alterations of soils and other ecosystem properties may also indirectly affect these species. Across 3.03 million ha of the western Great Basin, horse-occupied sites exhibited lower grass, shrub, and overall plant cover; higher cover of unpalatable forbs and abundance of cheatgrass; 2.2–10.0 times lower densities of ant mounds; and 2.9–17.4 times greater penetration resistance in soil surfaces, compared to sites from which horses had been removed for 10–14 years. As is true for all herbivores, equid effects on ecosystems vary markedly with elevation, stocking density, and season and duration of use. However, they may be especially pronounced in periods of drought, which are forecasted to occur with increasing frequency in the southwestern United States under climate change, and when they interact synergistically with livestock-grazing effects. Equids' use of sagebrush landscapes will have different ecological consequences than will livestock grazing, at both local and landscape scales. Spatially, the addition of horses to sagebrush landscapes means more of the landscape receives use by nonnative grazers than if domestic cattle alone were present. In spite of recent advances in ecological understanding of equid synecology, much remains to be learned. Life-history characteristics of Greater Sage-Grouse and other sagebrush-obligate species suggest the great value in evaluating equid effects more broadly than through a horses-vs.-livestock perspective, and in monitoring ecosystem components such as soil-surface hardness

Beever, E. A., and C. L. Aldridge. 2011. Influences of free-roaming equids on sagebrush ecosystems, with a focus on Greater Sage-Grouse. Pp. 273–290 *in* S. T. Knick and J. W. Connelly (editors). Greater Sage-Grouse: ecology and conservation of a landscape species and its habitats. Studies in Avian Biology (vol. 38), University of California Press, Berkeley, CA.

and ant-mound density that have ecological and management relevance yet for which data are relatively inexpensive to collect.

Key Words: burros, *Centrocercus urophasianus,* direct and indirect effects, *Equus asinus, Equus caballus,* grazing, horses, sagebrush, sage-grouse.

Efecto de los Équidos de Vida Libre en los Ecosistemas de *Artemisia* spp., con Énfasis en el Greater Sage-Grouse

Resumen. Équidos de movimiento libre en los Estados Unidos (caballos [*Equus caballus*] y burros [*E. asinus*]) fueron introducidos a Norteamérica a fines del siglo XVI, y ellos han tenido un manejo único entre los ungulados. La legislación requiere que estos animales no sean cazados, ni manejados con cercas, ni tampoco dejados en un tipo de pastoreo rotacional pero en vez éstos constituyen una parte esencial del sistema natural de tierras públicas. Investigaciones previas han demostrado que estos caballos de movimiento libre pueden tener influencias directas sobre la estructura y composición de los suelos y vegetación en comunidades dominadas por *Artemisia* spp., así como también influencias indirectas en varios grupos de animales en donde su abundancia colectiva puede indicar la integridad ecológica de estas comunidades. Alteraciones a las características de la vegetación e invertebrados pueden afectar en forma directa al éxito reproductivo del Greater Sage-Grouse (*Centrocercus urophasianus*) y de otras especies obligadamente relacionadas a comunidades de *Artemisia* spp. También alteraciones a los suelos y a otras propiedades del ecosistema pueden afectar indirectamente a estas especies. Dentro de 3,030,00 hectáreas de la Gran Cuenca Occidental, áreas ocupadas por caballos han exhibido una menor cobertura de gramíneas, arbustos, y vegetación en general; una mayor cobertura de hierbas no comestibles y mayor abundancia de la gramínea exótica *Bromus tectorum*; una reducción de entre 2.2 a 10 veces en la densidad de los nidos de hormigas en montículo; y un aumento de entre 2.9 y 17.4 veces en los niveles de resistencia a la penetración de la superficie en los suelos, todo esto si se compara con áreas en donde caballos han sido removidos por períodos de entre 10 a 14 años. Así como sucede con todos los herbívoros, los efectos de los équidos en ecosistemas varían de acuerdo a la densidad del ganado, duración y estación de pastoreo, y la elevación en el sitio. Sin embargo, estos efectos pueden ser especialmente dramáticos durante períodos de sequía, los cuales son previstos que ocurrirán en una mayor frecuencia en el suroeste de los Estados Unidos debido al cambio climático, y más aun cuando éstos interaccionan sinergísticamente con los efectos del pastoreo de la ganadería. El uso del paisaje en comunidades de *Artemisia* spp. por équidos tendrá consecuencias ecológicas muy distintas a las del pastoreo por medio del uso de ganado tanto a nivel local y del paisaje. En términos espaciales, la introducción de caballos a estos ecosistemas dominados por *Artemisia* significa que una mayor proporción de este ecosistema sea usado por herbívoros no nativos si este mismo fuese únicamente usado por el ganado doméstico. A pesar de los avances recientes en el entendimiento ecológico de la sinecología de équidos, todavía queda mucho que aprender. Las características naturales del *C. urophasianus* y de otras especies en estas comunidades sugieren que pueden ser de gran valor: 1) la evaluación del efecto de los caballos y que ésta debe ser tomada en forma más amplia que la simple perspectiva caballos vs. ganadería, y 2) el monitoreo de componentes del ecosistema tales como la dureza en la superficie de los suelos y la densidad de los nidos de hormigas en montículo, los cuales tienen relevancia ecológica y de manejo ya que la colección de datos es relativamente de bajo costo y factible de implementar en el campo.

Palabras Clave: Artemisia, burros, caballos, *Centrocercus urophasianus,* efectos directos e indirectos, *Equus asinus, Equus caballus,* pastoreo, sage-grouse.

Sagebrush (*Artemisia* spp.) habitats have undergone significant change during the last century because of conversion, degradation, and fragmentation (Knick et al. 2003, Connelly et al. 2004). Remaining sagebrush habitats continue to be threatened by direct loss of habitat due to agricultural conversion (West and Young 2000, Connelly et al. 2004), degradation or loss from energy exploration and extraction (Braun et al. 2002, Knick et al. 2003, Lyon and Anderson 2003, Aldridge and Boyce 2007), invasions of exotic plants (Knick et al. 2003, Connelly et al. 2004), intensive grazing practices (Beck and Mitchell 2000, Hayes and Holl 2003, Crawford et al. 2004), fire (Connelly et al. 2000c, 2004), and climate change (Neilson et al. 2005). Grazing is often implicated in reducing sagebrush habitat quality (Beck and Mitchell 2000, Hayes and Holl 2003, Crawford et al. 2004) resulting in consequences for sagebrush-obligate species; however, these effects are rarely addressed quantitatively. This is due to the difficulties in accounting for legacy effects related to past management (Beever et al. 2006, Rowe 2007) and developing defined experimental treatments integrated with management actions that allow for testing of hypotheses within an adaptive-management framework (Aldridge et al. 2004).

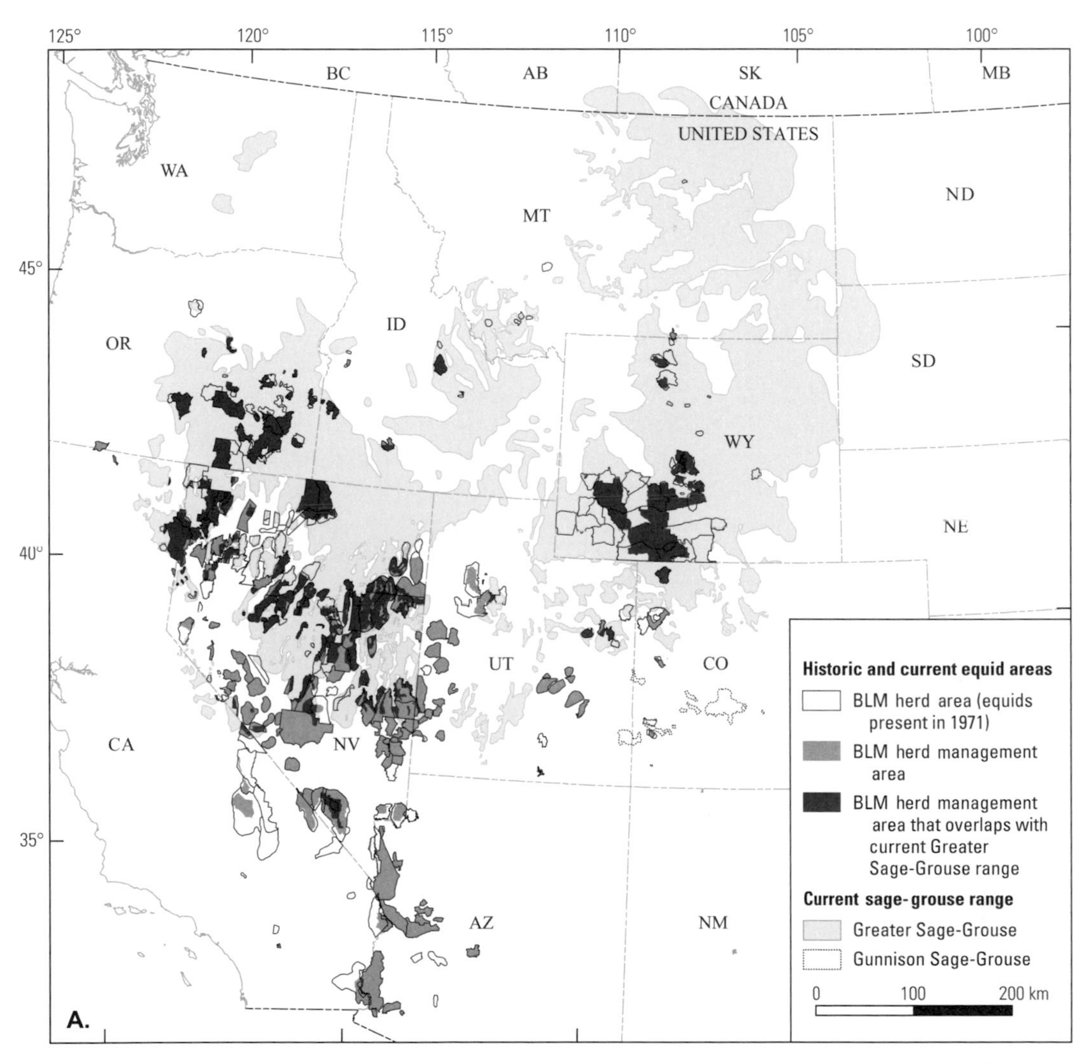

Figure 14.1. A) Distribution of Bureau of Land Management (BLM) free-roaming equid herd areas (equids historically present) and herd management areas (equids currently present), as well as B) U.S. Forest Service (USFS) wild horse and burro territories (equids currently present) across western North America. The current distributions (Schroeder et al. 2004) for both Greater Sage-Grouse and Gunnison Sage-Grouse are shown to illustrate the spatial overlap of equid landscape use with sage-grouse. Spatial data for equid ranges were compiled from state and local BLM and USFS offices.

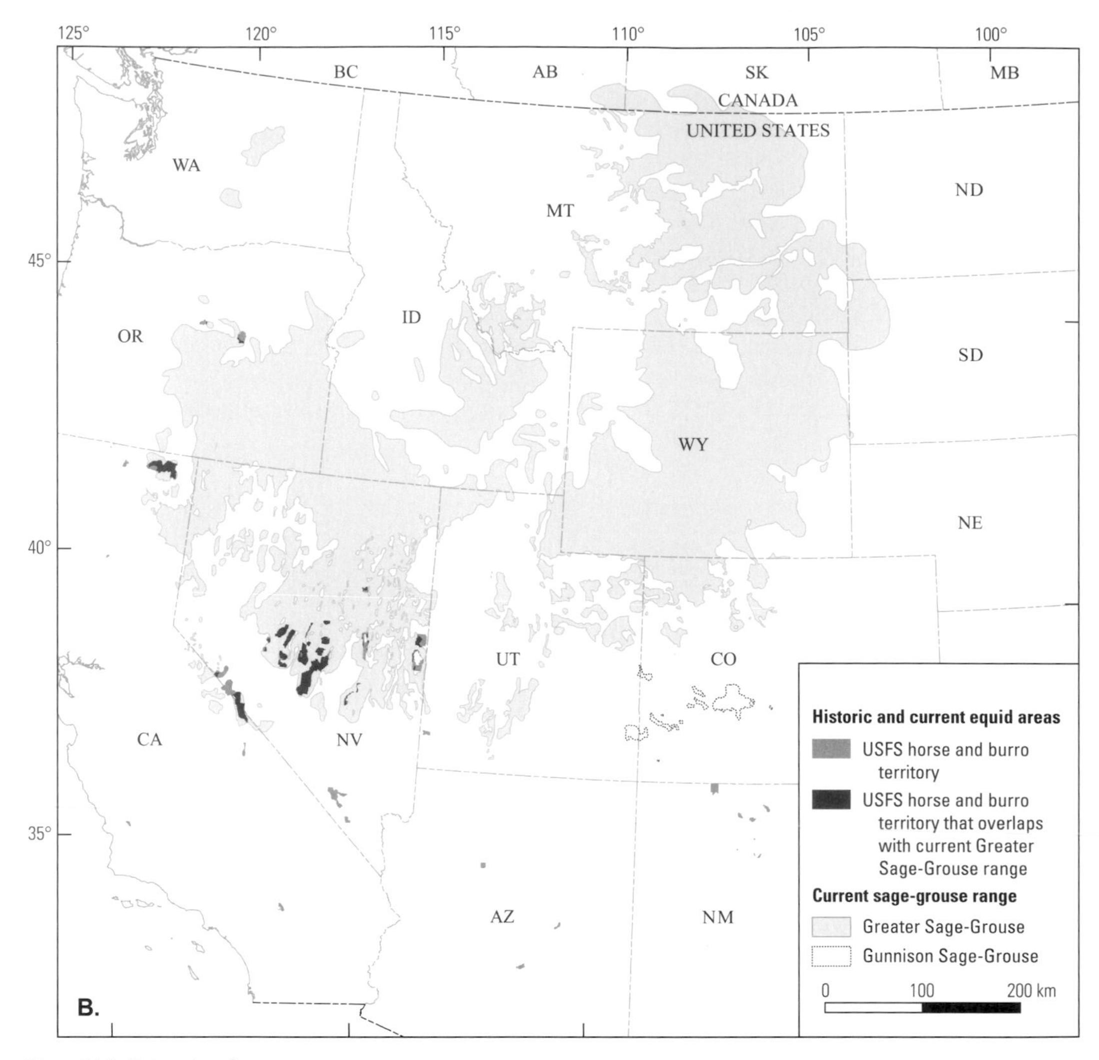

Figure 14.1. B. (*continued*)

Furthermore, relatively little consideration has been given to understanding links between the effects that free-roaming equids (hereafter, equids refer to horses [*Equus caballus*] and burros [*E. asinus*] in western North America; see distribution in Fig. 14.1) may have on sagebrush ecosystems and sagebrush-obligate species that these systems support.

Many wildlife species have been negatively affected by changes to sagebrush ecosystems. For example, many sagebrush-obligate birds have experienced population declines and range contractions over the past 40 years (Sauer et al. 2003, Connelly et al. 2004). Most monitored populations of Greater Sage-Grouse (*Centrocercus urophasianus*), for instance, have declined by ~2% or more per year since 1965 (Connelly and Braun 1997, Braun 1998, Connelly et al. 2004). Greater Sage-Grouse use a diversity of habitats across life stages and encompass annual home ranges as large as 2,975 km^2 (Connelly et al. 2004). Large tracts of contiguous sagebrush habitat are required for their persistence (Aldridge et al. 2008), although heterogeneity within those patches may also be important (Aldridge and Boyce 2007).

A larger body of knowledge exists on Greater Sage-Grouse habitat requirements at local scales (Connelly et al. 2000c, Crawford et al. 2004, Hagen et al. 2007) compared to landscape-scale understanding (Connelly et al., this volume, chapter 4). Greater Sage-Grouse require habitat patches with extensive cover of sagebrush during breeding and nesting (15–25% cover), brood-rearing (10–25%), and winter (10–30%) (Connelly et al. 2000c, Hagen et al. 2007). In addition to the vegetation structure that Greater Sage-Grouse require from woody plants, grass cover is important for nest concealment and as escape cover, and increased grass cover results in increased nest success (Connelly et al. 2000c, Crawford et al. 2004, Aldridge 2005, Hagen et al. 2007). Forb-rich mesic areas constitute important habitats during summer and brood-rearing; these areas provide forbs and insects as food resources that are required for chick survival (Johnson and Boyce 1990, Drut et al. 1994b, Sveum et al. 1998a). Greater Sage-Grouse exhibit strong selection for greater cover (i.e., 14–40%) and greater diversity of forbs (Klebenow and Gray 1968, Peterson 1970, Drut et al. 1994a, Sveum et al. 1998a).

Free-roaming equids within these sagebrush ecosystems, as is true for all large-bodied herbivores, can alter ecosystem components directly through any of several processes, including selective plant consumption, trampling of plants and surface soil horizons, and spatial redistribution of nutrients via ingestion and subsequent excretion (Fig. 14.2; Beever et al. 2003, Beever and Herrick 2006). These

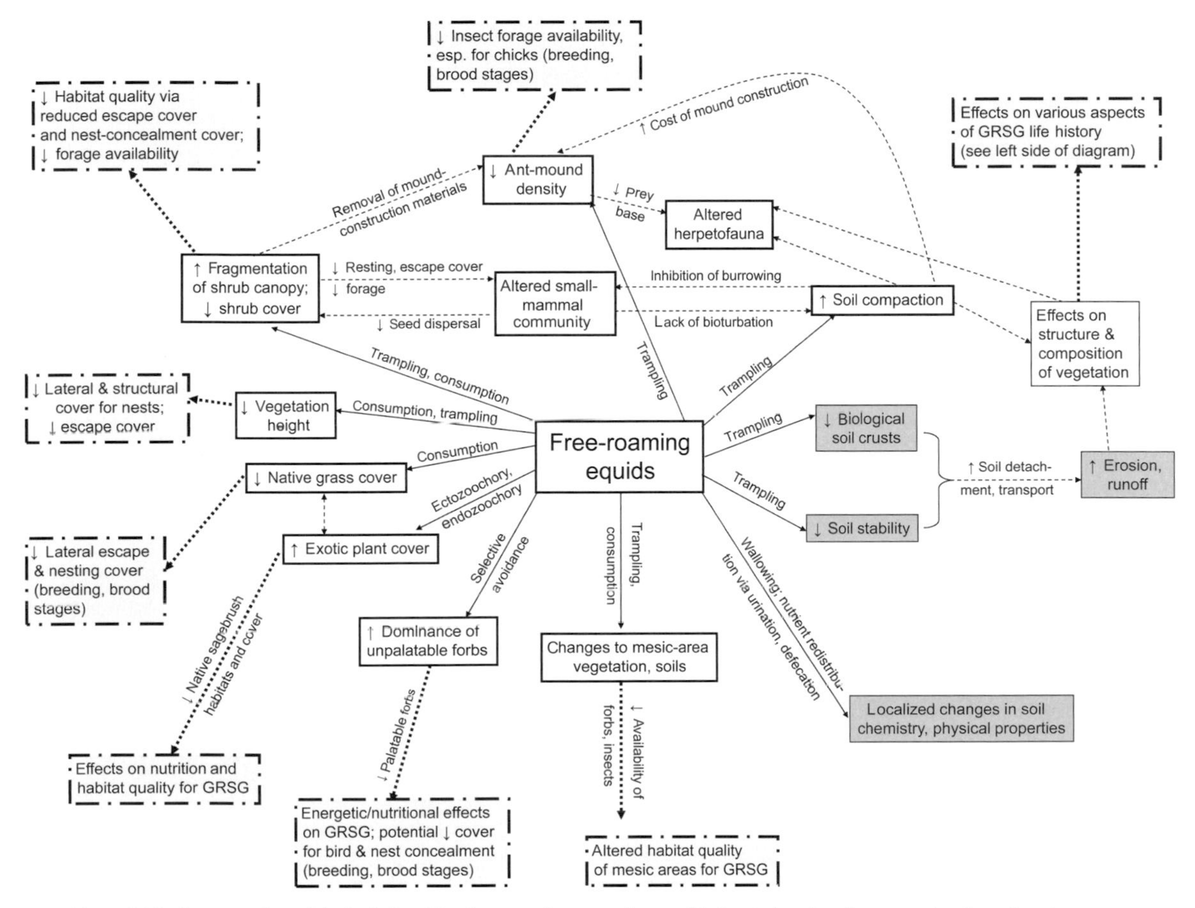

Figure 14.2. Conceptual model of relationships between free-roaming equids (center) and various aspects of sagebrush communities (inner ring of boxes) and, in turn, between those aspects and life-history characteristics of Greater Sage-Grouse (GRSG) (outer ring of boxes: dash-dot border). Direct effects of equids are denoted by solid lines, whereas potential ecological cascades and indirect effects are denoted with dashed lines. Unless otherwise noted, alterations to habitat may affect Greater Sage-Grouse fitness during multiple important life stages, including at winter, brood-rearing, and breeding and nesting habitats. Thick inner boxes denote that horse-occupied sites have been found to differ significantly from horse-removed sites for these aspects of ecosystems (see text). Labels along arrows describe the likely mechanisms underlying these relationships. Arrows inside the boxes can be interpreted as increased or decreased. Equid influences likely interact with other drivers in their relationships with ecosystem components; such drivers include rapid climate change, wildfire and prescribed fire, western juniper dominance, off-highway vehicle use, native and domestic ungulates, other herbivores, and pests and disease invasions. For model clarity, potentially beneficial effects of free-roaming equids (e.g., reduction of site vulnerability to catastrophic fires due to brief removal of fine fuels, increase of pore spaces via chipping of soils at low equid densities, and creation of germination sites for new plants) were not included graphically. Note that whereas the relationships of equids to ecosystem components and the relationships of Greater Sage-Grouse to various habitat conditions have both been documented in some systems, the entire link of equids to Greater Sage-Grouse has not been demonstrated experimentally to date.

processes all occur at local through landscape scales, and thus can have relevance for life processes of Greater Sage-Grouse and other sagebrush-obligate species. Equids' rubbing against plant parts and creation of wallows are two examples of these disturbance processes at local scales. In addition to exerting these direct effects, equids can also affect sagebrush-inhabiting vertebrates indirectly through their influences on structure, composition, and chemistry of vegetation, as well as on the structure and local chemistry of soils (Beever and Herrick 2006, Beever et al. 2008). In turn, effects on vegetation can affect its value as hiding cover for terrestrial species. Collectively, these influences on both vegetation and soils can alter the competitive interactions among plants, alter soil-erosion patterns, create niches for colonizing or ruderal (early-successional, often disturbance-tolerant) species, and create feedback loops (Hobbs 1996, Beever and Herrick 2006) that could ultimately reduce the functionality of habitats for sagebrush-obligate species such as sage-grouse (includes Greater Sage-Grouse and Gunnison Sage-Grouse [*Centrocercus minimus*]).

Greater Sage-Grouse currently occupy ~668,412 km^2 of habitat across 11 states and two Canadian provinces, roughly 56% of their historic pre-settlement range in North America (Schroeder et al. 2004: Fig. 1). Based on extents of Bureau of Land Management (BLM) wild horse and burro herd areas (HAs) and herd management areas (HMAs), and United States Forest Service (USFS) wild horse and burro territories, we estimate that during at least some of the years since the establishment of the Wild Free-Roaming Horses and Burros Act of 1971, these equids have occupied 366,690 km^2 of semiarid ecosystems in western North America (Table 14.1). Areas that have been managed for horses and/or burros from 1971 to 2007 constitute ~18% (119,703 km^2) of the currently occupied Greater Sage-Grouse range. This estimate excludes dispersal and extra-limital movements by equids (i.e., outside of management areas), which are difficult to quantify but may be widespread; considering these would appreciably increase the percentage of Greater Sage-Grouse habitats affected by equid grazing. About 12% (78,380 km^2) of the current range of Greater Sage-Grouse is now managed for free-roaming equids (Table 14.1). Thus, there may be unmeasured consequences for a significant portion of sage-grouse habitat throughout the species' range because of the aforementioned ways in which free-roaming equids can directly or indirectly impact sagebrush habitats.

Our goal is to critically consider the relationship between free-roaming equids and sage-grouse, focusing on Greater Sage-Grouse. Our objectives were to (1) synthesize the alterations to sagebrush and other communities that equids have been shown to exert in past research and (2) interpret these findings in view of the current body of knowledge of life-history and conservation requirements of Greater Sage-Grouse, as they pertain to continued persistence of the species.

A HISTORICAL REVIEW OF EQUID DYNAMICS IN NORTH AMERICA

Understanding relationships between free-roaming equids and sagebrush-obligate species such as sage-grouse requires a review over evolutionary and historical time. During the Eocene, equids (specifically, *Hyracotherium*) arose and began to diversify in North America (Simpson 1951, Zarn et al. 1977). Wild horses have occupied a variety of habitats worldwide throughout their evolutionary existence, but have generally favored cooler, more mesic habitats (e.g., grassland, open forest, and tundra), in contrast to the dry-steppe and desert habitats of burros and zebras such as *Equus burchellii* (Bennett 1992). After their long history in North America, however, horses were extirpated from the continent between 10,500 and >13,200 years ago, hypothesized to have resulted from overhunting by primitive peoples, climatic influences, or both (Martin 1984, Bennett and Hoffmann 1999). Equids were absent from North America until the end of the 15th century, when horses and burros were introduced by Spanish conquistadors into what is now the southwestern United States. Free-roaming horses spread rapidly across the Intermountain West, and populations reportedly peaked in the United States at 2,000,000–7,000,000 animals in the late 1700s to early 1800s (Ryden 1978, Thomas 1979). However, competing interests dictated that horses were captured in increasing numbers, a process that was accelerated by passage of the Taylor Grazing Act of 1934. These interests included using horses as saddle animals, breeding stock, or both; selling horses entrepreneurially for profit; and reducing suspected competition with livestock grazing. By 1934, numbers of free-roaming horses were reduced to ~150,000 animals (Thomas 1979).

TABLE 14.1

Area (km^2) affected by free-roaming equids across western United States, by individual states overlapping the current range of Greater Sage-Grouse.

				Area historically occupied by free-roaming equids[e]		Current area occupied by free-roaming equids	
State[a]	BLM herd area[b]	BLM herd management area[c, d]	USFS wild horse and burro territories[d]	Within western North America	Within Greater Sage-Grouse current range	Within western North America	Within Greater Sage-Grouse current range
Arizona	14,722	12,480	303	27,505	—	12,783	—
California	28,184	9,787	2,640	40,611	8,720	12,427	6,981
Colorado	2,973	1,638	—	4,611	816	1,638	634
Idaho	1,933	1,692	—	3,625	1,594	1,692	1,440
Montana[f]	905	150	18	1,073	362	168	8
New Mexico	512	116	641	1,269	—	757	—
Nevada	1,788	70,716	6,881	169,384	48,281	77,596	38,229
Oregon[f]	17,584	11,991	414	29,990	16,904	12,405	11,775
Utah	15,885	11,482	133	27,501	1,870	11,616	1,023
Wyoming	41,795	19,326	—	61,121	41,156	19,326	19,290
Totals	216,282	139,378	11,029	366,690	119,703	150,408	78,380

SOURCE: Spatial data for equid ranges were compiled from a variety of state and local BLM and USFS offices; current area and equid numbers represent data as of 2007.

[a] Indicates state that administers the federal land (not all land area reported is within the boundary of that state).

[b] BLM wild herd areas represent areas historically occupied by free-roaming equids, but that are not currently managed to contain horses, unless currently contained within a BLM herd management area.

[c] BLM herd management areas are contained within BLM herd areas.

[d] BLM herd management areas and USFS wild horse and burro territories both historically were, and currently are, occupied by free-roaming equids.

[e] During at least some period between 1971–2007.

[f] Equid areas managed by the BLM within South Dakota and North Dakota are administered by Montana; areas in the state of Washington are administered by Oregon. Numbers are inclusive within those states.

Given burros' value to both Native Americans and Euro-Americans as work animals and as food after their introduction to western North America, burros probably did not become feral until the 19th century, likely the result of their escaping from loose confinement or being deliberately released and not rounded up (McKnight 1958, Carothers et al. 1976, Wagner 1983). Based on mail-survey and field estimates, McKnight (1958) reported that feral burros occurred in 10 western states in 1957. Burro movements overlap sage-grouse habitats in multiple areas across the southwestern United States, and although the overlap is less extensive than is the overlap with horse habitats, burros tend to spend more time in lower-elevation habitats, as do sage-grouse. Free-roaming burros are descended from the African wild ass (*Equus africanus*) native to northeastern Africa, and are more abundant in the warmer deserts south of the sagebrush distribution. Historically, horses have outnumbered burros in the historic range of Greater Sage-Grouse by 50 to 1 to well over 80 to 1. In addition, much of the scientific literature on burros in the western United States (1) comes from study areas outside the historic range of Greater Sage-Grouse; (2) focuses on potential competition between burros and bighorn sheep (*Ovis canadensis*), and only indirectly considers effects of burros on other ecosystem components; and (3) occurs in proceedings and other gray literature not subject to the same level of scrutiny as other peer-reviewed sources (McKnight 1958, Carothers et al. 1976, Wagner 1983). We thus focus most of the remaining discussion on free-roaming horses and direct comparatively little attention to burros.

Legislative and Legal Mandates That Affect the Equid-Sagebrush Ecosystem Relationship

The legislation that established management policies for equids in western North America (Wild Free-Roaming Horses and Burros Act of 1971) deemed free-roaming equids an integral part of the natural system of public lands. Horse numbers rose sharply after enactment of the 1971 act that afforded horses and burros strong protection on BLM and USFS lands (Wagner 1983, United States Departments of Agriculture and the Interior 1997c). In the absence of any widespread, effective predator (but see Turner and Morrison 2001, Jeffress and Roush 2010 for a localized exception—mountain lions [*Felis concolor*]) and removal of any population constraint due to aridity because of human development of remote water sources such as guzzlers, humans became the strongest force regulating size of horse populations (Tables 14.2, 14.3). The Federal Land Policy and Management Act of 1976 and the Public Rangelands Improvement Act (PRIA) of 1978 amended the 1971 act and required record

TABLE 14.2

Criteria used by wild horse managers to set population goals reported in survey questionnaire from six states, administered in 1999.

Managers may have used more than one technique in Herd Management Areas (HMAs).

Technique used	No. of HMAs using technique	% of HMAs using technique
Grazing levels	31	55
Forage consumption rates	28	50
AUMs	16	29
Birth rates	3	5
Trend studies	3	5
Court order	2	4
Multiple-use decision (MUD)	2	4
Water availability	1	2
Genetics/N_e	1	2

TABLE 14.3

Removal strategies used by managers in six states to maintain population numbers of wild horses, reported in survey questionnaires in 1999.

Some Herd Management Areas (HMAs) use more than one strategy.

Removal strategy	No. of HMAs using technique	% of HMAs using technique
Selective by age	48	86
Selective by phenotype	13	23
Emergency removal due to drought, mortality, or both	6	11
Entire harems removed	2	4
Contraception	2	4
No removals conducted	2	4
Selective by breeding needs	1	2
Nuisance animals only	1	2
Ranchers remove without legal authorization	1	2

SOURCE: K. A. Schoenecker (unpubl. data).

keeping of equid population sizes and distribution; required removal of excess animals from public lands once excess was identified, as well as disposition of animals through adoption; and allowed use of helicopters for capture and removals (Boyles 1986). The Public Rangelands Improvement Act also required research on the interrelationships with wildlife, water, and forage resources, and analysis of what constitutes "excess" animals. Currently, free-roaming equids have a unique management status in North America (Beever 2003). They must be managed according to the 1971 act at the minimal feasible level; thus, many free-roaming herds of horses and burros are neither fenced nor rotated among grazing areas. The act also stipulates that, in contrast to other wild ungulates, hunting of horses or burros is not permitted. Free-roaming horses on other jurisdictions—national wildlife refuges and national parks—are not governed by the same legislation as are lands administered by the BLM or the USFS. Collectively, these evolutionary, cultural, and management histories have set the stage for critically considering the current relationship of equids to Greater Sage-Grouse and other wildlife species dependent on sagebrush habitats.

REVIEW OF PAST RESEARCH

Overview

Influences of large-bodied herbivores on sage-grouse and other sagebrush obligates can occur directly, indirectly, and via feedback loops (Fig. 14.2). Equids can directly affect these by disturbance of lekking behavior and trampling of nests or altricial young, but these interactions have not been investigated. However, direct interactions have been anecdotally observed in the field for other native and nonnative herbivores, such as nest trampling and nest abandonment due to cattle activity <30 cm from the nest (C. L. Aldridge, unpubl. data). We thus discuss studies of equids' effects on each of several components of sagebrush ecosystems and relate those effects to requirements of sagebrush-obligate species, focusing on Greater Sage-Grouse. The magnitude of grazing impacts will vary widely across the landscape, from areas minimally affected to areas used heavily (Coughenour 1999, Beever 2003), because of the nonuniform patterns of habitat use by horses and other ungulates (Hobbs 1996). However, the potential exists for these impacts to have consequences for Greater Sage-Grouse populations range-wide, given the broad distribution of equids within the range of Greater Sage-Grouse (Fig. 14.1).

Possible Effects of Equids on Plants Within Sagebrush Ecosystems

Horse diets may contain >95% grasses in some seasons (Hubbard and Hansen 1976, Hanley and Hanley 1982, McInnis 1985). It is thus not surprising that across nine mountain ranges of the western Great Basin, horse-removed sites averaged 1.9–2.9 times more grass cover and significantly higher grass density than did horse-occupied sites in years that received 104–120% of the long-term average of precipitation (Beever et al. 2003, 2008). As was true for most other vegetative variables in that study, the magnitude of difference in grass cover and density was most pronounced on the more-productive and species-rich high-elevation sites (Beever et al. 2008). Grasses were similarly heavily used in Utah pastures grazed moderately by domestic horses, regardless of their phenology and thus fiber content (Reiner and Urness 1982).

Grass cover is important for Greater Sage-Grouse throughout the year (Crawford et al. 2004, Hagen et al. 2007). Grass provides escape cover and is particularly important for concealing nests, which are often in heterogeneous sagebrush habitats (Aldridge and Boyce 2007) that have greater residual and live cover of tall grasses (Sveum et al. 1998b, Connelly et al. 2004, Crawford et al. 2004, Hagen et al. 2007). Thus, removal of tall-grass cover in horse-occupied sites could negatively impact quality of habitat for Greater Sage-Grouse (Fig. 14.2).

Sites where horses were removed across the western Great Basin exhibited 1.1–1.9 times greater shrub cover than did horse-occupied sites ($P > 0.027$), although stem density of shrubs did not differ between the two types of sites ($P > 0.60$) (Beever et al. 2003, 2008). In particular, cover of sagebrush (*Artemisia* spp.) among high-elevation sites was 1.47–1.57 times higher at horse-removed than horse-occupied sites, but the difference among all sites was statistically significant ($P < 0.05$) in only the wetter of two years (Beever et al. 2008). This trend in sagebrush cover occurred in spite of a lack of difference in the stem count of sagebrush between the two types of sites (Beever et al. 2008). Horse-occupied sites also had a more fragmented shrub canopy, thought to be due to the combined effects of trampling, rubbing, and consumption by horses. Specifically, randomly placed 50-m transects within horse-removed sites had significantly longer maximum intercepts of shrubs than did similar transects within horse-occupied sites (Beever et al. 2008). The relatively minor contribution of consumption to reductions in shrub cover was suggested by two studies from the Great Basin: (1) a study of the effects of domestic horses at different grazing intensities in northern Utah, which found that utilization of shrubs was usually ≤4.2% (Reiner and Urness 1982), and (2) a study of food-resource partitioning among five sympatric ungulates in northeastern California and northwestern Nevada, which found that horse diets included an average of only 5.0% (range 1.0–11.8%) of browse—shrubs and trees combined—across seasons (Hanley and Hanley 1982).

Loss of sagebrush cover can directly affect food resources for Greater Sage-Grouse, given that they consume sagebrush throughout the year (Patterson 1952; Connelly et al. 2000c, 2004). In addition, loss of sagebrush and other shrubs could directly reduce the quantity and quality of nesting and winter habitat. Reduction in escape cover can result in increased predation pressure on both nests and birds (Connelly et al. 1991, Schroeder and Baydack 2001), which could be exacerbated in the presence of free-roaming equids.

Domestic horses in heavy-grazing regimes in Utah pastures primarily consumed grasses, but consumption of grass was increasingly replaced by forb consumption as duration of grazing lengthened. However, rates of forage removal per animal were lower than in moderately grazed pastures. The authors hypothesized that this diet-switching may have been due to the increased search time for preferred species, lower intake, or more frequent animal interactions (Reiner and Urness 1982). Horse diets in the northwestern portion of the Great Basin contained an average of only 5.5% forbs across seasons (range 2.1–9.4%; Hanley and Hanley 1982).

Sites across the western Great Basin occupied by free-roaming horses at higher elevations possessed greater cover and abundance of grazing-tolerant and unpalatable (to ungulate grazers) forbs such as mules' ears (*Wyethia* spp.) and lupine (*Lupinus* spp.) than did horse-removed sites (Beever et al. 2008), which Ruthven (2007) also reported for sites grazed by domestic livestock in south Texas. These forbs are not food items ingested by Greater Sage-Grouse. This trend was generally true for all forbs combined, but was inconsistently and less strongly so than for other plant metrics. Furthermore, information-theoretic analyses suggested that variability in forb cover and frequency more strongly reflected precipitation, year, elevation, and site effects than the presence of horses (Beever et al. 2008). In spite of their greater (unpalatable) forb cover, horse-occupied sites were more depauperate floristically, averaging 4–12 fewer plant species per site; differences primarily reflected herbaceous species, particularly forbs (Beever et al. 2003). These site-level differences in plant species richness were evidenced in spite of a lack of statistical difference in average number of plants intercepted per transect, underscoring the greater heterogeneity within horse-removed sites (Beever et al. 2008) with which many vertebrates in sagebrush ecoregions are associated (Aldridge and Boyce 2007).

Cheatgrass (*Bromus tectorum*) tends to be more abundant at horse-occupied than horse-removed sites (1.6–2.6 times greater abundance; Beever et al. 2003, 2008). Dung piles of feral horses in montane natural grasslands of Argentina act as invasion windows for exotic plants: an invasive thistle was consistently associated with dung piles, and nonnative plant cover in abandoned dung piles was higher than in control plots (Loydi and Zalba 2009). From field-collected fecal samples in two research rangeland areas of the interior northwestern United States and subsequent studies of seed germination in greenhouses, Bartuszevige and Endress (2008) estimated that nonnative ungulates (domestic cattle) disperse 1,200,000 germinable exotic seeds/km^2—about two orders of magnitude greater than that estimated for the native ungulates in the same area (elk [*Cervus elaphus*] and mule deer [*Odocoileus hemionus*]). Invasion of cheatgrass into lower-elevation, xeric sites and alteration of fire frequencies result in reduced plant diversity and habitat structure (Connelly et al. 2004, Crawford et al. 2004). Herbaceous components of the system are lost, resulting in reduced food resources for sage-grouse, and fire frequency is increased, directly eliminating native forbs, shrubs, and perennial grasses within sagebrush habitats (d'Antonio and Vitousek 1992, Miller et al. 1994, Crawford et al. 2004). All of these ecological cascades reduce habitat suitability for sage-grouse (Fig. 14.2; Connelly et al. 2004). Given the drastic changes in stable-state dynamics and management alternatives when cheatgrass invasions occur, possible changes to sagebrush ecosystems as a result of the presence of wild horses—changes in vulnerability or resilience of communities to fire, mechanisms of spread or increase in nonnative plant density—may merit more consideration in light of concerns over the viability of Greater Sage-Grouse populations and other sagebrush-obligate bird species.

Within horse-occupied areas of the western Great Basin, transects adjacent to horse trails had less grass and shrub cover, equivalent or higher forb cover (primarily unpalatable species), and lower species richness than did randomly selected transects within the same site (Beever et al. 2008). That horse grazing affected ecosystem components similarly at both local and landscape scales is important because reports of home-range size in horses vary from 0.9 km^2 on Sable Island, Nova Scotia, to 303 km^2 in Wyoming (Pellegrini 1971, Welsh 1975, Green and Green 1977). We expect the most appropriate scale for analyses of interactions between equids and sage-grouse (or other species dependent on sagebrush habitats) will vary greatly across the geographic range of these sagebrush-associated species.

Insights from the Great Plains

Many of the longest-term and most intensive studies of free-roaming horses have occurred in the far northeastern corner of the current geographical extent of horses in the United States. The Pryor Mountain Wild Horse Range at the Montana-Wyoming border (Fig. 14.1) makes up <0.5% of the total extent of HMAs. Paleoevidence, coevolutionary arguments, and recent data all suggest that ecosystem response to equid grazing should be fundamentally different in grasslands and steppes of the Great Plains than in the lower-productivity communities of the Great Basin (Mack and Thompson 1982, Milchunas and Lauenroth 1993, Wilsey et al. 1997, Grayson 2006). Thus, extrapolating results from Pryor Mountain studies across the domain of free-roaming horses in North America should generally be avoided. However, long-term research in the Pryor Mountains suggests horse-grazing impacts on vegetation may vary by year (largely due to interannual variability in precipitation, especially precipitation during the growing season), short-term and especially long-term grazing history, life-form and species of plant, and, most commonly, site elevation and soil type (Coughenour 1999, Gerhardt and Detling 2000).

Insights from Research in Mesic Systems

Within sagebrush landscapes, streams and springs receive disproportionately heavy use by horses (Fig. 14.2; Crane et al. 1997, Beever and Brussard 2000). The extent to which these resources are negatively affected depends on timing, duration, and intensity of equid grazing; spring density in the area; and productivity and resilience of the habitats and soils adjacent to the water (Beever and Brussard 2000). Degradation of spring-associated resources in steeper terrain and higher-elevation areas by equids is of particular concern, given that these springs would less likely be accessible if domestic cattle were the only grazer in the area (Ganskopp and Vavra 1986, 1987; Beever 2003).

Research on ecosystem response to free-roaming horses in more mesic habitats such as salt marshes, oligotrophic bogs, pastures, and peatlands has a longer history and greater spatial replication at a global scale, but may contribute in only a limited manner to ecologists' understanding of grazing dynamics in semiarid landscapes and sagebrush communities. These ecosystems often have horses present in only limited numbers and over a limited area, and their response to horse grazing is likely to be fundamentally different from that expected in upland areas of semiarid ecosystems where thresholds and nonlinear dynamics dominate (Mack and Thompson 1982, Laycock 1991).

However, these studies may reveal how areas with wetted or hydric soils will respond to horse grazing, such as at springs and margins of riparian areas in the western United States; what plant attributes, portions of the plant community, or areas on the landscape may be particularly vulnerable to effects of horse grazing; and the mechanisms by which indirect effects are brought about by horse grazing. Collectively, these limitations in extrapolating results from other ecoregions to the Intermountain West, combined with the importance of springs and seeps for Greater Sage-Grouse and other sagebrush-obligate species, highlight equid relationships with mesic areas as a fruitful avenue for future research on conservation implications.

Possible Effects of Equids on Soils in Sagebrush Ecosystems

Soil hardness at horse-occupied sites averaged 2.9 times greater than at horse-removed high-elevation sites in the western Great Basin, and 17.4 times greater among low-elevation sites (Beever et al. 2003, Beever and Herrick 2006). Strong within-year correlations ($r = 0.69$, $P = 0.001$) in these studies between average soil-surface penetration resistance and the number of horse defecations at each site suggested a direct effect of horses on penetration resistance (Beever and Herrick 2006). The correlation was stronger within each of the two elevational strata, including the low-elevation ($r = 0.77$) and high-elevation plots ($r = 0.71$). Alterations of soil properties such as bulk density, thickness of surface horizons, and soil chemistry have been implicated in catastrophic degradation of vegetation, invasion of exotic plant species, and desertification (Thurow 1991, van de Koppel et al. 1997), all of which could reduce habitat availability and quality for Greater Sage-Grouse across all life stages (Fig. 14.2).

These alterations have been demonstrated for domestic-livestock grazing via compaction, hoof action and consequent erosion, and redistribution of nutrients such as nitrogen (Archer and Smeins 1991, van de Koppel et al. 1997). Surface soils are involved with numerous biotic and abiotic pathways (Thurow 1991, Belsky and Blumenthal 1997, Beever and Herrick 2006). Consequently, further work on equid-soil relationships would increase ecological understanding and identify implications of and identify management strategies to minimize the influences of free-roaming equids on sagebrush habitats and, thus, on Greater Sage-Grouse populations. For example, two questions that have management significance are to what depths do increases in penetration resistance extend below the soil surface, and what factors (e.g., soil texture or percent clays, consistence, and strength of effervescence) most strongly modify these relationships and pathways?

Whole-Community Perspectives in Analyzing Effects of Equid Grazing

The question of whether horse-occupied and horse-removed sites differ across a broad region in a community or whole ecosystem sense needs to be addressed, in addition to examining grazing effects on individual ecosystem components. Beever et al. (2003) investigated the relative performance of five different data sets, all collected or estimated for long-term abiotic properties across the same 19 sites and two years of field work, in detecting disturbance by free-roaming horses. Horse-occupied and horse-removed sites could not be clearly discriminated in multivariate ordination space using cover of key plant species consumed by horses (and measured by BLM horse specialists in horse-effects monitoring), cover or frequency of all plant species, or a full suite of abiotic site characteristics (Fig. 2 in Beever et al. 2003).

In contrast, the two types of sites were clearly distinguished by a suite of variables that were sensitive to grazing disturbance in prior ecological research. These diverse grazing-sensitive ecosystem attributes included density of ant (Formicidae) mounds, penetration resistance of the soil surface, percentage of hypothetically possible granivorous small mammals that were actually trapped, species richness of plants, grass cover, forb cover, shrub cover, and abundance of cheatgrass (Beever et al. 2003). This may be important for Greater Sage-Grouse and other sagebrush-obligate species because although neither cover of the species that free-roaming horses eat nor abiotic site properties differed between horse-removed and horse-occupied sites in a collective (multivariate) sense, community-level differences between the two types of sites clearly existed (Beever et al. 2003). These differences may be important to differing degrees, depending on the species, but effects of these ecosystem alterations have not been addressed for Greater Sage-Grouse. Greater Sage-Grouse have been shown to choose more complex vegetation structure (Crawford et al. 2004, Aldridge 2005). Changes at the community level could result in a reduction in visual obstruction barriers for Greater Sage-Grouse and increase predation on nests and individual birds (Martin 1993, Schroeder and Baydack 2001, Crawford et al. 2004, Aldridge 2005).

Habitat-Use Patterns by Equids and Potential Synergies with Livestock-Grazing Effects

One of the more important challenges in interpreting equids' role in the ecological integrity of sagebrush communities for the future of sage-grouse and other species of concern involves understanding how equids' effects are distributed across the landscape. Spatially explicit quantitative assessment of habitat-use patterns does not translate directly into equivalent understanding of how shorter- and longer-term grazing effects are distributed across semiarid landscapes. This is because of the differential vulnerability of various soil types and plant communities to equid grazing. However, habitat-use analyses provide a foundation that can be focused with emerging scientific information.

Several authors have noted particularly heavy use of the more mesic portions of semiarid landscapes—meadows, grasslands, springs, and riparian zones—by free-roaming horses, especially during the warmest, driest portions of the year (Feist 1971, Pellegrini 1971, Ganskopp and Vavra 1986). Beever and Brussard (2000) concluded local spring density was one of the strongest factors explaining differences in magnitude of impacts at springs across multiple mountain ranges. McKnight (1958) reported that feral burros may remain at watering locations for several hours, and roll and defecate in the water.

Mesic habitats are important for Greater Sage-Grouse survival, providing an abundance of forbs and insects as food resources from pre-laying in early spring (Crawford et al. 2004) through brood-rearing and into the fall (Klebenow and Gray 1968, Drut et al. 1994b, Sveum et al. 1998a, Aldridge 2005). Loss of productive mesic habitats can result in low productivity for Greater Sage-Grouse, which has been linked to population declines throughout the species' range (Connelly et al. 2004, Aldridge and Boyce 2007). Heavy use of these sites by equids could increase conflicts and reduce the availability and quality of important Greater Sage-Grouse habitats. Especially in arid and semiarid ecosystems, springs and riparian ecosystems are of paramount importance to maintaining regional biodiversity because of their facultative or obligate use by ≥70% of a region's species (Naiman et al. 1992). The extent to which free-roaming equids in sagebrush communities directly or indirectly alter the availability of insects important to Greater Sage-Grouse broods at these mesic sites is currently unknown.

Most ungulate species are known to move in response to local and seasonal shifts in distribution of forage abundance, availability, and nutrient quality (Talbot and Talbot 1963, Kitchen 1974). Horses in many of the Great Basin HMAs (Fig. 14.1) tend to move to higher elevations from late spring until early fall, and descend to lower habitats during colder months, when snow may persist at higher elevations (Pellegrini 1971, McInnis 1985). Seasonal shifts in habitat use by Greater Sage-Grouse occur during the late brood-rearing period (July and August) and are driven by the availability of forbs, which are required by chicks and adults (Patterson 1952, Klebenow and Gray 1968, Peterson 1970). Similar to the shift in habitat use by equids, Greater Sage-Grouse often move to higher-elevation mountain-shrub communities, spatially tracking the phenology of plant communities to ensure access to more mesic habitats with forbs throughout summer (Fischer et al. 1996a, Connelly et al. 2004). The seasonality of equids' use of the landscape may have important implications for Greater Sage-Grouse habitats and populations, given the varying requirements of Greater Sage-Grouse throughout the year.

Horses use plant communities and landscapes differently than do cattle and native ungulates, across both time and space. Within seasons, horses often segregate by elevation from cattle by using steeper slopes and occupying higher-elevation habitats (Pellegrini 1971; Ganskopp and Vavra 1986, 1987). Both cattle and horses most commonly used the sagebrush-grass vegetation type when averaged across the year in the Red Desert of Wyoming, but peak use by horses occurred in fall and winter, whereas use by cattle peaked in summer and fall (Miller 1980). Horses also spend considerably less time at watering locations than do cattle (Meeker 1979, Ganskopp and Vavra 1986) and range farther from watering sources throughout the year (up to 49.2 km in eastern Nevada; Feist 1971, Pellegrini 1971, Welsh 1975, Green and Green 1977). Horses graze an average of six hours per day longer than cows (Arnold and Dudzinski 1978, Rittenhouse et al. 1982, Duncan et al. 1990) and consume 20–65% more forage than would a cow of equivalent body mass because of their cecal digestion and rapid passage rate of food (Hanley 1982, Wagner 1983, Duncan et al. 1990, Menard et al. 2002). Horses can crop vegetation closer to the ground than can other ungulates due to their sensitive, flexible lips and possession of both upper and lower incisors (Stoddart et al. 1975, Symanski 1994). This cropping sometimes delays the recovery of plants (Symanski 1994, Menard et al. 2002). Furthermore, native ungulates such as mule deer and pronghorn (*Antilocapra americana*) can range farther from water than can horses, are browsers, and have much smaller body mass and hoof area relative to horses (Symanski 1994).

Taken collectively, these broader-scale differences in grazing ecology indicate that a larger proportion of the landscape experiences nonnative grazing when horses and cattle are present sympatrically, compared to when only cattle are present (Symanski 1994, Beever 2003). Domestic livestock consume an estimated 7,100,000 animal-unit months (AUMs) of forage annually (Table 14.4) within the current range of Greater Sage-Grouse. We estimate that free-roaming equids consume an additional 315,000–433,000 AUMs annually within the current range of sage-grouse (Table 14.4). It is unknown whether effects of cattle grazing, horse grazing, and native-ungulate browsing are synergistic or simply additive. However, the significance of the aforementioned differences is underscored by the importance that both patch dynamics and landscape patterns have been shown to have for several species, including Greater Sage-Grouse, which use the landscape at multiple spatial scales (Connelly et al. 2004, Aldridge 2005, Aldridge and Boyce 2007). The peak spatial overlap in use of sagebrush habitat with sage-grouse may thus occur during the breeding or late brood-rearing periods.

In addition to these contemporary additive or synergistic effects, large-mammal researchers and managers often struggle to separate the legacies of past mismanagement or poor implementation of livestock grazing strategies (especially during 1890–1920) from effects of current disturbances (Foster et al. 2003, Rowe 2007). Domestic livestock grazing is nearly ubiquitous across public lands of western North America, and the ratio of livestock to equid AUMs in this region averaged 23 to 1 in 1982 (Wagner 1983). However, within Nevada—the state with over half of the continent's free-roaming equids—this ratio averaged only 4.8 to 1 livestock-to-equid AUMs in 1982, and varied from 23 to 1 in the Elko BLM District to 2 to 1 in the Las Vegas BLM District (Wagner 1983).

FUTURE APPROACHES TO INVESTIGATE EQUID–SAGE-GROUSE RELATIONSHIPS

Several methods can quantitatively assess the influence of equids on the biotic integrity of sagebrush communities within the historic range of sage-grouse. The most informative for managers and perhaps most powerful statistically would be to regress an index of biotic integrity (Angermeier and Karr 1986) obtained from remotely sensed data against equid density using points from across the region. Unfortunately, this approach is currently untenable for several reasons. First, although the use of light detection and ranging (LiDAR) and satellite-based imagery to remotely

TABLE 14.4

Animal unit months (AUMs) permitted for domestic livestock and free-roaming equids by state, across the current distribution of Greater Sage-Grouse within the United States.

Domestic livestock and equid estimates are based on BLM-administered lands within the current range of Greater Sage-Grouse.

State	Domestic livestock AUMs (2007)[a]	Range of equid AUMs[b]	
		Lower (1.2 times)	Higher (1.65 times)
California	150,228	30,874	42,451
Colorado	326,393	5,875	8,078
Idaho	855,131	10,541	14,494
Montana	1,334,208	2,218	3,049
North Dakota	9,268	0	0
Nevada	1,436,771	159,869	219,820
Oregon	866,194	29,722	40,867
South Dakota	36,958	0	0
Utah	261,075	19,224	26,433
Washington	10,307	0	0
Wyoming	1,894,034	56,693	77,953
Totals	7,118,989	315,014	433,145

[a] Livestock estimates are based on 2007 allocations, as summarized by the BLM Rangeland Administration System, http://www.blm.gov/ras/, 24 October 2009.

[b] Permitted AUMs does not necessarily mean all AUMs were used, but the BLM system does not track used AUMs. Range of AUMs used by equids was estimated using numbers obtained from the National Wild Horse and Burro Program (http://www.wildhorseandburro.blm.gov/statistics/2007/index.htm, 24 October 2009), assuming 12 mo of use at 1.2–1.65 times greater forage consumption for equids (both horses and burros) compared to cattle (Hanley 1982, Wagner 1983, Menard et al. 2002).

map and monitor habitats across large landscapes is developing rapidly, indices that would be meaningful for Greater Sage-Grouse and other sagebrush obligates are not yet broadly available at sufficiently fine resolution to permit quantitative analysis.

Second, it is unclear whether maximum equid density or average density within AUMs would be a more logical predictor of current sagebrush conditions (Ellison 1960, van de Koppel et al. 1997). Third, except in the Pryor Mountain Wild Horse Range in Montana, the finest scale for which equid density estimates are available is the entire HMA, which is often an entire mountain range(s). This is critical because, as evidenced by the example that ~40% of the Pryor Mountain Wild Horse Range was unsuitable for horse grazing due to steep topography or large extents of rock or bare ground (Fahnestock and Detling 1999b), estimates of entire HMA equid density can be off by a factor of at least 1.7. This error factor for any given location could be even higher, because the conservative assumption that every remaining "suitable" location is grazed at the same intensity is clearly not true.

Fourth, high increase rates observed in populations of wild horses (to >20% annually; Eberhardt et al. 1982) indicate that the number of horses in a given fall or spring census period will not necessarily correspond either to the population size in the management unit at a given time in the future or to an average number of AUMs on the landscape during the entire year. This would only be problematic if error magnitudes were distributed nonrandomly, but this seems feasible (e.g., higher error in larger populations). Fifth, differences in precision of equid censuses across the sagebrush range due to differences in aircraft, observer experience, tree cover that reduces animal detectability, and other factors would reduce the power of the analysis (Caughley 1974, Ohmart et al. 1978).

Finally, and perhaps most seriously, the way in which equid, cattle, sheep, and native-ungulate AUMs are allocated to HMAs would confound the analysis if lower levels of equid AUMs are correspondingly or randomly replaced by increased cattle AUMs. Livestock grazing intensity could be incorporated as a covariate, but intensity is known only at the spatially coarse level of allotments, and historical records of density are not widely available, thus crippling this potential analysis.

A second analytical approach would compare a metric of sagebrush biotic integrity—such as cover of sagebrush from remotely sensed data—in management areas with and without equids. In addition to the aforementioned lack of alternative indicators, a problem for this analysis is the fact that because HMAs often correspond to entire mountain ranges, lower sagebrush cover within an HMA may be a consequence of that range's vicariance history rather than a reflection of greater disturbance. A new challenge created by this approach is that HMAs where equids have been removed are not allocated in a random, dispersed manner, but instead are driven by sociopolitical concerns (e.g., conflicts with railroad property rights). Thus, equid-removed herd areas are at times spatially clumped, a hindrance to landscape-scale analyses.

This second approach circumvents the second through fifth problems in the first analytical approach, but with the sixth, effects of other ungulates would remain—a serious confounding factor. Horse specialists realize it is difficult to separate effects of other sympatric ungulates, such as domestic livestock and native ungulates such as mule deer, elk, and pronghorn, from equid effects.

A final analytical approach would involve assessment of the possible direct interactions between equids and Greater Sage-Grouse or other sagebrush-obligate species. In this approach, radio transmitters placed on individuals of both grouse and equids would provide the raw data needed to develop habitat-selection maps for both species. In turn, these maps would indicate the factors driving habitat selection for the two taxa in regions where they are sympatric and allow investigation of overlap of use for the two species in different seasons that correspond to critical life-stages for Greater Sage-Grouse.

CONCLUSIONS

Conservation challenges posed by sage-grouse and other sagebrush-obligate species provide a timely impetus to consider the role free-roaming equids might play in ecoregional-scale conservation strategies. From a practical standpoint, such challenges may be an avenue by which the limiting horses versus cows debate that has dominated discussions and popular perception of equid management for decades may be broadened to include other resource values. Free-roaming burros and especially horses are undeniably charismatic and enigmatic, and have been used to symbolize power, freedom, wildness, and toughness. Given the multiple stresses that interact to influence ecosystem dynamics across western North America, however, the benefits these nonnative herbivores provide for various publics within society must be weighed against actual and potential ecological costs.

As with any large-bodied herbivore, the magnitude and permanence of impact that free-roaming horses and burros will have on sagebrush ecosystems depend upon a host of factors, including the plant-animal coevolutionary history, soil development, climate, recent weather, effects of other sympatric herbivores, site elevation, amount of recent and long-term precipitation, and the seasonality, duration, and intensity of grazing (Milchunas et al. 1988, Fahnestock and Detling 1999a, Beever et al. 2008). In turn, intensity of landscape use by equids is influenced by a host of biotic factors, including vegetation and soil types, vegetation physiognomy, proximity to water, presence and recency of wildfire, spatial and temporal distribution of plant defenses against herbivory, behavior of conspecifics or competitors, and past, current, and future nutritional states of animals, among other factors. Abiotic factors such as elevation, aspect and slope, snow depth, thermoregulatory constraints, and proximity to roads and other disturbances also affect habitat use of free-roaming equids.

Equid-induced changes in sagebrush and other communities that may most strongly affect sagebrush-obligate birds and have been observed to date include reduction in grass abundance and cover, alterations to the structure and composition of the shrub mosaic (e.g., lower cover and greater fragmentation of shrubs, and lower sagebrush cover), increases in compaction in surface soil horizons, and increased dominance of forbs unpalatable to livestock and Greater Sage-Grouse (Fig. 14.2). Collectively, these effects represent a nontrivial contribution to ecosystem integrity across the sagebrush range. All of these ecosystem alterations may increase the vulnerability of Greater Sage-Grouse and other species to predation, parasites, or disease; increase energetic costs and stress levels required to locate suitable habitat or resources; and negatively affect nest success, chick survival, or other aspects of fitness and survival, all of which could ultimately affect the viability of some populations.

CONSERVATION IMPLICATIONS

Multidisciplinary teams of managers and researchers comprising leading experts on various aspects of equid biology have long considered investigation of grazing impacts to be of highest priority (Blaisdell and Thomas 1977, Kitchen et al. 1977, Zarn et al. 1977, Wagner et al. 1980, Singer 2004), but understanding how feral horses may affect the nonungulate components of sagebrush ecosystems is only beginning (Crane et al. 1997, Fahnestock 1998, Peterson 1999, Beever and Brussard 2004, Beever and Herrick 2006). Recent research suggests that equid use of a landscape can indirectly affect a wide array of taxa in a community through numerous pathways (Fig. 14.2) (Levin et al. 2002). For example, compared to ecologically similar horse-removed sites in the western Great Basin, horse-occupied sites differ in numerous aspects of ecosystem structure, such as lower shrub cover, higher compaction of soil surfaces, and a more fragmented shrub canopy, composition (e.g., lower grass cover, lower total vegetative cover, lower plant species richness), and function (e.g., lower density of ant mounds). In spite of the increasing attention in recent research to synecological relationships of equid grazing in arid and semiarid landscapes of western North America, many basic questions remain unanswered.

Nearly all of the research on equids in western North America has been performed at spatially limited scales due to the advantages in terms of logistics, costs, and research design. Unfortunately, these focused investigations are difficult to amalgamate into a landscape- or regional-level understanding of equid biology. This is especially true given that horse behavior can be plastic across ecoregional scales (Bennett and Hoffmann 1999). Furthermore, most studies have usually used only two treatment levels for equid grazing—present and absent. If, instead, investigators were to select equid-occupied sites across a range of grazing intensities, this would allow clarification of the shape of the dose-response curves of various ecosystem indicators. This has importance for monitoring, because monitored indicators should ideally be sufficiently sensitive to grazing disturbance to act as early-warning indicators of system degradation.

Results from past research across 3,030,000 ha of the western Great Basin may help in forecasting how Greater Sage-Grouse and other sagebrush-obligate species may be affected by free-roaming equids. For example, community completeness may be lower in horse-grazed than in horse-free areas (e.g., for granivorous small mammals) as a result of the absence of rarer species from horse-grazed sites (Beever and Brussard 2004). In that work, community completeness was defined as the percentage of species detected during a standardized trapping effort, compared to the full suite of species predicted to exist at a site given the elevation, geographic location, and microhabitats of the site. Differences in grass cover are likely to have a biologically meaningful effect for some grass-specialist animals, as was illustrated by two to four times higher capture rates of the western harvest mouse (*Reithrodontomys megalotis*) at horse-removed compared to horse-occupied sites (Beever and Brussard 2004). Equid-influenced areas may be more dominated by disturbance-tolerant generalists like deer mice (*Peromyscus maniculatus*) and species that specialize in open areas without vegetation, such as western whiptails (*Cnemidophorus tigris*) (Beever and Brussard 2004). Greater density of ant mounds at horse-free sites than at horse-occupied sites suggests that at least a portion of the invertebrate community is more robust at horse-removed sites, and may also reflect differences in level of ecological function (Beever and Herrick 2006). Collectively, these responses not only illustrate the types of community restructuring that can occur in areas heavily influenced by equids, but also suggest indicators and mechanisms that may be important for further investigating equids' effects on Greater Sage-Grouse and other sagebrush-obligate species.

Given current knowledge, however, and the fact that recovery can be protracted and uncertain in arid and semiarid ecosystems within the geographic range of Greater Sage-Grouse—due to their low-productivity, event-driven, threshold state-and-transition dynamics (Laycock 1991, Tausch et al. 1993)—several conservation considerations appear warranted. First, arid land health and ecosystem integrity are important considerations for long-term conservation of sagebrush systems, along with minimum viable population sizes and other genetic concerns for free-roaming equid populations. Long-term conservation objectives should consider the appropriate management levels of horses and burros that can be maintained (Tables 14.2, 14.3) because free-roaming equids can influence the structure and function of sagebrush ecosystems. Fencing and rotational grazing can be used to limit spatial and temporal effects of domestic livestock grazing, but few options outside of controlling population levels are available to land managers for managing horse and burro distributions and grazing effects. Consequently, the primary mechanisms for managing equids will center on approaches that reduce herd levels or limit population growth.

Second, it remains important within this context to have explicit management goals to justify and reinforce decisions of where and how many free-roaming equids will occupy public lands. Third, extensive research into immunocontraception as a potential method to limit equid populations should be accompanied by synecological investigations that not only support adaptive management of equid herds but also provide insights into additional ecological indicators of grazing disturbance as it relates to ecosystem integrity. Finally, herd reductions during drought periods that managing agencies currently use may be important in sustaining sagebrush ecosystems during ecological stresses. Such emergency reductions not only avoid the pronounced grazing-induced ecosystem degradation that can accompany drought (Jardine and Forsling 1922, Archer and Smeins 1991) but also avoid conditions that can be stressful or lethal for equids.

ACKNOWLEDGMENTS

Previous versions of the manuscript were read by F. H. Wagner, S. T. Knick, C. E. Braun, and two anonymous reviewers. This manuscript benefited from fruitful discussions with numerous BLM horse specialists; many are cited in the text, but others include T. M. Pogacnik, J. M. Gianola, G. McFadden, and T. J. Seley. T. J. Assal compiled tabular and spatial data on BLM and USFS wild horse and burro data.

PART FOUR

Population Trends and Habitat Relationships

CHAPTER FIFTEEN

Greater Sage-Grouse Population Dynamics and Probability of Persistence

Edward O. Garton, John W. Connelly, Jon S. Horne, Christian A. Hagen, Ann Moser, and Michael A. Schroeder

Abstract. We conducted a comprehensive analysis of Greater Sage-Grouse (*Centrocercus urophasianus*) populations throughout the species' range by accumulating and analyzing counts of males at 9,870 leks identified since 1965. A substantial number of leks are censused each year throughout North America providing a combined total of 75,598 counts through 2007, with many leks having >30 years of information. These data sets represent the only long-term database available for Greater Sage-Grouse. We conducted our analyses for 30 Greater Sage-Grouse populations and for all leks surveyed in seven Sage-Grouse Management Zones (SMZs) identified in the Greater Sage-Grouse Comprehensive Conservation Strategy. This approach allowed grouping of leks into biologically meaningful populations, of which 23 offered sufficient data to model annual rates of population change. The best models for describing changes in growth rates of populations and SMZs, using information-theoretic criteria, were dominated by Gompertz-type models assuming density dependence on log abundance. Thirty-eight percent of the total were best described by a Gompertz model with no time lag, 32% with a one-year time lag, and 12% with a two-year time lag. These three types of Gompertz models best portrayed a total of 82% of the populations and SMZs. A Ricker-type model assuming linear density dependence on abundance in the current year was selected for 9% of the cases (SMZs or populations), while an exponential growth model with no density dependence was the best model for the remaining 9% of the cases. The best model in 44% of the cases included declining carrying capacity through time of −1.8% to −11.6% per year and in 18% incorporated lower carrying capacity in the last 20 years (1987–2007) than in the first 20 years (1967–1987). We forecast future population viability across 24 populations, seven SMZs, and the range-wide metapopulation using a hierarchy of best models applied to a starting range-wide minimum of 88,816 male sage-grouse counted on 5,042 leks in 2007 throughout western North America. Model forecasts suggest that at least 13% of the populations but none of the SMZs may decline below effective population sizes of 50 within the next 30 years, while at least 75% of the populations and 29% of the SMZs are likely to decline below effective population sizes of 500 within 100 years if current conditions and trends persist. Preventing high probabilities of extinction in many populations and in some SMZs in the long term will require concerted efforts to decrease continuing loss and

Garton, E. O., J. W. Connelly, J. S. Horne, C. A. Hagen, A. Moser, and M. A. Schroeder. 2011. Greater Sage-Grouse population dynamics and probability of persistence. Pp. 293–381 *in* S. T. Knick and J. W. Connelly (editors). Greater Sage-Grouse: ecology and conservation of a landscape species and its habitats. Studies in Avian Biology (vol. 38), University of California Press, Berkeley, CA.

degradation of habitat as well as addressing other factors (including West Nile virus) that may negatively affect Greater Sage-Grouse at local scales.

Key Words: carrying capacity, *Centrocercus urophasianus*, density dependence, effective population size, Greater Sage-Grouse, lek counts, management zones, models, N_e, probability of extinction, quasi-equilibrium, time lags.

Dinámicas De Población Y Probabilidad De Persistencia Del Greater Sage-Grouse

Resumen. Condujimos un análisis comprensivo de las poblaciones del Greater Sage-Grouse (*Centrocercus urophasianus*) en el rango de distribución de esta especie por medio de la acumulación y análisis de conteos de machos en 9,870 leks (asambleas de cortejo) identificados desde 1965. Un número considerable de leks es censado cada año en Norteamérica, lo que provee un total combinado de 75,598 conteos hasta el 2007, con muchos leks que poseen >30 años de información. Estos conjuntos de datos representan la única base de datos de largo plazo disponible para el Greater Sage-Grouse. Condujimos nuestros análisis sobre 30 poblaciones del Greater Sage-Grouse y para todos los leks examinados en siete zonas de manejo del sage-grouse (SMZs o Sage-Grouse Management Zones) que fueron identificadas en la Estrategia de Conservación Comprensiva del Greater Sage-Grouse (Greater Sage-Grouse Comprehensive Conservation Strategy). Este enfoque permitió agrupar a los leks en poblaciones biológicamente significativas de las cuales 24 ofrecieron suficientes datos para modelar tasas anuales de cambio de la población. Los mejores modelos para describir cambios en las tasas de crecimiento de poblaciones y SMZs, usando criterios informático-teóricos, fueron dominados por modelos del tipo Gompertz asumiendo dependencia de la densidad en la abundancia del registro. El 38% del total fue mejor descrito por un modelo de Gompertz sin acción diferida del tiempo, el 32% con una acción diferida del tiempo de 1 año, y el 12% con una acción diferida del tiempo de 2 años. Estos tres tipos de modelos Gompertz representaron mejor un total del 82% de las poblaciones y de SMZs. Un modelo de tipo Ricker asumiendo dependencia linear de densidad sobre la abundancia en el corriente año fue seleccionado para el 9% de los casos (SMZs o poblaciones), mientras que un modelo de crecimiento exponencial sin dependencia de densidad fue el mejor modelo para el restante 9% de los casos. El mejor modelo en el 44% de los casos incluyó capacidad de carga decreciente a través del tiempo de −1.8% a −11.6% por año y en el 18% incorporó capacidad de carga menor en los últimos 20 años (1987–2007) que en los primeros 20 años (1967–1987). Pronosticamos la viabilidad futura de la población en 24 poblaciones, siete SMZs, y la metapoblación del rango de distribución utilizando una jerarquía de los mejores modelos aplicados, comenzando con un mínimo a nivel de rango de distribución de 88,816 machos de sage-grouse contados en 5,042 leks en el 2007 en Norteamérica occidental. Los pronósticos del modelo sugieren que al menos 13% de las poblaciones pero ninguna de las SMZs podrán disminuir por debajo del tamaño efectivo de la población de 50 individuos en el plazo de los próximos 30 años, mientras que es probable que el 75% de las poblaciones y 29% de las SMZs disminuyan por debajo del tamaño efectivo de la población de 500 en el plazo de 100 años si las actuales condiciones y tendencias persisten. Prevenir las altas probabilidades de extinción a largo plazo en muchas poblaciones y en algunas SMZs requerirá rigurosos esfuerzos para disminuir la continua pérdida y degradación del hábitat así como también atender a otros factores (incluyendo el virus del Nilo occidental) que puedan afectar negativamente al Greater Sage-Grouse en escalas locales.

Palabras Clave: capacidad de carga, *Centrocercus urophasianus*, conteos de leks, cuasi-equilibrio, dependencia de la densidad, Greater Sage-Grouse, modelos, N_e, probabilidad de extinción, lapsos de tiempo, tamaño efectivo de la población, zonas de manejo.

Concerns about Greater Sage-Grouse (*Centrocercus urophasianus*; hereafter, sage-grouse) populations have been expressed for >90 years (Hornaday 1916, Patterson 1952, Crawford and Lutz 1985, Connelly and Braun 1997). Numerous investigators have assessed sage-grouse population trends since the mid-1990s in various states and Canadian provinces (Braun 1995, Schroeder et al. 2000, Aldridge and Brigham 2003, Beck et al. 2003, McAdam 2003, Smith 2003). In addition, Connelly and Braun (1997) synthesized available data for nine western states and one province and concluded that sage-grouse breeding populations have declined by 17–47%. They also examined sage-grouse production data for six states (Colorado, Idaho, Montana, Oregon, Utah, and Wyoming) and reported that production declined by an overall rate of 25%, comparing long-term averages to 1985–1994 data. Sage-grouse populations in five states were classified as secure and populations in six states and two provinces were considered at risk (Connelly and Braun 1997).

More recently, changes in the range-wide distribution of sage-grouse were analyzed by Schroeder et al. (2004), and they concluded this species now occupies about 56% of its likely pre-European settlement distribution. Connelly et al. (2004) analyzed lek data collected by states and provinces and concluded that sage-grouse populations declined at an overall rate of 2.0% per year from 1965 to 2003. Sage-grouse declined at an average annual rate of 3.5% from 1965 to 1985, and from 1986 to 2003 the population declined at a lower rate of 0.4% per year (Connelly et al. 2004). Recent trend analyses by the Sage- and Columbian Sharp-tailed Grouse Technical Committee (Anonymous 2008) suggest a long-term decline in Greater Sage-Grouse maximum male counts, with the greatest declines from the mid-1960s to the mid-1980s. The range-wide analysis showed quadratic, declining trends for the 1965–2007 and 1965–1985 time frames.

Connelly et al. (2004) also provided information on changes in sage-grouse populations by floristic province (Miller and Eddleman 2001). Stiver et al. (2006) suggested that sage-grouse populations should be assessed over broad scales without regard to political boundaries and indicated that floristic provinces could be slightly modified to provide Sage-Grouse Management Zones (SMZs) that would reflect ecological and biological issues and similarities. Our objectives were three-fold: (1) assess long-term changes (1965–2007) in sage-grouse populations by SMZ (Stiver et al. 2006) and population (Connelly et al. 2004) using information obtained from lek counts, (2) use information from these lek counts to reconstruct population abundance with an index to the minimum number of males observed, and (3) evaluate the likelihood of a variety of biologically significant models and their predictions concerning long-term probability of persistence of sage-grouse populations.

METHODS

Study Area

We analyzed lek data from within the Sage-Grouse Conservation Area first delineated in Connelly et al. (2004). This area included the pre-settlement distribution of sage-grouse (Schroeder et al. 2004) buffered by 50 km. The total assessment area comprised all or parts of 14 states and three provinces and encompassed approximately 2,063,000 km^2 (Connelly et al. 2004).

This area has been divided into seven SMZs that are similar to floristic regions and reflect ecological and biological similarities (Miller and Eddleman 2001). All areas occupied by sage-grouse within these floristic provinces are dominated by sagebrush (*Artemisia* spp.). These zones were developed by grouping sage-grouse populations within floristic regions (Stiver et al. 2006). Great Plains, Wyoming Basin, Snake River Plain, and Northern Great Basin SMZs encompassed core populations of sage-grouse (Connelly et al. 2004), while the Southern Great Basin SMZ included scattered populations in the southern part of the Great Basin. The Columbia Basin SMZ included sage-grouse in the state of Washington. The Colorado Plateau SMZ encompassed relatively small and isolated populations in Utah and Colorado.

Population Data

Lek counts are widely used to monitor sage-grouse populations, but a report for the Western Association of Fish and Wildlife Agencies questioned their usefulness (Beck and Braun 1980). Ideally populations threatened by extinction should be monitored by censusing breeding males and females and their progeny annually, yet the extensive spatial distribution of sage-grouse in regions with poor access, and the cryptic coloration and behavior of hens and their offspring preclude such an ideal approach. Counting breeding male sage-grouse provides a useful alternative index to the minimum number of breeding males within a local area because of their breeding behavior of concentrating and displaying at open or sparsely vegetated lek sites. Further complicating the use of this index, counts over the course of a single breeding season vary from a low at the beginning of the season, to peak in the middle, followed by a decline to the end, which necessitates using the maximum count from multiple counts across the entire season as the index. Nevertheless, techniques for correctly conducting lek counts have been described (Jenni and Hartzler 1978, Emmons and Braun 1984) and problems generally seem to be related to disregarding accepted techniques. All lek-monitoring procedures are supposed to be conducted during early morning (1/2 hour before to 1 hour after sunrise) with reasonably clear and calm weather (light or no wind, partly cloudy to clear) from early March to early May (Connelly et al. 2003b). Recent and ongoing investigations in southern Idaho revealed that lek counts ($N = 12$) collected using established guidelines (Connelly et al. 2003b) based on the maximum count from ≥4 surveys produced a highly repeatable index with maximum and second-highest counts in a season rarely differing by >4% over multiple years (J. A. Baumgardt, unpubl. data). Timing of lek monitoring is dependent on elevation of breeding habitat and persistence of winter conditions. We examined all lek data prior to analysis to ensure they were obtained following these procedures, and in some cases we had to assume that they were collected properly.

The same leks, or leks within the same area, have been counted by agency biologists for many years (Connelly et al. 2004). These leks were likely selected because they held many males, because of their accessibility, or for both reasons. Although some states and provinces attempt to monitor all known leks, leks surveyed in most states and provinces are not a random sample of those available, yet may provide unbiased and precise measures of the rate of change of populations when analyzed in a repeated measures framework. Connelly et al. (2004:Appendix 3) tested the lek count procedure because of potential biases in size of leks sampled and random changes in detection rates using simulated populations, and reported that average annual rate of population change estimated from 20 years of data collection at 20 leks sampled per population for 10,000 simulated populations provided unbiased estimates of the rate of change. The estimated rates of change deviated from the true simulated rate (using simulated surveys of each population) by an average of 0.04 (SD = 0.03). Precision of the estimates, measured by coefficient of determination of estimates with true simulated rates of change, increased with the simulated rate of population change from >80% for populations with an observed annual rate of change of at least 0.03 and >95% with rates of a least 0.07. Thus, while use of lek counts to assess change over a relatively large scale appears sound, we make no attempt to assess population dynamics at relatively small scales (e.g., harvest units, allotments) or estimate true population abundance using lek counts.

We used three time periods for analyses. The assessment period refers to the length of time that population dynamics for a given population or SMZ is assessed; in most cases, this ranges from 1965 through 2007. An analysis period is a five-year block of time over which data are averaged and corresponds with typical planning and assessment periods for management agencies. The final analysis period (2000–2007) contains eight years. The previous assessment of a portion of these data indicated that populations declined more steeply during the first 20 years evaluated (1965–1985) than during the last two decades (Connelly et al. 2004). Thus, we also evaluated models incorporating an early (1967–1987) and late (1987–2007) time period. We did not use the first two years of data (1965 and 1966 for most populations) to calculate rates of change so that models built with one- and two-year delays could be assessed in an information-theoretic framework on the basis of the same set of growth rate responses (e.g., rates calculated from 1967–2007).

We define a lek, for the purposes of this chapter, as a traditional display site with two or more males that has been recorded during the assessment period or within five years of that period. Substantial variation may exist among agencies with regard to the definition of a lek, because little published research documents the fluidity of lek establishment, formation, and extinction (Connelly et al. 2004). Although we assumed all lek data used in this analysis were obtained following established procedures (Connelly et al. 2003b), our review of state and provincial databases indicated there were some exceptions and that, in a few cases, the same lek had two or more somewhat different locations. Additionally, some agencies surveyed leks from the air in addition to using ground counts. Therefore, we carefully examined each state's and province's database and removed questionable data, e.g., leks for which no count data could be provided, and replicate locations (≥2 separate but nearby locations that represented the same lek). Many states had spatial data for leks but were lacking count data associated with them and thus no way of confirming that they actually were leks. We eliminated these data as well as leks when there was only a single count in a season for that lek and we eliminated data collected from the air regardless of the number of replicate counts in a year.

All information relating to population dynamics refers to changes in breeding populations. Delineating boundaries between local concentrations of breeding individuals (demes), populations, and metapopulations requires information on genetics, movements, habitat boundaries, and correlations in demographic rates (Garton 2002) that is sparsely available for sage-grouse across their extensive distribution (Fig. 15.1). Connelly et al. (1988) suggested that sage-grouse populations be defined on a temporal and spatial basis. A breeding population can be defined as a group of sage-grouse associated with one or more occupied leks in the same geographic area separated from other leks by >20 km (Connelly et al. 2003b). We followed these definitions for this analysis, and further defined sage-grouse populations throughout their North American distribution based on the known locations of leks. Concentrated areas of leks were considered breeding populations if they were separated from the nearest adjacent concentration of leks by at least 30 km and/or separated by unsuitable habitat such as mountain ranges, desert, or large areas of cropland (Connelly et al. 2004). These were grouped into SMZs including the Great Plains, Wyoming Basin, Snake River Plain, Columbia Basin, Northern Great Basin, Southern Great Basin, and Colorado Plateau (Fig. 15.1) (Miller and Eddleman 2001, Connelly et al. 2004, Stiver et al. 2006). Although individual SMZs consisting of multiple populations could be treated as metapopulations, three factors led us to combine data for all leks within SMZs into large single populations and only treat combinations of SMZs as a metapopulation: (1) our preliminary analysis indicated high correlations in growth rates among adjacent populations; (2) genetic studies suggest little genetic differentiation among populations

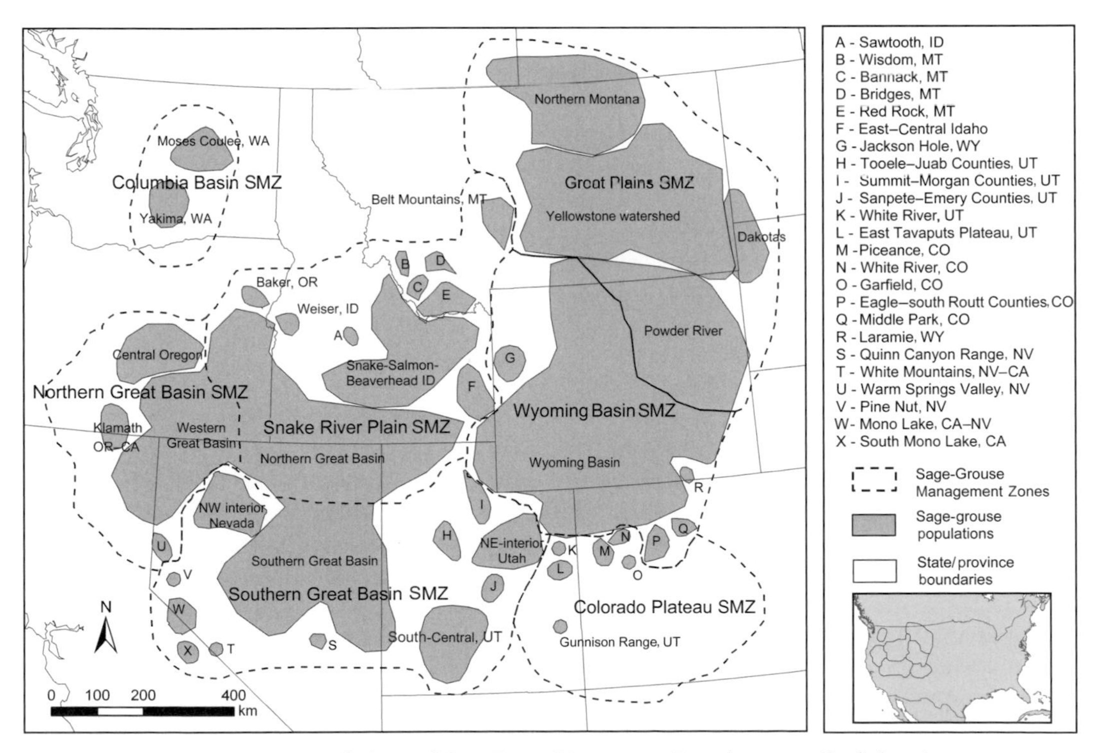

Figure 15.1. Greater Sage-Grouse populations and Sage-Grouse Management Zones in western North America.

(Oyler-McCance and Quinn, this volume, chapter 5); and (3) the large sample sizes of leks increase precision of estimates of abundance.

Forty-one distinct populations have been identified throughout the range of sage-grouse (Fig. 15.1) (Connelly et al. 2004). We were able to use 30 of the populations with sufficient data to allow some level of analysis (Table 15.1) and tended to include populations even if the data only included a handful of leks and ≥10 years of successive counts. Two large populations (Great Basin core and Wyoming Basin) were split by SMZ boundaries, and we split each into three and two smaller populations, respectively, to allow more meaningful analysis. We present findings from analyses of 30 populations, seven SMZs and range-wide. We organized our findings by presenting analyses for populations within SMZs, and then the results for the SMZ. We combined all lek counts within each SMZ even if some of them came from leks within populations for which the data were too sparse to perform an individual population analysis. This allowed us to use all lek counts meeting our standards for quality within each SMZ. Thus, sample sizes for SMZs in a particular year are often larger than the sum of the sample sizes for populations reported within that SMZ. We conclude with findings from a range-wide (metapopulation) analysis.

Monitoring Effort

We assessed monitoring effort within individual SMZs and populations by examining the average number of leks and number of active leks censused over five-year periods. This allowed evaluation of overall monitoring effort—the number of leks counted. We calculated the change in number of leks censused to describe the manner in which monitoring effort grew exponentially over time. Methods were developed to estimate trend and annual rates of change (see below) that would not be biased by this increasing monitoring effort.

Population Trends

Lek attendance data were obtained by counting the number of males attending leks during late March and April. In some cases, counts were made over a relatively short time frame or not made in consecutive years (Aldridge and Brigham 2003). For instance, Alberta conducted lek counts every other year for many years, while North Dakota conducted lek counts only during the third week of April, but has used this approach for >30 years.

Changes in sage-grouse breeding populations can be related to changes in number of leks, changes in lek size, or both. Ability to detect changes depends on monitoring effort. Different numbers of leks were often sampled annually in all states and provinces, so total counts of males (simple sums) provide almost meaningless information (Connelly et al. 2004). We used 1965 as a baseline for descriptive statistics in most cases because monitoring efforts by agencies were most consistent thereafter, and assumed that detection rates varied stochastically among years in assessing population dynamics.

We calculated mean lek size for all leks counted in a year based on the maximum count out of four or more counts in the year and averaged yearly means within periods to assess population change in each SMZ and population. We calculated λ (annual finite rate of change) from population reconstruction and summarized it by presenting its mean and standard error for each five-year period typical of agency planning periods. We also calculated mean lek size for active leks, defined as leks counted in a year with one or more males present on any count, because if a lek moved and was not detected, or if habitat changes from fires or development ended activity at a lek, counts would continue for a limited, but variable, number of years until the lek was deleted from annual surveys (Connelly et al. 2004). We averaged these values over five-year intervals (analysis periods) to provide a broader perspective of change in sage-grouse abundance and monitoring effort.

Population Reconstruction

Sage-grouse lek counts reported by individual states and provinces were summarized within SMZs and populations, and used to reconstruct an index to the historical abundance of the population within each SMZ and population. We treated the number of males counted at leks in the final year (2007) as an index to the minimum number of males attending leks because monitoring effort has grown exponentially in the last 10 years. In a few regions (e.g., Washington), counting every lek was attempted in 2007, making this index equal to the minimum known number

TABLE 15.1

Greater Sage-Grouse breeding populations in North America.

Population by management zone[a]	Brief description of population and justification for its delineation
	Great Plains SMZ I
Dakotas	Small population centered in southwest North Dakota and northwest South Dakota separated from adjacent populations by ~30–40 km and habitat features.
Northern Montana	Large population north of Missouri River in north central Montana, southeast Alberta, and southwest Saskatchewan separated from adjacent populations by ~20 km and Missouri River.
Powder River Basin, Montana	Large population in southeast Montana and northeast Wyoming separated from adjacent populations by ~20 km and habitat features.
Yellowstone watershed	Large population in central and southeast Montana separated from adjacent populations by 20–30 km and topography.
	Wyoming Basin SMZ II
Eagle–south Routt Counties, Colorado	Small population north of the Colorado River separated from adjacent populations by 20–30 km and topography.
Jackson Hole, Wyoming	Small population near Jackson Hole, Wyoming, separated from adjacent populations by ~50 km and topography.
Middle Park, Colorado	Small population in Middle Park, Colorado, separated from adjacent populations by 20–30 km and terrain.
Wyoming basin	Large population centered in Wyoming separated from adjacent populations by 20–40 km and topography.
	Southern Great Basin SMZ III
Mono Lake, California–Nevada	Small population on north side of Mono Lake area in California and Nevada isolated from adjacent populations by 20–40 km and topography.
South Mono Lake, California	Small population on south side of Mono Lake area in California separated from adjacent populations by 20–50 km and topography.
Northeast interior Utah	Small population in northeast interior Utah separated from adjacent populations by 30–50 km and topography.
Sanpete–Emery Counties, Utah	Small population in central Utah separated from adjacent populations by 50–60 km and topography.
South central Utah	Small population in south central Utah separated from adjacent populations by 50–70 km and topography.
Summit–Morgan Counties, Utah	Small population in northeast Utah separated from adjacent populations by 20–40 km and topography.
Tooele–Juab Counties, Utah	Small population in central Utah separated from adjacent populations by 20–40 km.
Southern Great Basin	A large population occupying much of central and eastern Nevada and a small portion of western Utah separated from adjacent populations by habitat and topographic features.

TABLE 15.1 *(continued)*

TABLE 15.1 (CONTINUED)

Population by management zone[a]	Brief description of population and justification for its delineation
	Snake River Plain SMZ IV
Baker, Oregon	Small population in Baker County, Oregon, that appears to be separated by topography and a mountain range from the nearest population by ~30 km.
Bannack, Montana	Small population near Bannack, Montana, separated from adjacent populations by 30–50 km and the Continental Divide.
Red Rocks, Montana	Small population in southwest Montana separated from adjacent populations by ~20–40 km and topography.
Wisdom, Montana	Small population in southwest Montana separated from adjacent populations by 4–60 km and terrain.
East central Idaho	Small population east of Snake River in east central Idaho separated from adjacent populations by 30–50 km, topography, and habitat.
Snake–Salmon–Beaverhead, Idaho	Large population along upper Snake, Salmon, and Beaverhead watersheds separated from adjacent populations by ~20–40 km and topography.
Northern Great Basin	Large population in Nevada, southeast Oregon, southwest Idaho, and northwest Utah separated from adjacent populations by 20–60 km and topography.
	Northern Great Basin SMZ V
Central Oregon	Relatively large population in central Oregon separated by topography from adjacent populations by ~30 km.
Klamath, Oregon–California	Small population along Oregon and California border separated from adjacent populations by ~50 km and topography.
Northwest interior Nevada	Small population in interior Nevada isolated from adjacent populations by ~20–30 km.
Western Great Basin	Large population in southeast Oregon, northwest Nevada, and northeast California separated from adjacent populations by ~25 km and unsuitable habitat.
	Columbia Basin SMZ VI
Moses Coulee, Washington	Small population along Moses Coulee in north central Washington separated from Yakima Washington population by ~50 km and Columbia River.
Yakima, Washington	Small population in south central Washington isolated by ~50 km and Columbia River from Moses Coulee Washington population.
	Colorado Plateau SMZ VII
Piceance Basin, Colorado	Small population in the Piceance Basin, Colorado, separated from adjacent populations by ~30–40 km and topography.

SOURCE: Adapted from Connelly et al. 2004.

[a] Sage-Grouse Management Zones from Stiver et al. 2006.

of males attending leks. In a few populations, the largest number of leks was counted in 2005 or 2006, and that year was taken as the basis for the index to the minimum males. Lek counts in each year were considered a cluster sample of male grouse and treated by standard finite population sampling procedures for cluster samples (Scheaffer et al. 1996:297). We estimated total number of males ($\hat{\tau}$) observed at all leks within a SMZ at time t as:

$$\hat{\tau}(t) = n\overline{M}(t) \qquad \text{(Eqn. 15.1)}$$

where an average of $\overline{M}(t)$ males are counted at n leks in year t. If counts were conducted at every lek within the region (e.g., Alberta and Washington), $\hat{\tau}$ would represent a census of all males attending leks rather than an index to the minimum number of males attending leks in year t. We estimated the precision of $\hat{\tau}(t)$ as:

$$SE = \sqrt{n(fpc)\,S_M^2} \qquad \text{(Eqn. 15.2)}$$

where *fpc* is a finite population correction which we assumed to equal 1.0 and S_M^2 is the sample variance of males counted per lek. We assumed the finite population correction (*fpc* = 1.0 − proportion of population sampled) is equal to 1.0 because only regions where agencies attempt a complete count of leks are sampled under an approximation to true probability sampling, but even in those regions detecting new or moving leks is problematic. Thus, the true *fpc* is unknown, and assuming it is equal to 1.0 prevents the estimated precision of the estimators in population reconstruction from overstating their true precision.

Sampling effort devoted to counting leks has varied from year to year and has grown appreciably in the last 10 years. We standardized estimates and removed bias due to variable sample sizes of leks by applying a ratio estimator (Scheaffer et al. 1996) to estimate the finite rate of change (λ_t) for the population between successive years at leks counted in both years. Beginning with the penultimate year (2006), males counted at each lek censused in both 2006 and 2007 were treated as cluster samples of individual males in successive years. The ratio of males counted in a pair of successive years estimates the finite rate of change (λ_t) at each lek site in that one-year interval. These ratios were combined across leks within a population for each year to estimate λ_t for the entire population or combined across all leks within a zone to estimate λ_t for that SMZ between successive years as:

$$\lambda(t) = \frac{\sum_{i=1}^{n} M_i(t+1)}{\sum_{i=1}^{n} M_i(t)} \qquad \text{(Eqn. 15.3)}$$

where $M_i(t)$ = number of males counted at lek i in year t, across n leks counted in both years t and $t + 1$. Ratio estimation under classic probability sampling designs—simple random, stratified, cluster, and probability proportional to size (PPS)—assumes the sample units (leks counted in two successive years in this case) are drawn according to some random process, but the strict requirement to obtain unbiased estimates is that the ratios measured represent an unbiased sample of the ratios (i.e., finite rates of change) from the population or SMZ sampled. This assumption seems appropriate for leks, and the possible tendency to detect larger leks than smaller leks does not bias the estimate of finite rate of change across a population or region, but makes it analogous to a PPS sample showing dramatically increased precision over simple random samples (Scheaffer et al. 1996). Precision (variance and standard error) of finite rates of change were estimated by treating $\lambda(t)$ as a standard ratio estimator (Scheaffer et al. 1996):

$$Var(\lambda_t) = \frac{fpc}{n\overline{M}(t)^2} \frac{\sum_{i=1}^{n}[M_i(t+1) - \lambda(t)M(t)]^2}{n-1} \qquad \text{(Eqn. 15.4)}$$

where *fpc* is again assumed to be 1.0 to avoid overestimating precision.

An index to the relative size of the previous year population ($\theta(t)$) was calculated in an analogous manner from the paired samples as the reciprocal of $\lambda(t)$:

$$\theta(t) = \frac{\sum_{i=1}^{n} M_i(t)}{\sum_{i=1}^{n} M_i(t+1)} \qquad \text{(Eqn. 15.5)}$$

with analogous variance:

$$Var(\theta_t) = \frac{fpc}{n\overline{M}(t+1)^2} \frac{\sum_{i=1}^{n}[M_i(t) - \theta_t M_i(t+1)]^2}{n-1} \qquad \text{(Eqn. 15.6)}$$

We used these ratios to calculate an index to population size by taking the number of males counted in the final year (2007, or another base year if 2007 sample sizes of leks were inadequate) as a minimum estimate of breeding male population size within that SMZ or population. We reconstructed the previous year's minimum male abundance index by multiplying the 2007 abundance by the ratio estimator of $\theta(2006)$, the relative number of males attending the same leks in 2006 compared to 2007. For example, $\lambda_t = 0.81$ between 2006 and 2007 corresponded to a $\theta(2006)$ of 1.23, suggesting the 15,761 males counted at 1,393 leks in 2007 were preceded by 19,461 males attending leks in 2006. This process was repeated for the change from 2005 to 2006 ($\lambda_t = 1.015$ indicating a $\theta(2005)$ of 0.985), yielding a minimum breeding male population index of 19,180 in 2005 and so on back to 1965. Repeating this process for each population and each SMZ yielded a population index for each population and zone stretching from 1965 to 2007 for populations in all SMZs except Columbia Basin and Colorado Plateau, for which valid indices were only estimable to 1967 and 1984, respectively. Application of this approach to individual populations yielded reconstructed population indices for variable but generally shorter periods of time. Sample sizes of leks for SMZs were much larger than the sum of leks analyzed for individual populations within each SMZ because some populations had too few leks counted over too few years to make modeling at the population level feasible. These small counts were feasible to include in SMZ analyses and made SMZ indices most representative of population changes within the entire zone. These population indices provided the basis for all further analyses and modeling.

The variance of previous years' population indices clearly involve the variance of a product of θs, with the product and therefore the variance growing progressively larger as the population reconstruction is extended back further and further. We estimated the variance by following Goodman (1962), who proved the validity of a straightforward approach to estimating the variance of these products as:

$$\hat{Var}\left(\prod_{i=1}^{k} \hat{\theta}_i\right) = \prod_{i=1}^{k}(\hat{Var}(\hat{\theta}_i) + \hat{\theta}_i^2) - \prod_{i=1}^{k} \hat{\theta}_i^2 \quad \text{(Eqn. 15.7)}$$

Fitting Population Growth Models

We fit a suite of stochastic population growth models, including two density-independent models, to time series of reconstructed population indices for each SMZ and population: (1) exponential growth with process error (EGPE; Dennis et al. 1991) and (2) exponential growth with differing mean rates of change between the two time periods (period 1 = 1967–1987, period 0 = 1987–2007). We also fit 24 density-dependent models consisting of all combinations of four factors: (1) Ricker-type density dependence in population growth (Dennis and Taper 1994) or Gompertz-type density dependence in population growth (Dennis et al. 2006), (2) presence or absence of a time delay in the effect density has on population growth rate (no delay, one-year delay, or two-year delay), (3) a period effect (period, as described above), and (4) time trend in population carrying capacity (year, see below). In an earlier analysis of lek data through 2003 (Connelly et al. 2004), we tested an additional model (the Gompertz state space model [GSS] Dennis et al. 2006) incorporating observation error into the Gompertz-type density-dependent model, which consistently indicated that observation error in our estimates of rate of change, $r(t)$, of populations and SMZs across large numbers of leks was close to 0. Thus, we did not include state-space models in our 26-model set.

Specifically, let $N(t)$ be the observed population index at time t, $Y(t) = \log[N(t)]$, and the annual growth rate be $r(t) = Y(t + 1) - Y(t)$. Note that $r(t)$ is estimated as $\log(\lambda_t)$ as described above. The global stochastic model incorporating Ricker-type density dependence was:

$$r(t) = a + b \times N(t - \Delta) + c \times Year + d \times Period + E(t) \quad \text{(Eqn. 15.8)}$$

where $\Delta = 0$ for no time-delay, $\Delta = 1$ for a one-year delay, or $\Delta = 2$ for a two-year delay; *Year* is the calendar year at time t; and *Period* = 1 if *Year* = 1967–1986 and *Period* = 0 if *Year* = 1987–2007. $E(t)$ represents environmental (i.e., process) variation in realized growth rates and is a normally distributed random deviate with mean = 0 and variance = σ^2. The analogous model for Gompertz-type density dependence was:

$$r(t) = a + b \times \ln(N(t - \Delta)) + c \times Year + d \times Period + E(t) \quad \text{(Eqn. 15.9)}$$

These models have five parameters (a, b, c, d, and σ^2) that can be estimated via maximum likelihood using the indices to past abundance data estimated from population reconstruction.

The only difference between the Ricker and Gompertz models is that Ricker assumes growth rates are a linear function of population size and Gompertz assumes growth rates are a linear function of the log of population size. Density-dependent models such as Gompertz and Ricker provide an objective approach to estimating a carrying capacity or quasi-equilibrium, which is defined as the population size at which the growth rate is 0. This quasi-equilibrium (hereafter, carrying capacity) represents a threshold in abundance below which population size tends to increase and above which population size tends to decrease.

Several plausible scenarios for population growth can be realized from these base models. Models involving time trends (+ *Year*) or period differences can be interpreted as inferring that carrying capacity is changing through time (e.g., negative slopes imply declines through time) or differs between time periods. For example, the parameter estimates from the Ricker model with a time trend (*Year*) and period effect (*Period*) can be used to estimate a carrying capacity as:

$$\hat{K} = -\hat{b}^{-1}(\hat{a} + \hat{c}Year + \hat{d}Period) \quad \text{(Eqn. 15.10)}$$

The *hat* notation over a parameter (e.g., $\hat{a}$) indicates this value was the maximum likelihood estimate for that parameter when fit to past abundance data. When parameters b and c are set to 0, these models reduce to the exponential growth with process error (EGPE) model (Dennis et al. 1991). Including *Period* simply allows for differing trends between the two time periods.

We fit 26 models to each set of estimated rates of change and observed abundance index data, using the statistical computing program R Version 2.6.1 (R Development Core Team 2008) and PROC MIXED and PROC REG in SAS (SAS Institute 2006). These stochastic growth models treat annual rates of change (r_t) as mixed effects of fixed effects (year and period) and random effects (reconstructed population index with or without log transformation and time lag). Annual rates of change (r_t) were consistently described well by a normal distribution. We used Akaike's information criterion corrected for small sample size (AIC_c) to compare the relative performance (i.e., predictive ability) of each model (Burnham and Anderson 2002). Likewise, we followed Akaike (1978, 1979, 1981, and 1983), Buckland et al. (1997) and Burnham and Anderson (2002) in calculating Akaike weights (w_i), which we treat as relative likelihoods for a model given the data:

$$w_i = \frac{exp(-0.5 \times \Delta_i)}{\sum_{j=1}^{R} exp(-0.5 \times \Delta_j)} \quad \text{(Eqn. 15.11)}$$

where Δ_i is the difference between the AIC_c for model i and the lowest AIC_c of all R models. We report a 95% confidence set of models based on the best model using the sum of model weights ≥ 0.95 for a given analysis unit (Burnham and Anderson 2002). This approach reduces the number of models reported for all analysis units to those with some potential of explaining the data but does not necessarily drop all models with ΔAIC_c less than 2 or 3.

Stochastic Population Projections

We performed parametric bootstraps (Efron and Tibshirani 1998) on minimum population size by projecting 100,000 replicate abundance trajectories for 30 and 100 years into the future (post-2007) for each population and SMZ using:

$$N(t+1) = N(t) \times e^{\hat{r}(t)} \quad \text{(Eqn. 15.12)}$$

where $\hat{r}(t)$ was the stochastic growth rate calculated using maximum likelihood parameter estimates for the given model. For example, to project based on the Gompertz model with no time lag, a time trend in carrying capacity and a difference between periods, we used:

$$N(t+1) = N(t) \times e^{\hat{a} + \hat{b}ln\, N(t) + \hat{c}Year + \hat{d}Period + E(t)}, \quad \text{(Eqn. 15.13)}$$

where $N(0)$, the initial abundance for the projections, was the final observed population size index (e.g., population size in 2007); *Period* = 0, indicating that future growth would be analogous to what occurred from 1987 to 2007; and $E(t)$ was a random deviate drawn from a normal distribution with mean 0 and standard deviation equal to $\hat{\sigma}$ (square root of maximum likelihood estimate of mean squared error remaining from mixed model). These replicate time series were used to calculate the probability that the population or SMZ would decline below a quasi-extinction threshold corresponding to minimum counts of 20 and 200 males at leks. Probability of quasi-extinction was the proportion of replications in which population abundance declined below the quasi-extinction threshold at some point during the time horizon (30 or 100 years). Thresholds of 20 and 200 were chosen to correspond approximately to the standard 50/500 rule for effective population sizes (N_e; Franklin 1980, Soulé 1980) expressed in terms of breeding males counted at leks and mean adult sex ratio at leks (2.5 adult females per adult male, Patterson 1952, Schroeder et al. 1999). N_e was formally defined by Sewall Wright (1938) as:

$$N_e = \frac{1}{\frac{1}{N_m} + \frac{1}{N_f}} \qquad \text{(Eqn. 15.14)}$$

where N_m = number of males successfully breeding and N_f = female breeders.

Aldridge (2001) estimated N_e for the population of sage-grouse in Alberta (part of the northern Montana population) by applying previous estimates of male and female breeding success to his counts of 140 males and 280 females attending eight leks to estimate an N_e of 88. However, Bush (2009) recently used genetic tools to estimate that 46% of the males at the same leks surveyed by Aldridge successfully breed yielding N_e = 228 from Wright's formula. This implies that N_e = 50 requires 30 males present at the lek. When Bush (2009) identified males present at the leks from individual genotypes extracted from feathers left at the lek sites during the lekking period, she found that 50% more males actually attended the site than were counted in surveys. Thus a maximum count of 20 males during the lekking period is required to have 30 males present at the lek, resulting in an N_e of 50. Likewise, a minimum count of 200 males at leks in a region is required to ensure N_e = 500.

In other words, forecasting future probability of a local population or SMZ declining below effective population size of 50 breeding adults (N_e = 50, corresponding to an index based on minimum males counted at leks of 20 or less) identifies populations or SMZs at short-term risk for extinction (Franklin 1980, Soulé 1980), while a local population or SMZ declining below effective population size of 500 breeding adults (N_e = 500, corresponding to an index based on minimum males counted at leks of 200 or less) identifies populations or SMZs at long-term risk for extinction (Franklin 1980, Soulé 1980).

Most populations and SMZs, based on our comparison of AIC_c values, had >1 model that could be considered a competing best model by scoring within the 95% set. This generally meant ΔAIC_c <3. We projected future population abundances using each of the 26 models and used model averaging to incorporate model selection uncertainty into forecasts of population viability (Burnham and Anderson 2002) to generate an overall estimate of the probability of quasi-extinction, based on all fitted models. Generally, a model-averaged prediction can be obtained by calculating the predicted value of a parameter of interest (e.g., probability of quasi-extinction) for each model and taking a weighted average of the predictions where the weights are the relative likelihoods of each model:

$$\hat{Pr}(Extinction) = \sum_{i=1}^{R} \left(\hat{Pr}(Extinction \mid Model_i) \times w_i\right) \qquad \text{(Eqn. 15.15)}$$

Probability of extinction under a particular model is conditional on that model and its maximum likelihood parameter estimates. We calculated a weighted variance for these probabilities of extinction to assess the precision of these model-averaged probabilities of quasi-extinction (Krebs 1998) similar to the variance of a mean for grouped data (Remington and Schork 1970:46):

$$\hat{Var}[\hat{Pr}(Extinction)] = \sum_{i=1}^{R} w_i^2 \times \left[\hat{Pr}(Extinction) - \hat{Pr}(Extinction \mid Model_i)\right]^2 \qquad \text{(Eqn. 15.16)}$$

Metapopulation Analyses

We analyzed viability of the metapopulation of sage-grouse SMZs similarly to the analysis for individual SMZs with three exceptions. First, instead of basing population projections on all 26 models, we used only the information-theoretic best models for Ricker- and Gompertz-type density dependence. Second, the metapopulation model required estimated dispersal rates among SMZs. Last, correlated dynamics among SMZs were modeled by including a covariance in the random deviates used to portray environmental stochasticity.

Specifically, the metapopulation was projected through time using:

$$N_{Meta}(t+1) = \sum_{j=1}^{7} N_j(t+1) \qquad \text{(Eqn. 15.17)}$$

where N_j is the abundance of SMZ j. Abundance of each SMZ was projected using

$$N_j(t+1) = N_j(t) \times e^{r_j(t)} + \sum_{i=1 \neq j}^{7} N_i(t) \times D_{ij} - \sum_{i=1 \neq j}^{7} N_j(t) \times D_{ji}$$

(Eqn. 15.18)

where D_{ij} is the dispersal rate between SMZ i and j. We followed the approach developed by Knick and Hanser (this volume, chapter 16) to estimate dispersal rates between populations within SMZs. The probability of connectivity between every pair of leks was estimated using graph theory, based on distance between known leks, the difference in size between adjacent leks, and the product of all probable steps (dispersal limited to 27 km) between the pair of leks (Knick and Hanser, this volume, chapter 16). We expressed the estimated number of probable connective links between leks in adjacent SMZs, based on graph theory, as a proportion of all the links shown between any pair of SMZs ($N = 112$). These proportions were standardized to an estimated maximum dispersal rate at a distance of 27 km of 0.05 (Knick and Hanser, this volume, chapter 16). The random deviate, $E_j(t)$, for the growth rate of the jth SMZ, $r_j(t)$, was drawn from a multivariate normal distribution with mean = 0 and the seven by seven variance/covariance matrix estimated from past abundance trajectories. We obtained estimates of covariance by correlating the residuals of the information-theoretic best models for each SMZ pair. We used a program written in Visual Basic (MetaPVA; J. S. Horne and E. O. Garton, unpubl.) for metapopulation projections.

RESULTS

Great Plains Management Zone

This SMZ represents sage-grouse populations in parts of Alberta, Saskatchewan, Montana, North Dakota, South Dakota, and Wyoming. Most of the sage-grouse in Montana and all sage-grouse in Alberta, Saskatchewan, and the Dakotas occur in this SMZ (Fig. 15.1).

Dakotas Population

This population occupies the western portions of North and South Dakota and small parts of southeastern Montana and northeastern Wyoming (Table 15.1). It occurs on the far eastern edge of the range of sage-grouse and is separated from other populations by distance and habitat features. The average number of leks counted per five-year period increased substantially from 1965–1969 to 2000–2007 (Table 15.2).

The proportion of active leks decreased 17% over the assessment period (Table 15.2). Population trends indicated by average number of males per lek declined 46% from 1965–1969 to 1995–1999 but then recovered during 2000–2007. Average number of males per active lek demonstrated the same pattern as males/lek (Table 15.2). Average rates of change were <1.0 for three of the eight

TABLE 15.2

Sage-grouse lek monitoring effort, lek size, and trends averaged over 5-yr periods for the Dakotas population, 1965–2007.

Parameter	2000–2007[a]	1995–1999	1990–1994	1985–1989	1980–1984	1975–1979	1970–1974	1965–1969
Leks counted	56	34	37	26	27	18	19	20
Males/lek	11	7	12	11	14	10	15	13
Active leks	39	24	31	22	24	16	16	17
% active leks	69	72	84	85	92	87	89	86
Males/active lek	16	10	14	13	16	11	17	15
λ	1.004	1.148	0.913	1.013	0.965	1.116	0.883	1.128
SE (λ)[b]	0.091	0.108	0.060	0.077	0.099	0.155	0.078	0.114

[a] Eight yr of data in this period.

[b] Standard error for annual rate of change.

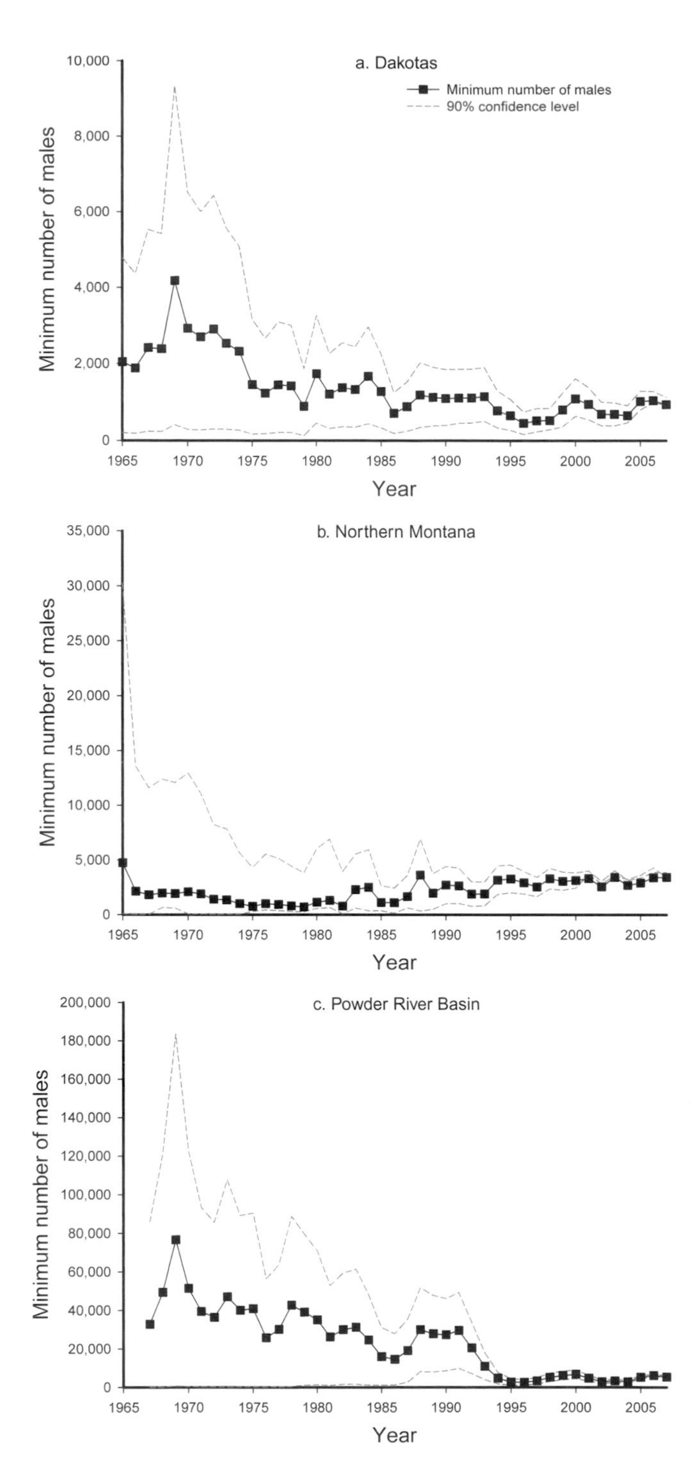

Figure 15.2. Population reconstructions for Great Plains populations and Great Plains Management Zone: (a) Dakotas; (b) northern Montana; (c) Powder River Basin; (d) Yellowstone watershed; (e) Great Plains Management Zone.

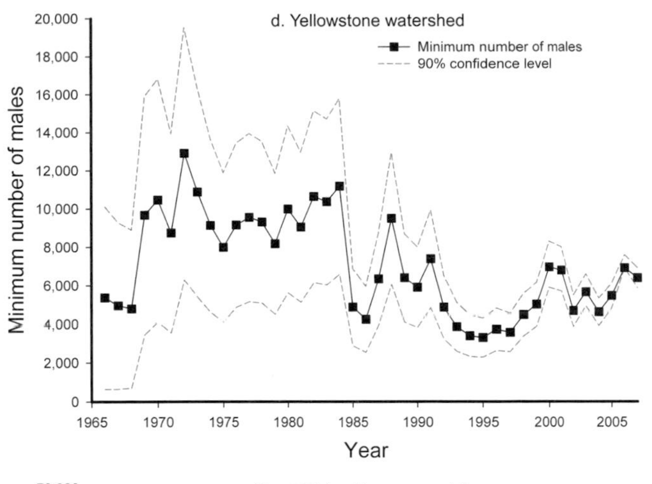

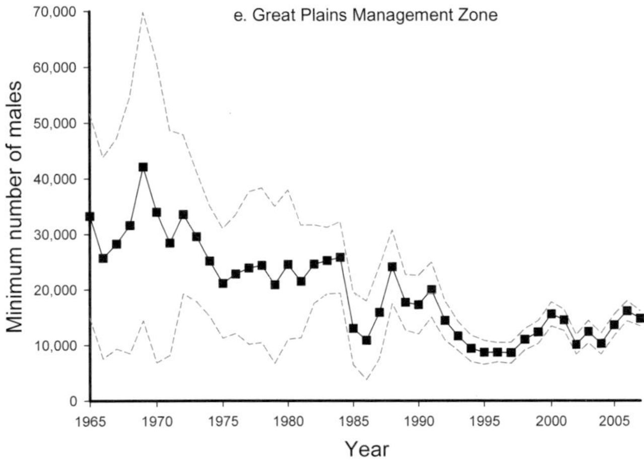

Figure 15.2. (*Continued*)

analysis periods. Contrary to lek size information, rate of change decreased 12.5% from the 1995–1999 analysis period to the 2000–2007 period, although both values remained at or above 1.0 for both of those periods (Table 15.2).

We used our 2007 minimum population estimate of 939 males (SE = 120) from 120 leks to reconstruct minimum population estimates for males back to 1965 (Fig. 15.2a). The population increased from about 2,000 males in 1965 to peak above 4,000 males in 1969, followed by a continuous decline through 2007.

The best stochastic model for the annual rates of change of the Dakotas population of sage-grouse was a Gompertz model with no time lags and a declining time trend of –3.2% per year ($r_t = 28.601 - 0.400 \ln N_t - 0.013$ year, $\sigma = 0.2503$, $r^2 = 0.190$; Table 15.3).

The Gompertz model with declining time trend implies the Dakotas population of sage-grouse will fluctuate around carrying capacities, which will decline from 587 males attending leks in 2007 to 222 attending leks in 2037 and only 23 males in 2107 if this trend continues at the same rate in the future. The 2007 count of 939 males was 50% higher than this estimated carrying capacity. A parametric bootstrap based on the Gompertz model with declining time trend (29% relative likelihood) infers there is virtually no chance of the population declining below $N_e = 50$, but declining below $N_e = 500$ is likely (72% relative probability) within 30 years. If this trend continues for 100 years there is a 67% chance of the population declining below $N_e = 50$ and a 100% probability of declining below $N_e = 500$.

TABLE 15.3

Candidate model set (contains 95% of model weight) and model statistics for estimating population trends and persistence probabilities for Greater Sage-Grouse in the Dakotas population, 1965–2007.

Model	Model statistics[a]			
	r^2	K	ΔAIC_c	w_i
Gompertz + year	0.190	4	0.0[b]	0.288
Gompertz	0.094	3	2.0	0.106
Gompertz + year, period	0.196	5	2.3	0.092
Ricker	0.070	3	3.0	0.063
Gompertz + period	0.121	4	3.3	0.056
Ricker + year	0.119	4	3.3	0.054
EGPE	0.000	3	3.6	0.048
Gompertz t − 1	0.047	4	4.0	0.039
Gompertz t − 2	0.038	4	4.4	0.032
Gompertz t − 1 + year	0.087	5	4.8	0.026
Ricker t − 1	0.026	4	4.9	0.025
Ricker t − 2	0.022	4	5.0	0.023
Ricker + period	0.079	4	5.1	0.022
Ricker + year, period	0.132	5	5.4	0.019
Gompertz t − 2 + year	0.069	5	5.5	0.018
Period	0.009	3	5.6	0.018
Gompertz t − 1 + period	0.058	5	6.0	0.014
Gompertz t − 2 + period	0.044	5	6.6	0.011

[a] Model fit described by coefficient of determination (r^2), the number of parameters (K), the difference in Akaike's information criterion corrected for small sample size (ΔAIC_c), and the AIC_c weights (w_i).
[b] $AIC_c = 11.9$ for best selected model.

Northern Montana Population

This population occupies parts of north-central Montana, southeast Alberta, and southwest Saskatchewan and is separated from adjacent populations by about 20 km and the Missouri River (Table 15.1). The average number of leks counted per five-year period increased substantially from 1965–1969 to 2000–2007 (Table 15.4).

The proportion of active leks declined somewhat over the assessment period, but in part this may be due to the relatively few leks counted until the mid-1990s (Table 15.4). Population trends indicated by average number of males per lek declined by 61% from 1965–1969 to 1995–1999 but increased by 91% from 1995–1999 to 2000–2007. Average number of males per active lek fluctuated but remained relatively constant over the assessment period (Table 15.4). Average rates of change were <1.0 for two of the eight analysis periods, and generally suggested a stable to increasing population during the 1995–1999 and 2000–2007 periods (Table 15.4).

From a minimum population estimate of 3,435 males (SE = 274) in 2007 based on counts at 156 leks, we reconstructed a minimum population estimate for males from 2007 back to 1965 (Fig. 15.2b) using 1,437 lek counts reported for this period. The first population estimate of >4,700 males in 1965 was the largest for the entire time period, but it and other estimates in the late 1960s were based on only two leks counted per year, yielding standard errors as large as 15,000 males.

TABLE 15.4

Sage-grouse lek monitoring effort, lek size, and trends averaged over 5-yr periods for the Northern Montana population, 1965–2007.

Parameter	2000–2007[a]	1995–1999	1990–1994	1985–1989	1980–1984	1975–1979	1970–1974	1965–1969
Leks counted	162	56	19	18	17	19	2	10
Males/lek	21	11	18	22	27	17	29	28
Active leks	123	31	15	17	16	19	2	10
% active leks	76	61	88	98	93	99	100	98
Males/active lek	28	18	20	22	28	18	29	28
λ	1.031	1.002	1.079	1.319	1.241	1.118	0.823	0.890
SE (λ)[b]	0.042	0.083	0.118	0.266	0.486	0.147	0.168	0.104

[a] Eight yr of data in this period.
[b] Standard error for annual rate of change.

TABLE 15.5

Candidate model set (contains 95% of model weight) and model statistics for estimating population trends and persistence probabilities for greater sage-grouse in the Northern Montana population, 1965–2007.

	Model statistic[a]			
Model	r^2	k	ΔAIC_c	w_i
Ricker + period	0.357	4	0.0[b]	0.470
Gompertz + period	0.331	4	1.6	0.216
Ricker + year, period	0.366	5	2.0	0.171
Gompertz + year, period	0.332	5	4.1	0.060

[a] Model fit described by coefficient of determination (r^2), the number of parameters (k), the difference in Akaike's information criterion corrected for small sample size (ΔAIC_c), and the AIC_c wt (w_i).
[b] $AIC_c = 19.2$ for best selected model.

Counts of ≥12 leks in the mid-1970s produced more precise estimates, with standard errors declining to be no larger than the estimates by the mid-1980s. Counts of 24 to 36 leks beginning in the mid-1990s provided more precise estimates, fluctuating in the range of 3,000–3,500 males attending leks from 1995–2007 (Fig. 15.2b).

The best stochastic model for the annual rates of change of the northern Montana population of sage-grouse was a Ricker model with no time lags and a period effect, suggesting that the carrying capacity in 2007 of 2,744 was 1,519 breeding males lower than in 1965–1987 (r_t = 1.067 – 0.000367 N_t – 0.556 Period, σ = 0. 2745, r^2 = 0.357; Table 15.5). The analogous Gompertz model had a ΔAIC_c of 1.6, an r^2 of 0.331 and a relative likelihood of 22% (w_i = 0.22).

The Ricker model with a period effect implies the northern Montana population of sage-grouse will fluctuate around a carrying capacity of 2,908 males attending leks if the pattern of change observed in the past 20 years remains for 30 or 100 years in the future. The Gompertz model with period effect gives virtually identical predictions. A parametric bootstrap based on the Ricker model with period effect, which has a 47% relative likelihood, infers there is virtually no chance of the population declining below $N_e = 50$ or $N_e = 500$ within 30 years. It is unlikely the population will decline below $N_e = 50$ or $N_e = 500$ if conditions remain

TABLE 15.6

Multimodel forecasts of probability (weighted mean percentage and standard error) of Greater Sage-Grouse populations (Fig. 15.1) and Sage-Grouse Management Zones declining below $N_e = 50$ and $N_e = 500$ in 30 and in 100 yr.

	Pr ($<N_e$) in 30 yr		Pr ($<N_e$) in 100 yr	
Populations by management zone	$N_e = 50$	$N_e = 500$	$N_e = 50$	$N_e = 500$
Great Plains SMZ I+				
Dakotas	4.6	39.5	44.6	66.3
Northern Montana	0.0	0.0	0.2	2.0
Powder River Basin, Montana	2.9	16.5	85.7	86.2
Yellowstone watershed	0.0	8.1	55.6	59.8
Overall[a]	9.5 (5.9)	11.1 (5.8)	22.8 (8.4)	24.0 (8.3)
Wyoming Basin SMZ II				
Jackson Hole, Wyoming	11.2	100	27.3	100
Middle Park, Colorado	2.5	100	7.1	100
Wyoming Basin	0.0	0.0	9.9	10.7
Overall[a]	0.1 (0.3)	0.3 (1.1)	16.1 (7.4)	16.2 (7.4)
Southern Great Basin SMZ III				
Mono Lake, California–Nevada	15.4	100.0	37.9	100.0
South Mono Lake, California	0.1	81.5	0.6	99.9
Northeast interior, Utah	0.8	51.8	8.8	78.6
Sanpete–Emery Counties, Utah	77.7	100.0	99.2	100.0
South central Utah	0.0	3.2	1.1	21.0
Summit–Morgan Counties, Utah	20.6	100.0	41.8	100.0
Tooele–Juab Counties, Utah	56.5	100.0	100.0	100.0
Southern Great Basin	0.0	2.0	4.2	78.0
Overall[a]	0.0 (0.0)	0.0 (0.1)	6.5 (4.9)	7.8 (5.3)
Snake River Plain SMZ IV				
Baker, Oregon	61.9	100.0	66.8	100.0
Bannack, Montana	6.4	70.2	32.7	97.7
Northern Great Basin	2.1	2.5	2.5	99.7
Red Rocks, Montana	0.1	55.3	2.5	91.9
Snake–Salmon–Beaverhead, Idaho	4.2	10.2	19.3	26.8
Overall[a]	2.3 (1.4)	10.5 (6.1)	19.4 (7.9)	39.7 (9.6)
Northern Great Basin SMZ V				
Central Oregon	4.2	15.2	74.9	91.3
Western Great Basin	5.5	6.4	6.4	99.1
Overall[a]	1.0 (2.0)	2.1 (2.3)	7.2 (5.0)	29.0 (8.1)

TABLE 15.6 (*continued*)

TABLE 15.6 (CONTINUED)

Populations by management zone	Pr (<N_e) in 30 yr		Pr (<N_e) in 100 yr	
	$N_e = 50$	$N_e = 500$	$N_e = 50$	$N_e = 500$
Columbia Basin SMZ VI				
Moses Coulee, Washington	9.8	87.6	62.4	99.8
Yakima, Washington	26.1	100	50.4	100.0
Overall[a]	12.4 (6.0)	76.2 (6.5)	62.1 (9.1)	86.3 (5.8)
Colorado Plateau SMZ VII	0.0 (0.0)	95.6 (3.7)	5.1 (2.3)	98.4 (3.7)
Summary[b]				
Popns < N_e = 50,500%	3	13	8	18
	13%	54%	33%	75%
SMZs < N_e = 50,500%	0	2	1	2
	0%	29%	14%	29%

[a] Overall estimates (SE) are based on all leks surveyed within an SMZ including small populations not listed in table because of small sample size of leks and/or years of data collection.

[b] Summary values are the number and percentage of populations and management zones with >50% likelihood of declining below $N_e = 50$ and $N_e = 500$.

the same for 100 years. Across all 26 models of population growth, there is only a 2% relative probability of the population declining below $N_e = 500$ within 100 years if population changes observed in the last 43 years continue unchanged (Table 15.6).

Powder River Basin, Montana, Population

This population occupies parts of southeastern Montana and northeastern Wyoming (Table 15.1). The average number of leks counted per five-year period increased substantially from 1965–1969 to 2000–2007 (Table 15.7). The proportion of active leks declined over the assessment period (Table 15.7). Population trends, indicated by average number of males per lek, declined by 45% from 1970–1974 to 2000–2007. The average number of males per active lek declined by 24% over the assessment period (Table 15.7). Average rates of change were <1.0 for three of the seven analysis periods, and decreased by 15.7% from the 1995–1999 to the 2000–2007 period, but both values remained at or above 1.0 for both periods (Table 15.7).

We reconstructed a minimum population estimate for males from 2007 back to 1967 (Fig. 15.2c) using 2,358 lek counts reported for this period from a minimum population estimate of 5,397 males (SE = 401) in 2007, based on counts at 344 leks. The estimated population peaked at more than 76,000 males (SE = 66,799) in 1969, with irregular short-duration fluctuations or cycles (four to five years between peaks) overlaid on a strongly declining trend through 1996. Counts at leks (range = 70–350) beginning in the mid-1990s provided relatively precise estimates fluctuating in the range of 3,000–6,000 males attending leks from 1996 to 2007 (Fig. 15.2c).

The best stochastic model for annual rates of change of the Powder River Basin population of sage-grouse was a Gompertz model with a one-year time lag and a rapidly declining time trend of −7.3% per year ($r_t = 60.417 - 0.377 \ln(N_{t-1}) - 0.0286$ year, $\sigma = 0.2618$, $r^2 = 0.315$), this model was supported by the data with a relative likelihood of 55% (Table 15.8).

The Gompertz model with declining time trend implies the Powder River Basin population of sage-grouse will fluctuate around a carrying capacity that will decline from 3,042 males attending leks in 2007 to only 312 males attending leks in 2037, to going extinct with only two males attending leks in 2107 if this trend continues at the same rate in the future. The 2007 count of 5,397 males is estimated to be about 2,000 males higher than the carrying capacity of

TABLE 15.7

Sage-grouse lek monitoring effort, lek size, and trends averaged over 5-yr periods for the Powder River Basin population, 1970–2007.

Parameter	2000–2007[a]	1995–1999	1990–1994	1985–1989	1980–1984	1975–1979	1970–1974
Leks counted	239	84	66	63	62	20	14
Males/lek	12	7	12	15	22	25	22
Active leks	158	46	48	49	56	17	13
% active leks	66	54	72	78	90	81	90
Males/active lek	19	13	16	19	24	30	25
λ	1.027	1.218	0.662	1.140	0.874	1.006	0.971
SE (λ)[b]	0.067	0.134	0.078	0.148	0.087	0.147	0.135

[a] Eight yr of data in this period.

[b] Standard error for annual rate of change.

TABLE 15.8

Candidate model set (contains 95% of model weight) and model statistics for estimating population trends and persistence probabilities for Greater Sage-Grouse in the Powder River Basin population, 1970–2007.

Model	Model statistic[a]			
	r^2	k	ΔAIC_c	w_i
Gompertz t − 1 + year	0.315	5	0.0[b]	0.553
Gompertz t − 1 + year, period	0.318	6	2.5	0.159
Gompertz t − 2 + year	0.242	5	3.9	0.081
Gompertz t − 1 + period	0.228	5	4.5	0.057
Gompertz t − 2 + year, period	0.249	6	6.1	0.026
Gompertz t − 2 + period	0.197	5	6.1	0.027
Ricker t − 1 + year	0.181	5	6.8	0.019
Gompertz t − 1	0.112	4	7.3	0.014
Gompertz t − 2	0.097	4	8.0	0.010
Ricker t − 1 + period	0.142	5	8.5	0.008

[a] Model fit described by coefficient of determination (r^2), the number of parameters (k), the difference in Akaike's information criterion corrected for small sample size (ΔAIC_c), and the AIC_c wt (w_i).

[b] AIC_c = 15.2 for best selected model.

the region. A parametric bootstrap based on the Gompertz model with declining time trend, which has a 29% relative likelihood, infers that there is little chance (3%) of the population declining below $N_e = 50$ but that declining below $N_e = 500$ is more likely (17% relative probability) within 30 years. Multimodel projections across all 26 models forecast that if this trend continues for 100 years there is an 86% probability of the population declining below $N_e = 50$ and $N_e = 500$ (Table 15.6).

Yellowstone Watershed Population

This population occupies much of southeastern Montana and northeastern Wyoming. It is separated

TABLE 15.9
Sage-grouse lek monitoring effort, lek size, and trends averaged over 5-yr periods for the Yellowstone watershed population, 1965–2007.

Parameter	2000–2007[a]	1995–1999	1990–1994	1985–1989	1980–1984	1975–1979	1970–1974	1965–1969
Leks counted	346	132	133	141	130	86	53	8
Males/lek	15	12	14	15	24	22	21	17
Active leks	231	89	96	111	118	79	48	8
% active leks	68	67	74	79	91	92	89	96
Males/active lek	21	18	19	19	26	23	24	18
λ	1.009	1.170	0.911	1.092	0.914	1.053	0.974	1.247
SE (λ)[b]	0.052	0.084	0.061	0.068	0.050	0.071	0.104	0.191

[a] Eight yr of data in this period.
[b] Standard error for annual rate of change.

from other populations by distance and topographic features (Table 15.1). The average number of leks counted per five-year period increased substantially from 1965–1969 to 2000–2007 (Table 15.9).

The proportion of active leks declined over the assessment period (Table 15.9). Population trends, as indicated by average number of males per lek, declined slightly from 1965–1969 to 2000–2007. Lek size increased by >41% from 1965–1969 to 1980–1984 and then decreased by 37% from 1980–1984 to 2000–2007. Average number of males per active lek also had the same pattern over the assessment period (Table 15.9). Average rates of change were <1.0 for three of the eight analysis periods and declined by 14% between the last two analysis periods (Table 15.9). Nevertheless, both values remained at or above 1.0 for both of the periods.

From a minimum population estimate of 6,385 males (SE = 327) in 2007 based on counts at 286 leks, we reconstructed a minimum population estimate for males from 2007 back to 1966 (Fig. 15.2d) using counts at 1,169 leks reported for this period. The estimated population peaked at just below 13,000 males (SE = 940) in 1972 during a period of relative high numbers (8,000–13,000) from 1969–1984, followed by fluctuations of 3,000–9,000 until present. Counts at >100 leks beginning in 1985 provided precise minimum estimates of number of males attending leks.

The best stochastic model for the annual rates of change of the Yellowstone watershed population of sage-grouse was a Ricker model with no time lags and a declining time trend of –4.5% per year (r_t = 27.938 – 0.00010421 ln(N_t) – 0.0138 year, σ = 0.2204, r^2 = 0.338). The analogous Gompertz model was not competitive, with only a 4% relative likelihood, while other Ricker models with Period or time + Period had high relative likelihoods (Table 15.10).

The Ricker model with declining time trend implies the Yellowstone watershed population of sage-grouse will fluctuate around a carrying capacity that will decline from 2,948 males in 2007 to extinction in 2037 if this trend continues at the same rate in the future. The 2007 count was more than twice as large as the estimated carrying capacity. The carrying capacity in 2037 was below 0. A parametric bootstrap based on the Ricker model with declining time trend infers there is virtually no chance of the population declining below a N_e = 50, but declining below N_e = 500 is more likely (21% relative probability) within 30 years. If this trend continues for 100 years, there is a 100% probability of the population declining below N_e = 50 and N_e = 500, though multimodel forecasts across all models predict lower (56% and 60%, respectively) probabilities.

Comprehensive Analysis of All Leks in the Management Zone

In 1965–1969, an average of 45 leks per year was censused. By 2005–2007, an average of 830 leks per year was counted, an increase of 1,744%

TABLE 15.10

Candidate model set (contains 95% of model weight) and model statistics for estimating population trends and persistence probabilities for Greater Sage-Grouse in the Yellowstone watershed population, 1965–2007.

Model	Model statistic[a]			
	r^2	k	ΔAIC_c	w_i
Ricker + year	0.338	4	0.0[b]	0.385
Ricker + period	0.317	4	1.2	0.211
Ricker + year, period	0.353	5	1.8	0.160
Gompertz + year	0.279	4	3.3	0.074
Gompertz + period	0.261	4	4.3	0.045
Gompertz + year, period	0.289	5	5.4	0.026
Gompertz t − 1 + period	0.225	5	6.1	0.018
Ricker t − 1 + period	0.225	5	6.1	0.018
Gompertz t − 1 + year	0.205	5	7.1	0.011
Ricker	0.153	3	7.1	0.011

[a] Model fit described by coefficient of determination (r^2), the number of parameters (k), the difference in Akaike's information criterion corrected for small sample size (ΔAIC_c), and the AIC_c wt (w_i).

[b] $AIC_c = 1.9$ for best selected model.

TABLE 15.11

Sage-grouse lek monitoring effort, lek size, and trends averaged over 5-yr periods for the Great Plains Management Zone, 1965–2007.

Parameter	2000–2007[a]	1995–1999	1990–1994	1985–1989	1980–1984	1975–1979	1970–1974	1965–1969
Leks counted	830	307	261	255	243	145	87	45
Males/lek	15	10	13	15	22	20	21	18
Active leks	564	191	194	206	221	133	79	41
% active leks	68	62	75	81	91	92	90	91
Males/active lek	22	16	18	19	24	22	23	20
λ	1.016	1.130	0.884	1.105	0.915	1.036	0.918	1.026
SE (λ)[b]	0.030	0.056	0.043	0.055	0.040	0.057	0.062	0.092

[a] Eight yr of data in this period.

[b] Standard error for annual rate of change.

(Table 15.11). The proportion of active leks decreased over the assessment period, averaging between 90% and 92% from 1965–1984, but declining to 68% by 2005–2007 (Table 15.11). Population trends, as indicated by average number of males per lek, decreased over the assessment period by 17% while average number of males per active lek increased by 10% (Table 15.11). Average annual rates of change were <1.0 for three of the eight analysis periods. Average annual rates of change declined by 10% from 1995–1999 to 2000–2007, but values remained at or above 1.0 for both of these periods.

From a minimum population estimate of 14,814 males (SE = 609) in 2007 based on counts at 905 leks, we reconstructed a minimum population estimate for males from 2007 back to 1967 (Fig. 15.2e), using counts at 1,977 leks reported for this period. The estimated population peaked at more than 42,000 males (SE = 13,702) in 1969 and showed irregular declines until the mid-1970s, followed by a plateau lasting until the mid-1980s, when the population fluctuated dramatically until it stabilized in the mid-1990s and began a slow increase until 2006. Counts at >200 leks beginning in 1980 provided precise minimum estimates of number of males attending leks relative to the earlier period.

The best stochastic model for annual rates of change of the population of sage-grouse in the entire Great Plains SMZ is a Gompertz model with a one-year time lag and a declining time trend of −2.9% per year (r_t = 29.245 − 0.430 $\ln(N_{t-1})$ − 0.013 year, σ = 0.197, r^2 = 0.315), and had a relative likelihood of 19% (Table 15.12). The analogous Ricker model has a relative likelihood of 8.6%, a ΔAIC_c of 1.7, an r^2 of 0.170, and a high annual rate of decline of −7.3% (r_t = 23.864 − 0.00002116 N_{t-1} − 0.012 year, σ = 0.201).

The Gompertz model with declining time trend implies the Great Plains SMZ population of sage-grouse will fluctuate around a carrying capacity that will decline from 9,579 males attending leks in 2007 to 3,974 males attending leks in 2037 to 510 in 2107 if this trend continues at the same rate in the future. The 2007 count of almost 15,000 males exceeded the

TABLE 15.12

Candidate model set (contains 95% of model weight) and model statistics for estimating population trends and persistence probabilities for Greater Sage-Grouse in the Great Plains Management Zone, 1965–2007.

	Model statistic[a]			
Model	r^2	k	ΔAIC_c	w_i
Gompertz t − 1 + year	0.203	5	0.0[b]	0.196
Gompertz + year	0.180	4	1.1	0.111
Ricker + year	0.176	4	1.4	0.099
Ricker t − 1 + year	0.170	5	1.7	0.086
Gompertz t − 2	0.161	4	2.0	0.071
Gompertz t − 2 + year	0.212	5	2.2	0.067
Gompertz t − 1	0.081	4	3.2	0.039
Gompertz	0.079	3	3.3	0.037
Ricker	0.072	3	3.6	0.032
Gompertz + period	0.127	4	3.7	0.032
Gompertz + year, period	0.180	5	3.8	0.030
Ricker t − 1	0.067	4	3.8	0.029
Ricker t − 2	0.124	4	3.8	0.030
Ricker + year, period	0.176	5	4.0	0.027
Ricker t − 2 + year	0.173	5	4.1	0.025
EGPE	0.000	3	4.3	0.023
Ricker + period	0.113	4	4.3	0.023

[a] Model fit described by coefficient of determination (r^2), the number of parameters (k), the difference in Akaike's information criterion corrected for small sample size (ΔAIC_c), and the AIC_c wt (w_i).

[b] AIC_c = −7.2 for best selected model.

estimated carrying capacity by 50%. Parametric bootstraps under this model also imply virtually no probability of the population declining below N_e = 50 or N_e = 500 if these rates are maintained indefinitely. The Ricker model analogous to the best Gompertz model predicts a carrying capacity of 7,647 males in 2007 that rapidly declines to extinction by 2037, with parametric bootstraps predicting 20% likelihood of the population declining below N_e = 500 in 30 years and 100% likelihood of numbers below N_e = 50 in 100 years. Multimodel forecasts across all 26 models predict 10% and 11% probabilities of declining below N_e = 50 and N_e = 500 in 30 years (Table 15.6), with standard errors of 5.9% and 5.8%, respectively, and higher (23% and 24%, respectively) probabilities in 100 years (SE = 8.4% and 8.3%, respectively).

Wyoming Basin Management Zone

This SMZ represents sage-grouse populations in parts of Montana, Colorado, Utah, and Wyoming (Fig. 15.1). Most of the sage-grouse in Wyoming and Colorado occur in this SMZ. Four of the five populations delineated within this management zone had data sufficient for analysis.

Eagle–South Routt Counties, Colorado, Population

The Eagle–South Routt Counties, Colorado, population is in north-central Colorado and is separated from nearby populations by distance and mountainous terrain (Table 15.1). The average number of leks counted per five-year period increased substantially from 1965–1969 to 2000–2007 (Table 15.13).

The proportion of active leks declined from 1985–1989 to 2000–2007 and declined substantially over the entire analysis period (Table 15.13). Population trends indicated by average number of males per lek declined from a high of 21 males/lek during 1965–1969 to 3 during 2000–2007, a decrease of 86%. Average numbers of males per active lek also decreased by 52% over the analysis period (Table 15.13). Average rates of change indicated a substantial decrease from 1995–1999 to 2000–2007, and a declining population during 2000–2007 (Table 15.13).

Population reconstruction, modeling, and persistence estimation were not conducted for the Eagle–South Routt population because of the large number of intervals of ≥3 successive years during which no or few leks were counted. It is unlikely the population will persist for 20 years (C. E. Braun, pers. comm.).

Jackson Hole, Wyoming, Population

This is a relatively isolated population in western Wyoming, separated from other populations by distance and mountainous terrain (Table 15.1). The average number of leks counted per five-year period increased from 1965–1969 to 2000–2007 (Table 15.14). In 1965–1969, an average of one lek

TABLE 15.13

Sage-grouse lek monitoring effort, lek size, and trends averaged over 5-yr periods for the Eagle–south Routt Counties, Colorado, population, 1965–2007.

Parameter	2000–2007[a]	1995–1999	1990–1994	1985–1989	1980–1984	1975–1979	1970–1974	1965–1969
Leks counted	21	15	20	16	3	0	5	9
Males/lek	3	3	4	3	—	—	—	21
Active leks	5	4	4	6	1	0	4	8
% active leks	34	24	23	41	—	—	—	84
Males/active lek	13	12	16	9	—	—	—	27
λ	0.906	1.840	—	—	—	—	—	1.092
SE (λ)[b]	0.137	—	0.000	0.155	—	—	—	0.247

[a] Eight yr of data in this period.
[b] Standard error for annual rate of change.

TABLE 15.14

Sage-grouse lek monitoring effort, lek size, and trends averaged over 5-yr periods for the Jackson Hole, Wyoming, population, 1965–2007.

Parameter	2000–2007[a]	1995–1999	1990–1994	1985–1989	1980–1984	1975–1979	1970–1974	1965–1969
Leks counted	6	8	7	5	1	—	—	1
Males/lek	16	10	28	19	—	—	—	—
Active leks	6	6	7	4	0	—	—	0
% active leks	92	73	100	77	—	—	—	—
Males/active lek	18	14	28	25	—	—	—	—
λ	1.075	0.851	1.029	1.217	—	—	—	—
SE (λ)[b]	0.158	0.346	0.223	0.371	—	—	—	—

[a] Eight yr of data in this period.
[b] Standard error for annual rate of change.

per five-year period was censused, but by 2005–2007, an average of six leks were counted.

The proportion of active leks fluctuated over the assessment period (Table 15.14). Population trends, as indicated by average number of males per lek, declined from 28 in 1990–1994 to 16 in 2000–2007. Average number of males per active lek also declined over the assessment period (Table 15.14). Average rates of change were <1.0 for one of the four periods for which data could be analyzed. However, the rate of change was <1.0 in the 1995–1999 analysis period and >1.0 in the 2000–2007 period, suggesting an increasing population over this period (Table 15.14).

Population reconstruction was only feasible for 1986–2007 for this population, but the estimated annual rates of change implied that the 2007 count of ≥129 males (SE = 44) attending nine leks was typical of counts during this time, with the maximum count of 200 occurring around 1990. Standard errors prior to 1998 were quite large (Fig. 15.3a), but the population had a pattern of declining counts with a count of 200 breeding males from 1988–1995 followed by a substantial decline to half that number in the late 1990s and early 2000s.

None of the 26 models garnered >10% relative likelihood, with the simplest models (exponential growth with process error, Gompertz, and Ricker) all showing similar relative likelihoods of 6–8% (Table 15.15).

The simple exponential growth with process error model estimated the annual rate of change for this population averaged −2.2%, which leads to relatively high likelihoods of populations declining below $N_e = 50$. Estimated male counts were already below 200, and probability of long-term persistence for $N_e = 500$ was 0%. Multimodel forecasts across all 26 models estimated 11% and 27% probabilities of declining below $N_e = 50$ in 30 and 100 years, respectively (Table 15.6).

Middle Park, Colorado, Population

This population occurs in north-central Colorado and is separated from adjacent populations by distance and mountainous terrain (Table 15.1). The average number of leks counted per five-year period increased from 7 to 35 from 1965–1969 to 1990–1994, and then decreased to 17 by 2000–2007 (Table 15.16).

The proportion of active leks declined from 1965–1969 to 1995–1999, and then increased in 2000–2007 (Table 15.16). Population trends indicated by average number of males per lek declined by 40% from 1965–1969 to 1995–1999, and then increased by 78% from 1995–1999 to 2000–2007. Average number of males per active lek remained relatively constant over the assessment period (Table 15.16). In contrast to males/lek data, average rates of change were <1.0 for five of the eight analysis periods and

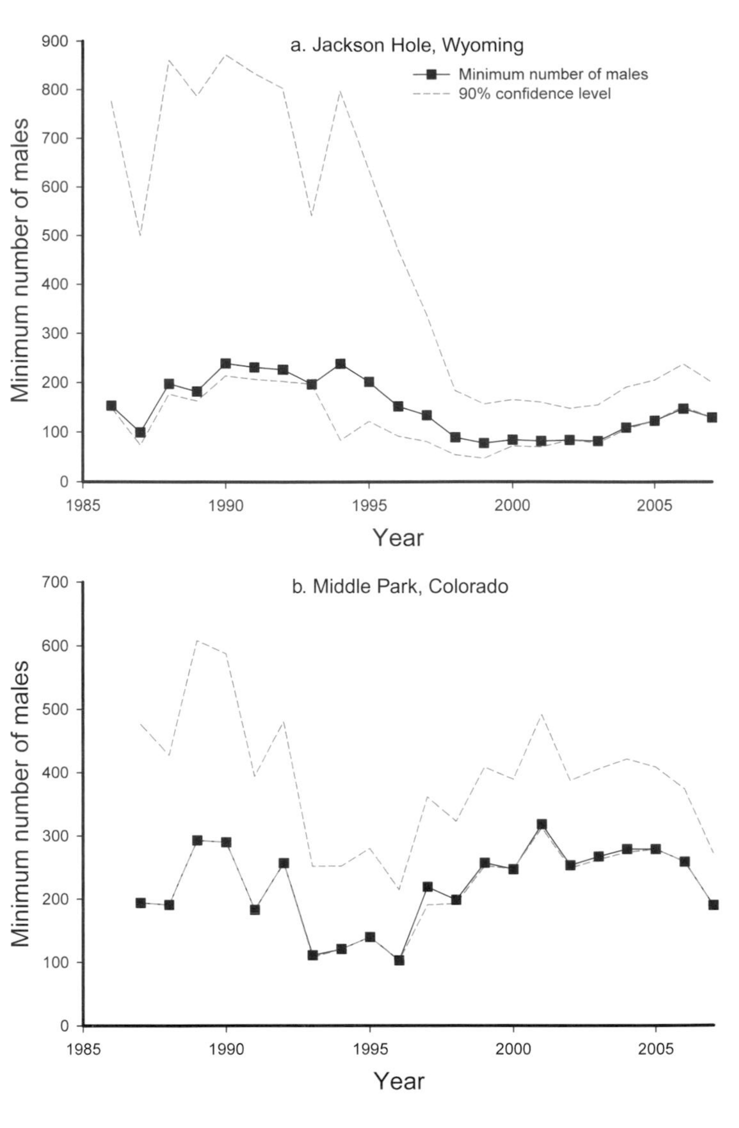

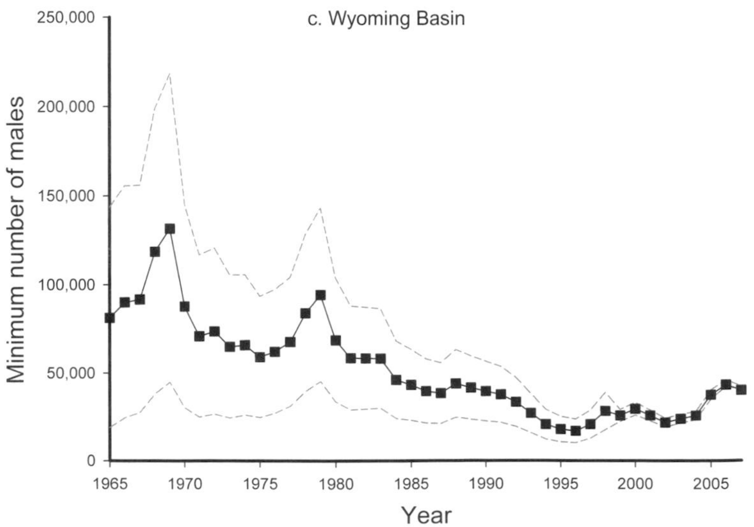

Figure 15.3. Population reconstructions for Wyoming Basin populations and Wyoming Basin Management Zone: (a) Jackson Hole, Wyoming; (b) Middle Park, Colorado; (c) Wyoming basin; (d) Wyoming Basin Management Zone.

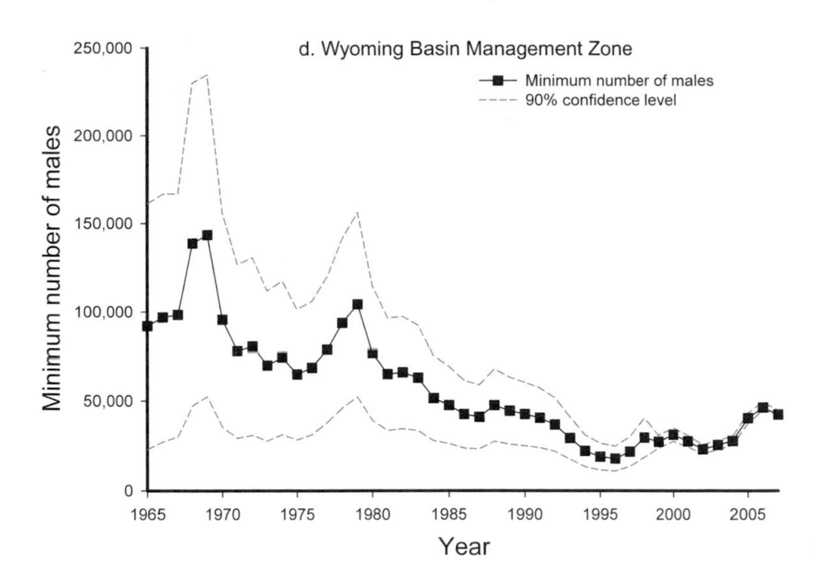

Figure 15.3. (*Continued*)

TABLE 15.15

Candidate model set (contains 95% of model weight) and model statistics for estimating population trends and persistence probabilities for Greater Sage-Grouse in the Jackson Hole, Wyoming, population, 1965–2007.

	Model statistic[a]			
Model	r^2	k	ΔAIC	w_i
EGPE	0.000	3	0.0[b]	0.170
Gompertz t − 2	0.126	4	0.3	0.148
Ricker	0.125	4	0.3	0.145
Ricker t − 2	0.112	4	0.6	0.126
Gompertz	0.111	4	0.6	0.125
Gompertz t − 1	0.049	4	1.9	0.066
Ricker t − 1	0.034	4	2.2	0.057
Ricker + year	0.162	5	2.7	0.043
Gompertz t − 2 + year	0.148	5	3.1	0.037
Gompertz + year	0.136	5	3.3	0.032
Ricker t − 2 + year	0.123	5	3.6	0.028

[a] Model fit described by coefficient of determination (r^2), the number of parameters (k), the difference in Akaike's information criterion corrected for small sample size (ΔAIC_c), and the AIC_c wt (w_i).
[b] $AIC_c = 5.4$ for best selected model.

indicated a generally decreasing population in 2000–2007 (Table 15.16). Moreover, rate of change declined by 21% between the 1995–1999 and 2000–2007 analysis periods and was <1.0 for 2000–2007.

Population reconstruction was only feasible for 1987–2007 for this population, but estimated annual rates of change implied the 2007 count of a minimum of 190 males (SE = 52) attending nine leks was typical of counts during this time, with maximum counts of 300 occurring around 1990 and 2001 separated by a decline to 100 in mid-1990s. Standard errors grew from 27% of the estimate in 2007 to approximately equal to it in 1987 (Fig. 15.3b), but the population showed only a simple pattern

TABLE 15.16

Sage-grouse lek monitoring effort, lek size, and trends averaged over 5-yr periods for the Middle Park, Colorado, population, 1965–2007.

Parameter	2000–2007[a]	1995–1999	1990–1994	1985–1989	1980–1984	1975–1979	1970–1974	1965–1969
Leks counted	17	21	35	26	9	8	5	7
Males/lek	16	9	5	7	14	23	18	15
Active leks	15	13	17	15	6	8	4	6
% active leks	90	66	49	58	71	96	96	84
Males/active lek	17	14	11	12	20	24	19	17
λ	0.978	1.241	0.916	0.729	1.412	0.834	0.714	1.095
SE (λ)[b]	0.080	0.197	0.149	0.122	1.217	0.271	0.182	0.259

[a] Eight yr of data in this period.
[b] Standard error for annual rate of change.

TABLE 15.17

Candidate model set (contains 95% of model weight) and model statistics for estimating population trends and persistence probabilities for Greater Sage-Grouse in the Middle Park, Colorado, population, 1986–2007.

Model	Model statistic[a]			
	r^2	k	ΔAIC	w_i
Gompertz	0.156	4	0.0[b]	0.228
EGPE	0.000	3	0.4	0.190
Ricker	0.134	4	0.5	0.179
Gompertz + year	0.195	5	2.3	0.070
Gompertz t − 2	0.041	4	2.4	0.068
Ricker + year	0.183	5	2.6	0.061
Ricker t − 1	0.018	4	2.9	0.054
Gompertz t − 1	0.017	4	2.9	0.054
Ricker t − 2	0.009	4	3.0	0.050

[a] Model fit described by coefficient of determination (r^2), the number of parameters (k), the difference in Akaike's information criterion corrected for small sample size (ΔAIC_c), and the AIC_c wt (w_i).
[b] $AIC_c = 9.1$ for best selected model.

of increasing counts from around 200 breeding males to 300, followed by a substantial decline to half that number in the mid-1990s and a general increase through the 2000s.

The Middle Park sage-grouse population was one of three in which six models had about 10% relative likelihoods (exponential growth with process error, Gompertz, and Ricker; Table 15.17). The simple exponential growth with process error model estimated the annual rate of change for this population is close to 0, which leads to moderate likelihoods (9–30%) of declining below $N_e = 50$ in 30 or 100 years. Estimated male counts were already below 200, and probability of long-term persistence unlikely given the best fit model. Multimodel

TABLE 15.18

Sage-grouse lek monitoring effort, lek size, and trends averaged over 5-yr periods for the Wyoming basin population, 1965–2007.

Parameter	2000–2007[a]	1995–1999	1990–1994	1985–1989	1980–1984	1975–1979	1970–1974	1965–1969
Leks counted	1149	752	610	515	330	184	137	130
Males/lek	24	15	16	19	23	25	24	29
Active leks	807	501	431	377	248	137	97	91
% active leks	70	67	71	74	75	74	72	69
Males/active lek	33	22	23	25	30	33	34	42
λ	1.061	1.118	0.856	0.986	0.915	1.046	0.928	1.039
SE (λ)[b]	0.025	0.082	0.038	0.049	0.055	0.072	0.076	0.098

[a] Eight yr of data in this period.

[b] Standard error for annual rate of change.

forecast across all 26 models projected 3% and 7% probabilities, respectively, of the population declining below $N_e = 50$ in 30 and 100 years (Table 15.6).

Wyoming Basin Population

This population occupies much of Wyoming as well as part of southern Montana, northeastern Utah, and northern Colorado. It is separated from other populations by distance and topographic features (Table 15.1). The average number of leks counted per five-year period increased substantially from 1965–1969 to 2000–2007 (Table 15.18).

The proportion of active leks fluctuated slightly, but remained stable over the assessment period (Table 15.18). Population trends, as indicated by average number of males per lek, declined by 17% from 1965–1969 to 2000–2007. However, lek size decreased by 48% from 1965–1969 to 1995–1999 and then increased by 60% from 1995–1999 to 2000–2007. Average number of males per active lek also had the same pattern and an overall decline of 21% over the assessment period (Table 15.18). Average rates of change were <1.0 for four of the eight analysis periods. Average rates of change were <1.0 during the 1980s and early 1990s, and then increased substantially during 1995–1999. However, average rates of change declined by 6% between the last two analysis periods, although both values remained at or above 1.0 for these periods (Table 15.18).

From a minimum population estimate of 40,166 males (SE = 1,401) in 2007 based on counts at 1,298 leks, we reconstructed a minimum population estimate for males from 2007 back to 1965 (Fig. 15.3c), using counts at 1,670 leks reported for this 43-year period. The estimated minimum population grew from 1965 to peak at more than 130,000 males in 1969, and showed regular declines and peaks at 9- to 10-year intervals until the present, with an overall declining trend and the relative magnitude of the difference between peaks and troughs decreasing. Counts at more than 300 leks beginning in 1980 provided more precise minimum estimates of number of males attending leks than the earlier periods.

The best stochastic model for the annual rates of change of the Wyoming Basin population is a Gompertz model with a one-year time lag and a declining time trend of −3.4% per year ($r_t = 23.017 - 0.294 \ln(N_{t-1}) - 0.010$ year, $\sigma = 0.152$, $r^2 = 0.188$); this model has a relative likelihood of 16% (Table 15.19). The analogous Ricker model has a relative likelihood of 6.4%, a ΔAIC_c of 1.8, an r^2 of 0.150, and a high annual rate of decline of −10.5% ($r_t = 14.255 - 0.0000445\ N_{t-1} - 0.00707$ year, $\sigma = 0.1554$).

TABLE 15.19

Candidate model set (contains 95% of model weight) and model statistics for estimating population trends and persistence probabilities for Greater Sage-Grouse in the Wyoming basin population, 1965–2007.

Model	Model statistic[a]			
	r^2	k	ΔAIC_c	w_i
Gompertz t − 1 + year	0.188	5	0.0[b]	0.162
Gompertz t − 1 + period	0.171	4	0.8	0.108
Gompertz t − 1	0.102	4	1.5	0.075
Gompertz t − 1 + year, period	0.211	5	1.5	0.078
Gompertz t − 2 + year	0.153	6	1.7	0.070
Ricker t − 1	0.097	4	1.8	0.066
Ricker t − 1 + year	0.150	5	1.8	0.064
Gompertz t − 2	0.083	5	2.4	0.049
Gompertz t − 2 + period	0.135	5	2.5	0.046
Ricker t − 1 + period	0.134	5	2.6	0.045
Gompertz t − 2 + year, period	0.174	6	3.3	0.031
EGPE	0.000	3	3.5	0.028
Gompertz	0.055	3	3.6	0.027
Ricker	0.054	3	3.6	0.027
Ricker t − 2	0.055	4	3.6	0.027
Ricker t − 1 + year, period	0.156	6	4.2	0.020
Period	0.018	3	5.1	0.013
Gompertz + period	0.065	4	5.6	0.010

[a] Model fit described by coefficient of determination (r^2), the number of parameters (k), the difference in Akaike's information criterion corrected for small sample size (ΔAIC_c), and the AIC_c wt (w_i).
[b] $AIC_c = -28.1$ for best selected model.

The Gompertz model with declining time trend implies the Wyoming Basin population of sage-grouse will fluctuate around an estimated carrying capacity that will decline from 20,980 males attending leks in 2007 to 7,545 males attending leks in 2037 to 2 in 2107 if this trend continues at the same rate in the future. The Ricker model analogous to the best Gompertz model predicts a carrying capacity of 15,079 males in 2007 that declines to extinction by 2037. Multimodel forecasts across all 26 models predict 10% and 11% probabilities of the Wyoming Basin minimum population index declining below $N_e = 50$ and $N_e = 500$, respectively, in 100 years (Table 15.6).

Comprehensive Analysis of All Leks in the Management Zone

The average number of leks counted per five-year period increased substantially over the assessment period (Table 15.20). In 1965–1969, an average of 138 leks per year was censused, but by 2005–2007, an average of 1,321 leks per year was counted, an increase of 857%.

The proportion of active leks decreased slightly over the assessment period, averaging between 70% and 72% from 1965–1989 but declining to 65% by 2005–2007 (Table 15.20). Population trends, as indicated by average number of males per lek, decreased over the assessment period by

TABLE 15.20

Sage-grouse lek monitoring effort, lek size, and trends averaged over 5-yr periods for the Wyoming Basin Management Zone, 1965–2007.

Parameter	2000–2007[a]	1995–1999	1990–1994	1985–1989	1980–1984	1975–1979	1970–1974	1965–1969
Leks counted	1321	852	701	603	394	214	145	138
Males/lek	21	13	15	17	21	23	24	30
Active leks	858	531	471	423	286	151	101	97
% active leks	65	62	67	70	72	71	71	70
Males/active lek	33	22	23	24	29	33	34	42
λ	1.062	1.118	0.853	0.982	0.912	1.049	0.930	1.036
SE (λ)[b]	0.025	0.081	0.037	0.047	0.053	0.073	0.073	0.102

[a] Eight yr of data in this period.

[b] Standard error for annual rate of change.

30%, and average number of males per active lek decreased by 21% (Table 15.20). Average annual rates of change were <1.0 in four of the eight analysis periods. The average annual rate of change declined by 5% from 1995–1999 to 2000–2007, but values remained at or above 1.0 for both of these periods (Table 15.20).

From a minimum population estimate of 42,429 males (SE = 1,494) in 2007 based on counts at 1,467 leks, we reconstructed a minimum population estimate for males from 2007 back to 1965 (Fig. 15.3d), using 18,701 counts at 2,080 leks reported for this 43-year period. The overall pattern for the SMZ is dominated by the core Wyoming population showing 9- to 10-year intervals between peaks overlaid upon a continually declining trend (Fig. 15.3d). The estimated minimum population grew from 1965 to peak at more than 140,000 males in 1969. Counts at >350 leks beginning in 1980 provided precise minimum estimates of number of males attending leks.

The best stochastic model for annual rates of change of the population of sage-grouse in the Wyoming Basin SMZ is a Gompertz model with a one-year time lag and a declining time trend of −3.5% per year ($r_t = 28.634 - 0.3443 \ln(N_{t-1}) - 0.01254$ year, $\sigma = 0.1511$, $r^2 = 0.192$). This has a relative likelihood of 18% (Table 15.21). The analogous Ricker model has a relative likelihood of 7.5%, a ΔAIC_c of 1.7, an r^2 of 0.156, and a high annual rate of decline of −12.9% ($r_t = 15.515 - 0.00004162\ N_{t-1} - 0.008$ year, $\sigma = 0.158$).

The one-year delayed Gompertz model with declining time trend implies the population of sage-grouse in the Wyoming Basin SMZ will fluctuate around an estimated carrying capacity that will decline from 21,954 males attending leks in 2007 to 7,452 males attending leks in 2037 to 600 in 2107 if this trend continues at the same rate in the future. Parametric bootstraps under this model imply virtually no likelihood of the sage-grouse population declining below $N_e = 50$ or $N_e = 500$ within 100 years. The Ricker model analogous to the best Gompertz model predicts a carrying capacity of 14,350 males in 2007 that rapidly declines to extinction by 2037. Parametric bootstraps under the one-year delayed Ricker model with declining trend through time implies little chance the population will decline below $N_e = 50$ or $N_e = 500$ in 30 years, but virtual certainty that it will in 100 years if this trend continues. The probability of declining below indices of $N_e = 50$ and $N_e = 500$ in 100 years are 16% (SE = 7.4%) under multimodel forecasts across all 26 models (Table 15.6).

Southern Great Basin Management Zone

This SMZ represents sage-grouse populations in parts of Utah, Nevada, and California. Nine of the 12 populations delineated within this zone had data sufficient for analysis.

TABLE 15.21

Candidate model set (contains 95% of model weight) and model statistics for estimating population trends and persistence probabilities for Greater Sage-Grouse in the Wyoming Basin Management Zone, 1965–2007.

	Model statistic[a]			
Model	r^2	k	ΔAIC_c	w_i
Gompertz t − 1 + year	0.192	5	0.0[b]	0.177
Gompertz t − 2	0.168	4	1.2	0.099
Gompertz t − 2 + year	0.215	5	1.5	0.085
Gompertz t − 1 + year, period	0.156	6	1.7	0.075
Ricker t − 1 + year	0.156	5	1.7	0.075
Gompertz t − 1	0.097	4	1.9	0.068
Ricker t − 1	0.095	4	2.1	0.063
Ricker t − 2	0.134	4	2.7	0.045
Gompertz t − 1 + period	0.078	5	2.8	0.044
Gompertz t − 2 + period	0.129	5	3.0	0.040
Gompertz t − 2 + year, period	0.175	6	3.4	0.032
EGPE	0.000	3	3.7	0.028
Ricker t − 1 + period	0.052	5	3.9	0.025
Gompertz	0.050	3	4.0	0.024
Ricker	0.050	3	4.0	0.024
Ricker t − 2 + year	0.161	5	4.1	0.023
Period	0.017	3	5.3	0.012
Ricker t − 1 + year, period	0.065	6	5.8	0.010

[a] Model fit described by coefficient of determination (r^2), the number of parameters (k), the difference in Akaike's information criterion corrected for small sample size (ΔAIC_c), and the AIC_c wt (w_i).
[b] $AIC_c = -26.5$ for best selected model.

Mono Lake, California–Nevada, Population

This population straddles the California and Nevada border and is separated from other populations by distance and topography (Table 15.1). The average number of leks counted per five-year period increased substantially from 1965–1969 to 2000–2007 (Table 15.22). The proportion of active leks fluctuated but remained relatively stable over the assessment period (Table 15.22). Population trends, as indicated by average number of males per lek, declined 35% from 1965–1969 to 2000–2007. The average number of males per active lek also declined by 41% over the assessment period (Table 15.22). Average rates of change were <1.0 for three of the eight analysis periods (Table 15.22). Average rate of change declined by 45% between the last two analysis periods, but values remained at or above 1.0 for both of these periods (Table 15.22).

From a minimum population estimate of 274 males (SE = 101) in 2007 based on counts at 11 of 19 leks, we reconstructed a minimum population estimate for males from 2007 back to 1965 (Fig. 15.4a), using 361 counts at leks reported for this 43-year period. The overall pattern for the population shows irregular fluctuations between peaks in 1970 and 1987 of 520–670 males with lows above 100 (Fig. 15.4a) and no consistent long-term trend over the 40-year period. Since 1991, minimum counts have been trending upward. Counts at almost all the leks in

TABLE 15.22

Sage-grouse lek monitoring effort, lek size, and trends averaged over 5-yr periods for the Mono Lake, California–Nevada, population, 1965–2007.

Parameter	2000–2007[a]	1995–1999	1990–1994	1985–1989	1980–1984	1975–1979	1970–1974	1965–1969
Leks counted	24	13	17	18	14	13	12	5
Males/lek	15	9	17	16	17	20	24	23
Active leks	19	7	12	12	8	10	10	4
% active leks	81	56	70	69	58	75	87	75
Males/active lek	19	17	24	23	29	27	27	32
λ	1.050	1.912	0.923	1.058	0.964	1.134	0.968	1.486
SE (λ)[b]	0.143	0.677	0.169	0.160	0.256	0.390	0.263	0.838

[a] Eight yr of data in this period.
[b] Standard error for annual rate of change.

recent years have provided more precise minimum estimates of number of males attending leks than earlier counts.

The best population growth models for the Mono Lake population were a simple Gompertz ($r_t = 3.545 - 0.677 \ln(N_t)$, $\sigma = 0.447$, $r^2 = 0.332$, 29% relative likelihood) or a simple Ricker (15% relative likelihood) or simple Gompertz with time, period, or time + period (14% to 6% relative likelihoods, Table 15.23). The Gompertz model explained 33% of the variation in growth rates, as did the Gompertz model with declining time trend ($r_t = 4.935 - 0.68 \ln(N_{t-1}) - 0.00069$ year, $\sigma = 0.4646$, $r^2 = 0.332$; Table 15.23), and both forecast carrying capacities under 200 males (187 for Gompertz and declining K in 2007, 2037, and 2107 and 183, 178, and 165, respectively, under Gompertz with time trend). Long-term persistence above $N_e = 500$ is clearly unlikely, but short-term probability of declining below $N_e = 50$ is 15% in 30 years and 38% in 100 years across multiple models (Table 15.6).

South Mono Lake, California, Population

This population occurs in eastern California and is separated from other populations by distance and topography (Table 15.1). The average number of leks counted per five-year period increased somewhat from 1965–1969 to 2000–2007 (Table 15.24).

The proportion of active leks increased from 1965–1969 to 1985–1989 and declined slightly thereafter (Table 15.24). Population trends, as indicated by average number of males per lek, increased by 218% from 1965–1969 to 1985–1989 but declined by 49% from 1985–1989 to 2000–2007. Average number of males per active lek followed the same pattern over the assessment period (Table 15.24). Average rates of change were <1.0 for three of the eight analysis periods and indicated a decreasing population during the mid- to late 1980s and early 1990s (Table 15.24). Average rate of change was relatively stable over the last two analysis periods, and values remained at or above 1.0 for both of these periods (Table 15.24).

We used 2005 as the base year to reconstruct this population because the most leks ($N = 32$) were counted in that year of any available (1965–2007). The estimated minimum number of males attending leks in the population was 459 (SE = 61), and the estimated rates of increase were used to reconstruct the population back to 1965 and forward to 2007 based on the estimated annual rates of change. Standard errors were large prior to 1985 (Fig. 15.4b), and the population showed no obvious pattern through time except a tendency to remain between 200 and 600 males attending leks.

South Mono Lake grouse population was modeled best by a Gompertz model with no time lags (50% relative likelihood), with two other related Gompertz models strongly supported by the data (Table 15.25). Male counts at

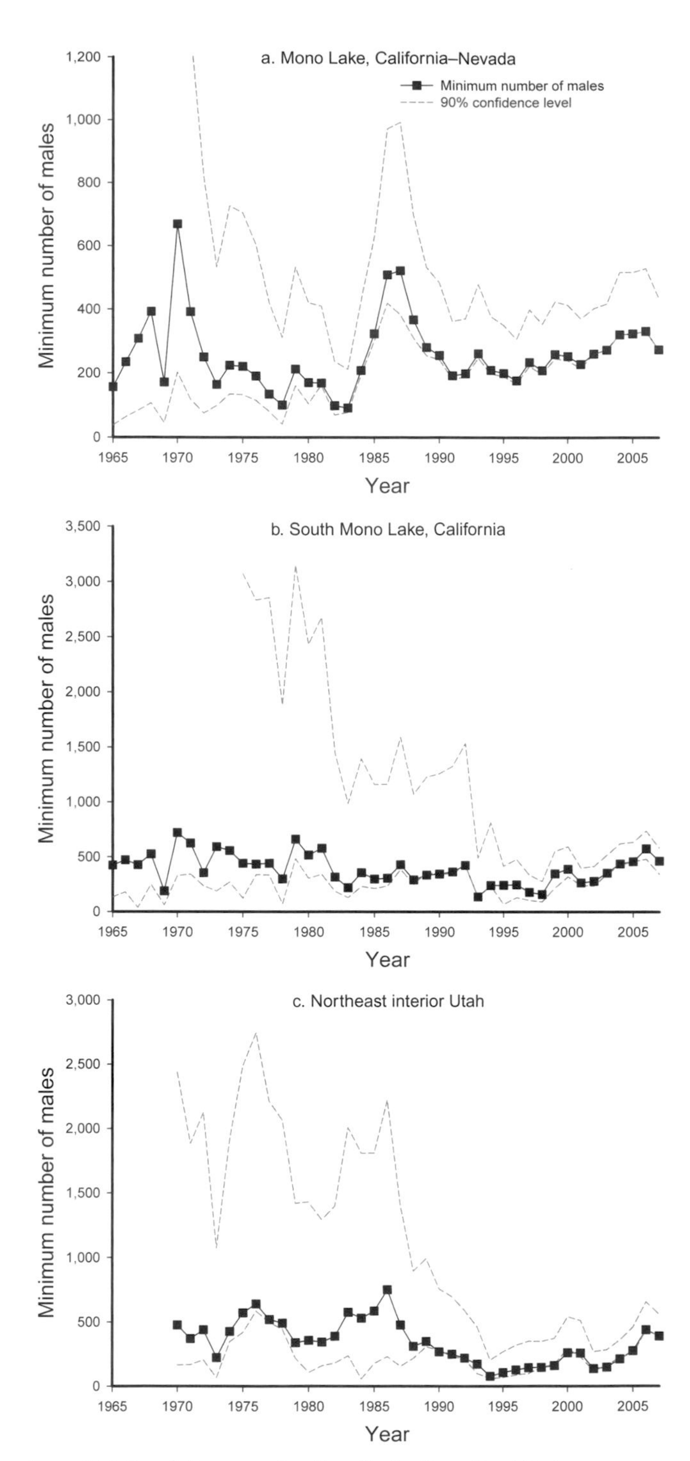

Figure 15.4. Population reconstructions for Southern Great Basin populations and Southern Great Basin Management Zone: (a) Mono Lake, California-Nevada; (b) south Mono Lake, California; (c) northeast interior Utah; (d) Sanpete–Emery Counties, Utah; (e) south-central Utah; (f) Summit–Morgan Counties, Utah; (g) Toole-Juab Counties, Utah; (h) Southern Great Basin; (i) Southern Great Basin Management Zone.

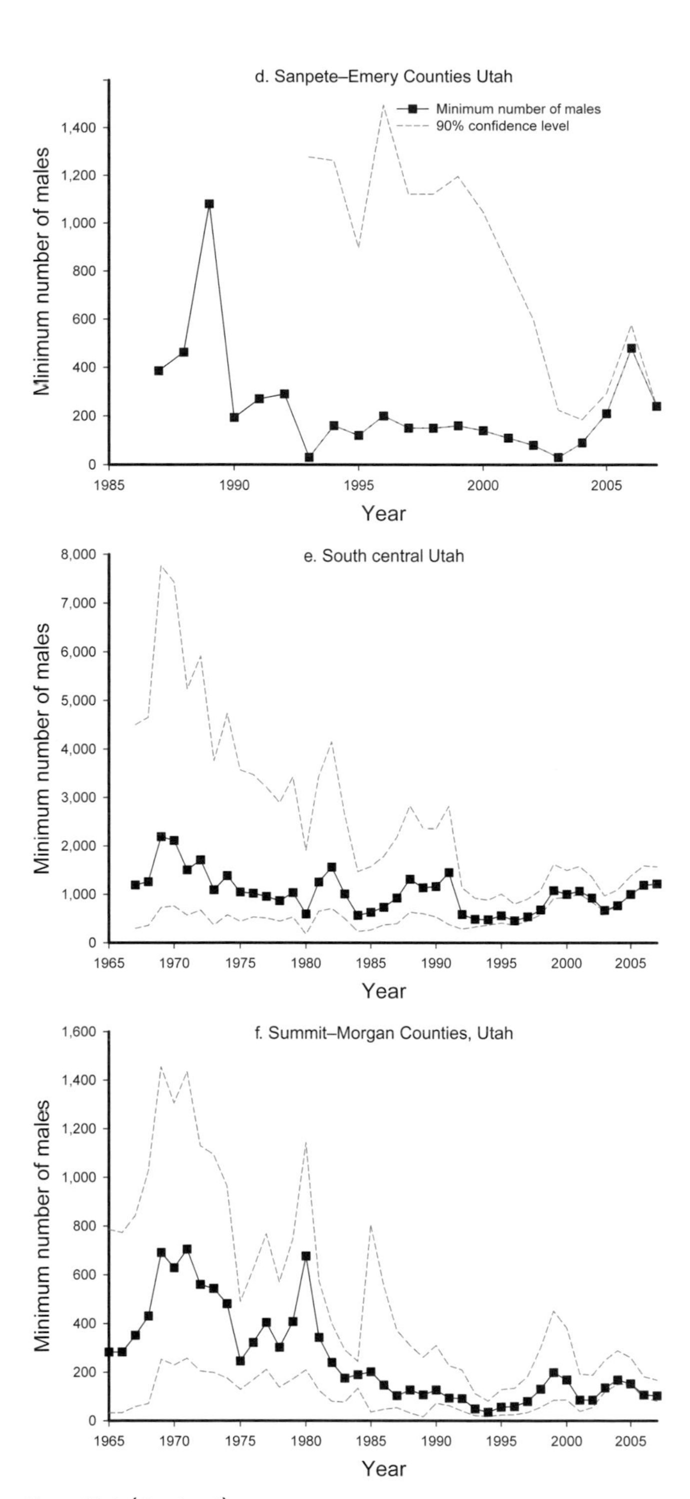

Figure 15.4. (*Continued*)

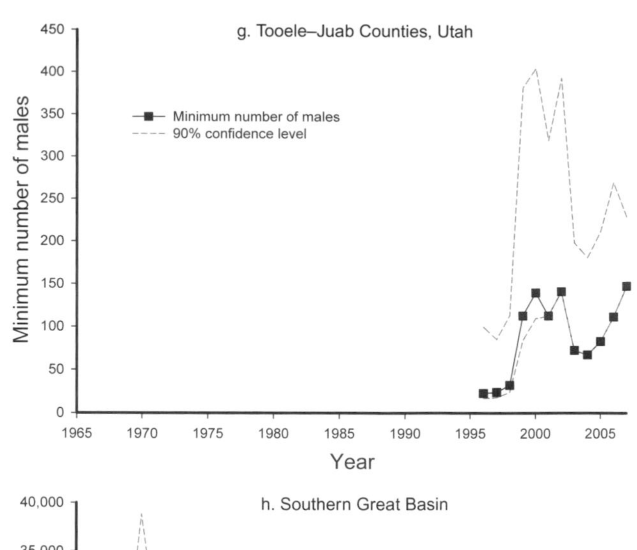

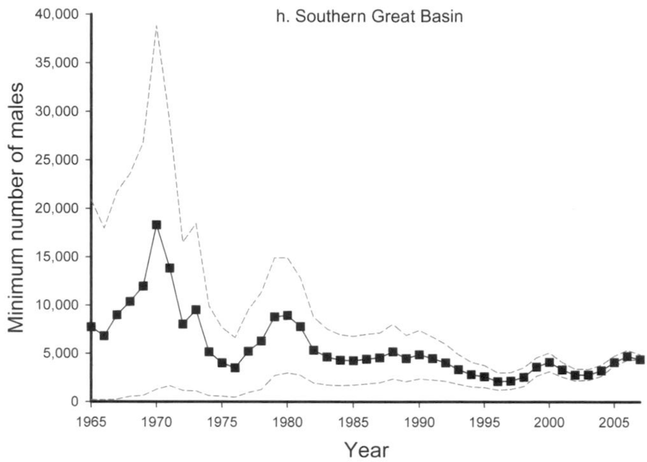

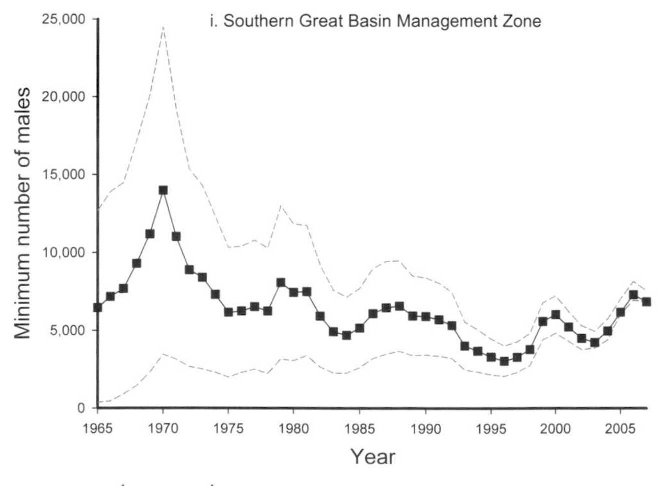

Figure 15.4. (*Continued*)

TABLE 15.23

Candidate model set (contains 95% of model weight) and model statistics for estimating population trends and persistence probabilities for Greater Sage-Grouse in the Mono Lake, California–Nevada, population, 1965–2007.

Model	Model statistic[a]			
	r^2	k	ΔAIC_c	w_i
Gompertz	0.332	3	0.0[b]	0.292
Ricker	0.229	3	1.3	0.153
Gompertz + year	0.332	4	1.4	0.143
Gompertz + period	0.342	4	2.3	0.092
Gompertz + year, period	0.364	5	3.1	0.062
Ricker + year	0.229	4	3.2	0.058
Ricker + period	0.233	4	3.8	0.044
Ricker + year, period	0.254	5	4.3	0.035
Gompertz t − 1	0.069	4	5.7	0.017
Ricker t − 1	0.062	4	6.0	0.015
Gompertz t − 2	0.059	4	6.1	0.014
EGPE	0.000	3	6.2	0.013
Ricker t − 2	0.057	4	6.2	0.013

[a] Model fit described by coefficient of determination (r^2), the number of parameters (k), the difference in Akaike's information criterion corrected for small sample size (ΔAIC_c), and the AIC_c wt (w_i).
[b] $AIC_c = 38.0$ for best selected model.

TABLE 15.24

Sage-grouse lek monitoring effort, lek size, and trends averaged over 5-yr periods for the South Mono Lake, California, population, 1965–2007.

Parameter	2000–2007[a]	1995–1999	1990–1994	1985–1989	1980–1984	1975–1979	1970–1974	1965–1969
Leks counted	10	11	11	10	9	9	8	6
Males/lek	18	19	19	35	14	12	16	11
Active leks	8	9	10	9	6	7	6	4
% active leks	83	83	84	90	74	78	72	74
Males/active lek	22	22	22	39	19	16	21	16
λ	1.080	1.063	0.971	0.995	1.267	1.048	0.846	1.683
SE (λ)[b]	0.191	0.170	0.219	0.097	0.367	0.321	0.155	0.831

[a] Eight yr of data in this period.
[b] Standard error for annual rate of change.

TABLE 15.25

Candidate model set (contains 95% of model weight) and model statistics for estimating population trends and persistence probabilities for Greater Sage-Grouse in the South Mono Lake, California, population, 1965–2007.

Model	Model statistic[a]			
	r^2	k	ΔAIC_c	w_i
Gompertz	0.331	3	0.0[b]	0.499
Gompertz + period	0.342	4	1.8	0.198
Gompertz + year	0.332	4	2.5	0.146
Gompertz + year, period	0.364	5	3.1	0.106

[a] Model fit described by coefficient of determination (r^2), the number of parameters (k), the difference in Akaike's information criterion corrected for small sample size (ΔAIC_c), and the AIC_c wt (w_i).
[b] $AIC_c = 55.7$ for best selected model.

TABLE 15.26

Sage-grouse lek monitoring effort, lek size, and trends averaged over 5-yr periods for the northeast interior Utah population, 1965–2007.

Parameter	2000–2007[a]	1995–1999	1990–1994	1985–1989	1980–1984	1975–1979	1970–1974	1965–1969
Leks counted	25	16	18	14	15	20	13	2
Males/lek	10	8	9	15	11	22	19	—
Active leks	15	9	13	11	10	19	9	1
% active leks	60	65	75	79	62	93	76	—
Males/active lek	17	12	13	19	18	23	23	—
λ	1.116	1.211	0.885	0.892	1.120	0.924	1.144	—
SE (λ)[b]	0.195	0.213	0.190	0.170	0.320	0.169	0.240	—

[a] Eight yr of data in this period.
[b] Standard error for annual rate of change.

south Mono Lake have already been below 200, and probability of long-term persistence is low, but the multimodel inference for short-term ($N_e = 50$) persistence is more likely (i.e., <1% probability of declining below $N_e = 50$; Table 15.6).

Northeast Interior Utah Population

This population is in northeast Utah and is separated from adjacent populations by distance and topography (Table 15.1). The average number of leks counted per five-year period increased substantially from 1965–1969 to 2000–2007 (Table 15.26). The proportion of active leks declined by 21% over the assessment period (Table 15.26). Population trends, as indicated by average number of males per lek, declined by 47% from 1970–1974 to 2000–2007. The average number of males per active lek also declined by 26% over the same period (Table 15.26). Average rates of change were <1.0 for three of eight analysis periods and indicated a decreasing population during the mid- to late 1980s and early 1990s (Table 15.26). Average rate of change increased during the 1990s but declined 8% between the last two analysis

TABLE 15.27

Candidate model set (contains 95% of model weight) and model statistics for estimating population trends and persistence probabilities for Greater Sage-Grouse in the northeast interior Utah population.

	Model statistic[a]			
Model	r^2	k	ΔAIC_c	w_i
Ricker + year, period	0.280	5	0.0[b]	0.185
Ricker + period	0.204	4	0.8	0.126
Gompertz + period	0.200	4	0.9	0.116
Gompertz + year, period	0.261	5	0.9	0.117
Gompertz t − 1 + period	0.162	5	2.6	0.051
Gompertz	0.090	3	2.9	0.043
Gompertz t − 1	0.085	4	3.1	0.040
Ricker	0.085	3	3.1	0.040
Ricker t − 1	0.076	4	3.4	0.033
Ricker t − 1 + period	0.135	5	3.7	0.029
EGPE	0.000	3	3.8	0.028
Gompertz t − 2	0.059	4	4.1	0.024
Ricker t − 2	0.059	4	4.1	0.024
Gompertz t − 1 + year, period	0.180	6	4.5	0.019
Gompertz t − 2 + period	0.102	5	5.0	0.015
Gompertz + year	0.098	4	5.1	0.014
Gompertz t − 1 + year	0.099	5	5.1	0.014
Ricker + year	0.093	4	5.4	0.013
Ricker t − 2 + period	0.092	5	5.4	0.013
Ricker t − 1 + year, period	0.156	6	5.6	0.011

[a] Model fit described by coefficient of determination (r^2), the number of parameters (k), the difference in Akaike's information criterion corrected for small sample size (ΔAIC_c), and the AIC_c wt (w_i).
[b] $AIC_c = 22.5$ for best selected model.

periods, but values remained at or above 1.0 for both of these periods (Table 15.26).

Starting from a minimum population estimate of 338 males (SE = 108) in 2007 based on counts at 32 leks, we reconstructed a minimum population estimate for males from 2007 back to 1970 (Fig. 15.4c). The population increased from about 476 males in 1970 and peaked in 1976 at 600 and again in 1986 at 750 males before reaching a low of 77 males in 1994 followed by a steady increase through 2007.

The best stochastic model for the annual rates of change of this population of sage-grouse was a Ricker model with no time lags and both a period effect and an increasing time trend of 3.7% per year ($r_t = -34.817 - 0.001322\ N_t + 0.0176\ \text{year} + 0.6558\ \text{Period}$, $\sigma = 0.2809$, $r^2 = 0.280$; Table 15.27).

The Ricker model with increasing time trend implies the northeast interior Utah population of sage-grouse will fluctuate around an estimated carrying capacity that will increase from 358 males attending leks in 2007 to 757 attending leks in 2037 to 1,688 males in 2107 if this trend continues at the same rate in the future, but a shortage of habitat in the region raises questions about the potential for this forecast to be realized. A parametric bootstrap based on the Ricker model with increasing time trend, which has a 19% relative likelihood, infers that there is virtually no chance of the population declining below $N_e = 50$ but declining below $N_e = 500$ is possible within 100 years (37% relative probability). Multimodel forecasts across all 26 models predict low probabilities of declining below population indices of $N_e = 50$ in 30 and 100 years (1% and 9%, respectively; Table 15.6) but higher probabilities of declining below $N_e = 500$ (52% and 79% in 30 and 100 years, respectively; Table 15.6).

Sanpete–Emery Counties, Utah, Population

This is an isolated population in central Utah and is separated from other populations by distance and topography (Table 15.1). The average number of leks counted per five-year period was relatively stable from 1965–1969 to 2000–2007 (Table 15.28). In 1965–1969 an average of one lek per five-year period was censused, and by 2000–2007 an average of three leks were counted. The average number of active leks counted per five-year period was also relatively stable.

The proportion of active leks declined over the assessment period but total leks counted was quite low (Table 15.28). Population trends, as indicated by average number of males per lek, declined by 30% from 1985–1989 to 2000–2007. Average number of males per active lek fluctuated considerably over the assessment period (Table 15.28). Average rates of change were >1.0 for all four analysis periods and generally suggested a stable to increasing population (Table 15.28). The average rate of change increased by 33% between the last two analysis periods.

A simple Gompertz model without time lags and a Gompertz with period effect were the most descriptive models for this population (Table 15.29, Fig. 15.4d). These two models together represented half of the likelihood based on model weights, but comparable Ricker models were second most likely. Multimodel forecasts predict high probabilities of this population declining below $N_e = 50$ in 30 and 100 years (78% and 99%, respectively; Table 15.6).

South Central Utah Population

This is an isolated population in south central Utah and is separated from other populations by distance and mountainous terrain (Table 15.1). The average number of leks counted per five-year period increased from 1965–1969 to 2000–2007 (Table 15.30).

The proportion of active leks remained relatively stable over the assessment period (Table 15.30). Population trends, as indicated by average number of males per lek, fluctuated but remained relatively stable from 1965–1969 to 2000–2007. Average number of males per active lek followed the same

TABLE 15.28

Sage-grouse lek monitoring effort, lek size, and trends averaged over 5-yr periods for the Sanpete–Emery Counties, Utah, population, 1965–2007.

Parameter	2000–2007[a]	1995–1999	1990–1994	1985–1989	1980–1984	1975–1979	1970–1974	1965–1969
Leks counted	3	3	3	2	0	1	3	1
Males/lek	7	5	5	10	—	—	—	—
Active leks	2	1	3	2	—	—	—	—
% active leks	60	33	85	100	—	—	—	—
Males/active lek	11	16	6	10	—	—	—	—
λ	1.430	1.072	1.732	1.089	—	—	—	—
SE (λ)[b]	0.609	0.000	1.438	0.580	—	—	—	—

[a] Eight yr of data in this period.
[b] Standard error for annual rate of change.

TABLE 15.29

Candidate model set (contains 95% of model weight) and model statistics for estimating population trends and persistence probabilities for Greater Sage-Grouse in the Sanpete–Emery Counties, Utah, population, 1965–2007.

	Model statistic[a]			
Model	r^2	k	ΔAIC_c	w_i
Gompertz	0.311	4	0.0	0.499
Ricker	0.247	4	1.8	0.204
Gompertz + year	0.329	5	2.6	0.133
Ricker + year	0.258	5	4.6	0.049
EGPE	0.000	3	4.7	0.049
Gompertz t − 1	0.022	4	7.0	0.015

[a] Model fit described by coefficient of determination (r^2), the number of parameters (k), the difference in Akaike's information criterion corrected for small sample size (ΔAIC_c), and the AIC_c wt (w_i).

[b] $AIC_c = 52.9$ for best selected model.

TABLE 15.30

Sage-grouse lek monitoring effort, lek size, and trends averaged over 5-yr periods for the south central Utah population, 1965–2007.

Parameter	2000–2007[a]	1995–1999	1990–1994	1985–1989	1980–1984	1975–1979	1970–1974	1965–1969
Leks counted	38	29	18	22	23	27	25	16
Males/lek	25	18	24	20	19	18	24	27
Active leks	29	22	13	16	18	20	20	14
% active leks	76	75	75	72	79	75	80	85
Males/active lek	33	24	32	28	23	24	29	32
λ	1.045	1.154	0.928	1.147	1.135	0.918	0.902	1.254
SE (λ)[b]	0.090	0.154	0.172	0.238	0.267	0.144	0.104	0.214

[a] Eight yr of data in this period.

[b] Standard error for annual rate of change.

pattern over the assessment period (Table 15.30). Average rates of change were <1.0 for three of the eight analysis periods but generally suggested a stable to increasing population (Table 15.30). Average rate of change declined 9% between the last two analysis periods, but values remained at or above 1.0 for both of these periods (Table 15.30).

We reconstructed a minimum population estimate for males from a minimum 2007 population estimate of 1,219 males (SE = 220) based on counts at 42 leks to 1967 (Fig. 15.4e), using 57 lek counts reported for this period. Counts were highest in 1969 and 1970 at just over 2,000 males, but the population has generally fluctuated around 1,000 over the period from 1971 to 2007.

The best stochastic model for the annual rates of change of the south central Utah population of sage-grouse was a Gompertz model with no time lags, period effects, or time trend ($r_t = 2.4099 - 0.3542\ln(N_t)$, $\sigma = 0.2776$, $r^2 = 0.209$; Table 15.31). Based on this model, the estimated carrying capacity for this population was 901 males.

TABLE 15.31

Candidate model set (contains 95% of model weight) and model statistics for estimating population trends and persistence probabilities for Greater Sage-Grouse in the south central Utah population, 1965–2007.

	Model statistic[a]			
Model	r^2	k	ΔAIC_c	w_i
Gompertz	0.209	3	0.0[b]	0.240
Ricker	0.205	3	0.2	0.222
Gompertz + period	0.213	4	2.3	0.077
Gompertz + year	0.213	4	2.3	0.075
Ricker + year	0.213	4	2.3	0.077
Ricker + period	0.211	4	2.4	0.072
Gompertz t − 1	0.143	4	3.0	0.053
Ricker t − 1	0.122	4	4.0	0.033
Gompertz + year, period	0.213	5	4.9	0.020
Ricker + year, period	0.213	5	4.9	0.020
Gompertz t − 1 + period	0.148	5	5.3	0.017
Gompertz t − 1 + year	0.146	5	5.4	0.016
Gompertz t − 2	0.077	4	5.8	0.013
Ricker t − 1 + period	0.125	5	6.3	0.010
Ricker t − 2	0.065	4	6.3	0.010

[a] Model fit described by coefficient of determination (r^2), the number of parameters (k), the difference in Akaike's information criterion corrected for small sample size (ΔAIC_c), and the AIC_c wt (w_i).
[b] AIC_c = 17.1 for best selected model.

The Gompertz model implies that in the future, the south-central Utah population will fluctuate around a carrying capacity of 901 males. A parametric bootstrap based on this model infers there is little chance (<1%) of this population declining below an effective population size of 50 or 500 within 30 or 100 years. There is a 0% chance, across all 26 models of population growth, of this population declining below $N_e = 50$ within 30 years, a 3% chance of declining below $N_e = 500$ within 30 years, a 1% chance of declining below $N_e = 50$ within 100 years, and a 21% chance of declining below $N_e = 500$ within 100 years (Table 15.6).

Summit–Morgan Counties, Utah, Population

This small population occurs in northeastern Utah and is separated from other populations by distance and mountainous terrain (Table 15.1). The average number of leks counted per five-year period increased from 1965–1969 to 2000–2007 (Table 15.32). The proportion of active leks fluctuated over the assessment period but generally decreased (Table 15.32). Population trends, as indicated by average number of males per lek, declined by 35% from 1965–1969 to 2000–2007. Average number of males per active lek fluctuated over the assessment period (Table 15.32). Average rates of change were <1.0 for four of the eight analysis periods, and generally indicated a fluctuating population from 1965–1969 to 2000–2007 (Table 15.32). Average rate of change declined by 23% between the last two analysis periods and was <1.0 for the 2000–2007 period.

The minimum population estimate was 81 males (SE = 40) in 2007 based on counts at seven leks. We reconstructed a minimum population estimate for males from the 2007 estimate back to 1965 (Fig. 15.4f), using 22 lek counts

TABLE 15.32

Sage-grouse lek monitoring effort, lek size, and trends averaged over 5-yr periods for the Summit–Morgan Counties, Utah, population, 1965–2007.

Parameter	2000–2007[a]	1995–1999	1990–1994	1985–1989	1980–1984	1975–1979	1970–1974	1965–1969
Leks counted	9	3	4	7	7	9	10	5
Males/lek	11	16	14	6	14	19	24	17
Active leks	5	1	3	3	5	7	8	4
% active leks	50	43	86	49	76	84	87	77
Males/active lek	23	36	15	13	19	22	28	21
λ	0.987	1.283	0.913	0.938	2.700	1.264	0.857	1.196
SE (λ)[b]	0.217	0.037	0.157	0.455	0.216	0.357	0.167	0.407

[a] Eight yr of data in this period.
[b] Standard error for annual rate of change.

TABLE 15.33

Candidate model set (contains 95% of model weight) and model statistics for estimating population trends and persistence probabilities for Greater Sage-Grouse in the Summit–Morgan Counties, Utah, population, 1965–2007.

	Model statistic[a]			
Model	r^2	k	ΔAIC_c	w_i
Gompertz	0.256	3	0.0[b]	0.437
Gompertz + period	0.267	4	1.9	0.172
Gompertz + year	0.266	4	1.9	0.166
Gompertz + year, period	0.268	5	4.5	0.047
Ricker	0.166	3	4.6	0.044
Gompertz t − 1	0.139	4	5.8	0.024
Ricker t − 1	0.139	4	5.8	0.024
Ricker + year	0.176	4	6.5	0.017
Ricker + period	0.171	4	6.8	0.015
Ricker t − 1 + year	0.151	5	7.8	0.009

[a] Model fit described by coefficient of determination (r^2), the number of parameters (k), the difference in Akaike's information criterion corrected for small sample size (ΔAIC_c), and the AIC_c wt (w_i).
[b] $AIC_c = 55.5$ for best selected model.

reported for this period. The population has generally fluctuated around 100 males during this period.

The best stochastic model for annual rates of change of this population was a Gompertz model with no time lag in density dependence, no time trend, and no period effect ($r_t = 2.1646 - 0.5092 \ln(N_t)$, $\sigma = 0.4455$, $r^2 = 0.256$; Table 15.33).

The Gompertz model implies that sage-grouse in the Summit–Morgan Counties population will fluctuate around an estimated carrying capacity of 70 males. A parametric bootstrap based on this model infers this population has a 19% chance of declining below $N_e = 50$ within the next 30 years and a 51% chance of declining below $N_e = 50$ within 100 years, though it is already below $N_e = 500$. Multimodel forecasts across all 26 models predict 21% and 42% probabilities of declining below $N_e = 50$ in 30 and 100 years, respectively (Table 15.6).

Toole–Juab Counties, Utah, Population

This isolated population occurs in central Utah and is separated from other populations by distance and topography (Table 15.1). The average number of leks counted per five-year period increased from 1965–1969 to 2000–2007 (Table 15.34). The proportion of active leks declined somewhat over the assessment period (Table 15.34). Population trends, as indicated by average number of males per lek, declined by 53% from 1965–1969 to 1995–1999 but increased by 122% in 2000–2007. Average number of males per active lek followed the same pattern as males per lek over the assessment period (Table 15.34). Average rates of change were <1.0 for two of the eight analysis periods (Table 15.34). However, average rate of change declined by 41% between the last two analysis periods, but values remained at or above 1.0 for both of these periods (Table 15.34).

A minimum population estimate of 148 males (SE = 51) in 2007 was based on counts at seven leks. We reconstructed a minimum population estimate for males from the 2007 value back to 1996 (Fig. 15.4g), using 13 lek counts reported for this period. The population generally increased during this period.

The best stochastic model for annual rates of change of this population was a Gompertz model with no time lag in density dependence, no time trend, and no period effect ($r_t = 4.2663 - 0.9109 \ln(N_t)$, $\sigma = 0.2727$, $r^2 = 0.682$; Table 15.35).

The Gompertz model implies sage-grouse in the Toole–Juab Counties population will fluctuate around an estimated carrying capacity of 108 males. A parametric bootstrap based on this model infers that this population has a 0% chance of declining below an effective population size of 50 within the next 100 years, though it is already below $N_e = 500$. Across all 26 models of population growth there is a 7% chance of this population declining below $N_e = 50$ within 30 years and a 13% chance of declining below $N_e = 50$ within 100 years (Table 15.6).

Southern Great Basin Population

This population occupies much of central and eastern Nevada and a small portion of western Utah (Table 15.1). The average number of leks counted per five-year period increased considerably from 1965–1969 to 2000–2007 (Table 15.36). The proportion of active leks increased from 1965–1969 to 1985–1989 and declined thereafter (Table 15.36). Population trends, as indicated by average number of males per lek, declined by 19% from 1965–1969 to 2000–2007. Average number of males per active lek followed the same pattern over the assessment period (Table 15.36). Average rates of change were <1.0 for three of the eight analysis periods (Table 15.36). Average rate of change declined by 8% between the last two analysis periods, but values remained at or above 1.0 for both of these periods (Table 15.36).

TABLE 15.34

Sage-grouse lek monitoring effort, lek size, and trends averaged over 5-yr periods for the Tooele–Juab Counties, Utah, population, 1965–2007.

Parameter	2000–2007[a]	1995–1999	1990–1994	1985–1989	1980–1984	1975–1979	1970–1974	1965–1969
Leks counted	6	3	4	5	4	5	5	1
Males/lek	20	9	10	13	6	15	19	—
Active leks	4	2	3	4	2	4	4	—
% active leks	76	67	88	86	62	74	88	—
Males/active lek	28	13	12	16	11	19	22	—
λ	1.057	1.783	0.935	1.067	1.033	1.269	0.913	—
SE (λ)[b]	0.296	0.425	0.287	0.224	0.326	0.390	0.190	—

[a] Eight yr of data in this period.

[b] Standard error for annual rate of change.

TABLE 15.35

Candidate model set (contains 95% of model weight) and model statistics for estimating population trends and persistence probabilities for Greater Sage-Grouse in the Tooele–Juab Counties, Utah, population, 1996–2007.

	Model statistic[a]			
Model	r^2	k	ΔAIC_c	w_i
Gompertz	0.682	3	0.0[b]	0.254
Ricker	0.635	3	1.3	0.135
Gompertz t − 1	0.537	4	3.4	0.087
EGPE	0	2	3.6	0.078
Ricker t − 1	0.434	4	5.2	0.035

[a] Model fit described by coefficient of determination (r^2), the number of parameters (k), the difference in Akaike's information criterion corrected for small sample size (ΔAIC_c), and the AIC_c wt (w_i).

[b] $AIC_c = 12.9$ for best selected model.

TABLE 15.36

Sage-grouse lek monitoring effort, lek size, and trends averaged over 5-yr periods for the southern Great Basin population, 1965–2007.

Parameter	2000–2007[a]	1995–1999	1990–1994	1985–1989	1980–1984	1975–1979	1970–1974	1965–1969
Leks counted	233	104	84	70	75	38	55	23
Males/lek	13	10	14	15	18	20	21	16
Active leks	159	80	76	65	64	31	46	16
% active leks	69	80	90	93	86	81	83	67
Males/active lek	19	13	16	16	21	25	25	23
λ	1.023	1.113	0.884	1.032	0.868	1.196	0.769	1.207
SE (λ)[b]	0.048	0.093	0.072	0.076	0.079	0.144	0.144	0.260

[a] Eight yr of data in this period.

[b] Standard error for annual rate of change.

We reconstructed a minimum population estimate for males from 2007 back to 1965 (Fig. 15.4h) using 675 lek counts reported for this period beginning from a minimum population estimate of 4,400 males (SE = 318) in 2007, based on counts at 219 leks. The highest estimate for this population was 18,310 males in 1970. Since 1983, the population has undergone a gradual decline, with recent fluctuations between 2,000 and 5,000 males.

The best stochastic model for annual rates of change of this population of sage-grouse was a Gompertz model with a two-year time lag in density dependence and a declining year trend ($r_t = 30.768 - 0.4342 \ln(N_{t-2}) - 0.01365$ year, $\sigma = 0.1875$, $r^2 = 0.325$; Table 15.37).

The Gompertz model with a declining trend implies that in the future, sage-grouse in the southern Great Basin population will fluctuate around an estimated carrying capacity of 2,446 males in 2007, 977 males in 2037, and 107 males in 2107. A parametric bootstrap based on this model infers this population has a 0% chance of declining below $N_e = 500$ within the next 30 years, a 0% chance of declining below $N_e = 50$ within 100 years, and a 100% chance of declining below $N_e = 500$ within the next 100 years. This population has a 0% chance

TABLE 15.37

Candidate model set (contains 95% of model weight) and model statistics for estimating population trends and persistence probabilities for Greater Sage-Grouse in the southern Great Basin population, 1965–2007.

Model	Model statistic[a]			
	r^2	k	ΔAIC_c	w_i
Gompertz t − 2 + year	0.325	5	0.0[b]	0.462
Gompertz t − 2 + year, period	0.329	6	2.4	0.142
Gompertz t − 1 + year	0.272	5	3.0	0.103
Gompertz t − 2 + period	0.270	5	3.1	0.096
Gompertz t − 1 + period	0.229	5	5.3	0.032
Gompertz t − 1 + year, period	0.275	6	5.5	0.029
Gompertz t − 2	0.164	4	6.1	0.022
Ricker t − 1 + year	0.204	5	6.6	0.017
Gompertz t − 1	0.149	4	6.8	0.015
Ricker t − 2 + year	0.197	5	7.0	0.014
Ricker t − 1	0.140	4	7.2	0.013
Ricker t − 2	0.135	4	7.5	0.011

[a] Model fit described by coefficient of determination (r^2), the number of parameters (k), the difference in Akaike's information criterion corrected for small sample size (ΔAIC_c), and the AIC_c wt (w_i).

[b] AIC_c = −11.3 for best selected model.

of declining below an effective population size of 50 within 30 years across all 26 models of population growth, a 2% chance of declining below $N_e = 500$ within 30 years, a 4% chance of declining below $N_e = 50$ within 100 years, and a 78% chance of declining below $N_e = 500$ within 100 years (Table 15.6).

Comprehensive Analysis of All Leks in the Management Zone

The average number of leks counted per five-year period increased substantially over the assessment period (Table 15.38). In 1965–1969, an average of 51 leks per year was censused, but by 2005–2007, an average of 387 leks per year was counted, an increase of 659%. The proportion of active leks decreased over the assessment period, declining from a high of 83% in 1985–1994 to a low of 67% by 2005–2007 (Table 15.38). Population trends, as indicated by average number of males per lek, decreased over the assessment period by 24%, and average number of males per active lek decreased by 9% (Table 15.38). Average annual rates of change were <1.0 in three of the eight analysis periods. The average annual rate of change from 1995–1999 to 2000–2007 declined by 10%, but values remained at or above 1.0 for both of these periods (Table 15.38).

Starting from a minimum population estimate of 6,851 males (SE = 435) in 2007 based on counts at 471 leks, we reconstructed a minimum population estimate for males from 2007 back to 1965 (Fig. 15.4i). The population increased from about 6,500 males in 1965 to peak at 14,000 males in 1970, followed by cycles of declines and peaks at 9- to 12-year intervals overlaid on a continuous long-term decline through 2007.

The best stochastic model for annual rates of change of the Southern Great Basin SMZ population of sage-grouse detected a cyclic nature and identified one- or two-year time-delayed Gompertz-type models with declining time trends of −2.6% per year as most representative (e.g., best model with relative likelihood of 28% was $r_t = 24.334 - 0.391\ \ln(N_{t-1}) - 0.010\ \text{year} + 0.156\ \text{period}$, $\sigma = 0.133$, $r^2 = 0.333$; Table 15.39).

TABLE 15.38

Sage-grouse lek monitoring effort, lek size, and trends averaged over 5-yr periods for the Southern Great Basin Management Zone, 1965–2007.

Parameter	2000–2007[a]	1995–1999	1990–1994	1985–1989	1980–1984	1975–1979	1970–1974	1965–1969
Leks counted	387	190	170	148	146	121	131	51
Males/lek	13	11	14	16	17	19	20	17
Active leks	257	136	142	123	114	98	103	39
% active leks	67	73	83	83	78	81	80	74
Males/active lek	20	15	17	20	21	24	25	22
λ	1.029	1.141	0.894	1.031	0.937	1.046	0.851	1.169
SE (λ)[b]	0.040	0.076	0.064	0.069	0.085	0.084	0.083	0.149

[a] Eight yr of data in this period.
[b] Standard error for annual rate of change.

TABLE 15.39

Candidate model set (contains 95% of model weight) and model statistics for estimating population trends and persistence probabilities for Greater Sage-Grouse in the Southern Great Basin Management Zone, 1965–2007.

	Model statistic[a]			
Model	r^2	k	ΔAIC_c	w_i
Gompertz t − 1 + year, period	0.333	6	0.0[b]	0.283
Gompertz t − 2+ period	0.326	5	0.4	0.230
Gompertz t − 2 + year, period	0.347	6	1.7	0.119
Gompertz t − 1 + year	0.285	5	2.7	0.072
Gompertz t − 2	0.276	4	3.3	0.056
Ricker t − 2 + period	0.262	5	4.0	0.038
Ricker t − 1 + year, period	0.260	6	4.1	0.036
Gompertz t − 1 + period	0.204	5	4.5	0.029
Gompertz t − 2 + year	0.294	5	4.9	0.025
Ricker t − 1 + year	0.244	5	5.0	0.023
Ricker t − 2	0.236	4	5.4	0.019
Gompertz t − 1	0.185	4	5.5	0.018
Ricker t − 1 + period	0.177	5	5.9	0.015

[a] Model fit described by coefficient of determination (r^2), the number of parameters (k), the difference in Akaike's information criterion corrected for small sample size (ΔAIC_c), and the AIC_c wt (w_i).
[b] $AIC_c = -43.5$ for best selected model.

The one-year delayed Gompertz model with declining time trend and period implies the Southern Great Basin SMZ population of sage-grouse will fluctuate around carrying capacities that will decline from 12,165 males attending leks in 2007 to 5,517 attending leks in 2037 and 872 males in 2107 if this trend continues at the same rate in the future. A parametric bootstrap based on the one-year time-delayed Gompertz model with declining time trend and period effect, which has a 28% relative likelihood, infers virtually no chance of the population declining below an effective population size of 50 or 500 within 100 years. Multimodel inference indicated that the probability of extinction is 0% in 30 years and only 6% to 8% (SE = 4.9–5.3%) in 100 years (Table 15.6).

Snake River Plain Management Zone

This SMZ represents sage-grouse populations in parts of Montana, Idaho, Utah, Nevada, and Oregon. Almost all sage-grouse in Idaho occur in this SMZ (Fig. 15.1). Seven of the 11 populations delineated within this management zone had data sufficient for analysis.

Baker, Oregon, Population

This population is in eastern Oregon and is separated by topography from the northern Great Basin population (Table 15.1). Routine monitoring did not start until the mid- to late 1980s (Table 15.40). The average number of leks counted per five-year period increased substantially from 1985–1989 to 2000–2007 (Table 15.40). An average of 1 lek per year was counted in 1985–1989, but by 2005–2007, an average of 15 leks per year was counted. The proportion of active leks increased over the assessment period, although this could be related to the relatively few leks counted when monitoring began (Table 15.40). Population trends indicated by average number of males per lek increased over the assessment period by 27%, and average number of males per active lek decreased by 14% (Table 15.40). Contrary to lek size data, rate of change declined by 11.9% from 1995–1999 to 2000–2007, with an average rate of change <1.0 for 2000–2007 (Table 15.40).

The minimum population estimate of 137 males in 2007 was based on counts at 13 leks. We reconstructed a minimum population estimate for males from 2007 back to 1993 (Fig. 15.5a) using 40 lek counts reported for this period. The population has generally fluctuated below 200 males during this period.

The best stochastic model for the annual rates of change of Baker, Oregon, population of sage-grouse was an Exponential Growth with Process Error (EGPE, $r_t = -0.00218$, $\sigma = 0.1838$, SE = 0.0510; Table 15.41).

The EPGE model implies this population will decline at an annual rate of −0.22% per year. A parametric bootstrap based on this model infers

TABLE 15.40

Sage-grouse lek monitoring effort, lek size, and trends averaged over 5-yr periods for the Baker, Oregon, population, 1965–2007.

Parameter	2000–2007[a]	1995–1999	1990–1994	1985–1989	1980–1984	1975–1979	1970–1974	1965–1969
Leks counted	15	10	4	1	0	0	0	0
Males/lek	14	13	11	—	—	—	—	—
Active leks	13	8	3	1	0	0	0	0
% active leks	89	75	78	—	—	—	—	—
Males/active lek	16	18	14	—	—	—	—	—
λ	0.951	1.079		—	—	—	—	—
SE (λ)[b]	0.126	0.214		—	—	—	—	—

[a] Eight yr of data in this period.

[b] Standard error for annual rate of change.

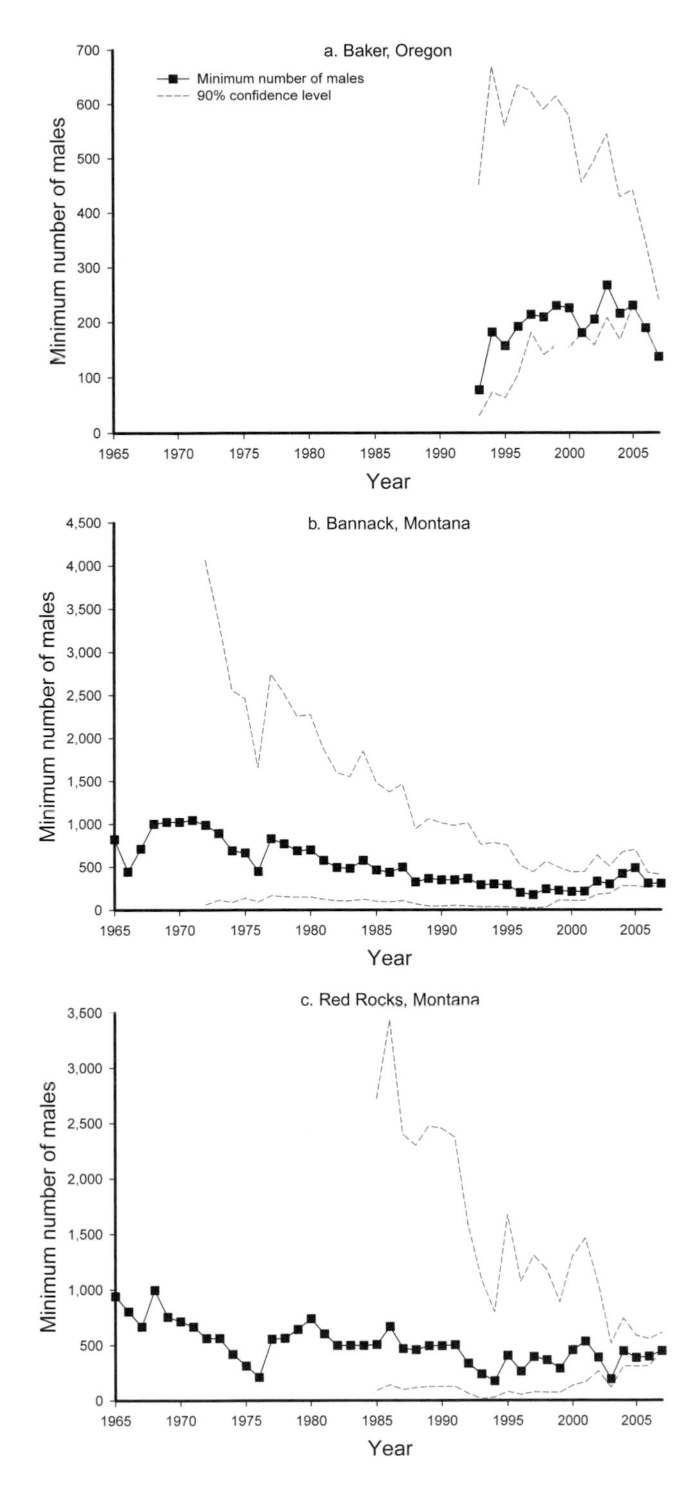

Figure 15.5. Population reconstructions for Snake River Plain populations and Snake River Plain Management Zone: (a) Baker, Oregon; (b) Bannack, Montana; (c) Red Rocks, Montana; (d) east-central Idaho; (e) Snake–Salmon–Beaverhead, Idaho; (f) Northern Great Basin; (g) Snake River Plain Management Zone.

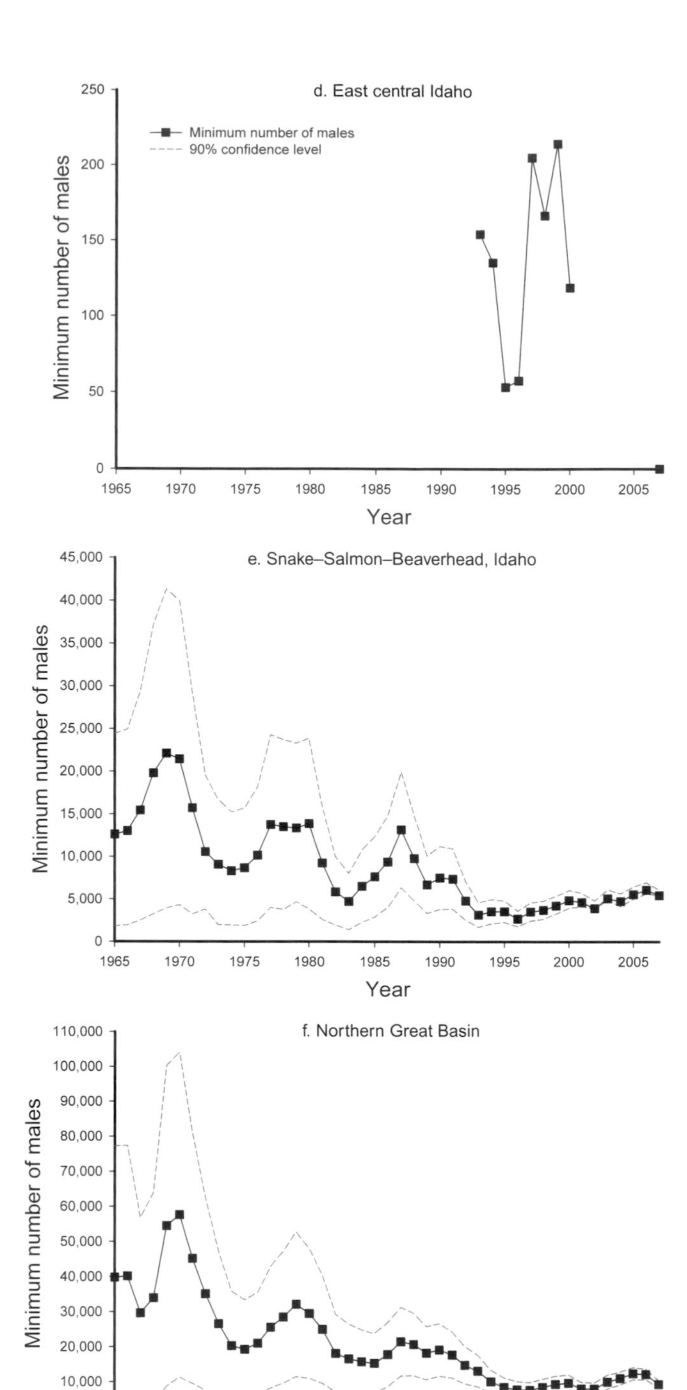

Figure 15.5. (*Continued*)

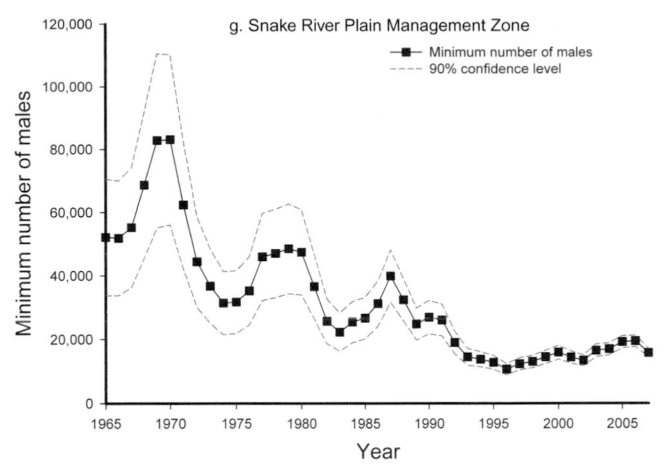

Figure 15.5. (*Continued*)

TABLE 15.41

Candidate model set (contains 95% of model weight) and model statistics for estimating population trends and persistence probabilities for Greater Sage-Grouse in the Baker, Oregon, population, 1985–2007.

	Model statistic[a]			
Model	r^2	k	ΔAIC_c	w_i
EGPE	0.000	3	0.0[b]	0.361
Ricker	0.278	4	1.9	0.138
Gompertz	0.271	3	2.0	0.131
Gompertz t − 1	0.183	4	3.4	0.066
Ricker t − 1	0.160	4	3.7	0.056
Ricker + year	0.418	5	4.0	0.050
Gompertz + year	0.410	5	4.2	0.045
Gompertz t − 2	0.123	4	4.2	0.040
Ricker t − 2	0.112	4	4.4	0.018

[a] Model fit described by coefficient of determination (r^2), the number of parameters (k), the difference in Akaike's information criterion corrected for small sample size (ΔAIC_c), and the AIC_c wt (w_i).
[b] $AIC_c = -2.0$ for best selected model.

this population has an 8% chance of declining below $N_e = 50$ and a 100% chance of declining below 500 within the next 30 years. There is a 62% chance, across all 9 models of population growth, of this population declining below an $N_e = 50$ within 30 yr; a 100% chance of declining below 500 within 30 yr; and a 67% chance of declining below 50 within 100 yr (Table 15.41).

Bannack, Montana, Population

The Bannack, Montana, population is a small population in southwestern Montana separated from nearby populations by distance and mountainous terrain (Table 15.1). The average number of leks counted per five-year period was relatively stable through most of the analysis period but

TABLE 15.42

Sage-grouse lek monitoring effort, lek size, and trends averaged over 5-yr periods for the Bannack, Montana, population, 1965–2007.

Parameter	2000–2007[a]	1995–1999	1990–1994	1985–1989	1980–1984	1975–1979	1970–1974	1965–1969
Leks counted	15	3	2	4	4	4	2	2
Males/lek	14	13	19	23	30	40	29	18
Active leks	13	3	2	4	4	4	2	2
% active leks	81	98	100	100	100	100	100	100
Males/active lek	17	14	19	23	30	40	29	18
λ	1.093	0.962	0.969	0.963	0.932	1.072	0.921	1.114
SE (λ)[b]	0.201	0.229	0.172	0.142	0.078	0.176	0.220	0.084

[a] Eight yr of data in this period.
[b] Standard error for annual rate of change.

increased substantially from 1995–1999 to 2000–2007 (Table 15.42). The proportion of active leks declined somewhat over the assessment period, although this could be because only two leks were counted when monitoring began (Table 15.42). Population trends, as indicated by average number of males per lek, declined from a high of 40 males/lek during 1975–1979 to 14 during 2000–2007, a decrease of 65%. Similarly, average number of males per active lek decreased by 58% (Table 15.42). Average rates of change were <1.0 for five of the eight analysis periods but, contrary to lek size data, rates of change increased by 13.6% from the 1995–1999 analysis period to the 2000–2007 period, and rate of change remained at or above 1.0 for the last period (Table 15.42).

From a minimum population estimate of 304 males (SE = 65) in 2007 based on counts at 24 leks, we reconstructed a minimum population estimate for males from 2007 back to 1965 (Fig. 15.5b), using 34 lek counts reported for this period. Less than two leks were counted prior to 1972, which was too few to construct valid confidence intervals.

The best stochastic model for the annual rates of change of this population was a Gompertz model with no time lag in density dependence and a period effect ($r_t = 1.8192 - 0.3233 \ln(N_t) + 0.2788$ period, $\sigma = 0.2015$, $r^2 = 0.175$; Table 15.43).

The Gompertz model implies that in the future, sage-grouse in the Bannack, Montana, population will fluctuate around an estimated carrying capacity of 278 males. A parametric bootstrap based on this model infers that this population has a 0% chance of declining below $N_e = 50$ within the next 30 years, an 86% chance of declining below $N_e = 50$ within 100 years, and a 99.9% chance of declining below $N_e = 500$ within the next 100 years. This population has a 6% chance of declining below an effective population size of 50 within 30 years based on multimodel projections across all 26 models of population growth, a 70% chance of declining below $N_e = 500$ within 30 years, a 33% chance of declining below $N_e = 50$ within 100 years, and a 98% chance of declining below $N_e = 500$ within 100 years (Table 15.6).

Red Rocks, Montana, Population

This population occurs in southwestern Montana just north of the Idaho border and is separated from adjacent populations by distance and mountainous terrain (Table 15.1). Few data were available for analysis until the mid-1980s. The average number of leks counted per five-year period from 1980–1984 to 2000–2007 increased by 1,900% (Table 15.44). The proportion of active leks declined slightly from 1985–1989 to 2000–2007 (Table 15.44). Population trends, as indicated by average number of males per lek, declined substantially from 1980–1984 to 2000–2007, but in part this is likely due to the low number of leks

TABLE 15.43

Candidate model set (contains 95% of model weight) and model statistics for estimating population trends and persistence probabilities for Greater Sage-Grouse in the Bannack, Montana, population, 1965–2007.

Model	Model statistic[a]			
	r^2	k	ΔAIC_c	w_i
Gompertz + period	0.175	4	0.0[b]	0.209
Gompertz + year	0.172	4	0.2	0.192
Gompertz + year, period	0.210	5	0.9	0.133
Ricker + year	0.118	4	2.7	0.054
EGPE	0.000	3	2.9	0.049
Gompertz	0.050	3	3.2	0.043
Ricker + period	0.104	4	3.3	0.040
Ricker	0.034	3	3.8	0.031
Gompertz t − 1 + period	0.090	5	3.9	0.029
Gompertz t − 1	0.026	4	4.2	0.026
Ricker t − 1	0.020	4	4.4	0.023
Gompertz t − 1 + year	0.072	5	4.7	0.020
Ricker + year, period	0.131	5	4.7	0.020
Gompertz t − 2	0.008	4	4.9	0.018
Ricker t − 2	0.008	4	4.9	0.018
Ricker t − 1 + period	0.063	5	5.1	0.016
Period	0.000	3	5.2	0.015
Ricker t − 1 + year	0.054	5	5.5	0.014
Gompertz t − 1 + year, period	0.097	6	6.3	0.009

[a] Model fit described by coefficient of determination (r^2), the number of parameters (k), the difference in Akaike's information criterion corrected for small sample size (ΔAIC_c), and the AIC_c wt (w_i).
[b] $AIC_c = -5.5$ for best selected model.

surveyed from 1965–1969 through 1980–1984. Average number of males per active lek followed a similar pattern over the same period (Table 15.44). Average rates of change were <1.0 for three of the eight analysis periods, and generally indicated a stable to increasing population during 1995–1999 and 2000–2007 (Table 15.44).

From a minimum population estimate of 448 males (SE = 103) in 2007 based on counts at 30 leks, we reconstructed a minimum population estimate for males from 2007 back to 1965 (Fig. 15.5c), using 39 lek counts reported for this period. However, too few leks were counted prior to 1985 to calculate valid confidence intervals. The population has generally fluctuated around 500 males.

The best stochastic model for annual rates of change of Red Rocks, Montana, population of sage-grouse was a Gompertz model with a period effect ($r_t = 3.596 - 0.6089 \ln(N_t) + 0.2348$ period, $\sigma = 0.2925$, $r^2 = 0.307$; Table 15.45).

The Gompertz model implies this population will fluctuate around an estimated carrying capacity of 367 males. A parametric bootstrap based on this model infers that this population has a 0% chance of declining below $N_e = 50$ within the next

TABLE 15.44

Sage-grouse lek monitoring effort, lek size, and trends averaged over 5-yr periods for the Red Rocks, Montana, population, 1965–2007.

Parameter	2000–2007[a]	1995–1999	1990–1994	1985–1989	1980–1984	1975–1979	1970–1974	1965–1969
Leks counted	20	5	4	4	1	1	1	1
Males/lek	13	15	18	31	71	57	73	104
Active leks	15	4	3	4	1	1	1	1
% active leks	75	76	79	100	100	100	100	100
Males/active lek	18	20	21	31	71	57	73	104
λ	1.105	1.088	1.088	1.016	0.932	1.326	0.854	0.977
SE (λ)[b]	0.282	0.360	0.224	0.169	—	—	—	—

[a] Eight yr of data in this period.

[b] Standard error for annual rate of change.

TABLE 15.45

Candidate model set (contains 95% of model weight) and model statistics for estimating population trends and persistence probabilities for Greater Sage-Grouse in the Red Rocks, Montana, population, 1965–2007.

Model	Model statistic[a]			
	r^2	k	ΔAIC_c	w_i
Gompertz + period	0.307	4	0.0[b]	0.382
Gompertz + year	0.295	4	0.7	0.273
Gompertz	0.228	3	1.8	0.152
Gompertz + year, period	0.310	5	2.5	0.112
Ricker + period	0.201	4	5.7	0.022
Ricker + year	0.197	4	5.9	0.020

[a] Model fit described by coefficient of determination (r^2), the number of parameters (k), the difference in Akaike's information criterion corrected for small sample size (ΔAIC_c), and the AIC_c wt (w_i).

[b] AIC_c = 24.3 for best selected model.

100 years, a 56.8% chance of declining below N_e = 500 within the next 30 years, and a 94.2% chance of declining below N_e = 500 within 100 years. There is a 0.1% chance, across all 26 models of population growth, of this population declining below N_e = 50 within 30 years, a 55% chance of declining below N_e = 500 within 30 years, a 2.5% chance of declining below N_e = 50 within 100 years, and a 92% chance of declining below N_e = 500 within 100 years (Table 15.6).

Wisdom, Montana, Population

This small, isolated population occurs in southwestern Montana and is separated from other populations by distance and mountainous terrain (Table 15.1). Data are only available for the 2000–2007 analysis period (Table 15.46). Estimates of rates of change were based on an average of six leks counted during this period, and the average rate of change was <1.0 for the 2000–2007 period.

TABLE 15.46

Sage-grouse lek monitoring effort, lek size, and trends for the Wisdom, Montana, population, 1965–2007.

Parameter	2000–2007	1965–1999
Leks counted	6	0
Males/lek	17	—
Active leks	5	—
% active leks	78	—
Males/active lek	22	—
λ	0.926	—
SE (λ)[a]	0.169	—

[a] Standard error for annual rate of change.

Population trends indicated by average number of males per lek declined from a high of 15 males/lek during 1980–1984 to 5 during 2000–2007, a decrease of 67%. Average number of males per active lek decreased from a high of 17 in 1985–1989 to a low of 7 in 1995–1999, but then increased to 11 males per active lek by 2000–2007 (Table 15.44). Average rates of change were <1.0 for two of the five periods that provided data for analysis, and decreased by 42.5% from 1995–1999 to 2000–2007; the values were <1.0 during both the last periods (Table 15.47).

Only eight years' worth of data were available to reconstruct this population's history, and population modeling or persistence analyses were not feasible (Fig. 15.5d).

East central Idaho Population

The east central Idaho population lies between the Snake River and the Wyoming border and is separated from nearby populations by distance and mountainous terrain (Table 15.1). The average number of leks counted per five-year period ranged from two to five and remained relatively stable from 1965–1969 to 2000–2007 (Table 15.44). Although 18 leks were counted in 2000, none were counted by 2003. The average number of active leks counted per five-year period was also relatively stable.

The proportion of active leks declined considerably over the assessment period (Table 15.47).

Snake–Salmon–Beaverhead, Idaho, Population

This population occupies much of central and eastern Idaho and is separated from other populations by habitat and mountainous terrain (Table 15.1). The average number of leks counted per five-year period increased substantially from 1965–1969 to 2000–2007 (Table 15.48). The proportion of active leks declined over the assessment period (Table 15.48). Population trends, as indicated by average number of males per lek, declined by 57% from 1965–1969 to 2000–2007. Similarly, average number of males per active lek declined by 41% over the assessment period (Table 15.48). Average rates of change were <1.0

TABLE 15.47

Sage-grouse lek monitoring effort, lek size, and trends averaged over 5-yr periods for the east central Idaho population, 1965–2007.

Parameter	2000–2007[a]	1995–1999	1990–1994	1985–1989	1980–1984	1975–1979	1970–1974	1965–1969
Leks counted	4	5	3	4	5	3	3	2
Males/lek	5	6	12	14	15	14	—	11
Active leks	3	4	3	3	4	3	3	2
% active leks	51	86	100	80	92	100	—	80
Males/active lek	11	7	12	17	16	14	—	—
λ	0.838	1.458		1.018	1.034	0.988	—	—
SE (λ)[b]	0.198	0.864		0.093	0.153	—	—	—

[a] Eight yr of data in this period.

[b] Standard error for annual rate of change.

TABLE 15.48

Sage-grouse lek monitoring effort, lek size, and trends averaged over 5-yr periods for the Snake–Salmon–Beaverhead, Idaho, population, 1965–2007.

Parameter	2000–2007[a]	1995–1999	1990–1994	1985–1989	1980–1984	1975–1979	1970–1974	1965–1969
Leks counted	323	234	165	131	187	138	112	76
Males/lek	15	10	13	20	12	25	30	35
Active leks	207	131	103	108	126	123	94	67
% active leks	65	56	63	81	69	87	85	89
Males/active lek	23	17	20	24	17	28	34	39
λ	1.028	1.082	0.883	1.036	0.932	1.107	0.845	1.117
SE (λ)[b]	0.050	0.071	0.087	0.089	0.088	0.090	0.070	0.089

[a] Eight yr of data in this period.
[b] Standard error for annual rate of change.

for three of the eight analysis periods and declined by 55% between the last two analysis periods, but values remained at or above 1.0 for both of these periods (Table 15.48).

The minimum population estimate of 5,457 males (SE = 397) in 2007 was based on counts at 340 leks. We reconstructed a minimum population estimate for males from 2007 back to 1965 (Fig. 15.5e), using 964 lek counts reported for this period. There was a general decrease in males attending leks in this population through 1992 following a high estimated population count of just over 22,000 in 1969. The population appears to be fluctuating at around 5,000 males since 1992, with an approximate eight-year cycle in population abundance from 1965–1990.

The best stochastic model for the annual rates of change of this population was a Gompertz model with a one-year time lag in density dependence and a period effect ($r_t = 2.757 - 0.3281 \ln(N_{t-1}) + 0.2616 \times \text{period}$, $\sigma = 0.1856$, $r^2 = 0.351$; Table 15.49). The Gompertz model implies sage-grouse in the Snake–Salmon–Beaverhead, Idaho, population will fluctuate around an estimated carrying capacity of 4,468 males. A parametric bootstrap based on this model infers that this population has a 0% chance of declining below $N_e = 500$ within the next 100 years. Across all 26 models of population growth, there is a 4% chance of this population declining below $N_e = 50$ within 30 years, a 10% chance of declining below $N_e = 500$ within 30 years, a 19% chance of declining below $N_e = 50$ within 100 years, and a 27% chance of declining below $N_e = 500$ within 100 years (Table 15.6).

Northern Great Basin Population

This population occupies portions of Nevada, southeastern Oregon, southwestern Idaho, and northwestern Utah (Table 15.1). The average number of leks counted per five-year period increased from 1965–1969 to 2000–2007 (Table 15.50). The proportion of active leks decreased during the assessment period (Table 15.50). Population trends, as indicated by average number of males per lek, declined by 37% from 1965–1969 to 2000–2007. Average number of males per active lek followed the same pattern over the assessment period and declined by 17% (Table 15.50). Average rates of change were <1.0 for three of the eight analysis periods (Table 15.50). Average rate of change declined by 2% between the last two analysis periods, but values remained at or above 1.0 for both of these periods.

We reconstructed a minimum population estimate for males back to 1965 starting from a minimum population estimate of 9,114 males (SE = 520) in 2007 based on counts at 952 leks (Fig. 15.5f), using 4,919 counts at 2,037 leks. The population increased from about 40,000 males in 1965 to peak at 57,655 males (SE = 28,345) in 1969, followed by a series of declines and peaks at

TABLE 15.49

Candidate model set (contains 95% of model weight) and model statistics for estimating population trends and persistence probabilities for Greater Sage-Grouse in the Snake–Salmon–Beaverhead, Idaho, population, 1965–2007.

Model	Model statistic[a]			
	r^2	k	ΔAIC_c	w_i
Gompertz t − 1 + period	0.351	5	0.0[b]	0.270
Ricker t − 1 + period	0.349	5	0.1	0.252
Gompertz t − 1 + year, period	0.359	6	2.1	0.095
Ricker t − 1 + year	0.314	5	2.2	0.089
Ricker t − 1 + year, period	0.357	6	2.2	0.088
Gompertz t − 1 + year	0.298	5	3.1	0.057
Ricker t − 2 + period	0.299	5	3.1	0.058
Gompertz t − 2 + period	0.273	5	4.5	0.028
Ricker t − 2 + year, period	0.301	6	5.6	0.017

[a] Model fit described by coefficient of determination (r^2), the number of parameters (k), the difference in Akaike's information criterion corrected for small sample size (ΔAIC_c), and the AIC_c wt (w_i).
[b] $AIC_c = -12.1$ for best selected model.

TABLE 15.50

Sage-grouse lek monitoring effort, lek size, and trends averaged over 5-yr periods for the northern Great Basin population, 1965–2007.

Parameter	2000–2007[a]	1995–1999	1990–1994	1985–1989	1980–1984	1975–1979	1970–1974	1965–1969
Leks counted	595	283	150	126	158	91	96	64
Males/lek	12	9	15	16	16	21	18	19
Active leks	366	196	128	94	127	80	72	52
% active leks	63	71	86	75	78	88	78	81
Males/active lek	19	13	18	22	20	24	23	23
λ	1.007	1.029	0.850	1.054	0.881	1.093	0.806	1.111
SE (λ)[b]	0.040	0.045	0.050	0.084	0.061	0.078	0.064	0.096

[a] Eight yr of data in this period.
[b] Standard error for annual rate of change.

irregular intervals of 8 to 13 years overlaid on a continuous decline through 2007.

The best stochastic model for the annual rates of change of the northern Great Basin population of sage-grouse was a Gompertz model with a one-year time lag and a declining time trend of −4.3% per year ($r_t = 49.971 - 0.4694 \ln(N_{t-1}) - 0.0208$ year, $\sigma = 0.1245$, $r^2 = 0.466$; Table 15.51).

The one-year delayed Gompertz model with declining time trend implies the northern Great Basin population of sage-grouse will fluctuate around an estimated carrying capacity that will decline from 6,770 males attending leks in 2007 to 1,787 attending leks in 2037 and only 80 males in 2107 if this trend continues at the same rate in the future. This model implies the population

TABLE 15.51

Candidate model set (contains 95% of model weight) and model statistics for estimating population trends and persistence probabilities for Greater Sage-Grouse in the northern Great Basin population, 1965–2007.

Model	Model statistic[a]			
	r^2	k	ΔAIC_c	w_i
Gompertz t − 1 + year	0.466	4	0.0[b]	0.680
Gomperz t − 1 + year, period	0.467	5	2.6	0.189
Gompertz t − 2 + year	0.406	4	4.3	0.079

[a] Model fit described by coefficient of determination (r^2), the number of parameters (k), the difference in Akaike's information criterion corrected for small sample size (ΔAIC_c), and the AIC_c wt (w_i).
[b] $AIC_c = 44.0$ for best selected model.

TABLE 15.52

Sage-grouse lek monitoring effort, lek size, and trends averaged over 5-yr periods for the Snake River Plain Management Zone, 1965–2007.

Parameter	2000–2007[a]	1995–1999	1990–1994	1985–1989	1980–1984	1975–1979	1970–1974	1965–1969
Leks counted	1,012	556	335	278	364	245	222	146
Males/lek	13	10	14	18	14	23	24	28
Active leks	643	356	249	220	271	219	180	125
% active leks	64	64	75	79	73	88	81	86
Males/active lek	20	15	19	23	19	26	29	33
λ	1.007	1.052	0.867	1.023	0.906	1.089	0.831	1.102
SE (λ)[b]	0.019	0.022	0.022	0.030	0.031	0.028	0.021	0.029

[a] Eight yr of data in this period.
[b] Standard error for annual rate of change.

estimate of 9,114 males attending leks in 2007 was 50% higher than the estimated current carrying capacity under this model. A parametric bootstrap based on the time-delayed Gompertz model with declining time trend, which has a 68% relative likelihood, infers that there is virtually no chance of the population declining below $N_e = 50$ in 30 or 100 years but declining below $N_e = 500$ is likely (100% relative probability) within 100 years. Multimodel forecasts imply this population has less than a 3% chance of declining below $N_e = 50$ or $N_e = 500$ in the short term (30 years) but declining below $N_e = 500$ in 100 years is certain if the carrying capacity continues to decline (Table 15.6).

Comprehensive Analysis of All Leks in the Management Zone

An average of 146 leks per year was censused in 1965–1969, but by 2005–2007, an average of 1,012 leks per year was counted, an increase of 593%. The proportion of active leks decreased over the assessment period, declining from 88% in 1975–1979 to 64% by 2005–2007 (Table 15.52). Population trends, as indicated by average number of males per lek, decreased over the assessment period by 54%, and average number of males per active lek decreased by 39% (Table 15.52). Average annual rates of change were <1.0 in three of the eight analysis periods. The average annual rate of

TABLE 15.53

Candidate model set (contains 95% of model weight) and model statistics for estimating population trends and persistence probabilities for Greater Sage-Grouse in the Snake River Plain Management Zone, 1965–2007.

Model	Model statistic[a]			
	r^2	k	ΔAIC_c	w_i
Gompertz t − 1 + year, period	0.413	3	0.0[b]	0.207
Gompertz t − 1 + year	0.363	3	0.7	0.146
Gompertz t − 1 + period	0.356	4	1.1	0.119
Gompertz t − 2 + year, period	0.385	4	1.9	0.080
Ricker t − 2 + period	0.339	5	2.2	0.069
Ricker t − 1 + period	0.336	4	2.3	0.066
Gompertz t − 2 + year	0.335	5	2.4	0.062
Ricker t − 1 + year	0.333	5	2.5	0.059
Gompertz t − 2 + period	0.328	6	2.8	0.051
Ricker t − 1 + year, period	0.367	4	3.0	0.046
Ricker t − 2 + year, period	0.368	5	3.0	0.046

[a] Model fit described by coefficient of determination (r^2), the number of parameters (k), the difference in Akaike's information criterion corrected for small sample size (ΔAIC_c), and the ΔIC_c wt (w_i).
[b] $AIC_c = -36.3$ for best selected model.

change declined by 4% from 1995–1999 to 2000–2007, but values remained at or above 1.0 for both of these periods.

The minimum population estimate of 15,761 males (SE = 676) in 2007 for the Snake River Plain SMZ was based on counts at 1,393 leks. We reconstructed minimum population estimates for males from 2007 back to 1965 (Fig. 15.5g) using 3,250 lek counts reported for this period. Since a high estimated population count of just over 82,000 in 1969 and 1970, the number of males attending leks in this SMZ has decreased. This population appears to have had an approximately eight-year cycle in population abundance from 1965 to about 1990.

The best stochastic model for the annual rates of change of the Snake River Plain SMZ population of sage-grouse was a Gompertz model with a one-year time lag in density dependence, a negative time trend, and a period effect ($r_t = 24.334 - 0.3910 \ln(N_{t-1}) - 0.0103$ year $+ 0.156$ period, $\sigma = 0.1327$, $r^2 = 0.413$; Table 15.53). The Gompertz model with a time trend implies sage-grouse in the Snake River Plain SMZ will fluctuate around a decreasing carrying capacity of 12,165 males in 2007, 5,517 males in 2037, and 872 males in 2107, representing a 2.6% decrease in carrying capacity per year. A parametric bootstrap based on this model infers that this population has a 0% chance of declining below $N_e = 50$ within the next 100 years. This population has a 2% chance (SE = 1.4%), across all 26 models of population growth, of declining below $N_e = 50$ within 30 years, a 10% (SE = 6.1%) chance of declining below $N_e = 500$ within 30 years, a 19% (SE = 7.9%) chance of declining below $N_e = 50$ within 100 years, and a 40% chance (SE = 9.6%) of declining below $N_e = 500$ within 100 years (Table 15.6).

Northern Great Basin Management Zone

This SMZ represents sage-grouse populations in parts of Oregon, Nevada, and California (Fig. 15.1). Four populations have been delineated within this management zone.

Central Oregon Population

The central Oregon population is separated from nearby populations by distance and topography (Table 15.1). The average number of leks counted

TABLE 15.54

Sage-grouse lek monitoring effort, lek size, and trends averaged over 5-yr periods for the central Oregon population, 1965–2007.

Parameter	2000–2007[a]	1995–1999	1990–1994	1985–1989	1980–1984	1975–1979	1970–1974	1965–1969
Leks counted	96	61	39	27	20	11	12	16
Males/lek	9	10	11	13	10	11	11	21
Active leks	62	48	32	23	14	7	10	13
% active leks	66	79	80	86	71	64	74	85
Males/active lek	13	13	14	15	15	18	15	25
λ	0.958	0.969	0.974	1.053	0.884	1.138	0.846	0.975
SE (λ)[b]	0.073	0.075	0.115	0.139	0.136	0.207	0.208	0.181

[a] Eight yr of data in this period.

[b] Standard error for annual rate of change.

per five-year period increased substantially from 1965–1969 to 2000–2007. The proportion of active leks declined over the assessment period (Table 15.54). Population trends indicated by average number of males per lek declined from a high of 21 males/lek during 1965–1969 to 9 during 2000–2007, a decrease of 48%. Similarly, average number of males per active lek decreased by 58% (Table 15.54). Average rates of change were <1.0 for six of the eight analysis periods, including the last three analysis periods (Table 15.54). The minimum population estimate was 835 males (SE = 106) in 2007 based on counts at 97 leks. We reconstructed minimum population estimates for males from 2007 back to 1965 (Fig. 15.6a), using 169 lek counts reported for this period.

The best stochastic model for the annual rates of change of the central Oregon population of sage-grouse was a Gompertz model with no time lags and a declining time trend ($r_t = 38.8227 - 0.4841 \ln(N_t) - 0.0177$ year, $\sigma = 0.1797$, $r^2 = 0.297$; Table 15.55). The Gompertz model with a declining time trend implies that the Central Oregon population will fluctuate around a decreasing carrying capacity of 783 males in 2007, 261 males in 2037, and 20 males in 2107, representing a 3.6% decrease in the carrying capacity per year. A parametric bootstrap based on this model infers that there is a 0% chance of this population declining below an effective population size of 50 within 30 years, a 14.2% chance of declining below $N_e = 500$ within 30 years, a 72.1% chance of declining below $N_e = 50$ within 100 years, and a 100% chance of declining below $N_e = 500$ within 100 years. This population has a 4% chance, across all 26 models of population growth, of declining below an effective population size of 50 within 30 years, a 15% chance of declining below $N_e = 500$ within 30 years, a 75% chance of declining below $N_e = 50$ within 100 years, and a 91% chance of declining below $N_e = 500$ within 100 years (Table 15.6).

Klamath, Oregon–California, Population

This is an isolated population straddling the Oregon and California border (Table 15.1). Monitoring started in the early 1970s and was inconsistent until the early 1990s. The average number of leks counted per five-year period increased from 1970–1974 to 2000–2007 (Table 15.56). Average number of males per lek fluctuated from 1990–1994 to 2000–2007, but the information is too sparse to allow calculations of annual rates of change or fit population models.

Northwest-Interior Nevada Population

This population occurs in north-central Nevada and leks are highly scattered (Table 15.1). Few data were available for analysis until the 1990s. The average number of leks counted per five-year period increased by 264% (Table 15.57) from 1990–1994 to 2000–2007. The proportion of active leks declined by 53% from 1990–1994 to 2000–2007 (Table 15.57).

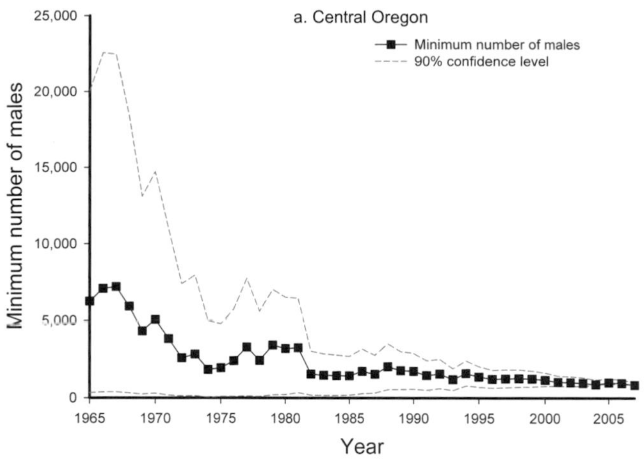

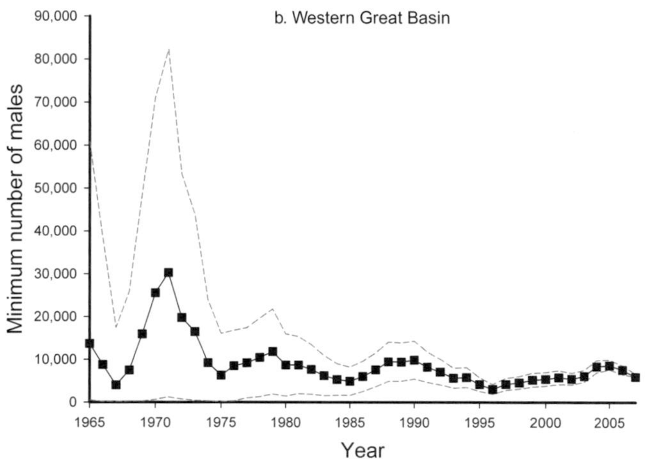

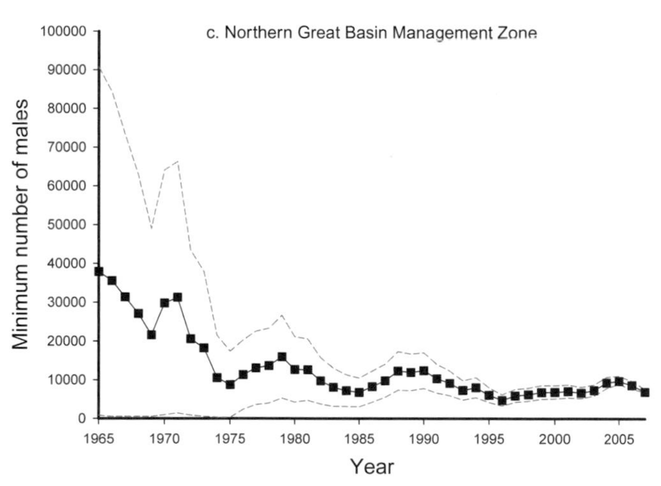

Figure 15.6. Population reconstructions for Northern Great Basin populations and Northern Great Basin Management Zone: (a) central Oregon; (b) western Great Basin; (c) Northern Great Basin Management Zone.

TABLE 15.55

Candidate model set (contains 95% of model weight) and model statistics for estimating population trends and persistence probabilities for Greater Sage-Grouse in the central Oregon population, 1965–2007.

Model	Model statistic[a]			
	r^2	k	ΔAIC_c	w_i
Gompertz + year	0.297	4	0.0[b]	0.523
Gompertz + year, period	0.323	5	1.2	0.293
Ricker + year	0.199	4	5.3	0.038
Ricker	0.128	3	6.2	0.024
Gompertz	0.119	3	6.6	0.020
Gompertz + period	0.158	4	7.2	0.014
Ricker + year, period	0.211	5	7.3	0.014
Gompertz t − 2 + year	0.148	5	7.7	0.011
Ricker + period	0.148	4	7.7	0.011
Gompertz t − 2	0.071	4	8.7	0.007

[a] Model fit described by coefficient of determination (r^2), the number of parameters (k), the difference in Akaike's information criterion corrected for small sample size (ΔAIC_c), and the AIC_c wt (w_i).
[b] $AIC_c = -14.7$ for best selected model.

TABLE 15.56

Sage-grouse lek monitoring effort, lek size, and trends averaged over 5-yr periods for the Klamath, Oregon, population, 1965–2007.

Parameter	2000–2007[a]	1995–1999	1990–1994	1985–1989	1980–1984	1975–1979	1970–1974	1965–1969
Leks counted	6	6	5	0	4	2	1	6
Males/lek	4	3	0	—	—	—	—	4
Active leks	1	1	0	—	—	—	—	1
% active leks	58	31	6	—	—	—	—	58
Males/active lek	7	11	3	—	—	—	—	7
λ	—	—	—	—	—	—	—	—
SE (λ)[b]	—	—	—	—	—	—	—	—

[a] Eight yr of data in this period.
[b] Standard error for annual rate of change.

Population trends, as indicated by average number of males per lek, remained largely unchanged from 1990–1994 to 2000–2007. Average number of males per active lek also remained largely unchanged over the same period (Table 15.57). Average rate of change could only be calculated for the 2000–2007 period and was <1.0 for this period (Table 15.57).

Western Great Basin Population

This population occupies portions of southeastern Oregon, northwestern Nevada, and northeastern California (Table 15.1). The average number of leks counted per five-year period increased considerably from 1965–1969 to 2000–2007 (Table 15.58). The proportion of active leks

TABLE 15.57

Sage-grouse lek monitoring effort, lek size, and trends averaged over 5-yr periods for the northwest interior Nevada population, 1965–2007.

Parameter	2000–2007[a]	1995–1999	1990–1994	1965–1989	1980–1984	1975–1979	1970–1974	1965–1969
Leks counted	40	9	11	0	0	0	0	0
Males/lek	3	4	4	—	—	—	—	—
Active leks	16	5	9	—	—	—	—	—
% active leks	41	77	88	—	—	—	—	—
Males/active lek	7	6	6	—	—	—	—	—
λ	0.936	—	—	—	—	—	—	—
SE (λ)[b]	0.330	0.244	—	—	—	—	—	—

[a] Eight yr of data in this period.
[b] Standard error for annual rate of change.

TABLE 15.58

Sage-grouse lek monitoring effort, lek size, and trends averaged over 5-yr periods for the western Great Basin population, 1965–2007.

Parameter	2000–2007[a]	1995–1999	1990–1994	1985–1989	1980–1984	1975–1979	1970–1974	1965–1969
Leks counted	285	96	88	45	34	19	34	17
Males/lek	18	18	29	31	23	24	20	16
Active leks	175	73	79	38	25	14	23	13
% active leks	62	76	88	84	76	76	68	78
Males/active lek	28	23	33	36	30	31	28	20
λ	1.028	1.075	0.847	1.155	0.895	1.088	0.782	1.339
SE (λ)[b]	0.062	0.065	0.060	0.124	0.116	0.224	0.165	0.401

[a] Eight yr of data in this period.
[b] Standard error for annual rate of change.

fluctuated throughout the assessment period (Table 15.58). Population trends, as indicated by average number of males per lek, increased 94% from 1965–1969 to 1985–1989 and decreased by 42% from 1985–1989 to 2000–2007. Average number of males per active lek followed the same pattern over the assessment period (Table 15.58). Average rates of change were <1.0 for three of the eight analysis periods (Table 15.58). Average rate of change declined by 4% between the last two analysis periods, but values remained at or above 1.0 for both of these periods (Table 15.58).

The minimum population estimate of 5,904 males (SE = 438) in 2007 was based on counts at 393 leks. We reconstructed minimum population estimates for males from 2007 back to 1965 (Fig. 15.6b), using 899 lek counts reported for this period. The highest recorded estimate for this population was 30,291 males in 1971. The population declined sharply to 6,277 males in 1975. The population has generally fluctuated between 3,000 and 10,000 males since 1975.

The best stochastic model for the annual rates of change of this population was a Gompertz model with a one-year time lag in density dependence

TABLE 15.59

Candidate model set (contains 95% of model weight) and model statistics for estimating population trends and persistence probabilities for Greater Sage-Grouse in the western Great Basin population, 1965–2007.

Model	Model statistic[a]			
	r^2	k	ΔAIC_c	w_i
Gompertz t − 1 + year	0.498	5	0.0[b]	0.620
Gompertz t − 1 + year, period	0.512	6	1.5	0.293
Ricker t − 1 + year	0.422	5	5.6	0.037

[a] Model fit described by coefficient of determination (r^2), the number of parameters (k), the difference in Akaike's information criterion corrected for small sample size (ΔAIC_c), and the AIC_c wt (w_i).
[b] $AIC_c = -8.4$ for best selected model.

TABLE 15.60

Sage-grouse lek monitoring effort, lek size, and trends averaged over 5-yr periods for the Northern Great Basin Management Zone, 1965–2007.

Parameter	2000–2007[a]	1995–1999	1990–1994	1985–1989	1980–1984	1975–1979	1970–1974	1965–1969
Leks counted	390	164	134	73	60	33	47	33
Males/lek	15	14	23	24	16	18	17	18
Active leks	240	123	112	63	41	23	33	26
% active leks	63	75	83	85	70	71	70	80
Males/active lek	24	19	27	28	24	25	24	23
λ	1.011	1.036	0.874	1.134	0.886	1.092	0.800	0.972
SE (λ)[b]	0.052	0.051	0.054	0.102	0.095	0.156	0.134	0.162

[a] Eight yr of data in this period.
[b] Standard error for annual rate of change.

and a declining year trend ($r_t = 32.6172 - 0.4682 \ln(N_{t-1}) - 0.0143$ year, $\sigma = 0.1943$, $r^2 = 0.498$; Table 15.59). The Gompertz model with a declining trend implies the sage-grouse in the Western Great Basin population will fluctuate around an estimated carrying capacity of 4,111 males in 2007, 1,695 males in 2037, and 200 males in 2107. A parametric bootstrap based on this model infers this population has a 0% chance of declining below $N_e = 500$ within the next 30 years, a 0% chance of declining below $N_e = 50$ within 100 years, and a 100% chance of declining below $N_e = 500$ within the next 100 years. A multimodel parametric bootstrap predicts that this population has a 5.5% chance of declining below $N_e = 50$ within 30 years, a 6% chance of declining below $N_e = 500$ within 30 years, a 6% chance of declining below $N_e = 50$ within 100 years, and a 99% chance of declining below $N_e = 500$ within 100 years (Table 15.6) across all 26 models of population growth.

Comprehensive Analysis of All Leks in the Management Zone

The average number of leks counted per five-year period increased over the assessment period (Table 15.60). An average of 33 leks per year was censused in 1965–1969, but by 2000–2007, an average of 390 leks per year was counted, an increase of 1,082%. The proportion of active leks decreased over the assessment period, declining from 85% in 1985–1989 to 63% in 2000–2007 (Table 15.60). Population trends, as indicated by average number of males per lek, decreased over the assessment period by 17%, but average number of males per active lek increased by 4% (Table 15.60). Average annual rates of change were <1.0 in four of the eight analysis periods. From 1995–1999 to 2000–2007, average annual rate of change declined by 2%, but values remained at or above 1.0 for both of these periods (Table 15.60).

A minimum population estimate of 6,925 males (SE = 464) in 2007 was estimated from counts at 495 leks. We reconstructed minimum population estimates for males from 2007 back to 1965 (Fig. 15.6c) using 1,122 lek counts reported for this period. Since a high estimated population count of 37,915 in 1965, males attending leks have decreased in this SMZ.

The best stochastic model for the annual rates of change of the Northern Great Basin SMZ population was a Gompertz model with a one-year time lag in density dependence and a negative time trend (r_t = 19.157 − 0.2990 $\ln(N_{t-1})$ − 0.0083 year, σ = 0.1683, r^2 = 0.240; Table 15.61). The Gompertz model with a time trend implies that sage-grouse in the Northern Great Basin SMZ will fluctuate around a decreasing carrying capacity of 5,529 males in 2007, 2,413 males in 2037, and 349 males in 2107, representing a 2.7%

TABLE 15.61

Candidate model set (contains 95% of model weight) and model statistics for estimating population trends and persistence probabilities for Greater Sage-Grouse in the Northern Great Basin Management Zone, 1965–2007.

	Model statistic[a]			
Model	r^2	k	ΔAIC_c	w_i
Gompertz t − 1 + year	0.240	3	0.0[b]	0.218
Gompertz t − 2 + year	0.221	3	1.0	0.133
Gompertz t − 1	0.157	4	1.7	0.094
Gompertz t − 1 + year, period	0.248	4	2.2	0.072
Gompertz t − 1 + period	0.188	5	2.6	0.058
Gompertz t − 2	0.135	4	2.7	0.056
Ricker t − 1	0.122	5	3.3	0.042
Gompertz t − 2 + year, period	0.223	5	3.5	0.038
Gompertz t − 2 + period	0.171	6	3.5	0.038
Gompertz	0.116	4	3.6	0.036
Ricker t − 2	0.104	5	4.1	0.028
Ricker t − 1 + year	0.157	6	4.2	0.027
Ricker	0.101	5	4.3	0.026
Gompertz + year	0.148	3	4.6	0.022
Ricker t − 1 + period	0.135	3	5.2	0.016
Ricker t − 2 + year	0.135	4	5.2	0.016
Gompertz + period	0.128	4	5.5	0.014
Ricker + year	0.117	5	6.0	0.011
Ricker t − 2 + period	0.117	4	6.0	0.011

[a] Model fit described by coefficient of determination (r^2), the number of parameters (k), the difference in Akaike's information criterion corrected for small sample size (ΔAIC_c), and the AIC_c wt (w_i).

[b] AIC_c = −19.9 for best selected model.

decrease in carrying capacity per year. A parametric bootstrap based on this model infers this population has a 0% chance of declining below N_e = 500 within the next 30 years, a 0% chance of declining below N_e = 50 within 100 years, and a 4.7% chance of declining below N_e = 500 within 100 years. There is a 1% chance (SE = 2.0%), across all 26 models of population growth, of this population declining below N_e = 50 within 30 years, a 2% (SE = 2.3%) chance of declining below N_e = 500 within 30 years, a 7% (SE = 5.0%) chance of declining below N_e = 50 within 100 years, and a 29% chance (SE = 8.1%) of declining below N_e = 500 within 100 years (Table 15.6).

Columbia Basin Management Zone

This SMZ represents sage-grouse populations in Washington. Two populations, Moses Coulee and Yakima, are delineated within this zone (Fig. 15.1). Genetic diversity in these populations has been lost due to population declines and isolation from core regions of the sage-grouse range (Oyler-McCance and Quinn, this volume, chapter 5).

Moses Coulee, Washington, Population

This population occurs in north-central Washington and is separated from the other Washington population (Yakima) by distance and topography (Table 15.1). The average number of leks counted per five-year period increased from 1965–1969 to 2000–2007 (Table 15.62). The proportion of active leks declined 60% over the assessment period (Table 15.62). Population trends, as indicated by average number of males per lek, declined by 79% from 1965–1969 to 2000–2007. Similarly, average number of males per active lek declined by 50% over the assessment period (Table 15.62). Average rates of change were <1.0 for four of the eight analysis periods. From 1995–1999 to 2000–2007, average rate of change decreased by 4.7%, but values remained at or above 1.0 for both of these periods (Table 15.62).

We estimated a minimum population of 230 males (SE = 54) in 2007 based on counts at 32 leks. We reconstructed minimum population estimates for males from 2007 back to 1965 (Fig. 15.7a), using 42 lek counts reported for this period. The highest estimated count was in 1965 with 2,433 males. Since then, the population dramatically declined until about 1993, with population counts fluctuating around 220 males between 1994 and 2007.

The best stochastic model for the annual rates of change of this population was a Gompertz model with a one-year time lag and a declining time trend of −4.3% per year (r_t = 34.7636 − 0.3689 ln(N_{t-1}) − 0.01637 year, σ = 0.2789, r^2 = 0.194; Table 15.63).

The Gompertz model with a declining time trend implies that in the future, the Moses Coulee, Washington, population will fluctuate around an estimated carrying capacity that will decline from 168 males attending leks in 2007 to 44 in 2037 and only 2 in 2107. A parametric bootstrap based

TABLE 15.62

Sage-grouse lek monitoring effort, lek size, and trends averaged over 5-yr periods for the Moses Coulee, Washington, population, 1965–2007.

Parameter	2000–2007[a]	1995–1999	1990–1994	1985–1989	1980–1984	1975–1979	1970–1974	1965–1969
Leks counted	33	24	19	16	15	14	13	3
Males/lek	7	10	12	13	21	13	27	34
Active leks	12	12	11	10	13	11	12	3
% active leks	38	48	60	66	89	88	95	95
Males/active lek	18	20	19	21	24	14	28	36
λ	1.002	1.051	0.902	1.327	0.766	1.122	0.811	0.929
SE (λ)[b]	0.124	0.108	0.151	0.241	0.073	0.210	0.124	0.343

[a] Eight yr of data in this period.
[b] Standard error for annual rate of change.

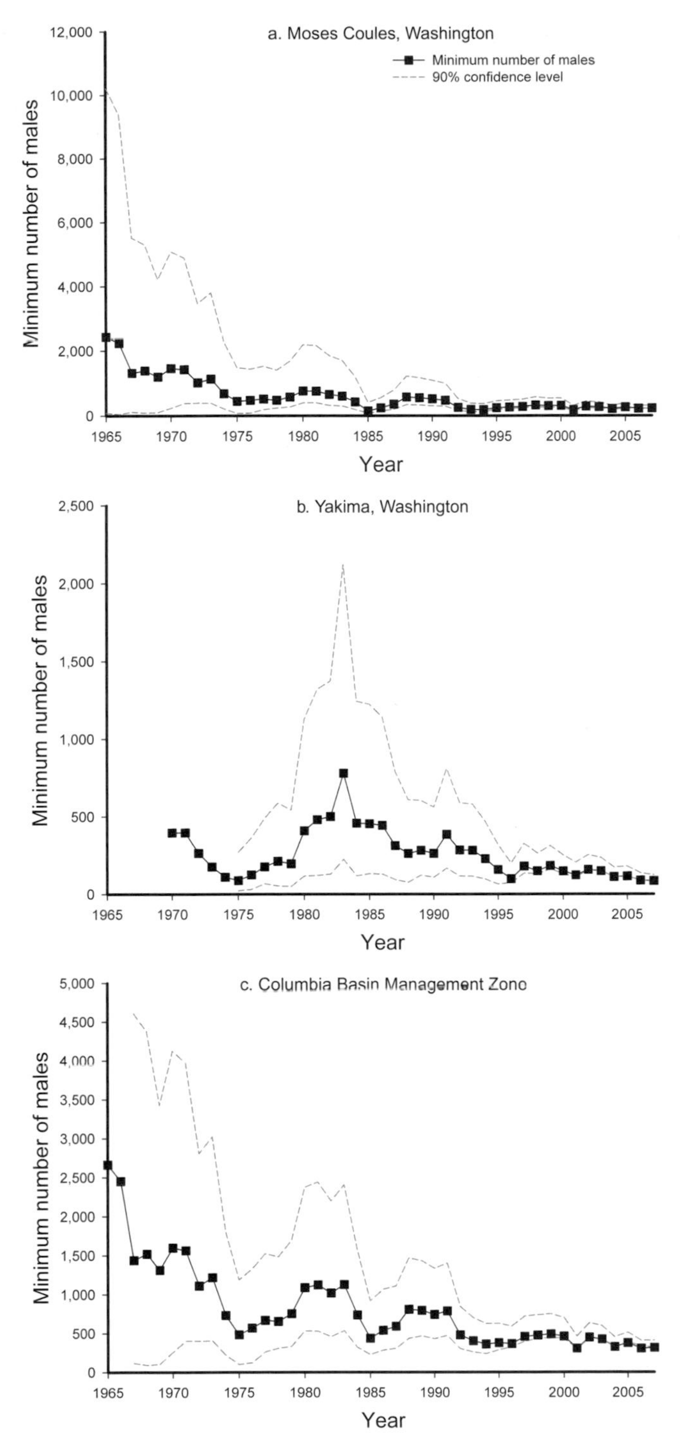

Figure 15.7. Population reconstructions for Columbia Basin populations and Columbia Basin Management Zone: (a) Moses-Coulee, Washington; (b) Yakima, Washington; (c) Columbia Basin Management Zone.

TABLE 15.63

Candidate model set (contains 95% of model weight) and model statistics for estimating population trends and persistence probabilities for Greater Sage-Grouse in the Moses Coulee, Washington, population, 1965–2007.

Model	Model statistic[a]			
	r^2	k	ΔAIC_c	w_i
Gompertz t − 1 + year	0.194	5	0.0[b]	0.207
Gompertz + year	0.192	4	0.1	0.197
Gompertz + year, period	0.213	5	1.7	0.090
Gompertz t − 1 + year, period	0.201	6	2.2	0.067
Gompertz	0.087	3	2.5	0.059
Gompertz t − 1	0.085	4	2.6	0.056
Gompertz t − 2 + year	0.132	5	3.0	0.047
Gompertz t − 1 + period	0.118	5	3.6	0.034
Gompertz t − 2	0.059	4	3.7	0.032
EGPE	0.000	3	3.8	0.031
Gompertz + period	0.112	4	3.9	0.030
Ricker	0.041	3	4.5	0.022
Gompertz t − 2 + period	0.092	5	4.8	0.019
Ricker t − 1	0.033	4	4.8	0.019
Ricker t − 2	0.024	4	5.2	0.016
Gompertz t − 2 + year, period	0.132	6	5.6	0.013
Ricker + year	0.072	4	5.7	0.012

[a] Model fit described by coefficient of determination (r^2), the number of parameters (k), the difference in Akaike's information criterion corrected for small sample size (ΔAIC_c), and the AIC_c wt (w_i).

[b] $AIC_c = 20.5$ for best selected model.

on this model infers a 9% chance of this population declining below $N_e = 50$ within 30 years, a 100% chance of this population declining below $N_e = 50$ within 100 years, and a 100% chance of declining below $N_e = 500$ within 30 years. Across all 26 models of population growth, this population has a 10% chance of declining below $N_e = 50$ within 30 years, an 88% chance of declining below $N_e = 500$ within 30 years, a 62% chance of declining below $N_e = 50$ within 100 years, and a 99.8% chance of declining below $N_e = 500$ within 100 years (Table 15.6).

Yakima, Washington, Population

This population occurs in south-central Washington and is separated from other populations by distance and the Columbia River (Table 15.1). The average number of leks counted per five-year period increased from 1970–1974 to 2000–2007 (Table 15.64). An average of 1 lek per five-year period was censused in 1970–1974, but by 2005–2007, an average of 10 leks was counted. The proportion of active leks declined over the assessment period (Table 15.64). Population trends, as indicated by average number of males per lek, increased 300% from 1975–1979 to 1980–1984 but declined by 73% from 1980–1984 to 2000–2007. However, relatively few leks existed to count throughout the assessment period. Average number of males per active lek also showed the same pattern over the assessment period (Table 15.64). Average rates of change were <1.0 for four of the eight analysis periods. Average rate of change declined by 12%

TABLE 15.64

Sage-grouse lek monitoring effort, lek size, and trends averaged over 5-yr periods for the Yakima, Washington, population, 1965–2007.

Parameter	2000–2007[a]	1995–1999	1990–1994	1985–1989	1980–1984	1975–1979	1970–1974	1965–1969
Leks counted	10	7	5	4	3	2	1	0
Males/lek	13	16	23	29	48	19	12	—
Active leks	8	6	5	4	3	2	1	—
% active leks	81	89	93	87	97	100	100	—
Males/active lek	16	18	26	34	49	19	12	—
λ	0.939	1.065	0.939	0.905	1.071	1.408	0.752	—
SE (λ)[b]	0.110	0.158	0.107	0.238	0.080	0.151	—	—

[a] Eight yr of data in this period.

[b] Standard error for annual rate of change.

between the last two analysis periods and was <1.0 during 2000–2007 (Table 15.64).

The minimum population estimate was 85 males (SE = 24) in 2007 based on counts at seven leks. We reconstructed minimum population estimates for males from 2007 back to 1970 (Fig. 15.7b), using 24 lek counts reported for this period. Because only one lek was surveyed from 1970–1974, confidence intervals on the population estimate could only be calculated to 1975. The population increased from 396 in 1970 to peak at 779 in 1983, followed by a decline through 2007.

The best stochastic model for the annual rates of change of this population was a Gompertz model with a two-year time lag and no period effects or time trend (r_t = 0.8762 – 0.1667 ln(N_{t-2}), $\sigma = 0.2934$, $r^2 = 0.085$; Table 15.65). Based on this model, the estimated carrying capacity for this population was 192 males. A parametric bootstrap based on this model infers this population has a 0.4% and 1.6% chance of declining below $N_e = 50$ within 30 and 100 years, respectively. The population is now below $N_e = 500$ and has a 26% chance, across all 26 models of population growth, of declining below $N_e = 50$ within 30 years and a 50% chance of declining below $N_e = 50$ within 100 years (Table 15.6).

Comprehensive Analysis of All Leks in the Management Zone

The average number of leks counted per five-year period increased over the assessment period (Table 15.65). An average of 3 leks per year was censused in 1965–1969, but by 2005–2007, an average of 42 leks per year was counted, an increase of 1,300%. The proportion of active leks decreased over the assessment period, averaging between 92% and 100% from 1965 to 1984 but decreased to 47% by 2000–2007 (Table 15.66). Population trends, as indicated by average number of males per lek and average number of males per active lek, also decreased over the assessment period by 76 and 53%, respectively (Table 15.66). Average annual rates of change were <1.0 in five of the eight analysis periods. The average annual rate of change from the 1995–1999 to 2000–2007 declined by 6.5% and the value was <1.0 for the last period (Table 15.66).

A minimum population estimate of 315 males (SE = 59) in 2007 was estimated from counts at 39 leks. We reconstructed minimum population estimates for males from 2007 back to 1965 (Fig. 15.7c) using 64 lek counts reported for this period. Because only one lek was surveyed in 1965 and 1966, confidence intervals on the population estimate could only be calculated to 1975. The number of males attending leks in this SMZ has decreased since an estimated high of 2,665 in 1965.

The best stochastic model for the annual rates of change in the Columbia Basin SMZ sage-grouse population was a Gompertz model with no time lag and a negative time trend (r_t = 28.817 – 0.3842

TABLE 15.65

Candidate model set (contains 95% of model weight) and model statistics for estimating population trends and persistence probabilities for Greater Sage-Grouse in the Yakima, Washington, population, 1965–2007.

Model	Model statistic[a]			
	r^2	k	ΔAIC_c	w_i
Gompertz t − 2	0.085	4	0.0[b]	0.106
Ricker t − 2	0.072	4	0.5	0.082
EGPE	0.000	3	0.7	0.074
Gompertz t − 2 + year	0.132	5	0.7	0.073
Gompertz t − 2 + period	0.121	5	1.2	0.059
Gompertz t − 1	0.049	4	1.4	0.053
Ricker t − 2 + year	0.114	5	1.4	0.052
Ricker t − 2 + period	0.114	5	1.4	0.052
Gompertz	0.044	3	1.5	0.049
Ricker	0.040	3	1.7	0.046
Ricker t − 1	0.033	4	1.9	0.040
Ricker + period	0.092	4	2.3	0.034
Gompertz + period	0.089	4	2.4	0.032
Gompertz t − 1 + period	0.086	5	2.5	0.030
Gompertz t − 1 + year	0.086	5	2.5	0.030
Gompertz + year	0.083	4	2.6	0.028
Period	0.015	3	2.6	0.029
Ricker + year	0.076	4	2.9	0.025
Ricker t − 1 + period	0.073	5	3.0	0.023
Ricker t − 1 + year	0.065	5	3.3	0.020
Gompertz t − 2 + year, period	0.132	6	3.5	0.019

[a] Model fit described by coefficient of determination (r^2), the number of parameters (k), the difference in Akaike's information criterion corrected for small sample size (ΔAIC_c), and the AIC_c wt (w_i).
[b] $AIC_c = 20.3$ for best selected model.

$\ln(N_t) - 0.0133 \times \text{year}$, $\sigma = 0.211$, $r^2 = 0.193$; Table 15.67). The Gompertz model with a time trend implies sage-grouse in the Columbia Basin SMZ will fluctuate around a decreasing carrying capacity of 192 males in 2007, 103 males in 2037, and nine males in 2107. A parametric bootstrap based on this model infers that this population has a 0% chance of declining below $N_e = 50$ within the next 30 years, a 100% chance of declining below $N_e = 50$ within 100 years. This population has a 12% (SE = 6.0%) chance across all 26 models of population growth of declining below $N_e = 50$ within 30 years, a 62% (SE = 9.1%) chance of declining below 50 within 100 years, a 76% (SE = 6.5%) chance of declining below $N_e = 500$ within 30 years, and an 86% (SE = 5.8%) of declining below $N_e = 500$ in 100 years (Table 15.6).

Colorado Plateau Management Zone

This SMZ represents sage-grouse populations in parts of Utah and Colorado. Of the five populations

TABLE 15.66

Sage-grouse lek monitoring effort, lek size, and trends averaged over 5-yr periods for the Columbia Basin Management Zone, 1965–2007.

Parameter	2000–2007[a]	1995–1999	1990–1994	1985–1989	1980–1984	1975–1979	1970–1974	1965–1969
Leks counted	42	31	24	20	18	16	16	3
Males/lek	8	11	14	16	25	13	24	34
Active leks	20	18	16	14	16	14	13	3
% active leks	47	57	67	70	89	90	90	95
Males/active lek	17	20	20	23	28	15	26	36
λ	0.977	1.045	0.891	1.123	0.858	1.185	0.809	0.929
SE (λ)[b]	0.088	0.102	0.110	0.163	0.091	0.170	0.117	—

[a] Eight yr of data in this period.

[b] Standard error for annual rate of change.

delineated within this zone, only the Piceance Basin had sufficient data for analysis.

Piceance Basin, Colorado, Population

This population occurs in the Piceance Basin, Colorado and is separated from adjacent populations by distance and topography (Table 15.1). Few data were available for analysis until the 2000–2007 analysis period (Fig. 15.8a). The proportion of active leks and lek size was low in 2000–2007 (Table 15.68). Average rate of change could only be calculated for the 2000–2007 period and was >1.0 for this period.

Comprehensive Analysis of All Leks in the Management Zone

The average number of leks counted per five-year period increased over the assessment period (Table 15.69). An average of 2 leks per year was censused in 1965–1969 but by 2005–2007, an average of 37 leks per year was counted, an increase of 1,750%. The proportion of active leks decreased over the assessment period, declining from 84% from 1980–1984 to 45% by 2000–2007 (Table 15.69). Population trends, as indicated by average number of males per lek, decreased from 1980–1984 to 2000–2007, but average number of males per active lek increased by 44% (Table 15.69). Average annual rates of change were <1.0 in three of the four analysis periods. The average annual rate of change increased by 34% from 1995–1999 to 2000–2007 (Table 15.69).

The minimum population estimate of 241 males (SE = 52) in 2007 was based on counts at 73 leks. We reconstructed minimum population estimates for males from 2007 back to 1984 (Fig. 15.8b) using 100 lek counts reported for this period.

The best stochastic model for the annual rates of change of the Colorado Plateau SMZ was a Gompertz model with no time lag (r_t = 2.788 – 0.5071 N_t, σ = 0.1454, r^2 = 0.298; Table 15.70) which was matched by a Ricker model with similar performance. The Gompertz model implies that sage-grouse populations in the Colorado Plateau SMZ will fluctuate around a carrying capacity of 244 males with no change through time. A parametric bootstrap based on this model infers that this SMZ has virtually no chance of declining below N_e = 50 within 30 or 100 years. This population of grouse in this SMZ has no chance, across all 26 models of population growth, of declining below N_e = 50 within 30 years, a 96% (SE = 3.7%) of declining below N_e = 500 in 30 years, a 5% (SE = 2.3%) chance of declining below N_e = 50 within 100 years, and a 98% (SE = 3.7%) chance of declining below N_e = 500 within 100 years (Table 15.6).

Metapopulation Analysis

Estimated dispersal rates among SMZs were generally low, never exceeding 5% of the SMZ's abundance dispersing to any other SMZ (Table 15.71).

TABLE 15.67

Candidate model set (contains 95% of model weight) and model statistics for estimating population trends and persistence probabilities for Greater Sage-Grouse in the Columbia Basin Management Zone, 1965–2007.

Model	Model statistic[a]			
	r^2	k	ΔAIC_c	w_i
Gompertz + year	0.193	3	0.0[b]	0.257
Gompertz t − 1 + year	0.160	3	1.6	0.113
Gompertz + year, period	0.201	4	2.2	0.084
Ricker + year	0.137	4	2.7	0.066
Gompertz	0.066	5	3.4	0.047
EGPE	0.000	4	3.8	0.039
Ricker	0.054	5	3.9	0.037
Gompertz t − 1	0.055	5	3.9	0.037
Gompertz t − 1 + year, Period	0.164	6	4.0	0.034
Gompertz + period	0.107	4	4.1	0.034
Gompertz t − 2 + year	0.103	5	4.3	0.031
Ricker t − 1	0.037	6	4.6	0.026
Gompertz t − 2	0.035	5	4.7	0.025
Gompertz t − 1 + period	0.090	3	4.8	0.023
Ricker + year, period	0.143	3	5.0	0.021
Ricker + period	0.081	4	5.2	0.019
Ricker t − 1 + year	0.082	4	5.2	0.020
Ricker t − 2	0.023	5	5.2	0.019
Period	0.001	4	6.1	0.012
Gompertz t − 2 + period	0.061	5	6.1	0.012

[a] Model fit described by coefficient of determination (r^2), the number of parameters (k), the difference in Akaike's information criterion corrected for small sample size (ΔAIC_c), and the AIC_c wt (w_i).

[b] $AIC_c = -1.9$ for best selected model.

Highest estimated dispersal occurred between the Great Plains and Wyoming Basin SMZs, between Snake River Plain and Northern Great Basin SMZs, and between Southern Great Basin and Snake River Plain SMZs with rates of 5%, 3.5%, and 2.4%, respectively. Correlated population dynamics were prevalent among many of the SMZs (Table 15.72). Highest correlations were between Southern Great Basin and Snake River Plain SMZs, between Southern Great Basin and Northern Great Basin SMZs, and between Snake River Plain and Northern Great Basin SMZs, with correlation coefficients of 0.58, 0.50, and 0.52, respectively. A slight negative correlation (−0.04) was found between Great Plains and Northern Great Basin SMZs.

Metapopulation projections were based on dispersal and correlations among SMZ populations and individual SMZ growth models based on Gompertz and Ricker types of density dependence (Table 15.73). Projections based on the information-theoretic (IT) best Gompertz models, which were also the best models overall, suggested a low probability

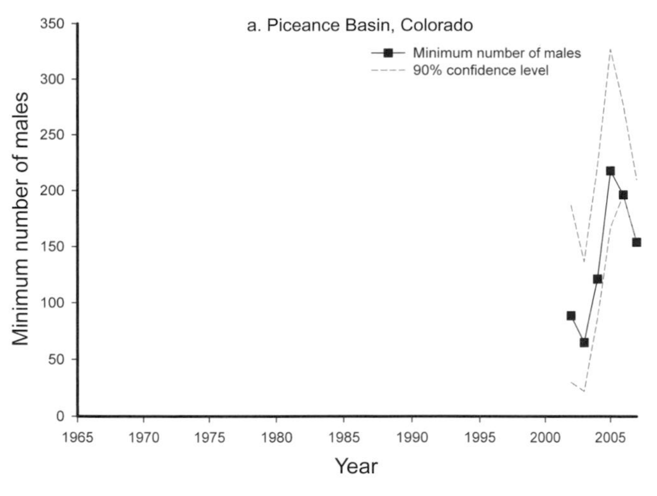

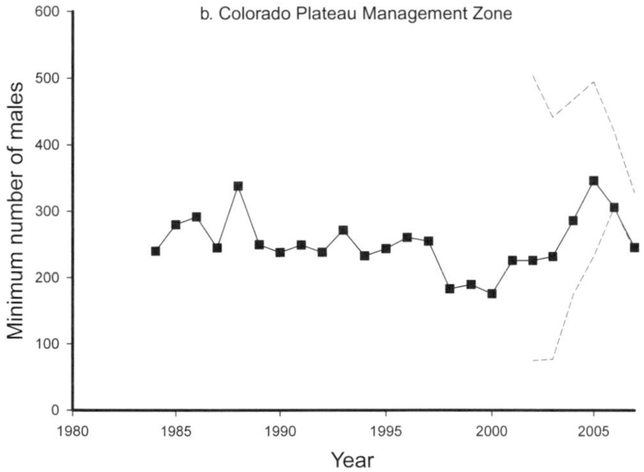

Figure 15.8. Population reconstructions for Colorado Plateau populations and Colorado Plateau Management Zone: (a) Piceance Basin, Colorado; (b) Colorado Plateau Management Zone.

(i.e., <0.1) that sage-grouse would fall <30,000 males within the next 30 years (Fig. 15.9a) or <5,000 males within 100 years (Fig. 15.9b). Mean final abundance was 45,870 and 39,817 males after 30 and 100 years, respectively. Mean minimum abundance was 6,965 and 5,998 males after 30 and 100 years, respectively. Projections based on the IT best Ricker models suggested much lower viability, with a low probability (i.e., <0.1) that sage-grouse would decline below 3,000 males within 30 years, but a 100% chance of extinction in the next 100 years (Fig. 15.9b). Mean final abundance was 5,652 and 0 males after 30 and 100 years, respectively. Mean minimum abundance was 5,577 and 0 males after 30 and 100 years, respectively.

DISCUSSION

Data Limitations

We based our analyses on all available attempted censuses of males on leks that met our standards for quality in SMZs and in 30 relatively discrete populations of sage-grouse in western North America (Table 15.1). Many of the populations appear to be spatially isolated, while narrow corridors connect other populations (Connelly et al. 2004; Knick and Hanser, this volume, chapter 16). The most isolated populations occur at the southern and western extremes of the range in the Southern Great Basin, Columbia Basin, and Colorado Plateau SMZs.

TABLE 15.68

Sage-grouse lek monitoring effort, lek size, and trends averaged over 5-yr periods for the the Piceance Basin, Colorado, population, 1965–2007.

Parameter	2000–2007[a]	1995–1999	1990–1994	1985–1989	1980–1984	1975–1979	1970–1974	1965–1969
Leks counted	28	1	7	13	6	0	0	0
Males/lek	4	—	—	—	—	—	—	—
Active leks	13	—	—	—	—	—	—	—
% active leks	51	—	—	—	—	—	—	—
Males/active lek	8	—	—	—	—	—	—	—
λ	1.216	—	—	—	—	—	—	—
SE (λ)[b]	0.229	—	—	—	—	—	—	—

[a] Eight yr of data in this period.
[b] Standard error for annual rate of change.

TABLE 15.69

Sage-grouse lek monitoring effort, lek size, and trends averaged over 5-yr periods for the Colorado Plateau Management Zone, 1965–2007.

Parameter	2000–2007[a]	1995–1999	1990–1994	1985–1989	1980–1984	1975–1979	1970–1974	1965–1969
Leks counted	37	4	12	19	10	2	2	2
Males/lek	6	9	6	9	8	—	—	3
Active leks	16	3	4	8	8	1	1	1
% active leks	45	64	54	68	84	—	—	63
Males/active lek	13	17	10	12	9	—	—	—
λ	1.063	0.793	0.938	0.991	—	—	—	—
SE (λ)[b]	0.396	—	—	—	—	—	—	—

[a] Eight yr of data in this period.
[b] Standard error for annual rate of change.

Not all leks are currently active, and many have been inactive for several years. The total number of leks that has been extirpated is unknown and hinders our attempts to fully understand the magnitude of change in sage-grouse populations. Although leks have become inactive throughout the species' range, the distribution of inactive leks appears to be clustered rather than widespread (Connelly et al. 2004). Proportionally, the largest number of inactive leks appears to occur in the Southern Great Basin, Columbia Basin, and Colorado Plateau SMZs. Applying a three-stage probability sampling approach, described below, in the future would provide unbiased estimates of the proportion of leks disappearing and new leks established, as well as facilitate modeling impacts of habitat changes and threat factors on sage-grouse abundance and population dynamics.

An examination of all trend data from the mid-1940s to 2003 suggests a substantial decline in the overall sage-grouse population in North America (Connelly et al. 2004). However, because

TABLE 15.70

Candidate model set (contains 95% of model weight) and model statistics in the estimating population trends and persistence probabilities for Greater Sage-Grouse in the Colorado Plateau Management Zone, 1965–2007.

Model	Model statistic[a]			
	r^2	k	ΔAIC_c	w_i
Ricker	0.264	4	0.0[b]	0.313
Gompertz	0.246	4	0.5	0.244
EGPE	0.000	3	1.5	0.148
Ricker + year	0.264	5	3.2	0.063
Gompertz + year	0.000	5	3.7	0.049
Ricker t − 1	0.247	4	4	0.042
Gompertz t − 1	0.108	4	4.1	0.040
Gompertz t − 2	0.103	4	4.3	0.037
Ricker t − 2	0.098	4	4.6	0.031

[a] Model fit described by coefficient of determination (r^2), the number of parameters (k), the difference in Akaike's information criterion corrected for small sample size (ΔAIC_c), and the AIC_c wt (w_i).

[b] $AIC_c = 15.0$ for best selected model.

TABLE 15.71

Dispersal rates among sage-grouse management zones representing the proportion of the population dispersing to another management zone each year.

	Wyoming Basin	Southern Great Basin	Snake River Plain	Northern Great Basin	Colorado Plateau
Great Plains	0.050				
Wyoming Basin		0.020	0.011		0.009
Southern Great Basin			0.024	0.004	0.005
Snake River Plain				0.035	

Connections between management zones not presented are assumed to be zero.

data collected in the 1940s and 1950s are highly variable and may have been collected in a somewhat haphazard fashion, they permit no means of assessing the true magnitude of the population change. Confidence intervals for population reconstructions for all populations and SMZs clearly show that precisions of recent population indices are dramatically smaller than the earlier ones based on smaller samples of leks in the decades from 1960 to 1980.

All states and provinces monitor sage-grouse breeding populations by counting males attending leks during the spring breeding season. Standard techniques for censusing leks have been available for a number of years (Patterson 1952, Eng 1963, Jenni and Hartzler 1978, Emmons and Braun 1984) and were recently summarized (Connelly et al. 2003b). Despite available information, methods differ among agencies and even among years within agencies (Connelly et al. 2004). These inconsistencies confound attempts to make comparisons of population trends across states and provinces. Nevertheless, long-term lek counts make up the largest range-wide database available for sage-grouse populations and provide the basis for reconstructing a remarkably precise index to minimum male abundance at a relatively broad spatial scale (Connelly et al. 2004). Without efforts to take a probability sample of leks in each spatial region and apply intense methods in local areas to convert this index into a valid estimator of both male and female breeding sage-grouse, data on populations of sage-grouse will face challenges to their validity.

Lek counts focus on attendance of males. Some male sage-grouse may not attend a lek or may attend two or more leks (Jenni and Hartzler 1978, Emmons and Braun 1984, Walsh et al. 2004). Lek data used to track populations have an implied assumption that probability of detection of birds does not change among years (i.e., the proportion missed because of nonattendance or attendance at a lek that is not counted remains about the same) or that it varies randomly. Even if the detection probability is unknown, which is usually the case, the problem can be minimized to more precise counts if leks counted on a single morning are relatively close and represent all or a significant part of a given local breeding population, a deme. Repeating these counts along a lek route at weekly intervals centered around the peak in male attendance will ensure their reliability and comparability across regions. Combining these extensive counts with intensive mark-resight or

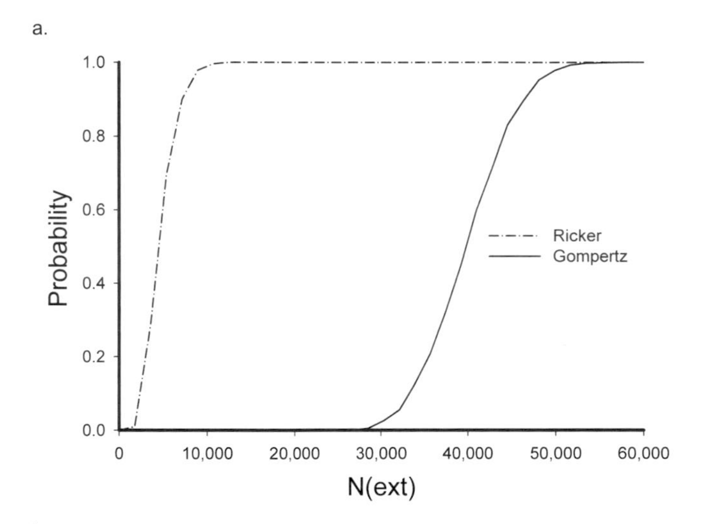

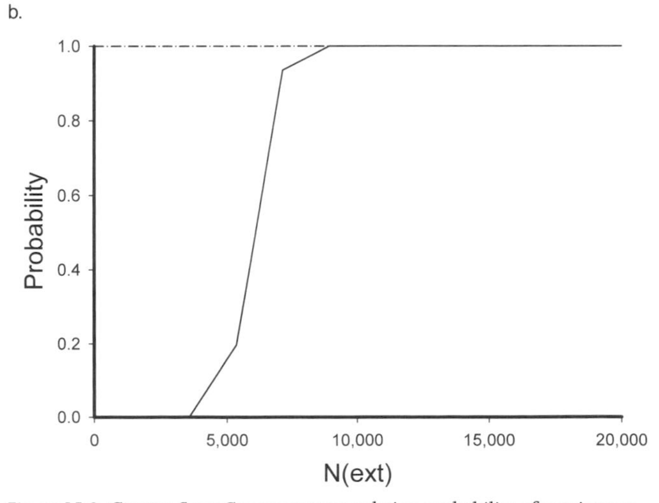

Figure 15.9. Greater Sage-Grouse metapopulation probability of persistence at: (a) 30 years and (b) 100 years. N(ext) is population abundance at which species is declared extinct.

TABLE 15.72

Correlation in model residuals among sage-grouse management zones for the AIC_c best Gompertz and Ricker (in parentheses) models.

	Great Plains	Wyoming Basin	Southern Great Basin	Snake River Plain	Northern Great Basin	Columbia Basin	Colorado Plateau
Great Plains	1	0.46 (0.46)	0.05 (0.08)	0.41 (0.41)	−0.04 (−0.02)	0.31 (0.31)	0.13 (0.09)
Wyoming Basin		1	0.32 (0.35)	0.34 (0.38)	0.09 (0.19)	0.10 (0.17)	0.27 (0.18)
Southern Great Basin			1	0.58 (0.58)	0.50 (0.50)	0.12 (0.14)	0.19 (0.10)
Snake River Plain				1	0.52 (0.57)	0.31 (0.32)	0.22 (0.07)
Northern Great Basin					1	0.36 (0.38)	0.53 (0.42)
Columbia Basin						1	0.03 (0.02)
Colorado Plateau							1

telemetry studies in a few areas would provide better data to verify their reliability as an unbiased index to grouse population abundance.

Unfortunately, counting males attending some leks in a region provides little more than an index to the minimum number of males present in a region. We developed an approach to analyzing this index that treats lek counts as a cluster sample of males within leks and applies ratio estimators to paired counts of males at leks in succeeding years to obtain unbiased estimators of $\lambda(t)$, the finite rate of change from the previous year to the present year, and $\theta(t)$, its reciprocal. Population reconstruction using these unbiased estimators provides remarkably precise estimates of the rates of change for reconstructing the index in previous years and is not biased by changes in the number of leks counted in different years. These rates of change are the basis for our modeling efforts. Unfortunately, the final year count of males attending leks is not based on a probability sample and cannot be used to infer the true number of males attending leks within the spatial region sampled, the true number of males present within the region, or the breeding population of both males and females present within the spatial region sampled. Methods to replace this weak foundation of lek counts representing an unknown proportion of leks in a spatial region by a true probability sample of leks and breeding males and females in defined spatial areas have been proposed but not widely adopted at this time (Garton et al. 2007).

Analytical Approach and Inference

We had sufficient data on 24 populations and six SMZs to reconstruct populations back to at least 1967. Forty-six percent of these populations peaked in 1969, and 77% peaked from 1969–1972. Fifty percent of these SMZs peaked in 1969. Clearly, our population reconstruction approach demonstrated a pattern of relatively high numbers of sage-grouse during the late 1960s and early 1970s. Independent data—population data not based on lek counts—could help validate or refute our approach to population reconstruction. Unfortunately, little published information is available on sage-grouse population change that is not based on lek counts, but some information on production and harvest is available. Sage-grouse production in six states, measured by age ratios in the harvest, was considerably higher from the late 1960s to the early 1980s compared to 1986–1996 (Connelly and Braun 1997). Between 1965 and 1979 in Idaho, the average number of chicks per hen peaked in 1969 (Autenrieth 1981). Between 1965 and 1992, sage-grouse production in Oregon peaked in 1969, while total birds counted during summer surveys peaked in 1971 (Willis et al. 1993). Similarly, from 1965 to 1990, chicks per adult, percent adults with broods, and mean brood size in Oregon were greatest in the late 1960s and early 1970s (Drut 1994). Additionally, estimated peak harvest occurred in Idaho in 1969 (Autenrieth 1981) and in Oregon and Washington in 1970 (Drut 1994). During the late

1960s, Montana substantially increased their season length for sage-grouse (Wallestad 1975a), which presumably was at least in part related to grouse abundance. Thus, several independent sources of data also strongly suggest relatively high populations of sage-grouse in the late 1960s and early 1970s and lend further support to our approach to population reconstruction.

Previous suggestions of the possibility of cyclic patterns in abundance of sage-grouse (Rich 1985, Connelly et al. 2004) led us to apply information-theoretic methods to assessment of first-order, second-order, and third-order models of density dependence. Zeng et al. (1998) earlier demonstrated their success in detecting complex patterns of density dependence using Schwartz's information criterion. Our use of AIC_c led to selection of one-year or two-year time delayed density-dependent models for 33% of the populations and SMZs, but in most cases no single model was selected as the single best model. Therefore, we chose to take a multimodel inferential approach to ensure adequate incorporation of model uncertainty and its implications for forecasting future viability and persistence of the populations.

Our findings generally agree with conclusions of other recent analyses (Connelly et al. 2004, Anonymous 2008) that also documented declining sage-grouse populations. However, our multimodel predictions of the likelihood of individual populations of sage-grouse declining below $N_e = 50$ (13%, Table 15.6) and $N_e = 500$ (54%) within 30 years are clearly underestimates of the true percentages because they are based solely on the 24 populations for which we had sufficient data to build stochastic growth models. Many smaller populations could not be analyzed and modeled because of lack of sufficient data. Our analyses suggest that smaller populations have suffered greater declines and tend to be at greater risk than larger populations.

Another recent approach to analyzing lek counts (Anonymous 2008) ignored serial correlation in successive lek counts by estimating the trend in male counts at individual leks with a log-linear regression of log(male count + 1) on year using mixed models cast in a hierarchical framework with year and year^2 as fixed effects and individual leks as random effects. Treating leks as random effects requires assuming that leks surveyed are a random sample from all leks rather than assuming rates of change in successive counts are a random sample of rates of change at leks throughout the population or SMZ, as our approach assumed. Recent work indicates that trend estimation based on regressions of log-linear abundance on time provides unbiased estimates of rates of change (Humbert et al. 2009), but inferential statistics such as confidence intervals are only correct when there is no process error in annual counts and all error is associated with sampling error, an unlikely assumption for sage-grouse. Our approach treating observation error as minimal also gave unbiased estimates of trend and was most efficient at detecting declining trends when sampling error was small. We evaluated the model incorporating both observation and process error (Exponential Growth State Space model; Staples et al. 2004, Humbert et al. 2009) early in our analysis and consistently found that this model's estimated sampling error was 0 or close to 0. Thus, we used the approach referred to as Exponential Growth with Process Error (Humbert et al. 2009) and applied inferential statistics using bootstraps under this model only. The approach taken by Anonymous (2008), referred to as Exponential Growth with Observation Error, provided unbiased estimates of trend, but inferential statistics yielded excessive levels of type 1 error. Moreover, this approach (Anonymous 2008:11) did not yield any estimates or indices of population size and ignored relative size of individual leks treating trend in large leks with more than 50 or 100 males present as equal in importance to leks with less than one dozen males attending. The modeling approach of Anonymous (2008) makes it impossible to forecast future population size or probability of a population or SMZ declining below a quasi-extinction threshold, as we did.

Even the best stochastic growth models in our analyses did not explain 50% of the variation in annual rates of change. Consequently, standard errors of the best estimates of model parameters were large (Appendix 15.1). Reducing this unexplained variation in growth rates will require efforts to decrease error in lek counts using probability sampling approaches and to incorporate key predictive factors into growth models describing environmental characteristics of lek sites (Johnson et al., this volume, chapter 17). If the recent expansion of monitoring efforts to include more leks within each geographic region is maintained or expanded, models of annual growth rates should improve as the temporal scale of the more precise estimates increases, making it feasible to increase the length of time lags testable. Nevertheless, the inherently stochastic nature of population changes of sage-grouse will require use of stochastic growth models to forecast future potential for persistence of the species.

Population indices for many of the distinct populations and SMZs of sage-grouse were 20–80% larger than estimated carrying capacities for those populations. How can those populations exceed their estimated carrying capacity? There are three potential explanations for these observations. The first is that a carrying capacity or quasi-equilibrium estimated from a stochastic growth model is not an upper boundary to population size but rather a diffuse area or cloud of points above which annual rates of change tend to switch from positive to negative, indicating a tendency to decline (Dennis and Taper 1994). Thus, populations will be above this carrying capacity as often as they are below it. Second, distributions of rates of change are typically log normal because they are skewed to large values. Carrying capacity value, as calculated, characterizes the median abundance (Dennis et al. 1991) rather than the mean. The third reason that many observed population indices in 2007 were higher than the carrying capacity may be due to the cyclic nature of many populations as indicated by one-year and two-year delayed density-dependent models being identified as the information-theoretic best models. For example, 2007 indices of abundance for the Southern Great Basin and Snake River Plain SMZs populations exceeded their estimated carrying capacities by 67% and 30%, respectively. The information-theoretic best models for both of these populations were Gompertz models with two-year and one-year time-delays in each case. The pattern of population change through time for each of these populations (Figs. 15.4i and 15.5g) suggests both populations are declining from cyclic highs that may lead to declines below their respective carrying capacities in the near future, if past patterns repeat.

Both basic density-dependent models based on a Ricker-type model in which annual log growth rates decline linearly with density and a Gompertz-type model in which log growth rates decline linearly with log abundance describe the data on growth rates comparably well, making it difficult to identify the best model. Overall, across 24 populations and seven SMZs, Gompertz-type models were selected in most cases (82%; Table 15.73), but Ricker-type models were often strongly competitive and vice versa when Ricker was chosen as best. Both models had similar r^2 values, indicating comparable abilities to describe the observed rates of change. The two basic types of models gave different projections of long-term viability and carrying capacities whether for populations, SMZs, or the metapopulation in spite of these strong similarities. Gompertz-type models usually estimated higher carrying capacities or quasi-equilibria and lower probabilities of extinction when the same populations were projected 30 or 100 years into the future. Plots of residuals versus abundance (N_t for Ricker or $\ln(N_t)$ for Gompertz) and plots of residuals versus predicted rates of change all appeared reasonable and provided no basis for rejecting either of the models for density dependence. We could find no reason for rejecting either model on a conceptual or statistical basis, but caution should be exercised in accepting the projections of the best model alone. Multimodel inference for assessing future forecasts of probability of extinction produce overall probabilities intermediate between either Gompertz-type or Ricker-type models alone (Table 15.6).

Population viability analysis is inherently problematic from the classic statistical perspective, as a model cannot be proven to provide reliable predictions for conditions outside of the range of the variables used to develop the model. For example, predicting the probability of declining below a quasi-extinction of $N_e = 50$ individuals necessitates modeling growth rates for population sizes well below the observed levels for many populations and SMZs. Forecasting future viability requires the assumption that future conditions will continue the same trajectory or trend observed in the past. We reiterated this assumption repeatedly in our presentation of results. Many of the dominant influences on sage-grouse populations, such as habitat changes associated with development, are under resource management agency control, and it is possible that future trajectories could be altered to benefit sage-grouse populations.

TABLE 15.73

Strongest models identified for populations and management zones of Greater Sage-Grouse on basis of type of density dependence (Gompertz versus Ricker), presence of a time trend, and presence of a period difference between 1967–1987 and 1987–2007.

Model	Populations		Management zones		Combined populations and zones	
	N	%	N	%	N	%
Exponential growth (EGPE)	3	11	0	0	3	9
Gompertz	12	44	1	14	13	38
Gompertz t − 1	3	11	4	57	7	21
Gompertz t − 2	2	7	2	29	4	12
All Gompertz	21	78	7	100	28	82
Ricker	3	11	—	0	3	9
Ricker t − 1	—	—	—	—	—	—
Ricker t − 2	—	—	—	—	—	—
All Ricker	3	11		0	3	9
Total	27	—	7	—	34	—
Temporal factors						
Year	10	37	6	86	16	47
Period	5	19	1	14	19	56

CONSERVATION IMPLICATIONS

Future Monitoring

Counts of males attending leks are not based on a probability sample of leks or spatial units, and they cannot be used to infer the true number of males attending leks within the spatial region sampled, the true number of males present within the region, or the breeding population of both males and females present within the spatial region sampled. Methods to replace this weak foundation of lek counts representing an unknown proportion of leks in a spatial region by a true probability sample of leks and breeding males and females in defined spatial areas have been proposed but not widely adopted at this time (Garton et al. 2007).

Obtaining unbiased estimates of breeding males and females could employ a three-stage sampling approach to counting males attending leks throughout the spatial region of interest combined with intensive methods applied at a small sample of sentinel leks within the region (Garton et al. 2007). The first stage sample requires drawing a stratified random sample of spatial units based on habitat (abundance and quality of sagebrush communities, seral stages, and disturbance threats such as human-footprint indices) as well as known or suspected lek sites (Garton et al. 2007). The second stage draws a stratified random sample of leks within these spatial units configured in the form of lek routes such that a complete count of core (large) leks is augmented by a random sample of new/satellite (small) leks in a way that allows inference to all the leks within the entire spatial unit sampled. The third stage sample consists of counts of males attending each lek sampled along the lek route according to established lek-counting protocols (Connelly et al. 2003b, 2004). The final component requires establishing a limited number of sentinel lek routes throughout the species' range for intensive studies of the proportion of breeding males counted on leks and ratio of breeding females to counted males using a combination of time-series, mark-resight, and/or sightability methods (Garton et al. 2007). These sentinel-lek routes and radio-marked grouse also provide ideal situations for ancillary work such as estimating survival rates, reproductive rates, harvest rates, and seasonal habitat use and requirements, as well as detailed demographic estimates by age and gender. Intensive research on these sentinel-lek routes need not occur every year but should sample the range of ecological and environmental conditions occurring within the area of interest over a period of time and also incorporate changes occurring there.

A related method involves a dual-frame sampling approach (Haines and Pollock 1998) that can be applied to long-term data sets collected at specific locations that lack a probability-based sample design. The dual-frame sampling consists of a list frame comprised of known lek sites and an area frame consisting of all other potential sage-grouse habitat where leks are not currently known to occur; the list and area frames should not overlap (Anonymous 2008). The list frame helps maintain continuity with historic data but is placed in a probability-based design, while data from the area frame allow inference to be made to the entire sage-grouse population. This dual-frame sampling approach could also be termed a stratified random sample of spatial units with only two strata, one defining spatial units containing known leks and a second containing all the remaining potential habitat. Both strata are sampled under a probability sampling design with appropriate allocating of sample units to minimize variances. Redesigning lek surveys using one of these probability sampling strategies would allow estimation of the true number of males and possibly females for a final year and thereby place population reconstruction on a firm foundation, rather than necessitating treating reconstruction simply as an index to minimum breeding male numbers.

Potential Trajectories

Cheatgrass (*Bromus tectorum*) has invaded many of the lower-elevation, more xeric sagebrush landscapes across the western portion of the range of Greater Sage-Grouse. Additionally, conifer woodlands have expanded into sagebrush habitats at higher elevations, creating stress on the sagebrush ecosystem from both extremes (Miller et al., this volume, chapter 10). Fire has also increased since 1980 throughout many portions of the species' range (Miller et al., this volume, chapter 10; Baker, this volume, chapter 11). Other areas have been impacted by energy development (Naugle et al., this volume, chapter 20) and West Nile virus (Walker and Naugle, this volume, chapter 9). The rapidity with which an entire sagebrush landscape can now be transformed through land use and changing environments (e.g., energy development and fire) is much greater than the natural disturbances that previously influenced sagebrush ecosystems (Knick et al., this volume, chapter 12). The ultimate influence of these unprecedented landscape changes is not well understood for sage-grouse populations. Our results (Appendices 15.1 and 15.2) indicate that in 44% of the cases the best model included a declining carrying capacity for sage-grouse through time and 18% incorporated a lower carrying capacity in the interval 1987–2007 than from 1967–1987. These lower carrying capacities provide supporting evidence for recent findings indicating a continuing decline in quality and quantity of habitat for sage-grouse (Miller et al., this volume, chapter 10; Baker, this volume, chapter 11; Knick et al., this volume, chapter 12; Leu and Hanser, this volume, chapter 13).

Theoretical work has questioned the precision of projections into the future beyond 20% of the length of the time series of counts unless populations are rapidly growing or declining or have low variances in population growth rate (Ludwig 1999, Fieberg and Ellner 2000). In contrast, Holmes et al. (2007) suggested that quasi-extinction thresholds can be estimated relatively precisely because variances in growth rate were not as large as had been previously supposed. Ellner and Holmes (2008) resolved this debate by deriving from general theory the combinations of projection interval into the future and quasi-extinction thresholds where estimates of quasi-extinction are certain and uncertain. They concluded that long-range (25 to 100 years) projections based solely on environmental stochasticity are sometimes possible, but usually problematic. Forecasting the future is always problematic, especially for population viability analysis (PVA), because it rests upon our analysis of past, incomplete information. We have attempted to improve upon the classic approaches by including models that are based upon estimates of both long-term changes (time or year effects) in carrying capacity (our terminology for the quasi-equilibrium), recent changes in rates of change in the last 20 years (period effects) and a variety of forms of density dependence (linear vs. log-linear and zero- to two-year time lags) that have increased the coefficients of determination of the models dramatically, thereby improving our confidence that these forecasts will be useful in guiding decisions concerning the future of sage-grouse and the sagebrush communities upon which they depend.

Our multimodel predictions of the likelihood of individual populations of sage-grouse declining below N_e = 50 (13%, Table 15.6) and N_e = 500 (54%) within 30 years are clearly underestimates of the true percentages because they are based solely on the 24 populations for which we had sufficient data to build stochastic growth models. Many smaller populations could not be analyzed and modeled because of lack of sufficient data. Our analyses suggest that smaller populations have suffered greater declines and tend to be at greater risk than larger populations. The percentage of individual populations forecast using multimodel inference to decline below N_e = 50 (33%; Table 15.6) and N_e = 500 (75%) within 100 years raise more concern for the long-term persistence of this species in local areas given continuing declines and degradation of habitat (Knick et al., this volume, chapter 12). By contrast, multimodel inferences for SMZs suggest that only two of seven are likely to decline below N_e = 500 in the next 30 years. Continuing loss and degradation of habitat will likely result in declines in carrying capacity for sage-grouse at the SMZ scale and will place higher percentages at risk of declining below N_e = 50 (14%) and N_e = 500 (29%) if declines continue for 100 years. Concerted effort will be necessary to maintain smaller populations of sage-grouse with continued declines in their habitats as well as increases in obvious threat factors such as West Nile virus. Populations distributed across broader scales such as SMZs and the continent-wide metapopulation seem to be following trajectories similar to many populations, but their larger size will extend the time before effective population sizes shrink to levels that are unlikely to persist.

ACKNOWLEDGMENTS

We thank the many state and federal biologists and volunteers that spent countless hours collecting and compiling lek attendance data. We thank members of the Western Sage and Columbian Sharp-tailed Grouse Technical Committee for insights and advice they provided on lek counts. We especially appreciate T. J. Christiansen's help with the Wyoming database. Drafts of this manuscript were greatly improved by reviews from K. M. Dugger, S. R. Beissinger, S. T. Knick, and D. K. Dahlgren. Four anonymous reviewers critical of our approach and analysis methods improved the manuscript by encouraging us to explain the approach and details of the analysis in greater detail. C. E. Braun provided helpful editorial suggestions.

Appendix 15.1

Maximum Likelihood Estimates of Parameters for AIC_c Best Models for Sage-Grouse Populations with Standard Errors, Coefficients of Determination (r^2), and Estimated Carrying Capacity (Quasi-Equilibrium) in Year τ (K_τ)

APPENDIX 15.1A

Maximum likelihood estimates of parameters for AIC_c best models for Sage-Grouse Populations.

Population	Model	a	$b_1 N_t$	$b_2 \ln N_t$	$c_2 \ln N_{t-1}$
Baker, Oregon	Ricker	0.721	−0.003493		
Bannack, Montana	Gompertz + Period	1.819		−0.323	
Wisdom, Montana	Gompertz	0.839		−0.195	
Central Oregon	Gompertz + Year	38.823		−0.484	
Eagle–South Routt Counties, Colorado	EGPE	−0.094			
Northern Great Basin	Gompertz $t-1$ + Year	45.970			−0.469
Western Great Basin	Gompertz $t-1$ + Year	32.617			−0.468
Southern Great Basin	Gompertz $t-2$ + Year	30.768			
Jackson Hole, Wyoming	EGPE	−0.023			
Dakotas	Gompertz + Year	28.601		−0.400	
Middle Park, Colorado	Gompertz	1.609		−0.303	
Moses Coulee, Washington	Gompertz $t-1$ + Year	34.764			−0.369
South Mono Lake, California	Gompertz	3.545		−0.678	
Northeast Interior Utah	Ricker + Year, Period	−34.817	−0.001322		
Northern Montana	Ricker + Period	1.067	−0.000367		
Red Rocks, Montana	Gompertz + Period	3.596		−0.609	
North Mono Lake	Gompertz	1.628		−0.396	
South-central Utah	Gompertz	2.410		−0.354	
Sanpete–Emery Counties, Utah	Gompertz	1.680		−0.601	
Snake–Salmon–Beaverhead, Idaho	Gompertz $t-1$ + Period	2.757			−0.328
Summit–Morgan Counties, Utah	Gompertz	2.165		−0.509	
Toole–Juab Counties, Utah	Gompertz	4.266		−0.911	
East-central Idaho	EGPE	−0.067			
Powder River, Wyoming	Gompertz $t-1$ + Year	60.417			−0.377
Wyoming Basin	Gompertz $t-1$ + Year	23.017			−0.294
Yakima, Washington	Gompertz $t-2$	0.876			
Yellowstone Watershed	Ricker + Year	27.938	−0.000104		

$d_2 \ln N_{t-2}$	e(year)	f(period)	s	r^2	K_{2007}	K_{2037}	K_{2107}
			0.153	0.277	206	206	206
		0.279	0.202	0.175	278	278	278
			0.260	0.166	75	75	75
	−0.0177		0.180	0.297	783	261	20
			0.142	0.000			
	−0.0208		0.125	0.466	6770	1787	80
	−0.0143		0.194	0.498	4239	1695	200
−0.431	−0.0136		0.187	0.325	2524	977	107
			0.246	0.000			
	−0.0130		0.250	0.190	587	222	23
			0.252	0.156	201	201	201
	−0.0164		0.279	0.194	168	44	2
			0.447	0.331	187	187	187
	0.0176	0.656	0.281	0.280	358	757	1688
		−0.556	0.275	0.357	2908	2908	2908
		0.235	0.292	0.307	367	367	367
			0.358	0.192	61	61	61
			0.278	0.209	901	901	901
			0.753	0.311	16	16	16
		0.292	0.186	0.351	4468	4468	4468
			0.446	0.256	70	70	70
			0.273	0.682	108	108	108
			0.294	0.000			
	−0.0286		0.262	0.315	3042	312	2
	−0.0100		0.152	0.188	20980	7545	694
−0.167			0.293	0.085	192	192	192
	−0.0138		0.220	0.338	2948	−1015	−10263

APPENDIX 15.1B

Standard errors of model parameters for Sage-Grouse Populations.

Population	Model	a	b_1N_t	$b_2\ln N_t$	$c_2\ln N_{t-1}$	$d_2\ln N_{t-2}$	e(year)	f(period)
Baker, Oregon	Ricker	0.377	0.002					
Bannack, Montana	Gompertz + Period	0.660		0.115				0.118
Wisdom, Montana	Gompertz	0.366		0.080				
Central Oregon	Gompertz + Year	12.360		0.126			0.006	
Eagle–South Routt Counties, Colorado	EGPE	0.050						
Northern Great Basin	Gompertz $t-1$ + Year	8.728			0.083		0.004	
Western Great Basin	Gompertz $t-1$ + Year	6.994			0.079		0.003	
Southern Great Basin	Gompertz $t-2$ + Year	9.868				0.104	0.005	
Jackson Hole, Wyoming	EGPE	0.058						
Dakotas	Gompertz + Year	13.121		0.138			0.006	
Middle Park, Colorado	Gompertz	0.907		0.171				
Moses Coulee, Washington	Gompertz $t-1$ + Year	15.197		0.125			0.007	
South Mono Lake, California	Gompertz	0.818		0.156				
Northeast Interior Utah	Ricker + Year, Period	0.194	0.000				0.010	0.231
Northern Montana	Ricker + Period	0.237	0.000					0.146
Red Rocks, Montana	Gompertz + Period	0.893		0.151				0.114
Mono Lake, California–Nevada	Gompertz	0.543		0.132				
South-central Utah	Gompertz	0.789		0.115				
Sanpete–Emery Counties, Utah	Gompertz	0.624		0.211				
Snake–Salmon–Beaverhead, Idaho	Gompertz $t-1$ + Period	0.633			0.074			0.084
Summit–Morgan Counties, Utah	Gompertz	0.600		0.141				
Toole–Juab Counties, Utah	Gompertz	1.061		0.235				
East-central Idaho	EGPE	0.098						
Powder River, Wyoming	Gompertz $t-1$ + Year	18.482			0.095		0.009	
Wyoming Basin	Gompertz $t-1$ + Year	11.129			0.108		0.005	
Yakima, Washington	Gompertz $t-2$	0.520				0.095		
Yellowstone Watershed	Ricker + Year	8.734	0.000				0.004	

Appendix 15.2

Maximum Likelihood Estimates of Parameters for AIC_C Best Gompertz and Ricker Models for Sage-Grouse Management Zones with Standard Errors, Coefficients of Determination (r^2), and Estimated Carrying Capacity (Quasi-Equilibrium) in Year τ (K_τ)

APPENDIX 15.2A

Maximum likelihood estimates of parameters for AIC_c best Gompertz and Ricker models for Sage-Grouse Management Zones.

Sage-Grouse Management Zone	Model	a	b_1N_t	$b_2\ln N_t$
Great Plains	Gompertz $t-1$ + Year	29.245		
	Ricker $t-1$ + Year	23.864		
Wyoming Basin	Gompertz $t-1$ + Year	24.388		
	Ricker $t-1$ + Year	15.515		
Southern Great Basin	Gompertz $t-2$ + Year	17.913		
	Ricker $t-2$ + Year	11.429		
Snake River Plain	Gompertz $t-1$ + Year, Period	24.334		
	Ricker $t-1$ + Year, Period	14.540		
Northern Great Basin	Gompertz $t-1$ + Year	19.157		
	Ricker $t-1$ + Year	9.810		
Columbia Basin	Gompertz + Year	28.817		−0.3842
	Ricker + Year	19.957	−0.000389	
Colorado	Ricker	0.516	−0.002083	
Plateau	Gompertz	2.788		−0.5071

APPENDIX 15.2B

Standard errors of model parameters for Sage-Grouse Management Zones.

Sage-Grouse Management Zone	Model	a	b_1N_t	$b_2\ln N_t$
Great Plains	Gompertz $t-1$ + Year	11.700		
	Ricker $t-1$ + Year	11.106		
Wyoming Basin	Gompertz $t-1$ + Year	11.272		
	Ricker $t-1$ + Year	9.405		
Southern Great	Gompertz $t-2$ + Year	6.069		
Basin	Ricker $t-2$ + Year	5.510		
Northern	Gompertz $t-1$ + Year, Period	12.264		
Great Basin	Ricker $t-1$ + Year, Period	10.694		
Northern Great	Gompertz $t-1$ + Year	8.866		
Basin	Ricker $t-1$ + Year	7.965		
Columbia Basin	Gompertz + Year	11.605		
	Ricker + Year	10.572		
Colorado Plateau	Ricker	0.194	0.000761	
	Gompertz	1.068		0.1936303

c_1N_{t-1}	$c_2\ln N_{t-1}$	d_1N_{t-2}	$d_2\ln N_{t-2}$	e(year)	f(period)	s	K_{2007}	K_{2037}	K_{2107}
	−0.430			−0.01261		0.19727	9579	3974	510
−0.0000212				−0.01181		0.20139	7647	−9096	−48164
	−0.296			−0.01067		0.15502	21954	7452	599
−0.0000042				−0.00770		0.15839	14350	−41159	−170680
			−0.3976	−0.00728		0.12523	4094	2364	657
		−0.0000498		−0.00560		0.13184	3990	617	−7251
	−0.391			−0.01029	0.156	0.13274	12165	5517	872
−0.0000094				−0.00721	0.130	0.13781	6542	−16479	−70194
	−0.299			−0.00826		0.16831	5529	2413	349
−0.0000141				−0.00487		0.17733	3031	−7355	−31589
				−0.01327		0.21069	291	103	9
				−0.00992		0.21795	117	−648	−2434
						0.14367	248	248	248
						0.14541	244	244	244

c_1N_{t-1}	$c_2\ln N_{t-1}$	d_1N_{t-2}	$d_2\ln N_{t-2}$	e(year)	f(period)	r^2
	0.141			0.00529		0.203
0.0000078				0.00552		0.176
	0.107			0.00515		0.192
0.0000017				0.00469		0.156
			0.0943	0.00273		0.333
		0.0000141		0.00274		0.262
			0.0908	0.00581	0.091	0.413
		0.0000023		0.00534	0.092	0.339
	0.092			0.00411		0.240
0.0000058				0.00398		0.221
	0.129			0.00548		0.193
0.0001609				0.00527		0.137
						0.265
						0.283

CHAPTER SIXTEEN

Connecting Pattern and Process in Greater Sage-Grouse Populations and Sagebrush Landscapes

Steven T. Knick and Steven E. Hanser

Abstract. Spatial patterns influence the processes that maintain Greater Sage-Grouse (*Centrocercus urophasianus*) populations and sagebrush (*Artemisia* spp.) landscapes on which they depend. We used connectivity analyses to: (1) delineate the dominant pattern of sagebrush landscapes; (2) identify regions of the current range-wide distribution of Greater Sage-Grouse important for conservation; (3) estimate distance thresholds that potentially isolate populations; and (4) understand how landscape pattern, environmental disturbance, or location within the spatial network influenced lek persistence during a population decline. Long-term viability of sagebrush, assessed from its dominance in relatively unfragmented landscapes, likely is greatest in south-central Oregon and northwest Nevada; the Owyhee region of southeast Oregon, southwest Idaho, and northern Nevada; southwest Wyoming; and south-central Wyoming. The most important leks (breeding locations) for maintaining connectivity, characterized by higher counts of sage-grouse and connections with other leks, were within the core regions of the sage-grouse range. Sage-grouse populations presently have the highest levels of connectivity in the Wyoming Basin and lowest in the Columbia Basin Sage-Grouse management zones (SMZs). Leks separated by distances >13–18 km could be isolated due to decreased probability of dispersals from neighboring leks. The range-wide distribution of sage-grouse was clustered into 209 separate components (units in which leks were interconnected within but not among) when dispersal was limited to distances <18 km. The most important components for maintaining connectivity were distributed across the central and eastern regions of the range-wide distribution. Connectivity among sage-grouse populations was lost during population declines from 1965–1979 to 1998–2007, most dramatically in the Columbia Basin SMZ. Leks that persisted during this period were larger in size, were more highly connected, and had lower levels of broad-scale fire and human disturbance. Protecting core regions and maintaining connectivity with more isolated sage-grouse populations may help reverse or stabilize the processes of range contraction and isolation that have resulted in long-term population declines.

Key Words: Artemisia, Centrocercus urophasianus, connectivity, graph analysis, Greater Sage-Grouse, landscape configuration, sagebrush.

Knick, S. T., and S. E. Hanser. 2011. Connecting pattern and process in Greater Sage-Grouse populations and sagebrush landscapes. Pp. 383–405 *in* S. T. Knick and J. W. Connelly (editors). Greater Sage-Grouse: ecology and conservation of a landscape species and its habitats. Studies in Avian Biology (vol. 38), University of California Press, Berkeley, CA.

Conectando Patrones y Procesos en Poblaciones de Greater Sage-Grouse y en Paisajes de Artemisa

Resumen. Los patrones espaciales influyen los procesos que mantienen a las poblaciones del Greater Sage-Grouse (*Centrocercus urophasianus*) y a los paisajes de artemisa (*Artemisia* spp.) de los que dependen. Utilizamos análisis de conectividad para: (1) delinear el patrón dominante de paisajes de artemisa; (2) identificar regiones actuales del rango de distribución del Greater Sage-Grouse que son importantes para la conservación; (3) estimar umbrales de distancia que potencialmente aíslan a las poblaciones; y (4) comprender cómo patrones de paisaje, disturbios ambientales, o la ubicación dentro de la red espacial influyó la persistencia del lek (asamblea de cortejo) durante un descenso de población. La viabilidad a largo plazo de artemisa, evaluado a partir de su dominancia en paisajes relativamente no fragmentados, probablemente sea mayor en Oregon central del sur y Nevada del noroeste; la región de Owyhee del sudeste de Oregon, el sudoeste de Idaho, y del norte de Nevada; el sudoeste de Wyoming; y Wyoming central del sur. Los leks más importantes (sitios de reproducción) para mantener conectividad, caracterizados por conteos más elevados de sage-grouse y conexiones con otros leks, se encontraron dentro de las regiones núcleo del territorio del sage-grouse. Actualmente las poblaciones de sage-grouse con los niveles más altos de conectividad se encuentran en el Wyoming Basin, y con los niveles más bajos en las zonas de manejo del Columbia Basin. Los leks separados por distancias de >13–18 km podrían aislarse debido a la disminuida probabilidad de dispersiones de leks vecinos. El rango de distribución del sage-grouse fue agrupado en 209 componentes separados (unidades en las que los leks se encontraban interconectados dentro pero no entre sí) cuando la dispersión estuvo limitada a distancias <18 km. Los componentes más importantes para mantener conectividad entre los leks se encontró distribuida a través de las regiones centrales y orientales de su rango de distribución. La conectividad entre poblaciones de sage-grouse fue perdida durante descensos de población entre 1965–1979 y 1998–2007, observándose la pérdida más dramática en las zonas de manejo del Columbia Basin. Los leks que persistieron durante este período eran de mayor tamaño, estaban altamente conectados, y tuvieron niveles más bajos de incendios y disturbios humanos de gran escala. Proteger las regiones del núcleo y mantener conectividad con poblaciones más aisladas de sage-grouse puede ayudar a revertir o estabilizar los procesos de contracción de territorio y aislamiento que han tenido como resultado descensos de población a largo plazo.

Palabras Clave: análisis gráfico, artemisa, *Artemisia*, *Centrocercus urophasianus*, conectividad, configuración del paisaje, Greater Sage-Grouse.

Greater Sage-Grouse (*Centrocercus urophasianus*) are wide-ranging, highly mobile birds that depend on sagebrush (*Artemisia* spp.) for most of their life requirements (Patterson 1952, Schroeder et al. 1999, Connelly et al. 2000c, Crawford et al. 2004). Extensive loss and alteration of sagebrush communities have resulted in regional and range-wide declines of Greater Sage-Grouse populations (Connelly and Braun 1997, Braun 1998, Connelly et al. 2004). The species currently occupies approximately half of its pre–Euro-American settlement range, and small populations at the edge are increasingly disjunct from larger populations at the core of the occupied range (Schroeder et al. 1999, 2004). The processes of range reduction, fragmentation, and isolation reduce connectivity among existing populations, which increases the probability of loss of genetic diversity (Benedict et al. 2003, Oyler-McCance et al. 2005b) and extirpation from stochastic events (Reese and Connelly 1997, Aldridge et al. 2008).

Relationships between spatial patterns of Greater Sage-Grouse populations and sagebrush landscapes can provide insights into underlying processes critical for managing populations and maintaining viability of the species. Knowledge of sage-grouse response to structural features of sagebrush landscapes, such as quantity, composition, and configuration, can be used to identify conservation strategies that necessarily rely on maintaining or restoring sage-grouse habitat across broad regional extents (Wisdom et al. 2002c, Meinke et al. 2009).

Connectivity analysis provides a framework to understand the way in which spatial pattern of a species' habitat influences individuals and populations (Taylor et al. 1993). Patches of habitat within a regional mosaic vary in quality and permeability to animal movements. Analysis of the pattern of habitat, or resources, or structural connectivity within a landscape is relatively common (Collinge and Forman 1998, Tischendorf and Fahrig 2000). Landscape metrics describing physical attributes of habitats or resources of presumed importance to a species are readily integrated into spatial data layers using geographic information system (GIS) technology (Schumaker 1996, Calabrese and Fagan 2004, Taylor et al. 2006). A transferable measure of connectivity has been difficult to define because species differ in habitat use and mobility. Patterns of land cover that influence one species may be transparent to others (O'Neill et al. 1988, Keitt et al. 1997). Translating landscape structure into species response has proven elusive (Milne 1992, Wiens and Milne 1989, Wiens 2002, Calabrese and Fagan 2004), and relies on our ability to understand how a species perceives its environment (Wiens et al. 1993, With et al. 1997, Baguette and Van Dyck 2007). If animals are sensitive to arrangement of their habitat and if we can describe that relationship, we can understand how landscapes influence individuals or populations in daily and seasonal movements, how they disperse to new locations from natal ranges, and gene flow. Analysis of connectivity thus merges complementary evaluation of landscape pattern, resource selection, and population characteristics to identify core regions, key areas or locations that link core regions, and pathways important for conserving a species (Schultz 1998, Crooks and Sanjayan 2006, Noss and Daly 2006).

We described structural connectivity of sagebrush-dominated landscapes across the Sage-Grouse Conservation Area (SGCA), which is the maximum extent encompassing historical and currently occupied ranges of Greater Sage-Grouse (Connelly et al. 2004). Management actions for Greater Sage-Grouse are focused primarily on conserving and restoring its sagebrush habitat (Stiver et al. 2006). However, maintaining landscapes dominated by sagebrush is a major challenge because changes in fire regimes, widespread invasion by nonnative plants, and increases in destructive land use are likely to accelerate the trajectory of fragmentation and loss (Knick et al. 2003, Wisdom et al. 2005b). Our objectives were to characterize the hierarchical pattern of sagebrush landscapes that results from natural and human disturbance and identify spatial scales perceived by Greater Sage-Grouse and other wildlife (Rotenberry and Wiens 1980a, Wiens et al. 1987, Knick et al. 2008).

We used graph theory (Horary 1969, Cantwell and Forman 1993) to delineate spatial and temporal patterns in Greater Sage-Grouse populations. Graph theory, as applied to ecological phenomena, represents spatial configurations by a set of nodes that describes arrangement of habitat patches or population centers and a corresponding set of linkages among nodes (Cantwell and Forman 1993, Ricotta et al. 2000, Urban and Keitt 2001, Minor and Urban 2008). Relative importance of nodes is a combined function of their size and location within the spatial network. Linkages represent real or implied pathways that facilitate functional processes, such as dispersal or migration, that maintain individual nodes and the population. The effect of loss or addition of nodes and linkages on a species' persistence can be assessed based on their contribution to overall connectivity (Pascual-Hortal and Saura 2006, Saura and Pascual-Hortal 2007, Pascual-Hortal and Saura 2008). Greater Sage-Grouse are well-suited for graph analyses because populations are distributed around leks that represent spatial foci during the breeding season (Patterson 1952, Dalke et al. 1960). We analyzed nodes and linkages based on the spatial pattern of more than 5,000 leks surveyed across the SGCA from 2003 to 2007. Our objectives were to identify lek locations and regions whose size and position within the sage-grouse range make them critical for maintaining connectivity and to determine thresholds in spatial distribution that limit dispersal and potentially isolate populations. We also sought to determine if lek persistence during long-term population declines was related to their connectivity within the sage-grouse network, structure of sagebrush landscapes, or environmental and human disturbance. The process of range contraction of sage-grouse then might be understood in terms of interactions among a spatial network of populations (Hanski and Gilpin 1991, Hanski 1994) superimposed on spatial and temporal patterns of habitat patches. Approaches such as ours, which combines landscape ecology and metapopulation theory (Wiens 1996, 1997), might provide the strongest insight into range-wide and regional dynamics of Greater Sage-Grouse populations.

STUDY AREA

The current range of Greater Sage-Grouse encompasses 670,000 km^2 and includes 11 states and two Canadian provinces (Schroeder et al. 2004). Seven individual sage-grouse SMZs have been delineated for monitoring and conservation actions (Stiver et al. 2006).

Our analysis included all mapped sagebrush habitat and currently surveyed Greater Sage-Grouse leks within the SGCA (Connelly et al. 2004). The SGCA covers 2,063,000 km^2, of which sagebrush is the dominant land cover on approximately 530,000 km^2 (Knick et al., this volume, chapter 12). Patches of sagebrush within this region comprise 20 taxa in 11 major species and subspecies of *Artemisia* (McArthur and Plummer 1978, Miller and Eddleman 2001). Few areas dominated by sagebrush remain unchanged since pre–Euro-American settlement (Braun et al. 1976, West and Young 2000), although the total amount of habitat that has been lost or altered is difficult to determine (Miller et al., this volume, chapter 10).

METHODS

Landscape Structure of Sagebrush

We used the LANDFIRE Existing Vegetation Map (LANDFIRE 2006) as the base GIS layer of land cover types for describing spatial structure of sagebrush landscapes. Land cover was classified at 30-m spatial resolution from Landsat Thematic Mapper images taken between 2000 and 2004. The map contained 210 cover types, which we collapsed into two classes, sagebrush presence or absence. Sagebrush taxa have different environmental optima (West and Young 2000) and are not used equally by Greater Sage-Grouse (Crawford et al. 2004, Connelly et al., this volume, chapter 4). We grouped sagebrush into a single class because of map inaccuracies in delineating different *Artemisia* species and because rangewide similarities in spatial structure may offset site-specific preferences in taxa used by Greater Sage-Grouse.

We changed the resolution of the original land cover data by resampling to 540-m grid-cells. We still were able to detect relatively fine-scale patterns at this resolution when considered at the spatial extent of the SGCA. Increasing minimum cell size to 540-m resolution would not influence detection of larger-scale disturbances, such as fire, and land cover changes of sufficient magnitude to influence ecological patterns and processes that affect sage-grouse and sagebrush distribution across a landscape. Our spatial resolution of 540 m precludes identification of understory species composition within the shrub community or small openings within a shrub community. These characteristics can influence sagebrush and sage-grouse dynamics but are not readily and accurately detected by remote sensing across broad regions (Bradley and Mustard 2005, Peterson 2005).

Analysis of structural connectivity in landscapes is based on metrics describing physical availability and arrangement of resources (Li and Reynolds 1994, Turner et al. 2001). Landscape quantity and composition were estimated from percent area dominated by sagebrush or other land cover types. We measured configuration as the total amount of edge between patches of sagebrush and nonsagebrush land cover. We divided total edge (km) by the area of sagebrush (km^2) to standardize the metric relative to proportion of sagebrush because landscapes dominated by single cover types have little edge.

Leks are important for breeding although habitat characteristics at leks are less important than the surrounding landscape used for nesting and brood-rearing (Connelly et al. 1981, Connelly et al. 2000c). Conservation of sagebrush within 5 km of leks has been recommended to maintain the most locations used for nesting and early brood-rearing by nonmigratory populations, whereas 18-km radii have been recommended for migratory populations (Wakkinen et al. 1992, Connelly et al. 2000c, Holloran and Anderson 2005). Characteristics at 54-km radii may influence seasonal movements and also incorporate habitats used outside the breeding season (Swenson et al. 1987, Leonard et al. 2000). We modeled landscape patterns within 5-, 18-, and 54-km radii surrounding each grid cell in the habitat coverage using a moving window analysis (ArcMap 9.1, ESRI Redlands, CA). Thus, we changed spatial extent to examine how structure of environmental attributes changed in the hierarchical landscape within which the focal point was embedded without confounding our results due to changing variable and resolution across scales (Doak et al. 1992, O'Neill et al.

1992). We captured a range of spatial scales by varying spatial radii that might influence movement and resource use by Greater Sage-Grouse (Aldridge and Boyce 2007; Hanser and Knick, this volume, chapter 19).

Spatial Structure of Greater Sage-Grouse Populations

We graphed the spatial structure of Greater Sage-Grouse populations based on lek locations because populations are focused on these sites for breeding (Patterson 1952). We assumed the distribution of leks represented the spatial structure of sage-grouse populations and that our analysis of connectivity captured probability of exchange among leks. Adult sage-grouse exhibit strong fidelity to lek locations each year, although subadults may disperse to neighboring leks (Gibson 1992). Broad-scale spatial arrangement of leks is relatively consistent, and individual lek locations can be maintained >80 years (Dalke et al. 1963, Smith et al. 2005).

We conducted separate graph analyses to assess connectivity of the current distribution of Greater Sage-Grouse and detect changes in connectivity from 1965 to 2007. The range-wide database on location and size of Greater Sage-Grouse leks was developed in 2004 (Connelly et al. 2004) and updated by individual states and provinces for surveys conducted through 2007 (Garton et al., this volume, chapter 15). Lek locations may remain stable over long periods, but leks may be relocated or new leks may be formed in response to population and environmental changes. We do not know what proportion of all leks was included in our range-wide sample.

Current Greater Sage-Grouse Populations

We measured connectivity for the network of leks surveyed from 2003 to 2007 across the SGCA and within SMZs. We assigned leks to SMZs based on geographical location and assumed that primary movements among leks were intrazone rather than interzone.

Landscape Probability of Connectivity (*PC*) is an index of connectivity negatively correlated with distance between nodes and isolation of populations (Saura and Pascual-Hortal 2007). Higher values of connectivity indicate larger nodes or greater probabilities of exchanging individuals with other nodes. We adapted the index

$$PC = \frac{\sum_{i=1}^{N} \sum_{j=1}^{N} a_i a_j p_{ij}^*}{A_L^2} \quad \text{(Eqn. 16.1)}$$

where sage-grouse leks (*N*) were the analysis unit in the range-wide or SMZ network, *i* and *j* were individual leks, a_i and a_j were sizes of leks (measured as the average of the yearly maximum of males counted at leks), and A_L was the total number of sage-grouse summed for all leks. The maximum product probability of moving between leks *i* and *j* (p_{ij}^*) represents all direct and intermediate steps (p_{ij}) between a given pair of leks. The index was standardized by the range-wide A_L, which permitted comparison among SMZs. We used Conefor Sensinode 2.2 software to calculate connectivity indices (*PC*)(Pascual-Hortal and Saura 2006, Saura and Pascual-Hortal 2007).

Sage-grouse populations can be connected directly by individuals moving between neighboring leks or indirectly through intermediary leks that serve as stepping-stones. The probability that two leks can be connected is a function of dispersal distance. We modeled probability of dispersal between leks (p_{ij}) as an exponential decay function (Bunn et al. 2000, Verheyen et al. 2004):

$$p_{ij} = e^{-kd_{ij}} \quad \text{(Eqn. 16.2)}$$

where d_{ij} is the distance (km) between leks *i* and *j*, and *k* is a constant that is set so the function returns a probability of dispersing a given distance. We used a decay function rather than a binary response to incorporate an increasing cost associated with moving longer distances. Dispersal also includes potential costs due to crossing inhospitable terrain, but a cost surface was not feasible in our study (Fall et al. 2007). Limited data on dispersal characteristics suggest that individuals were likely to move between leks *i* and *j* in equal proportions rather than as an asymmetrical flux from one density to another and that the proportion of dispersing males and females was similar (Dunn and Braun 1985). The probability of dispersal, and connectivity, between leks ranged from 0 for completely isolated leks to 1 for leks that can be reached with certainty.

Modeling dispersal in Greater Sage-Grouse is difficult because data are lacking. We defined dispersal as movements by yearlings to leks different from their maternal lek. Movement of some minimum number of animals must exceed the average distance between leks for potential genetic exchange to occur among leks and to effectively connect populations at different levels. Median dispersal distances in Colorado were 8.8 km for 12 yearling females and 7.4 km for 12 yearling males (Dunn and Braun 1985). Average distances traveled by Greater Sage-Grouse that visited more than one lek in Washington were 10.6 km for 5 males and 13.1 km for 14 females; the upper standard deviation ($P = 0.05$) of distances was 27.6 km (Schroeder and Robb 2003). We used a conservative approach to estimate an ~0.50 probability of dispersal beyond median distances reported for sage-grouse as well as to capture the upper standard deviation of distances in Washington. Setting $k = 0.1085$ yielded dispersal probabilities of 0.24 for 13.1 km, 0.38 for 8.8 km, and 0.58 for 5 km.

We calculated the importance of an individual lek for maintaining connectivity, dPC, as the difference in landscape connectivity (PC):

$$dPC(\%) = 100 \cdot \frac{PC - PC'}{PC} \quad \text{(Eqn. 16.3)}$$

when that lek was removed from the SGCA or SMZ network (Pascual-Hortal and Saura 2006). Relative importance (higher dPC) of individual leks was a function of lek size (number of male sage-grouse) and position within the network (number and strength of connections to other high-ranking leks). Number of male sage-grouse counted at a lek is the primary factor in maintaining breeding populations when leks are far apart and probability is low of individuals dispersing from other leks. Location within the network becomes more important when leks are closely spaced (Keitt et al. 1997, Saura and Pascual-Horton 2007). We calculated dPC for each lek using range-wide connectivity (overall A_L held constant) to provide a comparable measure across the entire SGCA.

We estimated number of components in the current range-wide network of Greater Sage-Grouse leks relative to dispersal distance. A component is a spatial unit in which all leks are connected, with no connections between separate components (Keitt et al. 1997, Pascual-Hortal and Saura 2006, Minor and Urban 2008). At one extreme, each individual lek represents a separate component when dispersal or exchange of individuals among leks is 0. Longer dispersal movements relative to interlek distances increase connectivity and decrease number of components. The network contains one component when all leks are connected. We assumed that distance between leks had behavioral or ecological significance relative to dispersal characteristics (Keitt et al. 1997). We varied dispersal distance as a binary response (rather than using a decay function) between 0 km (no exchange) and 100 km. We also estimated the relative importance (dPC) for each component to rank their conservation significance within the sage-grouse distribution.

Temporal Changes in Connectivity in Greater Sage-Grouse Populations

Connectivity in Greater Sage-Grouse populations (PC) should decrease with loss of larger leks, highly connected leks, or leks that serve as key steppingstones by connecting core regions. We used a subset of known lek locations that had been surveyed at least once within each interval from 1965–1974, 1980–1989, and 1998–2007 to avoid confounding analyses caused by increases in sampling effort that added new lek locations. Intervals were similar to periods used to estimate population trends (Connelly et al. 2004) and maximized the number of leks in the sample. Lek locations were surveyed in each interval, even though sage-grouse were not uniformly observed and recorded. Therefore, connectivity could increase between intervals.

We expected that changes in connectivity (PC) would mirror changes in abundance of Greater Sage-Grouse across their range and within SMZs (Connelly et al. 2004, Garton et al., this volume, chapter 15). Our results may be applicable to this cohort but we caution against their extrapolation to the entire Greater Sage-Grouse population because our sample was neither a random subset of surveyed leks nor stratified across the SGCA. We excluded the Colorado Plateau SMZ from analyses because of low sample sizes.

We estimated average number of sage-grouse counted, number of linkages, and average distance of linkages for each lek within each temporal period and SMZ. We also calculated the gamma index (γ):

$$\gamma = \frac{L}{3(V - 2)} \quad \text{(Eqn. 16.4)}$$

from the ratio of number of linkages (L) relative to maximum number of possible nonredundant pair-wise linkages for a given number of nodes

(V)(Forman 1995) to evaluate if linkages were lost at a higher rate than expected with leks that were abandoned.

Factors Associated with Lek Abandonment

Analysis of the current network of leks could improve conservation strategies if fate of individual leks was related to connectivity within the network, landscape structure of sagebrush, or environmental factors related to disturbance. We assumed numbers of male sage-grouse counted and persistence of leks were directly related to landscape features surrounding the lek (Connelly et al. 2000c, Holloran and Anderson 2005). We assumed the surrounding landscape no longer supported sage-grouse if a lek was abandoned or birds had moved to an alternate lek (Wallestad 1975b, Emmons and Braun 1984). For this analysis, we followed the fate through subsequent survey intervals only for those leks where sage-grouse were known to be present in 1965–1974.

We used a discrete-time hazard model (Singer and Willett 2003) to evaluate why leks may have been abandoned or persisted through declines in Greater Sage-Grouse populations since 1965. Hazard models are based on rate or timing of an occurrence and can be used to identify the effect of predictor variables on probability of an event (Singer and Willett 2003). Hazard ratios provide a comparison of rate of change in probability relative to a unit change in predictor variables. We defined a hazard event as lek abandonment, which occurred if no sage-grouse were counted at a lek within an interval; lek persistence was based on sage-grouse presence at leks for one (1965–1974), two (1965–1974, 1980–1989), or three intervals (1965–1974, 1980–1989, 1998–2007). Leks were right-censored (an event had not occurred) if sage-grouse were present in the last period (1998–2007).

We used likelihood ratio tests to identify the combination of connectivity and environmental variables and most appropriate spatial scales that best fit the hazard function. Likelihood ratio tests are appropriate when alternate models are fit to identical data and reduced models are nested within the full model (Singer and Willett 2003). Competing models are evaluated based on comparison of deviance statistics (-2 log likelihood; -2LL) to decide if additional predictor(s) improve the fit to the hazard function. Significant differences between deviance statistics are distributed as a χ^2 distribution with difference in number of estimated parameters as degrees of freedom.

We used a multistep process for model building. We first tested the lag effect of lek connectivity (dPC) calculated for the previous interval on the intercept-only (no covariate) hazard model as a time-varying predictor. We used this hazard function that included lek dPC as the base model to further identify environmental variables and spatial scales that influenced lek persistence.

Environmental variables characterizing habitat (landscape proportion of sagebrush), configuration (amount of habitat edge), and disturbance (proportion of burned area from 1965 to 2007; human-footprint score [Leu et al. 2008]) were time-invariant predictors because estimates concurrent with each interval (other than area burned) were not available. Values of environmental variables were estimated for the 540-m grid cell in which the lek was located as well as at 5-, 18-, and 54-km radii. We identified the most appropriate spatial scale for sagebrush, habitat edge, burned area, and human footprint based on deviance statistics that indicated an improved model fit (χ^2, $P < 0.1$) and by excluding variables and spatial scales whose odds ratios included 1. Last, we used the single variable within each environmental class to identify the best combination fitting the hazard function by comparing deviance statistics of reduced models nested within the full environmental model.

RESULTS

Structural Pattern of Sagebrush Landscapes

Different patterns of clustering within a landscape emerged with changes in the analysis radii (Fig. 16.1). Local patterns of sagebrush land cover, when mapped using a 5-km radius, were widely distributed and present across the Greater Sage-Grouse range. However, when using the larger 54-km radii, four primary regions with landscapes dominated by sagebrush land cover were evident: south-central Oregon and northwest Nevada; the Owyhee region of southeast Oregon, southwest Idaho, and northern Nevada; southwest Wyoming; and south-central Wyoming. Patterns of landscape fragmentation, delineated by calculating total distance of edge between sagebrush and other habitats, showed similar perspectives (Fig. 16.2).

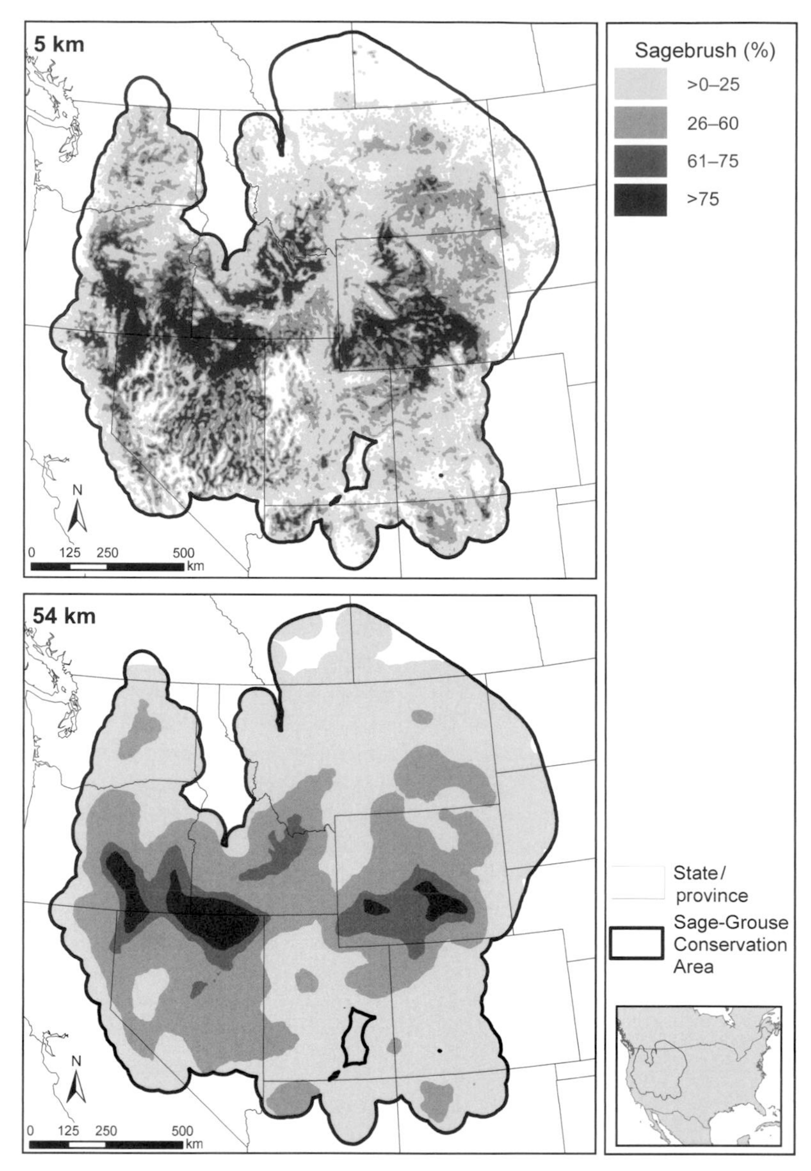

Figure 16.1. Percent of the landscape dominated by sagebrush within a 5-km (top) and 54-km (bottom) radius of each 0.5 km grid cell.

Connectivity in Greater Sage-Grouse Populations

The current geographic range of Greater Sage-Grouse was represented by 5,232 active leks surveyed from 2003 to 2007. Eighty percent (N = 4,143) of the leks were surveyed in ≥2 years and 50% (N = 2,596) in ≥4 years. Number of leks surveyed within SMZs ranged from 23 in the Columbia Basin to 1,495 in the Snake River Plain (Table 16.1). Average number of sage-grouse per lek was 20.3 ± 23.4 SD and ranged from 0.2 to 243.0 for the survey period. Average number of males counted per lek was highest in the Wyoming Basin SMZ and lowest in the Colorado Plateau SMZ (Table 16.1).

The graph of Greater Sage-Grouse leks contained 37,989 potential linkages when dispersal distance was <18 km (based on results from the component analysis)(Fig. 16.3). Average length (km) of linkages range-wide was 16.6 ± 7.3 SD compared to average straight-line distance between nearest neighbor lek pairs of 5.9 ± 5.2 SD. Linkages were primarily within SMZs; only 34 links (<0.1%) were between zones (Great Plains–Wyoming Basin, 22; Snake River Plain–Southern

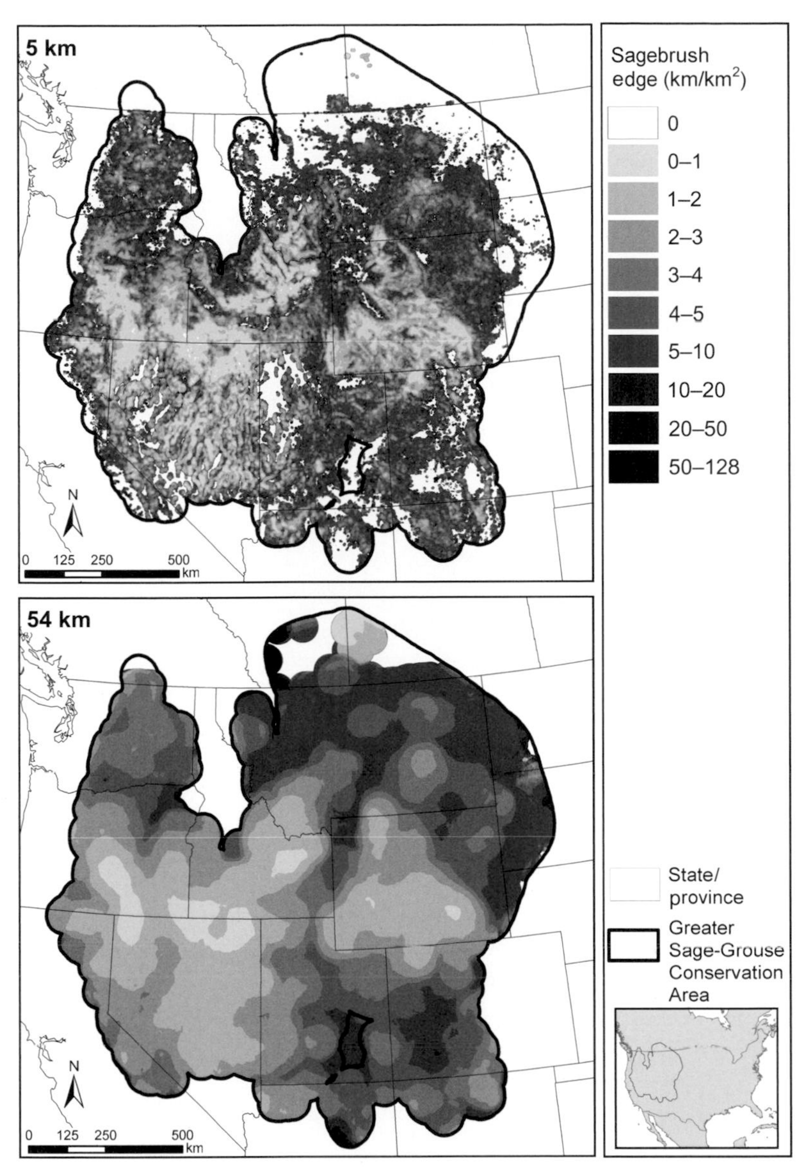

Figure 16.2. Small- and large-scale fragmentation of sagebrush habitats represented by the total distance of edge between sagebrush and other land cover types within a 5-km (top) and 54-km (bottom) radius of each 0.5 km grid cell.

Great Basin, 3; Snake River Plain–Northern Great Basin, 9).

The most important leks (*dPC*) were within core regions of the sage-grouse range (Fig. 16.4). The low probability of dispersal at long pair-wise distances resulted in isolation of leks and low relative importance, particularly in outlying regions of the range-wide distribution.

Relative measures of connectivity within SMZs (standardized by total number of sage-grouse range-wide) were a function of number of leks, average number of male sage-grouse counted, and number of linkages. Average number of male sage-grouse counted accounted for a large portion of the lek's importance (Pearson's correlation $r = 0.83$). The Wyoming Basin had the most highly connected network of sage-grouse leks (landscape *PC*), followed by the Snake River Plain and Great Plains SMZs (Table 16.1). Leks in the Colorado Plateau and Columbia Basin were the least connected of the SMZs (Table 16.1).

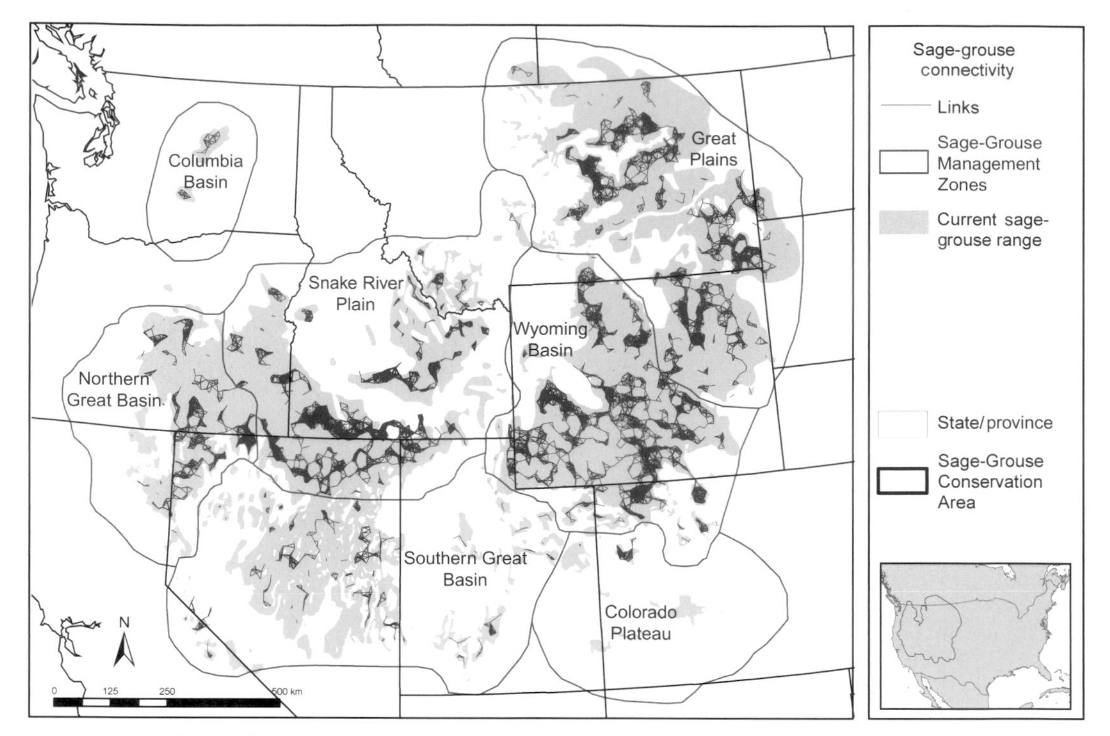

Figure 16.3. Distribution of current (2003–2007) Greater Sage-Grouse leks. Populations at leks were connected if the straight-line distance to neighbors was <18 km.

TABLE 16.1

Average number of male sage-grouse counted at leks (based on surveys from 2003–2007) and connectivity indices (dPC) for Greater Sage-Grouse range-wide and within Sage-Grouse Management Zones.

	Sage-grouse leks			Links among leks		
Sage-Grouse Management Zone	N	Average male count[a]	Average lek dPC[a, b]	N	Average distance (km)[a]	Average minimum distance (km)[a]
Great Plains	1,252	17.3B	0.021B	7,759	11.1B	5.8B
Wyoming Basin	1,397	29.5A	0.062A	9,046	11.2B	5.7B
Southern Great Basin	448	16.2B	0.011B	1,358	9.2C	5.9B
Snake River Plain	1,495	16.9B	0.029AB	14,430	10.4B	4.3C
Northern Great Basin	565	18.1B	0.031B	4,814	10.5B	4.5C
Columbia Basin	23	15.3B	0.005C	56	12.5A	9.5A
Colorado Plateau	52	5.7C	0.002D	492	8.6C	2.9C
Range-wide	5,232	20.3	0.034	37,955	10.7	5.2

[a] Management zones having different letters have significantly different ($P < 0.05$) values in a univariate analysis of variance: average number of individuals counted ($F_{6,5225} = 54.1$, $P < 0.001$), average lek *dpc* ($F_{6,5216} = 109.0$, $P < 0.001$), and average pair-wise distance between leks ($F_{6,\ 37,948} = 81.0$, $P < 0.001$).

[b] Lek *dPC* measures the change in landscape connectivity when individual leks were removed from the network.

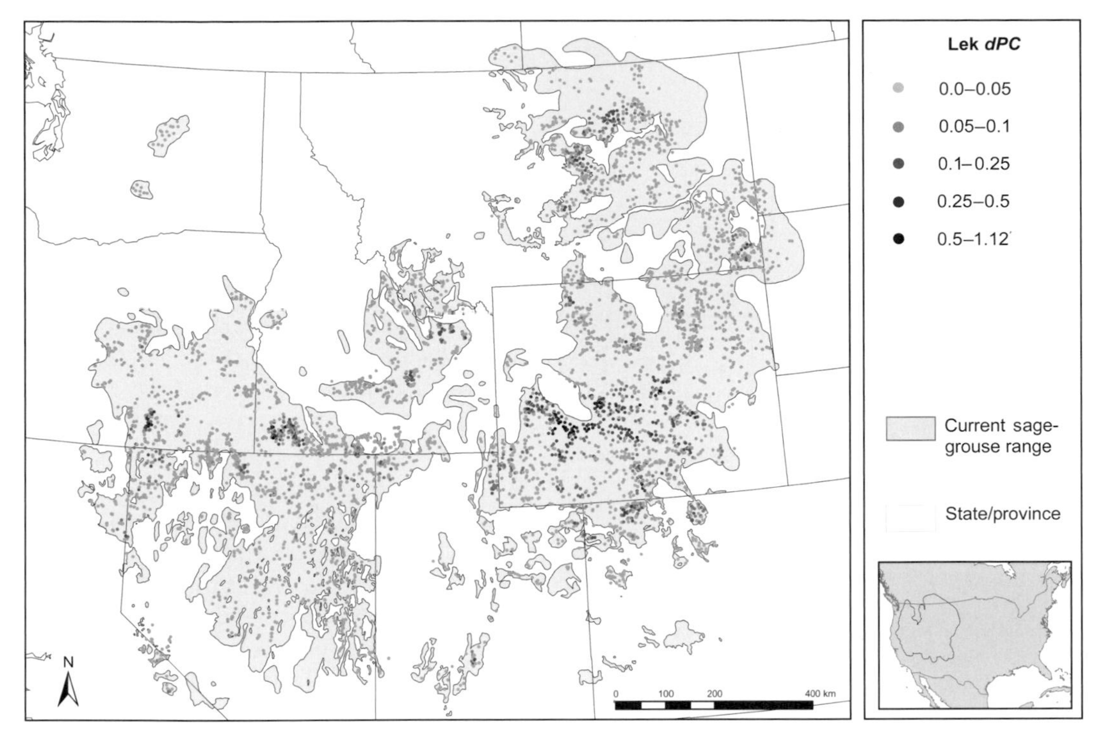

Figure 16.4. Importance of individual leks in maintaining connectivity in the range-wide distribution of Greater Sage-Grouse. Lek *dPC* measures change in landscape connectivity that results when a lek is removed from the network. Higher *dPC* values reflect larger numbers of sage-grouse at a lek and greater connectivity within the network.

The interaction of lek size and number of linkages in lek connectivity (*dPC*) was evident in comparisons among the Great Plains, Southern Great Basin, Snake River Plain, Northern Great Basin, and Columbia Basin. Average number of male sage-grouse counted per lek was similar among these zones and ranged from 15.3 (Columbia Basin) to 18.1 (Northern Great Basin) (Table 16.1). The small number of leks, longer average distance between leks, and few linkages in the Columbia Basin SMZ resulted in low lek connectivity (average $dPC = 0.005$). Connectivity also was low in the Southern Great Basin (Table 16.1); even though average distance between leks was short, leks were arranged in widely dispersed clusters separated by long distances.

Number of separate components (units encompassing leks connected within components but unconnected with others) in the range-wide distribution decreased when potential dispersal distance was increased from 0 km (no exchange) to 100 km (Fig. 16.5). Shorter dispersal distances limited linkages among leks by decreasing exchanges and resulted in larger numbers of separate components. An inflection point in the exponential relationship between a binary response in dispersal distance and number of components existed at distances between 13 and 18 km. Number of separate components increased rapidly when dispersal distances were <13 km because leks increasingly became more isolated. Most leks were connected when maximum dispersal distance was >18 km; further increases in dispersal distance resulted in proportionately less change in number of components. We used an 18-km dispersal distance in subsequent analyses for estimating connectivity among components.

Greater Sage-Grouse leks were clustered within 209 components when neighboring leks were connected by dispersal distances up to 18 km (Fig. 16.6). Relative importance of individual components (*dPC*) primarily reflected number of leks and total number of male sage-grouse counted within the component. Components with the highest relative importance within the sage-grouse range were distributed across the central and eastern parts of the SGCA (Fig. 16.6). Most components were small geographic units: 47% ($N = 98$) were <100 km^2 and 76% ($N = 160$) contained ≤ 10 leks (Fig. 16.7). Ten components were >5,000 km^2

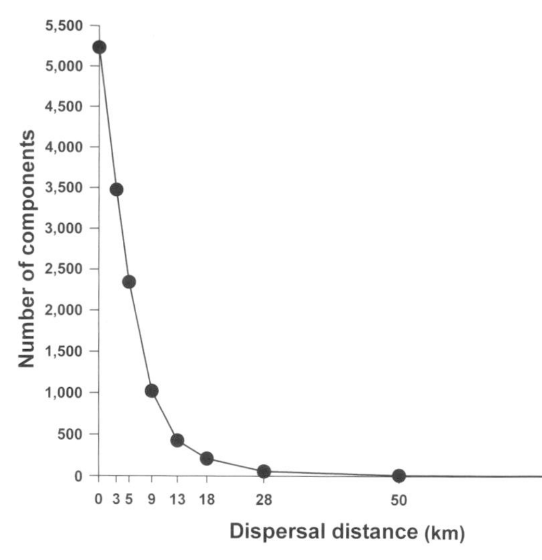

Figure 16.5. Number of components within the Greater Sage-Grouse range relative to dispersal distances. Components are spatially separated units in which leks are connected within but not among components. Distances between 13 and 18 km represent a threshold at which decreasing potential dispersal distance decreases the connections among leks and increases the number of components.

(range: 5,395–100,288 km^2), and eight components contained >100 leks (range: 143–1,139).

Temporal Changes in Connectivity in Greater Sage-Grouse Populations

Our sample included 907 lek locations at which surveys were conducted at least once in each interval. Proportion of leks at which no sage-grouse were recorded increased from 13% in 1965–1974 to 22% in 1980–1989 and 36% in 1998–2007. Average number of sage-grouse counted at leks where sage-grouse were present was 25.0 (N = 789 leks) in 1965–1974, 19.5 (N = 711) in 1980–1989, and 20.2 (N = 580) in 1998–2007. Relative landscape connectivity (*PC*) declined range-wide from 0.0054 in 1965–1974 to 0.0029 in 1980–1989 and 0.0025 in 1998–2007. Trends within individual SMZs mirrored range-wide trends in this cohort (Table 16.2).

Landscape connectivity (*PC*) declined within SMZs, although at different rates among zones and between temporal intervals (Fig. 16.8). Greatest rate of declines were between 1965–1974 and

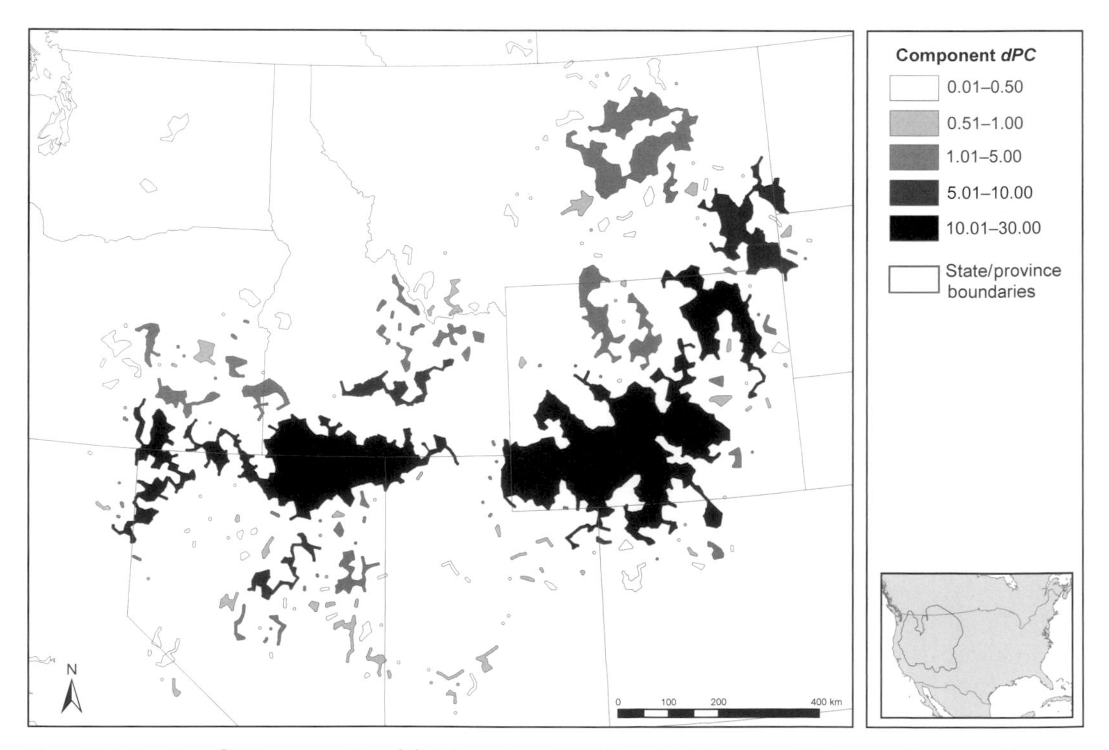

Figure 16.6. Location of 209 components and their importance (*dPC*) in maintaining connectivity across the range-wide distribution of Greater Sage-Grouse. Number and spatial arrangement of components was evaluated for a dispersal distance of 18 km.

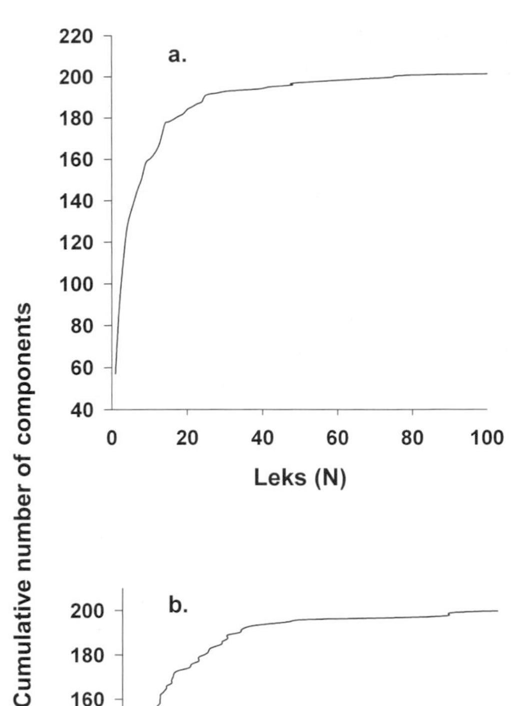

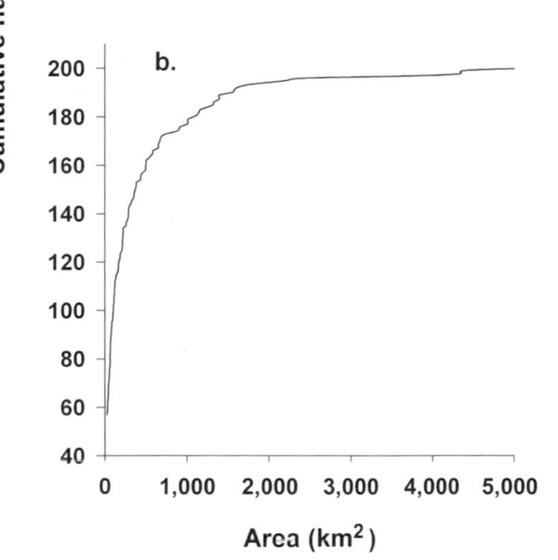

Figure 16.7. Size distribution of components (spatially separated units in which leks are connected within but not among components) relative to (a) number of leks and (b) component area. Eight components contained >100 leks and 10 were >5,000 km^2.

1980–1989. Relative stability in connectivity from 1980–1989 to 1998–2007 was reflected in landscape *PC* estimated for Snake River Plain, Southern Great Basin, and Wyoming Basin. Connectivity increased slightly in the Southern Great Basin and Wyoming Basin. Connectivity continued to decline between 1980–1989 and 1998–2007 in the Columbia Basin, Northern Great Basin, and Great Plains (Fig. 16.8). We did not evaluate changes in number of population components because most leks in our sample were widely dispersed.

Small leks with low connectivity were lost as abundance of sage-grouse declined (Fig. 16.9), which also changed the relative importance of persisting leks within the new network configuration (Fig. 16.10). Lek connectivity (*dPC*) was lower for 159 leks that were abandoned between 1965–1975 and 1980–1989 compared to 630 leks that persisted (Fig. 16.9). Similarly, 455 leks that persisted through all three sampling intervals had higher average abundance of sage-grouse and connectivity than 175 leks at which sage-grouse were present in 1965–1975 and 1980–1989 but were abandoned by 1998–2007 (Fig. 16.9). The slight decrease in average distance of links across intervals indicated that more-distant leks were abandoned. Decreased γ across intervals represented loss of a disproportionately higher number of linkages with leks that were abandoned.

Factors Associated with Lek Abandonment

Proportion of the landscape dominated by sagebrush and amount of habitat edge were similar between leks at which sage-grouse were present in 1998–2007 and those that had been abandoned (Table 16.3). Amount of burned area and the human footprint was higher for leks that were abandoned by 1998–2007.

Lek connectivity (*dPC*) improved the fit ($\Delta_{-2LL} = 13.6$, χ^2 1 df, $P < 0.001$) to the hazard function without covariates describing 271 leks that were abandoned between 1965–1975 and 1998–2007 and 546 leks that persisted (Table 16.4). This time-varying function was used as a base model for evaluating the subsequent contribution of environmental variables.

The most significant spatial scales for environmental predictors were proportion of sagebrush within 54 km of the lek ($P < 0.05$), proportion of burned area within 54 km of the lek ($P < 0.01$) and level of human footprint within 5 km ($P < 0.01$) (Table 16.4). Edge variables did not improve model fit ($P_{selection} = 0.1$) at any spatial scale and were not considered in subsequent model development.

The best model describing probability that a lek with sage-grouse present in 1965–1974 would be abandoned by 1998–2007 included proportion of area burned and level of human footprint (Table 16.5). The full environmental model including sagebrush did not further improve model fit ($\Delta_{-2LL} = 0.85$; $P > 0.1$). In the final hazard model (Table 16.6), probability of abandonment increased by 30% for each unit decrease in lek connectivity (*dPC*) during the previous interval [100 × (hazard

TABLE 16.2

Change in average number of male Greater Sage-Grouse counted at leks and connectivity (dPC) among leks during surveys conducted in 1965–1974, 1980–1989, and 1998–2007.

Sage-Grouse Management Zone	Years	Greater Sage-Grouse: Leks (N)	Greater Sage-Grouse: Average male count	Average lek *dPC*	Links among leks: N	Links among leks: γ[a]
Great Plains	1965–1974	118	24.3	1.27	484	1.39
	1980–1989	109	18.4	1.26	414	1.29
	1998–2007	69	9.6	1.20	196	0.98
Wyoming Basin	1965–1974	213	23.7	0.59	1,122	1.77
	1980–1989	194	14.1	0.59	828	1.44
	1998–2007	182	15.5	0.58	734	1.36
Southern Great Basin	1965–1974	139	18.8	0.94	500	1.22
	1980–1989	129	15.2	0.89	420	1.10
	1998–2007	107	14.0	0.88	258	0.82
Snake River Plain	1965–1974	259	20.5	0.55	1,488	1.93
	1980–1989	234	15.0	0.53	1,434	2.08
	1998–2007	182	11.5	0.50	838	1.55
Northern Great Basin	1965–1974	41	22.2	2.96	100	0.85
	1980–1989	29	16.7	3.13	56	0.69
	1998–2007	30	13.8	2.79	62	0.74
Columbia Basin	1965–1974	18	24.8	8.68	60	1.25
	1980–1989	16	16.0	8.44	56	1.33
	1998–2007	10	10.4	7.87	22	0.92

[a] The gamma function (γ) is the ratio of number of linkages relative to maximum number of possible non-redundant pair-wise linkages for a given number of leks.

ratio − 1)], by 30% for each unit increase in human footprint within 5 km of a lek, and by 800% for each unit increase in fire within 54 km of a lek (Table 16.5).

DISCUSSION

Connectivity analysis provides a conceptual framework for understanding dynamics of landscapes and wildlife (Schumaker 1996, Keitt et al. 1997, Ricotta et al. 2000, Crooks and Sanjayan 2006) but has not been applied previously to sagebrush systems. Spatial structure of sagebrush landscapes and Greater Sage-Grouse populations can reveal underlying processes that have led to long-term trajectories of habitat loss and population declines. These analyses also can provide a foundation for future conservation strategies by identifying critical locations necessary to maintain range-wide and regional networks of interacting populations of Greater Sage-Grouse.

Structural Pattern of Sagebrush Landscapes

Primary structural characteristics of landscapes that interact with underlying processes can be measured by the quantity, composition, and configuration of land cover types (Urban et al. 1987, Turner 1989, Turner et al. 2001). Sagebrush dominates >500,000 km^2 of the SGCA, and landscapes

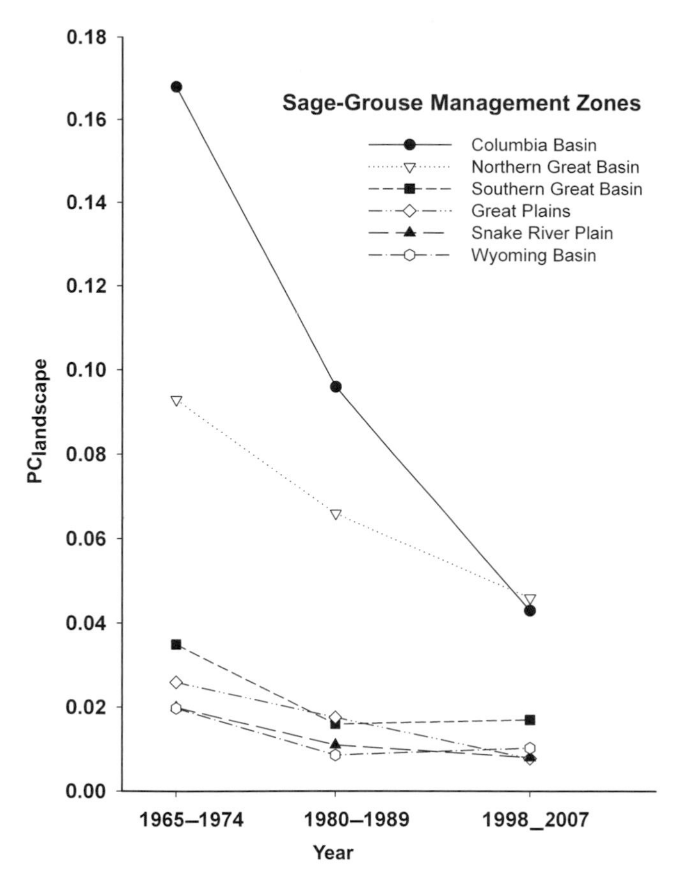

Figure 16.8. Change in landscape connectivity (*PC*) within Sage-Grouse Management Zones between 1965 and 1974, 1980 and 1989, and 1998 and 2007. Landscape *PC* is a relative index. Relative pattern but not absolute relationships are comparable among management zones.

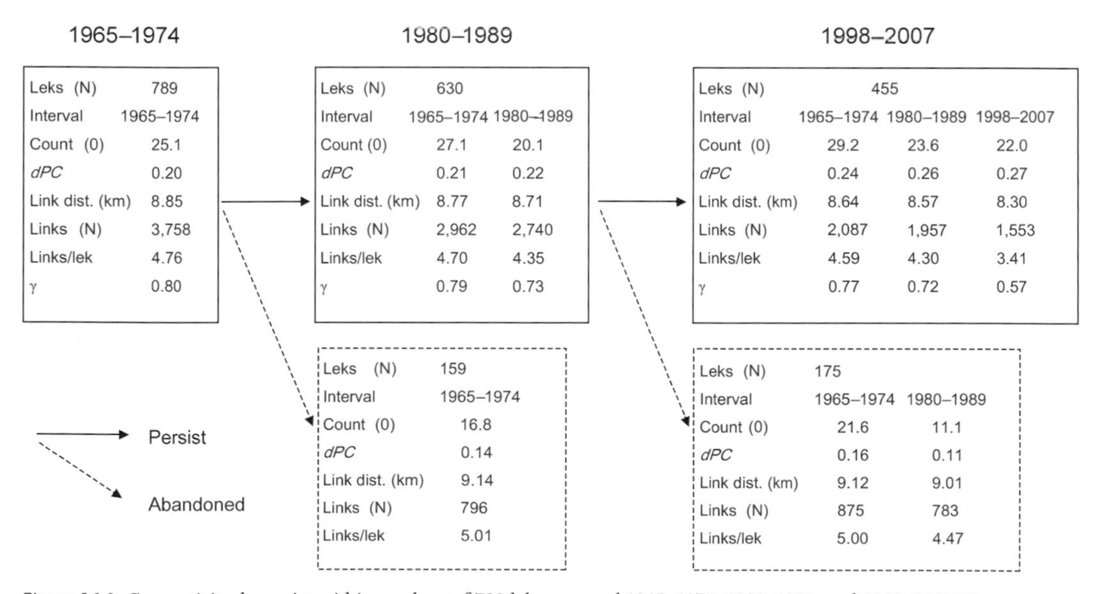

Figure 16.9. Connectivity dynamics within a cohort of 789 leks surveyed 1965–1974, 1980–1989, and 1998–2007. Sage-grouse were present at all leks in 1965–1974.

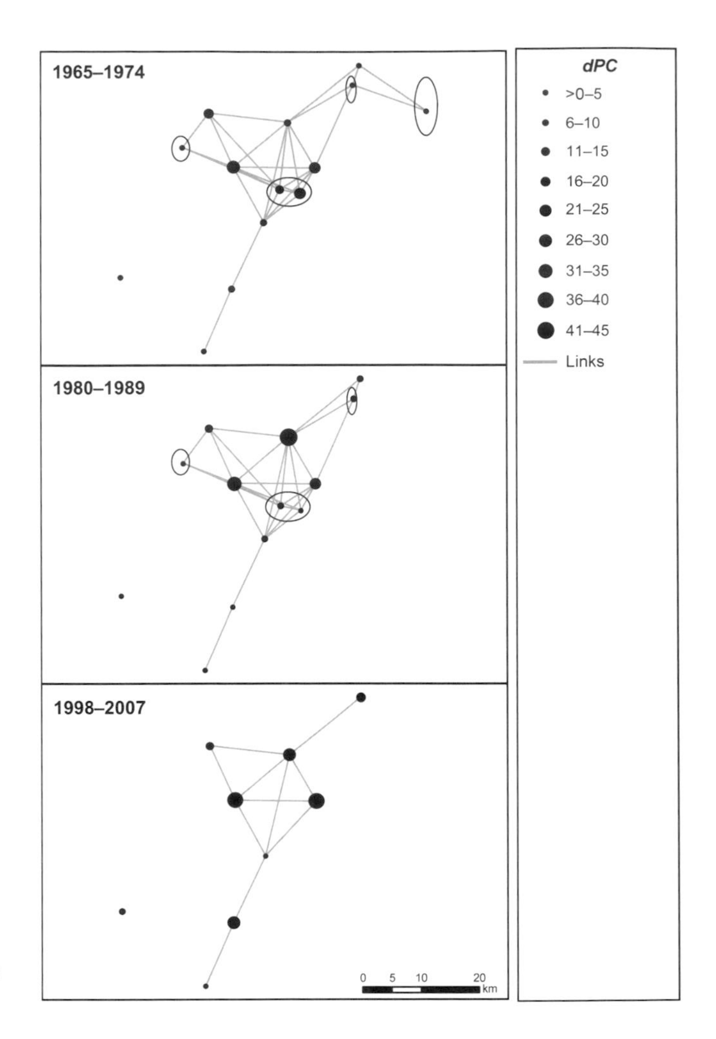

Figure 16.10. Change in connectivity (*dPC*) of individual leks and connections within a network of leks in central Washington over survey intervals from 1965 to 1974, 1980 to 1989, and 1998 to 2007. Ellipses enclose leks that will be abandoned by 1998–2007.

are arranged in a diverse array of patterns that vary with spatial scale (Leu and Hanser, this volume, chapter 13). The primary challenge to modeling sagebrush landscapes was to reduce this variation to metrics that: (1) capture the primary characteristics of landscape structure, (2) identify similar or repetitive patterns, and (3) provide meaningful measures to relate to the dynamics of wildlife populations (Cantwell and Forman 1993). Percent sagebrush (quantity and composition) and amount of edge with other land cover types (configuration) are related to disturbance and presence of the invasive annual cheatgrass (*Bromus tectorum*)(Knick and Rotenberry 1997), to distribution and population dynamics of passerine birds (Vander Haegen et al. 2000, Hamer et al. 2006, Vander Haegen 2007, Knick et al. 2008), and to presence and persistence of Greater Sage-Grouse (Aldridge and Boyce 2007; Doherty et al. 2008; Hanser and Knick, this volume, chapter 19). Numerous metrics have been developed (Gustafson 1998, Li and Wu 2004), but these basic attributes structuring sagebrush landscapes may be the primary characteristics that affect vegetation and disturbance dynamics and wildlife responses.

Management and monitoring in sagebrush systems have focused on successional or state-and-transition dynamics that occur over time within a location but not across space (Bestelmeyer et al. 2003, West 2003a, Crawford et al. 2004). Spatial variation traditionally has been regarded as a consequence of different states of individual locations along similar vegetational trajectories with the assumption that locations function independently rather than interacting as a mosaic within a landscape (Allen-Diaz and Bartolomé 1998, West 2003b, Briske et al. 2005). Yet many processes, such as fire or invasion of nonnative plant

TABLE 16.3

Univariate comparisons (two-sample t-test) of environmental variables at known leks where Greater Sage-Grouse were present (N = 580) compared to leks at which male sage-grouse were present in either 1965–1974 or 1980–1989 but absent by 1998–2007 (N = 326).

Values of environmental variables were estimated for the grid cell in which leks were located, and for 5-, 18-, and 54-km buffer zones surrounding the lek.

		Greater Sage-Grouse leks			
Environmental variable	Buffer (km)	Present $\bar{x} \pm SE$	Abandoned $\bar{x} \pm SE$	t	P
Sagebrush (%)[a]	Lek	74.3 ± 0.02	73.0 ± 0.02	0.43	0.67
	5	69.4 ± 0.01	66.7 ± 0.02	1.19	0.23
	18	57.0 ± 0.01	57.9 ± 0.01	0.83	0.41
	54	46.7 ± 0.01	48.8 ± 0.01	1.60	0.11
Burned (%)[a]	Lek	10.0 ± 0.01	16.3 ± 0.02	2.77	0.006
	5	10.1 ± 0.01	14.4 ± 0.02	2.70	0.007
	18	9.3 ± 0.01	12.3 ± 0.01	2.52	0.01
	54	8.9 ± 0.01	11.3 ± 0.01	2.66	0.008
Edge[b]	5	1.9 ± 0.08	2.0 ± 0.12	0.93	0.35
	18	2.1 ± 0.06	2.1 ± 0.09	0.12	0.90
	54	2.2 ± 0.05	2.2 ± 0.08	0.49	0.62
Human footprint[c]	Lek	4.3 ± 0.06	4.4 ± 0.08	1.10	0.27
	5	4.1 ± 0.04	4.3 ± 0.06	2.70	0.007
	18	4.0 ± 0.04	4.2 ± 0.06	2.31	0.02
	54	4.0 ± 0.03	4.1 ± 0.04	2.72	0.007

[a] Data were arcsine-transformed.

[b] Amount of edge was transformed relative to % sagebrush in the landscape. Edge was not calculated for the cell in which a lek was located.

[c] The human footprint was a cumulative score ranging from 0 (no effect) to 10 (maximum effect) derived from multiple submodels quantifying anthropogenic effects (Leu et al. 2008).

species, are inherently spatial, and they influence structure of current sagebrush landscapes as well as future dynamics of these systems (Peters et al. 2006, Davies and Sheley 2007).

Spatial attributes of land cover influence spatial and temporal stability and contribute to a system's resilience and resistance to change (Shugart 1998). Sagebrush systems in the western portion of the SGCA, including the Columbia Basin, Northern Great Basin, Southern Great Basin, and Snake River Plain SMZs, have two primary endpoints that are resistant to further change (Hemstrom et al. 2002; Miller et al., this volume, chapter 10). A landscape dominated by sagebrush communities containing an understory of native grasses and forbs represents one stable endpoint, because disturbance historically has been at small spatial scales and occurred at longer intervals than the period required for recovery (Laycock 1991; Baker, this volume, chapter 11). Cheatgrass-dominated grasslands without sagebrush represent an undesirable endpoint that remains stable because recurrent fires prevent recolonization by sagebrush and other native forbs and grasses (Young and Evans 1973, d'Antonio and Vitousek 1992, Chambers et al. 2007). A cheatgrass-dominated landscape persists because small islands of sagebrush are unlikely to remain as fire frequency increases (Brooks et al. 2004a, Link et al. 2006). An intermediate mosaic of patches dominated by

TABLE 16.4

Candidate predictor variables, spatial scale, and likelihood ratio tests to calculate fit to hazard models predicting lek abandonment.

Lek connectivity was first tested against the model without covariates. The model including lek connectivity was then used to test subsequent models to identify most significant spatial scale within environment categories.

Environmental variable	Buffer (km)[a]	−2LL	Likelihood ratio[b]	Hazard ratio[c]	95% CI
Model without covariates	—	1,387.90	—	—	—
Lek connectivity *dPC*	—	1,374.30	0.00	0.35	0.19−0.66
Sagebrush (%)	Lek	1,374.25	−0.05	0.97	0.71−1.31
	5	1,374.29	−0.06	0.98	0.59−1.63
	18	1,371.94	−2.36	1.55	0.88−2.71
	54	1,370.21	−4.089*	1.94	1.02−3.71
Fire (%)	Lek	1,366.40	−7.90**	1.71	1.19−2.46
	5	1,366.61	−7.68**	2.06	1.26−3.38
	18	1,366.90	−7.40**	2.61	1.34−5.11
	54	1,364.43	−9.87**	4.82	1.84−12.60
Edge[d]	5	1,374.24	−0.06	0.99	0.93−1.06
	18	1,373.89	−0.41	0.97	0.89−1.06
	54	1,373.02	−1.28	0.94	0.84−1.05
Human footprint	Lek	1,371.85	−2.49	1.08	0.98−1.19
	5	1,365.84	−8.46**	1.21	1.06−1.37
	18	1,368.47	−5.83*	1.18	1.03−1.35
	54	1,368.69	−5.61*	1.23	1.04−1.47

[a] Values of environmental variables were estimated for the grid cell in which leks were located, and for 5-, 18-, and 54-km buffer zones surrounding the lek.

[b] Likelihood ratios were the difference in deviance statistics (−2LL) between competing models. * = $P < 0.05$, ** = $P < 0.01$; significance level of likelihood ratio test for $H_o{:}\beta_{dPC} = \beta_{dPC} + \beta w_1 = 0$.

[c] Hazard ratios (presented with 95% CI) estimate the relative change in an event (lek persistence) for each unit change in value of a predictor.

[d] Edge was not calculated for the cell in which leks were located.

sagebrush and cheatgrass is inherently unstable. Small, dispersed patches of sagebrush within a larger landscape dominated by sagebrush can provide seed sources important for natural recolonization (Longland and Bateman 2002) or as building blocks for restoration (Wisdom et al. 2005a, Meinke et al. 2009). Alternatively, increases in frequency, intensity, or spatial extent of disturbance can prevent extensive recovery and dominance of sagebrush. In many landscapes, cheatgrass becomes the dominant land cover and return to sagebrush is unlikely (Billings 1990). Thus, basic structural attributes of composition, quantity, and arrangement expressed at multiple spatial scales are primary factors that affect future trajectories in sagebrush landscapes.

Landscape structure also influences the ability of an animal to move across a landscape. A single habitat patch can spread across an entire landscape when the proportion exceeds 60% in randomly generated landscapes (O'Neill et al. 1988). The mapped distribution of sagebrush land cover >60% indicates how movements restricted by habitat configuration might be facilitated at one spatial scale but constrained at others by the hierarchical organization of sagebrush landscapes.

TABLE 16.5

Evaluation of competing models combining environmental variables that best fit the hazard function describing probability of lek abandonment from 1965–1974 to 1998–2007.

Environmental variables (Table 16.4) used in candidate models were % sagebrush and % burned area within a 54-km buffer of a lek and human-footprint score within 5 km.

Candidate model	−2LL	Likelihood ratio[a]
Lek dPC + $sagebrush_{54\ km}$ + $fire_{54\ km}$	1,364.18	–10.12
Lek dPC + $sagebrush_{54\ km}$ + human $footprint_{5\ km}$	1,358.11	–17.24
Lek dPC + $fire_{54\ km}$ + human $footprint_{5\ km}$	1,348.74	–25.56
Lek dPC + $fire_{54\ km}$ + human $footprint_{5\ km}$ + $sagebrush_{54\ km}$	1,347.88	–26.42

[a] Likelihood ratios were the difference in deviance statistics (−2 LL) between candidate models and the base hazard function including the time-varying predictor for lek connectivity (dPC).

TABLE 16.6

Final hazard function identified from candidate models (Table 16.5) describing the probability of lek abandonment by 1980–1988 or 1998–2007.

Environmental variables were % fire within 54 km of a lek and human-footprint score within 5 km.

Variable	Coefficient	SE	Hazard ratio[b]	95% CI
Lek connectivity dPC[a]	−1.18	0.43	0.31	0.16−0.58
$Fire_{54\ km}$	2.12	0.52	9.00	3.25−24.95
Human $footprint_{5\ km}$	0.27	0.07	1.31	1.15−1.50

[a] Lek connectivity (dPC) was a time-varying predictor included in the base model (Table 16.4).

[b] Hazard ratios (presented with 95% CI) estimate the relative change in an event (lek persistence) for each unit change in value of a predictor.

Critical thresholds where habitat amount becomes less important than habitat arrangement vary with dispersal, habitat requirement, mobility, and vagility characteristics of a species (Andrén 1994, With and Crist 1995, Flather and Bevers 2002). Configuration of land cover may be less important to mobile species, such as sage-grouse, because small amounts of clumped habitat distributed across a landscape can be exploited and serve as population sources (With and King 2001). Agriculture is the primary factor influencing sagebrush landscapes in the Great Plains SMZ (Connelly et al. 2004), and edge between land cover types, rather than proportion of sagebrush in the landscape, is the dominant feature. Yet landscape connectivity (PC) of current sage-grouse populations in this matrix of agriculture, grassland, and sagebrush in the Great Plains was similar to other SMZs. In contrast, connectivity of sage-grouse populations was lower and populations have become increasingly isolated in the agricultural landscape of the Columbia Basin, where quantity of sagebrush may be lower, available patches more widely dispersed, and the human footprint most intensive (Leu and Hanser, this volume, chapter 13).

Connectivity in Greater Sage-Grouse Populations

Modeling functional relationships of Greater Sage-Grouse to their sagebrush habitat is challenging because different characteristics are important at different spatial scales (Aldridge and Boyce 2007, Aldridge et al. 2008). Habitat availability also varies by season and across regions (Connelly et al. 2000c, Doherty et al. 2008).

Fine-scale attributes of sagebrush habitats may influence within-season movements (such as choice of nesting or brood-rearing locations) and vital rates, including survival and productivity (Connelly et al. 2000c, Crawford et al. 2004). Highly mobile individuals tend to collapse landscape heterogeneity (Kotliar and Wiens 1990, With 1994); sage-grouse moving between seasonal ranges may be more sensitive to the broader-scale matrix and traverse areas that do not contain suitable habitats (Connelly et al. 1988, Leonard et al. 2000). A broad diversity of habitat configurations and environmental stressors are encompassed within annual ranges of individual sage-grouse populations as well as within the full geographic range of the species (Schroeder et al. 1999, Leonard et al. 2000). Thus, sagebrush landscapes may have multiple structures that vary in importance from the perspective of sage-grouse.

Identification of appropriate spatial scale(s) is important to effectively model habitat associations or track species' responses to changes in their habitat (Wiens 2002). Many shrubland birds are sensitive to landscape components at spatial scales much larger than individual home ranges (Vander Haegen et al. 2000, Knick and Rotenberry 2002). Large-scale characteristics within surrounding landscapes influenced locations selected by Greater Sage-Grouse in Alberta (1 km^2; Aldridge and Boyce 2007) and Wyoming (>4 km^2; Doherty et al. 2008). Probability of persistence of sage-grouse populations in a range-wide comparison of historical and current distributions was greatest in areas containing >30% sagebrush within a 30-km radius of a given point and with a human density <4/km^2 (Aldridge et al. 2008). In our study, fire within a 54-km radius and human activity within 5 km of a lek influenced the probability of persistence over 40 years.

Connectivity in Current Populations of Greater Sage-Grouse

We delineated the spatial arrangement of Greater Sage-Grouse populations as a network of connected leks. We modeled connections between lek pairs as straight-line distances, although in reality, movements are directed by patch characteristics and permeability of boundaries between patches (Wiens et al. 1993). We do not know how sage-grouse move through or over a landscape because radiotelemetry studies have emphasized daily or seasonal point locations of individuals rather than continuous movements during dispersal or seasonal migration. Our linkages estimated the minimum distance that individuals would have to traverse between leks or population components.

The range-wide distribution of Greater Sage-Grouse was dominated by a small set of core components, each containing a large number of leks and encompassing >5,000 km^2. Numerous small components were interspersed in between core regions and at the edges of the range. Populations of sage-grouse may have a spatial structure similar to Capercaillie (*Tetrao urogallus*), a forest grouse that is broadly distributed throughout central Europe and associated with late successional coniferous forests. Population dynamics of Capercaillie in northern Europe are primarily affected by amount of nonforested area and differences in forest quality at large (100-km^2) spatial scales. In contrast, fragmented forest distributions throughout central Europe have resulted in a metapopulation structure in which connectivity is important to maintaining individual Capercaillie populations (Storch and Segelbacher 2000, Storch 2003). Population dynamics of sage-grouse within core components similarly may depend on amount and quality of sagebrush or level of disturbance. Sage-grouse populations distributed in more isolated components at the edge of the range-wide distribution may depend on dispersal from connecting leks.

Metapopulation theory (Gilpin and Hanski 1991, Hanski 1994) suggests that long-term persistence of sage-grouse could be maintained by equilibria between colonizations and extinctions among independently functioning components. Alternately, source-sink models predict that individuals dispersing from larger, more productive regions would maintain satellite populations that are not self-sustaining because of low recruitment (Pulliam 1988). Under either hypothetical population model, connectivity and ability for spatially structured populations to exchange individuals are important conservation concerns; little is known about either phenomenon in Greater Sage-Grouse. If distance separating leks resulted from ecological forces (Keitt et al. 1997), the inflection of component number relative to dispersal distance suggests that dispersals up to 18 km occur with sufficient frequency to maintain exchanges of individuals within average constellations of leks within the network.

We do not know the extent to which gene flow is captured in the linkages (Oyler-McCance et al. 2005b), and modeling distance or multigenerational dispersal derived from genetic analysis may delineate components differently. Genetic evidence indicates that exchange of individuals has not been inhibited through the core regions (Oyler-McCance and Quinn, this volume, chapter 5). However, increasing isolation of components at the fringes of the sage-grouse range decreases the probability of dispersal to these regions. Sage-grouse in the state of Washington and in the Mono Lake region, which straddles the Nevada–California border, have been genetically isolated from other populations (Benedict et al. 2003, Oyler-McCance et al. 2005b). Our connectivity analysis suggests that isolation by distance, impermeable land cover, or topographical barriers could further increase the potential for loss of other population components.

Temporal Changes in Connectivity in Greater Sage-Grouse Populations

Contraction of the geographic range occupied by Greater Sage-Grouse (Schroeder et al. 2004) reflected decreasing connectivity within the spatial network of leks, leading to their isolation and loss. Range-wide, Greater Sage-Grouse populations declined at an annual rate of 3.5% from 1965 to 1985 and 0.4% from 1986 to 2003 (Connelly et al. 2004). Connectivity among sage-grouse populations in our lek cohort declined across intervals (1965–1974, 1980–1989, and 1998–2007) primarily because a lower proportion of surveyed leks were active and fewer male sage-grouse were counted at remaining leks. Sage-grouse in some locations may have moved to alternate leks following disturbance (Wallestad 1975b, Emmons and Braun 1984, Remington and Braun 1991), although this is unlikely for a large proportion of all leks abandoned within our cohort and in range-wide surveys (Connelly et al. 2004). Decreased indices of connectivity also reflected a greater loss of linkages between leks than expected on the basis of changes in maximum number of linkages possible in each interval.

Connectivity within SMZs tracked population trends and changes in lek characteristics, which generally were similar to range-wide estimates, although magnitude of change and temporal pattern varied (Connelly et al. 2004). However, decreases in connectivity despite more stable trend estimates (Connelly et al. 2004) indicate that isolation is continuing and is most severe within the Columbia Basin SMZ.

Lek connectivity (*dPC*) was a strong predictor of persistence to the next survey interval. Small decreases in lek connectivity resulted in large increases in probability of lek abandonment. Abandoned leks had fewer male sage-grouse than leks that persisted and had lower importance, reflecting position within the network, in the interval(s) prior to abandonment. However, connectivity among persisting leks declined because abandoned leks also tended to have more linkages with other leks.

Lek persistence was not strongly related to land cover of sagebrush or configuration. All active leks from 1965–1974 were in areas currently dominated by sagebrush. Sagebrush land cover dominated 70% of the surrounding landscape within 5 km of the lek and almost 50% within 54 km. Area burned and human-footprint score, which may estimate landscape quality from the perspective of sage-grouse, were the primary factors influencing fate of leks. Sage-grouse avoid burned areas in sagebrush landscapes because habitat characteristics important for nesting, brood concealment, and food are destroyed by fire and have slow recovery rates (Connelly et al. 2000b, Beck et al. 2009). Fire also facilitates invasion by cheatgrass and other nonnative plant species (Brooks et al. 2004a). Fires, prescribed and natural, have long-term effects (>10 years), and sage-grouse may continue to avoid burned areas even after sagebrush has recovered (Nelle et al. 2000). Frequent, large fires that resulted in loss of sagebrush were more likely to lead to extinction of simulated sage-grouse populations than small fires occurring at low frequencies across the landscape (Pedersen et al. 2003). In our analysis, small increases in the amount of area burned, particularly in the 54-km region surrounding a lek, had a large influence on the probability of lek abandonment.

Extensive conversion of sagebrush to agriculture within a landscape has decreased abundance of sage-grouse in many portions of their range (Swenson et al. 1987, Leonard et al. 2000, Smith et al. 2005, Aldridge and Boyce 2007, Aldridge et al. 2008). Negative influences on sage-grouse populations of other human disturbances, such as energy development, have also been documented

(Doherty et al. 2008; Naugle et al., this volume, chapter 20). The human-footprint score in our study assessed the physical and ecological effect of urbanization, infrastructure development (roads and power lines), agriculture, and energy development (Leu et al. 2008). We were unable to identify a specific source of human disturbance because the score represented a summed influence of all anthropogenic features. The cumulative effect of human activities may have a greater influence on persistence of sage-grouse populations than single land uses.

CONSERVATION IMPLICATIONS

Connectivity analysis provided a framework for quantifying the range-wide pattern of sage-grouse populations that integrated landscape arrangement of habitat and populations, population dynamics within components, and exchange of individuals among leks and components. Our analysis of spatial patterns in sage-grouse populations reflects processes such as dispersal and response to changes in their environment that can be incorporated into range-wide and regional conservation strategies.

The environmental matrix on which the network of sage-grouse leks were superimposed was also undergoing fragmentation, loss, and altered disturbance regimes (Knick et al. 2003, Wisdom et al. 2005b). Environmental factors, rather than stochastic events within the population, likely are the influences on population trends for Greater Sage-Grouse. Extinction currently is more probable than colonization for many sage-grouse components because of their low abundance and isolation coupled with fire (Pedersen et al. 2003; Miller et al., this volume, chapter 10; Baker, this volume, chapter 11) and human influence (Leu and Hanser, this volume, chapter 13). Population declines will track habitat loss or environmental changes, and extinctions occur when a species is unable to find suitable habitat within its dispersal distance (Thomas 1994). Thus, conservation strategies for species like sage-grouse should focus on conserving existing habitats, preserving large areas or connected networks of populations or habitat patches, and creating or restoring habitat within the species dispersal capabilities (Thomas 1994). Conserving smaller, more isolated components might depend on identifying or creating suitable habitat or connecting leks within 18 km that could function as intermediary islands or stepping-stones for migrating individuals.

Conservation of declining or endangered populations with limited resources involves assessing which populations or regions are critical to range-wide persistence. A primary concern in reviewing whether listing Greater Sage-Grouse under the United States' Endangered Species Act was warranted involved evaluating whether the species could persist 100 to 200 years into the future (United States Department of the Interior 2005b). Thus, it may be important to identify regions where Greater Sage-Grouse are likely to persist and whether we can focus conservation actions on specific regions or components to avoid global extinction. Our hierarchical analytic structure delineated leks nested within components, and components within the range-wide distribution. We then ranked each lek or component by a connectivity index (*dPC*) to prioritize its importance based on abundance of sage-grouse and location within the network. These rankings can suggest allocation of resources based on a relative measure of importance for maintaining a lek or component within the network. A strategy of no net loss (Stiver et al. 2006) may not be possible because altered fire regimes, spread of nonnative plants, climate change, and human land use present challenges to maintaining and restoring sagebrush habitats (Miller et al., this volume, chapter 10; Pyke, this volume, chapter 23).

The highest-ranked sage-grouse components for maintaining connectivity generally aligned with dominant patterns of sagebrush distribution. However, the variables that we used to represent sagebrush cover and fragmentation in the landscape did not influence lek persistence. Rather, landscape disturbance, measured by amount of fire since 1965 and level of human activity, was the primary factor affecting lek persistence. Of the major components delineated by connectivity, human land use in the form of energy development is high for the Powder River, south-central, and southwestern Wyoming. Fire and conversion of sagebrush to cheatgrass-dominated landscapes is a concern for important sage-grouse components in the western part of their range. Our results suggest that restoration of sagebrush will not be as successful in increasing the viability of sage-grouse populations long-term if those areas also are heavily influenced by human activities or fire.

We do not fully understand whether results from this cohort of leks are transferable to all leks within the sage-grouse range. We recommend that connectivity be monitored not only by counting sage-grouse on currently identified leks but also by conducting spatially extensive surveys to detect newly established leks, reuse of traditional locations, or relocation to new sites (Connelly et al. 2003b).

Currently, more than 5,000 leks are surveyed each year across the range of Greater Sage-Grouse. Tracking changes in connectivity of these leks can complement trend estimates and provide valuable information for conserving sage-grouse. The number of populations and active leks may further decrease in many of the SMZs because fire continues to be a dominant disturbance across much of the sage-grouse range (Baker, this volume, chapter 11). Human land use is also pervasive, and broad-scale creation of infrastructure (highways, transmission corridors, etc.) coupled with local-scale energy development (Leu et al. 2008; Knick et al., this volume, chapter 12) are likely to have strong influence at multiple scales on sagebrush landscapes that will be reflected by changes in sage-grouse populations and their connectivity.

ACKNOWLEDGMENTS

The Wilburforce Foundation and the project *Analysis and conservation prioritization of landscape connectivity in Nevada* provided the impetus for our application of connectivity concepts to Greater Sage-Grouse populations. We appreciate insights by C. W. Carroll, D. S. Dobkin, E. Fleishman, M. Leu, B. H. McRae, and N. H. Schumaker in developing this study. We consulted with B. S. Cade and L. Bond for statistical assistance. We thank the Western Association of Fish and Wildlife Agencies and members of the Western Sage and Sharp-tailed Grouse Technical Committee for sharing data on lek counts and location. J. W. Connelly critiqued our presentation of sage-grouse ecology. J. A. Bissonette, D. S. Dobkin, E. Fleishman, M. Leu, N. H. Schumaker; students and faculty in the College of Natural Resources, University of Idaho; and the Colorado Division of Wildlife reviewed the manuscript. Any use of trade, product, or firm names is for descriptive purposes only and does not imply endorsement by the United States Government.

CHAPTER SEVENTEEN

Influences of Environmental and Anthropogenic Features on Greater Sage-Grouse Populations, 1997–2007

Douglas H. Johnson, Matthew J. Holloran, John W. Connelly, Steven E. Hanser, Courtney L. Amundson, and Steven T. Knick

Abstract. The Greater Sage-Grouse (*Centrocercus urophasianus*), endemic to western North America, is of great conservation interest. Its populations are tracked by spring counts of males at lek sites. We explored the relations between trends of Greater Sage-Grouse lek counts from 1997 to 2007 and a variety of natural and anthropogenic features. We found that trends were correlated with several habitat features, but not always similarly throughout the range. Lek trends were positively associated with proportion of sagebrush (*Artemisia* spp.) cover, within 5 km and 18 km. Lek trends had negative associations with the coverage of agriculture and exotic plant species. Trends also tended to be lower for leks where a greater proportion of their surrounding landscape had been burned. Few leks were located within 5 km of developed land and trends were lower for those leks with more developed land within 5 km or 18 km. Lek trends were reduced where communication towers were nearby, whereas no effect of power lines was detected. Active oil or natural gas wells and highways, but not secondary roads, were associated with lower trends. Effects of some anthropogenic features may have already been manifested before our study period and thus not have been detected in this analysis. Results of this range-wide analysis complement those from more intensive studies on smaller areas. Our findings are important for identifying features that could threaten Greater Sage-Grouse populations.

Key Words: Centrocercus urophasianus, communication towers, Greater Sage-Grouse, lek, oil and gas wells, population, roads, sagebrush.

Influencias de Características Ambientales y Antropogénicas en Poblaciones de *Centrocercus urophasianus*, 1997–2007

Resumen. El Greater Sage-Grouse (*Centrocercus urophasianus*) es endémico a Norteamérica occidental y de gran interés para la conservación. Sus poblaciones son monitoreadas mediante conteos de los machos en los sitios donde se encuentran los leks (asambleas de cortejo) en la primavera. Exploramos las relaciones entre las tendencias de conteos de leks de Greater Sage-Grouse durante 1997–2007 y una variedad de características naturales y antropogénicas. Encontramos que las tendencias estaban correlacionadas con varias características del hábitat, pero no siempre de forma similar en todo el territorio. Las tendencias de los leks estuvieron positivamente asociadas a la

Johnson, D. H., M. J. Holloran, J. W. Connelly, S. E. Hanser, C. L. Amundson, and S. T. Knick. 2011. Influences of environmental and anthropogenic features on Greater Sage-Grouse populations, 1997–2007. Pp. 407–450 *in* S. T. Knick and J. W. Connelly (editors). Greater Sage-Grouse: ecology and conservation of a landscape species and its habitats. Studies in Avian Biology (vol. 38), University of California Press, Berkeley, CA.

proporción de cubierta de sagebrush (*Artemisia* spp.), entre los 5 km y 18 km. Las tendencias de los leks tuvieron asociaciones negativas con la cubierta vegetal agrícola y de especies de plantas exóticas. Las tendencias también tendieron a ser más bajas para los leks donde una mayor proporción del paisaje circundante había sido quemado. Se encontraron pocos leks situados a menos de 5 km de áreas desarrolladas, y las tendencias fueron más bajas para esos leks con áreas desarrolladas dentro de los 5 km y 18 km. Las tendencias de los leks fueron reducidas en sitios próximos a las torres de comunicación, mientras que no se detectó ningún efecto de las líneas eléctricas. Los pozos y carreteras activos utilizados para el transporte de petróleo y gas natural estuvieron asociados a tendencias más bajas, pero no se observó esta tendencia en carreteras o caminos secundarios. Los efectos de algunas características antropogénicas pudieron haberse manifestado en forma previa a nuestro período de estudio, y por ende no haber sido detectados en este análisis. Los resultados de este análisis a escala regional complementan ésos de estudios más intensivos en áreas más pequeñas. Nuestros resultados son importantes para identificar las características que podrían amenazar a poblaciones del Greater Sage-Grouse.

Palabras Clave: artemisa, carreteras, *Centrocercus urophasianus,* Greater Sage-Grouse, lek, población, pozos de petróleo y gas, torres de comunicación.

The Greater Sage-Grouse (*Centrocercus urophasianus*) is a species of major conservation interest. It is endemic to sagebrush (*Artemisia* spp.) ecosystems of western North America, is highly reliant on sagebrush for food and cover, and is declining in numbers in many areas. It is considered an umbrella species for several other sagebrush-obligate species (Rowland et al. 2006; Hanser and Knick, this volume, chapter 19). Many populations and habitats of Greater Sage-Grouse have been exposed in recent years to increasing levels of anthropogenic activities, such as oil and gas development, construction of communication towers, traffic on roads and highways, invasion by exotic plant species such as cheatgrass (*Bromus tectorum*), altered fire frequency (Young and Evans 1973, d'Antonio and Vitousek 1992), and invasion by pinyon (*Pinus* spp.) and juniper (*Juniperus* spp.) into sagebrush habitats (Miller and Tausch 2001; Weisberg et al. 2007; Miller et al., this volume, chapter 10).

Our goal was to gain insight into how Greater Sage-Grouse populations have responded recently to these potential stressors. Specifically, our objective was to relate recent trends in counts of sage-grouse males at leks to a variety of natural and anthropogenic features at the leks and in their vicinity (within 5- and 18-km radii). We examined counts of males at leks throughout the United States recorded from 1997 to 2007. This time period was chosen because of the higher quality and greater consistency of recent surveys compared with earlier surveys. We had information on 9,844 leks in Idaho, Washington, Oregon, Nevada, Utah, California, Colorado, Wyoming, Montana, North Dakota, and South Dakota, but counts on many of the leks were not made in some years.

The term lek is regularly used with two different meanings, as either a group of displaying birds or the sites occupied by those birds (Johnson and Rowland 2007). Here we mean a lek to be the site, not the group. Otherwise, a count of zero would be impossible, and extinctions of leks could not happen.

STUDY AREA

We analyzed lek data from within the Sage-Grouse Conservation Area (Connelly et al. 2004). This area includes the pre-settlement distribution of Greater Sage-Grouse (Schroeder et al. 2004) buffered by 50 km. The total study area comprises all or parts of 14 states and three provinces and encompasses about 2,063,000 km^2 (Connelly et al. 2004).

The study area includes Küchler's (1970) three sagebrush vegetation types as well as a portion of the northern Great Plains, which also supports stands of sagebrush. Sagebrush once was dominant across much of the nonforested parts of this region, but many areas now contain only islands of sagebrush within larger expanses of highly altered landscapes (Leu and Hanser, this volume, chapter 13). The study area has been divided into seven Sage-Grouse Management Zones (SMZs; Fig. 17.1), which had been developed by grouping sage-grouse populations within corresponding

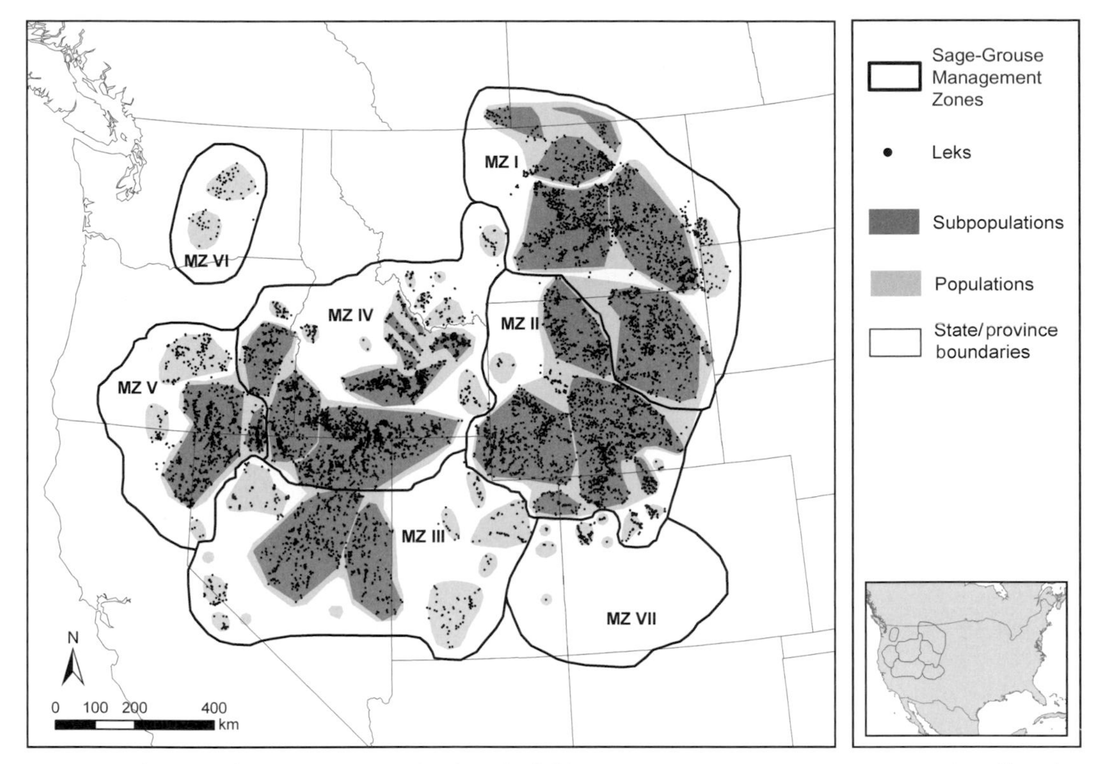

Figure 17.1. The range of Greater Sage-Grouse has been divided into seven Sage-Grouse Management Zones (Connelly et al. 2004, Stiver et al. 2006). Names of management zones are given in Table 17.1.

floristic provinces (Connelly et al. 2004, Stiver et al. 2006). Floristic provinces encompass contiguous areas with similar ecological and biological features. Areas occupied by sage-grouse within these floristic provinces are dominated by sagebrush. The Great Plains, Wyoming Basin, Snake River Plain, and Northern Great Basin SMZs encompass core populations of sage-grouse (Connelly et al. 2004). The Southern Great Basin SMZ includes scattered and often isolated populations. The Columbia Basin SMZ includes sage-grouse in the state of Washington, and the Colorado Plateau SMZ includes small, isolated populations of Greater Sage-Grouse and Gunnison Sage-Grouse (*Centrocercus minimus*) in Utah and Colorado.

METHODS

Counts of male Greater Sage-Grouse at leks were obtained from state wildlife agencies. Methods for gathering these data vary somewhat among agencies and sometimes within agencies among years (Connelly et al. 2004), but generally follow the same basic protocol. In brief, leks are visited multiple times during the breeding season (March–May) between 0.5 hour before sunrise and 1.5 hours after sunrise. The maximum number of males counted in a single visit is used as the count for the year (Connelly et al. 2003b).

Response Variable

We used the Pearson correlation coefficient between lek count (maximum number of males observed) and year for the period 1997–2007 as a measure of change in the number of males observed on leks. We refer to this correlation coefficient as lek trend, which can range from −1, indicating a consistent decline in lek count, to +1, indicating a consistent increase in lek count. We seek to examine how lek trend is related to several explanatory variables. Only the 3,679 leks with at least four annual counts during the 11-year period were included. To account for differing numbers of years among leks and corresponding differences in the precision of trend estimates, observations in the analysis were weighted by the square root of the number of lek counts minus one.

The lek trend is affected both by the magnitude of the trend and by the variation around that trend. Both characteristics are useful in this analysis, the former because it reflects the number of birds involved in the change, and the latter because it reflects the consistency of the change. For other analyses, different metrics, such as slope of the regression line, would be more appropriate. The lek trend captures only the linear component of the change in lek count. However, lek counts did not vary linearly, or even monotonically, with year. If an independently identified cyclicity in lek counts existed, other metrics might better describe trends. Although speculation has been made about such cycles in sage-grouse, peer-reviewed literature does not support it and no causal mechanisms have been identified.

Log transformations often confer some benefits when analyzing data, such as counts whose values are always nonnegative (Weisberg 2005). To determine if log transformations would be advantageous, we compared the strength of association between year and lek count and between year and log of lek count on all leks having at least seven years with counts. The correlation between count and year was stronger than one between log count and year, so the log-transformed variable was not used as the response variable.

Explanatory Variables

A large number of geospatial variables could potentially explain sage-grouse trends at each lek location. Certain variables were measured at the lek site itself; others reflected conditions in circles of 5- and 18-km radii centered at the lek. These two radii were selected because they correspond to the distances around lek locations recommended for management of nonmigratory (5 km) and migratory (18 km) populations of Greater Sage-Grouse (Connelly et al. 2000c).

Among these variables were measures of physical attributes, vegetation (land cover), fire history, and anthropogenic features. The physical attributes included the latitude and longitude, slope, elevation in meters, and average total annual precipitation in millimeters from 1971 to 2000 (Doggett et al. 2004). Land cover information was derived from the LANDFIRE Existing Vegetation Map (LANDFIRE 2006). Vegetation in LANDFIRE was classified from Landsat Thematic Mapper satellite imagery taken in 2002 (± 2 years). We developed eight land cover classes from the original 210 ecological systems mapped by LANDFIRE: tall sage (includes big sagebrush shrubland, big sagebrush steppe, and mountain sagebrush classes), low sage (includes low sagebrush shrubland and steppe), all sage (all sagebrush classes combined), grassland, uncultivated exotic vegetation, riparian, agriculture (cultivated crops, pasture, and hayland); and developed (urban and suburban areas, and interstate and state highways) (Leu and Hanser, this volume, chapter 13). For each land-cover class, we calculated the proportional area within the 5- and 18-km radii using Fragstats (McGarigal et al. 2002).

We analyzed the fire history of each lek using a spatial data set of annual fire perimeters from 1980 to 2007. We calculated the area burned each year within 5 km and 18 km of each lek. Thus, we needed to determine how to incorporate fire, which was recorded annually, in an analysis of population trend, which had but one value for the entire 1997–2007 period. After a separate analysis, we concluded that effects of fire could be expressed as the area burned (within each 5- and 18-km radius) and the number of years in the period that had been preceded by a fire. Because virtually all fires occurred later in the year than when lek counts were made, a fire was associated with the year after it actually occurred. For example, a fire occurring in 1999 would be assumed to first affect a lek count in 2000, and the number of years attributed to the fire would be eight (2000, 2001, . . . 2007). We did not know which specific land cover types were involved in the burns.

We calculated several measures associated with anthropogenic features in the area surrounding each lek location. For three road categories (interstate highways, state or federal highways, and secondary roads), pipelines, and power lines (transmission and distribution lines combined), we computed the lengths within 5 km and 18 km of the lek and the distance from the lek to the nearest feature. Road categories were investigated separately, and we further combined interstate, state, and federal highways and all road types as two additional categories. For point features such as communication towers and oil and gas wells, we computed the number of towers and wells within 5 km and 18 km of the lek and the distance from the lek to the nearest tower or well. Oil and gas wells were categorized by their status (active, pending, temporarily abandoned, abandoned, other, or unknown), as reported by individual states. We investigated active wells and the sum of all wells. A final anthropomorphic metric was the human footprint (Leu et al. 2008), a derived index that combines 14 landscape structural and anthropogenic features: human habitation, interstate highways, federal and state highways, secondary roads, railroads, irrigation canals, power lines, linear feature densities, agricultural lands, campgrounds, highway rest stops, landfills, oil and gas developments, and human-induced fires. Values of human footprint ranged from 0 to 10 and were a cumulative measure of anthropogenic features. Human footprint was measured at a lek site; in addition, median values within 5 km and 18 km of the lek were included.

Many of the factors we considered were in place well before our study period, so any immediate effects they may have had on sage-grouse had occurred earlier. Accordingly, our analysis might not show that these factors currently have an effect. Also, estimated effects might pertain only to recent changes associated with a factor, such as increased traffic on roads.

Many of the explanatory variables had extremely wide ranges in value. We therefore employed a logarithmic transformation ($\log_{10} x + 1$) for length and distance measurements (Weisberg 2005:150).

We took two approaches to analyzing the data: single variable and multivariable. The single-variable methodology was mainly descriptive. Because of the large data set and strong possibility of nonlinear responses to explanatory variables, we used a locally weighted scatterplot smoother (loess) to examine relations between lek trend and the explanatory variables. Unless noted otherwise, we used the default smoothing options in SAS PROC LOESS (Cohen 1999), which is based on an underlying model selected to provide an optimal fit. Each graph shows the loess best-fitting line and upper and lower 95% confidence bounds, as well as the actual data. The value of the smoothing parameter is given in the legend. We recognize, however, that if explanatory variables (e.g., x_1 and x_2) are correlated among themselves, an apparent association between the response variable and x_1 might actually be due to the association between the response variable and x_2. Also, to the extent that data from individual leks are spatially autocorrelated, effective sample sizes would be less than indicated, and the confidence intervals shown might have less than the nominal 95% coverage. Regardless, we focused on the pattern shown rather than the confidence limits. Space limitations preclude presentation of all loess curves examined. We did not include these figures if no relation was apparent, sample sizes were small, or the relation was similar to others that were presented.

Patterns may be stronger than they appear from some of the loess curves. The figures show a wide range in observed values of lek trend; most figures have values from nearly −1 to nearly +1. The figures do not, however, indicate the weights that observations were given when constructing the loess curves. Most of the extreme values of lek trend arose from leks with few years of surveys. In fact, 19% of leks for which trends were based on only four years had extreme trend values, either greater than 0.9 or less than −0.9. In contrast, fewer than 4% of lek trends based on seven or more years of counts were extreme. And the latter leks were weighted more heavily in the analysis than the former.

For the multivariable approach, we applied three methods for selecting variables in a regression setting: stepwise regression, least angle regression (Efron et al. 2004), and LASSO regression (Tibshirani 1996). Analyses were done by individual SMZ as well as across all SMZs. As it turned out, each method selected at most one explanatory variable other than the intercept. For this reason and because of the simplicity of interpreting results, we report single-variable results only.

RESULTS

Patterns in Lek Counts

Somewhat more leks (56%) had increasing population trends from 1997 to 2007 than decreasing (44%). The mean trend (correlation between lek count and year) was 0.065 (SE = 0.010); the median trend was 0.141. The average count increased for the 854 leks with complete counts from 1997 to 2007, although not monotonically (Fig. 17.2). Average lek trends by SMZ ranged from −0.279 (Columbia Basin, only 24 leks) to 0.173 (Wyoming Basin, 1,321 leks; Table 17.1).

Physical Attributes

Average elevation within 5 km of leks varied among management zones (Table 17.1), as did observed relations between lek trends and elevation

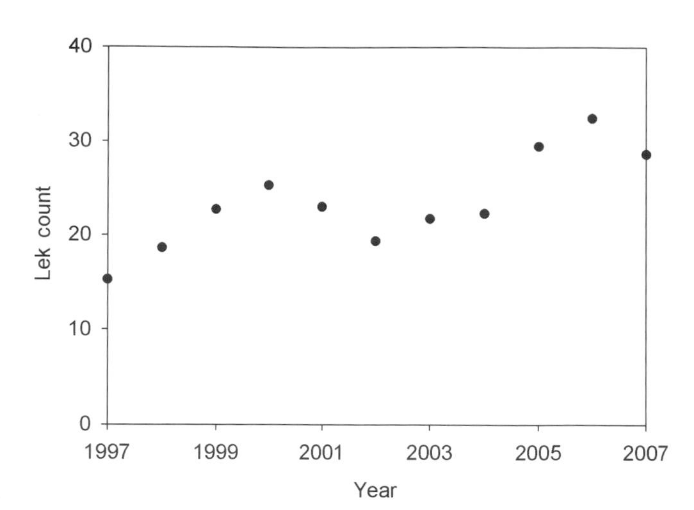

Figure 17.2. Average counts during 1997–2007 for 854 leks that had complete records for that period, across all management zones.

(Table 17.2). Relations in all SMZs but the Great Plains were positive. The Wyoming Basin appeared to have a peak near 2,100 m (Fig. 17.3), but in the other SMZs patterns continued upward (e.g., Northern Great Basin; Fig. 17.4).

We found no relation between lek trends and average annual precipitation (Table 17.2). Our metric was too general to capture any short-term effects due to either drought or deluge. Extremes of drought and deluge, especially on a more local scale, likely have more important consequences to sage-grouse reproduction and survival than does a simple long-term average.

Land Cover

Sagebrush

Sagebrush was common in all SMZs, of course, but least common in the Colorado Plateau, Great Plains, and Columbia Basin and most common in the Northern Great Basin (Table 17.1). The type of sagebrush cover also varied by SMZ. Of note is that the average percentage cover of sagebrush within 5 km exceeded the average within 18 km (Table 17.1), suggesting that sage-grouse leks are preferentially located within sagebrush. Lek trends across all SMZs increased modestly but steadily with the cover of tall sagebrush (combined categories for big sagebrush shrubland, big sagebrush steppe, mountain big sagebrush) at both 5-km (Fig. 17.5) and 18-km (Fig. 17.6) radii (Table 17.2). Patterns within individual SMZs generally were similar, except in the Columbia Basin and the Colorado Plateau, where data were minimal. Similarly, lek trends across all SMZs increased modestly but steadily with the cover of all sagebrush (combined categories for tall sage and low sagebrush) at both 5-km (Fig. 17.7) and 18-km (Fig. 17.8) radii (Table 17.2).

Low sagebrush was present in all SMZs but the Great Plains. There was no consistent pattern for all SMZs combined (Fig. 17.9), and results for 5- and 18-km radii were similar. Low sagebrush was fairly common in the Northern Great Basin, where trends in lek counts generally increased until the cover of low sagebrush was about 40% of the landscape, after which trends were roughly steady (Fig. 17.10).

Grassland

Grassland was common only in the Great Plains, where a slight positive relation could be seen (Fig. 17.11). The Columbia Basin also contained large areas of grassland but had too few leks to draw any conclusions.

Exotic Vegetation

Average cover of exotic vegetation was about 2–3% for all SMZs except the Great Plains and Colorado Plateau, where it was lower (Table 17.1). Lek trends tended to be lower as the cover of exotic species increased at both 5-km (Fig. 17.12) and 18-km scales, although few leks had >8% coverage within either 5 km or 18 km.

TABLE 17.1

Average values of lek count trend, environmental features, and anthropogenic variables, by Sage-Grouse Management Zone (SMZ).

Lek characteristics	Buffer[a]	Great Plains SMZ I	Wyoming Basin SMZ II	Southern Great Basin SMZ III	Snake River Plain SMZ IV	Northern Great Basin SMZ V	Columbia Basin SMZ VI	Colorado Plateau SMZ VII
Number of leks		804	1,321	374	813	310	24	33
Trend		−0.065	0.173	0.050	0.064	−0.018	−0.279	0.115
Survey yr (N)		6.9	8.3	7.5	7.6	7.4	9.8	5.0
				Physical setting				
Elevation (m)	5	1,075.2	2,025.8	2,055.9	1,653.5	1,631.5	684.8	2,303.4
	18	1,086.3	2,042.7	2,078.8	1,696.4	1,594.2	660.0	2,257.0
Slope (degrees)	Lek	0.0	2.7	4.6	3.2	2.6	1.9	8.4
Precipitation (mm)	Lek	347.5	339.6	360.9	379.0	335.9	268.8	471.2
Longitude (degrees)	Lek	−106.8	−108.4	−115.4	−114.4	−119.6	−119.8	−108.5
Latitude (degrees)	Lek	46.0	42.2	39.4	43.0	42.4	47.3	39.7
				Land cover (%)				
Big sagebrush shrubland	5	2.0	52.2	31.1	29.3	20.7	3.9	22.7
	18	1.8	44.4	21.9	24.2	26.1	5.2	17.1
Big sagebrush steppe	5	23.0	12.3	4.6	27.8	13.3	16.3	0.1
	18	20.7	10.8	3.5	21.1	12.4	14.8	0.1
Low sagebrush	5	0.0	1.7	24.9	11.5	45.6	7.7	7.8
	18	0.0	1.5	16.1	9.0	30.2	6.4	4.1

TABLE 17.1 *(continued)*

TABLE 17.1 (CONTINUED)

Lek characteristics	Buffer[a]	Great Plains SMZ I	Wyoming Basin SMZ II	Southern Great Basin SMZ III	Snake River Plain SMZ IV	Northern Great Basin SMZ V	Columbia Basin SMZ VI	Colorado Plateau SMZ VII
Exotic	5	0.4	1.9	1.7	2.4	1.2	1.5	0.7
	18	0.6	2.4	3.1	2.3	3.1	2.6	0.8
Grassland	5	52.8	2.6	0.5	2.7	1.2	42.4	0.5
	18	49.5	3.3	0.8	3.0	1.4	34.5	0.3
Mountain big sagebrush	5	0.9	11.1	5.8	11.7	6.2	0.3	5.7
	18	0.8	10.4	5.1	15.8	5.8	0.3	4.6
Riparian	5	2.9	3.4	1.0	2.5	1.2	0.2	2.5
	18	3.3	3.7	1.3	2.6	1.2	0.3	2.5
Agriculture	5	5.9	2.8	0.6	3.6	0.6	24.1	0.5
	18	8.5	4.1	1.1	6.0	1.4	30.3	0.5
Developed	5	0.3	0.4	0.7	0.3	0.2	1.7	0.1
	18	0.6	0.7	0.7	0.6	0.4	2.8	0.2
Fire characteristics								
Fires (N)	5	0.1	0.2	0.5	0.7	0.7	0.2	0.4
	18	0.9	1.3	2.5	3.5	3.2	1.4	2.8
Fire yr (N)	5	0.8	1.4	3.3	7.3	3.8	3.0	1.5
	18	4.4	5.2	8.3	10.1	9.6	5.2	7.6
Burned area (ha)	5	0.9	1.2	4.2	21.7	5.6	4.2	0.6
	18	16.2	16.3	68.8	248.1	82.6	34.4	12.2

Anthropogenic features								
Distance to interstate hwy (km)		89.3	72.6	101.3	51.6	174.8	49.3	39.6
Length interstate hwy (km)	5	0.2	0.1	0.1	0.1	0.0	0.6	0.0
	18	5.8	3.0	2.9	4.5	0.1	8.7	0.9
Distance to state/federal hwy (km)		11.1	8.9	11.9	10.5	15.5	6.8	23.4
Length state/federal hwy (km)	5	2.2	2.9	3.3	3.2	1.8	3.9	0.0
	18	38.5	49.5	40.3	38.7	26.5	77.3	9.1
Distance to secondary (km)		0.7	0.3	0.4	0.5	0.9	0.5	0.3
Length secondary (km)	5	51.7	75.4	55.1	56.2	43.0	72.0	59.5
	18	703.8	954.4	666.4	669.3	624.9	927.4	741.8
Minimum distance to interstate hwy (km)		11.0	8.7	11.8	10.1	15.5	6.3	21.8
Length of interstate and other hwys (km)	5	2.4	2.9	3.4	3.3	1.8	4.6	0.0
	18	44.3	52.5	43.1	43.1	26.6	86.0	10.0
Minimum distance to any road (km)		0.7	0.3	0.4	0.5	0.9	0.5	0.3
Length of all roads (km)	5	54.1	78.4	58.5	59.5	44.8	76.5	59.5
	18	748.1	1,006.9	709.6	712.4	651.4	1,013.4	751.7
Distance to pipeline (km)		19.6	8.3	78.3	21.6	86.4	42.8	8.3
Length pipeline (km)	5	7.0	14.4	1.5	1.2	0.1	0.0	7.0
	18	101.3	199.7	16.3	18.1	3.2	3.8	99.2
Distance to power line (km)		26.7	9.5	19.9	12.5	22.4	8.2	9.2
Length power line (km)	5	1.9	4.2	1.5	2.3	1.1	3.3	3.3
	18	29.4	65.3	26.9	33.9	18.3	71.5	51.5

TABLE 17.1 *(continued)*

TABLE 17.1 (CONTINUED)

Lek characteristics	Buffer[a]	Great Plains SMZ I	Wyoming Basin SMZ II	Southern Great Basin SMZ III	Snake River Plain SMZ IV	Northern Great Basin SMZ V	Columbia Basin SMZ VI	Colorado Plateau SMZ VII
Distance to oil/gas well (km)		2.7	3.2	78.9	73.8	50.0	233.2	1.4
Oil/gas well (N)	5	47.4	25.1	2.4	0.0	0.0	0.0	47.1
	18	676.6	293.7	28.9	0.3	0.2	0.0	894.3
Distance to producing well (km)		9.0	10.4	98.2	149.4	73.8	233.2	2.5
Producing wells (N)	5	19.4	11.9	1.2	0.0	0.0	0.0	30.3
	18	313.2	136.9	14.1	0.0	0.1	0.0	631.3
Distance to communication towers (km)		19.6	15.6	25.3	19.3	34.0	13.7	25.1
Communication towers (N)	5	0.1	0.1	0.0	0.1	0.0	0.0	0.0
	18	3.1	3.4	1.9	2.4	0.4	9.8	0.4
Human footprint[b]	Lek	3.7	4.3	3.9	4.4	3.0	6.3	4.2
	5	3.3	3.8	3.4	3.9	2.7	6.4	3.2
	18	3.4	3.9	3.3	3.8	2.9	6.6	3.1

[a] Values of environmental variables were estimated for the grid cell in which leks were located, and for 5- and 18-km buffer zones surrounding the lek.

[b] The human footprint is a cumulative score ranging from 0 (no effect) to 10 (maximum effect) derived from multiple submodels quantifying anthropogenic effects (Leu et al. 2008).

TABLE 17.2

Pearson correlation coefficients of lek count trend with environmental features and anthropogenic variables, by Sage-Grouse Management Zone (SMZ).

Lek characteristics	Buffer[a]	Great Plains SMZ I	Wyoming Basin SMZ II	Southern Great Basin SMZ III	Snake River Plain SMZ IV	Northern Great Basin SMZ V	Columbia Basin SMZ VI	Colorado Plateau SMZ VII
Number of leks		804	1,321	374	813	310	24	33
Physical Setting								
Elevation (m)	5	0.044	0.052	0.110	0.134	0.225	0.081	0.397
	18	0.053	0.047	0.121	0.152	0.222	0.292	0.404
Slope (degrees)	Lek	0.067	0.022	0.027	0.054	0.095	−0.122	−0.029
Precipitation (mm)	Lek	0.180	−0.026	0.086	0.001	−0.057	0.223	0.390
Longitude (degrees)	Lek	0.111	0.066	0.023	0.091	0.133	0.108	0.284
Latitude (degrees)	Lek	−0.103	0.011	−0.044	0.042	−0.033	0.259	−0.372
Land cover (%)								
Big sagebrush shrubland	5	−0.133	−0.034	0.062	0.035	−0.053	−0.183	−0.479
	18	−0.146	−0.007	0.054	−0.009	0.162	−0.287	−0.410
Big sagebrush steppe	5	0.102	0.097	0.073	0.006	−0.146	−0.053	−0.063
	18	0.112	0.134	0.084	0.001	−0.157	−0.181	0.452
Low sagebrush	5	0.010	0.034	−0.010	0.067	0.170	−0.167	−0.114
	18	0.052	0.067	0.053	0.062	0.122	−0.310	−0.032
Exotic	5	0.036	−0.064	−0.182	−0.113	−0.142	−0.264	−0.358
	18	0.078	−0.037	−0.148	−0.191	−0.031	−0.254	−0.366

TABLE 17.2 *(continued)*

TABLE 17.2 (CONTINUED)

Lek characteristics	Buffer[a]	Great Plains SMZ I	Wyoming Basin SMZ II	Southern Great Basin SMZ III	Snake River Plain SMZ IV	Northern Great Basin SMZ V	Columbia Basin SMZ VI	Colorado Plateau SMZ VII
Grassland	5	−0.074	−0.014	−0.087	−0.076	−0.085	−0.187	0.224
	18	−0.102	0.002	0.000	−0.072	−0.031	−0.214	0.112
Mountain big sagebrush	5	−0.025	0.047	0.028	0.019	0.121	−0.112	0.334
	18	−0.102	0.002	0.000	−0.072	−0.031	−0.214	0.112
Riparian	5	−0.059	0.053	0.004	−0.056	0.045	0.064	0.458
	18	−0.022	0.044	0.085	−0.068	−0.012	−0.025	0.330
Agriculture	5	−0.012	−0.096	0.009	−0.112	−0.113	0.387	−0.139
	18	−0.030	−0.105	−0.008	−0.103	−0.105	0.362	−0.233
Developed	5	0.128	−0.125	0.042	−0.106	−0.093	0.378	−0.062
	18	0.169	−0.112	0.004	−0.085	−0.147	0.128	−0.192
Fire characteristics								
Fires (N)	5	0.067	0.022	0.027	0.054	0.095	−0.122	−0.029
	18	0.130	−0.016	−0.002	−0.089	−0.028	−0.229	0.099
Fire yr (N)	5	0.045	−0.042	−0.032	−0.042	0.078	−0.267	−0.048
	18	0.097	−0.038	−0.031	0.016	0.055	−0.342	0.185
Burned area (ha)	5	0.051	−0.007	−0.131	−0.042	−0.020	−0.006	0.144
	18	0.173	0.031	−0.112	−0.022	−0.014	−0.219	0.346

Anthropogenic features								
Distance to interstate hwy (km)		−0.201	−0.054	−0.054	0.043	0.142	0.246	−0.399
Length interstate hwy (km)	5	0.045	−0.016	−0.011	−0.030	0.000	0.080	0.000
	18	0.136	0.011	−0.092	0.020	−0.088	−0.199	−0.011
Distance to state/federal hwy (km)		−0.134	0.154	−0.155	−0.023	0.119	−0.262	0.015
Length state/federal hwy (km)	5	0.106	−0.121	0.063	−0.008	−0.045	0.267	0.000
	18	0.212	−0.160	0.040	−0.039	−0.097	0.168	−0.285
Distance to secondary (km)		−0.076	−0.031	0.046	0.029	0.067	0.206	−0.324
Length secondary (km)	5	0.113	0.035	−0.124	0.066	−0.167	−0.054	0.183
	18	0.122	0.083	−0.143	−0.023	−0.215	0.033	0.207
Distance to interstate or other hwy (km)		−0.133	0.153	−0.153	−0.031	0.119	−0.196	−0.025
Length of interstate and other hwys (km)	5	0.111	−0.122	0.060	−0.013	−0.045	0.291	0.000
	18	0.225	−0.150	0.013	−0.029	−0.101	0.052	−0.287
Minimum distance to any road (km)		−0.079	−0.031	0.034	0.027	0.067	0.206	−0.324
Length of all roads (km)	5	0.130	0.011	−0.103	0.064	−0.171	−0.006	0.183
	18	0.144	0.059	−0.130	−0.026	−0.219	0.038	0.159
Distance to pipeline (km)		−0.127	0.049	−0.044	−0.033	0.343	0.115	0.172
Length pipeline (km)	5	0.103	−0.096	−0.010	−0.047	−0.020	0.000	−0.103
	18	0.106	−0.073	−0.050	−0.068	−0.156	−0.104	−0.290
Distance to power line (km)		−0.135	0.009	−0.021	0.018	0.166	0.075	0.107
Length power line (km)	5	0.127	−0.018	0.029	0.042	−0.001	0.107	−0.043
	18	0.136	−0.027	−0.066	−0.035	−0.168	−0.205	−0.155

TABLE 17.2 *(continued)*

TABLE 17.2 (CONTINUED)

Lek characteristics	Buffer[a]	Great Plains SMZ I	Wyoming Basin SMZ II	Southern Great Basin SMZ III	Snake River Plain SMZ IV	Northern Great Basin SMZ V	Columbia Basin SMZ VI	Colorado Plateau SMZ VII
Distance to oil/gas well (km)		−0.017	0.048	−0.021	−0.029	0.254	0.251	0.532
Oil/gas well (N)	5	0.111	−0.165	−0.054	−0.001	−0.085	0.000	−0.280
	18	0.090	−0.182	−0.042	0.020	0.040	0.000	−0.198
Distance to producing well (km)		−0.058	0.052	0.000	0.007	0.060	0.251	0.253
Producing wells (N)	5	0.091	−0.172	−0.058	0.001	−0.085	0.000	−0.161
	18	0.076	−0.170	−0.038	−0.025	0.020	0.000	−0.058
Distance to communication towers (km)		−0.145	0.044	0.011	0.010	0.207	0.111	0.357
Communication towers (N)	5	0.111	−0.096	0.008	0.012	−0.066	−0.125	0.000
	18	0.167	−0.046	−0.077	−0.035	−0.103	−0.105	−0.412
Human footprint[b]	Lek	0.125	−0.059	−0.025	−0.042	−0.177	0.218	0.210
	5	0.161	−0.098	0.014	−0.057	−0.133	0.316	0.007
	18	0.141	−0.083	0.026	−0.033	−0.126	0.393	0.076

[a] Values of environmental variables were estimated for the grid cell in which leks were located, and for 5- and 18-km buffer zones surrounding the lek.

[b] The human footprint is a cumulative score ranging from 0 (no effect) to 10 (maximum) derived from multiple submodels quantifying anthropogenic effects (Leu et al. 2008).

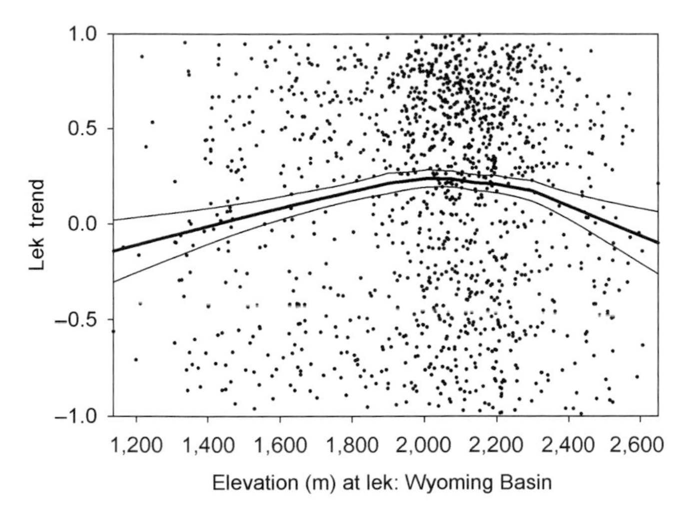

Figure 17.3. Relation between trend of lek counts and elevation for Wyoming Basin Sage-Grouse Management Zone (smooth = 0.625).

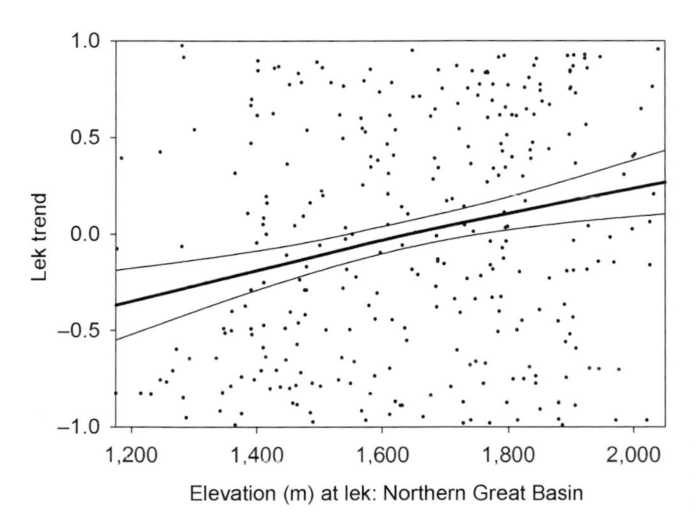

Figure 17.4. Relation between trend of lek counts and elevation for Northern Great Basin Sage-Grouse Management Zone (smooth = 1.00).

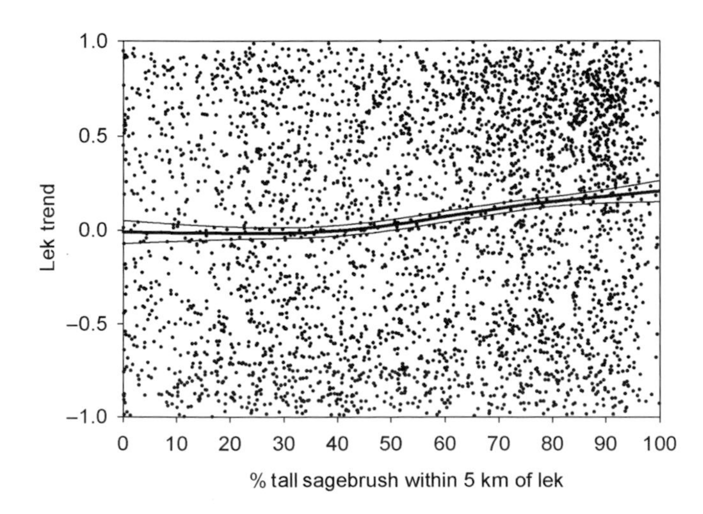

Figure 17.5. Relation between trend of lek counts and tall sagebrush within 5 km, all management zones combined (smooth = 0.623).

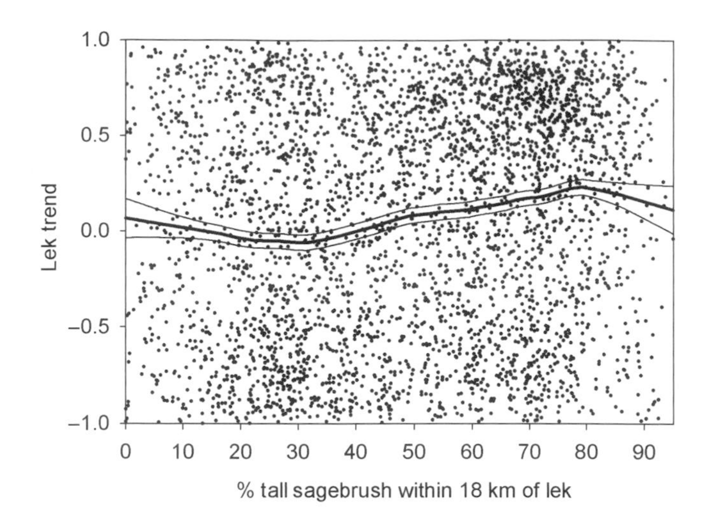

Figure 17.6. Relation between trend of lek counts and tall sagebrush within 18 km, all management zones combined (smooth = 0.311).

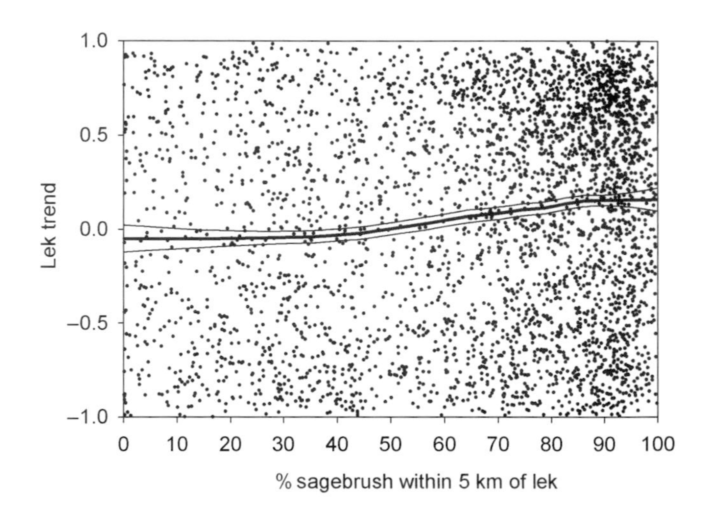

Figure 17.7. Relation between trend of lek counts and all sagebrush within 5 km, all management zones combined (smooth = 0.511).

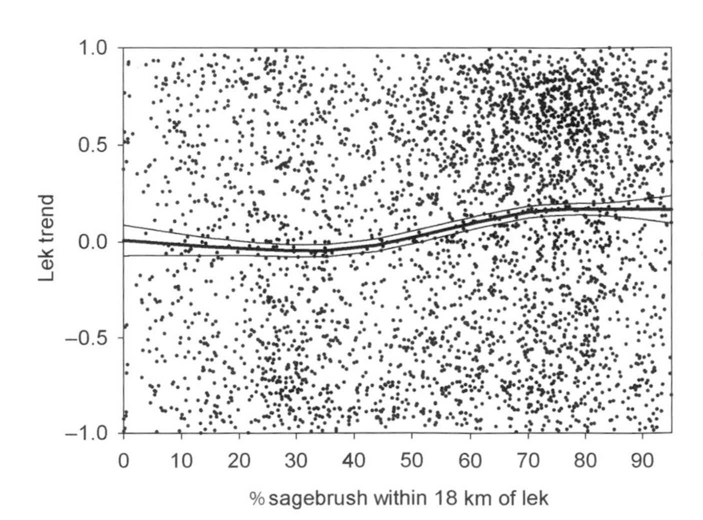

Figure 17.8. Relation between trend of lek counts and all sagebrush within 18 km, all management zones combined (smooth = 0.485).

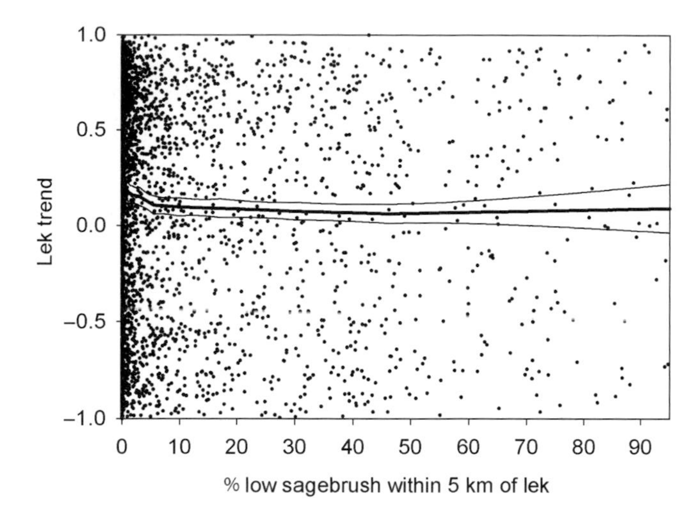

Figure 17.9. Relation between trend of lek counts and low sagebrush within 5 km, all management zones combined (smooth = 0.452).

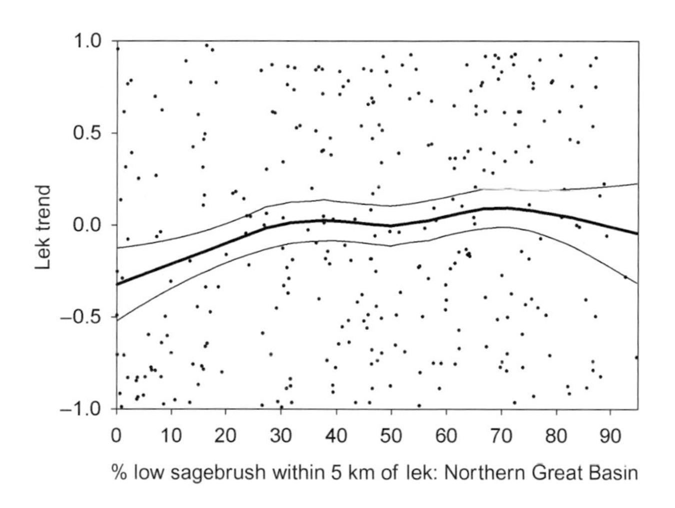

Figure 17.10. Relation between trend of lek counts and low sagebrush within 5 km, Northern Great Basin Sage-Grouse Management Zone (smooth = 0.550).

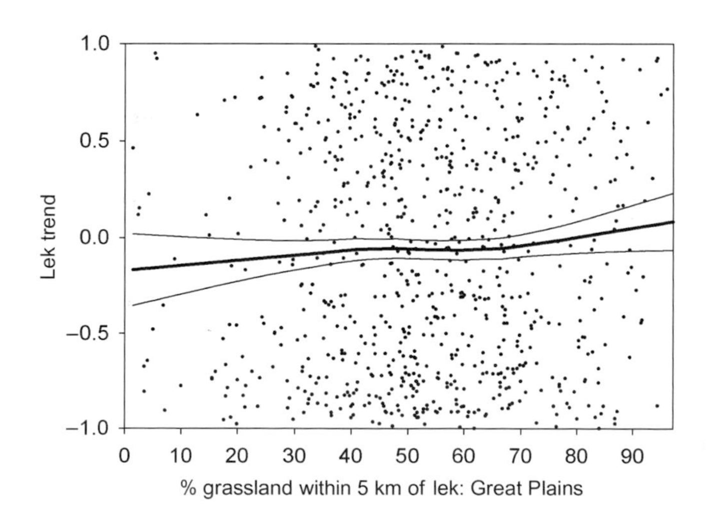

Figure 17.11. Relation between trend of lek counts and grassland within 5 km, Great Plains Sage-Grouse Management Zone (smooth = 0.864).

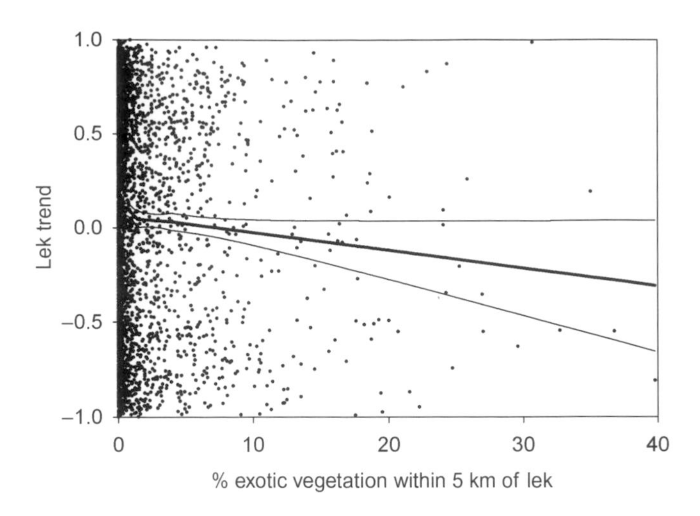

Figure 17.12. Relation between trend of lek counts and exotic vegetation within 5 km, all management zones combined (smooth = 0.702).

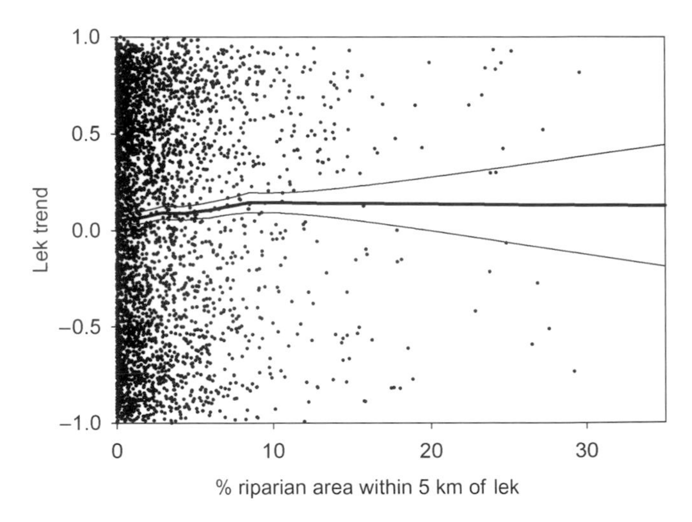

Figure 17.13. Relation between trend of lek counts and riparian cover within 5 km, all management zones combined (smooth = 0.624).

Riparian Cover

Riparian cover was low in all SMZs, especially in the Columbia Basin. Overall, trends in lek counts tended to be slightly lower for leks with little riparian coverage within either 5 km or 18 km (Figs. 17.13, 17.14) (Table 17.2).

Fire History

The area burned within 5 km (Fig. 17.15) and within 18 km was greatest in the Snake River Plain and lowest in the Colorado Plateau, Great Plains, and Wyoming Basin (Table 17.1). Relations between trends in lek counts and area burned were also variable among SMZs (Table 17.2). Within 5 km, the loess curve was increasingly downward in the Southern Great Basin (Fig. 17.16) and mostly downward in Snake River Plain. Within 18 km (Fig. 17.17), the curve was steadily downward in the Great Plains (Fig. 17.18) and declining after about 40 km^2 in the Southern Great Basin. There was no detectable relation between trends in lek counts and the number of post-burn years at either 5 km or 18 km (Fig. 17.19).

Anthropogenic Features

Communication Towers

Communication towers were most common in the Columbia Basin, least common in the Northern Great Basin and Colorado Plateau, and intermediate in the other SMZs (Table 17.1). Despite the generally small numbers of towers, across all SMZs lek trends generally increased with distance to nearest tower (Fig. 17.20). Within SMZs, patterns were similar for the Great Plains,

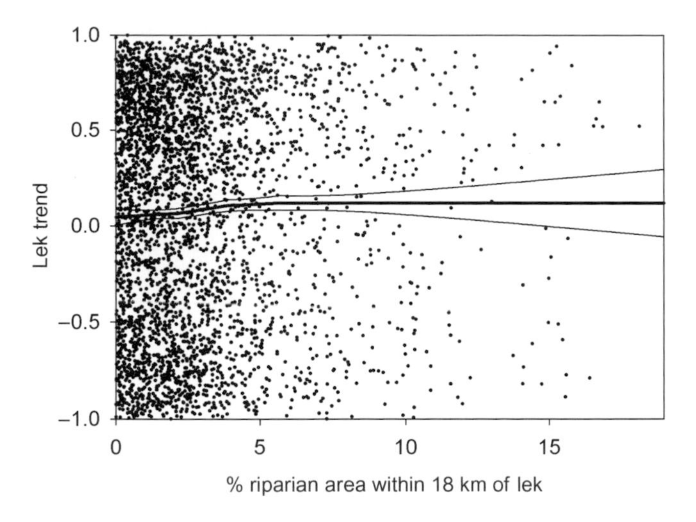

Figure 17.14. Relation between trend of lek counts and riparian cover within 18 km, all management zones combined (smooth = 0.790).

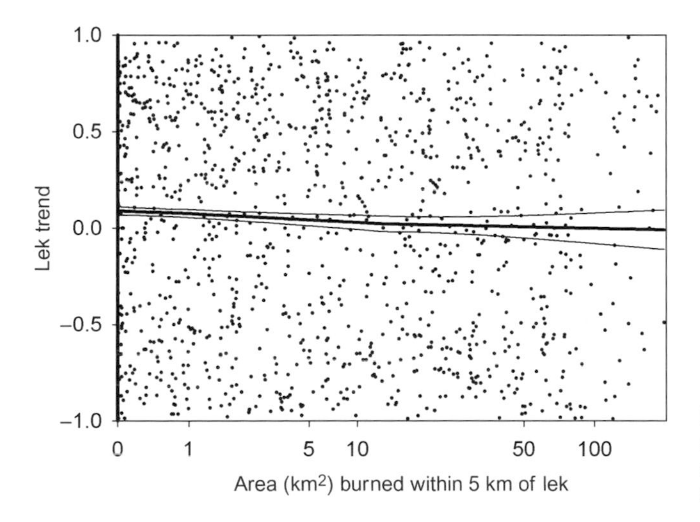

Figure 17.15. Relation between trend of lek counts and area burned within 5 km, all management zones combined (smooth = 0.973).

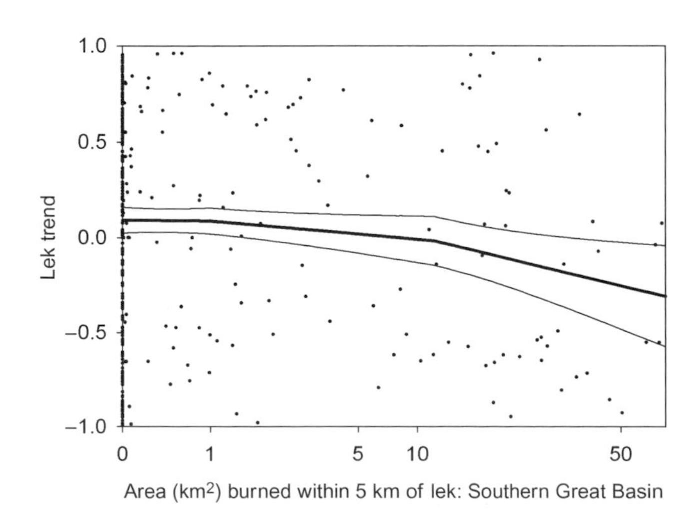

Figure 17.16. Relation between trend of lek counts and area burned within 5 km, Southern Great Basin Sage-Grouse Management Zone (smooth = 0.977).

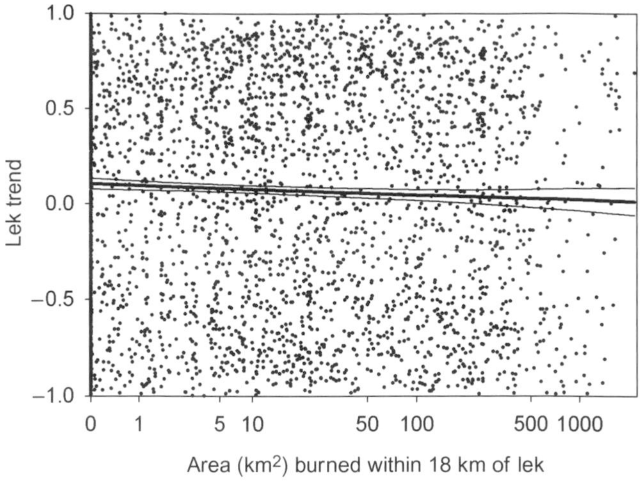

1.0
0.5
0.0
−0.5
−1.0
Lek trend
0
1
5
10
50
100
500
1000
Area (km2) burned within 18 km of lek

Figure 17.17. Relation between trend of lek counts and area burned within 18 km, all management zones combined (smooth = 1.00).

1.0
0.5
0.0
−0.5
−1.0
Lek trend
0
1
5
10
50
100
Area (km2) burned within 18 km of lek: Great Plains

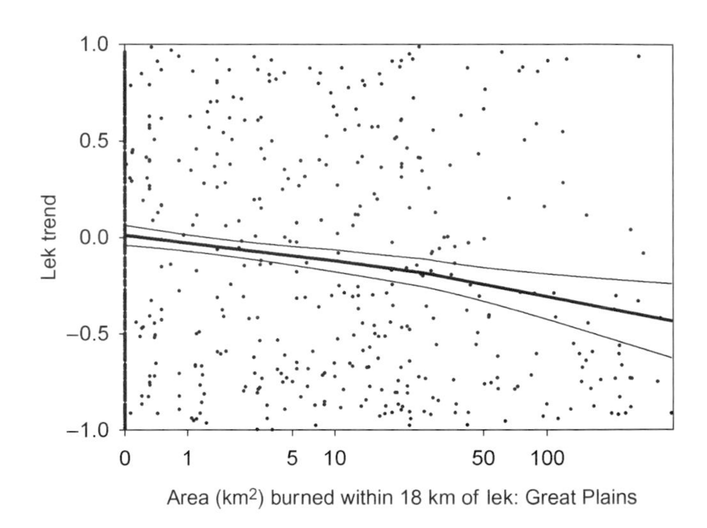

Figure 17.18. Relation between trend of lek counts and area burned within 18 km, Great Plains Sage-Grouse Management Zone (smooth = 1.00).

1.0
0.5
0.0
−0.5
−1.0
Lek trend
0
1
5
10
Fire-years within 18 km of lek

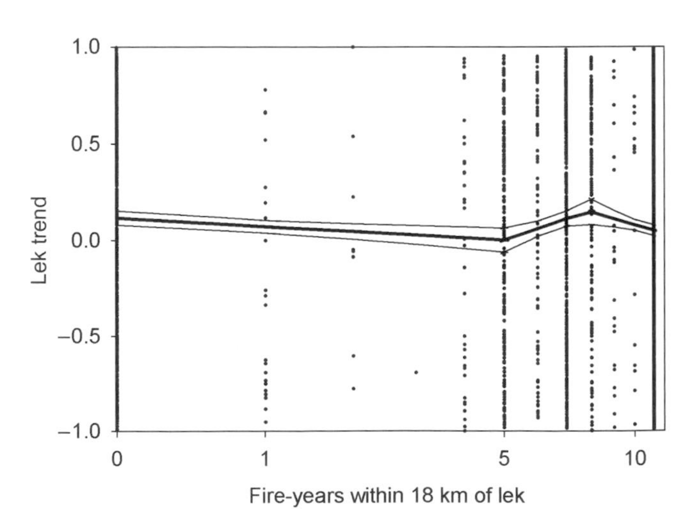

Figure 17.19. Relation between trend of lek counts and number of post-fire years within 18 km, all management zones combined (smooth = 0.624).

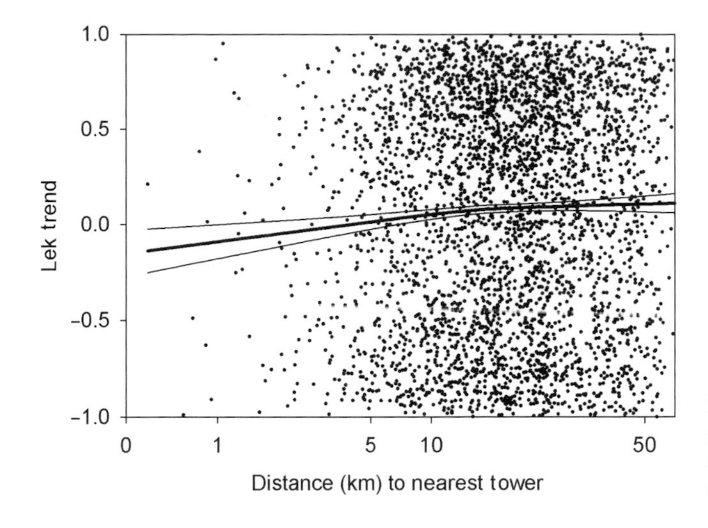

Figure 17.20. Relation between trend of lek counts and distance to nearest tower, all management zones combined (smooth = 0.985).

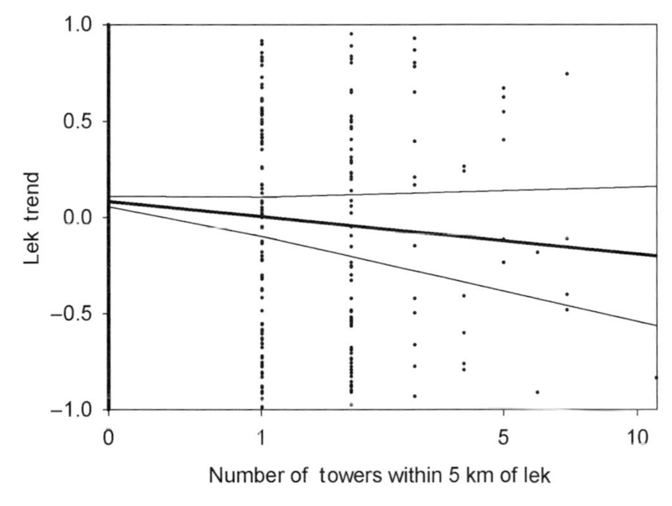

Figure 17.21. Relation between trend of lek counts and number of towers within 5 km, all management zones combined (smooth = 0.494).

Wyoming Basin (leveling off at about 10 km), Northern Great Basin, and Colorado Plateau, but not evident in the Southern Great Basin or Snake River Plain. Analogously, across all SMZs there was a steady downward pattern of trends of lek counts as the number of towers increased, either within 5 km (Fig. 17.21) or within 18 km (Fig. 17.22). This pattern was especially noticeable in the Wyoming Basin (Fig. 17.23).

Oil and Gas Wells

Active oil and gas wells were most common in the Colorado Plateau, followed by the Great Plains and Wyoming Basin, with few occurring in the other SMZs. Analyses of producing wells indicated somewhat stronger effects than did all wells (not shown). Leks tended to have more positive trends if they were farther from producing wells. Across all SMZs, there was an increasing pattern up to about 5 km (Fig. 17.24). Within SMZs, lek trends appeared to increase to about 20 km in the Great Plains and Wyoming Basin (Fig. 17.25), 90 km in the Northern Great Basin, and steadily in the Colorado Plateau. There were no noticeable patterns or data were limited in other SMZs. For the count of producing wells within 5 km, there was no overall effect until about 10 wells, after which the curve declined (Fig. 17.26). Within SMZs, declines occurred after about 20 wells in the Great Plains and 8 wells in the Wyoming Basin. Declines were steady in the Southern Great Basin, which had few wells, and remaining SMZs had too few leks or too few wells to detect a pattern. For the count of producing wells within 18 km, the pattern across all SMZs suggested lower trends beginning with about 160 wells (Fig. 17.27). The only individual SMZ showing a pattern at the 18-km radius was

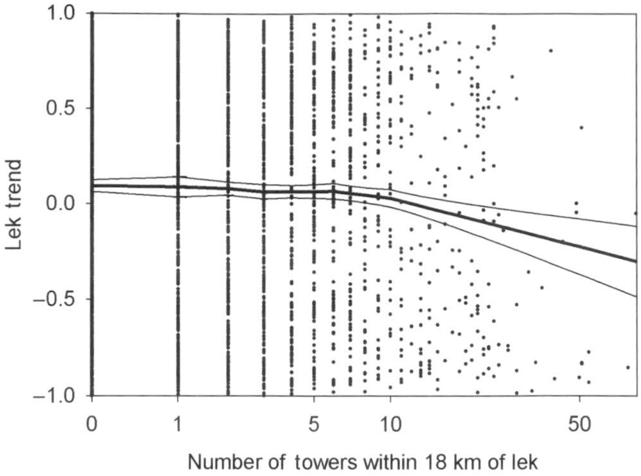

Figure 17.22. Relation between trend of lek counts and number of towers within 18 km, all management zones combined (smooth = 0.372).

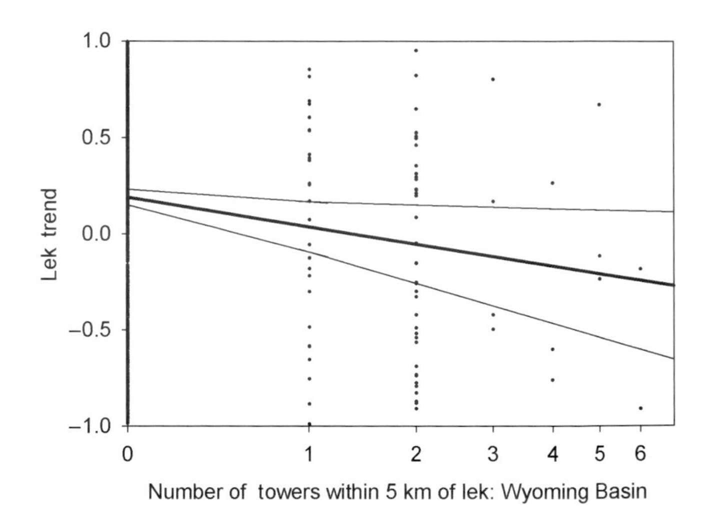

Figure 17.23. Relation between trend of lek counts and number of towers within 5 km, Wyoming Basin Sage-Grouse Management Zone (smooth = 0.541).

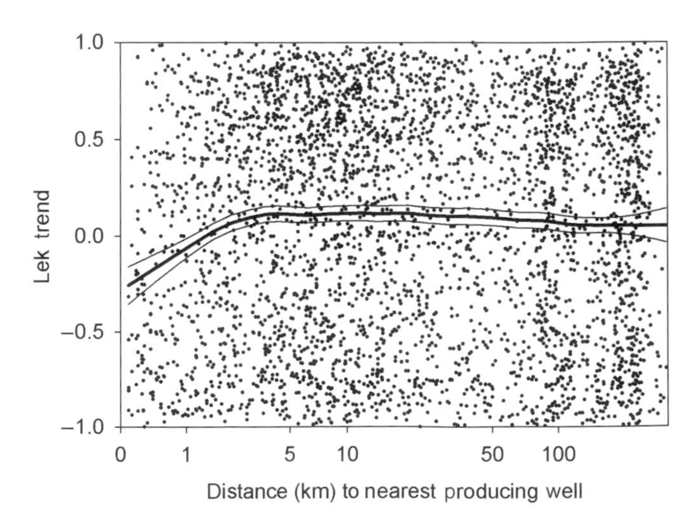

Figure 17.24. Relation between trend of lek counts and distance to producing well, all management zones combined (smooth = 0.304).

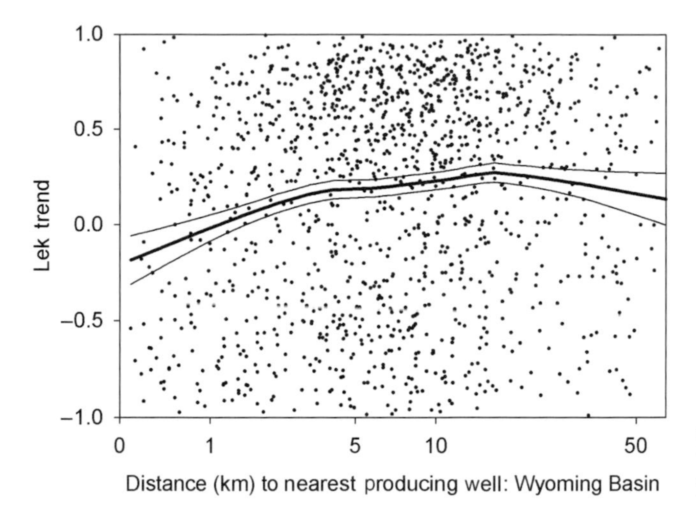

Figure 17.25. Relation between trend of lek counts and distance to producing well, Wyoming Basin Sage-Grouse Management Zone (smooth = 0.555).

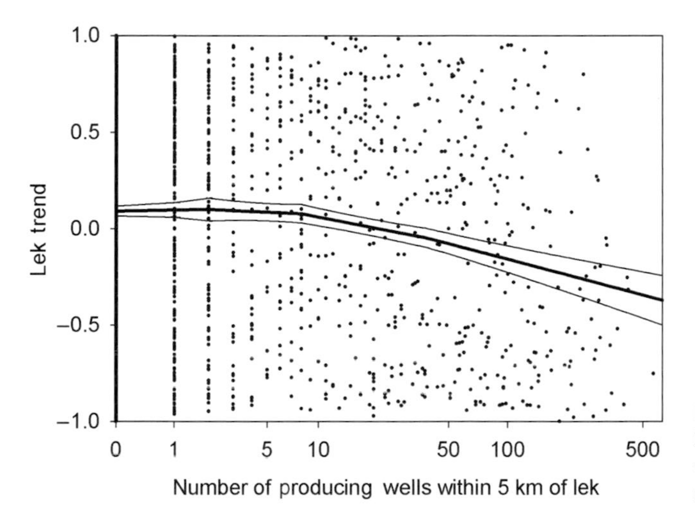

Figure 17.26. Relation between trend of lek counts and number of producing wells within 5 km, all management zones combined (smooth = 0.528).

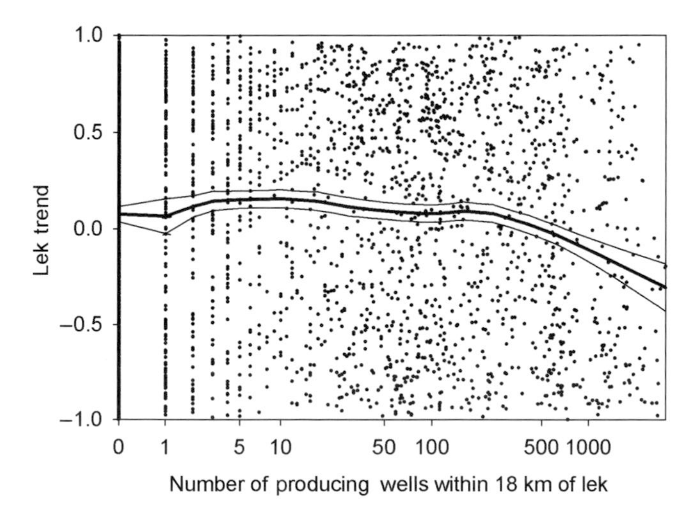

Figure 17.27. Relation between trend of lek counts and number of producing wells within 18 km, all management zones combined (smooth = 0.230).

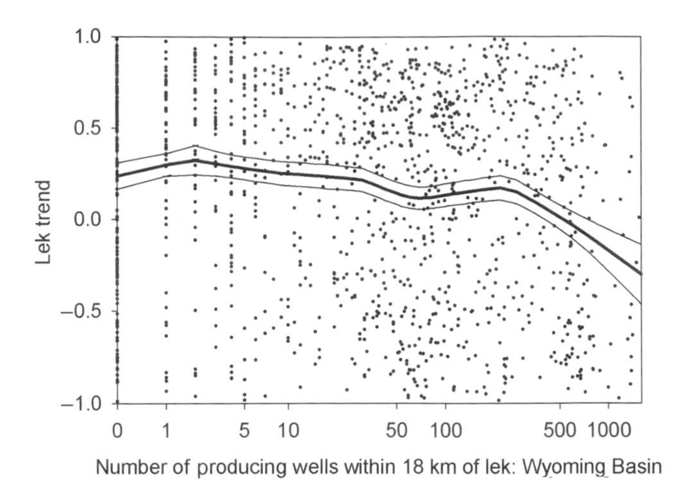

Figure 17.28. Relation between trend of lek counts and number of producing wells within 18 km, Wyoming Basin Sage-Grouse Management Zone (smooth = 0.310).

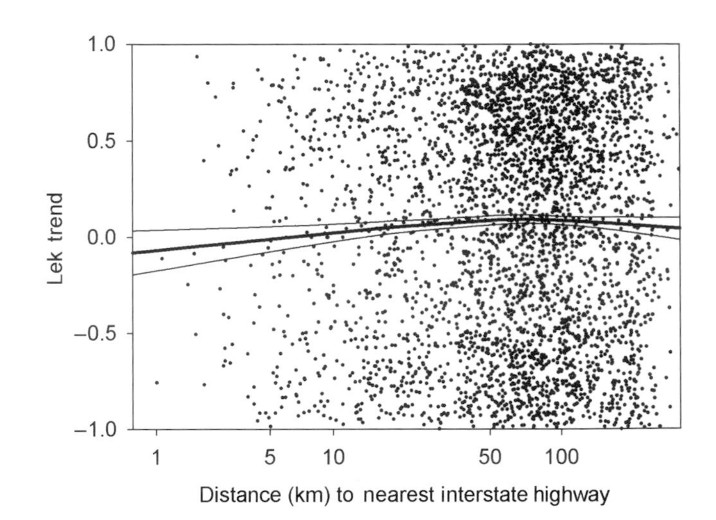

Figure 17.29. Relation between trend of lek counts and distance to nearest interstate highway, all management zones combined (smooth = 0.878).

Wyoming Basin, where a modest decline became a steep decline after about 250 wells (Fig. 17.28).

Highways and Roads

Relatively few leks had interstate highways nearby. The Columbia Basin had the greatest length of interstate highway near leks; the Northern Great Basin and Colorado Plateau had none within 5 km, and the others were intermediate (Table 17.1). It appears from the limited data that the few leks near interstate highways generally fared less well than those farther away (Fig. 17.29 for all SMZs; patterns within SMZs with sufficient data were overall similar). Loess curves relating trend to the length of interstate highways within 5 km showed a steady downward pattern across all SMZs (Fig. 17.30) and in all SMZs with sufficient data. The length of interstate highways within 18 km showed a less-strong relation (Fig. 17.31). Across all SMZs, and within the Great Plains, the loess curves appeared to decline especially with greater than about 30 km of interstate; other individual SMZs displayed no pattern.

Other federal or state highways seemed less related to lek count trends than did interstate highways (Table 17.2). The distance to nearest other federal or state highway seemed only slightly associated with higher trends across all SMZs (Fig. 17.32) and more so in the Great Plains (Fig. 17.33). The loess curves for length of other federal or state highway showed slight declines, to about 6 km of highway within 5 km (Fig. 17.34) and beyond 16 km of highway within 18 km (Fig. 17.35).

Trend in lek counts was positively associated with distance to nearest interstate or other highways combined, both across all SMZs (Fig. 17.36) and especially in the Great Plains (Fig. 17.37) (Table 17.2). The length of interstate and other highways

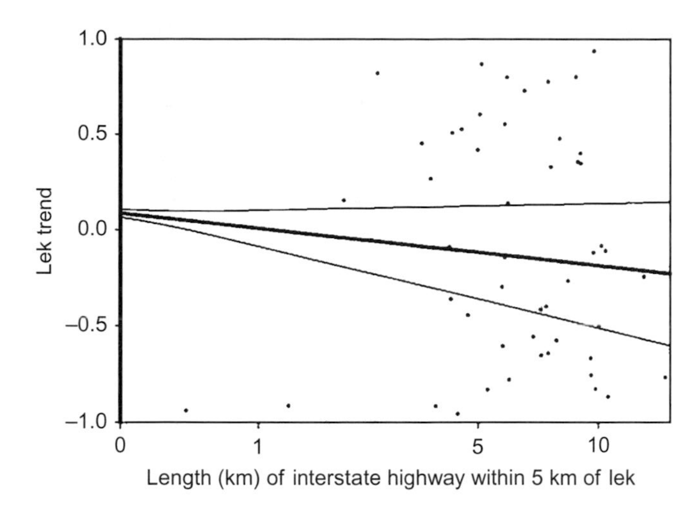

Figure 17.30. Relation between trend of lek counts and length of interstate highways within 5 km, all management zones combined (smooth = 0.850).

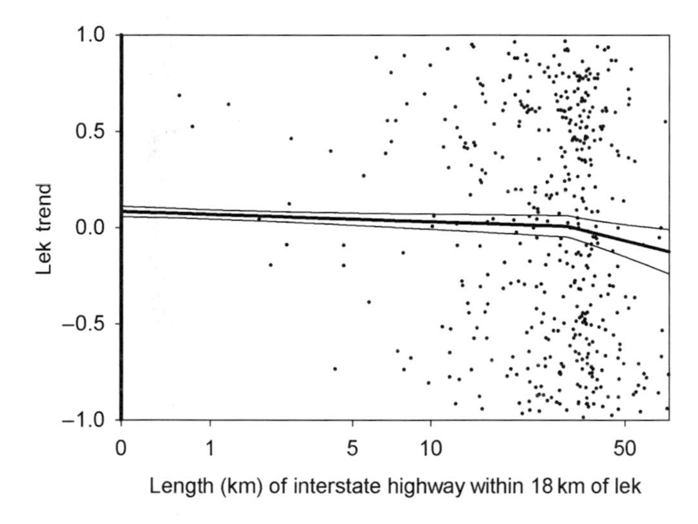

Figure 17.31. Relation between trend of lek counts and length of interstate highways within 18 km, all management zones combined (smooth = 0.494).

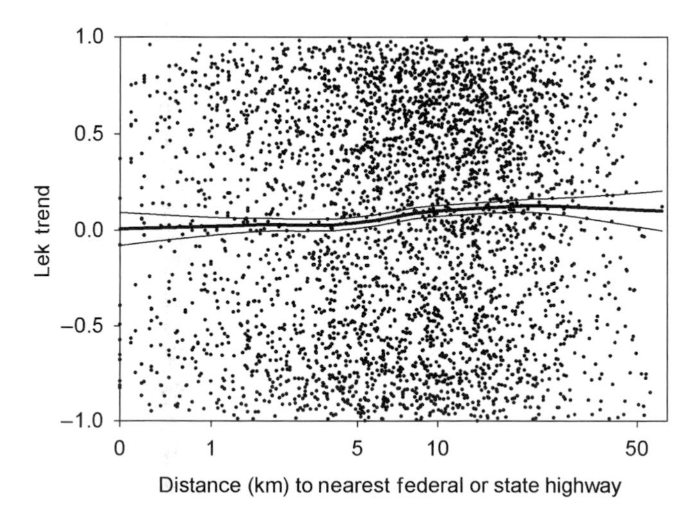

Figure 17.32. Relation between trend of lek counts and distance to nearest other federal or state highway, all management zones combined (smooth = 0.585).

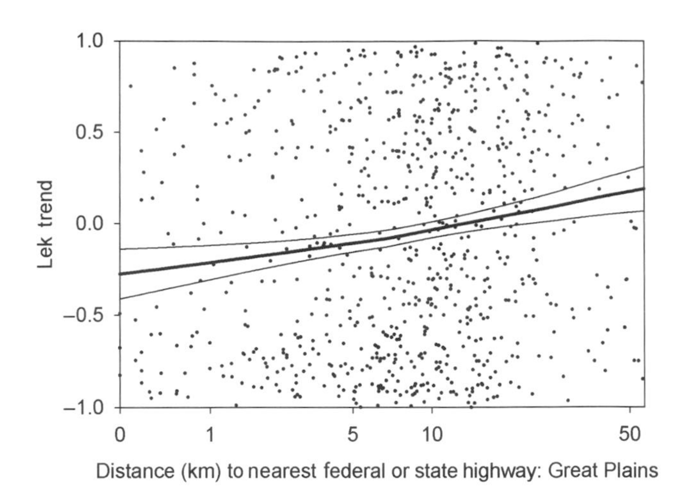

Figure 17.33. Relation between trend of lek counts and distance to nearest other federal or state highway, Great Plains Sage-Grouse Management Zone (smooth = 0.991).

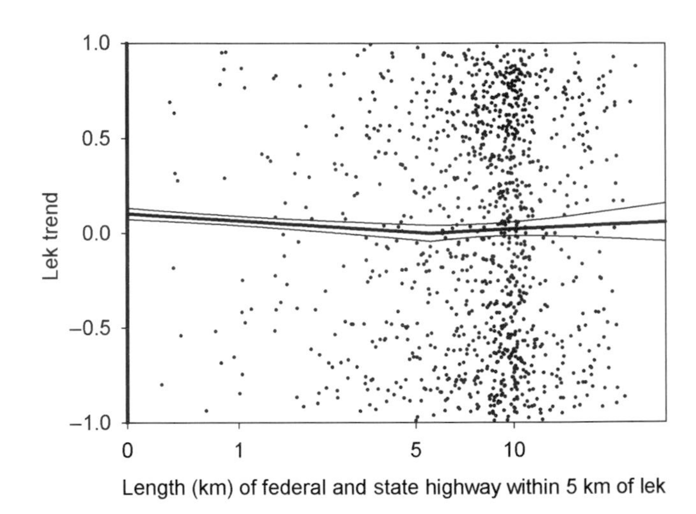

Figure 17.34. Relation between trend of lek counts and length of other federal or state highways within 5 km, all management zones combined (smooth = 0.399).

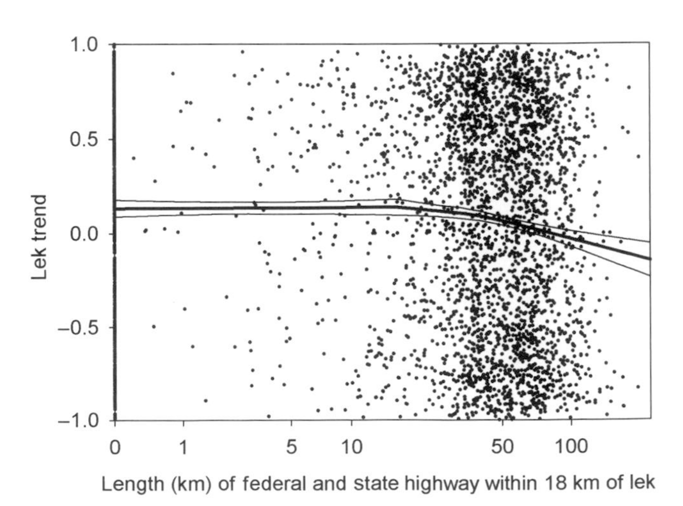

Figure 17.35. Relation between trend of lek counts and length of other federal or state highways within 18 km, all management zones combined (smooth = 0.759).

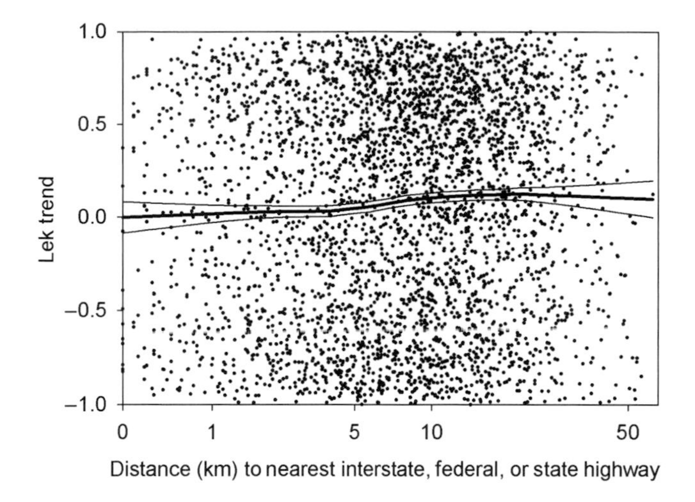

Figure 17.36. Relation between trend of lek counts and distance to nearest interstate, other federal, or state highway, all management zones combined (smooth = 0.627).

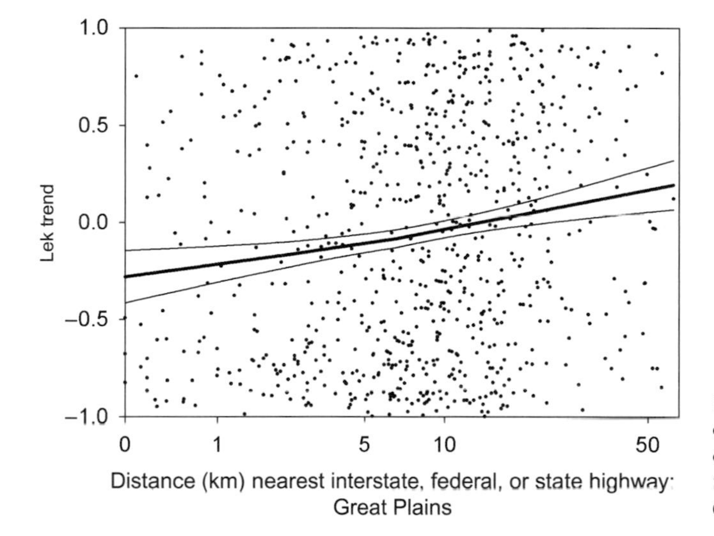

Figure 17.37. Relation between trend of lek counts and distance to nearest interstate, other federal, or state highway, Great Plains Sage-Grouse Management Zone (smooth = 0.988).

combined within 5 km did not seem associated with trends of lek counts (Fig. 17.38), except possibly in the Great Plains (Fig. 17.39). For the length of interstate and other highways within 18 km, lek trends decreased for values beyond 20 km, all SMZs combined (Fig. 17.40). Roughly similar patterns occurred for the Columbia Basin, Wyoming Basin, Southern Great Basin, and Snake River Plain.

The presence of secondary roads appeared not to influence lek trends. The distance to secondary roads displayed no consistent effects (Fig. 17.41). Similarly, the loess curves involving length of secondary road within 5 km (Fig. 17.42) or within 18 km (Fig. 17.43) exhibited no pattern across SMZs, and patterns within individual SMZs were highly variable, some increasing, some decreasing.

Distance to nearest road of any type appeared related to lek trends only in the Northern Great Basin SMZ (Table 17.2), where higher trends were associated with greater distances to a road (Fig. 17.44). Loess curves relating trends in lek counts to the length of all roads combined within 5 km showed no pattern overall (Fig. 17.45) but suggested declines beginning at about 20 km in the Columbia Basin and Northern Great Basin (Fig. 17.46) and 60 km in the Southern Great Basin. For length of all roads within 18 km, again, there was no overall pattern (Fig. 17.47), but declines beginning at about 300 km in the Great Plains (Fig. 17.48), 600 km in the Southern Great Basin, and 400 km in the Northern Great Basin were evident.

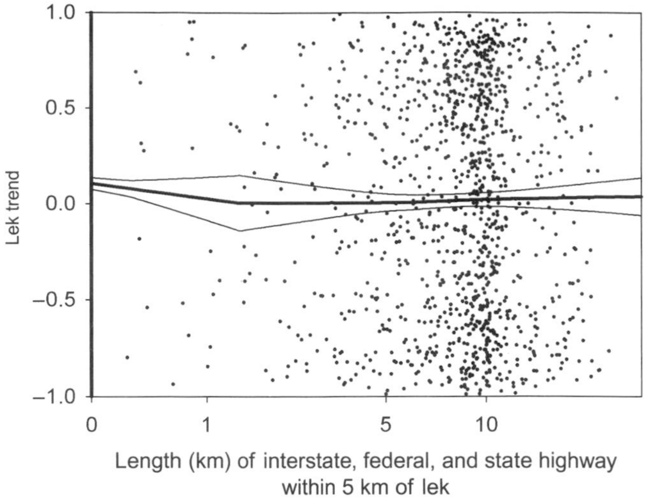

1.0
0.5
0.0
−0.5
−1.0
Lek trend
0 1 5 10
Length (km) of interstate, federal, and state highway within 5 km of lek

Figure 17.38. Relation between trend of lek counts and length of interstate, other federal, or state highways within 5 km, all management zones combined (smooth = 0.397).

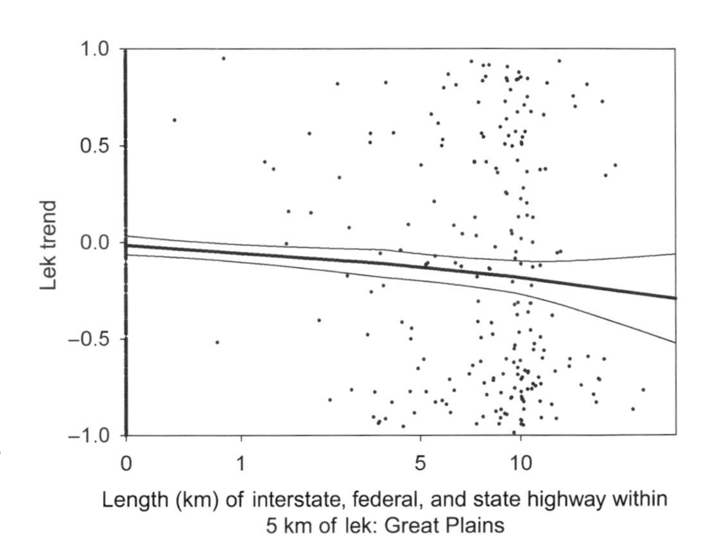

Figure 17.39. Relation between trend of lek counts and length of interstate, other federal, or state highways within 5 km, Great Plains Sage-Grouse Management Zone (smooth = 1.00).

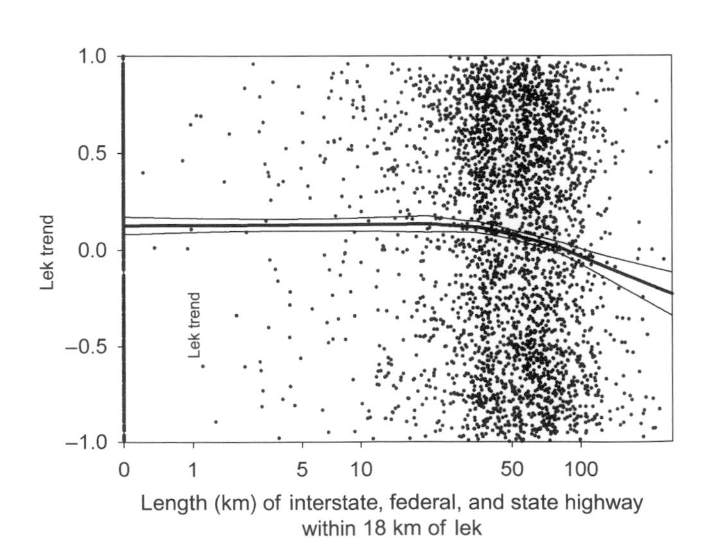

Figure 17.40. Relation between trend of lek counts and length of interstate, other federal, or state highways within 18 km, all management zones combined (smooth = 0.717).

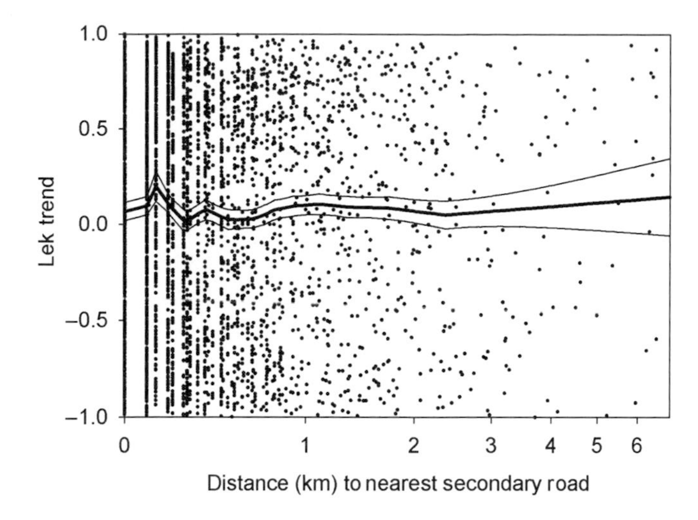

Figure 17.41. Relation between trend of lek counts and distance to nearest secondary road, all management zones combined (smooth = 0.214).

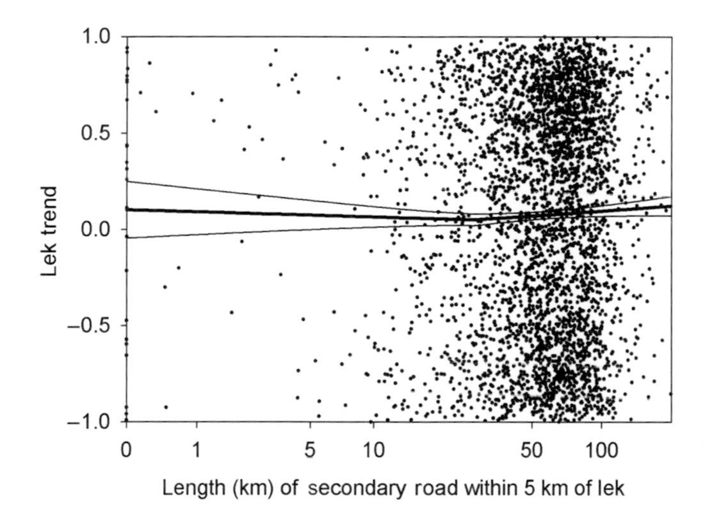

Figure 17.42. Relation between trend of lek counts and length of secondary roads within 5 km, all management zones combined (smooth = 0.999).

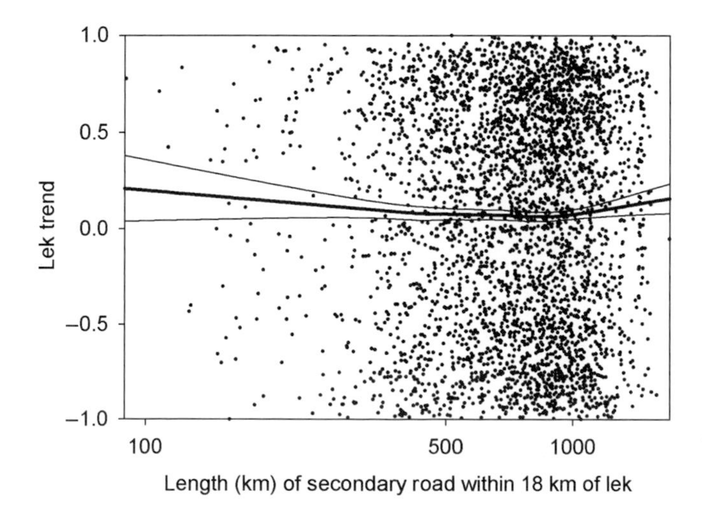

Figure 17.43. Relation between trend of lek counts and length of secondary roads within 18 km, all management zones combined (smooth = 0.779).

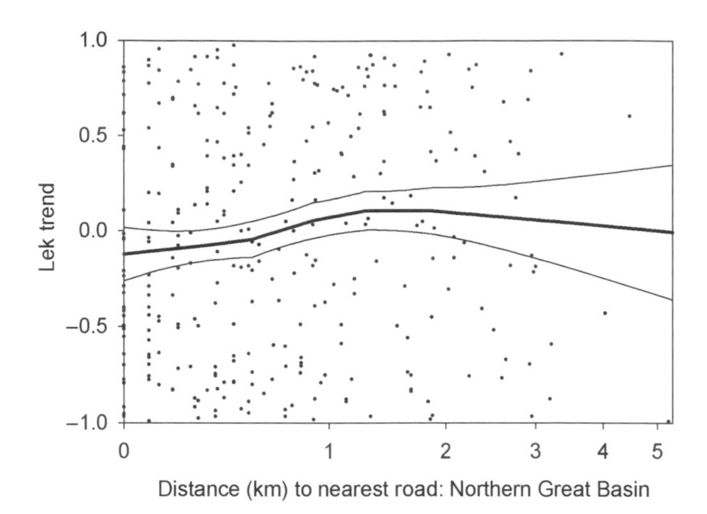

Figure 17.44. Relation between trend of lek counts and distance to nearest road of any type, Northern Great Basin Sage-Grouse Management Zone (smooth = 0.779).

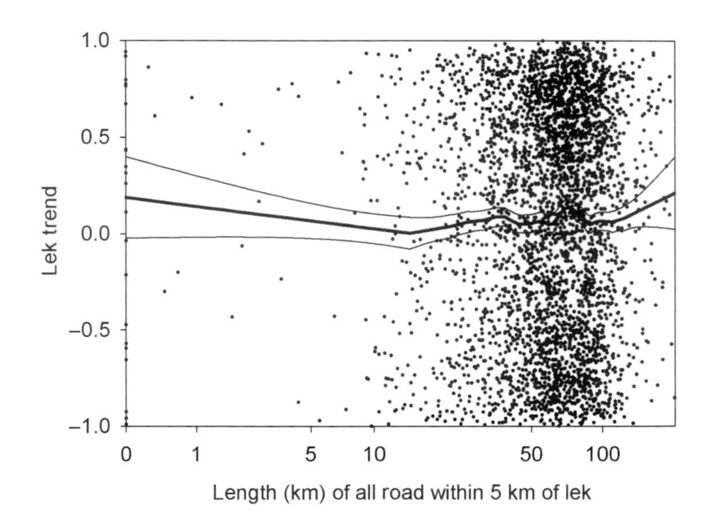

Figure 17.45. Relation between trend of lek counts and length of all roads within 5 km, all management zones combined (smooth = 0.245).

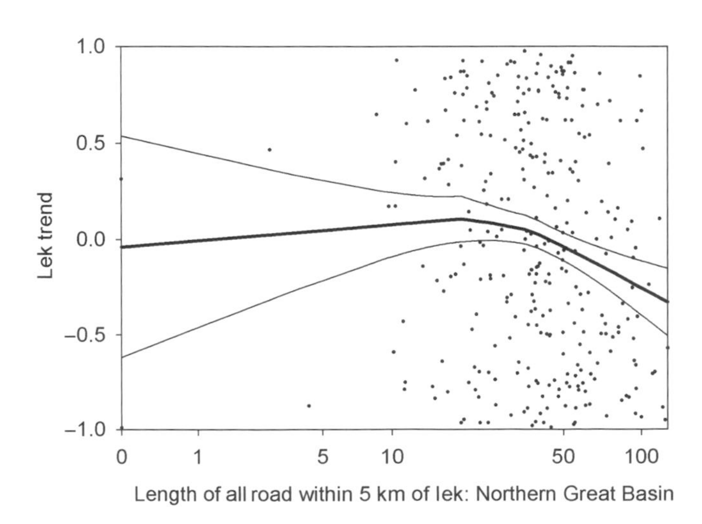

Figure 17.46. Relation between trend of lek counts and length of all roads within 5 km, Northern Great Basin Sage-Grouse Management Zone (smooth = 0.973).

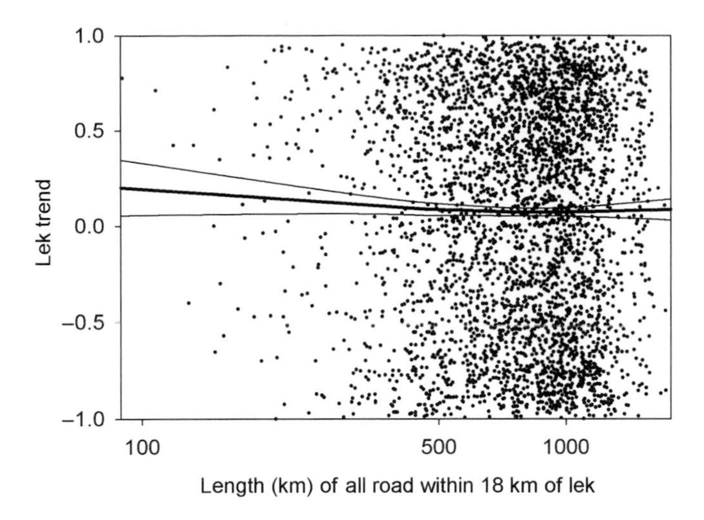

Figure 17.47. Relation between trend of lek counts and length of all roads within 18 km, all management zones combined (smooth = 0.972).

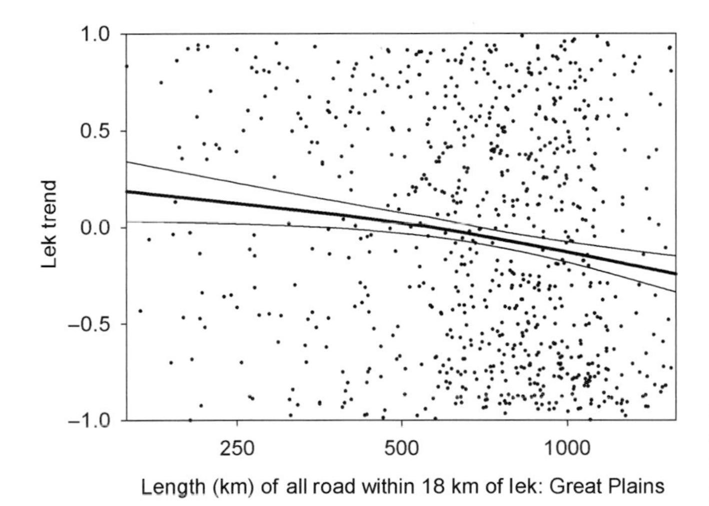

Figure 17.48. Relation between trend of lek counts and length of all roads within 18 km, Great Plains Sage-Grouse Management Zone (smooth = 1.00).

Pipelines

Pipelines were most prevalent in the Wyoming Basin, with fewest in the Northern Great Basin and Columbia Basin (Table 17.1). Trends in lek counts seemed unrelated to distance to the nearest pipeline over all SMZs (Fig. 17.49), but positively related in the Great Plains (Fig. 17.50) and Wyoming Basin. Over all SMZs, the loess curve for the 5-km radius indicated a slight decline once length of pipeline exceeded about 6 km (Fig. 17.51). Some individual SMZs showed similar patterns, with declines beginning at about 16 km in the Great Plains (Fig. 17.52), 0 km in the Wyoming Basin, and 10 km in the Snake River Plain. For the 18-km radius, no pattern emerged for combined data from all SMZs (Fig. 17.53), but decreases were evident beginning with about 200 km in Great Plains and from about 3 km in the Snake River Plain (Fig. 17.54).

Power Lines

Power lines occurred regularly in all SMZs, but somewhat more abundantly in the Wyoming Basin and less commonly in the Northern Great Basin (Table 17.1). Distance from lek to nearest power line suggested no relationship across all SMZs (Fig. 17.55), but a possible increasing relationship in the Great Plains individually (Fig. 17.56). No general pattern across all SMZs emerged between trends in lek counts and length of power line, within either 5 km (Fig. 17.57) or 18 km (Fig. 17.58) (Table 17.2). Possible declines

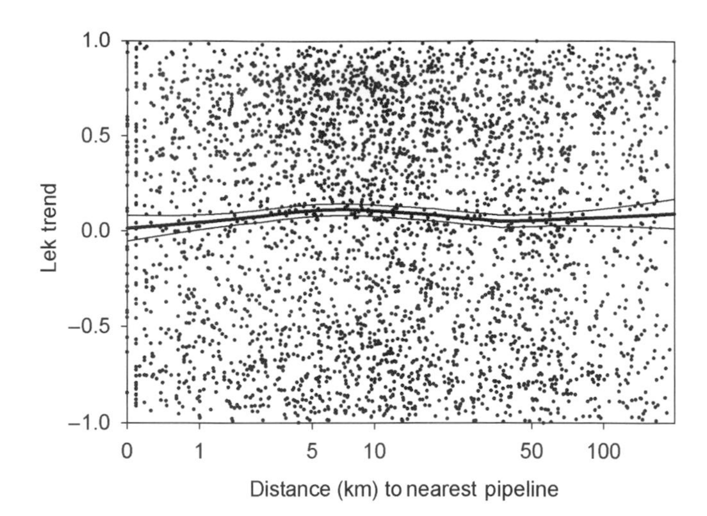

Figure 17.49. Relation between trend of lek counts and distance to nearest pipeline, all management zones combined (smooth = 0.507).

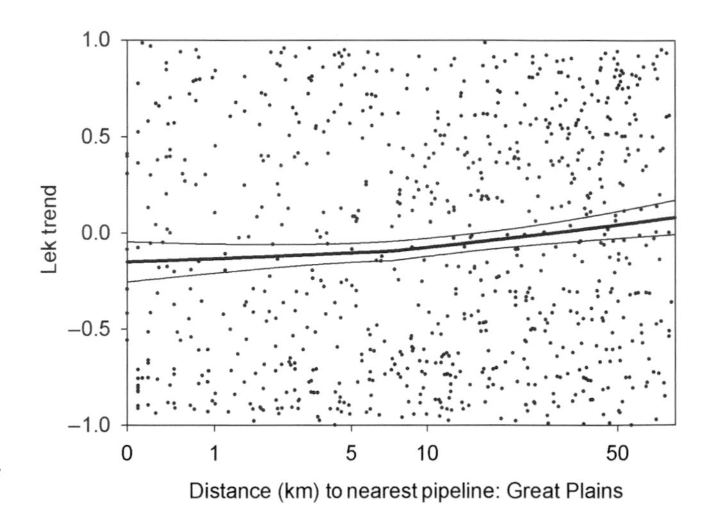

Figure 17.50. Relation between trend of lek counts and distance to nearest pipeline, Great Plains Sage-Grouse Management Zone (smooth = 1.00).

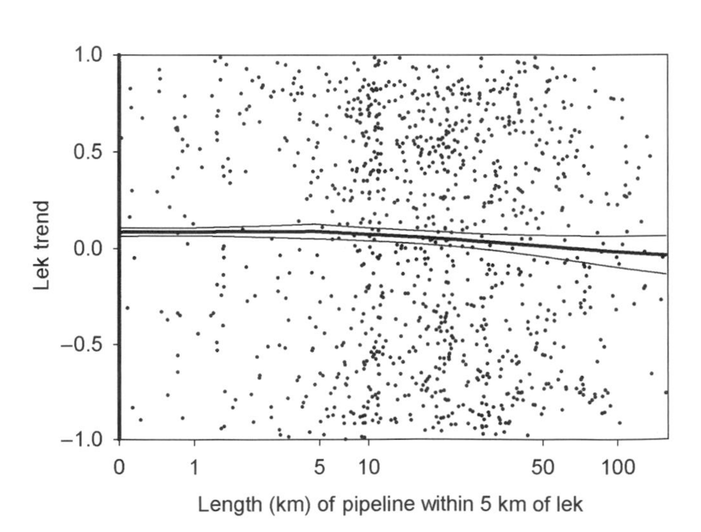

Figure 17.51. Relation between trend of lek counts and length of pipelines within 5 km, all management zones combined (smooth = 0.955).

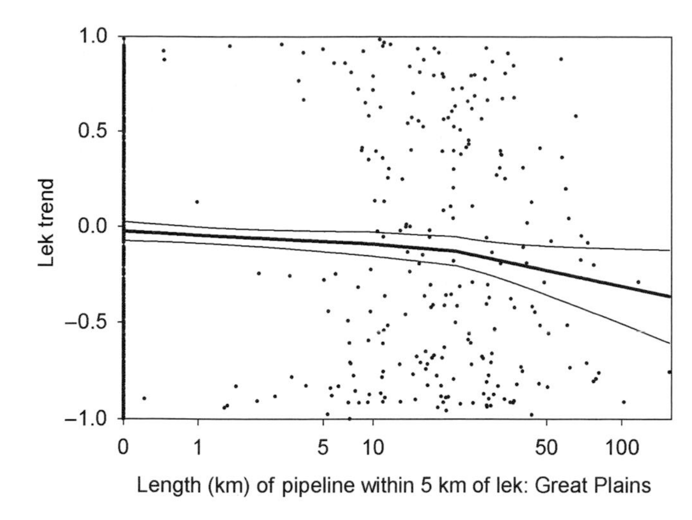

Figure 17.52. Relation between trend of lek counts and length of pipelines within 5 km, Great Plains Sage-Grouse Management Zone (smooth = 1.00).

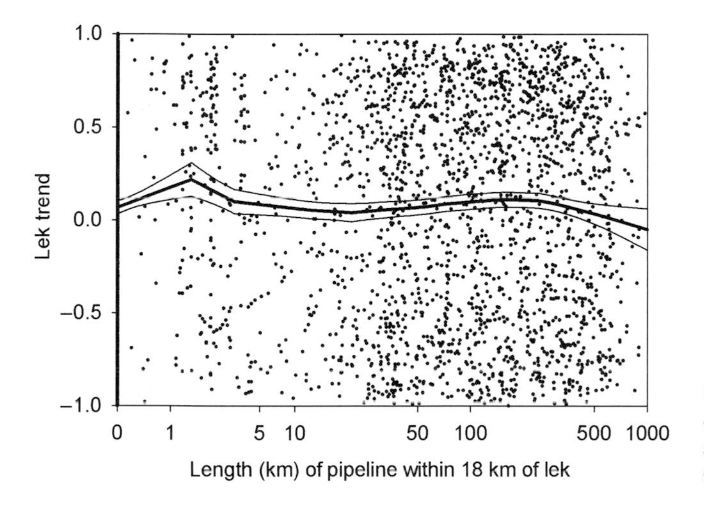

Figure 17.53. Relation between trend of lek counts and length of pipelines within 18 km, all management zones combined (smooth = 0.297).

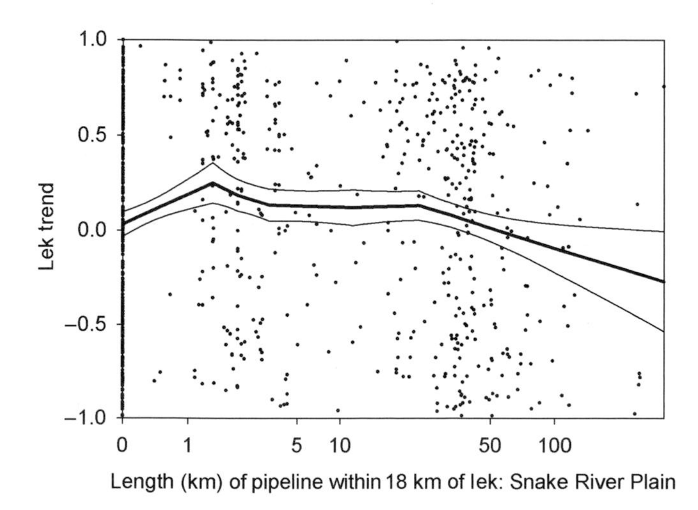

Figure 17.54. Relation between trend of lek counts and length of pipelines within 18 km, Snake River Plain Sage-Grouse Management Zone (smooth = 0.650).

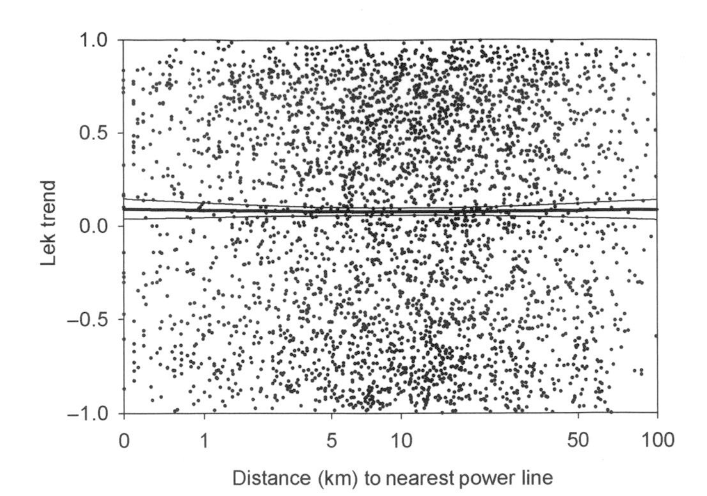

Figure 17.55. Relation between trend of lek counts and distance to nearest power line, all management zones combined (smooth = 1.00).

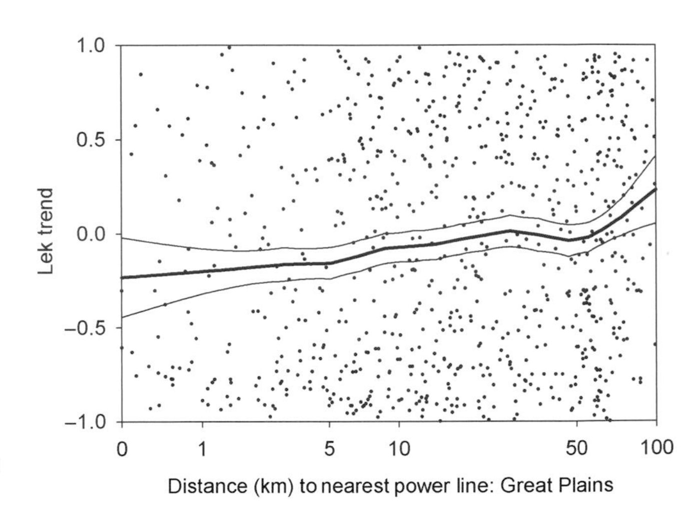

Figure 17.56. Relation between trend of lek counts and distance to nearest power line, Great Plains Sage-Grouse Management Zone (smooth = 0.364).

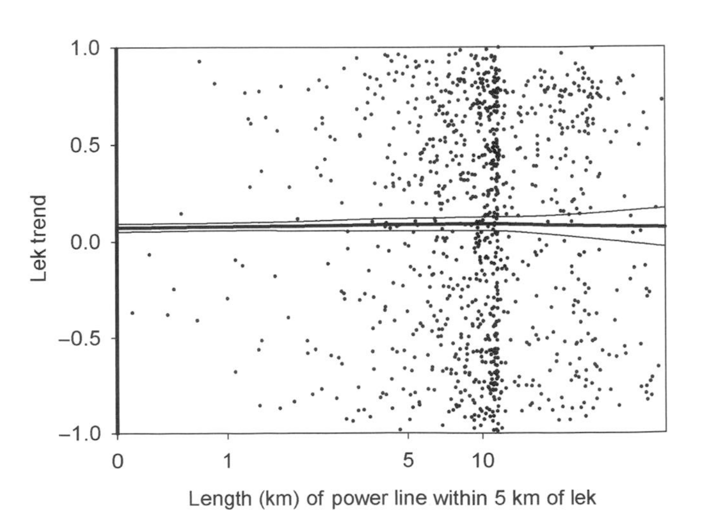

Figure 17.57. Relation between trend of lek counts and length of power lines within 5 km, all management zones combined (smooth = 1.00).

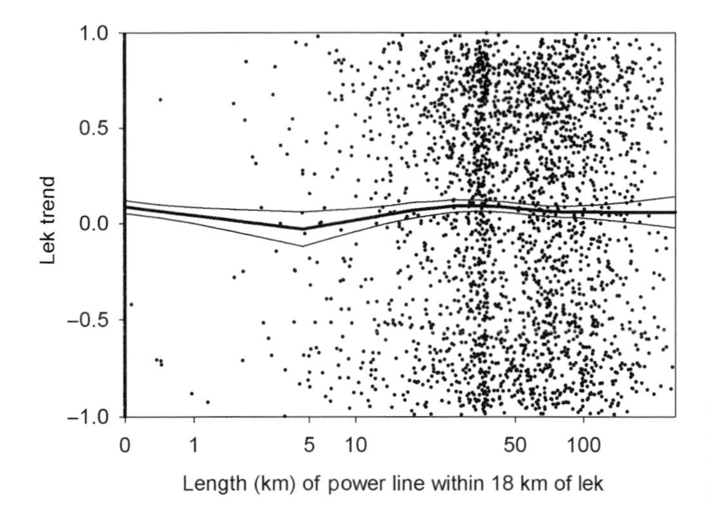

Figure 17.58. Relation between trend of lek counts and length of power lines within 18 km, all management zones combined (smooth = 0.585).

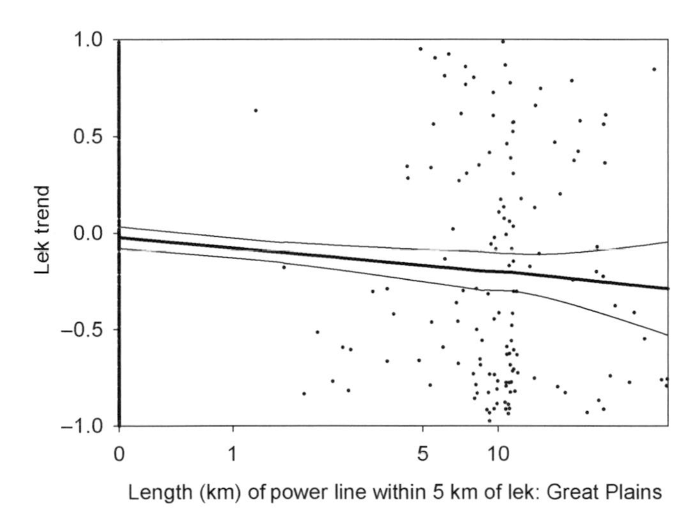

Figure 17.59. Relation between trend of lek counts and length of power lines within 5 km, Great Plains Sage-Grouse Management Zone (smooth = 0.556).

occurred within SMZs when length of power line within 5 km exceeded about 0 km in the Great Plains (Fig. 17.59) and 10 km in the Wyoming Basin and Southern Great Basin, or when length of power line within 18 km exceeded about 25 km in the Great Plains, 15 km in the Southern Great Basin (Fig. 17.60), and 30 km in the Northern Great Basin.

Agriculture

Agricultural (cultivated) land was most common by far in the Columbia Basin and least common in the Southern Great Basin, Northern Great Basin, and Colorado Plateau (Table 17.1). Trends were higher for leks with no agricultural land within either 5 km (Fig. 17.61) or 18 km (Fig. 17.62). They tended to decline steeply with increasing agriculture until about 2.5% cover within 5 km and 1.5% cover within 18 km; further declines were minimal at higher values.

Development

Responses to developed land were negative. Most leks had no developed land within 5 km, but those that did tended to have lower trends than those that did not (Fig. 17.63). At 18 km, a strong and consistent negative relation was evident (Fig. 17.64).

Human Footprint

Trends in lek counts tended to be somewhat reduced if human-footprint scores at the lek location exceeded 2 across all SMZs (Fig. 17.65) and within the Great Plains (Fig. 17.66) and Wyoming

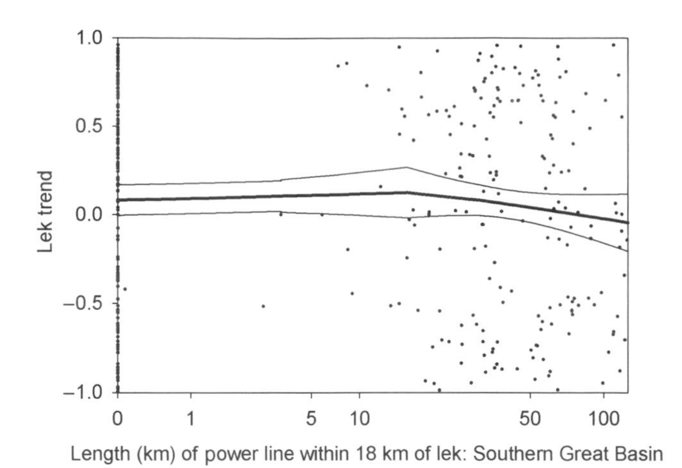

Figure 17.60. Relation between trend of lek counts and length of power lines within 18 km, Southern Great Basin Sage-Grouse Management Zone (smooth = 1.00).

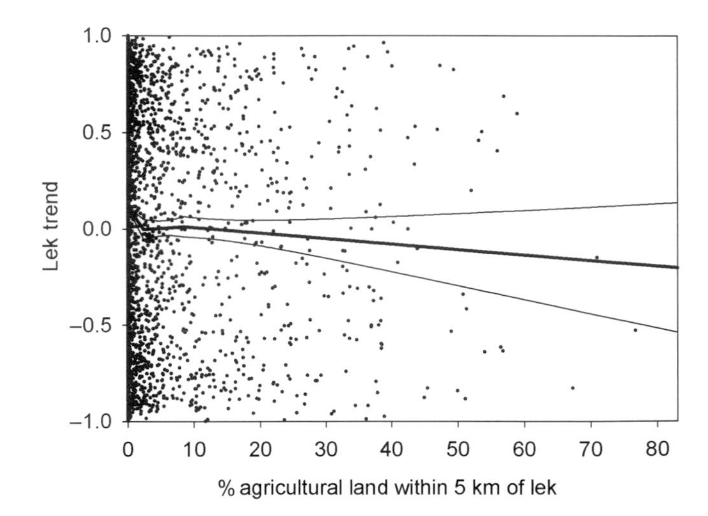

Figure 17.61. Relation between trend of lek counts and agricultural land within 5 km, all management zones combined (smooth = 0.820).

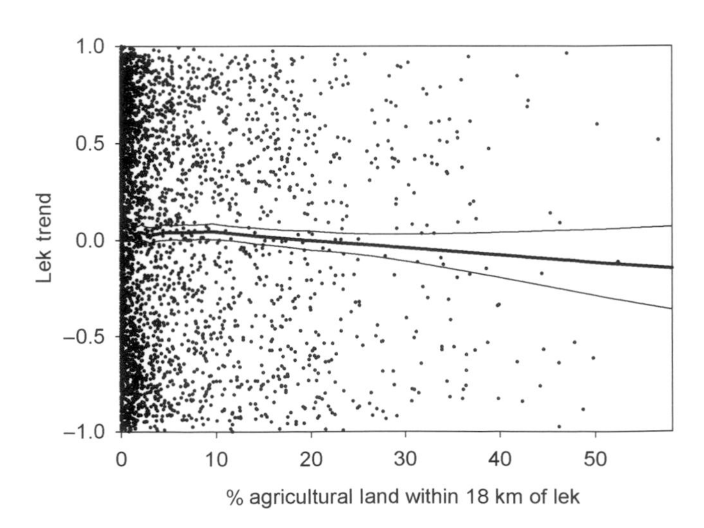

Figure 17.62. Relation between trend of lek counts and agricultural land within 18 km, all management zones combined (smooth = 0.478).

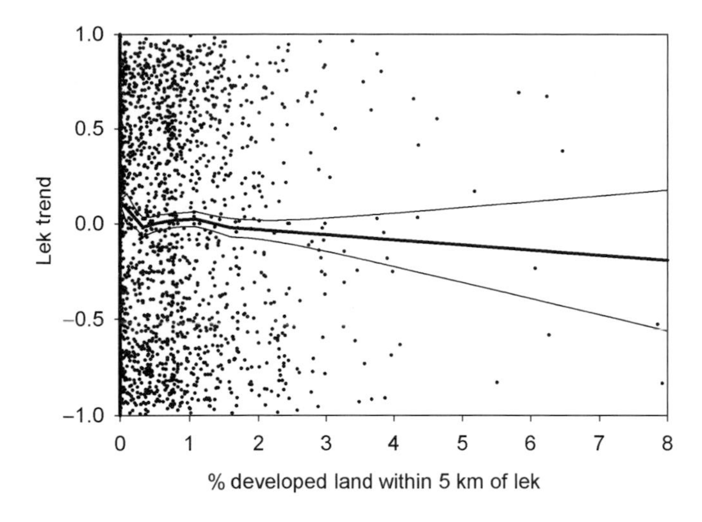

Figure 17.63. Relation between trend of lek counts and developed land within 5 km, all management zones combined (smooth = 0.624).

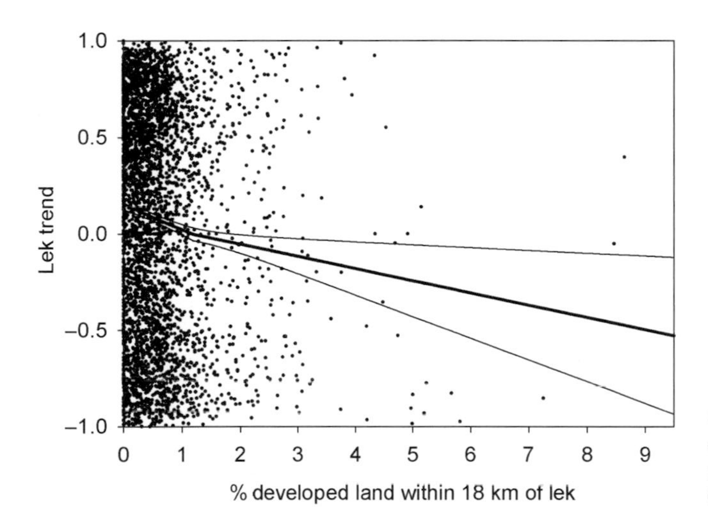

Figure 17.64. Relation between trend of lek counts and developed land within 18 km, all management zones combined (smooth = 0.982).

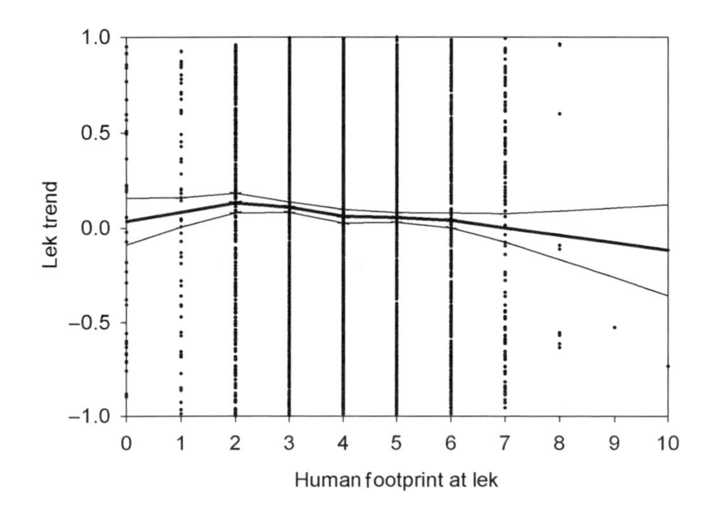

Figure 17.65. Relation between trend of lek counts and human-footprint score at the lek site, all management zones combined (smooth = 0.625).

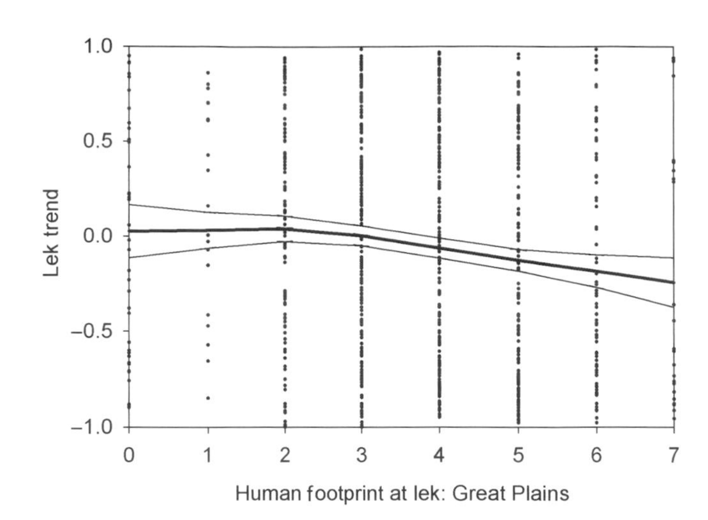

Figure 17.66. Relation between trend of lek counts and human-footprint score at the lek site, Great Plains Sage-Grouse Management Zone (smooth = 0.862).

Basin. Declines in lek trends related to human-footprint scores occurred in other SMZs as well, but the patterns varied somewhat.

For the median human footprint within either 5 km (Fig. 17.67) or 18 km (Fig. 17.68), declines occurred across all SMZs for values >3. Also, the Great Plains had a similar pattern (Fig. 17.69), and the Wyoming Basin showed a minor downward trend. For other SMZs, either no pattern emerged or data were limited.

DISCUSSION

Some Caveats

Our results should be viewed with a number of caveats in mind. For one, lek counts are fraught with problems (Beck and Braun 1980, Applegate 2000, Walsh et al. 2004). Among these are imprecise definitions of a lek, possibility that surveyed leks are not representative of the entire population of leks in certain areas, inconsistent number and timing of surveys conducted at different leks, and incompleteness of a count compared with the number of birds actually using a lek (Johnson and Rowland 2007).

In addition, the response variable, correlation between lek count and year, is only an index to whether lek numbers are decreasing, staying roughly similar, or increasing. More sophisticated metrics could be used if lek counts followed certain monotonic patterns or had a periodicity that was understood.

The patterns of lek counts shown in Fig. 17.3 may suggest that populations fluctuate in a cyclic manner, but no peer-reviewed publications have provided compelling evidence that sage-grouse populations are cyclic (but see Garton et al., this volume, chapter 15). Most populations of animals display irregular fluctuations. To be considered cyclic, these fluctuations need to occur regularly and with a period differing from that of obvious environmental drivers (Nisbet and Gurney 1976). Intervals between sage-grouse population peaks were not consistent; further, the clear fluctuations evident through the 1980s (Connelly et al. 2004; Garton et al., this volume, chapter 15) have become less apparent in the last few decades (Garton et al., this volume, chapter 15). Regardless of whether or not sage-grouse populations fluctuate in a cyclic pattern, the time period we investigated would cover at least one full cycle, so our use of a measure of linear change seems appropriate.

A potential bias is that agencies may have discontinued surveys of leks that had become inactive. In that event, inactive leks would tend to have fewer years in which surveys were conducted and would not be included in the analysis if that number of years (during 1997–2007) was less than four, and would be down-weighted even if they were included. We investigated the possibility of a discontinuation bias by calculating the percentage of occasions from 1997 to 2007 for which a zero count was followed by a missing and the percentage of occasions for which a nonzero count was followed by a missing count. If the two

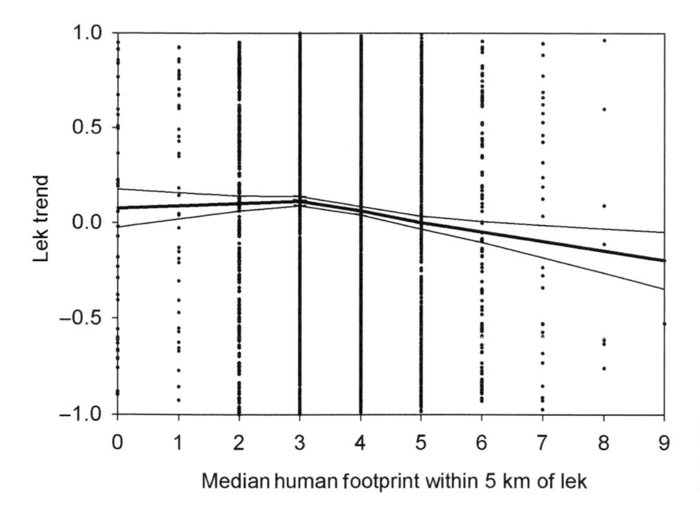

Figure 17.67. Relation between trend of lek counts and median human-footprint score within 5 km, all management zones combined (smooth = 0.947).

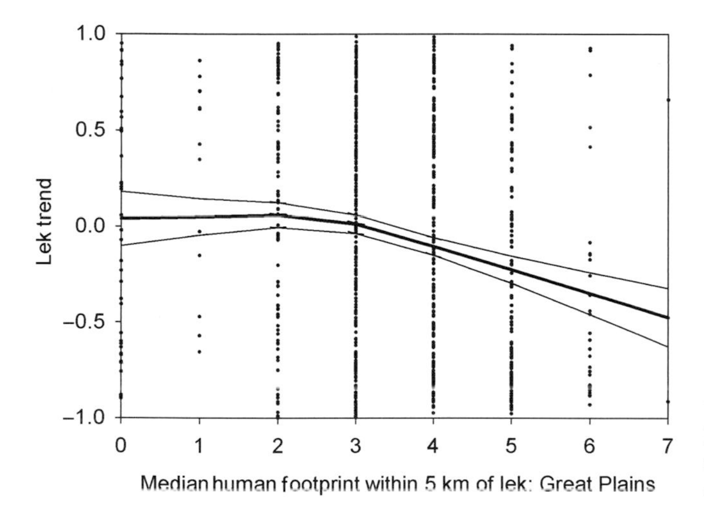

Figure 17.68. Relation between trend of lek counts and median human-footprint score within 18 km, all management zones combined (smooth = 0.801).

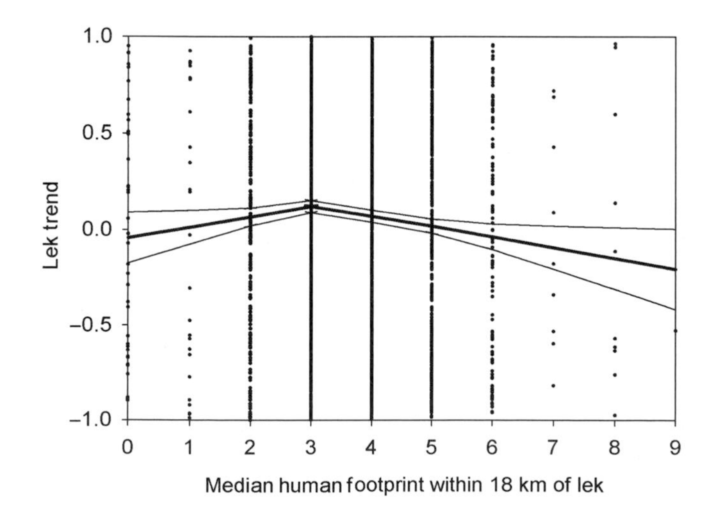

Figure 17.69. Relation between trend of lek counts and median human-footprint score within 5 km, Great Plains Sage-Grouse Management Zone (smooth = 0.958).

percentages are roughly similar, there is no evidence of a discontinuation bias. In fact, nonzero counts were following by missing counts 19% (4,226/22,517) of the time, whereas zero counts were followed by missing counts 36% (4,467/12,532) of the time. So the data provide strong evidence of a discontinuation bias: leks on which no sage-grouse were observed in one year were less likely to be surveyed the following year. As an aside, a missing count was followed by another missing count 82% of the time, indicating a rather low probability that a missing count would be resumed the following year. The net effect of the discontinuation bias is that the data set includes disproportionately fewer abandoned leks than would be representative, and our average lek trend estimates may be biased high.

Our analysis is but a snapshot of the time period selected because it had the highest-quality data. In many instances, the lek count data we used may not temporally relate to when the anthropogenic stressors examined were added to the landscape. As examples, most of the conversion of fertile soils supporting sagebrush to agriculture occurred during the first half of the 20th century, cheatgrass was well established throughout much of the Intermountain West by the 1920s, and the majority of paved interstate highways were opened to traffic in the 1960s and 1970s (Connelly et al. 2004). These developments may have caused the extirpation of populations from a region (Aldridge et al. 2008), or leks that remained active may have been isolated from effects of the disturbances (Braun et al. 2002). The results we report here therefore may not accurately reflect the total response of populations to the addition of these factors.

A final caveat is that ours is an observational, rather than experimental, study (Shaffer and Johnson 2008). We were not able to assign treatments (reflected by the various explanatory variables: elevation, landscape composition, roads, towers, etc.) in a balanced, random manner. Of the three cornerstones of inference, we lack control and randomization, but have fair to excellent replication both within SMZs and among them. One consequence of this is confounding of the explanatory variables, which makes it difficult, at best, to determine which of them is responsible for any effect on the response variable. A more severe consequence is a greater risk incurred by presuming that associations observed in the data reflect causation.

Biological Implications

Greater Sage-Grouse populations across the range of the species are influenced by a myriad of natural and anthropogenic factors. A stressor may affect populations throughout the range, or it may be unique to a single population. For example, sage-grouse populations in the upper Green River Basin of western Wyoming (Wyoming Basin SMZ) are exposed to influences from energy development (Lyon and Anderson 2003, Holloran 2005), populations in the Powder River Basin of northern Wyoming and southern Montana (Great Plains SMZ) must deal with energy development and West Nile virus (Walker et al. 2007a,b; Doherty 2008), and populations in the upper Snake River Plain of southeastern Idaho (Snake River Plain SMZ) and Yakima–Kittitas Counties in central Washington (Columbia Basin SMZ) face agricultural conversion and the effects of fire (Leonard et al. 2000, Nelle et al. 2000, Schroeder et al. 2000, Wambolt et al. 2002). Trends of populations in individual areas may additionally be related to stressors acting at larger spatial scales, such as livestock grazing or drought (Connelly and Braun 1997, Connelly et al. 2000c). Although we interpreted only single-variable analyses here because multivariable analyses failed to reveal any additional patterns in the data, effects of individual stressors on sage-grouse populations should be considered in the cumulative context of the suite of stressors affecting populations.

Patterns in Lek Counts

Consistent patterns in sage-grouse population trends as estimated from lek counts across SMZs (except the Columbia Basin, where populations generally declined) suggest the effects of prevailing conditions that influence populations at broad spatial scales. Our study period (1997–2007) began when populations were generally at low levels (Connelly et al. 2004; Garton et al., this volume, chapter 15) and included two periods of increasing populations and one period of decreasing populations. Two population peaks occurred, in 1999–2001 and in 2006, along with one period of low population in 2002–2004. This fact may explain why there were more leks with increasing versus decreasing trends during our study period, even though most populations generally had been declining range-wide since the 1960s (Connelly et al. 2004; Garton et al., this volume, chapter 15).

Regardless of whether sage-grouse populations increased or decreased during our study period, the analytical techniques we used compared relative patterns, for example, how numbers at leks with towers nearby changed in comparison to leks without nearby towers. Given the consistent effects of prevailing conditions, the relative patterns established across the large number of and broad spatial distribution of leks investigated suggest relationships that were essentially standardized for these prevailing conditions.

Physical Attributes, Land Cover, and Fire History

The distribution of sage-grouse is closely aligned with the distribution of sagebrush-dominated landscapes (Schroeder et al. 2004). The well-documented dependence of the species on sagebrush cannot be overemphasized (Patterson 1952, Connelly et al. 2000c, Hagen et al. 2007). Lek trends across the range of the species were positively associated with the proportions of tall-stature sagebrush and all sagebrush land covers within 5- and 18-km radii of the lek location. Also, the coverage of sagebrush was greater within 5 km than within 18 km. Low sagebrush was common in only the Northern Great Basin SMZ, where it was positively associated with sage-grouse. Walker et al. (2007a) found strong support for models relating lek persistence in southern portions of the Great Plains SMZ with the proportion of sagebrush habitat within 6.4 km. Aldridge et al. (2008) predicted that across the range of the species areas where sage-grouse persisted, compared with areas where populations were extirpated, were those containing at least 25% sagebrush cover within 30 km. Clearly, sagebrush at both local and landscape scales is a necessary, if not sufficient, requirement for viable sage-grouse populations.

Sage-grouse depend on sagebrush through all seasonal periods. Females consistently select areas with more sagebrush canopy cover than other available locations during nesting (Sveum et al. 1998b, Holloran et al. 2005, Hagen et al. 2007), and increased shrub cover at nests has been linked to increased probability of a successful hatch (Wallestad and Pyrah 1974, Gregg et al. 1994). Sage-grouse broods throughout Wyoming selected relatively dense sagebrush stands between hatching and 2 weeks post-hatching, presumably for thermal and predator protection of young chicks (Thompson et al. 2006). During summer, sage-grouse select mesic areas closely associated with sagebrush that are used for loafing and security cover (Klebenow 1982). And the primary requirement of wintering sage-grouse is sagebrush exposure above the snow (Patterson 1952, Hupp and Braun 1989, Crawford et al. 2004); winter ranges in general are characterized by large expanses of dense sagebrush (Eng and Schladweiler 1972, Beck 1977).

Lek trends were positively related to elevation in the northwestern portion of sage-grouse range. The positive relation was consistent over all elevations in the Northern Great Basin and at elevations greater than average in the Snake River Plain and Columbia Basin SMZs. These SMZs have lower mean lek elevations (Table 17.1). The risk of invasion by exotic annual grasses such as cheatgrass in big sagebrush communities is greatest at lower-elevation (<1,500 m), drier sites (Hemstrom et al. 2002, Crawford et al. 2004). Because of shortened fire-return intervals and the inability of sagebrush-steppe plant communities to recover from fire quickly, invasion of a sagebrush-dominated site by annual grasses may result in the permanent conversion of that site into one that is unsuitable for sage-grouse (d'Antonio and Vitousek 1992, Brooks et al. 2004a, Baker 2006a). Although SMZs differed little in the average proportional area dominated by exotic vegetation (Table 17.1), few leks across the range of the species had >8% cover within 5 km or 18 km, suggesting that when the extent of the landscape dominated by exotic vegetation becomes relatively high, leks become inactive (Enyeart 1956; Connelly et al. 2000c; Miller et al., this volume, chapter 10).

The functional conversion of relatively small amounts of landscape either through exotic species invasion or agricultural conversion had pronounced negative effects on sage-grouse populations. Lek count trends tended to be more negative for leks with greater proportions of either exotic species or agricultural land cover, even at low proportions. Extirpation of sage-grouse throughout the species' range was found to be most likely in areas that had >25% cultivated cropland within 30 km (Aldridge et al. 2008). Inactive sage-grouse leks were noted to have a higher percentage of tillage agriculture within 4 km and within 6.4 km compared with active leks in eastern (Smith 2003) and southern portions of the Great Plains (Walker et al. 2007a). Additionally, replacement of native understory vegetation by exotic annual grasses such as cheatgrass has altered fire regimes, resulting in losses of sagebrush over large expanses of the Southern Great Basin, Snake River Plain, Northern Great Basin, and the Columbia Basin SMZs (Mack 1981, Knick and Rotenberry 1997, Wambolt et al. 2002, Crawford et al. 2004, Beck et al. 2009).

Greater Sage-Grouse populations have consistently responded negatively to habitat management practices that manipulated large proportions of sagebrush on a landscape (Connelly et al. 2000c). We found that lek count trends across the range of the species generally were lower the greater the area burned within 5 km and 18 km. No relationship was evident between number of post-fire years and lek count trends, however. Baker (2006a) reports that the time required to burn once through an area in big sagebrush-dominated habitats is >70 years, indicating that the 11 years of our analysis was inadequate to determine long-term effects of fire on sage-grouse populations.

In most SMZs, sage-grouse lek trends tended to be higher at higher elevations. Exceptions were the Great Plains, with no consistent trend, and the Wyoming Basin, where a peak at about 2,000 m was evident. This association may reflect the greater human development and disturbances associated with valleys in mountainous areas.

We found no consistent relationship between lek count trends and mean precipitation across SMZs. However, our study period may be too short to identify potential long-term effects of precipitation on populations. Severe, persistent drought conditions have coincided with declining sage-grouse populations throughout the range of the species both historically (1930s) and recently (mid-1980s to early 1990s) (Patterson 1952, Connelly and Braun 1997, Aldridge et al. 2008), and researchers have surmised that grouse numbers were reduced during periods of below-average precipitation (Slater 2003, Holloran et al. 2005, Moynahan et al. 2007). Mean annual precipitation by states throughout the species' range suggests that below-average conditions have persisted for one to four ($\bar{x}$ = 1.7) consecutive years since 1960. Our use of mean precipitation during the 11-year period did not allow us to explore any short-term effects of annual precipitation on sage-grouse populations.

Anthropogenic Features

Sage-grouse populations and distributions in many areas have diminished as a result of modifications to sagebrush-dominated landscapes from anthropogenic activities (Connelly et al. 2000c, Knick et al. 2003). Few sage-grouse leks were located within 5 km of developed land, and a strong negative relationship was found between lek count trends and the proportion of the landscape developed within 5 km or 18 km. Many of the anthropogenic stressors associated with urban and suburban areas may act synergistically to cause sagebrush habitat fragmentation (Knick et al. 2003, Leu et al. 2008). As a landscape becomes bisected into smaller patches of habitat, functional connectivity among patches may be lost (With and Crist 1995). A patch of habitat that by itself is too small to support sage-grouse may nonetheless be valuable to the species if it is connected to other patches of suitable habitat. If that connection is lost, the patch may no longer have value (Fahrig 2003). Viable sage-grouse populations require large landscapes (Patterson 1952, Connelly et al. 2000c), so fragmentation resulting even from scattered disturbances may lead to disproportionate population declines or regional extinctions.

Lek count trends tended to be lower on leks that had >20 km of interstate, federal, or state highway within 18 km and on leks nearer such highways, but no apparent relationship occurred between lek count trends and the presence of secondary roads. Avoidance of roads by Greater Sage-Grouse (Lyon and Anderson 2003), Gunnison Sage-Grouse (Oyler-McCance 1999), and Lesser Prairie-Chickens (*Tympanuchus pallidicinctus*; Crawford and Bolen 1976, Hagen 2003) has been documented. Given that most major highways were constructed before 1997, their major effects likely occurred before our study period, and leks near highways that have persisted may be situated where topography has isolated them from these sources of disturbance (Braun et al. 2002). However, declines in count trends during the study period suggest a continuing disturbance associated with highways, possibly due to increased traffic levels, which have been identified as reducing numbers of sage-grouse occupying leks (Remington and Braun 1991, Holloran 2005).

The apparent reaction of sage-grouse to vertical structures at leks varied by the type of structure. Lek count trends were negatively related to proximity to the closest communication tower and to the number of towers within 18 km. However, trends were not consistently related to the distance to nearest power line or to the length of power line within 5 km or 18 km. Towers typically indicate high human-use areas, whereas power lines (especially transmission lines) are more uniformly distributed across the landscape. The lower trends at sage-grouse leks near towers may be in response to these spatially associated activities and not the towers themselves. However, towers themselves may be stressors, and differences in relations between lek trends and the two types of vertical structures may be due to the different times they were erected. Most power lines were in place prior to our study period, and any effects they had may have already occurred. In contrast, communication towers have only recently become common in the area, and sage-grouse populations may have responded to them during our study period. Other research has suggested that sage-grouse avoid transmission lines in general and during the breeding season (Ellis 1985, Braun 1998), and Lesser Prairie-Chickens also avoid them in general and when nesting (Hagen 2003, Pitman et al. 2005).

Evidence from studies conducted in the Great Plains and Wyoming Basin suggests that high-density development of energy resources excludes sage-grouse from developed areas (Holloran 2005, Walker et al. 2007a, Doherty 2008). Our analysis indicates that, across the range of the species, trends on leks within 5 km of a producing oil or natural gas well were depressed. Trends also were lower on leks with more than 10 producing wells within 5 km or more than 160 wells within 18 km. The effect-distance of the infrastructure of gas or oil developments we found was similar to those from other studies; the well density results differed, however. Naugle et al. (this volume, chapter 20) suggested that impacts to sage-grouse leks from energy development remained discernible out to distances of 6.2–6.4 km at study sites in the western Wyoming Basin and southern Great Plains SMZs. The authors further stated that breeding populations of sage-grouse were influenced by development in excess of one pad/2.6 km^2 and impacts at well densities of eight pads/2.6 km^2 exceeded the species' threshold of tolerance (Naugle et al., this volume, chapter 20). We cannot assume a uniform distribution of wells within the buffers examined and thus cannot conclusively estimate thresholds of well density at which sage-grouse are affected, but our results conservatively suggest that a density of more than one producing well/6.4 km^2 within 18 km of leks negatively influences lek count trends.

Portions of most pipelines are inherently near oil or gas wells, suggesting that, because of the effects on sage-grouse populations of producing wells described above, the magnitude of the effect of pipelines is confounded by the relative proximity of other forms of disturbance. We found slightly lower count trends on leks near pipelines in only two regions, and a slight negative relation between trend in lek counts and length of pipeline within 5 km or 18 km. The difference in the magnitude of apparent effects of pipelines and producing wells (slight vs. relatively severe, respectively) suggests that pipelines by themselves do not relate to lower population trends.

Although assemblages of stressors varied by SMZ, the cumulative effects of increased human activity appear to relate negatively to sage-grouse population trends across the species' range. Many of the individual stressors we investigated were prevalent in a few SMZs; in these instances, relationships between trends in lek counts across all SMZs can be generalized to those zones contributing the bulk of the data. The Columbia Basin had high proportions of agricultural and developed land and high values for towers, power lines, highways, and the human-footprint index relative to other SMZs. This disparity may explain why lek counts from 1997 to 2007 generally declined in the Columbia Basin while they fluctuated in the other SMZs. Energy development influenced populations predominantly in the Great Plains, Wyoming Basin, and Colorado Plateau. The Great Plains and Wyoming Basin also had relatively high values for agricultural conversion, towers, and highways, whereas the Wyoming Basin and Colorado Plateau had relatively high values for power lines. The area burned around leks was high in the Snake River Plain, and this zone also had relatively high values for exotic species invasion, towers, and highways. The Southern Great Basin and Northern Great Basin did not have high values for any of the directly anthropogenic stressors examined, but did have relatively high values for exotic species invasion and area burned; the Southern Great Basin also had relatively high values for towers and highways.

We stress that our results are correlative and do not establish causation. For example, human developments, roads, and a prevalence of agricultural land may not generally cause direct mortalities to sage-grouse, but these stressors do provide habitats and often reliable and abundant food sources for predators such as corvids and red foxes (*Vulpes vulpes*), which may reduce the productivity and survival of sage-grouse (Connelly et al. 2000a, Storch et al. 2005, Marzluff and

Neatherlin 2006). Additionally, mechanistic explanations of responses of populations to different stressors may not be readily apparent. For example, avoidance of natural gas fields by sage-grouse appears to be dictated by the response of yearling cohorts to development (Holloran et al. 2010), and fidelity to breeding sites results in temporal lags that differ depending on adult survival probability differences between males and females (Zablan et al. 2003). Focused research on the stressors discussed is needed to determine causation and to identify mechanisms behind population-level responses. This information is necessary to establish effective management options. For example, raptor perch deterrents on vertical structures may not mitigate the effects of these structures if sage-grouse population declines result from avoidance of habitats in close proximity and not reduced survival due to changes in predator distributions (Avian Power Line Interaction Committee 2006, Lammers and Collopy 2007).

The cumulative effects of relatively low levels of anthropogenic features in landscapes surrounding leks were associated with reduced sage-grouse population trends. Consistent patterns in population trends across SMZs regardless of differences in the suite of stressors in each zone suggest functional similarities between stressors. Lek count trends were lower when human-footprint scores exceeded 2 at leks, or when median scores exceeded 3 within either 5 km or 18 km of a lek. The human-footprint index is a measure of the totality of direct anthropogenic features on a landscape expressed on a one to 10 scale. We have discussed individually the relations between range-wide population trends and several of the stressors included in this index. Our results from a large number and broad spatial distribution of leks complement those from more intensive studies on smaller areas. Findings from both kinds of studies are important for identifying site-specific features that could threaten Greater Sage-Grouse populations, and for effectively managing the species throughout its range.

CONSERVATION IMPLICATIONS

We found that the most pronounced effects on sage-grouse populations were from factors that recently have become more pronounced on western landscapes—gas and oil wells and communication towers. However, several of the stressors that first occurred on landscapes prior to our study period (e.g., conversion of range to cultivation and paved highways) appeared to still be influencing populations, even if their major effects occurred before our study period.

Managers responsible for sage-grouse should consider the cumulative nature of anthropogenic activity when developing management plans for the species. For example, the development of an energy field involves more than just the infrastructure needed to extract resources. Urban areas near the developing field will expand, communication towers and transmission lines will be erected, and traffic on highways will increase, to name a few of the additional factors that accompany energy development. Unintended consequences to development will occur, as evidenced by the proliferation of West Nile virus in the coal-bed methane fields of the Great Plains (Walker et al. 2007b). These stressors generally will be added to those affecting populations prior to development, such as livestock grazing, drought, and exotic species (Connelly et al. 2000c). No single factor is responsible for declining sage-grouse populations, and no single action may be sufficient to restore them. Conservation of the species will initially require a recognition of the intrinsic value of sagebrush-dominated landscapes, followed by the development of a comprehensive approach to sagebrush habitat conservation that involves commitments and partnerships among state and federal agencies, academia, industry, private organizations, and landowners; Knick et al. (2003:627) affirm that only through this concerted effort and commitment can we afford to be optimistic about the future of sagebrush ecosystems and their avifauna.

ACKNOWLEDGMENTS

We thank the state and federal biologists and volunteers who spent many hours collecting and compiling lek attendance data. We also thank members of the Western Sage and Columbian Sharp-tailed Grouse Technical Committee for insights they provided on lek counts. We appreciate T. J. Christiansen's help with the Wyoming database. Earlier drafts of this manuscript were greatly improved by reviews from C. E. Braun, B. S. Cade, M. M. Rowland, M. A. Schroeder, and several anonymous reviewers.

CHAPTER EIGHTEEN

Factors Associated with Extirpation of Sage-Grouse

Michael J. Wisdom, Cara W. Meinke, Steven T. Knick, and Michael A. Schroeder

Abstract. Geographic ranges of Greater Sage-Grouse (*Centrocercus urophasianus*) and Gunnison Sage-Grouse (*C. minimus*) have contracted across large areas in response to habitat loss and detrimental land uses. However, quantitative analyses of the environmental factors most closely associated with range contraction have been lacking, results of which could be highly relevant to conservation planning. Consequently, we analyzed differences in 22 environmental variables between areas of former range (extirpated range), and areas still occupied by the two species (occupied range). Fifteen of the 22 variables, representing a broad spectrum of biotic, abiotic, and anthropogenic conditions, had mean values that were significantly different between extirpated and occupied ranges. Best discrimination between extirpated and occupied ranges, using discriminant function analysis (DFA), was provided by five of these variables: sagebrush area (*Artemisia* spp.); elevation; distance to transmission lines; distance to cellular towers; and land ownership. A DFA model containing these five variables correctly classified >80% of sage-grouse historical locations to extirpated and occupied ranges. We used this model to estimate the similarity between areas of occupied range with areas where extirpation has occurred. Areas currently occupied by sage-grouse, but with high similarity to extirpated range, may not support persistent populations. Model estimates showed that areas of highest similarity were concentrated in the smallest, disjunct portions of occupied range and along range peripheries. Large areas in the eastern portion of occupied range also had high similarity with extirpated range. By contrast, areas of lowest similarity with extirpated range were concentrated in the largest, most contiguous portions of occupied range that dominate Oregon, Idaho, Nevada, and western Wyoming. Our results have direct relevance to conservation planning. We describe how results can be used to identify strongholds and spatial priorities for effective landscape management of sage-grouse.

Key Words: Centrocercus minimus, Centrocercus urophasianus, extirpated range, extirpation, Greater Sage-Grouse, Gunnison Sage-Grouse, range contraction, sagebrush.

Factores Asociados a la Extirpación del Sage-Grouse

Resumen. Las distribuciones geográficas del Greater Sage-Grouse (*Centrocercus urophasianus*) y el Gunnison Sage-Grouse (*C. minimus*) se han

Wisdom, M. J., C. W. Meinke, S. T. Knick, and M. A. Schroeder. 2011. Factors associated with extirpation of Sage-Grouse. Pp. 451–472 *in* S. T. Knick and J. W. Connelly (editors). Greater Sage-Grouse: ecology and conservation of a landscape species and its habitats. Studies in Avian Biology (vol. 38), University of California Press, Berkeley, CA.

contraído a través de extensas áreas en respuesta a la pérdida de hábitat y a usos perjudiciales del suelo. Sin embargo, se carece de análisis cuantitativos de los factores ambientales que más se asocian a la contracción del territorio, cuyos resultados podrían ser altamente relevantes al planeamiento de la conservación. Por lo tanto, analizamos diferencias en 22 variables ambientales entre las áreas del territorio original (territorio extirpado), y las áreas todavía ocupadas por las dos especies (territorio ocupado). Quince de las 22 variables, representando un amplio espectro de condiciones bióticas, abióticas, y antropogénicas, tuvieron valores medios que resultaron significativamente diferentes entre los territorios extirpados y ocupados. La mejor discriminación entre los territorios extirpados y ocupados, usando el análisis de función discriminante (DFA), fue proporcionada por cinco de estas variables: área del sagebrush (*Artemisia* spp.); elevación; distancia a las líneas de transmisión; distancia a las torres celulares; y propiedad del terreno. Un modelo de DFA que contenía estas cinco variables clasificó correctamente >80% de las ubicaciones históricas del sage-grouse como territorios extirpados y ocupados. Utilizamos este modelo para estimar la semejanza entre las áreas del territorio ocupado con las áreas donde ha ocurrido la extirpación. Las áreas ocupadas actualmente por sage-grouse, pero con alta semejanza al territorio extirpado, pueden no ser capaces de sostener a las poblaciones persistentes. Las estimaciones del modelo demostraron que las áreas de mayor semejanza están concentradas en las porciones más pequeñas y divididas del territorio ocupado, y a lo largo de las periferias del territorio. Extensas áreas en la porción este del territorio ocupado también tuvieron gran semejanza con el territorio extirpado. Por el contrario, las áreas de menor semejanza con el territorio extirpado están concentradas en las porciones más grandes y más contiguas del territorio ocupado que dominan Oregon, Idaho, Nevada, y Wyoming occidental. Nuestros resultados tienen relevancia directa al planeamiento de la conservación. Describimos cómo los resultados pueden utilizarse para identificar baluartes y prioridades espaciales para el eficaz manejo del paisaje de sage-grouse.

Palabras Clave: artemisa, *Centrocercus minimus, Centrocercus urophasianus,* contracción del rango geográfico, extirpación, Greater Sage-Grouse, Gunnison Sage-Grouse, rango geográfico extirpado.

Species across the world are threatened by human activities that degrade and eliminate habitats at a massive scale. The World Conservation Union estimates that >12,000 species are at risk of extinction from the pervasive and accelerating effects of human-associated causes of habitat loss (Baillie et al. 2004). Habitat loss is reflected in range contraction for many widely distributed species. Large, contiguous ranges of many terrestrial species have become smaller and fragmented, resulting in population isolation and increased vulnerability to extirpation and extinction. In western North America, a myriad of widely distributed birds and mammals have experienced large contractions in their historical ranges in response to habitat loss and detrimental human activities (Wisdom et al. 2000a, Laliberte and Ripple 2004).

Range contraction for many species is well documented and the causes generally accepted. However, the specific changes in environmental conditions associated with contraction often are not well studied and thus poorly quantified. Consequently, specific factors and their threshold values associated with range contraction, or regional extirpation of a species, have rarely been documented (see Laliberte and Ripple 2004 as an exception). The advent of continuous coverage spatial data now allows environmental conditions to be summarized across vast areas, encompassing extirpated and occupied portions of a species historical range. These spatial data provide novel and compelling opportunities for formal analysis of conditions associated with extirpation in areas where species ranges have contracted (Aldridge et al. 2008). Differences in environmental conditions between extirpated and occupied portions of a species' historical range could provide important insights for conservation planning and recovery. This is particularly true for many species whose populations are declining and considered imperiled, yet data are insufficient to conduct a formal population viability analysis (Morris and Doak 2002).

Greater Sage-Grouse (*Centrocercus urophasianus*) and Gunnison Sage-Grouse (*C. minimus*) (collectively referred to as sage-grouse) are typical of many widely distributed species whose ranges have contracted in response to habitat loss and detrimental land uses. Habitats and populations have declined steadily over long periods and across large areas (Connelly and Braun 1997, Braun 1998, Schroeder et al. 1999, Connelly et al. 2004, Aldridge et al. 2008) resulting in widespread range contraction (Schroeder et al. 2004). Notably, sage-grouse are strongly associated with sagebrush (*Artemisia* spp.), and like many other sagebrush-associated vertebrates, are highly vulnerable to regional extirpation because of extensive habitat loss and degradation (Raphael et al. 2001).

Our goal was to identify environmental factors associated with regional extirpation of sage-grouse. Our specific objectives were to: (1) identify spatially explicit environmental factors most strongly associated with, and providing the best discrimination between, currently occupied versus extirpated ranges of sage-grouse; (2) use these factors in a spatially explicit model to estimate the similarity of remaining areas of occupied range with areas where extirpation has occurred as a means of identifying areas where sage-grouse may be vulnerable to extirpation; (3) interpret results for conservation planning at regional and range-wide spatial extents, and (4) describe data deficiencies and research needs to enhance knowledge about environmental conditions that potentially contribute to sage-grouse extirpation at regional extents.

METHODS

We used six steps to meet our objectives: (1) delineate boundaries of currently occupied versus extirpated portions of sage-grouse historical range; (2) obtain or derive continuous-coverage spatial layers for all environmental variables likely to differ between occupied and extirpated ranges based on known or hypothesized environmental associations with sage-grouse at landscape scales; (3) develop an unbiased system of sampling or census of these environmental variables in occupied versus extirpated ranges at a spatial extent compatible with that used by sage-grouse populations to meet year-round needs, and consequently, the extent at which regional extirpation may occur; (4) use the system to analyze patterns and differences in environmental variables between occupied and extirpated ranges; (5) build and validate spatial models based on these patterns and differences that best discriminate between occupied and extirpated ranges; and (6) apply the best-performing model to different regions of occupied range to estimate each region's similarity with areas where extirpation has occurred.

Step 1: Range Delineation

We used the range map for Greater and Gunnison Sage-Grouse as the basis for identifying their occupied and extirpated ranges (Schroeder et al. 2004). The historical ranges of the two species could not always be distinguished. Until recently, the two species were considered one, and historical records often were identified simply as sage-grouse (Schroeder et al. 2004). As a result, our analysis combines both species, recognizing that most areas of their collective ranges were and continue to be dominated by Greater Sage-Grouse (Schroeder et al. 2004). Both species have similar environmental requirements and respond similarly to habitat loss from human activities, and both have undergone substantial range contractions in response to habitat loss (Oyler-McCance et al. 2001, Rowland 2004).

The range map of Schroeder et al. (2004) depicts the potential pre-settlement and current range of sage-grouse. Potential pre-settlement was defined as the range before 1800, when settlement of western North America by large numbers of Euro-Americans had not yet occurred. We assumed that the potential pre-settlement range not currently occupied represented areas where sage-grouse once existed but now are extirpated. This assumption is supported by the large number of sage-grouse collected or observed during the latter phases of Euro-American settlement (late 1800s and early 1900s) in areas where sage-grouse no longer exist. Collected specimens or unambiguous observations of sage-grouse provided clear evidence of areas where sage-grouse occurred historically, although collections and observations were not systematic across the range and exact locations not always documented. Given this background information, we assumed that potential pre-settlement range, minus the current range, represented the best estimate of areas where sage-grouse have been extirpated. We refer to current range as occupied and to potential pre-settlement range, excluding current range, as extirpated.

Step 2: Environmental Variables

We identified 22 environmental variables relevant to sage-grouse or sagebrush landscapes whose values likely differed between occupied and extirpated ranges (Table 18.1). Most variables were identified from earlier research as being associated with sage-grouse extirpation at large spatial extents (>100,000 ha; Oyler-McCance et al. 2001, Wisdom et al. 2002c, Aldridge and Boyce 2007, Aldridge et al. 2008), or that have modified sagebrush habitats across large areas of sage-grouse range (Schroeder et al. 1999, Rowland 2004). Other variables represented common landscape features potentially helpful for accurate discrimination between occupied and extirpated ranges. Inclusion of these additional variables was important because of the paucity of prior landscape research on sage-grouse–environmental relations and our objective to identify the best discriminators between occupied and extirpated ranges, regardless of whether such variables had previously been evaluated as causal factors of extirpation.

Nine of the 22 variables were biological measures such as area, patch size, and fragmentation of sagebrush. Five variables were abiotic measures including precipitation, elevation, and soil characteristics. Eight variables were anthropogenic measures such as distance to roads, area in agriculture, and human population density. Of the 22 variables, 16 were raster-based and 6 were vector-based (polygon- or contour-based) estimates (Table 18.1).

Map resolution (cell size, polygon size, or contour interval) differed by variable, but most raster-based estimates used a 90-m cell size, and contour-based estimates used a resolution as fine as 10 m (Table 18.1). Variables also had to be available as continuous-coverage layers in a geographic information system (GIS) and encompass most areas of pre-settlement range. Some fringes of pre-settlement range in the United States and in Canada could not be analyzed because variables were not available in continuous coverage or in compatible GIS formats. These small areas not included in our analysis composed <2% of sage-grouse pre-settlement range. Estimates of variables were made for 2000–2004, and thus were compatible with the time frame in which sage-grouse ranges were delineated (Schroeder et al. 2004).

Variables used in our analysis were assumed to affect or be associated with changes in sage-grouse habitats or populations at regional spatial extents (≥100,000 ha). Analysis at regional extents was purposefully different than more typical analyses conducted at smaller spatial extents (<100,000 ha), such as evaluation of factors within a seasonal range or a specific use area (e.g., evaluating a lekking, nesting, brood-rearing, or wintering area used by individual sage-grouse or a subpopulation). Consequently, variables included in our analysis did not include all factors associated with smaller areas of fine-scale habitat use or subpopulation dynamics (Connelly et al. 2000c; Connelly et al., this volume, chapter 4). In addition, some variables potentially associated with population dynamics of sage-grouse at regional extents, such as livestock stocking rates and grazing systems, were not available in continuous coverage formats, and thus could not be considered for analysis.

Step 3: Sampling Design

We used historical locations of sage-grouse for analyzing differences in environmental variables between occupied and extirpated ranges. Historical locations came from two sources (Schroeder et al. 2004): museum specimens collected mostly during the early 1900s and published observations documented for this period. Historical locations represent documented areas of occurrence in pre-settlement range (Schroeder et al. 2004).

We used 375 of >1,300 historical locations after eliminating multiple collections or observations from the same locations and excluding locations or observations clearly outside the established pre-settlement range where individual birds may have occasionally occurred (Schroeder et al. 2004). Use of historical locations focused our analysis on documented areas of species occurrence before and during European settlement, in contrast to an analysis of randomly selected areas within pre-settlement range that might include regions not having direct physical evidence of species occurrence.

Each historical location was classified as occupied or extirpated range. A circle with an 18-km radius, encompassing an area of 101,740 ha, was then centered on each historical location (Fig. 18.1). Of the 375 historical locations, 239 were in occupied range and 136 were in extirpated

TABLE 18.1

Descriptions of the 22 environmental variables used in discriminant function analysis.

Estimates of the variables were made for the time period 2000–2004, and thus are compatible with the time period in which sage-grouse ranges were estimated (Schroeder et al. 2004). Estimates of the 22 variables were based on conditions within the circles of 18-km radius that encompassed each of the 375 historical locations of sage-grouse. Raster-based variables were derived or estimated using a 90 × 90-m cell size unless stated otherwise.

Variable	Type	Definition and estimation method
Sagebrush area (%)	Raster	Percentage of 18-km radius composed of sagebrush cover types[a].
Patch size	Raster	Mean size (ha) of sagebrush patches, where a patch is defined as the cells of sagebrush cover types that are contiguous with one another (touching on at least one side)[b].
Patch density	Raster	Number of sagebrush patches divided by the area[b].
Edge density 1	Raster	Number of edges between sagebrush patches and non-sagebrush cover types, weighted by sagebrush area. Weighting by sagebrush area differentiates between a low number of edges when little sagebrush is present versus a low number of edges when sagebrush occupies most or all of the area. Resulting values were transformed as 1/n, such that high edge density indicates a high amount of edge, and low edge density indicates low edge[b].
Edge density 2	Raster	Total length (m) of all edges between sagebrush patches and non-sagebrush cover types divided by area[b].
Nearest neighbor	Raster	The mean distance (m) between sagebrush patches, where distance between each patch is measured as the shortest distance (edge to edge) to another patch within the circle[b, c].
Proximity index	Raster	The mean proximity (unitless scale) among sagebrush patches. Mean proximity is calculated as the area of each sagebrush patch divided by the squared mean distance of all distances between the patch and all other patches in the circle, with these values summed for all patches in the circle and divided by the total number of patches[b].
Core area	Raster	The mean size (ha) of core areas of sagebrush. A core area is defined as a sagebrush patch plus all additional cells of sagebrush within 100 m of the edge of each patch (i.e., all additional sagebrush within the distance of two cells from the edge of each sagebrush patch).
Distance to occupied–extirpated boundary	Vector	Distance (m) from the sage-grouse historical location (the center of each circle) to the boundary between occupied and extirpated range[b].
Precipitation	Raster	Mean annual precipitation (cm) within each 18-km circle for the period 1961–2004. Precipitation estimates were derived from parameter-elevation regression on independent slopes model (PRISM), which uses point data and a digital elevation model (DEM) to generate grid-based estimates of annual, monthly, and event-based climatic parameters[d].
Elevation	Raster	Mean elevation (m) among all cells, using a 1:24,000-scale digital elevation model downloaded from the United States Geological Survey National Elevation Dataset[d].
Soil water capacity	Raster	The total amount of water available in all soil profiles (cm of water/cm of soil) for each cell, averaged over all cells. Estimates were derived from the USDA Natural Resources Conservation Service[d].

TABLE 18.1 *(continued)*

TABLE 18.1 (CONTINUED)

Variable	Type	Definition and estimation method
Soil rock depth	Raster	The mean depth (cm) to bedrock, or soil depth, as averaged over all cells. Estimates were derived from the USDA Natural Resources Conservation Service[d].
Soil salinity	Raster	The mean salinity (mmhos/cm) of soil, as averaged over all cells. Estimates were derived from the USDA Natural Resources Conservation Service[d].
Agriculture area	Raster	Percentage of cells of agricultural cover types[d].
Human density	Raster	Number of humans/ha in 2000[e].
Distance to roads	Vector	Distance (m) to the nearest road. All roads identified in the 2000 United States Census Bureau 1:100,000 scale line files[d].
Road density	Vector	Linear km of road per unit area. All roads identified in the 2000 United States Census Bureau 1:100,000 scale line files[d].
Distance to highways	Vector	Distance (m) to the nearest major highway[f].
Distance to transmission lines	Vector	Distance (m) to the nearest electrical transmission line.
Distance to cellular towers	Vector	Distance (m) to the nearest cellular tower, based on locations of towers registered with the Federal Communications Commission.
Land ownership	Raster	Dominant land ownership, either public or private, based on state and federal sources of ownership data[d]. These data were summarized as the percentage of circles dominated by public land.

[a] Sagebrush cover types were defined and estimated by Comer et al. (2002) and further described by Wisdom et al. (2005b).
[b] Landscape statistics estimated using FRAGSTATS (McGarigal et al. 2002).
[c] Gustafson and Parker (1994).
[d] Data available at http://sagemap.wr.usgs.gov.
[e] United States census block data (United States Census Bureau 2001).
[f] Major highways documented in the National Atlas (http://nationalatlas.gov/) (Comer et al. 2002, Wisdom et al. 2005b).

range. Portions of some of the associated circles overlapped the boundary between occupied and extirpated ranges. We retained these locations for analysis because the majority of area in the circle was always in the same portion of range as its historical location, and we wanted to include the full spectrum of environmental conditions across areas far from, and close to, the occupied–extirpated range boundary.

Step 4: Environmental Conditions

We used each historical location and associated 18-km radius as our unit of observation to analyze conditions for each environmental variable in occupied versus extirpated range (Table 18.1). For this analysis, we first calculated the composite value of each environmental variable within each circle. The composite value was the average of all values for a variable that composed the cells, polygons, or contours within the circle. We then calculated the mean and associated 95% confidence interval (CI) for the composite values among all circles associated with occupied ($N = 239$) and extirpated range ($N = 136$)(Fig. 18.1).

We treated each circle as a sample unit, although most (279 of 375) circles overlapped one another on their outer edges. However, most of the area within circles did not overlap other circles ($\bar{x}$ overlap = 22%). Moreover, circle overlap occurred most often along the occupied–extirpated range boundary, where we chose to retain circles because of their contribution to the occupied–extirpated gradient.

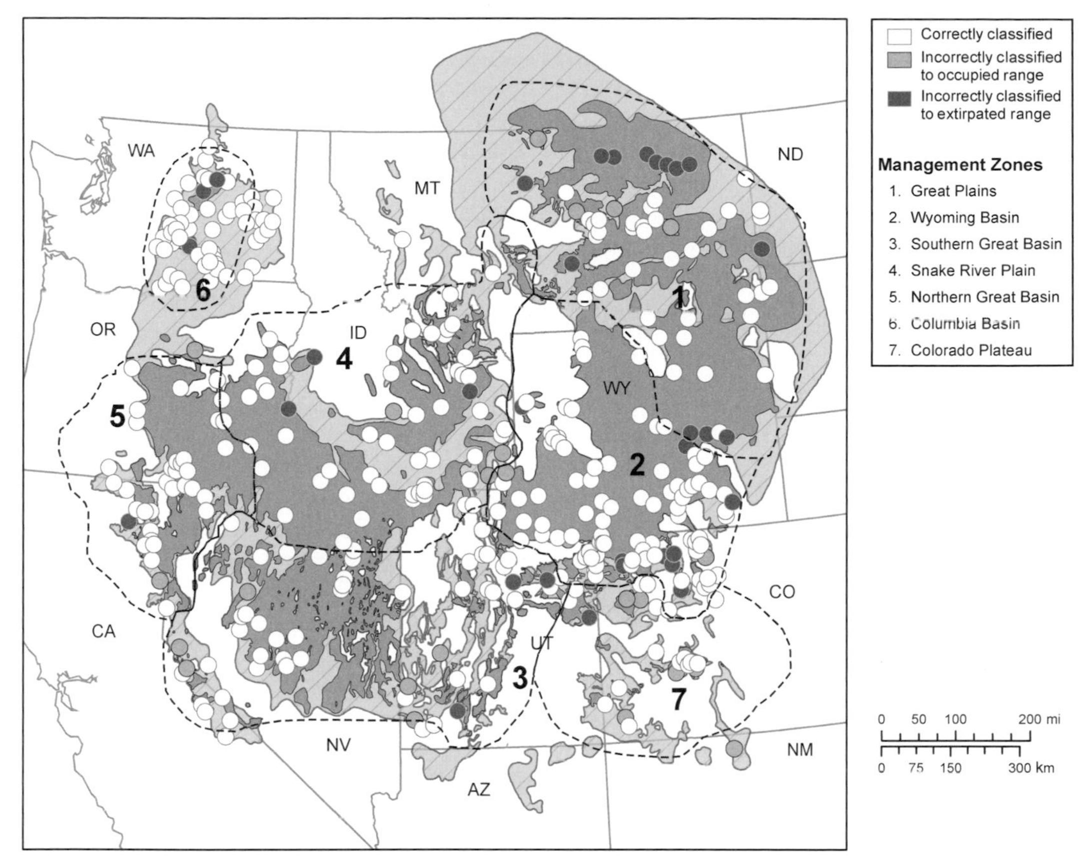

Figure 18.1. Distribution of 375 historical locations (circles) of sage-grouse (Schroeder et al. 2004). Locations are overlaid on occupied (dark grey) and extirpated (light grey) ranges of sage-grouse. Shaded locations represent the classification accuracy of discriminant function analysis (model 2, Table 18.3). Black locations are those present in occupied range but incorrectly classified to extirpated range. Grey locations are those present in extirpated range but incorrectly classified to occupied range. White locations were correctly classified to occupied and extirpated ranges.

Step 5: Discriminant Analysis

We used discriminant function analysis (DFA; SAS Institute 1990) to identify which environmental variables discriminated best between historical locations in occupied versus extirpated range. Discriminant function analysis is an appropriate method for discriminating between two or more groups when variables used for discrimination are quantitative and normally distributed (Hair et al. 1992). When these assumptions are met, DFA generally has more discriminatory power than analogues such as logistic regression (Efron 1975). Prior probabilities of classification were set proportional to sample sizes in occupied and extirpated ranges. Variance-covariance structure for the two classification groups were not pooled (i.e., we used quadratic discriminant functions), as recommended when equal variances between groups is not assured (SAS Institute 1990, Hair et al. 1992). Examination of the frequency distributions of each variable showed that data were normally distributed for all variables within both classification groups, thus meeting this assumption. Discriminatory variables also should not be highly correlated if stepwise procedures are used. Correlation coefficients among all discriminatory variables were <0.35, positive or negative, indicating that stepwise procedures could be used.

Results from the discriminant function analysis were used in cross-validation analysis by withholding data for a different circle for each run to jackknife the assessment of classification accuracy of each combination of discriminatory variables in a given model (SAS Institute 1990, Hair et al. 1992). Results were expressed as the percentage of locations correctly classified to occupied range, to extirpated range, and incorrectly classified to each (SAS Institute 1990).

We used cross-validation results to rank model performance. First, we summed the percentage of historical locations correctly classified to occupied or extirpated range to obtain a cumulative percentage of correct classifications (Table 18.2). For a model to perform perfectly, the cumulative percentage would be 200%–100% of locations correctly classified to occupied range and to extirpated range. Second, we subtracted the percentage of locations correctly classified to occupied range from the percentage correctly classified to extirpated range. This absolute difference measured the evenness of correct classifications between occupied and extirpated ranges. The best evenness value would be 0, indicating that a model was equally consistent in correct classifications between occupied and extirpated ranges. Third, we subtracted the evenness value from the cumulative percentage of correct classifications. This difference, or performance value, provided an overall measure of model performance, considering both accuracy and evenness of classifications (Table 18.2). For example, a given model might correctly classify 100% of locations associated with occupied range but only 75% of locations associated with extirpated range, yielding a cumulative percentage of 175, an evenness of 25, and an overall performance value of 150. By contrast, a second model that correctly classified 90% of locations to occupied range and 85% of locations to extirpated range also results in a cumulative percentage of 175, but an evenness of 5, and an overall performance value of 170. The second model has a higher performance value, owing to its superior capability to correctly classify locations to both occupied and extirpated ranges.

We used this process to evaluate DFA models containing different combinations of the 22 discriminatory variables. The combinations included evaluation of: (1) each environmental variable individually; (2) biotic variables as a group; (3) abiotic variables as a group; (4) anthropogenic variables as a group; (5) all combinations of the three groups of biotic, abiotic, and anthropogenic variables; (6) all variables that had nonoverlapping 95% confidence intervals between their mean values for occupied versus extirpated ranges; (7) all groups of variables whose individual performance values were ≥75 and ≥100; and (8) all variables identified in forward stepwise DFA (Hair et al. 1992) as statistically significant ($P < 0.05$) discriminators. All of these DFA models were identified a priori of any modeling results.

Step 6: Spatial Modeling

The combination of variables with highest performance value in discriminating between extirpated and occupied ranges was used in a predictive DFA to estimate the probability that different regions of occupied range had environmental conditions similar to conditions in extirpated range. Our purpose was to identify and map areas of occupied range where environmental conditions indicated that sage-grouse may be at higher risk of regional extirpation, versus areas with conditions likely to serve as regional strongholds for population persistence.

We first subdivided the occupied range into 100,000-ha blocks. These blocks were compatible in size with the circular areas used to evaluate performance of various models at historical locations, and likewise compatible with regional effects on sage-grouse. We then applied the model to each of 2,661 blocks that encompassed occupied range. Results were expressed as the probability of environmental similarity of a given block of occupied range with conditions in extirpated range.

The probability of similarity for each block was placed in one of six categories: 0.0–0.10, >0.10–0.25, >0.25–0.50, >0.50–0.75, >0.75–<0.90, and 0.90–1.0. These categories were most narrow for the lowest and highest probabilities because these values represent extreme conditions where similarity to extirpated range is either highly probable or improbable. Categories for intermediate probability values were wider, reflecting higher uncertainty about environmental differences between occupied and extirpated ranges. We summarized results by these categories across occupied range within each Sage-Grouse Management Zone (SMZ; Stiver et al. 2006). We also mapped similarity values as a continuous variable by state and SMZ to compare and contrast these results with values summarized by categories.

TABLE 18.2

Classification accuracy and resulting performance of biotic, abiotic, and anthropogenic variables contained in discriminant function models that were used to discriminate between historical locations of sage-grouse in occupied versus extirpated ranges under cross-validation.

See methods for details regarding cross-validation.

Discriminatory variables	Correctly classified to occupied range (%)[a]	Correctly classified to extirpated range (%)[a]	Total % correctly classified[b]	Evenness of correctly classified[c]	Performance value (rank)[d]
Sagebrush area (%)[e, f]	76	65	141	11	130 (2)
Patch density	100	0	100	100	0
Patch size[f]	41	96	137	55	82 (8)
Edge density 1	98	6	104	92	12 (17)
Edge density 2[e]	96	4	100	92	8 (18)
Proximity index[f]	35	86	121	51	70 (12)
Nearest neighbor	99	0	99	99	0
Mean core area[f]	39	95	134	56	78 (11)
Distance to occupied–extirpated boundary[e, f]	92	24	116	68	48 (16)
All biotic variables	52	92	144	40	104 (6)
Precipitation	98	0	98	98	0
Elevation[e, f]	85	50	135	35	100 (8)
Soil water capacity[f]	90	29	119	61	58 (13)
Soil rock depth	100	0	100	100	0
Soil salinity[e, f]	100	0	100	100	0

TABLE 18.2 *(continued)*

TABLE 18.2 (CONTINUED)

Discriminatory variables	Correctly classified to occupied range (%)[a]	Correctly classified to extirpated range (%)[a]	Total % correctly classified[b]	Evenness of correctly classified[c]	Performance value (rank)[d]
All abiotic variables	89	47	136	42	94 (9)
All biotic and abiotic variables	54	92	146	38	108 (5)
Agriculture area (%)[f]	92	40	132	52	80 (10)
Distance to roads	100	0	100	0	0
Human density[f]	99	25	124	74	50 (15)
Road density[e, f]	93	28	121	65	56 (14)
Distance to highways[f]	100	0	100	0	0
Distance to transmission lines[e, f]	64	63	127	1	126 (4)
Distance to cellular towers[e, f]	81	51	132	30	102 (7)
Land ownership[f]	64	74	138	10	128 (3)
All anthropogenic variables	96	42	138	54	84 (10)
All anthropogenic and abiotic variables	94	52	146	42	104 (6)
All anthropogenic and biotic variables	75	81	156	6	150 (1)

[a] Percentage of historical locations in currently occupied or extirpated range correctly classified to that range based on the associated discriminatory variable or variables using cross-validation.

[b] Sum of correct classification percentages for occupied and extirpated ranges based on the associated discriminatory variable or variables.

[c] Absolute difference between percentages of locations correctly classified to occupied versus extirpated ranges.

[d] Performance value is the evenness subtracted from total percent correctly classified. A value of 200 represents highest performance, which is possible if all locations are correctly classified to occupied (100%) and to extirpated (100%) ranges, for a total percent of 200 and an evenness of 0 (100% minus 100%). Variables are ranked, shown in parentheses, according to their performance values, with a rank of 1 representing the best performance considering all discriminant function models listed in Table 18.2. Each line of the table represents a discriminant function model that was evaluated.

[e] Variables with significant discriminatory value ($P < 0.05$) as estimated by forward stepwise discriminant analysis.

[f] Variables with non-overlapping 95% confidence intervals between mean values in occupied versus extirpated ranges (Figs. 18.2–18.4).

RESULTS

Differences Between Extirpated and Occupied Ranges

Fifteen of the 22 environmental variables had mean values with nonoverlapping 95% CIs between extirpated and occupied ranges (Figs. 18.2–18.4). These variables included five biotic, three abiotic, and seven anthropogenic variables.

The five significant biotic variables were sagebrush area, patch size of sagebrush, proximity of sagebrush patches, size of sagebrush core areas, and distance to the boundary between occupied and extirpated ranges. Historical locations in occupied range contained almost twice as much area in sagebrush as those in extirpated range (Fig. 18.2). Mean patch size of sagebrush was >9 times larger, and mean core area >11 times larger, in occupied versus extirpated range (Fig. 18.2). Sagebrush patches also were substantially closer to one another in occupied range (Fig. 18.2). In addition, historical locations in occupied range were closer to the boundary between occupied and extirpated ranges than locations in extirpated range (Fig. 18.2).

The three significant abiotic variables were elevation, soil water capacity, and soil salinity. Elevation was almost 50% higher in occupied range than in extirpated range (Fig. 18.3). Occupied range had lower soil water capacity and higher soil salinity (Fig. 18.3).

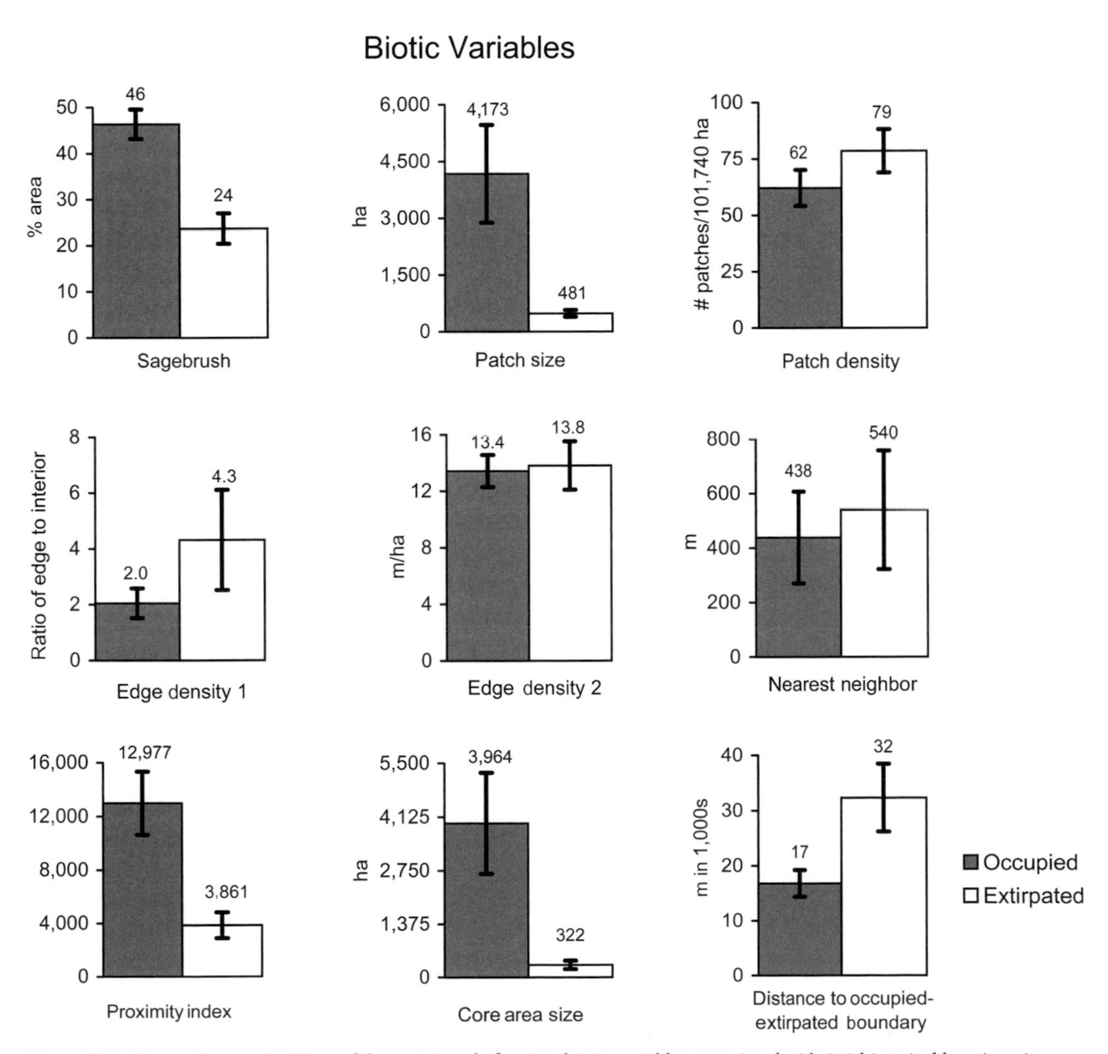

Figure 18.2. Mean values and 95% confidence intervals for nine biotic variables associated with 239 historical locations in occupied range and 136 historical locations in extirpated range for sage-grouse.

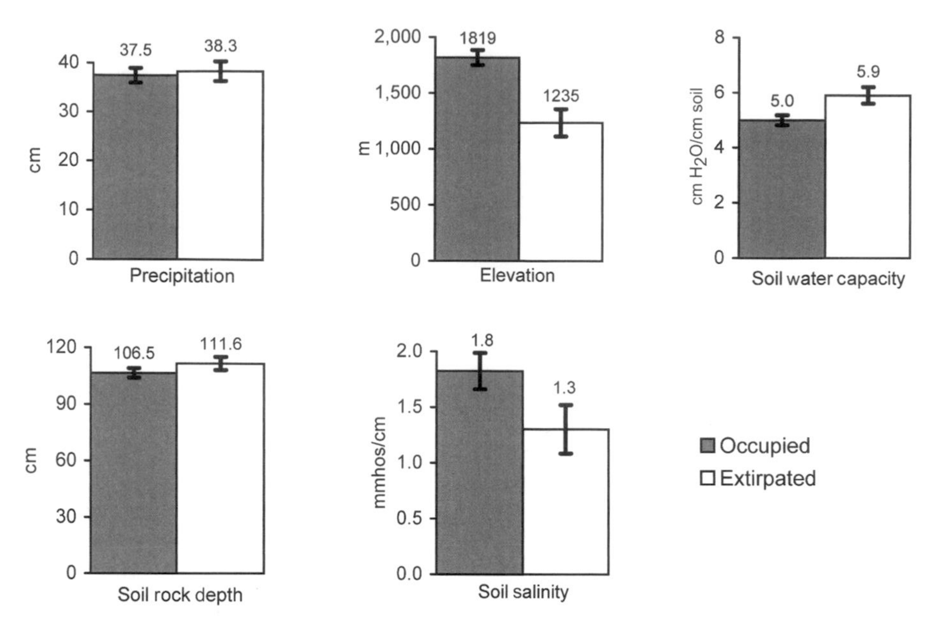

Figure 18.3. Mean values and 95% confidence intervals for five abiotic variables associated with 239 historical locations in occupied range and 136 historical locations in extirpated range for sage-grouse.

The seven significant anthropogenic variables were area in agriculture, human density, road density, distance to highways, distance to electric transmission lines, distance to cellular towers, and landownership. Area in agriculture was almost three times lower and mean human density was 26 times lower in occupied than in extirpated range (Fig. 18.4). Road density also was lower and highways substantially farther from historical locations in occupied range (Fig. 18.4). Mean distance to electric transmission lines was >2 times farther in occupied range than in extirpated range (Fig. 18.4). The distance to cellular towers averaged almost twice as far in occupied range (Fig. 18.4). Occupied range also had substantially more public ownership (Fig. 18.4); 64% of circles encompassing historical locations in occupied range were dominated by public ownership compared to 26% in extirpated range.

Environmental Discrimination Between Extirpated and Occupied Ranges

Individual Variables and Biotic, Abiotic, and Anthropogenic Groups

We first evaluated performance of DFA models containing individual discriminatory variables and those containing all combinations of biotic, abiotic, and anthropogenic groups of variables (Table 18.1). The best-performing of these models contained all biotic and anthropogenic variables, which correctly classified 72% of historical locations to occupied range and 80% to extirpated range (Table 18.2). The second-best model contained just one variable, sagebrush area, which correctly classified 76% of historical locations to occupied range and 65% to extirpated range. The landownership model had third-best performance, followed by models containing distance to transmission lines, all biotic and abiotic variables, distance to cellular towers, elevation, all biotic variables, and all anthropogenic and abiotic variables (Table 18.2). Additional models containing the remaining individual variables performed poorly as discriminators between occupied and extirpated ranges (Table 18.2).

Best-Performing Combinations of Variables

We evaluated four additional models that contained combinations of variables with potential for high classification accuracy (Table 18.3), based on our a priori modeling approaches described in step 5 of Methods. The best-performing model, model 2, contained just five variables: sagebrush area, elevation, distance to transmission lines,

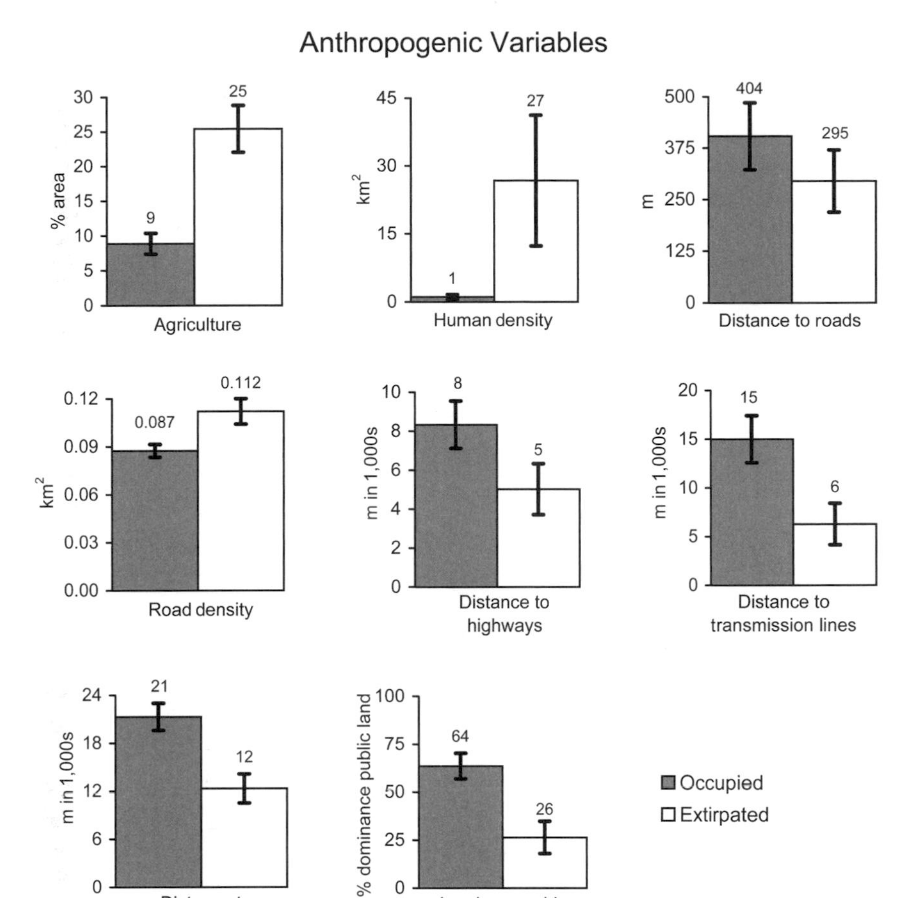

Figure 18.4. Mean values and 95% confidence intervals for eight anthropogenic variables associated with 239 historical locations in occupied range and 136 historical locations in extirpated range for sage-grouse.

distance to cellular towers, and landownership (Table 18.3). This model correctly classified 85% of locations to occupied range and 83% to extirpated range (performance value 166; Table 18.3). Model 4, which contained the 15 variables with nonoverlapping confidence intervals between mean values in occupied and extirpated ranges, performed slightly worse than model 2 (performance value 154) and substantially better than models 1 and 3 (Table 18.3). Both models 2 and 4 outperformed all single-variable models and all models based on biotic, abiotic, and anthropogenic groups of variables (Tables 18.2, 18.3).

Nearly all errors in correctly classifying historical locations to occupied and extirpated ranges with model 2, our best-performing model, occurred under two conditions: they were located in the Great Plains SMZ (N = 17), or they were substantially closer to the boundary between occupied and extirpated ranges (N = 41)(Fig. 18.1). Locations incorrectly classified to occupied and extirpated ranges and not within the Great Plains SMZ were <10 km from the boundary between occupied and extirpated ranges. By contrast, >80% of correctly classified locations were >20 km from the boundary between occupied and extirpated ranges. Incorrectly classified locations close to the occupied–extirpated range boundary had large portions of the associated circles that overlapped both occupied and extirpated ranges. Thus, locations associated with these circles represented a mix of conditions from both ranges. As a result, performance of model 2 was diminished with the inclusion of these circles that overlapped both classification groups (occupied versus

TABLE 18.3

Classification accuracy and performance of four models used to discriminate between historical locations of sage-grouse in occupied versus extirpated range using cross-validation.

Discriminatory variables in each model were selected using different criteria (see table notes).

Discriminatory models	Correctly classified to occupied range (%)	Correctly classified to extirpated range (%)	Total % correctly classified	Evenness of correctly classified	Performance value[a]
Model 1[b]					
SB, PS, MCA, E, AA, DL, CT, LO	54	93	147	39	106
Model 2[b]					
SB, E, TL, CT, LO	85	83	168	2	166
Model 3[c]					
SB, ED, RB, E, S, RD, TL, CT	90	70	160	20	140
Model 4[d]					
SB, PS, PI, MCA, RB, E, SWC, S, AA, HD, RD, DH, TL, CL, LO	77	88	165	11	154

ABBREVIATIONS: Variables used in one or more of the models included sagebrush area (SB), patch size (PS), edge density 2 (ED2), proximity index (PI), mean core area (MCA), distance to occupied–extirpated range boundary (RB), elevation (E), soil salinity (S), soil water capacity (SWC), agriculture area (AA), road density (RD), human density (HD), distance to highways (DH), distance to transmission lines (TL), distance to cellular towers (CL), and land ownership (LO).

[a] Evenness subtracted from total percent correctly classified.

[b] Models 1 and 2 included variables with individual performance values ≥75 and ≥100 (Table 18.1).

[c] Model 3 contained variables selected under stepwise discriminant analysis.

[d] Model 4 included the 15 variables with non-overlapping 95% confidence intervals between mean values in occupied versus extirpated ranges (Figs. 18.2–18.4).

extirpated ranges). However, classification accuracy was high for model 2 (>80%; Table 18.3) despite the inclusion of these circles along the occupied–extirpated range boundary.

Similarity of Occupied Range with Extirpated Range

Estimates based on application of model 2 to all 100,000-ha blocks of occupied range showed that similarity to extirpated range was highest along most range peripheries (Fig. 18.5). Similarity to extirpated range also was highest in the smaller, disjunct areas of occupied range in Washington, southwest Oregon, northeast California, Idaho, northeast Utah, southern Montana, and in larger areas of east-central Montana and eastern and north-central Wyoming (Fig. 18.5).

Environmental similarity to extirpated range was lowest in the expansive area of occupied range in southeast Oregon, southwest Idaho, northern and east-central Nevada, and west-central and southwest Wyoming (Fig. 18.5); these areas compose the largest, most contiguous blocks of occupied range of Greater Sage-Grouse. By contrast, the small, disjunct areas occupied by Gunnison Sage-Grouse in southeast Utah and southern Colorado had similarity values that were mostly intermediate with those of extirpated range (Fig. 18.5).

The Columbia Basin had the highest percentage of environmental similarity with extirpated range: 65% of the zone was in the two highest similarity classes (probabilities >0.75) and mostly in the highest class (0.90–1.0; Fig. 18.5, Table 18.4). The Great Plains had the next-highest percentage of occupied range in the two highest similarity

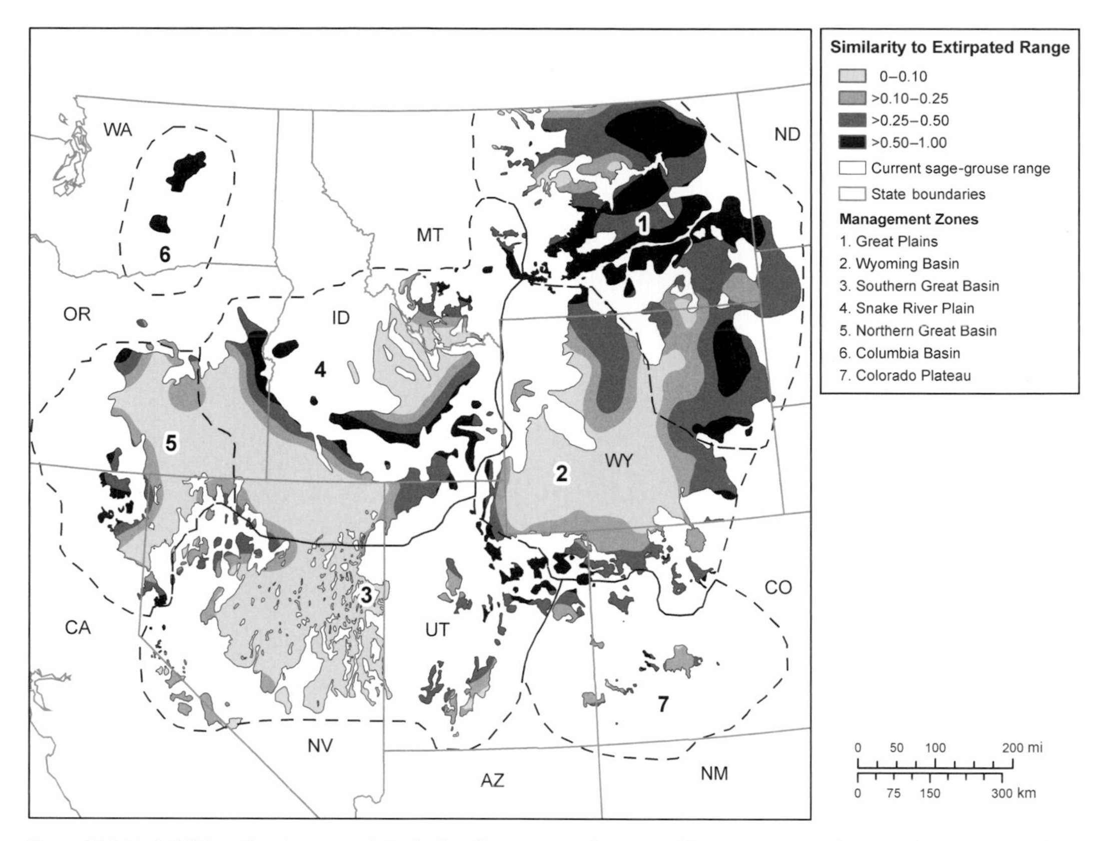

Figure 18.5. Probabilities of environmental similarity of areas currently occupied by sage-grouse with areas where extirpation has occurred, based on estimates from model 2 discriminant function analysis. Probabilities range from 0.0–1.0 and are mapped as a continuous variable. Areas in black show high similarity with extirpated range. Areas in light grey show low similarity.

classes (37%), followed by Colorado Plateau at 10% (Table 18.4). SMZs with lowest similarity to extirpated range were Northern Great Basin, Southern Great Basin, Snake River Plain, and Wyoming Basin. The large majority of occupied range in these four SMZs had probabilities of similarity of ≤0.10. All four, however, had high similarity with extirpated range along range peripheries or in smaller, disjunct areas (Fig. 18.5).

DISCUSSION

Factors Associated with Extirpation

Biotic Variables

Sage-grouse occupation versus extirpation was strongly associated with measures of sagebrush abundance and distribution, including sagebrush area, patch size, proximity of patches, and size of core areas. These results support past studies that identified sage-grouse as a sagebrush obligate, dependent on sagebrush for persistence (Braun et al. 1976, Schroeder et al. 1999, Rowland 2004).

Sagebrush area was the single best discriminator between occupied and extirpated ranges among the 22 variables evaluated. The DFA model containing this single variable was one of the top-performing models. These results agree with recent findings that sagebrush area is one of the best landscape predictors of sage-grouse persistence (Wisdom et al. 2002c; Walker et al. 2007a, Doherty et al. 2008, Aldridge et al. 2008).

The upper 95% CI for sagebrush area in extirpated range was 27%. Landscapes occupied by sage-grouse with sagebrush <27% would thus have a >97.5% probability of being no different than a random sample of extirpated ranges, suggesting that associated populations in these occupied ranges could be more vulnerable to extirpation. Similarly, the lower 95% CI for sagebrush area in occupied range was 50%. Landscapes occupied by sage-grouse with values above this lower bound thus have a >97.5% probability of being no different than a

TABLE 18.4

Percent area of occupied range by categories of the probability of similarity with extirpated range, summarized by Sage-Grouse Management Zone.

Probabilities of similarity are summarized in six categories: (1) 0.0–0.10; (2) >0.10–0.25; (3) >0.25–0.50; (4) >0.50–0.75; (5) >0.75–<0.90; and (6) 0.90–1.0. Probabilities were estimated for each of 2,661 100,000-ha blocks that encompass the occupied range of sage-grouse.

	Categories of similarity to extirpated range												
	1 (0.0–0.10)		2 (>0.10–0.25)		3 (>0.25–0.50)		4 (>0.50–0.75)		5 (>0.75–<0.90)		6 (0.90–1.0)		
Sage-Grouse Management Zone	ha	% area	ha	% area	ha	% area	ha	% area	ha	% area	ha	% area	Total area
Great Plains	9,783,456	49.9	876,405	4.5	962,934	4.7	778,925	4.0	1,068,834	5.4	6,176,855	31.5	19,611,209
Wyoming Basin	11,176,049	76.4	1,088,868	7.4	781,287	5.3	359,034	2.5	502,765	3.4	715,584	4.9	14,623,587
Southern Great Basin	8,426,483	82.2	756,360	7.4	336,628	3.3	454,331	4.4	108,158	1.1	175,149	1.7	10,257,109
Snake River Plain	11,531,252	79.7	451,982	3.1	555,622	3.8	434,949	3.0	443,068	3.1	1,044,541	7.2	14,461,414
Northern Great Basin	5,978,359	86.4	211,164	3.1	203,219	2.9	102,251	1.5	134,475	1.9	286,249	4.1	6,915,717
Columbia Basin	69,720	14.7	0	0.0	0	0.0	97,724	20.6	84,004	17.7	222,305	46.9	473,753
Colorado Plateau	507,907	56.2	174,546	19.3	36,991	4.1	94,962	10.5	52,531	5.8	36,666	4.1	903,603

random sample of occupied ranges, suggesting a higher capability to support persistent populations.

Recent landscape studies of Greater Sage-Grouse identified similar threshold values for sagebrush area to maintain population persistence. Aldridge et al. (2008:990), using a 30.77-km radius around sampling locations, estimated that at least 25% and preferably 65% of the landscape needed to be dominated by sagebrush for long-term sage-grouse persistence. These estimates mirror our values of 27% and 50% for sagebrush area, with values <27% indicating a high risk of extirpation, and values above 50% indicating a high probability of persistence. Our estimates also are for large landscapes, based on the 18-km radius circles that we analyzed. Similarly, Walker et al. (2007a) estimated that the lowest probability of lek persistence, approximately 40–50%, occurred for landscapes with <30% area in sagebrush within 6.4 km of a lek center. These probabilities declined even more for landscapes with <30% sagebrush that were subjected to energy development (see fig. 5 in Walker et al. 2007a).

Abiotic Variables

Three abiotic variables—elevation, soil salinity, and soil water capacity—also differed between occupied and extirpated ranges. Elevation was a good discriminator, probably because most sagebrush loss has occurred disproportionately at lower elevations where human activities and developments have been concentrated (Hann et al. 1997; Knick et al., this volume, chapter 12; Leu and Hanser, this volume, chapter 13), and where invasive grasses have displaced large areas of sagebrush (Suring et al. 2005b, Meinke et al. 2009). Lower soil salinity and higher soil water capacity in extirpated range also indicate a higher suitability for agricultural development (Knick, this volume, chapter 1), which also was associated with sage-grouse extirpation.

Anthropogenic Variables

Seven of the eight anthropogenic variables differed between occupied and extirpated ranges. The number of these variables, their diversity, and the strength of differences between occupied and extirpated ranges suggest that a variety of human activities and land uses have contributed to or been associated with sage-grouse extirpation. This inference agrees with findings from recent landscape studies that documented negative effects of anthropogenic variables on sage-grouse populations, including human density and percent agriculture (Aldridge et al. 2008), roads and traffic (Lyon and Anderson 2003, Holloran 2005), and energy development (Holloran 2005; Aldridge et al. 2007; Walker et al. 2007a; Doherty et al. 2008; Naugle et al., this volume, chapter 20). We did not specifically evaluate energy development. However, extirpated range contained almost 27 times the human density, had almost three times more area in agriculture, was 60% closer to highways, and had 25% higher density of roads, in contrast to occupied range. These patterns agree with research cited above that evaluated these or similar variables. Moreover, the four variables of human density, area in agriculture, distance to highways, and road density were part of model 4, which outperformed all models except the top-ranked model 2.

Three additional anthropogenic variables—distance to transmission lines, distance to cellular towers, and landownership—also differed between occupied and extirpated ranges. These variables were the best discriminators among the eight anthropogenic variables considered and ranked among the best of all individual variables. These variables have received little attention in landscape research on sage-grouse—only distance to transmission lines has been formally evaluated (Connelly et al. 2000a, Aldridge and Boyce 2007, Walker et al. 2007a). Transmission lines can cause sage-grouse mortality via bird collisions with lines (Beck et al. 2006, Aldridge and Boyce 2007) and facilitate raptor predation of sage-grouse (Connelly et al. 2000a). In addition, the electromagnetic radiation emitted from transmission lines has a variety of negative effects on other bird species using areas on or near lines (Fernie and Reynolds 2005). Moreover, transmission lines convert habitat to nonhabitat and fragment the remaining habitat, similar to roads (Naugle et al., this volume, chapter 20).

The strong association between distance to cellular towers and sage-grouse extirpation was an especially intriguing result, given that no previous studies of sage-grouse have evaluated this variable. Whether cellular towers function in a cause-effect manner or simply are aligned with other detrimental factors cannot be addressed without additional research. Recent studies, however, suggest possible cause-effect relationships between high levels of electromagnetic radiation within 500 m of cellular towers and reduced population or reproductive performance of a limited number of bird and amphibian species (Balmori 2005, 2006; Balmori and Hallberg 2007; Everaert and Bauwens 2007). These negative effects are similar to those documented for bird species exposed to electromagnetic radiation generated by power lines (Fernie and Reynolds 2005). Cellular towers also are likely to cause sage-grouse mortality via collisions with these structures or influence movements by visual obstruction, but no research has investigated these issues.

Distance to cellular towers may also indicate the most intensive human developments and uses, given that cellular towers are concentrated along major highways and within and near larger towns and cities across the range of sage-grouse. Although correlation coefficients between this and the other environmental variables were low, cellular towers represent discrete points within areas of high human use. Consequently, distance to cellular towers may serve as a finely measured indicator of more concentrated human uses, in contrast to other anthropogenic variables that reflect more general landscape measures of human uses. This pattern would explain the variable's low correlation with other anthropogenic variables yet high discriminatory performance.

Landownership also was an ideal indicator of underlying causes of sage-grouse extirpation, given that many private lands have been converted from sagebrush to other land uses (Vander Haegen 2007; Knick et al., this volume, chapter 12). In addition, the conversion of private lands to nonsagebrush land uses has fragmented remaining sagebrush habitats nearby (Vander Haegen et al. 2000) and facilitated the spread of exotic plants in sagebrush habitats near such conversions (Hann et al. 1997; Wisdom et al. 2005a,c).

Combinations of Biotic, Abiotic, and Anthropogenic Variables

Performance of the many discriminant function models, each containing different combinations of environmental variables, largely reflected differences in individual variables between occupied and extirpated ranges. Models that performed best either contained all 15 variables whose mean values had nonoverlapping confidence intervals between occupied and extirpated ranges—model 4—or contained a subset of five of those variables (sagebrush area, elevation, distance to transmission lines, distance to cellular towers, and landownership) that provided highly distinct and precise differences between ranges—model 2. The superior performance of models 2 and 4 suggests that different combinations of the 15 environmental variables could be used as effective predictors of sage-grouse vulnerability to extirpation for current or projected landscape conditions. These results also clearly demonstrate that sage-grouse extirpation is associated with a varied combination of biotic, abiotic, and anthropogenic influences, and that holistic consideration of these many environmental factors in land management appears important to maintain persistent populations at large landscape extents like those studied here.

Geographic Patterns of Environmental Similarity with Extirpated Range

Our estimates of environmental similarity of areas occupied by sage-grouse to areas where extirpation has occurred have direct implications for rangewide conservation planning. First, populations along the peripheries of occupied range may have a higher risk of extirpation. This higher risk is an expected extension of past extirpation patterns that have largely occurred from the outside inward. That is, sage-grouse extirpation has occurred mostly along the outer portions of pre-settlement range and contracted inward (Schroeder et al. 2004, Aldridge et al. 2008). Most areas along the outer portion of pre-settlement range are at lower elevations where land uses and habitat conversions have been concentrated, particularly on private lands. Moreover, this pattern is expected because populations on the periphery of their range immediately adjacent to areas where extirpation has occurred often are more vulnerable to extirpation than populations closer to the center when anthropogenic factors disproportionately affect the periphery (Brown et al. 1996, Laliberte and Ripple 2004). This is the case for sage-grouse. By contrast, this may not be the case for declining populations of other species when peripheral areas provide refuge from habitat degradation occurring in core areas (Lomolino and Channell 1995, 2000).

Populations of many species at high risk along range peripheries may undergo extirpation during periods of high environmental variation, such as during a severe and prolonged drought. Extirpation also may occur in such areas when a combination of environmental, genetic, stochastic, and demographic sources of variation manifest over time in ways not easily predicted (Mills 2007). Populations in the periphery of a species range typically experience high temporal variation in abundance in contrast to core populations (Vucetich and Waite 2003); this variation may reflect the many sources of variation described above that contribute to extirpation in small populations of sage-grouse.

Second, populations in small, disjunct areas of occupied range may have a high risk of extirpation. This pattern also is expected, given principles of population viability, which have consistently shown that extinction probability increases for populations that become increasingly small and isolated (Purvis et al. 2000). Populations of Greater Sage-Grouse occupying small, disjunct areas in Washington, northeast California, southwest Oregon, north-central Idaho, eastern Idaho, northeast Utah, and southern Montana, which are separated from larger core populations, fit these conditions.

Third, populations in many areas of occupied range in the Great Plains may have a higher risk of extirpation. This result is not unexpected, given the relatively low sagebrush area in the Great Plains (Knick, this volume, chapter 1), which is dominated more by grasslands (Küchler 1964, 1970; McArthur and Ott 1996). In addition, the southern part of the Great Plains has been altered by extensive energy development, resulting in extensive sagebrush loss and concomitant development of roads, power lines, and other infrastructure (Walker et al. 2007a; Walker 2008; Naugle et al., this volume, chapter 20). Energy development in Wyoming has progressed, at varying rates in relation to varying energy prices, for many decades (Braun et al. 2002; Naugle et al., this volume, chapter 20). Consequently, long-term changes in sage-grouse environments based on energy development in the Wyoming portion of the Great Plains were reflected through the early 2000s in our estimates of sagebrush area and distance to transmission lines—two of the five discriminatory variables included in model 2 that we used to estimate environmental similarity with areas where extirpation has occurred.

Given that sagebrush is substantially less common in the Great Plains in contrast to other areas of sage-grouse range (Knick, this volume, chapter 1), our analyses suggest that sage-grouse in this zone may be vulnerable to further reductions in sagebrush area. Additional loss of sagebrush in the Great Plains would approach potential thresholds for sage-grouse extirpation faster than in other areas where sagebrush dominates a larger proportion of the landscape. Our results also indicate that other detrimental factors are at play in the Great Plains.

Finally, our mapped estimates of similarity could be used to identify strongholds for sage-grouse, that is, areas of occupied range where the risk of extirpation appears low (e.g., areas with similarity values ≤ 0.10; Fig. 18.5) and those that compose the largest areas of contiguous range. Two large strongholds for Greater Sage-Grouse are evident. One, a western stronghold, is the extensive, contiguous area encompassing southeast Oregon, northwest Nevada, southwest Idaho, northeast Nevada, and east-central Nevada that includes most areas in the Northern Great Basin, Southern Great Basin, and Snake River Plain SMZs—the lightest grey areas within these zones in Fig. 18.5. The other, an eastern stronghold, is the area encompassing south-central and southwest Wyoming in the Wyoming Basin SMZ. This second stronghold is approximately one-half the size of the western stronghold. In addition, an area in east-central Idaho has low similarity to extirpated range (Fig. 18.5) but is smaller than either of the two primary strongholds.

No strongholds are evident for Gunnison Sage-Grouse that consist of expansive, contiguous areas where similarity with extirpated range is ≤ 0.10 (Fig. 18.5). Intensive management to conserve existing habitats and populations of the species, combined with efforts to restore habitats, are obvious needs for Gunnison Sage-Grouse (Oyler-McCance et al. 2001, 2005a; Lupis et al. 2006).

Our documented spatial patterns of environmental similarity with extirpated range are similar to recent range-wide estimates of sage-grouse persistence (Aldridge et al. 2008). Similarities between these separate analyses are particularly compelling, given that different methods and variables were used. In that regard, our spatial estimates of environmental similarity with extirpated range, and those of persistence by Aldridge et al. (2008), are mutually reinforcing, thus providing a stronger basis for inferences made from each study (Johnson et al. 2002).

Spatial Priorities for Management

Our mapped estimates of environmental similarity of areas currently occupied by sage-grouse with areas where extirpation has occurred could be used to help establish management priorities across existing sage-grouse range. Strongholds identified from our analysis are potential areas of focus for maintenance and improvement over time. Management emphasis on strongholds is more effective and efficient than devoting limited resources to restoration of areas where populations are at high risk of extirpation because of widespread habitat deficiencies (Wisdom et al. 2005c; Meinke et al. 2009; Doherty et al., this volume, chapter 21). In the latter situation, it is highly uncertain as to whether populations can persist, or how effective it would be to use limited resources in an attempt to improve a myriad of challenging environmental conditions to assure population persistence. This uncertainty revolves around three related issues: (1) areas with high similarity to extirpated range could be population sinks, given that these areas are mostly along the boundary with extirpated range, and range contraction along this boundary appears to be an ongoing process for sage-grouse; (2) areas with high similarity to extirpated range are associated with a variety of anthropogenic management challenges that may be difficult or impossible to mitigate (e.g., minimizing current infrastructure of roads, highways, transmission lines, cellular towers, and agricultural and urban areas that dominate these areas), thus negating benefits to restore sagebrush, which also is deficient in these areas; and (3) areas with high similarity to extirpated range are mostly at lower elevations characterized by warmer conditions that have low resistance to exotic plant invasions and low resiliency for returning to native vegetation states following any natural or human-caused disturbances, including restoration treatments (Wisdom et al. 2005c, Meinke et al. 2009).

Despite these challenges, the presence of sage-grouse populations in areas with high similarity to extirpated range may help maintain a lower risk of extirpation for populations in strongholds, by maintaining a larger population size overall and thus helping buffer the negative effects of environmental stochasticity and loss of genetic variation. More isolated or disjunct populations, especially at the range periphery, may have different genetic, phenotypic, and behavioral characteristics important to the species. Understanding the role of these high-risk populations in relation to those in strongholds warrants immediate research attention (Nielson et al. 2001).

Regardless of the role of high-risk populations, effective management of strongholds is important because detrimental anthropogenic factors in strongholds are less common and extensive areas of sagebrush remain. Thus, the management challenge in strongholds is one of maintaining or improving current conditions, which largely translates to prevention of detrimental land uses and minimizing undesirable ecological processes (Wisdom et al. 2005c). In many cases, this combination of passive management and passive restoration involves modifications to existing land uses that maintain or improve conditions (McIver and Starr 2001). This contrasts with active restoration, requiring intensive management and large inputs to restore or rehabilitate conditions in areas where extensive degradation and loss of habitat has occurred, and which may be difficult or impossible to reverse for many sites formerly dominated by sagebrush (McIver and Starr 2001; Pyke, this volume, chapter 23).

If management emphasis is placed on strongholds, a comprehensive and detailed assessment of threats to habitats and populations within these areas is appropriate (Wisdom et al. 2005a,b,c). Most areas of sagebrush in the western stronghold are threatened by large-scale invasion of exotic plants, particularly cheatgrass (*Bromus tectorum*)(Suring et al. 2005b; Miller et al., this volume, chapter 10). Minimizing this threat warrants comprehensive management of all human activities that act as vectors for spread and establishment of exotic plants, and that increase their competitive edge over native vegetation. More than 25 different human-associated disturbances would need to be effectively managed to reduce this threat (Wisdom et al. 2005b,c). Among these disturbances are obvious factors such as high densities of roads open to motorized travel and expansive areas of public land open to off-road motorized travel (Barton and Holmes 2007). A myriad of less obvious human-associated disturbances also are prevalent and warrant management attention (Wisdom et al. 2005b).

Another common threat in the western stronghold is displacement of sagebrush by highly invasive pinyon pine (*Pinus* spp.) and juniper (*Juniperus* spp.) woodlands (Suring et al. 2005b; Miller et al., this volume, chapter 10). Woodland control can be achieved through aggressive mechanical or burning treatments; which treatments, if any, are appropriate and effective depends on local site conditions, the potential interaction with exotic plants, and the anticipated responses of affected sagebrush community types (Suring et al. 2005b; Miller et al. 2007; this volume, chapter 10). Comprehensive assessment of risks posed by this threat, mapped across the western stronghold, would provide a basis for developing and implementing effective management controls (Suring et al. 2005b).

The eastern stronghold continues to be a focal area of large-scale energy development, and attempts to mitigate the associated negative effects on sage-grouse populations have been ineffective (Holloran 2005, Walker et al. 2007a, Walker 2008). If the eastern stronghold is to be maintained, a holistic redesign of mitigation practices for energy development is needed (Kiesecker et al. 2009). For mitigations to be effective, they must be implemented over substantially larger areas than current practices, which focus on small areas around leks at a scale too small to sustain year-round needs of sage-grouse populations (Walker et al. 2007a; Walker 2008; Naugle et al., this volume, chapter 20; Doherty et al., this volume, chapter 21).

Research and Information Needs

Our analysis was one of the first to associate a diverse set of environmental factors with sage-grouse extirpation. As part of this process, we noted a number of deficiencies in spatial data. One was the lack of spatial data available for livestock grazing, which constitutes the most pervasive land use across the range of sage-grouse (Knick et al. 2003). Federal agencies are required to closely manage and monitor livestock grazing. However, associated data are not available in consistent, spatially explicit formats across the range of sage-grouse, or even for smaller areas that span multiple administrative boundaries within or among federal agencies. This deficiency precluded our analysis of livestock grazing. Likewise, no other studies of potential effects of livestock grazing on sage-grouse have been conducted at regional or range-wide spatial extents because of this data deficiency (Crawford et al. 2004).

Primitive and secondary roads also may be underestimated in current spatial layers. Our distance- and density-based road analyses might have changed with a more accurate inventory. In addition, exotic plant occurrence, another extensive landscape feature, has not been mapped accurately across the range of sage-grouse, and this factor may have substantial effects on habitat (Hemstrom et al. 2002, Rowland et al. 2005). Regional models of cheatgrass occurrence recently were developed and validated for western areas of sage-grouse range (Peterson 2005, Bradley and Mustard 2006, Meinke et al. 2009). Ultimately, such an approach is needed to estimate and map range-wide occurrence of the more common exotic plants, such that potential effects on sage-grouse extirpation can be evaluated range-wide. Similarly, more accurate, range-wide maps of the occurrence of pinyon pine and juniper woodlands would provide a basis for analyzing this variable in relation to range-wide patterns of sage-grouse extirpation.

Another data challenge for range-wide analyses is that some factors may be common or pervasive in specific parts of sage-grouse range, but uncommon or absent in most areas. Energy development is concentrated in Wyoming and adjacent areas but is spatially uncommon, concentrated in small areas, or absent elsewhere. Consequently, we could not evaluate energy development as a range-wide variable because we had insufficient sampling coverage, using historical sage-grouse locations in occupied and extirpated ranges, to evaluate different levels of energy development (but see earlier discussion regarding variables evaluated that are associated with energy development).

Other factors that may affect sage-grouse populations are just emerging, such as West Nile virus (Naugle et al. 2005; Walker 2008; Walker and Naugle, this volume, chapter 9). Such variables are not related to past extirpations, and the range-wide prevalence of West Nile virus within all populations is uncertain. Consequently, an evaluation was not possible.

Finally, identifying which environmental factors are operating in a cause-effect manner in relation to extirpation and which may simply be correlative is a challenge not easily addressed except through consideration of our results in relation to the larger body of sage-grouse literature. Our results confirm prior research documenting sage-grouse as a species whose persistence depends on adequate areas of sagebrush. This inference extends to other sagebrush variables—patch size, proximity among patches, and size of core areas—that also were associated with extirpation. These results illustrate the strong effect of sagebrush abundance and distribution on sage-grouse persistence; without large areas of contiguous sagebrush, sage-grouse cannot persist.

A cause-effect relationship of anthropogenic variables such as area in agriculture, human density, road density, and distance to highways is indicated by past research documenting the widespread conversion of sagebrush habitat to these land uses (Braun 1998; Vander Haegen et al. 2000; Knick et al., this volume, chapter 12); by the facilitation of exotic plant invasions into sagebrush habitats adjacent to these land uses (Hann et al. 1997), especially adjacent to roads (Gelbard and Belnap 2003); and by mortality of sage-grouse along roads and highways (Lyon and Anderson 2003, Aldridge and Boyce 2007).

The strong associations of elevation and landownership with sage-grouse extirpation represent the widespread conversion of low-elevation, private lands to nonsagebrush land uses, such as agricultural and urban development (Vander Haegen et al. 2000; Knick et al., this volume, chapter 12), as well as the substantial loss of sagebrush from widespread exotic plant invasions at lower elevations (Hann et al. 1997, Meinke et al. 2009). In that context, elevation and landownership are ideal indicators of underlying causes of extirpation.

Finally, two variables strongly associated with sage-grouse extirpation, distance to transmission lines and distance to cellular towers, have unknown relations with sage-grouse population dynamics at regional extents. New, mechanistic research is needed to understand the potential relation between these variables and sage-grouse extirpation. Until then, our results suggest that transmission lines and cellular towers warrant consideration as part of holistic conservation strategies for sage-grouse.

CONSERVATION IMPLICATIONS

A variety of biotic, abiotic, and anthropogenic factors are associated with extirpation of sage-grouse. Consequently, holistic attention to a diverse set of environmental factors—beyond those considered for sage-grouse in current guidelines and management—appears necessary to help maintain population persistence.

Maintenance of desired conditions in areas identified as strongholds for Greater Sage-Grouse appears critical to the species' future persistence. Strongholds provide extensive areas of contiguous sagebrush that can be maintained and improved with less active management and fewer resource inputs. Restoration and rehabilitation of areas within and near the small, disjunct areas of Gunnison Sage-Grouse range likewise is important to recovery and viability of this species. The future of sage-grouse will depend on new, holistic management approaches that are strategically designed and effectively implemented range-wide, and that minimize all forms of detrimental factors and maximize best uses of limited resources.

ACKNOWLEDGMENTS

K. E. Doherty and C. L. Aldridge provided valuable comments that substantially improved this chapter. Comments from L. Bond and D. B. Marx improved our statistical methods and their description. J. M. Boyd assisted with data summaries, spatial analysis, and editing. Our work was supported by the Pacific Northwest Research Station, United States Forest Service; the Forest and Rangeland Ecosystem Science Center, United States Geological Survey; and the Washington Department of Fish and Wildlife. Any use of trade, product, or firm names is for descriptive purposes only and does not imply endorsement by the United States Government.

PART FIVE

Conservation and Management

CHAPTER NINETEEN

Greater Sage-Grouse as an Umbrella Species for Shrubland Passerine Birds

A MULTISCALE ASSESSMENT

Steven E. Hanser and Steven T. Knick

Abstract. Working groups and government agencies are planning and conducting land actions in sagebrush (*Artemisia* spp.) habitats to benefit Greater Sage-Grouse (*Centrocercus urophasianus*) populations. Managers have adopted an umbrella concept, creating habitat characteristics specific to sage-grouse requirements, in the belief that other wildlife species dependent on sagebrush will benefit. We tested the efficacy of this approach by first identifying the primary environmental gradients underlying sagebrush steppe bird communities (including Greater Sage-Grouse). We integrated field sampling for birds and vegetation with geographic information system (GIS) data to characterize 305 sites sampled throughout the current range of Greater Sage-Grouse in the Intermountain West, United States. The primary environmental axis defining the bird community represented a gradient from local-scale Wyoming/basin big sagebrush (*A. t.* ssp. *wyomingensis/A. t.* ssp. *tridentata*), and bare ground cover to local and regional grassland cover; the second axis represented a transition from low-elevation Wyoming/basin big sagebrush and bare ground to mountain big sagebrush (*A. t.* ssp. *vaseyana*) and habitat edge. We identified the relative overlap of sage-grouse with 13 species of passerine birds along the multiscale gradients and estimated the width of the umbrella when applying management guidelines specific to sage-grouse. Passerine birds associated with sagebrush steppe habitats had high levels of overlap with Greater Sage-Grouse along the multiscale environmental gradients. However, the overlap of the umbrella was primarily a function of the broad range of sagebrush habitats used by sage-grouse. Management that focuses on creating a narrow set of plot-scale conditions will likely be less effective than restoration efforts that recognize landscape scale heterogeneity and multiscale organization of habitats. These multiscale efforts may improve some sage-grouse habitats and strengthen the management umbrella for shrub steppe passerine birds.

Key Words: canonical correspondence analysis, *Centrocercus urophasianus*, Greater Sage-Grouse, habitat management, passerine birds, sagebrush ecosystem, umbrella species.

Hanser, S. E., and S. T. Knick. 2011. Greater Sage-Grouse as an umbrella species for shrubland passerine birds: a multiscale assessment. Pp. 473–487 *in* S. T. Knick and J. W. Connelly (editors). Greater Sage-Grouse: ecology and conservation of a landscape species and its habitats. Studies in Avian Biology (vol. 38), University of California Press, Berkeley, CA.

Greater Sage-Grouse Como Especie Sombrilla Para los Pájaros Paseriformes de los Arbustos: Evaluación Multi-escala

Resumen. Los grupos de trabajo y agencias de gobierno están planeando y realizando acciones en los hábitats de artemisa (*Artemisia* spp.) para favorecer a las poblaciones de Greater Sage-Grouse (*Centrocercus urophasianus*). Los administradores han adoptado un concepto sombrilla, creando características de hábitat específicas para las necesidades del Greater Sage-Grouse, en la creencia de que otras especies de vida silvestre que dependen de artemisa se beneficiarán. Probamos la eficacia de este enfoque, primero identificando los principales gradientes subyacentes a las comunidades de pájaros de la artemisa de estepa (incluyendo Greater Sage-Grouse). Integramos el muestreode campo para pájaros y vegetación con los datos del sistema de información geográfica (GIS) para caracterizar 305 sitios muestreados en todo el territorio actual del Greater Sage-Grouse en el Intermountain West, EE.UU. El principal eje ambiental que define la comunidad de pájaros representa un gradiente desde una escala local de Wyoming/basin big sagebrush (*A. t.* ssp. *wyomingensis/A. t.* ssp. *tridentata*), y una cubierta de terreno desnuda, hasta cubierta de grama local y regional; el segundo eje representó una transición de artemisa de baja altitud de Wyoming/basin big sagebrush y terreno desnudo hasta mountain big sagebrush (*A. t.* ssp. *vaseyana*) en el límite del hábitat. Identificamos la superposición relativa del Greater Sage-Grouse con 13 especies de aves paseriformes junto con los gradientes multi-escala y estimamos el ancho de la sombrilla cuando se aplicaron lineamientos específicos para Greater Sage-Grouse. Los pájaros paseriformes asociados a hábitats de artemisa de estepa tenían alto nivel de superposición con Greater Sage-Grouse junto con los gradientes ambientales multi-escala. Sin embargo, el solapamiento de la sombrilla era principalmente una función del amplio rango de los hábitats de artemisa usados por Greater Sage-Grouse. El manejo que se centra en crear un conjunto estrecho de condiciones a escala de parcela probablemente será menos efectivo que los esfuerzos de restauración que reconocen laheterogeneidad a nivel de territorio y la organización multi-escala de los hábitats. Estos esfuerzos multi-escala pueden mejorar algunos hábitats de artemisa y fortalecer la administración sombrilla para las aves paseriformes de los arbustos de estepa.

Palabras Clave: análisis de correspondencia canónica, aves paseriformes, *Centrocercus urophasianus*, ecosistemas de artemisa (sagebrush), especie sombrilla, Greater Sage-Grouse, manejo del hábitat.

Sagebrush (*Artemisia* spp.) ecosystems, particularly those in xeric regions at low elevations, are among the most imperiled in North America (Noss and Peters 1995, Mac et al. 1998). Few native sagebrush ecosystems remain intact; most are disturbed, altered, fragmented, or lost due to numerous factors including agriculture; improper livestock grazing; energy and natural resource development; urbanization; invasive plant and animal species; and natural, prescribed, or other human-caused fires (Noss et al. 1995, Hann et al. 1997, Miller and Eddleman 2001, Connelly et al. 2004). Consequently, restoration and management of sagebrush habitats are important conservation concerns.

More than 350 species, including Greater Sage-Grouse (*Centrocercus urophasianus*), depend on sagebrush habitats for all or part of their life cycle (Wisdom et al. 2005b); many have declined numerically and spatially from their historical status concomitant with changes to their habitats (Braun et al. 1976, Connelly and Braun 1997, Vander Haegen et al. 2000, Connelly et al. 2004, Dobkin and Sauder 2004, Schroeder et al. 2004). These declines have led to the classification of many sagebrush obligates as species of conservation concern by state and federal agencies and nongovernmental conservation organizations (Knick and Rotenberry 2002, Suring et al. 2005a). Threatened or endangered species status has been petitioned for for Gunnison Sage-Grouse (*C. minimus*) and Greater Sage-Grouse, and granted for pygmy rabbits (*Brachylagus idahoensis*) in Washington state (United States Department of the Interior 2003a, 2005a, 2005b, 2006b). The 2010 decision by the United States Department of the Interior that listing Greater Sage-Grouse under the Endangered Species Act was warranted but precluded (United States Department of the Interior 2010), emphasized the need for management planning and actions by local working groups and state and federal management agencies to provide critical sage-grouse habitat (United States Department of the Interior 2004b, Stiver et al. 2006).

Land management to provide habitat based on needs of Greater Sage-Grouse explicitly espouses an umbrella concept, the actions from which will benefit other wildlife species in sagebrush communities (Rich and Altman 2002, Braun 2005, Rich et al. 2005, Rowland et al. 2006). Criteria for an effective umbrella species include having a broad geographic range, an overlap of resource requirements with target species, similar responses to management actions, a known life history, and legal protection (Caro and O'Doherty 1999, Fleishman et al. 2000, Rubinoff 2001, Rowland et al. 2006). Sage-grouse meet these criteria, with the exception of legal status. Previous studies have used delineations of species ranges, coarse habitat requirements, or relative abundances on Breeding Bird Survey routes to evaluate effectiveness of a sage-grouse umbrella (Rich et al. 2005, Rowland et al. 2006). These approaches primarily focused on identifying coarse-resolution regional patterns of overlap and found that Greater Sage-Grouse were only a moderately effective umbrella species (Wisdom et al. 2005b). Species have different requirements and use resources at differing spatial and temporal resolutions (Holling 1977, Southwood 1977, Wiens 1989a, Kolasa and Waltho 1998). The umbrella concept as applied to Greater Sage-Grouse has not been tested using multiscale data to characterize the hierarchical environments important to the sagebrush steppe bird community.

We tested the effectiveness of using Greater Sage-Grouse as an umbrella species for other birds dependent on sagebrush steppe habitats using field- and geographic information system (GIS)–derived data to capture the spatial and temporal scales at which birds in this community select habitats (Rotenberry and Wiens 1980b, Wiens 1981, Wiens et al. 1987, Wiens 1989a, Rotenberry and Knick 1999, Knick et al. 2008). We designed our study to identify the primary environmental gradients underlying the bird community that included Greater Sage-Grouse. We then estimated overlap between Greater Sage-Grouse and 13 passerine bird species along these environmental gradients. Finally, we estimated the width of the umbrella when habitat is managed to provide characteristics recommended in the management guidelines for Greater Sage-Grouse (Connelly et al. 2000c).

Testing this concept presented challenges because habitat use by species may overlap spatially but not temporally (Suter et al. 2002). Most passerine birds are present in sagebrush habitats only during the breeding season, migrating to wintering ranges in the southwestern United States and Mexico (Wiens 1981, Wiens and Rotenberry 1981, Knick et al. 2003). Sage-grouse may move >150 km to seasonal ranges within areas >2,500 km^2 but remain within sagebrush habitats to meet their yearly requirements (Schroeder et al. 1999, Connelly et al. 2000c). Because of this seasonal differential in habitat use, we used methods to detect current use by passerines and prior use by sage-grouse.

We conducted our study across the range currently occupied by Greater Sage-Grouse in the Intermountain West. We measured habitat variables commonly manipulated by land management agencies and listed as important sage-grouse habitat components (Connelly et al. 2000c, Connelly et al. 2004, Braun et al. 2005) or used to assess habitat configuration (McGarigal et al. 2002). Our results are important for evaluating the prevailing management paradigm focused on meeting requirements of Greater Sage-Grouse as an umbrella to benefit passerine bird communities.

METHODS

We tested the potential for sage-grouse to function as an umbrella for 13 species of passerine birds associated with sagebrush steppe habitats (Table 19.1). We selected these 13 passerine species based on their primary habitat associations (shrubland or grassland) and conservation status (Braun et al. 1976, Paige and Ritter 1999, Dobkin and Sauder 2004). Of the 13 species, nine are of conservation interest in one or more states within our study area (Dobkin and Sauder 2004) and six are undergoing significant population declines (Sauer et al. 2008).

We surveyed passerine bird communities, examined sage-grouse use, and sampled habitats at 305 randomly located sites within the current distribution of Greater Sage-Grouse (Schroeder et al. 2004) across Washington (N = 11), Idaho (N = 81), Oregon (N = 80), Nevada (N = 80), and Utah (N = 53). Our sample sites were within sagebrush habitat at the plot scale and accessible on public land within 1 km of a two-track unpaved road to reduce travel time (Fig. 19.1).

TABLE 19.1

Population trend and conservation status of passerine species associated with sagebrush steppe used to examine the Greater Sage-Grouse umbrella species concept.

Common name	Scientific name	Trend[a]	Status[b]
Gray Flycatcher	*Empidonax wrightii*	4.61*	X
Loggerhead Shrike	*Lanius ludovicianus*	−2.70***	X
Horned Lark	*Eremophila alpestris*	−2.39***	
Sage Thrasher	*Oreoscoptes montanus*	ns	X
Green-tailed Towhee	*Pipilo chlorurus*	ns	X
Brewer's Sparrow	*Spizella breweri*	−2.01**	X
Vesper Sparrow	*Pooecetes gramineus*	ns	X
Lark Sparrow	*Chondestes grammacus*	ns	
Black-throated Sparrow	*Amphispiza bilineata*	−1.73***	X
Sage Sparrow	*Amphispiza belli*	ns	X
Savannah Sparrow	*Passerculus sandwichensis*	ns	
Grasshopper Sparrow	*Ammodramus savannarum*	−5.31***	X
Western Meadowlark	*Sturnella neglecta*	−1.23***	

[a] Population trend (% change/year) in the western United States calculated by standard BBS analysis (1966–2007, Sauer et al. 2008).

[b] Federal or state species of concern (Dobkin and Sauder 2004).

* $P \leq 0.05$

** $P \leq 0.01$

*** $P \leq 0.001$

ns = not significant ($P > 0.05$).

Characteristics of sagebrush habitats differ by stewardship, but most sagebrush habitat in this region is public land and managed for multiple uses (Knick et al. 2003, this volume, chapter 12; Knick, this volume, chapter 1). The majority of roads in sage-grouse range are two-track dirt roads, and >95% of sagebrush habitats are <2.5 km from a road (Connelly et al. 2004).

Bird Surveys

Plot size used for surveys at each of the 305 sites was 180 × 180 m, which approximates the home-range size of the majority of the passerine birds in our study. We combined point counts and walking surveys within plots to detect passerine bird abundance at each location (Thompson et al. 1992, Diefenbach et al. 2003). Two sampling techniques were used to maximize the strengths of each method and increase the probability of detecting the passerine bird species present on plots. Point counts were conducted during May and June 2006 from 30 min before sunrise to 1100 H (~6-hour span), on mornings with little wind and no rain. We recorded presence of all birds seen or heard within a 10-min interval after a 3-min waiting period and recorded distance to each individual using a laser range finder. The point count data set was truncated to include only individuals found within the study plot. Walking surveys were conducted immediately following the point-count period by walking three parallel transects spaced 36 m apart to cover the entire area of the sample plot (Fig. 19.1). Observers again recorded all species, independent of point counts, detected by sight or sound during the walking survey period.

We used pellet-count surveys to identify plot use by sage-grouse (Boyce 1981, Schroeder and Vander Haegen 2006). Pellets are easily identified and can persist for up to three years (Boyce 1981), but begin to breakdown after one year (Dahlgren

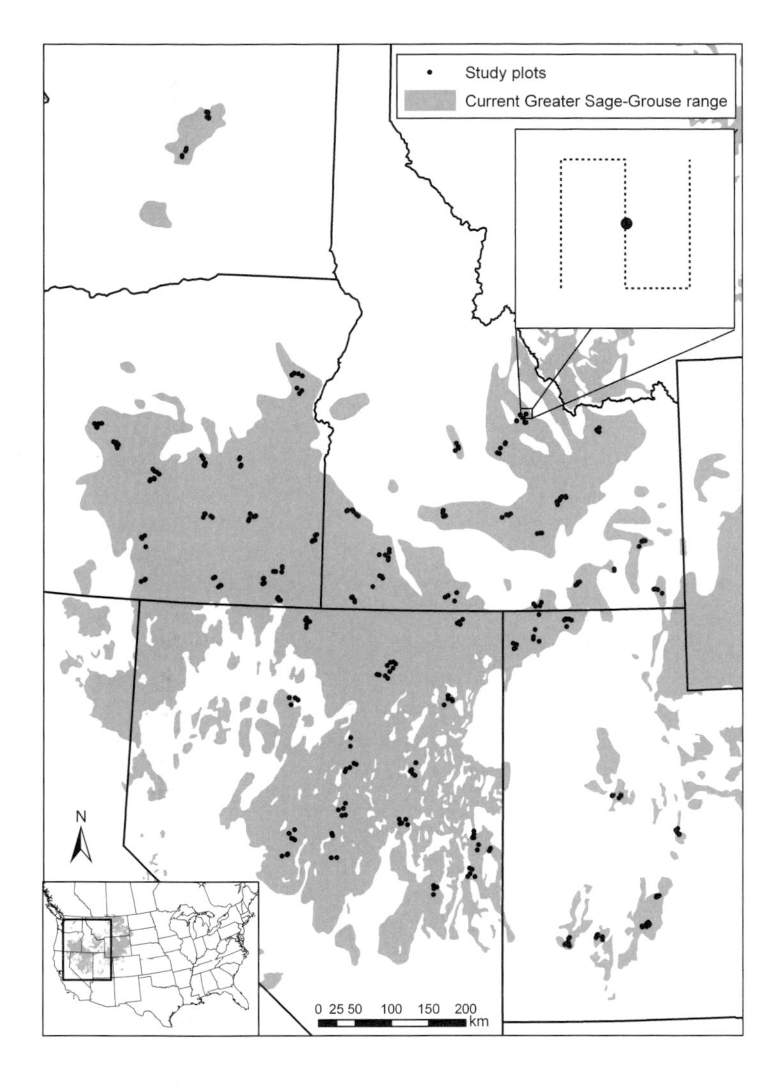

Figure 19.1. Distribution of 305 study plots throughout the western United States. Plots (inset) were 180 × 0180 m with a passerine point-count station (point) at the center and a walking transect for passerine bird and sage-grouse pellet surveys (dashed line).

et al. 2006). We searched within 2 m of transects used for passerine bird surveys to detect sage-grouse pellets. The majority of pellets are detected within this distance, and detection of pellets does not vary dramatically between areas of different vegetation cover (Dahlgren et al. 2006). We counted both single pellets and pellet clusters. Sage-grouse were considered present if pellets were found.

Habitat Surveys

We characterized habitat at three spatial scales: plot, 1-, and 5-km landscape. Habitat characteristics within 1 km influenced fitness and probability of occurrence of sage-grouse (Aldridge and Boyce 2007), and probability of occurrence of passerine birds in sagebrush habitats (Knick and Rotenberry 1995, Vander Haegen et al. 2000). We also estimated habitat characteristics within 5 km of each plot based on recommended distance for habitat management surrounding leks of nonmigratory sage-grouse populations (Connelly et al. 2000c).

We measured vegetation within the 180 × 180-m plot using line-point intercept (Heady et al. 1959) on two parallel 50-m transects spaced 20 m apart. Cover values were recorded at 0.5-m intervals along each transect totaling 200 points/plot. Sagebrush shrubs were recorded as Wyoming/basin big sagebrush (*A. t.* ssp. *wyomingensis*/*A. t.* ssp. *tridentata*), mountain big sagebrush (*A. t.* ssp. *vaseyana*), or low/black sagebrush (*A. arbuscula* or *A. nova*), along with their maximum height, excluding inflorescence. The understory at each point was classified as native grass, forb, litter, bare ground including biological soil crust, or exotic grass, with the possibility of multiple values per point. Bare ground and litter classes were only available in the absence of herbaceous vegetation. The exotic grass category included two species: cheatgrass (*Bromus tectorum*), which has invaded sagebrush communities, resulting in altered vegetation dynamics and increased fire frequencies (d'Antonio and Vitousek 1992); and crested wheatgrass (*Agropyron cristatum*), which has been planted extensively by management agencies for range improvement and soil stabilization (Reynolds and Trost 1981, Connelly et al. 2000c).

We estimated landscape composition and configuration for 1- and 5-km radii around plot centroids in a GIS using the Shrubmap land cover map (Hanser et al. 2005) to characterize the surrounding landscape. Spatial resolution of the Shrubmap coverage was 30-m pixels, which we resampled to 90 m to facilitate computation of landscape composition and configuration metrics. Landscape composition was percent of Wyoming/basin big sagebrush, mountain big sagebrush, low/black sagebrush, grassland, and exotic grass/forb habitat within each buffer. We described landscape configuration from mean patch size and mean nearest neighbor using an aggregation of all sagebrush types to delineate habitat patches. Mean patch size was the average size (ha) of sagebrush patches and mean nearest neighbor estimated patch isolation from the mean Euclidean distance between all sagebrush patches (McGarigal et al. 2002). We also calculated contrast weighted edge density (CWED) to assess influence of habitat edges (McGarigal et al. 2002). CWED combines amount of edge and dissimilarity of adjacent habitats to estimate complexity of the landscape. The relative dissimilarity of land cover types at each buffer size was calculated from the Euclidian distance between environmental centroid scores between pair-wise combinations in a preliminary correspondence analysis. Values for CWED range from 0 (all edges between habitats are equal) to infinity (many different habitats adjacent to one another). We used Shannon's diversity index (SDI) to estimate relative landscape diversity among plot locations. Shannon's diversity index increases from 0 to infinity with increasing number of habitat patches and relative proportions of habitats within the landscape. All GIS analyses were conducted using ArcMap 9.1 (ESRI 2006) and Fragstats (McGarigal et al. 2002).

Statistical Analysis

Passerine species were considered present if an individual was detected during either point-count or walking surveys. We explored occupancy modeling to estimate detection probability and occupancy at each sample location, but lack of temporal independence and closure between the two passerine surveys led us to use naïve presence (MacKenzie et al. 2006). We also refrained from estimating abundance of passerine species due to inability to scale sage-grouse pellet counts as an indicator of abundance. Consequently, our analysis was based on a binary presence-absence response.

We estimated the primary environmental gradients underlying the shrub steppe bird community using canonical correspondence analysis (CCA), which identifies the linear combination of environmental variables that maximizes the dispersion of species (ter Braak 1995, ter Braak and Smilauer 1997). We combined the multiscale data set (plot, 1-, and 5-km radii) in the ordinations. The relative importance of variables contributing to each gradient was identified from the interset correlations between environmental variables and a CCA axis (ter Braak 1995). We used a Monte Carlo test with 1,000 iterations to assess significance of the first canonical axis and the combination of all canonical axes.

We conducted a variance decomposition to separate the influence of plot-, 1-, and 5-km scale factors on bird community structure (Cushman and McGarigal 2004). Variance decomposition quantifies relative importance and redundancy of each spatial scale by providing a measure of the unique contribution of each spatial scale; variance shared among scales quantifies the level of redundancy.

Use of presence-absence data is appropriate for these analyses; the distribution of each species along environmental axes are Gaussian curves centered on the species mean (score) with a dispersion (tolerance) equal to two standard deviations (ter Braak 1986). The range of values on each axis provides context for the species arrangement along the axis and the range of variability within the explanatory data set. We used gradient analyses instead of species-specific modeling because of our interest in arrangement of the bird community in relation to sage-grouse along a consistent set of habitat parameters. Species-specific modeling is appropriate for examining resource selection of individual species, but comparisons among a group of species are best examined using a gradient approach (Guisan et al. 1999).

We estimated amount of overlap between sage-grouse and passerine birds along the primary environmental gradients (CCA axes I and II). We also calculated niche breadth of each study species along CCA axes I–IV, which is the species score ± species tolerance (ter Braak and Smilauer 1997, Ferrero et al. 2002). We computed phi correlation coefficients (Zar 1998) for pair-wise comparisons of passerine birds and sage-grouse using overlap between estimates of niche breadth. Phi values range from −1 to 1 and indicate strength and direction of the association between species of interest. Phi values $>|0.3|$ indicate moderate associations between the species, and phi values $>|0.7|$ indicate strong associations (Rowland et al. 2006). We also calculated the weighted mean phi using eigenvalues of the four CCA axes as weights.

We calculated the niche breadth representing the sage-grouse management guidelines (Connelly et al. 2000c) by adding study plots that met recommended site conditions for either breeding or brood-rearing habitat as a supplementary species in the CCA analysis. Supplementary species are used to examine the relation of a species or grouping of samples to other species in the ordination without influencing existing ordination axes (ter Braak and Smilauer 1997). We aggregated all sagebrush to facilitate selection of study plots that met management guidelines, which do not distinguish among sagebrush sub- or species levels, to calculate overall plot-scale sagebrush cover and 5-km scale proportion of sagebrush, and combined native grass and forb components for an overall plot-scale herbaceous cover. We compared the niche breadth for management guidelines to sage-grouse and shrub steppe passerine birds. We did not collect grass or forb height to estimate primary environmental gradients, and this factor was not used for guideline site designation. Wintering habitat conditions were excluded due to lack of snow depth data at study plots. The exclusion of these characteristics may lead to an increased overlap of the guideline criteria, with species having grass height restrictions or limited overlap with species that inhabit wintering habitats that may not be optimal habitat during the rest of the annual cycle.

RESULTS

We detected 2,344 individuals of 13 passerine species on surveys at 305 plots. Sage-grouse pellets were detected on 93 plots, including 52 plots with both single pellets and clusters, 17 with only singles, and 24 with only clusters. Vegetation characteristics measured at plot locations and within 1- and 5-km radii indicated that our sampling effort captured a large array of environmental conditions (Table 19.2).

A significant portion of the species data (25.7%) was explained by the CCA axes ($F = 3.55$, $P < 0.001$, Monte Carlo method, 1,000 random permutations). The primary environmental axis (CCA I) explained a significant ($F = 20.91$, $P < 0.001$, Monte Carlo method, 1,000 random permutations) portion of the species-environment relationship (27.3%) and was 8.332 units in length (range = –3.51–4.81). CCA I was correlated with amount of bare ground, and cover of grass and Wyoming/basin big sagebrush at the plot-scale and proportion of grassland within 5 km (Fig. 19.2, Table 19.2). The secondary environmental axis (CCA II) explained 24.8% of the species-environment relationship. The axis was 14.58 units in length (range = –2.63–11.95) and correlated with plot-scale mountain big sagebrush cover, proportion of Wyoming/basin big sagebrush within 1- and 5-km, and CWED within 1-km radii (Fig. 19.2, Table 19.2). The centroid for Greater Sage-Grouse was at 0.19 with a tolerance of 1.05 on CCA I and was −0.03 with a tolerance of 0.86 on CCA II.

Variance decomposition indicated a high amount of shared variance between spatial scales (Fig. 19.3). All three spatial scales (plot, 1-, and 5-km) explained similar amounts of the species data when considered separately (13.0%, 11.2%, and 13.5%, respectively). Plot-scale variables had the largest independent effect (7.9%); 3.5% of the variance was shared among the three spatial scales; and 1- and 5-km scale variables had the largest amount of shared variance (3.4%).

The majority of the 13 passerine species had at least a moderate level of overlap with Greater Sage-Grouse on the first two canonical axes (Table 19.3; Fig. 19.4). Eleven species had positive associations (five species >0.70), and two species (Grasshopper Sparrow [*Ammodramus savannarum*] and Savannah Sparrow [*Passerculus sandwichensis*]) had negative associations (one

TABLE 19.2

Summary statistics for habitat variables measured at study plot locations and interset correlations between environmental variables and canonical ordination axes.

Percent (%) is the proportion of ground cover for plot-level variables and landscape cover at 1- and 5-km scales. Bolded values indicate variables that define each canonical axis.

			Correlation coefficient, r			
Habitat variable	Mean ± SD	Range	CCA 1	CCA 2	CCA 3	CCA 4
		Plot				
Low/black sagebrush, %	5.80 ± 10.71	0–49	−0.06	−0.04	0.17	0.00
Mountain big sagebrush, %	2.53 ± 8.03	0–53	−0.05	**0.51**	0.01	0.10
Wyoming/basin big sagebrush, %	10.48 ± 10.89	0–42	**−0.35**	−0.14	0.01	0.13
Total sagebrush, %	18.81 ± 11.61	0–53	na	na	na	na
Sagebrush height, cm	49.85 ± 25.98	0–120	−0.13	0.17	−0.02	0.07
Grass, %	18.08 ± 16.01	0–85.5	**0.36**	0.21	0.14	0.01
Forb, %	6.48 ± 9.02	0–86.5	0.23	0.29	0.04	−0.01
Total native grass/forb, %	24.56 ± 20.42	0–93.5	na	na	na	na
Exotic grass, %	14.41 ± 21.79	0–94	−0.01	−0.14	**−0.28**	**−0.17**
Bare ground, %	35.50 ± 16.86	0–75	**−0.23**	−0.19	0.13	0.03
Litter, %	19.97 ± 10.66	0.5–65	−0.17	0.13	0.07	**0.13**
		1 km				
Low/black sagebrush, %	14.68 ± 23.89	0–99.2	0.03	−0.04	0.12	−0.09
Mountain big sagebrush, %	10.80 ± 25.14	0–100	0.08	0.45	0.15	−0.04
Wyoming/basin big sagebrush, %	57.79 ± 34.07	0–100	−0.16	**−0.36**	0.02	0.05
Total sagebrush, %	83.27 ± 19.30	2.6–100	na	na	na	na
Sagebrush mean nearest neighbor, m	27.72 ± 38.66	0–276.6	0.10	0.16	−0.23	0.08
Sagebrush mean patch size, ha	186.33 ± 121.86	1–314.1	−0.07	−0.15	**0.42**	−0.06
Grassland, %	2.06 ± 4.87	0–33.5	0.16	−0.12	−0.14	−0.01
Exotic grass/forb, %	2.88 ± 8.79	0–67.2	0.04	−0.14	−0.21	−0.02
Contrast weighted edge density	36.17 ± 13.94	25.6–89.5	−0.02	**0.53**	−0.17	**−0.11**
Shannon's diversity index	0.66 ± 0.41	0–1.77	0.09	0.17	**−0.32**	−0.05
		5 km				
Low/black sagebrush, %	12.61 ± 15.29	0–62.2	−0.05	−0.07	0.15	−0.09
Mountain big sagebrush, %	10.47 ± 21.28	0–99.9	0.12	0.41	0.21	−0.07
Wyoming/basin big sagebrush, %	47.17 ± 27.85	0–99.5	−0.14	**−0.40**	0.05	0.03
Total sagebrush, %	70.25 ± 21.77	12.4–99.9	na	na	na	na
Sagebrush mean nearest neighbor, m	65.32 ± 30.63	0–224.9	0.18	0.06	−0.13	**0.24**
Sagebrush mean patch size, ha	648.89 ± 1,466.02	4.9–7,848.5	0.24	−0.06	**0.27**	−0.07
Grassland, %	3.32 ± 5.20	0–37.1	**0.28**	−0.16	−0.14	0.10

Habitat variable	Mean ± SD	Range	Correlation coefficient, r CCA 1	CCA 2	CCA 3	CCA 4
Exotic grass/forb, %	3.45 ± 9.65	0–77.2	0.02	−0.16	−0.17	−0.07
Contrast weighted edge density	22.58 ± 11.88	5.1–59	−0.15	0.29	0.06	−0.11
Shannon's diversity index	1.10 ± 0.42	0.01–2.1	−0.03	0.17	−0.25	−0.03

NOTE: na indicates that variable was not used in the canonical correspondence analysis.

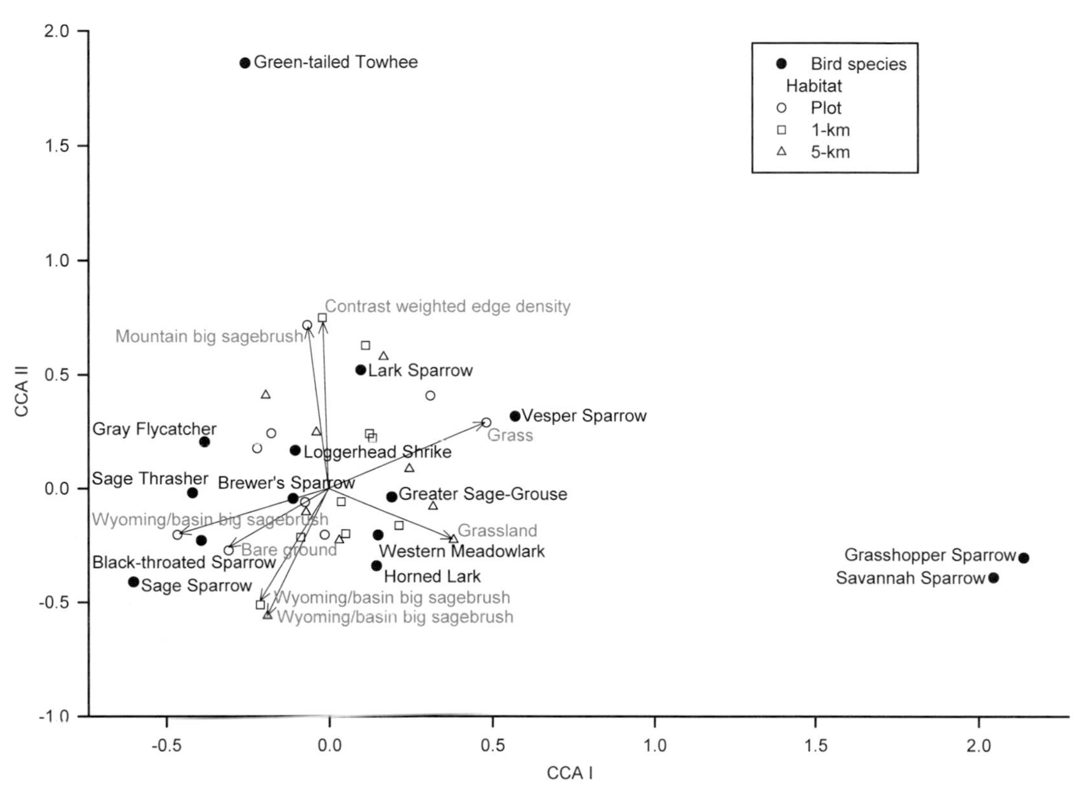

Figure 19.2. Species-environment biplot of species occurrence along the first 2 CCA axes. Vectors representing the environmental variables that define the ends of each gradient are indicated by arrows. Arrow direction and length indicate orientation relative to canonical axes and explanatory strength of the variable.

species ≤0.70) along CCA I. Twelve species had positive associations (seven species >0.70), and one species (Green-tailed Towhee [*Pipilo chlorurus*]) had a strong negative association on CCA II (Table 19.3, Fig. 19.4).

Niche breadths across all CCA axes for 10 of 13 passerine species had weighted mean phi scores positively associated with sage-grouse, including all sagebrush obligates (Brewer's Sparrow [*Spizella breweri*], Sage Sparrow [*Amphispiza belli*], and Sage Thrasher [*Oreoscoptes montanus*]). Three species exhibited no overall association with Greater Sage-Grouse: Grasshopper Sparrows, Savannah Sparrows, and Green-tailed Towhees (Table 19.3).

The niche breadth of plots meeting the sage-grouse management guideline criteria on CCA I contained species centroids of 11 of 14 species, including Greater Sage-Grouse, and overlapped with the niche breadth of 12 of 14 species. On CCA II, 12 of 14 species centroids were within the niche breadth of plots meeting sage-grouse management guideline criteria, and all 14 species' niche breadths overlapped with the guidelines (Fig. 19.4).

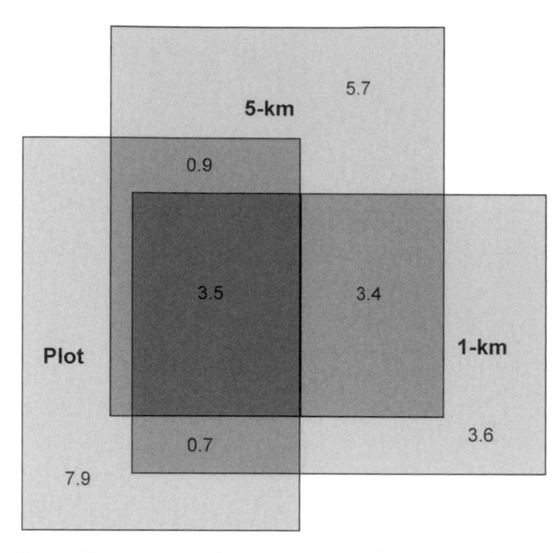

Figure 19.3. Variance decomposition of plot, 1-, and 5-km scales on bird community structure. The area of each rectangular cell is proportional to the variance explained by that component. Numbers list the percent variance explained by each component.

DISCUSSION

The broad diversity of sagebrush habitats used by Greater Sage-Grouse may provide an effective umbrella for the broader community of passerine birds associated with this ecosystem. Passerine birds generally had positive associations with sage-grouse along the environmental gradients defining the community. However, the strength and direction of association between sage-grouse and individual passerine species varied at differing scales as well as across primary environmental gradients. Small-scale heterogeneity in sagebrush ecosystems is composed of local variations in topography, soil type, and cover of grasses, forbs, and shrubs (West and Young 2000). Landscape heterogeneity at larger scales is characterized by the mosaic of habitats, including sagebrush patches of different age-class and species/subspecies and the juxtaposition of these patches to other habitats types, such as juniper (*Juniperus* spp.) woodlands, aspen (*Populus* spp.) groves, and

TABLE 19.3

Overlap between Greater Sage-Grouse and passerine birds in species niche breadth along environmental gradients measured by phi correlation coefficient.

Phi values range from −1 to 1 and indicate both the strength and direction of the relationship. Phi values >|0.3| indicate moderate associations between the passerine species and sage-grouse. Values of phi >|0.7| are indicated in bold and correspond to strong positive or negative associations. Species are sorted by weighted mean phi in ascending order. Weighted mean phi was calculated using eigenvalues of the four CCA axes as weights.

Species	Habitat affinity	CCA 1	CCA 2	CCA 3	CCA 4	Weighted mean
Savannah Sparrow	Grassland	−0.59	0.68	−0.52	−0.05	−0.08
Grasshopper Sparrow	Grassland	**−0.70**	0.68	−0.05	−0.09	−0.05
Green-tailed Towhee	Shrubland	0.62	**−0.71**	−0.17	0.64	0.04
Sage Sparrow	Sagebrush obligate	0.31	0.60	0.16	**0.72**	0.43
Lark Sparrow	Shrubland	**0.70**	0.59	−0.46	0.65	0.46
Black–throated Sparrow	Shrubland	0.51	**0.82**	−0.32	**0.8**	0.51
Loggerhead Shrike	Shrubland	**0.74**	**0.96**	−0.07	0.56	0.66
Sage Thrasher	Sagebrush obligate	0.47	**0.88**	0.58	**0.82**	0.67
Gray Flycatcher	Dry woodland, sagebrush	0.54	**0.78**	0.68	**0.72**	0.67
Vesper Sparrow	Grassland	0.67	0.69	0.58	**0.89**	0.69
Brewer's Sparrow	Sagebrush obligate	**0.74**	**0.78**	0.28	**0.86**	0.69
Western Meadowlark	Grassland, shrubland	**0.88**	**0.85**	−0.09	**0.92**	**0.71**
Horned Lark	Grassland	**0.96**	**0.72**	0.16	**0.91**	**0.74**

grasslands (Wiens et al. 1987, Miller and Eddleman 2001, Knick and Rotenberry 2002). The primary environmental gradients were a function of both scales of heterogeneity. Similar amounts of explained variance and large shared variance among local, 1-, and 5-km spatial scales illustrate the integral nature of scale in defining environmental gradients and community structure. Assessment of the potential effectiveness for Greater Sage-Grouse to be an umbrella species should consider not only the organizational level of habitats but also differences among species due to hierarchical selection of those habitats (Hildén 1965, Cushman and McGarigal 2004, Knick et al. 2008).

Our results indicate that Greater Sage-Grouse may be a more effective umbrella species than shown in studies that used species ranges and coarse habitat requirements as evaluation metrics for analyzing species associations (Rowland et al. 2006). Our field-collected occurrence data for sage-grouse and the passerine species provided a better resolution on the habitats used by these species, leading to a clearer picture of species overlap. In addition, our development of multiscale environmental gradients allowed us to interpret the

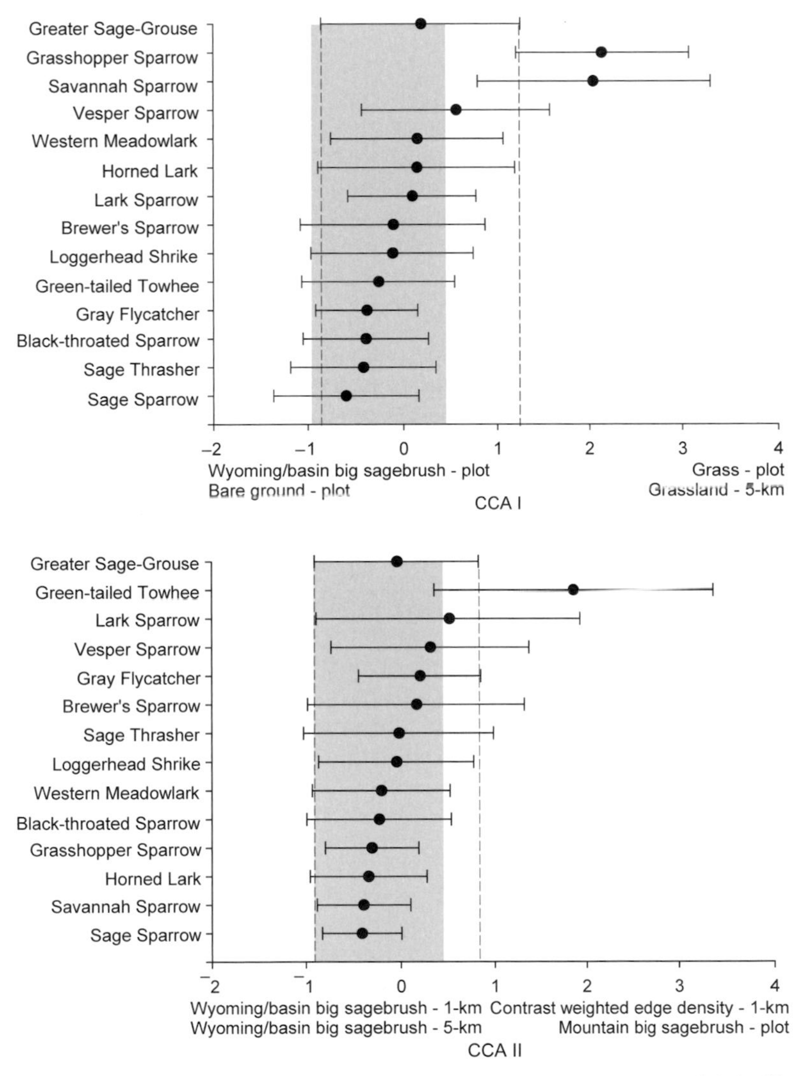

Figure 19.4. Species niche breadth along CCA axes 1 and 2 relative to Greater Sage-Grouse (vertical dashed lines) and the sage-grouse management guidelines (Connelly et al. 2000c) (shaded area). Niche breadth is the dispersion of each species along each axis within multivariate space. Species are sorted by score along the corresponding axis. Labels on the CCA axes indicate the variables correlated with the corresponding ends of the axes.

importance of individual environmental factors to organization of this suite of species.

The primary environmental gradient (CCA I) underlying the sagebrush steppe bird community represented a transition from habitats dominated at local scales by Wyoming/basin big sagebrush and bare ground to increasing cover of local-scale grass embedded within larger regional grasslands. Passerine species, including sagebrush obligates, positively associated with sage-grouse were clustered near the sagebrush end of this ordination axis. Centroids of grassland species, such as Grasshopper and Savannah Sparrows, that were negatively associated with sage-grouse were at the grassland end of this gradient. These grassland species primarily occupy grassland patches and mesic micro-habitats within the sagebrush matrix and largely occur outside the range of sage-grouse (Wheelwright and Rising 1993, Vickery 1996).

The second environmental gradient (CCA 2) represented a transition from regional dominance by Wyoming/basin big sagebrush habitats at lower elevations to higher elevations containing local-scale mountain big sagebrush and regional high-contrast habitat edge. The vegetation dynamics captured by this gradient include low-elevation, xeric sagebrush habitats susceptible to exotic plant invasion to more heterogeneous landscapes of mountain big sagebrush intermixed with pinyon-juniper, aspen, and conifer forest and woodlands at higher elevations (West and Young 2000; Miller and Eddleman 2001; Miller et al., this volume, chapter 10). The strong association of passerine species and sage-grouse along this axis was consistent with previous research in which species associated with sage-grouse were found in mid-elevation sagebrush habitats (Medin et al. 2000). Green-tailed Towhees were negatively associated with sage-grouse on this axis. Green-tailed Towhees primarily select ecotones between sagebrush, other shrublands, or open woodlands at higher elevations (Knopf et al. 1990), which are generally avoided by sage-grouse (Commons et al. 1999).

CONSERVATION IMPLICATIONS

Sagebrush-obligate passerines, Brewer's Sparrow, Sage Sparrow, and Sage Thrasher had moderate to strong positive associations with sage-grouse. This group of species has been declining throughout the western United States over several decades (Rich et al. 2005), and restoration efforts developed to benefit sage-grouse may have the greatest influence on this suite of species. The pattern of association between Sage Sparrows and sage-grouse illustrates the importance of scale in maintaining the environmental heterogeneity in the sagebrush ecosystem as restoration efforts to benefit sage-grouse are implemented. Sage Sparrows select shrubland habitats with sparse understory and a high bare ground component (Rotenberry and Wiens 1980a, Martin and Carlson 1998, Misenhelter and Rotenberry 2000); the species centroid in our study was on the region of primary gradient characterized by sagebrush habitats with high proportion of bare ground. Projects focused on reducing sagebrush canopy cover to improve grass and forb conditions for sage-grouse brood habitat could negatively affect Sage Sparrows. However, efforts to change the landscape pattern at larger scales to those dominated by large contiguous patches of sagebrush will benefit both sage-grouse and Sage Sparrows. Maintaining existing large patches of sagebrush while restoring areas of degraded or disturbed sagebrush habitat to create a mosaic of sagebrush habitats can be an important step toward reversing population declines of sage-grouse and sagebrush-obligate passerines, such as Sage Sparrows.

Management to benefit Greater Sage-Grouse may benefit the broader community of birds that use sagebrush steppe habitats. The diversity of habitats used by sage-grouse on their annual home ranges encompassed a large portion of the more specific habitat characteristics used by passerines for breeding. Although the management guidelines for sage-grouse prescribe a relatively narrow range of plot-level criteria and regional habitat parameters (Connelly et al. 2000c), sites in our study meeting these guideline criteria also encompassed portions of the primary and secondary environmental gradients used by Greater Sage-Grouse and the majority of the shrub steppe passerine community. Thus, the sage-grouse management guidelines can provide a reference to manage and restore habitat for sage-grouse and shrub steppe passerine birds.

The strength of a Greater Sage-Grouse umbrella depends on maintenance of the natural environmental and landscape heterogeneity within the sagebrush ecosystem. Historically, the sagebrush ecosystem was dominated by large contiguous patches of sagebrush interspersed with patches of native bunchgrasses in recently burned areas (Frémont 1845, Young 1989). The niche breadth of the majority of passerine species along the environmental gradients was associated with areas of moderate to high local-scale cover of sagebrush and dominance of sagebrush shrublands at larger spatial scales. Sagebrush obligates have adapted to this multiscale habitat and landscape structure, and changes to the environmental and landscape heterogeneity of the ecosystem may have decreased habitat suitability for the suite of species (Knick and Rotenberry 2002, Knick et al. 2008). Therefore, if Greater Sage-Grouse are acknowledged as an umbrella species in future management planning, restoration efforts to improve or create additional sage-grouse habitat (Wisdom et al. 2002a, 2005b; Meinke et al. 2009) should be designed to preserve environmental and landscape heterogeneity. The sage-grouse management guidelines (Connelly et al. 2000c) explicitly describe plot-level conditions but do not include landscape characteristics or multiscale habitat conditions. Restoration efforts focused on maintaining or achieving a narrow set of plot-scale conditions, without recognition of the historical contiguous nature of sagebrush habitats and environmental and landscape heterogeneity, would likely be less effective.

ACKNOWLEDGMENTS

We thank K. J. VanGunst, K. A. Fesenmyer, and J. Matthews for their assistance with field work. We appreciate manuscript reviews by M. M. Rowland, K. E. Doherty, W. M. Vander Haegen, and one anonymous reviewer. This project was funded by the United States Geological Survey project Environmental Correlates of Greater Sage-Grouse. Any use of trade, product, or firm names is for descriptive purposes only and does not imply endorsement by the United States Government.

CHAPTER TWENTY

Energy Development and Greater Sage-Grouse

David E. Naugle, Kevin E. Doherty, Brett L. Walker, Matthew J. Holloran, and Holly E. Copeland

Abstract. Rapidly expanding energy development in western North America poses a major new challenge for conservation of Greater Sage-Grouse (*Centrocercus urophasianus*). We reviewed the scientific literature documenting biological responses of sage-grouse to development, quantified changes in landscape features detrimental to sage-grouse that result from development, examined the potential for landscape-level expansion of energy development within sage-grouse range, and outlined recommended landscape-scale conservation strategies. Shrublands developed for energy production contained twice as many roads and power lines, and where ranching, energy development, and tillage agriculture coincided, human features were so dense that every 1 km^2 could be bounded by a road and bisected by a power line. Sage-grouse respond negatively to three different types of energy development, and conventional densities of oil and gas wells far exceed the species' threshold of tolerance. These patterns were consistent among studies regardless of whether they examined lek dynamics or demographic rates of specific cohorts within populations. Severity of current and projected impacts indicates the need to shift from local to landscape conservation. The immediate need is for planning tools that overlay the best remaining areas for sage-grouse with the extent of current and anticipated development. This will allow stakeholders to consider a hierarchy of set-aside areas, lease consolidations, and more effective best-management practices as creative solutions to reduce losses. Multiple stressors, including energy development, must be managed collectively to maintain sage-grouse populations over time in priority landscapes.

Key Words: Centrocercus urophasianus, conservation planning, cumulative impacts, energy development, landscapes, natural gas, oil, sagebrush, sage-grouse, wells.

Desarrollo Energético y el Greater Sage-Grouse

Resumen. El desarrollo energético en rápida expansión en el oeste de Norte América, presenta un nuevo desafío importante para la conservación del Greater Sage-Grouse (*Centrocercus urophasianus*). Revisamos la literatura científica que documenta las respuestas biológicas del Greater Sage-Grouse al desarrollo, cuantificamos los cambios en los caracteres del territorio del Greater Sage-Grouse que le son perjudiciales, examinamos el potencial de expansión, a nivel de territorio, del desarrollo energético dentro de las extensiones que ocupa el

Naugle, D. E., K. E. Doherty, B. L. Walker, M. J. Holloran, and H. E. Copeland. 2011. Energy development and Greater Sage-Grouse. Pp. 489–503 *in* S. T. Knick and J. W. Connelly (editors). Greater Sage-Grouse: ecology and conservation of a landscape species and its habitats. Studies in Avian Biology (vol. 38), University of California Press, Berkeley, CA.

Greater Sage-Grouse, y reseñamos las estrategias recomendadas para la conservación del territorio. Los territorios con arbustos desarrollados para la producción de energía contenían el doble de carreteras y líneas de energía, y donde coincidían el desarrollo de ranchos, el desarrollo energético, y la agricultura tillage, las características humanas eran tan densas que cada 1 km cuadrado puede estar unido por una carretera y ser atravesado por una línea de energía. El Greater Sage-Grouse responde negativamente a tres tipos diferentes de desarrollo y las densidades convencionales de pozos de petróleo y de gas excedían ampliamente los umbrales de tolerancia de la especie. Estas pautas fueron consistentes entre los estudios, independientemente de si examinaron la dinámica del lek o las tasas demográficas de los cohortes específicos al interior de las poblaciones. La severidad de los impactos actuales y proyectados indican la necesidad de cambiar de la conservación local a la de territorios. La necesidad inmediata es de herramientas de planificación que cubran las mejores áreas que queden para el Greater Sage-Grouse, con el alcance para el desarrollo actual y el esperado. Esto permitirá a las partes interesadas considerar una jerarquía de áreas reservadas, consolidación de arrendamientos, y buenas prácticas de manejo más efectivas como soluciones creativas que ayuden a reducir las pérdidas. Múltiples estresantes, incluyendo el desarrollo energético, deben ser manejados colectivamente para mantener las poblaciones de Greater Sage-Grouse en paisajes con prioridad a lo largo del tiempo.

Palabras Clave: artemisa, *Centrocercus urophasianus*, desarrollo energético, gas natural, impactos acumulativos, paisajes, petróleo, planificación de la conservación, pozos, sage-grouse.

World demand for energy increased by >50% in the last half-century, and a similar increase is projected between now and 2030 (National Petroleum Council 2007). Fossil fuels will likely remain the largest source of energy worldwide, with oil, natural gas, and coal accounting for 83–87% of total world demand. A primary focus of the 2005 amendments to the National Energy Policy and Conservation Act in the United States is to expedite the leasing and permitting process on public lands to increase domestic production of fossil fuels (American Gas Association 2005). Of 320, 192 federal applications to drill in 13 western states from 1929–2007, 85.7% were authorized, 1.0% were pending, 6.7% were withdrawn and 6.6% were rejected (table 12.17 in Knick et al., this volume, chapter 12). Projected growth in United States energy demand is 0.5–1.3% annually (National Petroleum Council 2007), and trends suggest development of domestic fossil fuel reserves will expand through the first half of the 21st century.

The Greater Sage-Grouse (*Centrocercus urophasianus*; hereafter, sage-grouse) is a galliform endemic to western semiarid sagebrush (*Artemisia* spp.) habitats in North America (Schroeder et al. 1999). Previously widespread, loss and degradation of sagebrush habitat have resulted in extirpation of the species from almost half of its original range (Schroeder et al. 2004). Energy development has emerged as a major issue in conservation because areas currently under development contain some of the highest densities of sage-grouse and other sagebrush-obligate species in western North America (Knick et al. 2003, Connelly et al. 2004). An understanding of the biological response of sage-grouse to energy development will inform decision makers as to whether current lease stipulations are adequate or if landscape conservation is required to maintain populations.

Early studies evaluating potential impacts to sage-grouse are few (Braun 1986, 1987, 1998; Remington and Braun 1991), but interest in research has followed the pace and extent of energy development (Fig. 20.1). The science is evolving from small-scale reactive studies (Rost and Bailey 1979, Van Dyke and Klein 1996) into broad-scale and comprehensive evaluations of cumulative impacts (Johnson et al. 2005, Sorensen et al. 2008) that use before-after-control-impact designs (Underwood 1997) and viability models capable of quantifying population-level impacts (Haight et al. 2002, Carroll et al. 2003). Past major reviews conclude (Schroeder et al. 1999, Connelly et al. 2000c, Crawford et al. 2004) and recent studies reaffirm (Holloran and Anderson 2005, Walker et al. 2007a, Doherty et al. 2008) that sage-grouse are landscape specialists that require large and intact sagebrush habitats to maintain populations.

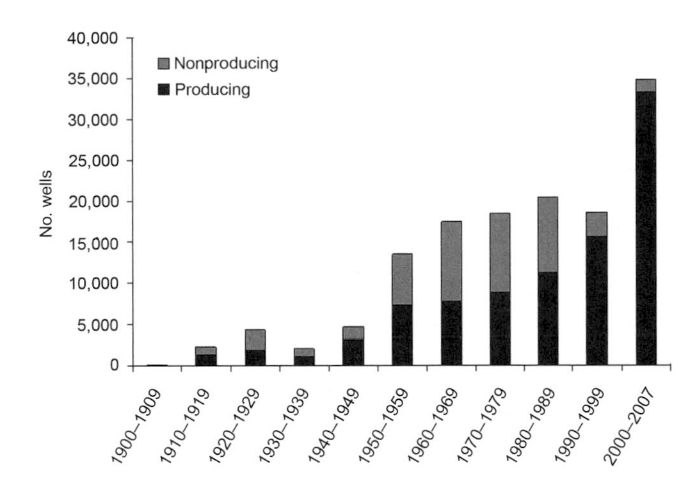

Figure 20.1. Number of producing and nonproducing oil and gas wells through time on public and private lands within the range of Greater Sage-Grouse (Schroeder et al. 2004; gray shaded area in Fig. 20.2) within the United States portion of the Great Plains and Wyoming Basin SMZs (Connelly et al. 2004). Number of wells in each time interval is based on the decade during which the permit to drill was issued.

Recent studies also show that both direct and indirect impacts can result from synergistic effects of energy development. Direct impacts result when animals avoid human infrastructure (Sawyer et al. 2006, Doherty et al. 2008) or when development negatively affects survival (Holloran 2005) or reproduction (Aldridge and Boyce 2007). Indirect impacts include changes in habitat quality (Bergquist et al. 2007), predator communities (Hebblewhite et al. 2005), or disease dynamics (Daszak et al. 2000) and can be equally deleterious if cascading effects negatively influence sensitive species.

The potential for energy development to impact sagebrush-obligate species is high (Holloran 2005, Sawyer et al. 2006, Walker et al. 2007a) because the five geologic basins that contain most of the onshore oil and gas reserves in the Intermountain West are within the sagebrush ecosystem (Connelly et al. 2004). Loss and degradation of native sagebrush habitats have already imperiled much of this ecosystem and its associated wildlife (Knick et al. 2003, Connelly et al. 2004, Wisdom et al. 2005b). Conservation of public lands is vital, because while 70% of remaining sagebrush habitat is publicly owned, almost none of it is protected within a federal reserve system (Knick et al. 2003). Federal policy largely dictates the fate of imperiled species because the Bureau of Land Management (BLM) is both the primary steward of public shrubland and the lead agency responsible for administering the federal mineral estate. Scientific studies that evaluate potential impacts, test sufficiency of mitigation measures, and provide conservation planning solutions help decision makers formulate policy that enables the BLM to carry out its multiple-use mandate through the Federal Land Policy and Management Act of 1976 (United States Bureau of Land Management and Office of the Solicitor 2001).

Our goal was to synthesize the biological response of sage-grouse to energy development and evaluate whether mitigation at the local scale can sustain populations as cumulative impacts from energy development increase at the landscape scale. We address this question by using coal-bed natural gas development in the Powder River Basin in northeast Wyoming as a case study to quantify changes in landscape features detrimental to sage-grouse that result from energy development. We also provide a critical review of the scientific literature to synthesize the biological response of sage-grouse to energy development. Further, we depict the current extent of development and leasing of the federal mineral estate within the eastern range of the species. Finally, we recommend a paradigm shift from local to landscape conservation and discuss the implications of this change.

ENERGY DEVELOPMENT AND LAND-USE CHANGE

Energy development and its infrastructure may negatively affect sage-grouse populations via several different mechanisms. Mechanisms responsible for cumulative impacts that lead to population declines depend in part on the magnitude and extent of human disturbance. We quantified changes in landscape features detrimental to sage-grouse that result from energy development. Males and females may abandon leks if repeatedly disturbed by raptors perching on power lines near leks (Ellis 1984), by vehicle traffic on nearby roads (Lyon and Anderson 2003), or by noise and human activity associated with energy development (Braun et al. 2002, Holloran 2005, Kaiser 2006). Collisions with power lines and vehicles and increased predation by raptors may increase mortality of birds at leks (Connelly et al. 2000a, Lammers and Collopy 2007). Roads and power lines may also indirectly affect lek persistence by altering productivity of local populations or survival at other times of the year. Sage-grouse mortality associated with power lines and roads occurs year-round (Aldridge and Boyce 2007), and artificial ponds created by development (Zou et al. 2006b) that support breeding mosquitoes known to vector West Nile virus (Walker et al. 2007b) elevate risk of mortality from disease in late summer (Walker and Naugle, this volume, chapter 9). Sage-grouse may also avoid otherwise suitable habitat as development increases (Lyon and Anderson 2003, Holloran 2005, Kaiser 2006, Doherty et al. 2008).

Methods

We quantified changes that accompany ranching, energy development, and tillage agriculture in the Powder River Basin in northeast Wyoming and southeast Montana, a landscape where intensive energy development and irrigated agriculture are intermixed with sagebrush-dominated ranchlands. The traditional land use is cattle and sheep ranching, but tillage agriculture is prevalent, with most fields planted to alfalfa. Coal-bed natural gas is the new land use relative to energy development in the Powder River Basin, the largest coal-producing basin in the United States and one of the largest natural gas fields in North America. Energy development covers much of the basin in northeast Wyoming (Fig. 20.2), with ~35,000 gas wells drilled since 1997 and 68,000 authorized on public lands. Each additional group of 2–10 wells increases the number of new roads, power lines, artificial ponds, pipelines, and compressor stations.

We quantified the cumulative change in human disturbance by classifying land cover using SPOT-5 satellite imagery acquired in 2003 for a 9,081-km^2 area of the Powder River Basin north of Sheridan, Wyoming, near Decker, Montana. Vegetation was predominantly Wyoming big sagebrush (*A. tridentata* ssp. *wyomingensis*) intermixed with native bluebunch wheatgrass (*Pseudoroegneria spicata*), western wheatgrass (*Pascopyrum smithii*), and blue grama (*Bouteloua gracilis*). Rocky Mountain juniper (*Juniperus scopulorum*) and ponderosa pine (*Pinus ponderosa*) occurred along slopes and at higher elevations. We used SPOT-5 imagery to classify five cover types as sagebrush, conifer, grassland, riparian, and bare ground. We combined the SPOT-5 panchromatic and multispectral images into a single panchromatic, multispectral file. We then used the panchromatic 25-m^2-pixel image to perform pan-sharpening to reduce the multispectral image pixel size from 100 m^2 to 25 m^2, greatly increasing the resolution of our analysis. Classification accuracy was 83% for sagebrush, 77% for conifer, 76% for grassland, 70% for riparian, and 80% for barren, with an overall accuracy of 78% (Doherty et al. 2008).

We overlaid a grid of 9-km^2 cells onto classified land cover and randomly selected 20 cells of each of four land use types: (1) ranchlands, (2) ranchlands with energy development, (3) ranchlands with tillage agriculture, and (4) ranchlands with energy development and tillage agriculture. Cells with >10% of area in cropland defined land use types with tillage agriculture. Cells with ≥4 wells defined land use types with energy development. We obtained locations of coal-bed natural gas wells from the Montana Board of Oil and Gas Conservation and the Wyoming Oil and Gas Conservation Commission. The three companies that supply electricity to this region provided spatially referenced locations of power lines. Analysts in the Spatial Analysis Laboratory at the University of Montana manually digitized roads and boundaries of tillage agriculture using SPOT-5 imagery and 1-m digital orthophotography. We also used SPOT-5 imagery to quantify the number of ponds in each 9-km^2 cell. Some ponds were stock water for cattle, but most were retention ponds to hold groundwater pumped to the surface as part of the energy extraction process. We estimated average density (linear km or number/km^2; ± SE) of each human feature in each of four land use types. We buffered collectively around all human features to estimate the area (%) of the landscape within 50, 100, and 200 m of a road, power line, pond, or tillage agriculture in each of four land use types.

Analysis

Ranching was the most environmentally benign land use that accumulated fewer human features

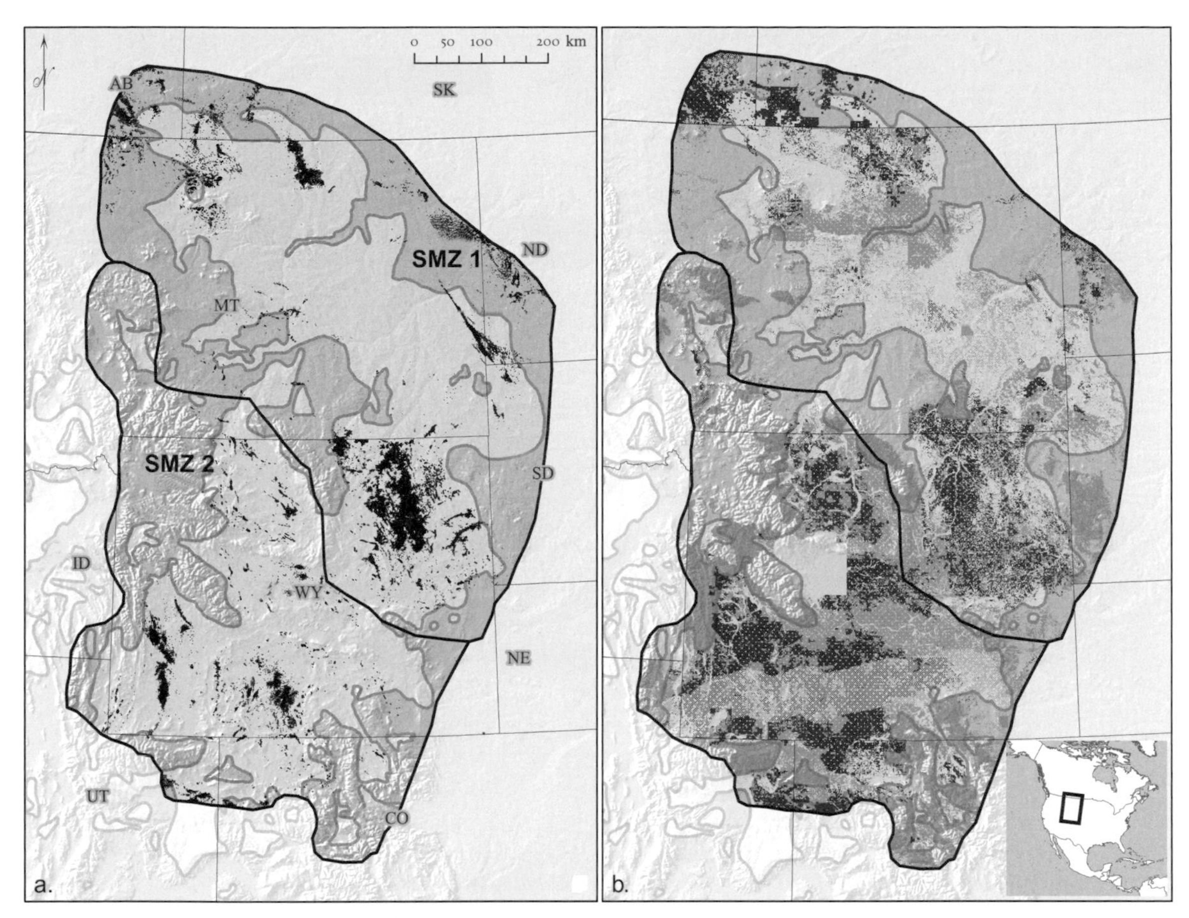

Figure 20.2. (a) Locations of producing oil and gas wells within the Great Plains (SMZ I) and Wyoming Basin (SMZ II) (Connelly et al. 2004). Range of Greater Sage-Grouse (Schroeder et al. 2004) within management zones is shown in gray. (b) Federal mineral estate is shown in gray. Authorized leases from the federal mineral estate in the United States and Canada are shaded black. Leases were authorized for exploration and development on or before 1 June 2007 for each state except Utah (which was authorized 1 May 2007). Leases in Canada were authorized for development on or before 29 January 2008 in Saskatchewan and 4 April 2008 in Alberta. A swath of authorized leases across southern Wyoming appears lighter in color because mineral ownership is mixed.

than landscapes that also contained tillage agriculture, energy development, or both (Fig. 20.3, Table 20.1; Holechek 2007). A moderate addition of tillage agriculture into ranchlands (5–10% of area tilled in 9-km^2 cells) removed sagebrush habitat and increased densities of roads (33%), power lines (59%), and water sources (167%; Table 20.1). Ranchlands with tillage agriculture had fewer human features than those with energy development (Table 20.1), but the area of the landscape juxtaposed to disturbance was similar in both (~70% within 200 m; Fig. 20.3) because tilled fields resulted in more direct habitat conversion. Ranchlands with energy development contained twice the density of roads (1.57 vs. 3.13 km/km^2) and power lines (0.27 vs. 0.58 km/km^2) and five times as many ponds (0.12 vs. 0.62 per km^2) as those where ranching was the primary land use. Human features had the highest density where ranching, tillage agriculture, and energy development coincided (Table 20.1). At this intensity of land use, 70% of the landscape was within 100 m and 85% was within 200 m of a human feature (Fig. 20.3), and densities were sufficiently high that every 1 km^2 of land could be bounded by a road (4.10 km/km^2) and bisected by a power line (0.86 km/km^2).

Quantitative analyses provide the baseline for describing the magnitude and extent of change that accompanies energy development. Impacts of energy development have been documented for a few species in sagebrush ecosystems, including mule deer (*Odocoileus hemionus*), which avoided otherwise suitable habitats within

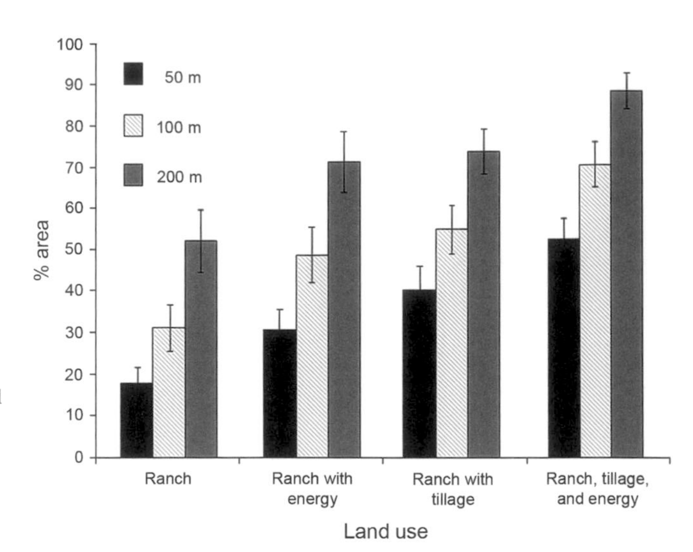

Figure 20.3. Percent area of a 9-km^2 landscape (N = 20 each category) within 50, 100, and 200 m of a road, power line, artificial pond, or agricultural tillage. Estimates of area (%) calculated for sagebrush landscapes where primary use is ranchland either with or without agricultural tillage or energy development. Error bars represent 95% CIs.

TABLE 20.1

Density (linear km or number/km^2; ± SE) of human features in sagebrush landscapes where the primary land use is livestock grazing with or without agricultural tillage or coal-bed natural gas development.

Estimates based on a sample of 80 9-km^2 cells stratified by status of agricultural and energy development (N = 20 cells in each category), Powder River Basin, Wyoming and Montana.

	Land use			
Density of human feature	Ranch land	Ranch land with tillage	Ranch land with energy	Ranch land with tillage with energy
Number of wells	0.04 ± 0.02	0.02 ± 0.01	2.82 ± 0.58	4.82 ± 0.73
Km of roads	1.57 ± 0.15	2.10 ± 0.16	3.13 ± 0.29	4.10 ± 0.36
Km of power lines	0.27 ± 0.08	0.43 ± 0.09	0.58 ± 0.09	0.86 ± 0.08
Number of ponds	0.12 ± 0.04	0.32 ± 0.12	0.51 ± 0.09	0.62 ± 0.09

2.7–3.7 km of gas wells (Sawyer et al. 2006); and Brewer's Sparrow (*Spizella breweri*) and Sage Sparrow (*Amphispiza belli*), for which breeding densities declined 36–57% within 100 m of roads in gas fields (Ingelfinger and Anderson 2004). Until energy-specific research is available for more species, the magnitude and extent of change can be used in land-use planning to anticipate thresholds of disturbance that trigger biological responses in species such as elk (*Cervus elaphus*), which either alter their habitat use (Hurley and Sargeant 1991), avoid roads altogether (Lyon 1979, Frair et al. 2007, Sawyer et al. 2007), or are affected by increased rates of poaching or legal harvest (Leptich and Zager 1991, Unsworth and Kuck 1991). Most new research on this topic has focused on sage-grouse in particular, and we further evaluate the importance of cumulative impacts by synthesizing the biological response of sage-grouse to energy development.

BIOLOGICAL RESPONSE OF SAGE-GROUSE TO ENERGY DEVELOPMENT

Methods

We conducted a literature review (Pullin and Stewart 2006) for studies that investigated relationships between sage-grouse and energy development by searching from 1980 to present in the databases of ISI Web of Science, Google Scholar, Agricola, Biological Abstracts, CAB Abstracts,

TABLE 20.2

Research citations on effects of energy development (oil and gas) to Greater Sage-Grouse.

Citation and study location	Research outlet	Pretreatment design	Length or control	Years	Sample size
Walker et al. (2007a), Powder River Basin, NE Wyoming and SE Montana	Scientific journal	Correlative	Y/Y	8	97–154 leks/year for trends, 276 leks in status analysis
Doherty et al. (2008), Powder River Basin, NE Wyoming and SE Montana	Scientific journal	Correlative	N/N	4	435 locations to build model, 74 new locations from different years to test it
Aldridge and Boyce (2007), SE Alberta, Canada	Scientific journal	Correlative	N/N	4	113 nests, 669 locations on 35 broods, 41 chicks from 22 broods
Lyon and Anderson (2003), Pinedale Mesa, SW Wyoming	Scientific journal	Observational	N/Y	2	48 females from 6 leks
Holloran (2005), Pinedale Anticline Project Area and Jonah II gas field, SW Wyoming	Ph.D. dissertation	Correlative and observational	N/Y	7	Counts of 21 leks, 209 females from 14 leks, 162 nests
Kaiser (2006), Pinedale Anticline Project Area and Jonah II gas field, SW Wyoming	M.S. thesis	Correlative and observational	N/Y	1	18 leks, 83 females (23 yearlings), plus 20 yearling males
Holloran et al. (2007), Pinedale Anticline Project Area and Jonah II gas field, SW Wyoming	U.S. Geological Survey	Correlative and observational	N/Y	2	86 yearlings (52 females), 23 yearlings (17 females) with known maternity

CSA Biological Sciences, Wildlife and Ecology Studies Worldwide, Dissertation Abstracts, and Zoological Record. We searched databases using combinations of the following key words: sage-grouse; sage grouse; *Centrocercus urophasianus*; sagebrush; habitat; land use change; resource selection function; energy development; oil development; gas development; coal-bed natural gas development; four seasons in the life cycle of a sage-grouse (breeding, brood-rearing, fall, and winter); six states, Montana, North Dakota, South Dakota, Wyoming, Colorado, and Utah; and two Canadian provinces, Alberta and Saskatchewan. Annotation for each citation used in the synthesis indicates the study length and location, rigor of peer review, scientific outlet, general research design, presence of pretreatment or control data, and sample size (Table 20.2). We provide context by describing each of the three oil or gas fields studied, giving the status of their associated sage-grouse populations (Table 20.3), and summarizing how leasing of the federal mineral estate works (Appendix 20.1).

Analysis

The search yielded 32 documents for screening from which we identified seven scientific investigations that form the foundation for this review (Table 20.2). Four investigations are published in

TABLE 20.3

Characteristics of oil and gas fields where research was conducted to evaluate response of Greater Sage-Grouse to energy development.

	Study areas		
Description	Powder River Basin	Pinedale Anticline Project Area	Manyberries Oil Field
Location	A 24,000-km^2 area (1,230 m elevation) of the basin in northeast Wyoming–southeast Montana (44°13′46″ N, 106°6′21″ W). Part of largest coal-producing basin in United States and one of the largest gas fields in North America.	A 2,550-km^2 area located within a high-desert ecosystem (2,100–2,350 m elevation) near the town of Pinedale in southwest Wyoming (42°37′33″ N, 109°53′7″ W). Pinedale is 120 km south of Jackson Hole in the northern reach of the Green River basin, an area rich in oil and gas resources.	A 150-km^2 area in southeast Alberta (49°23′60″ N, 110°42′1″ W) just outside the hamlet of Manyberries (903 m elevation) and 85 km south of Medicine Hat. This field lies within the northwest quarter of remaining occupied range of sage-grouse in the province.[b]
Development	Coal-bed natural gas is most recent and extensive development. First wells drilled in 1990s with rapid expansion in 1997. By 2007, 35,000 producing wells had been drilled. 68,000 wells are authorized for development on federal lands in Wyoming (50,000) and Montana. All but 3% of federal leases have been authorized for development.	Natural gas wells were first drilled near Pinedale in 1930s, but activity was limited until 1990s when renewed interest was prompted by production in nearby Jonah gas field. Existing development is 700 producing well pads with 645 km of pipeline and 445 km of road. Development is expected for ≥10 yr. Life of field is 50–100 yr. On federal lands, an anticipated 242 new wells will be drilled annually from 2001–2020, of which 60% will be in the anticline and Jonah field.	Renewed interest in fluid minerals in the late 1970s resulted in increased oil drilling in southern Alberta. About 1,500 wells have been drilled in and around this field, of which 30% are still producing.[c]
Ownership	Mix of federal, state, and private land in split estate where subsurface mineral rights have legal precedence over surface rights. Most of surface is privately owned (85%), but 75% of mineral rights and 10% of surface are federally owned and administered by BLM. Producing wells are on federal (30%), state (10%), and private (60%) holdings.	Ownership in the anticline is straightforward when compared to the Powder River basin. In the anticline, the BLM owns surface and mineral rights, administers permits to drill, and approves construction.	Mix of public and private land but heavily weighted to provincial surface ownership. As in the United States, mineral rights in Canada hold legal precedence over surface rights. Mineral rights are leased to industry by the province.

Stipulations and restrictions	Drilling authorized at 1 well/32 ha on federal lands. Federal lease stipulations prohibit surface infrastructure within 0.4 km of lek and restrict timing of drilling and construction activities within 3.2 km of documented lek during the breeding season (15 March–15 June) and within crucial winter habitat (1 Dec–31 March, Montana only). Restrictions (excludes operational phase), can be modified or waived, or other conditions of approval applied, on a case-by-case basis. Private and state minerals are largely developed with no restrictions.	Drilling authorized at 1 pad/16 ha. Federal stipulations prohibit surface disturbance with 0.4 km of lek, no activity within 0.8-km radius of active lek between 0000–0900 H in breeding season, and no construction or drilling during breeding within 1.6 km of active lek. BLM restricts construction activities in breeding and nesting seasons in suitable habitat within 3.2 km of active lek; restrictions only apply during initial construction. Suitable habitat designation requires that an active nest be located during on-site review. No additional restrictions placed on well field activities in operational phase.	Few limitations are placed on spacing or density of wells in Alberta. Alberta Fish and Wildlife Division provides a set of voluntary guidelines that recommend timing and setback distances around leks, but no provincial or federal legislation commits Alberta Public Lands or Alberta Energy Utility Board to implement recommendations.
Population status	Component of larger Wyoming Basin population that represents 25% of sage-grouse in species' range.[a] Supports an important regional population, with >500 known leks since 2005[a] and is a link to fringe populations in eastern Wyoming and western South Dakota and between the Wyoming Basin and central Montana.	Population stronghold for sage-grouse with some of the highest densities of males per km^2 anywhere in remaining range of species. Component of larger Wyoming Basin population that represents 25% of sage-grouse in species' range. Part of southwest Wyoming–northwest Colorado subpopulation with >800 known leks.[a]	Endangered status since 2003 under Canadian Species at Risk Act. Extirpated from British Columbia, sage-grouse remain in Alberta and Saskatchewan but occupy 6% of former range, and constitute <1% of range-wide population.[d] Population in Canada, which totals <700 individuals, declined 82–92% in <20 yr.[e] Population in Alberta declined 77–84% since 1968 and was estimated at 285–422 birds in 2005.

[a] Aldridge and Boyce (2007).
[b] Braun et al. (2002).
[c] Connelly et al. (2004).
[d] Aldridge and Brigham (2003).
[e] Lungle and Pruss (2007).

peer-reviewed journals and three are available as a dissertation, thesis, or agency report from the United States Geological Survey (Table 20.2). Nine additional documents excluded from further review include either cautionary statements regarding the potential for impacts (Braun 1987, Remington and Braun 1991, Braun 1998, Connelly et al. 2000c, Crompton and Mitchell 2005, Hanson and Wright 2006) or anecdotal evidence of lek abandonment following development (Braun 1986, Braun et al. 2002, Aldridge and Brigham 2003). The remaining 18 documents were mostly state and federal management plans that discuss sage-grouse and energy development but included no new or additional research.

Seven studies reported negative impacts of energy development on sage-grouse (Table 20.2). No study reported any positive influence of development on populations or habitats. Findings suggested that development in excess of one pad/2.6 km^2 resulted in impacts to breeding populations (Holloran 2005), and impacts at conventional well densities (8 pads/2.6 km^2) exceeded the species' threshold of tolerance (Holloran 2005, Walker et al. 2007a, Doherty et al. 2008). Negative impacts are known for three different sage-grouse populations in three different types of development (Table 20.3), including shallow coal-bed natural gas in the Powder River Basin of northeast Wyoming and extreme southeast Montana (Walker et al. 2007a, Doherty et al. 2008), deep gas in the Pinedale Anticline Project Area in southwest Wyoming (Lyon and Anderson 2003, Holloran 2005, Kaiser 2006, Holloran et al. 2007), and oil extraction in the Manyberries oil field in southeast Alberta (Aldridge and Boyce 2007). Population trends in the Powder River Basin indicated that from 2001 to 2005 lek-count indices inside gas fields declined by 82%, whereas indices outside development declined by 12% (Fig. 20.4). By 2004–2005, 38% of leks inside gas fields remained active, whereas 84% of leks outside of development remained active (Walker et al. 2007a). Male lek attendance in the Pinedale Anticline decreased with distance to the nearest active drilling rig (Fig. 20.5), producing gas well, and main haul road, and declines were most severe (40–100%) at breeding sites within 5 km of an active drilling rig or within 3 km of a producing gas well or main haul road (Holloran 2005). In an endangered population in Alberta, where low chick survival (12% to 56 days) limits population growth, risk of chick mortality in the Manyberries oil field was 1.5 times higher for each additional well site visible within 1 km of a brood location (Aldridge and Boyce 2007).

Studies have also quantified the distance from leks at which impacts of development become

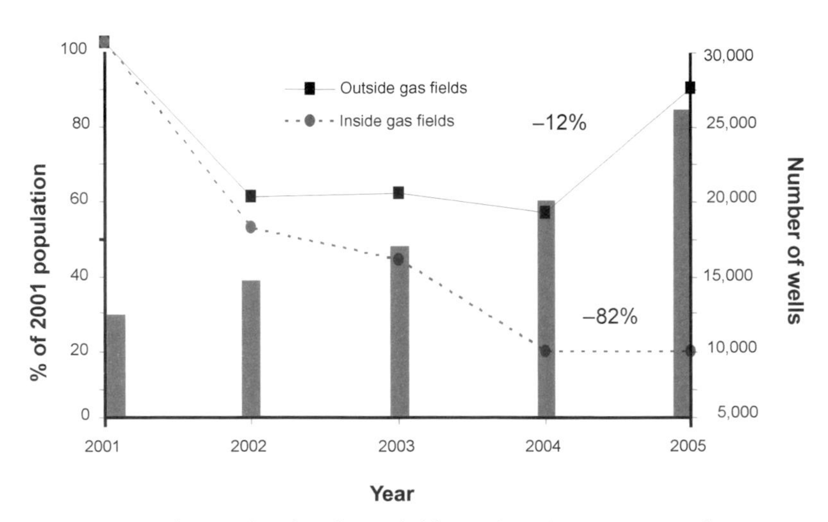

Figure 20.4. Population indices based on male lek attendance for sage-grouse in the Powder River Basin, Montana and Wyoming, 2001–2005, for leks categorized as inside or outside of coal-bed natural gas fields on a year-by-year basis (modified from Walker et al. 2007a). Leks defined as inside gas fields had ≥40% energy development within 3.2 km or >25% development within 3.2 km and ≥1 well within 350 m of the lek center. Number of producing gas wells in the Powder River Basin documents the overall increase in development coincident with declines in sage-grouse population indices.

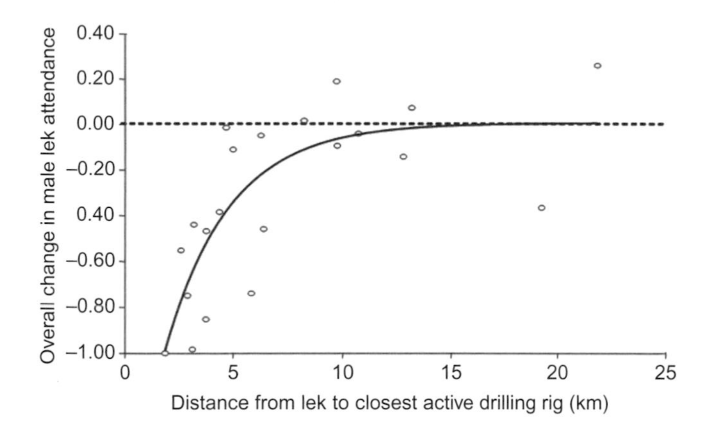

Figure 20.5. Relationship between number of male sage-grouse attending leks and average distance from leks to closest active drilling rig, Pinedale Anticline Project Area, southwest Wyoming, 1998–2004 (modified from Holloran 2005). Each point along the regression line represents one lek (N = 21).

negligible and assessed the efficacy of the current BLM stipulation of no surface infrastructure within 0.4 km of a lek (United States Department of the Interior 1992, 1994, 2004e). Impacts to leks from energy development were most severe near the lek, remained discernible out to distances >6 km (Holloran 2005, Walker et al. 2007a), and have resulted in the extirpation of leks within gas fields (Holloran 2005, Walker et al. 2007a). Curvilinear relationships in Holloran (2005) showed that lek counts decreased with distance to the nearest active drilling rig, producing well, or main haul road, and that development influenced counts of displaying males to a distance of between 4.7 and 6.2 km (Fig. 20.5). All well-supported models in Walker et al. (2007a) indicated a strong negative effect of energy development, estimated as proportion of development within either 0.8 km or 3.2 km, on lek persistence. Models with development at 6.4 km had considerably less support (5–7 ΔAIC_c units lower), but the regression coefficient ($\beta = -5.11$, SE = 2.04) indicated that negative impacts were still apparent out to 6.4 km. Walker et al. (2007a) used the resulting model to demonstrate that the 0.4-km lease stipulation was insufficient to conserve breeding sage-grouse populations in fully developed gas fields because this buffer distance leaves 98% of the landscape within 3.2 km open to full-scale development (Table 20.3). Full-field development of 98% of the landscape within 3.2 km of leks in a typical landscape in the Powder River Basin reduced the average probability of lek persistence from 87% to 5% (Walker et al. 2007a).

Negative responses of sage-grouse to energy development were consistent among studies, regardless of whether they examined lek dynamics or demographic rates of specific cohorts within populations. Recent research demonstrated that sage-grouse populations declined when birds behaviorally avoid infrastructure in one or more seasons (Doherty et al. 2008) and when cumulative impacts of development negatively affect reproduction or survival (Aldridge and Boyce 2007) or both (Lyon and Anderson 2003, Holloran 2005, Kaiser 2006, Holloran et al. 2007). Avoidance of energy development reduces the distribution of sage-grouse and may result in population declines if density dependence, competition, or displacement into poor-quality habitats lowers survival or reproduction among displaced birds (Holloran and Anderson 2005, Aldridge and Boyce 2007). Sage-grouse in the Powder River Basin were 1.3 times less likely to use otherwise suitable winter habitats that have been developed for energy (12 wells/4 km^2), and avoidance was most pronounced in high-quality winter habitat with abundant sagebrush (Doherty et al. 2008).

These studies provide compelling evidence of impacts, and long-term studies in the Pinedale Anticline Project Area in southwest Wyoming present the most complete picture of cumulative impacts of energy development on sage-grouse populations (Table 20.3). Lyon and Anderson (2003) showed that early in the development process, nest sites were farther from disturbed leks than from undisturbed leks, that nest initiation rate for females from disturbed leks was 24% lower than for birds breeding on undisturbed leks, and that 26% fewer females from disturbed leks initiated nests in consecutive years. As development of the anticline progressed, Holloran (2005) reported that adult female sage-grouse remained in traditional nesting areas regardless of increasing levels of development, but yearling females that had not yet imprinted on habitats inside the gas field avoided development by nesting farther from main haul roads. Kaiser (2006) and Holloran et al. (2007) later confirmed that yearling females avoided infrastructure when selecting nest sites and that yearling males that avoided leks inside of development were displaced to the periphery of the gas field. Recruitment of males to leks also declined as distance within the external limit of development increased, indicating greater likelihood of lek loss near the center of development.

Perhaps the most important finding from studies in the Pinedale Anticline was that sage-grouse declines are at least partially explained by lower annual survival of female sage-grouse, and that impacts to survival resulted in a population-level decline (Holloran 2005). The population decline observed in sage-grouse is similar to that observed in Kansas for the Lesser Prairie-Chicken (*Tympanuchus pallidicinctus*; Hagen 2003), a federally threatened species that also avoided otherwise suitable sand sagebrush (*Artemisia filifolia*) habitats close to oil and gas development (Pitman et al. 2005, Johnson et al. 2006). High site fidelity but low survival of adult sage-grouse combined with lek avoidance by younger birds (Kaiser 2006, Holloran et al. 2007) resulted in a time lag of 3–4 years between the onset of development activities and lek loss (Holloran 2005). The time lag observed by Holloran (2005) in the anticline matched that for leks that became inactive 3–4 years following intensive coal-bed natural gas development in the Powder River Basin (Walker et al. 2007a).

The scientific evidence from 1998 to the present that energy development is impacting sage-grouse populations has become apparent. However, questions remain concerning the exact mechanisms responsible for population declines, and manipulative experiments are needed to test the efficacy of mitigation policies and practices (Table 20.3). Burying power lines (Connelly et al. 2000a); minimizing road and well pad construction, vehicle traffic, and industrial noise (Lyon and Anderson 2003, Holloran 2005); and managing produced water to prevent the spread of West Nile virus (Zou et al. 2006b, Walker et al. 2007b) may reduce impacts. Rigorous testing is needed to know whether these or other modifications will allow sage-grouse to persist in developed areas. The severity of population-level impacts is a major concern because attempts to translocate birds for reintroduction or to supplement existing populations are rarely successful (Reese and Connelly 1997, Baxter et al. 2008).

CURRENT AND POTENTIAL EXTENT OF DEVELOPMENT

The pace and extent of oil and gas development has emerged as a major issue because areas being developed in southwest Wyoming and northwest Colorado are some of the largest and most ecologically intact sagebrush landscapes with the highest densities of sage-grouse remaining in North America (Connelly et al. 2004). Documented negative impacts suggest the pace and extent of future development will have a large role in the future status of region-wide sage-grouse populations. The current pace and extent of drilling in sagebrush habitats (Figs. 20.1, 20.2), and continued leasing of the federal mineral estate with inadequate stipulations for conservation of wildlife (Table 20.3, Appendix 20.1), increase risk of further declines in sage-grouse distribution and abundance. We depict the extent of existing development and uncertainty about the scale of future development, demonstrating the need for a fundamental shift from local to landscape conservation.

Methods

Most current energy exploration and development occurs within the remaining eastern range of sage-grouse (Fig. 20.2). IHS Incorporated provided geo-referenced data layers depicting locations of producing oil and gas wells as of 1 September 2007 on public and private lands in Montana, North Dakota, South Dakota, Wyoming, Colorado, and Utah. State offices of the BLM provided information on the extent of the federal mineral estate and locations of authorized leases within these states. Leases were authorized for exploration and development on or before 1 June 2007 for all states except Utah (1 May 2007). From Saskatchewan Industry and Resources via the *Geological Atlas of Saskatchewan*, we acquired locations of producing oil and gas wells on or before August 2007; data on authorized leases on crown lands were obtained from the same source on or before 29 January 2008. Producing well data for

Alberta were acquired from the Resource Data Branch through IHS Energy (Canada) Limited; Alberta Energy provided data on leases authorized on or before 4 April 2008.

We overlaid locations of producing wells within the range of sage-grouse (Schroeder et al. 2004) in the Great Plains and Wyoming Basin Sage-Grouse Management Zones (SMZs; Stiver et al. 2006) to illustrate the scope of energy development within sage-grouse habitats. We then overlaid authorized leases within the same geographic areas to quantify the proportion of the federal mineral estate that has been authorized for future oil and gas exploration and potential development. We overlaid wells within authorized leases to estimate the proportion of leases that contain ≥1 producing wells and to estimate the proportion of leases that are held by production (Appendix 20.1). We excluded federally protected national parks and wilderness areas from calculations; lands with other federal designations that might exclude oil and gas development—wilderness study areas and areas of critical environmental concern—were included in our analysis.

Analysis

The number of producing wells has tripled from 11,231 in the 1980s to 33,280 in 2007 within the eastern range of sage-grouse in the United States (Fig. 20.1). The United States government has authorized exploration and development (Appendix 20.1) in 7,000,000 of 16,000,000 ha (44%) of the federal mineral estate within the range of sage-grouse in the Great Plains and Wyoming Basin SMZs (Table 20.4; Fig. 20.2). Almost two-thirds of the federal mineral estate in the Great Plains and Wyoming Basin SMZs is within Wyoming (Table 20.4). Wyoming also has the highest proportion of federal leases (52%) authorized for exploration and development (Table 20.4). Lease sales are conducted under a market-driven process whereby most lands offered for lease are nominated by industry (Appendix 20.1). Opportunities for conservation remain because 4,300,000 ha of authorized leases have not yet been developed, but options are limited once leases are authorized (Appendix 20.1) and no comprehensive plan is in place to reduce impacts to sage-grouse populations. To date, 25% of the 7,000,000 ha of authorized leases contain ≥1 producing wells and 39% are held by production. The proportion of authorized leases that will be developed remains uncertain.

TABLE 20.4

Proportion of federal mineral estate within the range of Greater Sage-Grouse in the Great Plains and Wyoming Basin Sage-Grouse Management Zones with authorized leases Range is from (Schroeder et al. (2004); grey shaded area in Fig. 20.2) and management zones are from Stiver et al. (2006).

State	Proportion authorized	Area of federal mineral estate (1,000 ha)
Wyoming	52%	10,600
Colorado	50%	915
Montana	27%	3,700
Utah	25%	405
Dakotas	14%	365
Totals	44%	15,985

CONSERVATION IMPLICATIONS

Severity of impacts and continued leasing of the public mineral estate dictate the need to shift from local to landscape conservation. The scientific basis of this shift should transcend state and other political boundaries to develop and implement a plan for conservation of sage-grouse populations across the western United States and Canada. The immediate need is for planning tools that overlay the best existing areas for sage-grouse with the extent of current and projected development for all of the Great Plains and Wyoming Basin SMZs (Stiver et al. 2006). Maps that depict locations of the best remaining sage-grouse populations and their relative risk of loss will provide decision makers with the information they need to implement a conservation strategy (Doherty et al., this volume, chapter 21). Following initial implementation, site-specific information, including that of seasonal habitat use, will be necessary to test and refine the strategy. Ultimately, multiple stressors—not just energy development—must be managed collectively to maintain populations over time in priority landscapes. Integrated analyses should consider how additional factors such as habitat loss (Knick et al. 2003), restoration (Wisdom et al. 2002c), range management (Crawford et al. 2004), disease (Naugle et al. 2004), invasive weeds

(Bergquist et al. 2007), and other ecological threats such as climate change will cumulatively affect sage-grouse populations over time.

A scientifically defensible strategy can be constructed, and the most reliable measure of success will be long-term maintenance of sage-grouse populations in their natural habitats. The challenge will be for federal and state governments and the energy industry to implement solutions at a sufficiently large scale across multiple jurisdictions to meet the biological requirements of sage-grouse. One approach to conserving large populations may be to forgo development in priority landscapes until new best-management practices proven to safeguard populations are implemented. New best-management practices can be applied and rigorously tested in landscapes less critical to conservation. Practices to reduce impacts may include a combination of unitization, phased development, consolidation of well pads per unit area, and remote instrumentation to reduce traffic volume. Accelerated restoration programs may increase the probability of reestablishing populations in landscapes that are developed for energy production. We have the capability and opportunity to reduce future losses, but time is becoming critical, and the need for interjurisdictional cooperation is paramount.

ACKNOWLEDGMENTS

We thank the numerous state and federal decision makers who have pressed us to provide biological data on this important conservation issue. Comments by T. J. Christiansen, C. A. Hagen, J. M. Kiesecker, J. T. Nicholson, E. T. Rinkes, and four anonymous reviewers greatly improved this manuscript. Funding for this work came from the state offices of the Bureau of Land Management in Montana and Wyoming, Wolf Creek Charitable Foundation, the Liz Claiborne and Art Ortenberg Foundation, and the University of Montana. Well location data were supplied by IHS Incorporated, its subsidiary, and affiliated companies (copyright 2008 with all rights reserved).

APPENDIX 20.1

Laws and Processes Governing Mineral Leasing with the United States Federal Government.

Regulations governing leasing are in Title 43, Part 3100, of the Code of Federal Regulations. The Mineral Leasing Act of 1920 and the Mineral Leasing Act for Acquired Lands of 1947 give the Bureau of Land Management (BLM) leasing authority on 230,000,000 ha of BLM, National Forest, and other federal lands, and on private lands where mineral rights are retained by the federal government. Qualifications to hold a Federal lease are identified in 43 CFR 3102. Leases can be held by U.S. citizens, associations, municipalities, and companies that are incorporated in the United States. The Federal Onshore Oil and Gas Leasing Reform Act of 1987 stipulates that publicly owned lands available for leasing first be offered by competitive leasing. If no bid is received on a parcel at the competitive sale, lands are available for filing of a noncompetitive offer for two years. Maximum lease sizes are 1,032 ha for competitive (except Alaska) and 4,129 ha for noncompetitive parcels. Before land can be offered for leasing, it must be identified in a land-use plan as suitable. Title 43, Section 1712, of the Federal Land Policy and Management Act stipulates the Secretary of the Interior shall for land-use plans use a systematic and interdisciplinary approach to achieve integrated consideration of physical, biological, economic, and other sciences, and give priority to the designation and protection of areas of critical environmental concern. The Reform Act requires the BLM to hold a minimum of four oral auctions annually; because the leasing process is market driven, most lands offered for lease are nominated by industry. Values of leases vary and are determined by marketplace at auction. The BLM does not make a separate evaluation for oil and gas potential. A notice of competitive lease sale is posted for at least 45 days prior to the auction date. Leases grant lessees rights to explore and drill for, extract, remove, and dispose of oil and gas deposits found in leased lands. Granted rights are subject to terms of lease, stipulations attached to lease, applicable laws, the Secretary of the Interior's regulations, and formal orders in effect as of lease issuance and to regulations and formal orders subsequently placed into effect that are not inconsistent with specific lease provisions. Mitigation measures identified in the planning document are attached to leases. Typical mitigation measures for sage-grouse are described (Table 20.3). Lessees submit application to the BLM for a permit to drill, at which time restrictions can be modified or waived by the BLM, or additional conditions of approval applied, on a site-by-site basis. Most state and private minerals are developed with few or no requirements to mitigate impacts to wildlife. Prior to approval of a drilling permit, bonding must be filed. Minimum amounts for drilling bonds are found in 43 CFR 3104. The Energy Policy Act of 1992 directs competitive and noncompetitive leases to be issued for a 10-year period. Annual rental rates for leases paid to the Federal Minerals Management Service are $3.75/ha or fraction thereof the first five years and $5.00/ha each year thereafter. For leases issued under the 1987 Reform Act, royalty rate for produced oil or gas is 12.5% and payable to the Minerals Management Service. Leases are subject to extension privileges outlined in 43 CFR 3107 and may be held by production as long as there is oil or gas being produced in paying quantities on the lease or within a unit agreement (e.g., communitization) to which the lease is committed. Lessee may surrender a lease in whole or part by filing a written relinquishment with the BLM.

CHAPTER TWENTY-ONE

Energy Development and Conservation Tradeoffs

SYSTEMATIC PLANNING FOR GREATER SAGE-GROUSE IN THEIR EASTERN RANGE

Kevin E. Doherty, David E. Naugle, Holly E. Copeland, Amy Pocewicz, and Joseph M. Kiesecker

Abstract. We developed a framework for conservation planning to evaluate options for reducing development impacts on Greater Sage-Grouse (*Centrocercus urophasianus*) in Wyoming, Montana, Colorado, Utah, North Dakota, and South Dakota that contained some of the largest populations and highest risk of energy development. We used lek-count data (N = 2,336 leks) to delineate high-abundance population centers, which we termed core regions, that contained 25%, 50%, 75%, and 100% of the known breeding population. We assessed vulnerability of these areas by examining risk of future land transforming uses from energy development. Sage-grouse abundance varied by state. Core regions contain a disproportionately large segment of the breeding population, and core regions vary dramatically by risk of future energy development. Wyoming contains 64% of the known sage-grouse population and more active leks than all the other states combined within our study area. Conservation success in Wyoming will depend on leasing and permitting policy decisions because this state has the highest risk of development. Montana contains fewer sage-grouse (24%) than Wyoming, but actions that reduce sagebrush (*Artemisia* spp.) tillage by providing private landowners incentives to maintain sagebrush-dominated landscapes would provide lasting benefits because core regions in Montana are at comparatively low development risk. Habitat restoration in areas with low risk of development but containing fewer sage-grouse fit into the overall conservation strategy by targeting populations that promote connectivity of core regions. This vulnerability assessment illustrates the tradeoffs between conservation and energy development, and provides a framework for maintaining populations across the species' eastern range.

Key Words: Centrocercus urophasianus, conservation planning, core regions, energy development, lek counts, prioritization, risk assessment, sage-grouse, Wyoming.

El Equilibrio Entre el Desarrollo de Energía y la Conservación: Planeamiento Sistemático para el Greater Sage-Grouse en la Extensión Oriental de Su Territorio

Resumen. Desarrollamos un marco para el planeamiento de la conservación para evaluar opciones que ayuden a reducir los impactos del desarrollo sobre el Greater Sage-Grouse (*Centrocercus urophasianus*) en Wyoming, Montana, Colorado, Utah,

Doherty, K. E., D. E. Naugle, H. E. Copeland, A. Pocewicz, and J. M. Kiesecker. 2011. Energy development and conservation tradeoffs: systematic planning for Greater Sage-Grouse in their eastern range. Pp. 505–516 *in* S. T. Knick and J. W. Connelly (editors). Greater Sage-Grouse: ecology and conservation of a landscape species and its habitats. Studies in Avian Biology (vol. 38), University of California Press, Berkeley, CA.

y Dakota del Norte y del Sur, siendo que estos estados contuvieron algunas de las poblaciones más grandes y de riesgo más alto de desarrollo de energía. Utilizamos datos de conteo de los leks (asambleas de cortejo) ($N = 2{,}336$ leks) para delinear centros de población de gran abundancia a las que denominamos regiones núcleo, que contuvieran 25%, 50%, 75%, y 100% de la población de cría conocida. Determinamos la vulnerabilidad de estas áreas examinando el riesgo de futuros cambios en el uso de la tierra para el desarrollo de energía. La abundancia del sage-grouse varía según el estado, las regiones núcleo contienen un segmento desproporcionadamente grande de la población de cría, y las regiones núcleo varían dramáticamente por el riesgo de desarrollo de energía futuro. Wyoming contiene el 64% de la población conocida del sage-grouse, y la mayor cantidad de leks activos que el resto de los estados combinados dentro de nuestra área de estudio. El éxito de conservación en Wyoming dependerá del leasing (alquiler con opción a compra) y de permitir decisiones políticas ya que este estado tiene el riesgo más alto de desarrollo. Montana contiene menos sage-grouse (24%) que Wyoming, pero las acciones que reduzcan la labranza del sagebrush (*Artemisia* spp.) ofreciendo incentivos a los terratenientes privados para mantener los paisajes dominados por artemisa, proporcionarían ventajas duraderas porque las regiones núcleo en Montana están en riesgo de desarrollo comparativamente bajo. La restauración del hábitat en áreas con poco riesgo de desarrollo pero que contengan menos sage-grouse cabe dentro de la estrategia general de conservación al fijar como objetivo poblaciones que promuevan la conectividad de las regiones núcleo. Esta evaluación de la vulnerabilidad ilustra el equilibrio (o tradeoffs) entre la conservación y el desarrollo de energía, y proporciona un marco para proteger a las poblaciones en la extensión oriental del territorio de esta especie.

Palabras Clave: Centrocercus urophasianus, conteos de lek, desarrollo de la energía, evaluación de riesgos, planificación de la conservación, priorización, regiones centrales, sage-grouse, Wyoming.

World demand for energy is predicted to increase by ≥50% in the next 20 years (International Energy Agency 2007, National Petroleum Council 2007). The Rocky Mountain West will be one of the most heavily affected landscapes in the continental United States, as it has 7% of proven onshore oil reserves and 26% of natural gas reserves (United States Departments of the Interior, Agriculture, and Energy 2006). Meeting 20% of United States energy demand with wind power could impact 50,000 km^2, a significant portion of which would be in the Rocky Mountain West (United States Department of Energy 2008). The increasing energy demand of an expanding human population poses a challenge to conservation of wildlife populations in North America (Sawyer et al. 2006, Walker et al. 2007a). Energy development is known to impact wildlife directly by altering habitat use (Doherty et al. 2008) and population dynamics (Sorensen et al. 2008), and indirectly by facilitating spread of nonnative invasive plants (Bergquist et al. 2007) and new diseases such as West Nile virus in North America (Naugle et al. 2004, Zou et al. 2006b). The ability to identify areas of high biological value and assess the potential for adverse habitat alteration is a component of a proactive rather than a reactive approach to conservation (Groves et al. 2002). Not all wildlife areas are of equal value, and mapping high-abundance population centers for a priority species can help frame regional plans. Realization of conservation goals requires plans be constructed at broad spatial scales to provide for effective management (Soulé and Terborgh 1999, Margules and Pressey 2000). Plans that explicitly examine tradeoffs between wildlife conservation and energy development will need to be broad in scale to be effective, given the scale of anticipated energy development in the western United States.

Loss and degradation of native vegetation has impacted much of the sagebrush (*Artemisia* spp.) ecosystem and associated wildlife (Knick et al. 2003, Connelly et al. 2004). Greater Sage-Grouse (*Centrocercus urophasianus*; hereafter, sage-grouse) is a gallinaceous species native only to western semiarid sagebrush landscapes (Schroeder et al. 1999). Previously widespread, sage-grouse have been extirpated from nearly half of their original range in western North America (Schroeder et al.

2004), with a range-wide population decline of 45–80% and local declines of 17–92% (Connelly and Braun 1997, Braun 1998, Connelly et al. 2004). Energy development has emerged as a key issue in sage-grouse conservation for three reasons: (1) sage-grouse populations decline with oil and gas development (Holloran 2005, Aldridge and Boyce 2007, Walker et al. 2007a); (2) landscapes being developed contain some of the highest abundance estimates for sage-grouse in North America; and (3) 44% of the lands that the federal government has authority to control for oil and gas development in the eastern range of sage-grouse (7,000,000 of 16,000,000 ha) have already been authorized for exploration and development (Naugle et al., this volume, chapter 20).

It is urgent that we identify areas of high biological value and areas of potential future development to evaluate options for reducing impacts (Abbitt et al. 2000, Balmford et al. 2001, Wilson et al. 2005), given sage-grouse sensitivity to oil and gas development and the projected rate of increased development. We focused on identifying core regions of sage-grouse abundance to illustrate the process of risk assessment and to contrast opposing conservation strategies. Lek-count data provided an opportunity to spatially identify the distribution and abundance of core regions of habitat that support breeding populations. Our goal was to develop a conservation planning framework (Pressey and Bottrill 2008) to address the following questions using readily available spatial data: (1) Where are landscapes with the highest biological value for sage-grouse? (2) How do these landscapes differ with respect to risk from future energy development? and (3) How does variation and juxtaposition in risk and biological values of areas affect the potential to develop a successful conservation strategy for sage-grouse?

STUDY AREA

Our study area included landscapes within the eastern distribution of sage-grouse including portions of Colorado, Montana, North Dakota, South Dakota, Utah, and Wyoming (Fig. 21.1; Schroeder et al. 2004). Schroeder et al. (2004) used a combination of lek-survey data; geographic information system (GIS) habitat layers to exclude barren areas, alpine areas, and forest habitats; and locations of radio-marked sage-grouse to delineate the current occupied distribution for sage-grouse in all of North America. We modified this boundary to include 27 additional known lek locations in Montana, South Dakota, Wyoming, and Colorado outside the boundaries suggested by Schroeder et al. (2004). We adopted a spatial organizational framework based on the Western Association of Fish and Wildlife Agencies Sage-Grouse Management Zones (SMZs; Stiver et al. 2006), which are delineated by floristic provinces and used to group sage-grouse populations for management actions. We restricted analyses to areas within the eastern distribution that were within the Great Plains SMZ (portions of Montana, Wyoming, North Dakota, South Dakota, Saskatchewan, and Alberta) and the Wyoming Basin SMZ (portions of Idaho, Wyoming, Utah, Montana, and Colorado) (Fig. 21.1; Stiver et al. 2006) because these populations are experiencing the highest risk of energy development. All analyses presented evaluate the relative importance of an individual breeding area to all other breeding areas within these management zones (Fig. 21.1).

METHODS

Sage-Grouse Abundance Data

Knowledge of high-abundance population centers for priority species represents a starting point to frame regional conservation initiatives and can direct management actions to landscapes where they will have the largest benefit to regional populations (Groves et al. 2002, Sanderson et al. 2002b). Techniques such as resource selection functions have been widely used in the absence of large-scale survey data to identify critical habitat needs and to map those habitats at appropriate scales for a wide range of species (McLoughlin et al. 2002, Boyce et al. 2003, Johnson et al. 2006) including sage-grouse (Aldridge and Boyce 2007, Doherty et al. 2008). No seamless habitat coverage is available for sage-grouse to build seasonal models that could form the comparison of the relative biological value of different landscapes. Fortunately, sage-grouse are one of the few species in which extensive data sets exist on distribution and relative abundance across their entire breeding distribution, making an analysis of this scale possible (Connelly et al. 2004, Schroeder et al. 2004). The concept of using high-abundance population centers to define the size, shape, connectivity,

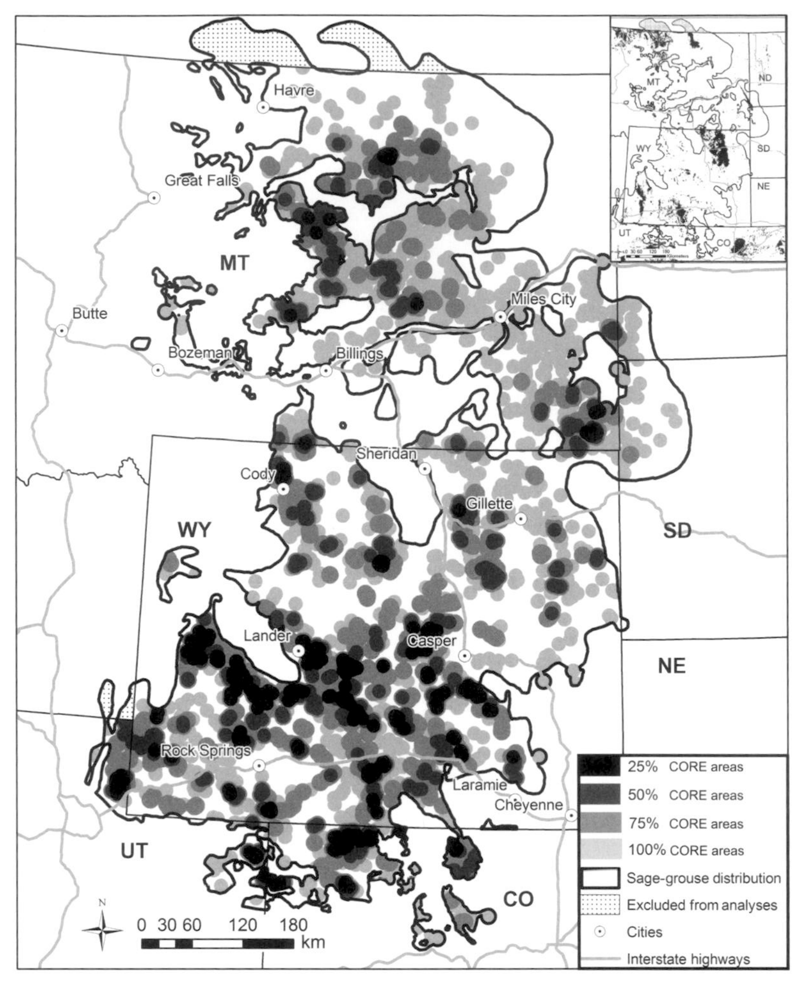

Figure 21.1. Core regions that contain 25%, 50%, 75%, and 100% of the known breeding population of Greater Sage-Grouse in their eastern range. Distribution boundaries are the combined areas of the Great Plains and Wyoming Basin Greater Sage-Grouse Sage-Grouse Management Zones (Stiver et al. 2006). Inset depicts locations of producing oil and gas wells (black triangles) as of September 2007.

replication, and spacing of conservation areas is well documented in other systems (Myers et al. 2000, Groves et al. 2002, Sanderson et al. 2002b).

Breeding ground (lek) data have been widely used by agencies to monitor sage-grouse population trends and are considered a reasonable index to relative abundance (Walsh et al. 2004, Reese and Bowyer 2007). Each spring, displaying males are counted within each state on sage-grouse leks in a large, coordinated effort by state, federal, and contract employees across the entire distribution of the species. Typically, leks are visually surveyed at least three times each spring from the air or ground, and displaying males are counted during the early morning. Protocols for counting males at leks were almost identical among states following the recommendations of Connelly et al. (2003b), which allowed for comparisons between state populations.

We used the maximum count of male sage-grouse to identify high-abundance areas. Each state wildlife agency assembled and provided us a maximum lek count for each year the lek was surveyed over the past 11 years, along with spatial coordinates of lek locations. This maximum count database provided us the ability to map relative abundance of sage-grouse breeding areas. We analyzed 2,336 active leks to delineate breeding core regions. We defined active leks as those on which ≥2 males were counted in the last year the lek was surveyed. We used the highest count during 2005–2007 because not all leks are counted each year, but most are counted within a three-year interval. However,

249 leks in Montana, primarily in Rosebud, Custer, and Garfield Counties, were not counted during this interval, and we used the most recent survey within the 11-year interval to assign abundance values to these leks. We also included the last count of five leks in Colorado after consultation with regional biologists indicated that counts of 0 males recorded in 2007 were likely a result of no survey effort.

Mapping Core Sage-Grouse Breeding Areas

Kernel density functions have been commonly used in ecology to delineate home ranges of individual animals and to map concentrated areas of use by populations (Silverman 1986, Worton 1989). A kernel is a mathematical density function that groups cells of concentrated use by attributing a grid placed over top of a study site with animal use or count data (Silverman 1986, Worton 1989). We populated a 1-km^2 grid of cells with counts of sage-grouse males at leks across the eastern range of the species. We used this grid to select individual leks for conservation priority groupings. We modified the kernel function because choice of smoothing bandwidth is known to drastically affect area estimates and outer boundaries of home ranges and concentrated areas of use by populations (Seaman et al. 1999, Kernohan et al. 2001, Horne and Garton 2006). We circumvented the bandwidth choice problem and used known distributions of nesting females around leks to delineate the outer boundaries of core regions (Holloran and Anderson 2005, table B-1 of Colorado Division of Wildlife 2008).

The value of each grid cell is a function of the number and proximity of leks in the surrounding landscape. We attributed each cell with counts of males at leks within a radius of 6.4 km (4.0 mi). We chose this distance because nesting females distribute their nests spatially in relation to lek location, with 79% of nests located within a 6.4-km radius from lek-of-capture (table B-1 of Colorado Division of Wildlife 2008). We ranked leks by abundance values and placed each into four groups that contained 25%, 50%, 75%, or 100% of the known breeding population, and buffered these leks by 6.4 km to delineate nesting areas. We extended the radius from 6.4 to 8.5 km for leks in 75% and 100% core regions (Holloran and Anderson 2005), because a post-hoc analysis indicated that 6.4 km was too small an area to contain simulated nest densities in lower population density areas and fragmented habitats where a few leks were far apart (e.g., North and South Dakota; Table 21.1). Increasing the radius in

TABLE 21.1

Characteristics of Greater Sage-Grouse leks used to delineate core regions.

State	No. of leks with ≥2 males 2005–2007	Average (±SD) maximum male count 2005–2007	Relative abundance (%)	Average (±SD) distance (km) to nearest lek	Median distance (km) to nearest lek	Current distribution
Colorado[a]	200	33.4 ± 32.4	8	4.6 ± 4.2	3.5	17,061[b]
Montana	869	23.6 ± 20.5	24	4.6 ± 4.2	3.5	127,242
North Dakota	14	15.4 ± 14.5	<1	8.6 ± 2.6	8.3	2,829
South Dakota	21	28.2 ± 13.0	1	10.4 ± 5.2	9.8	10,074
Utah[c]	71	37.3 ± 34.6	3	4.4 ± 4.6	2.7	7,046
Wyoming	1,190	47.3 ± 45.2	64	5.0 ± 3.6	4.3	176,424
Great Plains and Wyoming Basin SMZs[d]	2,336	37.2 ± 40.0	100	4.8 ± 3.7	4.1	338,789[e]

[a] Total included 29 leks in the Colorado Plateau Sage-Grouse Management Zone (SMZ).

[b] Area estimate included portions of the Colorado Plateau SMZ.

[c] Included leks in the Colorado Plateau SMZ.

[d] Leks do not sum because Colorado includes 29 leks from the Colorado Plateau SMZ.

[e] Area estimate excludes Idaho, Canada, and the Colorado Plateau SMZ.

75% and 100% core regions provided more realistic estimates of the area needed to support breeding populations in low-abundance or fragmented landscapes. Our model output is a grouping of leks in four shades that represent the smallest area necessary to contain 25%, 50%, 75%, and 100% of the nesting sage-grouse population. Area estimates are inclusive; 25% core regions are included within the boundaries of 50% core regions.

Mapping Energy Potential

We used readily available spatial data to rapidly assess the potential for energy development in sage-grouse core regions. Our risk assessment included indicators for two major forms of energy development in the eastern range: oil and gas, and more recently, wind power (Fig. 21.2). We acquired information on oil and gas development by compiling locations of authorized oil and gas leases within Montana, North Dakota, South Dakota, Wyoming, Colorado, and Utah from Bureau of Land Management state offices within the Great Plains and Wyoming Basin SMZs. Leases were authorized for exploration and development on or before 1 June 2007 for all states except Utah (1 May 2007). We obtained geo-referenced data layers depicting locations of producing oil and gas wells as of 1 September 2007 on public and private lands in Montana, North Dakota, South Dakota, Wyoming, Colorado, and Utah from IHS Incorporated, Englewood, CO. We used data from the National

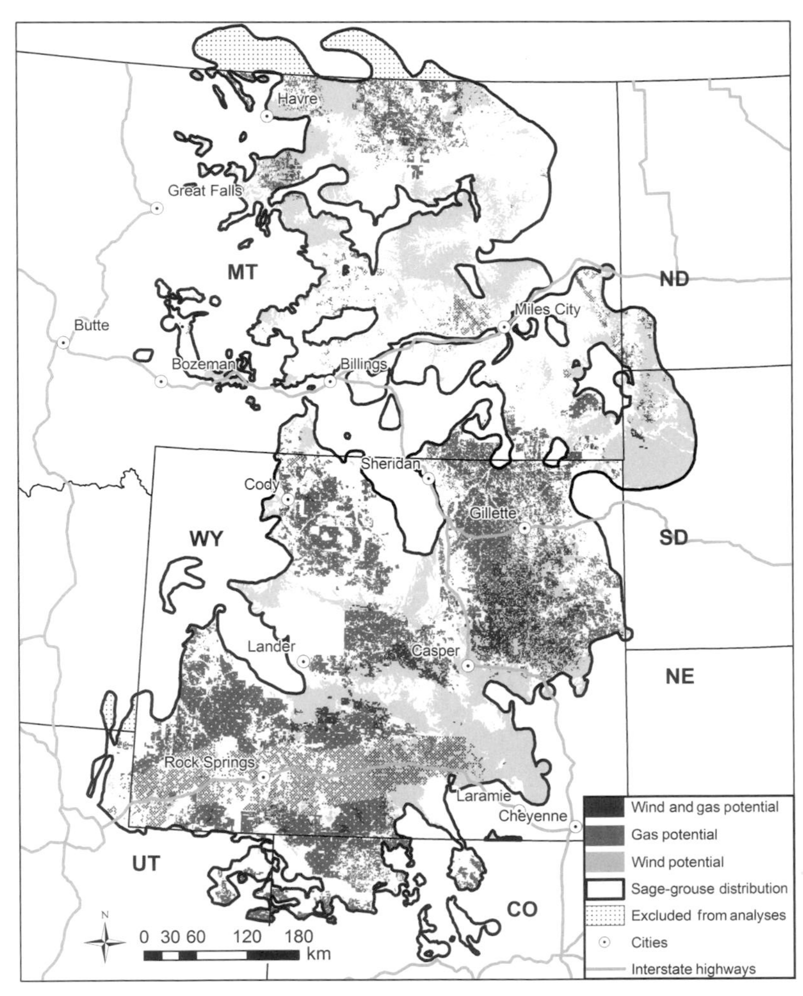

Figure 21.2. Potential for oil and gas and wind development in the eastern range of Greater Sage-Grouse (Great Plains and Wyoming Basin Sage-Grouse Management Zones; Stiver et al. 2004). A swath of authorized leases across southern Wyoming following the interstate between Laramie and Rock Springs appears lighter in color because of the checkerboard pattern of mineral ownership.

Renewable Energy Laboratory to represent the potential for commercial wind potential (National Renewable Energy Laboratory 2008). Wind classes are grouped from 1–7 with all wind classes ≥4 having potential for commercial energy production.

Conservation Planning Analyses

Systematic conservation planning requires identification of areas to achieve specific goals (Pressey et al. 2007). Our core-areas analyses delineate specific landscapes that differ markedly in biological value and offer a means to rank their relative importance. Conservation planning also requires that areas identified with high value have the ability to persist over time (Groves et al. 2002). We conducted a series of GIS overlays of biological values of sage-grouse with the potential for energy development to frame the opportunities and challenges facing sage-grouse in relation to energy development. The intersection of high biological value with high energy potential frames the risk of development to sage-grouse populations. We first quantified the proportion of 25%, 50%, 75%, and 100% core regions at risk from oil and gas development, wind power development, or both. We quantified the risk of development of oil and gas and of wind power to 75% core regions by state and quantified the proportion of land with federal management to document how risk varies by state. We mapped the location of current oil and gas wells in relation to core regions to highlight the importance of core regions next to development to promote resilience of areas disturbed by energy development (Groves et al. 2002, Lindenmeyer et al. 2008). We used a factorial analysis to categorically define biological value and energy potential into four categories that show opportunities for both conservation and energy development across the landscape based on all possible combinations of biological value (low or high) and energy potential (low or high). We defined an area as having high biological value if it was in the top three groupings of breeding densities (25%, 50%, and 75% core regions), as these groups contained 75% of the regional breeding population in only 30% of the total eastern sage-grouse distribution. We included 100% core regions as high biological value in North Dakota and South Dakota because these fringe populations experience the highest risk of extirpation (Aldridge et al. 2008). We defined our 100% core area group as low biological value elsewhere. If an area did not have a lek within 8.5 km (Holloran and Andersen 2005), it was not assigned a biological value because we did not have information on other seasonal habitats. We considered an area to have high potential for energy development if it had either an authorized oil or gas lease from the federal government or showed potential for commercial wind production (Fig. 21.2). Areas excluded from the high potential category were classified as having low potential for energy development. The result was four categorical and spatially explicit groups mapped in a GIS (Fig. 21.3).

RESULTS

Sage-grouse abundance regionally exhibited a clumped distribution, making it possible to identify core regions that contained a large proportion of the breeding population within a relatively small proportion of the species' eastern range (Fig. 21.1). Core regions contained 25%, 50%, 75%, and 100% of the breeding population within 5%, 12%, 30%, and 60% of the eastern sage-grouse range. Bird abundance varied within core regions. Wyoming contained the highest proportion of high-density areas (Fig. 21.1), largest number of leks, highest male sage-grouse abundance at leks, and the broadest species distribution among the six states within our study area (Table 21.1). Wyoming provides habitat for nearly two-thirds of all known sage-grouse within our study area, while Montana, having the second-largest and most expansive population, provides habitat for an additional quarter of the sage-grouse in our study area (Table 21.1). A small area of northwest Colorado also supports an especially high abundance of breeding birds per unit area, relative to the entire eastern range of sage-grouse (Fig. 21.1).

Risk of energy development to sage-grouse core regions increased as the relative biological value increased across the entire eastern range (Table 21.2). Half (51%) of 25% core regions are at risk from either wind or oil and gas development, whereas 39% of the 100% core regions are at risk. This is a function of the locations of oil and gas leases. Over one-third of the 25% core regions have been leased for oil and gas development, whereas one-fifth of the eastern distribution is leased (Table 21.2). Potential for wind energy development is also widespread across the eastern range; however, core regions did not exhibit increasing risks as biological value increased (19–21% risk; Table 21.2). Development risk is highly noncomplementary with <5% spatial overlap of potential oil/gas and wind development, which increased the total land area at risk (Table 21.2, Fig. 21.2).

Energy development risks differed by state (Table 21.3, Fig. 21.2) and are highest in Wyoming, intermediate in Colorado, and lowest in Montana, the three states with 95% of the sage-grouse (Table 21.3). Wyoming has the highest proportion of 75% core regions at risk from both oil and gas and wind development (Table 21.3). Oil and gas development is the primary threat in Colorado and Utah, while wind development poses a greater risk to sage-grouse core regions in Montana, North Dakota, and South Dakota (Table 21.3). Overall, threats from energy development to 75% core regions ranged from 9% to 73% of breeding areas (Table 21.3).

Factorial analysis documented large landscapes within each category (Fig. 21.3). Analyses classified 84,896 km^2 of land as low biological value with high potential for energy development (25% of range; Fig. 21.3) and 64,641 km^2 as low potential for energy development (19% of range; Fig. 21.3). The inclusion of 100% core regions in North Dakota and South Dakota brought the total area classified as high biological value to 31%. Analyses classified 46,419 km^2 of land as high biological value for sage-grouse with high potential for energy development (14% of range; Fig. 21.3) and 59,237 km^2 as low energy potential (17%; Fig. 21.3). The proportion of areas with high biological value and low energy potential varied greatly by state, as did federal surface and mineral ownership (Table 21.3). Montana had 72% of its high value core regions with low potential for development and had 31% federal surface ownership and 45% federal subsurface ownership (Fig. 21.3; Table 21.3). Wyoming had 49% of areas with high biological value and low energy potential but was 57% federally owned on the surface and 69% controlled by federal subsurface ownership. Large-scale

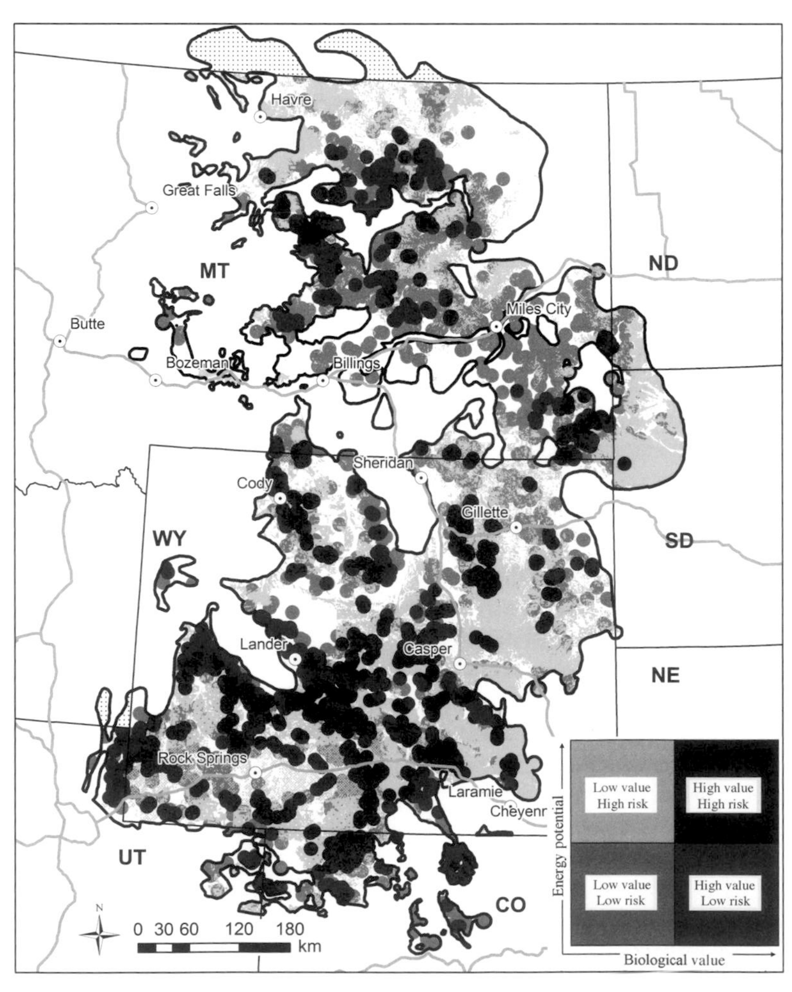

Figure 21.3. Overlay of biological value (25–75% core regions = high value) with energy potential for oil and gas or wind development to assess risk of development to Greater Sage-Grouse core regions.

TABLE 21.2

Percent of Greater Sage-Grouse core regions at risk of wind and/or oil and gas development in Wyoming, Montana, Colorado, Utah, South Dakota, and North Dakota (through September 2007).

	High wind potential[a]	Authorized oil and gas leases[b]	Both	Either
25% core regions	20.2	34.3	3.7	50.8
50% core regions	19.4	31.5	4.1	46.9
75% core regions	19.0	28.0	3.9	43.1
100% core regions	18.7	23.4	3.2	38.8
Eastern distribution	21.4	20.8	3.3	38.8

[a] We defined high wind potential as a wind class rating ≥4 (National Renewable Energy Laboratory 2008).

[b] Authorized leases include federal oil and gas leases authorized for exploration and development on or before 1 June 2007 for each state except Utah (1 May 2007).

TABLE 21.3

Greater Sage-Grouse 75% core regions at risk of wind and/or oil and gas development by state (through September 2007).

State	High wind potential[a] ownership	Authorized oil and gas leases[b]	Both	Either	Federal surface
Wyoming	21.2	35.7	5.7	51.2	57.1
Montana	19.8	8.5	0.6	27.7	31.4
Colorado	0.3	33.7	0.1	33.8	43.3
Utah	0.4	8.6	0.0	9.1	46.5
South Dakota	72.3	3.7	3.2	72.9	11.5
North Dakota	28.9	10.0	2.2	36.6	56.8

[a] We defined high wind potential as a wind class rating ≥4 (National Renewable Energy Laboratory 2008).

[b] Authorized leases include federal oil and gas leases authorized for exploration and development on or before 1 June 2007 for each state except Utah (1 May 2007).

development has already occurred next to core regions, especially in Wyoming (Fig. 21.1).

DISCUSSION

Landscape planning to balance wildlife conservation with resource development must be analogous in scale to be effective given the spatial extent of anticipated impacts. Successful planning must embrace the social and political realities of the region (Lindenmeyer et al. 2008). Our analysis is sufficiently broad in scale to allow a relevant examination of the necessary tradeoffs, and by assessing the potential impacts of energy development, we bring recognition of the political reality of energy development in the West. The framework presented provides the necessary structure to illustrate the tradeoffs between sage-grouse conservation and energy development. The next generation of analyses to direct conservation action should be twofold. First, there is a need to support implementation of core regions with studies that document seasonal habitat use and migration patterns of radio-marked sage-grouse

(Aldridge and Boyce 2007, Doherty et al. 2008) to ensure identified priority landscapes meet all seasonal habitat needs. Second, incorporation of future modeling of other relevant risks, such as cheatgrass (*Bromus tectorum*) invasion, to core regions will ensure gains in conservation will not be offset by unevaluated risks.

Resources available to implement landscape conservation invariably are in short supply relative to need. Setting priorities for conservation action is a necessary and major task for agencies and organizations concerned with conservation of species and ecosystems (Groves et al. 2002, Newburn et al. 2005). Core regions enable decision makers to spatially prioritize their targets for sage-grouse conservation. Our results suggest that, given the nature of sage-grouse distribution, a large portion of the breeding population can be conserved within core regions. For example, 75% of the breeding population can be captured within only 30% of the area. However, distribution of core regions and their value vary. Wyoming contains 64% of the known breeding population in this region and more active leks than all other states combined. Risks to core regions vary dramatically in concordance with variation in value of these regions. Wyoming has the greatest combined risk from both wind energy and oil and gas development, but also has the greatest potential for conservation in terms of the value of core regions. The intersection of the value of the core regions and the risks to which they are exposed (Figs. 21.2, 21.3) suggests a series of strategies needed to ensure long-term persistence of sage-grouse: (1) policy changes are needed in areas of high biological value and high risk of energy development to manage leasing and permitting of oil and gas development on federal lands and to proactively site future wind developments; (2) rapid implementation of conservation is needed to enhance populations in high value biological areas without energy potential; and (3) restoration of fringe habitats and low-density areas with limited risk is needed to promote connectivity. We explore each of these strategies in detail.

Landscapes with high biological value for sage-grouse and high risk for development represent the greatest challenge facing land use managers. This is a concern because 44% of areas with high biological value are at risk for energy development (darkest gray areas, Fig. 21.3). The rapid pace and scale of oil and gas drilling has emerged as a major issue because areas being developed (i.e., southwest Wyoming and northwest Colorado) include some of the largest remaining sagebrush landscapes with the highest densities of sage-grouse in North America (Fig. 21.1; Connelly et al. 2004). The future of sage-grouse conservation is in question in the eastern range in part because 44% of the lands that the federal government has authority to control for oil and gas development (7,000,000 of 16,000,000 ha) have been authorized for exploration and development (Naugle et al., this volume, chapter 20). Lease sales continue, despite concerns, because no policy is in place that would permit an environmental assessment of risk at the scale at which impacts occur. Severity of impacts (Holloran 2005, Aldridge and Boyce 2007, Walker et al. 2007a) and the unprecedented leasing of the public mineral estate dictate the need for a shift from piecemeal to landscape-scale conservation. Our analyses will enable policymakers to consider a portfolio of set-aside areas, priority conservation areas, lease consolidations, and more stringent, spatially based, best-management practices as creative solutions to balance energy development with sage-grouse conservation.

Wind power is an emerging issue contributing to the overall risk of energy development to sage-grouse populations (Figs. 21.2, 21.3). There is an urgent need for policies that promote landscape-scale considerations when siting wind facilities, as well as for replicated research to quantify potential impacts (Stewart et al. 2007). The low overlap between wind potential and oil and gas leasing highlights the need to incorporate multiple stressors in planning efforts, because unconsidered stressors could negate conservation actions. Lands with federal surface ownership are being leased at increasing rates for wind development, and a similar portfolio of tools could be considered to reduce impacts on these lands. However, much of the future wind energy development is anticipated to occur on private lands with little or no regulatory oversight. The lack of a landscape-planning paradigm is especially of concern for populations in Montana, North Dakota, and South Dakota, where the primary risk is unplanned large-scale wind development on private lands. Private lands with high value sage-grouse habitat might be considered for purchased conservation easement agreements with landowners that limit surface development. Yet the high purchase cost of easements and even higher profitability of wind development for private landowners require broader strategies to minimize wind development footprints. Ultimately, policy decisions on placement of new energy transmission corridors built to carry electricity from new wind developments will be a major factor in wind development and may be used to further refine risk assessment.

High biological value and low energy potential identify low-conflict areas to immediately focus conservation actions. Currently, 17% of the eastern sage-grouse range has high biological value and low risk from energy development (Fig. 21.3). Maintaining these quality sage-grouse habitats, especially in areas adjacent to development (Fig. 21.1) or where development is anticipated (Figs. 21.2, 21.3), will be critical to ensure genetic connectivity (Oyler-McCance et al. 2005a,b) and natural recolonization after oil and gas development activities have ceased (Gonzalez et al. 1998). Strategies in these high value and low energy potential areas should further focus on reducing risks from other stressors to sagebrush habitats (Klebenow 1970; Connelly et al. 2000b,c; Leonard et al. 2000; Smith et al. 2005; Walker et al. 2007a) such as tillage (Farrell et al. 2006, United States Government Accounting Office 2007), residential development (Theobald 2003, 2005), and invasive plants such as cheatgrass (Bergquist et al. 2007). Rural areas with desirable natural amenities and recreational opportunities throughout the United States have experienced a surge in rural development since the 1970s (Brown et al. 2005), with growth in the Intermountain West during the 1990s occurring faster than in any other region of the country (Hansen et al. 2002). Conservation easements are one tool to reduce residential development and agricultural conversion on private lands (Kiesecker et al. 2007). Opportunities also exist to target existing federal and state incentive programs in these areas, focusing on compatible grazing practices and habitat enhancement activities. A preponderance of private surface ownership in Montana and Utah coupled with low risk of development make core regions in many parts of these states ideal places to develop incentives for ranching and rural lifestyles through long-term easement programs such as the Conservation Reserve Program, which reduces habitat loss by conversion to agriculture. Opportunities for easements and management programs are available in Wyoming because of the sheer size of this population, but long-term viability is more of a public policy decision.

Areas of low biological value and low energy potential (19% of eastern range; Fig. 21.3) represent low-conflict opportunities for sage-grouse. Our analyses document the importance of these areas in maintaining connectivity to high value core regions in Montana (Fig. 21.3). Core regions with low biological value and low energy potential will be important in this regard, with restoration being one of the key strategies. Recent experience has shown the difficulty of maintaining numbers of Gunnison Sage-Grouse (*Centrocercus minimus*; Oyler-McCance et al. 2005a) and Lesser Prairie-Chicken (*Tympanuchus pallidicinctus*; Hagen et al. 2004) when only small and fragmented populations remain. Sage-grouse have already been lost from half of their former range (Schroeder et al. 2004), and many of the low value and low potential areas identified in this analysis are the same areas where continued range contraction is expected to be most severe (Aldridge et al. 2008). Fringe populations in North Dakota, South Dakota, Montana, and Canada need to pursue aggressive habitat restoration programs if they hope to maintain their biological value. Programs should focus on restoring adjacent lands presently in tillage agriculture to sagebrush-dominated grasslands in addition to enhancing existing native habitats.

CONSERVATION IMPLICATIONS

Explicitly combining information about the vulnerability of landscapes to anthropogenic risk enables conservation planners to consider aspects of urgency as well the probability for success of a given conservation strategy (Wilson et al. 2005, Copeland et al. 2007, Pressey and Bottrill 2008). Core regions and assessment of the potential future impacts they may experience represents a starting point to initiate conservation of landscapes where results will have the largest benefit to populations. Prioritization of landscapes is an admission that threats are large, resources are limited, and conservation action targeting every remaining population is improbable. Core regions represent a proactive attempt to identify a set of conservation targets to maintain a viable and connected set of populations before the opportunity to do so is lost.

Strategies must be integrated among all states and provinces involved for landscape-scale conservation to be successful. Each state and province will need to do its part to maintain sage-grouse distribution and abundance. Successful implementation in one state, such as Montana, will not be sufficient to compensate for losses in important places like Wyoming. Conservation concerns related to sage-grouse will remain at the forefront until collaborative landscape planning and conservation are demonstrated. Analyses reported here provide a framework for planning across political boundaries and a currency for measuring the success of implementation.

ACKNOWLEDGMENTS

We thank state and federal wildlife managers throughout the Great Plains and Wyoming Basin Sage-Grouse Management Zones for helping us envision this project and its end products. We thank the agencies for providing lek-count data, and the following people that helped make it happen: A. D. Apa and T. E. Remington (Colorado), R. D. Northrup (Montana), D. W. Olsen (Utah), T. J. Christiansen (Wyoming), T. R. Kirschenmann (South Dakota), and A. C. Robinson (North Dakota). A. A. Messer (Montana), K. J. Eichhoff (Colorado), and N. I. Whitford (Wyoming) provided GIS support and helped answer questions about lek databases. We replicated core area analyses at the level of individual states; the resulting areas and statistics are available upon request from K. E. Doherty. Funding for this work was provided by the state offices of the Bureau of Land Management in Montana and Wyoming, Wolf Creek Charitable Foundation, the Hewlett Foundation, the Liz Claiborne and Art Ortenberg Foundation, Google Earth, and the University of Montana. Comments by S. T. Knick and three anonymous reviewers greatly improved this manuscript.

CHAPTER TWENTY-TWO

Response of Greater Sage-Grouse to the Conservation Reserve Program in Washington State

Michael A. Schroeder and W. Matthew Vander Haegen

Abstract. We examined the relationship between the Conservation Reserve Program (CRP) lands and Greater Sage-Grouse (*Centrocercus urophasianus*) in Washington state including an assessment of population change, nest-site selection, and general habitat use. We monitored nest-site selection of 89 female sage-grouse between 1992 and 1997 with the aid of radiotelemetry. The proportion of nests in CRP lands significantly increased from 31% in 1992–1994 to 50% in 1995–1997, although more nests were detected in shrub steppe (59% vs. 41% of 202 nests). The increase appeared to be associated with maturation of CRP fields, which were characterized by increased cover of perennial grass and big sagebrush (*Artemisia tridentata*). Nest success was similar ($P = 0.38$) for nests placed in the two cover types (45% in CRP and 39% in shrub steppe). Counts of fecal pellets indicated that sage-grouse selected areas with greater sagebrush cover, especially in relatively new CRP in a shrub steppe landscape. Analysis of male lek attendance prior to implementation of CRP (1970–1988) illustrated similar rates of declines in two separate populations of sage-grouse in north-central and south-central Washington. Data from 1992 to 2007 following establishment of the CRP revealed a reversal of the population decline in north-central Washington while the south-central population continued a long-term decline (~17% vs. 2% of the occupied areas were in the CRP, respectively). These results indicate that lands enrolled in the CRP can have a positive impact on Greater Sage-Grouse, especially if they include big sagebrush and are focused in landscapes with substantial extant shrub steppe. The CRP for sage-grouse and other sage-dependent species should be considered a long-term investment because of the time required for sagebrush plants to develop.

Key Words: Artemisia tridentata, big sagebrush, *Centrocercus urophasianus,* Conservation Reserve Program, Greater Sage-Grouse, habitat restoration, population dynamics, shrub steppe, Washington.

Respuesta del Greater Sage-Grouse al Programa de Reservas para la Conservación en el Estado de Washington

Resumen. Examinamos la relación entre las tierras del Conservation Reserve Program (CRP, Programa de Reservas para la Conservación) y el

Schroeder, M. A., and W. M. Vander Haegen. 2011. Response of Greater Sage-Grouse to the conservation reserve program in Washington State. Pp. 517–529 *in* S. T. Knick and J. W. Connelly (editors). Greater Sage-Grouse: ecology and conservation of a landscape species and its habitats. Studies in Avian Biology (vol. 38), University of California Press, Berkeley, CA.

Greater Sage-Grouse (*Centrocercus urophasianus*) en el estado de Washington, incluyendo una evaluación de cambios de la población, de selección de sitios de anidación, y del uso general del hábitat. Monitoreamos la selección de sitios de anidación de 89 hembras de sage-grouse entre 1992 y 1997 con la ayuda de radio telemetría. La proporción de nidos en tierras del CRP aumentó perceptiblemente de un 31% en 1992-1994 a un 50% en 1995–1997, aunque más nidos (59 contra el 41% de 202 nidos) fueron detectados en el hábitat de shrub steppe (arbustos de estepa). El aumento parecería estar asociado a la maduración de los campos de CRP, que estuvieron caracterizados por la cubierta creciente de hierbas perennes y de sagebrush grande (*Artemisia tridentata*). El éxito de anidación fue similar ($P = 0.38$) para nidos ubicados en los dos tipos de cubierta vegetal (el 45% en CRP y el 39% en hábitats de shrub steppe). Los conteos de heces fecales indicaron que el sage-grouse seleccionó las áreas con mayor cubierta de artemisa, especialmente en áreas nuevas del CRP en paisajes de shrub steppe. El análisis de la asistencia de machos a los leks (sitios de cortejo) antes de la puesta en marcha del CRP (1970–1988) ilustró índices similares de descenso en los números de dos poblaciones separadas de sage-grouse en Washington central del norte y central del sur. Los datos que surgieron entre 1992–2007 a partir del establecimiento del CRP revelaron una revocación leve de la disminución de la población en Washington central del norte, mientras que la población central del sur continuó un descenso de población de largo plazo (~el 17% contra el 2% de las áreas ocupadas estaba en el CRP, respectivamente). Estos resultados indican que las tierras alistadas en el CRP pueden tener un impacto positivo en el Greater Sage-Grouse, especialmente si incluyen a la *A. tridentata* y se centran en paisajes que han permanecido con shrub steppe abundante. El CRP debe considerarse como una inversión a largo plazo tanto para el sage-grouse como para otras especies dependientes de la artemisa debido al tiempo requerido para el desarrollo de la vegetación en este tipo de hábitat.

Palabras Clave: Artemisia tridentata, big sagebrush (artemisa), *Centrocercus urophasianus*, dinámica poblacional, estepa arbustiva, Greater Sage-Grouse, Programa Reserva para la Conservación (Conservation Reserve Program), restauración del hábitat, Washington.

Shrub steppe communities historically dominated the landscape of eastern Washington (Daubenmire 1970). Today, <50% of Washington's historical shrub steppe remains, and much of it is degraded, fragmented, and/or isolated from other similar habitats (Jacobson and Snyder 2000, Vander Haegen et al. 2000). Conversion to cropland has resulted in the greatest loss of shrub steppe in Washington, particularly among deep-soil communities (Dobler et al. 1996, Vander Haegen et al. 2000). Similar large-scale conversion of shrub steppe to cropland has occurred in north-central Oregon, southern Idaho, and eastern Montana (Wisdom et al. 2000a, Knick et al. 2003). Shrub steppe communities across the Intermountain West also have been lost or degraded by extensive energy extraction, inappropriate livestock grazing, invasion by exotic plants, and changes in fire frequency (Yensen et al. 1992, Pashley et al. 2000, Knick et al. 2003).

Loss and degradation of extensive shrub steppe communities has greatly reduced habitat available for a wide range of shrub steppe–associated wildlife (Quigley and Arbelbide 1997, Saab and Rich 1997, Vander Haegen et al. 2000). Most species identified as having a high management concern in an analysis of species at risk within the interior Columbia River Basin were those associated with shrub steppe (Quigley and Arbelbide 1997). Moreover, according to the Breeding Bird Survey, half of these shrub steppe–associated species have experienced long-term population declines (Saab and Rich 1997).

The Greater Sage-Grouse (*Centrocercus urophasianus*) illustrates the problems associated with shrub steppe–dependent wildlife. Populations of Greater Sage-Grouse have declined in recent decades throughout much of their range (Connelly and Braun 1997, Braun 1998, Connelly et al. 2004, Schroeder et al. 2004). These declines have been particularly dramatic in Washington, where sage-grouse have been reduced to two separate populations, one in north-central and the other in south-central Washington (Schroeder et al. 2000). The reduction in distribution of sage-grouse has been caused by numerous factors, but foremost is the conversion of native sagebrush (*Artemisia* spp.)–dominated shrub steppe to cropland (Yocom 1956, Swenson et al. 1987, Dobler et al. 1996, Schroeder et al. 2000). Degradation of remaining habitats, particularly those used for nesting and brood-rearing, has also had negative impacts (Connelly et al. 1991, Gregg et al. 1994, Schroeder 1997, Connelly et al. 2000c, Connelly et al. 2004). Declines of sage-grouse populations throughout their range have resulted in numerous efforts to list this species as either threatened or endangered (United States Department of the Interior 2005b, 2008a). The Greater Sage-Grouse is currently considered warranted for federal listing in the state of Washington, but precluded from listing by higher priorities (United States Department of the Interior 2001a). The species is presently listed by the state of Washington as threatened (Hays et al. 1998).

Habitat restoration is a fundamental component of the recovery plan for Greater Sage-Grouse in Washington (Stinson et al. 2004). Shrub steppe is currently being restored through the Conservation Reserve Program (CRP), both by design and by happenstance. This voluntary program (administered by the United States Department of Agriculture) pays farmers to take agricultural lands out of production to achieve conservation objectives, including reduced soil erosion and improvement of wildlife habitat. The program allows farmers to periodically enroll lands for intervals of at least 10 years. In Washington as of July 2006, 599,314 ha of converted farmland had been planted to perennial grasses, forbs, and shrubs under the CRP (Schroeder and Vander Haegen 2006). The vast majority of CRP land in Washington occurs on land that was historically shrub steppe. The current acreage of CRP land in eastern Washington equals roughly 10% of the region's total agricultural lands. This program is not the ideal solution to the problem of declining native habitat, but CRP has enormous potential to provide cover and food for many species associated with shrub steppe habitat.

CRP fields have historically been planted with a variety of nonnative grasses; more recently (late 1990s), an increasing number of fields have been planted to native grasses, forbs, and arid-land shrubs. Native shrubs (particularly big sagebrush [*Artemisia tridentata*]) frequently seed-in from adjacent shrub steppe, making some fields potentially usable by shrub-nesting species. Despite the potential of CRP land as wildlife habitat, no studies have examined use of these lands by wildlife, and specifically Greater Sage-Grouse, in Washington.

The purpose of our research was to examine the behavioral and population response of Greater Sage-Grouse to the presence of CRP fields in the state of Washington. In addressing this goal, some basic questions were considered. Do Greater Sage-Grouse use CRP land for nesting, and is this use proportional to its availability? Is the use of CRP land a function of its characteristics, such as age and configuration relative to native shrub steppe? Does CRP positively impact populations of Greater Sage-Grouse?

METHODS

Telemetry

We studied Greater Sage-Grouse on a 3,000-km^2 area near Mansfield, Washington (Schroeder 1997). Female sage-grouse were trapped on seven leks with walk-in traps (Schroeder and Braun 1991) during March and April, 1992–1996. Sex and age were ascertained for all captured birds (Beck et al. 1975); all females were fitted with battery-powered radio transmitters attached by poncho-like collars (Amstrup 1980) or necklaces.

Females were located either visually or by triangulation at least once every three days with a portable receiver and four-element Yagi antenna to monitor location and success of nests. Variation in intensity of transmitter signals was also used as an indication of female behavior; radio transmitters emitted a constant signal when a female was on her nest and a variable signal when she was walking or flying. Fixed-wing aircraft were used to locate lost birds.

Observations of females on nests were made by triangulating from a distance of about 30 m from the nest site to minimize disturbance while allowing nests to be located following hatch or failure. We considered females to have nested successfully if at least one egg hatched. Analyses of nest success and habitat selection were conducted with logistic regression (PROC CATMOD, SAS Institute 2006). Most nests were located during the laying stage or early in incubation, and exposure period differed little among nests.

After a female ceased her nesting effort, two 18-m perpendicular transects were established, centered on the nest, with orientation of the initial transect chosen randomly. Ten point-intercept locations were positioned 2 m apart along each transect (total of 20 points). All plant species intercepted at each point were identified and recorded (or the site was recorded as bare ground). Data were simplified to shrub cover, grass cover, forb cover, and bare ground for purposes of our analysis. The majority of shrubs in CRP land were big sagebrush (~75% near nest sites); the other shrubs consisted primarily of threetip sagebrush (*Artemisia tripartita*), antelope bitterbrush (*Purshia tridentata*), and rabbitbrush (*Chrysothamnus* spp.). No effort was made to differentiate between the subspecies of big sagebrush (Frisina and Wambolt 2004) due to ambiguity in their characteristics in Washington. A visual obstruction reading was recorded with the aid of a Robel pole (Robel et al. 1970) at each of the 20 points. Visual obstruction readings were taken from a distance of 4 m from each point perpendicular to the transect direction and at a height of 1 m. Height of the tallest shrub (shrub height) within 9 m of the nest site (length of each transect outward from the nest site) was recorded to the nearest centimeter. Species diversity was recorded as the number of different plant species identified within 9 m of the nest site in a 10-min period.

We used logistic regression (Hosmer and Lemeshow 2000) to examine the likelihood of a nest occurring in CRP land versus other vegetation types, primarily shrub steppe or wheat fields. The outcome variable was CRP land or shrub steppe with three explanatory variables: female age (adult or yearling), order of the nest (first nest or renest), and year (1992 through 1997). Females tend to display spatial fidelity to nesting sites (Schroeder and Robb 2003), and the relative likelihood of consecutive nests being in the same cover type was also examined.

We used logistic regression to assess the influence of cover type on nest success, with nest fate (successful or failed) as the outcome variable and three explanatory variables: cover type (CRP land or shrub steppe); female age (adult or yearling); and order of nest (first nest or renest). Nests for females that were killed by predators while off the nest were excluded from this analysis. Nests for which knowledge of success was ambiguous also were excluded. Variables describing vegetation structure were not included in this analysis; measurements were not taken at all nests because some nests could not be examined in a timely fashion. We tested for model fit using the Hosmer-Lemeshow test (Hosmer and Lemeshow 2000). Percentage data were arcsin-transformed prior to analysis to improve normality and converted back to percentages for presentation. We also considered the influence of multiple comparisons when evaluating significance values.

Pellet Counts

Greater Sage-Grouse pellet counts (a pellet is an individual unit of fecal material) were conducted as part of a larger study designed to evaluate the potential role of CRP in the long-term conservation of species associated with shrub steppe habitats in the Columbia Basin (Schroeder and Vander Haegen 2006). Sage-grouse deposit pellets throughout the year, but the vast majority of those detected were deposited in autumn, winter, and early spring, when the sage-grouse diet is primarily sagebrush. Pellets deposited during late spring and summer tend to be less durable because of the relatively moist diet of sage-grouse during this time of year (Schroeder et al. 1999). Most observed pellets likely represent distinct seasons—the fresh pellets (light brown with a hint of dull yellow) from the recent autumn-spring period and the old pellets (grayish-brown) from previous autumn-spring periods.

Pellet count research was divided into two phases. Forty-eight study sites were selected in Douglas, Grant, Lincoln, and Adams Counties for phase 1 in 2004–2005 (Fig. 22.1). The 48 study sites were divided into eight spatially separated clusters, each with 6 study sites with a different treatment. The six treatments included three cover types (shrub steppe, old CRP, new CRP) and two landscapes (surrounding landscape dominated by shrub steppe or cropland). Shrub steppe cover was not grazed by livestock and was dominated by native vegetation, with an overstory of big sagebrush and an understory of bunchgrasses and forbs. Old CRP cover was former cropland planted to nonnative bunchgrasses in the late 1980s; shrubs were present if big sagebrush encroached from adjacent shrub steppe. New CRP cover was former cropland planted to a mix of nonnative and native species including big sagebrush, generally in the late 1990s. Pellet count research was conducted in the four

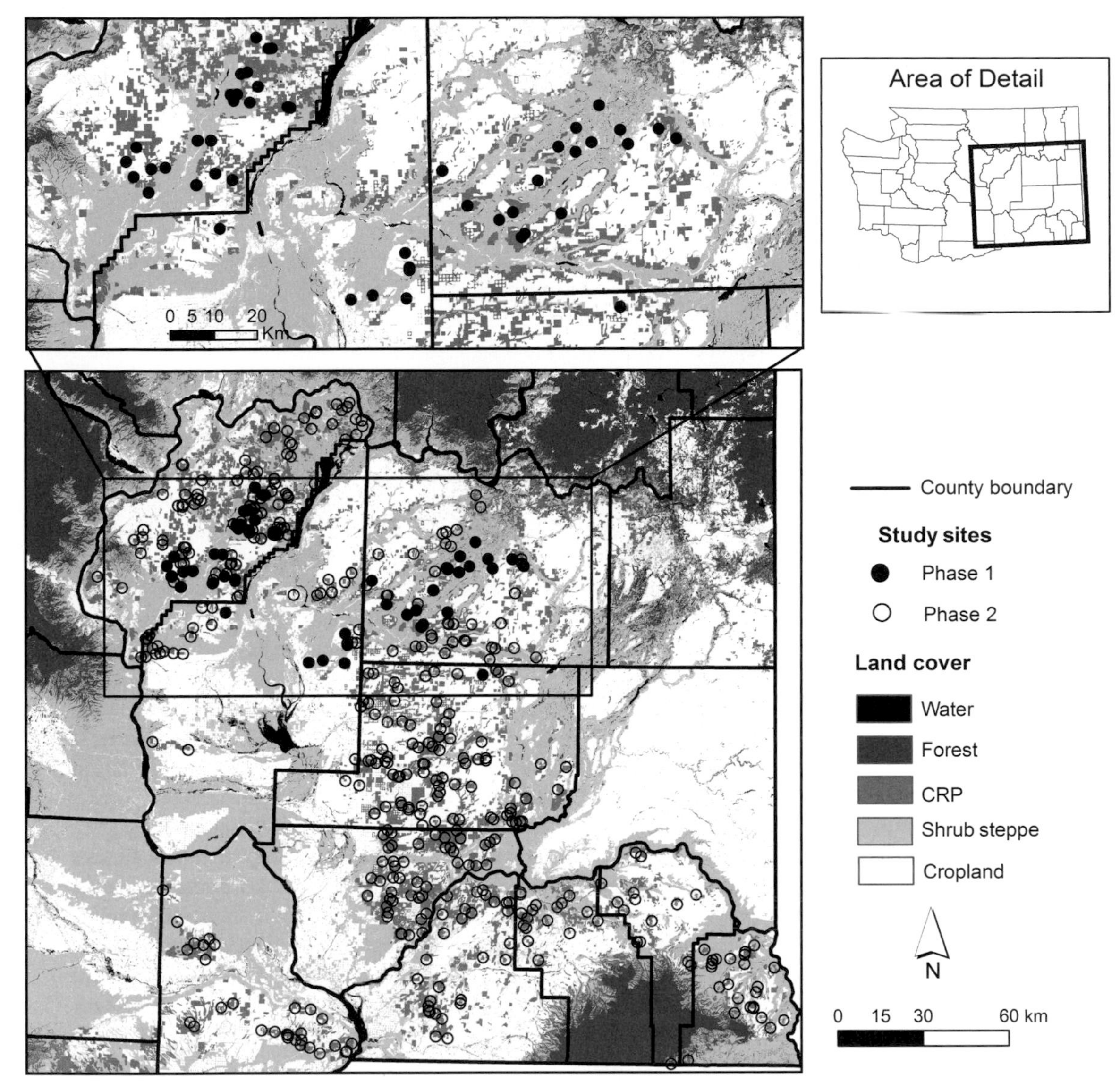

Figure 22.1. Phase 1 (enlarged map at top) and phase 2 (large map at bottom) study sites for Greater Sage-Grouse pellet surveys in eastern Washington, 2004–2006. During phase 1, pellets were only counted in the western 24 study sites in Douglas and Grant Counties. Land cover was derived from Landsat imagery (1993–1994; Jacobson and Snyder 2000) and aerial photography in Conservation Reserve Program archives (1996 photos; USDA Farm Services Agency).

westernmost clusters of study sites (six sites/cluster equally divided by habitat and landscape). These 24 westernmost sites were within the distribution of Greater Sage-Grouse, whereas the easternmost sites were outside the distribution (Schroeder et al. 2000).

Each study site in phase 1 was 25 ha, buffered by at least 100 m of similar vegetation to prevent edge effects. Each study site contained four 100-m fixed-radius circles spaced 300 m apart, providing a 100-m buffer between each circle perimeter. Pellet counts were conducted within circular 50-m^2 plots at cardinal directions 50 m from the center of each fixed-radius circle (16 plots/study site). Each circular 50-m^2 plot was delineated with the aid of a 4-m string looped over a permanent center stake. By walking the perimeter at the end of the string, an observer was able to identify pellets that were in the circle. Once the perimeter was established, pellets clearly within the circle were identified, counted, and removed. Surveys were conducted in October 2004 and April 2005.

Vegetation on each study site was assessed within 15 × 6.67 m (100 m^2) rectangular plots randomly located and oriented within each of the four 100-m fixed-radius circles (Fig. 22.2). Vegetation plots were stratified by plant community; if more than one plant community was present within the circle, an additional plot was randomly placed within the additional community. Thus, from four to eight vegetation plots were measured on each study site.

Colored flags were placed at set distances along the plot boundary to create "subplots" and assist with cover estimation (Fig. 22.2). Percent cover of shrubs, perennial grasses, and forbs was visually estimated as one of nine values: (1) ≤1%, (2) >1–5%, (3) >5–15%, (4) >15–25%, (5) >25–35%, (6) >35–50%, (7) >50–75%, (8) >75–95%, and (9) >95%. The midpoint of each cover category was used for analysis. A visual obstruction reading (Robel et al. 1970) was recorded for each of 10 fixed points along the perimeter of the sampling plot. Readings were taken at a height of 1 m and at a distance of 4 m from the inside of the plot looking toward the outside. The average visual obstruction reading was used in subsequent analyses. All sampling was completed in June and July of 2004.

Phase 2 of the pellet count research was conducted in 2006 on 410 study sites scattered throughout much of eastern Washington in CRP lands of differing ages, conditions, and landscape configurations (Fig. 22.1). Study sites were randomly placed in CRP fields similar to those in phase 1 except that only one 100-m fixed-radius circle was used within each study site, rather than four circles. Fifty-four of the 410 study sites were eliminated from analyses due to lack of established perennial vegetation. The same basic technique used to record sage-grouse pellets in phase 1 was also used in phase 2, with the following exceptions: four plots were examined for each of the 356 study sites in April–July 2006 (rather than 16 plots for each study site in phase 1); plots in phase 2 were 100 m from the center of the fixed-radius circles, rather than 50 m; pellets were not removed; and pellets detected while in transit between plots were also recorded.

Habitat data were collected the same way in phase 2 as in phase 1, except the visual obstruction reading was not recorded and rectangular plots were positioned with a corner in the center of the circle and oriented randomly. Study sites in phase 2 of the pellet research were positioned randomly relative to the general landscape, and three additional variables were quantified for the area within 1 km of the center of each circle: (1) proportion of cover in CRP land, (2) proportion of cover in shrub steppe, and (3) ratio of CRP land to nonshrub steppe cover. We also recorded distance to the nearest active sage-grouse lek and maximum attendance of males at the nearest lek in 2006.

We used a general linear model to examine pellet counts at the level of the fixed-radius circles. The outcome variable was the number of pellets, and the explanatory variables were cover type, landscape, and year. We used logistic regression to examine presence and absence of pellets in relation to cover type and specific vegetation characteristics. We tested for model fit using the Hosmer-Lemeshow test (Hosmer and Lemeshow 2000).

Populations

Greater Sage-Grouse historically were found throughout much of eastern Washington, but surveys conducted between 1955 and 2007 suggest that only three populations existed during that

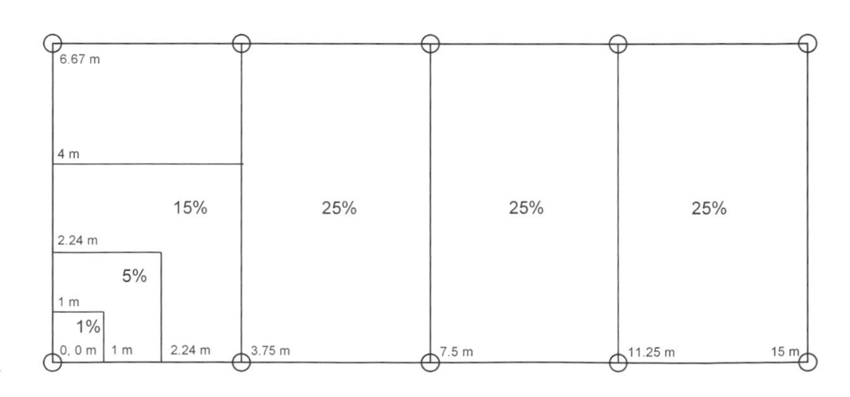

Figure 22.2. Sample vegetation plot showing corner stakes and distances along plot edges for placement of pin flags to create subplots for percent cover estimation. Circles represent locations for placement of the Robel pole.

time interval (Schroeder et al. 2000). These include a population primarily in the Moses-Coulee area of Douglas County, north-central Washington; a population in south-central Washington, primarily on the United States Army's Yakima Training Center in Yakima and Kittitas Counties; and a population primarily in Lincoln County. We did not consider the population in Lincoln County in our analysis of CRP because it has been extinct since the mid-1980s. The north-central Washington population has been monitored regularly since 1955, and the south-central Washington population has been monitored since 1970. Only data collected since 1970 were used in our analysis of CRP to make comparisons between areas consistent.

We defined a lek complex as a group of leks <3 km from one another. Lek complexes were spatially separated from adjacent lek complexes by >6 km. We surveyed lek complexes between 1970 and 2007 to obtain information on sage-grouse populations and annual rates of change (Schroeder et al. 2000). The survey protocol included searches for new and/or previously unknown complexes and multiple (≥3) visits to all known complexes. Some original data from the 1970s were lost so that only single high counts remained, despite some complexes having been observed on more than one occasion.

Numbers of males attending lek complexes were analyzed using the greatest number of males observed on a single day for each complex for each year. This technique is well established for Greater Sage-Grouse (Connelly et al. 2004), but it may have biases (Jenni and Hartzler 1978, Emmons and Braun 1984, Walsh 2002, Walsh et al. 2004). Despite these potential biases, lek counts presently provide the only assessment of a population's long-term trend (Connelly et al. 2004). We estimated annual rates of population change by comparing total number of males counted at lek complexes in consecutive years. Sampling was occasionally affected by effort and/or size and accessibility of leks, and those not counted in consecutive years were excluded from the sample for the applicable intervals (Connelly et al. 2004). Annual instantaneous rates of change for each population were estimated as the natural logs of the males counted on leks in one year divided by the males counted on the same leks the previous year. We also estimated rates of change for individual leks in each population. In this analysis, lek counts were transformed by adding one male to each count; this avoided undefined calculations. Annual rates of change were only started when a lek was first discovered.

The CRP was authorized in 1986, and analysis of population data in Washington permitted a comparison of pretreatment data (before CRP) with treatment data (after CRP). However, because implementation of CRP was not instantaneous and planted fields took time to develop, we eliminated transition years to avoid confusion. CRP lands were not usable by grouse in 1987, the year most of the fields were first planted, thus we did not consider data for population changes between 1988 and 1992 in subsequent analyses because 1992 was the first year nests were documented in CRP. These five years of data represent four annual intervals of population change. We used the 1987–1988 interval as the last pretreatment interval because CRP fields resembled wheat fields during their first year. Thus, pretreatment years included 1970 through 1988. The treatment years included 1992 through 2007. We treated the population in south-central Washington as a control because of the small amount of CRP land (Table 22.1).

RESULTS

Telemetry

We documented 204 nests from 89 females monitored between 1992 and 1998. However, only one nest was found in 1998 and it was eliminated from the sample; a second nest was eliminated from the analysis because it was in a wheat field. The remaining 202 nests were either in CRP land or shrub steppe. Females nesting in shrub steppe were usually in vegetation dominated by big sagebrush, but were occasionally in areas dominated by other shrub species or perennial grass. CRP land was also variable, with some fields containing a mixture of shrubs and perennial grasses and other fields largely dominated by grasses.

Eighty-three (41.1%) of the 202 nests were in CRP fields, and 119 nests (58.9%) were in shrub steppe. Neither age ($\chi^2 = 0.19$, $P = 0.67$) nor nest order ($\chi^2 = 0.99$, $P = 0.32$) was significant in the logistic regression, but year offered a significant explanation ($\chi^2 = 6.60$, $P = 0.01$) for the observed variation in nest placement between CRP land and shrub steppe cover (Hosmer and Lemeshow test; $\chi^2 = 3.48$, $P = 0.84$). Nests were more likely to be

TABLE 22.1

Potential habitat quantity in relation to current and historical distribution of Greater Sage-Grouse in Washington (adapted from Table 1 in Schroeder et al. 2000).

	Area dominated by each cover type (%)				
Range or population	Shrub steppe[a]	Cropland[a]	CRP[b]	Other[b]	Total area (km^2)
North-central Washington	44.3	35.1	16.7	3.9	3,529
South-central Washington	95.6	0.5	1.9	1.9	1,154
Total occupied range	57.0	26.6	13.0	3.4	4,683
Unoccupied range	42.3	42.8	5.5	9.4	53,058
Total historical range	43.5	41.5	6.1	8.9	57,741

[a] Landsat Thematic Mapper, 1993.

[b] Conservation Reserve Program (CRP) was measured from aerial photos dated 1996.

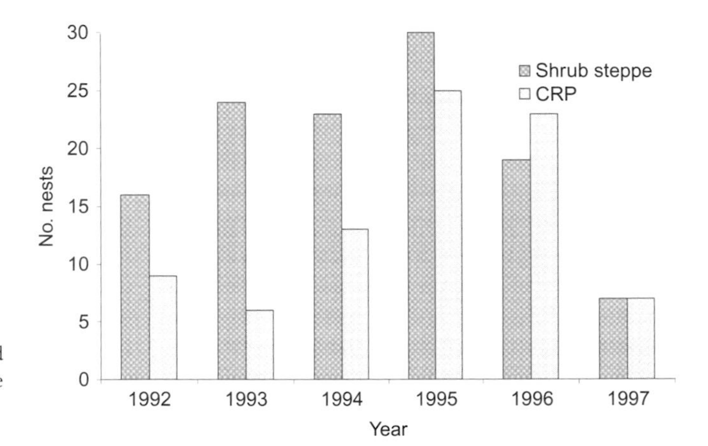

Figure 22.3. Number of nests of radio-marked Greater Sage-Grouse in Conservation Reserve Program and shrub steppe cover in north-central Washington between 1992 and 1997.

in CRP land later in the study (Fig. 22.3), perhaps in response to the maturation of CRP fields, most of which were planted in the late 1980s. Some variability in this trend was noted, particularly in 1997, as that year also had the smallest sample size of nests (14). During the first half of the study (1992–1994), 30.8% of nests were in CRP land, and during the last half of the study (1995–1997), 49.5% of nests were in CRP land.

The shift in selection of nesting cover was also noted when comparing individual females. Between 1992 and 1997, 121 of the 202 documented nests for radio-marked females were at least a female's second nest. Seventy-seven of the 121 nests (63.6%) were in the same cover as the previous nest. Of the 44 changes in cover type, 26 (59.0%) were shifts from shrub steppe to CRP land and 18 (41.0%) were shifts from CRP land to shrub steppe.

Sage-grouse nested in CRP land in proportions substantially greater (41.1% of nests were in CRP land) than availability would suggest (16.7% of the sage-grouse range in north-central Washington was CRP land; Table 22.1). Sage-grouse also nested in shrub steppe more often than its availability would suggest (58.9% use vs. 44.3% availability). These observations are due to the almost complete absence of nests in cropland, which is an abundant cover type (35.1%). If nonnest cover types are removed from consideration, 27.4% of the potential nesting habitat was CRP land and 72.6% was shrub steppe. Thus, CRP land and shrub steppe were used in similar proportions to availability during 1992–1994, but CRP land was used more during 1995–1997.

We also examined nest success in relation to habitat selection for 192 nests (excluding 10 nests in which females were killed by predators while

TABLE 22.2

Logistic regression analysis of Conservation Reserve Program versus shrub steppe vegetation characteristics at Greater Sage-Grouse nest sites in north-central Washington, 1992–1997.

Parameter	χ^2	P
Visual obstruction reading	0.05	0.82
Species diversity	15.40	<0.001
Shrub height	0.55	0.46
Shrub cover	8.01	0.005
Grass cover	4.74	0.03
Forb cover	0.11	0.74
Bare ground	1.88	0.17
Hosmer and Lemeshow goodness of fit	15.10	0.06

TABLE 22.3

General linear model analysis of study site configuration in relation to number of Greater Sage-Grouse pellets in phase 1 of pellet research in north-central Washington, 2004–2005.

Parameter	F	P
Area (four different clusters of study sites)	3.87	0.01
Cover type (shrub steppe, old CRP[a], or new CRP)	4.83	0.01
Landscape (shrub steppe or cropland)	5.70	0.02
Cover type–landscape interaction	4.16	0.02
Year (2004 or 2005)	1.61	0.21

[a] CRP refers to the Conservation Reserve Program.

off the nest or when the nest's outcome was ambiguous). Seventy-nine nests were in CRP land and 113 nests in shrub steppe. Nest success did not significantly differ by age of female (χ^2 = 0.15, P = 0.70), nest order (first nests vs. renests; χ^2 = 0.24, P = 0.62), year (χ^2 = 0.24, P = 0.62), or cover type (χ^2 = 0.77, P = 0.38; Hosmer and Lemeshow test; χ^2 = 9.33, P = 0.32). Overall nest success for this sample was 37.0%. Apparent nest success was estimated to be 40.5% in CRP land and 34.5% in shrub steppe; these values were not significantly different (χ^2 = 0.72, P = 0.40).

We used 161 nests for analysis of specific vegetation characteristics. The difference in sample sizes from the previous analysis represented nests that could not be examined in a timely fashion or for which the actual nest bowl was not located. Vegetation characteristics differed between nests in CRP land and nests in shrub steppe (Table 22.2). Nests in CRP land had lower average (± SE) plant diversity (10.47 ± 0.57, CRP; 17.95 ± 0.60, shrub steppe), lower shrub cover (1.85 ± 0.20%, CRP; 3.74 ± 0.06%, shrub steppe), and higher grass cover (4.74 ± 0.03%, CRP; 4.43 ± 0.03%, shrub steppe).

Large differences in specific vegetation characteristics occurred between general cover types (CRP and shrub steppe), and specific characteristics were examined in relation to nest success for each cover type separately. None of the characteristics was apparently related to nest success in a logistic regression with vegetation characteristics as independent variables in either shrub steppe (overall χ^2 = 7.12, df = 7, P = 0.42; Hosmer and Lemeshow test, χ^2 = 11.48, P = 0.18) or CRP land (overall χ^2 = 3.64, df = 7, P = 0.82; Hosmer and Lemeshow test, χ^2 = 2.67, P = 0.95).

Pellet Counts

We counted 1,839 individual Greater Sage-Grouse pellets on 24 study sites in phase 1 of the pellet research. Pellets were found on 12 of the 24 study sites, 32 of 96 fixed-radius circles, and 60 of 384 plots. The number of pellets detected on each study site was strongly correlated with sampling period (r^2 = 0.90, P < 0.01). A general linear model detected significant effects of area, cover type, landscape, and interaction of cover and landscape on number of pellets (Table 22.3). The largest number of pellets was observed in new CRP land in the shrub steppe landscape (Fig. 22.4). The only significant variables detected in a logistic regression were cover type and percent shrub cover (Table 22.4). New CRP land was 53.1% likely to have pellets, while the likelihood was only 21.9% in old CRP land and 25.0% in shrub steppe. Average (± SE) shrub cover was 12.0 ± 2.4% at points with sage-grouse pellets and 8.6 ± 1.2% at points without pellets. In addition, 48.4% of the points without pellets had <5% shrub cover, while 71.9% of the points with pellets had >5% shrub cover (Fig. 22.5).

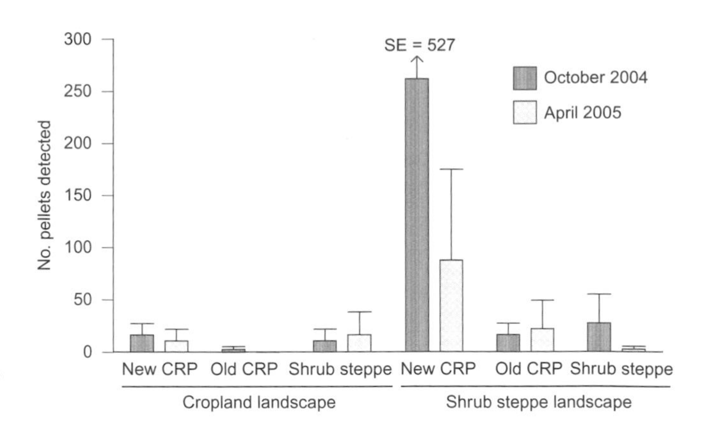

Figure 22.4. Number of Greater Sage-Grouse pellets counted in study sites in native shrub steppe, old Conservation Reserve Program (CRP), and new CRP in landscapes dominated by shrub steppe or cropland. Bars show mean and standard error.

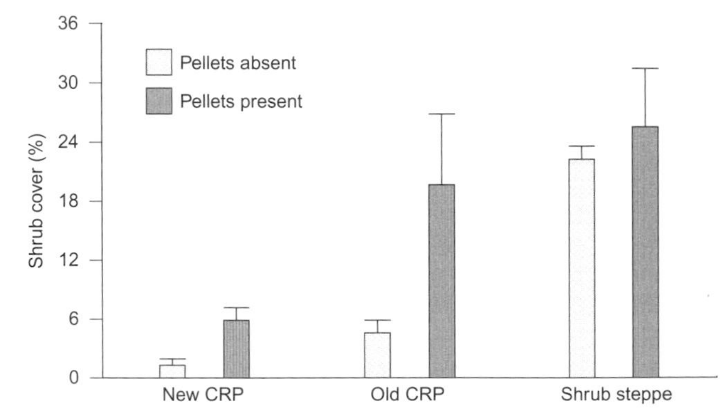

Figure 22.5. Percent shrub cover in relation to cover type and presence or absence of Greater Sage-Grouse pellets in north-central Washington, 2004–2005. Bars show mean and standard error.

During phase 2 of this study, 65 Greater Sage-Grouse pellets were counted on 356 study sites (single fixed-radius circles similar to phase 1). Presence and absence of pellets was considered in a logistic regression analysis, and pellets found in transit between the four standard pellet count plots were also considered. This increased the sample size of study sites with sage-grouse pellets from four to nine. No pellets of Greater Sage-Grouse were found in the 267 study sites outside their known distribution in Washington, and the logistic regression was only used for the remaining 89 study sites.

Pellets were found in the same areas in phase 2 as in phase 1, despite the substantially greater distribution of study sites in phase 2. Landscape and vegetation parameters were analyzed in a logistic regression to identify which characteristics of study areas were correlated with presence of sage-grouse pellets (Table 22.5). The only significant variables were percent shrub cover and maximum attendance of males at the nearest active lek. Percent shrub cover (± SE) was higher on study sites with pellets (14.67 ± 5.33%) than on areas without pellets (1.66 ± 0.55%). The maximum (± SE) number of males attending the nearest lek was greater for study sites with pellets (24.11 ± 4.89 males) than for sites without pellets (13.66 ± 1.23 males).

Populations

Sixty-five active leks were documented in Washington between 1955 and 2007; 32 in north-central Washington, 23 in south-central Washington, and 10 in the Lincoln County population that is now extinct. Leks in north-central Washington and south-central Washington, except one, were active at least as recently as 1970. As of 2007, 15 active leks occurred in north-central Washington and seven active leks were in south-central Washington.

The average (± SE) annual instantaneous rate of change for populations during the pretreatment period was −0.016 ± 0.073 in north-central Washington and −0.012 ± 0.063 in south-central Washington. The variances in annual rates of change were large, but the declines were comparable. Overall, the population in north-central Washington declined 25% and the population in south-central Washington declined 19% between 1970 and 1988. The average (± SE) annual

TABLE 22.4

Logistic regression analysis of study site characteristics in relation to presence or absence of Greater Sage-Grouse pellets in phase 1 of pellet research in north-central Washington, 2004–2005.

Parameter	χ^2	P
Cover type (shrub steppe, old CRP[a], or new CRP)	12.43	<0.001
Landscape (shrub steppe or cropland)	0.01	0.90
Cover type–landscape interaction	0.18	0.67
Visual obstruction reading	0.09	0.77
Shrub cover	9.54	0.002
Perennial grass cover	1.95	0.16
Forb cover	0.59	0.44
Bare ground	0.01	0.94
Hosmer and Lemeshow goodness of fit	12.80	0.12

[a] CRP refers to the Conservation Reserve Program.

TABLE 22.5

Logistic regression analysis of study area characteristics in relation to presence or absence of Greater Sage-Grouse pellets in phase 2 of pellet research in north-central Washington, 2006.

Parameter	χ^2	P
CRP[a] within 1 km (%)	1.72	0.19
Shrub steppe within 1 km (%)	1.01	0.31
Ratio of CRP to non-shrub steppe within 1 km	1.36	0.24
Shrub cover	10.04	0.002
Perennial grass cover	0.16	0.69
Forb cover	0.002	0.96
Bare ground	0.10	0.75
Maximum attendance of males at the nearest active lek in 2006	3.72	0.05
Distance to nearest active lek	2.07	0.15
Hosmer and Lemeshow goodness of fit	4.45	0.81

[a] CRP refers to the Conservation Reserve Program.

instantaneous rate of change for the treatment period (addition of CRP in north-central Washington) was 0.011 ± 0.065 in north-central Washington and −0.055 ± 0.060 in south-central Washington. Overall, the population in north-central Washington increased 19%, while the population in south-central Washington decreased 56% between 1992 and 2007.

Observations for the populations were similar when each lek complex was considered individually. The average (± SE) annual instantaneous rate of change for the pretreatment period was −0.038 ± 0.019 in north-central Washington and −0.145 ± 0.065 in south-central Washington. When leks were considered individually, variances were smaller, but results were still not significant. The average (± SE) annual instantaneous rate of change for the treatment period was 0.001 ± 0.025 in north-central Washington and −0.074 ± 0.038 in south-central Washington.

DISCUSSION

CRP lands were extremely variable in north-central Washington with regard to landscape and vegetative characteristics, primarily related to shrub steppe and sagebrush. This variation clearly influenced habitat use, as illustrated by pellet data. Pellet research indicated substantial use of CRP land by sage-grouse, particularly CRP land in shrub steppe landscapes and CRP land with relatively abundant shrub cover, especially sagebrush. Results from telemetry research were consistent with the pellet data, despite the different seasons involved in the research. The likelihood of sage-grouse nesting in CRP land increased with age of the CRP field—at the same time shrubs were becoming established. Nest success was at least as high in CRP land as in native shrub steppe during this study.

We considered the potential for biases in this research. The telemetry data included 89 females and 204 nests; many of the females nested more than once in a year and in more than one year. However, little variation was observed in the likelihood of nesting (100%) or renesting (87%). Nothing indicated that observations were influenced by an unusual effect of age, year, weather, or capture location. We also considered the possibility that pellets may be more observable in new

CRP fields, where use appeared to be highest, when compared to other habitat types. Examination of old pellets that were detected during the 2005 portion of the phase 1, and missed during the 2004 portion of the survey, indicated the potential bias in detection was negligible. For example, the proportion of old pellets detected in 2005 was 11% in new CRP land (41 pellets), 2% in old CRP land (2 pellets), and 9% in shrub steppe (6 pellets).

Habitat use data would have little value if habitat selection were not related, at least in part, to long-term changes in populations. Greater Sage-Grouse populations have declined substantially in Washington (Schroeder et al. 2000) across years and regions, with one exception. The population in north-central Washington that was 17% CRP land appeared to be at least stable since 1992, while the population in south-central Washington (2% CRP) substantially declined. This is particularly noteworthy given the widespread declines of most populations of sage-grouse in North America during the same time interval (Connelly et al. 2004). Population data from Washington are consistent with the habitat use information gathered with telemetry and pellet data; thus, we believe that CRP is benefiting sage-grouse in north-central Washington.

No evidence suggested that other factors such as potential regional variation in weather affected differences in rates of population change. However, the sage-grouse population in south-central Washington is primarily on the Yakima Training Center, which is a focal point for anthropogenic disturbance associated with military training activities (Stinson et al. 2004). The north-central Washington population, in contrast, is a focal point for anthropogenic disturbance associated with many other activities, such as farming, residential activity, and recreation, by a population of about 400 people within the sage-grouse population perimeter. Additional research is needed to evaluate the impacts of these disturbances.

Concurrent efforts have been made to restore shrub steppe within both populations (Stinson et al. 2004). The primary management tool has been removal of livestock from public lands (public lands are more common in the south-central Washington population), but direct enhancement of native vegetation has also been implemented. Reduced livestock grazing has resulted in an apparent increase in herbaceous cover, at least in localized areas (Schroeder et al. 2009), but it is not clear whether these efforts have impacted sage-grouse. Our research did not focus on the tremendous variation in characteristics of shrub steppe. We considered shrub steppe one cover type for comparison with CRP land. However, little doubt exists that shrub steppe represents substantial diversity in soil type, moisture regime, dominant plants, habitat condition, and historical management (Dobler et al. 1996, Vander Haegen et al. 2000). More research is needed to illustrate why some areas of shrub steppe support sage-grouse and others do not.

CRP land and shrub steppe are not independent cover types in north-central Washington. The region has a historic mix of cropland and shrub steppe, with close spatial association between the different cover types. For example, 10 of 15 leks in north-central Washington in 2007 were in cropland, while the other five were in shrub steppe. It was not unusual for sage-grouse to nest on the edge of habitats and have access to multiple vegetation types. It seemed clear that presence of CRP land adjacent to shrub steppe improved the value of each cover type for sage-grouse, but it was difficult to quantify this interaction. The majority of sites used by sage-grouse in north-central Washington, regardless of cover type, are in a shrub steppe–dominated landscape. Many areas in eastern Washington had extensive areas of CRP land, but little or no shrub steppe in the surrounding landscape; none of these areas support sage-grouse.

We were limited in our assessment of new CRP land in that sagebrush had been planted in these cover types only since 1997. They will likely be used more by sage-grouse as vegetation in these areas matures and shrub height increases. CRP is clearly benefiting sage-grouse in north-central Washington, but it is possible these observations are peculiar to habitat types and landscape configurations in this area and may not be applicable to other areas. For instance, north-central Washington had a relatively large component of cropland prior to implementation of the CRP (>50% of the area); consequently, the quantity of potential nesting habitat may have been limiting. Replacement of cropland with CRP land may make little difference in regions with a smaller proportion of cropland. Sage-grouse in north-central Washington have larger clutch sizes and higher rates of nest initiation and renesting when compared with sage-grouse elsewhere (Schroeder

1997). These life-history characteristics may be related to the unique configuration of cropland (and CRP land) with shrub steppe in north-central Washington (Schroeder et al. 2000).

The best way to examine the significance of sage-grouse use of CRP land in north-central Washington would be to compare these observations to different locations throughout the range of sage-grouse (data used in Connelly et al. 2004). These data could be compared for multiple areas using controls with both pretreatment and treatment data, as was done for Washington. Areas of particular interest include portions of the current distribution of Greater Sage-Grouse with abundant CRP land such as northern Utah, southeastern Idaho, western Colorado, and eastern Montana.

CONSERVATION IMPLICATIONS

CRP land was of greatest benefit to sage-grouse when it contained sagebrush and when it was in a shrub steppe landscape. CRP land with a sagebrush component is increasingly providing suitable nesting cover for sage-grouse. Use of CRP land by sage-grouse in north-central Washington appears to be correlated with slight increases in population size. Nesting and early brood-rearing are often identified as the most important time periods in the annual life cycle of sage-grouse (Connelly et al. 2004). CRP is supporting a substantial portion of the sage-grouse breeding population in north-central Washington, and it is likely the population would be severely impacted if CRP cover were removed.

The use of CRP as a conservation tool has potential benefits beyond sage-grouse. CRP may help connect fragmented patches of shrub steppe, thereby creating a relatively continuous vegetative community. CRP land in these shrub steppe landscapes supports many species that normally depend on sagebrush-dominated habitats such as the Brewer's Sparrow (*Spizella breweri*) and Sage Thrasher (*Oreoscoptes montanus*; Schroeder and Vander Haegen 2006). CRP in north-central Washington is concurrent with restoration efforts in intact shrub steppe, leading to the following question: would it be more beneficial and/or efficient to restore existing shrub steppe or to convert cropland to CRP? This question is fundamentally difficult to answer. The historical conversion of shrub steppe to cropland in Washington was heavily focused on areas with relatively deep soil (Vander Haegen et al. 2000). It is likely that soils supporting croplands are more productive than soils supporting shrub steppe, which may explain, at least in part, why CRP fields have been so successful in supporting wildlife.

Improvement of native shrub steppe habitat can increase its usefulness for sage-grouse (Bunting et al. 2003). However, cropland in a shrub steppe landscape offers at least two encouraging opportunities. First, CRP converts areas that are unlikely to support sage-grouse to habitat that has the potential to support this species, particularly in a shrub steppe landscape. CRP fields planted to benefit sage-grouse should include big sagebrush and be focused in landscapes with substantial extant shrub steppe. Second, croplands typically have deeper soil than unconverted native habitats (Vander Haegen et al. 2000) and provide substantial opportunities for establishing suitable plant communities to benefit wildlife. CRP for sage-grouse and other sage-dependent species should be considered a long-term investment because of the time required for sagebrush plants to mature.

ACKNOWLEDGMENTS

The fieldwork represented in this chapter, in particular the lek surveys, was accomplished with the aid of many more biologists than can possibly be mentioned. The primary fieldwork was provided by Devon Anderson, Ron Bassar, Scott Downes, Catherine Engleman, Gabrielle Gareau, Melissa Hill, Wendy Jessop, Marie-France Julien, Sherri Kies, James Lawrence, Luke R. Lillquist, Susan Lundsten, Ann Manning, James Mason, Joanne McDonald, Ann Peterka, Leslie A. Robb, Dina Roberts, Darina Roediger, Audrey Sanfacon, Jeff Scales, Anna Schmidt, John Slotterback, and Ashley Spenceley. We thank Wan-Ying Chang for help with statistical analyses. We also received a great deal of cooperation from many landowners in the region who graciously permitted access to their lands. Substantial financial support was provided by the United States Department of Agriculture Farm Service Agency, the United States Fish and Wildlife Service, the National Fish and Wildlife Foundation, and the Washington Department of Fish and Wildlife through the Federal Aid in Wildlife Restoration Project W-96-R. We thank C. A. Hagen, J. W. Connelly, L. A. Robb, S. T. Knick, and R. D. Northrup for their thoughtful reviews.

CHAPTER TWENTY-THREE

Restoring and Rehabilitating Sagebrush Habitats

David A. Pyke

Abstract. Less than half of the original habitat of the Greater Sage-Grouse (*Centrocercus urophasianus*) currently exists. Some has been permanently lost to farms and urban areas, but the remaining varies in condition from high quality to no longer adequate. Restoration of sagebrush (*Artemisia* spp.) grassland ecosystems may be possible for resilient lands. However, Greater Sage-Grouse require a wide variety of habitats over large areas to complete their life cycle. Effective restoration will require a regional approach for prioritizing and identifying appropriate options across the landscape. A landscape triage method is recommended for prioritizing lands for restoration. Spatial models can indicate where to protect and connect intact quality habitat with other similar habitat via restoration. The ecological site concept of land classification is recommended for characterizing potential habitat across the region along with their accompanying state and transition models of plant community dynamics. These models assist in identifying if passive, management-based or active, vegetation manipulation–based restoration might accomplish the goals of improved Greater Sage-Grouse habitat. A series of guidelines help formulate questions that managers might consider when developing restoration plans: (1) site prioritization through a landscape triage; (2) soil verification and the implications of soil features on plant establishment success; (3) a comparison of the existing plant community to the potential for the site using ecological site descriptions; (4) a determination of the current successional status of the site using state and transition models to aid in predicting if passive or active restoration is necessary; and (5) implementation of post-treatment monitoring to evaluate restoration effectiveness and post-treatment management implications to restoration success.

Key Words: Artemisia, Centrocercus urophasianus, Greater Sage-Grouse, habitat restoration, landscape triage, restoration guidelines, sagebrush grassland.

Restaurando y Rehabilitando Hábitats de Artemisa

Resumen. Menos de la mitad del hábitat original del Greater Sage-Grouse (*Centrocercus urophasianus*) existe actualmente. Parte del hábitat se ha perdido permanentemente debido a las granjas y a las zonas urbanas, pero el remanente varía en su condición, con áreas de hábitat de alta calidad a

Pyke, D. A. 2011. Restoring and rehabilitating sagebrush habitats. Pp. 531–548 *in* S. T. Knick and J. W. Connelly (editors). Greater Sage-Grouse: ecology and conservation of a landscape species and its habitats. Studies in Avian Biology (vol. 38), University of California Press, Berkeley, CA.

hábitat no adecuado. La restauración de los ecosistemas de pastizales de sagebrush (*Artemisia* spp.) puede ser posible en áreas de tierras resistentes. Sin embargo, el Greater Sage-Grouse requiere una gran variedad de hábitats sobre extensas áreas para completar su ciclo de vida. La restauración eficaz requerirá un enfoque regional para dar prioridad e identificar opciones apropiadas a lo largo de su territorio. Se recomienda utilizar un método de clasificación del paisaje para dar prioridad a las tierras para la restauración. Los modelos espaciales pueden indicar dónde proteger y conectar hábitats intactos de calidad con otros hábitats similares por medio de la restauración. Se recomienda el concepto del sitio ecológico de clasificación de tierras para caracterizar hábitats potenciales a través de la región, acompañados por sus respectivos modelos del estado y de la transición de las dinámicas de comunidades vegetales. Estos modelos asisten en identificar qué tipo de restauración (pasiva, activa o basada en el manejo, o manipulación de la vegetación) puede lograr la meta de un hábitat mejorado para el Greater Sage-Grouse. Una serie de pautas ayuda a formular preguntas que los directores pueden considerar al momento de desarrollar planes de restauración: (1) localización de sitios de prioridad a través de una clasificación del paisaje, (2) verificación del suelo y las implicaciones de las características del suelo en el éxito del establecimiento de la vegetación, (3) una comparación de la comunidad vegetal existente con el potencial del sitio mediante el uso de descripciones ecológicas del sitio, (4) una determinación del estado actual de sucesión del sitio usando modelos de estado y de transición para ayudar a predecir si es necesaria una restauración pasiva o activa, y (5) puesta en práctica de un monitoreo post-tratamiento para evaluar la eficacia de la restauración y las implicaciones del manejo post-tratamiento sobre el éxito de la restauración.

Palabras Clave: Artemisia, Centrocercus urophasianus, clasificación del paisaje, Greater Sage-Grouse, pastizales de artemisa, pautas de restauración, restauración del hábitat.

The sagebrush (*Artemisia* spp.) ecosystem is in jeopardy from increasing dominance of exotic annual grasses and native trees, altered fire regimes, inappropriate livestock-grazing practices and off-road vehicle activity, increasing development of energy sources, and climate change (Miller et al., this volume, chapter 10; Knick et al., this volume, chapter 12). These disturbances will likely result in temporary changes in relative dominance of plants if ecosystems are sufficiently resilient, yet all life-forms and species that make up native plant communities will be maintained. Ecosystems lacking resilience may cross ecological thresholds leading them to alternative stable communities; alternative communities differ considerably in structure and function from the original. Returning to original communities will not likely occur without human intervention, including control of undesirable species or reintroduction of previously dominant species (Briske et al. 2006). Severe alterations to original ecosystems, ranging from soil erosion to dominance of competitive invasive plants, may require introduction of new plants that provide similar structure and function, resulting in an alternative yet desirable ecosystem (Aronson et al. 1993). Changes in plant communities can result in simultaneous changes in animal communities as a result of habitat changes.

Greater Sage-Grouse (*Centrocercus urophasianus*) depend on characteristics of sagebrush ecosystems for their survival. Locations with co-dominance of a subspecies of big sagebrush (*A. tridentata*) and mid to tall perennial bunchgrasses during spring nesting and brood-rearing generally provide the most important habitat. Summer and autumn habitats vary from farmland to wet meadows to sagebrush rangelands. Greater Sage-Grouse require big sagebrush for cover and food in winter, but can use little sagebrush (*A. arbuscula*), black sagebrush (*A. nova*), scabland sagebrush (*A. rigida*), or silver sagebrush (*A. cana*) for food (Connelly et al. 2000c).

Dominance of each sagebrush species in a specific location is dependent on the suite of soil characteristics, climate, and natural disturbances that result in a dynamic set of plant species (associations) that may change in relative dominance depending on time since disturbances. The ecological site concept as defined by the United States Department of Agriculture (2003) is a land classification system that describes this set of soil-climate-plant associations across the United States. Ecological site descriptions attempt to depict the variation in plant community dynamics and natural disturbances for specific land areas. Ecological site descriptions use state (a relatively stable set of plant communities that are resilient to disturbances) and transition (the drivers of change among alternative states) as two successional concepts to describe the natural range in variation of plant communities (Westoby et al. 1989a,b; Bestelmeyer et al. 2003; Stringham et al. 2003). The reference state often includes multiple plant communities that differ in their dominant plant species relative to time since disturbance. Alternative states describe new sets of communities where relatively irreversible transitions (thresholds) may maintain these new plant communities in their own stable states with their unique set of dominant plants (Fig. 23.1).

The reference state of big sagebrush ecosystems is a suite of dynamic community phases changing from shrub-dominated to grass-dominated when fire removes fire-intolerant big sagebrush. Recovery of big sagebrush in burned locations requires

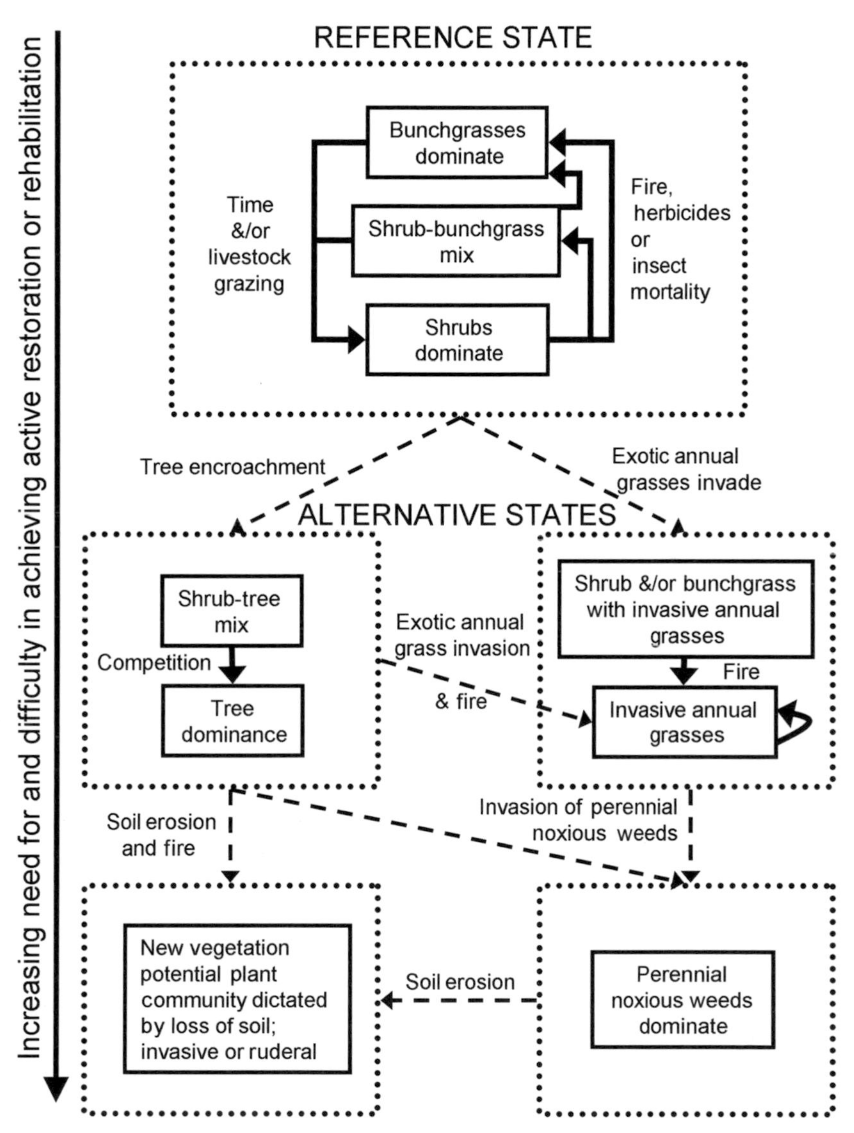

Figure 23.1. Generalized conceptual model for sagebrush ecosystems showing plant dynamics using state and transition (dotted boxes and dashed lines) models within Greater Sage-Grouse distribution. The uppermost dotted box represents the reference state for a site, while the lower dotted boxes represent various alternative states. Solid boxes and arrows within states are plant communities and pathways.

seedling establishment. This may be accomplished by seeds surviving in the soil or being dispersed from big sagebrush plants that escaped the fire. These sagebrush seedlings grow slowly, increasing in size and dominance over time and eventually leading to late successional communities represented by a combination of sagebrush and perennial grasses (20–45 years; Watts and Wambolt 1996). Mature sagebrush, once dominant, may remain dominant beyond 45 years provided livestock grazing has been removed (Robertson 1971, Sanders and Voth 1983, West et al. 1984, Allen-Diaz and Bartolomé 1998, Anderson and Inouye 2001). No evidence supports the belief that sagebrush dominance will continue at the expense of perennial grass cover or survival. Fires reset succession to perennial grass-dominated communities, and the cycle continues.

Animals that depend on habitats dominated by certain plant species may optimize their demographics within a certain plant community phase in the reference state. For example, grassland-dependent animals are favored in grass-dominated community phases developing after fires, whereas shrub-dependent species require shrub dominance. Greater Sage-Grouse tend to reach this optimum when sagebrush species co-dominate with mid-statured perennial bunchgrasses (Connelly et al. 2000c, Crawford et al. 2004). Managers who wish to optimize habitat for Greater Sage-Grouse populations might first identify the land's potential to support a specific sagebrush plant community. This potential is known as the ecological site. The current plant community that exists on the land can then be identified, as can management options for achieving the desired sage-grouse habitat.

This chapter describes a process for managing, restoring, or rehabilitating sagebrush ecosystems to achieve desired plant communities for Greater Sage-Grouse habitat. In doing this, managers might consider not only the land area where the desired plant community is the objective but also the temporal and spatial dynamics of the multitude of plant communities that currently exist and will develop across the surrounding landscape used by Greater Sage-Grouse. Some areas will require management to sustain current vegetation; some may require restoration or rehabilitation. Ecosystem restoration is the recovery of an ecosystem that has been degraded, damaged, or destroyed. The ecosystem contains a self-sustaining biotic and abiotic system through an assemblage of native species and community structures. These restored systems will maintain a suite of natural disturbances and their associated ranges of environmental conditions given the soils and climate of the location (Society for Ecological Restoration International Science and Policy Working Group 2004). Restored Greater Sage-grouse habitat will favor those plants that support self-sustaining populations of sage-grouse. Rehabilitation has the same goal of repairing ecosystem processes, productivity, and services as restoration; however, rehabilitation tends to achieve this goal using nonnative plants (Society for Ecological Restoration International Science and Policy Working Group 2004).

RESTORATION AND REHABILITATION OPTIONS

Past vegetation manipulations reflected land settlement patterns and mandated federal policies in the Greater Sage-Grouse range. Homesteading and irrigation development aided the development of farms where large expanses of sagebrush once grew (Knick, this volume, chapter 1). The Public Rangelands Improvement Act of 1978 (United States Code 43, Chapter 37 Public Rangelands Improvement Section 1901–1908) recognized the continued need to improve rangeland conditions for multiple uses of public lands. The major source of measuring land condition was based on a technique that organized plants into three categories: those that increase, decrease, or invade in response to livestock grazing (Pyke and Herrick 2003, West 2003a). Methods used to implement these improvements tended to rely on the science of the day. That science was reflected in the principal textbook of that time (Vallentine 1971), which focused on rangeland improvements as special treatments, developments, and structures used to improve range forage resources or to facilitate their use by grazing animals. The focus of many revegetation efforts was to increase forage production for livestock and to decrease the abundance of undesirable plants that interfered with livestock forage production and invasive annuals that provided unreliable forage. Undesirable plants included the major invasive plant, cheatgrass (*Bromus tectorum*; Young et al. 1972), and sagebrush. Sagebrush is still treated as a weed in some books (Whitson 1996). Restoration options may require some reductions or temporary eliminations of sagebrush if desired outcomes include increases in herbaceous life-forms in the community.

Effective restoration takes a regional perspective when considering when and where to restore lands for Greater Sage-Grouse habitat. Greater Sage-Grouse use large land areas and multiple plant communities, and have a variety of habitat needs, depending on their life stage (Crawford et al. 2004). Restoration decisions become challenging in view of economical considerations, restoration potentials, status of existing habitat, and logistical concerns such as landownership or topography. A prioritization process for selecting sites is required as the first stage in a successful restoration plan (Wisdom et al. 2005c, Meinke et al. 2009).

Landscape Triage

In ecosystem restoration terminology, triage is an initial prioritization technique where ecosystems are grouped into three categories, one that receives immediate care and two for which no immediate care is necessary. The group receiving immediate care and intervention has significant damage and benefits from aid. The other two groups are at opposite ends of the care spectrum. One needs no immediate intervention and will recover through later treatment, whereas the other has terminal damage and will not recover even with intervention (Kennedy et al. 1996, Samways 2000). Assessments ascertain ecosystem status (Groves 2003, Pyke and Herrick 2003, Wisdom et al. 2005a). Assessments of land status or ecosystem health are based on a manager's knowledge of the ecosystem's current status relative to the level of ecosystem threats—the drivers of ecosystem change—and the probability of ecosystem recovery from those threats (Hobbs and Kristjanson 2003). The ecological site land classification system, along with its accompanying ecological site description, provides a baseline in the United States for assessing land status (Briske et al. 2006). Techniques are currently being applied that use indicators of ecological processes to determine land status at multiple scales (Pyke et al. 2002, Spaeth et al. 2003, Pellant et al. 2005).

Land assessments can aid in developing ecosystem intervention grids (Hobbs and Kristjanson 2003). These grids provide decision levels for prioritizing management actions and restoration options. A potential grid for sagebrush grassland ecosystems could involve extent of departure from the reference state (Fig. 23.1), and potential for land to recover—referred to as resilience—after management changes or restoration activities (Table 23.1). Land assessments provide an approximation of departure from the reference state. Ecological site descriptions provide information necessary for predicting level of resilience of ecological sites. Areas with higher annual precipitation and greater soil depth provide approximations of increasing resiliency for most sagebrush grassland ecosystems. Intervention grids may contain additional axes such as cost-benefit ratios for proposed actions (Hobbs et al. 2003) that may assist managers in deciding if investment in restoration within an area is worth the risk. These grids simplify relationships into decision groups, but many of these decisions could be represented in continuous probability scales and entered into models to formulate decision tools.

When considering type and level of restoration intervention to use for improving Greater Sage-Grouse habitat, managers might consider the status of habitats adjacent to and surrounding potential restoration projects, since these areas, in combination with restoration areas, will encompass Greater Sage-Grouse habitat. Reasons for considering larger areas than the restoration site alone are based on criteria relating to Greater Sage-Grouse biology as well as the probability of restoration success. Greater Sage-Grouse have a large home range (Connelly et al. 2000c) that generally exceeds the size of most restoration projects. Thus, it is useful to provide land status evaluations spatially over a larger landscape, ideally for the entire region where birds exist. Assessments done in a spatially balanced and consistent manner can be placed into a Geographic Information System (GIS), where supporting data layers for the same spatial location are useful in accessing probability of restoration success. These data layers may include current vegetation and its appropriateness as Greater Sage-Grouse habitat, climate, soils, topography, and Greater Sage-Grouse habitat use information. Meinke et al. (2009) demonstrated a prioritization model for sagebrush restoration that used data layers derived from environmental conditions for growth of two subspecies of big sagebrush, potential for connecting existing stands of sagebrush, locations of viable Greater Sage-Grouse populations, and potential for invasive cheatgrass to impede success. This approach would ensure that local restoration projects were considering regional

TABLE 23.1

Potential sagebrush grassland intervention grid for identifying appropriate restoration interventions (modified from Hobbs and Kristjanson 2003).

Departure from the reference state is assigned using a land status assessment similar to Interpreting Indicators of Rangeland Health (Pyke et al. 2002, Pellant et al. 2005). Information from state and transition models is employed to identify probability of recovery (Fig. 23.1).

	Departure from the reference state		
	None to slight	Moderate	State change occurred
Probability of recovery or restoration	All plant functional and structural plant groups are present, but may not be in desired composition.	Some functional or structural plant groups are missing or under represented; invasive species common, but not dominant.	Invasive plants dominate; sagebrush or tall grasses are rare; soil stability and hydrologic functioning may be impaired.
High	**No Action.** Maintain status; monitor to prevent changes. Adjust management as necessary.	**Attempt Passive Restoration** if feasible: If unsuccessful use active restoration.	**Active Restoration.** Potential for successful restoration is high because of deep soils and higher precipitation. Potential for invasive plant control is high.
Medium	**No Action.** Monitor frequently to ensure that management is adjusted before habitat quality is impaired.	**Attempt Passive Restoration** if feasible. If unsuccessful use active restoration.	**Active Restoration,** but lower priorty because of lower probability of success.
Low	**No Action.** Monitor frequently to ensure that management is adjusted before habitat quality is impaired.	**No Action.**	**Conduct Inventory** and adjust management to fit new site and conditions.

factors for success of achieving both restoration and improved habitat for Greater Sage-Grouse.

Restoration of sagebrush habitats can take two forms—passive and active (McIver and Starr 2001, Hemstrom et al. 2002). Passive forms of restoration generally do not require human-aided revegetation, nor do they require applications of herbicide to modify the habitat, because desired species exist at the site as plants or seeds. Passive restoration of desired plant communities, including factors such as community structure (plant height and cover) and ecosystem processes (e.g., nutrient cycling), may be achieved by changing current management practices. Recovery of desired species or vegetation structure in the community occurs through normal successional processes and through drivers of change via new management. Active restoration (e.g., revegetation and severe modifications of plant communities using a variety of techniques) may be necessary if desired species are eliminated from sites or are too far from locations for successful dispersal and recovery to occur.

Passive Restoration

Passive restoration may achieve desired habitat changes provided that degradation of habitat quality has not been too severe and the community has remained within the reference state (Fig. 23.1; Stringham et al. 2003). Loss of dominant species such as tall bunchgrasses or sagebrush from a community, even if they are not replaced by invasive species, may require active restoration because the community no longer has an adequate density of those species or an adequate seed bank to draw upon to reestablish them in the community. The plant composition, or relative proportions of each plant species in the community, that defines these thresholds among states is largely unknown and is an active area of research. Major shifts in relative dominance (proportion of the total cover or production) among plant structural groups from a balanced mix of grasses and shrubs may require active restoration (e.g., thinning of shrubs and seeding of additional grasses) to achieve a balanced mix. For example, <5% tall perennial grass relative canopy cover with >80% shrub relative canopy cover may keep the community within the reference state, but the community is at risk of crossing a threshold because of the lack of adequate tall perennial grasses to provide recovery. A fire in a similar at-risk community may leave a void to be filled by invasive species (Sheley et al. 1996).

Common forms of passive restoration are removals or reductions of land uses. Changes in season of use may at times be adequate to achieve desired responses. If the goal is achieving increases in tall perennial grass composition, and these plants currently exist on a site but in small numbers, ensuring reproduction of existing grasses is paramount for providing propagules of these grasses. Several bottlenecks to restoration may exist that can severely hinder recovery of species. Increases in desirable plants will rely on a combination of seed production, longevity, and dispersal in conjunction with adequate safe sites for establishment and growth of desirable species if their dominance and density are inadequate (Archer and Pyke 1991, Pyke and Archer 1991, Bakker and van Diggelen 2006, van Andel 2006a,b).

Livestock Grazing Modifications

The greatest land-use adjustment within the Greater Sage-Grouse region that might bring about passive restoration is to change livestock management, largely because of the prevalence of livestock grazing as a land use. Simple modifications, such as shifting to no livestock use, may not provide desired outcomes, such as increases in herbaceous components of the plant community (West et al. 1984). Increases in herbaceous cover occur along with increases in shrub cover (Anderson and Holte 1981). Beck and Mitchell (2000) reviewed the literature and presented evidence for both positive and negative impacts of livestock grazing on Greater Sage-Grouse habitats. Modifications to grazing management might be considered as prescriptive techniques in conjunction with other ecosystem and management options available to achieve desired habitat conditions.

Past rangeland improvements through grazing modifications (adjustments in grazing seasons, period of grazing, or numbers of animals) sought improved amounts and composition of grasses and forbs (Dyksterhuis 1949). Adjustments were achieved by constructing new fences or developing additional water sources, which spread livestock use over larger areas. The greatest change was the shift from growing-season-long grazing to seasonal-rotational-grazing practices for livestock throughout the western United States (United States General Accounting Office 1977). Season of use by livestock often differs between intermountain and Great Plains regions in sage-grouse habitat. The seasons are somewhat intermediate in Wyoming and the Colorado Plateau. Season of use often reflects differences in types of grasses growing in each region.

Herbaceous vegetation in the Intermountain West is exclusively cool-season plants, whereas the Great Plains has both cool- and warm-season grasses (Sage et al. 1999). These two mechanisms for capturing solar energy and converting it into plant growth mainly differ in their optimal leaf temperatures for growth. Warm-season grasses tend to grow at optimal temperatures between 30°C and 45°C, while cool-season plants grow optimally between 20°C and 35°C. Optimal growth requires adequate moisture during the time when temperatures reach these levels.

Most of sagebrush grassland is a winter-dominated precipitation region, and cool-season plants typically dominate herbaceous layers (Miller and Eddleman 2001). Cool-season plants generally grow fastest in late spring and early summer (April–June). They reproduce and mature seeds at the end of this growth period and enter a summer dormancy period as soil moisture becomes limited and rainfall becomes less predictable during summer, or as temperatures exceed their optimum growth level. Exceptions occur in the Colorado Plateau of southern Utah, the remainder of eastern Utah, northeastern Colorado, eastern Wyoming, and eastern Montana, where tropical moisture from North American monsoons moves across the area, creating a second peak of predictable moisture in July and August. Warm-season plants may co-dominate with cool-season plants in herbaceous layers in regions with monsoonal rains, but season of active growth differs between these two plant functional types. Warm-season plants will begin to actively grow as cool-season plants reproduce and become dormant due to higher temperatures. Warm-season plants generally reproduce in mid- to late summer and become dormant as temperatures cool and moisture becomes limited in fall or winter. These periods of active growth can be important in sustaining and recovering these plants from herbivory by large grazers.

Cool-season plants typically tolerate moderate grazing (40–60% utilization) from mid-summer through early spring, when they are typically dormant or just beginning growth (Mueggler 1950, Laycock 1967, Laycock and Conrad 1981, Bork et al. 1998). Grazing at this level may not provide adequate hiding cover for sage-grouse, so lower levels of utilization may be necessary to achieve sage-grouse habitat. Reproduction normally occurs during late spring and early summer, the time of favorable active growth. These plants are highly susceptible to defoliation when reproduction occurs (Mueggler 1950, Laycock 1967, Bork et al. 1998), and defoliation of reproductive stems will reduce propagules needed for new grasses and forbs. Late spring grazing was deleterious or caused smaller increases in forb composition in threetip sagebrush (*Artemisia tripartita*) and big sagebrush ecosystems of southern Idaho and Oregon (Hyder and Sawyer 1951, Bork et al. 1998).

Resting pastures from grazing during periods of fastest growth of dominant grasses and forbs in sagebrush grasslands tends to enhance herbaceous plant growth and reproduction (Hyder and Sawyer 1951, Briske and Richards 1995, Bork et al. 1998). Pasture rest during this same period generally increases culm height, tiller production over the long term, and flower and seed production within the intermountain sagebrush steppe (Miller et al. 1994). Managers may wish to consider maintaining livestock stocking at a low enough level to achieve an average stubble height (Holechek and Galt 2000) of 18 cm, similar to recommendations of Gregg et al. (1994) and Connelly et al. (2000c), if sage-grouse nesting and hiding cover is a management goal and herbaceous plant height can potentially achieve this goal. This can also be achieved by removing livestock when the apical meristem is beginning to elevate in the culm of the grass, about one month before flowering of tall perennial grasses, so maximum leaf and inflorescence development and height may be achieved before the end of the growing season (Mueller and Richards 1986, Briske and Richards 1995). This may be as early as mid-April in arid, lower-elevation sites or as late as mid-June in mesic, higher-elevation sites, depending on the phenology of these grasses. Livestock must not graze during the dormant season, or they will likely remove this material and will not allow adequate regrowth before sage-grouse need the hiding cover in the following spring.

Grazing of the herbaceous layer late in the growing season in sagebrush grasslands favors plants avoided by grazing (Anderson and Briske 1995), such as sagebrush (Mueggler 1950, Laycock 1967). Repeated grazing during late spring and early summer, when grasses grow actively immediately before reproduction, tends to favor sagebrush growth until sagebrush becomes so dense that competition from it restricts recovery of herbaceous plants (Reichenberger and Pyke 1990). Once this level of sagebrush density and cover is achieved on a site, passive restoration may no longer be an option for improving sagebrush rangelands (Rice and Westoby 1978, West et al. 1984, Wambolt and Payne 1986).

Passive restoration through adjustments in grazing seasons or reductions in livestock numbers may improve Greater Sage-Grouse habitat quality if the vegetation community consists of adequate densities of sagebrush and perennial grass (Fig. 23.1). This community retains both sagebrush and tall bunchgrass densities necessary for quality habitat (Connelly et al. 2000c, Crawford et al. 2004), but cover or height of grasses may be inadequate. Release from livestock grazing during the later portion of the growing season should allow full expression of vegetation height for hiding cover and nest protection. Improvements in cover and height may not be expressed fully in the next growing season but may take three to five years for preexisting plants to fully express their height. Livestock grazing, when it occurs during dormant or early growing season, must be at low enough stocking levels to maintain adequate standing dead tiller density and culm height to provide cover and protection. Stubble height monitoring may provide a measure to adjust livestock stocking levels to attain adequate tiller densities with adequate height for sage-grouse. This form of passive restoration may take time and adequate weather, if seedling establishment of sagebrush or perennial grasses are required to increase proportional cover of either group. Studies tracking vegetation change after removal of livestock in big sagebrush ecosystems generally retained their initial proportions (Anderson and Holte 1981, West et al. 1984, Anderson and Inouye 2001) and took a minimum of 10–15 years for seed production, seedling establishment, and growth to occur, since these events may be episodic (Call and Roundy 1991, Pyke 1995).

Active Restoration Versus Rehabilitation

A common goal shared between restoration and rehabilitation is renewal of ecosystem processes, productivity, structure, and function (Society for Ecological Restoration International Science and Policy Working Group 2004). Restoration accomplishes this goal using native species, while rehabilitation may use species introduced to the site that may have similar structure and function. Active restoration or rehabilitation is warranted when desired species or structural groups are poorly represented in communities. Desired species are often replaced by undesirable, frequently invasive, species that can eventually dominate the site. These species include, but are not limited to, cheatgrass, noxious weeds, or native species including juniper (*Juniperus* spp.) or pinyon pine (*Pinus edulis* and *P. monophylla*; Miller et al., this volume, chapter 10). When left unchecked, these species can become dominant and lead to positive feedbacks that maintain their existence on the site and negatively impact desirable species such as sagebrush, perennial bunchgrasses, and forbs (Briske et al. 2006). A sagebrush site can progress along a transition into an alternative vegetation state, but transitions between states are generally unidirectional, without resilience, and are not likely to return to the previous state (Fig. 23.1).

Some state changes retain soils and hydrologic processes and may still retain the capability of supporting original plant communities; thus, restoration is possible if biological constraints such as weedy competitors can be reduced. Other state changes, in contrast, can result in sufficient soil loss or changes in hydrologic function so the site is no longer capable of supporting former plant communities found in reference states (Davenport et al. 1998, Briske et al. 2006). The ecological site changed because of soil loss. This new form of ecological site will eventually come to a dynamic equilibrium and will likely support a different ecological site with a new type and/or amount of plants. Restoration is no longer possible, and rehabilitation, defined as an alternative to the historic native plant community that provides similar structure and function without allowing further degradation of the site, may be the only remaining alternative that might make the site usable by Greater Sage-Grouse (Bradshaw 1983, Aronson et al. 1993).

Once lands degrade to the extent that the ecological site changes, repair of the former productivity along with its structure and function may become more difficult, if not impossible. A hypothetical example would be a mountain big sagebrush (*Artemisia tridentata* ssp. *vaseyana*) community that once supported an understory plant association with a wide diversity of forbs and mid to tall perennial grasses. Once pinyon and juniper trees increase in this ecosystem, sagebrush dies, the herb layer declines, and seed banks of former dominant plants in the community become depleted (Koniak and Everett 1982, Miller et al. 2000). Declines in shrub and herbaceous components of communities can leave soil susceptible to erosion (studies cited in Miller et al. 2005). Severe erosion could change the soil depth from a deep to a shallow soil. This leads to a site changing from a mountain big sagebrush community into a more arid Wyoming big sagebrush (*A. tridentata* ssp. *wyomingensis*) community with less herbaceous cover. Subsequent rehabilitation might lead to a less-productive site with different species and less structural diversity.

Another scenario using the Wyoming big sagebrush community might include natural disturbances, such as fire. Fire normally eliminates or reduces trees, but as trees age and they dominate sagebrush sites, fires become rare (Miller and Tausch 2001; Miller et al., this volume, chapter 10). Fires that occur after tree dominance tend to be severe crown fires of high intensity. High-intensity fires on warm, dry sites often dominated by Wyoming big sagebrush are capable of causing shifts from woodlands to introduced annual plant communities (Tausch 1999a,b).

Invasions of exotic annual grasses often make communities more susceptible to frequent fires because of the increase in fuel continuity caused by the annual grasses filling the interspaces between perennials (Whisenant 1990). Fine fuels in the pre-invasion community represented by the reference state would be distributed in patches represented by perennial bunchgrasses (Fig. 23.1). Cheatgrass is known to be a successful competitor against native plants for resources necessary to establish and grow (Harris 1967, Melgoza et al. 1990, Booth et al. 2003, Chambers et al. 2007). This alternative state for the ecosystem may require rehabilitation rather than restoration to successfully renew ecosystem structure and function.

Greater Sage-Grouse Habitat Goals Through Restoration or Rehabilitation

Greater Sage-Grouse use a diverse set of plant communities within a year, and their population success requires specific habitat needs for each stage of their seasonal life cycle (Connelly et al., this volume, chapter 4). The key for overall sage-grouse population sustainability and improvement, especially for successful reproduction and winter survival, is expanses of big sagebrush or silver sagebrush >4,000 ha (Leonard et al. 2000, Walker et al. 2007a). Woodward (2006) recommended not sacrificing preexisting stands of sagebrush, even if the herbaceous community is depleted and not ideal for Greater Sage-Grouse habitat. Fires will burn these stands eventually, allowing an opportunity to restore the herbaceous component. Connelly et al. (2000c) recommended altering a maximum of 20% of large sagebrush stands only if managers deem that the alteration is necessary.

Grass cover, in addition to sagebrush, is important during nesting because it provides horizontal cover to reduce depredation of sage-grouse eggs or young (Connelly et al., this volume, chapter 4). Forbs provide important food for hens and chicks in the spring and early summer. This combination of herbaceous plants with sagebrush is a habitat goal in nesting and brood-rearing areas. Diverse mixtures of plant species in communities should provide a diverse mixture of invertebrates, another critical sage-grouse food during fledging (Connelly et al., this volume, chapter 4).

Restoration and rehabilitation of Greater Sage-Grouse habitat should focus on maintaining or improving key habitat components for survival and reproductive success. Passive restoration goals focus on maintaining sagebrush cover while increasing grass cover and height and increasing forb cover and reproduction. This could be achieved by setting appropriate stocking levels for livestock while shifting grazing seasons to periods when active growth is slow and plant reproduction has not been initiated. Active restoration goals attempt to reestablish a sagebrush overstory with an understory mixture of native short, mid, and tall grasses and forbs. Rehabilitation should seek to achieve these same goals even though they may require introduced species to achieve them. The shrub component of these rehabilitated communities should include the appropriate native sagebrush species or subspecies for the site. Sage-grouse may nest under shrubs other than sagebrush, but most records of successful nesting are associated with one of the larger sagebrush species (Connelly et al. 1991; Schroeder and Vander Haegen, this volume, chapter 22). Studies have shown sage-grouse nesting or using lands sown with some introduced grasses (Connelly et al., this volume, chapter 4; Schroeder and Vander Haegen, this volume, chapter 22). Thus, rehabilitation using introduced perennial grasses may benefit sage-grouse populations. These studies also warn that the benefit may be related to proximity of large areas with mature sagebrush; an isolated rehabilitation within a large expanse of farmed land may not benefit sage-grouse. Most of these studies have not focused on the amount of cover of introduced grasses; this is a future research need.

PAST AND CURRENT VEGETATION MANIPULATION APPROACHES

Types of active revegetation and rehabilitation used within Greater Sage-Grouse habitats vary. Some involve revegetation, while others only control for invasive or undesirable species. Combinations of vegetation control followed by revegetation are common (Monsen et al. 2004). Each of these approaches has advantages and disadvantages that should be considered before applying a particular set of techniques.

Woody Plant Removal

Removal of woody plants (trees and shrubs) to increase herbaceous forage and allow grasses and forbs to dominate has been a common habitat treatment (Vale 1974, Olson and Whitson 2002). The original goal of eliminating sagebrush because it was considered a weed that competes with forage for livestock has been replaced in many locations with a goal to achieve a balance between shrubs and herbaceous plants to provide not only forage for livestock but also habitat for wildlife (Whitson 1996, Olson and Whitson 2002). Several techniques have been used to accomplish this balance with differing impacts on spatial and temporal development of structure and function of the ecosystem.

Prescribed fires may kill, eliminate, or reduce the density of most woody species and provide a temporary flush of nutrients that may result in increases in herbaceous plants, but may also leave sites susceptible to soil erosion during the first years after the fire (Blank et al. 1994, Stubbs 2000, Wrobleski and Kaufmann 2003). This tool is currently being used on sagebrush rangelands where pinyon or juniper have increased. Tree increases, if left unchecked, may decrease species diversity, increase soil loss, and reduce the potential for ecosystem recovery of former sagebrush grasslands (Miller and Tausch 2001). A good example of when to use fire as a restoration tool was presented in Miller et al. (2007). Managers considering using fire as a tool for controlling woody plants should consider that sagebrush dominance may be, in some ecosystems, lost for 25–45 years (Watts and Wambolt 1996, Wambolt et al. 2001) depending on distance from seed sources. However, sagebrush dominance in more mesic mountain big sagebrush communities may recover in less time (Miller et al., this volume, chapter 10).

Herbicide applications of 2,4-D or tebuthiuron have been used to kill large expanses of sagebrush, leaving standing dead skeletons of shrubs (Crawford et al. 2004). Western juniper (*Juniperus occidentalis*) was controlled successfully with tebuthiuron and picloram (Britton and Sneva 1981, Evans and Young 1985). These chemicals have an advantage over fire in that they lower the risk of soil erosion, as grasses and grass-like plants in the community generally remain unharmed and will likely increase as a result of decreasing competition from woody plants.

Complete elimination of sagebrush is not a goal for sage-grouse habitat; thus, partial reduction may be preferred because the understory is anticipated to respond and increase. Tebuthiuron used at low rates is a technique for thinning dense sagebrush and opening the community for herbaceous plants, including forbs, to respond (Olson and Whitson 2002), but effectiveness is highly dependent on soil type and depth. This technique might yield improvements to habitat quality, provided herbaceous perennial plants exist in the understory. Unfortunately, if exotic annual grasses exist in the community, expansion and spread of these invasive plants may result. Herbicides such as tebuthiuron and 2,4-D, if used at strengths recommended for killing all sagebrush, may kill or injure many forbs, since active ingredients that kill sagebrush also kill forbs.

Mechanical techniques are designed to remove all or a portion of the aboveground plant growth (e.g., mowing, roller chopping, rotobeating, and harrowing) or to uproot some or all of the plants from soil (e.g., grubbing, plowing, bulldozing, anchor chaining, cabling, railing, raking, and plowing)(Scifres 1980, Stevens and Monsen 2004). Some techniques such as sawing or mastication focus on cutting or grinding individual plants. Indiscriminate techniques tend to remove the more upright and stiff woody plants, while shorter, younger, or more pliable woody species survive being pushed over. Uprooting techniques create the greatest soil disturbance, adding to the risk of post-treatment soil erosion. They may tend to harm the herbaceous community, at least initially, by uprooting plants, but strong evidence for this in chaining is lacking (Ott et al. 2003). Control of pinyon and juniper through removal of individual trees can have minor impacts on the shrub community because they tend to be selective (Miller et al. 2005). Removal of tree competition should also facilitate rapid recovery of the shrub and herb understory if adequate densities are present prior to treatment. Mowing, roller chopping, rotobeating, and plowing will have a greater and longer-lasting impact on the shrub layer. Critical for success of these techniques is that the community remains in the reference state and that invasive annual grasses do not exist within the community (Fig. 23.1). Some mechanical removal techniques are capable of preparing a seedbed for revegetation if communities entered an alternative state dominated by trees (Fig. 23.1; Stevens and Monsen 2004).

Livestock may also be used to reduce woody plants in a targeted grazing approach, but generally plants need to be young, the browsing animal must be appropriate, and browsing must occur at a time when the animal's preference favors woody plants. Mueggler (1950), Laycock (1967), and Bork et al. (1998) cited long-term declines in threetip sagebrush with recovery of herbaceous vegetation at high-elevation sites in Idaho. Declines of Wyoming and mountain big sagebrush densities due to browsing by deer (*Odocoileus* spp.) or elk (*Cervus elaphus*) have been noted in Utah and Montana (Smith 1949, Austin et al. 1986, McArthur et al. 1988, Patten 1993, Wambolt 1996). The potential exists for goat browsing to target and reduce juniper density, provided trees are small (Fuhlendorf et al. 1997, Pritz et al. 1997). However, direct findings within the sagebrush steppe region have proved promising only with threetip sagebrush, not with big sagebrush or with western junipers (Fajemisin et al. 1996). Further research on use of woody versus herbaceous plants needs to be evaluated before any recommendation can be given.

Revegetation

Historic revegetation on most sagebrush grasslands had the goal of improving livestock forage, including replacing invasive forbs and annual grasses such as *Halogeton glomeratus* and cheatgrass with perennial grasses while protecting soils from erosion. Early trials comparing native versus introduced grasses in several locations within the Greater Sage-Grouse distribution found that native species often did not establish or produced less forage than introduced species. Recommendations during early history of rangeland revegetation favored use of introduced grasses, such as crested wheatgrass (*Agropyron cristatum* or *A. desertorum*) to meet combined goals of livestock forage production and erosion control (Asay et al. 2001).

Wildfire rehabilitation is a major source of revegetation in the Great Basin. The mandated goals of these projects are to: (1) reduce soil loss, (2) provide species palatable to livestock, and (3) reduce spread of invasive species. Total restoration of ecosystems with a complete suite of plant life-forms is not a designated objective for expenditure of funds. Thus, comparisons with restoration projects often show use of fewer plant species and an emphasis on introduced grasses that establish quickly (Richards et al. 1998). Only modest increases in use of native plants were reported in recent evaluations, although federal policies have advocated use of native plants in revegetation efforts when natives are available (Clinton 1999). The average number of species used on a rehabilitation project has remained between four and five, while the number of native species in the mixture has increased from one before 1996 to two after 1996, and the proportional increase in weight of native bulk seeds has been from 20% to 40% (Pyke et al. 2003). Land managers cited poor competitiveness and poor establishment of natives and high seed cost compared with introduced grasses as main reasons why they elected to use introduced species (McArthur 2004).

Currently, the prevalence and continued spread of exotic annual grasses, specifically cheatgrass and medusahead (*Taeniatherum caput-medusae*), throughout most of the sagebrush biome (Miller et al., this volume, chapter 10) has created the desire for revegetation projects to stop this trend (Monsen and Kitchen 1994). In addition, federal legislation encouraged use of grasses that could quickly stabilize soils, effectively compete with weedy or poisonous plants, and provide ample forage for livestock (Young and McKenzie 1982, Young and Evans 1986). Research during the mid-1900s pointed directly to the use of introduced forage grasses to meet these goals (Young and McKenzie 1982, Young and Evans 1986, Pellant and Lysne 2005). Characteristics that made these species effective also created communities dominated by near monocultures of introduced grasses that are less diverse (e.g., lacking sagebrush or forbs) and created poor habitat quality for Greater Sage-Grouse (Crawford et al. 2004). Methods for improving these sites have been proposed and are currently being tested in expanded trials (Cox and Anderson 2004, Pellant and Lysne 2005). One proposal is for revegetation of annual grass-dominated lands using competitive introduced forage grasses. An assisted succession approach may be used to reintroduce native plants into communities once introduced forage grasses dominate the site (Cox and Anderson 2004). Pellant and Lysne (2005) provide details for this process, which includes: (1) reduction in density and thus competition of introduced forage grasses, (2) seeding or transplanting of desired native plants, and (3) adaptive management to encourage establishment and reproduction of desired plants. Caution is advised in using these techniques. Timing of reductions in introduced forage grasses is critical for success (Fansler 2007). Sage-grouse may use introduced grasses if they include sagebrush and are near large stands of sagebrush

(Connelly et al., this volume, chapter 4; Schroeder and Vander Haegen, this volume, chapter 22). Thus, this technique should focus on reestablishing sagebrush in these communities to improve these lands for Greater Sage-Grouse habitat.

Replacing Annual Grasslands with Native Perennials

Rehabilitation and restoration techniques to transform lands currently dominated by invasive annual grasses into quality Greater Sage-Grouse habitat are largely unproven and experimental. Several components of the process are being investigated with varying success. The first of the process is reduction in the competition that invasive annual grasses provide against native seedlings during the establishment phase. Methods to reduce annual grass densities are therefore necessary. Techniques often mentioned are herbicides (Ogg 1994), defoliation via livestock grazing (Hulbert 1955, Finnerty and Klingman 1961, Mosley 1996, but see limitations in Hempy-Mayer and Pyke 2008), pathogenic bacteria (Kennedy et al. 1991), and fungi (Meyer et al. 2001, Beckstead and Augspurger 2004). Prescribed fire may be an effective technique if applied in combination with an herbicide treatment and if fire is conducted in either late spring or autumn. Prescribed fire alone is not recommended (Mosley et al. 1999, DiTomaso et al. 2006).

Herbicides have been widely applied throughout the Intermountain West (Vallentine 2004). At least 21 herbicides are labeled for use in controlling cheatgrass (Ogg 1994), but not all are registered for rangelands. Paraquat and atrazine were early herbicides that showed promise in controlling annual grasses, but environmental concerns led to their elimination as rangeland chemicals (Young and Clements 2000). Two herbicide groups currently used to control invasive annual grasses are broad-spectrum contact herbicides that kill or injure most plants they contact and preemergent herbicides that kill plants as they germinate but are less damaging to those plants already established.

Glyphosate (including Roundup®) is a contact systemic herbicide that kills most plants growing actively at time of application. It has no soil residual activity, and any plants emerging after application will survive. It kills plants late in the growing season and can prepare a fuel bed for fire that can reduce residual seeds of cheatgrass in litter seed banks. Follow-up applications of glyphosate the next spring may be necessary to ensure that cheatgrass populations are decreased sufficiently to reduce competition with any seeded desirable plants. Applications of carbon in a form readily available for soil microbial uptake may increase soil microbial content and cause microbes to reduce available soil nitrogen, reducing growth and potentially reducing competition with cheatgrass (McLendon and Redente 1990, 1992; Young et al. 1997; Blumenthal et al. 2003).

Imazapic (Plateau®, Panoramic 2SL®) and sulfometuron-methyl (Oust®, SFM 75®, Spyder®) are preemergent as well as contact herbicides. Sulfometuron-methyl showed promise in reducing the continuity of cheatgrass fuels in stands of crested wheatgrass (Pellant et al. 1999). Some agricultural crops are highly sensitive to this herbicide, and caution is paramount when applying near crops. Imazapic has recently been tested successfully within the sagebrush biome for the control of cheatgrass (Shinn and Thill 2002). It is an amino-acid–inhibiting herbicide that can operate as a preemergent or a contact herbicide. Annual plants are generally more susceptible than perennials to this herbicide, but some perennials such as antelope bitterbrush (*Purshia tridentata*) can vary in their susceptibility from being killed to having reproduction reduced during year of application (e.g., Wyoming big sagebrush; Eddington 2006, Vollmer and Vollmer 2008). Fall applications followed by sowing of six native species, including Wyoming big sagebrush, successfully controlled cheatgrass and medusahead while providing mixed results for native plant establishment (Bekedam 2004). Susceptibility of native perennial plants as adults or seedlings is unknown for many species and soil types; thus, care should be taken when managers use this herbicide as a selective herbicide for annual plants with the hope of retaining native perennials or revegetating immediately after herbicide applications. Imazapic applied to reduce cheatgrass fuel continuity has been successful and has not reduced some perennial grasses (Shinn and Thill 2004, Miller 2006, Davison and Smith 2007). Native annual plants, if they emerge at the same time as invasive annual grasses, may also be susceptible and harmed by imazapic applications. This herbicide has shown considerable promise, but continued monitoring and interpreting the impacts of its application are needed.

Immediate revegetation is advised after use of any of these density-reduction techniques; otherwise, invasive annual grasses that escape control treatments will likely grow unabated and quickly dominate sites by producing large numbers of seeds (Mack and Pyke 1983). No evidence for complete eradication of invasive annual grasses with control techniques and revegetation has been noted. However, successful revegetation efforts that have controlled invasive annual grass populations and have maintained perennial plants are generally rehabilitation projects sown with introduced forage grasses (Asay et al. 2001). Some evidence from wildfire rehabilitation studies shows that native plants can be sown and eventually coexist with invasive annuals, but these were generally sown in combination with introduced grasses (Pyke et al. 2003, Cox and Anderson 2004). Theoretical frameworks hypothesize that multiple native species representing a variety of growth and life-forms may successfully compete with invasive plants where any one species would be unsuccessful (Sheley et al. 1996). Invasive annual grasses can germinate in fall or early winter, and an appropriate mixture of plants would require perennials with shallow and deep roots and with early, middle, and late phenological development.

Restoration of Greater Sage-Grouse habitat will require time for sagebrush to establish and mature. It is critical when revegetating big sagebrush that appropriate subspecies are selected for the site. Big sagebrush has a number of subspecies, however three are most common—basin big sagebrush (*A. tridentata* ssp. *tridentata*), Wyoming big sagebrush, and mountain big sagebrush. These subspecies dominate in distinctly different environments (West 1983a). Mountain big sagebrush occurs most often in cooler and moister sites, while Wyoming big sagebrush dominates warmer and drier sites. Basin big sagebrush grows on deep soils, many of which are now farmed. A common problem associated with seeding big sagebrush has been that purchased seed often included more than one subspecies, even when only a single subspecies was requested (Dalzal 2004). Matching subspecies to the site is critical for establishment and growth of sagebrush and can be associated with seeding failure (Lysne 2005, Shaw et al. 2005a). Surface sowing of big sagebrush followed by soil-surface compaction may be necessary for establishment (Shaw et al. 2005a). Broadcasting seeds of Wyoming big sagebrush without covering the seed or pressing it into soil was unsuccessful in southern Idaho (Dalzal 2004) and should be used cautiously elsewhere in the region.

Locations that have been dominated by invasive annual grasses often have few forbs remaining, and forb species should be considered as part of seed mixtures. Establishment of forbs important to Greater Sage-Grouse has also shown promise (Wirth and Pyke 2003), but availability of seed tends to limit widespread use in rangeland restoration and rehabilitation projects (McArthur 2004). That limitation is being addressed, with more seed becoming available each year (Walker and Shaw 2005).

BOTTLENECKS TO SUCCESS

Availability and cost of native seed are major obstructions to use of native seeds in revegetation projects (McArthur 2004). The difficulties and vagaries of collecting, growing, and selling native seeds that have not been used historically within sagebrush ecosystems tends to raise prices and increase risks to both sellers and buyers (Bermant and Spackeen 1997, Currans et al. 1997, Roundy et al. 1997, Dunne 1999) relative to tested and released plants that are widely available (Currans et al. 1997).

Equipment for sowing native seeds is not widely available (Wiedemann 2005). Most revegetation projects in sagebrush habitats use rangeland drills that were developed for the rough terrain of wildland environments and for ease of seeding introduced forage grasses. Many native seeds, because of their differing sizes and appendages, require mixing within seed boxes on drills to ensure equal proportions of all seeds are sown on a site or will require separate seed boxes with effective depth bands to allow seeds of different sizes to be buried at different optimal depths (Boltz 1994, Stevens and Monsen 2004). These requirements will either require purchases of new seed drills or retrofitting of old drills to accommodate these needs.

GUIDELINES FOR RESTORATION PROJECTS

Success is not guaranteed when conducting Greater Sage-Grouse habitat restoration projects in semiarid environments. The only guarantee is that annual weather conditions can vary widely and that these often dictate success of restoration projects. Managers cannot influence immediate weather in a region to assist in restoration, and it is necessary to follow useful guidelines in preparing and implementing a restoration project. Goals and objectives should be explicitly stated and should represent both management and sampling objectives for projects. Wirth and Pyke (2007) provide examples of how to state these objectives and outline a potential monitoring protocol. These are important for monitoring and ultimately for adaptive management. The steps and questions in Table 23.2 are modified from those developed for management of western juniper on sagebrush grasslands (Miller et al. 2007), and are intended to aid managers in making restoration decisions. The initial step is to examine the region and prioritize lands into those that provide adequate Greater Sage-Grouse habitat and those that do not. Sites that do not provide adequate habitat but have the potential to provide it are affected by the soils and climate at the site. This process leads to identification of the ecological site for each soil unit. The decision of which areas to choose for restoration or rehabilitation of sage-grouse habitat is made during this stage using geospatial tools described in Wisdom et al. (2005c) or Meinke et al. (2009).

The second step involves ascertaining if the plant community currently existing on the site is one of the community phases within the reference state for the community dynamics model of that site (Fig. 23.1). Managers can refer to ecological site descriptions to make this assessment. Conducting a rangeland health assessment (Pyke et al. 2002, Pellant et al. 2005) may be helpful in identifying the status of the site relative to state and transition model.

The appropriate action for restoring or rehabilitating sage-grouse habitat is the third step. This involves estimating the relative cover or production that sagebrush, grasses, and forbs make to the overall dominance of the plant community. This informs managers if the site has the potential to restore itself through changes in the current management practice of passive restoration or if it may require greater intervention to achieve adequate sage-grouse habitat. Managers will decide if plant control techniques will be necessary and how restoration or rehabilitation will be conducted. Managers might consult resources for selecting appropriate plants if active restoration or rehabilitation is necessary. An initial resource is the local United States Department of Agriculture (USDA) Natural Resources Conservation Service rangeland management specialist who is trained and has access to plant materials information appropriate for local sites. Other general resources are ecological site descriptions for the soil. These can be found at the Ecological Site Information System on the USDA Natural Resources Conservation Service website (http://plants.usda.gov). This source will have lists of native species that typically occur on the site. Several publications (Vallentine 1989, Whisenant 1999, Monsen et al. 2004, Shaw et al. 2005b) provide recommendations for developing revegetation plans including plant control techniques, species recommendations, and development of seed mixtures, seeding techniques, and rates.

Decisions regarding protecting the site from disturbances must be made if active restoration or rehabilitation has been used. Often sites are closed to livestock grazing and recreational uses to provide seedlings the best potential for establishment and growth. Unfortunately, most recommendations are based on expert observations rather than replicated studies. Stevens and Monsen (2004) and Shaw et al. (2005a) offer recommendations of two or more years of protection. Stevens (2004) recommends a base time for protection ranging from two years (mountain big sagebrush in sites with >36 cm annual precipitation) to four years (Wyoming big sagebrush in sites with <36 cm annual precipitation). Managers may need to add one to six additional years, depending on type of restoration project and environmental conditions before and after project implementation.

A critical element is post-project effectiveness monitoring. Monitoring provides knowledge regarding where, when, and with what species successful restoration and rehabilitation projects occur. The United States General Accounting Office (2003) conducted an intensive analysis of emergency stabilization and rehabilitation projects in the United States Departments of Agriculture and the Interior and found that neither department could report on the effectiveness of their projects. These projects represent the largest set of revegetation treatments conducted on federal lands, most of which fall within the habitat range of Greater Sage-Grouse. The United States General Accounting Office (2003)

TABLE 23.2

Guidelines for conducting a restoration project for improving Greater Sage-Grouse habitat

Steps in the process	Questions to be asked	How to answer the question
I. Identify landscape priorities and ecological sites	1. Where are priority sites for restoration?	Conduct a landscape triage.
	2. What kind of soils are on the site?	Verify soils mapped to the location and provide further detail regarding the distribution of soil components at the site. This will require collecting information on soil texture and depth and some basic soil chemistry (pH, calcium carbonate presence).
	3. How will soils and physical features affect vegetation establishment and erosion?	Erosion is a major concern with any restoration project, especially if it is necessary to remove vegetation or disturb soils to conduct the project. Finer soils and steeper slopes generally have an increased risk of erosion. Soil descriptions will provide a guide regarding erosion risks on sites. Caution should be used in conducting soil disturbances on highly erosive sites. If revegetation is attempted, use fast-growing plants to protect and stabilize soils quickly. Generally, revegetation to protect soils from erosion takes many years and often does not provide adequate protection if high rainfall occurs (Robichaud et al. 2000).
	4. What is the native plant community for this site?	Match soil components on the site to their correlated ecological site description (ESD). Generally, there is only one ecological site mapped to a single soil component. The ESD will provide details on plant species and relative composition of these species in the community. This will provide an initial list of potential species for the site.
	5. Is old-growth juniper growing?	If yes, site may be a juniper site. Refer to Miller et al. (2007) for guidance. This site may not be appropriate for restoration. If no, proceed onward.
II. Determine current state of the site	6. Is site still within the reference state for the state and transition (S&T) model of this ecological site?	Compare current plant community on the site to those described in the S&T model. If plant community appears to fit in the reference state, and soil and hydrology of the site appear intact, then attempt passive restoration to improve habitat.
III. Select appropriate action	7. Does sagebrush dominate, yet herbaceous life-forms that should be co-dominant are missing from the site and annual invasive plants are rare?	This is a difficult situation. A need exists to reintroduce the herbaceous component of the habitat, but sagebrush competition may make revegetation difficult (Reichenberger and Pyke 1990). Consider restoring other higher-priority sites and wait to restore this site until fire burns sagebrush on the site.

Steps in the process	Questions to be asked	How to answer the question
	8. Is sagebrush missing, but native herbaceous life-forms are present and dominant?	Although sagebrush seed could be added to this site, it might be more cost-effective to introduce small patches of sagebrush transplants. As those plants mature, they will reproduce and spread seed naturally.
	9. Do invasive annual grasses co-dominate with native plants on the site?	Consider passive restoration first to attempt to increase competitive ability of native plants. Otherwise, wait for a fire to occur and attempt active restoration with herbicide to control annual grasses.
	10. Do invasive annual grasses dominate the site while native life-forms are missing or severely underrepresented?	Active restoration is necessary to restore habitat.
IV. Determine post-treatment management	11. How long should the site be protected before land uses begin?	Although some authors believe that only a minimum of two years of protection is necessary (Stevens 1994), most believe that two years is too short when native plants are being used in the restoration (Stevens 2004, Shaw et al. 2005a). A good rule of thumb is to continue protection until two-thirds of the restored plants become reproductive. Stevens (2004) provides some guidelines for increasing the time of protection depending on the ecosystem and precipitation after seeding. Uses should aim to minimize defoliation and trampling during the most active growing period (from just before reproduction until after seed dispersal).
	12. How will monitoring occur?	Monitoring of effectiveness of restoration treatments requires that a complete set of monitoring elements be completed such that an analysis and report are completed.
	13. Are adjustments to the restoration recommended?	Adaptive management is complete when lessons learned from the previous project can be applied in future projects. This requires completion of reports and meta-analyses of these reports to provide spatial recommendations based on consistent findings in multiple locations. This can be expedited through a common database for restoration monitoring reports.

SOURCE: Modified from Miller et al. 2007.

recommended that projects be monitored using similar techniques and that data be stored and made available for future query in a common database. Wirth and Pyke (2007) provide an example of a monitoring system with methods and a database that meets these goals. Monitoring data should reflect the quantitative objectives the manager wants to achieve with restoration or rehabilitation projects. Data analysis is directed at learning if projects were successful in achieving management objectives using simple statistical methods with graphical interpretations (Wirth and Pyke 2007). Consideration of a similar monitoring storage and retrieval database and analysis tool for sage-grouse restoration and rehabilitation projects would be useful to provide region-wide information for adaptive management of Greater Sage-Grouse habitat restoration and rehabilitation.

CONSERVATION IMPLICATIONS

Dramatic changes within sagebrush grassland ecosystems are a major contributor to population changes of Greater Sage-Grouse. However, the large spatial area that represents the distribution of Greater Sage-Grouse and variety of types of plant communities that are optimal for population growth and sustenance (Connelly et al. 2000c) require planning and prioritization to accomplish needed changes. Restoration, whether passive or active, often carries economic costs that are borne by private or government entities. Risks of not succeeding due to factors out of the manager's control, such as weather, may add to costs of these projects.

Restoration of habitat for Greater Sage-Grouse is more complex than the typical restoration project, which generally is site-specific, with goals and objectives dependent on a single site, and often smaller than the home range of a Greater Sage-Grouse. Successful restoration of Greater Sage-Grouse habitat will require not only vegetation changes in a single area but also connectivity among patches of currently intact vegetation. Many partnerships and working groups throughout the region have begun to implement efforts to assist in conservation of Greater Sage-Grouse, including some restoration projects (Western Governors' Association 2004). Coordination of these efforts might improve these projects and increase their effectiveness in providing habitat for Greater Sage-Grouse where it is most needed.

The most effective restoration for Greater Sage-Grouse habitat will require regional assessments of current status of sagebrush grasslands. It will require protection and proper management for the maintenance of intact sagebrush grasslands while identifying those lands where modifications to management may improve and restore quality habitat for this native and endemic bird. Active restoration will be needed to ultimately improve areas where current vegetation has already crossed a threshold and management alone will not achieve habitat improvement. For these projects, strategic placement will be critical for enhancing the likelihood of restoration success while keeping economic costs reasonable.

ACKNOWLEDGMENTS

This manuscript benefited from comments by anonymous reviewers from the Ecological Society of America, who reviewed previous material in the Range-wide Conservation Assessment for Greater Sage-Grouse and Sagebrush Habitats, as well as R. F. Miller, E. B. Allen, D. D. Musil, C. E. Braun, S. T. Knick, and C. D. Marti, who aided in editing and improvements. This is contribution Number 26 of the Sagebrush Steppe Treatment Evaluation Project (SageSTEP Proj #05-S-08), funded by the United States Joint Fire Science Program. Additional financial support came from U.S. Geological Survey, Forest and Rangeland Ecosystem Science Center, Coordinated Intermountain Restoration Project (9354-AKFC1). Any use of trade names is for descriptive purposes only and does not imply endorsement by the U.S. Government.

CHAPTER TWENTY-FOUR

Conservation of Greater Sage-Grouse

A SYNTHESIS OF CURRENT TRENDS AND FUTURE MANAGEMENT

J. W. Connelly, S. T. Knick, C. E. Braun, W. L. Baker, E. A. Beever, T. Christiansen, K. E. Doherty, E. O. Garton, S. E. Hanser, D. H. Johnson, M. Leu, R. F. Miller, D. E. Naugle, S. J. Oyler-McCance, D. A. Pyke, K. P. Reese, M. A. Schroeder, S. J. Stiver, B. L. Walker, and M. J. Wisdom

Abstract. Recent analyses of Greater Sage-Grouse (*Centrocercus urophasianus*) populations indicate substantial declines in many areas but relatively stable populations in other portions of the species' range. Sagebrush (*Artemisia* spp.) habitats necessary to support sage-grouse are being burned by large wildfires, invaded by nonnative plants, and developed for energy resources (gas, oil, and wind). Management on public lands, which contain 70% of sagebrush habitats, has changed over the last 30 years from large sagebrush control projects directed at enhancing livestock grazing to a greater emphasis on projects that often attempt to improve or restore ecological integrity. Nevertheless, the mandate to manage public lands to provide traditional consumptive uses as well as recreation and wilderness values is not likely to change in the near future. Consequently, demand and use of resources contained in sagebrush landscapes plus the associated infrastructure to support increasing human populations in the western United States will continue to challenge efforts to conserve Greater Sage-Grouse. The continued widespread distribution of sage-grouse, albeit at very low densities in some areas, coupled with large areas of important sagebrush habitat that are relatively unaffected by the human footprint, suggest that Greater Sage-Grouse populations may be able to persist into the future. We summarize the status of sage-grouse populations and habitats, provide a synthesis of major threats and challenges to conservation of sage-grouse, and suggest a roadmap to attaining conservation goals.

Key Words: Centrocercus urophasianus, Greater Sage-Grouse, habitats, management, populations, restoration, sagebrush.

Conservación del Greater Sage-Grouse: Una Síntesis de las Tendencias Actuales y del Manejo Futuro

Resumen. Los análisis recientes de poblaciones de Greater Sage-Grouse (*Centrocercus urophasianus*) indican declinaciones substanciales en muchas áreas, pero con poblaciones relativamente estables en otras porciones de la distribución de esta especie. Los hábitats de artemisa (*Artemisia* spp.)

Connelly, J. W., S. T. Knick, C. E. Braun, W. L. Baker, E. A. Beever, T. Christiansen, K. E. Doherty, E. O. Garton, S. E. Hanser, D. H. Johnson, M. Leu, R. F. Miller, D. E. Naugle, S. J. Oyler-McCance, D. A. Pyke, K. P. Reese, M. A. Schroeder, S. J. Stiver, B. L. Walker, and M. J. Wisdom. 2011. Conservation of Greater Sage-Grouse: a synthesis of current trends and future management. Pp. 549–563 *in* S. T. Knick and J. W. Connelly (editors). Greater Sage-Grouse: ecology and conservation of a landscape species and habitats. Studies in Avian Biology (vol. 38), University of California Press, Berkeley, CA.

necesarios para sustentar al sage-grouse están siendo quemados por grandes incendios naturales, invadidos por plantas introducidas, y desarrollados para recursos energéticos (gas, petróleo, y energía eólica). El manejo de tierras públicas, las cuales contienen el 70% del hábitat de sagebrush, ha cambiado durante los últimos 30 años: desde grandes proyectos de control del sagebrush dirigidos a aumentar el pastoreo de ganado, a un mayor énfasis en los proyectos que intentan a menudo mejorar o restaurar la integridad ecológica. Sin embargo, el mandato que incita a manejar tierras públicas para proporcionar aplicaciones de consumo tradicionales, así como valores de recreación y de áreas naturales, probablemente no vaya a cambiar en un futuro cercano. Por lo tanto, la demanda y el uso de los recursos contenidos en paisajes de artemisa, más la infraestructura asociada al soporte de las crecientes poblaciones humanas en el oeste de los Estados Unidos, continuarán desafiando los esfuerzos para conservar al Greater Sage-Grouse. La incesante extensa distribución del sage-grouse, no obstante sus bajas densidades en algunas áreas, junto con grandes áreas del importante hábitat de artemisa que se encuentran relativamente inafectadas por la mano del hombre, sugieren que las poblaciones del Greater Sage-Grouse podrán persistir en el futuro. Resumimos el estado de las poblaciones y de los hábitats del sage-grouse, proporcionamos una síntesis de amenazas y de desafíos importantes a la conservación del sage-grouse, y sugerimos un mapa para lograr metas de conservación.

Palabras Clave: artemisa (sagebrush), *Centrocercus urophasianus*, gestión, Greater Sage-Grouse, hábitats, poblaciones, restauración.

The Greater Sage-Grouse (*Centrocercus urophasianus*; hereafter, sage-grouse), now occupies only 56% of its likely distribution prior to European settlement (Schroeder et al. 2004). Range-wide, populations have been declining at an average of 2.0% per year from 1965 to 2003 (Connelly et al. 2004). Concerns about declining sage-grouse populations (Braun 1995, Connelly and Braun 1997, Connelly et al. 2004, Schroeder et al. 2004) coupled with information on habitat loss (Connelly et al. 2004) have prompted multiple petitions to list the species under the Endangered Species Act (Stiver, this volume, chapter 2).

The United States Fish and Wildlife Service determined in 2010 that listing Greater Sage-Grouse under the Endangered Species Act was biologically warranted but was precluded by other higher priorities (United States Department of the Interior 2010). During the four years since the first detailed range-wide analysis of sage-grouse populations and sagebrush habitats (Connelly et al. 2004), negative impacts of energy development and West Nile virus on Greater Sage-Grouse were documented (Naugle et al. 2004, 2005; Holloran et al. 2005; Aldridge and Boyce 2007; Doherty et al. 2008; Walker 2008). Hundreds of thousands of hectares of sagebrush (*Artemisia* spp.) steppe were also burned by wildfire (Miller et al., this volume, chapter 10; Baker, this volume, chapter 11). Large-scale conversion of sagebrush-dominated landscapes to exotic annual grasslands following these fires further increases the likelihood of future fire (Miller et al., this volume, chapter 10) and decreases any potential for recovery or restoration (Pyke, this volume, chapter 23). Along with these habitat changes, sage-grouse populations in some portions of the species' range have continued to decline (Garton et al., this volume, chapter 15) despite the collaborative efforts of many local working groups (Stiver, this volume, chapter 2).

We do not expect land uses to decrease, because growing human populations will increase demand for traditional consumptive resources and recreation. Thus, the human footprint (Leu and Hanser, this volume, chapter 13) is likely to continue to influence sagebrush-dominated landscapes (Knick et al., this volume, chapter 12). Nevertheless, the continued widespread distribution of sage-grouse (although some areas have very low densities) and relatively large areas providing key sagebrush habitats suggest that long-term conservation of sage-grouse populations should be possible. This chapter summarizes information on Greater Sage-Grouse populations and habitats presented in this volume, provides a synthesis of major threats and challenges to conservation of Greater Sage-Grouse, and suggests a roadmap to attaining conservation goals.

CURRENT KNOWLEDGE OF POPULATIONS

The Greater Sage-Grouse is genetically distinct from the congeneric Gunnison Sage-Grouse (*Centrocercus minimus*). Greater Sage-Grouse populations in Washington and the Lyon-Mono population, spanning the border between Nevada and California, also have unique genetic characteristics (Oyler-McCance and Quinn, this volume, chapter 5) but have not been described as separate species. The distribution of genetic variation has shifted gradually across the range, suggesting movement among neighboring populations is not yet likely across the species' range (Oyler-McCance et al. 2005b). Most populations have similar levels of genetic diversity even at the periphery of the range. With declining populations and habitat as well as increased threats from anthropogenic sources, however, current connectivity among populations may become eroded.

Although Moynahan et al. (2006) reported relatively high mortality during one winter of their study, sage-grouse generally have low over-winter mortality (<20%) and relatively high annual survival (30–78%). The average likelihood of a female nesting in a given year varies from 63% to 100% and averages 82% in the eastern part of the species' range and 78% in the western portion of the range (Connelly et al., this volume, chapter 3). Clutch size of sage-grouse averages six to nine eggs and nest success rates average 52% in relatively nonaltered habitats, while those in altered habitats average 37% (Connelly et al., this volume, chapter 3). Adult female sage-grouse survival is greater than adult male survival and adults have lower survival than yearlings, but not all estimates of survival rates are directly comparable (Zablan et al. 2003; Connelly et al., this volume, chapter 3). These relatively high survival rates and low reproductive rates suggest that sage-grouse populations may be slow to respond to improved habitat conditions.

Many populations are migratory (Connelly et al., this volume, chapter 3). Lengthy migration between separate seasonal ranges is one of the more distinctive characteristics of many sage-grouse populations (Connelly et al. 1988, 2000b). These migratory movements (>20 km) and large annual home ranges (>600 km^2) help integrate sage-grouse populations across vast landscapes of sagebrush-dominated habitats (Connelly et al., this volume, chapter 3; Knick and Hanser, this volume, chapter 16).

All state and provincial fish and wildlife agencies monitor sage-grouse breeding populations annually, but monitoring techniques have varied somewhat among areas and years both within and among agencies. This methodological variation complicates attempts to understand grouse population trends and make comparisons among areas (Connelly et al. 2004). Population monitoring efforts increased substantially between 1965 and 2007 throughout the range of sage-grouse (Garton et al., this volume, chapter 15). The largest increases in effort occurred in the Great Plains Sage-Grouse Management Zone (SMZ)(parts of Alberta, Saskatchewan, Montana, North Dakota, South Dakota, and Wyoming) and Colorado Plateau SMZ (representing parts of Utah and Colorado). In 2007, a minimum of 88,816 male sage-grouse were counted on 5,042 leks throughout western North America (Garton et al., this volume, chapter 15).

CURRENT KNOWLEDGE OF HABITATS

Invasive plant species, wildfires, weather, and climate change are major influences on sagebrush habitats and present significant challenges to long-term conservation (Miller et al., this volume, chapter 10; Baker, this volume, chapter 11). All of these factors are spatially pervasive and have considerable potential to influence processes within sagebrush communities. In addition, habitat loss or degradation can have a significant influence on sage-grouse populations by increasing the role of predation and disease (Hagen, this volume, chapter 6; Walker and Naugle, this volume, chapter 9).

Cheatgrass (*Bromus tectorum*) has invaded many of the lower-elevation, more-xeric sagebrush landscapes in the western United States. A large proportion of the remaining sagebrush communities is at moderate to high risk of invasion by cheatgrass (Connelly et al. 2004; Wisdom et al. 2005a; Miller et al., this volume, chapter 10). Moreover, juniper (*Juniperus* spp.) and pinyon (*Pinus* spp.) woodlands have expanded into sagebrush habitats at higher elevations (Miller et al., this volume, chapter 10). Numbers of fires and total area burned have increased since 1980 throughout most sagebrush-dominated habitats.

Sage-grouse have been eliminated from many former areas of their likely distribution prior to Euro-American settlement (Schroeder et al. 2004, Aldridge et al. 2008). Extirpated ranges had a lower percent area of sagebrush compared to those currently occupied by sage-grouse. Extirpated ranges also were at lower elevation, contained greater levels of human infrastructure such as transmission lines and communication towers, and had more private landownership relative to occupied regions (Wisdom et al., this volume, chapter 18). Moreover, this analysis identified those areas currently occupied by sage-grouse but characterized by environmental features most similar to extirpated range. These areas generally were concentrated in small, disjunct portions of occupied range and along peripheries of the current sage-grouse distribution (Wisdom et al., this volume, chapter 18). These regions will likely not support populations far into the future without active restoration or management that improves habitat conditions. In contrast, areas characterized by environmental factors where sage-grouse were most likely to persist were concentrated in the largest, most contiguous portions of occupied range in Oregon, Idaho, Nevada, and western Wyoming (Wisdom et al., this volume, chapter 18).

Urbanization and increasing human populations throughout much of the sage-grouse distribution have resulted in an extensive system of roads, power lines, railroads, and communication towers with an expanding influence on sagebrush habitats (Knick et al., this volume, chapter 12). Less than 5% of current sagebrush habitats was >2.5 km from a mapped road (Knick et al., this volume, chapter 12). Roads and other corridors promote invasion of exotic plants, provide travel routes for predators, and facilitate human access into sagebrush habitats. Human-caused fires also were closely related to existing roads.

Wildfire dynamics under the historic range of variation were likely characterized in all sagebrush landscapes by infrequent episodes of large, high-severity fires followed by long interludes with smaller, patchier fires, allowing mature sagebrush to dominate for extended periods (Baker 2006). Fire rotation, estimated from recent fire records, suggests fire exclusion had little effect on fire in sagebrush ecosystems, especially in more xeric areas. Instead, cheatgrass invasion, increases in number of human-set fires, and global warming have resulted in greatly increased amounts of fire relative to the historic variation in the Columbia Basin, Northern Great Basin, Southern Great Basin, and Snake River Plain SMZs (Baker, this volume, chapter 11). In addition, global climate change is likely to further promote cheatgrass and increase frequency of fire (Miller et al., this volume, chapter 10).

Additional fire created by widespread prescribed burning of sagebrush is unnecessary and exacerbates this increasing dominance of fire, particularly in lower-elevation landscapes dominated by Wyoming big sagebrush (*Artemisia tridentata* spp. *wyomingensis*) (Baker 2006a; Baker, this volume, chapter 11). Sagebrush ecosystems in these low-productivity regions characterized by low resilience and resistance to disturbance would benefit from rest, rather than the increased levels of disturbance that prescribed fire contributes to the natural regime. Thus, fire suppression is appropriate where cheatgrass invasion or expansion is likely to impede restoration treatments or natural recovery of native plant communities (Baker, this volume, chapter 11; Pyke, this volume, chapter 23).

Energy development for oil and gas influences sagebrush habitats by physical removal of habitat to construct well pads, roads, power lines, and pipelines (Naugle et al., this volume, chapter 20; Doherty et al., this volume, chapter 21). Indirect effects include habitat fragmentation and soil disturbance along roads, spread of exotic plants, and increased predation from raptors that have access to new perches for nesting and hunting (Knick et al., this volume, chapter 12; Naugle et al., this volume, chapter 20). Available evidence clearly supports the conclusion that conserving large landscapes with suitable habitat is important for conservation of sage-grouse, but that doing so involves overcoming numerous environmental challenges (Miller et al., this volume, chapter 10).

By creating habitat characteristics specific to sage-grouse requirements (Connelly et al., this volume, chapter 4), managers have adopted an umbrella concept that should similarly benefit other wildlife species dependent on sagebrush (Hanser and Knick, this volume, chapter 19). Passerine birds associated with sagebrush steppe habitats had high levels of overlap with sage-grouse along multiscale environmental gradients. However, this overlap was primarily a function of the broad range of sagebrush habitats used by sage-grouse (Hanser and Knick, this volume, chapter 19). Management that focuses on creating a narrow set of plot-scale conditions for a single species or site restoration will likely be less effective in addressing the needs of multiple species than restoration efforts that recognize landscape heterogeneity and multiscale organization of habitats (Hanser and Knick, this volume, chapter 19).

THREATS

Predation is often identified as a potential threat to sage-grouse (Schroeder and Baydack 2001; Hagen, this volume, chapter 6). However, predator management studies have not provided sufficient evidence to support implementation of predator control to improve sage-grouse populations over broad geographic or temporal scales. The limited information available suggests predator management may provide short-term relief for a sage-grouse population sink in the few cases where this situation has been documented (Hagen, this volume, chapter 6).

Hunting has also been identified as a management concern for sage-grouse populations (Connelly et al. 2003a; Reese and Connelly, this volume, chapter 7). Nine of 11 states with sage-grouse presently have hunting seasons for this species. Sage-grouse normally experience high survival over winter (Wik 2002, Hausleitner 2003, Beck et al. 2006, Battazo 2007); thus, mortality from hunter harvest in September and October may not be totally compensatory. Nevertheless, harvest mortality is low on most populations of sage-grouse, and no studies have demonstrated that hunting is a primary cause reducing populations (Reese and Connelly, this volume, chapter 7).

Despite the prevalence of organisms that may infect individual birds, population-level effects of parasites and disease have rarely been documented in sage-grouse (Christiansen and Tate, this volume, chapter 8). However, West Nile virus has shown greater impact on sage-grouse populations than any other infectious agent detected to date. This virus was an important new source of mortality in low- and mid-elevation sage-grouse populations range-wide from 2003 to 2007 (Naugle et al. 2004; Walker et al. 2007b; Walker 2008; Walker and Naugle, this volume, chapter 9). West Nile virus can significantly reduce survival and may lead to local and regional population declines. Simulations of West Nile virus mortality projected reduced growth of susceptible sage-grouse populations by an average of 0.06% to 0.09% per year. However, marked spatial and annual fluctuations in nest success, chick survival, and other sources of adult mortality may mask population-level impacts in most years. Resistance to West Nile virus–related disease appears to be low but is expected to increase slowly over time (Walker et al. 2007b; Walker and Naugle, this volume, chapter 9).

Livestock grazing is the most widespread use of sage-grouse habitats, but data used by agencies (e.g., permitted animal unit months) do not provide information on management regime, habitat condition, or type of livestock that allows the assessment of direct effects of grazing at large spatial scales (Milchunas and Lauenroth 1993; Jones 2000; Knick et al., this volume, chapter 12). These data may be collected for individual allotments. However, they often are subjective estimates or are not collected systematically across a region or through time in a way that permits an evaluation of grazing levels and intensity relative to habitat condition. Consequently, the significance of decreased numbers of livestock on public lands (Mitchell 2000) cannot be interpreted without corresponding information on changes in habitat productivity. Thus, the direct effect of livestock grazing expressed through habitat changes to population-level responses of sage-grouse cannot be addressed using existing information.

The effects of livestock grazing management, however, can have significant influences on landscape patterns and processes (Freilich et al. 2003; Miller et al., this volume, chapter 10; Knick et al., this volume, chapter 12). Large treatments designed to remove sagebrush and increase forage for livestock may no longer be the primary emphasis by agencies for management of public lands. Nevertheless, habitat manipulations, water developments, and fencing are still widely implemented to manage livestock grazing, and large-scale treatments still occur on some private lands. More than 1,000 km of fences were constructed annually on public lands from 1996 to 2002; linear density of fences exceeded 2 km/km^2 in some regions of the sagebrush biome (Knick et al., this volume, chapter 12). Fences provide perches for raptors and modify access and movements by humans and livestock, thus exerting a new mosaic of disturbance and use on the landscape (Freilich et al. 2003).

Development of oil and gas resources will continue to be a major influence on sagebrush habitats and sage-grouse because advanced technology allows access to reserves, high demand for these resources will continue, and a large number of applications have been approved and are still being submitted and approved annually. Future oil and gas development is projected to cause a 7–19% decline from 2007 sage-grouse lek population counts and impact 3,700,000 ha of sagebrush shrublands and 1,100,000 ha of grasslands throughout much of the current and likely historical range of sage-grouse (Copeland et al. 2009). Sagebrush landscapes developed for energy production contained twice as many roads and power lines, and in some areas where ranching, energy development, and tillage agriculture coincided, human features were so dense that every 1 km^2 could be bounded by a road and bisected by a power line (Naugle et al., this volume, chapter 20). Sage-grouse respond negatively to different types of development, and conventional densities of oil and gas wells likely far exceed the species' threshold of tolerance (Naugle et al., this volume, chapter 20). Noise disturbance from construction activities and vehicles may also disrupt sage-grouse breeding and nesting (Lyon and Anderson 2003).

Highly productive regions with deeper soils throughout the sagebrush biome have been converted to agriculture, in contrast to relatively xeric areas with rather shallow soils that characterize the larger landscapes still dominated by sagebrush. Agriculture currently influences 49% of sagebrush habitats within the sage-grouse range through habitat loss or by large-scale fragmentation of remaining sagebrush. Potential predators on sage-grouse nests, such as Common Ravens (*Corvus corax*; Coates 2007), are subsidized by agriculture and associated practices. In addition, insecticides can be a major cause of mortality for sage-grouse attracted to lush croplands during summer brood-rearing (Blus et al. 1989).

The human footprint is defined as the cumulative extent to which anthropogenic resources and actions influence sagebrush ecosystems within the range of sage-grouse (Leu and Hanser, this volume, chapter 13). The levels and broad-scale effects of the human footprint across the sage-grouse distribution strongly support the importance of managing and maintaining sagebrush habitats at larger spatial scales than currently recognized by land management agencies (Leu and Hanser, this volume, chapter 13). The greatest influence of the human footprint was within the Columbia Basin SMZ, followed by the Wyoming Basin, Great Plains, Colorado Plateau, Snake River Plain, Southern Great Basin, and Northern Great Basin SMZs (Leu and Hanser, this volume, chapter 13). Populations within the Columbia Basin, which had the highest levels of human footprint, are decreasing and have a reasonably high likelihood of declining to <50 sage-grouse within 100 years (Garton et al., this volume, chapter 15).

The cumulative and interactive impact of multiple disturbances, continued spread and dominance of invasive species, and increased impacts of land use have the most significant influence on the trajectory of sagebrush ecosystems, rather than any single source (Knick et al., this volume, chapter 12). Sage-grouse populations and sagebrush habitats that once were continuous now are separated by agriculture, urbanization, and development. Thus, understanding how to conserve sage-grouse involves multiscale patterns and dynamics in sagebrush ecosystems as well as population trends, behavior, and ecology of sage-grouse (Knick et al., this volume, chapter 12).

Fifteen major threats (Table 24.1) have been identified in recent syntheses of sage-grouse conservation issues (Connelly and Braun 1997; Braun 1998; Connelly et al. 2004; Knick and Connelly, this volume). These reports generally agreed that energy development, drought, and wildfire posed a serious risk to sage-grouse conservation. Drought was listed in all reports, while energy development and wildfire were listed in three of four reports. Invasive species, grazing management, and urban development were listed in two of the three reports (Table 24.1). In addition, one federal agency and two state agencies convened expert panels to assess threats to sage-grouse populations (Table 24.2). Together, these panels listed 15 threats to sage-grouse and collectively identified energy development, wildfire, urban development, West Nile virus, conifer encroachment, and invasive species as the most serious threats to sage-grouse conservation. Considered as a whole, these seven different assessments of threats identified two levels of risk. Energy development, invasive species, drought, grazing management, and wildfire, listed on five threat assessments, constitute the first level and could be judged as the most significant range-wide threats to sage-grouse conservation. Urbanization and West Nile virus, listed on three or four assessments, represent the second level, suggesting a broad concern about these issues as well. Infrastructure was listed on two assessments and fences, roads, and reservoirs (all potential energy-related infrastructures) were listed separately on a third assessment. In summary, these efforts to identify threats

TABLE 24.1

Threats to sage-grouse identified by recent reviews.

Threat	Connelly and Braun (1997)	Braun (1998)	Connelly et al. (2004)	This volume
Agriculture		X		
Drought[a]	X	X	X	X
Energy development		X	X	X
Fences		X		
Grazing management	X			X
Hunting		X		
Invasive species			X	X
Predation		X		
Power lines		X		
Reservoirs		X		
Roads		X		
Urban development[b]		X		X
Vegetation treatments		X		
West Nile virus[c]				X
Wildfire	X		X	X

[a] Includes climate change induced drought.

[b] Includes factors associated with the human footprint.

[c] West Nile virus was first detected within Greater Sage-grouse range in 2002 after completion of the 1997 and 1998 assessments (Naugle et al. 2004).

suggest that energy development, invasive species, wildfire, grazing management, urbanization, West Nile virus, and infrastructure pose the greatest risk to long-term conservation of sage-grouse. The relative importance of each of these threats undoubtedly varies throughout the range of sage-grouse.

POPULATION AND HABITAT TRAJECTORIES

Lek size declined over the assessment period (1965–2007) for 20 of 28 (71%) populations that had sufficient data for analysis (Garton et al., this volume, chapter 15). Average rates of change declined between the 1995–1999 and 2000–2007 analysis periods for 20 of 26 (77%) populations (Garton et al., this volume, chapter 15). Nevertheless, 20 of 29 (69%) populations had an average rate of change ≥1 while nine of 29 (31%) populations had an average rate of change ≤1.0 for the 2000–2007 analysis period. Although lek size and average rates of change declined for six of seven management zones, all but one had an average rate of change ≥1.0 during the 2000–2007 analysis period. Only the Columbia Basin management zone had an average rate of change ≤1.0 during the last analysis period (Garton et al., this volume, chapter 15).

For 86% of management zones and 50% of populations, the best statistical model indicated a declining carrying capacity through time of −1.8% to −11.6% per year, and 18% of models for all populations and management zones indicated a lower carrying capacity in the last 20 years (1987–2007) compared to the first 20 years (1967–1987) of analysis (Garton et al., this volume, chapter 15). These lower carrying capacities support other findings in this volume suggesting that declines in quality and quantity of habitat for sage-grouse are continuing across regional and range-wide scales (Miller et al., this volume, chapter 10; Baker, this volume, chapter 11; Knick et al., this volume, chapter 12; Leu and Hanser, this volume, chapter 13). Forecasts of future population

TABLE 24.2

Threats to Greater Sage-Grouse identified by expert panels.

Threat	USFWS[a] panel	IDFG[b] panel	WGF[c] panel
Agriculture	X		
Climate change		X	
Conifer encroachment	X	X	
Energy development	X		X
Grazing management	X	X	X
Infrastructure	X	X	
Invasive species	X	X	X
Human disturbance		X	
Prescribed fire		X	
Seeded grassland		X	
Strip/coal mining	X		
Urbanization	X		X
West Nile virus		X	X
Wildfire	X	X	
Weather	X		

[a] U.S. Fish and Wildlife Service.

[b] Idaho Department of Fish and Game.

[c] Wyoming Game and Fish Department.

viability across 27 populations and all management zones suggest that 96% of populations and all management zones will likely remain above effective population sizes of 50 within the next 30 years. However, 78% of populations and 29% of management zones are likely to decline below effective population sizes of 500 within 100 years if current conditions and trends persist (Garton et al., this volume, chapter 15). Sage-grouse populations in the Colorado Plateau, Columbia Basin, and Snake River Plain management zones appear to be at higher risk than populations in core regions enclosed within the Great Plains, Northern Great Basin, Southern Great Basin, and Wyoming Basin management zones.

Trends in number of male sage-grouse counted at leks were correlated with several habitat features, although the relationships differed across the sage-grouse range (Johnson et al., this volume, chapter 17). In low-elevation regions, trends tended to be greater at higher elevations (i.e., positive correlations with elevation); the reverse was true in higher-elevation areas. Lek trends across all management zones increased steadily with cover of tall sagebrush at 5- and 18-km radii. Similarly, lek trends across all management zones increased with cover of all sagebrush (combined categories for tall sage and low sagebrush) at both radii (Johnson et al., this volume, chapter 17). In contrast, associations were negative with the coverage of agriculture and exotic plant species. Trends also tended to be lower for leks at which a greater proportion of the surrounding landscape had been burned (Johnson et al., this volume, chapter 17). Few leks were within 5 km of developed land, and trends were lower for those leks with more developed land within 5 or 18 km of the lek. Lek counts were reduced where communication towers were nearby, whereas no effects of power lines were detected. Producing oil or natural gas wells and paved highways, but not secondary roads, were also associated with lower counts (Johnson et al., this volume, chapter 17). Roads, power lines and other disturbances that have been in place for many years may have affected lek attendance in years prior to this analysis period (1997–2007), while other disturbances, such as communication towers, are relatively new; their effects may be expressed in the current data or may not have been detected due to lags in population response. Conversion of sagebrush habitats to cultivation and paved highways that occurred before the 1997–2007 study period likely continues to influence sage-grouse populations (Johnson et al., this volume, chapter 17).

Sage-grouse now occupy <60% of their probable historical range prior to European settlement (Connelly and Braun 1997, Schroeder et al. 2004). Moreover, synergistic feedbacks among invasive plant species, fire, and climate change coupled with current trajectories of habitat changes and rates of disturbance, both natural and human-caused, likely will continue to change sagebrush communities and create challenges for future conservation and management of sage-grouse populations and habitat.

CHALLENGES TO SAGE-GROUSE CONSERVATION

Conservation programs for sage-grouse populations and habitat can be developed to address threats (Stiver, this volume, chapter 2), but administrative or natural impediments to development and implementation of successful programs may still exist (Forbis et al. 2006). Land management agencies continually make decisions regarding land use actions and vegetation management (Knick et al., this volume, chapter 12). These agencies also develop programs to address potential or actual environmental issues including wildfire, invasive species, and vegetation restoration or rehabilitation efforts (Miller et al., this volume, chapter 10; Baker, this volume, chapter 11; Pyke, this volume, chapter 23). The continued interest in prescribed burning and other forms of sagebrush reduction in sagebrush-dominated landscapes (Wyoming Interagency Vegetation Committee 2002; Davies et al. 2008, 2009), despite a large body of evidence documenting the negative effects of these actions on sage-grouse, may continue to degrade and fragment sage-grouse habitats. Similarly, development of energy-related projects in key habitats will continue to negatively affect important sage-grouse habitat (Knick et al., this volume, chapter 12; Naugle et al., this volume, chapter 20).

Natural phenomena may act to degrade or eliminate sage-grouse habitat. Wildfire (Baker et al. 2006, this volume, chapter 11; Miller et al., this volume, chapter 10) and drought (Patterson 1952, Connelly and Braun 1997, Connelly et al. 2000a) can negatively affect sage-grouse populations. The incidence of wildfire may be reduced by suppression efforts, but fire will never be eliminated as a threat to sagebrush-dominated landscapes. Periodic drought will also be part of the arid west and pose a threat to sage-grouse productivity by reducing nest and chick survival (Connelly et al. 2000a). In addition, restoration following treatments, such as prescribed fire, often is severely hindered or is unsuccessful because of unpredictable weather and lack of precipitation necessary for plant establishment (Pyke, this volume, chapter 23).

Climate change also has an important influence on sagebrush landscapes (Miller et al., this volume, chapter 10). Climate change scenarios for the sagebrush region predict increasing temperature, atmospheric carbon dioxide, and severe weather events, all of which favor cheatgrass expansion and increased wildfire (Miller et al., this volume, chapter 10). Approximately 12% of the current distribution of sagebrush is predicted to be replaced by expansion of other woody vegetation for each 1°C increase in temperature (Miller et al., this volume, chapter 10). All of these factors are likely to result in a loss of sagebrush and decline of sage-grouse.

A broad array of invasive plants is widely distributed across the range of sage-grouse, has a major influence on the structure and function of sagebrush habitats, and presents significant challenges to the long-term conservation of sagebrush-dominated landscapes (Miller et al., this volume, chapter 10). Many sagebrush communities at low elevations are at moderate to high risk of invasion by cheatgrass (Wisdom et al. 2005b; Miller et al., this volume, chapter 10). At higher elevations, woodland expansion has altered the fire regime and resulted in loss of sagebrush and the understory of grasses and forbs (Miller et al., this volume, chapter 10).

Invasions into native plant communities may be sequential as initial invaders are replaced by a series of new exotics or by species adapting to new habitats within their range (Young and Longland 1996). For example, areas that were once dominated by cheatgrass in some locations in southwestern Idaho are now characterized by medusahead (*Taeniatherum caput-medusae*; Miller et al., this volume, chapter 10). Rush skeletonweed (*Chondrilla juncea*), which originally was localized to disturbed areas in drier sagebrush grassland communities, is now invading areas previously dominated by medusahead (Sheley et al. 1999) and following wildfire (Kinter et al. 2007).

Free-roaming equids (horses [*Equus caballus*] and burros [*E. asinus*]) in the United States were introduced to North America near the end of the 16th century. These species could be considered invasive, but they have unique management status and by law are neither hunted nor as intensively managed as livestock (Beever and Aldridge, this volume, chapter 14). Free-roaming horses can exert direct influences on structure and composition of vegetation and soils in sagebrush communities, as well as indirectly affect numerous animal groups whose abundance collectively may indicate the ecological integrity of such communities (Beever and Aldridge, this volume, chapter 14). Compared to ecologically similar sites in which horses were removed in the western Great Basin, sites that still supported wild horses had lower shrub cover, higher compaction of soil surfaces, more fragmented shrub canopy, lower grass cover, lower total vegetative cover, lower plant species richness, and lower density of ant mounds (Beever and Aldridge, this volume, chapter 14). Greater density of ant mounds at horse-free sites than at horse-occupied sites suggests that at least a portion of the invertebrate community is more robust at horse-removed sites, and may also reflect differences in level of ecological function (Beever and Herrick 2006).

Restoration of sage-grouse habitat is more complex than typical restoration projects, which often focus on individual sites and have objectives specific to that location (Pyke, this volume, chapter 23). Successful restoration of sage-grouse habitat will not only necessitate vegetation changes in a single area but will also require connectivity among patches of currently intact vegetation (Wisdom et al. 2005b; Meinke et al. 2009; Knick and Hanser, this volume, chapter 16; Pyke, this volume, chapter 23). Additionally, availability and cost are major obstructions to the use of native seeds in revegetation projects (McArthur 2004), and equipment for planting native seeds is not widely available (Wiedemann 2005).

Many partnerships and working groups throughout the West have begun to initiate efforts to assist in conservation of sage-grouse, including some restoration projects (Western Governors' Association 2004). Unfortunately, to the best of our knowledge, the effectiveness of these actions in stabilizing or increasing sage-grouse populations has yet to be documented. In part, this is because some projects are too recent to demonstrate positive effects, while others may have had competing interests or lacked a complete understanding of the ecological challenges during planning and implementation.

A ROADMAP TO CONSERVATION

Realistic approaches to issues, understanding threats, and implementing levels of effort appropriate to combat inherent challenges are important considerations in developing long-term conservation plans. We discuss many of the key issues presented in this volume and, based on the chapters within this volume, attempt to provide some insight and guidance to addressing these issues, threats, and challenges within the broad context of sage-grouse conservation.

Population Management

Harvest Management

Hunting opportunity for sage-grouse has been reduced where data suggested a negative impact from hunting and in response to general population declines of known and unknown origin. Seasons may need to be adjusted or reduced as necessary in those regions where sage-grouse continue to decline or are at risk of extirpation from other causes of mortality (Reese and Connelly, this volume, chapter 7). A risk-sensitive harvest strategy (Williams et al. 2004a) that avoids reducing individual populations of sage-grouse will require new research and continued routine population monitoring. We suggest social implications, as well as biological effects, are important considerations for management in areas where harvest is strictly controlled or altered to better conserve sage-grouse (Reese and Connelly, this volume, chapter 7).

Predation Management

Thus far, little information suggests that predator management should be routinely applied to conserve sage-grouse populations (Schroeder and Baydack 2001; Hagen, this volume, chapter 6). Where predator management is necessary, both lethal and nonlethal methods might be needed to buffer population sinks to increase survival and recruitment of grouse in these areas in the short-term (two to three years) from adverse effects of predation rates. The relatively broad financial and political costs to removing predators at a scale and extent that may be effective is no longer likely to be socially or ecologically viable (Messmer et al. 1999). Because of these considerations, predator management for sage-grouse has generally been accomplished most efficiently by manipulating habitat rather than by predator removal to enhance populations (Schroeder and Baydack 2001). For future sage-grouse conservation efforts, we recommend quantifying predator communities as they relate to demographic rates and habitat variables so the predator-cover complex as it pertains to sage-grouse life history can be better understood (Hagen, this volume, chapter 6). Additionally, information is needed on how species that prey on sage-grouse respond to anthropogenic changes on sagebrush-dominated landscapes (Coates 2007).

Disease Management

Documentation of population-level effects of parasites, infectious diseases, and noninfectious diseases related to toxicants is rare (Christiansen and Tate, this volume, chapter 8). Thus, little recent emphasis has been placed on managing this aspect of sage-grouse biology. Within the last few years, West Nile virus has had severe effects on some sage-grouse populations (Walker and Naugle, this volume, chapter 9). The severity of the potential impact and the need for more information require future studies to better document effects and relate outbreaks to environmental variables. The potential implications of climate change further underscore the need to effectively monitor disease impacts on sage-grouse (Christiansen and Tate, this volume, chapter 8; Miller et al., this volume, chapter 10). Many pathogens are sensitive to temperature, rainfall, and humidity (Harvell et al. 2002). Warmer climates can increase pathogen development and survival rates, disease transmission, and host susceptibility. Most host-parasite systems are likely to experience more frequent or severe disease impacts with warming climates (Harvell et al. 2002).

Habitat Management

Habitat Protection

Much sage-grouse habitat has been lost or altered, but substantial habitat still exists to support this species in many parts of its range (Connelly et al. 2004; Schroeder et al. 2004; Leu and Hanser, this volume, chapter 13). Characteristics of important habitats and general guidelines for protecting and managing these habitats are well known (Connelly et al. 2000b, Crawford et al. 2004, Hagen et al. 2007). We suggest the most effective strategy to stabilize or recover many sage-grouse populations will be protecting existing sagebrush habitat (Stiver et al. 2006). Energy development and other anthropogenic change represent substantial challenges to protecting existing habitat, and will require development and implementation of broad-scale and long-term conservation plans (Stiver et al. 2006; Stiver, this volume, chapter 2) that are carefully developed using the best available data. A wide range of local and regional concerns may need to be considered, including urban development, fire, grazing (livestock, equid, and wildlife), fragmentation, roads, structures, invasive species, West Nile virus, and habitat quality and quantity. The importance of each of these issues varies spatially and temporally.

Landscapes with high biological value for sage-grouse and high risk for development represent the greatest challenge facing land use managers. This is a concern because 44% of areas with high biological value are at risk for energy development (Doherty et al., this volume, chapter 21). The rapid pace and scale of energy development is a major issue, because areas being developed include some of the largest remaining sagebrush landscapes with the highest densities of sage-grouse in North America (Connelly et al. 2004; Doherty et al., this volume, chapter 21). Sage-grouse conservation faces major challenges in the eastern portion of the species' range, where 44% of the lands that the federal government has authority to control for oil and gas development has been authorized for exploration and development (Naugle et al., this volume, chapter 20; Doherty et al., this volume, chapter 21). Severity of impacts and extensive leasing of the public mineral estate suggest a need for landscape-scale conservation (Holloran 2005, Aldridge and Boyce 2007, Walker et al. 2007a). Lease sales continue, despite concerns, because no policy is in place that would permit an environmental assessment of risk at the scale at which impacts occur.

Areas of high biological value combined with low energy potential represent regions where conservation actions can be immediately implemented (Doherty et al., this volume, chapter 21). Currently, 17% of the eastern sage-grouse range has high biological value and low risk from energy development (Doherty et al., this volume, chapter 21). Maintaining these quality sage-grouse habitats, especially in areas adjacent to development or where development is planned, will be critical to ensure genetic connectivity (Oyler-McCance et al. 2005a,b) and persistence of source populations for natural recolonization after energy development activities have ceased (Gonzalez et al. 1998). Reducing risks from other stressors to sagebrush habitats will be an important component of conservation strategies in high value and low energy potential areas (Klebenow 1970; Connelly et al. 2000a,b; Leonard et al. 2000; Smith et al. 2005; Walker et al. 2007a). Habitat loss to agricultural development (Farrell et al. 2006, United States Government Accounting Office 2007), urban and exurban expansion (Theobald 2003, 2005), and conversion to communities dominated by invasive plants (e.g., cheatgrass; Bergquist et al. 2007) are significant concerns in many of these regions.

Conservation easements are one tool to reduce residential development and agricultural conversion on private lands (Kiesecker et al. 2007). A preponderance of private surface ownership in Montana and Utah coupled with low risks of development make core regions in many parts of these states ideal places to develop incentives for ranching and rural lifestyles through long-term programs such as the Conservation Reserve Program (CRP; Schroeder and Vander Haegen, this volume, chapter 22). Opportunities for easements and management programs are available in other states, but long-term viability of them is a public policy decision (Doherty et al., this volume, chapter 21).

Areas of low biological value and low energy potential represent low-conflict opportunities for sage-grouse and could be important in maintaining connectivity to high value core regions (Doherty et al., this volume, chapter 21). Restoration of these linkage habitats will be a key strategy in some areas. Many of the low value and low potential areas identified by Doherty et al. (this volume, chapter 21) are the same areas where continued range contraction is expected to be most severe (Aldridge et al. 2008; Garton et al., this volume, chapter 15). Aggressive habitat protection and restoration programs may be necessary to maintain the biological integrity of fringe populations in North Dakota, South Dakota, northern Montana, and Canada. Explicitly combining information about vulnerability of landscapes to anthropogenic risk allows planners to consider the relative urgency and likelihood of success of a given conservation strategy (Wilson et al. 2005, Copeland et al. 2007, Pressey and Bottrill 2008). Core regions and assessment of potential impacts these regions may experience represent a starting point to begin conservation of landscapes where results will have the largest benefit to populations. Prioritizing landscapes simply reflects the reality that threats are large, resources are limited, and conservation actions targeting all remaining populations are not feasible (Wisdom et al. 2005c, Meinke et al. 2009). Identification of core regions represents a proactive attempt to maintain a viable and connected set of populations before the opportunity to do so is lost (Knick and Hanser, this volume, chapter 16; Doherty et al., this volume, chapter 21).

Strategies that are integrated among all states and provinces involved for landscape-scale conservation are most likely to be successful. Successful implementation of conservation strategies in one state or province may not be sufficient to compensate for losses in other areas. Conservation concerns related to sage-grouse will present challenges until collaborative landscape planning and conservation are implemented. Doherty et al. (this volume, chapter 21) provide a framework for planning across political boundaries and suggestions for measuring success.

Habitat Restoration

Much of the original sage-grouse habitat has been permanently lost to agricultural development and urban areas, and the remaining habitat ranges in condition from high quality to inadequate (Pyke, this volume, chapter 23). Sage-grouse require somewhat different seasonal habitats distributed over large areas to complete their life cycle. Thus, restoration that incorporates a broad perspective when considering when and where to restore lands is likely to be the most effective for improving sage-grouse habitat. Restoration decisions are often difficult because of economics, restoration potentials, status of existing habitat, and logistics such as landownership or topography (Knick, this volume, chapter 1).

Prioritization is an important first step in a successful restoration plan for selecting sites when resources are limited (Wisdom et al. 2005c, Meinke et al. 2009). The triage approach is an initial prioritization technique where ecosystems are grouped into three categories, one that receives immediate care and two others where no urgent care is warranted (Pyke, this volume, chapter 23). The category provided immediate care and intervention has significant damage requiring immediate intervention to aid likely recovery. The second category needs no immediate intervention and, with some later treatment, will likely recover, whereas the third category represents areas so severely damaged they could not recover even with intervention (Kennedy et al. 1996, Samways 2000). A framework was presented (Doherty et al., this volume, chapter 21) that demonstrated tradeoffs between sage-grouse conservation and energy development. However, landscape planning for sage-grouse is likely to be most successful if it includes restoration and identifies core regions (Doherty et al., this volume, chapter 21) that reflect seasonal habitats and migration of radio-marked sage-grouse (Connelly et al. 1988, Aldridge and Boyce 2007, Doherty et al. 2008) to ensure priority landscapes meet, or with restoration will contribute to, all habitat needs. Moreover, future modeling of other relevant risks, such as invasive species, will help ensure that gains in conservation will not be offset by unknown risks (Doherty et al., this volume, chapter 21).

Functioning landscapes that consist of an integrated mosaic of individual sites are important objectives when considering type and level of restoration intervention for improving sage-grouse habitat (Pyke, this volume, chapter 23). Reasons for considering larger areas than the restoration site alone are based on criteria relating to sage-grouse biology as well as the likelihood of restoration success. Sage-grouse have large annual and seasonal home ranges (Connelly et al. 2000b) that often exceed the size of restoration projects. In addition to enhancing existing native habitats, restoring adjacent lands presently in tillage agriculture to sagebrush-dominated grasslands could facilitate the larger goal of landscape restoration (Schroeder and Vander Haegen, this volume, chapter 22).

Effective restoration and rehabilitation of sage-grouse habitat focuses on maintaining or improving key habitat components necessary for survival and reproductive success. We caution that simply replacing vegetation components may not produce the intended benefit to sage-grouse populations. The negative influence of fire and the human footprint, not sagebrush quantity or configuration, were the significant factors in persistence of sage-grouse leks (Knick and Hanser, this volume, chapter 16). Reestablishing suitable vegetation will be difficult because of increasing fire frequencies throughout much of the sage-grouse range coupled with long periods for vegetation recovery (Baker, this volume, chapter 11). Increasing levels of all land uses for traditional commodity development as well as for recreation and exurban living by a growing human population also indicate that the human footprint will continue to be a primary impediment to successful restoration.

Passive restoration goals focus on maintaining sagebrush cover while increasing grass cover and height and increasing forb cover and reproduction (Pyke, this volume, chapter 23). This could be achieved through setting appropriate livestock stocking levels while shifting grazing seasons to periods when active growth is slow and plant reproduction has not been initiated (Kirby and Grosz 1995, Norton 2005, Sidle 2005). Active restoration is necessary in some situations to reestablish a sagebrush overstory with an understory mixture of native forbs and short, mid, and tall grasses (Pyke, this volume, chapter 23). Appropriate native sagebrush species and subspecies for the site are significant factors in successful restoration for sage-grouse. Nevertheless, we recognize that some efforts may require introduced species such as palatable forbs and bunchgrasses to quickly stabilize soils as well as different techniques to achieve similar goals.

Effective restoration will require protection and proper management for maintenance of intact, healthy sagebrush grasslands, while identifying those lands where modifications to management might improve quality habitat for sage-grouse (Pyke, this volume, chapter 23). Strategic placement will be critical for enhancing the likelihood of restoration success while minimizing costs. Unfortunately, sagebrush grassland restoration is largely in its infancy. Large acreages are still being affected by invasive species and wildfire, while funding and resources necessary for rehabilitating these areas are often severely limited. Farm programs such as the CRP have the potential to affect large portions of the landscape and positively influence sage-grouse populations in some parts of the species' range (Schroeder and Vander Haegen, this volume, chapter 22). However, these programs can only be applied to private lands; comparable programs to affect public land at a similar scale with effective restoration are needed. We are concerned that many lands currently in the CRP and benefiting sage-grouse populations are increasingly being converted to other uses, such as production of biofuels (Fargione et al. 2009).

Monitoring and Assessment

Throughout the sagebrush biome, various natural and anthropogenic actions are and will be occurring that may have positive (e.g., restoration work) or negative (e.g., energy development, wildfire) effects on sage-grouse. Monitoring and assessment activities are necessary to provide an objective appraisal of the effects of potentially positive activities and assess the relative damage to sage-grouse populations or habitats of potentially negative actions.

Protocols that include statistically sound sampling and analysis designs are necessary to obtain unbiased information. Casual field surveys, ocular assessment, and other forms of subjective evaluation provide unreliable information. For proposed projects that occupy spatially discrete (as opposed to dispersed) areas, a before-after-control-impact (BACI) design may provide the most powerful statistical approach.

To assess population effects, we recommend that BACI include marking sage-grouse at each impact and control site. Required sample sizes of marked birds will vary depending on size and extent of the grouse population being considered, questions being asked, and marking technology employed. We recommend capturing and marking birds in a manner that allows sampling of the entire project area, focusing on leks most proximate to the proposed impact site(s). We also recommend marking additional female grouse in an 18-km buffer zone to characterize the migratory status of the population, but this sample will not allow evaluation of avoidance behavior. Because of the effect of lag periods on population response, a minimum of at least three years pre-construction and four years post-construction may be required in addition to the year of construction to fully assess project effects on grouse populations. Given the lifespan of sage-grouse, strong fidelity to breeding areas, and lag effects in population dynamics, some longer-term (8–12 years), less-intensive monitoring will be necessary to fully assess impacts.

Unbiased characterization of habitat use or habitat change requires a random sampling approach and often a stratified random sample. Strata will depend on vegetation, treatment, and topographic characteristics of the area. Most habitat assessments will include measurements of one or more of the following: cover, height, density, frequency, and visual obstruction for individual plant species or groups of species (Connelly et al. 2003b). Density, height, and frequency are direct measurements or counts, but canopy or foliar cover can be estimated by several techniques. Well-recognized techniques that are largely free of observer bias and that can be easily replicated in other studies are important in ensuring widespread application and interpretation of results.

We have emphasized throughout this volume that the Greater Sage-Grouse is a landscape species. Although regional and range-wide dynamics of sage-grouse populations are monitored (Reese and Bowyer 2007), we have yet to develop protocols to assess landscape change in sagebrush habitats (West 2003a,b). Recent analyses suggest that >25–30% sagebrush and <25% agriculture are threshold levels at a landscape scale important to maintaining sage-grouse populations (Aldridge et al. 2008; Wisdom et al., this volume, chapter 18). Other studies have emphasized the importance of the landscape surrounding sage-grouse leks for distances up to 54 km (Holloran and Anderson 2005; Walker et al. 2007a; Knick and Hanser, this volume, chapter 16). Landscape effects also were significant in winter habitat selection by sage-grouse (Doherty et al. 2008). Thus, monitoring approaches that detect changes in quantity, composition, and configuration in regional and range-wide landscapes would significantly improve our ability to relate environmental features at the primary scales driving population dynamics.

Well-planned and carefully implemented monitoring and assessment will allow an objective evaluation of conservation measures over varying temporal and spatial frames. It will also provide an unbiased assessment of impacts that can be used to guide appropriate mitigation efforts.

CONSERVATION IMPLICATIONS

Much is known about the biology of sage-grouse and its response to various management actions as well as natural and anthropogenic disturbance. Despite this knowledge, many threats to sage-grouse and numerous constraints to successful conservation for this species and its habitats remain. Rigorously and objectively addressing these threats and constraints should result in sound management practices and decisions that perpetuate sage-grouse populations.

A minimum of 88,816 male sage-grouse were counted on 5,042 leks in 2007 (Garton et al., this volume, chapter 15), and sagebrush is the dominant land cover on approximately 530,000 km^2 within sage-grouse range (Knick, this volume, chapter 1). Therefore, even though some populations are declining and a few have a relatively low likelihood of persistence, opportunities to conserve sage-grouse throughout much of the species' current range still exist.

Land and wildlife managers, as well as policymakers, face many challenges and difficult decisions. We have attempted to assemble a volume that presents unbiased, current information spanning multiple facets of Greater Sage-Grouse and their habitats. The information, presented from an ecological perspective, is intended to aid sage-grouse conservation efforts, including those currently undertaken for the very similar Gunnison Sage-Grouse. We hope that this volume on sage-grouse populations and their habitats will be used to inform these decisions and guide policies in a manner that will allow future generations to enjoy this icon of the West.

ACKNOWLEDGMENTS

Earlier drafts of this manuscript were improved by reviews from T. P. Hemker, F. L. Knopf, J. D. Brittell, and B. Everitt. We also appreciate the thoughts and insight that members of the Western Sage and Columbian Sharp-tailed Grouse Technical Committee shared with us. Finally, we thank the numerous technicians, graduate students, and field biologists who have labored throughout western North America to collect data on sage-grouse populations and habitats. Many sage-grouse conservation efforts will be based on their diligent, but often unrecognized, work.

LITERATURE CITED

Abbitt, R. J. F., J. M. Scott, and D. S. Wilcove. 2000. The geography of vulnerability: incorporating species geography and human development patterns into conservation planning. Biological Conservation 96:169–175.

Adams, A. W. 1975. A brief history of juniper and shrub populations in southern Oregon. Oregon State Wildlife Commission Research Division. Wildlife Research Report No. 6, Corvallis, OR.

Addison, E. M., and R. C. Anderson. 1969. *Oxyspirura lumsdeni;* n. sp. (Nematoda:Thelaziidae) from Tetraonidae in North America. Canadian Journal of Zoology 47:1223–1227.

Agee, J. K. 1993. Fire ecology of Pacific Northwest forests. Island Press, Washington, DC.

Akaike, H. 1978. A Bayesian analysis of the minimum AIC procedure. Annals of the Institute of Statistical Mathematics 30:9–14.

Akaike, H. 1979. A Bayesian extension of the minimum AIC procedure of autoregressive model fitting. Biometrika 66:37–242.

Akaike, H. 1981. Likelihood of a model and information criteria. Journal of Econometrics 16:3–14.

Akaike, H. 1983. Information measures and model selection. International Statistical Institute 44:277–291.

Alatalo, R. V., T. Burke, J. Dann, O. Hanotte, J. Höglund, A. Lundberg, R. Moss, and P. T. Rintamäki. 1996. Paternity, copulation disturbance and female choice in lekking Black Grouse. Animal Behavior 52:861–873.

Aldous, A. E., and H. L. Shantz. 1924. Types of vegetation in the semiarid portion of the United States and their economic significance. Journal of Agricultural Research 28:99–128.

Aldrich, J. W. 1946. New subspecies of birds from western North America. Proceedings of the Biological Society of Washington 59:129–136.

Aldrich, J. W. 1963. Geographic orientation of American Tetraonidae. Journal of Wildlife Management 27:529–545.

Aldridge, C. L. 2000. Reproduction and habitat use by Sage Grouse (*Centrocercus urophasianus*) in a northern fringe population. M.S. thesis, University of Regina, Regina, SK, Canada.

Aldridge, C. L. 2001. Do Sage Grouse have a future in Canada? Population dynamics and management suggestions. Proceedings of the 6th Prairie Conservation and Endangered Species Conference, Winnipeg, MB, Canada.

Aldridge, C. L. 2005. Identifying habitats for persistence of Greater Sage-Grouse (*Centrocercus urophasianus*) in Alberta, Canada. Ph.D. dissertation, University of Alberta, Edmonton, AB, Canada.

Aldridge, C. L., and M. S. Boyce. 2007. Linking occurence and fitness to persistence: habitat-based approach for endangered Greater Sage-Grouse. Ecological Applications 17:508–526.

Aldridge, C. L., and R. M. Brigham. 2001. Nesting and reproductive activities of Greater Sage-Grouse in a declining northern fringe population. Condor 103:537–543.

Aldridge, C. L., and R. M. Brigham. 2002. Sage-grouse nesting and brood habitat use in southern Canada. Journal of Wildlife Management 66:433–444.

Aldridge, C. L., and R. M. Brigham. 2003. Distribution, abundance, and status of the Greater Sage-Grouse, *Centrocercus urophasianus,* in Canada. Canadian Field-Naturalist 117:25–34.

Aldridge, C. L., M. S. Boyce, and R. K. Baydack. 2004. Adaptive management of prairie grouse: how do we get there? Wildlife Society Bulletin 32:92–103.

Aldridge, C. L., S. E. Nielsen, H. L. Beyer, M. S. Boyce, J. W. Connelly, S. T. Knick, and M. A. Schroeder. 2008. Range-wide patterns of Greater

Sage-Grouse persistence. Diversity and Distributions 17:983–994.
Allen, D. L. 1962. Our wildlife legacy. Funk and Wagnall's, New York, NY.
Allen, T. F. H., and T. B. Starr. 1982. Hierarchy: perspectives for ecological complexity. University of Chicago Press, Chicago, IL.
Allen-Diaz, B., and J. W. Bartolome. 1998. Sagebrush-grass vegetation dynamics: comparing classical and state-transition models. Ecological Applications 8:795–804.
American Bird Conservancy. 2007. Top 20 most threatened bird habitats in the U.S.http://www.abcbirds.org/newsandreports/habitatreport.pdf (1 August 2009).
American Gas Association. 2005. Natural gas: balancing supply, demand and the environment. White paper delivered to the Natural Gas: Balancing Supply, Demand and the Environment Forum, Washington, DC.
Amstrup, S. C. 1980. A radio-collar for game birds. Journal of Wildlife Management 44:214–217.
Anderson, D. R., and K. P. Burnham. 1976. Population ecology of the Mallard: VI. The effect of exploitation on survival. USDI, Fish and Wildlife Service, Resource Publication 128, Washington, DC.
Anderson, J. E., and K. E. Holte. 1981. Vegetation development over 25 years without grazing on sagebrush-dominated rangeland in southeastern Idaho. Journal of Range Management 34:25–29.
Anderson, J. E., and R. S. Inouye. 2001. Landscape-scale changes in plant species abundance and biodiversity of a sagebrush steppe over 45 years. Ecological Monographs 71:531–556.
Anderson, J. E., and M. L. Shumar. 1986. Impacts of black-tailed jackrabbits at peak population densities on sagebrush steppe vegetation. Journal of Range Management 39:152–156.
Anderson, J. E., M. L. Shumar, N. L. Toft, and R. S. Nowak. 1987. Control of soil water balance by sagebrush and three perennial grasses in a cold-desert environment. Arid Soil Research and Rehabilitation 1:229–244.
Anderson, R. M., and R. M. May. 1978. Regulation and stability of host parasite population interactions: I. Regulatory processes. Journal of Animal Ecology 47:219–247.
Anderson, R. M., and R. M. May. 1979. Population biology of infectious diseases: part I. Nature 280:361–367.
Anderson, R. M., and R. M. May. 1981. The population dynamics of microparasites and their invertebrate hosts. Philosophical Transactions of the Royal Society of London, Series B, Biological Sciences 291:451–524.
Anderson, S. H. 2002. Managing our wildlife resources. Fourth edition. Prentice Hall, Upper Saddle River, NJ.
Anderson, V. J., and D. D. Briske. 1995. Herbivore-induced species replacement in grasslands: is it driven by herbivory tolerance or avoidance? Ecological Applications 5:1014–1024.
Andreadis, K. M., and D. P. Lettenmaier. 2006. Trends in 20th century drought over the continental United States. Geophysical Research Letters 33:L10403.
Andrén, H. 1992. Corvid density and nest predation in relation to forest fragmentation: a landscape perspective. Ecology 73:794–804.
Andrén, H. 1994. Effects of habitat fragmentation on birds and mammals in landscapes with different proportions of suitable habitat: a review. Oikos 71:355–366.
Andrén, H., and P. Angelstam. 1988. Elevated predation rates as an edge effect in habitat islands: experimental evidence. Ecology 69:544–547.
Andrén, H., P. Angelstam, E. Lindström, and P. Widén. 1985. Differences in predation pressure in relation to habitat fragmentation—an experiment. Oikos 45:273–277.
Angelstam, P. 1986. Population dynamics in tetraonids: the role of extrinsic factors. Proceedings of the International Ornithological Congress 19:2458–2477.
Angermeier, P. L., and J. R. Karr. 1986. Applying an index of biotic integrity based on stream-fish communities: considerations in sampling and interpretation. North American Journal of Fisheries Management 6:418–429.
Anonymous. 2008. Greater sage-grouse population trends: an analysis of lek count databases 1965–2007. Sage- and Columbian Sharp-tailed Grouse Technical Committee, Western Association of Fish and Wildlife Agencies, Cheyenne, WY.
Antevs, E. 1938. Rainfall and tree growth in the Great Basin. Carnegie Institute of Washington, American Geographical Society, Special Publication 21. New York, NY.
Antevs, E. 1948. Climatic changes and pre-white man. The Great Basin; with emphasis on glacial and postglacial times. University of Utah Bulletin 38:168–191.
Antos, J. A., B. McCune, and C. Bara. 1983. The effect of fire on an ungrazed western Montana grassland. American Midland Naturalist 110:354–364.
Apa, A. D. 1998. Habitat use and movements of sympatric Sage and Columbian Sharp-Tailed Grouse in southeastern Idaho. Ph.D. dissertation, University of Idaho, Moscow, ID.
Applegate, R. D. 2000. Use and misuse of prairie chicken lek surveys. Wildlife Society Bulletin 28:457–459.
Archer, S., and D. A. Pyke. 1991. Plant-animal interactions affecting plant establishment and persistence on revegetated rangeland. Journal of Range Management 44:558–565.
Archer, S., and F. E. Smeins. 1991. Ecosystem-level processes. Pp. 109–139 *in* R. K. Heitschmidt and J. W. Stuth (editors). Grazing management: an ecological perspective. Timber Press, Portland, OR.

Arno, S. F., and G. E. Gruell. 1983. Fire history at the forest-grassland ecotone in southwestern Montana. Journal of Range Management 36:332–336.

Arnold, G. W., and M. L. Dudzinski. 1978. Ethology of free-living domestic animals. Elsevier, Amsterdam, The Netherlands.

Aronson, J., C. Floret, E. Le Floc'h, C. Ovalle, and R. Pontanier. 1993. Restoration and rehabilitation of degraded ecosystems in arid and semiarid lands. I. A view from the south. Restoration Ecology 1:8–17.

Asay, K. H., W. H. Horton, K. B. Jensen, and A. J. Palazzo. 2001. Merits of native and introduced Triticeae grasses on semiarid rangelands. Canadian Journal of Plant Science 81:45–52.

Askins, R. A., F. Chávez-Ramírez, B. C. Dale, C. A. Haas, J. R. Herkert, F. L. Knopf, and P. D. Vickery. 2007. Conservation of grassland birds in North America: understanding ecological processes in different regions. Ornithological Monographs 64:1–46.

Atamian, M. T. 2007. Brood ecology and sex ratio of Greater Sage-Grouse in East-Central Nevada. M.S. Thesis, University of Nevada, Reno, NV.

Atkinson, C. T. 1999. Hemosporidiosis. Pp. 193–199 *in* M. Friend and J. C. Franson (editors). Field manual of wildlife diseases: general field procedures and diseases of birds. U.S. Geological Survey Biological Resources Division Information and Technical Report 1999-001. USGS, Madison, WI.

Austin, D. D., P. J. Urness, and R. A. Riggs. 1986. Vegetal change in the absence of livestock grazing, mountain brush zone, Utah. Journal of Range Management 39:514–517.

Autenrieth, R. E. 1981. Sage Grouse management in Idaho. Wildlife Bulletin 9. Idaho Department of Fish and Game, Boise, ID.

Autenrieth, R. E., W. Molini, and C. E. Braun. 1982. Sage Grouse management practices. Technical Bulletin 1. Western States Sage Grouse Committee, Twin Falls, ID.

Avian Power Line Interaction Committee. 2006. Suggested practices for avian protection on power lines: the state of the art in 2006. Edison Electric Institute, APLIC, and the California Energy Commission. Washington, DC, and Sacramento, CA.

Avise, J. C. 1989. A role for molecular genetics in the recognition and conservation of endangered species. Trends in Ecology and Evolution 4:279–281.

Avise, J. C. 1994. Molecular markers, natural history, and evolution. Chapman and Hall, New York, NY.

Avise, J. C., and W. S. Nelson. 1989. Molecular genetic relationship of the extinct Dusky Seaside Sparrow. Science 243:646–648.

Babero, B. B. 1953. Studies on the helminth fauna of Alaska. XVI. A survey of the helminth parasites of ptarmigan (*Lagopus* spp.). Journal of Parasitology 39:538–546.

Bachelet, D., R. P. Neilson, J. M. Lenihan, and R. J. Drapek. 2001. Climate change effects on vegetation distribution and carbon budget in the United States. Ecosystems 4:164–185.

Back, G. N., M. R. Barrington, and J. K. McAdoo. 1987. Sage Grouse use of snow burrows in northeastern Nevada. Wilson Bulletin 99:488–490.

Baguette, M., and H. Van Dyck. 2007. Landscape connectivity and animal behavior: functional grain as a key determinant for dispersal. Landscape Ecology 22:1117–1129.

Bai, Y., and J. T. Romo. 1996. Fringed sagebrush response to sward disturbances: seedling dynamics and plant growth. Journal of Range Management 49:228–233.

Bailey, S. F., D. A. Eliason, and B. L. Hoffman. 1965. Flight and dispersal patterns of the mosquito *Culex tarsalis* Coquillett in the Sacramento Valley of California. Hilgardia 37:73–113.

Baillie, J. E. M., C. Hilton-Taylor, and S. N. Stuart (editors). 2004. 2004 IUCN red list of threatened species: a global species assessment. IUCN, Gland, Switzerland, and Cambridge, UK.

Baines, D. 1996. The implications of grazing and predator management on the habitats and breeding success of Black Grouse *Tetrao tetrix*. Journal of Applied Ecology 33:54–62.

Baines, D., R. Moss, and D. Dugan. 2004. Capercaillie breeding success in relation to forest habitat and predator abundance. Journal of Applied Ecology 41:59–71.

Baisan, C. H., and T. W. Swetnam. 1997. Interactions of fire regimes and land use in the Central Rio Grande Valley. USDA Forest Service Research Paper RM-RP-330. USDA Forest Service, Rocky Mountain Research Station, Ft. Collins, CO.

Baker, A. J., and H. D. Marshall. 1997. Mitochondrial control region sequences as tools for understanding evolution. Pp. 51–82 *in* D. P. Mindell (editor). Avian molecular evolution and systematics. Academic Press, San Diego, CA.

Baker, W. L. 1989. Effect of scale and spatial heterogeneity on fire-interval distributions. Canadian Journal of Forest Research 19:700–706.

Baker, W. L. 1993. Spatially heterogenous multi-scale response of landscapes to fire suppression. Oikos 66:66–71.

Baker, W. L. 2002. Indians and fire in the Rocky Mountains: the wilderness hypothesis renewed. Pp. 41–76 *in* T. R. Vale (editor). Fire, native peoples and the natural landscape. Island Press, Washington, DC.

Baker, W. L. 2006a. Fire and restoration of sagebrush ecosystems. Wildlife Society Bulletin 34:177–185.

Baker, W. L. 2006b. Fire history in ponderosa pine landscapes of Grand Canyon National Park: is it reliable enough for management and restoration? International Journal of Wildland Fire 15:433–437.

Baker, W. L., 2009. Fire ecology in Rocky Mountain landscapes. Island Press, Washington, DC.

Baker, W. L., and D. Ehle. 2001. Uncertainty in surface-fire history: the case of ponderosa pine forests in the western United States. Canadian Journal of Forest Research 31:1205–1226.

Baker, W. L., and D. J. Shinneman. 2004. Fire and restoration of piñon-juniper woodlands in the western United States: a review. Forest Ecology and Management 189:1–21.

Bakker, J. P., and R. van Diggelen. 2006. Restoration of dry grasslands and heathlands. Pp. 95–110 *in* J. van Andel and J. Aronson (editors). Restoration ecology: the new frontier. Blackwell Publishing, Malden, MA.

Balmford, A., J. L. Moore, T. Brooks, N. Burgess, L. A. Hansen, P. Williams, and C. Rahbek. 2001. Conservation conflicts across Africa. Science 291:2616–2619.

Balmori, A. 2005. Possible effects of electromagnetic fields from phone masts on a population of White Stork (*Ciconia ciconia*). Electromagnetic Biology and Medicine 24:109–119.

Balmori, A. 2006. The incidence of electromagnetic pollution on the amphibian decline: is this an important piece of the puzzle? Toxicological and Environmental Chemistry 88:287–299.

Balmori, A., and Ö. Hallberg. 2007. The urban decline of the House Sparrow (*Passer domesticus*): a possible link with electromagnetic radiation. Electromagnetic Biology and Medicine 26:141–151.

Barbarika, A., S. Hyberg, J. Williams, and J. Agapoff. 2004. Conservation Reserve Program overview. CRP: planting for the future. United States Department of Agriculture, Farm Services Agency, Washington, DC.

Barker, J. R., and C. M. McKell. 1983. Habitat differences between basin and Wyoming big sagebrush in contiguous populations. Journal of Range Management 36:450–456.

Barnett, J. K., and J. A. Crawford. 1994. Pre-laying nutrition of Sage Grouse hens in Oregon. Journal of Range Management 47:114–118.

Barney, M. A., and N. C. Frischknecht. 1974. Vegetation changes following fire in the pinyon-juniper type of west-central Utah. Journal of Range Management 27:91–96.

Barrowclough, G. F., and R. J. Gutiérrez. 1990. Genetic variation and differentiation in the Spotted Owl (*Strix occidentalis*). Auk 107:737–744.

Barrowclough, G. F., J. G. Groth, L. A. Mertz, and R. J. Gutiérrez. 2004. Phylogeographic structure, gene flow and species status in Blue Grouse (*Dendragapus obscurus*). Molecular Ecology 13:1911–1922.

Barrowclough, G. F., J. G. Groth, L. A. Mertz, and R. J. Gutiérrez. 2005. Genetic structure, introgression, and a narrow hybrid zone between northern and California Spotted Owls (*Strix occidentalis*). Molecular Ecology 14:1109–1120.

Barton, D. C., and A. L. Holmes. 2007. Off-highway vehicle trail impacts on breeding songbirds in northeastern California. Journal of Wildlife Management 71:1617–1620.

Bartuszevige, A. M., and B. A. Endress. 2008. Do ungulates facilitate native and exotic plant spread? Seed dispersal by cattle, elk and deer in northeastern Oregon. Journal of Arid Environments 72:904–913.

Battazzo, A. M. 2007. Winter survival and habitat use by female Greater Sage-Grouse (*Centrocercus urophasianus*) in south Phillips County, Montana. M.S. thesis, University of Montana, Missoula, MT.

Batterson, W. M., and W. B. Morse. 1948. Oregon Sage Grouse. Oregon Fauna Series Number 1. Oregon State Game Commission, Portland, OR.

Bauer, J. 2006. Fire history and stand structure of a central Nevada pinyon-juniper woodland. M.S. thesis, University of Nevada, Reno, NV.

Baxter, R. J., K. D. Bunnell, J. T. Flinders, and D. L. Mitchell. 2007. Impacts of predation on Greater Sage-Grouse in Strawberry Valley, Utah. Transactions of the 72nd North American Wildlife and Natural Resources Conference 72:258–269.

Baxter, R. J., J. T. Flinders, and D. L. Mitchell. 2008. Survival, movements, and reproduction of translocated Greater Sage-Grouse in Strawberry Valley, Utah. Journal of Wildlife Management 72:179–186.

Bazzaz, F. A., S. L. Bassow, G. M. Berntson, and S. C. Thomas. 1996. Elevated CO_2 and terrestrial vegetation: implications for and beyond the global carbon budget. Pp. 43–76 *in* B. Walker and W. Steffen (editors). Global change and terrestrial ecosystems. Cambridge University Press, Cambridge, UK.

Bean, M. J., and M. J. Rowland. 1997. The evolution of national wildlife law. Third edition. Environmental Defense Fund and World Wildlife Fund. Praeger Publishers, Westport, CT.

Bean, R. W. 1941. Life history studies of the Sage Grouse (*Centrocercus urophasianus*) in Clark County, Idaho. M. S. thesis, Utah State Agricultural College, Logan, UT.

Beck, J. L., and D. L. Mitchell. 2000. Influences of livestock grazing on Sage Grouse habitat. Wildlife Society Bulletin 28:993–1002.

Beck, J. L., J. W. Connelly, and K. P. Reese. 2009. Recovery of Greater Sage-Grouse habitat features in Wyoming big sagebrush following prescribed fire. Restoration Ecology 17:393–403.

Beck, J. L., D. L. Mitchell, and B. D. Maxfield. 2003. Changes in the distribution and status of Sage-Grouse in Utah. Western North American Naturalist 63:203–214.

Beck, J. L., K. P. Reese, J. W. Connelly, and M. B. Lucia. 2006. Movements and survival of juvenile Greater Sage-Grouse in southeastern Idaho. Wildlife Society Bulletin 34:1070–1078.

Beck, T. D. I. 1975. Attributes of a wintering population of Sage Grouse, North Park, Colorado. M.S. thesis, Colorado State University, Ft. Collins, CO.

Beck, T. D. I. 1977. Sage Grouse flock characteristics and habitat selection in winter. Journal of Wildlife Management 41:18–26.

Beck, T. D. I., and C. E. Braun. 1978. Weights of Colorado Sage Grouse. Condor 80:241–243.

Beck, T. D. I., and C. E. Braun. 1980. The strutting ground count: variation, traditionalism, management needs. Proceedings of the Western Association of Fish and Wildlife Agencies 60:558–566.

Beck, T. D. I., R. B. Gill, and C. E. Braun. 1975. Sex and age determination of Sage Grouse from wing characteristics. Game Information Leaflet 49 (revised). Colorado Department of Natural Resources, Denver, CO.

Beckstead, J., and C. K. Augspurger. 2004. An experimental test of resistance to cheatgrass invasion: limiting resources at different life stages. Biological Invasions 6:417–432.

Beehler, B. M., and M. S. Foster. 1988. Hotshots, hotspots, and female preference in the organization of lek mating systems. American Naturalist 131:203–219.

Beehler, J. W., and M. S. Mulla. 1995. Effects of organic enrichment on temporal distribution and abundance of culicine egg rafts. Journal of the American Mosquito Control Association 11:167–171.

Beever, E. A. 2003. Management implications of the ecology of free-roaming horses in semiarid ecosystems of the western United States. Wildlife Society Bulletin 31:887–895.

Beever, E. A., and P. F. Brussard. 2000. Examining ecological consequences of feral horse grazing using exclosures. Western North American Naturalist 60:236–254.

Beever, E. A., and P. F. Brussard. 2004. Community- and landscape-level responses of reptiles and small mammals to feral-horse grazing in the Great Basin. Journal of Arid Environments 59:271–297.

Beever, E. A., and J. E. Herrick. 2006. Effects of feral horses in Great Basin landscapes on soils and ants: direct and indirect mechanisms. Journal of Arid Environments 66:96–112.

Beever, E. A., M. Huso, and D. A. Pyke. 2006. Multiscale responses of soil stability and invasive plants to removal of non-native grazers from an arid conservation reserve. Diversity and Distributions 12:258–268.

Beever, E. A., R. J. Tausch, and P. F. Brussard. 2003. Characterizing grazing disturbance in semiarid ecosystems across broad spatial scales, using multiple indices. Ecological Applications 13:119–136.

Beever, E. A., R. J. Tausch, and P. F. Brussard. 2008. Multi-scale responses of vegetation to removal of horse grazing from Great Basin (USA) mountain ranges. Plant Ecology 197:163–184.

Bekedam, S. 2004. Establishment tolerance of six native sagebrush steppe species to imazapic (Plateau®) herbicide: implications for restoration and recovery. M.S. thesis, Oregon State University, Corvallis, OR.

Belcher, J. W., and S. D. Wilson. 1989. Leafy spurge and the species composition of a mixed-grass prairie. Journal of Range Management 42:172–175.

Bell, J. A., C. M. Brewer, N. J. Mickelson, G. W. Garman, and J. A. Vaughan. 2006. West Nile virus epizootiology, central Red River valley, North Dakota and Minnesota, 2002–2005. Emerging Infectious Diseases 12:1245–1247.

Belnap, J., and O. Lange (editors). 2001. Biological soil crusts: structure, function, and management. Springer-Verlag, New York, NY.

Belnap, J., S. L. Phillips, and T. Troxler. 2006. Soil lichen and moss cover and species richness can be highly dynamic: the effects of invasion by the annual exotic grass *Bromus tectorum*, precipitation, and temperature on biological soil crusts in SE Utah. Applied Soil Ecology 32:63–76.

Belsky, A. J., and D. M. Blumenthal. 1997. Effects of livestock grazing on stand dynamics and soils in upland forests of the Interior West. Conservation Biology 11:315–327.

Bender, E. A., T. J. Case, and M. E. Gilpin. 1984. Perturbation experiments in community ecology: theory and practice. Ecology 65:1–13.

Benedict, N. G., S. J. Oyler-McCance, S. E. Taylor, C. E. Braun, and T. W. Quinn. 2003. Evaluation of the eastern (*Centrocercus urophasianus urophasianus*) and western (*Centrocercus urophasianus phaios*) subspecies of sage-grouse using mitochondrial control-region sequence data. Conservation Genetics 4:301–310.

Bennett, D. K. 1992. Origin and distribution of living breeds of the domestic horse. Pp. 41–61 *in* J.W. Evans (editor). Horse breeding and management. World Animal Science, C7. Elsevier, Amsterdam, The Netherlands.

Bennett, D. K., and R. S. Hoffmann. 1999. *Equus caballus*. Mammalian Species 628:1–14.

Bennett, G. G., M. A. Peirce, and R. W. Ashford. 1993. Avian hemotozoa: mortality and pathogenicity. Journal of Natural History 27:993–1001.

Bergerud, A. T. 1985. The additive effect of hunting mortality on the natural mortality rates of grouse. Pp. 345–366 *in* S. L. Beasom and S. F. Roberson (editors). Game harvest management. Caesar Kleberg Wildlife Research Institute, Kingsville, TX.

Bergerud, A. T. 1988a. Mating systems in grouse. Pp. 439–472 *in* A. T. Bergerud and M. W. Gratson (editors). Adaptive strategies and population ecology of northern grouse. University of Minnesota Press, Minneapolis, MN.

Bergerud, A. T. 1988b. Population ecology of North American grouse. Pp. 578–685 *in* A. T. Bergerud and M. W. Gratson (editors). Adaptive strategies and population ecology of northern grouse. University of Minnesota, Minneapolis, MN.

Bergquist, E., P. Evangelista, T. J. Stohlgren, and N. Alley. 2007. Invasive species and coal bed methane development in the Powder River Basin, Wyoming. Environmental Monitoring and Assessment 128:381–394.

Bergstrom, R. C. 1982. Nematode parasites of game birds. Pp. 222–224 *in* E. T. Thorne, N. Kingston, W. R. Jolley, and R. C. Bergstrom (editors). Diseases of wildlife in Wyoming. Second edition. Wyoming Game and Fish Department, Cheyenne, WY.

Bermant, D., and S. Spackeen. 1997. Native species in the commercial seed industry. Pp. 26–29 *in* N. L. Shaw and B.A. Roundy (compilers). Using seeds of native species on rangelands. USDA Forest Service General Technical Report INT-GTR-372. USDA Forest Service, Intermountain Research Station, Ogden, UT.

Bernard-Laurent, A., and Y. Magnani. 1994. Statut, évolution et facteurs limitant les populations de Gelinotte des Bois (*Bonasa bonasia*) en France: synthèse bibliographique. Gibier Faune Sauvage 11(Part 1):5–40.

Berry, J. D., and R. L. Eng. 1985. Interseasonal movements and fidelity to seasonal use areas by female Sage Grouse. Journal of Wildlife Management 49:237–240.

Berryman, A. A. 2002. Population: a central concept for ecology? Oikos 97:439–442.

Bertelsen, M. F., R.-A. Ølberg, G. J. Crawshaw, A. Dibernardo, L. R. Lindsay, M. Drebot, and I. K. Barker. 2004. West Nile virus infection in the eastern Loggerhead Shrike (*Lanius ludovicianus migrans*): pathology, epidemiology, and immunization. Journal of Wildlife Diseases 40:538–542.

Bestelmeyer, B. T., J. R. Brown, K. M. Havstad, R. Alexander, G. Chavez, and J. E. Herrick. 2003. Development and use of state-and-transition models for rangelands. Journal of Range Management 56:114–126.

Bestelmeyer, B. T., A. J. Tugel, G. L. Peacock, Jr., D. G. Robinett, P. L. Shaver, J. R. Brown, J. E. Herrick, H. Sanchez, and K. M. Havstad. 2009. State-and-transition models for heterogeneous landscapes: a strategy for development and application. Rangeland Ecology and Management 62:1–15.

Betancourt, J. 1987. Paleoecology of pinyon juniper woodlands: summary. Pp. 129–139 *in* R. L. Everett (compiler). Proceedings—Pinyon-Juniper Conference. USDA Forest Service General Technical Bulletin Report INT-215. USDA Forest Service, Intermountain Research Station, Ogden, UT.

Beveroth, T. A., M. P. Ward, R. L. Lampman, A. M. Ringia, and R. J. Novak. 2006. Changes in seroprevalence of West Nile virus across Illinois in free-ranging birds from 2001 through 2004. American Journal of Tropical Medicine and Hygiene 74:174–179.

Beyers, J. L. 2004. Postfire seeding for erosion control: effectiveness and impacts on native plant communities. Conservation Biology 18:947–956.

Billings, W. D. 1990. *Bromus tectorum*, a biotic cause of ecosystem impoverishment in the Great Basin. Pp. 301–322 *in* G. M. Woodwell (editor). The earth in transition: patterns and processes of biotic impoverishment. Cambridge University Press, Cambridge, UK.

Blaisdell, J. P. 1953. Ecological effects of planned burning of sagebrush-grass range on the upper Snake River Plains. USDA Technical Bulletin 1075. United States Government Printing Office, Washington, DC.

Blaisdell, J. P., and J. W. Thomas. 1977. Prospectus for research related to management of wild and free-roaming horses and burros. Pp. 49–52 *in* J. L. Artz (forum chairman). Proceedings, National Wild Horse Forum. Cooperative Extension Service, University of Nevada, Reno, NV.

Blank, R. R., R. Allen, and J. A. Young. 1994. Extractable anions following wildfire in a sagebrush grass community. Soil Science Society of America Journal 58:564–570.

Blumenthal, D. M., N. R. Jordan, and M. P. Russelle. 2003. Soil carbon addition controls weeds and facilitates prairie restoration. Ecological Applications 13:605–615.

Blus, L. J., C. S. Staley, C. J. Henny, G. W. Pendleton, T. H. Craig, E. H. Craig, and D. K. Halford. 1989. Effects of organophosphorus insecticides on Sage Grouse in southeastern Idaho. Journal of Wildlife Management 53:1139–1146.

Boarman, W. I., and B. Heinrich. 1999. *Corvus corax*: Common Raven. Pp. 1–32 *in* A. Poole and F. Gill (editors). The birds of North America, No. 476. The Academy of Natural Sciences, Philadelphia, PA, and The American Ornithologists' Union, Washington, DC.

Bock, C. E., J. H. Bock, and H. M. Smith. 1993. Proposal for a system of federal livestock exclosures on public rangelands in the western United States. Conservation Biology 7:731–733.

Bohls, R. L., E. W. Collisson, S. L. Gross, N. J. Silvy, and D. N. Phalen. 2006. Experimental infection of Attwater's/Greater Prairie Chicken hybrids with the reticuloendotheliosis virus. Avian Diseases 50:613–619.

Bolton, Jr., H., J. L. Smith, and S. L. Link. 1993. Soil microbial biomass and activity of a disturbed and undisturbed shrub-steppe ecosystem. Soil Biology and Biochemistry 25:545–552.

Bolton, Jr., H., J. L. Smith, and R. E. Widung. 1990. Nitrogen mineralization potentials of shrub-steppe soils with different disturbance histories. Soil Science Society of America Journal 54:887–891.

Boltz, M. 1994. Factors influencing postfire sagebrush regeneration in south-central Idaho. Pp. 281–290 *in* S. B. Monsen and S. G. Kitchen (compilers). Ecology and management of annual rangelands. USDA Forest Service General Technical Report INT-GTR-313.

USDA Forest Service, Intermountain Research Station, Ogden, UT.

Booth, D. T., and K. P. Vogel. 2006. Revegetation priorities. Rangelands 28(5):24–30.

Booth, M. S., M. M. Caldwell, and J. M. Stark. 2003. Overlapping resource use in three Great Basin species: implications for community invasibility and vegetation dynamics. Journal of Ecology 91:36–48.

Bork, E. W., N. E. West, and J. W. Walker. 1998. Cover components on long-term seasonal sheep grazing treatments in three-tip sagebrush steppe. Journal of Range Management 51:293–300.

Boughton, R. V. 1937. Endoparasitic infestations in grouse, their pathogenicity and correlation with meteoro-topographical conditions. Agricultural Experiment Station Technical Bulletin 121. University of Minnesota, Minneapolis, MN.

Bouzat, J. L., and K. Johnson. 2003. Genetic structure among closely spaced leks in a peripheral population of Lesser Prairie-Chickens. Molecular Ecology 13:499–505.

Bouzat, J. L., H. H. Cheng, H. A. Lewin, R. L. Westemeier, J. D. Brawn, and K. N. Paige. 1998. Genetic evaluation of a demographic bottleneck in the Greater Prairie Chicken. Conservation Biology 12:836–843.

Bowen, B. W., A. B. Meylen, and J. C. Avise. 1991. Evolutionary distinctiveness of the endangered Kemp's Ridley sea turtle. Nature 352:709–711.

Box, T. W. 1990. Rangelands. Pp. 101–120 *in* R. N. Sampson and D. Hair (editors). Natural resources in the 21st century. Island Press, Covelo, CA.

Boyce, M. S. 1981. Robust canonical correlation of Sage Grouse habitat. Pp. 152–159 *in* D. Capen (editor). The use of multivariate statistics in studies of wildlife habitat. USDA Forest Service General Technical Report RM-87. USDA Forest Service, Rocky Mountain Research Station, Ft. Collins, CO.

Boyce, M. S. 1990. The red queen visits Sage Grouse leks. American Zoologist 30:263–270.

Boyce, M. S. 1992. Population viability analysis. Annual Review of Ecology and Systematics 23:481–506.

Boyce, M. S., J. S. Mao, E. H. Merrill, D. Fortin, M. G. Turner, J. Fryxell, and P. Turchin. 2003. Scale and heterogeneity in habitat selection by elk in Yellowstone National Park. Ecoscience 10:421–431.

Boyko, A. R., R. M. Gibson, and J. R. Lucas. 2004. How predation risk affects the temporal dynamics of avian leks: Greater Sage-Grouse versus Golden Eagles. American Naturalist 163:154–165.

Boyles, J. S. 1986. Managing America's wild horses and burros. Journal of Equine Veterinary Science 6:261–265.

Bradbury, J. W., and R. M. Gibson. 1983. Leks and mate choice. Pp. 109–138 *in* P. Bateson (editor). Mate choice. Cambridge University Press, Cambridge, MA.

Bradbury, J. W., S. L. Vehrencamp, and R. M. Gibson. 1989. Dispersion of displaying male Sage Grouse. I. Patterns of temporal variation. Behavioral Ecology and Sociobiology 24:1–14.

Bradley, B. A., and J. F. Mustard. 2005. Identifying land cover variability distinct from land cover change: cheatgrass in the Great Basin. Remote Sensing of Environment 94:204–213.

Bradley, B. A., and J. F. Mustard. 2006. Characterizing the landscape dynamics of an invasive plant and risk of invasion using remote sensing. Ecological Applications 16:1132–1147.

Bradshaw, A. D. 1983. The reconstruction of ecosystems. Journal of Applied Ecology 20:1–17.

Braun, C. E. 1984. Attributes of a hunted Sage Grouse population in Colorado, U.S.A. International Grouse Symposium 3:148–162.

Braun, C. E. 1986. Changes in Sage Grouse lek counts with advent of surface coal mining. Proceedings: Issues and Technology in the Management of Impacted Western Wildlife, Thorne Ecological Institute 2:227–231.

Braun, C. E. 1987. Current issues in Sage Grouse management. Proceedings of the Western Association of State Fish and Wildlife Agencies 67:134–144.

Braun, C. E. 1995. Distribution and status of Sage Grouse in Colorado. Prairie Naturalist 27:1–9.

Braun, C. E. 1998. Sage Grouse declines in western North America: what are the problems? Proceedings of the Western Association of State Fish and Wildlife Agencies 78:139–156.

Braun, C. E. 2005. Multi-species benefits of the proposed North American Sage-Grouse management plan. Pp. 1162–1164 *in* Proceedings of the Third International Partners in Flight Symposium. General Technical Report PSW-GTR-191. USDA Forest Service, Pacific Southwest Research Station, Albany, CA.

Braun, C. E., and T. D. I. Beck. 1985. Effects of changes in hunting regulations on Sage Grouse harvest and populations. Pp. 335–343 *in* S. L. Beasom and S. F. Roberson (editors). Game harvest management. Caesar Kleberg Wildlife Research Institute, Kingsville, TX.

Braun, C. E., and T. D. I. Beck. 1996. Effects of research on Sage Grouse management. Transactions of the North American Wildlife and Natural Resources Conference 61:429–436.

Braun, C. E., and W. B. Willers. 1967. The helminth and protozoan parasites of North American grouse (family: Tetraonidae): a checklist. Avian Diseases 11:170–187.

Braun, C. E., M. F. Baker, R. L. Eng, J. S. Gashwiler, and M. H. Schroeder. 1976. Conservation committee report on effects of alteration of sagebrush communities on the associated avifauna. Wilson Bulletin 88:165–171.

Braun, C. E., T. E. Britt, and R. O. Wallestad. 1977. Guidelines for maintenance of Sage Grouse habitats. Wildlife Society Bulletin 5:99–106.

Braun, C. E., J. W. Connelly, and M. A. Schroeder. 2005. Seasonal habitat requirements for Sage-grouse: spring, summer, fall, and winter. Pp. 38–42 *in* N. L. Shaw, M. Pellant, and S. B. Monson (compilers). Sage-grouse Habitat Restoration Symposium proceedings. USDA Forest Service Research Report RMRS-P-38, USDA Forest Service, Rocky Mountain Research Station, Ft. Collins, CO.

Braun, C. E., K. Martin, and L. A. Robb. 1993. White-tailed Ptarmigan (*Lagopus leucurus*). Pp. 1–22 *in* A. Poole and F. Gill (editors). The birds of North America, No. 68. The Academy of Natural Sciences, Philadelphia, PA, and The American Ornithologists' Union, Washington, DC.

Braun, C. E., O. O. Oedekoven, and C. L. Aldridge. 2002. Oil and gas development in western North America: effects on sagebrush steppe avifauna with particular emphasis on Sage Grouse. Transactions of the North American Wildlife and Natural Resources Conference 67:337–349.

Brennan, L. A. 1994. Introductory remarks: do we need a national upland gamebird plan? North American Wildlife and Natural Resources Conference 59:411–414.

Brennan, L. A., and W. P. Kuvlesky, Jr. 2005. North American grassland birds: an unfolding conservation crisis? Journal of Wildlife Management 69:1–13.

Bright, R. C., and O. K. Davis. 1982. Quaternary paleoecology of the Idaho National Engineering Laboratory, Snake River Plain, Idaho. American Midland Naturalist 108:21–33.

Briske, D. D., and J. H. Richards. 1995. Plant responses to defoliation: a morphological, physiological and demographic evaluation. Pp. 635–710 *in* D. J. Bedunah and R. E. Sosebee (editors). Wildland plants: physiological ecology and developmental morphology. Society for Range Management, Denver, CO.

Briske, D. D., S. D. Fuhlendorf, and F. E. Smeins. 2003. Vegetation dynamics on rangelands: a critique of the current paradigms. Journal of Applied Ecology 40:601–614.

Briske, D. D., S. D. Fuhlendorf, and F. E. Smeins. 2005. State-and-transition models, thresholds, and rangeland health: a synthesis of ecological concepts and perspectives. Rangeland Ecology and Management 58:1–10.

Briske, D. D., S. D. Fuhlendorf, and F. E. Smeins. 2006. A united framework for assessment and application of ecological thresholds. Rangeland Ecology and Management 59:225–236.

Britton, C. M., and R. G. Clark. 1985. Effects of fire on sagebrush and bitterbrush. Pp. 101–109 *in* K. Sanders and J. Durham (editors). The sagebrush ecosystem: a symposium. Utah State University, College of Natural Resources, Logan, UT.

Britton, C. M., and F. A. Sneva. 1981. Effects of tebuthiuron on western juniper. Journal of Range Management 34:30–32.

Broms, K. 2007. Small game population reconstruction: model development and applications. M.S. thesis, University of Washington, Seattle, WA.

Brooks, M. L., C. M. d'Antonio, D. M. Richardson, J. B. Grace, J. E. Keeley, J. M. DiTomaso, R. J. Hobbs, M. Pellant, and D. Pyke. 2004a. Effects of invasive alien plants on fire regimes. BioScience 54:677–688.

Brooks, M. L., J. R. Matchett, and K. H. Berry. 2006. Effects of livestock watering sites on alien and native plants in the Mojave Desert, USA. Journal of Arid Environments 67:125–147.

Brooks, T. M., M. I. Bakarr, T. Boucher, G. A. B. Da Fonseca, C. Hilton-Taylor, J. M. Hoekstra, T. Moritz, S. Olivieri, J. Parrish, R. L. Pressey, A. S. L. Rodrigues, W. Sechrest, A. Stattersfield, W. Strahm, and S. N. Stuart. 2004b. Coverage provided by the global protected-area system: is it enough? BioScience 54:1081–1091.

Browers, H. W. 1983. Dispersal and harvest of Sage Grouse utilizing the Test Reactor Area on the Idaho National Engineering Laboratory. M.S. thesis, South Dakota State University, Brookings, SD.

Browers, H. W., and L. D. Flake. 1985. Breakup and sibling dispersal of two Sage Grouse broods. Prairie Naturalist 17:249–249.

Brown, D. G., K. M. Johnson, T. R. Loveland, and D. M. Theobald. 2005. Rural land-use trends in the conterminous United States, 1950–2000. Ecological Applications 15:1851–1863.

Brown, J. H., G. C. Stevens, and D. M. Kaufman. 1996. The geographic range: size, shape, boundaries, and internal structure. Annual Review of Ecology and Systematics 27:597–623.

Brown, J. R., T. Svejcar, M. Brunson, J. Dobrowolski, E. Fredrickson, U. Krueter, K. Launchbaugh, J. Southworth, and T. Thurow. 2002. Range sites: are they the appropriate spatial unit for measuring and managing rangelands? Rangelands 24(6):7–12.

Brussard, P. F., D. D. Murphy, and C. R. Tracy. 1994. Cattle and conservation biology—another view. Conservation Biology 8:919–921.

Brust, R. A. 1991. Environmental regulation of autogeny in *Culex tarsalis* (Diptera Culicidae) from Manitoba, Canada. Journal of Medical Entomology 28:847–853.

Buckland, S. T., Burnham, K. P., and N. H. Augustin. 1997. Model selection: an integral part of inference. Biometrics 53:603–618.

Buechling, A., and W. L. Baker. 2004. A fire history from tree rings in a high-elevation forest of Rocky Mountain National Park. Canadian Journal of Forest Research 34:1259–1273.

Bump, G., R. W. Darrow, F. C. Edminster, and W. F. Crissey. 1947. The Ruffed Grouse: life history, propagation, management. New York State Conservation Department, Albany, NY.

Bunn, A. G., D. L. Urban, and T. H. Keitt. 2000. Landscape connectivity: a conservation application of graph theory. Journal of Environmental Management 59:265–278.

Bunnell, K. D. 2000. Ecological factors limiting Sage Grouse recovery and expansion in Strawberry Valley, Utah. M.S. thesis, Brigham Young University, Provo, UT.

Bunning, M. L., P. E. Fox, R. A. Bowen, N. Komar, G.-J. J. Chang, T. J. Speaker, M. R. Stephens, N. M. Nemeth, N. A. Panella, S. A. Langevin, P. Gordy, M. Teehee, P. R. Bright, and M. J. Turell. 2007. DNA vaccination of the American Crow (*Corvus brachyrhynchos*) provides partial protection against lethal challenge with West Nile virus. Avian Diseases 51:573–577.

Bunting, S. C. 1984. Prescribed burning of live standing western juniper and post-burning succession. Pp. 69–73 *in* T. E. Bedell (compiler). Western Juniper Short Course. Oregon State University Extension Service, Bend, OR.

Bunting, S. C., J. L. Kingery, M. A. Hemstrom, M. A. Schroeder, R. A. Gravenmier, and W. J. Hann. 2002. Altered rangeland ecosystems in the Interior Columbia Basin. USDA Forest Service General Technical Report PNW-GTR-553. Pacific Northwest Research Station, Portland, OR.

Bunting, S. C., J. L. Kingery, and M. A. Schroeder. 2003. Assessing the restoration potential of altered rangeland ecosystems in the Interior Columbia Basin. Ecological Restoration 21:77–86.

Burkhardt, J. W., and E. W. Tisdale. 1976. Causes of juniper invasion in southwestern Idaho. Ecology 57:472–484.

Burnham, K. P., and D. R. Anderson. 2002. Model selection and multimodel inference: a practical information-theorectic approach. Second edition. Springer-Verlag, New York, NY.

Bush, G. W. 2001. Actions to expedite energy-related projects. Executive Order 13212. Federal Register 66:28357.

Bush, K. 2009. Genetic diversity and paternity analysis of endangered Canadian Greater Sage-Grouse (*Centrocercus urophasianus*). Ph.D. dissertation, University of Alberta, Edmonton, AB, Canada.

Buss, I. O., and E. S. Dziedic. 1955. Relation of cultivation to the disappearance of the Columbian Sharp-tailed Grouse from southeastern Washington. Condor 57:185–187.

Buth, J. L., R. A. Brust, and R. A. Ellis. 1990. Development time, oviposition activity and onset of diapause in *Culex tarsalis, Culex restuans* and *Culiseta inornata* in southern Manitoba. Journal of the American Mosquito Control Association 6:55–63.

Butler, B. W., and T. D. Reynolds. 1997. Wildfire case study: Butte City fire, southeastern Idaho, July 1, 1994. USDA Forest Service General Technical Report INT-GTR-351. USDA Forest Service, Rocky Mountain Research Station, Ft. Collins, CO.

Byrne, M. W. 2002. Habitat use by female Greater Sage Grouse in relation to fire at Hart Mountain National Antelope Refuge, Oregon. M.S. thesis, Oregon State University, Corvallis, OR.

Caffrey, C., S. C. R. Smith, and T. J. Weston. 2005. West Nile virus devastates an American Crow population. Condor 107:128–132.

Caffrey, C., T. J. Weston, and S. C. R. Smith. 2003. High mortality among marked crows subsequent to the arrival of West Nile virus. Wildlife Society Bulletin 31:870–872.

Caizergues, A., A. Berndard-Laurent, J.-F. Brenot, L. Ellison, and J. Y. Rasplus. 2003a. Population genetic structure of Rock Ptarmigan *Lagopus mutus* in Northern and Western Europe. Molecular Ecology 12:2267–2274.

Caizergues, A., O. Rätti, P. Helle, L. Rotelli, L. Ellison, and J.-Y. Rasplus. 2003b. Population genetic structure of male Black Grouse (*Tetrao tetrix* L.) in fragmented vs. continuous landscapes. Molecular Ecology 12:2297–2305.

Calabrese, J. M., and W. F. Fagan. 2004. A comparison-shopper's guide to connectivity metrics. Frontiers in Ecology and Environment 2:529–536.

Call, C. A., and B. A. Roundy. 1991. Perspectives and processes in revegetation of arid and semiarid rangelands. Journal of Range Management 44:543–549.

Campbell, G. L., A. A. Marfin, R. S. Lanciotti, and D. J. Gubler. 2002. West Nile virus. Lancet: Infectious Diseases 2:519–529.

Canada Gazette. 2003. Chapter 29. Species at Risk Act. Queen's Printer for Canada, Ottawa, Canada.

Canadian Sage-Grouse Recovery Team. 2001. Canadian Sage-Grouse recovery strategy. Alberta Sustainable Resource Development, Edmonton, Canada. http://www.srd.gov.ab.ca/fishwildlife/speciesatrisk/pdf/SageGrousePlan.pdf (18 March 2009).

Cannon, R. W., and F. L. Knopf. 1981. Lesser Prairie Chicken densities on shinnery oak and sand sagebrush rangelands in Oklahoma. Journal of Wildlife Management 45:521–524.

Cantwell, M. D., and R. T. T. Forman. 1993. Landscape graphs: ecological modeling with graph theory to determine configurations common to diverse landscapes. Landscape Ecology 8:239–255.

Carhart, A. H. 1943. Disease studies in Moffat County, summer of 1943. Sage Grouse survey. Special Memo, Colorado Game and Fish Department, Denver, CO.

Caro, T. M., and G. O'Doherty. 1999. On the use of surrogate species in conservation biology. Conservation Biology 13:805–814.

Carothers, S. W., M. E. Stitt, and R. R. Johnson. 1976. Feral asses on public lands: an analysis of biotic impact, legal considerations and management alternatives. Transactions of the North American Wildlife and Natural Resources Conference 41:396–405.

Carpenter, F. R. 1981. Establishing management under the Taylor Grazing Act. Rangelands 3(3):105–115.

Carroll, C., R. F. Noss, P. C. Paquet, and N. H. Schumaker. 2003. Use of population viability analysis and reserve selection algorithms in regional conservation plans. Ecological Applications 13:1773–1789.

Carroll, J. P. 1993. Gray Partridge (*Perdix perdix*). *In* A. Poole and F. Gill (editors.) The birds of North America, No. 58. The Academy of Natural Sciences, Philadelphia, PA, and The American Ornithologists' Union, Washington, DC.

Caughley, G. 1974. Bias in aerial survey. Journal of Wildlife Management 38:921–933.

Caughley, G., and A. R. E. Sinclair. 1994. Wildlife ecology and management. Blackwell Scientific Publications, Cambridge, MA.

Cawley, R. M., and J. Freemuth. 2007. Science, politics and the federal lands. Pp. 69–88 *in* Z. Smith and J. Freemuth (editors). Environmental politics and policy in the West. Second edition. University of Colorado Press, Boulder, CO.

Cayan, D. R., S. A. Kammerdiener, M. D. Dettinger, J. M. Caprio, and D. H. Peterson. 2001. Changes in the onset of spring in the western United States. Bulletin of the American Meterological Society 82:399–415.

Center for Science, Economics, and Environment. 2002. The state of the nation's ecosystems: measuring the lands, waters, and living resources of the United States. Cambridge University Press, Cambridge, UK.

Centers for Disease Control and Prevention. 2007. West Nile virus. http://www.cdc.gov/ncidod/dvbid/westnile/index.htm (13 April 2009).

Centers for Disease Control and Prevention. 2008. West Nile virus. http://www.cdc.gov/ncidod/dvbid/westnile/ (15 October 2008).

Chambers, J. C., B. A. Roundy, R. R. Blank, S. E. Meyer, and A. Whittaker. 2007. What makes Great Basin sagebrush ecosystems invasible by *Bromus tectorum*? Ecological Monographs 77:117–145.

Chambers, J. C., E. W. Schupp, and S. B. Vander Wall. 1999a. Seed dispersal and seedling establishment of piñon and juniper species within the piñon-juniper woodland. Pp. 29–34 *in* S. B. Monson, R. Stevens, R. J. Tausch, R. Miller, and S. Goodrich (editors). Ecology and management of pinyon-juniper communities within the interior West. USDA Forest Service General Technical Report RMRS-P-9. USDA Forest Service, Rocky Mountain Research Station, Ogden, UT.

Chambers, J. C., S. B. Vander Wall, and E. W. Schupp. 1999b. Seed and seedling ecology of piñon and juniper species in the pygmy woodlands of western North America. Botanical Review 65:1–38.

Chaplin, S. J., R. A. Gerrard, H. M. Watson, L. L. Master, and S. R. Flack. 2000. The geography of imperilment: targeting conservation toward critical biodiversity areas. Pp. 159–199 *in* B. A. Stein, L. S. Kutner, and J. S. Adams (editors). Precious heritage: the status of biodiversity in the United States. Oxford University Press, New York, NY.

Charley J. L., and N. E. West. 1977. Micropatterns of nitrogen mineralization activity in soils of some shrub dominated semi-desert ecosystems of Utah. Soil Biogeochemistry 9:357–365.

Chavez-Ramirez, F., and R. D. Slack. 1994. Effects of avian foraging and post-foraging behavior on seed dispersal patterns of Ashe juniper. Oikos 71:40–46.

Chi, R. Y. 2004. Greater Sage-Grouse on Parker Mountain, Utah. M.S. thesis, Utah State University, Logan, UT.

Chomel, B. B., A. Belotto, and F. X. Meslin. 2007. Wildlife, exotic pets, and emerging zoonoses. Emerging Infectious Diseases 13:6–11.

Christiansen, E. M., and H. B. Johnson. 1964. Presettlement vegetation and vegetational change in three valleys in central Utah. Brigham Young University Science Bulletin 4:1–16.

Christensen, G. C. 1996. Chukar (*Alectoris chukar*). *In* A. Poole and F. Gill (editors.) The birds of North America, No. 258. The Academy of Natural Sciences, Philadelphia, PA, and The American Ornithologists' Union, Washington, DC.

Christensen, J. H., B. Hewitson, A. Busuioc, A. Chen, X. Gao, I. Held, R. Jones, R. K. Kolli, W.-T. Kwon, R. Laprise, V. Magaña Rueda, L. Mearns, C. G. Menéndez, J. Räisänen, A. Rinke, A. Sarr, and P. Whetton. 2007. Regional climate projections. Pp. 848–940 *in* S. Solomon, D. Qin, M. Manning, Z. Chen, M. Marquis, K. B. Avery, M. Tignor, and H. L. Miller (editors). Climate change 2007: the physical science basis. Contribution of Working Group I to the fourth assessment report of the Intergovernmental Panel on Climate Change. Cambridge University Press, Cambridge, UK.

Christiansen, T. 2008. Hunting and sage-grouse: a technical review of harvest management on a species of concern in Wyoming. Wyoming Game and Fish Department, Cheyenne, WY.

Church, K. E., and W. F. Porter. 1990. Winter and spring habitat use by Gray Partridge in New York. Journal of Wildlife Management 54:653–657.

Clark, L., J. Hall, R. McLean, M. Dunbar, K. Klenk, R. Bowen, and C. A. Smeraski. 2006. Susceptibility of Greater Sage-Grouse to experimental infection with West Nile virus. Journal of Wildlife Diseases 42:14–22.

Clawson, M., and B. Held. 1957. The federal lands: their use and management. John Hopkins Press, Baltimore, MD.

Clayton, D. H. 1990. Mate choice in experimentally parasitized Rock Doves: lousy males lose. American Zoologist 30:251–262.

Clinton, W. J. 1999. Invasive species. Executive Order 13112. Federal Register 64:6183–6186.

Cluff, G. J., J. A. Young, and R. A. Evans. 1983. Edaphic factors influencing the control of Wyoming big sagebrush and seedling establishment of crested wheatgrass. Journal of Range Management 36:786–792.

Coates, P. S. 2007. Greater Sage-Grouse (*Centrocercus urophasianus*) nest predation and incubation behavior. Ph.D. dissertation, Idaho State University, Pocatello, ID.

Coates, P. S., and D. J. Delehanty. 2008. Effects of environmental factors on incubation patterns of Greater Sage-Grouse. Condor 110:627–638.

Coates, P. S., J. W. Connelly, and D. J. Delehanty. 2008. Predators of Greater Sage-Grouse nests identified by video monitoring. Journal of Field Ornithology 79:421–428.

Coggins, K. A. 1998. Relationship between habitat changes and productivity of Sage Grouse at Hart Mountain National Antelope Refuge, Oregon. M.S. thesis, Oregon State University, Corvallis, OR.

Cohen, R. A. 1999. An introduction to PROC LOESS for local regression. http://support.sas.com/rnd/app/papers/loesssugi.pdf (21 August 2009).

Cole, R. A. 1999. Trichomoniasis. Pp. 201–206 *in* M. Friend and J. C. Franson (editors). Field manual of wildlife diseases: general field procedures and diseases of birds. U.S. Geological Survey Biological Resources Division Information and Technical Report 1999-001. USGS, Madison, WI.

Cole, R. A., and M. Friend. 1999. Miscellaneous parasitic diseases. Pp. 249–258 *in* M. Friend and J. C. Franson (editors). Field manual of wildlife diseases: general field procedures and diseases of birds. U.S. Geological Survey Biological Resources Division Information and Technical Report 1999-001. USGS, Madison, WI.

Colket, E. C. 2003. Long-term vegetation dynamics and post-fire establishment patterns of sagebrush steppe. M.S. thesis, University of Idaho, Moscow, ID.

Collinge, S. K., and R. T. T. Forman. 1998. A conceptual model of land conversion processes: predictions and evidence from a microlandscape experiment with grassland insects. Oikos 82:66–84.

Colorado Division of Wildlife. 2008. Colorado Greater Sage-Grouse conservation plan. http://wildlife.state.co.us/WildlifeSpecies/SpeciesOfConcern/Birds/GreaterSagegrouseConservationPlan.htm (31 August 2008).

Colorado Greater Sage-Grouse Steering Committee. 2008. Colorado Greater Sage-Grouse conservation plan. Colorado Division of Wildlife, Denver, CO. http://wildlife.state.co.us/wildlifespecies/speciesofconcern/birds/greatersagegrouseconsplan2.htm (18 March 2009).

Coltman, D. W., D.R. Bancroft, A. Robertson, J. A. Smith, T. H. Clutton-Brock, and J. M. Pemberton. 1999. Male reproductive success in a promiscuous mammal: behavioural estimates compared with genetic paternity. Molecular Ecology 8:1199–1209.

Comer, P., J. Kagan, M. Heiner, and C. Tobalske. 2002. Current distribution of sagebrush and associated vegetation in the western United States (excluding NM and AZ). USGS Forest and Rangeland Ecosystems Science Center, Boise, ID. http://sagemap.wr.usgs.gov/ (4 June 2009).

Commons, M. L., R. K. Baydack, and C. E. Braun. 1999. Sage Grouse response to pinyon-juniper management. Pp. 238–239 *in* S. B. Monson and R. Stevens (compilers). Ecology and management of pinyon-juniper communities within the interior West. USDA Forest Service RMRS-P-9. USDA Forest Service, Rocky Mountain Research Station, Ft. Collins, CO.

Connelly, Jr., J. W. 1982. An ecological study of Sage Grouse in southeastern Idaho. Ph. D. dissertation, Washington State University, Pullman, WA.

Connelly, J. W., and C. E. Braun. 1997. Long-term changes in Sage Grouse *Centrocercus urophasianus* populations in western North America. Wildlife Biology 3:229–234.

Connelly, J. W., and L. A. Doughty. 1989. Sage Grouse use of wildlife water developments in southeastern Idaho. Pp. 167–172 *in* G. K. Tsukamoto and S. J. Stiver (editors). Wildlife water development: a proceedings of the wildlife water development symposium. Nevada Department of Fish and Game, Reno, NV.

Connelly, J. W., and O. D. Markham. 1983. Movements and radionuclide concentrations of Sage Grouse in southeastern Idaho. Journal of Wildlife Management 47:169–177.

Connelly, J. W., and M. A. Schroeder. 2007. Historical and current approaches to monitoring Greater Sage-Grouse. Pp. 3–9 *in* K. P. Reese and R. T. Bowyer (editors). Monitoring populations of sage-grouse. University of Idaho College of Natural Resources Experiment Station Bulletin 88. University of Idaho, Moscow, ID.

Connelly, J. W., A. D. Apa, R. B. Smith, and K. P. Reese. 2000a. Effects of predation and hunting on adult Sage Grouse *Centrocercus urophasianus* in Idaho. Wildlife Biology 6:227–232.

Connelly, J. W., W. J. Arthur, and O. D. Markham. 1981. Sage Grouse leks on recently disturbed sites. Journal of Range Management 34:153–154.

Connelly, J. W., H. W. Browers, and R. J. Gates. 1988. Seasonal movements of Sage Grouse in southeastern Idaho. Journal of Wildlife Management 52:116–122.

Connelly, J. W., R. A. Fischer, A. D. Apa, K. P. Reese, and W. L. Wakkinen. 1993. Renesting of Sage Grouse in southeastern Idaho. Condor 95:1041–1043.

Connelly, J. W., J. Gammonley, and J. M. Peek. 2005. Harvest management. Pp. 658–690 *in* C. E. Braun (editor). Techniques for wildlife investigation and management. The Wildlife Society, Bethesda, MD.

Connelly, J. W., M. W. Gratson, and K. P. Reese. 1998. Sharp-tailed Grouse (*Tympanuchus phasianellus*). *In* A. Poole and F. Gill (editors). The birds of North America, No. 354. The Academy of Natural Sciences, Philadelphia, PA, and The American Ornithologists' Union, Washington, DC.

Connelly, J. W., S. T. Knick, M. A. Schroeder, and S. J. Stiver. 2004. Conservation assessment of Greater Sage-Grouse and sagebrush habitats.

Western Association of Fish and Wildlife Agencies, Cheyenne, WY.

Connelly, J. W., K. P. Reese, R. A. Fischer, and W. L. Wakkinen. 2000b. Response of a Sage Grouse breeding population to fire in southeastern Idaho. Wildlife Society Bulletin 28:90–96.

Connelly, J. W., K. P. Reese, E. O. Garton, and M. L. Commons-Kemner. 2003a. Response of Greater Sage-Grouse *Centrocercus urophasianus* populations to different levels of exploitation in Idaho, USA. Wildlife Biology 9:335–340.

Connelly, J. W., K. P. Reese, and M. A. Schroeder. 2003b. Monitoring of Greater Sage-Grouse habitats and populations. University of Idaho College of Natural Resources Experiment Station Bulletin Bulletin 80. University of Idaho, Moscow, ID.

Connelly, J. W., K. P. Reese, W. L. Wakkinen, M. D. Robertson, and R. A. Fischer. 1994. Sage Grouse ecology. Study I: Sage Grouse response to a controlled burn. Pitmann-Robertson Project W-160-R-21. Idaho Department of Fish and Game, Boise, ID.

Connelly, J. W., M. A. Schroeder, A. R. Sands, and C. E. Braun. 2000c. Guidelines to manage Sage Grouse populations and their habitats. Wildlife Society Bulletin 28:967–985.

Connelly, J. W., W. L. Wakkinen, A. D. Apa, and K. P. Reese. 1991. Sage Grouse use of nest sites in southeastern Idaho. Journal of Wildlife Management 55:521–524.

Cook, J. G., T. J. Hershey, and L. L. Irwin. 1994. Vegetative response to burning on Wyoming mountain shrub big game ranges. Journal of Range Management 47:296–302.

Cooper, S. V., P. Lesica, and G. M. Kudray. 2007. Post-fire recovery of Wyoming big sagebrush shrub-steppe in central and southeast Montana. The Montana Natural Heritage Program, United States Bureau of Land Management, Montana State Office, Helena, MT.

Copeland, H. E., K. E. Doherty, D. E. Naugle, A. Pocewicz, and J. M. Kiesecker. 2009. Mapping oil and gas development potential in the US Intermountain West and estimating impacts to species. PLoS ONE 4(10):e7400.

Copeland, H. E., J. M. Ward, and J. M. Kiesecker. 2007. Assessing tradeoffs in biodiversity, vulnerability and cost when prioritizing conservation sites. Journal of Conservation Planning 3:1–16.

Cordell, H. K., C. J. Betz, G. Green, and M. Owens. 2005. Off-highway vehicle recreation in the United States, regions and states: a national report from the National Survey on Recreation and the Environment (NSRE). USDA Forest Service Southern Research Station, Athens, GA. http://www.fs.fed.us/recreation/programs/ohv/OHV_final_report.pdf (15 August 2009).

Costigan, G. P. 1912. Cases on the American law of mining. Bobbs-Merrill, Indianapolis, IN.

Côté, I. M., and W. J. Sutherland. 1997. The effectiveness of removing predators to protect bird populations. Conservation Biology 11:395–405.

Cottam, W. P., and G. Stewart. 1940. Plant succession as a result of grazing and of meadow desiccation by erosion since settlement in 1862. Journal of Forestry 38:613–626.

Coughenour, M. B. 1999. Ecosystem modeling of the Pryor Mountain Wild Horse Range. Final report to USGS-BRD and USDI-BLM, Natural Resource Ecology Laboratory, Colorado State University, Ft. Collins, CO.

Cox, R. D., and V. J. Anderson. 2004. Increasing native diversity of cheatgrass-dominated rangeland through assisted succession. Journal of Range Management 57:203–210.

Crane, K. K., M. A. Smith, and D. Reynolds. 1997. Habitat selection patterns of feral horses in southcentral Wyoming. Journal of Range Management 50:374–380.

Crawford, J. A. 1982. Factors affecting Sage Grouse harvest in Oregon. Wildlife Society Bulletin 10:374–377.

Crawford, J. A., and E. G. Bolen. 1976. Effects of lek disturbances on Lesser Prairie-Chickens. Southwestern Naturalist 21:238–240.

Crawford, J. A., and R. S. Lutz. 1985. Sage Grouse population trends in Oregon, 1941–1983. Murrelet 66:69–74.

Crawford, J. A., R. A. Olson, N. E. West, J. C. Mosley, M. A. Schroeder, T. D. Witson, R. F. Miller, M. A. Gregg, and C. S. Boyd. 2004. Ecology and management of sage-grouse and sage-grouse habitat. Journal of Range Management 57:2–19.

Crompton, B., and D. Mitchell. 2005. The sage-grouse of Emma Park—survival, production, and habitat use in relation to coalbed methane development. Completion report. Southeastern Region, Utah Division of Wildlife Resources, Price, UT.

Cronin, M. A. 2006. A proposal to eliminate redundant terminology for intra-species groups. Wildlife Society Bulletin 34:237–241.

Cronquist, A., A. H. Holmgren, N. H. Holmgren, and J. L. Reveal. 1994. Intermountain flora: vascular plants of the Intermountain West, USA. Vol. 5. Asterales. Hafner Publishing Company, New York, NY.

Crooks, K. R., and M. Sanjayan. 2006. Connectivity conservation: maintaining connections for nature. Pp. 1–19 *in* K. R. Crooks and M. Sanjayan (editors). Connectivity conservation. Cambridge University Press, New York, NY.

Crosbie, S. P., W. D. Koenig, W. K. Reisen, V. L. Kramer, L. Marcus, R. Carney, E. Pandolfino, G. M. Bolen, L. R. Crosbie, D. A. Bell, and H. B. Ernest. 2008. Early impact of West Nile virus on the Yellow-billed Magpie (*Pica nuttalli*). Auk 125:542–550.

Cumming, S. G. 2001. A parametric model of the fire-size distribution. Canadian Journal of Forest Research 31:1297–1303.

Currans, S. P., S. G. Kitchen, and S. M. Lambert. 1997. Ensuring identity and quality of native seeds. Pp. 17–20 *in* N. L. Shaw and B. A. Roundy (compilers). Using seeds of native species on rangelands. USDA Forest Service General Technical Report INT-GTR-372. USDA Forest Service, Intermountain Research Station, Ogden, UT.

Curtin, C. G. 2002. Livestock grazing, rest, and restoration in arid landscapes. Conservation Biology 16:840–842.

Cushman, S. A., and K. McGarigal. 2004. Hierarchical analysis of forest bird species–environment relationships in the Oregon Coast Range. Ecological Applications 14:1090–1105.

Cushman, S. A., K. McGarigal, and M. C. Neel. 2008. Parsimony in landscape metrics: strength, universality, and consistency. Ecological Indicators 8:691–703.

Czech, B., P. R. Krausman, and P. K. Devers. 2000. Economic associations among causes of species endangerment in the United States. BioScience 50:593–601.

Dahl, B. E., and E. W. Tisdale. 1975. Environmental factors related to medusahead distribution. Journal of Range Management 28:463–468.

Dahlgren, D. K. 2006. Greater Sage-Grouse reproductive ecology and response to experimental management of mountain big sagebrush on Parker Mountain, Utah. M.S. thesis, Utah State University, Logan, UT.

Dahlgren, D. K., R. Chi, and T. A. Messmer. 2006. Greater Sage-Grouse response to sagebrush management in Utah. Wildlife Society Bulletin 34:975–985.

Dalebout, M. L., J. G. Mead, C. S. Baker, A. N. Baker, and A. L. Van Helden. 2002. A new species of beaked whale *Mesoplodon perrini* sp. n. (Cetacea: Ziphiidae) discovered through phylogenetic analyses of mitochondrial DNA sequences. Marine Mammal Science 18:577–608.

Dalke, P. D., D. B. Pyrah, D. C. Stanton, J. E. Crawford, and E. F. Schlatterer. 1960. Seasonal movements and breeding behavior of Sage Grouse in Idaho. Transactions of the North American Wildlife and Natural Resources Conference 25:396–406.

Dalke, P. D., D. B. Pyrah, D. C. Stanton, J. E. Crawford, and E. F. Schlatterer. 1963. Ecology, productivity, and management of Sage Grouse in Idaho. Journal of Wildlife Management 27:811–841.

Daly, C., R. P. Neilson, and D. L. Phillips. 1994. A statistical-topographic model for mapping climatological precipitation over mountainous terrain. Journal of Applied Meteorology 33:140–158.

Dalzal, C. R. 2004. Post-fire establishment of vegetation communities following reseeding on southern Idaho's Snake River Plain. M.S. thesis, Boise State University, Boise, ID.

Dando, L. M., and K. J. Hansen. 1990. Tree invasion into a range environment near Butte, Montana. Great Plains-Rocky Mountain Geographer 18:65–76.

d'Antonio, C. M. 2000. Fire, plant invasions, and global changes. Pp. 65–93 *in* H. A. Mooney and R. J. Hobbs (editors). Invasive species in a changing world. Island Press, Washington, DC.

d'Antonio, C. M., and P. M. Vitousek. 1992. Biological invasions by exotic grasses, the grass/fire cycle, and global change. Annual Review of Ecology and Systematics 23:63–87.

Dargan, L.M., R. J. Keller, H. R. Shepherd, and R. N. Randall. 1942. Sage Grouse survey, Colorado, Volume. 4: survey of 1941–42, food studies, parasite relations, habitat requirements. Including preliminary data on sharp-tail grouse in Moffatt and Routt Counties. Pittman-Robertson project, Colorado 4-R, season of 1941–1942. Colorado Game and Fish Commission, Denver, CO.

Dasmann, R. F. 1964. Wildlife biology. John Wiley and Sons, New York, NY.

Daszak, P., A. A. Cunningham, and A. D. Hyatt. 2000. Emerging infectious diseases of wildlife—threats to biodiversity and human health. Science 287:443–449.

Daubenmire, R. 1970. Steppe vegetation of Washington. Washington Agricultural Experiment Station Technical Bulletin 62. Washington State University, Pullman, WA.

Daubenmire, R. 1975a. An analysis of structural and functional characters along a steppe-forest catena. Ecology 53:419–424.

Daubenmire, R. F. 1975b. Ecology of *Artemisia tridentata* subsp. *tridentata* in the state of Washington. Northwest Science 49:24–35.

Daubenmire, R. F. 1978. Plant geography with special reference to North America. Academic Press, New York, NY.

Davenport, D. W., D. D. Breshears, B. P. Wilcox, and C. D. Allen. 1998. Viewpoint: sustainability of piñon-juniper ecosystems—a unifying perspective of soil erosion thresholds. Journal of Range Management 51:231–240.

Davidson, J. W. 1956. A crossroads of freedom: the 1912 campaign speeches of Woodrow Wilson. Yale University Press, New Haven, CT.

Davies, K. W., and R. L. Sheley. 2007. A conceptual framework for preventing the spatial dispersal of invasive plants. Weed Science 55:178–184.

Davies, K. W., J. D. Bates, D. D. Johnson, and A. M. Nafus. 2009. Influence of mowing *Artemisia tridentata* ssp. *wyomingensis* on winter habitat for wildlife. Environmental Management. http://www.springerlink.com/content/5j40l7t2024612p1/ (20 December 2009).

Davies, K. W., J. D. Bates, and R. F. Miller. 2007a. Environmental and vegetation relationships of *Artemisia tridentata* spp. *wyomingensis* alliance. Journal of Arid Environments 70:478–494.

Davies, K. W., J. D. Bates, and R. F. Miller. 2007b. The influence of *Artemisia tridentata* spp. *wyomingensis* on microsite and herbaceous vegetation heterogeneity. Journal of Arid Environments 69:441–457.

Davies, K. W., J. D. Bates, and R. F. Miller. 2007c. Short-term effects of burning Wyoming big sagebrush steppe in southeast Oregon. Journal of Range Ecology and Management 60:515–522.

Davies, K. W., R. L. Sheley, and J. D. Bates. 2008. Does fall prescribed burning *Artemisia tridentata* steppe promote invasion or resistance to invasion after a recovery period? Journal of Arid Environments 72:1076–1085.

Davis, C. T., D. W. C. Beasley, H. Guzman, M. Siirin, R. E. Parsons, R. B. Tesh, and A. D. T. Barrett. 2004. Emergence of attenuated West Nile virus variants in Texas, 2003. Virology 330:342–350.

Davis, C. T., G. D. Ebel, R. S. Lanciotti, A. C. Brault, H. Guzman, M. Siirin, A. Lambert, R. E. Parsons, D. W. C. Beasley, R. J. Novak, D. Elizondo-Quirogag, E. N. Green, D. S. Young, L. M. Stark, M. A. Drebot, H. Artsob, R. B. Tesh, L. D. Kramer, and A. D. T. Barrett. 2005. Phylogenetic analysis of North American West Nile virus isolates, 2001–2004: evidence for the emergence of a dominant genotype. Virology 342:252–265.

Davis, J. O. 1982. Bits and pieces: the last 35 ka years in the Lahontan area. Society of American Archeology Papers 2:53–75.

Davison, J. C., and E. G. Smith. 2007. Imazapic provides 2-year control of weedy annuals in a seeded Great Basin fuelbreak. Native Plants Journal 8:91–95.

Dawson, J. R., W. B. Stone, G. D. Ebel, D. S. Young, D. S. Galinski, J. P. Pensabene, M. A. Franke, M. Eidson, and L. D. Kramer. 2008. Crow deaths caused by West Nile virus during winter. Emerging Infectious Diseases 13:1912–1914.

Deacon, J. E., A. E. Williams, C. D. Williams, and J. E. Williams. 2007. Fueling population growth in Las Vegas: how large-scale groundwater withdrawal could burn regional biodiversity. BioScience 57:688–698.

Deem, S. L., W. B. Karesh, and W. Weisman. 2001. Putting theory in practice: wildlife health in conservation. Conservation Biology 15:1224–1233.

Deibert, P. A. 1995. Effect of parasites on Sage Grouse (*Centrocercus urophasianus*) mate selection. Ph.D. dissertation, University of Wyoming, Laramie, WY.

DeLap, J. H., and R. L. Knight. 2004. Wildlife response to human food. Natural Areas Journal 24:112–118.

Delcourt, H. R., P. A. Delcourt, and T. Webb, III. 1983. Dynamic plant ecology: the spectrum of vegetational change in space and time. Quaternary Science Reviews 1:153–175.

DeLong, A. K., J. A. Crawford, and D. C. DeLong, Jr. 1995. Relationships between vegetational structure and predation of artificial Sage Grouse nests. Journal of Wildlife Management 59:88–92.

Dennis, B., and M. L. Taper. 1994. Density dependence in time series observations of natural populations: estimation and testing. Ecological Monographs 64:205–224.

Dennis, B., P. L. Munholland, and J. M. Scott. 1991. Estimation of growth and extinction parameters for endangered species. Ecological Monographs 61:115–143.

Dennis, B., J. M. Ponciano, S. R. Lele, and M. Taper. 2006. Estimating density dependence, process noise and observation error. Ecological Monographs 76:323–341.

Derner, J. D., and X. B. Wu. 2001. Light distribution in mesic grasslands: spatial patterns and temporal dynamics. Applied Vegetation Science 3:189–196.

DeYoung, R. W., and R. L. Honeycutt. 2005. The molecular toolbox: genetic techniques in wildlife ecology and management. Journal of Wildlife Management 69:1362–1384.

Diefenbach, D. R., D. W. Brauning, and J. A. Mattice. 2003. Variability in grassland bird counts related to observer difference and species detection rates. Auk 120:1168–1179.

Diersing, V. E., R. B. Shaw, and D. J. Tazik. 1992. United States Army land condition–trend analysis (LCTA) program. Environmental Management 16:405–414.

Dimcheff, D. E., S. V. Drovetski, M. Krishnan, and D. P. Mindell. 2000. Cospeciation and horizontal transmission of avian sarcoma and leukosis virus *gag* genes in galliform birds. Journal of Virology 74:3984–3995.

DiMenna, M. A., R. Bueno, Jr., R. R. Parmenter, D. E. Norris, J. M. Sheyka, J. L. Molina, E. M. Labeau, E. S. Hatton, and G. E. Glass. 2006. Emergence of West Nile virus in mosquito (Diptera: Culicidae) communities of the New Mexico Rio Grande Valley. Journal of Medical Entomology 43:594–599.

Dingman, J. D. 1980. Characteristics of Sage Grouse leks, North Park, Colorado. M.S. thesis, University of Denver, Denver, CO.

DiTomaso, J. M., M. L. Brooks, E. B. Allen, R. Minnich, P. M. Rice, and G. B. Kyser. 2006. Control of invasive weeds with prescribed burning. Weed Technology 20:535–548.

Dixon, K. R., M. A. Horner, S. R. Anderson, W. D. Henriques, D. Durham, and R. J. Kendall. 1996. Northern Bobwhite habitat use and survival on a South Carolina plantation during winter. Wildlife Society Bulletin 24:627–635.

Doak, D. F., P. C. Marino, and P. M. Kareiva. 1992. Spatial scale mediates the influence of habitat fragmentation on dispersal success: implications for conservation. Theoretical Population Biology 41:315–336.

Dobkin, D. S. 1995. Management and conservation of Sage Grouse, denominative species for the ecological health of shrubsteppe ecosystems. USDI Bureau of Land Management, Portland, OR.

Dobkin, D. S., and J. D. Sauder. 2004. Shrubsteppe landscapes in jeopardy. Distributions, abundances, and the uncertain future of birds and small mammals in the Intermountain West. High Desert Ecological Research Institute, Bend, OR.

Dobler, F. C., J. Eby, C. Perry, S. Richardson, and M. Vander Haegen. 1996. Status of Washington's shrub-steppe ecosystem: extent, ownership, and wildlife/vegetation relationships. Research Report. Washington Department of Fish and Wildlife, Olympia, WA.

Dobson, A., and J. Foufopoulos. 2001. Emerging infectious pathogens of wildlife. Philosophical Transactions of the Royal Society of London Series B Biological Sciences 356:1001–1012.

Doelger, M. J., and J. A. Barlow, Jr. 1989. Wyoming and the central Rocky Mountain area natural gas supply: a United States perspective. Pp. 21–29 *in* J. Eisert (editor). Gas Resources of Wyoming. 40th Field Conference guidebook. Wyoming Geological Association, Casper, WY.

Doescher, P. S., R. F. Miller, and A. H. Winward. 1984. Soil chemical patterns under eastern Oregon plant communities dominated by big sagebrush. Soil Science Society of America Journal 48:659–663.

Doggett, M., C. Daly, J. Smith, W. Gibson, G. Taylor, G. Johnson, and P. Pasteris. 2004. High-resolution 1971–2000 mean monthly temperature maps for the western United States. *In* Proceedings of the 14th American Meteorological Society Conference on Applied Climatology. Paper 4.3, CD-ROM. Seattle, WA.

Doherty, K. E. 2008. Sage-grouse and energy development: integrating science with conservation planning to reduce impacts. Ph.D. dissertation, University of Montana, Missoula, MT.

Doherty, K. E., D. E. Naugle, B. L. Walker, and J. M. Graham. 2008. Greater Sage-Grouse winter habitat selection and energy development. Journal of Wildlife Management 72:187–195.

Doherty, M. K. 2007. Mosquito populations in the Powder River Basin, Wyoming: a comparison of natural, agricultural and effluent coal-bed natural gas aquatic habitats. M.S. thesis, Montana State University, Bozeman, MT.

Dohm, D. J., M. L. O'Guinn, and M. J. Turell. 2002. Effect of environmental temperature on the ability of *Culex pipiens* (Diptera: Culicidae) to transmit West Nile virus. Journal of Medical Entomology 39:221–225.

Dombeck, M. P., C. A. Wood, and J. E. Williams. 2003. From conquest to conservation: our public lands legacy. Island Press, Washington, DC.

Donahue, D. L. 1999. The western range revisited: removing livestock from public lands to conserve native biodiversity. University of Oklahoma Press, Norman, OK.

Drew, M. L. 2007. Retroviral infections. Pp. 216–235 *in* N. J. Thomas, D. B. Hunter, and C. T. Atkinson (editors). Infectious diseases of wild birds. Blackwell Publishing, Ames, IA.

Drew, M. L., W. L. Wigle, D. L. Graham, C. P. Griffin, N. J. Silvy, A. M. Fadly, and R. L. Witter. 1998. Reticuloendotheliosis in captive Greater and Attwater's Prairie-Chickens. Journal of Wildlife Disease 34:783–791.

Drovetski, S. V. 2002. Molecular phylogeny of grouse: individual and combined performance of W-linked, autosomal, and mitochondrial loci. Systematic Biology 5:1930–945.

Drut, M. S. 1994. Status of Sage Grouse with emphasis on populations in Oregon and Washington. Audubon Society of Portland, Portland, OR.

Drut, M. S., J. A. Crawford, and M. A. Gregg. 1994a. Brood habitat use by Sage Grouse in Oregon. Great Basin Naturalist 54:170–176.

Drut, M. S., W. H. Pyle, and J. A. Crawford. 1994b. Diets and food selection of Sage Grouse chicks in Oregon. Journal of Range Management 47:90–93.

Dubose, R. T. 1965. Pox in the Sage Grouse. Bulletin of the Wildlife Disease Association 1:6.

Dunbar, M. R., and M. A. Gregg. 2004. Infectious disease survey of Sage-Grouse in Nevada and Oregon. Abstract only *in* C. K. Baer (editor). American Association of Zoo Veterinarians, American Association of Wildlife Veterinarians, Wildlife Disease Association Joint Conference. San Diego, CA.

Dunbar, M. R., S. Tornquist, and M. R. Giordano. 2003. Blood parasites in Sage-Grouse from Nevada and Oregon. Journal of Wildlife Diseases 39:203–208.

Duncan, P., T. J. Foose, I. J. Gordon, C. G. Gakahu, and M. Lloyd. 1990. Comparative nutrient extraction from forages by grazing bovids and equids: a test of the nutritional model of equid/bovid competition and coexistence. Oecologia 84:411–418.

Dunkle, S. W. 1977. Swainson's Hawks on the Laramie Plains, Wyoming. Auk 94:65–71.

Dunn, P. O., and C. E. Braun. 1985. Natal dispersal and lek fidelity of Sage Grouse. Auk 102:621–627.

Dunn, P. O., and C. E. Braun. 1986a. Late summer–spring movements of juvenile Sage Grouse. Wilson Bulletin 98:83–92.

Dunn, P. O., and C. E. Braun. 1986b. Summer habitat use by adult female and juvenile Sage Grouse. Journal of Wildlife Management 50:228–235.

Dunne, R. 1999. Common difficulties encountered in collecting native seed. Pp. 28–29 *in* L. K. Holzworth and R.W. Brown (compilers). Revegetation with native species. USDA Forest Service RMRS-P-8. USDA Forest Service, Rocky Mountain Research Station, Ogden, UT.

Dyksterhuis, E. J. 1949. Condition and management of range land based on quantitative ecology. Journal of Range Management 2:104–115.

Eberhardt, L. L., and J. M. Thomas. 1991. Designing environmental field studies. Ecological Monographs 61:53–73.

Eberhardt, L. L., A. K. Majorowicz, and J. A. Wilcox. 1982. Apparent rates of increase for two feral horse herds. Journal of Wildlife Management 46:367–374.

Eckert, Jr., R. E., 1957. Vegetation-soil relationships in some *Artemisia* types in northern Harney and Lake Counties, Oregon. Ph.D. dissertation, Oregon State University, Corvallis, OR.

Eddington, D. B. 2006. Effects of cheatgrass control on Wyoming big sagebrush in southeastern Utah. M.S. thesis, Brigham Young University, Provo, UT.

Edelmann, F. B., M. J. Ulliman, M. J. Wisdom, K. P. Reese, and J. W. Connelly. 1998. Assessing habitat quality using population fitness parameters: a remote sensing/GIS- based habitat-explicit population model for Sage Grouse (*Centrocercus urophasianus*). Technical Report 25. Idaho Forest, Wildlife, and Range Experiment Station, Moscow, ID.

Edminster, F. C. 1954. American game birds of field and forest. Charles Scribner's Sons, New York, NY.

Efron, B. 1975. The efficiency of logistic regression compared to normal discriminant analysis. Journal of the American Statistical Association 70:892–898.

Efron, B., and R. J. Tibshirani. 1998. An introduction to the bootstrap. Monographs on Statistics and Applied Probability No. 57. Chapman and Hall, New York, NY.

Efron, B., T. Hastie, and R. Tibshirani. 2004. Least angle regression. Annals of Statistics 32:407–451.

Eichhorn, L. C., and C. R. Watts. 1984. Plant succession of burns in the River Breaks of central Montana. Proceedings of the Montana Academy of Sciences 43:21–34.

Eisenhart, K. S. 2004. Historic range of variability and stand development in piñon-juniper woodlands of western Colorado. Ph.D. dissertation, University of Colorado, Boulder, CO.

Eiswerth, M. E., and J. S. Shonkwiler. 2006. Examining post-wildfire reseeding on arid rangeland: a multivariate tobit modelling approach. Ecological Modelling 192:286–298.

Elkie, P. C., and R. S. Rempel. 2001. Detecting scales of pattern in boreal forest landscapes. Forest Ecology and Management 147:253–261.

Ellis, K. L. 1984. Behavior of lekking sage grouse in response to a perched Golden Eagle. Western Birds 15:37–38.

Ellis, K. L. 1985. Distribution and habitat selection of breeding male Sage Grouse in northeastern Utah. M.S. thesis, Brigham Young University, Provo, UT.

Ellis, K. L., J. R. Murphy, and G. H. Richins. 1987. Distribution of breeding male Sage Grouse in northeastern Utah. Western Birds 18:117–121.

Ellis, K. L., J. R. Parish, J. R. Murphy, and G. H. Richins. 1989. Habitat use by breeding male Sage Grouse: a management approach. Great Basin Naturalist 49:404–407.

Ellison, L. 1960. Influence of grazing on plant succession of rangelands. Botanical Review 26:1–35.

Ellison, L. N. 1991. Shooting and compensatory mortality in tetraonids. Ornis Scandinavica 22:229–240.

Ellner, S. P., and E. E. Holmes. 2008. Commentary on Holmes et al. (2007): resolving the debate on when extinction risk is predictable. Ecology Letters 11:E1–E5.

Ellsworth, D. L., R. L. Honeycutt, and N. J. Silvy. 1996. Systematics of grouse and ptarmigan determined by nucleotide sequences of the mitochondrial cytochrome-*B* gene. Auk 113:811–822.

Ely, J. B., A. W. Jensen. L. R. Chatten, and H. W. Jori. 1957. Air tankers—a new tool for forest fire fighting. Fire Control Notes 18:103–109.

Emmons, S. R., and C. E. Braun. 1984. Lek attendance of male Sage Grouse. Journal of Wildlife Management 48:1023–1028.

Eng, R. L. 1963. Observations of the breeding biology of male Sage Grouse. Journal of Wildlife Management 27:841–846.

Eng, R. L., and P. Schladweiler. 1972. Sage Grouse winter movements and habitat use in central Montana. Journal of Wildlife Management 36:141–146.

Enyeart, G. W. 1956. Responses of Sage Grouse to grass reseeding in the Pines Area, Garfield County, Utah. M.S. thesis, Utah State University, Logan, UT.

Epstein, P. R. 2001. West Nile virus and the climate. Journal of Urban Health: Bulletin of the New York Academy of Medicine 78:367–371.

Epstein, P. R., and C. Defilippo. 2001. West Nile virus and drought. Global Change and Human Health 2:105–107.

Errington, P. L. 1945. Some contributions of a fifteen-year local study of the Northern Bobwhite to a knowledge of population phenomena. Ecological Monographs 15:1–34.

Errington, P. L. 1956. Factors limiting higher vertebrate populations. Science 124:304–307.

ESRI. 1998. ArcGIS version 9.1. Environmental Systems Research Institute, Redlands, CA.

ESRI. 2006. ARC/INFO version 9.2. Environmental Systems Research Institute, Redlands, CA.

Evans, C. C. 1986. The relationship of cattle grazing to Sage Grouse use of meadow habitat on the Sheldon National Wildlife Refuge. M.S. thesis, University of Nevada, Reno, NV.

Evans, K. L. 2004. A review of the potential for interactions between predation and habitat change to cause population declines of farmland birds. Ibis 146:1–13.

Evans, R. A., and J. A. Young. 1975. Aerial applications of 2,4-D plus picloram for green rabbitbrush control. Journal of Range Management 28:315–318.

Evans, R. A., and J. A. Young. 1977. Weed control–revegetation systems for big sagebrush–downy brome rangelands. Journal of Range Management 30:331–336.

Evans, R. A., and J. A. Young. 1978. Effectiveness of rehabilitation practices following wildfire in a degraded big sagebrush–downy brome community. Journal of Range Management 31:185–188.

Evans, R. A., and J. A. Young. 1985. Plant succession following control of western juniper (*Juniperus occidentalis*) with picloram. Weed Science 33:63–68.

Evans, R. D., R. Rimer, L. Sperry, and J. Belnap. 2001. Exotic plant invasion alters nitrogen dynamics in an arid grassland. Ecological Applications 11:1301–1310.

Everaert, J., and D. Bauwens. 2007. A possible effect of electromagnetic radiation from mobile phone base stations on the number of breeding house sparrows (*Passer domesticus*). Electromagnetic Biology and Medicine 26:63–72.

Ewers, B. E., and E. Pendall. 2008. Spatial patterns in leaf area and plant functional type cover across chronosequences of sagebrush ecosystems. Plant Ecology 194:67–83.

Fahnestock, J. T. 1998. Vegetation responses to herbivory and resource supplementation in the Pryor Mountain Wild Horse Range. Ph.D. dissertation, Colorado State University, Ft. Collins, CO.

Fahnestock, J. T., and J. K. Detling. 1999a. The influence of herbivory on plant cover and species composition in the Pryor Mountain Wild Horse Range, USA. Plant Ecology 144:145–157.

Fahnestock, J. T., and J. K. Detling. 1999b. Plant responses to defoliation and resource supplementation in the Pryor Mountains. Journal of Range Management 52:263–270.

Fahrig, L. 2003. Effects of habitat fragmentation on biodiversity. Annual Review Ecology and Evolutionary Systems 34:487–515.

Fajemisin, B., D. Ganskopp, R. Cruz, and M. Vavra. 1996. Potential for woody plant control by Spanish goats in the sagebrush steppe. Small Ruminant Research 20:229–238.

Fall, A., M.-J. Fortin, M. Manseau, and D. O'Brien. 2007. Spatial graphs: principles and applications for habitat connectivity. Ecosystems 10:448–461.

Fang, Y., and W. K. Reisen. 2006. Previous infection with West Nile or St. Louis encephalitis viruses provides cross protection during reinfection in House Finches. American Journal of Tropical Medicine and Hygiene 75:480–485.

Fansler, V. A. 2007. Establishing native plants in crested wheatgrass stands using successional management. M.S. thesis, Oregon State University, Corvallis, OR.

Farajollahi, A., W. J. Crans, D. Nickerson, P. Bryant, B. Wolf, A. Glaser, and T. G. Andreadis. 2005. Detection of West Nile virus RNA from the louse fly *Icosta americana* (Diptera: Hippoboscidae). Journal of the American Mosquito Control Association 21:474–476.

Fargione, J. E., T. R. Cooper, D. J. Flaspohler, J. Hill, C. Lehman, T. McCoy, S. McLeod, E. J. Nelson, K. S. Oberhauser, and D. Tilman. 2009. Bioenergy and wildlife: threats and opportunities for grassland conservation. BioScience 59:767–777.

Farrell, A. E., R. J. Plevin, B. T. Turner, A. D. Jones, M. O'Hare, and D. M. Kammen. 2006. Ethanol can contribute to energy and environmental goals. Science 311:506–508.

Fechner, G. H., and J. S. Barrows. 1976. Aspen stands as wildfire fuel breaks. Eisenhower Consortium Bulletin 4. College of Natural Resources, Department of Forestry and Wood Science, Colorado State University, Ft. Collins, CO.

Fedy, B. C., K. Martin, C. Ritland, and J. Young. 2008. Genetic and ecological data provide incongruent interpretations of population structure and dispersal in naturally subdivided populations of White-tailed Ptarmigan (*Lagopus leucura*). Molecular Ecology 17:1905–1917.

Feist, J. D. 1971. Behavior of feral horses in the Pryor Mountain Wild Horse Range. M.S. thesis, University of Michigan, Ann Arbor, MI.

Fernie, K. J., and S. J. Reynolds. 2005. The effects of electromagnetic fields from power lines on avian reproductive biology and physiology: a review. Journal of Toxicology and Environmental Health, Part B 8:127–140.

Ferrero, R. C., R. C. Hobbs, and G. R. Vanblaricom. 2002. Indications of habitat use patterns among small cetaceans in the central North Pacific based on fisheries observer data. Journal of Cetacean Research and Management 4:311–321.

Fichter, E., and R. Williams. 1967. Distribution and status of the red fox in Idaho. Journal of Mammalogy 48:219–230.

Fieberg, J., and S. P. Ellner. 2000. When is it meaningful to estimate an extinction probability? Ecology 81:2040–2047.

Fiero, B. 1986. Geology of the Great Basin; a natural history. University of Nevada Press, Las Vegas, NV.

Finnerty, D. W., and D. L. Klingman. 1961. Life cycles and control studies of some weed bromegrass. Weeds 10:40–47.

Fischer, R. A. 1994. The effects of prescribed fire on the ecology of migratory Sage Grouse in southeastern Idaho. Ph.D. dissertation, University of Idaho, Moscow, ID.

Fischer, R. A., A. D. Apa, W. L. Wakkinen, K. P. Reese, and J. W. Connelly. 1993. Nesting-area fidelity of Sage Grouse in southeastern Idaho. Condor 95:1038–1041.

Fischer, R. A., K. P. Reese, and J. W. Connelly. 1996a. Influence of vegetal moisture content and nest fate on timing of female Sage Grouse migration. Condor 98:868–872.

Fischer, R. A., K. P. Reese, and J. W. Connelly. 1996b. An investigation on fire effects within xeric Sage Grouse brood habitat. Journal of Range Management 49:194–198.

Fischer, R. A., W. L. Wakkinen, K. P. Reese, and J. W. Connelly. 1997. Effects of prescribed fire on movements of female Sage Grouse from breeding to summer ranges. Wilson Bulletin 109:82–91.

Fiske, P., and J. A. Kålås. 1995. Mate sampling and copulation behaviour of Great Snipe females. Animal Behaviour 49:209–219.

Flather, C. H., and M. Bevers. 2002. Patchy reaction-diffusion and population abundance: the relative importance of habitat amount and arrangement. American Naturalist 159:40–56.

Fleishman, E., D. D. Murphy, and P. F. Brussard. 2000. A new method for selection of umbrella species for conservation planning. Ecological Applications 10:569–579.

Flores, D. 2001. The natural West: environmental history in the Great Plains and Rocky Mountains. University of Oklahoma Press, Norman, OK.

Floyd, M. L., D. D. Hanna, and W. H. Romme. 2000. Fire history and vegetation pattern in Mesa Verde National Park, Colorado, USA. Ecological Applications 10:1666–1680.

Floyd, M. L., D. D. Hanna, and W. H. Romme. 2004. Historical and recent fire regimes in piñon-juniper woodlands on Mesa Verde, Colorado, USA. Forest Ecology and Management 198:269–289.

Floyd, M. L., W. H. Romme, D. D. Hanna, M. Winterowd, D. Hanna, and J. Spence. 2008. Fire history of piñon-juniper woodlands on Navajo Point, Glen Canyon National Recreation Area. Natural Areas Journal 28:26–36.

Foley, J. A., R. DeFries, G. P. Asner, C. Barford, G. Bonan, S. R. Carpenter, F. S. Chapin, M. T. Coe, G. C. Daily, H. K. Gibbs, J. H. Helkowski, T. Holloway, E. A. Howard, C. J. Kucharik, C. Monfreda, J. A. Patz, I. C. Prentice, N. Ramankutty and P. K. Snyder. 2005. Global consequences of land use. Science 309:570–574.

Forbis, T. A., L. Provencher, L. Frid, and G. Medlyn. 2006. Great Basin land management planning using ecological modeling. Environmental Management 38:62–83.

Forman, R. T. T. 1995. Land mosaics. The ecology of landscapes and regions. Cambridge University Press, Cambridge, UK.

Forman, R. T. T. 2000. Estimate of the area affected ecologically by the road system in the United States. Conservation Biology 14:31–35.

Forman, R. T. T., and L. E. Alexander. 1998. Roads and their major ecological effects. Annual Review of Ecology and Systematics 29:207–231.

Fosberg, M. A., and M. Hironaka. 1964. Soil properties affecting the distribution of big and low sagebrush communities in southern Idaho. Forage plant physiology and soil-range relationships. American Society Agronomy Special Publication 5:230–236.

Foster, D., F. Swanson, J. Aber, I. Burke, N. Brokaw, D. Tilman, and A. Knapp. 2003. The importance of land-use legacies to ecology and conservation. BioScience 53:77–88.

Fox, A., and P. J. Hudson. 2001. Parasites reduce territorial behaviour in Red Grouse (*Lagopus lagopus scoticus*). Ecology Letters 4:139–143.

Fraas, W. W., C. L. Wambolt, and M. R. Frisina. 1992. Prescribed fire effects on a bitterbrush-mountain big sagebrush-bluebunch wheatgrass community. Pp. 212–216 *in* W. Clary, E. D. McArthur, D. Bedunah, and C. L. Wambolt (editors). Symposium on ecology and management of riparian shrub communities. USDA, Forest Service General Technical Report INT-289. USDA Forest Service, Intermountain Forest and Range Experiment Station, Ogden, UT.

Frair, J. L., E. H. Merrill, J. R. Allen, and M. S. Boyce. 2007. Know thy enemy: experience affects elk translocation success in risky landscapes. Journal of Wildlife Management 71:541–554.

Francis, M., and D. A. Pyke. 1996. Crested wheatgrass–cheatgrass seedling competition using a mixed-density design. Journal of Range Management 49:432–438.

Franklin, I. R. 1980. Evolutionary change in small populations. Pp. 135–139 *in* M. E. Soulé and B. A. Wilcox (editors). Conservation biology: an ecological-evolutionary perspective. Sinauer Associates, Sunderland, MA.

Freilich, J. E., J. M. Emlen, J. J. Duda, D. C. Freeman, and P. J. Cafaro. 2003. Ecological effects of ranching: a six-point critique. BioScience 53:759–765.

Frémont, J. C. 1845. Report of the exploring expedition to the Rocky Mountains in the year 1842, and to Oregon and Northern California in the years 1843–44. Gales and Seaton, Washington, DC.

Friedel, M. H. 1991. Range condition assessment and the concept of thresholds: a viewpoint. Journal of Range Management 44:422–426.

Friedman, M., and D. C. Carlton. 1999. Petition for a rule to list the Washington population of western Sage-grouse, *Centrocercus urophasianus phaios*, as "threatened" or "endangered" under the Endangered Species Act, 16 U.S.C. Sec. 1531 et seq (1973) as amended. USDI Fish and Wildlife Service, Spokane, WA.

Friend, M. 1999a. Aspergillosis. Pp. 129–133 *in* M. Friend and J. C. Franson (editors). Field manual of wildlife diseases: general field procedures and diseases of birds. U.S. Geological Survey Biological Resources Division Information and Technical Report 1999-001. USDI Geological Survey, Madison, WI.

Friend, M. 1999b. Miscellaneous bacterial diseases. Pp. 121–126 *in* M. Friend and J. C. Franson (editors). Field manual of wildlife diseases: general field procedures and diseases of birds. U.S. Geological Survey Biological Resources Division Information and Technical Report 1999-001. USDI Geological Survey, Madison, WI.

Friend, M. 1999c. Salmonellosis. Pp. 99–109 *in* M. Friend and J. C. Franson (editors). Field manual of wildlife diseases: general field procedures and diseases of birds. U.S. Geological Survey Biological Resources Division Information and Technical Report 1999-001. USDI Geological Survey, Madison, WI.

Friend, M., R. G. McLean, and F. J. Dein. 2001. Disease emergence in birds: challenges for the twenty-first century. Auk 118:290–303.

Frischknecht, N. C. 1978. Effects of grazing, climate, fire, and other disturbances on long-term productivity of sagebrush-grass ranges. Pp. 633–635 *in* D. N. Hyder (editor). Proceedings of the First International Rangeland Congress. Society for Range Management, Denver, CO.

Frisina, M. R., and C. L. Wambolt. 2004. A guide for identifying the four subspecies of big sagebrush. Rangelands 26(1):12–16.

Frost, W. E., E. L. Smith, and P. R. Ogden. 1994. Utilization guidelines. Rangelands 16(6):256–259.

Fuhlendorf, S. D., F. E. Smeins, and C. A. Taylor. 1997. Browsing and tree size influences on Ashe juniper understory. Journal of Range Management 50:507–512.

Gaines, W. L., P. H. Singleton, and R. C. Ross. 2003. Assessing the cumulative effects of linear recreation routes on wildlife habitats on the Okanogan and Wenatchee National Forests. USDA Forest Service General Technical Report PNW-GTR-586. USDA Forest Service, Pacific Northwest Research Station, Portland, OR.

Ganskopp, D., and M. Vavra. 1986. Habitat use by feral horses in the northern sagebrush steppe. Journal of Range Management 39:207–212.

Ganskopp, D., and M. Vavra. 1987. Slope use by cattle, feral horses, deer, and bighorn sheep. Northwest Science 61:74–81.

Garton, E.O. 2002. Mapping a chimera? Pp. 663–666 *in* J. M. Scott, P. J. Heglund, J. B. Hauffler, M. G. Raphael, W. A. Wall, and F. B. Samson (editors). Predicting species occurrence: issues of accuracy and scale. Island Press, Covelo, CA.

Garton, E. O., D. D. Musil, K. P. Reese, J. W. Connelly, and C. L. Anderson. 2007. Sentinel lek-routes: an integrated sampling approach to estimate Greater Sage-Grouse population characteristics. Pp. 31–41 in K. P. Reese and R. T. Bowyer (editors). Monitoring populations of sage-grouse. College of Natural Resources Experiment Station Bulletin 88. University of Idaho, Moscow, ID.

Gates, R. J. 1983. Sage Grouse, lagomorph, and pronghorn use of a sagebrush grassland burn site on the Idaho National Engineering Laboratory. M.S. thesis, Montana State University, Bozeman, MT.

Gelbard, J. L., and J. Belnap. 2003. Roads as conduits for exotic plant invasions in a semiarid landscape. Conservation Biology 17:420–432.

Gerding, M. (editor). 1986. Fundamentals of petroleum. Petroleum Extension Service, University of Texas, Austin, TX.

Gerhardt, T., and J. K. Detling. 2000. Summary of vegetation dynamics at the Pryor Mountain Wild Horse Range, 1992–1996. Pp. 3–36 *in* F. J. Singer and K. A. Schoenecker (compilers). Managers' summary—ecological studies of the Pryor Mountain Wild Horse Range, 1992–1997. USDI Geological Survey, Ft. Collins Science Center, Ft. Collins, CO.

Ghil, M., and R. Vautgard. 1991. Interdecadal oscillations and the warming trend in global temperature time series. Nature 350:324–327.

Gibbs, S. E. J., D. M. Hoffman, L. M. Stark, N. L. Marlenee, B. J. Blitvich, B. J. Beaty, and D. E. Stallknecht. 2005. Persistence of antibodies to West Nile virus in naturally infected Rock Pigeons (*Columba livia*). Clinical and Diagnostic Laboratory Immunology 12:665–667.

Gibson, R. M. 1990. Relationship between blood parasites, mating success and phenotypic cues in male Sage Grouse *Centrocercus urophasianus*. American Zoologist 30:271–278.

Gibson, R. M. 1992. Lek formation in Sage Grouse: the effect of female choice on male territorial settlement. Animal Behaviour 43:443–450.

Gibson, R. M. 1996. A re-evaluation of hotspot settlement in lekking Sage Grouse. Animal Behaviour 52:993–1005.

Gibson, R. 1998. Effects of hunting on the population dynamics of two Sage Grouse populations in Mono County, California. Western Sage and Columbian Sharp-tailed Grouse Workshop 21:15.

Gibson, R. M., and J. W. Bradbury. 1986. Male and female mating strategies on Sage Grouse leks. Pp. 379–398 *in* D. I. Rubenstein and R. W. Wrangham (editors). Ecological aspects of social evolution. Princeton University Press, Princeton, NJ.

Gibson, R. M., and J. W. Bradbury. 1987. Lek organization in Sage Grouse: variations on a territorial theme. Auk 104:77–84.

Gibson, R. M., A. S. Aspbury, and L. L. McDaniel. 2002. Active formation of mixed-species grouse leks: a role of predation in lek evolution? Proceedings of the Royal Society of London B 269:2503–2507.

Gibson, R. M., J. W. Bradbury, and S. L. Vehrencamp. 1991. Mate choice in lekking Sage Grouse revisited: the roles of vocal display, female site fidelity and copying. Behavioral Ecology 2:165–180.

Gibson, R. M., D. Pires, K. S. Delaney, and R. K. Wayne. 2005. Microsatellite DNA analysis shows that Greater Sage-Grouse leks are not kin groups. Molecular Ecology 14:4453–4459.

Giezentanner, K. I., and W. H. Clark. 1974. The use of western harvester ant mounds as strutting locations by Sage Grouse. Condor 76:218–219.

Gilbert, L., L. D. Jones, M. K. Laurenson, E. A. Gould, H. W. Reed, and P. J. Hudson. 2004. Ticks need not bite their Red Grouse hosts to infect them with louping ill virus. Proceedings of the Royal Society of London B (Supplement) 271:S202–S205.

Gill, R. B. 1965. Distribution and abundance of a population of Sage Grouse in North Park, Colorado. M.S. thesis, Colorado State University, Ft. Collins, CO.

Gill, R. B., and F. A. Glover. 1965. Daily and seasonal movements of Sage Grouse. Technical Paper 3. Colorado Cooperative Wildlife Research Unit, Ft. Collins, CO.

Gilpin, M., and I. Hanski. 1991. Metapopulation dynamics: empirical and theoretical investigations. Academic Press, London, UK.

Giudice, J. H., and J. T. Ratti. 2001. Ring-necked Pheasant (*Phasianus colchicus*). *In* A. Poole, and F. Gill (editors). The birds of North America, No. 572. The Academy of Natural Sciences, Philadelphia, PA, and The American Ornithologists' Union, Washington, DC.

Goddard, L. B., A. E. Roth, W. K. Reisen, and T. W. Scott. 2002. Vector competence of California mosquitoes for West Nile virus. Emerging Infectious Diseases 8:1385–1391.

Goddard, L. B., A. B. Roth, W. K. Reisen, and T. W. Scott. 2003. Vertical transmission of West Nile virus by three California *Culex* (Diptera: Culicidae) species. Journal of Medical Entomology 40:743–747.

Gonzalez, A., J. H. Lawton, F. S. Gilbert, T. M. Blackburn, and I. Evans-Freke. 1998. Metapopulation dynamics, abundance, and distribution of a microecosystem. Science 281:2045–2047.

Goodman, L. A. 1962. The variance of the product of *K* random variables. Journal of American Statistical Association 57:54–60.

Gosnell, H., and W. R. Travis. 2005. Ranchland ownership dynamics in the Rocky Mountain west. Rangeland Ecology and Management 58:191–198.

Gottschalk, J. S. 1972. The German hunting system, West Germany, 1968. Journal of Wildlife Management 36:110–118.

Gray, R. L., and B. M. Teels. 2006. Wildlife and fish conservation through the farm bill. Wildlife Society Bulletin 34:906–913.

Grayson, D. K. 1993. The desert's past: a natural prehistory of the Great Basin. Smithsonian Institution, Washington, DC.

Grayson, D. K. 2006. The late Quaternary biogeographic histories of some Great Basin mammals (western USA). Quaternary Science Reviews 25:2964–2991.

Green, N. F., and H. D. Green. 1977. The wild horse population of Stone Cabin Valley, Nevada: a preliminary report. Pp. 59–65 *in* J. L. Artz (forum chairman). National Wild Horse Forum. University of Nevada Cooperative Extension Service, Reno, NV.

Greenwood, P. J., and P. H. Harvey. 1982. The natal and breeding dispersal of birds. Annual Review of Ecology and Systematics 13:1–21.

Greenwood, R. J., and M. A. Sovada. 1996. Prairie duck population management. Transactions of the North American Wildlife and Natural Resources Conference 61:31–42.

Gregg, L. E. 1990. Harvest rates of Sharp-tailed Grouse on managed areas in Wisconsin. Research Report 152. Wisconsin Department of Natural Resources, Park Falls, WI.

Gregg, M. A. 1991. Use and selection of nesting habitat by Sage Grouse in Oregon. M.S. thesis, Oregon State University, Corvallis, OR.

Gregg, M. A. 2006. Greater Sage-Grouse reproductive ecology: linkages among habitat resources, maternal nutrition, and chick survival. Ph.D. dissertation, Oregon State University, Corvallis, OR.

Gregg, M. A., J. A. Crawford, and M. S. Drut. 1993. Summer habitat use and selection by female Sage Grouse (*Centrocercus urophasianus*) in Oregon. Great Basin Naturalist 53:293–298.

Gregg, M. A., J. A. Crawford, M. S. Drut, and A. K. DeLong. 1994. Vegetational cover and predation of Sage Grouse nests in Oregon. Journal of Wildlife Management 58:162–166.

Gregg, M. A., M. R. Dunbar, J. A. Crawford, and M. D. Pope. 2006. Total plasma protein and renesting by Greater Sage-Grouse. Journal of Wildlife Management 70:472–478.

Griffin, D. 2002. Prehistoric human impacts on fire regimes and vegetation in the northern Intermountain West. Pp. 77–100 *in* T. R. Vale (editor). Fire, native peoples, and the natural landscape. Island Press, Washington, DC.

Griffiths, D. 1902. Forage conditions on the northern border of the Great Basin. United States Department of Agriculture Bureau of Plant Industry Bulletin No. 15. U.S. Government Printing Office, Washington, DC.

Griner, L. A. 1939. A study of the Sage Grouse (*Centrocercus urophasianus*), with special reference to life history, habitat requirements, and numbers and distribution. M.S thesis, Utah State Agricultural College, Logan, UT.

Grossman, D. H., D. Gaver-Langendoen, A. S. Weakley, M. Anderson, P. Bourgeron, R. Crawford, K. Goodin, S. Landaal, K. Metzler, K. Patterson, M. Pyne, M. Reid, and L. Sneddon. 1998. International classification of ecological communities: terrestrial vegetation of the United States. Vol. 1: the national vegetation classification system: development, status, and applications. The Nature Conservancy, Arlington, VA.

Grover, G. 1944. Disease among Sage Grouse. Colorado Conservation Comments 1:14–15.

Groves, C. R. 2003. Drafting a conservation blueprint—a practitioner's guide to planning for biodiversity. Island Press, Washington, DC.

Groves, C. R., D. B. Jensen, L. L. Valutis, K. H. Redford, M. L. Shaffer, J. M. Scott, J. V. Baumgartner, J. V. Higgins, M. W. Beck, and M. G. Anderson. 2002. Planning for biodiversity conservation: putting conservation science into practice. BioScience 52:499–512.

Gruell, G. E. 1985. Fire on the early western landscape: an annotated list of recorded wildfires in presettlement times. Northwest Science 59:97–107.

Gubler, D. J. 2007. The continuing spread of West Nile virus in the western hemisphere. Emerging Infections 45:1039–1046.

Gubler, D. J., G. L. Campbell, R. Nasci, N. Komar, L. Petersen, and J. T. Roehrig. 2000. West Nile virus in the United States: guidelines for detection, prevention, and control. Viral Immunology 13:469–475.

Guisan, A., S. B. Weiss, and A. D. Weiss. 1999. GLM versus CCA spatial modeling of plant species distribution. Plant Ecology 143:107–122.

Gullion, G. W. 1984. Ruffed Grouse management—where do we stand in the eighties? Pp. 169–181 *in* W. L. Robinson (editor). Ruffed Grouse management: state of the art in the early 1980's. North Central Section of The Wildlife Society, BookCrafters, Chelsea, MI.

Gustafson, E. J. 1998. Quantifying landscape spatial pattern: what is the state of the art? Ecosystems 1:143–156.

Gustafson, E. J., and G. R. Parker. 1994. Using an index of habitat patch proximity for landscape design. Landscape and Urban Planning 29:117–130.

Guthery, F. S. 2008. Statistical ritual versus knowledge accrual in wildlife science. Journal of Wildlife Management 72:1872–1875.

Gutiérrez, R. J. 1994. North American upland gamebird management at crossroads: which road will we take? North American Wildlife and Natural Resources Conference 59:494–497.

Gutiérrez, R. J., G. F. Barrowclough, and J. G. Groth. 2000. A classification of the grouse (Aves: Tetraoninae) based on mitochondrial DNA sequences. Wildlife Biology 6:205–211.

Hagen, C. A. 1999. Sage Grouse habitat use and seasonal movements in a naturally fragmented landscape, northwestern Colorado. M.S. thesis, University of Manitoba, Winnepeg, MB, Canada.

Hagen, C. A. 2003. A demographic analysis of Lesser Prairie-Chicken populations in southwestern Kansas: survival, population viability, and habitat use. Ph.D. dissertation, Kansas State University, Manhattan, KS.

Hagen, C. A. 2005. Greater Sage-Grouse conservation assessment and strategy for Oregon: a plan to maintain and enhance populations and habitat. Oregon Department of Fish and Wildlife, Salem, OR.

Hagen, C. A., and R. J. Bildfell. 2007. An observation of *Clostridium perfringens* in Greater Sage-Grouse. Journal of Wildlife Diseases 43:545–547.

Hagen, C. A., and K. M. Giesen. 2005. Lesser Prairie-Chicken (*Tympanuchus pallidicinctus*). *In* A. Poole (editor). The birds of North America online. http://bna.birds.cornell.edu/BNA/account/Lesser_Prairie-Chicken (10 March 2009).

Hagen, C. A., and T. M. Loughin. 2008. Productivity estimates from upland bird harvests: estimating variance and necessary sample sizes. Journal of Wildlife Management 72:1369–1375.

Hagen, C. A., J. W. Connelly, and M. A. Schroeder. 2007. A meta-analysis of Greater Sage-Grouse *Centrocercus urophasianus* nesting and brood-rearing habitats. Wildlife Biology 13 (Supplement 1):42–50.

Hagen, C. A., S. S. Crupper, R. D. Applegate, and R. J. Robel. 2002. Prevalence of *Mycoplasma* antibodies in Lesser Prairie Chicken sera. Avian Diseases 46:708–712.

Hagen, C. A., B. E. Jamison, K. M. Giesen, and T. Z. Riley. 2004. Guidelines for managing Lesser Prairie-Chicken populations and their habitats. Wildlife Society Bulletin 32:69–82.

Hagen, C. A., N. C. Kenkel, D. J. Walker, R. K. Baydack, and C. E. Braun. 2001. Fractal-based spatial analysis of radiotelemetry data. Pp. 167–187 *in* J. Millspaugh and J. Marzluff (editors). Radio tracking and animal populations. Academic Press, San Diego, CA.

Hagen, C. A., J. C. Pitman, B. K. Sandercock, R. J. Robel, and R. D. Applegate. 2005. Age-specific variation in apparent survival rates of male Lesser Prairie-Chickens. Condor 107:78–86.

Hahn, D. C., N. M. Nemeth, E. Edwards, P. R. Bright, and N. Komar. 2006. Passive West Nile virus antibody transfer from maternal Eastern Screech-Owls (*Megascops asio*) to progeny. Avian Diseases 50:454–455.

Haig, S. M. 1998. Molecular contributions to conservation. Ecology 79:413–425.

Haig, S. M., E. A. Beever, S. M. Chambers, H. M. Draheim, B. D. Dugger, S. Dunham, E. Elliott-Smith, J. B. Fontaine, D. C. Kesler, B. J. Knaus, I. F. Lopes, P. Loschl, T. D. Mullins, and L. M. Sheffield. 2006. Taxonomic considerations in listing subspecies under the U.S. Endangered Species Act. Conservation Biology 20:1584–1594.

Haig, S. M., T. D. Mullins, and E. D. Forsman. 2004. Subspecific relationships and genetic structure in the Spotted Owl. Conservation Genetics 5:683–705.

Haig, S. M., R. S. Wagner, E. D. Forsman, and T. D. Mullins. 2001. Geographic variation and genetic structure in Spotted Owls. Conservation Genetics 2:25–40.

Haight, R. G., B. Cypher, P. A. Kelly, S. Phillips, H. P. Possingham, K. Ralls, A. M. Starfield, P. J. White, and D. Williams. 2002. Optimizing habitat protection using demographic models of population viability. Conservation Biology 16:1386–1397.

Haines, A. M., M. Leu, L. K. Svancara, J. M. Scott, and K. P. Reese. 2008. A theoretical approach to using human footprint models to measure landscape level conservation success. Conservation Letters 1:165–172.

Haines, D. E., and K. H. Pollack. 1998. Estimating the number of active and successful Bald Eagle nests: an application of the dual frame method. Environmental and Ecological Statistics 5:245–256.

Hair, Jr., J. F., R. E. Anderson, R. L. Tatham, and W. C. Black. 1992. Multivariate analysis with readings. Third edition. Macmillan Publishing Company, New York, NY.

Halvorson, J. J., H. Bolton, Jr., J. L. Smith, and R. E. Rossi. 1995. Evaluating shrub associated spatial patterns of soil properties in a shrub steppe ecosystem using multiple variable geostatistics. Soil Science Society of America Journal 59:1476–1478.

Hamer, T. L., C. H. Flather, and B. R. Noon. 2006. Factors associated with grassland bird species richness: the relative roles of grassland area, landscape structure, and prey. Landscape Ecology 21:569–583.

Hamilton, J. R., and R. L. Gardner. 1986. Value added and secondary benefits in regional projection evaluation: irrigation development in the Snake River Basin. Annals of Regional Science 20:1–11.

Hanf, J. M., P. A. Schmidt, and E. B. Groshens. 1994. Sage Grouse in the high desert of central Oregon: results of a study, 1988–1993. Series P-SG-01. USDI Bureau of Land Management, Prineville, OR.

Hanley, T. A. 1982. The nutritional basis for food selection by ungulates. Journal of Range Management 35:146–151.

Hanley, T. A., and K. A. Hanley. 1982. Food resource partitioning by sympatric ungulates on Great Basin rangeland. Journal of Range Management 35:152–158.

Hann, W. J., J. L. Jones, M. G. Karl, P. F. Hessburg, R. E. Keane, D. G. Long, J. P. Menakis, C. H. McNicoll, S. G. Leonard, R. A. Gravenmier, and B. G. Smith. 1997. Landscape dynamics of the basin. Pp. 334–1056 *in* T. M. Quigley and S. J. Arbelbide (technical editors). An assessment of ecosystem components in the interior Columbia Basin and portions of the Klamath and Great Basins. USDA Forest Service General Technical Report PNW-GTR-405. USDA Forest Service, Pacific Northwest Research Station, Portland, OR.

Hansen, A. J., R. L. Knight, J. M. Marzluff, S. Powell, K. Brown, P. H. Gude, and K. Jones. 2005. Effects of exurban development on biodiversity: patterns, mechanisms, and research needs. Ecological Applications 15:1893–1905.

Hansen, A. J., R. Rasker, B. Maxwell, J. J. Rotella, J. D. Johnson, A. W. Parmenter, U. Langner, W. B. Cohen, R. L. Lawrence, and M. P. V. Kraska. 2002. Ecological causes and consequences of demographic change in the new West. BioScience 52:151–162.

Hansen, K., W. Wyckoff, and J. Banfield. 1995. Shifting forests: historical grazing and forest invasion in southwestern Montana. Forest and Conservation History 39:66–76.

Hanser, S. E., S. T. Knick, J. Hak, and J. Kagan. 2005. Current distribution of sagebrush and associated vegetation in the Columbia Basin and southwestern regions. http://sagemap.wr.usgs.gov (27 October 2008).

Hanski, I. 1994. A practical model of metapopulation dynamics. Journal of Animal Ecology 63:151–162.

Hanski, I., and M. Gilpin. 1991. Metapopulation dynamics: brief history and conceptual domain. Biological Journal of the Linnean Society 42:3–16.

Hanson, P., and A. Wright. 2006. West Tavaputs sage-grouse research project. Progress Report. Southeastern Region, Utah Division of Wildlife Resources, Price, UT.

Harniss, R. O., and R. B. Murray. 1973. 30 years of vegetal change following burning of sagebrush-grass range. Journal of Range Management 26:322–325.

Haroldson, K. J., R. O. Kimmel, M. R. Riggs, and A. H. Berner. 2006. Association of Ring-necked Pheasant, Gray Partridge, and meadowlark abundance to Conservation Reserve Program grasslands. Journal of Wildlife Management 70:1276–1284.

Harris, G. A. 1967. Some competitive relationships between *Agropyron spicatum* and *Bromus tectorum*. Ecological Monographs 37:89–111.

Hartzler, J. E. 1974. Predation and the daily timing of Sage Grouse leks. Auk 91:532–536.

Harvell, C. D., C. E. Mitchell, J. R. Ward, S. Altizer, A. P. Dobson, R. S. Ostfeld, and M. D. Samuel. 2002. Climate warming and disease risks for terrestrial and marine biota. Science 296:2158–2162.

Hausleitner, D. 2003. Population dynamics, habitat use and movements of Greater Sage-Grouse in Moffat County, Colorado. M.S. thesis, University of Idaho, Moscow, ID.

Havlick, D. G. 2002. No place distant: roads and motorized recreation on America's public lands. Island Press, Washington, DC.

Hayes, E. B., N. Komar, R. S. Nasci, S. P. Montgomery, D. R. O'Leary, and G. L. Campbell. 2005a. Epidemiology and dynamics of transmission of West Nile virus disease. Emerging Infectious Diseases 11:1167–1173.

Hayes, E. B., J. J. Sejvar. S. R. Zaki, R. S. Lanciotti, A. V. Bode, and G. L. Campbell. 2005b. Virology, pathology, and clinical manifestations of West Nile virus disease. Emerging Infectious Diseases 11:1174–1179.

Hayes, G. F., and K. D. Holl. 2003. Cattle grazing impacts on annual forbs and vegetation composition of mesic grasslands in California. Conservation Biology 17:1694–1702.

Hays, D. W., M. J. Tirhi, and D. W. Stinson. 1998. Washington state status report for the Sage Grouse. Washington Department of Fish and Wildlife, Olympia, WA.

Heady, H. F., R. P. Gibbens, and R. W. Powell. 1959. A comparison of charting, line intercept, and line point methods of sampling shrub types of vegetation. Journal of Range Management 12:180–188.

Heath, B. J., R. Straw, S. H. Anderson, and J. Lawson. 1997. Sage-grouse productivity, survival, and seasonal habitat use near Farson, Wyoming. Pitmann-Robertson Project Completion Report. Wyoming Game and Fish Department, Cheyenne, WY.

Heath, B. J., R. Straw, S. H. Anderson, J. Lawson, and M. J. Holloran. 1998. Sage-Grouse productivity, survival, and seasonal habitat use among three ranches with different livestock grazing, predator control, and harvest management practices. Pitmann-Robertson Project Completion Report. Wyoming Game and Fish Department, Cheyenne, WY.

Hebblewhite, M., C. A. White, C. G. Nietvelt, J. A. Mckenzie, T. E. Hurd, J. M. Fryxell, S. E. Bayley, and P. C. Paquet. 2005. Human activity mediates a trophic cascade caused by wolves. Ecology 86:2135–2144.

Heckel, G., C. C. Voigt, F. Mayer, and O. Von Helversen. 1999. Extra-harem paternity in the white-lined bat *Saccopteryx bilineata* (Emballonuridae). Behaviour 136:1173–1185.

Heinselman, M. L. 1973. Fire in the virgin forests of the Boundary Waters Canoe Area, Minnesota. Quaternary Research 3:329–382.

Heitschmidt, R. K., K. D. Klement, and M. R. Haferkamp. 2005. Interactive effects of drought and grazing on Northern Great Plains rangelands. Rangeland Ecology and Management 58:11–19.

Hemker, T. P., and C. E. Braun. 2001. Innovative approaches for the development of conservation plans for Sage Grouse: examples from Idaho and Colorado. Transactions of the North American Wildlife and Natural Resource Conference 66:456–463.

Hempy-Mayer, K., and D. A. Pyke. 2008. Defoliation effects on *Bromus tectorum* seed production: implications for grazing. Rangeland Ecology and Management 61:116–123.

Hemstrom, M. A., M. J. Wisdom, W. J. Hann, M. M. Rowland, B. C. Wales, and R. A. Gravenmier. 2002. Sagebrush-steppe vegetation dynamics and restoration potential in the interior Columbia Basin, U.S.A. Conservation Biology 16:1243–1255.

Hepworth, W. G. 1962. Diagnosis of disease in mammals and birds. P-R Project FW-3-R-8, Job 1W. Wyoming Game and Fish Commission, Cheyenne, WY.

Herman, C. M. 1963. Disease and infection in the Tetraonidae. Journal of Wildlife Management 27:850–855.

Herman-Brunson, K. M. 2007. Nesting and brood-rearing success and habitat selection of Greater Sage-Grouse and associated survival of hens and broods at the edge of their historic distribution. M.S. thesis, South Dakota State University, Brookings, SD.

Heyerdahl, E. K., R. F. Miller, and R. A. Parsons. 2006. History of fire and Douglas-fir establishment in a savanna and sagebrush-grassland mosaic, southwestern Montana, USA. Forest Ecology and Management 230:107–118.

Higgs, S., B. S. Schneider, D. L. Vanlandingham, K. A. Klingler, and E. A. Gould. 2005. Nonviremic transmission of West Nile virus. Proceedings of the National Academy of Sciences 102:8871–8874.

Hildén, O. 1965. Habitat selection in birds. Annales Zoologici Fennici 2:53–75.

Hill, D., and P. Robertson. 1988. The pheasant: ecology, management and conservation. BSP Professional Books, Cambridge, MA.

Hironaka, M., M. A. Fosberg, and A. H. Winward. 1983. Sagebrush-grass habitat types in southern Idaho. Bulletin No. 35. University of Idaho Forest, Wildlife, and Range Experiment Station, Moscow, ID.

Hjorth, I. 1970. Reproductive behavior in Tetraonidae with special reference to males. Viltrevy 7:381–357.

Hobbs, N. T. 1996. Modification of ecosystems by ungulates. Journal of Wildlife Management 60:695–713.

Hobbs, R. J., and L. J. Kristjanson. 2003. Triage: how do we prioritize health care for landscapes? Ecological Management and Restoration 4 (Supplement):S39–S45.

Hobbs, R. J., V. A. Cramer, and L. J. Kristjanson. 2003. What happens if we cannot fix it? Triage, palliative care and setting priorities in salinising landscapes. Australian Journal of Botany 51:647–653.

Hoberg, E. P., L. P. Polley, E. J. Jenkins, S. K. Kutz, A. M. Veitch, and B. T. Elkin. 2008. Integrated approaches and empirical models for investigation of parasitic diseases in northern wildlife. Emerging Infectious Diseases 14:10–17.

Hockett, G. A. 2002. Livestock impacts on the herbaceous components of Sage Grouse habitat: a review. Intermountain Journal of Sciences 8:105–114.

Hofmann, L. A. 1991. The western Sage Grouse (*Centrocercus urophasianus phaios*) on the Yakima Training Center in central Washington: a case study of a declining species and the military. M.S. thesis, Central Washington University, Ellensburg, WA.

Höglund, J., and L. Shorey. 2003. Local genetic structure in a White-bearded Manakin population. Molecular Ecology 12:2457–2463.

Höglund, J., R. V. Alatalo, A. Lundberg, P. T. Rintamäkil, and J. Lindell. 1999. Microsatellite markers reveal the potential for kin selection on Black Grouse leks. Proceeding of the Royal Society of London B 266:813–816.

Holechek, J. L. 1981. A brief history of range management in the United States. Rangelands 3(1):16–18.

Holechek, J. L. 2001. Western ranching at the crossroads. Rangelands 23(1):17–21.

Holechek, J. L. 2006. Changing western landscapes, debt, and oil: a perspective. Rangelands 28(4):28–32.

Holechek, J. L. 2007. National security and rangelands. Rangelands 29(5):33–38.

Holechek, J. L., and D. Galt. 2000. Grazing intensity guidelines. Rangelands 22(3):11–14.

Holechek, J. L., and T. Stephenson. 1983. Comparison of big sagebrush vegetation in northcentral New Mexico under moderately grazed and grazing excluded conditions. Journal of Range Management 36:455–456.

Holechek, J. L., T. T. Baker, J. C. Boren, and D. Galt. 2006. Grazing impacts on rangeland vegetation: what we have learned. Rangelands 28(1):7–13.

Holechek, J. L., R. D. Piper, and C. H. Herbel. 1998. Range management: principles and practices. Third edition. Prentice-Hall, Upper Saddle River, NJ.

Holechek, J. L., M. Thomas, F. Molinar, and D. Galt. 1999. Stocking desert rangelands: what we've learned. Rangelands 21(6):8–12.

Holling, C. S. 1992. Cross-scale morphology, geometry, and dynamics of ecosystems. Ecological Monographs 62:447–502.

Holloran, M. J. 1999. Sage Grouse (*Centrocercus urophasianus*) seasonal habitat use near Casper, Wyoming. M.S. thesis, University of Wyoming, Laramie, WY.

Holloran, M. J. 2005. Greater Sage-Grouse (*Centrocercus urophasianus*) population response to natural gas field development in western Wyoming. Ph.D. dissertation, University of Wyoming, Laramie, WY.

Holloran, M. J., and S. H. Anderson. 2003. Direct identification of Northern Sage-Grouse, *Centrocercus urophasianus*, nest predators using remote sensing cameras. Canadian Field-Naturalist 117:308–310.

Holloran, M. J., and S. H. Anderson. 2005. Spatial distribution of Greater Sage-Grouse nests in relatively contiguous sagebrush habitats. Condor 107:742–752.

Holloran, M. J., B. J. Heath, A. G. Lyon, S. J. Slater, J. L. Kuipers, and S. H. Anderson. 2005. Greater Sage-Grouse nesting habitat selection and success in Wyoming. Journal of Wildlife Management 69:638–649.

Holloran, M. J., R. C. Kaiser, and W. A. Hubert. 2007. Population response of yearling Greater Sage-Grouse to the infrastructure of natural gas fields in southwestern Wyoming. Completion Report. USDI Geological Survey, Laramie, WY.

Holloran, M. J., R. C. Kaiser, and W. A. Hubert. 2010. Yearling Greater Sage-Grouse response to energy development in Wyoming. Journal of Wildlife Management 74:65–72.

Holmes, E. E., J. L. Sabo, S. V. Viscido, and W. Fagan. 2007. A statistical approach to quasi-extinction forecasting. Ecology Letters 10:1182–1198.

Homer, C. G., T. C. Edwards, Jr., R. D. Ramsey, and K. P. Price. 1993. Use of remote sensing methods in modelling Sage Grouse winter habitat. Journal of Wildlife Management 57:78–84.

Honess, R. F. 1942. Sage Grouse coccidiosis not transmissible to chickens. Poultry Science 21:560.

Honess, R. F. 1955. A new flagellate, *Tritrichomonas simony*, from the Sage Grouse, *Centrocercus urophasianus*. Bulletin 1:1–3. Wyoming Game and Fish Commission, Cheyenne, WY.

Honess, R. F. 1968. Part 2: coccidial species infecting the Sage Grouse. Sage Grouse coccidiosis. Pp. 23–33 *in* History of an epizootic in sage grouse. Science Monograph 14. University of Wyoming Agricultural Experiment Station, Laramie, WY.

Honess, R. F. 1982a. Cestodes of grouse. Pp. 161–164 *in* E. T. Thorne, N. Kingston, W. R. Jolley, and R. C. Bergstrom (editors). Diseases of wildlife in Wyoming. Second edition. Wyoming Game and Fish Department, Cheyenne, WY.

Honess, R. F. 1982b. Order: Mallophaga (chewing lice). Pp. 252–254 *in* E. T. Thorne, N. Kingston, W. R. Jolley, and R. C. Bergstrom (editors). Diseases of wildlife in Wyoming. Second edition. Wyoming Game and Fish Department, Cheyenne, WY.

Honess, R. F., and R. C. Bergstrom. 1982. Ectoparasites. Pp. 231–237 *in* E. T. Thorne, N. Kingston, W. R. Jolley, and R. C. Bergstrom (editors). Diseases of wildlife in Wyoming. Second edition. Wyoming Game and Fish Department, Cheyenne, WY.

Honess, R. F., and G. Post. 1955. Eimeria of grouse (family Tetraonidae), with a description of *Eimeria pattersoni* n. sp. from the Sage Grouse. Bulletin 2:5–11. Wyoming Game and Fish Commission, Cheyenne, WY.

Honess, R. F., and G. Post. 1968. History of an epizootic in Sage Grouse. Science Monograph 14. University of Wyoming Agricultural Experiment Station, Laramie, WY.

Honess, R. F., and K. B. Winter. 1956. Diseases of wildlife in Wyoming. Bulletin 9. Wyoming Game and Fish Commission, Cheyenne, WY.

Hopla, C. E. 1974. The ecology of tularemia. Advances in Veterinary Science and Comparative Medicine 18:25–53.

Horary, F. 1969. Graph theory. Addison-Wesley, Reading, MA.

Hornaday, W. T. 1916. Save the Sage Grouse from extinction, a demand from civilization to the western states. New York Zoological Park Bulletin 5:179–219.

Horne, J. S., and E. O. Garton. 2006. Likelihood cross-validation versus least squares cross-validation for choosing the smoothing parameter in kernel home-range analysis. Journal of Wildlife Management 70:641–648.

Hosmer, D. W., and S. Lemeshow. 2000. Applied logistic regression. Second edition. John Wiley and Sons, New York, NY.

Houghton, J. G. 1969. Characteristics of rainfall in the Great Basin. Desert Research Institute, University of Nevada System, Reno, NV.

Houston, D. B. 1973. Wildfires in northern Yellowstone National Park. Ecology 54:1111–1117.

Houston, D. B. 1982. The northern Yellowstone elk: ecology and management. Macmillan, New York, NY.

Hubbard, R., and R. Hansen. 1976. Diets of wild horses, cattle, and mule deer in the Piceance Basin, Colorado. Journal of Range Management 29:389–392.

Huber-Sannwald, E., and D. A. Pyke. 2005. Establishing native grasses in a big sagebrush–dominated site: an intermediate restoration step. Restoration Ecology 13:292–301.

Hudson, P. J. 1986. The effect of a parasitic nematode on the breeding production of Red Grouse. Journal of Animal Ecology 55:85–92.

Hudson, P. J. 1992. Grouse in space and time: the population biology of a managed gamebird. The report of the Game Conservancy's Scottish grouse research project and north of England grouse research project. Game Conservancy Trust, Fordingbridge, UK.

Hudson, P. J., and A. P. Dobson. 2001. Harvesting unstable populations: Red Grouse *Lagopus lagopus scoticus* in the United Kingdom. Wildlife Biology 7:189–195.

Hudson, P. J., A. P. Dobson, and D. Newborn. 1992. Do parasites make prey vulnerable to predation? Red Grouse and parasites. Journal of Animal Ecology 61:681–692.

Hudson, P. J., A. P. Dobson, and D. Newborn. 1998. Prevention of population cycles by parasite removal. Science 282:2256–2258.

Hudson, P. J., A. P. Dobson, and D. Newborn. 1999. Population cycles and parasitism response. Science 286:2425.

Hulbert, L. C. 1955. Ecological studies of *Bromus tectorum* and other annual bromegrasses. Ecological Monographs 25:181–213.

Hulet, B. V. 1983. Selected responses of Sage Grouse to prescribed fire, predation and grazing by domestic sheep in southeastern Idaho. M.S. thesis, Brigham Young University, Provo, UT.

Hull, Jr., A. C. 1974. Species for seeding arid rangeland in southern Idaho. Journal of Range Management 27:216–218.

Hull, Jr., A. C. 1976. Rangeland use and management in the Mormon West. Symposium on Agriculture, Food and Man—A Century of Progress. Brigham Young University, Provo, UT.

Hull, Jr., A. C., and M. K. Hull. 1974. Presettlement vegetation of Cache Valley, Utah, and Idaho. Journal of Range Management 27:27–29.

Humbert, J.-Y., L. S. Mills, J. S. Horne, and B. Dennis. 2009. A better way to estimate population trend. Oikos 118:1940–1946.

Humphrey, L. D. 1984. Patterns and mechanisms of plant succession after fire on *Artemisia*-grass sites in southeastern Idaho. Vegetatio 57:91–101.

Humphrey, L. D., and E. W. Schupp. 2004. Competition as a barrier to establishment of a native perennial grass (*Elymus elymoides*) in alien annual grass (*Bromus tectorum*) communities. Journal of Arid Environments 58:405–422.

Hunter, R. 1991. *Bromus* invasions on the Nevada test site: present status of *B. rubens* and *B. tectorum* with notes on their relationship to disturbance and altitude. Great Basin Naturalist 51:176–182.

Hupp, J. W., and C. E. Braun. 1989. Topographic distribution of Sage Grouse foraging in winter. Journal of Wildlife Management 53:823–829.

Hupp, J. W., and C. E. Braun. 1991. Geographic variation among Sage Grouse in Colorado. Wilson Bulletin 103:255–261.

Hurley, M. A., and G. A. Sargeant. 1991. Effects of hunting and land management on elk habitat use, movement patterns, and mortality in western Montana. Pp. 94–98 *in* A. G. Christensen, L. J. Lyon, and T. N. Nooner (editors). Elk Vulnerability Symposium. Montana State University, Bozeman, MT.

Hutcheson, H. J., C. H. Gorham, C. Machain-Williams, M. A. Lorono-Pino, A. M. James, N. L. Marlenee, B. Winn, B. J. Beaty, and C. D. Blair. 2005. Experimental transmission of West Nile virus (Flaviviridae: *Flavivirus*) by *Carios capensis* ticks from North America. Vector Borne Zoonotic Diseases 5:293–295.

Huwer, S. L. 2004. Evaluating Greater Sage-Grouse brood habitat using human-imprinted chicks. M.S. thesis, Colorado State University, Ft. Collins, CO.

Hyder, D. N., and W. A. Sawyer. 1951. Rotation-deferred grazing as compared to season-long grazing on sagebrush-bunchgrass ranges in Oregon. Journal of Range Management 4:30–34.

Idaho Department of Fish and Game. 1997. Idaho sage grouse management plan. Idaho Department of Fish and Game, Boise, ID.

Idaho Sage-Grouse Advisory Committee. 2006. Conservation plan for the Greater Sage-Grouse in Idaho. Idaho Department of Fish and Game, Boise, ID. http://fishandgame.idaho.gov/cms/hunt/grouse/conserve_plan/Sage-grousePlan.pdf (18 March 2009).

Ingelfinger, F., and S. Anderson. 2004. Passerine response to roads associated with natural gas extraction in a sagebrush steppe habitat. Western North American Naturalist 64:385–395.

Innis, H. A. 1923. A history of the Canadian Pacific Railway. McClelland and Steward, Toronto, ON Canada.

Intergovernmental Panel on Climate Change. 2007. Climate change 2007: synthesis report. http://www.ipcc.ch/ipccreports/ar4-syr.htm (12 August 2008)

International Energy Agency. 2007. World energy outlook 2007. Paris, France.

Jackson, R. B., and M. M. Caldwell. 1993. Geostatistical patterns of soil heterogeneity around individual perennial plants. Journal of Ecology 81:683–692.

Jacobs, K., and C. Whitlock. 2008. A 2000-year environmental history of Jackson Hole, Wyoming, inferred from lake-sediment records. Western North American Naturalist 68:350–364.

Jacobson, J. E., and M. C. Snyder. 2000. Shrubsteppe mapping of eastern Washington using Landsat Satellite Thematic Mapper data. Washington Department of Fish and Wildlife, Olympia, WA.

Jardine, J. T., and C. L. Forsling. 1922. Range and cattle management during drought. U.S. Department of Agriculture Bulletin 1031. Washington, DC. U.S. Government Printing Office. Washington, DC.

Jeffress, J., and P. Roush. 2010. Lethal hoof beats: the rising toll of feral horses and burros. The Wildlife Professional 4:50–55.

Jenni, D. A., and J. E. Hartzler. 1978. Attendance at a Sage Grouse lek: implications for spring censuses. Journal of Wildlife Management 42:46–52.

Jensen, B. M. 2006. Migration, transition range and landscape use by Greater Sage-Grouse (*Centrocercus urophasianus*). M.S. thesis, University of Wyoming, Laramie, WY.

Jensen, M. E. 1989a. Soil characteristics of mountainous northeastern Nevada sagebrush community types. Great Basin Naturalist 49:469–481.

Jensen, M. E. 1989b. Soil temperature and moisture regime relationships within some rangelands of the Great Basin. Soil Science 147:134–138.

Jensen, M. E. 1990. Interpretation of environmental gradients which influence sagebrush community distribution in northeastern Nevada. Journal of Range Management 43:161–167.

Jensen, M. E., G. H. Simonson, and M. Dosskey. 1990. Correlation between soils and sagebrush-dominated plant communities of northeastern Nevada. Soil Science Society of America Journal 54:902–910.

Johansen, J. R. 1993. Cryptogamic crusts of semiarid and arid lands of North America. Journal of Phycology 29:140–147.

Johnsen, Jr., T. N. 1962. One-seed juniper invasion of northern Arizona grasslands. Ecological Monographs 32:187–207.

Johnsgard, P. A. 1973. Grouse and quails of North America. University of Nebraska Press, Lincoln, NE.

Johnsgard, P. A. 1983. Grouse of the world. University of Nebraska Press, Lincoln, NE.

Johnson, C. J., M. S. Boyce, R. L. Case, H. D. Cluff, R. J. Gau, A. Gunn, and R. Mulders. 2005. Cumulative effects of human developments on arctic wildlife. Wildlife Monographs 160. The Wildlife Society, Bethesda, MD.

Johnson, C. J., S. E. Nielsen, E. H. Merrill, T. L. McDonald, and M. S. Boyce. 2006. Resource selection functions based on use-availability data: theoretical motivation and evaluation methods. Journal of Wildlife Management 70:347–357.

Johnson, D. D., and R. F. Miller. 2006. Structure and development of expanding western juniper woodlands as influenced by two topographic variables. Forest Ecology and Management 229:7–15.

Johnson, D. D., and R. F. Miller. 2008. Old-growth juniper: distribution, abundance, and influence on post-settlement expansion. Journal of Range Ecology and Management 61:82–92.

Johnson, D. H. 1980. The comparison of usage and availability measurements for evaluating resource preference. Ecology 61:65–71.

Johnson, D. H. 2002. The importance of replication in wildlife research. Journal of Wildlife Management 66:919–932.

Johnson, D. H., and M. M. Rowland. 2007. The utility of lek counts for monitoring Greater Sage-Grouse. Pp. 15–23 *in* K. P. Reese and R. T. Bowyer (editors). Monitoring populations of Sage-Grouse. Bulletin 88. Idaho Forest, Wildlife and Range Experiment Station, College of Natural Resources, University of Idaho, Moscow, ID.

Johnson, G. D., and M. S. Boyce. 1990. Feeding trials with insects in the diet of Sage Grouse chicks. Journal of Wildlife Management 54:89–91.

Johnson, J. A., J. E. Toepfer, and P. O. Dunn. 2003. Contrasting patterns of mitochondrial and microsatellite population structure in fragmented populations of Greater Prairie-Chickens. Molecular Ecology 12:3335–3347.

Johnson, J. R., and G. F. Payne. 1968. Sagebrush reinvasion as affected by some environmental influences. Journal of Range Management 21:209–213.

Johnson, K. H., and C. E. Braun. 1999. Viability and conservation of an exploited Sage Grouse population. Conservation Biology 13:77–84.

Johnson, K., T. B. Neville, and P. Neville. 2006. GIS habitat analysis for Lesser Prairie-Chickens in southeastern New Mexico. BMC Ecology 6:18.

Johnson, L. L., and M. S. Boyce. 1991. Female choice of males with low parasite loads in sage grouse. Pp. 377–388 *in* J. E. Loye and M. Zuk (editors). Bird-parasite interactions: ecology, evolution, and behaviour. Oxford Ornithological Series 2. Oxford University Press, Oxford, UK.

Jolley, W. R. 1982. Protozoa. Pp. 107–154 *in* E. T. Thorne, N. Kingston, W. R. Jolley, and R. C. Bergstrom (editors). Diseases of wildlife in Wyoming. Second edition. Wyoming Game and Fish Department, Cheyenne, WY.

Jones, A. 2000. Effects of cattle grazing on North American arid ecosystems: a quantitative review. Western North American Naturalist 60:155–164.

Joyce, L. A. 1993. The life cycle of the range condition concept. Journal of Range Management 46:132–138.

June, J. W. 1963. Wyoming Sage Grouse population measurement. Proceedings of the Western Associations of State Game and Fish Commission 43:206–211.

Kaczor, N. W. 2008. Nesting and brood-rearing success and resource selection of Greater Sage-Grouse in northwestern South Dakota. M.S. thesis, South Dakota State University, Brookings, SD.

Kahn, N. W., C. E. Braun, J. R. Young, S. Wood, D. R. Mata, and T. W. Quinn. 1999. Molecular analysis of genetic variation among large- and small-bodied sage-grouse using mitochondrial control-region sequences. Auk 116:819–824.

Kaiser, R. C. 2006. Recruitment by Greater Sage-Grouse in association with natural gas development in Western Wyoming. M.S. thesis, University of Wyoming, Laramie, WY.

Kauhala, K., P. Helle, and E. Helle. 2000. Predator control and the density and reproductive success of grouse populations in Finland. Ecography 23:161–168.

Kay, C. E. 2007. Are lightning fires unnatural? A comparison of aboriginal and lightning ignition rates in the United States. Tall Timbers Fire Ecology Conference 23:16–28.

Keeley, J. E. 2002. Native American impacts on fire regimes of the California coastal ranges. Journal of Biogeography 29:303–320.

Keeley, J. E., C. J. Fotheringham, and M. Morais. 1999. Reexamining fire suppression impacts on brushland fire regimes. Science 284:1829–1832.

Keitt, T. H., D. L. Urban, and B. T. Milne. 1997. Detecting critical scales in fragmented landscapes. Conservation Ecology [online] 1(1):4. http://www.consecol.org/vol1/iss1/art4 (4 March 2009).

Keller, R. J., H. R. Shepherd, and R. N. Randall. 1941. North Park, Jackson County, Moffat County, including comparative data of previous seasons. Sage Grouse survey 3. Colorado Game and Fish Commission, Ft. Collins, CO.

Kellogg, K. A., J. A. Markert, J. R. Stauffer, Jr., and T. D. Kocher. 1995. Microsatellite variation demonstrates multiple paternity in lekking cichlid fishes from Lake Malawi, Africa. Proceedings of the Royal Society of London B 260:79–84.

Kennedy, A. C., L. F. Elliott, F. L. Young, and C. L. Douglas. 1991. Rhizobacteria suppressive to the weed downy brome. Soil Science Society of America Journal 55:722–727.

Kennedy, J. T., B. L. Fox, and T. D. Osen. 1995. Changing social values and images of public rangeland management. Rangelands 17(4):127–132.

Kennedy, K., R. V. Aghababian, L. Gans, and C. P. Lewis. 1996. Triage: techniques and applications in decisionmaking. Annals of Emergency Medicine 28:136–144.

Kernohan, B. J., R. A. Gitzen, and J. J. Millspaugh. 2001. Analysis of animal space use and movements. Pp. 125–166 *in* J. J. Millspaugh and J. M. Marzluff (editors). Radio tracking and animal populations. Academic Press, San Diego, CA.

Kiesecker, J. M., T. Comendant, T. Grandmason, E. Gray, C. Hall, R. Hilsenbeck, P. Kareiva, L. Lozier, P. Naehu, A. Rissman, M. R. Shaw, and M. Zankel. 2007. Conservation easements in context: a quantitative analysis of their use by The Nature Conservancy. Frontiers in Ecology and the Environment 5:125–130.

Kiesecker, J. M., H. Copeland, A. Pocewicz, N. Nibbelink, B. McKenney, J. Dahlke, M. Holloran, and D. Stroud. 2009. A framework for implementing biodiversity offsets: selecting sites and determining scale. BioScience 59:77–84.

Kilpatrick, A. M., P. Daszak, M. J. Jones, P. P. Marra, and L. D. Kramer. 2006a. Host heterogeneity dominates West Nile virus transmission. Proceedings of the Royal Society of London B 273:2327–2333.

Kilpatrick, A. M., L. D. Kramer, M. J. Jones, P. P. Marra, and P. Daszak. 2006b. West Nile virus epidemics in North America are driven by shifts in mosquito feeding behavior. Public Library of Science—Biology 4:1–5.

Kilpatrick, A. M., S. L. LaDeau, and P. P. Marra. 2007. Ecology of West Nile virus transmission and its impact on birds in the western hemisphere. Auk 124:1121–1136.

King, T. L., J. F. Switzer, C. L. Morrison, M. S. Eackles, C. C. Young, B. A. Lubinski, and P. Cryan. 2006. Comprehensive genetic analyses reveal evolutionary distinction of a mouse (*Zapus hudsonius preblei*) proposed for delisting from the United States Endangered Species Act. Molecular Ecology 15:4331–4359.

Kingston, N., and R. J. Honess. 1982. Miscellaneous hard ticks. Pp. 237–239 *in* E. T. Thorne, N. Kingston, W. R. Jolley, and R. C. Bergstrom (editors). Diseases of wildlife in Wyoming. Second edition. Wyoming Game and Fish Department, Cheyenne, WY.

Kinter, C. L., and R. N. Mack. 2004. Comparing phenotype and fitness of native, naturalized and invasive populations of downy brome (Cheatgrass, *Bromus tectorum*). Pp. 18–23 *in* A. L. Hild, N. L. Shaw, S. E. Meyer, T. D. Booth, and E. D. McArthur (compilers). Seed and soil dynamics in shrubland ecosystems. USDA Forest Service Rocky Mountain Research Paper RMRS-P-31. USDA Forest Service Rocky Mountain Research Station, Ogden, UT.

Kinter, C. L., B. A. Mealor, N. L. Shaw, and A. L. Hild. 2007. Postfire invasion potential of rush skeletonweed (*Chondrilla juncea*). Rangeland Ecology and Management 60:386–394.

Kirby, D. R., and K. L. Grosz. 1995. Cattle grazing and Sharp-tailed Grouse nesting success. Rangelands 17(4):124–126.

Kitchen, D. W. 1974. Social behavior and ecology of the pronghorn. Wildlife Monographs 38. The Wildlife Society, Bethesda, MD.

Kitchen, D. W., N. Green, and H. Green. 1977. Research needed on wild horse ecology and behavior to develop adequate management plans for public lands. Pp. 54–59 *in* National Wild Horse Forum. University of Nevada Cooperative Extension Service, Reno, NV.

Klebenow, D. A. 1969. Sage Grouse nesting and brood habitat in Idaho. Journal of Wildlife Management 33:649–662.

Klebenow, D. A. 1970. Sage Grouse versus sagebrush control in Idaho. Journal of Range Management 23:396–400.

Klebenow, D. A. 1973. The habitat requirements of Sage Grouse and the role of fire in management. Tall Timbers Fire Ecology Conference 12:305–315.

Klebenow, D. A. 1982. Livestock grazing interactions with Sage Grouse. Pp. 113–123 *in* J. M. Peck and P. D. Dalke (editors). Wildlife-Livestock Relationships Symposium: Proceedings 10. University of Idaho, Moscow, ID.

Klebenow, D. A. 1985. Habitat management for Sage Grouse in Nevada. World Pheasant Association Journal 10:34–46.
Klebenow, D. A., and G. M. Gray. 1968. Food habits of juvenile Sage Grouse. Journal of Range Management 21:80–83.
Klemmedson, J. O., and J. G. Smith. 1964. Cheatgrass (*Bromus tectorum* L.). Botanical Review 30:226–262.
Klopatek, J. M., R. J. Olson, C. J. Emerson, and J. L. Joness. 1979. Land-use conflicts with natural vegetation in the United States. Environmental Conservation 6:191–199.
Klott, J. H., and F. G. Lindzey. 1990. Brood habitats of sympatric Sage Grouse and Columbian Sharp-tailed Grouse in Wyoming. Journal of Wildlife Management 54:84–88.
Knapp, P. A. 1995. Intermountain West lightning-caused fires: climatic predictors of area burned. Journal of Range Management 48:85–91.
Knapp, P. A. 1997. Spatial characteristics of regional wildfire frequencies in Intermountain West grass-dominated communities. Professional Geographer 49:1–13.
Knapp, P. A. 1998. Spatio-temporal patterns of large grassland fires in the Intermountain West, U.S.A. Global Ecology and Biogeography Letters 7:259–272.
Knapp, P. A., and P. T. Soulé. 1998. Recent *Juniper occidentalis* (western juniper) expansion on a protected site in central Oregon. Global Change Biology 4:347–357.
Knapp, P. A., and P. T. Soulé. 1999. Vegetation change and the role of atmospheric CO_2 enrichment on a relict site in central Oregon: 1960–1994. Annals of the Association of American Geographers 86:397–411.
Knapp, P. A., P. T. Soulé, and H. D. Grissino-Mayer. 2001. Detecting potential regional effects of increased atmospheric CO_2 on growth rates of western juniper. Global Change Biology 7:903–917.
Knick, S. T. 1999. Requiem for a sagebrush ecosystem? Northwest Science 73:53–57.
Knick, S. T., and J. T. Rotenberry. 1995. Landscape characteristics of fragmented shrubsteppe landscapes and breeding passerine birds. Conservation Biology 9:1059–1071.
Knick, S. T., and J. T. Rotenberry. 1997. Landscape characteristics of disturbed shrubsteppe habitats in southwestern Idaho (U.S.A.). Landscape Ecology 12:287–297.
Knick, S. T., and J. T. Rotenberry. 2000. Ghosts of habitats past: contribution of landscape change to current habitats used by shrubland birds. Ecology 81:220–227.
Knick, S. T., and J. T. Rotenberry. 2002. Effects of habitat fragmentation on passerine birds breeding in intermountain shrubsteppe. Studies in Avian Biology 25:130–140.
Knick, S. T., D. S. Dobkin, J. T. Rotenberry, M. A. Schroeder, W. M. Vander Haegen, and C. Van Riper, III. 2003. Teetering on the edge or too late? Conservation and research issues for avifauna of sagebrush habitats. Condor 105:611–634.
Knick, S. T., A. L. Holmes, and R. F. Miller. 2005. The role of fire in structuring sagebrush habitats and bird communities. Studies in Avian Biology 30:63–75.
Knick, S. T., J. T. Rotenberry, and M. Leu. 2008. Habitat, topographical, and geographical components structuring shrubsteppe bird communities. Ecography 31:389–400.
Knight, D. H. 1994. Mountains and plains: the ecology of Wyoming landscapes. Yale University Press, New Haven, CT.
Knight, R. L, and J. Y. Kawashima. 1993. Responses of raven and Red-tailed Hawk populations to linear right-of-ways. Journal of Wildlife Management 57:266–271.
Knight, R. L., G. N. Wallace, and W. E. Riebsame. 1995. Ranching the view: subdivisions versus agriculture. Conservation Biology 9:459–461.
Knopf, F. L., J. A. Sedgwick, and D. B. Inkley. 1990. Regional correspondence among shrubsteppe bird habitats. Condor 92:45–53.
Koenig, W. D., L. Marcus, T. W. Scott, and J. L. Dickinson. 2007. West Nile virus and California breeding bird declines. EcoHealth 4:18–24.
Kokko, H. 2001. Optimal and suboptimal use of compensatory responses to harvesting: timing of hunting as an example. Wildlife Biology 7:141–150.
Kokko, H., P. T. Rintamäki, R. V. Alatalo, J. Höglund, E. Karvonen, and A. Lundberg. 1999. Female choice selects for lifetime lekking performance in Black Grouse males. Proceedings of the Royal Society of London B 266:2109–2115.
Kolada, E. J. 2007. Nest site selection and nest success of Greater Sage-Grouse in Mono County, California. M.S. thesis, University of Nevada, Reno, NV.
Kolasa, J., and N. Waltho. 1998. A hierarchical view of habitat and its relationship to species abundance. Pp. 55–76 *in* D. L. Peterson and V. T. Parker (editors). Ecological scale: theory and applications. Columbia University Press, New York, NY.
Komar, N., N. Panella, S. Langevin, A. Brault, M. Amador, E. Edwards, and J. Owen. 2005. Avian hosts for West Nile virus in St. Tammany Parish, Louisiana. American Journal of Tropical Medicine and Hygiene 73:1031–1037.
Koniak, S., and R. L. Everett. 1982. Seed reserves in soils of successional stages of pinyon woodlands. American Midland Naturalist 108:295–303.
Körner, C. 1996. The response of complex multi-species systems to elevated CO_2. Pp. 20–42 *in* B. Walker and W. Steffen (editors). Global change and terrestrial ecosystems. Cambridge University Press, Cambridge, UK.
Kotliar, N. B., and J. A. Wiens. 1990. Multiple scales of patchiness and patch structure: a hierarchical

framework for the study of heterogeneity. Oikos 59:253–260.
Kou, X., and W. L. Baker. 2006a. Accurate estimation of mean fire interval for managing fire. International Journal of Wildland Fire 15:489–495.
Kou, X., and W. L. Baker. 2006b. A landscape model quantifies error in reconstructing fire history from scars. Landscape Ecology 21:735–745.
Kramer, L. D., L. M. Styer, and G. D. Ebel. 2008. A global perspective on the epidemiology of West Nile virus. Annual Review of Entomology 53:4.1–4.21.
Krebs, C. J. 1998. Ecological methodology. Second edition. Addison-Wesley Educational Publishers, Menlo Park, CA.
Kristan, III, W. B. 2006. Sources and expectations for hierarchical structure in bird-habitat associations. Condor 108:5–12.
Kristan, III, W. B., and W. I. Boarman. 2003. Spatial pattern of risk of Common Raven predation on desert tortoises. Ecology 84:2432–2443.
Küchler, A. W. 1964. Manual to accompany the map—potential natural vegetation of the conterminous United States. American Geographic Society Special Publication No. 36. American Geographical Society, New York, NY.
Küchler, A. W. 1970. The potential natural vegetation of the conterminous United States. The national atlas of the United States of America. USDI Geological Survey, Washington, DC.
Kurki, S., A. Nikula, P. Helle, and H. Lindén. 1997. Landscape-dependent breeding success of forest grouse in Fennoscandia. Wildlife Biology 3:295.
Kuvlesky, Jr., W. P., L. A. Brennan, M. L. Morrison, K. K. Boydston, B. M. Ballard, and F. C. Bryant. 2007. Wind energy development and wildlife conservation: challenges and opportunities. Journal of Wildlife Management 71:2487–2498.
Lackey, R. T. 2007. Science, scientists, and policy advocacy. Conservation Biology 21:12–17.
LaDeau, S. L., A. M. Kilpatrick, and P. P. Marra. 2007. West Nile virus emergence and large-scale declines of North American bird populations. http://nationalzoo.si.edu/Publications/ScientificPublications/pdfs/nature05829.pdf (15 October 2008).
Laerm, J., J. C. Avise, J. C. Patton, and R. A. Lansman. 1982. Genetic determination of the status of an endangered species of pocket gopher in Georgia. Journal of Wildlife Management 46:513–518.
Laliberte, A. S., and W. J. Ripple. 2004. Range contractions of North American carnivores and ungulates. BioScience 54:123–138.
Lammers, W. M., and M. W. Collopy. 2007. Effectiveness of avian predator perch deterrents on electric transmission lines. Journal of Wildlife Management 71:2752–2758.
Lanctot, R. B., K. T. Scribner, B. Kempenaers, and P. J. Weatherhead. 1997. Lekking without a paradox in the Buff-breasted Sandpiper. American Naturalist 6:1051–1070.
Lande, R., and G. W. Barrowclaugh. 1987. Effective population size, genetic variation, and their use in population management. Pp. 87–124 *in* M. E. Soulé (editor). Viable conservation for populations. Cambridge University Press, Cambridge, UK.
LANDFIRE. 2006. National existing vegetation type layer. USDI Geological Survey. http://gisdata.usgs.net/website/landfire/ (21 August 2009).
Landres, P. B., P. Morgan, and F. J. Swanson. 1999. Overview of the use of natural variability concepts in managing ecological systems. Ecological Applications 9:1179–1188.
Lank, D. B., C. M. Smith, O. Hanotte, A. Ohtonen, S. Baily, and T. Burke. 2002. High frequency of polyandry in a lek mating system. Behavioral Ecology 13:209–215.
Lawrence, J. S., and N. J. Silvy. 1995. Effect of predator control on reproductive success and hen survival of Attwater's Prairie-Chicken. Proceedings of Southeastern Association of Fish and Wildlife Agencies 49:275–282.
Laycock, W. A. 1967. How heavy grazing and protection affect sagebrush-grass ranges. Journal of Range Management 20:206–213.
Laycock, W. A. 1991. Stable states and thresholds of range condition on North American rangelands: a viewpoint. Journal of Range Management 44:427–433.
Laycock, W. A., and P. W. Conrad. 1981. Responses of vegetation and cattle to various systems of grazing on seeded and native mountain rangelands in eastern Utah. Journal of Range Management 34:52–58.
Le Page, S. L., R. A. Livermore, D. W. Cooper, and A. C. Taylor. 2000. Genetic analysis of a documented population bottleneck: introduced Bennett's wallabies (*Macropus rufogriseus rufogriseus*) in New Zealand. Molecular Ecology 9:753–763.
Leach, H. R., and A. L. Hensley. 1954. The Sage Grouse in California with special reference to food habitats. California Fish and Game 40:385–394.
Lebigre, C., R. V. Alatalo, H. Sitari, and S. Parri. 2007. Restrictive mating by females on Black Grouse leks. Molecular Ecology 16:4380–4389.
Lee, J. H., H. Hassan, G. Hill, E. W. Cupp, T. B. Higazi, C. J. Mitchell, M. S. Godsey, Jr., and T. R. Unnasch. 2002. Identification of mosquito avian-derived blood meals by polymerase chain reaction-heteroduplex analysis. American Journal of Tropical and Medicine and Hygiene 66:599–604.
Leidy, J. 1887. Tape-worms in birds. Journal of Comparative Medicine and Surgery 8:1–11.
Leonard, K. M., K. P. Reese, and J. W. Connelly. 2000. Distribution, movements and habitats of Sage Grouse *Centrocercus urophasianus* on the upper Snake River Plain of Idaho: changes from the 1950s to the 1990s. Wildlife Biology 6:265–270.

Leopold, A., T. M. Sperry, W. S. Feeney, and J. A. Catenhusen. 1943. Population turnover on a Wisconsin pheasant refuge. Journal of Wildlife Management 7:383–394.

Leptich, D. J., and P. Zager. 1991. Road access management effects on elk mortality and population dynamics. Pp. 126–131 *in* A. G. Christensen, L. J. Lyon, and T. N. Nooner (editors). Elk Vulnerability Symposium. Montana State University, Bozeman, MT.

Lesica, P., S. V. Cooper, and G. Kudray. 2005. Big sagebrush shrub-steppe postfire succession in southwest Montana. Montana Natural Heritage Program, Natural Resource Information System, Montana State Library, Helena, MT.

Lesica, P., S. V. Cooper, and G. Kudray. 2007. Recovery of big sagebrush following fire in southwest Montana. Rangeland Ecology and Management 60:261–269.

Leu, M., S. E. Hanser, and S. T. Knick. 2008. The human footprint in the West: a large-scale analysis of anthropogenic impacts. Ecological Applications 18:1119–1139.

Levin, P. S., J. Ellis, R. Petrik, and M. E. Hay. 2002. Indirect effects of feral horses on estuarine communities. Conservation Biology 16:1364–1371.

Levin, S. A. 1992. The problem of pattern and scale in ecology. Ecology 73:1943–1967.

Levine, N. D. 1953. A review of the coccidia from the avian orders Galliformes, Anseriformes, and Charadriiformes with descriptions of three new species. American Midland Naturalist 49:696–719.

Li, H., and J. F. Reynolds. 1994. A simulation experiment to quantify spatial heterogeneity in categorical maps. Ecology 75:2446–2455.

Li, H., and J. Wu. 2004. Use and misuse of landscape indices. Landscape Ecology 19:389–399.

Li, Y. C., A. B. Korol, T. Fahima, A. Beiles, and E. Novo. 2002. Microsatellites: genomic distribution, putative functions and mutational mechanisms: a review. Molecular Ecology 11:2453–65.

Lieurance, M. T. 1979. Grazing outlook on public lands managed by the Bureau of Land Management. Rangelands 1(2):50–51.

Linberg, M. S., J. S. Sedinger, D. V. Derksen, and R. F. Rockwell. 1998. Natal and breeding philopatry in a Black Brant, *Branta bernicla nigricans*, metapopulation. Ecology 79:1893–1904.

Lindenmeyer D., R. J. Hobbs, R. Montague-Drake, J. Alexandra, A. Bennett, M. Burgman, P. Cale, A. Calhoun, V. Cramer, P. Cullen, D. Driscoll, L. Fahrig, J. Fischer, J. Franklin, Y. Haila, M. Hunter, P. Gibbons, S. Lake, G. Luck, C. Macgregor, S. Mcintyre, R. Mac Nally, A. Manning, J. Miller, H. Mooney, R. Noss, H. Possingham, D. Saunders, F. Schmiegelow, M. Scott, D. Simberloff, T. Sisk, G. Tabor, B. Walker, J. Wiens, J. Woinarski, and E. Zavaleta. 2008. A checklist for ecological management of landscapes for conservation. Ecology Letters 11:78–91.

Link, S. O., C. W. Keeler, R. W. Hill, and E. Hagen. 2006. *Bromus tectorum* cover mapping and fire risk. International Journal of Wildland Fire 15:113–119.

Lockwood, J. 2008, June 9. Why the West needs mythic cowboys. (Paonia, CO) High Country News, 7 January 2011. http://www.hcn.org/issues/372/17740.

Lommasson, T. 1948. Succession in sagebrush. Journal of Range Management 1:19–21.

Lomolino, M. V., and R. Channell. 1995. Splendid isolation: patterns of geographic range collapse in endangered mammals. Journal of Mammalogy 76:335–347.

Lomolino, M. V., and R. Channell. 2000. Dynamic biogeography and conservation of endangered species. Nature 403:84–86.

Longland, W. S., and S. L. Bateman. 2002. Viewpoint: the ecological value of shrub islands on disturbed sagebrush rangelands. Journal of Range Management 55:571–575.

Lord, C. C., and J. F. Day. 2001. Simulation studies of St. Louis encephalitis and West Nile viruses: the impact of bird mortality. Vector Borne and Zoonotic Diseases 1:317–329.

Lord, C. C., R. Rutledge, and W. J. Tabachnick. 2006. Relationships between host viremia and vector susceptibility for arboviruses. Journal of Medical Entomology 43:623–630.

Lovich, J. E., and D. Bainbridge. 1999. Anthropogenic degradation of the southern California desert ecosystem and prospects for natural recovery and restoration. Environmental Management 24:309–326.

Lowe, B. S. 2006. Greater Sage-Grouse use of threetip sagebrush and seeded sagebrush steppe. M.S. thesis, Idaho State University, Pocatello, ID.

Loydi, A., and S. M. Zalba. 2009. Feral horse dung piles as potential invasion windows for alien plant species in natural grasslands. Plant Ecology 201:471–480.

Ludwig, D. 1999. Is it meaningful to estimate a probability of extinction? Ecology 80:298–310.

Lungle, K., and S. Pruss. 2007. Recovery strategy for the Greater Sage-Grouse (*Centrocercus urophasianus urophasianus*) in Canada [proposed]. Species at Risk Act Recovery Strategy Series. Parks Canada, Ottawa, Canada.

Lupis, S. G., T. A. Messmer, and T. Black. 2006. Gunnison Sage-Grouse use of Conservation Reserve Program fields in Utah and response to emergency grazing: a preliminary evaluation. Wildlife Society Bulletin 34:957–962.

Luttrell, M. P., L. H. Creekmore, and J. W. Mertins. 1996. Avian tick paralysis caused by *Ixodes brunneus* in the southeastern United States. Journal of Wildlife Diseases 32:133–136.

Lyon, A. G. 2000. The potential effects of natural gas development on Sage Grouse near Pinedale,

Wyoming. M.S. thesis, University of Wyoming, Laramie, WY.

Lyon, A. G., and S. H. Anderson. 2003. Potential gas development impacts on Sage Grouse nest initiation and movement. Wildlife Society Bulletin 31:486–491.

Lyon, L. J. 1979. Habitat effectiveness for elk as influenced by roads and cover. Journal of Forestry 77:658–660.

Lysne, C. R. 2005. Restoring Wyoming big sagebrush. Pp. 93–98 *in* N. L. Shaw, M. Pellant, and S. B. Monsen (compilers). Sage-Grouse Habitat Restoration Symposium proceedings. USDA Forest Service RMRS-P-38. USDA Forest Service, Rocky Mountain Research Station, Ft. Collins, CO.

Mac, M. J., P. A. Opler, E. P. Haecker, and P. D. Doran. 1998. Status and trends of the nation's biological resources. Volume 2. USDI Geological Survey, Reston, VA.

Mack, R. N. 1981. Invasion of *Bromus tectorum* L. into western North America: an ecological chronicle. Agro-ecosystems 7:145–165.

Mack, R. N., and D. A. Pyke. 1983. The demography of *Bromus tectorum*: variation in time and space. Journal of Ecology 71:69–93.

Mack, R. N., and J. N. Thompson. 1982. Evolution in steppe with few large, hooved mammals. American Naturalist 119:757–773.

MacKenzie, D. I., J. D. Nichols, J. A. Royle, K. H. Pollock, L. L. Bailey, and J. E. Hines. 2006. Occupancy estimation and modeling. Elsevier-Academic, San Diego, CA.

Maestas, J. D., R. L. Knight, and W. C. Gilgert. 2003. Biodiversity across a rural land-use gradient. Conservation Biology 17:1425–1434.

Mahalovich, M. F., and E. D. McArthur. 2004. Sagebrush (*Artemisia* spp.) seed and plant transfer guidelines. Native Plants (Fall):141–148.

Maier, A. M., B. L. Perryman, R. A. Olson, and A. L. Hild. 2001. Climatic influences on recruitment of 3 subspecies of *Artemisia tridentata*. Journal of Range Management 54:699–703.

Malcomson, R. O. 1960. Mallophaga from birds of North America. Wilson Bulletin 72:182–197.

Manier, D. J., and N. T. Hobbs. 2007. Large herbivores in sagebrush steppe ecosystems: livestock and wild ungulates influence structure and function. Oecologia 152:739–750.

Manson, C. 2008. For Your Information 16(12). http://www.mcgeorge.edu/documents/publications/insidemcgeorge/FYI-2008-06.pdf (25 September 2009). http://www.msnbc.msn.com/id/24761281/ns/us_news-environment/

Manzer, D. L., and S. J. Hannon. 2005. Relating grouse nest success and corvid density to habitat: a multiscale approach. Journal of Wildlife Management 69:110–123.

Marcström, V., R. E. Kenward, and E. Engren. 1988. The impact of predation on boreal tetraonids during vole cycles: an experimental study. Journal of Animal Ecology 57:859–872.

Margules, C. R., and R. L. Pressey. 2000. Systematic conservation planning. Nature 405:243–253.

Marra, P. P., S. Griffing, C. Caffrey, A. M. Kilpatrick, R. McLean, C. Brand, E. Saito, A. P. Dupuis, L. Kramer, and R. Novak. 2004. West Nile virus and wildlife. BioScience 54:393–402.

Martin, C. A., J. C. Alonso, J. Alonso, C. Pitra, and D. Lieckfeldt. 2001. Great Bustard population in central Spain: concordant results from genetic analysis and dispersal study. Proceedings of the Royal Society of London B 269:119–125.

Martin, J. W., and B. A. Carlson. 1998. Sage Sparrow (*Amphispiza belli*). *In* A. Poole and F. Gill (editors). The birds of North America, No. 326. The Academy of Natural Sciences, Philadelphia, PA, and The American Ornithologists' Union, Washington, DC.

Martin, N. S. 1970. Sagebrush control related to habitat and Sage Grouse occurrence. Journal of Wildlife Management 34:313–320.

Martin, P. S. 1984. Prehistoric overkill: the global model. Pp. 354–403 *in* P.S. Martin and R.G. Klein (editors). Quaternary extinctions: a prehistoric revolution. University of Arizona Press, Tucson, AZ.

Martin, T. E. 1993. Nest predation and nest sites—new perspectives on old patterns. Bioscience 43:523–532.

Marzluff, J. M., and E. Neatherlin. 2006. Corvid response to human settlements and campgrounds: causes, consequences, and challenges for conservation. Biological Conservation 130:301–314.

Mata-González, R., R. G. Hunter, C. L. Coldren, T. McLendon, and M. W. Paschke. 2008. A comparison of modeled and measured impacts of resource manipulations for control of *Bromus tectorum* in sagebrush steppe. Journal of Arid Environments 72:836–846.

MathWorks, Inc. 2007. Matlab. Version R2007a. Natick, MA.

May, R. M., and R. M. Anderson. 1978. Regulation and stability of host parasite population interactions. II: destabilizing processes. Journal of Animal Ecology 47:249–267.

May, R. M., and R. M. Anderson. 1979. Population biology of infectious diseases: part II. Nature 280:455–461.

McAdam, S. M. 2003. Lek occupancy by Greater Sage Grouse in relation to habitat in southwestern Saskatchewan. M.S. thesis, Royal Roads University, Victoria, BC, Canada.

McArdle, R. E., D. F. Costello, E. E. Birkmaier, E. Ewing, B. A. Hendricks, C. A. Kutzleb, A. A. Simpson, and A. R. Standing. 1936. III. The white man's toll. Letter from the Secretary of Agriculture to the U.S. Senate. A report on the western range—a great but neglected resource. Senate Document 199:81–116.

McArthur, E. D. 1983. Taxonomy, origin, and distribution of big sagebrush (*Artemisia tridentata*) and allies (subgenus Tridentatae). Pp. 3–11 *in* R. L. Johnson (editor). First Utah Shrub Ecology Workshop. College of Natural Resources, Utah State University, Logan, UT.

McArthur, E. D., and J. E. Ott. 1996. Potential natural vegetation in the 17 conterminous western United States. Pp. 16–28 *in* J. R. Barrow, E. D. McArthur, R. E. Sosebee, and R. J. Tausch (compilers). Shrubland ecosystem dynamics in a changing environment. USDA Forest Service General Technical Report INT-GTR-338. USDA Forest Service, Rocky Mountain Research Station, Ft. Collins, CO.

McArthur, E. D., and A. P. Plummer. 1978. Biogeography and management of native western shrubs: a case study, section Tridentatae of *Artemisia*. Great Basin Naturalist Memoirs 2:229–243.

McArthur, E. D., A. C. Blauer, and S. C. Sanderson. 1988. Mule deer–induced mortality of mountain sagebrush. Journal of Range Management 41:114–117.

McArthur, T. O. 2004. Emergency fire rehabilitation of BLM lands in the Great Basin: revegetation and monitoring. M.S. thesis, Oregon State University, Corvallis, OR.

McCarthy, J. J., and G. D. Kobriger. 2005. Management plan and conservation strategies for Greater Sage-Grouse in North Dakota. North Dakota Game and Fish Department, Bismark, ND. http://www.gf.nd.gov/conservation/sage-grouse-plan.html (18 March 2009).

McDaniel, K. C., D. L. Anderson, and J. F. Balliette. 1991. Wyoming big sagebrush control with metsulfuron and 2,4-D in northern New Mexico. Journal of Range Management 44:623–627.

McDonald, D. B., and W. K. Potts. 1994. Cooperative display and relatedness among males in a lek-mating bird. Science 266:1030–1032.

McDonald, M. W., and K. P. Reese. 1998. Landscape changes within the historical distribution of Columbian Sharp-tailed Grouse in eastern Washington: is there hope? Northwest Science 72:34–41.

McGarigal, K., S. A. Cushman, M. C. Neel, and E. Ene. 2002. Fragstats: spatial pattern analysis program for categorical maps. Computer software program produced by the authors. University of Massachusetts, Amherst, MA. http://www.umass.edu/landeco/research/fragstats/fragstats.html (21 August 2009).

McInnis, M. L. 1985. Ecological relationships among feral horses, cattle, and pronghorn in southeastern Oregon. Ph.D. dissertation, Oregon State University, Corvallis, OR.

McIntyre, N. E., and J. A. Wiens. 2000. A novel use of the lacunarity index to discern landscape function. Landscape Ecology 15:313–321.

McIver, J., and L. Starr. 2001. Restoration of degraded lands in the interior Columbia River basin: passive vs. active approaches. Forest Ecology and Management 153:15–28.

McKean, W. T. 1976. Winter guide to central Rocky Mountain shrubs (with summer key). Colorado Department of Natural Resources, Division of Wildlife Resources, Denver, CO.

McKinney, M. L. 2002. Urbanization, biodiversity, and conservation. BioScience 52:883–890.

McKnight, T. L. 1958. The feral burro in the United States: distribution and problems. Journal of Wildlife Management 22:163–179.

McLean, R. G. 2006. West Nile virus in North American birds. Ornithological Monographs 60:44–64.

McLendon, T., and E. F. Redente. 1990. Succession patterns following soil disturbance in a sagebrush steppe community. Oecologia 85:293–300.

McLendon, T., and E. F. Redente. 1992. Effects of nitrogen limitation on species replacement dynamics during early succession on a semiarid sagebrush site. Oecologia 91:312–317.

McLoughlin, P. D., R. L. Case, R. J. Gau, H. Dean Cluff, R. Mulders, and F. Messier. 2002. Hierarchical habitat selection by barren-ground grizzly bears in the central Canadian arctic. Oecologia 132:102–108.

McMaster, D. G., and S. K. Davis. 2001. An evaluation of Canada's permanent cover program: habitat for grassland birds? Journal of Field Ornithology 72:195–210.

McSweegan, E. 1996. The infectious diseases impact statement: a mechanism for addressing emerging diseases. Emerging Infectious Diseases 2:103–108.

Medin, D. E., B. L. Welch, and W. P. Clary. 2000. Bird habitat relationships along a Great Basin elevational gradient. USDA Forest Service Research Paper RMRS-RP-23. USDA Forest Service, Rocky Mountain Research Station, Ft. Collins, CO.

Meece, J. K., T. A. Kronenwetter-Koepel, M. F. Vandermause, and K. D. Reed. 2006. West Nile virus infection in commercial waterfowl operation, Wisconsin. Emerging Infectious Diseases 12:1451–1453.

Meeker, J. O. 1979. Interactions between pronghorn antelope and feral horses in northwestern Nevada. M.S. thesis, University of Nevada, Reno, NV.

Mehringer, P. J. 1985. Late-Quaternary pollen records from the interior Pacific Northwest and northern Great Basin of the United States. Pp. 167–189 *in* W. M. Bryan, Jr., and R. G. Holloway (editors). Pollen records of late-Quaternary North American sediments. American Association of Stratigraphic Palynoligists, Dallas, TX.

Mehringer, Jr., P. J., and P. E. Wigand. 1987. Western juniper in the Holocene. Pp. 13–16 *in* R. L. Everett (compiler). Pinyon-Juniper Conference. USDA Forest Service General Technical Report INT-215. USDA Forest Service, Intermountain Research Station, Ogden UT.

Mehus, C. A. 1995. Influences of browsing and fire on sagebrush taxa of the northern Yellowstone winter range. M.S. thesis, Montana State University, Bozeman, MT.

Meinke, C. W., S. T. Knick, and D. A. Pyke. 2009. A spatial model to prioritize sagebrush landscapes in the Intermountain West (U.S.A.) for restoration. Restoration Ecology 17:652–659.

Meisner, B. N., R. A. Chase, M. H. McCutchan, R. Mees, J. W. Benoit, B. Ly, D. Albright, D. Strauss, and T. Ferryman. 1994. A lightning fire ignition assessment model. Proceedings of the Conference on Fire and Forest Meteorology 12:172–178.

Melgoza, G., R. Nowak, and R. Tausch. 1990. Soil water exploitation after fire: competition between *Bromus tectorum* (cheatgrass) and two native species. Oecologia 83:7–13.

Menard, C., P. Duncan, G. Fleurance, J. Georges, and M. Lila. 2002. Comparative foraging and nutrition of horses and cattle in European wetlands. Journal of Applied Ecology 39:120–133.

Mensing, S., S. Livingston, and P. Barker. 2006. Long-term fire history in Great Basin sagebrush reconstructed from macroscopic charcoal in spring sediments, Newark Valley, Nevada. Western North American Naturalist 66:64–77.

Messmer, T. A., M. W. Brunson, D. Reiter, and D. G. Hewitt. 1999. United States public attitudes regarding predators and their management to enhance avian recruitment. Wildlife Society Bulletin 27:75–85.

Meyer, S. E., D. L Nelson, and S. Clement. 2001. Evidence for resistance polymorphism in the *Bromus tectorum–Ustilago bullata* pathosystem: implications for biocontrol. Canadian Journal of Plant Pathology 23:19–27.

Mezquida, E. T., S. J. Slater, and C. W. Benkman. 2006. Sage-Grouse and indirect interactions: potential implications of coyote control on Sage-Grouse populations. Condor 108:747–759.

Michener, G. R. 2005. Limits on egg predation by Richardson's ground squirrels. Canadian Journal of Zoology 83:1030–1037.

Milby, M. M., and R. P. Meyer. 1986. The influence of constant versus fluctuating water temperatures on the preimaginal development of *Culex tarsalis*. Journal of the American Mosquito Control Association 2:7–10.

Milchunas, D. G., and W. K. Lauenroth. 1993. Quantitative effects of grazing on vegetation and soils over a global range of environments. Ecological Monographs 63:327–366.

Milchunas, D. G., O. E. Sala, and W. K. Lauenroth. 1988. A generalized model of the effects of grazing by large herbivores on grassland community structure. American Naturalist 132:87–106.

Milchunas, D. G., K. A. Schulz, and R. B. Shaw. 2000. Plant community structure in relation to long-term disturbance by mechanized military maneuvers in a semiarid region. Environmental Management 25:525–539.

Miller, D. 2006. Controlling annual bromes: using rangeland "greenstrips" to create natural fire breaks. Rangelands 28(2):22–25.

Miller, H. C., D. Clausnitzer, and M. M. Borman. 1999. Medusahead. Pp. 271–281 *in* R. L. Sheley and J. K. Petroff (editors). Biology and management of noxious rangeland weeds. Oregon State University Press, Corvallis, OR.

Miller, J., D. Germanoski, K. Waltman, R. Tausch, and J. Chambers. 2001. Influence of late Holocene hillslope processes and landforms on modern channel dynamics in upland watersheds in central Nevada. Geomorphology 38:373–391.

Miller, R. 1980. The ecology of feral horses in Wyoming's Red Desert. Ph.D. dissertation, University of Wyoming, Laramie, WY.

Miller, R. F., and L. Eddleman. 2001. Spatial and temporal changes of Sage Grouse habitat in the sagebrush biome. Technical Bulletin 151. Oregon State University Agricultural Experiment Station, Corvallis, OR.

Miller, R. F., and E. K. Heyerdahl. 2008. Fine-scale variation of historical fire regimes in sagebrush-steppe and juniper woodland: an example from California, USA. International Journal of Wildland Fire 17:245–254.

Miller, R. F., and J. A. Rose. 1995. Historic expansion of *Juniperus occidentalis* (western juniper) in southeastern Oregon. Great Basin Naturalist 55:37–45.

Miller, R. F., and J. A. Rose. 1999. Fire history and western juniper encroachment in sagebrush steppe. Journal of Range Management 52:550–559.

Miller, R. F., and L. M. Shultz. 1987. Development and longevity of ephemeral and perennial leaves on *Artemisia tridentata* Nutt ssp. *wyomingensis*. Western North American Naturalist 47:227–230.

Miller, R. F., and R. J. Tausch. 2001. The role of fire in pinyon and juniper woodlands: a descriptive analysis. Tall Timbers Research Station Miscellaneous Publication No. 11:15–30.

Miller, R. F., and P. E. Wigand. 1994. Holocene changes in semiarid pinyon-juniper woodlands. BioScience 44:465–474.

Miller, R. F., J. D. Bates, T. J. Svejcar, F. B. Pierson, and L. E. Eddleman. 2005. Biology, ecology, and management of western juniper. Technical Bulletin 152. Oregon State University Agricultural Experiment Station, Corvallis, OR.

Miller, R. F., J. D. Bates, T. J. Svejcar, F. B. Pierson, and L. E. Eddleman. 2007. Western juniper field guide: asking the right questions to select appropriate management actions. USDI Geological Survey Circular 1321. USDI Geological Survey, Reston, VA.

Miller, R. F., T. Svejcar, and J. A. Rose. 2000. Impacts of western juniper on plant community composition and structure. Journal of Range Management 53:574–585.

Miller, R. F., T. J. Svejcar, and N. E. West. 1994. Implications of livestock grazing in the intermountain sagebrush region: plant composition. Pp. 101–146 *in* M. Vavra, W.A. Laycock, and R. D. Pieper

(editors). Ecological implications of livestock herbivory in the West. Society for Range Management, Denver, CO.
Miller, R. F., R. J. Tausch, E. D. MacArthur, D. D. Johnson, and S. C. Sanderson. 2008. Age structure and expansion of piñon-juniper woodlands: a regional perspective in the Intermountain West. USDA Forest Service Research Paper Report RMRS-RP-69. USDA Forest Service, Ft. Collins, CO.
Miller, S. G., R. L. Knight, and C. K. Miller. 1998. Influence of recreational trails on breeding bird communities. Ecological Applications 8:162–169.
Mills, L. S. 2007. Conservation of wildlife populations: demography, genetics, and management. Blackwell Publishing, Malden, MA.
Mills, T. J., and R. N. Clark. 2001. Roles of research scientists in natural resource decision-making. Forest Ecology and Management 153:189–198.
Milne, B. T. 1992. Spatial aggregation and neutral models in fractal landscapes. American Naturalist 139:32–57.
Milton, S. J., W. R. J. Dean, M. A. du Plessis, and W. R. Siegfried. 1994. A conceptual model of arid rangeland degradation. BioScience 44:70–76.
Minor, E. S., and D. L. Urban. 2008. A graph-theory framework for evaluating landscape connectivity and conservation planning. Conservation Biology 22:297–307.
Misenhelter, M. D., and J. T. Rotenberry. 2000. Choices and consequences of habitat occupancy and nest site selection in Sage Sparrows. Ecology 81:2892–2901.
Mitchell, J. E. 2000. Rangeland resource trends in the United States: a technical document supporting the 2000 USDA Forest Service RPA Assessment. USDA Forest Service General Technical Report RMRS-GTR-68. USDA Forest Service, Rocky Mountain Research Station, Ft. Collins, CO.
Mitchell, J. E., and R. H. Hart. 1987. Winter of 1886–87: the death knell of open range. Rangelands 9(1):3–8.
Mitchell, J. E., R. L. Knight, and R. J. Camp. 2002. Landscape attributes of subdivided ranches. Rangelands 24(1):3–9.
Mitchell, V. L. 1976. Regionalization of climate in the western United States. Journal of Applied Meteorology 5:920–927.
Monsen, S. B., and S. G. Kitchen (compilers). 1994. Ecology and management of annual rangelands. USDA Forest Service General Technical Report INT-GTR-313. USDA Forest Service, Intermountain Research Station, Ogden, UT.
Monsen, S. B., R. Stevens, and N. L. Shaw (compilers). 2004. Restoring western ranges and wildlands. USDA Forest Service RMRS-GTR-136. USDA Forest Service, Rocky Mountain Research Station, Ft. Collins, CO.
Montana Sage-Grouse Work Group. 2003. Management plan and conservation strategies for sage-grouse in Montana. Montana Fish, Wildlife and Parks Department, Helena, MT. http://fwp.mt.gov/fwppaperapps/wildthings/SGFinalPlan.pdf (18 March 2009).
Mooney, H. A., and R. J. Hobbs (editors). 2000. Invasive species in a changing world. Island Press, Washington, DC.
Moritz, W. E. 1988. Wildlife use of fire-disturbed areas in sagebrush steppe on the Idaho National Engineering Laboratory. M.S. thesis, Montana State University, Bozeman, MT.
Morris, M. S., R. G. Kelsey, and D. Griggs. 1976. The geographic and ecological distribution of big sagebrush and other woody *Artemisia* in Montana. Proceedings of the Montana Academy of Sciences 36:56–79.
Morris, W. F., and D. Doak. 2002. Quantitative conservation biology: theory and practice of population viability analysis. Sinauer Associates, Sunderland, MA.
Morrison, R. B. 1964. Lake Lahontan: geology of southern Carson Desert, Nevada. USGS Professional Paper 401. U.S. Government Printing Office, Washington, DC.
Mosley, J. C. 1996. Prescribed sheep grazing to suppress cheatgrass: a review. Sheep and Goat Research Journal 12:74–81.
Mosley, J. C., S. C. Bunting, and M. E. Manoukian. 1999. Cheatgrass. Pp. 175–188 *in* R. L. Sheley and J. K. Petroff (editors). Biology and management of noxious rangeland weeds. Oregon State University Press, Corvallis, OR.
Motha, M. X., J. R. Egerton, and A. W. Sweeney. 1984. Some evidence of mechanical transmission of reticuloendotheliosis virus by mosquitos. Avian Diseases 28:858–867.
Moudy, R. M., M. A. Meola, L. L. Morin, G. D. Ebel, and L. D. Kramer. 2007. A newly emergent genotype of West Nile virus is transmitted earlier and more efficiently by *Culex* mosquitos. American Journal of Tropical Medicine and Hygiene 77:365–370.
Moynahan, B. J. 2004. Landscape-scale factors affecting population dynamics of Greater Sage-Grouse (*Centrocercus urophasianus*) in north-central Montana, 2001–2004. Ph.D. dissertation, University of Montana, Missoula, MT.
Moynahan, B. J., M. S. Lindberg, J. J. Rotella, and J. W. Thomas. 2007. Factors affecting nest survival of Greater Sage-Grouse in northcentral Montana. Journal of Wildlife Management 71:1773–1783.
Moynahan, B. J., M. S. Lindberg, and J. W. Thomas. 2006. Factors contributing to process variance in annual survival of female Greater Sage-Grouse in Montana. Ecological Applications 16:1529–1538.
Mueggler, W. F. 1950. Effects of spring and fall grazing by sheep on vegetation of the upper Snake River plains. Journal of Range Management 3:308–315.
Mueggler, W. F. 1956. Is sagebrush seed residual in the soil of burns or is it wind-borne? USDA Forest Service Research Note RN-INT-35. USDA Forest Service, Intermountain Forest and Range Experiment Station, Ogden, UT.

Mueller, R. J., and J. H. Richards. 1986. Morphological analysis of tillering in *Agropyron spicatum* and *A. desertorum*. Annals of Botany 58:911–921.

Muscha, J. M., and A. L. Hild. 2006. Biological soil crusts in grazed and ungrazed Wyoming sagebrush steppe. Journal of Arid Environments 67:195–207.

Musil, D. D., J. W. Connelly, and K. P. Reese. 1993. Movements, survival and reproduction of Sage Grouse translocated into central Idaho. Journal of Wildlife Management 57:85–91.

Myers, N., R. A. Mittermeier, C. G. Mittermeier, G. A. B. Dafonseca, and J. Kent. 2000. Biodiversity hotspots for conservation priorities. Nature 403:853–858.

Naiman, R. J., H. Décamps, and M. Pollock. 1992. The role of riparian corridors in maintaining regional biodiversity. Ecological Applications 3:209–212.

Nasci, R. S., H. M. Savage, D. J. White, J. R. Miller, B. C. Cropp, M. S. Godsey, A. J. Kerst, P. Bennett, K. Gottfreid, and R. S. Lanciotti. 2001. West Nile virus in overwintering *Culex* mosquitos, New York City, 2000. Emerging Infectious Diseases 7:742–744.

National Energy Policy Development Group. 2001. National energy policy. http://www.pppl.gov/common_pics/national_energy_policy/national_energy_policy.pdf (15 August 2009).

National Petroleum Council. 1999. Natural gas: meeting the challenges of the nation's growing natural gas demand. Vol. 1: summary report. http://www.npc.org/reports/ReportVol1.pdf (15 August 2009)

National Petroleum Council. 2003. Balancing natural gas policy—fueling the demands of a growing economy. Vol 1: summary of findings and recommendations. Final Report. http://www.npc.org/reports/NG_Volume1.pdf (21 August 2009).

National Petroleum Council. 2007. Facing the hard truths about energy: a comprehensive view to 2030 of global oil and natural gas. http://www.npchardtruthsreport.org (7 April 2009).

National Renewable Energy Laboratory. 2008. Wind resource assessment. http://www.nrel.gov/wind/resource_assessment.html (9 April 2009).

National Research Council. 1989. Land use planning and oil and gas leasing on onshore federal lands. National Academy Press, Washington, DC.

National Research Council. 1994. Rangeland health: new methods to classify, inventory, and monitor rangelands. National Academy Press, Washington, DC.

Naugle, D. E., C. L. Aldridge, B. L. Walker, T. E. Cornish, B. J. Moynahan, M. J. Holloran, K. Brown, G. D. Johnson, E. T. Schmidtmann, R. T. Mayer, C. Y. Kato, M. R. Matchett, T. J. Christiansen, W. E. Cook, T. Creekmore, R. D. Falise, E. T. Rinkes, and M. S. Boyce. 2004. West Nile virus: pending crisis for Greater Sage-Grouse. Ecology Letters 7:704–713.

Naugle, D. E., C. L. Aldridge, B. L. Walker, K. E. Doherty, M. R. Matchett, J. McIntosh, T. E. Cornish, and M. S. Boyce. 2005. West Nile virus and Sage-Grouse: what more have we learned? Wildlife Society Bulletin 33:616–623.

Neilson, R. P., J. M. Lenihan, D. Bachelet, and R. J. Drapek. 2005. Climate change implications for sagebrush ecosystems. Transactions of the North American Wildlife and Natural Resources Conference 70:145–159.

Nelle, P. J., K. P. Reese, and J. W. Connelly. 2000. Long-term effects of fire on Sage Grouse habitat. Journal of Range Management 53:586–591.

Nelson, N. A., and J. Pierce. 2010. Late-Holocene relationships among fire, climate and vegetation in a forest-sagebrush ecotone of southwestern Idaho, USA. The Holocene 20:1179–1194.

Nemeth, N. M., and R. A. Bowen. 2007. Dynamics of passive immunity to West Nile virus in domestic chickens (*Gallus gallus domesticus*). American Journal of Tropical Medicine and Hygiene 76:310–317.

Nemeth, N. M., D. Gould, R. Bowen, and N. Komar. 2006a. Natural and experimental West Nile virus infection in five raptor species. Journal of Wildlife Diseases 42:1–13.

Nemeth, N. M., D. C. Hahn, D. H. Gould, and R. A. Bowen. 2006b. Experimental West Nile virus infection in Eastern Screech Owls (*Megascops asio*). Avian Diseases 50:252–258.

Nettles, V. F. 1984. Report of the Fish and Wildlife Health Committee. Proceedings of the International Association of Fish and Wildlife Agencies 74:89–101.

Nettles, V. F., and E. T. Thorne. 1982. Annual report of the Wildlife Diseases Committee. Proceedings of the 86th Annual Meeting of the United States Animal Health Association 86:64–65.

Nevada Governor's Sage-Grouse Conservation Team. 2004. Greater Sage-Grouse Conservation Plan for Nevada and Eastern California. First edition. Nevada Department of Wildlife, Reno, NV. http://ndow.org/wild/conservation/sg/plan/SGPlan063004.pdf (18 March 2009).

Newburn, D., S. Reed, P. Berck, and A. Merenlender. 2005. Economics and land-use changes in prioritizing private land conservation. Conservation Biology 19:1411–1420.

Nielson, J. L., J. M. Scott, and J. L. Aycrigg. 2001. Endangered species and peripheral populations: cause for conservation. Endangered Species Update 18:194–197.

Niemuth, N. D. 1992. Effects of nest predation on breeding biology of Sage Grouse (*Centrocercus urophasianus*). M.S. thesis, University of Wyoming, Laramie, WY.

Nisbet, R. M., and W. S. C. Gurney. 1976. A simple mechanism for population cycles. Nature 263:319–320.

Nisbet, R. A., S. H. Berwick, and K. L. Reed. 1983. A spatial model of Sage Grouse habitat quality. Developments in Environmental Modeling 5:267–276.

Nixon, R. M. 1972. Use of off-road vehicles on the public lands. Executive Order 11644. Federal Register 37:2877.

Noble, I. R. 1996. Linking the human dimension to landscape dynamics. Pp. 173–183 *in* B. Walker and W. Steffen (editors). Global change and terrestrial ecosystems. Cambridge University Press, Cambridge, UK.

Nokkentved, N. S. 2008. A forest of wormwood: sagebrush, water and Idaho's Twin Falls Canal Company. Caxton Printers, Caldwell, ID.

Nordling, D., M. Andersson, S. Zohari, and L. Gustafsson. 1998. Reproductive effort reduces specific immune response and parasite resistance. Proceedings of the Royal Society of London B 265:1291–1298.

Norrdahl, K., and E. Korpimäki. 2000. Do predators limit the abundance of alternative prey? Experiments with vole eating avian and mammalian predators. Oikos 91:528–540.

Norton, J. B., T. A. Monaco, J. M. Norton, D. A. Johnson, and T. A. Jones. 2004. Soil morphology and organic matter dynamics under cheatgrass and sagebrush-steppe plant communities. Journal of Arid Environments 57:445–466.

Norton, M. A. 2005. Reproductive success and brood habitat use of Greater Prairie Chickens and Sharp-tailed Grouse on the Fort Pierre National Grassland of central South Dakota. M.S. thesis, South Dakota State University, Brookings, SD.

Noss, R. F. 1994. Cows and conservation biology. Conservation Biology 8:613–616.

Noss, R. F., and K. M. Daly. 2006. Incorporating connectivity into broad-scale conservation planning. Pp. 587–619 *in* K. R. Crooks and M. Sanjayan (editors). Connectivity conservation. Cambridge University Press, New York, NY.

Noss, R. F., and R. L. Peters. 1995. Endangered ecosystems. A status report on America's vanishing habitat and wildlife. Defenders of Wildlife, Washington, DC.

Noss, R. F., J. F. Franklin, W. L. Baker, T. Schoenagel, and P. B. Moyle. 2006. Managing fire-prone forests in the western United States. Frontiers in Ecology and Environment 4:481–487.

Noss, R. F., E. T. LaRoe, III, and J. M. Scott. 1995. Endangered ecosystems of the United States: a preliminary assessment of loss and degradation. Biological Report 28. USDI National Biological Service, Washington, DC.

Notaro, M., Z. Liu, and J. W. Williams. 2006. Observed vegetation-climate feedbacks in the United States. Journal of Climate 19:763–86.

O'Brien, S. J., and E. Mayr. 1991. Bureaucratic mischief: recognizing endangered species and subspecies. Science 251:1187–1188.

Obrist, D., D. Yakir, and J. A. Arnone, III. 2004. Temporal and spatial patterns on soil water following wildfire-induced changes in plant communities in the Great Basin in Nevada, USA. Plant and Soil 262:1–12.

Odell, E. A., D. M. Theobald, and R. L. Knight. 2003. Incorporating ecology into land use planning. Journal of the American Planning Association 69:72–82.

Odum, E. P. 1985. Trends expected in stressed ecosystems. BioScience 35:419–422.

Ogg, A. G. 1994. A review of the chemical control of downy brome. Pp. 194–196 *in* S. B. Monsen and S. G. Kitchen (compilers). Symposium on Ecology and Management of Annual Rangelands. USDA Forest Service General Technical Report INT-GTR-313. USDA Forest Service, Intermountain Research Station, Ogden, UT.

Ohmart, R. D., J. E. Walters, R. R. Johnson, and E. J. Bicknell. 1978. On estimating burro numbers: a more reliable method. Transactions of the Desert Bighorn Council 22:45–46.

Oliphant, J. O. 1968. On the cattle ranges of the Oregon country. University of Washington Press, Seattle, WA.

Olson, R. A., and T. D. Whitson. 2002. Restoring structure in late-successional sagebrush communities by thinning with tebuthiuron. Restoration Ecology 10:146–155.

O'Neill, R. V., D. L. DeAngelis, J. B. Waide, and T. F. H. Allen. 1986. A hierarchical concept of ecosystems. Princeton University Press, Princeton, NJ.

O'Neill, R. V., R. H. Gardner, and M. G. Turner. 1992. A hierarchical neutral model for landscape analysis. Landscape Ecology 7:55–61.

O'Neill, R. V., B. T. Milne, M. G. Turner, and R. H. Gardner. 1988. Resource utilization scale and landscape pattern. Landscape Ecology 2:63–69.

Otis, D. L. 2002. Survival models for harvest management of Mourning Dove populations. Journal of Wildlife Management 66:1052–1063.

Ott, J. E., E. D. McArthur, and B. A. Roundy. 2003. Vegetation of chained and non-chained seedings after wildfire in Utah. Journal of Range Management 56:81–91.

Ouren, D. S., C. Hass, C. P. Melcher, S. C. Stewart, P. D. Ponds, N. R. Sexton, L. Burris, T. Fancher, and Z. H. Bowen. 2007. Environmental effects of off-highway vehicles on Bureau of Land Management lands: a literature synthesis, annotated bibliographies, and internet resources. USDI Geological Survey Open File Report 2007–1353. USDI Geological Survey, Reston, VA. http://webmesc.cr.usgs.gov/products/publications/22021/22021.pdf (15 August 2009).

Owen, J., F. Moore, N. Panella, E. Edwards, R. Bru, M. Hughes, and N. Komar. 2006. Migrating birds as dispersal vehicles for West Nile virus. EcoHealth 3:79–85.

Oyler-McCance, S. J. 1999. Genetic and habitat factors underlying conservation strategies for Gunnison Sage Grouse. Ph.D. dissertation, Colorado State University, Ft. Collins, CO.

Oyler-McCance, S. J., and P. L. Leberg. 2005. Conservation genetics in wildlife management. Pp. 632–657 *in* C. E. Braun (editor). Techniques for wildlife investigations and management. The Wildlife Society, Bethesda, MD.

Oyler-McCance, S. J., K. P. Burnham, and C. E. Braun. 2001. Influence of changes in sagebrush on Gunnison Sage Grouse in southwestern Colorado. Southwestern Naturalist 46:323–331.

Oyler-McCance, S. J., N. W. Kahn, K. P. Burnham, C. E. Braun, and T. W. Quinn. 1999. A population genetic comparison of large- and small-bodied sage-grouse in Colorado using microsatellite and mitochondrial DNA markers. Molecular Ecology 8:1457–1465.

Oyler-McCance, S. J., J. St. John, S. E. Taylor, A. D. Apa, and T. W. Quinn. 2005a. Population genetics of Gunnison Sage-Grouse: implications for management. Journal of Wildlife Management 69:630–637.

Oyler-McCance, S. J., S. E. Taylor, and T. W. Quinn. 2005b. A multilocus population genetic survey of the Greater Sage-Grouse across their range. Molecular Ecology 14:1293–1310.

Paige, C., and S. A. Ritter. 1999. Birds in a sagebrush sea: managing sagebrush habitats for bird communities. Partners in Flight Western Working Group, Boise, ID.

Palmer, W. C. 1965. Meterological drought. Research Paper No. 45. United States Department of Commerce, Weather Bureau, Washington, DC.

Parker, R. R., C. B. Phillip, and G. E. Davis. 1932. Tularemia: occurrence in the Sage Hen, *Centrocercus urophasianus*. Public Health Reports 47:479–487.

Parsons, R. A., E. K. Heyerdahl, R. E. Keane, B. Dorner, and J. Fall. 2007. Assessing accuracy of point fire intervals across landscapes with simulation modeling. Canadian Journal of Forest Research 37:1605–1614.

Pascual-Hortal, L., and S. Saura. 2006. Comparison and development of new graph-based landscape connectivity indices: towards the prioritization of habitat patch and corridors for conservation. Landscape Ecology 21:959–967.

Pascual-Hortal, L., and S. Saura. 2008. Integrating landscape connectivity in broad-scale forest planning through a new graph-based habitat availability methodology: application to Capercaille (*Tetrao urogallus*) in Catalona (NE Spain). European Journal of Forest Research 127:23–31.

Pashley, D. N., C. J. Beardmore, J. A. Fitzgerald, R. P. Ford, W. C. Hunter, M. S. Morrison, and K. V. Rosenberg. 2000. Partners in flight: conservation of the land birds of the United States. American Bird Conservancy, The Plains, VA.

Passey, H. B., V. K. Hugie, E. W. Williams, and D. E. Ball. 1982. Relationships between soil, plant community, and climate on rangelands of the Intermountain West. Technical Bulletin 1662. USDA Soil Conservation Service, Washington, DC.

Patten, D. T. 1993. Herbivore optimization and overcompensation: does native herbivory on western rangelands support these theories? Ecological Applications 3:35–26.

Patterson, R. L. 1952. The Sage Grouse in Wyoming. Sage Books, Denver, CO.

Pechanec, J. F., A. P. Plummer, J. H. Robertson, and A. C. Hull, Jr. 1965. Sagebrush control on rangelands. United States Department of Agriculture Handbook No. 277. US Department of Agriculture, Washington, DC.

Pedersen, E. K., J. W. Connelly, J. R. Hendrickson, and W. E. Grant. 2003. Effect of sheep grazing and fire on Sage Grouse populations in southeastern Idaho. Ecological Modelling 165:23–47.

Peek, J. M. 1986. A review of wildlife management. Prentice-Hall, Englewood Cliffs, NJ.

Pellant, M. 1990. The cheatgrass-wildfire cycle—are there any solutions? Pp. 11–18 *in* E. D. McArthur, E. M. Romney, S. D. Smith, and P. T. Tueller (editors). Symposium on Cheatgrass Invasion, Shrub Die-off, and Other Aspects of Shrub Biology and Management. USDA Forest Service General Technical Report INT-276. USDA Forest Service, Intermountain Forest and Range Experiment Station, Ogden, UT.

Pellant, M., and C. Hall. 1994. Distribution of two exotic grasses on public lands in the Great Basin: status in 1992. Pp. 109–112 *in* S. B. Monsen and S. G. Kitchen (compilers). Ecology and management of annual rangelands. USDA Forest Service General Technical Report INT-GTR-313. USDA Forest Service, Intermountain Research Station, Ogden, UT.

Pellant, M., and C. R. Lysne. 2005. Strategies to enhance plant structure and diversity in crested wheatgrass seedings. Pp. 81–92 *in* N. L. Shaw, M. Pellant, and S. B. Monsen (compilers). Sage-grouse habitat restoration symposium proceedings.USDA Forest Service RMRS-P-38. USDA Forest Service, Rocky Mountain Research Station, Ft. Collins, CO.

Pellant, M., J. Kaltenecker, and S. Jirik. 1999. Use of oust herbicide to control cheatgrass in the northern Great Basin. Pp. 322–326 *in* S. B. Monsen and R. Stevens (compilers). Ecology and management of pinyon-juniper communities within the interior West: sustaining and restoring a diverse ecosystem. USDA Forest Service RMRS-P-9. USDA Forest Service, Rocky Mountain Research Station, Ogden, UT.

Pellant, M., P. Shaver, D. A. Pyke, and J. E. Herrick. 2005. Interpreting indicators of rangeland health, version 4. Interagency Technical Reference 1734-6. National Science and Technology Center, Information and Communications Group, USDI, Bureau of Land Management, Denver, CO.

Pellegrini, S. W. 1971. Home range, territoriality and movement patterns of wild horses in the Wassuk

Range of western Nevada. M.S. thesis, University of Nevada, Reno, NV.

Perryman, B. L., A. M. Maier, A. L. Hild, and R. A. Olson. 2001. Demographic characteristics of 3 *Artemisia tridentata* Nutt. subspecies. Journal of Range Management 54:166–170.

Perryman, B. L., R. A. Olson, S. Petersburg, and T. Naumann. 2002. Vegetation response to prescribed fire in Dinosaur National Monument. Western North American Naturalist 62:414–422.

Pesole, G., C. Gissi, A. De Chirico, and C. Saccone. 1999. Nucleotide substitution rate of mammalian mitochondrial genomes. Journal of Molecular Evolution 48:427–434.

Peterjohn, B. G., and J. R. Sauer. 1999. Population status of North American grassland birds from the North American Breeding Bird Survey, 1966–1996. Studies in Avian Biology 19:27–44.

Peters, D. P. C., B. T. Bestelmeyer, J. E. Herrick, E. L. Fredrickson, H. C. Monger, and K. M. Havstad. 2006. Disentangling complex landscapes: new insights into arid and semiarid system dynamics. BioScience 56:491–501.

Petersen, B. E. 1980. Breeding and nesting ecology of female Sage Grouse in North Park, Colorado. M.S. thesis, Colorado State University, Ft. Collins, CO.

Petersen, L. R., R. T. Dumke, and J. M. Gates. 1988. Pheasant survival and the role of predation. Pp. 165–196 *in* D. L. Hallett, W. R. Edwards, and G. V. Burger (editors). Pheasants: symptoms of wildlife problems on agricultural lands. North Central Section of The Wildlife Society, Bloomington, IN.

Peterson, A. T., D. A. Vieglais, and J. K. Andreasen. 2003. Migratory birds modeled as critical transport agents for West Nile virus in North America. Vector-Borne and Zoonotic Diseases 3:27–37.

Peterson, D. L., and V. T. Parker (editors). 1998. Ecological scale: theory and applications. Columbia University Press, New York, NY.

Peterson, E. B. 2005. Estimating cover of an invasive grass (*Bromus tectorum*) using tobit regression and phenology derived from two dates of Landsat ETM+ data. International Journal of Remote Sensing 26:2491–2507.

Peterson, J. 1999. Ungulate/vegetation dynamics in the Pryor Mountain Wild Horse Range. Ph.D. dissertation, Colorado State University, Ft. Collins, CO.

Peterson, J. G. 1970. The food habits and summer distribution of juvenile Sage Grouse in central Montana. Journal of Wildlife Management 34:147–155.

Peterson, M. J. 2004. Parasites and infectious diseases of prairie grouse: should managers be concerned? Wildlife Society Bulletin 32:35–55.

Peterson, M. J., P. J. Ferro, M. N. Peterson, R. M. Sullivan, B. E. Toole, and N. J. Silvy. 2002. Infectious disease survey of Lesser Prairie Chickens in north Texas. Journal of Wildlife Diseases 38:834–839.

Petrie, M., M. Hall, T. Halliday, H. Budgey, and C. Pierpoint. 1992. Multiple mating in a lekking bird: why do peahens mate with more than one male and with the same male more than once? Behavioral Ecology and Sociobiology 31:349–358.

Petrie, M., A. Krupa, and T. Burke. 1999. Peacocks lek with relatives even in the absence of social and environmental cues. Nature 401:155–157.

Pickett, S. T. A., and P. S. White (editors). 1985. The ecology of natural disturbance and patch dynamics. Academic Press, San Diego, CA.

Pickford, G. D. 1932. The influence of continued heavy grazing and of promiscuous burning on spring-fall ranges in Utah. Ecology 13:159–171.

Piemeisel, R. L. 1951. Causes affecting change and rate of change in a vegetation of annuals in Idaho. Ecology 32:53–72.

Piertney, S. B., A. D. C. MacColl, P. J. Bacon, and J. F. Dallas. 1998. Local genetic structure in Red Grouse (*Lagopus lagopus scoticus*): evidence from microsatellite DNA markers. Molecular Ecology 7:1645–1654.

Pietz, P. J., and D. A. Granfors. 2000. Identifying predators and fates of grassland passerine nests using miniature video cameras. Journal of Wildlife Management 64:71–87.

Pitman, J. C., C. A. Hagen, B. E. Jamison, R. J. Robel, T. M. Loughin, and R. D. Applegate. 2006. Nesting ecology of Lesser Prairie-Chickens in sand sagebrush prairie of southwestern Kansas. Wilson Journal of Ornithology 118:23–35.

Pitman, J. C., C. A. Hagen, R. J. Robel, T. M. Loughin, and R. D. Applegate. 2005. Location and success of Lesser Prairie-Chicken nests in relation to vegetation and human disturbance. Journal of Wildlife Management 69:1259–1269.

Platt, K. B., B. J. Tucker, P. G. Halbur, S. Tiawsirisup, B. J. Blitvich, F. G. Fabiosa, L. C. Bartholomay, and W. A. Rowley. 2007. West Nile virus viremia in eastern chipmunks (*Tamias striatus*) sufficient for infecting different mosquitoes. Emerging Infectious Diseases 13:831–837.

Platt, K. B., B. J. Tucker, P. G. Halbur, S. Tiawsirisup, B. J. Blitvich, F. G. Fabiosa, K. Mullin, G. R. Parikh, P. Kitikoon, L. C. Bartholomay, and W. A. Rowley. 2008. Fox squirrels (*Sciurus niger*) develop West Nile virus viremias sufficient for infecting select mosquito species. Vector-Borne and Zoonotic Diseases 5:9.

Plotnick, R. E., R. H. Gardner, and R. V. O'Neil. 1993. Lacunarity indices as measures of landscape pattern. Landscape Ecology 8:201–211.

Poling, M. A. 1991. Legal milestones in range management. Renewable Resources Journal 9:7–10.

Polley, H. W. 1997. Implications of rising atmospheric carbon dioxide concentration for rangelands. Journal of Range Mangement 50:561–577.

Ponzetti, J. M., B. McCune, and D. A. Pyke. 2007. Biotic soil crusts in relation to topography, cheatgrass and fire in the Columbia Basin, Washington. Bryologist 110:706–722.

Popham, G. P. 2000. Sage Grouse nesting habitat in northeastern California. M.S. thesis, Humboldt State University, Arcata, CA.

Popham, G. P., and R. J. Gutiérrez. 2003. Greater Sage-grouse *Centrocercus urophasianus* nesting success and habitat use in northeastern California. Wildlife Biology 9:327–334.

Post, G. 1951. Effects of toxaphene and chlordane on certain game birds. Journal of Wildlife Management 15:381–386.

Post, G. 1960. Diagnosis of disease in mammals and birds. Pp. 5–27 *in* Federal Aid in Fish and Wildlife Restoration, Project FW-3-R-7, Work Plan 1, Job 1W. Wyoming Game and Fish Commission, Cheyenne, WY.

Postovit, B. C. 1981. Suggestions for Sage Grouse habitat reclamation on surface mines in northeastern Wyoming. M.S. thesis, University of Wyoming, Laramie, WY.

Potts, G. R. 1986. The partridge: pesticides, predation, and conservation. Collins, London, UK.

Prater, M. R., D. Obrist, J. A. Arnone, III, and E. H. DeLucia. 2006. Net carbon exchange and evapotranspiration in postfire and intact sagebrush communities in the Great Basin. Oecologia 146:595–607.

Pressey, R. L., and M. C. Bottrill. 2008. Opportunism, threats, and the evolution of systematic conservation planning. Conservation Biology 22:1340–1345.

Pressey, R. L., M. Cabeza, M. E. Watts, R. M. Cowling, and K. A. Wilson. 2007. Conservation planning in a changing world. Trends in Ecology and Evolution 22:583–592.

Pritchard, D., J. Anderson, C. Correll, J. Fogg, K. Gebhardt, R. Krapf, S. Leonard, B. Mitchell, and J. Staats. 1998. Riparian area management: a user's guide to assessing proper functioning condition and the supporting science for lotic areas. BLM/RS/ST-98/001+1737. United States Department of the Interior, Bureau of Land Management, National Applied Resources Sciences Center, Denver, CO.

Pritchard, J. K., M. Stephens, and P. J. Donnelly. 2000. Inference of population structure using multilocus genotype data. Genetics 155:945–959.

Pritz, R. K., K. L. Launchbaugh, and C. A. Taylor, Jr. 1997. Effects of breed and dietary experience on juniper consumption by goats. Journal of Range Management 50:600–606.

Prosser, C. W., K. M. Skinner, and K. K. Sedivec. 2003. Comparison of 2 techniques for monitoring vegetation on military lands. Journal of Range Management 56:446–454.

Provencher, L., T. A. Forbis, L. Frid, and G. Medlyn. 2007. Comparing alternative management strategies of fire, grazing, and weed control using spatial modeling. Ecological Modelling 209:249–263.

Pulliam, H. R. 1988. Sources, sinks and population regulation. American Naturalist 132:652–661.

Pullin, A. S., and G. B. Stewart. 2006. Guidelines for systematic review in conservation and environmental management. Conservation Biology 20:1647–1656.

Purvis, A., J. L. Gittleman, G. Cowlishaw, and G. M. Mace. 2000. Predicting extinction risk in declining species. Proceedings of the Royal Society of London B 267:1947–1952.

Pyke, D. A. 1995. Population diversity with special reference to rangeland plants. Pp. 21–32 *in* N. E. West (editor). Biodiversity of rangelands. Natural Resources and Environmental Issues, Vol. IV. College of Natural Resources, Utah State University, Logan, UT.

Pyke, D. A., and S. Archer. 1991. Plant-plant interactions affecting plant establishment and persistence on revegetated rangeland. Journal of Range Management 44:550–557.

Pyke, D. A., and J. E. Herrick. 2003. Transitions in rangeland evaluations: a review of the major transitions in rangeland evaluations during the last 25 years and speculation about future evaluations. Rangelands 25(6):22–30.

Pyke, D. A., J. E. Herrick, P. Shaver, and M. Pellant. 2002. Rangeland health attributes and indicators for qualitative assessment. Journal of Range Management 55:584–597.

Pyke, D. A., T. O. McArthur, K. S. Harrison, and M. Pellant. 2003. Coordinated intermountain restoration project: fire, decomposition and restoration. Pp. 1116–1124 *in* N. Allsopp, A. R. Palmer, S. J. Milton, K. P. Kirkman, G. I. H. Kerley, and C. R. Hurt (editors). Proceedings of the VIIth International Rangelands Congress, Durban, South Africa. Document Transformation Technologies, Irene, South Africa.

Pyle, W. H., and J. A. Crawford. 1996. Availability of foods of Sage Grouse chicks following prescribed fire in sagebrush-bitterbrush. Journal of Range Management 49:320–324.

Quigley, T. M. 2005. Evolving views of public land values and management of natural resources. Rangelands 27(3):37–44.

Quigley, T. M., and S. J. Arbelbide. 1997. An assessment of ecosystem components in the Interior Columbia Basin and portions of the Klamath and Great Basins. General Technical Report PNW-GTR-405. USDA Forest Service, Pacific Northwest Research Station, Portland, OR.

R Development Core Team. 2008. R: a language and environment for statistical computing. R Foundation for Statistical Computing. Vienna, Austria.

Ramey, R. R., H.-P. Liu, C. W. Epps, L. M. Carpenter, and J. D. Wehausen. 2005. Genetic relatedness of the Preble's meadow jumping mouse (*Zapus hudsonius preblei*) to nearby subspecies of *Z. hudsonius* as inferred from variation in cranial morphology, mitochondrial DNA and microsatellite DNA:

implications for taxonomy and conservation. Animal Conservation 8:329–346.
Ransom, B. H. 1909. The taenioid cestodes of North American Birds. Bulletin 69. United States National Museum, Washington, DC.
Raphael, M. G., M. J. Wisdom, M. M. Rowland, R. S. Holthausen, B. C. Wales, B. G. Marcot, and T. D. Rich. 2001. Status and trends of habitats of terrestrial vertebrates in relation to land management in the interior Columbia River Basin. Forest Ecology and Management 153:63–88.
Rappole, J. H., and Z. Hubálek. 2003. Migratory birds and West Nile virus. Journal of Applied Microbiology 94 (Supplement 1):47–58.
Rapport, D. J., and W. G. Whitford. 1999. How ecosystems respond to stress: common properties of arid and aquatic systems. BioScience 49:193–203.
Rasmussen, D. I., and L. A. Griner. 1938. Life history and management studies of the Sage Grouse in Utah, with special reference to nesting and feeding habits. Transactions of the North American Wildlife Conference 3:852–864.
Rebholz, J. L. 2007. Influence of habitat characteristics on Greater Sage-Grouse reproductive success in the Montana Mountains, Nevada. M.S. thesis, Oregon State University, Corvallis, OR.
Reed, J. M., L. S. Mills, J. B. Dunning, Jr., E. S. Menges, K. S. McKelvey, R. Frye, S. R. Beissinger, M.-C. Anstett, and P. Miller. 2002. Emerging issues in population viability analysis. Conservation Biology 16:7–19.
Reed, K. D., J. K. Meece, J. S. Henkel, and S. K. Shukla. 2003. Birds, migration, and emerging zoonoses: West Nile virus, Lyme disease, influenza A and enteropathogens. Clinical Medicine and Research 1:5–12.
Reed, W. J. 2006. A note on fire frequency concepts and definitions. Canadian Journal of Forest Research 36:1884–1888.
Reese, K. P., and R. T. Bowyer (editors). 2007. Monitoring populations of Sage-Grouse. College of Natural Resources Experiment Station Bulletin 88. University of Idaho, Moscow, ID.
Reese, K. P., and J. W. Connelly. 1997. Translocations of Sage Grouse *Centrocercus urophasianus* in North America. Wildlife Biology 3:235–241.
Reese, K. P., J. W. Connelly, E. O. Garton, and M. L. Commons-Kemner. 2005. Exploitation and Greater Sage-Grouse *Centrocercus urophasianus*: a response to Sedinger and Rotella. Wildlife Biology 11:377–381.
Reichenberger, G., and D. A. Pyke. 1990. Impact of early root competition on fitness components of four semiarid species. Oecologia 85:159–166.
Reid, M., P. Comer, H. Barrett, S. Caicco, R. Crawford, C. Jean, G. Jones, J. Kagan, M. Karl, G. Kittel, P. Lyon, M. Manning, E. Peterson, R. Rosentreter, S. Rust, D. Tart, C. Williams, and A. Winward. 2002. International classification of ecological communities: terrestrial vegetation of the United States. Sagebrush vegetation of the western United States. Final Report for the USGS. NatureServe, Arlington, VA.
Reiner, R. J., and P. J. Urness. 1982. Effect of grazing horses managed as manipulators of big game winter range. Journal of Range Management 35:567–571.
Reinert, S. E. 1984. Use of introduced perches by raptors: experimental results and management implications. Journal of Raptor Research 18:25–29.
Reisen, W. K., and A. C. Brault. 2007. West Nile virus in North America: perspectives on epidemiology and intervention. Pest Management Science 63:641–646.
Reisen, W. K., Y. Fang, H. D. Lothrop, V. M. Martinez, J. Wilson, P. O'Connor, R. Carney, B. Cahoon-Young, M. Shafii, and A. C. Brault. 2006a. Overwintering of West Nile virus in southern California. Journal of Medical Entomology 43:344–355.
Reisen, W. K., Y. Fang, and V. M. Martinez. 2006b. Effects of temperature on the transmission of West Nile virus by *Culex tarsalis* (Diptera: Culicidae). Journal of Medical Entomology 43:309–317.
Reisen, W. K., Y. Fang, and V. M. Martinez. 2007. Is nonviremic transmission of West Nile virus by *Culex* mosquitoes (Diptera: Culicidae) nonviremic? Journal of Medical Entomology 44:299–302.
Reisen, W. K., H. D. Lothrop, and B. Lothrop. 2003. Factors influencing the outcome of mark-release-recapture studies with *Culex tarsalis* (Diptera: Culicidae). Journal of Medical Entomology 40:820–829.
Reisen, W. K., V. M. Martinez, Y. Fang, S. Garcia, S. Ashtari, S. S. Wheeler, and B. D. Carroll. 2006c. Role of California (*Callipepla californica*) and Gambel's (*Callipepla gambelii*) Quail in the ecology of mosquito-borne encephalitis viruses in California, USA. Vector-borne and Zoonotic Diseases 6:248–260.
Remington, R. D., and M. A. Schork. 1970. Statistics with application to the biological and health sciences. Prentice-Hall, Englewood Cliffs, NJ.
Remington, T. E., and C. E. Braun. 1985. Sage Grouse food selection in winter, North Park, Colorado. Journal of Wildlife Management 49:1055–1061.
Remington, T. E., and C. E. Braun. 1991. How surface coal mining affects Sage Grouse, North Park, Colorado. Issues and Technology in the Management of Western Wildlife 5:128–132.
Restani, M., J. M. Marzluff, and R. E. Yates. 2001. Effects of human food sources on movements, survivorship, and sociality of Common Ravens in the Arctic. Condor 103:399–404.
Reynolds, J. F., R. A. Virginia, and W. H. Schlesinger. 1997. Defining functional types for models of desertification. Pp. 195–216 *in* T. M. Smith, H. H. Shugart, and F. I. Woodward (editors). Plant functional types: their relevance to ecosystem properties and global change. Cambridge University Press, Cambridge, UK.

Reynolds, T. D., and C. H. Trost. 1981. Grazing, crested wheatgrass, and bird populations in southeastern Idaho. Northwest Science 55:225–234.

Rice, B., and M. Westoby. 1978. Vegetative responses of some Great Basin shrub communities protected against jackrabbits or domestic stock. Journal of Range Management 31:28–34.

Rice, P. M. 2004. Invaders database system. University of Montana, Missoula, MT. http://invader.dbs.umt.edu (13 July 2008).

Rich, T. 1985. Sage Grouse population fluctuation: evidence for a 10-year cycle. BLM Technical Bulletin 85-1. USDI Bureau of Land Management, Boise, ID.

Rich, T. D., and B. Altman. 2002. Under the sage-grouse umbrella. Bird Conservation 14:10.

Rich, T. D., M. J. Wisdom, and V. A. Saab. 2005. Conservation of priority birds in sagebrush ecosystems. Pp. 598–606 *in* Proceedings of the Third International Partners in Flight Symposium. General Technical Report PSW-GTR-191. USDA Forest Service, Pacific Southwest Research Station, Albany, CA.

Richards, R. T., J. C. Chambers, and C. Ross. 1998. Use of native plants on federal lands: policy and practice. Journal of Range Management 51:625–632.

Ricotta, C., A Stanisci, G. C. Avena, and C. Blasi. 2000. Quantifying the network connectivity of landscape mosaics: a graph-theoretical approach. Community Ecology 1:89–94.

Riebsame, W. E., H. Gosnell, and D. M. Theobald. 1996. Land use and landscape change in the Colorado mountains I. Theory, scale, and pattern. Mountain Research and Development 16:395–405.

Riley, T. Z. 2004. Private-land habitat opportunities for prairie grouse through federal conservation programs. Wildlife Society Bulletin 32:83–91.

Rinella, M. J., B. D. Maxwell, P. K. Fay, T. Weaver, and R. L. Sheley. 2009. Control effort exacerbates invasive-species problem. Ecological Applications 19:155–162.

Rittenhouse, L. R., D. E. Johnson, and M. M. Borman. 1982. A study of food consumption rates and nutrition of horses and cattle. USDI Bureau of Land Management, Washington, DC.

Robel, R. J., J. N. Briggs, A. D. Dayton, and L. C. Hulbert. 1970. Relationships between visual obstruction measurements and weight of grassland vegetation. Journal of Range Management 23:295–297.

Roberts, C., and J. A. Jones. 2000. Soil patchiness in juniper-sagebrush communities of central Oregon. Plant and Soil 223:45–61.

Robertson, D. R., J. L. Nielsen, and N. H. Bare. 1966. Vegetation and soils of alkali sagebrush and adjacent big sagebrush ranges in North Park, Colorado. Journal of Range Management 19:17–20.

Robertson, J. H. 1971. Changes on a sagebrush-grass range in Nevada ungrazed for 30 years. Journal of Range Management 24:397–400.

Robertson, J. H., and P. B. Kennedy. 1954. Half-century changes on northern Nevada ranges. Journal of Range Management 7:117–121.

Robertson, J. H., and C. K. Pearse. 1945. Artificial reseeding and the closed community. Northwest Science 19:58–66.

Robertson, M. D. 1991. Winter ecology of migratory Sage Grouse and associated effects of prescribed fire in southern Idaho. M.S. thesis, University of Idaho, Moscow, ID.

Robichaud, P. R., J. L. Beyers, and D. G. Neary. 2000. Evaluating the effectiveness of postfire rehabilitation treatments. USDA Forest Service General Technical Report RMRS-GTR-63. USDA Forest Service, Rocky Mountain Research Station, Ft. Collins, CO.

Robinette, C. F., and P. D. Doerr. 1993. Survival of Northern Bobwhite on hunted and nonhunted study areas in the North Carolina sandhills. Pp. 74–78 *in* K. E. Church and T. V. Dailey (editors). Quail III: National Quail Symposium. Kansas Department of Wildlife and Parks, Pratt, KS.

Roché, Jr., B. F., and C. T. Roché. 1999. Diffuse knapweed. Pp. 217–230 *in* R. L. Sheley and J. K. Petroff (editors.). Biology and management of noxious rangeland weeds. Oregon State University Press, Corvallis, OR.

Rogers, G. E. 1964. Sage Grouse investigations in Colorado. Technical Publication Number 16. Colorado Game, Fish and Parks Department, Ft. Collins, CO.

Rollins, M. G., and C. K. Frame (editors). 2006. The LANDFIRE prototype project: nationally consistent and locally relevant geospatial data for wildland fire management. USDA Forest Service General Technical Report RMRS-GTR-175. USDA Forest Service, Rocky Mountain Research Station, Ft. Collins, CO.

Romme, W. H., C. D. Allen, J. D. Bailey, W. L. Baker, B. T. Bestelmeyer, P. M. Brown, K. S. Eisenhart, L. Floyd-Hanna, D. W. Huffman, B. Jacobs, R. F. Miller, E. H. Muldavin, T. W. Swetnam, R. J. Tausch, and P. J. Weisberg. 2008. Historical and modern disturbance regimes, stand structures, and landscape dynamics in piñon-juniper vegetation in the western U.S. Colorado Forest Restoration Institute, Colorado State University, Ft. Collins, CO. http://www.cfri.colostate.edu/reports.htm (13 November 2008).

Rosenbaum, H. C., R. L. Brownwell, Jr., M. W. Brown, C. Schaeff, V. Portway, B. N. White, S. Malik, L. A. Pastene, N. J. Patenaude, C. S. Baker, M. Goto, P. B. Best, P. J. Clapham, P. Hamilton, M. Moore, R. Payne, V. Rowntree, C. T. Tynan, J. L. Bannister, and R. DeSalle. 2000. World-wide genetic differentiation of *Eubalaena*: questioning the number of right whale species. Molecular Ecology 9:1793–1802.

Rosene, W. 1969. The Bobwhite Quail: its life and management. Rutgers University Press, New Brunswick, NJ.

Rosentreter, R. 2005. Sagebrush identification, ecology, and palatability relative to sage-grouse. Pp. 3–16 *in* N. L. Shaw, M. Pellant, and S. B. Monsen (compilers). Sage-grouse habitat restoration symposium proceedings. USDA Forest Service Proceedings RMRS-P-38. USDA Forest Service, Rocky Mountain Research Station, Ft. Collins, CO.

Ross, J. 2006. FLPMA turns 30. Rangelands 28(5):16–23.

Ross, J. V. H. 1984. Managing the public rangelands: 50 years since the Taylor Grazing Act. Rangelands 6(4):147–151.

Rost, G. R., and J. A. Bailey. 1979. Distribution of mule deer and elk in relation to roads. Journal of Wildlife Management 43:634–641.

Rotenberry, J. T. 1998. Avian conservation research needs in western shrublands: exotic invaders and the alteration of ecosystem processes. Pp. 261–272 *in* J. M. Marzluff and R. Sallabanks (editors). Avian conservation. Research and management. Island Press, Washington, DC.

Rotenberry, J. T., and S. T. Knick. 1999. Multiscale habitat associations of the Sage Sparrow: implications for conservation biology. Studies in Avian Biology 19:95–103.

Rotenberry, J. T., and J. A. Wiens. 1980a. Habitat structure, patchiness, and avian communities in North American steppe vegetation: a multivariate analysis. Ecology 61:1228–1250.

Rotenberry, J. T., and J. A. Wiens. 1980b. Temporal variation in habitat structure and shrubsteppe bird dynamics. Oecologia 47:1–9.

Rothenmaier, D. 1979. Sage Grouse reproductive ecology: breeding season movements, strutting ground attendance and site characteristics, and nesting. M.S. thesis, University of Wyoming, Laramie, WY.

Roundy, B. A., S. P. Hardegree, J. C. Chambers, and A. Whittaker. 2007. Prediction of cheatgrass field germination potential using wet thermal accumulation. Rangeland Ecology and Management 60:613–623.

Roundy, B. A., N. L. Shaw, and D. T. Booth. 1997. Using native seed on rangelands. Pp. 1–8 *in* N. L. Shaw and B. A. Roundy (compilers). Using seeds of native species on rangelands. USDA Forest Service General Technical Report INT-GTR-372. USDA Forest Service, Intermountain Research Station, Ogden, UT.

Rowe, R. J. 2007. Legacies of land use and recent climatic change: the small mammal fauna in the mountains of Utah. American Naturalist 170:242–257.

Rowland, M. M. 2004. Effects of management practices on grassland birds: Greater Sage-Grouse. USDI Geological Survey, Northern Prairie Wildlife Research Center, Jamestown, ND.

Rowland, M. M., L. H. Suring, M. J. Wisdom, C. W. Meinke, and L. Schueck. 2005. Habitats for vertebrate species of conservation concern. Pp. 163–204 *in* M. J. Wisdom, M. M. Rowland, and L. H. Suring (editors). Habitat threats in the sagebrush ecosystem: methods of regional assessment and applications in the Great Basin. Alliance Communications Group, Lawrence, KS.

Rowland, M. M., M. J. Wisdom, L. H. Suring, and C. W. Meinke. 2006. Greater Sage-Grouse as an umbrella species for sagebrush-associated vertebrates. Biological Conservation 129:323–335.

Roy, C., and A. Woolf. 2001. Effects of hunting and hunting-hour extension on Mourning Dove foraging and physiology. Journal of Wildlife Management 65:808–815.

Rubinoff, D. 2001. Evaluating the California Gnatcatcher as an umbrella species for conservation of southern California coastal sage scrub. Conservation Biology 15:1374–1383.

Ruthven, III, D. C. 2007. Grazing effects on forb diversity and abundance in a honey mesquite parkland. Journal of Arid Environments 68:668–677.

Ryden, H. 1978. America's last wild horses. E. P. Dutton, New York, NY.

Ryel, R. J., M. M. Caldwell, A. J. Leffler, and C. K. Yoder. 2003. Rapid soil moisture recharge to depth by roots in a stand of *Artemisia tridentata*. Ecology 84:757–764.

Ryel, R. J., M. M. Caldwell, and J. H. Manwaring. 1996. Temporal dynamics of soil spatial heterogeneity in sagebrush-wheatgrass steppe during a growing season. Plant and Soil 184:299–309.

Saab, V. A., and T. D. Rich. 1997. Large-scale conservation assessment for Neotropical migratory land birds in the Interior Columbia River Basin. USDA Forest Service General Technical Report PNW-GTR-399. USDA Forest Service, Pacific Northwest Research Station, Portland, OR.

Saab, V. A., C. E. Bock, T. D. Rich, and D. S. Dobkin. 1995. Livestock grazing effects in western North America. Pp. 311–353 *in* T. E. Martin and D. M. Finch (editors). Ecology and management of neotropical migratory birds. A synthesis and review of critical issues. Oxford University Press, New York, NY.

Sabinske, D. W., and D. H. Knight. 1978. Variation within the sagebrush vegetation of Grand Teton National Park, Wyoming. Northwest Science 52:195–204.

Sage, R. F., D. A. Wedin, and M. Li. 1999. The biogeography of C4 photosynthesis: patterns and controlling factors. Pp. 313–376 *in* R. F. Sage and R. K. Monson (editors). C4 plant biology. Academic Press, San Diego, CA.

Saito, E. K., L. Sileo, D. E. Green, C. U. Meteyer, G. S. McLaughlin, K. A. Converse, and D. E. Docherty. 2007. Raptor mortality due to West Nile virus in the United States, 2002. Journal of Wildlife Diseases 43:206–213.

Sallach, B. K. 1986. Vegetation changes in New Mexico documented by repeat photography. M.S. thesis, New Mexico State University, Las Cruces, NM.

Salt, W. R. 1958. *Sarcocystis rileyi* in Sage Grouse. Journal of Parasitology 44:511.

Samways, M. J. 2000. A conceptual model of ecosystem restoration triage based on experiences from three remote oceanic islands. Biodiversity and Conservation 9:1073–1083.

Sanders, K. D., and A. S. Voth. 1983. Ecological changes of grazed and ungrazed plant communities. Pp. 176–179 *in* S. B. Monsen and N. Shaw (compilers). Managing intermountain rangelands—improvement of range and wildlife habitats. USDA Forest Service General Technical Report INT-157. USDA Forest Service, Intermountain Research Station, Ogden, UT.

Sanderson, E. W., M. Jaiteh, M. A. Levy, K. H. Redford, A. V. Wannebo, and G. Woolmer. 2002a. The human footprint and the last wild. BioScience 52:891–904.

Sanderson, E. W., K. H. Redford, A. Vedder, P. B. Coppolillo, and S. E. Ward. 2002b. A conceptual model for conservation planning based on landscape species requirements. Landscape and Urban Planning 58:41–56.

Sapsis, D. B., and J. B. Kauffman. 1991. Fuel consumption and fire behavior associated with prescribed fires in sagebrush ecosystems. Northwest Science 65:173–179.

Sardelis, M. R., M. J. Turell, D. J. Dohm, and M. L. O'Guinn. 2001. Vector competence of selected North American *Culex* and *Coquillettidia* mosquitos for West Nile virus. Emerging Infectious Diseases 7:1018–1022.

SAS Institute. 1990. SAS/STAT user's guide, version 6. Fourth edition. SAS Institute, Cary, NC.

SAS Institute. 2006. SAS Version 9.1. SAS Institute, Cary, NC.

Sauer, J. R., J. E. Hines, and J. Fallon. 2003. The North American breeding bird survey, results and analysis 1966–2002. Version 2003.1. USDI Geological Survey, Patuxent Wildlife Research Center, Laurel, MD.

Sauer, J. R., J. E. Hines, and J. Fallon. 2008. The North American breeding bird survey, results and analysis 1966–2007. Version 5.15.2008. USDI Geological Survey, Patuxent Wildlife Research Center, Laurel, MD.http://www.mbr-pwrc.usgs.gov/bbs/trend/tf07.html (27 October 2008).

Saunders, D. A., R. J. Hobbs, and C. R. Margules. 1991. Biological consequences of ecosystem fragmentation: a review. Conservation Biology 5:18–32.

Saunders, L. M., D. M. Tompkins, and P. J. Hudson. 2002. Stochasticity accelerates nematode egg development. Journal of Parasitology 88:1271–1272.

Saura, S., and J. Martínez-Millán. 2000. Landscape patterns simulation with modified random clusters method. Landscape Ecology 15:661–678.

Saura, S., and L. Pascual-Hortal. 2007. A new habitat availability index to integrate connectivity in landscape conservation planning: comparison with existing indices and application to a case study. Landscape and Urban Planning 83:91–103.

Savage, D. E. 1969. Relation of Sage Grouse to upland meadows in Nevada. Nevada Fish and Game Commission, Job Completion Report, Project W-39-R-9, Job 12. Nevada Division of Wildlife, Reno, NV.

Savage, M., and T. W. Swetnam. 1990. Early 19th-century fire decline following sheep pasturing in a Navajo ponderosa pine forest. Ecology 71:2374–2378.

Sawyer, H., R. M. Nielson, F. G. Lindzey, L. Keith, J. H. Powell, and A. A. Abraham. 2007. Habitat selection of Rocky Mountain elk in a nonforested environment. Journal of Wildlife Management 71:868–874.

Sawyer, H., R. M. Nielson, F. Lindzey, and L. L. McDonald. 2006. Winter habitat selection of mule deer before and during development of a natural gas field. Journal of Wildlife Management 70:396–403.

Scarnecchia, D. L. 1995. Viewpoint: the rangeland condition concept and range sciences search for identity: a system viewpoint. Journal of Range Management 48:181–186.

Schacht, W. H. 1993. A new approach for range condition assessment is needed. Rangelands 15(6):245, 247.

Schaefer, R. J., D. J. Thayer, and T. S. Burton. 2003. Forty-one years of vegetation change on permanent transects in northeastern California: implications for wildlife. California Fish and Game 89:55–71.

Scheaffer, R. L., W. Mendenhall, III, and R. L. Ott. 1996. Elementary survey sampling. Wadsworth Publishing, Belmont, CA.

Schlesinger, W. H., J. F. Reynolds, G. L. Cunningham, L. F. Huenneke, W. M. Jarrell, R. A. Virginia, and W. G. Whitford. 1990. Biological feedbacks in global desertification. Science 247:1043–1047.

Schmidt, G. D., and L. S. Roberts. 1996. Foundations of parasitology. McGraw Hill Education, New York, NY.

Schmidtmann, E. T., G. D. Johnson, C. Y. Kato, B. L. Walker, D. E. Naugle, R. L. Mayer, and J. E. Lloyd. 2005. Coalbed methane impoundments, mosquitoes, and West Nile virus: an emerging problem? Pp. 87–89 *in* J. R. Stine (editor). Public Information Circular 43. Wyoming State Geological Survey, Laramie, WY.

Schneegas, E. R. 1967. Sage Grouse and sagebrush control. Transactions of the North American Wildlife and Natural Resources Conference 32:270–274.

Schneider, S. H. 1993. Scenarios of global warming. Pp. 9–23 *in* P. M. Kareiva, J. G. Kingsolver, and R. B. Huey (editors). Biotic interactions and global change. Sinauer Associates, Sunderland, MA.

Schoenberg, T. J. 1982. Sage Grouse movements and habitat selection in North Park, Colorado. M.S. thesis, Colorado State University, Ft. Collins, CO.

Schroeder, M. A. 1997. Unusually high reproductive effort by Sage Grouse in a fragmented habitat in north-central Washington. Condor 99:933–941.

Schroeder, M. A., and R. K. Baydack. 2001. Predation and the management of prairie grouse. Wildlife Society Bulletin 29:24–32.

Schroeder, M. A., and C. E. Braun. 1991. Walk-in traps for capturing Greater Prairie-Chickens on leks. Journal of Field Ornithology 62:378–385.

Schroeder, M. A., and L. A. Robb. 1993. Greater Prairie Chicken (*Tympanuchus cupido*). *In* A. Poole and F. Gill (editors). The birds of North America, No. 36. The Academy of Natural Sciences, Philadelphia, PA, and The American Ornithologists' Union, Washington, DC.

Schroeder, M. A., and L. A. Robb. 2003. Fidelity of Greater Sage-Grouse *Centrocercus urophasianus* to breeding areas in a fragmented landscape. Wildlife Biology 9:291–299.

Schroeder, M. A., and W. M. Vander Haegen. 2006. Use of Conservation Reserve Program fields by Greater Sage-Grouse and other shrubsteppe-associated wildlife in Washington State. Technical report for USDA Farm Service Agency. Washington Department of Fish and Wildlife, Olympia, WA.

Schroeder, M. A., C. L. Aldridge, A. D. Apa, J. R. Bohne, C. E. Braun, S. D. Bunnell, J. W. Connelly, P. A. Deibert, S. C. Garnder, M. A. Hilliand, G. D. Kobriger, S. M. McAdam, C. W. McCarthy, J. J. McCarthy, D. L. Mitchell, E. V. Rickerson, and S. J. Stiver. 2004. Distribution of sage-grouse in North America. Condor 106:363–376.

Schroeder, M. A., P. R. Ashley, and M. Vander Haegen. 2009. Terrestrial wildlife and habitat assessment on Bonneville Power Administration–funded wildlife areas in the state of Washington: monitoring and evaluation activities of the past and recommendations for the future. Washington Department of Fish and Wildlife, Olympia, WA.

Schroeder, M. A., D. W. Hays, M. F. Livingston, L. E. Stream, J. E. Jacobson, D. J. Pierce, and T. McCall. 2000. Changes in the distribution and abundance of Sage Grouse in Washington. Northwestern Naturalist 81:104–112.

Schroeder, M. A., J. R. Young, and C. E. Braun. 1999. Sage Grouse (*Centrocercus urophasianus*). *In* A. Poole and F. Gill (editors). The birds of North America, No. 425. The Academy of Natural Sciences, Philadelphia, PA, and The American Ornithologists' Union, Washington, DC.

Schultz, C. B. 1998. Dispersal behavior and its implications for reserve design in a rare Oregon butterfly. Conservation Biology 12:284–292.

Schumaker, N. H. 1996. Using landscape indices to predict habitat connectivity. Ecology 77:1210–1225.

Schupp, E. W. 1993. Quantity, quality and the effectiveness of seed dispersal by animals. Vegetatio 107/108:15–29.

Schupp, E. W., J. M. Gomez, J. E. Jimenez, and M. Fuentes. 1997. Dispersal of *Juniperus occidentalis* (western juniper) seeds by frugivorous mammals on Juniper Mountain, southeastern Oregon. Great Basin Naturalist 57:74–78.

Shultz, L. M. 2009. Monograph of Artemisia subgenus Tridentatae (Asteracea-Anthemideae). Systematic Botany Monographs 89.

Scifres, C. J. 1980. Brush management: principles and practices for Texas and the Southwest. Texas A & M University Press, College Station, TX.

Scott, J. M., F. Davis, B. Csuti, R. Noss, B. Butterfield, G. Groves, H. Anderson, S. Caicco, J. Ulliman, and R. G. Wright. 1993. Gap analysis: a geographic approach to protection of biological diversity. Wildlife Monographs 123:1–41.

Scott, J. M., M. Murray, R. G. Wright, B. Csuti, P. Morgan, and R. L. Pressey. 2001. Representation of natural vegetation in protected areas: capturing the geographic range. Biodiversity and Conservation 10:1297–1301.

Scott, J. W. 1940. The role of coccidia as parasites of wildlife. Colorado-Wyoming Academy of Science 2:45.

Scott, J. W. 1942. Mating behavior of the Sage Grouse. Auk 59:477–498.

Scott, J. W., and R. F. Honess. 1933. On a serious outbreak of coccidiosis among Sage Chickens. Colorado-Wyoming Academy of Science 1:87–88.

Scott, J. W., and R. F. Honess. 1937. A further investigation of coccidia in the Sage Hen. Colorado-Wyoming Academy of Science 2:48.

Seager, R., Y. Kushnir, C. Herweijer, N. Naik, and J. Velez. 2005. Modeling of tropical forcing of persistent droughts and pluvials over western North America: 1856–2000. Journal of Climate 18:4065–4088.

Seager, R., M. Ting, I. Held, Y. Kushnir, J. Lu, G. Vecchi, H.-P. Huang, N. Harnik, A. Leetmaa, N-C. Lau, C. Li, J. Velez, and N. Naik. 2007. Model projections of an imminent transition to a more arid climate in southwestern North America. Science 316:1181–1184.

Seaman, D. E., J. J. Millspaugh, B. J. Kernohan, G. C. Brundige, K. J. Raedeke, and R. A. Gitzen. 1999. Effects of sample size on kernel home range estimates. Journal of Wildlife Management 54:42–45.

Sedinger, J. S., and J. J. Rotella. 2005. Effect of harvest on Sage-Grouse *Centrocercus urophasianus* populations: what can we learn from the current data? Wildlife Biology 11:371–375.

Seefeldt, S. S., M. Germino, and K. DiCristina. 2007. Prescribed fires in *Artemisia tridentata* ssp. *vaseyana* steppe have minor and transient effects on vegetation cover and composition. Applied Vegetation Science 10:249–256.

Segelbacher, G., J. Höglund, and I. Storch. 2003. From connectivity to isolation: genetic consequences of population fragmentation in Capercaillie across Europe. Molecular Ecology 12:1773–1780.

Semple, K., R. K. Wayne, and R. M. Gibson. 2001. Microsatellite analysis of female mating behavior in lek-breeding Sage Grouse. Molecular Ecology 10:2043–2048.

Shafer, S. L., P. J. Bartlein, and R. S. Thompson. 2001. Potential changes in the distributions of western North America tree and shrub taxa under future climate scenarios. Ecosystems 4:200–215.

Shaffer, T. L., and D. H. Johnson. 2008. Ways of learning: observational studies versus experiments. Journal of Wildlife Management 72:4–13.

Shaman, J., J. F. Day, and M. Stieglitz. 2005. Drought-induced amplification and epidemic transmission of West Nile virus in southern Florida. Journal of Medical Entomology 42:134–141.

Shane, R. L., J. R. Garrett, and G. S. Lucier. 1983. Relationship between selected factors and internal rate of return from sagebrush removal and seeding crested wheatgrass. Journal of Range Management 36:782–786.

Shaw, N. L., and A. L. Hild. 2007. Rush skeletonweed in the northern Great Basin. USDA Forest Service RMRS Tech Transfer. Research Summary. http://www.fs.fed.us/rm/boise/research/shrub/projects/documents/shaw_rust_tech_transfer110107.pdf (13 July 2008).

Shaw, N. L., A. M. DeBolt, and R. Rosentreter. 2005a. Reseeding big sagebrush techniques and issues. Pp. 99–108 *in* N. L. Shaw, M. Pellant, and S. B. Monsen (compilers). Sage-grouse habitat restoration symposium proceedings. USDA Forest Service RMRS-P-38. USDA Forest Service, Rocky Mountain Research Station, Ft. Collins, CO.

Shaw, N. L., M. Pellant, and S. B. Monsen (compilers). 2005b. Sage-grouse habitat restoration symposium proceedings. USDA Forest Service RMRS-P-38. USDA Forest Service, Rocky Mountain Research Station, Ft. Collins, CO.

Shaw, R. B., and V. E. Diersing. 1989. Allowable use estimates for tracked vehicular training on Pinon Canyon maneuver site, Colorado, USA. Environmental Management 13:773–782.

Shaw, R. B., and V. E. Diersing. 1990. Tracked vehicle impacts on vegetation at the Pinon Canyon maneuver site, Colorado. Journal of Environmental Quality 19:234–243.

Shaw, R. B., and D. G. Kowalski. 1996. U.S. Army lands: a national survey. CEEML TPS 96-1. Center for Ecological Management of Military Lands, Colorado State University, Ft. Collins, CO.

Sheley, R. L., and J. K. Petroff (editors). 1999. Biology and management of noxious rangeland weeds. Oregon State University Press, Corvallis, OR.

Sheley, R. L., J. M. Hudak, and R. T. Grubb. 1999. Rush skeletonweed. Pp. 308–314 *in* R. L. Sheley and J. K. Petroff (editors). Biology and management of noxious rangeland weeds. Oregon State University Press, Corvallis, OR.

Sheley, R. L., T. J. Svejcar, and B. D. Maxwell. 1996. A theoretical framework for developing successional weed management strategies on rangeland. Weed Technology 10:766–773.

Shepherd, III, J. F. 2006. Landscape-scale habitat use by Greater Sage-Grouse (*Centrocercus urophasianus*) in southern Idaho. Ph.D. dissertation, University of Idaho, Moscow, ID.

Sherfy, M. H. 1992. The influence of season, temperature, and wind speed on Sage Grouse metabolism. M.S. thesis, University of New Hampshire, Durham, NH.

Sherfy, M. H., and P. J. Pekins. 1995. Influence of wind speed on Sage Grouse metabolism. Canadian Journal of Zoology 73:749–754.

Shiflet, T. N. (editor). 1994. Rangeland cover types of the United States. Society for Range Management, Denver, CO.

Shillinger, J. E., and L. C. Morley. 1937. Diseases of upland game birds. Farmer's Bulletin 1781. United States Department of Agriculture, Washington, DC.

Shinn, S. L., and D. C. Thill. 2002. The response of yellow starthistle (*Centaurea solstitialis*), annual grasses, and smooth brome (*Bromus inermus*) to Imazapic and Picloram. Weed Technology 16:366–370.

Shinn, S. L., and D. C. Thill. 2004. Tolerance of several perennial grasses to Imazapic. Weed Technology 18:60–65.

Shinneman, D. J. 2006. Determining restoration needs for piñon-juniper woodlands and adjacent ecosystems on the Uncompahgre Plateau, western Colorado. Ph.D. dissertation, University of Wyoming, Laramie, WY.

Shinneman, D. J., and W. L. Baker. 2009. Environmental and climatic variables as potential drivers of post-fire cover of cheatgrass (*Bromus tectorum*) in seeded and unseeded semi-arid ecosystems. International Journal of Wildland Fire 18:191–202.

Shorey, L., S. Peirtney, J. Stone, and J. Höglund. 2000. Fine-scale genetic structuring on *Manacus manacus* leks. Nature 408:352–353.

Shugart, H. H. 1998. Terrestrial ecosystems in changing environments. Cambridge University Press, Cambridge, UK.

Shumar, M. L., and J. E. Anderson. 1986. Gradient analyses of vegetation dominated by two subspecies of big sagebrush. Journal of Range Management 39:156–160.

Sidle, J. G. 2005. In my opinion: grousing and grazing on national grasslands. Wildlife Society Bulletin 33:1139–1144.

Sika, J. L. 2006. Breeding ecology, survival rates, and causes of mortality of hunted and nonhunted Greater Sage-Grouse in central Montana. M.S. thesis, Montana State University, Bozeman, MT.

Silverman, B. W. 1986. Density estimation for statistics and data analysis. Chapman and Hall, London, UK.

Sime, C. A. 1991. Sage Grouse use of burned, non-burned, and seeded vegetation on the Idaho National Engineering Laboratory, Idaho. M.S. thesis, Montana State University, Bozeman, MT.

SIMMAP. 2003. SIMMAP 2.0. Landscape categorical spatial patterns simulation software. http://web.udl.es/usuaris/saura/ (18 June 2009).

Simon, F. 1937. A new cestode, *Raillietina centrocerci*, from the Sage Grouse *Centrocercus urophasianus*. Transactions of the American Microscopical Society 56:340–343.

Simon, F. 1939a. *Cheilospirura certrocerci*, a new nematode from the Sage Grouse *Centrocercus urophasianus*. Transactions of the American Microsopical Society 58:78–80.

Simon, F. 1939b. *Eimeria centrocerci* n. sp. du *Centrocercus urophasianus* (coq de Bruyere). Annales de Parasitology, Humaine et Comparee 17:137–138.

Simon, F. 1940. The parasites of the Sage Grouse. Publication 7:77–100. University of Wyoming, Laramie, WY.

Simpson, G. G. 1951. Horses: the story of the horse family in the modern world and through sixty million years of history. Oxford University Press, New York, NY.

Sindelar, B. W. 1971. Douglas-fir invasion of western Montana grasslands. Ph. D. dissertation. University of Montana, Missoula, MT.

Singer, F. J. 2004. Strategic research plan: wild horse and burro management. Report produced by United States Geological Survey, Fort Collins Science Center and Natural Resource Ecology Laboratory, for the Bureau of Land Management, Ft. Collins, CO.

Singer, J. D., and J. B. Willett. 2003. Applied longitudinal data analysis. Oxford University Press, New York, NY.

Slater, S. J. 2003. Sage-grouse (*Centrocercus urophasianus*) use of different-aged burns and the effects of coyote control in southwestern Wyoming. M.S. thesis, University of Wyoming, Laramie, WY.

Small, R. J., J. C. Holzwart, C. James, and D. H. Rusch. 1991. Predation and hunting mortality of Ruffed Grouse in central Wisconsin. Journal of Wildlife Management 55:512–520.

Smith, A. D. 1949. Effects of mule deer and livestock upon a foothill range in northern Utah. Journal of Wildlife Management 13:21–23.

Smith, J. A. 1991. The idea brokers. Free Press, New York, NY.

Smith, J. T. 2003. Greater Sage Grouse on the edge of their range: leks and surrounding landscapes in the Dakotas. M.S. thesis, South Dakota State University, Brookings, SD.

Smith, J. T., L. D. Flake, K. F. Higgins, and G. D. Kobriger. 2006. Microhabitat characteristics relative to lek abandonment by Greater Sage-Grouse in the Dakotas. Intermountain Journal of Science 12:1–11.

Smith, J. T., L. D. Flake, K. F. Higgins, G. D. Kobriger, and C. G. Homer. 2005. Evaluating lek occupancy of Greater Sage-Grouse in relation to landscape cultivation in the Dakotas. Western North American Naturalist 65:310–320.

Smith, S. D., T. E. Huxman, S. F. Zitzer, T. N. Charlet, D. C. Housman, J. S. Coleman, L. K. Fenstermaker, J. R. Seeman, and R. S. Nowak. 2000. Elevated CO_2 increases productivity and invasive species success in an arid ecosystem. Nature 408:79–82.

Smith, S. D., B. R. Strain, and T. D. Sharkey. 1987. Effects of CO_2 enrichment on four Great Basin grasses. Functional Ecology 1:139–143.

Society for Ecological Restoration International Science and Policy Working Group. 2004. The SER international primer on ecological restoration, version 2. Society for Ecological Restoration International, Tucson AZ. http://www.ser.org/pdf/primer3.pdf (30 April 2009).

Society for Range Management. 1989. Assessment of rangeland condition and trend of the United States. Society for Range Management, Denver, CO.

Society for Range Management. 1995. New concepts for assessment of rangeland condition. Journal of Range Management 48:271–282.

Sorensen, T., P. D. Mcloughlin, D. Hervieux, E. Dzus, J. Nolan, B. Wynes, and S. Boutin. 2008. Determining sustainable levels of cumulative effects for boreal caribou. Journal of Wildlife Management 72:900–905.

Soulé, M. E. 1980. Thresholds for survival: maintaining fitness and evolutionary potential. Pp. 151–169 *in* M. E. Soulé and B. A. Wilcox (editors). Conservation biology: an ecological-evolutionary perspective. Sinauer Associates, Sunderland, MA.

Soulé, M. E., and M. A. Sanjayan. 1998. Conservation targets: do they help? Science 279:2060–2061.

Soulé, M. E., and J. Terborgh. 1999. Conserving nature at regional and continental scales—a scientific program for North America. BioScience 49:809–817.

Soulé, P. T., and P. A. Knapp. 1999. Western juniper expansion on adjacent disturbed and near relict sites. Journal of Range Management 52:525–533.

Soulé, P. T., and P. A. Knapp. 2000. *Juniperus occidentalis* (western juniper) establishment history on two minimally disturbed research natural areas in central Oregon. Western North American Naturalist 60:26–33.

Soulé, P. T., P. A. Knapp, and H. D. Grissino-Mayer. 2003. Comparative rates of western juniper afforestation in south-central Oregon and the role of anthropogenic disturbance. Professional Geographer 55:43–55.

Soulé, P. T., P. A. Knapp, and H. D. Grissino-Mayer. 2004. Human agency, environmental drivers, and western juniper establishment during the late Holocene. Ecological Applications 14:96–112.

South Dakota Game, Fish and Parks. 2008. Greater Sage-Grouse management plan, South Dakota—2008–2017. South Dakota Game, Fish and Parks, Pierre, SD. http://www.sdgfp.info/Publications/SageGrousePlanDRAFT.pdf (18 March 2009).

Southwood, T. R. E. 1977. Habitat, the template for ecological strategies? Journal of Animal Ecology 46:337–365.

Sovada, M. A., P. J. Pietz, K. A. Converse, D. T. King, E. K. Hofmeister, P. Scherr, and H. S. Ip. 2008. Impact of West Nile virus and other mortality factors on American White Pelicans at breeding colonies in the northern plains of North America. Biological Conservation 141:1021–1031.

Spaeth, K. E., F. B. Pierson, J. E. Herrick, P. L. Shaver, D. A. Pyke, M. Pellant, D. Thompson, and B. Dayton. 2003. New proposed national resources inventory protocols on nonfederal rangelands. Journal of Soil and Water Conservation 58:18–21.

Spaulding, A. W., K. E. Mock, M. A. Schroeder, and K. I. Warheit. 2006. Recency, range expansion, and unsorted lineages: implications for interpreting neutral genetic variation in the Sharp-tailed Grouse (*Tympanuchus phasianellus*). Molecular Ecology 9:2317–2332.

Species at Risk Act. 2002. Canada Gazette. Part III, chapter 29, volume 25, number 3. http://www.sararegistry.gc.ca/default_e.cfm (6 April 2009).

Spurrier, M. F., M. S. Boyce, and B. F. J. Manly. 1991. Effects of parasites on mate choice by captive Sage Grouse. Pp. 389–399 *in* J. E Loye and M. Zuk (editors). Ecology, behavior and evolution of bird-parasite interactions. Oxford Ornithological Series 2. Oxford University Press, Oxford, UK.

Stabler, R. M., and N. J. Kitzmiller. 1974. Hematozoa from Colorado birds. Journal of Parasitology 60:536–537.

Stabler, R. M., C. E. Braun, and T. D. I Beck. 1977. Hematozoa in Sage Grouse from Colorado. Journal of Wildlife Diseases 13:414–417.

Stabler, R. M., P. A. Holt, and N. J. Kitzmiller. 1966. *Trypanosoma avium* in the blood and bone marrow from 677 Colorado birds. Journal of Parasitology 52:1141–1144.

Staples, D. F., M. L. Taper, and B. Dennis. 2004. Estimating population trend and process variation for PVA in the presence of sampling error. Ecology 85:923–929.

Steenhof, K., M. N. Kochert, and J. A. Roppe. 1993. Nesting by raptors and Common Ravens on electrical transmission line towers. Journal of Wildlife Management 57:271–281.

Sterling, T. M., D. C. Thompson, and K. C. McDaniel. 1999. Snakeweeds. Pp. 323–335 *in* R. L. Sheley and J. K. Petroff (editors). Biology and management of noxious rangeland weeds. Oregon State University Press, Corvallis, OR.

Stevens, R. 1994. Interseeding and transplanting to enhance species composition. Pp. 300–306 *in* S. B. Monsen and S. G. Kitchen (compilers). Ecology and management of annual rangelands. USDA Forest Service General Technical Report INT-GTR-313. USDA Forest Service, Intermountain Research Station, Ogden, UT.

Stevens, R. 2004. Management of restored and revegetated sites. Pp. 193–198 *in* S. B. Monsen, R. Stevens, and N. L. Shaw (compilers). Restoring western ranges and wildlands, Vol. I. USDA Forest Service RMRS-GTR-136-vol-1. USDA Forest Service, Rocky Mountain Research Station, Ft. Collins, CO.

Stevens, R., and S. B. Monsen. 2004. Mechanical plant control. Pp. 65–87 *in* S. B. Monsen, R. Stevens, and N. L. Shaw (compilers). Restoring western ranges and wildlands, Vol. I. USDA Forest Service RMRS-GTR-136-vol-1. USDA Forest Service, Rocky Mountain Research Station Ft. Collins, CO.

Stewart, G. B., A. S. Pullin, and C. F. Coles. 2007. Poor evidence-base for assessment of wind farm impacts on birds. Environmental Conservation 34:1–11.

Stigar, M. S. 1989. Hunting low density Sage-Grouse populations. M.S. thesis, University of Nevada, Reno, NV.

Stinson, D. W., D. W. Hays, and M. A. Schroeder. 2004. Washington state recovery plan for the Greater Sage-Grouse. Washington Department of Fish and Wildlife, Olympia, WA. http://wdfw.wa.gov/wlm/diversty/soc/recovery/sage_grouse/final_sage_grouse_recovery.pdf (18 March 2009).

Stiver, S. J., A. D. Apa, J. R. Bohne, S. D. Bunnell, P. A. Deibert, S. C. Gardner, M. A. Hilliard, C. W. McCarthy, and M. A. Schroeder. 2006. Greater Sage-Grouse comprehensive conservation strategy. Western Association of Fish and Wildlife Agencies, Cheyenne, WY. http://www.wafwa.org/pdf/GreaterSage-grouseConservationStrategy2006.pdf (21 August 2009).

Stoddart, L. A., A. D. Smith, and T. W. Box. 1975. Range management. McGraw-Hill, New York, NY.

Stohlgren, T. J., L. D. Schell, and B. Vanden Heuvel. 1999. How grazing and soil quality affect native and exotic plant diversity in Rocky Mountain grasslands. Ecological Applications 9:45–64.

Stoms, D. M., F. W. Davis, K. L. Driese, K. M. Cassidy, and M. P. Murray. 1998. Gap analysis of the vegetation of the intermountain semi-desert ecoregion. Great Basin Naturalist 58:199–216.

Storch, I. 2000. Conservation status and threats to grouse worldwide: an overview. Wildlife Biology 6:195–204.

Storch, I. 2003. Linking a multiscale habitat concept to species conservation. Pp. 303–320 *in* J. A. Bissonette and I. Storch (editors). Landscape ecology and resource management: linking theory with practice. Island Press, Washington, DC.

Storch, I., and G. Segelbacher. 2000. Genetic correlates of spatial population structure in central European Capercaillie *Tetrao urogallus* and Black Grouse *T. tetrix*: a project in progress. Wildlife Biology 6:305–310.

Storch, I., E. Woitke, and S. Krieger. 2005. Landscape-scale edge effect in predation risk in forest-farmland mosaics of central Europe. Landscape Ecology 20:927–940.

Strauss, D., L. Bednar, and R. Mees. 1989. Do one percent of forest fires cause ninety-nine percent of the damage? Forest Science 35:319–328.

Strickland, M. D., H. J. Harju, K. R. McCaffery, H. W. Miller, L. M. Smith, and R. J. Stoll. 1994. Harvest management. Pp. 445–473 *in* T. A. Bookhout (editor). Research and management techniques for wildlife and habitats. The Wildlife Society, Bethesda, MD.

Stringham, T. K., W. C. Krueger, and P. L. Shaver. 2003. State and transition modeling: an ecological process approach. Journal of Range Management 56:106–113.

Stubbs, M. M. 2000. Time-based nitrogen availability and risk of *Taeniaterum caput-medusae* (L.) Nevski invasion in central Oregon. M.S. thesis, Oregon State University, Corvallis, OR.

Styer, L. M., K. A. Kent, R. G. Albright, C. J. Bennett, L. D. Kramer, and K. A. Bernard. 2007a. Mosquitos inoculate high doses of West Nile virus as they probe and feed on live hosts. PLoS Pathology 3:1262–1270.

Styer, L.M., M. A. Meola, and L. D. Kramer. 2007b. West Nile virus infection decreases fecundity of *Culex tarsalis* females. Journal of Medical Entomology 44:1074–1085.

Suring, L. H., M. M. Rowland, and M. J. Wisdom. 2005a. Identifying species of conservation concern. Pp. 150–162 *in* M. J. Wisdom, M. M. Rowland, and L. H. Suring (editors). Habitat threats in the sagebrush ecosystem: methods of regional assessment and applications in the Great Basin. Alliance Communications Group, Lawrence, KS.

Suring, L. H., M. J. Wisdom, R. J. Tausch, R. F. Miller, M. M. Rowland, L. Schueck, and C. W. Meinke. 2005b. Modeling threats to sagebrush and other shrubland communities. Pp. 114–149 *in* M. J. Wisdom, M. M. Rowland, and L. H. Suring (editors). Habitat threats in the sagebrush ecosystem: methods of regional assessment and applications in the Great Basin. Alliance Communications Group, Lawrence, KS.

Suter, W., R. F. Graf, and R. Hess. 2002. Capercaillie (*Tetrao urgallus*) and avian biodiversity: testing the umbrella-species concept. Conservation Biology 16:778–788.

Sutherland, W. J. 2001. Sustainable exploitation: a review of principles and methods. Wildlife Biology 7:131–140.

Svancara, L. K., R. Brannon, J. M. Scott, C. R. Groves, R. F. Noss, and R. L. Pressey. 2005. Policy-driven versus evidence-based conservation: a review of political targets and biological needs. BioScience 55:989–995.

Svejcar, T., and R. Sheley. 2001. Nitrogen dynamics in perennial- and annual-dominated arid rangeland. Journal of Arid Environments 47:33–46.

Sveum, C. M. 1995. Habitat selection by Sage Grouse hens during the breeding season in south-central Washington. M.S. thesis, Oregon State University, Corvallis, OR.

Sveum, C. M., J. A. Crawford, and W. D. Edge. 1998a. Use and selection of brood-rearing habitat by Sage Grouse in south central Washington. Great Basin Naturalist 58:344–351.

Sveum, C. M., W. D. Edge, and J. A. Crawford. 1998b. Nesting habitat selection by Sage Grouse in south-central Washington. Journal of Range Management 51:265–269.

Swenson, J. E. 1986. Differential survival by sex in juvenile Sage Grouse and Gray Partridge. Ornis Scandinavica 17:14–17.

Swenson, J. E., C. A. Simmons, and C. D. Eustace. 1987. Decrease of Sage Grouse *Centrocercus urophasianus* after ploughing of sagebrush steppe. Biological Conservation 41:125–132.

Swetnam, T. W., C. H. Baisan, and J. M. Kaib. 2001. Forest fire histories of the sky islands of La Frontera. Pp. 95–119 *in* G. L. Webster and C. J. Bahre (editors). Changing plant life of La Frontera: observations on vegetation in the U.S./Mexico borderlands. University of New Mexico Press, Albuquerque, NM.

Symanski, R. 1994. Contested realities: feral horses in outback Australia. Annals of the Association of American Geographers 84:251–269.

Szabó, P., and G. Meszéna. 2006. Spatial ecological hierarchies: coexistence on heterogeneous landscapes via scale niche diversification. Ecosystems 9:1009–1016.

Talbert, C. B., R. L. Knight, and J. E. Mitchell. 2007. Private ranchlands and public land grazing in the southern Rocky Mountains. Rangelands 29(3):5–8.

Talbot, L. M., and M. H. Talbot. 1963. The wildebeest in western Masailand, East Africa. Wildlife Monographs 12. The Wildlife Society, Washington, DC.

Tausch, R. J. 1999a. Historic woodland development. Pp. 12–19 *in* S. B. Monsen, R. Stevens, R. J. Tausch, R. Miller, and S. Goodrich (compilers). Ecology and management of pinyon-juniper communities within the Interior West. USDA Forest Service RMRS-P-9. USDA Forest Service, Rocky Mountain Research Station, Ogden, UT.

Tausch, R. J. 1999b. Transitions and thresholds: influences and implications for management in pinyon and Utah juniper woodlands. Pp. 61–65 *in* S. B. Monsen, R. Stevens, R. J. Tausch, R. Miller, and S. Goodrich (compilers). Ecology and management of pinyon-juniper communities within the Interior West. USDA Forest Service RMRS-P-9. USDA Forest Service, Rocky Mountain Research Station, Ogden, UT.

Tausch, R. J., and R. S. Nowak. 1999. Fifty years of ecotone change between shrub and tree dominance in the Jack Springs Pinyon Research Natural Area. Pp. 71–77 *in* E. D. McArthur, W. K. Ostler, and C. L. Wambolt (compilers). Shrubland ecotones. USDA Forest Service Research Paper RMRS-P-11. USDA Forest Service, Rocky Mountain Research Station, Ogden, UT.

Tausch, R. J., N. E. West, and A. A. Nabi. 1981. Tree age and dominance patterns in Great Basin pinyon-juniper woodlands. Journal of Range Management 34:259–264.

Tausch, R. J., P. E. Wigand, and J. W. Burkhardt. 1993. Viewpoint: plant community thresholds, multiple steady states, and multiple successional pathways: legacy of the Quaternary? Journal of Range Management 46:439–447.

Taylor, A. R., and R. L. Knight. 2003. Wildlife responses to recreation and associated visitor perceptions. Ecological Applications 13:951–963.

Taylor, P. D., L. Fahrig, K. Henein, and G. Merriam. 1993. Connectivity is a vital element of landscape structure. Oikos 68:571–573.

Taylor, P. D., L. Fahrig, and K. A. With. 2006. Landscape connectivity: a return to the basics. Pp. 29–43 *in* K. R. Crooks and M. Sanjayan (editors). Connectivity conservation. Cambridge University Press, New York, NY.

Taylor, S. E., and J. R. Young. 2006. A comparative behavioral study of three Greater Sage-Grouse populations. Wilson Journal of Ornithology 118:36–41.

Teich, A. H. 2008. Can administrative measures resolve a political conflict? Public Administration Review 68:19–22.

ter Braak, C. J. F. 1986. Correspondence analysis of incidence and abundance data: properties in terms of a unimodel response model. Biometrics 41: 859–873.

ter Braak, C. J. F. 1995. Ordination. Pp. 91–173 *in* R. H. G. Jongman, C. J. F. ter Braak, and O. F. R. van Tongeren (editors). Data analysis in community and landscape ecology. Cambridge University Press, Cambridge, UK.

ter Braak, C. J. F., and P. Smilauer. 1997. CANOCO. Version 4.02. Centre for Biometry Wageningen, CPRO-DLO, Wageningen, Netherlands.

Tewksbury, J. J., A. E. Black, N. Nur, V. A. Saab, B. D. Logan, and D. S. Dobkin. 2002. Effects of anthropogenic fragmentation and livestock grazing on western riparian bird communities. Studies in Avian Biology 25:158–202.

Theobald, D. M. 2001. Land use dynamics beyond the American urban fringe. Geographical Review 91:544–564.

Theobald, D. M. 2003. Targeting conservation action through assessment of protection and exurban threats. Conservation Biology 17:1624–1637.

Theobald, D. M. 2005. Landscape patterns of exurban growth in the USA from 1980 to 2020. Ecology and Society 10:32. http://www.ecologyandsociety.org/vol10/iss1/art32/ (7 September 2009).

Thomas, C. D. 1994. Extinction, colonization, and metapopulations: environmental tracking by rare species. Conservation Biology 8:373–378.

Thomas, D. H. 1973. An empirical test for Steward's model of Great Basin settlement patterns. American Antiquity 38:155–176.

Thomas, H. S. 1979. The wild horse controversy. A. S. Barnes, New York, NY.

Thompson, F. R., W. D. Dijak, T. G. Kulowiec, and D. A. Hamilton. 1992. Breeding bird populations in Missouri Ozark forests with and without clearcutting. Journal of Wildlife Management 56:23–30.

Thompson, K. M., M. J. Holloran, S. J. Slater, J. L. Kuipers, and S. H. Anderson. 2006. Early brood-rearing habitat use and productivity of Greater Sage-Grouse in Wyoming. Western North American Naturalist 66:332–342.

Thorne, E. T. 1969. Diseases in Wyoming Sage Grouse. Proceedings of the Western States Sage Grouse Workshop 6:192–198.

Thurow, T. L. 1991. Hydrology and erosion. Pp. 141–159 *in* R. K. Heitschmidt and J. W. Stuth (editors). Grazing management: an ecological perspective. Timber Press, Portland, OR.

Thurow, T. L., and C. A. Taylor, Jr. 1999. Viewpoint: the role of drought in range management. Journal of Range Management 52:413–419.

Tiawsirisup, S., J. R. Kinley, B. J. Tucker, R. B. Evans, W. A. Rowley, and K. B. Platt. 2008. Vector competence of *Aedes vexans* (Diptera: Culicidae) for West Nile virus and potential as an enzootic vector. Journal of Medical Entomoogy 45:452–457.

Tiawsirisup, S., K. B. Platt, B. J. Tucker, and W. A. Rowley. 2005. Eastern cottontail rabbits (*Sylvilagus floridanus*) develop West Nile virus viremias sufficient for infecting select mosquito species. Vector-Borne and Zoonotic Diseases 5:342–349.

Tibshirani, R. 1996. Regression shrinkage and selection via the lasso. Journal of the Royal Statistical Society B58:267–288.

Tirhi, M. J. 1995. Washington state management plan for sage-grouse. Washington Department of Fish and Wildlife, Olympia, WA.

Tischendorf, L., and L. Fahrig. 2000. On the usage and measurement of landscape connectivity. Oikos 90:7–19.

Tisdale, E. W. 1994. Great Basin region: sagebrush types. Pp. 40–46 *in* T. N. Shiflet (editor). Rangeland cover types. Society for Range Management, Denver, CO.

Tisdale, E. W., and M. Hironaka. 1981. The sagebrush-grass ecoregion: a review of the ecological literature. Forest, Wildlife, and Range Experiment Station Contribution Number 209. University of Idaho, Moscow, ID.

Tisdale, E. W., M. Hironaka, and F. A. Fosberg. 1965. An area of pristine vegetation in Craters of the Moon National Monument, Idaho. Ecology 46:349–352.

Tisdall, J. M., and J. M. Oades. 1982. Organic matter and water-stable aggregates in soils. Journal of Soil Science 33:141–163.

Toft, N. L., J. E. Anderson, and R. S. Nowak. 1989. Water use efficiency and carbon isotope composition of plants in a cold desert environment. Oecologia 80:11–18.

Tompkins, D. M., A. P. Dobson, P. Arneberg, M. E. Begon, I. M. Cattadori, J. V. Greenman, J. A. P. Heesterbeek, P. J. Hudson, D. Newborn, A. Pugliese, A. P. Rizzoli, R. Rosa, F. Rosso, and K. Wilson. 2002. Parasites and host population dynamics. Pp. 45–62 *in* P. J. Hudson, A. Rizzoli, B. T. Grenfell, H. Heesterbeek, and A. P. Dobson (editors). The ecology of wildlife diseases. Oxford University Press, Oxford, UK.

Tompkins, D. M., R. A. H. Draycott, and P. J. Hudson. 2000a. Field evidence for apparent competition mediated via the shared parasites of two gamebird species. Ecology Letters 3:10–14.

Tompkins, D. M., J. V. Greenman, P. A. Robertson, and P. J. Hudson. 2000b. The role of shared parasites in the exclusion of wildlife hosts: *Heterakis gallinarum* in the Ring-necked Pheasant and the Grey Partridge. Journal of Animal Ecology 69:829–840.

Torregrosa, A., and N. Devoe. 2008. Urbanization and changing land use in the Great Basin. Pp. 9–13 *in* J. C. Chambers, N. Devoe, and A. Evenden (editors). Collaborative management and research in the Great Basin—examining the issues and developing a framework for action. USDA Forest Service Technical Report RMRS-GTR-204. USDA Forest Service, Rocky Mountain Research Station General, Ft. Collins, CO.

Touchan, R., T. W. Swetnam, and H. D. Grissino-Mayer. 1995. Effects of livestock grazing on pre-settlement fire regimes in New Mexico. Pp. 268–272 *in* J. K. Brown, R. W. Mutch, C. W. Spoon, and R. H. Wakimoto (technical coordinators). Symposium on Fire in Wilderness and Park Management. USDA Forest Service General Technical Report INT-GTR-320. USDA Forest Service, Intermountain Research Station, Ogden, UT.

Trail, P. W. 1985. Courtship disruption modifies mate choice in a lek breeding bird. Science 227:778–780.

Trombulak, S. C., and C. A. Frissell. 2000. Review of ecological effects of roads on terrestrial and aquatic communities. Conservation Biology 14:18–30.

Trueblood, R. W. 1954. The effect of grass reseeding in sagebrush lands on Sage Grouse populations. M.S. thesis, Utah State University, Logan, UT.

Tueller, P. T., and R. A. Evans. 1969. Control of green rabbitbrush and big sagebrush with 2,4-D and picloram. Weed Science 17:233–235.

Turell, M. J., D. J. Dohm, M. R. Sardelis, M. L. O'Guinn, T. G. Andreadis, and J. A. Blow. 2005. An update on the potential of North American mosquitos (Diptera: Culicidae) to transmit West Nile virus. Journal of Medical Entomology 42:57–62.

Turell, M. J., M. L. O'Guinn, D. J. Dohm, and J. W. Jones. 2001. Vector competence of North American mosquitos (Diptera: Culicidae) for West Nile virus. Journal of Medical Entomology 38:130–134.

Turell, M. J., M. L. O'Guinn, and J. Oliver. 2000. Potential for New York mosquitoes to transmit West Nile virus. American Journal of Tropical Medicine and Hygiene 62:413–414.

Turner, Jr., J. W., and M. L. Morrison. 2001. Influence of predation by mountain lions on numbers and survivorship of a feral horse population. Southwestern Naturalist 46:183–190.

Turner, M. G. (editor). 1987. Landscape heterogeneity and disturbance. Springer-Verlag, NY.

Turner, M. G. 1989. Landscape ecology: the effect of pattern on process. Annual Review of Ecology and Systematics 20:171–197.

Turner, M. G., and S. P. Bratton. 1987. Fire, grazing, and the landscape heterogeneity of a Georgia barrier island. Pp. 85–101 *in* M. G. Turner (editor). Landscape heterogeneity and disturbance. Springer-Verlag, New York, NY.

Turner, M. G., and V. H. Dale. 1991. Modeling landscape disturbance. Pp. 323–351 *in* M. G. Turner and R. H. Gardner (editors). Quantitative methods in landscape ecology. Springer-Verlag, New York, NY.

Turner, M. G., R. H. Gardner, and R. V. O'Neill. 2001. Landscape ecology in theory and practice. Springer-Verlag, New York, NY.

Underwood, A. J. 1997. Experiments in ecology: their logical design and interpretation using analysis of variance. Cambridge University Press, Cambridge, UK.

United States Bureau of Land Management. 2000. Science strategy. http://www.blm.gov/nstc/pdf/scistrat.pdf (19 September 2009).

United States Bureau of Land Management. 2004. Fire history database. ftp://ftp.blm.gov/pub/gis/wildfire/firehistory2003 (13 November 2008).

United States Bureau of Land Management and Office of the Solicitor. 2001. The Federal Land Policy and Management Act, as amended. USDI Bureau of Land Management, Office of Public Affairs, Washington, DC.

United States Census Bureau. 2001. U.S. Census 2000 block data. United States Census Bureau Systems Support Division. http://www.census.gov/main/www/cen2000.html (4 June 2009).

United States Department of Agriculture. 1995. State Soil Geographic (STATSGO) data base. Data use information. USDA Natural Resources Conservation Service Miscellaneous Publication Number 1492. Natural Resources Conservation Service, Ft. Worth, TX. http://www.drought.unl.edu/whatis/palmer/riverbasin.htm (13 July 2008).

United States Department of Agriculture. 2003. National range and pasture handbook. 190-VI-NRPH. Grazing Lands Technology Institute, Washington, DC.

United States Department of Agriculture. 2009. Cumulative CRP enrollment by county, FY 1986 through FY 2007. http://www.fsa.usda.gov/Internet/FSA_File/public.xls (15 August 2009).

United States Department of Agriculture, Natural Resources Conservation Service. 2003. National range and pasture handbook, revision 1. 190-VI-NRPH. USDA NRCS, Grazing Lands Technology Institute, Ft. Worth, TX.

United States Department of Commerce. 2004. Historical graphs of the Palmer Drought Severity Index. USDC National Drought Mitigation Center. http://www.drought.unl.edu/whatis/palmer/riverbasin.htm (13 July 2008).

United States Department of Energy. 2008. 20% energy by 2030: increasing wind energy's contribution to U.S. electricity supply. http://www1.eere.energy.gov/windandhydro/pdfs/41869.pdf (1 September 2008).

United States Department of Energy. 2009. Annual energy outlook 2009 with projections to 2030. http://www.eia.doe.gov/oiaf/aeo/gas.html (15 August 2009).

United States Department of Health and Human Services. 1995. Chlordane CAS#57-74-9 ToxFAQs. Agency for Toxic Substances and Disease Registry, Atlanta, GA.

United States Department of Health and Human Services. 1997. Toxaphene CAS#8001-35-2 ToxFAQs. Agency for Toxic Substances and Disease Registry, Atlanta, GA.

United States Department of the Interior. 1989. Oil and gas surface operating standards for oil and gas exploration and development. USDI Bureau of Land Management, Washington, DC.

United States Department of the Interior. 1991. Final environmental statement: vegetation treatment on BLM lands in 13 western states. USDI Bureau of Land Management, Cheyenne, WY.

United States Department of the Interior. 1992. Final oil and gas resource management plan/environmental impact statement amendment for the Billings, Powder River and South Dakota resource areas. USDI Bureau of Land Management, Miles City District, MT.

United States Department of the Interior. 1994. Record of decision: Powder River, Billings, and South Dakota oil and gas resource management plan/environmental impact statement amendment. USDI Bureau of Land Management, Miles City District, MT.

United States Department of the Interior. 1996. Effects of military training and fire on habitats, prey and raptors in the Snake River Birds of Prey National Conservation Area. BLM/IDARNG Final Report. USDI Geological Survey, Boise, ID.

United States Department of the Interior. 1997. Public land statistics. USDI Bureau of Land Management. http://www.blm.gov/natacq/pls97 (16 August 2009).

United States Department of the Interior. 1998. Public land statistics. USDI Bureau of Land Management. http://www.blm.gov/natacq/pls98 (16 August 2009).

United States Department of the Interior. 1999. Public land statistics. USDI Bureau of Land Management. http://www.blm.gov/natacq/pls99/Pls99home.html (13 July 2008).

United States Department of the Interior. 2000a. Endangered and threatened wildlife and plants; 90-day finding and commencement of status review for a petition to list the western sage-grouse in Washington as threatened or endangered. Federal Register 65:51578–51584.

United States Department of the Interior. 2000b. Endangered and threatened wildlife and plants; notice of designation of Gunnison Sage-Grouse as a candidate species. Federal Register 65:82310–82312.

United States Department of the Interior. 2000c. Environmental assessment: the Lousiana Land and Exploration Company. WYO-050-EA-0139. USDI Bureau of Land Management, Lander Field Office, Lander, WY.

United States Department of the Interior. 2000d. Public land statistics. USDI Bureau of Land Management. http://www.blm.gov/natacq/pls00 (12 August 2008).

United States Department of the Interior. 2001a. Endangered and threatened wildlife and plants; 12-month finding for a petition to list the Washington population of Western Sage-Grouse (*Centrocercus urophasianus phaios*). Federal Register 66:22984–22986.

United States Department of the Interior. 2001b. Public land statistics. USDI Bureau of Land Management. http://www.blm.gov/natacq/pls01 (13 July 2008).

United States Department of the Interior. 2001c. SAGEMAP: a GIS database for sage-grouse and shrubsteppe management in the Intermountain West. USDI Geological Survey. http://sagemap.wr.usgs.gov (16 August 2009).

United States Department of the Interior. 2002a. Endangered and threatened wildlife and plants; 90-day finding on a petition to list the Mono Basin Area Sage-Grouse as endangered. Federal Register 67:78811–78815.

United States Department of the Interior. 2002b. Public land statistics. USDI Bureau of Land Management. http://www.blm.gov/natacq/pls02/ (13 July 2008).

United States Department of the Interior. 2003a. Endangered and threatened wildlife and plants; Final rule to list the Columbia Basin distinct population segment of the pygmy rabbit (*Brachylagus idahoensis*) as endangered. 50 CFR 17:10388–10409.

United States Department of the Interior. 2003b. Endangered and threatened wildlife and plants; 90-day finding on a petition to list the Western Sage-Grouse. Federal Register 68:6500–6504.

United States Department of the Interior. 2003c. Final environmental statement: Powder River Basin Oil and Gas Project. USDI Bureau of Land Management, Wyoming State Office, Cheyenne, WY.

United States Department of the Interior. 2003d. Grazing administration—exclusive of Alaska; proposed rule. USDI Bureau of Land Management. Federal Register 68:68474.

United States Department of the Interior. 2003e. Proposed revisions to grazing regulations for the public lands. Draft Environmental Impact Statement DES 03-62. USDI Bureau of Land Management, Washington, DC.

United States Department of the Interior. 2003f. Public land statistics. USDI Bureau of Land Management. http://www.blm.gov/natacq/pls03/ (13 July 2008).

United States Department of the Interior. 2004a. BLM sage-grouse habitat conservation strategy. Washington, DC. http://www.blm.gov/nhp/spotlight/sage_grouse/docs/Sage-Grouse_Strategy.pdf (18 March 2009).

United States Department of the Interior. 2004b. Bureau of Land Management national sage-grouse habitat conservation strategy. USDI Bureau of Land Management, Washington, DC.

United States Department of the Interior. 2004c. Endangered and threatened wildlife and plants; 90-day finding for a petition to list the eastern subspecies of the Greater Sage-Grouse as endangered. Federal Register 69:933–936.

United States Department of the Interior. 2004d. Endangered and threatened wildlife and plants; 90-day finding for petitions to list the Mono Basin area population of the sage-grouse as threatened or endangered. Federal Register 69:21484–21494.

United States Department of the Interior. 2004e. Public land statistics. USDI Bureau of Land Management. http://www.blm.gov/natacq/pls04/ (13 July 2008).

United States Department of the Interior. 2004f. Statement of policy regarding sage grouse management definitions, use of protective stipulations, and conditions of approval. Instructional Memorandum Number WY-2004-057. USDI Bureau of Land Management, Wyoming State Office, Cheyenne, WY.

United States Department of the Interior. 2005a. Endangered and threatened wildlife and plants; 90-day finding on a petition to list the pygmy rabbit as threatened or endangered. 50 CFR 17:29253–29265.

United States Department of the Interior. 2005b. Endangered and threatened wildlife and plants; 12-month finding for petitions to list the Greater Sage-Grouse as threatened or endangered; proposed rule. Federal Register 70:2244–2282.

United States Department of the Interior. 2005c. Energy facts. Onshore federal lands. USDI Bureau of Land Management. http://www.blm.gov/pgdata/etc/medialib/blm/wo/Communications_Directorate/general_publications/energy_facts.Par.76690.File.dat/energy_brochure_2005.pdf (16 August 2009).

United States Department of the Interior. 2005d. Public land statistics. USDI Bureau of Land Management. http://www.blm.gov/natacq/pls05/ (12 August 2008).

United States Department of the Interior. 2006a. Endangered and threatened wildlife and plants; Final listing determination for the Gunnison Sage-Grouse as threatened or endangered; final rule. 50 CFR 17:19953–19982.

United States Department of the Interior. 2006b. Endangered and threatened wildlife and plants; 90-day finding for petitions to list the Mono Basin area population of Greater Sage-Grouse as threatened or endangered. Federal Register 71:76058–76079.

United States Department of the Interior. 2006c. Public land statistics. USDI Bureau of Land Management. http://www.blm.gov/wo/st/en/res/Direct_Links_to_Publications/ann_rpt_and_pls/2006_pls_index.html (13 July 2008).

United States Department of the Interior. 2007a. Public land statistics. USDI Bureau of Land Management. http://www.blm.gov/public_land_statistics/pls07/index.htm (16 August 2009).

United States Department of the Interior. 2007b. Vegetation treatments on Bureau of Land Management lands in 17 western states: final programmatic environmental report. USDI Bureau of Land Management, Reno, NV. http://www.blm.gov/wo/st/en/prog/more/veg_eis.html (16 August 2009).

United States Department of the Interior. 2007c. Vegetation treatments using herbicides on Bureau of Land Management lands in 17 western states: final programmatic environmental impact statement. USDI Bureau of Land Management, Reno, NV. http://www.blm.gov/wo/st/en/prog/more/veg_eis.html (16 August 2009).

United States Department of the Interior. 2008a. Endangered and threatened wildlife and plants; Initiation of status review for the Greater Sage-Grouse (*Centrocercus urophasianus*) as threatened or endangered. USDI Fish and Wildlife Service. Federal Register 73:10218–10219.

United States Department of the Interior. 2008b. Endangered and threatened wildlife and plants; 90-day finding for petitions to list the Mono Basin area population of Greater Sage-Grouse (*Centrocercus urophasianus*) as threatened or endangered. Federal Register 73:23173–23175.

United States Department of the Interior. 2010. Endangered and Threatened Wildlife and Plants; 12-Month Findings for Petitions to List the Greater Sage-Grouse (*Centrocercus urophasianus*) as Threatened or Endangered. Federal Register 75:13910–13958.

United States Departments of Agriculture and the Interior. 1997a. Eastside draft environmental impact statement. Vol. 1. BLM-OR-WA-PL-96-037+1792. USDA Forest Service and USDI Bureau of Land Management, Portland, OR.

United States Departments of Agriculture and the Interior. 1997b. Eastside draft environmental impact statement. Vol. 2. BLM-OR-WA-PL-96-037+1792. USDA Forest Service and USDI Bureau of Land Management, Portland, OR.

United States Departments of Agriculture and the Interior. 1997c. The 10th and 11th report to Congress on the administration of the Wild Free-Roaming Horses and Burro Act for fiscal years 1992–1995. USDI Bureau of Land Management and the USDA Forest Service. United States Government Printing Office, Washington, DC.

United States Departments of Agriculture, Energy, and the Interior. 2003. Energy policy and conservation act report: scientific inventory of onshore federal lands' oil and gas resources and reserves and the extent and nature of restrictions or impediments to their development. BLM/WO/GI-03/002+3100. http://www.doi.gov (20 August 2008).

United States Departments of Agriculture, Energy, and the Interior. 2006. Scientific inventory of onshore federal lands' oil and gas resources and the extent and nature of restrictions or impediments to their development. http://www.blm.gov/epca/phase2/text/Executive_Summary.pdf/ (9 April 2009).

United States Departments of Commerce and the Interior. 2003. Policy for evaluation of conservation efforts when making listing secisions. Federal Register 68:15100–15115.

United States Departments of Energy and the Interior. 2008. Programmatic environmental impact statement, designation of energy corridors on federal land in the 11 western states. United States Department of Energy DOE/EIS-0386. http://www.gc.energy.gov/NEPA/draft-eis0386.htm (21 August 2009).

United States Departments of the Interior, Agriculture, and Energy. 2003. Energy policy and conservation act report: scientific inventory of onshore federal lands' oil and gas resources and reserves and the extent and nature of restrictions or impediments to their development. USDI Bureau of Land Management BLM/W0/GI-03/002+3100. http://www.blm.gov/epca/epcaI.htm (21 August 2009).

United States Departments of the Interior, Agriculture, and Energy. 2006. Inventory of onshore federal oil and natural gas resources and restrictions to their development. Phase II cumulative inventory: Northern Alaska, Montana Thrust Belt, Greater Green River Basin, Denver Basin, Uinta-Piceance Basin, Paradox/San Juan Basin, Appalachian Basin, Black Warrior Basin, Florida Peninsula. http://www.blm.gov/epca/phase2/EPCA06full72.pdf (21 August 2009).

United States Departments of the Interior, Agriculture, and Energy. 2008. Inventory of onshore federal oil and natural gas resources and restrictions to their development. Phase III inventory: onshore United States. http://www.blm.gov/wo/st/en/prog/energy/oil_and_gas/EPCA_III/EPCA_III_geodata.html (16 August 2009).

United States Endangered Species Act of 1973. 1973. 16 USC 1531–1543. http://www.fws.gov/endangered/ESA/content.html (6 April 2009).

United States General Accounting Office. 1977. Public rangelands continue to deteriorate. United States General Accounting Office Report CED-77-88. US General Accounting Office, Washington, DC.

United States General Accounting Office. 2003. Wildland fires: better information needed on effectiveness of emergency stabilization and rehabilitation treatments. United States General Accounting Office GAO-03-430. US General Accounting Office, Washington, DC.

United States Geological Survey. 2005a. Current distribution of sagebrush and associated vegetation in the Columbia Basin and southwestern regions. http://sagemap.wr.usgs.gov (18 June 2009).

United States Geological Survey. 2005b. Gap analysis data warehouse. http://gapanalysis.nbii.gov/portal/server.pt (13 November 2008).

United States Geological Survey. 2006. West Nile virus in Greater Sage-Grouse. Wildlife Health Bulletin 06-08. http://www.nwhc.usgs.gov/publications/wildlife_health_bulletins/WHB_06_08.jsp (15 October 2008).

United States Government Accounting Office. 2005. Wind power: impacts on wildlife and government responsibilities for regulating development and protecting wildlife. United States Government Accounting Office GAO-05-906. US Government Accounting Office, Washington, DC. http://www.gao.gov/new.items/d05906.pdf (21 August 2009).

United States Government Accounting Office. 2007. Farm program payments are an important factor in landowners' decisions to convert grassland to cropland. GAO-07-1054. US Government Accounting Office, Washington, DC. http://www.gao.gov/products/GAO-07-1054 (9 April 2009).

Unsworth, J. W., and L. Kuck. 1991. Bull elk vulnerability in the Clearwater Drainage of north-central Idaho. Pp. 85–93 *in* A. G. Christensen, L. J. Lyon, and T. N. Nooner (editors). Elk Vulnerability Symposium. Montana State University, Bozeman, MT.

Urban, D. L., and T. H. Keitt. 2001. Landscape connectedness: a graph theoretic perspective. Ecology 82:1205–1281.

Urban, D. L., R. V. O'Neill, and H. H. Shugart, Jr. 1987. Landscape ecology. BioScience 37:119–127.

Utah Department of Wildlife Resources. 2002. Strategic management plan for Sage-Grouse 2002. Publication 02-20. Utah Division of Wildlife Resources, Salt Lake City, UT. http://www.wildlife.utah.gov/uplandgame/pdf/2002manplan.pdf (18 March 2009).

Vale, T. R. 1974. Sagebrush conversion projects: an element of contemporary environmental change in the western United States. Biological Conservation 6:274–284.

Vale, T. R. 1975. Presettlement vegetation in the sagebrush-grass area of the Intermountain West. Journal of Range Management 28:32–36.

Vale, T. R. (editor). 2002a. Fire, native peoples, and the natural landscape. Island Press, Washington, DC.

Vale, T. R. 2002b. The pre-European landscape of the United States: pristine or humanized? Pp. 1–39 *in* T. R. Vale (editor). Fire, native peoples, and the natural landscape. Island Press, Washington, DC.

Valentine, K. A. 1947. Distance from water as a factor in grazing capacity of rangeland. Journal of Forestry 45:749–754.

Valkama, J., E. Korpimaki, B. Arroyo, P. Beja, V. Bretagnolle, E. Bro, R. Kenward, S. Manosa, S. M. Redpath, S. Thirgood, and J. Vinuela. 2005. Birds of prey as limiting factors of gamebird populations in Europe: a review. Biological Reviews 80:171–203.

Vallentine, J. F. 1971. Range development and improvements. Brigham Young University Press, Provo, UT.

Vallentine, J. F. 1989. Range development and improvements. Third edition. Academic Press, San Diego, CA.

Vallentine, J. F. 2004. Herbicides for plant control. Pp. 89–99 *in* S. B. Monsen, R. Stevens, and N. L. Shaw (compilers). Restoring western ranges and wildlands, Vol. I. USDA Forest Service RMRS-GTR-136-vol-1. USDA Forest Service, Rocky Mountain Research Station, Ft. Collins, CO.

van Andel, J. 2006a. Communities: interspecific interactions. Pp. 58–69 *in* J. van Andel and J. Aronson (editors). Restoration ecology: the new frontier. Blackwell Publishing, Malden, MA.

van Andel, J. 2006b. Populations: intraspecific interactions. Pp. 70–81 *in* J. van Andel and J. Aronson (editors). Restoration ecology: the new frontier. Blackwell Publishing, Malden, MA.

van de Koppel, J., M. Reitkerk, and F. J. Weissing. 1997. Catastrophic vegetation shifts and soil degradation in terrestrial grazing systems. Trends in Ecology and Evolution 12:352–356.

van der Meulen, K. M., M. B. Pensaert, and H. J. Nauwynck. 2005. West Nile virus in the vertebrate world. Archives of Virology 150:637–657.

Van Dyke, F., and W. C. Klein. 1996. Response of elk to installation of oil wells. Journal of Mammalogy 77:1028–1041.

Van Horne, M. L., and P. Z. Fulé. 2006. Comparing methods of reconstructing fire history using fire scars in a southwestern United States ponderosa pine forest. Canadian Journal of Forest Research 36:855–867.

Van Kooten, G. C., A. J. Eagle, and M. E. Eiswerth. 2007. Determinants of threatened Sage Grouse in northeastern Nevada. Human Dimensions of Wildlife 12:53–70.

Vander Haegen, W. M. 2007. Fragmentation by agriculture influences reproductive success of birds in a shrubsteppe landscape. Ecological Applications 17:934–947.

Vander Haegen, W. M., F. C. Dobler, and D. J. Pierce. 2000. Shrubsteppe bird response to habitat and landscape variables in eastern Washington, U.S.A. Conservation Biology 14:1145–1160.

Vander Haegen, W. M., S. M. McCorquodale, C. R. Peterson, G. A. Green, and E. Yensen. 2001. Wildlife of eastside shrubland and grassland habitats. Pp. 292–316 *in* D. H. Johnsen and T. A. O'Neil (editors). Wildlife-habitat relationships in Oregon and Washington. University of Oregon Press, Corvallis, OR.

Vavra, M., and J. Brown. 2006. Rangeland research: strategies for providing sustainability and stewardship to the rangelands of the world. Rangelands 28(6):7–14.

Verheyen, K., M. Vellend, H. Van Calster, G. Peterken, and M. Hermy. 2004. Metapopulation dynamics in changing landscapes: a new spatially realistic model for forest plants. Ecology 85:3302–3312.

Vickery, P. D. 1996. Grasshopper Sparrow (*Ammodramus savannarum*). *In* A. Poole and F. Gill (editors). The birds of North America, No. 239. The Academy of Natural Sciences, Philadelphia, PA, and The American Ornithologists' Union, Washington, DC.,

Vickery, P. D., P. L. Tubaro, J. M. Cardoso da Silva, B. G. Peterjohn, J. R. Herkert, and R. B. Cavalcanti. 1999. Conservation of grassland birds in the western hemisphere. Studies in Avian Biology 19:2–26.

Vignieri, S. N., E. M. Hallerman, B. J. Bergstrom, D. J. Hafner, A. P. Martin, P. Devers, P. Grobler, and N. Hitt. 2006. Mistaken view of taxonomic validity undermines conservation of an evolutionarily distinct mouse: a response to Ramey et al. (2005). Animal Conservation 9:237–243.

Vollmer, J. L., and J. G. Vollmer. 2008. Controlling cheatgrass in winter range to restore habitat and endemic fire. Pp. 57–60 *in* S. G. Kitchen, R. L. Pendleton, T. A. Monaco, and J. Vernon (compilers). Shrublands under fire: disturbance and recovery in a changing world. USDA Forest Service RMRS-P-52. USDA Forest Service, Rocky Mountain Research Station, Ft. Collins, CO.

Vucetich, J. A., and T. A. Waite. 2003. Spatial patterns of demography and genetic processes across the species' range: null hypotheses for landscape conservation genetics. Conservation Genetics 4:639–645.

Wagner, F. H. 1983. Status of wild horse and burro management on public rangelands. Transactions of the North American Wildlife and Natural Resources Conference 48:116–133.

Wagner, F. H., G. L. Achterman, J. L. Artz, F. J. Ayala, W. H. Blackburn, W. H. Conley, L. L. Eberhardt, S. K. Fairfax, W. E. Johnston, S. R. Kellert, J. C. Malechek, P. D. Moehlman, U. S. Seal, and J. W. Swan. 1980. Wild and free-roaming horses and burros: current knowledge and recommended research. Phase I, Final Report of the Committee on Wild and Free-Roaming Horses and Burros, National Research Council. National Academy Press, Washington, DC.

Waichler, W. S., R. F. Miller, and P. S. Doescher. 2001. Community characteristics of old-growth western juniper woodlands in the pumice zone of central Oregon. Journal of Range Management 54:518–527.

Wakkinen, W. L. 1990. Nest site characteristics and spring-summer movements of migratory Sage Grouse in southeastern Idaho. M.S. thesis, University of Idaho, Moscow, ID.

Wakkinen, W. L., K. P. Reese, and J. W. Connelly. 1992. Sage Grouse nest locations in relation to leks. Journal of Wildlife Management 56:381–383.

Waldo, D. 1984. The administrative state. Holmes and Meier, New York, NY.

Walhof, K. S. 1997. A comparison of burned and unburned big sagebrush communities in southwest Montana. M.S. thesis, Montana State University, Bozeman, MT.

Walker, B. H., and W. L. Steffen (editors). 1996. Global change and terrestrial ecosystems. Cambridge University Press, Cambridge, UK.

Walker, B. L. 2006. West Nile and sage-grouse update. SAGEMAP. http://sagemap.wr.usgs.gov/ (15 October 2008).

Walker, B. L. 2008. Greater Sage-Grouse response to coal-bed natural gas development and West Nile virus in the Powder River Basin, Montana and Wyoming, USA. Ph.D. dissertation, University of Montana, Missoula, MT.

Walker, B. L., D. E. Naugle, and K. E. Doherty. 2007a. Greater Sage-Grouse population response to energy development and habitat loss. Journal of Wildlife Management 71:2644–2654.

Walker, B. L., D. E. Naugle, K. E. Doherty, and T. E. Cornish. 2004. From the field: outbreak of West Nile virus in Greater Sage-Grouse and guidelines for monitoring, handling, and submitting dead birds. Wildlife Society Bulletin 32:1000–1006.

Walker, B. L., D. E. Naugle, K. E. Doherty, and T. E. Cornish. 2007b. West Nile virus and Greater Sage-Grouse: estimating infection rate in a wild bird population. Avian Diseases 51:691–696.

Walker, S. C., and N. L. Shaw. 2005. Current and potential use of broadleaf herbs for reestablishing native communities. Pp. 56–61 *in* N. L. Shaw, M. Pellant, and S. B. Monsen (compilers). Sage-grouse habitat restoration symposium proceedings. USDA Forest Service RMRS-P-38. USDA Forest Service, Rocky Mountain Research Station, Ft. Collins, CO.

Wallestad, R. O. 1971. Summer movements and habitat use by Sage Grouse broods in central Montana. Journal of Wildlife Management 35:129–136.

Wallestad, R. O. 1975a. Life history and habitat requirements of Sage Grouse in central Montana. Montana Department of Fish, Wildlife and Parks, Helena, MT.

Wallestad, R. 1975b. Male Sage Grouse responses to sagebrush treatment. Journal of Wildlife Management 39:482–484.

Wallestad, R. O., and D. B. Pyrah. 1974. Movement and nesting of Sage Grouse hens in central Montana. Journal of Wildlife Management 38:630–633.

Wallestad, R. O., and P. Schladweiler. 1974. Breeding season movements and habitat selection of male Sage Grouse. Journal of Wildlife Management 38:634–637.

Wallestad, R. O., J. G. Peterson, and R. L. Eng. 1975. Foods of adult Sage Grouse in central Montana. Journal of Wildlife Management 39:628–630.

Walsh, D. P. 2002. Population estimation techniques for Greater Sage-Grouse. M.S. thesis, Colorado State University, Ft. Collins, CO.

Walsh, D. P., G. C. White, T. E. Remington, and D. C. Bowden. 2004. Evaluation of the lek-count index for Greater Sage-Grouse. Wildlife Society Bulletin 32:56–68.

Wambolt, C. L. 1996. Mule deer and elk foraging preference for 4 sagebrush taxa. Journal of Range Management 49:499–503.

Wambolt, C. L., and M. R. Frisina. 2002. Montana sagebrush: a taxonomic key and habitat descriptions. Intermountain Journal of Sciences 8:46–59.

Wambolt, C. L., and G. F. Payne. 1986. An 18-year comparison of control methods for Wyoming big sagebrush in southwestern Montana. Journal of Range Management 39:314–319.

Wambolt, C. L., A. J. Harp, B. L. Welch, N. Shaw, J. W. Connelly, K. P. Reese, C. E. Braun, D. A. Klebenow, E. D. McArthur, J. G. Thompson, L. A. Torell, and J. A. Tanaka. 2002. Conservation of Greater Sage-Grouse on public lands in the western U. S.: implications of recovery and management policies. Policy Paper SG-02-02. Policy Analysis Center for Western Public Lands, Caldwell, ID.

Wambolt, C. L., T. L. Hoffman, and C. A. Mehus. 1999. Response of shrubs in big sagebrush habitats to fire on the northern Yellowstone winter range. Pp. 238–242 *in* E. D. McArthur, W. K. Ostler, and C. L. Wambolt (editors). Shrubland ecotones. USDA Forest Service Proceedings RMRS-P-11. USDA Forest Service, Rocky Mountain Research Station, Ft. Collins, CO.

Wambolt, C. L., K. S. Walhof, and M. R. Frisina. 2001. Recovery of big sagebrush communities after burning in south-western Montana. Journal of Environmental Management 61:243–252.

Wangler, M. J., and R. A. Minnich. 1996. Fire and succession in pinyon-juniper woodlands of the San Bernardino Mountains, California. Madroño 43:493–514.

Ward, M. R., D. E. Stallknecht, J. Willis, M. J. Conroy, and W. R. Davidson. 2006. Wild bird mortality and West Nile virus surveillance: biases associated with detection, reporting, and carcass persistence. Journal of Wildlife Diseases 42:92–106.

Washington-Allen, R. A., T. G. Van Niel, R. D. Ramsey, and N. E. West. 2004. Remote sensing-based biosphere analysis. GIScience and Remote Sensing 41:136–154.

Washington-Allen, R. A., N. E. West, R. D. Ramsey, and R. A. Efroymson. 2006. A protocol for retrospective remote sensing-based ecological monitoring of rangelands. Rangeland Ecology and Management 59:19–29.

Watts, M. J., and C. L. Wambolt. 1996. Long-term recovery of Wyoming big sagebrush after four treatments. Journal of Environmental Management 46:95–102.

Watts, S. E. 1998. Short-term influence of tank tracks on vegetation and microphytic crusts in shrubsteppe habitat. Environmental Management 22:611–616.

Webb, R. H., and H. G. Wilshire. 1983. Environmental effects of off-road vehicles: impacts and management in arid regions. Springer-Verlag, New York, NY.

Webb, W. C., W. I. Boarman, and J. T. Rotenberry. 2009. Movements of juvenile Common Ravens in an arid landscape. Journal of Wildlife Management 73:72–81.

Wehr, E. E. 1931. A new species of nematode worm from the Sage Grouse. United States National Museum Proceedings 79:1–3.

Wehr, E. E. 1933. Notes of the 148th meeting of the Helminthological Society of Washington. Journal of Parasitology 19:90–91.

Weigand, J. P. 1980. Ecology of the Hungarian Partridge in north-central Montana. Wildlife Monograph 74. The Wildlife Society, Bethesda, MD.

Weiguo, S., W. Jian, S. Kohyu, and F. Weicheng. 2006. Three types of power-law distribution of forest fires in Japan. Ecological Modelling 196:527–532.

Weisberg, P. J., E. Lingua, and R. B. Pillai. 2007. Spatial patterns of pinyon-juniper woodland expansion in central Nevada. Rangeland Ecology and Management 60:115–124.

Weisberg, S. 2005. Applied linear regression. Third edition. Wiley, Hoboken, NJ.

Welch, B. L., and C. Criddle. 2003. Countering misinformation concerning big sagebrush. USDA Forest Service Research Paper RMRS-RP-40. USDA Forest Service, Rocky Mountain Research Station, Ft. Collins, CO.

Welch, B. L., F. J. Wagstaff, and R. L. Williams. 1990. Sage Grouse status and recovery plan for Strawberry Valley, Utah. USDA Forest Service Research Paper INT-430. USDA Forest Service, Intermountain Research Station, Provo, UT.

Welsh, D. A. 1975. Population, behavioral, and grazing ecology of the horses of Sable Island, Nova Scotia. Ph.D. dissertation, Dalhousie University, Halifax, NS, Canada.

West, N. E. 1983a. Great Basin—Colorado Plateau sagebrush semi-desert. Pp. 331–349 *in* N. E. West (editor). Ecosystems of the world. Vol. 5: temperate deserts and semi-deserts. Elsevier Scientific Publishing Company, New York, NY.

West, N. E. 1983b. Western intermountain sagebrush steppe. Pp. 351–397 *in* N. E. West (editor). Ecosystems of the world. Vol. 5: temperate deserts and semi-deserts. Elsevier Scientific Publishing Company, New York, NY.

West, N. E. 1996. Strategies for maintenance and repair of biotic community diversity on rangelands. Pp. 326–346 *in* R. C. Szaro and D. W. Johnston (editors). Biodiversity in managed landscapes: theory and practice. Oxford University Press, New York, NY.

West, N. E. 1999. Managing for diversity of rangelands. Pp. 101–126 *in* W. W. Collins and C. O. Qualset (editors). Biodiversity in agrosystems. CRC Press, Boca Raton, FL.

West, N. E. 2000. Synecology and disturbance regimes of sagebrush steppe ecosystems. Pp. 15–26 *in* P. G. Entwistle, A. M. DeBolt, J. H. Kaltenecker, and K. Steenhof (compilers). Sagebrush Steppe Ecosystems Symposium. USDI Bureau of Land Management Publication No. BLM/ID/PT-001001+1150. USDI Bureau of Land Management, Boise, ID.

West, N. E. 2003a. History of rangeland monitoring in the U.S.A. Arid Land Research and Management 17:495–545.

West, N. E. 2003b. Theoretical underpinnings of rangeland monitoring. Arid Land Research and Management 17:333–346.

West, N. E., and N. S. Van Pelt. 1987. Successional patterns in pinyon-juniper woodlands. Pp. 43–52 *in* R. L. Everett (compiler). Pinyon-Juniper Conference. USDA Forest Service General Technical Report INT-215. USDA Forest Service, Intermountain Research Station, Ogden, UT.

West, N. E., and T. P. Yorks. 2002. Vegetation responses following wildfire on grazed and ungrazed sagebrush semi-desert. Journal of Range Management 55:171–181.

West, N. E., and J. A. Young. 2000. Intermountain valleys and lower mountain slopes. Pp. 255–284 *in* M. G. Barbour and W. D. Billings (editors). North American terrestrial vegetation. Second edition. Cambridge University Press, Cambridge, UK.

West, N. E., F. D. Provensa, P. S. Johnson, and D. K. Owens. 1984. Vegetation changes after 13 years of livestock grazing exclusion on sagebrush semidesert in west central Utah. Journal of Range Management 37:262–264.

Westerling, A. L., A. Gershunov, T. J. Brown, D. R. Cayan, and M. D. Dettinger. 2003. Climate and wildfire in the western United States. Bulletin of the American Meteorological Society 84:595–604.

Westerling, A. L., H. G. Hidalgo, D. R. Cayan, and T. W. Swetnam. 2006. Warming and earlier spring increase western U.S. forest wildfire activity. Science 313:940–943.

Western Governors' Association. 2004. Conserving the Greater Sage Grouse: examples of partnerships and strategies at work across the west. Western Governors' Association and USDA Natural Resources Conservation Service, Denver, CO.

Westoby, M., B. Walker, and I. Noy-Meir. 1989a. Opportunistic management for rangelands not at equilibrium. Journal of Range Management 42:266–274.

Westoby, M., B. Walker, and I. Noy-Meir. 1989b. Range management on the basis of a model which does not seek to establish equilibrium. Journal of Arid Environments 17:235–239.

Wheelwright, N. T., and J. D. Rising. 1993. Savannah Sparrow (*Passerculus sandwichensis*). *In* A. Poole and F. Gill (editors). The birds of North America, No. 45. The Academy of Natural Sciences, Philadelphia, PA, and The American Ornithologists' Union, Washington, DC.

Whisenant, S. G. 1990. Changing fire frequencies on Idaho's Snake River Plains: ecological and management implications. Pp. 4–10 *in* E. D. McArthur, E. M. Romney, S. D. Smith, and P. T. Tueller (compilers). Symposium on Cheatgrass Invasion, Shrub Die-Off, and Other Apects of Shrub Biology and Management. USDA Forest Service General Technical Report INT-276. USDA Forest Service, Intermountain Research Station, Ogden, UT.

Whisenant, S. G. 1999. Repairing damaged wildlands. A process-orientated, landscape-scale approach. Cambridge University Press, New York, NY.

White, G. C. 2000. Population viability analysis: data requirements and essential analyses. Pp. 288–331 *in* L. Boitani and T. K. Fuller (editors). Research techniques in animal ecology: controversies and consequences. Columbia University Press, New York, NY.

Whitford, W. G., A. G. de Soyza, J. W. Van Zee, J. E. Herrick, and K. M. Havstad. 1998. Vegetation, soil, and animal indicators of rangeland health. Environmental Monitoring and Assessment 51:179–200.

Whitson, T. D. (editor). 1996. Weeds of the west. Fifth edition. Western Society of Weed Science, Newark, CA.

Wiedemann, H. T. 2005. Revegetation equipment catalog. Revegetation Equipment Technology Committee, USDA Forest Service; USDI Bureau of Land Management. http://reveg-catalog.tamu.edu/ (30 April 2009).

Wiedenfeld, D. A., D. H. Wolfe, J. E. Toepfer, L. M. Mechlin, R. D. Applegate, and S. K. Sherrod. 2002. Survey for reticuloendotheliosis viruses in wild populations of Greater and Lesser Prairie Chickens. Wilson Bulletin 114:142–144.

Wiens, J. A. 1981. Scale problems in avian censusing. Studies in Avian Biology 6:513–521.

Wiens, J. A. 1989a. The ecology of bird communities. Vol. 1: foundations and patterns. Cambridge University Press. Cambridge, UK.

Wiens, J. A. 1989b. Spatial scaling in ecology. Functional Ecology 3:385–397.

Wiens, J. A. 1996. Wildlife in patchy environments: metapopulations, mosaics, and management. Pp. 53–84 *in* D. R. McCullough (editor). Metapopulations and wildlife conservation. Island Press, Washington, DC.

Wiens, J. A. 1997. Metapopulation dynamics and landscape ecology. Pp. 43–62 *in* I. Hanski and M. Gilpin (editors). Metapopulation dynamics: ecology, genetics, and evolution. Academic Press, New York, NY.

Wiens, J. A. 2002. Predicting species occurrences: progress, problems, and prospects. Pp. 739–749 *in* J. M. Scott, P. J. Heglund, M. L. Morrison, J. B. Haufler, M. G. Raphael, W. A. Wall, and F. B. Samson (editors). Predicting species occurrences: issues of accuracy and scale. Island Press, Washington, DC.

Wiens, J. A., and B. T. Milne. 1989. Scaling of "landscapes" in landscape ecology, or, landscape ecology from a beetle's perspective. Landscape Ecology 3:87–96.

Wiens, J. A., and K. R. Parker. 1995. Analyzing the effects of accidental environmental impacts: approaches and assumptions. Ecological Applications 5:1069–1083.

Wiens, J. A., and J. T. Rotenberry. 1981. Habitat associations and community structure of birds in shrubsteppe environments. Ecological Monographs 51:21–41.

Wiens, J. A., J. T. Rotenberry, and B. Van Horne. 1987. Habitat occupancy patterns of North American shrubsteppe birds: the effects of spatial scale. Oikos 48:132–147.

Wiens, J. A., N. C. Stenseth, B. Van Horne, and R. A. Ims. 1993. Ecological mechanisms and landscape ecology. Oikos 66:369–380.

Wigand, P. E. 1987. Diamond Pond, Harney County, Oregon: vegetation history and water table in the eastern Oregon desert. Great Basin Naturalist 47:427–458.

Wijayratne, U., and Pyke, D. A., 2009. Investigating seed longevity of big sagebrush (Artemisia tridentata). U.S. Geological Survey Open-File Report 2009–1146.

Wik, P. A. 2002. Ecology of Greater Sage-Grouse in south-central Owyhee County, Idaho. M.S. thesis, University of Idaho, Moscow, ID.

Wilcove, D. S., D. Rothstein, J. Dubow, A. Phillips, and E. Losos. 1998. Quantifying threats to imperiled species in the United States. BioScience 48:607–615.

Wiley, R. H. 1973a. The strut display of male Sage Grouse: a "fixed" action pattern. Behaviour 47:129–152.

Wiley, R. H. 1973b. Territoriality and non-random mating in the Sage Grouse *Centrocercus urophasianus*. Animal Behavior Monographs 6:87–169.

Wiley, R. H. 1974. Evolution of social organization and life history patterns among grouse. Quarterly Review of Biology 49:201–227.

Wiley, Jr., R. H. 1978. The lek mating system of the Sage Grouse. Scientific American 238:114–125.

Willebrand, T., and M. Hornell. 2001. Understanding the effects of harvesting Willow Ptarmigan *Lagopus lagopus* in Sweden. Wildlife Biology 7:205–212.

Williams, C. K., F. S. Guthery, R. D. Applegate, and M. J. Peterson. 2004a. The Northern Bobwhite decline: scaling our management for the twenty-first century. Wildlife Society Bulletin 32:861–869.

Williams, C. K., A. R. Ives, and R. D. Applegate. 2003. Population dynamics across geographical ranges: time series analyses of three small game species. Ecology 84:2654–2667.

Williams, C. K., R. S. Lutz, and R. D. Applegate. 2004b. Winter survival and additive harvest in Northern Bobwhite coveys in Kansas. Journal of Wildlife Management 68:94–100.

Willis, M. J., G. P. Keister, Jr., D. A. Immell, D. M. Jones, R. M. Powell, and K. R. Durbin. 1993. Sage Grouse in Oregon. Wildlife Research Report 15. Oregon Department of Fish and Wildlife, Portland, OR.

Wilsey, B. J., J. S. Coleman, and S. J. McNaughton. 1997. Effects of elevated CO_2 and defoliation on grasses: a comparative ecosystem approach. Ecological Applications 7:844–853.

Wilson, K., R. L. Pressey, A. Newton, M. Burgman, H. Possingham, and C. Weston. 2005. Measuring and incorporating vulnerability into conservation planning. Environmental Management 35:527–543.

Wilson, P. I. 2008. Preservation versus motorized recreation: institutions, history, and public lands management. Social Science Journal 45:194–202.

Wing, L. W. 1951. Practice of wildlife conservation. John Wiley and Sons, New York, NY.

Winward, A. H. 1991. A renewed commitment to management of sagebrush grasslands. Pp. 2–7 *in* Research in rangeland management. Special Report No. 880. Agricultural Experiment Station, Oregon State University, Corvallis, OR.

Winward, A. H. 2004. Sagebrush of Colorado. Taxonomy, distribution, ecology, and management. Colorado Division of Wildlife, Department of Natural Resources, Denver, CO.

Wirth, T. A., and D. A. Pyke. 2003. Restoring forbs for Sage Grouse habitat: fire, microsites, and establishment methods. Restoration Ecology 11:370–377.

Wirth, T. A., and D. A. Pyke. 2007. Monitoring post-fire vegetation rehabilitation projects—a common approach for non-forested ecosystems. United States Geological Survey, Scientific Investigations Report 2006-5048. USDI Geological Survey, Reston, VA.

Wisdom, M. J., R. S. Holthausen, B. C. Wales, C. D. Hargis, V. A. Saab, D. C. Lee, W. J. Hann, T. D. Rich, M. M. Rowland, W. J. Murphy, and M. R. Eames. 2000a. Source habitats for terrestrial vertebrates of focus in the interior Columbia Basin: broad-scale trends and management implications. General Technical Report PNW-GTR-485. USDA Forest Service, Portland, OR.

Wisdom, M. J., L. S. Mills, and D. F. Doak. 2000b. Life state simulation analysis: estimating vital rate effects on population growth for species conservation. Ecology 81:628–641.

Wisdom, M. J., M. M. Rowland, M. A. Hemstrom, and B. A. Wales. 2005a. Landscape restoration for Greater Sage-Grouse: implications for multiscale planning and management. Pp. 62–69 *in* N. L. Shaw, M. Pellant, and S. B. Monsen (compilers). Sage-grouse habitat restoration symposium proceedings. USDA Forest Service RMRS-P-38. USDA Forest Service, Rocky Mountain Research Station, Ft. Collins, CO.

Wisdom, M. J., M. M. Rowland, and L. H. Suring (editors). 2005b. Habitat threats in the sagebrush ecosystem: methods of regional assessment and applications in the Great Basin. Alliance Communications Group, Lawrence, KS.

Wisdom, M. J., M. M. Rowland, and R. J. Tausch. 2005c. Effective management strategies for sage-grouse and sagebrush: a question of triage? Transactions of the North American Wildlife and Natural Resources Conference 70:206–227.

Wisdom, M. J., M. M. Rowland, B. C. Wales, M. A. Hemstrom, W. J. Hann, M. G. Raphael, R. S. Holthausen, R. A. Gravenmier, and T. D. Rich. 2002a. Modeled effects of sagebrush-steppe restoration on Greater Sage-Grouse in the interior Columbia Basin, U.S.A. Conservation Biology 16:1223–1231.

Wisdom, M. J., M. Vavra, J. M. Boyd, M. A. Hemstrom, A. A. Ager, and B. K. Johnson. 2006. Understanding ungulate herbivory-episodic disturbance effects on vegetation dynamics: knowledge gaps and management needs. Wildlife Society Bulletin 34:283–292.

Wisdom, M. J., B. C. Wales, R. S. Holthausen, W. J. Hann, M. A. Hemstrom, and M. M. Rowland. 2002b. A habitat network for terrestrial wildlife in the interior Columbia Basin. Northwest Science 76:1–13.

Wisdom, M. J., B. C. Wales, M. M. Rowland, M. G. Raphael, R. S. Holthausen, T. D. Rich, and V. A. Saab. 2002c. Performance of Greater Sage-Grouse models for conservation assessment in the interior Columbia Basin, U.S.A. Conservation Biology 16:1232–1242.

With, K. A. 1994. Using fractal analysis to assess how species perceive landscape structure. Landscape Ecology 9:25–36.

With, K. A., and T. O. Crist. 1995. Critical thresholds in species' responses to landscape structure. Ecology 76:2446–2459.

With, K. A., and A. W. King. 1999. Dispersal success on fractal landscapes: a consequence of lacunarity thresholds. Landscape Ecology 14:73–82.

With, K. A., and A. W. King. 2001. Analysis of landscape sources and sinks: the effect of spatial pattern on avian demography. Biological Conservation 100:75–88.

With, K. A., R. H. Gardner, and M. G. Turner. 1997. Landscape connectivity and population distributions in heterogeneous landscapes. Oikos 78:151–169.

Withey, J. C., T. D. Bloxton, and J. M. Marzluff. 2001. Effects of tagging and location error in wildlife radiotelemetry studies. Pp. 43–75 *in* J. J. Millspaugh and J. M. Marzluff (editors). Radio tracking and animal populations. Academic Press, San Diego, CA.

Witmer, G. W., J. L. Bucknall, T. H. Fritts, and D. G. Moreno. 1996. Predator management to protect endangered avian species. Transactions of the North American Wildlife and Natural Resources Conference 61:102–108.

Wittenberger, J. F. 1978. The evolution of mating systems in grouse. Condor 80:126–137.

Woodward, J. K. 2006. Greater Sage-Grouse (*Centrocercus urophasianus*) habitat in central Montana. M.S. thesis, Montana State University, Bozeman, MT.

Worton, B. J. 1989. Kernel methods for estimating the utilization distribution in home-range studies. Ecology 70:164–168.

Wright, C. S., and S. J. Prichard. 2006. Biomass consumption during prescribed fires in big sagebrush ecosystems. Pp. 489–500 *in* P. L. Andrews and B. W. Butler (editors). Fuels management—how to measure success. USDA Forest Service Proceedings RMRS-P-41. USDA Forest Service, Rocky Mountain Research Station, Ft. Collins, CO.

Wright, H. A., and A. W. Bailey. 1982. Fire ecology, United States and southern Canada. John Wiley and Sons, New York, NY.

Wright, R. G., J. M. Scott, S. Mann, and M. Murray. 2001. Identifying unprotected and potentially at risk plant communities in the western USA. Biological Conservation 98:97–106.

Wright, S. 1938. Size of population and breeding structure in relation to evolution. Science 87:430–431.

Wrobleski, D. W., and J. B. Kauffman. 2003. Initial effects of prescribed fire on morphology, abundance, and phenology of forbs in big sagebrush communities in southeastern Oregon. Restoration Ecology 11:82–90.

Wu, X. B., and D. Z. Sui. 2001. An initial exploration of a lacunarity-based segregation measure. Environment and Planning B: Planning and Design 28:433–446.

Wu, X. B., T. L. Thurlow, and S. G. Whisenant. 2000. Fragmentation and changes in hydrologic function of tiger brush landscapes, south-west Niger. Journal of Ecology 88:790–800.

Wünschmann, A., and A. Ziegler. 2006. West Nile virus–associated mortality events in domestic Chukar Partridges (*Alectoris chukar*) and domestic Impeyan Pheasants (*Lophophorus impeyanus*). Avian Diseases 50:456–459.

Wyoming Interagency Vegetation Committee. 2002. Wyoming guidelines for managing sagebrush communities with an emphasis on fire management. USDI Bureau of Land Management, Wyoming State Office, Cheyenne, WY.

Wyoming Sage-Grouse Working Group. 2003. Wyoming Greater Sage-Grouse conservation plan. Wyoming Game and Fish Department, Cheyenne, WY. http://gf.state.wy.us/wildlife/wildlife_management/sagegrouse.asp (18 March 2009).

Yaremych, S. A., R. E. Warner, P. C. Mankin, J. D. Brawn, A. Raim, and R. Novak. 2004. West Nile virus and high death rate in American crows. Emerging Infectious Diseases 10:709–711.

Yensen, D. L. 1981. The 1900 invasion of alien plants into southern Idaho. Great Basin Naturalist 41:176–183.

Yensen, E., D. Quinney, K. Johnson, K. Timmerman, and K. Steenhof. 1992. Fire, vegetation changes, and population fluctuations of Townsend's ground squirrels. American Midland Naturalist 128:299–312.

Yocom, C. F. 1956. The Sage Hen in Washington state. Auk 73:540–550.

Yorks, T. P., N. E. West, and K. M. Capels. 1992. Vegetation differences in desert shrublands of western Utah's Pine Valley between 1933 and 1989. Journal of Range Management 45:569–578.

Young, J. A. 1989. Intermountain shrubsteppe plant communities—pristine and grazed. Pp. 3–14 *in* B. G. Pendleton (editor). Western Raptor Management Symposium and Workshop. National Wildlife Foundation, Washington, DC.

Young, J. A., and F. L. Allen. 1997. Cheatgrass and range science: 1930–1950. Journal of Range Management 50:530–535.

Young, J. A., and C. D. Clements. 2000. Cheatgrass control and seeding. Rangelands 22(4):3–6.

Young, J. A., and R. A. Evans. 1973. Downy brome—intruder in the plant succession of big sagebrush communities in the Great Basin. Journal of Range Management 26:410–415.

Young, J. A., and R. A. Evans. 1978. Population dynamics after wildfires in sagebrush grasslands. Journal of Range Management 31:283–289.

Young, J. A., and R. A. Evans. 1981. Demography and fire history of a western juniper stand. Journal of Range Management 34:501–506.

Young, J. A., and R. A. Evans. 1986. History of crested wheatgrass in the intermountain area. Pp. 21–25 *in* K. L. Johnson (editor). Crested wheatgrass: its values, problems, and myths: proceedings. Utah State University, Logan, UT.

Young, J. A., and R. A. Evans. 1989. Dispersal and germination of big sagebrush (*Artemisia tridentata*) seeds. Weed Science 37:201–206.

Young, J. A., and W. S. Longland. 1996. Impact of alien plants on Great Basin rangelands. Weed Technology 10:384–391.

Young, J. A., and D. McKenzie. 1982. Rangeland drill. Rangelands 4(3):108–113.

Young, J. A., and B. A. Sparks. 2002. Cattle in the cold desert. University of Nevada Press, Reno, NV.

Young, J. A., C. D. Clements, and R. R. Blank. 1997. Influence of nitrogen on antelope bitterbrush seedling establishment. Journal of Range Management 50:536–540.

Young, J. A., R. A. Evans, and R. E. Eckert, Jr. 1981. Environmental quality and the use of herbicides on *Artemisia*/grasslands of the U.S. intermountain area. Agriculture and Environment 6:53–61.

Young, J. A., R. A. Evans, and J. J. Major. 1972. Alien plants in the Great Basin. Journal of Range Management 25:194–201.

Young, J. R., C. E. Braun, S. J. Oyler-McCance, J. W. Hupp, and T. W. Quinn. 2000. A new species of sage-grouse (Phasianidae: *Centrocercus*) from southwestern Colorado. Wilson Bulletin 112:445–453.

Young, J. R., J. W. Hupp, J. W. Bradbury, and C. E. Braun. 1994. Phenotypic divergence of secondary sexual traits among Sage Grouse populations. Animal Behavior 47:1353–1362.

Zablan, M. A. 1993. Evaluation of Sage Grouse banding program in North Park, Colorado. M.S. thesis, Colorado State University, Ft. Collins, CO.

Zablan, M. A., C. E. Braun, and G. C. White. 2003. Estimation of Greater Sage-Grouse survival in North Park, Colorado. Journal of Wildlife Management 67:144–154.

Zamora, B., and P. T. Tueller. 1973. *Artemisia arbuscula, A. longiloba,* and *A. nova* habitat types in northern Nevada. Great Basin Naturalist 33:225–242.

Zar, J. H. 1998. Biostatistical analysis. Fourth edition. Prentice Hall, Englewood Cliffs, NJ.

Zarn, M., T. Heller, and K. Collins. 1977. Wild, free-roaming horses—status and present knowledge. USDI Bureau of Land Management Technical Note 294. USDI Bureau of Land Management, Washington, DC.

Zavala, G. S., S. Cheng, T. Barbosa, and H. Haefele. 2006. Enzootic reticuloendotheliosis in the endangered Attwater's and Greater Prairie Chickens. Avian Disease 50:520–525.

Zeng, Z., R. M. Nowierski, M. L. Taper, B. Dennis, and W. P. Kemp. 1998. Complex dynamics in the real world: modeling the influence of time-varying parameters and time lags. Ecology 79:2193–2209.

Ziegenhagen, L. L. 2004. Shrub reestablishment following fire in the mountain big sagebrush (*Artemisia tridentata* ssp. *vaseyana*) alliance. M.S. thesis, Oregon State University, Corvallis, OR.

Ziegenhagen, L. L., and R. F. Miller. 2009. Postfire recovery of two shrubs in the interiors of large burns in the Intermountain West, USA. Western North American Naturalist 69:195–205.

Zink, R. M. 2004. The role of subspecies in obscuring avian biological diversity and misleading conservation policy. Proceedings of the Royal Society of London B 271:561–564.

Ziska, L. H., J. B. Reeves, III, and B. Blank. 2005. The impact of recent increases in atmospheric CO_2 on biomass production and vegetative retention of cheatgrass (*Bromus tectorum*): implications for fire disturbance. Global Change Biology 11:1325–1332.

Zou, L., S. N. Miller, and E. T. Schmidtmann. 2006a. A GIS tool to estimate West Nile virus risk based on a degree-day model. Environmental Modeling and Assessment. Environmental Modeling and Assessment 129:413–420.

Zou, L., S. N. Miller, and E. T. Schmidtmann. 2006b. Mosquito larval habitat mapping using remote sensing and GIS: implications of coalbed methane development and West Nile virus. Journal of Medical Entomology 43:1034–1041.

Zunino, G. W. 1987. Harvest effects on Sage Grouse densities in northwest Nevada. M.S. thesis, University of Nevada, Reno, NV.

INDEX

Indexer: Leslie A. Robb
Composition: Michael Bass Associates
Text: 9.25/11.75 Scala
Display: Scala Sans, Scala Sans Caps
Printer and Binder: Thomson-Shore

STUDIES IN AVIAN BIOLOGY

1. Kessel, B., and D. D. Gibson. 1978.
Status and Distribution of Alaska Birds.

2. Pitelka, F. A., editor. 1979.
Shorebirds in Marine Environments.

3. Szaro, R. C., and R. P. Balda. 1979.
Bird Community Dynamics in a Ponderosa Pine Forest.

4. DeSante, D. F., and D. G. Ainley. 1980.
The Avifauna of the South Farallon Islands, California.

5. Mugaas, J. N., and J. R. King. 1981.
Annual Variation of Daily Energy Expenditure by the Black-billed Magpie: A Study of Thermal and Behavioral Energetics.

6. Ralph, C. J., and J. M. Scott, editors. 1981.
Estimating Numbers of Terrestrial Birds.

7. Price, F. E., and C. E. Bock. 1983.
*Population Ecology of the Dipper (*Cinclus mexicanus*) in the Front Range of Colorado.*

8. Schreiber, R. W., editor. 1984.
Tropical Seabird Biology.

9. Scott, J. M., S. Mountainspring, F. L. Ramsey, and C. B. Kepler. 1986.
Forest Bird Communities of the Hawaiian Islands: Their Dynamics, Ecology, and Conservation.

10. Hand, J. L., W. E. Southern, and K. Vermeer, editors. 1987.
Ecology and Behavior of Gulls.

11. Briggs, K. T., W. B. Tyler, D. B. Lewis, and D. R. Carlson. 1987.
Bird Communities at Sea off California: 1975 to 1983.

12. Jehl, J. R., Jr. 1988.
Biology of the Eared Grebe and Wilson's Phalarope in the Nonbreeding Season: A Study of Adaptations to Saline Lakes.

13. Morrison, M. L., C. J. Ralph, J. Verner, and J. R. Jehl, Jr., editors. 1990.
Avian Foraging: Theory, Methodology, and Applications.

14. Sealy, S. G., editor. 1990.
Auks at Sea.

15. Jehl, J. R., Jr., and N. K. Johnson, editors. 1994.
A Century of Avifaunal Change in Western North America.

16. Block, W. M., M. L. Morrison, and M. H. Reiser, editors. 1994.
The Northern Goshawk: Ecology and Management.

17. Forsman, E. D., S. DeStefano, M. G. Raphael, and R. J. Gutiérrez, editors. 1996.
Demography of the Northern Spotted Owl.

18. Morrison, M. L., L. S. Hall, S. K. Robinson, S. I. Rothstein, D. C. Hahn, and T. D. Rich, editors. 1999.
Research and Management of the Brown-headed Cowbird in Western Landscapes.

19. Vickery, P. D., and J. R. Herkert, editors. 1999.
Ecology and Conservation of Grassland Birds of the Western Hemisphere.

20. Moore, F. R., editor. 2000.
Stopover Ecology of Nearctic–Neotropical Landbird Migrants: Habitat Relations and Conservation Implications.

21. Dunning, J. B., Jr., and J. C. Kilgo, editors. 2000.
Avian Research at the Savannah River Site: A Model for Integrating Basic Research and Long-Term Management.

22. Scott, J. M., S. Conant, and C. van Riper, II, editors. 2001.
Evolution, Ecology, Conservation, and Management of Hawaiian Birds: A Vanishing Avifauna.

23. Rising, J. D. 2001.
*Geographic Variation in Size and Shape of Savannah Sparrows (*Passerculus sandwichensis*).*

24. Morton, M. L. 2002.
The Mountain White-crowned Sparrow: Migration and Reproduction at High Altitude.

25. George, T. L., and D. S. Dobkin, editors. 2002.
Effects of Habitat Fragmentation on Birds in Western Landscapes: Contrasts with Paradigms from the Eastern United States.

26. Sogge, M. K., B. E. Kus, S. J. Sferra, and M.J. Whitfield, editors. 2003. *Ecology and Conservation of the Willow Flycatcher.*

27. Shuford, W. D., and K. C. Molina, editors. 2004. *Ecology and Conservation of Birds of the Salton Sink: An Endangered Ecosystem.*

28. Carmen, W. J. 2004. *Noncooperative Breeding in the California Scrub-Jay.*

29. Ralph, C. J., and E. H. Dunn, editors. 2004. *Monitoring Bird Populations Using Mist Nets.*

30. Saab, V. A., and H. D. W. Powell, editors. 2005. *Fire and Avian Ecology in North America.*

31. Morrison, M. L., editor. 2006. *The Northern Goshawk: A Technical Assessment of Its Status, Ecology, and Management.*

32. Greenberg, R., J. E. Maldonado, S. Droege, and M. V. McDonald, editors. 2006. *Terrestrial Vertebrates of Tidal Marshes: Evolution, Ecology, and Conservation.*

33. Mason, J. W., G. J. McChesney, W. R. McIver, H. R. Carter, J. Y. Takekawa, R. T. Golightly, J. T. Ackerman, D. L. Orthmeyer, W. M. Perry, J. L. Yee, M. O. Pierson, and M. D. McCrary. 2007. *At-Sea Distribution and Abundance of Seabirds off Southern California: A 20-Year Comparison.*

34. Jones, S. L., and G. R. Geupel, editors. 2007. *Beyond Mayfield: Measurements of Nest-Survival Data.*

35. Spear, L. B., D. G. Ainley, and W. A. Walker. 2007. *Foraging Dynamics of Seabirds in the Eastern Tropical Pacific Ocean.*

36. Niles, L. J., H. P. Sitters, A. D. Dey, P. W. Atkinson, A. J. Baker, K. A. Bennett, R. Carmona, K. E. Clark, N. A. Clark, C. Espoz, P. M. González, B. A. Harrington, D. E. Hernández, K. S. Kalasz, R. G. Lathrop, R. N. Matus, C. D. T. Minton, R. I. G. Morrison, M. K. Peck, W. Pitts, R. A. Robinson, and I. L. Serrano. 2008. *Status of the Red Knot (*Calidris canutus rufa*) in the Western Hemisphere.*

37. Ruth, J. M., T. Brush, and D. J. Krueper, editors. 2008. *Birds of the US–Mexico Borderland: Distribution, Ecology, and Conservation.*

38. Knick, S. T., and J. W. Connelly, editors. 2011. *Greater Sage-Grouse: Ecology and Conservation of a Landscape Species and Its Habitats.*